SHELF SAND AND
SANDSTONE BODIES

# Shelf Sand and Sandstone Bodies
## GEOMETRY, FACIES AND SEQUENCE STRATIGRAPHY

Edited by D.J.P. Swift,
G.F. Oertel, R.W. Tillman
and J.A. Thorne

SPECIAL PUBLICATION NUMBER 14 OF THE
INTERNATIONAL ASSOCIATION OF SEDIMENTOLOGISTS
PUBLISHED BY BLACKWELL SCIENTIFIC PUBLICATIONS
OXFORD LONDON EDINBURGH BOSTON
MELBOURNE PARIS BERLIN VIENNA

© 1991 The International Association
of Sedimentologists
and published for them by
Blackwell Scientific Publications
Editorial offices:
Osney Mead, Oxford OX2 0EL
25 John Street, London WC1N 2BL
23 Ainslie Place, Edinburgh EH3 6AJ
3 Cambridge Center, Cambridge,
  Massachusetts 02142, USA
54 University Street, Carlton
  Victoria 3053, Australia

Other Editorial Offices:
Librairie Arnette SA
2, rue Casimir-Delavigne
75006 Paris
France

Blackwell Wissenschafts-Verlag
Meinekestrasse 4
D-1000 Berlin 15
Germany

Blackwell MZV
Feldgasse 13
A-1238 Wien
Austria

All rights reserved. No part of this
publication may be reproduced, stored
in a retrieval system, or transmitted,
in any form or by any means,
electronic, mechanical, photocopying,
recording or otherwise without the
prior permission of the copyright
owner.

First published 1991

Set by Excel Typesetters Co., Hong Kong
Printed and bound in Great Britain
by The Alden Press, Oxford

DISTRIBUTORS

Marston Book Services Ltd
PO Box 87
Oxford OX2 0DT
(*Orders*: Tel: 0865 791155
         Fax: 0865 791927
         Telex: 837515)

USA
  Blackwell Scientific Publications, Inc
  3 Cambridge Center
  Cambridge, MA 02142
  (*Orders*: Tel: 800 759-6102)

Canada
  Oxford University Press
  70 Wynford Drive
  Don Mills
  Ontario M3C 1J9
  (*Orders*: Tel: (416) 441-2941)

Australia
  Blackwell Scientific Publications
  (Australia) Pty Ltd
  54 University Street
  Carlton, Victoria 3053
  (*Orders*: Tel: (03) 347-0300)

British Library
Shelf sand and sandstone bodies.
  1. Sedimentary rocks
  I. Swift, D.J.P.  II. Series
  552.5
Cataloguing in Publication Data
ISBN 0-632-03237-5

Library of Congress
Cataloging-in-Publication Data

Shelf sand and sandstone bodies:
  geometry, facies, and sequence stratigraphy/
  edited by D.J.P. Swift . . . [*et al*.].
          p.      cm.
  (Special publication number 14 of the International Asssociation
  of Sedimentologists)
  'This volume is the outgrowth of a symposium . . .
  held at the 1987 midyear SEPM meeting
  in Raleigh, North Carolina'—Pref.
  ISBN 0-632-03237-5
  1. Sandstone.  2. Sand.
  3. Sedimentation and deposition.
  4. Continental margins.  5. Continental shelf.
  6. Geology, Stratigraphic.
  I. Swift, Donald J.P.  II. SEPM (Society of Economic
  Paleontologists and Mineralogists)
  III. Series: Special publication . . . of the
  International Association of Sedimentologists; no. 14.
  QE571.S438  1992
  552'.5—dc20

# Contents

vii Preface

**Concepts of Continental Margin Sedimentation**

3   Sedimentation on continental margins, I: a general model for shelf sedimentation
    *D.J.P. Swift and J.A. Thorne*

33  Sedimentation on continental margins, II: application of the regime concept
    *J.A. Thorne and D.J.P. Swift*

59  Sedimentation on continental margins, III: the depositional fabric—an analytical approach to stratification and facies identification
    *J.A. Thorne, E. Grace, D.J.P. Swift and A. Niedoroda*

89  Sedimentation on continental margins, IV: lithofacies and depositional systems
    *D.J.P. Swift, S. Phillips and J.A. Thorne*

153 Sedimentation on continental margins, V: parasequences
    *D.J.P. Swift, S. Phillips and J.A. Thorne*

189 Sedimentation on continental margins, VI: a regime model for depositional sequences, their component systems tracts, and bounding surfaces
    *J.A. Thorne and D.J.P. Swift*

**Regressive Shelf and Coastal Sandstones**

259 Facies architecture of a falling sea-level strandplain, Doce River coast, Brazil
    *J.M.L. Dominguez and H.R. Wanless*

283 Origin and geometry of storm-deposited sand beds in modern sediments of the Texas continental shelf
    *J.W. Snedden and D. Nummedal*

309 Model for genesis of shoreface and shelf sandstone sequences, southern Appalachians: palaeoenvironmental reconstruction of an Early Silurian shelf system
    *S.G. Driese, M.W. Fischer, K.A. Easthouse, G.T. Marks, A.R. Gogola and A.E. Schoner*

339 Prograding shoreline deposits in the Lower Silurian Medina Group, Ontario and New York: storm- and tide-influenced sedimentation in a shallow epicontinental sea, and the origin of enigmatic shore-normal channels encapsulated by open shallow-marine deposits
*W.L. Duke, P.J. Fawcett and W.C. Brusse*

**Transgressive Coastal and Shelf Sandstones**

379 Implications of tide-dominated lagoonal processes on the preservation of buried channels on a sediment-starved continental shelf
*G.F. Oertel, V.J. Henry and A.M. Foyle*

395 Generation of late Holocene sand ridges on the middle continental shelf of New Jersey, USA—evidence for formation in a mid-shelf setting based on comparisons with a nearshore ridge
*J.M. Rine, R.W. Tillman, S.J. Culver and D.J.P. Swift*

**Sandstones Deposited at Sequence Boundaries**

427 Evolution of an Upper Cretaceous (Turonian) shelf sandstone ridge: analysis of the Crossfield–Cardium pool, Alberta, Canada
*F.F. Krause and D.A. Nelson*

457 Truncated prograding strandplain or offshore sand body?—sedimentology and geometry of the Cardium (Turonian) sandstone and conglomerate at Willesden Green field, Alberta
*D.A.W. Keith*

489 Facies of the lower Ferron Sandstone and Blue Gate Shale Members of the Mancos Shale: lowstand and early transgressive facies architecture
*P.E. Riemersma and M.A. Chan*

511 Index

# Preface

The past few years have been exciting for the sedimentological community. A decade of study of the shallow-marine boundary layer has arrived at a synthesis of sediment dynamic principles that we can begin to transfer from the time-scales of observation (seconds to months) to the time-scales of the rock record (years to millenia). At the same time, the technology of petroleum exploration has led to a fundamentally new way of examining the deposits of sedimentary basins at large spatial scales. We and our fellow contributors apply these insights to continental shelf and continental margin deposits. In order to assess our problems, we have evaluated the ideas of fluid dynamicists in the context of the ideas of geophysicists and stratigraphers, and vice versa. The result has been profoundly synergistic. Our papers set forth a series of sedimentological principles that have not been apparent from either viewpoint.

This volume is the outgrowth of a symposium entitled 'Shelf Sand and Sandstone Bodies' held at the 1987 midyear meeting of the Society of Economic Paleontologists and Mineralogists in Raleigh, North Carolina. The symposium was convened by some of the editors and contributors of this volume because we thought we could see more order in shelf sand and sandstone bodies than had been appreciated previously, and because there had been no previous examination of the problem since the publication of *Geometry of Sandstone Bodies* by Peterson & Osmond in 1961 (American Association of Petroleum Geologists, Tulsa). The decision to seek publication for the papers was made jointly by the symposium contributors and the audience of the last session. We continued to see the character of shelf sand and sandstone bodies and their relationship to their depositional setting as our central theme. We invited the participation of other sedimentologists whom we felt could make contributions to this theme. As a consequence of this overriding concern, important regions and sediment types have been omitted. However, we were unwilling to take a more encyclopaedic approach at the expense of our focus. We see our study areas as laboratories in which we can develop generalizations that are more widely applicable in time and space.

The main body of the volume is preceded by a six-part synthesis of fluid dynamic and sequence stratigraphic insights as they apply to continental margins. Our first problem in this introductory portion was the problem of scale itself. How do we transfer the insights from the fluid dynamic time-scale of seconds and centimetres to the stratigraphic time-scale of hundreds of kilometres and millions of years? Our response has been a return to the regime concept of hydraulic engineers and geomorphologists. These people, confronted with the need to explain channel geometry from first principles of fluid dynamics, solved the problem by bypassing it. They recast river hydraulics in terms of a system of interdependent variables that can attain a state of equilibrium.

It turns out, however, that in adopting this solution we were reaching further back into the history of science. The thermodynamicist Prigogine and the science historian Stengers have described the growing dismay of eighteenth century chemists as they realized that the new Newtonian synthesis was *not* going to solve their problems. Specifying a chemical system by the position and velocity of its constituent parts was no longer a serious solution after Avogadro showed that there were $6 \times 10^{23}$ molecules in a mole of gas. Chemists responded with the new science of thermodynamics. In thermodynamics, the molecular system is specified by a set of macroscopic parameters, such as temperature, pressure and volume. Given boundary conditions it is possible to write equations of state and evaluate the system in terms of an equilibrium condition. The regime approach of hydraulic engineers and geomorphologists stands in similar relationship to sediment transport mechanics as does thermodynamics to physics.

Fortified by the regime approach, we could access the building blocks of sedimentary deposits at successive time and space scales. Event beds seemed the place to begin; we could apply the insights of boundary layer fluid dynamics directly to the problem. Beds could be related to facies, and facies to the 'systems' stratigraphy of the Texas Bureau of Economic Geology. Systems stratigraphy could be assimilated, not without some difficulty, into the rapidly growing concepts of sequence stratigraphy. At each level, the regime concept has allowed us to take a more analytical approach than would be possible otherwise.

The introduction and the contributions were written more or less simultaneously, and as a consequence, not all of the introductory concepts appear in the contributions. The volume should be viewed as the inverse of a textbook; we have no grand chapter of conclusions, but we do start with an executive summary, which draws together some of the themes that we have discerned among the contributions, and systematizes their concepts. Further attempts at organization would be contrary to the timeliness and plurality of approach that are the strengths of a symposium volume.

We would like to thank the authors for their cooperation throughout the period of preparation, and the reviewers for their time and effort. We would also like to thank Marcus Milling who encouraged us to begin the work, and Roger Slatt, for his support while the work was in progress. We would also like to thank Donald Emminger for the outstanding quality of the graphics in the introductory section. This book was edited for the IAS by Maurice Tucker.

DONALD J.P. SWIFT
*Norfolk, Virginia, USA*

GEORGE F. OERTEL
*Norfolk, Virginia, USA*

RODERICK W. TILLMAN
*Tulsa, Oklahoma, USA*

JULIAN A. THORNE
*Plano, Texas, USA*

# Concepts of Continental Margin Sedimentation

# Sedimentation on continental margins, I: a general model for shelf sedimentation

D.J.P. SWIFT[*] and J.A. THORNE[†]

[*] *Department of Oceanography, Old Dominion University, Norfolk, VA 23529, USA; and* [†]*Research and Technical Services, ARCO Oil and Gas Company, Plano, TX 75075, USA*

## ABSTRACT

The shelf surface is subjected to complexly varying hydraulic conditions over short time periods. However, over geological time, it can be viewed as a surface of dynamic equilibrium, controlled by such variables as the rate of relative sea-level rise ($R$), the rate and character of sediment input ($Q, M$), the fluid power ($P$), and the resulting rate of sediment transport ($D$). Sediment accumulating on the continental margin builds up toward 'wave base', the depth to which, on a time-averaged basis, there is sufficient wave power to bypass sediment at a rate equal to its prevailing rate of input. In this manner, continental margins evolve from diverse structural origins toward a characteristic shelf-slope configuration. During the evolutionary process, shifts in the values of variables occur with relative frequency, so that the sedimentary column beneath mature shelves tends to consist of multiple depositional sequences, rather than of a single clinoform sequence. Such variations in the character of shelf deposits over geological time are not examples of disequilibrium, but are rather homeostatic responses of the depositional regime, in which the equilibrium is maintained through mutual adjustments of the governing variables. This concept is akin to the 'regime' concept of hydraulic engineers, and the 'grade' concept of geomorphologists. In this paper, a theory of regime sedimentation is developed as a unifying idea for the analysis of continental margin deposits.

In this introductory paper, recent studies of sedimentation on modern shelves are synthesized to provide a general descriptive model for shelf sedimentary regimes. Shelf regimes can be characterized by an accommodation/supply ratio, $R \cdot D : Q \cdot M$. Values of the ratio greater than 1 identify *accommodation-dominated regimes*, characterized by high values of sea-level rise or sediment dispersal. Values less than 1 identify *supply-dominated regimes*, characterized by high values of sediment input or a coarse grain size. Shelf regimes experience an important feedback between the regime variables and coastal morphodynamics. Changes in the accommodation/supply ratio modify the coastal configuration, which in turn intensifies the change in the value of the accommodation/supply ratio. The feedback is effected through the coastal bypassing mechanism. Shoaling and breaking waves on oceanic coasts create a mean landward-directed bottom stress, which tends to drive sand towards the shore, resulting in a 'littoral energy fence'. Sediment passes through the fence in one of two ways. Supply-dominated regimes are characterized by regressions in which *river mouth bypassing* transports sediment to the shore face via the flood stage jet of a deltaic river mouth. It is then transported along the coast and offshore by coastal storm currents. Accommodation-dominated regimes are characterized by transgressions, in which river mouths become sediment-trapping estuaries. The shoreface undergoes erosional shoreface retreat so that *shoreface bypassing* becomes dominant. The two bypassing modes result in two basic classes of sediment provenance. River mouth bypassing results in *allochthonous* (of far-travelled origin) sedimentation, in which large volumes of fine, mobile, first-cycle, river-derived sediment spread rapidly to all parts of the shelf, and build thick, uniform deposits. Shoreface bypassing results in *autochthonous* (of *in situ* origin) sedimentation. Autochthonous sediment is derived primarily from erosional shoreface retreat. Autochthonous deposits are thin, variable and relatively coarse.

The general model provides a framework with which to consider the process of facies differentiation. In this context, shelf facies are best understood as the result of *progressive sorting* during episodic transport. In progressive sorting, successive storm or tidal flows carry sediment obliquely down the shoreface and over the deepening shelf floor, through regions of steadily decreasing bottom wave power. Competence decreases in a down-current direction, as a function of the decrease in fluid power. Consequently, coarser particles are less likely to be re-entrained during each transport event, so that the sediment load delivered to successive down-current points is increasingly fine-grained in nature. As relative sea-level rise continues, sequences of graded event beds tend to accumulate. Such sequences are records of successive sorting events. The beds become finer grained in the direction of sediment transport, as their coarser, basal zones pinch out. Episodic progressive sorting thus simultaneously creates both the vertical stratification patterns and the horizontal grain-size gradients that characterize sedimentary facies, and determine the sequence of facies tracts that succeed each other down the transport path. Progressive sorting and strata formation are the main subequilibrium (event scale) processes that underlie regime sedimentation.

## INTRODUCTION

Stratigraphy is a classical branch of the geological sciences. The emphasis has historically been on the 'graphic', or descriptive part of the science. In the years since Steno (1630–1687) first described the principles of superposition, geologists have mapped ('graphed') a good part of the sedimentary rock record. However, the science of stratigraphy is presently changing. It is becoming increasingly apparent that stratigraphic interpretation requires an analytical understanding of the physics of sediment accumulation at small time and space-scales. Stratigraphic interpolation between measured sections, seismic profiles and drill-holes can be reinforced by the application of quantitative techniques based on appropriate representation of dynamic systems.

Geologists are often faced, however, with the incomplete nature of inferences based on the sedimentary record. One aspect of this problem has been stated by Schwartzacher (1975) as follows:

> The measured stratigraphic section is the basis of historical geology and provides the raw material for the reconstruction of physical and biological conditions in the past ... The relationships in such a dynamic [reconstructed] system are always extraordinarily complex and there is no environmental variable which is not, in one way or another, related to some other factor.

We argue in this paper that the physical environment, though characterized by such complex interactions on a short time-scale, achieves a predictable steady-state condition or 'regime' when viewed on a geological time-scale. Depositional style, as interpreted by geologists working at different scales (outcrop to regional seismic scales), is therefore a direct reflection of this regime and the geological history of the forcing factors that change the physical environment.

The regime approach has many historical precedents. Its roots lie in the geomorphological concepts developed around the turn of the century (Gilbert, 1877; Davis, 1899, 1902; Barrell, 1917). It has achieved its most quantitative expression in the regime relationships of hydraulic engineering (e.g Lacey, 1929; Blench, 1957). Though the theory of steady-state sedimentation (regime theory) has been applied extensively to river system and landscape evolution (Melhorn & Flemal, 1975; Chorley *et al.*, 1984), a similar approach has not been adopted widely in the study of shelf sedimentation. This paper presents a preliminary attempt to apply regime theory to the study of shelf sedimentation in order that we may better understand the shallow-marine rock record.

In this section we present a general model for shelf sedimentation. We outline the general theoretical and conceptual groundwork for the regime approach to shelf sedimentation. We discuss the variables that control the geological history of sedimentation and the making of the stratigraphic record. We synthesize existing ideas concerning shelf sedimentation, with particular emphasis on what is known from modern sediment transport studies, and present a general, descriptive model for shelf depositional regimes.

Subsequent sections explore the implications of the general model. Thorne & Swift (this volume, pp. 35–58) are concerned with the application of the regime concept, and show that regime theory, as developed by geomorphologists, allows us to take a more analytical approach to shelf regimes. Two examples of regime sedimentation are then illustrated. The first, from a modern shelf, discusses the supply-dominated regime of inshore areas of the Louisiana prodelta shelf. The second example applies a simplification of the regime concept to the modelling and interpretation of large-scale sequence geometries, as are typically observed on seismic profiles of passive continental margins.

The regime approach, as presented by Thorne & Swift (this volume, pp. 33–58), is ideal for the integration of geological field studies on various scales. Thorne *et al.* (this volume, pp. 59–87) consider the depositional fabric in terms of an analytical approach to stratification and facies identification: examination of shallow marine deposits at subequilibrium time-scales, and the time-scales of the depositional events that create them. In this paper Thorne *et al.* establish parameters of strata formation and consider their interrelationships. Field observations of stratification thickness are integrated into the larger regime context inferred from the sequence architecture of the deposits. Swift *et al.* (this volume, pp. 89–152) examine lithofacies and depositional systems. They apply the principles of strata formation set forth in the previous paper to the problem of facies differentiation, and examine the families of shallow-marine depositional systems (facies arrays) that characterize supply-dominated and accommodation-dominated regimes. In their consideration of parasequences, Swift *et al.* (this volume, pp. 153–187) move up in spatial scale to the field study of depositional sequences, and examine the relationship between depositional systems and high-frequency depositional sequences on one hand, and the relationship between high-frequency and low-frequency sequences on the other. Finally, a regime model for depositional sequences, their component systems tracts, and bounding surfaces is presented by Thorne & Swift (this volume, pp. 189–255). This section demonstrates that the regime model presented here and the depositional sequence model used by stratigraphers are complementary concepts that work well in describing field observations.

## EVOLUTION OF CONTINENTAL SHELVES

In the broadest sense, shelves (continental, insular, or bank shelves) can be defined as shallow marine surfaces of regional extent. They may be floored by basement rock. However, they commonly occur on continental margins or in interior basins. These are subsidence-prone areas, and unless subsidence is very rapid relative to the rate of sediment input, the aggrading water–sediment interface will tend to keep pace with the rate of subsidence, so that most

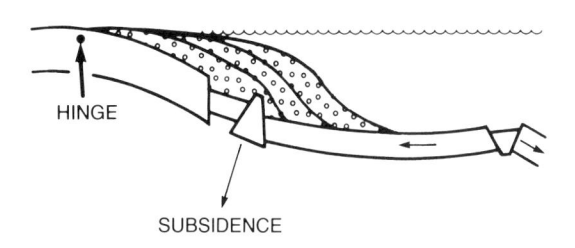

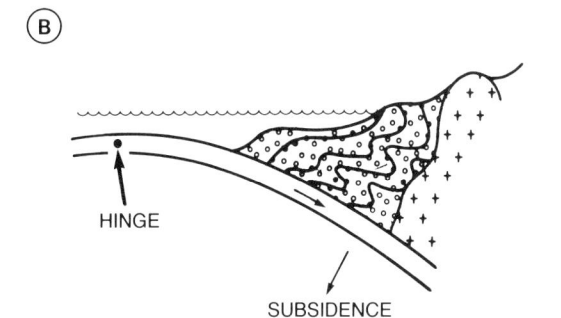

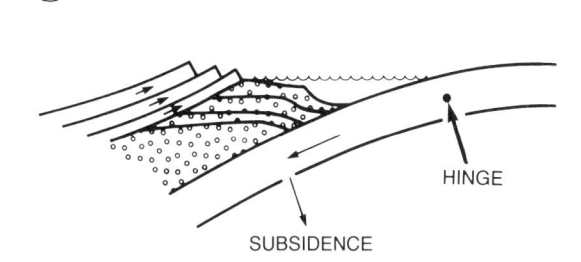

**Fig. 1.** Structural settings of continental shelves. (A) Passive margin (Atlantic continental margin); (B) convergent margin; (C) foreland basin.

mature shelves are underlain by relatively thick shallow-marine sedimentary sequences (Emery, 1968a).

Three basic shelf patterns exist on continental margins. Rifted and thermally subsiding continental margins (passive margins) are hinged on their landward side (Pitman, 1978; Fig. 1A). Sediment is delivered to them by large, integrated drainage systems (Inman & Nordstrom, 1971). On convergent margins, subsidence is the direct result of subduction. Convergent margin shelves tend to be narrow, wave-cut rock terraces, but if the sediment supply is sufficient, a seaward-thickening shelf prism may build seaward over the trench as an accretionary wedge formed by the subduction process (Makran coast, Harms et al., 1982; Fig. 1B). In the case of a foreland basin, subsidence is induced by the loading of the continental crust by successive thrust slices (Jordan, 1982; Fig. 1C). In this case the hinge is on the cratonic side of the basin, while the greatest subsidence is on the orogenic side. Since this is also the side of maximum sediment input, the clastic wedge, accumulating in the basin, may maintain a shelf-slope morphology as it does so (Swift et al., 1987).

The first two cases of shelf evolution result in *pericontinental shelves* on continental margins. However, the dynamics of plate motion frequently result in movements within continental crust, so that back-arc basins, or other kinds of cratonic basins, subside, and epeiric seas form. Most of these are shallow enough to be of shelf depth, resulting in *epicontinental shelves* with simple ramp profiles. In deeper cratonic basins, shelf-slope morphologies may form above neritic or shallow bathyal basin floors. Whether or not the shelf-slope morphology forms, the basic principles of fluid and sediment dynamics and facies formation are the same for pericontinental and epicontinental shelves.

## GEOHISTORICAL VARIABLES

### General

An understanding of the physical processes of modern sedimentation only indirectly aids in the analysis of the ancient rock record. The stratigraphic record reflects two additional dimensions: (1) it is a record of the factors that control the *geological history* of basin sedimentation; and (2) it is record of *preserved* sedimentation. The problem of relative preservation is examined by Thorne et al. (this volume, pp. 64-87) and Swift et al. (this volume, pp. 153-187). In this section we consider the geohistorical factors.

Analyses of continental margin sedimentation patterns have identified a small number of forcing factors as the dominant controls on the style of basin infill (Grabau, 1913; Sloss, 1962; Allen, 1964; Swift, 1976a). Following the notation of Sloss (1962) these variables can be grouped into four categories: the rate of sediment input from the landward source terrane ($Q$) in volume per unit width per unit time; the type of sediment input ($M$), which can be specified by grain-size distribution or median grain size; the rate (and sense) of relative sea-level variation ($R$) in unit length per unit time; and the rate of dispersive sediment transport ($D$) in volume per unit width per unit time. Sediment transport varies across the shelf with variations in fluid power ($P$). These variables can be referred to as *geohistorical variables*, and as governing variables they are distinguished by bold type. They will be described in the international system of units except where quoted directly from the literature.

### Sediment input rate

The first of the geohistorical variables, the sediment input rate ($Q$), has been studied mainly by geomorphologists (e.g. Schumm, 1977). Subaerial denudation rates (mm yr$^{-1}$) increase with increasing relief, and are also a function of climate. The sediment yield, or amount of sediment delivered by the drainage system (m$^3$ m$^{-2}$ yr$^{-1}$) is greatest in regions of moderate rainfall; plant cover suppresses erosion in areas of intense rainfall, while in arid areas there is insufficient discharge for the transport of much sediment (Chorley et al., 1984). The Mississippi, an example of a major river, discharges 2.4 × 10$^{11}$ kg yr$^{-1}$ to the northwestern Gulf of Mexico shelf (Coleman, 1981). The sediment is dispersed across the shelf, and also along the shelf as far southwest as the Mexican border (Holmes, 1982). One can thus compute sediment input per metre of shoreline for the sector supplied by a river. However, many modern shelves are undergoing transgression in response to glacio-eustatic sea-level rise. The mouths of these rivers are flooded by the sea, and the resulting estuaries trap the river sediment. Such shelves receive their sediment instead from shoreface erosion. In these cases, sediment input per metre of shoreline can be estimated by the rate of erosion of the shoreface from a time-series of bathymetric maps.

Delivered grain size ($M$) is a related sediment consideration. The grain size delivered is greatest on convergent margins where relief is greatest, and least on mature passive margins serviced by large, integrated drainage basins (Atlantic margins, Inman & Nordstrom, 1971), which tend to retain the coarser fraction of their received load and bypass only the finer fractions (progressive sorting; Russell, 1939).

**Relative sea-level change**

The third variable, the rate of relative sea-level change ($R$) is measured in mm yr$^{-1}$. It is the net effect of: (1) global (eustatic) changes in sea-level; (2) local or regional (diastrophic) changes in base level owing to tectonic basin subsidence or uplift; and (3) compaction-induced subsidence created by fluid escape.

Global variation in sea-level spans many orders of frequency. For example, two first-order cycles have occurred in the Phanerozoic owing to secular changes in the volume of the ocean basins (Heller & Angevine, 1985). At the other extreme, glacio-eustatic cycles modulated by the 21 000, 40 000 and 100 000-year Milankovich cycles have dominated the Pleistocene record of sea-level change (Hays et al., 1976).

Glacio-eustatic cycles, unlike low-order sea-level cycles, are not strictly synchronous on a global basis. For example, Holocene sea-level curves vary markedly from continent to continent; the record for eastern South America for instance, includes a mid-Holocene high stand (Parker et al., 1982). Part of this variation appears to be the result of differential isostatic effects caused by deformation of the geoid by the loading and unloading of the lithosphere by ice at high latitudes (Peltier, 1980, 1986).

Earlier workers have suggested that Quaternary shelf deposits are anomalous with respect to those of the rock record because shelf facies could not have maintained equilibrium with respect to their formative processes during the rapid glacio-eustatic fluctuations (Curray, 1965). However, as our knowledge of the stratigraphic record has increased, the discrepancies seem less significant. High-frequency sea-level oscillations may have been widespread in time and space. During the Ordovician and Permo-Pennsylvanian episodes of glaciation, sea-level oscillations must have had the amplitude as well as the frequency of Quaternary cycles. High-frequency climatic cycles seem to have persisted through geological time between glacial episodes, resulting in sea-level fluctuations of lower amplitude, or perhaps merely in variations in the sediment input rate (Barron et al., 1985; see Swift et al., this volume, pp. 153–187).

Regional changes in sea-level occur at conservative and convergent plate margins and on continental margins undergoing ice or sediment loading (Chorley et al., 1984). Relative sea-level changes feed back into sediment input, as base-level variations modify rates of denudation. Considerations of the possible relationships between diastrophism and erosion have played a surprisingly small part in the development of geomorphology of the English-speaking world (Chorley et al., 1984, p. 71). These authors note that the common assumption has been the Davisian, or interrupted equilibrium model, in which uplift is discontinuous and short-lived, and initiates a cycle of erosion, during which the landscape passes through a series of transient states of youth, maturity, and old age, as equilibrium is restored. In contrast, the Penckian model sees landscapes as equilibrium responses to slow, prolonged uplift. Chorley et al. point out that Davis' view is broadly supported by the more numerous measurements now available. Comparison between rates of orogeny ($c.7.62$ mm yr$^{-1}$) and maximum regional rates of denudation show that the former are about eight times the latter (Chorley et al., 1984).

Subsidence associated with sediment compaction results from the introduced weight of the sedimentary overburden. When sedimentary compaction proceeds as a direct response to ongoing sedimentation, it does not alter depositional style or affect the depositional topography. If, however, a significant time-lag occurs between the introduction of sedimentary overburden and the resulting compaction, then the time history of compaction will affect depositional style. An example of this second case occurs during foundering of sand bodies on the chenier plains of the Gulf Coast (Gould & McFarlan, 1959).

**Sediment transport**

The fourth variable, the dispersive sediment transport function, $D$, represents the time-averaged transport of sediments as suspended or bed load in response to the fluid power, $P$, of waves, wind-driven and density driven currents, tidal currents, and sediment density flows. The interaction of these depositional agents will vary in space and time across the basin, creating a series of flow-integrated depositional environments. Collectively these can be referred to as the transport system (Swift et al., 1987), or to stress the modification of the grain size that occurs during the transport process, the *dispersal system*.

The fluid power, $P$, expended to resuspend sediment, can be expressed as 'the fluid power of velocity', or the time-averaged product of bottom shear stress and the velocity ($P = \tau \cdot u$ in Newtons m$^{-1}$s$^{-1}$). The fluid power budget is complex and not easy to quantify. Major energy sources are waves, storm currents and tidal currents. Locally, intruding oceanic currents are also important (Curray, 1960; Flemming, 1988). Wave climate statistics are available for many continental margins and can be incorporated into models of continental margin fluid power budgets (for instance, McCave, 1971). Coasts can be classified from high to low wave energy depending on the incidence of waves on the shore (Bird & Schwartz, 1985). The variation of wave energy depends on several contributing factors. Foremost among these factors are those that control the propagation of deep-water oceanic swell into the area. Empirical relationships have been developed to predict the reduction of wave energy owing to refraction, bottom friction, destructive interference, reflection and wave transformation as swell propagates from deep to shallow water (Bretshneider, 1966). Wright & Coleman (1972), using these empirical formulations, demonstrate that the ratio of deep-water to nearshore wave power can approach 10 000:1 for river-dominated deltas such as the Mississippi.

During periods of coastal set-up, storm currents flow along the shelf in geostrophic adjustment to the wind-forced sea-level tilt (Csanady, 1982), and the sea-floor is stirred by high-frequency wave orbital motions. Sediment transport during these events is proportional to the shear stress caused on the sea-floor by the interaction of the wave orbital velocity field with the mean shelf current (Grant & Madsen, 1979). The intensity of wave-induced fluid motions decreases rapidly away from the water surface, at a rate depending on the period of the wave. This rapid decrease has led many authors to refer to the concept of a time-averaged 'wave base', below which the net effects of waves on sediment transport are minimal. On margins affected by long-period high-amplitude swell this 'wave-base' may be over 100 m of water depth (Vincent, 1986). Storm-dominated shelf regimes occur on about 80% of the world's continental margins. Mid-latitude storms (extra-tropical storms) have weaker winds than do low-latitude hurricanes, but are much wider, last longer, and are more frequent (Swift & Nummedal, 1987). However, the most striking contrasts are between storm-dominated shelves and tide-dominated shelves.

Tidal currents on continental margins are readily modelled and predicted (A.C. Davies, 1983). Tide-dominated shelf regimes occur on shelves with mid-tide tidal current velocities in excess of 25 cm s$^{-1}$ (Swift *et al.*, 1981) and account for about 15% of modern continental margins. These shelves are bordered by coasts whose tides display the high mesotidal or macrotidal ranges of J.L. Davies (1964). Mid-latitude shelves experience several important storms a year (Vincent *et al.*, 1981), but tide-dominated shelves with macrotidal ranges experience the equivalent of a storm twice a day.

The sediment transport rate, $D$, can be expressed in kg m$^{-1}$s$^{-1}$, or kilograms per metre transverse to the transport path per second. The rate of accumulation, $\bar{A}$, a related value, is expressed in mm yr$^{-1}$. These variables reach maximum values near the mouths of such major rivers as the Yellow River (China), where Wright *et al.* (1988) have reported density underflows into the low-salinity Bohai Sea with concentrations of 25 kg m$^{-3}$, moving seaward at 1.5 m s$^{-1}$. Transport rates here are of the order of 35 kg m$^{-1}$s$^{-1}$. Accumulation rates near river mouths are proportionately high; McKee (1983) reports short-term accumulation rates of 4.4 cm per month at the mouth of the Yangtze River, based on $^{234}$Th profiles. Accumulation rates based on $^{210}$Pb yield 100-year rates that are an order of magnitude less (about 5.4 cm yr$^{-1}$). Similarly, Adams *et al.* (1988), have computed accumulations of the order of a metre in 60 m of water during a single flood season, 15 km seaward of Southwest Pass on the Louisiana prodelta shelf, on the basis of transmissometer records.

As a shelf edge delta, the Mississippi Delta has developed sufficiently steep prodelta slopes that gravity transport of sediment has become significant (Coleman, 1988). Most movement occurs as submarine landslides. Redeposition by sediment density flows may occur on the prodelta slope in response to fluidization of underconsolidated delta-front muds by storm and earthquake activity, but there is in general little evidence for density driven underflows on the low gradient surfaces of most modern shelves.

The distal prodelta shelves, tens to hundreds of kilometres down-current from major river mouths, tend to accumulate muddy sediments at lower rates, as indicated by $^{210}$Pb analysis. Maldonado *et al.*

(1983) have reported accumulation rates of the order of 1.0–0.1 mm yr$^{-1}$, 100 km down-shelf from the Ebro Delta on the Spanish Mediterranean shelf. Similar rates prevail on the Texas shelf (Holmes, 1982), and on the distal Amazon shelf (Kuehl et al., 1982), but the prograding scarp that defines the front of the mud deposit on the outer Amazon shelf is accumulating at 10 cm yr$^{-1}$.

Shelves beyond the influence of major rivers receive their sediment primarily from shoreface erosion and are primarily sandy in nature. Vincent et al. (1981) have computed transport rates of $10^5$–$10^4$ gm cm$^{-1}$ yr$^{-1}$ on the inner Atlantic shelf, and $10^3$–$10^2$ gm cm$^{-1}$ yr$^{-1}$ on the outer Atlantic shelf. Most of the transport occurs in one or two major storms a year. Net accumulation occurs primarily on the inner shelf, where radiocarbon dates suggest accumulation rates of 2 mm yr$^{-1}$ over 1000 years (Swift et al., 1972).

## CONCEPT OF THE EQUILIBRIUM SHELF

### Evolution of the shelf surface

Although the shelf surface is subjected to complexly varying hydraulic conditions over short time periods, it can be viewed as a surface of dynamic equilibrium controlled by the stratigraphic process variables described in the preceding section (Watts & Thorne, 1984). The relationship of the rate of sediment input ($Q$) to the fluid power ($P$) available to remove it ($Q/P$) is particularly important. If $P$ is computed as the fluid power of velocity (Newtons m$^{-1}$ s$^{-1}$), and sediment input is computed in terms of the rate of application of force needed to provide it per unit length shoreline (also Newtons m$^{-1}$ s$^{-1}$), then $Q/P$ becomes a dimensionless ratio, which we will describe as the *sediment supply index*.

Numerical experiments (Harbaugh & Bonham-Carter, 1977) have shown that sediment accumulating on the continental margin builds up toward 'wave base', the depth to which, on a time-averaged basis, there is sufficient fluid power ($P$) to transport sediment, at a given rate of sediment input ($Q$) (Gulliver, 1899, pp. 176–177; Moore & Curray, 1964). The surface is inclined seaward, relatively steeply on the inner shelf (nearly 1°) and more gently on the outer shelf (a fraction of a degree). The seaward-deepening profile is in part an adjustment to the progressive landward loss of energy from waves to the bottom as the waves propagate landward. However, on a constructional shelf, it is mainly a response to supply; the equilibrium depth near land is shallower because the sediment input there is higher, and the bottom must become shallower before wave energy is sufficient for sediment bypassing to equal input (Wright & Coleman, 1972).

As wave base is approached by an aggrading sediment prism, the ratio of sediment deposited to sediment bypassed decreases, until, ideally, a surface of non-deposition results (continental shelf). During this process, the locus of deposition shifts seaward to the forward face of the sediment pile, to form a secondary surface dominated by gravity processes (continental slope). As this morphology evolves, the typical clinoform pattern of time surfaces develops inside the growing sediment pile (Fig. 2).

In a more realistic model, the continental margin sediment prism accumulates not *after* the margin subsides, but *as* it subsides. In this case, if the sediment supply is sufficient, the shelf surface will continuously aggrade so as to maintain the 'wave base' appropriate to the existing fluid power level and rate of sediment input. The ratio of deposition to bypassing remains fixed at a value appropriate to achieve this result. It does not change until one

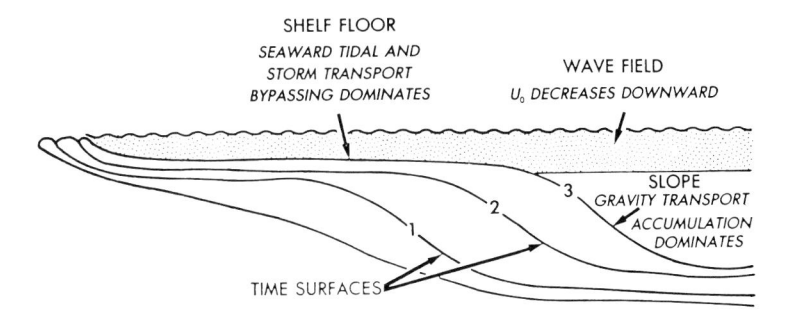

Fig. 2. Geometry of a growing sediment prism on the continental margin, with numbered time lines indicating successive positions of the depositional surface.

**(A) EQUILIBRIUM DEFINITION SKETCH**

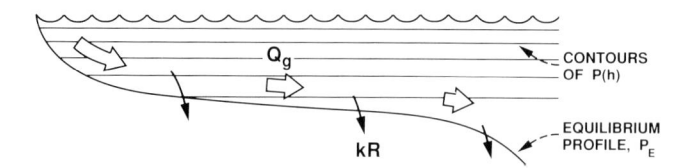

$Q_g$
CONTOURS OF P(h)
EQUILIBRIUM PROFILE, $P_E$
kR

**(B) DISTRIBUTION OF VARIABLES**

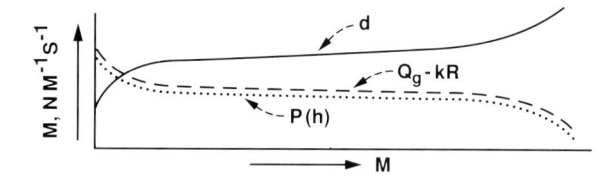

M, N M$^{-1}$ s$^{-1}$
d
$Q_g$ - kR
P(h)
M

**(C) INCREASED Q; SHELF SHALLOWS**

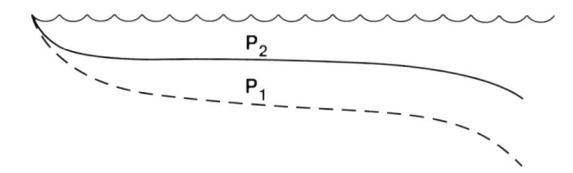

$P_2$
$P_1$

**(D) INCREASED P; SHELF DEEPENS**

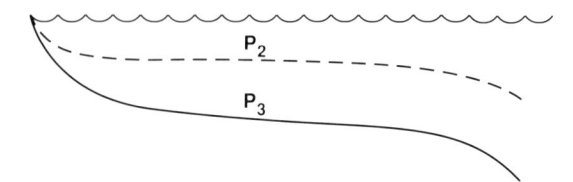

$P_2$
$P_3$

**Fig. 3.** The shelf as an equilibrium surface. See text for explanation.

or more of the geohistorical variables change in value, so that a new equilibrium surface must be constructed.

In this manner, continental margins evolve from diverse structural origins (as divergent, convergent, or conservative margins) toward a characteristic shelf-slope configuration (Fig. 2). During the evolutionary process, shifts in the values of geohistorical variables occur with relative frequency, so that mature shelves tend to consist of multiple depositional sequences, rather than of a single clinoform sequence. Changes in sedimentation thus induced are *not* examples of disequilibrium, but are rather homeostatic responses of the depositional regime, in which the equilibrium is maintained through mutual adjustments of the governing variables.

### Regime parameters and equilibrium

Regime behaviour is described conveniently in terms of parametric ratios. For instance, we can define, for a point on an evolving shelf surface, a dimensionless *equilibrium index*, $E(d)$ as an elaboration of the sediment supply index, $Q/P$. Because changes in sea-level, $R$, or sediment input, $Q$, cause the equilibrium surface to translate landward or seaward, as well as rise or fall, the equilibrium is best defined in terms of distance from shore, $d$, rather than in terms of absolute geodetic coordinates:

$$E(d) = \frac{Q_g - k\mathbf{R}}{\mathbf{P}_g(h)} \quad (1)$$

Where $\mathbf{Q}_g$ is sediment input of grain size $g$ in kilograms, in excess of the amount $k\mathbf{R}$ needed to keep pace with sea-level rise; the constant $k$ is a dimensional constant of conversion relating length to mass; $\mathbf{P}_g(h)$ is the fluid power expended to disperse sediment of that grain size at the depth $(h)$ associated with that point. Thus the equilibrium index is the ratio of the excess sediment input to the available fluid power. Equilibrium is achieved when for all points on the shelf, $E(d) = 1$.

Several important characteristics of regime sedimentation can be illustrated by equation (1) (see Fig. 3). Because of the homeostatic (self-adjusting) nature of the equilibrium, the equation defines not a single surface, but a family of surfaces. For instance, should $Q$ increase (Fig. 3C), the increment of sediment input increase, $\Delta Q$, will cause the surface to aggrade from depth $h_1$ to a new and shallower depth $h_2$, higher in the wave orbital velocity field, where a higher fluid power level, $P(h_2)$, is competent to bypass the enlarged surplus, $Q + \Delta Q - k\mathbf{R}$. The index $E(d)$ is then once more equal to 1. Similarly, if $Q$ is reduced or if $P$ is increased (Fig. 3D), bypassing will increase. The sea-floor will be eroded until the depth is increased and the fluid power level diminished to match the reduced surplus $Q - (\Delta Q + k\mathbf{R})$, so that $E(d)$ is once more equal to 1. The regime is 'locked on' to equilibrium. However, the equilibrium will fail if sediment input is less than sufficient to keep pace with subsidence, $Q - k\mathbf{R} < 0$. Models describing classes of equilibrium surfaces are presented by Thorne & Swift (this volume, pp. 38–39).

If we wish to examine the equilibrium in more detail, we consider a unit volume of the shelf water column and its associated unit area of sea-floor. An accumulation rate, $\mathring{A}$, can be defined in this space as the difference between the sediment input and sediment bypassed (the circle over the $A$ indicates a rate). In order to do so, we must consider the

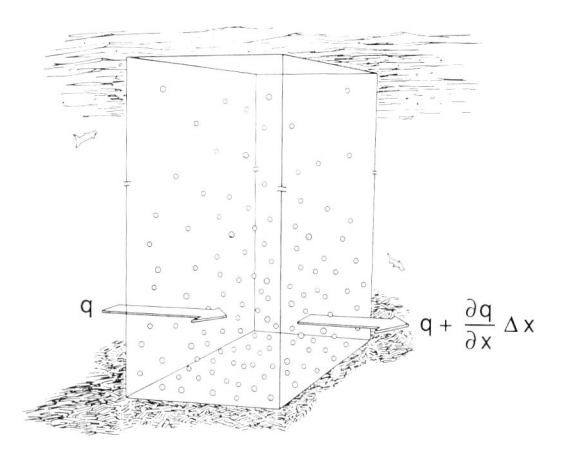

**Fig. 4.** Sketch of unit volume of shelf water column, illustrating sediment continuity equation. (After Niedoroda *et al.*, 1989.)

sediment continuity equation, which governs all aspects of sedimentation and erosion (Fig. 4). In the sediment continuity equation, the change in sea-floor elevation, $\Delta\eta$, is expressed as a function of the sum of time and space-dependent losses and gains of sediment. Figure 4 presents a case of two-dimensional transport. Sediment in suspension enters one side of the unit volume at a discharge $q$. The discharge leaves the box at the same rate, $q$, plus the product of the horizontal discharge gradient, $\partial q/\partial x$, and the width of the box, $\Delta x$. If the gradient is negative (sediment is falling out of suspension on to the floor of the box), then the added term, $(\partial q/\partial x)\Delta x$, is negative, and the exiting discharge, $q + (\partial q/\partial x)\Delta x$, is less than the entering discharge, $q$. Thus the rate of change of elevation of the sea-floor, $\partial\eta/\partial t$, can be related to the spatial discharge gradient, $\partial q/\partial x$. The temporal term is defined as the rate of change of sediment concentration in the water column $\partial C/\partial t$, as sediment is entrained into, or deposited from, the transported load. The two terms must be multiplied by a dimensional constant, $1/e$, so that the discharge in kg m$^{-1}$ s$^{-1}$ can be equated to $\partial\eta/\partial t$ in m s$^{-1}$:

$$\frac{\partial\eta}{\partial t} = \frac{1}{e}\left(\frac{\partial C}{\partial t} + \frac{\partial q}{\partial x}\right) \quad (2)$$

When considering regime sedimentation, the temporal term can be dropped (it will be used later to describe strata-forming processes at sub-equilibrium scales). We can substitute the parameter $\mathring{A}$ for $\partial\eta/\partial t$ and reduce equation (2) to:

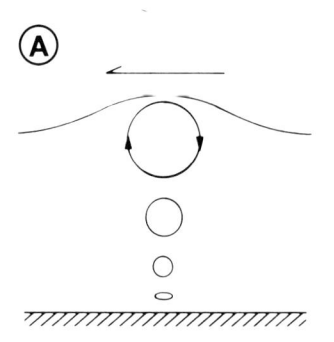

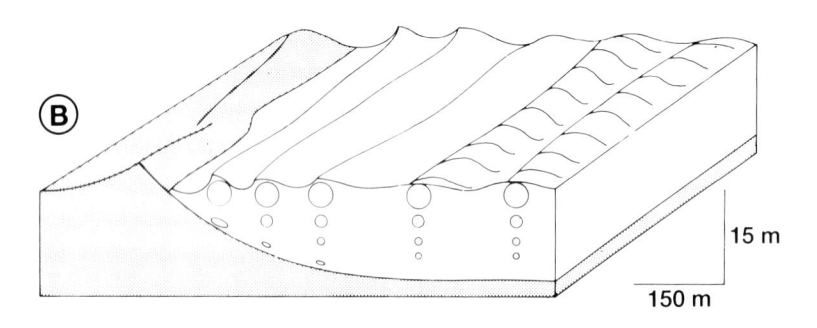

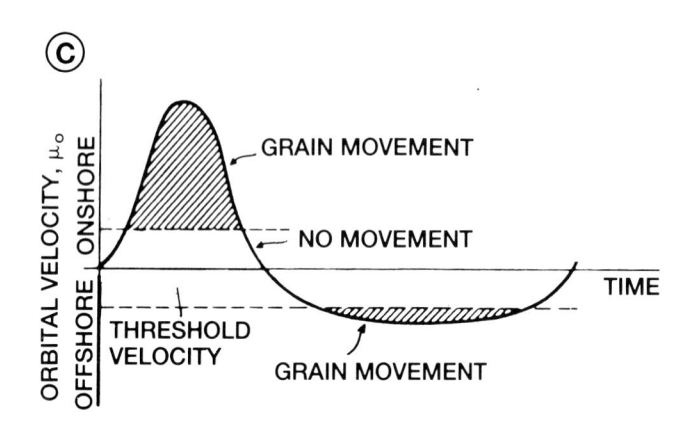

**Fig. 5.** The littoral energy fence. (A) As a wind-induced wave travels over the sea surface, water parcels within the wave travel in circular orbits. Orbital diameter decreases with depth, and becomes a to-and-fro motion near the bottom. (B) As waves enter shallow water, their orbitals are distorted by the frictional drag of the bottom. (C) Time–velocity record of bottom flow during the passage of a shoaling wave. The landward stroke as the crest passes is a higher velocity flow and carries more sediment than does the return stroke associated with the passage of the trough.

$$\mathring{A} = -\frac{1}{e}\left(\frac{\partial q}{\partial x}\right) \quad (3)$$

Because $q$, the sediment discharge rate, varies with grain size, so does $\mathring{A}$, the accumulation rate, and its complementary parameter, and equation (3) can be used to describe facies differentiation. Facies concepts are discussed in more detail in Part IV (Swift et al., this volume).

## A GENERAL MODEL FOR CONTINENTAL SHELF SEDIMENTATION

### Earlier ideas

Sand deposits occur in modern continental shelves in a variety of settings. The variety is such that Francis Shepard (1932), who first considered the problem, concluded that their distribution did not reflect the modern sedimentary regime at all, but was rather a relict distribution, a response to processes acting during glacially lowered sea level. The concept of relict sediments on continental shelves was given its fullest expression by K.O. Emery (1952, 1968b). However, recent work on modern shelf sediments has shown that such transgressive deposits *are* responding systematically to the modern hydraulic regime (Swift *et al.*, 1986a,b) and that, in addition, many shelves near major river mouths have reverted to regressive regimes (Nittrouer *et al.*, 1986). Meanwhile, seismic records of ancient sedimentary deposits have begun to be interpreted in terms of a model that sees sedimentary sequences as responses to sea-level fluctuation, sediment input, and subsidence (Vail, 1988). In this section, we will consolidate these earlier concepts into a general model of shelf sediment dispersal and facies differentiation.

### The littoral energy fence

Regime parameters and their relationships are important gauges of shelf behaviour. However, in order to develop a general model of shelf sedimentation, we need to look at the machinery itself. Grabau (1913), Sloss (1963) and Curray (1964) have stressed the response of the shoreline to sea-level changes as a key to the nature of the attendant depositional regime, and we follow these authors, expanding our model to include shoreline behaviour. In order to understand this behaviour, it is necessary to look at sediment dynamics at smaller time and space-scales, especially on the inner shelf and near the shoreline. This region has been described by J.R.L. Allen (1970, p. 169) as the '*littoral energy fence*'. Shoaling and breaking waves on oceanic coasts create a mean landward-directed bottom stress, which tends to drive sand towards the shore (Fig. 5). The coastal fine-sediment budget is more complex, but a variety of hydraulic mechanisms work together to move suspended fine sediment landward, against the concentration gradient (Postma, 1967; McCave, 1972; Swift *et al.*, 1985).

There are two ways that sediment can escape from the land to the shelf through the littoral energy fence (Fig. 6). During regressions or slow transgressions, *river mouth bypassing* transports sediment to the shoreface via the flood stage jet of a river mouth (Swift, 1976a). It is then transported along the coast and offshore by coastal storm currents. During transgressions, however, the entire shoreface may become a direct sediment source as a consequence of erosional shoreface retreat (*shoreface bypassing*).

### Dynamics of river mouth bypassing

River mouths have two fundamental geomorphological configurations. These geomorphological configurations are direct reflections of the relative values of the geohistorical variables $Q$, $M$, $D$ and $R$. Rivers that deliver more sediment to their mouths (a function of $Q \cdot M$) than rising sea-level ($R$) can accommodate or marine processes ($D$) can remove are, in geodynamic terms, *deltaic*. We can express this condition in terms of an accommodation–supply relationship $Q \cdot M > R \cdot D$. Such an inequality is not quantitative and is intended to indicate the qualitative sense of change owing to an increase or decrease of each variable. A similar heuristic expression was derived by Lane (1957) in his work on river morphology. In geomorphological terms, deltas are constructional landforms at river mouths that generally cause the shoreline to bulge seaward.

Rivers that deliver less sediment to their mouths than rising sea-level can accommodate or marine processes can disperse ($Q \cdot M < R \cdot D$) are, in morphodynamic terms, *estuarine*. Estuaries are ocean-flooded river mouths that may be either erosional or depositional in nature. In hydrodynamic terms, they are the portion of the river intruded by the salt wedge, and usually, by the oceanic tide (Pritchard, 1967). When the inequality is large ($Q \cdot M \ll R \cdot D$), then the estuary shifts landward much faster than it can fill. The submerging river valley is often dendritic in plan view, like a newly dammed reservoir (for instance, Chesapeake Bay). However, when the inequality is small ($Q \cdot M \leq R \cdot D$), then the estuary can fill the valley largely with estuarine sediment as it retreats up the valley's thalweg. In the limiting case, the area of each cross-section of the estuary closes down to an area just sufficient to service the tidal prism upstream of the cross-section (Swift, 1976a). The aggrading channel tends to be trumpet-shaped, with a flaring mouth, and a straight or meandering stem (Leopold *et al.*, 1964, p. 273). River mouth bypassing occurs mainly through regressive, deltaic coasts, although under conditions of very slow transgression or high tidal range, equilibrium estuaries may be able to bypass significant amounts of sediment.

River mouth bypassing begins with river flooding

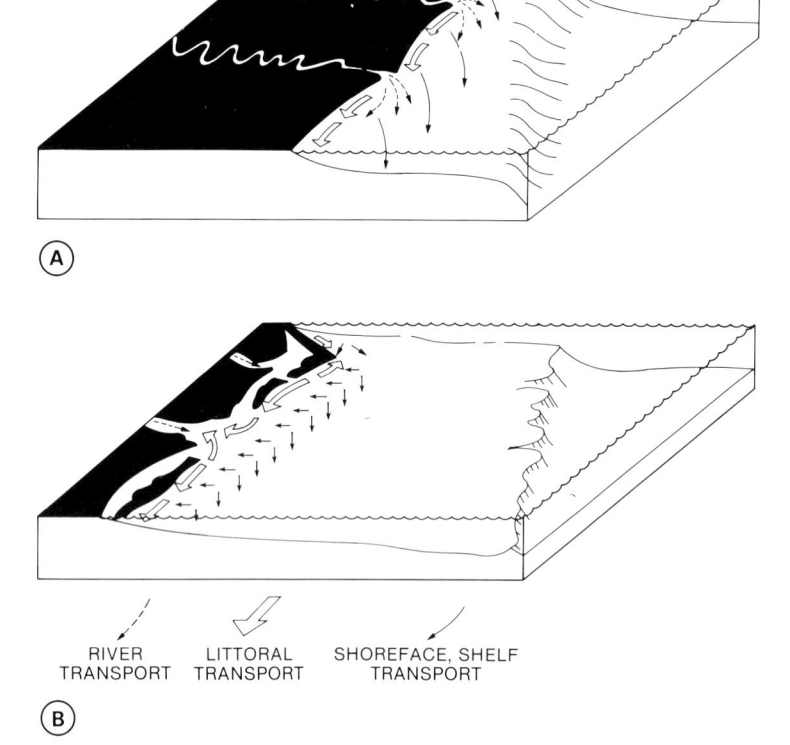

**Fig. 6.** Categories of coastal bypassing. (A) River mouth bypassing. Flood stage jets bypass fine sand into the littoral drift, and fine sand, silt and clay on to the inner shelf. (B) Shoreface bypassing. Downwelling, along-coast storm currents erode sand from the shoreface and bypass it on to the shelf floor.

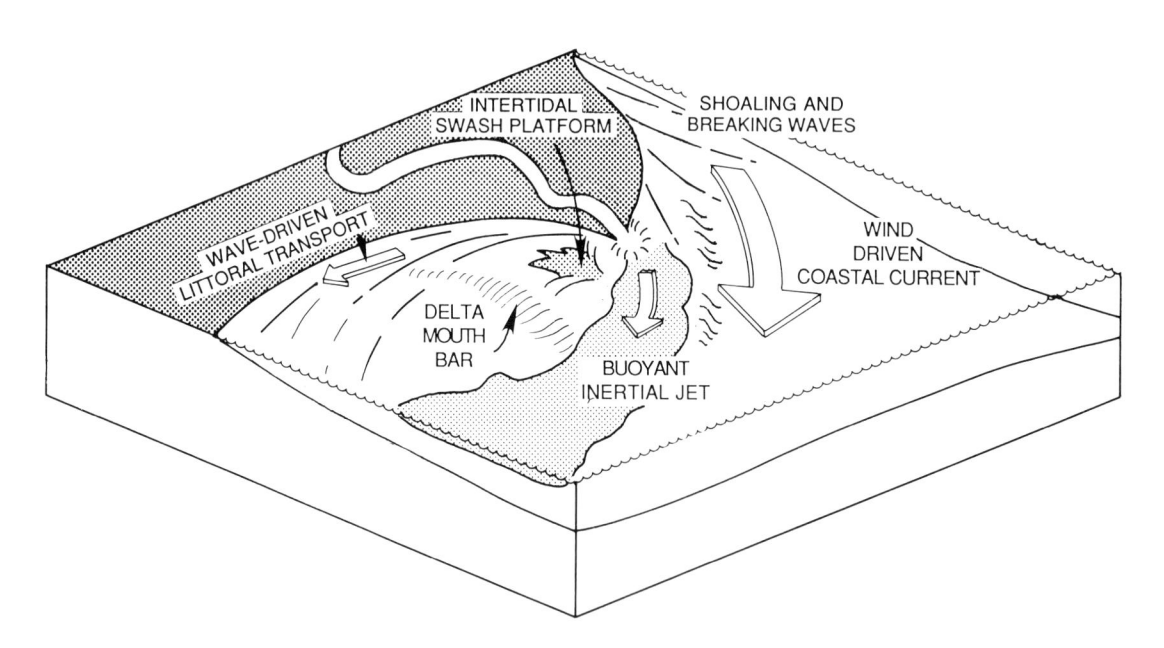

**Fig. 7.** Dynamics of proximal river mouth bypassing. Most of the flood-delivered sand is stored in the delta mouth bar, and slowly re-entrained in the wave-driven littoral sand stream. Fine sand, very fine sand, silt, and clay are carried in the flood stage river-mouth jet, and rained out on the inner shelf floor. (After Wright, 1977, and Wright & Coleman, 1974.)

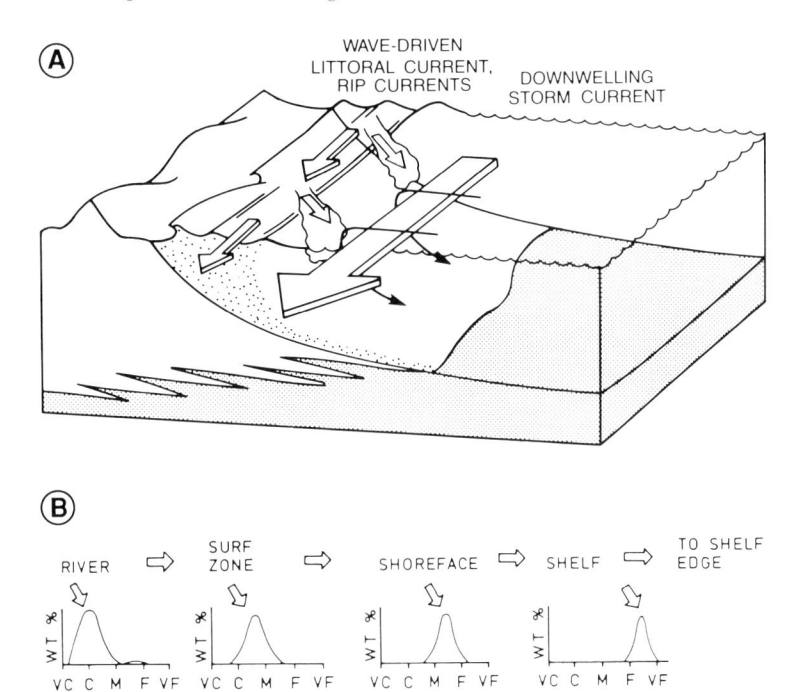

**Fig. 8.** Distal river mouth bypassing. During storms, rip currents form in the surf zone and carry suspended sand out to the landward edge of the main shelf current. The shelf flow is downwelling, and sand is swept obliquely down the shoreface. (A) Block diagram illustrating basic dynamic mechanisms; (B) schematic diagram of the sediment dispersal system illustrating the downstream shift in mean diameter owing to progressive sorting. (After Swift *et al.*, 1985.)

(Wright & Coleman, 1974; Fig. 7). Sand deposited by this process on the river mouth bar is re-entrained by the wave-driven littoral current in the surf zone. In this narrow zone (only several hundred metres wide), forces associated with shoaling and breaking waves propel a powerful along-shore current (Komar, 1983). Sand moves in the littoral current almost continuously, although the bulk of the transport is accomplished by storms. In the years following the flood, along-shore transport by wave-driven currents in the surf zone redistributes the sand into spits and strandplains that flank the delta (Galloway & Hobday, 1983).

During the more intense storms, a mechanism for the offshore transport of sand becomes operative. In the Northern Hemisphere, when the wind blows with the coast on its right, surface water dragged along by the wind is deflected landward by the Earth's rotation (Csanady, 1982). As shelf water piles up against the coast, the tilt in the sea surface causes a seaward-directed pressure gradient and drives a storm current. In the current, the pressure gradient force is balanced by the Coriolis force (geostrophic balance), so that water moves along-shelf, parallel to the contours of the tilted sea surface. Near the shoreline, however, the head of excess pressure drives downwelling, and induces an offshore component of bottom flow. Offshore veering also occurs at the bottom of the main geostrophic flow further out on the shelf, owing to frictional damping. Taken together, the vertical and horizontal components of storm flow in response to coastal set-up cause surface water to move obliquely landward, to downwell, and then to move obliquely seaward across the bottom (Swift *et al.*, 1985).

This mechanism is effective for offshore sediment transport because during peak storm flow the wave-driven littoral current of the breaker zone interacts with the main shelf current. Rip currents flow seaward from the breaker zone, and dump vast quantities of sand into the downwelling coastal margin of the shelf current (Fig. 8). Sand is swept obliquely down the shoreface on to the shelf floor (Swift *et al.*, 1985). If more sand is supplied to the littoral current than is carried on to the shelf floor, then the difference is deposited on the shoreface, which consequently progrades (Niedoroda *et al.*, 1985).

### Dynamics of shoreface bypassing

When relative sea-level rises rapidly with respect to the rate of sediment input, the coastal sediment budget described in the previous section is run in

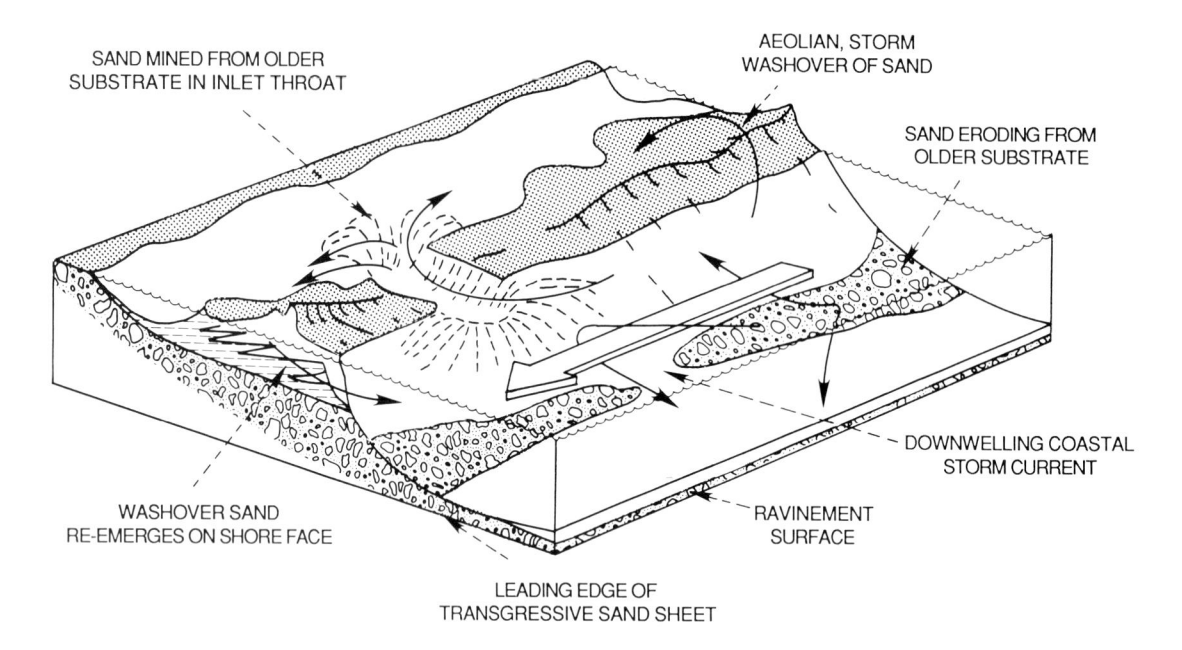

**Fig. 9.** Shoreface bypassing by erosional shoreface retreat. Rising sea-level causes an upward and landward translation of the shoreface profile. The barrier top cycles in tank-tread fashion by storm washover, burial, and re-emergence on the shoreface. Sand is eroded from the shoreface by downwelling along-coast storm currents. Storm-eroded sand is swept over the barrier and into the inlet. It also moves seaward to join the leading edges of the transgressive shelf sand sheet. (After Swift *et al.*, 1985, and Niedoroda *et al.*, 1985.)

reverse. River mouths become estuaries that not only trap river sediment but also sand from the littoral current of the coast on either side of the estuary (Meade, 1969). The beach, nourished with sand from the wave-driven littoral current, can grow upwards as sea-level rises, but the zone behind the dunes is sediment-starved and becomes flooded. The main oceanic shoreline detaches in this fashion and becomes a barrier island, backed by a lagoon (Hoyt, 1967). The shorefaces between estuaries are also sediment starved and can no longer build out faster than the downwelling storm currents can erode them. They undergo erosional retreat, and the resulting debris is swept out onto the shelf floor, or is transported along coast in the breaker zone into adjacent estuaries. Sand is also withdrawn from the along-coast littoral flux, to build back-barrier tidal deltas for storm-breached inlets, is swept over the barrier on to washover fans, or is blown on to the back-barrier flats by strong onshore winds (Fig. 9). When a back-barrier flat is built in this manner, the barrier rolls over on it (Leatherman, 1983), and coastal retreat resumes. In this scheme, the cycling barrier serves as temporary storage for sand during its journey out of the substrate and on to the shelf floor. Under conditions of rapid subsidence and high sediment input, the shoreface retreat process may re-entrain fluvial and estuarine deposits of the same sedimentary cycle (e.g. Mesaverde Sandstone of the Western interior; Swift *et al.*, 1987). Under conditions of slower subsidence, however, the shoreface may also be incised into, and may release deposits of a previous cycle (e.g. the Atlantic shelf of North America; Niedoroda *et al.*, 1985).

The geometry of shoreface translation in response to rising sea-level was first described by Bruun (1962). Bruun argued that if the coastal profile was accepted as an equilibrium response of the sea-floor to the hydraulic climate, the effect of sea-level rise must be a landward and upward translation of the profile. The geometric consequences of this conclusion are quite rigid. The shoreface must undergo erosional retreat, while accommodation space is created for the resulting debris beneath the rising seaward limb of the profile. Bruun (1962) suggested that if there was no along-shore gradient

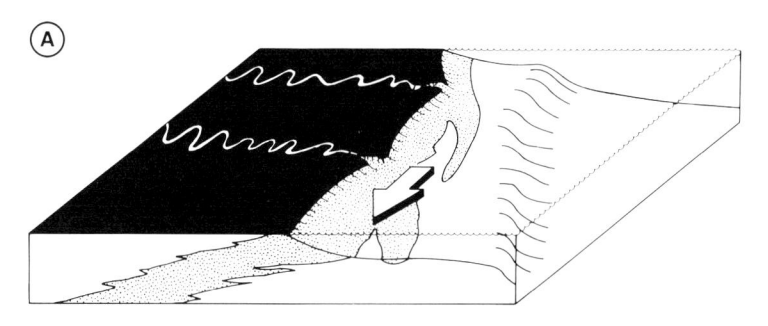

**Fig. 10.** Shelf sedimentary regimes. (A) Supply-dominated regime. River mouth bypassing releases mobile fine sediment that spreads rapidly to all parts of the shelf. (B) Accommodation-dominated regime. Shoreface bypassing leaves a thin, coarse lag deposit over ravinement surface, created by erosional shoreface retreat. The sediment-starved surface seaward of the maximum seaward position of the shoreline undergoes marine erosion. Sand bypassed in the flood stage jets of river mouths during the previous sedimentary cycle is winnowed out and swept into sand bodies. (From Swift *et al.*, 1987.)

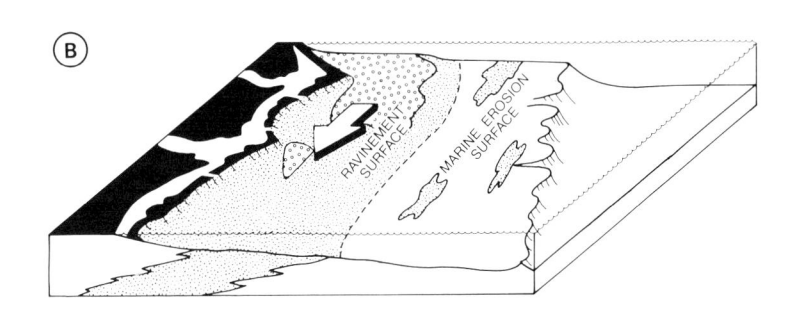

of sand discharge in the littoral current, then there should be an equal volume transfer of sediment from the eroding shoreface to the adjacent shelf floor. The erosional surface cut by the retreat process is the *ravinement surface* of Stamp, 1921 (Fig. 10B).

The two-dimensional Bruun model has been expanded recently to account for longshore discharge gradients and back-barrier sand storage (Dean & Maurmeyer, 1983; Everts, 1987; Pilkey & Davis, 1987). The middle Atlantic coast is responding in this fashion to a sea-level rise of $1\,\text{cm yr}^{-1}$ by retreating at a mean rate of $80\,\text{cm}-\text{yr}^{-1}$ (May *et al.*, 1982). The Bruun model is analysed in terms of regime sedimentation by Thorne & Swift (this volume, pp. 38–39).

## Shelf depositional regimes

River mouth and shoreface bypassing are thus two modes by which sediment can cross the coastal zone. The selection of one or the other of these modes results from the values assumed by the geohistorical variables listed above. Since the coastline is the gateway through which sediment must pass, coastal bypassing mechanisms control the depositional regime on the shelf. Two basic regimes are possible (Swift, 1976b; Fig. 10). River mouth bypassing results in allochthonous sedimentation (allochthonous: of far-travelled origin; Naumann, 1858) and an 'open' shoreline, through which large volumes of fine, mobile, river-derived sediment spread rapidly to all parts of the shelf. Allochthonous settings are characterized by a supply-dominated configuration of the values of the accommodation–supply inequality, $Q \cdot M > R \cdot D$. Thus regimes in allochthonous settings are *supply-dominated regimes*. The accumulation rate, $\dot{A}$, is high. There is consequently little opportunity for reworking, differential bypassing and facies differentiation. Textural gradients are gentle, and facies are weakly defined. Deposits are thick, fine-grained and homogeneous. Following convention, we refer to these deposits as regressive depositional systems. Regressive systems are considered in more detail by Swift *et al.* (this volume, pp. 107–118.)

In contrast, shoreface bypassing results in autochthonous sedimentation (autochthonous: of *in situ* origin; Naumann, 1858). Autochthonous settings are characterized by an accommodation-dominated configuration of the accommodation–supply inequality, $Q \cdot M < R \cdot D$, which creates a transgressive, 'closed' shoreline. In the resulting *accommodation-dominated regime*, the accumulation rate, $\dot{A}$, is low. Sediment is repeatedly resuspended before final burial, and differential bypassing is high. Textural gradients are steep, and facies are strongly defined. The resulting transgressive deposits are thin, coarse-grained and heterogeneous. Transgressive depositional systems are considered in more detail by Swift *et al.* (this volume, pp. 118–143).

Note that in both supply-dominated and accommodation-dominated regimes, the shoreline configuration serves as a 'flip-flop mechanism' that enhances the trend of the accommodation–supply ratio, $Q \cdot M : R \cdot D$, towards either high or low values. A reduction in the supply term results in a transgressive (barrier–estuarine) shoreline and a yet lower value of the ratio. An increase in the supply

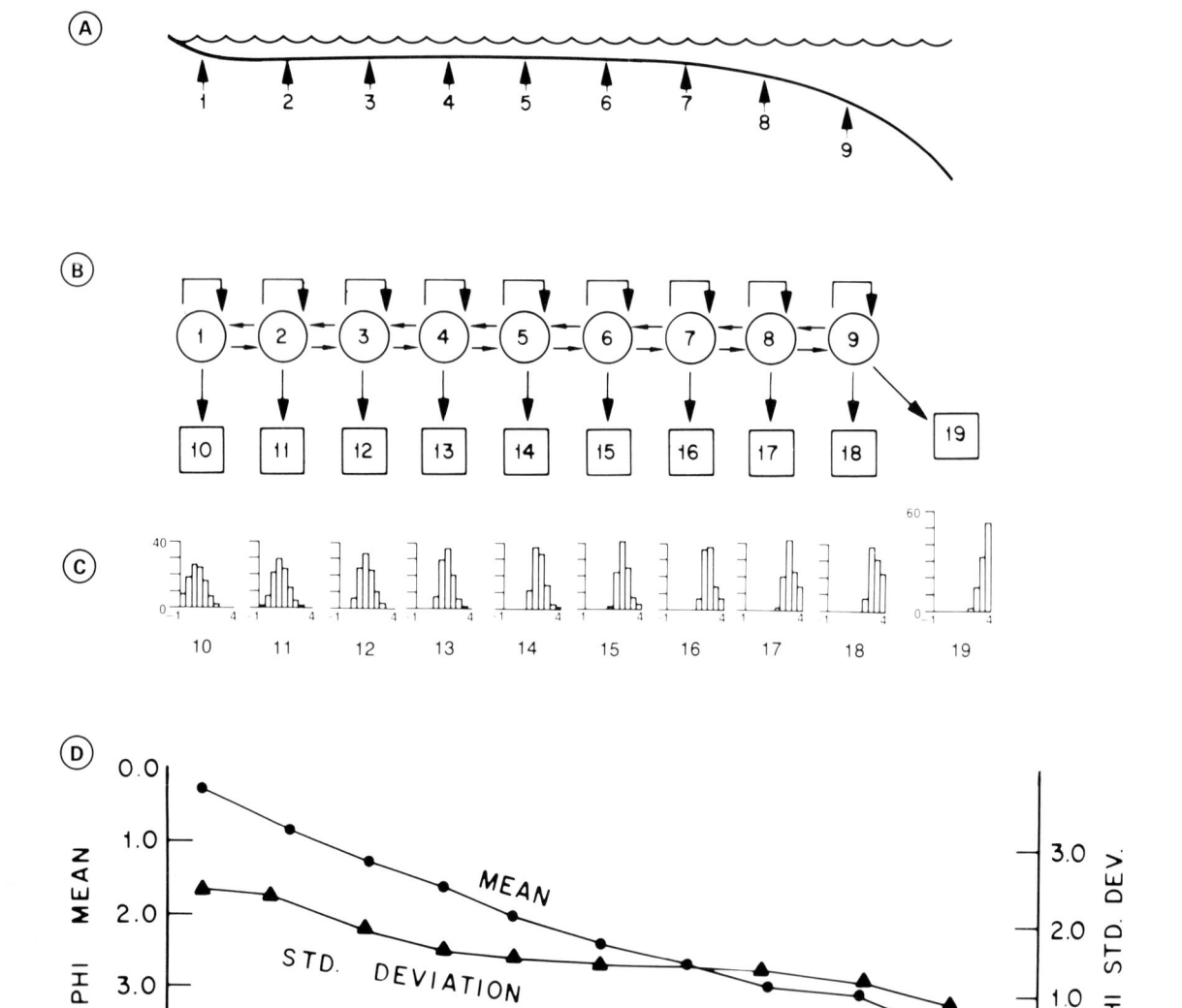

**Fig. 11.** Markov process model for progressive sorting. (A) Shelf cross-sections with nine stations. (B) Markov chain, showing transition states (circles), trapping states (squares), and transition pathways. (C) Limiting grain-size frequency distributions for trapping states. (D) Computed variation in mean and standard deviation. (From Swift *et al.*, 1972b.)

term results in a regressive (deltaic) shoreline and a yet higher value.

## SHELF FACIES PATTERNS

### Progressive sorting

Shelf facies, modified in this fashion by the regime variables, are best understood as the result of *progressive sorting* (Russell, 1939; Swift *et al.*, 1972b), operating through time on a *sediment dispersal system*, which we define as an assemblage of flow-integrated depositional environments (Swift *et al.*, 1987). A dispersal system can be modelled as consisting of three classes of elements; sources, reservoirs and sinks. The elements are connected by a transport pathway, along which operates a source to sink gradient of decreasing competence. In a shelf dispersal system, sediment moves down-shelf and offshore, in response to intermittent storm and tidal currents. The competence of the currents (the maximum grain size that they can carry) decreases in a seaward direction as the water deepens and bottom wave stresses weaken. As a result, the sediment load delivered to successive downstream points on the shelf is increasingly fine-grained in nature (Fig. 11). Progressive sorting is not just a matter of the fine particles outrunning the coarsest particles; it is instead a large-scale diffusion mechanism, and is dependent on the episodic nature of the transport. At the end of each transport step (river flood, shelf storm flow, etc.) the sediment load is redeposited. Each time that an event is followed by a less energetic event, the re-entrainment is only partial, and the coarsest portion of the load is subtracted permanently out of the transport system.

The progressive sorting process is a complex one, and is poorly understood. As the current intensifies over a depositional surface of mixed grain sizes, entrainment is biased towards coarser grains, which project further into the flow. Finer grains are thus intrinsically not as 'entrainable' in a multiple size class aggregate as they would be in single size-class aggregates because they experience shielding (Bridge, 1981; Mclaren & Bowles, 1985). On the other hand, the suspension load domain occupies a greater area of the flow cross-section than the bedload domain; in this sense, finer particles are more 'transportable'. Also the finer grains were the last to be deposited in the previous event, and are more abundant on the surface than they are within the body of the deposit. They have a higher probability of being re-entrained for this reason.

The effects of progressive sorting on shelf grain-size gradients have been modelled by Swift *et al.* (1972b), as a Markov process. In Fig. 11A, a shelf cross-section with nine stations is diagrammed. Figure 11B presents the equivalent Markov model. Here station 1 becomes, in Markov terminology, an *initial state*, and stations 2–9 become *transition states*. These stations may be thought of as unit volumes in the sense of Fig. 4, into which sediment particles can enter and leave. Stations 10–19 are *trapping states* into which particles can enter, but cannot leave. These are equivalent to sediment particles that enter a unit volume such as Fig. 4, and undergo permanent burial. A sediment particle, entering the Markov model, passes through a series of steps called *transitions*. A particle may undergo a transition to another transition state, to a trapping state, or it may remain where it is during the transition step (undergo autotransition). The model is probabilistic in nature, in that each transition is assigned a probability between 0 and 1. The selection of each transition for a given particle is made by means of a random number generator, but the choices are biased according to the transition probabilities. The Markov model is thus a sort of pinball machine through which sediment particles may pass. In order to simulate progressive sorting, it is necessary to consider a family of 10 transition matrices, one for each half-phi size class.

Limiting state probability distributions computed for a normally distributed entering population are presented in Fig. 11C. Sediment deposited along the cross-section becomes finer and better sorted down the transport path, as more and more of the coarser particles are trapped out (Fig. 11D): it is important to realize that the system has a 'memory'. The grain-size distribution at a station is the result of a time-averaged discharge gradient across the station, $\partial q/\partial x$, as modelled by the transition probabilities. However, it is also a *cumulative* effect, reflecting both the input distribution and the succession of hydraulic climates experienced on the way to the final depositional site. Thus two shelf floor sites with identical hydraulic characteristics may both be floored by equilibrium sediment—but may exhibit very different grain-size distributions if their transport histories are different.

Progressive sorting operates in different ways in supply-dominated and accommodation-dominated regimes. In supply-dominated regimes, much of the progressive sorting is accomplished before the transported load reaches the shelf. In the Mississippi River system, for instance, pebbles go no further than Cairo, Illinois. By the time the Mississippi Delta is reached, fine sand is the coarsest material being carried (Dunbar & Rodgers, 1957). J.R.L. Allen (1964) has described a similar situation from the Niger Delta, and its prodelta shelf. In contrast, the retrograding shorelines of accommodation-dominated regimes yield sediment directly from their eroding shorefaces. In this process, sediments from very different back-barrier environments are overrun by the retreating barrier and released on to the shelf by shoreface erosion. Underlying strata deposited during earlier sea-level cycles may be breached and may release their sediment also. The coarse channel lags of the subaerial portions of earlier progressive sorting systems are remixed with fine material that had been bypassed downstream. As a result, coarse materials emerge on the shelf floor, and progressive sorting must begin anew. Progressive sorting is more efficient on supply-dominated shelves than it is in the accommodation-dominated case: the lower the accumulation rate, the more time is available for sorting before the sediment is buried.

**Shelf flow and stratification**

The record of progressive sorting on the continental margins is contained in the stratification patterns that record the sequence of sorting events. On most shelves, sorting episodes are the result of wind-generated currents. Fair-weather flow is generally not capable of moving sediment of the continental shelf (Swift et al., 1986a). On those limited shelf areas where tidal currents are strong enough to be more important than storm currents in transporting sediment, the flow is still episodic, but at a much higher frequency; daily or twice daily instead of several times a year.

Because of the episodic nature of shelf flow, shelf deposits are initially stratified, although the stratification may be erased later by bioturbation. The scale of stratification ranges from lamina (strata less than a centimetre thick; Ingram, 1954) to very thick beds (over 3 m thick). Lamina, very thin and thin beds (thickness <1 m) are generally *event strata* (Seilacher, 1982), created by brief, high-frequency events such as storms and tidal flows. Event strata are to be distinguished from the thicker units that result from the slow shift of depositional environments attendant on change in the stratigraphic process variables (lithofacies and depositional systems; Fisher & McGowen, 1967). Event strata are the basic building blocks of all larger scale sedimentary units, including not only lithofacies, but also depositional systems and depositional sequences. Relationships between event strata and larger scale units are discussed by Swift et al. (this volume, pp. 145 & 158).

The manner in which shallow-marine currents create event strata is indicated in Fig. 12. In this model (Niedoroda et al., 1989), the curve represents the waxing and waning of a storm current over a period of several days. As the current intensifies, it can hold at each successive moment more than it did during the previous moment (has greater capacity). It erodes the bottom and is carried along in the bottom boundary layer. As soon as the boundary layer adjusts to the velocity increase, erosion ceases. The sediment load is in equilibrium with the level of fluid power, and as a result the bottom is dynamically armoured; for every grain that is swept up into the current, another one must fall out of the boundary layer on to the bottom. In the time-dependent model of Fig. 12, however, this equilibrium condition is momentary. As the current wanes, capacity decreases and sediment falls out. The thickness of the resulting deposit can be described by the continuity equation (equation (2)), but in most cases, the spatial gradient term may be omitted, since on the flat continental shelf, it is several orders of magnitude lower than the temporal term. The bed thus produced is a record of the flow event. The basal unconformity is the record of the earlier erosional period, while the graded bed above it is a record of the period of waning flow.

In event beds, coarser grains are stored in the basal portions, because they are the first to be deposited as the current wanes. During the rising phase of the next flow event, much of the finer, upper portion of the original bed may be re-entrained, but the coarse basal lag is more likely to remain buried. Sequences of event beds are therefore records of the progressive sorting process (Fig. 13). However, the event bed is more than a passive record; its formation is an essential part of the sorting process. Unless coarser particles are sequestered by the bed formation, horizontal grain-size gradients cannot develop. Sequences of event beds formed in this manner become finer grained in the direction of sediment transport, as their coarser lower layers successively pinch out (Aigner & Reineck, 1982).

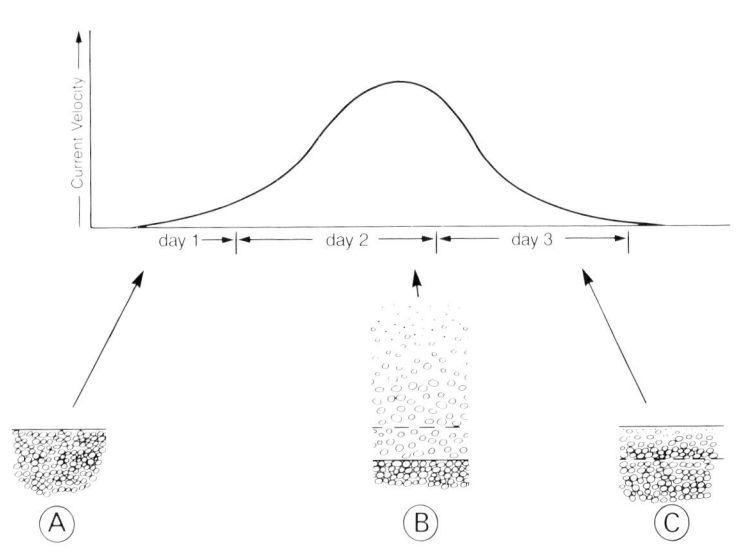

**Fig. 12.** Model for the deposition of a storm bed. As current increases, sediment is eroded from the bed and travels as a graded suspension in the benthic boundary layer. As the current wanes, the sediment is redeposited as a graded bed. (From Niedoroda *et al.*, 1989.)

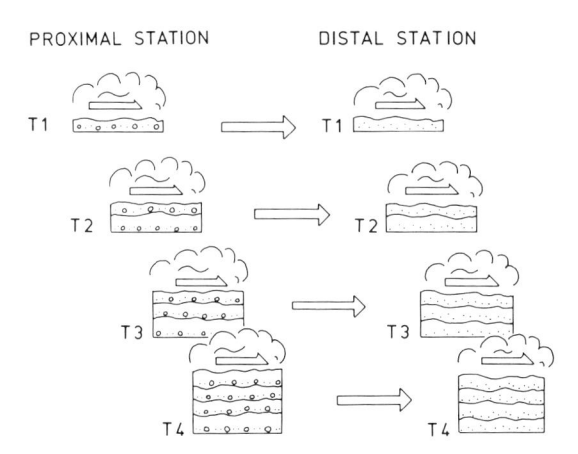

**Fig. 13.** Progressive sorting and facies differentiation. During each resuspension event, the coarsest grains deposited by the previous event have the least probability of being resuspended and transported down the dispersal system. A proximal facies accumulates through successive events, which differs in grain size and in stratification pattern from a distal facies.

Considerations of storm stratification patterns (Nittrouer & Sternberg, 1981), and more recent numerical simulations (Niedoroda *et al.*, 1989) indicate that the efficiency of the progressive sorting mechanism is a function of the sediment supply index, *Q/P*. Nittrouer & Sternberg (1981; see also Nittrouer *et al.*, 1986) have parameterized strata-forming processes in order to provide a conceptual framework for the analysis of $^{210}$Pb activity profiles in box cores. They have divided the uppermost metre of the sea-bed into a zone of resuspension and mixing, underlain by a zone of accumulation, or permanent burial. The zone of burial corresponds to the trapping states of Fig. 11B. The stratal pattern is a signal impressed on the sediment as it passes downwards through the zone of mixing. The zone of mixing is thus the 'recording head' for the unreeling tape of the stratigraphic record, as it develops in the zone of permanent burial (Fig. 14). Nittrouer & Sternberg (1981), following Goldberg & Koide (1962) and Guinasso & Schink (1975), have characterized this signal as a dimensionless mixing parameter: a record of the varying ratio of the sediment mixing rate to the rate of accumulation.

Following suggestions by Smith (1977), Nittrouer & Sternberg (1981) have introduced a physical mixing parameter, *h*, that describes the formation of strata by fluid dynamic processes:

$$h = \frac{l_p f_p}{\mathring{a}} \qquad (4)$$

where $l_p$ is the depth to which the sea-bed is eroded by storms in centimetres, $f_p$ is the storm frequency and $\mathring{a}$ is the accumulation rate in cm$^{-1}$ yr. The stratal parameter $\mathring{a}$ fulfils a function similar to that of the regime parameter, $\mathring{A}$, but is distinguished by a lower case letter to indicate that it is a subequilibrium parameter, measured over a shorter time-scale. Observations of contrasting short and long-term sedimentation rates on continental shelves have been reported by McKee (1983). The physical mix-

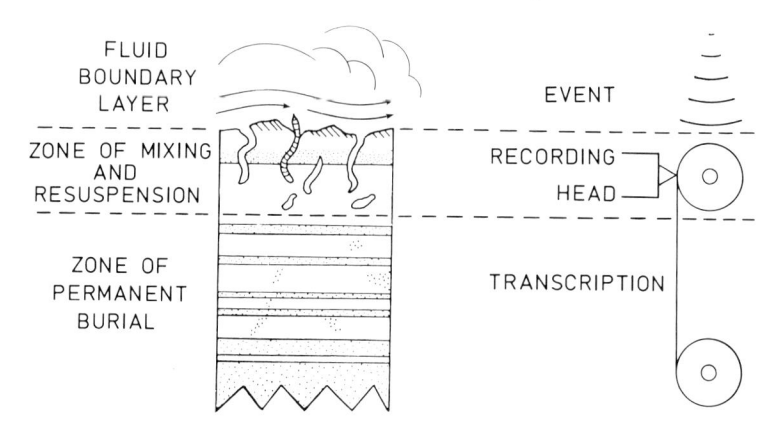

**Fig. 14.** Dynamic zones of the sea floor. The fluid dynamic and biogenic signals are transcribed in the zone of mixing, and pass downward into 'the unreeling tape of the rock record' in the zone of permanent burial. (After Nittrouer & Sternberg, 1981.)

ing ratio, which has broad applicability in analysis of facies and study of strata, is equivalent to the *reworking ratio*, $r$ (Thorne, 1988; see Thorne *et al.*, this volume, pp. 64–65).

Variations in stratal patterns for varying values of $h$ and $å$ are presented in Fig. 15. For higher values of the accumulation rate $å$ (lower values of $h$), more of the thinner beds are preserved (Fig. 15A). Sedimentary particles are buried after relatively few erosion–deposition events. The efficiency of progressive sorting is low, and the horizontal separation (in kilometres) of grain-size grades is large; horizontal grain-size gradients on the sea-floor are gentle. As $h$ increases, the deposits of milder events are increasingly cannibalized by more intense ones (Fig. 15B). The number of resuspension events participated in prior to final burial increases, and so, correspondingly, does the steepness of the horizontal grain size gradient. In the limit $å = 0$, and $h = \infty$, a single bed is repeatedly remade, and steadily bleeds its finer grains down the sediment transport pathway. As size class after size class is lost, the residuum (lag bed) progressively coarsens (Fig. 15C). The relationship $h = l_p f_p / å$ is thus a 'preservation filter' (Crowley, 1984), which controls the degree of condensation of section (extent to which each bed is cannibalized by the succeeding bed). These relationships are examined in greater detail by Thorne *et al.* (this volume, p. 64) and Swift *et al.* (this volume, pp. 99–101).

The relationships also can be observed in shoreface deposits (Niedoroda *et al.*, 1989; Fig. 16). The sediment accumulation rate, $å$, is high in the surf zone; but fluid power parameters, $f(P) = l_p f_p$, are even higher, so that the ratio $h = l_p f_p / å$ is high. Surf zone beds are thick and consist almost entirely of

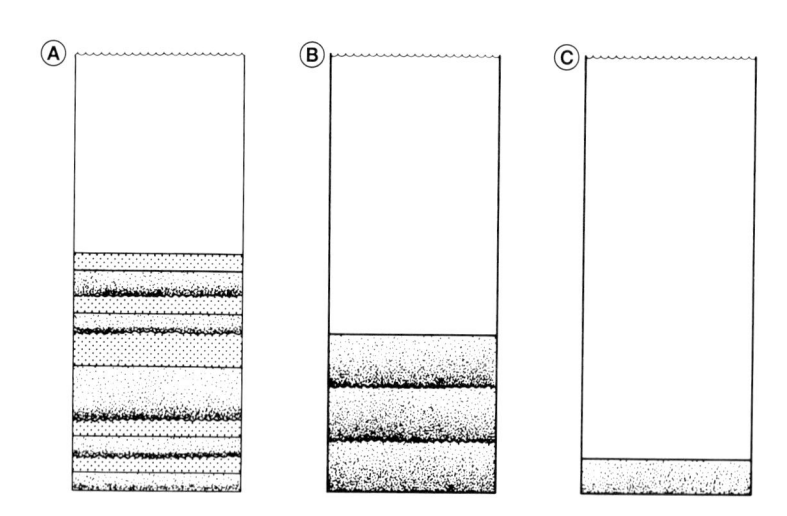

**Fig. 15.** Schematic diagram illustrating the role of sedimentation rate $å$, and the reworking ratio, $h$, in determining strata thickness. (A) High sedimentation rate, low reworking ratio. Degree of cannibalization of beds by later beds is minor. (B) High reworking ratio, low sedimentation rate, resulting in condensed record. (C) Zero sedimentation rate, resulting in lag bed. (From Niedoroda *et al.*, 1989.)

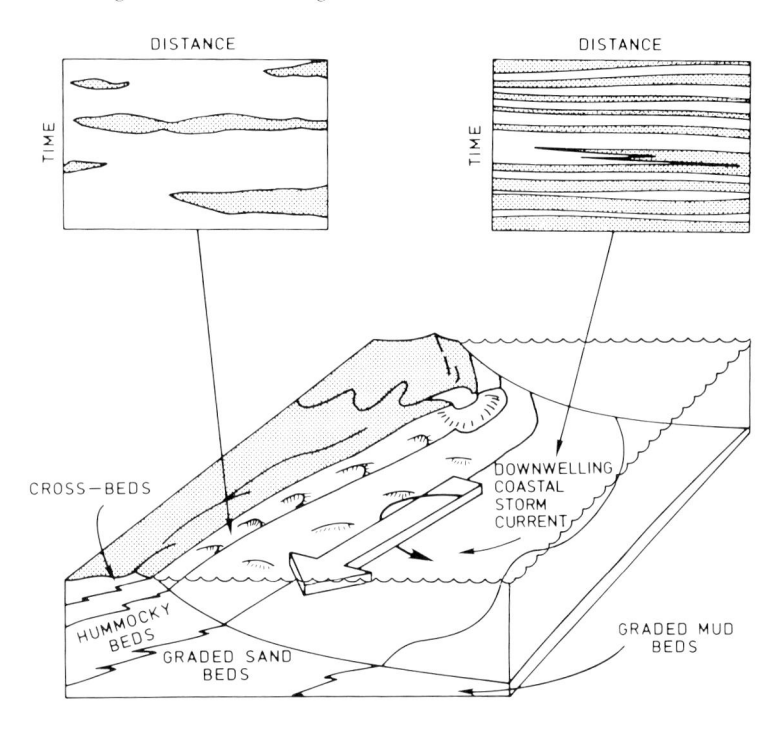

**Fig. 16.** Schematic diagram illustrating storm stratification patterns on a prograding shoreface. The changing stratification pattern, from the beach to the lower shoreface reflects the seaward decrease in the accumulation rate $å$, and the physical mixing parameter, $h$. (From Niedoroda *et al.*, 1989.)

the truncated bases of extreme event deposits (Niedoroda *et al.*, 1989). The parameter $å$ is lower on the lower shoreface, but $l_p f_p$ is much lower, hence $h = l_p f_p / å$ is low. Lower shoreface beds are consequently thin, and include much more of the depositional record. The shoreface is characterized by a seaward decrease in grain size, but most of this occurs on the upper shoreface where $h$ is large (see Thorne *et al.*, this volume, pp. 64–65).

The effects of the accumulation rate and fluid power are modified by the role of grain size in progressive sorting systems. In effect, the downstream drop in fluid power that characterizes such systems is buffered by a compensating decrease in grain size. These relationships can be observed in the shoreface system described in Fig. 16. On supply-dominated shelves, progressive sorting continues out across the shelf floor despite the weakening of the wave field with depth, because as the wave field weakens, the grain size is simultaneously reduced to silt size (Snedden & Nummedal, this volume, pp. 291–307).

Episodic progressive sorting thus simultaneously creates both the vertical stratification patterns and the horizontal grain-size gradients that characterize sedimentary facies, and these characteristics determine in turn the sequence of facies tracts that succeed each other down the transport path. The stratal patterns are modulated by the 'preservation filter', which controls the degree of condensation of the section and determines the efficiency of the progressive sorting mechanism. Progressive sorting and strata formation are thus the main subequilibrium (event scale) processes that underlie regime sedimentation.

### Shelf dispersal systems and facies assemblages

At the next larger spatial scale, variations in shelf flow create a lithofacies assemblage, or a *depositional system* (assemblage of process-related facies; Fisher & McGowen, 1967). The dynamic equivalent is a *dispersal system* or an assemblage of flow-integrated depositional environments (Swift *et al.*, 1987). We have noted above that a dispersal system can be modelled as consisting of three classes of elements; sources, reservoirs and sinks, connected by transport pathways, as has been done in Fig. 11B. A depositional system is thus a three-dimensional, stratigraphic concept that describes rock volumes, while a dispersal system is a two-dimensional dynamic and geomorphological concept that concerns itself with depositional surfaces. However, the two concepts are related in a fundamentally geological way; a depositional system consists of the fossilized 'hard parts' of the original dispersal system. Like other kinds of fossils, the depositional system is subject to taphonomic distortion; selective preservation modifies the information content of the original system.

Most dispersal systems consist of two classes of elements; an erosional source environment, and a series of depositional environments that serve as successive partial sinks, as in Fig. 11B. There is a progressive loss of competence through these environments, down the sediment transport pathway. Competence is lost because the currents become weaker in the down-current direction. They therefore lose successively finer fractions of the transported load, in the process described above as progressive sorting. Where currents become weaker in the down-current direction, it is usually a result of flow expansion and deceleration. In fluid dynamics, the continuity principle says that the cross-sectional area of the flow and its velocity must vary inversely so that a constant discharge is maintained. For shelf flows, the principle requires that they contract, accelerate and erode as they enter a constricted zone, and that they expand and decelerate as the shelf widens or deepens (Fig. 17).

Shelf width as well as shelf depth counts in this process. Because shelf flows tend to be geostrophic in nature and to conserve vorticity (Csanady, 1982), they follow shelf contours, and except under special circumstances, cannot simply pass over the shelf edge where the shelf narrows. As a consequence, fluid exchange with the slope water mass is inhibited and occurs only slowly. Shelf flow is therefore like a river, even though only one bank is solid. The flow must accelerate to get through the narrow zone, and will expand and decelerate in the zone of shelf widening beyond. In one common pattern of shelf-facies distribution, acceleration zones become sediment sources, and the downstream zones of flow expansion and deceleration become sediment sinks, marked by shelf-transverse facies tracts that become increasingly finer in the down-shelf direction (Fig. 18).

When the sediment load is predominantly fine-grained, a second mechanism of shelf sediment dispersal becomes important. It is controlled by the interaction of the wind-driven current field with the wave field. During storm periods of coastal set-up and geostrophic adjustment of flow, the sea-floor is stirred by high-frequency wave orbital motions. The geostrophic component of velocity is controlled by

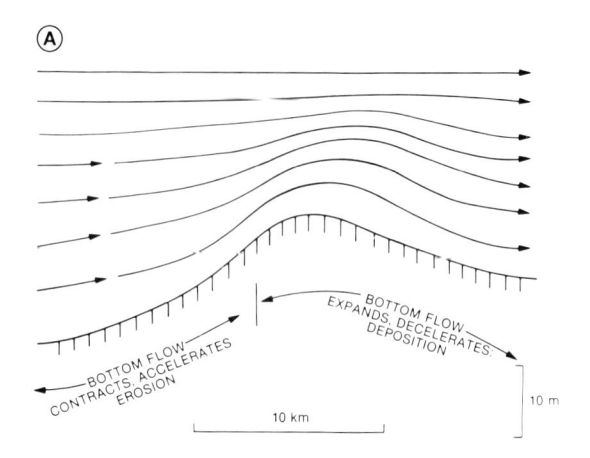

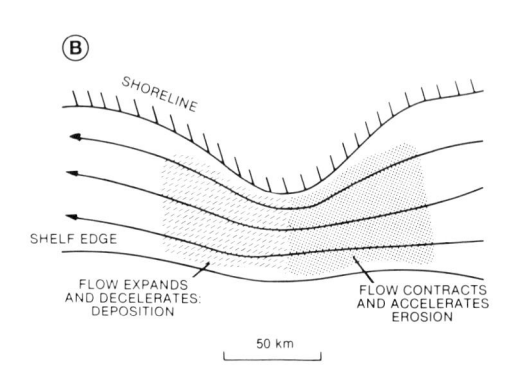

**Fig. 17.** Topographic control of sedimentation on the continental shelf. (A) Cross-section. Shelf flows must accelerate as the bottom shoals in a down-current direction. Erosion results. The flow expands and decelerates and deposition occurs in the deeper water beyond the shelf high. (B) Plan view. Shelf currents accelerate and erode shelf sectors that narrow down-current, and deposit the sediment in the zones of shelf widening beyond.

the slope of the sea surface, and is depth invariant. The wave motion, however, decreases with water depth. Shelf sediment is entrained and transported by the combined bottom stresses of wave and fluid motion. The capacity and competence of shelf transport must therefore decrease as wave orbital motion decreases, and with increasing water depth. There is thus a built-in mechanism for seaward sediment transport on shelves, and for offshore progressive sorting (McCave, 1972). In this class of dispersal system, facies boundaries are shelf-parallel.

Studies of the hydraulic climate of modern shelves

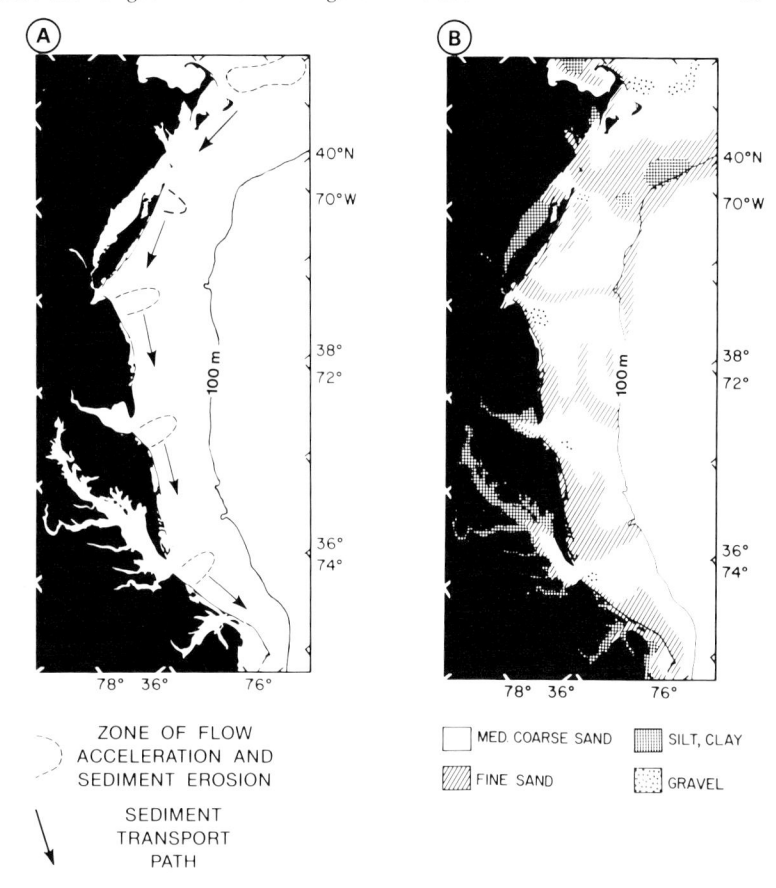

**Fig. 18.** Cross-shelf facies boundaries on a coarse-grained (accommodation-dominated) shelf, a response to advection-dominated sediment dispersal. Middle Atlantic shelf of North America. (A) Flow pattern; (B) sediment distribution. (From Swift *et al.*, 1986b.)

have shown that at large time and space-scales, sediment particles undergo trajectories that can be described by the mathematics of a random walk (Clarke *et al.*, 1982, 1983). Particles are resuspended for minutes, hours or days, transported metres or kilometres in the current direction of the moment, then buried again for months, years, or centuries. At these time-scales, sediment transport is a diffusion process, and transport gradients can be mapped by means of advection–diffusion equations, calibrated against data sets describing the wave, current and tidal climates (Clarke *et al.*, 1982, 1983). Shelves such as the Atlantic continental shelf, which is coarse grained, have relatively low diffusion coefficients, hence the shelf-transverse facies pattern prevails (Fig. 18). Such shelves are governed by *advective dispersal systems*. On muddy shelves, however, sediment particles are fine and diffusion coefficients are large. Facies boundaries are shelf-parallel (Fig. 18). Such shelves are governed by *diffusive dispersal systems*.

### Supply-dominated facies patterns

Thus the facies created by progressive sorting events are arranged in broader patterns controlled by the dispersal system. On supply-dominated shelves, significant facies differentiation is largely restricted to the inner shelf, where the relatively high fluid power expenditure $P$ reduces the value of the sediment supply index, $Q/P$, and increases the event scale variable $r$ (reworking ratio, equivalent to $h = l_p f_p / å$, see Thorne *et al.*, this volume, pp. 64–65). As a result, the efficiency of progressive sorting is high. Facies differentiation on supply-dominated shelves is described in more detail by Swift *et al.* (this volume, pp. 93–96).

Further seaward on supply-dominated shelves, the efficiency of progressive sorting decreases and grain-size gradients become gentle. Diffusion processes become dominant. Steep underwater slopes cannot be maintained, and supply-dominated shelves tend to be simple, seaward-inclined planes.

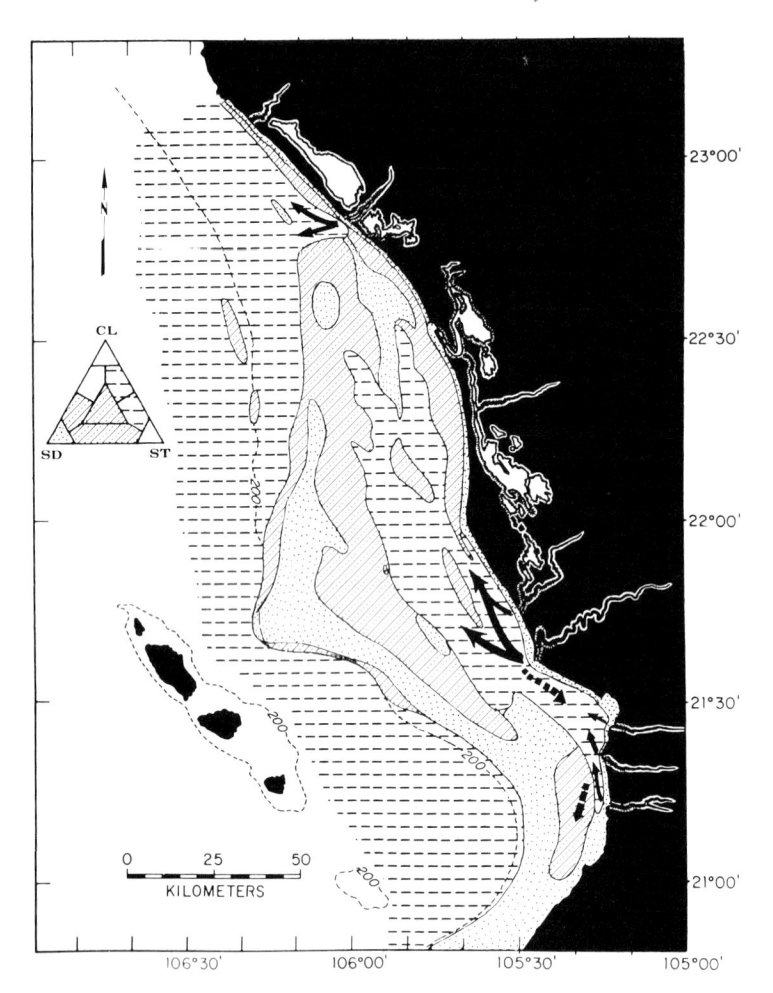

Fig. 19. Shelf-parallel facies boundaries on a fine-grained (supply-dominated) shelf, a response to diffusion-dominated sediment dispersal. Nayarit shelf, Pacific Mexico. Zone of coarse sand is a window in the allochthonous sediment blanket, exposing older transgressive sand. Such windows are characteristic of supply-dominated shelves. (From Curray, 1969.)

While fluid and sediment advection continues to be predominantly along shore, the cross-shelf diffusive component of transport is sufficiently strong to control facies patterns. Grain-size gradients are consequently shore normal, and facies boundaries are shore-parallel. Supply-dominated shelves are the classic 'graded' shelves, in which grain-size decreases offshore (Fig. 19; see concept of graded shelf, Thorne & Swift, this volume, pp. 33–35). Supply-dominated shelf facies patterns have been described in detail by Nelson (1982), Aigner & Reineck (1982) and Snedden & Nummedal (this volume). Supply-dominated facies patterns are described in greater detail by Swift et al. (this volume, pp. 116–118).

### Accommodation-dominated facies patterns

On accommodation-dominated shelves, the sediment input, $Q$, and the accumulation rate, $\mathring{A}$, are lower than on supply-dominated shelves. The event-scale parameter, $r$, is high (the reworking ratio, equivalent to $h = l_p f_p / \mathring{a}$; see Thorne et al., this volume, pp. 59–87). Sediment is resuspended and retransported many more times than on supply-dominated shelves before undergoing permanent burial. Progressive sorting is consequently a highly efficient process, and grain-size gradients are steep. Because the autochthonous sediments on accommodation-dominated shelves are composed of relatively coarse, heavy grains, their diffusion constants are low. They tend to be advected primarily along shelf in the direction of the prevailing mean flow (dispersal system is advective). As a result, grain-size gradients and facies boundaries tend to be oriented normal to the shoreline, and to the direction of flow.

Accommodation-dominated shelves tend to have considerable relief (tens of metres over tens of kilometres of horizontal spacing), because they are locally subject to erosion, and because where depositional, their coarser sediments can sustain steeper underwater slopes. As a consequence, regional patterns of flow acceleration and deceleration are common, and give rise to multiple sediment transport systems on the shelf. Localized sediment transport systems begin wherever topographic constrictions result in flow acceleration and erosion of the substrate, and continue through the sediment fallout shadow in the downstream zone of flow expansion.

Sediment dispersal systems also occur at larger spatial scales on accommodation-dominated shelves. A complex pattern has developed on the Atlantic shelf of North America during the Holocene transgression (Fig. 18). The shelf consists of a series of segments, separated by the retreat paths of estuaries (Swift et al., 1986b). The shorelines of the successive segments rotate counterclockwise, from north to south. As the prevailing southerly storm flows strike each coastal segment, flow on the inner shelf rotates its orientation to accommodate the coast, and accelerates in order to satisfy continuity. The inner shelf in these areas is eroded, and a lag gravel develops. As the current flows south along the shelf segment, the inner shelf flow (coastal boundary layer, Csanady, 1982) completes its adjustment to the new orientation. The flow expands, decelerates, and deposits successively finer grades of sand. The finest sand is deposited in the drowned cross-shelf river valley at the terminus of the segment. Accommodation-dominated facies patterns are described in greater detail by Swift et al. (this volume, pp. 89–152).

## CONCLUSIONS

Over geological time-scales, the shelf surface can be viewed as a surface of dynamic equilibrium controlled by the variables of the relative sea-level rise rate ($R$), the rate ($Q$) and character of sediment input ($M$), and rate of sediment transport ($D$). Sediment accumulating on the continental margin builds up toward 'wave base', the depth to which, on a time-averaged basis, there is sufficient wave power to transport sediment at its prevailing rate of input. As wave base is approached, the ratio of sediment deposited to sediment bypassed decreases, until, ideally, a surface of non-deposition results (continental shelf). The locus of deposition shifts seaward to the forward face of the sediment pile, to form a secondary surface dominated by gravity processes (continental slope). The shelf surface aggrades, at decreasing rates of net deposition, until deposition balances subsidence, so that a constant depth is maintained. At this point, the ratio of deposition to bypassing becomes fixed until such a time as one or more of the process variables changes in value.

In this manner, continental margins evolve from diverse structural origins toward a characteristic shelf-slope configuration. During the evolutionary process, shifts in the values of variables occur with relative frequency, so that mature shelves tend to consist of multiple depositional sequences, rather then of a single clinoform sequence.

Sedimentary facies on continental shelves are responses to the subequilibrium processes of progressive sorting and strata formation. In progressive sorting, a sediment load undergoing intermittent transport is progressively depleted of its coarse fraction, resulting in a deposit with a downstream-fining grain-size gradient. Progressive sorting is an episodic process. The stratification pattern of the deposit is a record of the successive sorting events that have created the grain-size gradients, and is also a measure of the degree of cannibalization that occurs as the beds pass through the 'preservation filter'. Progressive sorting operates through *dispersal systems*, or assemblages of flow-integrated depositional environments. Over time these create lithofacies assemblages, or *depositional systems* (assemblages of process-related facies). The shelf flows responsible are primarily along shelf in orientation. Since vorticity must be conserved, they cannot easily cross over the shelf edge. Flows must accelerate through narrow zones, and expand and decelerate in the zone of shelf widening beyond. Acceleration zones become sediment sources, At the same time, because bottom wave orbital currents weaken with depth, sediment transporting currents lose competence and capacity in an offshore direction, and there is an offshore diffusive component of sediment transport on the shelf.

On transgressive shelves, river mouths become sediment trapping estuaries, and the shelf is sourced by erosional shoreface retreat (shoreface bypassing), leading to *accommodation-dominated* shelf sedimentation. The accumulation rate, $\mathring{A}$, is low. Differential bypassing is favoured, and progressive sorting is efficient. Deposits are coarse, variable and thin, and transport is largely advective in nature. On such sandy shelves, sediment dispersal systems are marked by flow-transverse facies tracts that become increasingly finer in the down-current direction, so that facies boundaries tend to be shore normal. In autochthonous settings, sediments may have moved to the shelf site during an earlier depositional cycle, or are at least time-lagged contributions of the present cycle, released after prolonged back-barrier storage.

On shelves undergoing slow transgression or regression, river mouths become sediment sources (river mouth bypassing), and fine, mobile, first-cycle sediment is released in large quantities. The depositional regime is supply dominated. The accumulation rate is high. The differential bypassing ratio is reduced and progressive sorting is less efficient. Deposits are fine-grained and thick. Because they are subject to intensive wave resuspension, diffusive offshore transport dominates, and facies boundaries are more nearly shelf parallel.

## REFERENCES

ADAMS, J. JR., SWIFT, D.J.P. & COLEMAN, J. (1988) Bottom currents and fluviomarine sedimentation on the Mississippi Prodelta Shelf, February–May, 1984. *J. Geophys. Res.* **92**, 14585–14609.

AIGNER, T. & REINECK, H.E. (1982) Proximity trends in modern storm sands from the Heligoland Bight (North Sea) and their implications for basin analysis. *Senckenbergiana Maritima* **14**, 183–215.

ALLEN, J.R.L. (1964) The Quaternary Niger Delta and adjacent areas: sedimentary environments and lithofacies. *Bull., Am. Assoc. Petrol. Geol.* **49**, 547–600.

ALLEN, J.R.L. (1970) *Physical Processes of Sedimentation*. Elsevier, New York, 433 pp.

ALLEN, P. (1964) Sedimentological models. *J. Sediment. Petrol.* **34**, 289–293.

BARRELL, J. (1917) Rhythms and measurement of geological time. *Geol. Soc. Am. Bull.* **28**, 754–904.

BARRON, E.J., ARTHUR, M.A. & KAUFFMAN, E.G. (1985) Cretaceous rhythmic bedding sequences: a plausible link between orbital variations and climate. *Earth Planet. Sci. Lett.* **72**, 327–340.

BIRD, E.C.F. & SCHWARTZ, M.L. (1985) *The World's Coastline*. Van Nostrand Reinhold, New York, 1071 pp.

BLENCH, T. (1957) *Regime Behaviour of Canals and Rivers*. Butterworths, London, 138 pp.

BRETSHNEIDER, C.L. (1966) Wave generation by wind, deep and shallow water (Chapter 3), Engineering aspects of hurricane surge (Chapter 5). In: *Estuary and Coastline Hydrodynamics* (Ed. Ippen, A.T.) pp. 133–196 & 231–256. McGraw Hill, New York.

BRIDGE, J.R. (1981) Hydraulic interpretation of grain-size distributions using a physical model for bed load transport. *J. Sediment. Petrol.* **51**, 1109–1124.

BRUUN, P. (1962) Sea-level rise as a cause of shore erosion. *J. Waterways Harbors Div., Am. Soc. Civ. Eng.* **88**, 117–130.

CHORLEY, R.J., SCHUMM, S.A. & SUGDEN, D.E. (1984) *Geomorphology*. Methuen, New York, 498 pp.

CLARKE, T.L., SWIFT, D.J.P. & YOUNG R.A. (1982) A numerical model of fine sediment transport on the continental shelf. *Environ. Geol.* **4**, 117–129.

CLARKE, T.L., SWIFT, D.J.P. & YOUNG, R.A. (1983) A stochastic modeling approach to the fine sediment budget of the New York Bight. *J. Geophys. Res.* **88**, 9653–9660.

COLEMAN, J.M. (1981) *Deltas*. Burgess, Minneapolis, 124 pp.

COLEMAN, J.M. (1988) Dynamic changes and processes in the Mississippi River delta. *Geol. Soc. Am. Bull.* **100**, 999–1015.

CROWLEY, K.D. (1984) Filtering of depositional events and the completeness of the sedimentary record. *J. Sediment. Petrol.* **54**, 127–136.

CSANADY, G.T. (1982) *Circulation in the Coastal Ocean*. D. Reidel, Dordrecht, 280 pp.

CURRAY, J.R. (1960) Sediments and history of Holocene transgression, continental shelf, northwest Gulf of Mexico. In: *Recent Sediments, Northwestern Gulf of Mexico* (Eds Shepard, F.P., Phleger, F. & Van Andel, Tj.) pp. 221–226. American Association of Petroleum Geologists, Tulsa, OK.

CURRAY, J.R. (1964) Transgressions and regressions. In: *Papers in Marine Geology* (Ed. Miller, R.C.) pp. 175–203. McMillan, New York.

CURRAY, J.R. (1965) Late Quaternary history, continental shelves of the United States. In: *The Quaternary of the United States* (Eds Wright, H.E., Jr & Frey, D.G) pp. 723–735. Princeton University Press, Princeton, NJ.

CURRAY, J.R. (1969) Lectures 2, 3 and 6. In: *The New Concepts of Continental Margin Sedimentation* (Ed. Stanley, D.J.). Short Course Lecture Notes, 236 pp. American Geological Institute, Washington, DC.

CURRAY, J.R., EMMEL, F.J. & CRAMPTON, P.J.S. (1969) Holocene history of a strandplain. In: *Simposio Internacional Lagunas Costeras* (Eds Castañares, A.A. & Phleger, F.B.) pp. 63–100. Univ. Nac. Autonoma De Mexico, Mexico City.

DAVIES, A.C. (1983) Application of a three-dimensional shelf model to the calculation of North Sea currents. In: *North Sea Dynamics* (Eds Sundermann, J. & Lenz, W.) pp. 44–62. Springer Verlag, New York.

DAVIES, J.L. (1964) A morphogenic approach to world shorelines. *Z. Geomorphol.* **8**, 127–142.

DAVIS, W.M. (1899) The geographic cycle. *Geogr. J.* **14**, 481–504.

DAVIS, W.M. (1902) Base level, grade, and peneplain. *J. Geol.* **10**, 77–110.

DEAN, R.G. & MAURMEYER, E.M. (1983) Models of beach profile response. In: *CRC Handbook of Coastal Processes and Erosion* (Eds Komar, P. & Moore, J.) pp. 151–165. CRC Press, Boca Raton, FL.

DUNBAR, C.O. & RODGERS, J. (1957) *Principles of Stratigraphy*. Wiley, New York, 356 pp.

EMERY, K.O. (1952) Continental shelf sediments off Southern California. *Geol. Soc. Am. Bull.* **63**, 1105–1108.

EMERY, K.O. (1968a) Shallow structure of continental shelves and slopes. *Southeastern Geology*, **9**, 178–194.

EMERY, K.O. (1968b) Relict sediments on continental shelves of the world. *Bull., Am. Assoc. Petrol. Geol.* **52**, 445–464.

EVERTS, C.H. (1987) Continental shelf evolution in response to a rise in sea level. In: *Sea-level Fluctuation and Coastal Evolution* (Eds Nummedal, D., Pilkey, O. & Howard, J.). *Soc. Econ. Paleontol. Mineral.* 41, 49–58.

FISHER, W.L. & MCGOWEN, J.H. (1967) Depositional systems in the Wilcox Group of Texas and their relationship to the occurrence of oil and gas. *Trans. Gulf Coast Assoc. Geol. Soc.* 17, 105–125.

FLEMMING, B.W. (1988) Pseudo-tidal sedimentation in a non tidal shelf environment (southeastern African margin). In: *Tide-influenced Sedimentary Environments and Facies* (Eds Deboer, P.L., van der Gelder, A. & Nio, S.D.) pp. 167–180. D. Reidel, Dordrecht.

GALLOWAY, W.E. & HOBDAY, D.K. (1983) *Terrigenous Clastic Depositional Systems*. Springer Verlag, New York, 423 pp.

GILBERT, G.K. (1877) Report on the geology of the Henry Mountains. *U.S. Geog. Geol. Surv. Rocky Mtn. Region.* 160 pp.

GOLDBERG, E.D. & KOIDE, M. (1962) Geochronological studies of deep sea sediments by the ionium-thorium method. *Geochem. Cosmochim. Acta.* 26, 417–456.

GOULD, H.R. & MCFARLAN, E., JR. (1959) Geologic history of the chenier plain, southwestern Louisiana. *Trans. Gulf Coast Assoc. Geol. Soc.* 9, 261–270.

GRABAU, W.A. (1913) *Principles of Stratigraphy*. Facsimile edn, Dover, New York, 1185 pp.

GRANT, W.D. & MADSEN, O.S. (1979) Combined wave and current interaction with a rough bottom. *J. Geophys. Res.* 84, 1797–1808.

GUINASSO, N.L. & SCHINK, D.R. (1975) Qualitative estimates of biological mixing rates in abyssal sediments. *Geophys. Res.* 80, 3032–3043.

GULLIVER, F. (1899) Shoreline topography. *Am. Acad. Arts Sciences Proc.* 34, 151–258.

HARBAUGH, J. & BONHAM-CARTER, G. (1977) In: *The Sea*, Vol. 6, *Marine Modeling* (Eds Goldberg, E., McCave, I.N. O'Brien, J.J. & Steele, J.H.) pp. 623–674. Wiley Interscience, New York.

HARMS, J.C., SOUTHARD, J.B. & WALKER, R.G. (1982) Structures and sequence in clastic rocks. *Soc. Econ. Paleontol. Mineral. Short Course* 9.

HAYS, J.D., IMBRIE, J. & SHACKLETON, N.J. (1976) Variations in the Earth's orbit: pacemaker of the ice ages. *Science*, 194, 2212–2232.

HELLER, P.L. & ANGEVINE, C.L. (1985) Sea-level cycles during the growth of Atlantic-type oceans. *Earth Planet. Sci. Lett.* 75, 417–426.

HOLMES, C.W. (1982) Geochemical indices of fine sediment transport, northwest Gulf of Mexico. *J. Sediment. Petrol.* 52, 307–321.

HOYT, J.H. (1967) Barrier island formation. *Geol. Soc. Am. Bull.* 78, 1123–1136.

INGRAM, R.L. (1954) Terminology for the thickness of stratification and parting units in sedimentary rocks. *Geol. Soc. Am. Bull.* 65, 937–938.

INMAN, D.L. & NORDSTROM, C.E. (1971) On the tectonic and morphologic classification of coasts. *J. Geology* 709, 1–21.

JORDAN, T.E. (1982) Thrust loads and foreland basin evolution, Cretaceous Western United States. *Am. Assoc. Petrol. Bull.* 65, 2506–2520.

KOMAR, P.D. (1983) Nearshore currents and sand transport on beaches. In: *Physical Oceanography of Coastal and Shelf Seas*. Amsterdam, Elsevier, 234 pp.

KUEHL, S.A., NITTROUER, C.A. & DEMASTER, D.J. (1982) Modern sediment accumulation and strata formation on the Amazon continental shelf. *Mar. Geol.* 49, 279–300.

LACEY, G. (1929) Stable channels in alluvium. *Min. Proc. Inst. Civil Eng. London*, 229, 259–292.

LANE, E.W. (1957) *A Study of the Shape of Channels Formed by Natural Streams Flowing in Erodible Material* Sediment Series 9. US Army Corps of Engineers, Missouri River Division, 31 pp.

LEATHERMAN, S.P. (1983) Barrier dynamics and landward migration with Holocene sea-level rise. *Nature* 301, 415–418.

LEOPOLD, L.B., WOLMAN, M.G. & MILLER, J.P. (1964) *Fluvial Processes in Geomorphology*. W.H. Freeman, San Francisco, 522 pp.

MALDONADO, A., SWIFT, D.J.P., YOUNG, R.A., HAN, G., NITTROUER, C.E., DEMASTER, D., REY, J., PALOMO, C., ACOSTA, J., BALLESTER, C.E. & CASTELLVI, I. (1983) Sedimentation on the Valencia continental Shelf: preliminary results. *Continental Shelf Res.* 3, 195–211.

MAY, S., KIMBALL, W.H., GRANDY, N. & DOLAN, R. (1982) CEIS: coastal erosion information system. *Shore Beach* Jan. 19–25.

MCCAVE, I.N. (1971) Wave effectiveness at the sea bed and its relationship to bedforms and deposition of mud. *J. Sediment. Petrol.* 41, 89–96.

MCCAVE, I.N. (1972) Sediment transport and escape of fine-grained sediment from shelf areas. In: *Shelf Sediment Transport, Process and Pattern* (Eds Swift, D.J.P., Duane, D.B. & Pilkey, O.H.) pp. 215–248. Dowden Hutchinson & Ross, Stroudsbury, PA.

MCKEE, B.A. (1983) Concepts of sediment deposition and accumulation applied to continental shelf near the mouth of the Yangtze River. *Geology* 11, 631–633.

MCLAREN, P. & BOWLES, D. (1985) The effects of sediment transport on grain size distributions. *J. Sediment. Petrol.* 55, 457–470.

MEADE, R.H. (1969) Landward transport of bottom sediments in estuaries of the Atlantic coastal plain. *J. Sediment. Petrol.* 39, 222–234.

MELHORN, W.N. & FLEMAL, R.C. (eds) (1975) *Theories of Landform Development*. Publications in Geomorphology, State University of New York, Binghamton, New York, 306 pp.

MOORE, D.G. & CURRAY, J.R. (1964) Wave-base, marine profile of equilibrium and wave built terraces, discussion. *Geol. Soc. America. Bull.* 47, 1267–1274.

NAUMANN, C.F. (1858) *Lehrbuch der Geognosie*, Bd.1. 2d Ed. Wilhelm Engelmann, Leipzig, 960 pp.

NELSON, L.H. (1982) Modern shallow water graded sand layers from storm surges, Bering Shelf: a mimic of Bouma sequences and turbidite systems. *J. Sediment. Petrol.* 52, 537–545.

NIEDORODA, A.W., SWIFT, D.J.P., FIGUEIREDO, A.G. & FREELAND, G.L. (1985) Barrier island evolution, middle Atlantic shelf, USA. Part II: evidence from the shelf floor. *Mar. Geol.* 63, 363–396.

NIEDORODA, A.W., SWIFT, D.J.P. & THORNE, J.A. (1989) Modeling shelf storm beds: controls of bed thickness and bedding sequence. In: *Shelf Sandstones, Shelf Deposit-*

*ional Sequences and Petroleum Accumulation: a Symposium*, pp. 15–39. Gulf Coast Section, Seventh Annual Research Conference Proceedings, Society of Economic Paleontologists and Mineralogists, Tulsa, OK.

NITTROUER, C.A. & STERNBERG, R.W. (1981) The formation of sedimentary strata in an allochthonous shelf environment: the Washington Continental Shelf. *Mar. Geol.* **42**, 201–232.

NITTROUER, C.A., DEMASTER, D.J., KUEHL, S.A. & MCKEE, B.A. (1986) Association of sand with mud deposits accumulating on continental shelves. *Shelf Sands and Sandstones* (Eds Knight, R.J. & McLean, J.R.). Can. Soc. Petrol. Geol. Mem. 11, 17–25.

NUMMEDAL, D. & SWIFT, D.J.P. (1987) Transgressive stratigraphy at sequence bounding unconformities—some principles derived from Holocene and Cretaceous examples. In: *Sea-level Fluctuation and Coastal Evolution* (Eds Nummedahl, D., Pilkey, O.H. & Howard, J.D.). Soc. Econ. Paleon. Mineral. Spec. Publ. 41, 241–260.

PARKER, G., LANFREDI, N.W. & SWIFT, D.J.P. (1982) Substrate response to flow in a southern hemisphere ridge field: Argentina inner shelf. *Sediment. Geol.* **33**, 195–216.

PELTIER, W.R. (1980) Models of glacial isostasy and relative sea level. In: *Dynamics of Plate Interiors* (Eds Ball, A.W., Bender, P.L., McGretchin, T.R. & Walcott, R.I.) *Geodynamics Series*, pp. 111–127. American Geophysical Union, Vol. I, Washington, DC.

PELTIER, W.R. (1986) Deglaciation-induced vertical motion of the North American continent and transient lower mantle rheology. *J. Geophys. Res.* **91**, 9099–9123.

PILKEY, O.H. & DAVIS, T.W. (1987) An analysis of coastal recession models: North Carolina Coast. In: *Sea-level Fluctuation and Coastal Evolution* (Eds Nummedal, N., Pilkey, O.H. & Howard, J.D.). Soc. Econ. Paleontol. Mineral. Spec. Publ. 41, pp. 59–68.

POSTMA, H. (1967) Sediment transport and sedimentation in the estuarine environment. In: *Estuaries* (Ed. Lauff, G.H.) pp. 158–179. American Association for Advancement of Science, Washington, DC.

PRITCHARD, D.W. (1967) Observations of circulation in coastal plain estuaries. In: *Estuaries* (Ed. Lauff, G.H.) pp. 37–44. American Association for Advancement of Science, Washington, DC.

RUSSELL, R.D. (1939) Effects of transportation on sedimentary particles. In: *Recent Marine Sediments* (Ed. Trask, P.D.) pp. 32–47. Am. Assoc. Petrol. Geol., Tulsa, OK.

SCHUMM, S.A. (1977) *The Fluvial System*, Wiley Interscience, New York, 235 pp.

SCHWARTZACHER, W. (1975) *Sedimentation Models in Quantitative Stratigraphy*, Vol, 19, *Developments in Sdeimentology*. Elsevier, New York, 382 pp.

SEILACHER, A. (1982) General remarks about event deposits. In: *Cyclic and Event Stratification* (Eds Einsele, G. & Seilacher, A.) pp. 161–174. Springer Verlag, New York.

SHEPARD, F.P. (1932) Sediments of the continental shelves. *Geol. Assoc. Am. Bull.* **43**, 1017–1040.

SLOSS, L.L. (1962) Stratigraphical models in exploration. *J. Sediment. Petrol.* **32**, 415–422.

SLOSS, L.L. (1963) Sequences in the cratonic interior of North America. *Geol. Soc. Am. Bull.* **74**, 93–114.

SMITH, J.D. (1977) Modeling of sediment transport on continental shelves. In: *The Sea*, Vol. 6, *Marine Modeling* (Eds Goldberg, E.D., McCave, I.N., O'Brien, J.J. & Steele, J.H.) pp. 539–577. Wiley Interscience, New York.

STAMP, L.D. (1921) On cycles of sedimentation in the Eocene strata of the Anglo-Franco Belgian basin. *Geol. Mag.* **58**, 108–114, 194–200.

SWIFT, D.J.P. (1976a) Coastal shelf sedimentation. In: *Marine Sediment Transport and Environmental Management* (Eds Stanley, D.J. & Swift, D.J.P.) pp. 255–311. Wiley, New York.

SWIFT, D.J.P. (1976b) Continental shelf sedimentation. In: *Marine Sediment Transport and Environmental Management* (Eds Stanley, D.J. & Swift, D.J.P.) pp. 311–350. Wiley, New York.

SWIFT, D.J.P. & NUMMEDAL, D. (1987) Discussion: Hummocky cross-stratification, tropical hurricanes and intense winter storms. *Sedimentology*, **34**, 338–344.

SWIFT, D.J.P., HOLLIDAY, B., AVIGNONE, N. & SHIDELER, G. (1972a) Anatomy of a shoreface ridge system, False Cape, Virginia. *Mar. Geol.* **12**, 59–84.

SWIFT, D.J.P., LUDWICK, J.C. & BOEHMER, W.R. (1972b) Shelf sediment transport, a probability model. In: *Shelf Sediment Transport: Process and Pattern* (Eds Swift, D.J.P., Duane, D.B. & Pilkey, O.H.) pp. 195–223. Dowden Hutchinson & Ross, Stroudsburg, PA.

SWIFT, D.J.P., CLARKE, T., NIEDORODA, A.W., YOUNG, R.A. & VINCENT, C.E. (1981) Sediment transport in the Middle Atlantic Bight of North America, synopsis of recent observations. In: *Holocene Marine Sedimentation in the North Sea Basin* (Eds Nio, D.S., Schuttenholm, R.T.E. & van Weering, T.C.E.). Spec. Publ. Int. Assoc. Sediment. 5, 361–383.

SWIFT, D.J.P., NIEDORODA, A.W., VINCENT, C.E. & HOPKINS, T.S. (1985) Barrier island evolution, middle Atlantic shelf, USA, Part I: shoreface dynamics. *Mar. Geol.* **63**, 331–361.

SWIFT, D.J.P., HAN, G. & VINCENT, C.E. (1986a) Fluid process and sea floor response on a modern storm-dominated shelf: middle Atlantic shelf of North America. Part 1, the storm current regime. In: *Shelf Sands and Sandstone Reservoirs* (Eds Knight, R.J. & McLean, J.R.). Can. Soc. Petrol. Geol. Mem. 11, 99–119.

SWIFT, D.J.P., THORNE, J.A. & OERTEL, G.F. (1986b) Fluid process and sea floor response on a storm dominated shelf: middle Atlantic shelf of North America. Part II: response of the shelf floor. In: *Shelf Sands and Sandstone Reservoirs* (Eds Knight, R.J. & McLean, J.R.). Can. Soc. Petrol. Geol. Mem. 11, 191–211.

SWIFT, D.J.P., HUDELSON, P.M., BRENNER, R.L. & THOMPSON, P. (1987) Shelf construction in a foreland basin: storm beds, shelf sand bodies and shelf-slope sequences in the Upper Cretaceous Mesaverde group, Book Cliffs, Utah. *Sedimentology*, **34**, 423–457.

THORNE, J.A. (1988) Modeling storm bed genesis: oceanography or geology—which one holds the key? *Program, Workshop on Quantitative Dynamic Stratigraphy*, Society of Economic Paleontologists and Mineralogists, 14–18 February, Deckers, CO, 12 pp.

VAIL, P.R. (1988) Part 1: Seismic stratigraphy interpretation procedure. *Atlas of Seismic Stratigraphy* (Ed. Bally, A.W.). Am. Assoc. Petrol. Geol. Stud. 27, 1–9.

VINCENT, C.E. (1986) Processes affecting sand transport on a storm-dominated shelf. In: *Shelf Sands and Sandstones* (Eds Knight, R.J. & McLean, J.R.). Can. Soc. Petrol. Geol. Mem. 11, 121–132.

VINCENT, C.E., SWIFT, D.J.P. & HILLARD, B. (1981) Sediment transport in the New York Bight, North American Atlantic Shelf. Mar. Geol. 42, 369–398.

WATTS, A.B. & THORNE, J.A. (1984) Tectonics, global changes in sea level and their relationship to stratigraphical sequences at the U.S. Atlantic continental margin. Mar. Petrol. Geol. 1, 319–339.

WRIGHT, L.D. (1977) Sediment transport and deposition at river mouths: a synthesis. Geol. Soc. Am. Bull. 55, 857–868.

WRIGHT, L.D. & COLEMAN, J.M. (1972) River delta morphology: wave climate and the role of the subaqueous profile. Science, 172, 282–284.

WRIGHT, L.D. & COLEMAN, J.M. (1974) Mississippi River Mouth processes: effluent dynamics and morphologic development. J. Geology 82, 751–778.

WRIGHT, L.D., WISEMAN, W.J., BORNHOLD, D.B., PRIOR, D.B., SUHAYDA, J.N., KELLER, G.H., YANG, Z.-S. & FANG, Y.B. (1988) Marine dispersal and deposition of Yellow River silts by gravity driven underflows. Nature 332, 629–632.

# Sedimentation on continental margins, II: application of the regime concept

J.A. THORNE* and D.J.P. SWIFT[†]

*Research and Technical Services, ARCO Oil and Gas Company, Plano, TX 75075, USA; and
†Department of Oceanography, Old Dominion University, Norfolk, VA 23529, USA

## ABSTRACT

The theory of steady-state sedimentation (regime theory) has been applied extensively to the analysis of river systems and landscape evolution. In the regime concept, sedimentation systems are governed at geological scales by interdependent variables, whose mutual adjustments maintain a state of dynamic equilibrium. Regime models may be described in terms of equilibrium surfaces controlled by homeostatic responses of erosion or deposition. Regime theory offers geomorphologists a means to realize the consequences of the small-scale behaviour of sedimentary particles at large time and space-scales. The success of regime theory in geomorphological (denudation) modelling indicates that it is an appropriate tool for the complementary task of stratigraphic (basin accumulation) modelling.

The *depositional regime*, within the context of quantitative stratigraphic models, can thus be defined as the state of the geohistorical variables. These are sediment input rate ($Q$), sediment character ($M$), the sediment transport rate ($D$), and the rate of relative sea-level change ($R$). A change in any of the geohistorical variables produces, by definition, a change in depositional regime. Topographic profiles are one type of theoretical simplification of the equilibria governed by regime sedimentation. Such a topographic profile in defining the geometry of the resulting sedimentary deposit depends on its type, of which there are five: (1) *graded stream profile*; (2) *isostatic equilibrium profile*; (3) *coastal equilibrium profile, R-dependent case*; (4) *coastal equilibrium profile, Q-dependent, advective case*; and (5) *coastal equilibrium profile, Q-dependent, diffusive case*. The $R$-dependent coastal profile model describes autochthonous (transgressive, sediment starved) regimes, while the $Q$-dependent profile model describes allochthonous (regressive, sediment flooded) regimes. The isostatic equilibrium model is useful for investigating the behaviour of the continental margin depositional regime at larger time and space-scales.

Shelf depositional regimes, operating through time in response to shifting process variables, create in this fashion the depositional sequences of the rock record. Two examples of regime sedimentation are illustrated. The subaqueous topography of the inshore region of the Louisiana prodelta shelf is shown to form a $Q$-dependent advective profile that has adjusted to regime conditions of wave energy and sediment supply. In a second illustration of regime sedimentation, a hypothetical shelf is examined at the larger time and space-scale governing its coastal equilibrium profile. An algorithm is given, utilizing a simple, $R$-dependent, coastal equilibrium profile, that can be used to create computer simulations of the 'systems tracts' of sequence stratigraphy. This model is used to investigate the effect of sediment supply, rate of subsidence and sea-level fluctuation on these geometries.

## INTRODUCTION

The natural development of stratigraphic sequences involves major processes that occur over a range of time-scales. Sedimentary basins develop over periods of $10^8$ years. Changes in palaeoclimate, palaeoceanography, and sea-level are characterized by time intervals of $10^3$ to $10^7$ years. Sedimentary processes are driven by processes on the annual and subannual time-scales. The concept of regime sedimentation is one way to bridge the gap between these varying time-scales. Using this approach as a simplification of sedimentary dynamics allows the generation of quantitative stratigraphic models.

The roots of the regime concept are complex. Swift & Thorne (this volume, pp. 3–31) present a simplified regime model when they argue that the shelf surface is a surface of dynamic equilibrium, whose shape is controlled by the geohistorical variables of the rate of relative sea-level change (**R**), the rate and character of sediment input (**Q, M**) and the available fluid power (**P**), and resulting sediment transport rate (**D**). They stress the ratio of the rate of sediment input to the fluid power available to disperse it (**Q/P**) as a key to the equilibrium relationship. The defining regime variables will be distinguished from others in this paper by bold type.

We have been guided in our thinking by the work of L.L. Sloss (1962) who first defined these variables for purposes of basin analysis, and also by the work of D.W. Johnson (1919) who first applied Davisian ideas of geomorphic evolution to the continental shelf. However, regime ideas can be traced considerably further back, as far as the work of nineteenth century hydraulic engineers who were responsible for designing major irrigation systems in India, Pakistan and Egypt (Blench, 1957; Chorley *et al.*, 1984, p. 291). It was necessary for the movement of water through these channels that they have stable channels, or be 'in regime'. Empirical equations were developed that related velocity of flow to channel depth, and discharge to channel dimensions for 'regime' canals.

A related idea emerges in the writings of G.K. Gilbert (1877), who first introduced the idea of the 'graded' river, the transportation of whose load exactly consumed the potential energy released during flow, so that the channel neither aggraded nor eroded (degraded). Gilbert was applying thermodynamic concepts of equilibrium to physical geology as was his contemporary Willard Gibbs to chemistry (Pyne, 1975). Another contemporary, W.M. Davis, however, was more interested in incorporating the evolutionary concepts of Charles Darwin into geomorphological thought. Davis (1899) proposed the geomorphic cycle of youth, maturity and old age. The concept of grade was incorporated into the geomorphic cycle, but was seen as delayed until maturity. In the Davis model, rivers develop concave-upward longitudinal profiles. The profiles are considered an equilibrium response, although not one of constant form; thalweg slopes decrease as the erosion surface descends through maturity and old age to base level.

Davis' student, Douglas Johnson (1919), transferred the concept of the geomorphic cycle to the marine environment. At maturity, the shelf becomes graded as a 'profile of equilibrium' develops. It is

> steepest near the land where the debris is coarsest, and most abundant; and progressively more gentle further seaward where the debris has been ground finer and reduced in volume by a removal of a part in suspension. At every point the slope is precisely of the steepness required to enable the amount of wave energy there developed to dispose of the size and volume of debris there in transit ... A common error is to confuse ... profiles of equilibrium with [wave base] ... submarine platforms ... merely represent surfaces of equilibrium which are being slowly reduced toward a wave base ...

The concept of grade joined (or perhaps rejoined) the concept of regime in the 1950s (Knox, 1975), when geomorphologists, now better equipped with quantitative tools, returned to a consideration of hydraulic geometry (Leopold & Maddock, 1953; Blench, 1957). Channels were seen as complex systems of feedback relationships among variables, in which cause and effect are difficult to distinguish. The detailed response of such a system was considered to be indeterminate in that 'its complexity and richness of linkage make it difficult to specify the precise effects of a given input' (Chorley *et al.*, 1984).

In the ensuing spate of quantitative geomorphological studies, it became apparent that grade was only one aspect of the equilibria possible within such systems, and that 'mutual adjustments of the variables take place whether the stream is at grade, aggrading or degrading' (Wolman, 1955). Hack (1960) showed that the concave-upward profile that Gilbert (1877) had been calling the 'profile of equilibrium' was achieved rather early in the valley's history and was generally not delayed to maturity as Davis (1899) had maintained. Meanwhile, geological oceanographers, as yet without the equipment to precisely measure marine flows and its transported load, were rediscovering Douglas

Johnson, and were arguing over the critical time constants over which equilibrium (wave base) might be defined (Dietz, 1963, 1964; Moore & Curray, 1964)

We see as the common thread in this sequence of ideas the concept of a sedimentation system governed by interdependent variables. An equilibrium prevails, in that a change in one variable is compensated for by adjustments of the others. The equilibrium is a dynamic one, in that oscillations occur around a mean value that is itself changing through time (Chorley *et al.*, 1984). Such a concept might be called dynamic equilibrium theory, following Strahler (1952), Hack (1960), and Chorley *et al.* (1984, p. 11). We prefer, however, the more restricted term *regime theory*, historically used to denote a dynamic equilibrium occurring at geological time and space-scales. A regime, in this sense, may govern a sedimentation surface, for which homeostasis (the return, after disturbance, to the equilibrium state) involves erosion or deposition (see, for instance, Schumm, 1975).

As sedimentologists and stratigraphers, we find this geomorphological insight to be a valuable one. As Wheeler (1964) has demonstrated, the stratigraphic problem is the inverse of the geomorphological problem. The regime concept has been developed largely by geomorphologists concerned with the denudation of an emergent landscape. Stratigraphers are concerned with the blanketing and burial of a subsiding landscape, and regime theory can be applied usefully to this problem.

The divergence of approach, in geomorphological versus stratigraphic thought, to the problem of scale is particularly interesting. Stratigraphers have traditionally left dynamic analysis and the application of physical principles to sedimentologists, and their colleagues, the fluid dynamicists. The application of Newtonian physics to the small-scale problems of sediment dynamics has been a great success. However, the stratigraphic approach, at larger spatial and temporal scales, has been, until recently, a more descriptive one.

Sediment dynamics, at small time and space-scales, has been an important part of geomorphology too, but geomorphologists, perhaps because they could see and directly measure their landscapes, also have been concerned from the beginning with dynamics at larger scales. Here, Newtonian physics has not served as well. Geomorphologists, grappling with the problems of channel geometry have turned repeatedly to thermodynamic analogies (e.g. Leopold & Langbein, 1962), in which molecular behaviour is related to the macroscale as a system specified by equations of state. Modern thermodynamic thought has in part led into a more general paradigm of the relationship of disorder and indeterminacy to order called 'Chaos theory', hailed as a significant advance on classical physics (Gleick, 1987, p. 6). As quantitative geomorphology has developed over the past decades, its practitioners have been driven to take positions that, though lacking the new 'chaos' vocabulary, are the core concepts of chaos theory (Leopold *et al.*, 1964, p. 274; Chorley *et al.*, 1984. pp. 290–291).

We thus turn to the geomorphological concept of regime as a historically available and potentially unifying concept for analytical stratigraphy. To this end, the *depositional regime*, within the context of quantitative stratigraphic models, can be defined as the state of the geohistorical variables $Q$, $M$, $D$ and $R$ (defined by Swift & Thorne, this volume, pp. 3–31). A change in any of the geohistorical variables produces, by *definition*, a change in sedimentary regime. Sloss (1962) postulated that the geohistorical variables $Q$, $M$, $D$ and $R$ defined the shape, or external geometry, of a body of sedimentary rocks as a function of time, $t$, formally expressed as:

$$\text{shape}(t) = f[Q(t), M(t), D(t), R(t)] \qquad (1)$$

Numerical stratigraphic modellers have achieved realistic cross-sectional simulations of stratal geometries by specifying the shape function as a *topographic profile* (Horowitz, 1976; Pitman, 1978; Watts & Thorne, 1984; Helland-Hansen *et al.*, 1988; Jervey, 1988). By extension, a three-dimensional shape function can be specified as a *topographic surface*. The remainder of this paper, for simplicity, will refer to two-dimensional profiles. Profiles used in this manner are one type of theoretical simplification of regime sedimentation. The remainder of this paper will focus on the use and application of such profiles. Thorne *et al.* (this volume, pp. 59–87) and Swift *et al.* (this volume, pp. 89–152 & 153–187) discuss other aspects of regime sedimentation.

## KINDS OF REGIME SEDIMENTATION

### General

The role of a topographic profile in controlling the external geometry of sedimentary rocks depends on its type, of which there are five: (1) *graded stream profile*; (2) *isostatic equilibrium profile*; (3) *coastal equilibrium profile, **R**-dependent case*; (4) *coastal equilibrium profile, **Q**-dependent, advective case*; and (5) *coastal equilibrium profile, **Q**-dependent, diffusive case*.

**Graded stream profile**

A *graded profile* is achieved when the sediment transport system can transport all the sediment supplied to it through without loss. Mackin (1948), following Gilbert (1877; see above), described a *graded stream*, an example of this type of profile, as follows:

> A graded stream is one in which, over a period of years, slope is delicately adjusted to provide, with available discharge and with prevailing channel characteristics, just the velocity required for transportation of the load supplied from the drainage basin.

In a graded stream, therefore, the geohistorical variables $Q$, $M$ and $D$ have achieved an equilibrium. We can write, therefore:

$$\text{graded profile} = f_g(Q, M, D, x)$$
$$\text{such that } \mathring{A} = R = 0; Q = D \quad (2)$$

where $x$ is the horizontal coordinate along the profile and $\mathring{A}$ is the accumulation rate.

The Gilbert–Mackin concept of grade is an example of a class of geomorphological equilibrium described by Chorley *et al.* (1984) as a *steady state equilibrium* (Fig. 1A). The slope of the profile is fixed at each point along its course; any departure from the appropriate slope will result in either erosion or deposition sufficient to restore the equilibrium value. Therefore, the fourth variable, relative sea-level change, $R$, does not enter into the balance. The fundamental effect of relative sea-level change is to cause sediment accumulation or erosion; i.e. to cause the profile to depart from grade.

The graded stream profile was defined more broadly by Davis (1902) to include the effects of denudation and sea-level change. In the Davisian concept, profiles change during the course of landscape evolution. As the landscape matures, relief is reduced so that the profile through time becomes a family of successively flatter curves, each tied to the coastline (Fig. 1B). Ultimately, base level is attained; a surface of relief so subdued that streams can no longer release potential (gravitational)

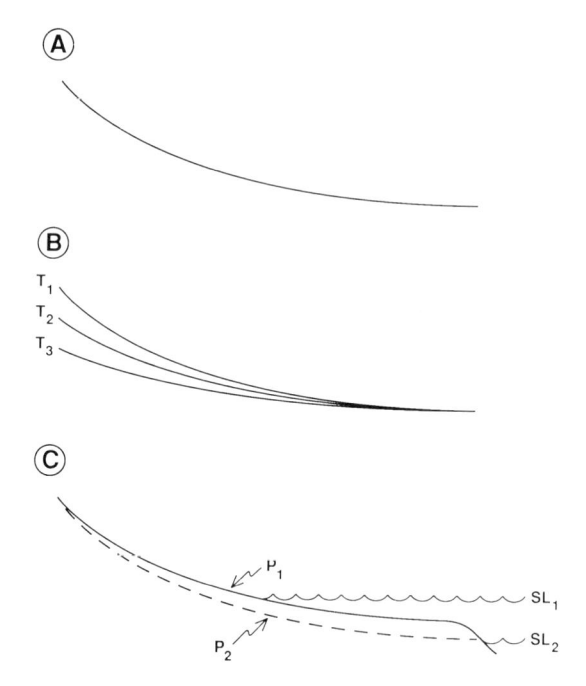

**Fig. 1.** Graded profiles. (A) the Gilbert–Mackin profile (steady-state equilibrium); (B) the Davis profile (dynamic equilibrium); (C) the Davis profile regraded in response to a sea-level fall.

energy in significant amounts, and no further reduction of the profile can occur. In this process, the slope at each point steadily decreases. Through the same period, drainage basins expand, and the discharge at a point on the profile increases, so that less slope is necessary to maintain the graded condition; thus equilibrium between discharge, velocity, and slope is maintained. The Davisian graded profile corresponds to the class of geomorphological equilibrium described by Chorley *et al.* (1984) as a *dynamic equilibrium*, in which the equilibrium value (in this case a slope) oscillates about a mean that is itself changing through time.

The steady-state and dynamic concepts of grade are not mutually exclusive, but rather are complementary (Brunsden & Thornes, 1979). Steady-state grade prevails at time-scales of a century, over which the effects of individual floods are compensated by adjustments of channel variables. Dynamic grade prevails over periods in excess of one million years, during which time there is compensation not only of hydraulic variation, but also of the changing geometry of the drainage basin.

In the Davisian model, the evolutionary clock is

periodically reset by base level (sea-level) changes, which are abrupt with respect to the rate of denudation. Relative sea-level falls attach a new stream segment to the old stream mouth. If the new segment is steeper or shallower than required for its discharge, the entire stream must ultimately regrade, by aggrading and prograding if too steep, or by downcutting if too flat (Fig. 1C).

Graded stream profiles are important in regulating sediment input ($Q$) for shelf regimes. Their role is discussed by Swift *et al.* (this volume, pp. 89–152).

## Isostatic equilibrium profile

A second type of regime profile occurs when sediment accumulation and relative sea-level changes are isostatically balanced so that sedimentation at each point along the profile maintains an equilibrium water depth or height above sea-level. It has long been recognized that this balance represents a combination of the effects of basin subsidence, isostatic compensation to sediment loading, compaction of underlying sediments, and secular eustatic variations (e.g. Barrell, 1917). Modern expression of these concepts is based on the flexural theory of isostasy as introduced by Vening-Meinesz (1931), and Gunn (1944). Following these ideas we can write for any time interval, $\Delta t$, of accumulated sediment, $\Delta A$:

$$\boldsymbol{R}\Delta t = [r_m/(r_m - r_w)] \Delta SL + \Delta T + \Delta E$$
$$= \Delta W + \Phi * \Delta A \qquad (3)$$

where $\Delta W$ is the rate of change of water depth, $*$ stands for spatial convolution and $\Phi$ is the isostatic response function for sediment accumulation in water. This modified value for sediment accumulation must be used because some of the shallowing owing to sediment accumulation is lost to load subsidence of the lithosphere. Spatial convolution of the isostatic response function is required because the effect of point load on the lithosphere extends for some distance beyond the point, the distance depending on the rigidity of the lithosphere. Loading at a point will depress that point, but will also depress neighbouring points to a lesser extent.

The symbols $r_m$ and $r_w$ are mantle and water densities. The variation in relative sea-level, $R$, is separated into three components: eustatic sea-level, $\Delta SL$, tectonic subsidence in water, $\Delta T$, and subsidence owing to expulsion of pore fluids by compaction, $\Delta E$. Note that $\Delta A$, the added accumulation,

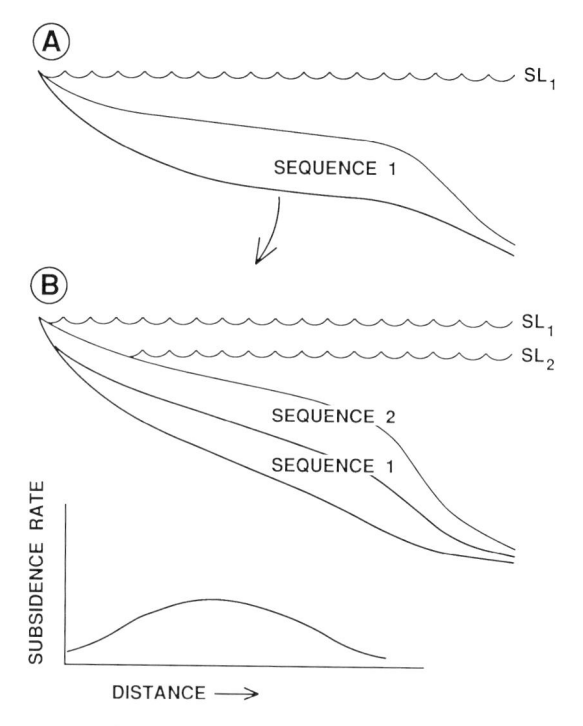

**Fig. 2.** The isostatic equilibrium profile.

does not include reworking of previous accumulation that does not increase the net thickness of the sediment column.

Accumulation along an *isostatic equilibrium topographic profile* occurs without a change in water depth (Fig. 2). Setting $\Delta W$ to 0 in equation (3) gives the defining criteria for such a profile:

isostatic equilibrium profile $= f_i\,(\boldsymbol{Q}, \boldsymbol{M}, \boldsymbol{D}, \boldsymbol{R}, x)$

such that $\quad \Phi * \mathring{A} = \boldsymbol{R} \qquad (4)$

Watts & Thorne (1984) were able to reproduce the salient features of USA Atlantic margin stratigraphy by combining a thermo-mechanical model of passive continental margin subsidence, eustatic sea-level changes derived from tectonic subsidence analysis and a sedimentation model using an isostatic equilibrium topographic profile of the shelf, slope and rise. Application of the isostatic equilibrium model to the analysis of continental margin sediment prisms is undertaken by Swift *et al.* (this volume, pp. 153–187).

## Coastal profile of equilibrium: *R*-dependent case

A third type of regime profile is the coastal profile of equilibrium, as described by D.W. Johnson (1919;

see above). The profile in question takes the form of a concave-upwards exponential curve, whose steepest limb (shoreface segment) is nearest the shore. The profile is thus attached to the shoreline, as is the graded stream profile. In fact, the original Johnsonian coastal profile model parallels in many respects the Davisian stream profile model described above. It is a response to wave power that diminishes with depth, and is limited by wave base, while the Davisian profile is a response to (gravitational) stream power that diminishes as the slope decreases, and is limited by base level. Like the Davisian profile, the Johnson coastal profile maintains grade throughout an evolutionary cycle, initiated by a sea-level change (see quote above).

We propose a more general form of the coastal evolutionary profile, in which relative sea-level change is one of the participating variables. The mathmatical description of the profile is based on two assumptions: (1) in the nearshore region the topographic profile adjusts quickly to grade in response to changes in the geohistorical variables; and (2) in other regions of the shelf, the topographic profile adjusts so as to maintain grade in the nearshore region. The defining criteria for such a profile is therefore:

equilibrium shoreline profile =
$f_c(Q, M, D, R, x - x_s)$

such that $Q = D(x_s)$      (5)

where $x_s$ is the position of the shoreline.

The coastal equilibrium profile can be referenced to a shoreline whose position can be controlled by the rate of sea-level variation, $R$, or by both $R$ and the sediment input rate, $Q$. In the $R$-dependent case, the profile must move with the shoreline as sea-level varies. Profile shifts in turn control the spatial distribution of sedimentation rates. An expression for sedimentation rate as a function of coordinate $x$ along the equilibrium shoreline profile under these conditions can be derived as follows: let $f_c(x - x_s)$ be the water depth or height above sea-level of the equilibrium shoreline profile as a function of distance from the shoreline. Then the accumulation rate as function of $x$ is given by:

$$\Phi Å(x) = R(x) - \frac{\partial}{\partial t} f_c(x - x_s) + C \quad (6)$$

where we have replaced, for simplicity, the convolution with the isostatic response function with a multiplication. Taking the derivative (assuming that the shape function $f_c$ does not change with time) results in the following expression:

$$\Phi Å(x) = R(x) - \frac{f'_c(x - x_s)}{f'_c(0)} R(x_s) + C \quad (7)$$

where $f'(x - x_s)$ is the slope of the equilibrium shoreline profile as a function of distance from the shoreline, and $f'_c(0) = f'_c(x - x_s)$ where $x = x_s$, or the slope at the shoreline. Thus, in equation (7), the load-adjusted sedimentation rate, $\Phi Å(x)$, is set equal to the sea-level rise rate $R(x)$ at point $x$, minus some fraction of sea-level rise at the shoreline, $R(x_s)$. The fraction is the ratio of the shoreline slope (maximum slope) to the slope at point $x$. It describes the extent to which the maximum slope has been achieved. A constant is added to the slope adjusted rise rate.

The mechanism described by the equation can be explained as follows. As sea-level rises, the exponentially shaped shoreface profile must slide landward up the inclined coastal plain surface (Fig. 3A). Sediment accumulation is occurring, yet the profile must be maintained. To do so, sedimentation must be inhibited at the shoreline, to keep the nearshore zone steep. This is accomplished in the

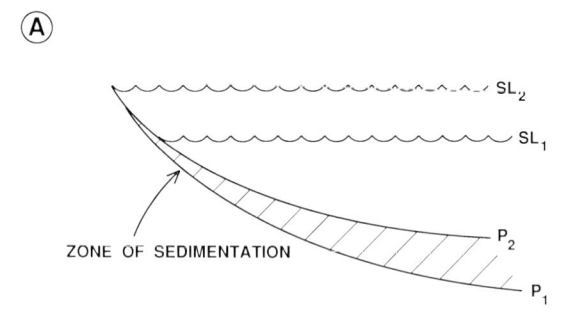

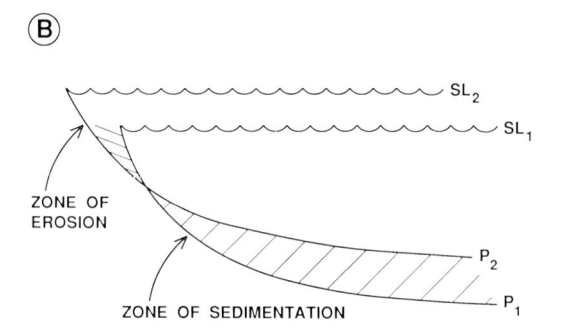

Fig. 3. The coastal equilibrium profile, $R$-dependent case. (A) Without shore erosion; (B) with shoreface erosion.

equation, because at the shoreline ($x = 0$) the slope fraction = 1, so (if subsidence is constant with $x$) the whole right-hand side of the equation goes to zero. The slope fraction decreases with increasing $x$, so the inhibition of sedimentation falls off with distance from shore until sedimentation becomes equal to $R$ ($x_s$). The constant $C$ adjusts the trajectory of the profile up the coastal plain surface, and determines whether its passage will be associated with deposition or erosion. If the constant is positive, sedimentation will extend into the shoreline. If it is negative, the zero point will be shifted seaward from the shoreline, and a shoreface zone of erosion will appear (Fig. 3B).

Thus if sea-level rise is positive and the shoreline is moving landward, the nearshore area will be sediment-starved and will retain the steep subaerial slope as it floods (Fig. 3A). If, on the other hand, $R$ is negative, then $\Phi\mathring{A}(x) = 0$ at the shoreline, but becomes increasingly negative seaward (the sea-floor erodes), until the erosion rate equals the sea-level fall rate.

The coastal profile model can be used to describe an *accommodation-dominated regime*, as defined by Swift & Thorne (this volume, pp. 3–31). In such a regime, as sea-level rises the lack of accumulation at the inshore end of the coastal profile leads to erosion, so that the profile is maintained primarily by the process of erosional shoreface retreat, and sediment is thus supplied to the shelf by shoreface bypassing in order to account for shoreface erosion. Thus in an autochthonous setting, the constant $C$ is a negative number, whose magnitude determines the intensity of erosion and the width of the erosional zone (Fig 3B). The role of $R$-dependent coastal profiles in generating stratigraphy on autochthonous shelves is described by Swift *et al.* (this volume, pp. 89–152 & 153–187).

The coastal profile of equilibrium has been used extensively in the literature to study sedimentation response at various time-scales. Studies of model beach profiles that emphasize yearly adjustment to hydraulic climate include Waters (1939), Johnson (1949), Iwagaki & Noda (1962), Johnson & Eagleson (1966), Noda (1972), Sunamara & Horikawa (1974) and Herbich (1977).

On a longer time-scale, models of shoreline retreat due to Holocene sea-level rise have been constructed by Bruun (1962; see discussion by Swift & Thorne, this volume, pp. 3–31). The Bruun concept has been applied by Schwartz (1965, 1967), Edelman (1968, 1970), Dubois (1977), Rosen (1978), Dean & Maurmeyer (1983), Everts (1987) and Pilkey & Davis (1987).

On a still longer time-scale, the concept of an equilibrium shoreline profile has been used to study the stratigraphic effects of cyclic relative sea-level changes. Pitman (1978) examined the case in which the equilibrium shoreline profile has a constant slope. In this case the slope ratio in the second term on the right side of equation (7) reduces to unity, hence:

equilibrium slope model $\quad \Phi\mathring{A}(x) = R(x) - R(x_s)$ \hfill (8)

Jervey (1988) and Posamentier *et al.* (1988) have used the concept of an equilibrium shoreline profile to provide a theoretical framework for the depositional sequence model of Vail (1987). In this model the equilibrium profile exhibits a relatively steeply dipping segment at the *depositional shoreline break* (shelf edge). In this case the slope ratio in the second term on the right side of equation (7) approaches zero, hence:

shoreface model $\quad \Phi\mathring{A}(x) \cong R(x)$ \hfill (9)

The relevance of the Jervey–Posamentier version is discussed by Swift *et al.* (this volume, pp. 153–187).

## Coastal profile of equilibrium: $Q$-dependent case

In the $Q$-dependent case, the input of sediment causes movement of the profile, as well as vice-versa. The general criteria for a $Q$-dependent profile is given by:

$Q$-dependent profile $= f_s(Q, M, D, R, x)$

such that $\quad \Phi\mathring{A}(x) - R = u_v + u_h f'_s(x)$ \hfill (10)

where $u_v$ and $u_h$ are two sedimentation rate coefficients specifying the vertical and horizontal translation rates of the profile, respectively.

Equation (10) says that the load-adjusted sedimentation rate $[\Phi\mathring{A}(x)]$, minus the sea-level rise rate, $R$, equals the rate of vertical profile rise, $u_v$, plus the slope-adjusted horizontal rise rate, $u_h f'_s(x)$. This mechanism of profile maintenance can be explained as follows. With no sea-level rise, the profile responds to the sediment input rate, $Q$, by moving up and out as sedimentation aggrades the sloping shelf floor. The accumulation rate at a point, $\Phi\mathring{A}(x)$, must include the entire vertical component of shift of the profile ($u_v$). If the profile is the

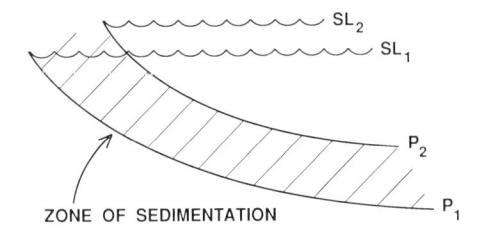

Fig. 4. The coastal equilibrium profile, $Q$-dependent case.

characteristic concave upward curve, then the accumulation rate also benefits from $u_h$, because horizontal translation of the curved profile moves a shallower portion of the profile on to the station in question, so a slope-adjusted fraction of the horizontal translation component, $u_h f'_s(x)$, must be added (Fig. 4). While shoreline migration controls sedimentation for the $R$-dependent coastal profile, the reverse is true for the $Q$-dependent profile. A positive value of $R$ works to decrease the apparent sedimentation rate, $\Phi \mathring{A}(x)$, for a $Q$-dependent profile, and in equation (10), must be subtracted from it.

The $Q$-dependent coastal profile model may thus be used to describe the sedimentary regime described by Swift & Thorne (this volume, pp. 3–31) as a *supply-dominated shelf regime*. In such a regime, the shoreface is nourished by littoral drift derived from river mouths (river mouth bypassing), and the shoreface progrades. The role of $Q$-dependent profiles in generating stratigraphy on allochthonous shelves is described by Swift *et al.* (pp. 89–152 & 153–187).

There are two types of $Q$-dependent profiles depending on whether sediment transport is primarily by the *diffusive* or *advective* components of sediment transport. Studies of the hydraulic climate of modern shelves have shown that at large time and space-scales, sediment particles undergo trajectories that can be described by the diffusion and advection equations of stochastic motion (e.g. Clarke *et al.*, 1983). Any individual particle is resuspended for minutes, hours or days, transported metres or kilometres in some current direction, and then buried again for months, years, or centuries. Though the path of any individual particle cannot be predicted, the flux of many particles, after sufficient time and spatial averaging, can be described by advective and diffusive components of sediment transport (see discussion by Swift & Thorne, this volume, pp. 3–31).

The advective component of transport results from a long-term flux of sediments from one area to another owing to a net drift. Advective sediment flux can be given as the product of net, time-averaged velocity of transport, $V$, and the load in the water column, $L$. Accumulation, the net sum of this flux in and out of the area, is the spatial gradient of the sediment flux along the direction of sediment transport. If the direction of sediment transport along which $V$ is measured is designated $x$, we can write:

$$\mathring{A} = \frac{\partial L(x) \cdot V(x)}{\partial x} \quad (11)$$

where $L$ is given in sediment volume per unit area.

Substituting into equation (10) from equation (11) and integrating with respect to $x$ we have:

$$R = -u_v$$

$$\Phi L(x) \cdot V(x) = u_h f_s(x) + \text{constant of integration} \quad (12)$$

where the constant of integration can be evaluated from boundary conditions at $x = 0$.

The diffusive component is a measure of sediment mobility along the sea-floor. Local concentrations of highly mobile sediments are quickly spread over large distances of the sea-floor by the diffusive component of sediment transport. For instance, progressive vector diagrams of sediment transport on the Atlantic continental shelf show that transport down-shelf in one storm event is nearly equalled by up-shelf transport during a subsequent storm (Vincent *et al.*, 1981). The net advective flux of sediment motion in this situation may be low. The diffusive mixing, however, will be high. Many authors have shown that the evolution of topography can be modelled as a diffusive process (e.g. Culling, 1960; Scheidegger, 1961; Hirano, 1968; Schwarzacher, 1975; Kenyon & Turcotte, 1985). Following Kenyon & Turcotte (1985) the form of a steady-state diffusive profile is given by:

diffusive profile ($\alpha$ constant)  $f_{sd}(x) =$

$$w_\infty \left[ 1 - \exp\left(-\frac{u_h x}{\Phi \alpha}\right) \right] \quad (13)$$

where $w_\infty$ is the water depth for large $x$ and $\alpha$ is a diffusion coefficient with units of length squared per unit time.

Equation (13) is derived by assuming that the diffusion coefficient, $\alpha$, remains constant with $x$. In general, the diffusion coefficient varies as the product of the time-averaged standard deviation of the velocity field, $V(x)$ and the percentage of time that the sea-bed is in a state of suspended transport (Clarke et al., 1983). The general criteria for a steady-state diffusive profile from equation (10) is given by:

diffusive profile (α variable)   $\Phi \dfrac{\partial}{\partial x} a(x) \dfrac{\partial}{\partial x} L(x) = u_h f'_{sd}(x)$   (14)

## REGIME CONDITIONS OF THE LOUISIANA PRODELTA SHELF

The Louisiana prodelta shelf is an ideal location in which to document regime conditions. Stratigraphic sections of the prodelta shelf indicate a rapid progradation of the prodelta profile since the late Quaternary (Frazier, 1974). This profile, including the shoreline, deltaic plain, delta front, continental shelf and slope, has prograded some 40 km during the Holocene (Fisk & McFarlan, 1955) owing to the large supply of dominantly fine sand and silt brought to the Gulf by the Mississippi River. Rates of progradation of the modern prodelta shelf are approximately 76 m y$^{-1}$ (Fisk et al., 1954; Kenyon & Turcotte, 1985). The Louisiana prodelta shelf is a classic example of an autochthonous shelf (see discussion of the autochthonous regime of the Louisiana shelf by Swift et al. (this volume, pp. 153–187)).

The Mississippi delta is a shelf edge rather than a platform delta (Galloway & Hobday, 1983). Platform deltas are, in terms of their relative dimensions, pancake-thin deposits, whose subaerial surfaces, nearly at sea-level, are separated from their shallow prodelta shelves by shorefaces of 10 m or less of relief (Wright & Coleman, 1972). The Mississippi delta has, however, prograded nearly to the shelf edge, and has developed a more complex profile in the process. The transition from a platform delta to a shelf edge delta occurs as deeper water is approached, the gradient in regime parameters become so steep that a discontinuity develops in the regime-maintained equilibrium surface. The shallow, rapidly aggrading prodelta shelf develops a prograding constructional scarp. Behind the scarp or 'delta front' (Coleman, 1988), the subaerial delta plain and the prodelta shelf lie at almost the same elevation, and are separated only by the shoreface, 10 m, or less, high. As progradation continues, the submarine scarp bridges the increasing discrepancy in elevation between the newly formed prodelta shelf and the old shelf surface being buried. Thus the profile of the Louisiana prodelta shelf between South Pass and Main Pass consists of four segments each of relatively constant slope (Fig. 5). The shelf between distributaries (interdistributary bay) extends 15 km from the bayline at a very low gradient (c.0.05°). The 'delta front' in this region now stands 60 m above the remnant of the old outer shelf surface. Delta front gradients are steeper and range from 0.2° to 0.5°. The outer shelf sector has the same low gradient as the higher, inner sector, and on the upper slope, the steep gradient of the 'delta front' is resumed.

We are concerned here with the equilibrium sur-

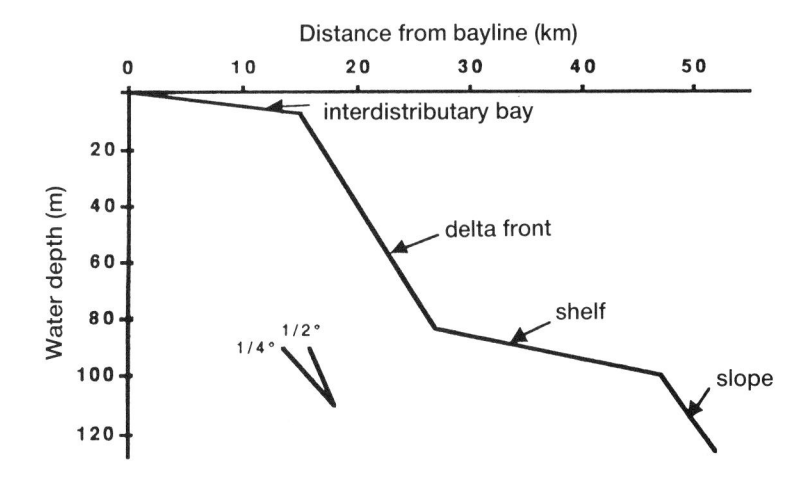

Fig. 5. Generalized modern subaqueous topographic profile of the Mississippi Delta. Bathymetric data from Wright & Coleman (1972), Kenyon & Turcotte (1985) and Coleman (1988).

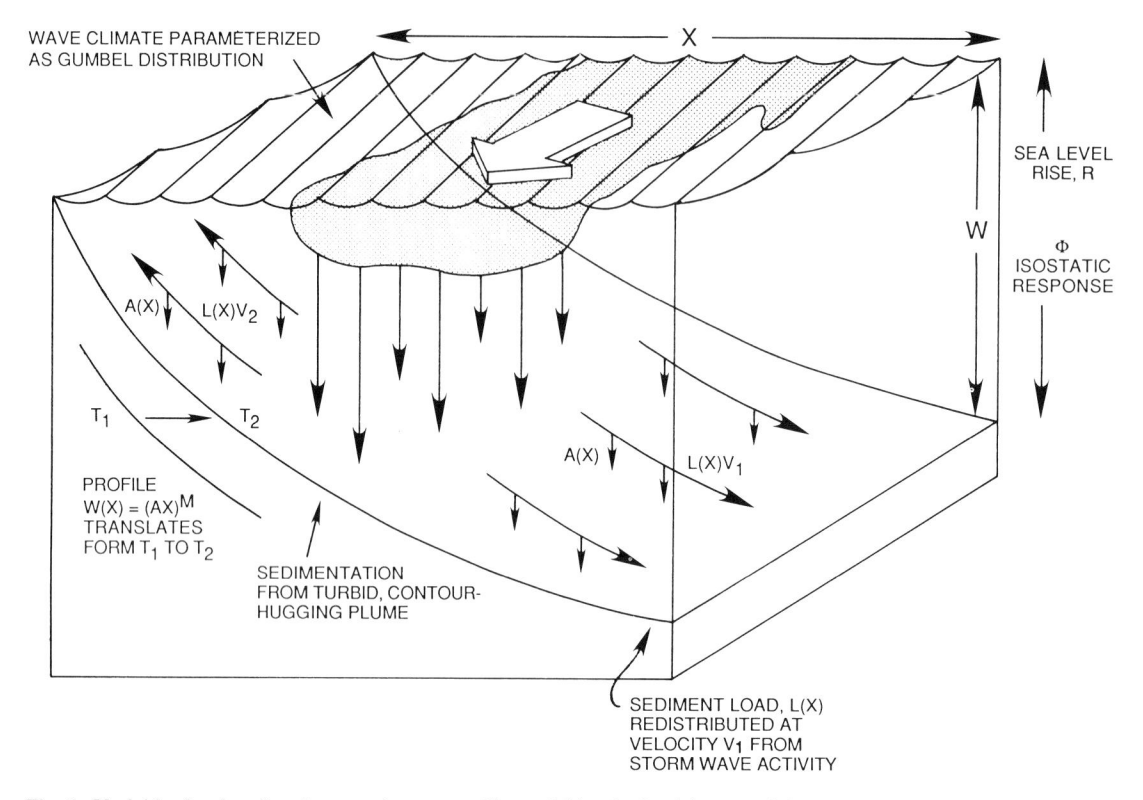

**Fig. 6.** Variables for the advective, steady-state profile model for the Louisiana prodelta shelf.

face maintained on the inner prodelta shelf. Wright & Coleman (1972), in a study of the wave climate of the inner sector, suggest that it is determined by a balance between the nearshore wave power and the rate at which the Mississippi River can supply sediment to the nearshore zone. Coleman & Prior (1982) and Lindsay et al. (1984) show that on the oversteepened 'delta front', the profile is controlled by subaqueous landslides and sediment creep.

To quantitatively test the hypothesis of Wright & Coleman (1972) concerning the shelf sector, we have applied semi-empirical coastal engineering methods to calculate the variation along the profile of the expected sediment accumulation rate as a function of oceanographic variables (see Appendix I). In order to make the problem tractable to quantitative analysis we have made two assumptions based on review of the oceanography of the Louisiana prodelta shelf (Scruton, 1956; Wells, 1983; Schroeder et al., 1986; Dinnel & Wiseman, 1986) as follows: (1) the inshore circulation system advects sediment from the delta mouth into bay areas as a contour-following, flood-stage jet of sediment as a result of the action of river-mouth bypassing; and (2) sediment is redistributed from this main axis of circulation by the action of storm currents.

In general, subaqueous topographic profiles can be expressed as a power law of the form:

$$w = (A \cdot x)^m \quad (15)$$

where $w$ is the water depth in metres, $x$ is the distance from land in metres, and $A$ and $m$ are empirical coefficients (Fig. 6). Variation along a profile of this form of sediment accumulation rate is a function of the coefficients $A$ and $m$. If the regime concept is applicable, then in order for a topographic profile to achieve steady state, the values of $A$ and $m$ must adjust to the wave and current climate so as to produce sedimentation that satisfies the criterion given in equation (10).

In Fig. 7 we show, as a function of distance along a profile with $A = 4 \times 10^{-4}$ and $m = 1.25$, the computed relative, time-averaged, suspended sediment load due to the exerted shear stress of waves on the sea-floor. Sediment concentration increases landward as a function of increasing along-shore

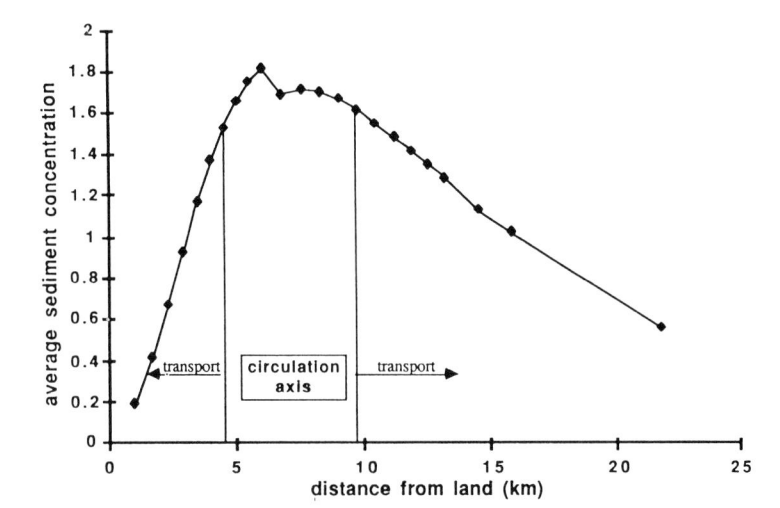

**Fig. 7.** Computed relative, time-averaged, suspended sediment load due to the exerted shear stress of waves on the sea-floor as a function of distance along a profile with $A = 4 \times 10^{-4}$ and $m = 1.25$.

mean velocity and increasing near-bottom wave orbital velocity. A maximum in suspended sediment concentration occurs where these parameters begin to decrease landward in response to the increasing role of bottom friction on the flow.

The time history of bed shear stress is computed by the methods described in Appendix I. The relative frequency of storms of different sizes and generated waves of different sizes have been calculated from deep-water wave climate statistics of the Gulf of Mexico (Bea, 1974; Quayle & Fulbright, 1975). The shallow-water wave climate is computed from deep-water statistics by accounting for the effects of coastal set-up, wave set-up, wave breaking, and wave attenuation by bottom friction.

The predicted average sediment load achieves a maximum value between c.5 and c.10 km offshore and falls off linearly in offshore and onshore directions. The zone of maximum sediment load acts as a sediment source for the rest of the profile. Sediment is advected into this zone during floods. During storms the decreasing concentration gradient to either side of this zone acts by diffusion transport to redistribute sediment towards the landward and seaward ends of the profile.

Accumulation is proportional to gradient of transport velocity times sediment load. We will assume that gradients of transport velocity do not contribute in the long-term average to net sedimentation. Given a net landward transport velocity given by $-v_l$ and, correspondingly, a seaward velocity of $v_s$, then equation (11) gives:

$$\mathring{A}(x < 5\,\text{km}) = -v_l \frac{\partial L(x)}{\partial x}$$

$$\mathring{A}(x > 10\,\text{km}) = v_s \frac{\partial L(x)}{\partial x} \qquad (16)$$

Although $v_s$ and $v_l$ are not known, the gradient of the sediment load in equation (16) can be determined from the model results of Appendix I.

For any given set of values for $A$ and $m$, the relative accumulation rates across the profile predicted from equation (16) can be compared with the accumulation rates from equation (10). In this comparison the shape of the two curves is known but the absolute magnitude is not. A measure of the error of fit, $\varepsilon$, is given by:

$$\varepsilon = \frac{1}{N}\sum_{i=1}^{N}\left|\frac{\mathring{A}_{\text{pr}}(i)}{\mathring{A}_{\text{ss}}(i)} - 1\right| \qquad (17)$$

where $\mathring{A}_{\text{pr}}(i)$ is the predicted accumulation rate from equation (15) assessed at sampling point $i$ along the profile, $\mathring{A}_{\text{ss}}(i)$ is the accumulation rate from equation (10), and $N$ is the total number of sampling points chosen. In this comparison, the velocities $v_l$ and $v_s$ have been chosen to minimize the net error. Table 1 gives the percentage error determined in this way for 12 combinations of coefficients $A$ and $m$. The steps taken to arrive at this comparison are indicated in Fig. 8.

As can be seen in Table 1, coefficients $A$ and $m$ are narrowly constrained by the condition that the

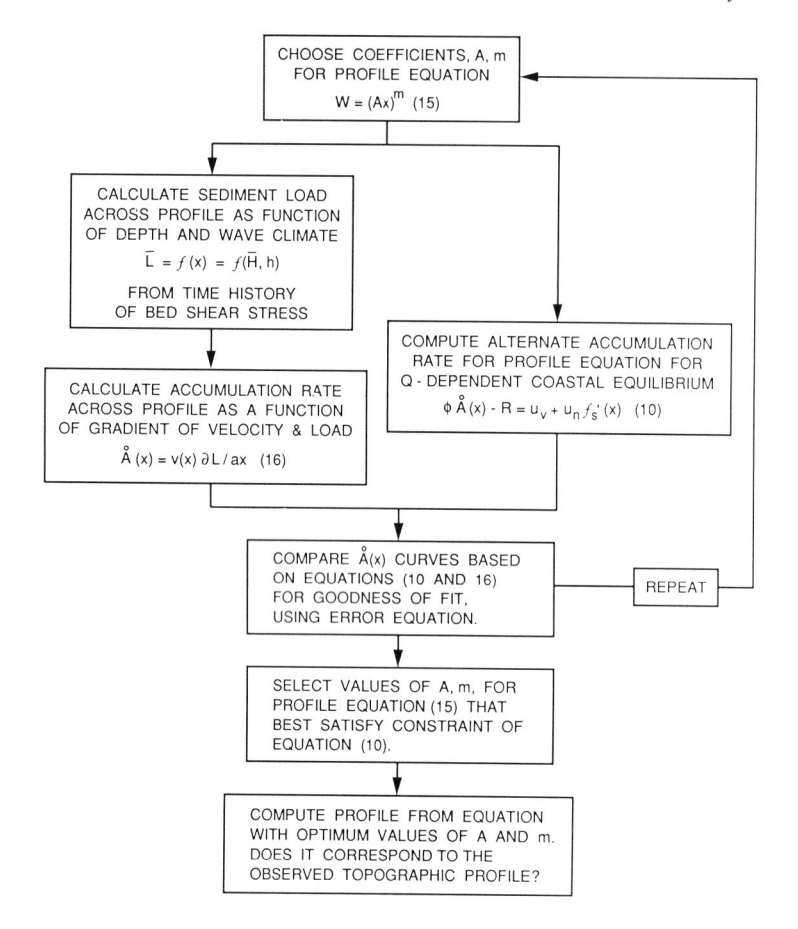

**Fig. 8.** Computational scheme for evaluation of Louisiana shelf for $Q$-dependent equilibrium condition.

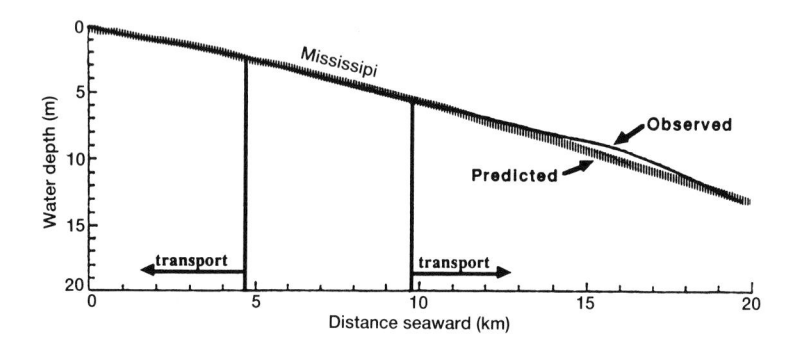

**Fig. 9.** Comparison of observed topographic profile (from Wright & Coleman, 1972) of the Louisiana prodelta shelf and predicted profile based on regime theory.

wave and current climate produces sedimentation that satisfies the criterion given in equation (10). Figure 9 shows the predicted $Q$-dependent advective profile for the best-fitting values of $A = 4 \times 10^{-4}$ and $m = 1.25$ compared with a representative topographic profile of the South Pass and Main Pass inshore areas. Most importantly, the figure shows that the present-day topographic profile clearly has adjusted to the predicted profile. We conclude that the present subaqueous topography of the inshore delta region forms a coastal profile of equilibrium that has adjusted to the prevailing allochthonous regime conditions of wave energy and sediment supply.

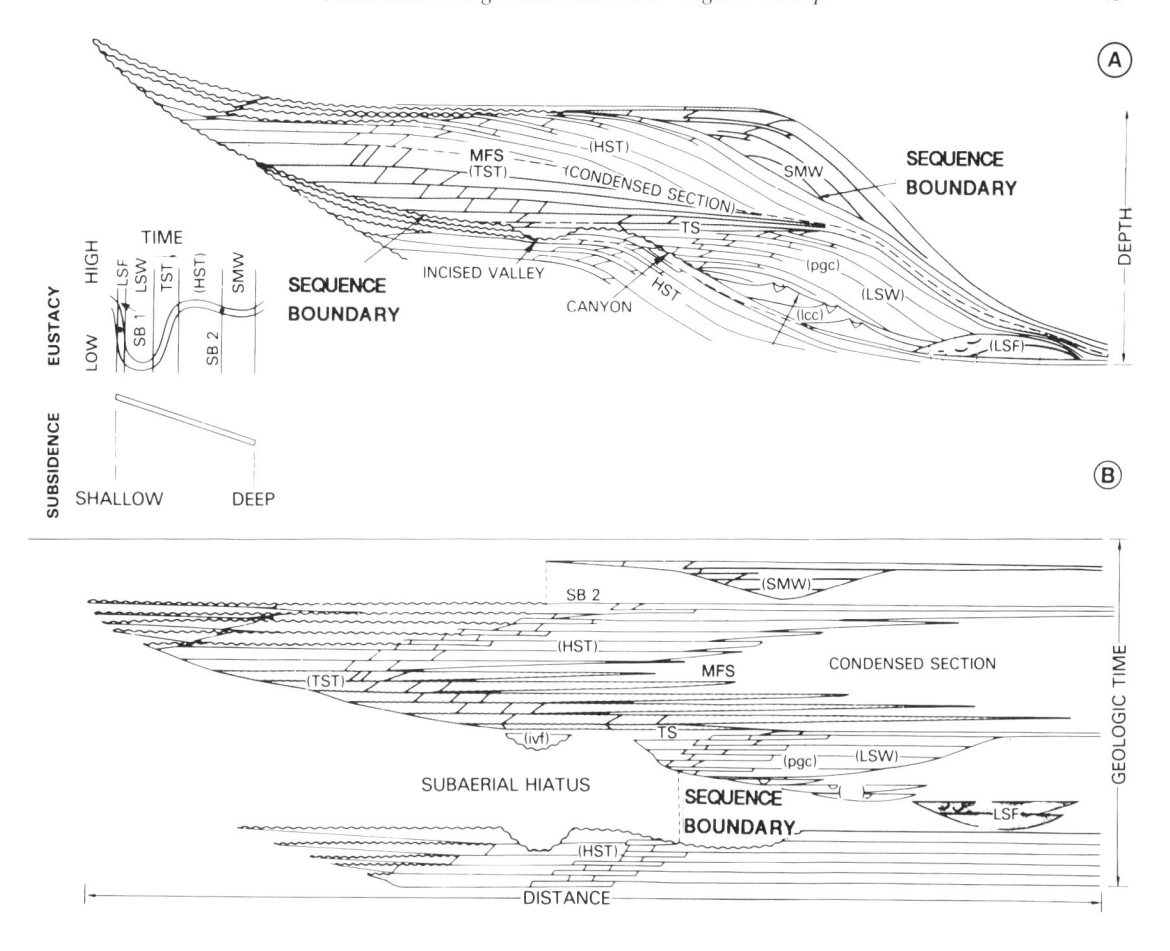

**Fig. 10.** The Vail–Haq model for a depositional sequence. (A) Lithostratigraphic diagram; (B) chronostratigraphic diagram. HST = highstand systems tract. TST = transgressive systems tract. Lowstand systems tract elements include SMW (shelf-margin wedge), LSW (lowstand wedge) and LSF (lowstand fan). SB = sequence boundary. MFS = maximum flooding surface. ivf = incised valley fill. pgc = prograding complex. lcc = levee channel complex. (From Haq *et al.*, 1987.)

## APPLICATIONS OF REGIME MODELS TO SEQUENCE STRATIGRAPHY

The application of the regime approach to geological interpretation of the ancient rock record has been used extensively in sequence stratigraphic interpretations of seismic profiles. The depositional sequence model of Vail (1987) and Haq *et al.* (1987) is used to interpret the geohistory of sea-level, subsidence, and sediment supply, on the basis of observed patterns of stratal geometries (Fig. 10). Computer simulations of the basic stratal patterns shown in this depositional sequence model have been carried out by Helland-Hansen *et al.* (1988) and Jervey (1988) based on a simplification of the concept of regime sedimentation. These models

**Table 1.** Average percentage deviation from steady-state accumulation rates for predicted accumulation rates based on a wave energy model for various values of coefficients $A$ and $m$. The best fit is achieved when $A = 4 \times 10^{-4}$ and $m = 1.25$

| $A$ | $m$ | | |
|---|---|---|---|
| | 1.20 | 1.25 | 1.30 |
| $2 \times 10^{-4}$ | 338.0 | 189.0 | 128.0 |
| $3 \times 10^{-4}$ | 63.9 | 45.0 | 47.0 |
| $4 \times 10^{-4}$ | 48.0 | 23.4 | 47.1 |
| $5 \times 10^{-4}$ | 57.0 | 61.8 | 68.1 |

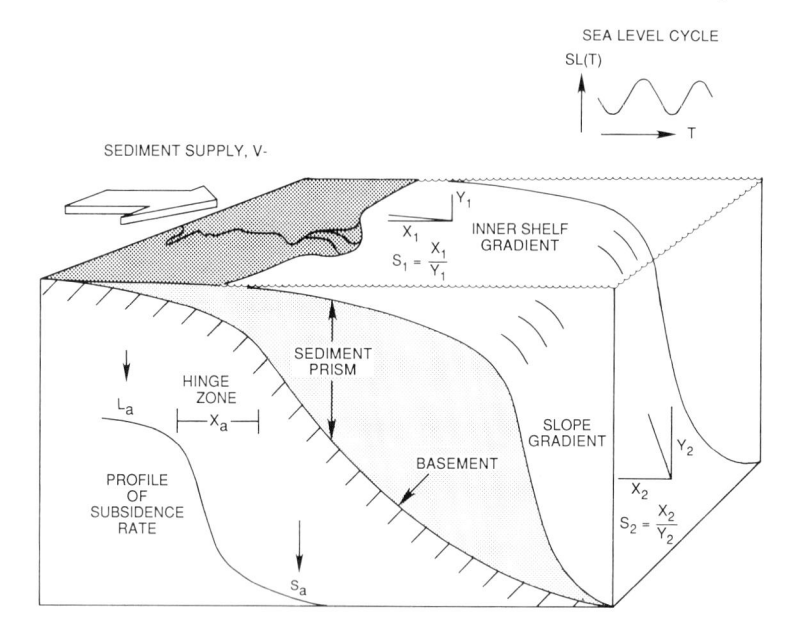

**Fig. 11.** Variables for the *R*-dependent coastal equilibrium profile model for an Atlantic-type margin.

create synthetic stratigraphy by assuming that an *R*-dependent coastal profile controls the shelf topography, while the continental slope is characterized by a *Q*-dependent profile of constant gradient. In general, the gradient of the coastal profile is much less than the gradient of the steady-state slope. The position where these two profiles meet has been called variously the *depositional shelf edge* and the *depositional shoreline break* (Vail, 1987). In this paper, the profile position will be referred to as the *depositional break in slope*.

Helland-Hansen *et al.* (1988) and Jervey (1988), unfortunately, do not document the sensitivity of model results to variations in assumed model parameters. The control of geohistorical variables, *Q*, *M*, *D* and *R*, on stratal geometry patterns depends in models of this type on: (1) the time history of eustatic sea-level; (2) the form of basin subsidence; (3) the rate of sediment supply; and (4) the assumed form of the equilibrium shoreface and steady-state slope profile.

To illustrate the relative control of each of these factors a numerical algorithm has been constructed to simulate regime sedimentation of the isostatic equilibrium type (Appendix II). The algorithm is constructed to create synthetic depositional sequences on typical Atlantic-type passive margins. A sigmoidal form of subsidence is assumed, which is described by three zones: (1) a landward zone of low subsidence rate; (2) a seaward zone of greatest subsidence rate; and (3) a transitional 'hinge' zone that smoothly connects these two previous zones. The geological input to this simple algorithm is specified by: (1) a eustatic sea-level curve, $SL(t)$; (2) the rate of sediment supply (assumed constant), $V$; (3) the rate of subsidence in the seaward zone of greatest subsidence, $S_a$; (4) the width of the transitional hinge zone, $x_a$; (5) the gradient of the shoreface equilibrium profile (assumed constant), $s_1$; and (6) the gradient of the profile of the slope, $s_2$ (see Fig. 11).

Vail *et al.* (1977) and Vail (1987) describe two types of sequence geometries (Fig. 10): a *type 1* sequence displays: (1) onlap of *lowstand wedge* deposits against the slope profile at a point well below the depositional break in slope; and (2) downlap against basin-floor deposits. Whereas the *shelf-margin deposits* of a *type 2* sequence: (1) onlap above the depositional break in slope; and (2) downlap against the slope profile. To model these two types of sequences depositional geometries are assumed to take two forms depending on whether the shoreline is landward or seaward of the depositional break in slope of the immediately preceding deposition. If the shoreline is landward of this break in slope, sediment is deposited as a wedge with topset beds inclined at the gradient $s_1$ and foreset beds inclined at gradient $s_2$. If the shoreline is at or seaward of this break in slope then sediments are bypassed to the basin. It is assumed that base-of-

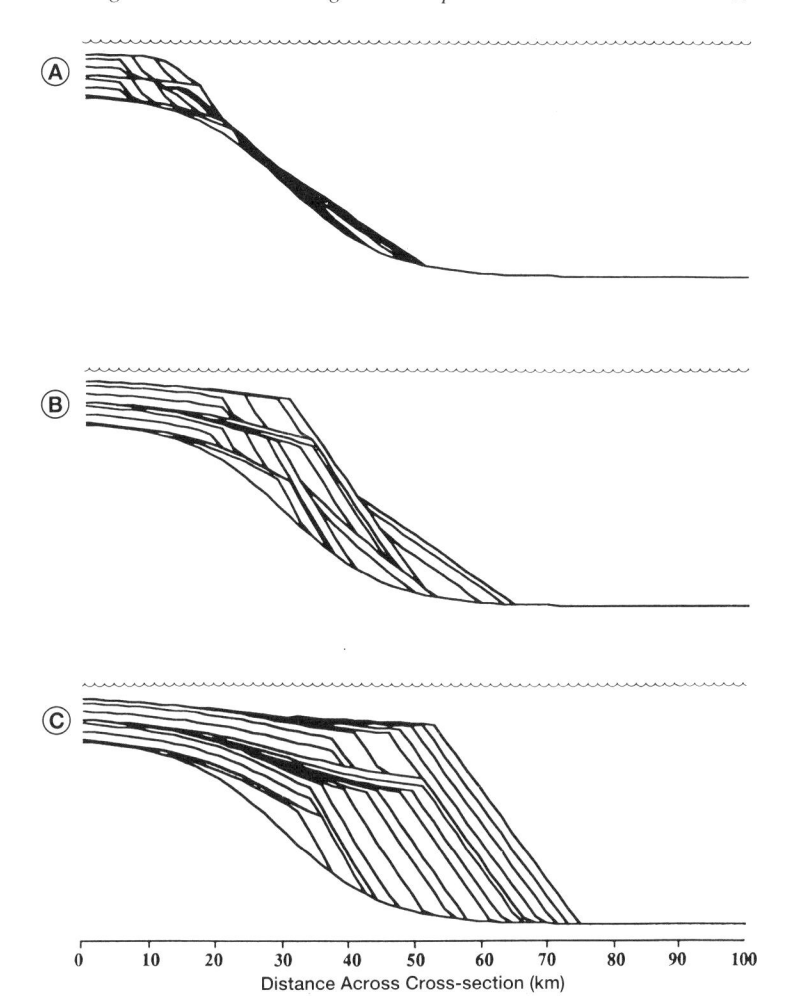

**Fig. 12.** *R*-dependent coastal equilibrium model. Effect of varying sediment supply on resulting geometry is shown for simulations created with the eustatic sea-level curve of Fig. 17. Case B is a standard reference run created with a sediment supply of 3.75 km² my⁻¹, a shelf gradient of 0.08°, a slope gradient of 2°, a basin subsidence rate of 50 m my⁻¹ and a hinge zone width of 30 km. The model cross-section shows lines (time lines) of synchronous deposition for every simulated 200 000 years. The cross-sections are shown at the end of 4 my of deposition. The water depth of the sea-floor on the far right side of model B is 600 m. Case A, low sediment supply (1.25 km² my). Case C, high sediment supply (7.5 km² my).

slope deposition occurs with a bedding inclination of $s_2/2$. The point of downlap of base-of-slope deposits against the basin floor is determined by requiring that the rate of deposition equal $V$.

The control of geohistorical variables $Q$, $D$ and $R$ on simulated stratal geometry patterns using the algorithm described in Appendix II is shown in Figs 12–16. Variable $Q$ is specified by the sediment input rate, $V$; variable $D$ by the gradient $s_1$ and $s_2$; and variable $R$ by the addition of basin subsidence specified by $S_a$ and $x_a$, and by eustatic sea-level. The effects of geohistorical variable $M$ and the component of $R$ owing to sediment compaction are not included in this simple model. Figures 12–16 show three simulations each. Model case B, in each figure, is a standard reference run using 'average' conditions for model parameters. Model cases A and C show the effect of varying one model parameter while keeping all other parameters identical to the standard case. Simulations shown in Figs 12–16 use the eustatic sea-level curve given in Fig. 17. This sea-level curve was constructed by superimposing a sinusoidal 2 my period, 100 m amplitude cycle on the rising limb of a 16 my period, 100 m amplitude cycle.

The pattern of stratal geometries seen in Figs 12–16 is similar to the stratal geometries of the depositional sequence model of Haq (1987) and Haq *et al.* (1987). The simulation results create model 'look-alikes' for the lowstand wedge, transgressive systems tract, highstand systems tract, and shelf-margin systems tract. The model does not produce features of a lowstand fan, incised valley, or canyon. Jervey (1988) and Helland-Hansen *et al.* (1988) have

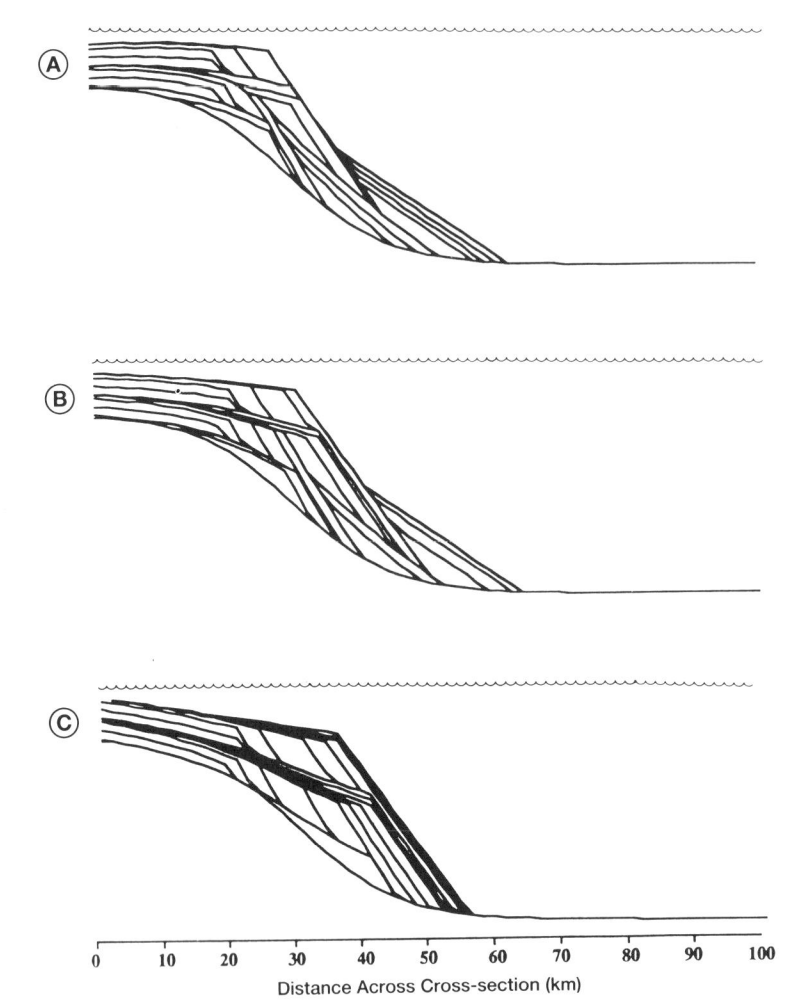

**Fig. 13.** A, low (0.02°); B, medium (0.08°); C, high (0.1°). Other *R*-dependent coastal equilibrium model—effect of varying shelf gradient: parameters as in Fig. 12.

noted previously the apparent similarity between numerical model results based on algorithms similar to the one used here and the seismic sequence stratigraphy model of Vail (1987) and Haq *et al.* (1987).

Figures 12–16 show that the conditions leading to the formation of a type 1 versus type 2 sequence is sensitive to model parameters $V$, $s_1$ and $s_2$. Conditions of high sediment supply (Fig. 12C), high shelf gradient (Fig. 13C) and high slope gradient (Fig. 14C) all prevent the shoreline from falling below the depositional slope break. Falling sea-level fails to produce a type 1 sequence when either: (1) the water depth of the slope break during highstand time is too deep to allow a relative sea-level fall to expose the shelf; or (2) the rapid subsidence rate at the slope break effectively cancels the eustatic component of sea-level fall.

When sediment supply is high (Fig. 12C), the width of the shelf is large, producing a relatively deep water depth at the depositional break in slope. Furthermore, the position of the slope break is located relatively far seaward where subsidence rates approach their maximum. Thus, a type 2 sequence is to be expected for conditions of high sediment supply. High shelf gradient (Fig. 13C) and high slope gradient conditions (Fig. 14C) also lead to a relatively deep, rapidly subsiding slope break.

Model simulations for low, average, and high rates of basin subsidence (Fig. 15C) all show type 1 sequence geometries. For the case of high basin subsidence, the position of the depositional break in

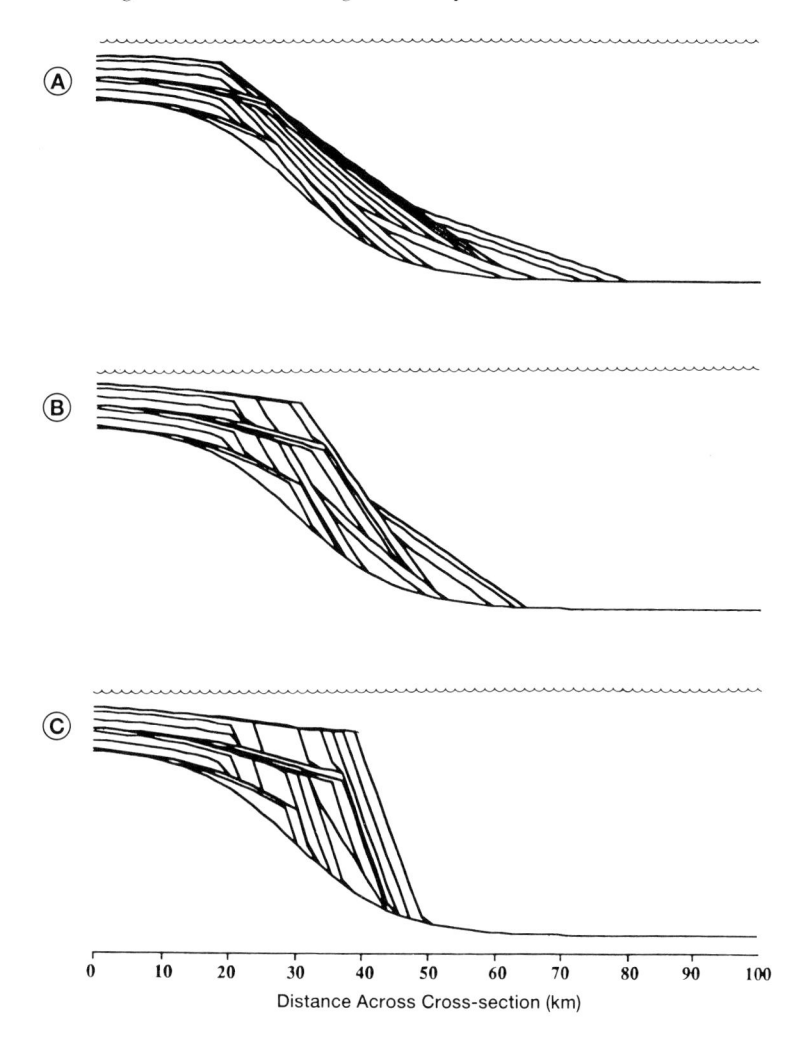

**Fig. 14.** *R*-dependent coastal equilibrium model—effect of varying slope gradient: A, low (1°) B, average (2°); and C, high (4°). Other parameters as in Fig. 12.

slope is located in the landward zone of low margin subsidence. For the case of average basin subsidence, the position of the break in slope moves to a more seaward position. Thus the effect of basin subsidence rate, $S_a$, on the rate of subsidence at the break in slope is buffered by this change in position.

The effect of model parameter $x_a$, the width of the hinge zone, is shown in Fig. 16. The greatest effect of this parameter is on the aspect ratio of sedimentary units. A wide hinge zone (Fig. 16A) produces relatively long units, while a narrow hinge zone (Fig. 16C) produces relatively thick units. This is particularly apparent for the transgressive deposit, which are quite thin in Fig. 16A and appreciably thicker in Fig. 16C.

Aside from the variations in geometry just mentioned, Figs 12–16 show a similar overall pattern of deposition that reflects the eustatic sea-level history of Fig. 17. The dominating effect of eustatic sea-level on resulting stratal geometries is shown also in Fig. 18. In this figure an additional higher order sea-level cycle has been added to the reference sea-level curve. The amplitude of this higher order cycle varies from 20 m in Fig. 18A to 100 m in Fig. 18C. In Fig. 18A the stratal geometries can be broken out into lowstand, highstand and transgressive system tracts. However, the highstand systems no longer show continuous progradation but exhibit a paracycle-scale transgressive event. The depositional history of Fig. 18(B, C) becomes very hard to unravel. The transgressive units overlying highstand units are no longer present on the shelf. Slope

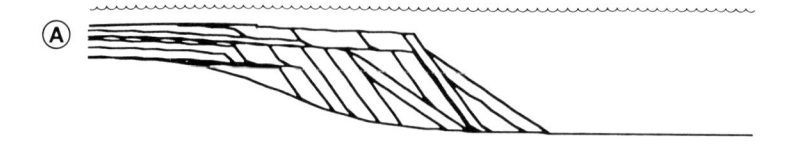

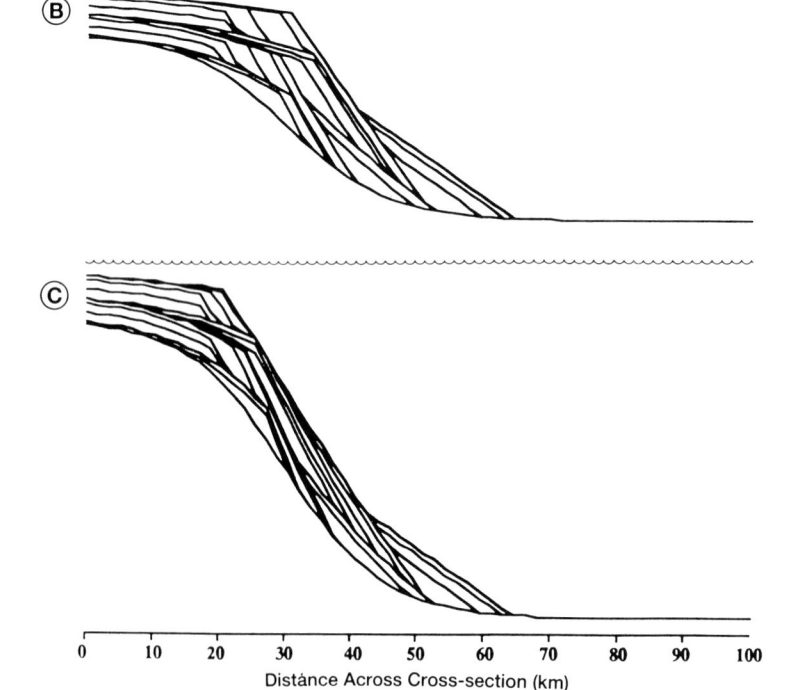

Fig. 15. *R*-dependent coastal equilibrium model—effect of varying subsidence rate: A, low (0 m my$^{-1}$); B, average (50 m my$^{-1}$); and C, high (100 m my$^{-1}$). Other parameters as in Fig. 12.

bypass, basin-floor, lowstand deposits have increased in importance in these two cases. In Fig. 18C, all ten 100 m cycles in the simulation have produced basin-floor deposits. In addition the 100 m cycle rises that occur during the falling stages of the lower order sea-level cycle are not sufficient to move the shoreline landward of the depositional break in slope. The resulting deposition is dominated by basin-floor sedimentation.

## CONCLUSION: AN APPROACH TO INTEGRATED STUDIES

Swift & Thorne (this volume, pp. 3–31) have presented a descriptive model of shelf sedimentation, which is organized about two basic ideas: (1) it is useful to think of the shelf surface as an equilibrium surface, adjusted to geohistorical variables; and (2) we can distinguish, in our descriptive model, between an allochthonous regime, with regression, high sedimentation rates and a low bypassing ratio, and an autochthonous regime, with transgression, low sedimentation rates and a high bypassing ratio. The two regimes are characterized by distinctively different deposits; autochthonous deposits are thin and relatively coarse-grained, while allochthonous deposits are thicker and finer. We briefly discussed the subequilibrium processes of progressive sorting and strata formation that control facies differentiation in allochthonous and autochthonous regimes.

In this paper we attempt a more analytical approach to these ideas. We argue that the term

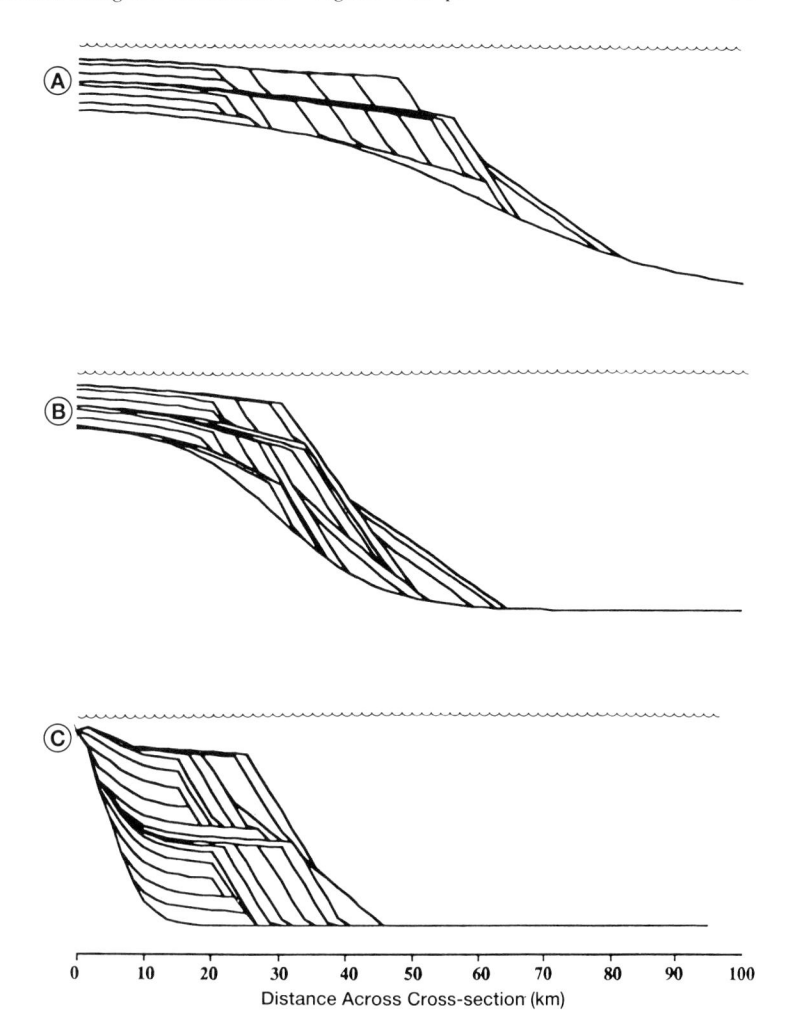

**Fig. 16.** *R*-dependent coastal equilibrium model—effect of position of hinge zone: A, wide (60 km); B, average (30 km); and C, narrow (10 km). Other parameters as in Fig. 12.

*regime*, used descriptively by Swift & Thorne (this volume, pp. 3–31), is appropriate in a more analytical context, in that it has been historically used to describe sedimentation systems occurring at geological time and space-scales. Such systems are typically governed by interdependent variables whose mutual adjustments constitute a state of dynamic equilibrium. Regimes can be described usefully in terms of sedimentation surfaces controlled by homeostatic responses of erosion or deposition.

We have reviewed such applications of regime theory to geomorphological (denudation) modelling, and have suggested that it is a similarly appropriate tool for stratigraphic (basin accumulation) modelling. Shelf sedimentation thus can be interpreted as a response to the geohistory of sea-level change, sediment input, oceanographic conditions, and basin subsidence, which, taken together, define the *depositional regime*. One theoretical characteristic of the depositional regime is that sediment accumulating on a continental margin, given sufficient time under constant conditions, builds up towards a steady-state topographic profile. The role of a such a topographic profile in controlling the external geometry of sedimentary rocks depends on its type, of which there are five: (1) graded stream profile; (2) isostatic equilibrium profile; (3) coastal equilibrium profile, *R*-dependent case; (4) coastal equilibrium profile, *Q*-dependent, advective case; and (5) coastal equilibrium profile, *Q*-dependent, diffusive case. The use of each of these profiles presents a different hypothesis as to the nature of long-term sedimentary equilibrium. The use of any

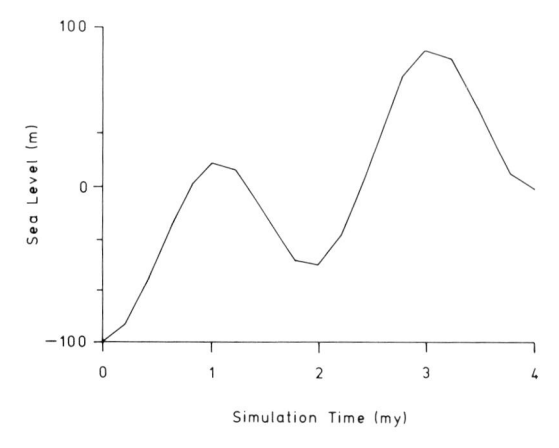

**Fig. 17.** Eustatic sea-level curve used for the generation of simulations shown in Figs 12–16. This sea-level curve was constructed by superimposing a sinusoidal 2 my period, 100-m amplitude cycle on the rising limb of a 16 my period, 100-m amplitude cycle.

of these profiles, in any specific geological setting, requires an understanding of the important physical processes that control regime sedimentation. *R*-dependent and *Q*-dependent coastal profiles describe autochthonous and allochthonous regimes, respectively, while the isostatic equilibrium profile is useful in describing conditions at longer time-scales.

The Louisiana prodelta shelf is an ideal location in which to document regime conditions. The shelf is a classic example of an allochthonous depositional regime. Stratigraphic sections of the delta indicate a rapid progradation of a topographic deltaic profile continuing since the late Quaternary. Wright & Coleman (1972), in a study of the wave climate of the inshore parts of the delta, suggest that the morphology of the subaqueous profile off the Louisiana prodelta shelf represents a balance between the nearshore wave power and the rate at which the Mississippi River can supply sediment to the nearshore zone. To quantitatively test the hypothesis of Wright & Coleman (1972) we have applied semi-empirical coastal engineering methods to calculate the variation along the profile of the expected sediment accumulation rate as a function of oceanographic variables. The computations show that the present-day topographic profile clearly has adjusted to the profile predicted by the coastal model for the *Q*-dependent, advective case. We conclude that the present subaqueous topography of the Louisiana shelf forms an equilibrium that has adjusted to regime conditions of wave energy and sediment supply.

The application of the regime approach to geological interpretation of the ancient rock record has been used extensively in sequence stratigraphic interpretations of seismic profiles. Regime models of this type serve as a basis for the depositional sequence model of Vail (1987) and Haq *et al.* (1987). To illustrate the relative control of each of these factors a numerical algorithm has been constructed to simulate regime sedimentation and the role of sea-level oscillations in shaping the architecture of the continental margin sediment prism. The simulation creates model look-alikes for the 'systems tracts' of sequence stratigraphy, and it is possible to establish the effect of sediment supply and rates of subsidence and sea-level fluctuation on these geometries. Conditions leading to the formation of a type 1 versus type 2 sequence are sensitive to model parameters. Conditions of high sediment supply (Fig. 12C), high shelf gradient (Fig. 13C) and high slope gradient (Fig. 14C) all prevent the shoreline from falling below the depositional slope break. Falling sea-level fails to produce a type 1 sequence when either: (1) the water depth of the slope break during highstand time is too deep to allow a relative sea-level fall to expose the shelf; or (2) the rapid subsidence rate at the slope break effectively cancels the eustatic component of sea-level fall.

The addition of higher order sea-level cycles renders it very hard to unravel the depositional history. The transgressive units overlying highstand units are no longer present on the shelf. Slope bypass, basin-floor, lowstand deposits have increased in importance in these two cases. As the frequency and amplitude of such high-order cycles increases, deposition is eventually dominated by basin-floor sedimentation.

## APPENDIX I: COASTAL ENGINEERING APPROACH TO SHALLOW-WATER REGIME

In this appendix we use empirical relationships derived by coastal engineers to derive the time-average sediment load function $L(x)$ along a topographic profile. The approach taken is mainly from Bretshneider (1966). Recent advances in parametric modelling of winds, waves and currents (e.g. Bea *et al.*, 1983; Gordon *et al.*, 1987; Cooper, 1988) have not been adopted here in order to maintain the relative simplicity of the analytical approach.

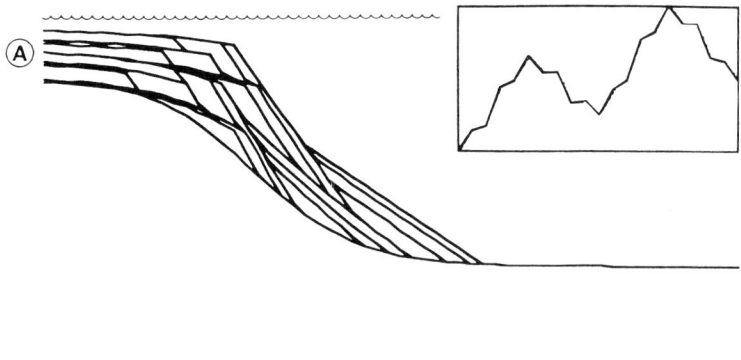

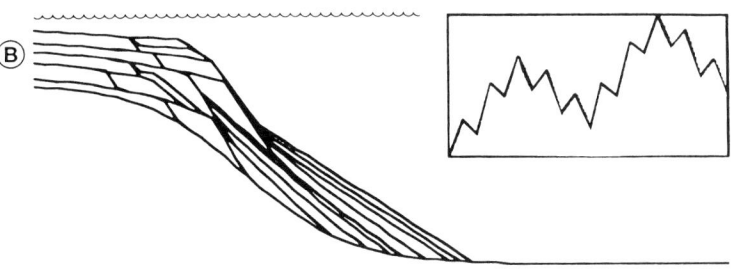

**Fig. 18.** Effect on stratal geometries of superimposing high-order variations on lower order sea-level curve of Fig. 17. The multicomponent sea-level curve is shown as an inset for each simulation: A, 20 m cycles; B, 50 m cycles; and C 100 m cycles. All other model parameters are identical to the standard reference run.

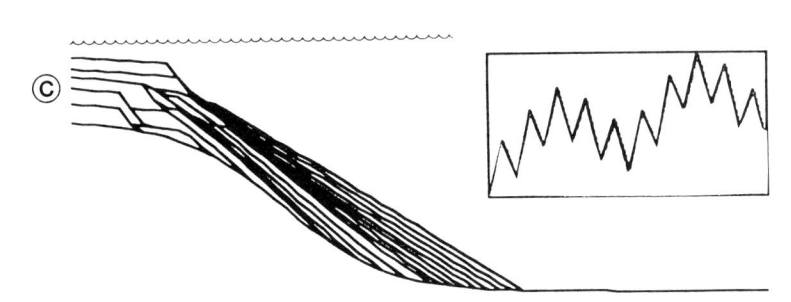

The fair-weather water depth, $w$, is given as a power function of distance from the shoreline $x$:

$$w(x) = (A \cdot x)^m \qquad (AI.1)$$

where $A$ and $m$ are empirical coefficients.

The classical theory of extreme values provides three different statistical distributions for the maxima of stochastic processes (Gumbel, 1958). The probability $p$, of a significant wave height less than or equal to $H_{sig}$ is most often (e.g. Augustine *et al.*, 1978; Silveria & Brillinger, 1978; Ward *et al.*, 1978) taken to follow the extreme value distribution of Type 1 (Gumbel, 1958):

$$H_{sig}(p) = \alpha - \beta \ln[-\ln(p)] \qquad (AI.2)$$

where $\alpha$ and $\beta$ are empirical coefficients.

The return period, $T$, is related to the probability $p$ by (Gumbel, 1958):

$$T \equiv 1/(1-p) \qquad (AI.3)$$

The probability, $p'$, of a wave of less than, or equal to, height $H$, during a storm with significant wave height $H_{sig}$ is given by the Rayleigh distribution for wave heights (Bretshneider, 1966):

$$H(p, p') = \bar{H}(p) \left[ 1 - \exp\left\{ -\frac{\pi}{4}\left(\frac{H(p')}{\bar{H}(p)}\right)^2 \right\} \right]$$
$$(AI.4)$$

where $\bar{H}$ is the average wave height given by:

$$\bar{H}(p) = 0.63 H_{sig}(p) \qquad (AI.5)$$

The deep-water wave period, $T$, associated with a wave of a given height is given by:

$$T = 3.893\sqrt{H} \quad (AI.6)$$

Where $H$ is in metres and $T$ is in seconds. The proportionality constant is from the empirical data of Bretshneider (1966) and is based on short fetch, high wind-speed observations.

Deep-water wave height, $H$, is affected by the combined effects of internal viscous damping, shoaling wave transformation, wave breaking, and friction with the sediment bottom. Of these factors the latter two are more important than the first two. We do not consider the effects of refraction, reflection and defraction created by irregular sea-floor topography.

The effect of bottom friction can be computed using the method of Bretshneider (1966) in which the wave height reduced by friction $H_f$ is given by:

$$H_f(p, x) = \frac{H(p, \infty)}{(1 + K_f)} \quad (AI.7)$$

where the frictional reduction factor $k_f$ can be computed from:

$$k_f(p, x) = -\int_{\infty}^{x} \frac{32\pi^3 f H(p, \infty)}{3g^2 T^4 [\sinh(kh)]^3} dx \quad (AI.8)$$

where $f$ is an empirical wave friction coefficient from Jonsson (1966), $g$ is the acceleration of gravity, and $k$ is the wave frequency for the wave of probability $p$ at the depth of the sea surface from the sea bottom $h$. The wave frequency $k$ must be determined iteratively from:

$$k(h) = \frac{4\pi^2}{gT^2 \tanh(kh)} \quad (AI.9)$$

The friction coefficient, $f$, is, in general, a function of the sediment surface boundary roughness. For sediments of mixed grain size, in the absence of vegetation, the particles in the bedload determine this roughness. The partitioning between bedload and suspended load of each grain-size fraction depends on the flow conditions. The coarsest particles that can be transported in suspension are those whose settling velocity, $w$, is equal to the shear velocity exerted on the bed by passing waves (Bagnold, 1966; Middleton, 1977). An analytical solution to the effective grain size $D$ meeting this criteria can be found by solving:

$$\frac{1}{18}\frac{(\sigma - \rho)gD}{\eta} = \sqrt{\frac{1}{4}(D/A_b)^{0.434} u_b^2} \quad (AI.10)$$

where $\sigma$ is grain density, $\rho$ is fluid density, $\eta$ is fluid viscosity, $D$ is grain size diameter, $u_b$ is the maximum near-bottom velocity and $A_b$ is the excursion amplitude of water particles relative to the sea-bed. We have used Stokes law for the settling velocity and a simple power law approximation of the shear velocity, as computed by Jonsson (1966) for rough turbulent conditions. Using $7.5 \times 10^{-4}\,\mathrm{N\,s\,m^{-2}}$ for viscosity, 2650 and 1025 kg m$^{-3}$ for grain and fluid densities, solution of equation (AI.10) gives:

$$f = 0.014 \left(\frac{2\pi}{T \cdot A_b}\right)^{0.243} \quad (AI.11)$$

The depth $h$ is the sum of the water depth $w$ from equation (AI.1) and the set-up, $S$, of the sea surface caused by wind stress. Wind stress can be related to maximum wave height by (Bretshneider, 1966):

$$U^2 F = \int_{x_1}^{x_2} U_x^2 dx = 0.470 \times 10^6 H_{\max}^2 \quad (AI.12)$$

where $U$ is the wind speed (m s$^{-1}$), $F$ is the effective fetch (m) over which the winds act, and the integral is computed over two distances $x_1$ and $x_2$ that span the circulating wind field.

An expression for set-up, $S(p, x)$, created by storm surge on a continental shelf bottom of constant slope of width $sw$, created by a storm moving perpendicular to the coast with a triangular wind speed distribution has been derived by Bretshneider (1966). Set-up created on smoothly varying topographic profiles given by equation (AI.1) can be approximated iteratively by:

$$S(p, x) = \frac{\kappa U^2 (sw - x)}{g(w(sw) - w(x) - S(p, x))} \ln\left[\frac{w(sw)}{(w(x) + S(p, x))}\right] \quad (AI.13)$$

where $\kappa$ is $3.0 \times 10^{-6}$.

An additional elevation of the sea is due to wave set-up. Wave set-up, $WS(p, x)$ can be approximated by (Longuet-Higgins & Stewart, 1963):

$$WS = 0.19H[1 - 2.82(H/gT^2)^{1/2}] \quad (AI.14)$$

where $H$ and $WS$ are functions of $p$ and $x$.

The depth $h$ in equations (AI.8) and (AI.9) is then given by:

$$h(p, x) = w(x) + S(p, x) + WS(p, x) \quad (AI.15)$$

The previous equations determine the frictional wave height $H_f$. The effect of wave breaking can be determined from the widely used breaking criteria first derived by McCowan (1894):

$$\left(\frac{H_{fb}}{h}\right)_{max} = 0.78 \quad (AI.16)$$

where $H_{fb}$ is the wave height accounting for frictional and breaking effects.

The maximum wave orbital velocity at the sea-bed using first-order wave theory is then given by:

$$u(p, x) = \frac{\pi H_{fb}(p, x)}{T(p) \sinh[k(p, x) \cdot h(p, x)]} \quad (AI.17)$$

Vincent et al. (1981) found that the entrained sediment load $L$ is linearly proportional to the excess Shield's number, i.e.:

$$L(t) = 0.09[\psi(t) - \psi(\text{threshhold})]$$

where

$$\psi(t) = \frac{0.5 f_w \rho u^2(t)}{(\rho_s - \rho)gD} \quad (AI.18)$$

and $t$ is time, $\rho_s$ the sediment density, $\rho$ the fluid density, $f_w$ the coefficient of friction and $D$ grain diameter.

For the large wave heights considered here the Shield's number $\psi(t)$ in most cases will be significantly greater than the threshold Shields parameter for fine sand and smaller grain sizes. Making this simplification we can write a proportionality between sediment load, $L$, and maximum wave orbital velocity, $u$:

$$L(p, x) \propto u^2(p, p', x) \quad (AI.19)$$

The net load $L(x)$ averaged over all waves of probability $p'$ during events of maximum wave height probability $p$ is found by integrating equation (AI.18):

$$L(x) \propto \int_0^1 \int_0^1 u^2(p, p', x) dp' dp \quad (AI.20)$$

## APPENDIX II: ALGORITHM FOR NUMERICAL SEQUENCE MODEL

The following notation is used:

| Symbol | Description |
|---|---|
| $SL_n$ | Input sea-level during the $n$th timestep |
| $V_n$ | Input volume of sediment supply/unit width during $n$th timestep |
| $sub(x)$ | Input subsidence at a position $x$ during any one timestep |
| $b_n$ | Position of the beach during the $n$th timestep |
| $d_{a,b}(x)$ | Depth at position $x$ at timestep $b$ of a horizon deposited at timestep $a$ |
| $bs_n$ | Position of the break in slope at the $n$th timestep |
| $dlap_n$ | Downlap position of base of slope deposits at the $n$th timestep |
| $\Delta x$ | Functions in $x$ are defined for discrete intervals $\Delta x, 2\Delta x \ldots n$ col $\cdot \Delta x$ |
| $s_1$ | Topographic slope for $b < x < bs$ |
| $s_2$ | Topographic slope for $x > bs$ when $b < bs$ |
| $\frac{s_2}{2}$ | Topographic slope for $x > bs$ when $b \geq bs$ (base of slope deposits) |
| $x_a$ | Parameter determining the width of the tectonic hinge zone |
| $S_a$ | Parameter determining the rate of tectonic subsidence |
| $T_n(x)$ | Thickness at a position $x$ of deposition during the $n$th timestep |

### Initial conditions

1 Given $b_0$, $d_{0,0}(x)$, and $bs_0$ as initial conditions
2 Assume a 'passive margin' sigmoidal form of subsidence:
   compute subsidence array for $i = 1$, ncol

$$sub(i\Delta x) = S_a + \frac{S_a}{\tanh(2)} \tanh\left(2\frac{i\Delta x}{x_a} - 2\right)$$

### Main loop

For each timestep $n = 1$, total steps do steps 3–7 as follows:

3 $\Delta SL_n = SL_n - SL_{n-1}$
4 Find new beach position:
   if $\Delta SL_n + sub(b_{n-1}) > 0$ then
     (i) solve for $j$ to minimize the absolute value $|\varepsilon|$
     where $\varepsilon$ is given by:

$$\varepsilon = d_{n,n}(j\Delta x) + \Delta SL_n + sub(j\Delta x)$$

   (ii) $b_n = j\Delta x$
   else
     (i) solve for $j$ to minimize the absolute value $|\varepsilon|$
     where $\varepsilon$ is given by:

$$\varepsilon = d_{n,n}(j\Delta x) + \Delta SL_n + [sub(b_{n-1}) + sub(j\Delta x)]/2$$

   (ii) $b_n = j\Delta x$
   endif

**5** If $b_n < bs_{n-1}$ then

find the new break in slope position:

solve for $bs_n$ to minimize the absolute value $|\varepsilon|$ where $\varepsilon$ is given by:

$$\varepsilon = V_n - \Delta x \sum_{i=1}^{ncol} [d_{n,n}(i\Delta x) - d_{n-1,n}(i\Delta x)]$$

where $d_{n,n}$ is given by:

$$d_{n,n}(x) = \begin{cases} d_{n-1,n}(x) & x < b_n \\ s_1(x - b_{n-1}) & b_n \leq x \leq bs_n \\ s_2(x - bs_{n-1}) + sd & x > bs_n \text{ and} \\ & d_{n,n} < d_{n-1,n} \\ d_{n-1,n} & x > bs_n \text{ and} \\ & d_{n,n} \geq d_{n-1,n} \end{cases}$$

where $sd = s_1(bs_n - b_{n-1})$.

Else

compute position of downlap of base of slope deposits, $dlap_n$

solve for $dlap_n$ to minimize the absolute value $|\varepsilon|$ where $\varepsilon$ is given by:

$$\varepsilon = V_n - \Delta x \sum_{i=1}^{ncol} [d_{n,n}(i\Delta x) - d_{n-1,n}(i\Delta x)]$$

where $d_{n,n}$ is given by:

$$d_{n,n}(x) = \begin{cases} sd - s_2/2 \cdot (dlap_n - x) & d_{n,n} < d_{n-1,n} \\ d_{n-1,n} & d_{n,n} \geq d_{n-1,n} \end{cases}$$

where $sd = d_{n-1}(bs_{n-1})$

**6** Recompute $d_{n,n}(x)$ with the value of $bs_n$ or $dlap_n$ from previous step

**7** Compute deposited thickness from:

$$T_n(x) = d_{n,n}(x) - d_{n-1,n}(x)$$

**End main loop**

## REFERENCES

AUGUSTINE, F.E., MAXWELL, F.D. & LAZANOFF, S.M. (1978) Extreme wave heights in the Gulf of Alaska. *Offshore Tech. Conf.*, pp. 1551–1557. Soc. Expl. Geop., Richardson, TX.

BAGNOLD, R.A. (1966) An approach to the sediment transport problem from general physics. *U.S. Geol. Surv. Prof. Pap.* **422-1**.

BARRELL, J. (1917) Rhythms and measurement of geological time. *Bull. Geol. Soc. Am.* **28**, 754–904.

BEA, R.G. (1974) Gulf of Mexico hurricane wave heights. *Offshore Tech. Conf. Pap.* **422-1**, 40 pp. Soc. Expl. Geop., Richardson, TX.

BEA, R.G., LAI, N.W., NIEDORODA, A.W. & MOORE, G.H. (1983) Gulf of Mexico shallow-water wave heights and forces. *Offshore Tech. Conf.* **3**, 49–62.

BLENCH, T. (1957) *Regime Behaviour of Canals and Rivers*. Butterworths, London, 138 pp.

BRETSHNEIDER, C.L. (1966) Wave generation by wind, deep and shallow water (Chapter 3), Engineering aspects of hurricane surge (Chapter 5). In: *Estuary and Coastline Hydrodynamics* (Ed. Ippen, A.T.) pp. 133–196, 231–256. McGraw Hill, New York.

BRUNSDEN, D. & THORNES., J.B. (1980) Landscape sensitivity and change. *Trans. Inst. Brit. Geogr.* **4**, 463–484.

BRUUN, P. (1962) Sea-level rise as a cause of shore erosion. *J. Waterways Harbors Div., Am. Soc. Civ. Eng.* **88**, 117–130.

CHORLEY, R.J., SCHUMM, S.A. & SUGDEN, D.E. (1984) *Geomorphology*. Methuen, New York, 498 pp.

CLARKE, T.L., SWIFT, D.J.P. & YOUNG, R.A. (1983) A stochastic modeling approach to the fine sediment budget of the New York Bight. *J. Geophys. Res.* **88**, 9653–9660.

COLEMAN, J.M. (1988) Dynamic changes and processes in the Mississippi River delta. *Geol. Soc. Am. Bull.* **100**, 999–1015.

COLEMAN, J.M. & PRIOR, D.B. (1982). Deltaic environments of deposition. In: *Sandstone Depositional Environments* (Eds Scholle, P. & Spearing, D.). Mem. Am. Assoc. Petrol. Geol. 31, 139–178.

COOPER, C.K (1988) Parametric models of hurricane-generated winds, waves and currents in deep water. *Offshore Tech. Conf.* **2**, 475–483.

CULLING, W.E.H. (1960) Analytical theory of erosion. *J. Geology*, **68**, 336–344.

DAVIS, W.M. (1899) The geographic cycle. *Geogr. J.*, **14**, 481–504.

DAVIS, W.M. (1902) Base level, grade, and peneplain. *J. Geol.* **10**, 77–111.

DEAN, R.G. & MAURMEYER, E.M. (1983) Models of beach profile response. In: *CRC Handbook of Coastal Processes and Erosion* (Eds Komar, P. & Moore, J.) pp. 151–165. CRC Press, Boca Raton, FL.

DIETZ, R.S. (1963) Wave base, marine profile of equilibrium, and wave built terraces: a critical appraisal. *Geol. Soc. Am. Bull.* **74**, 971–990.

DIETZ, R.S. (1964) Wave base, marine profile of equilibrium, and wave built terraces: reply. *Geol. Soc. Am. Bull.* **75**, 1275–1282.

DINNEL, S.P. & WISEMAN, W.J., JR. (1986) *Fresh Water on the West Louisiana and Texas Shelves*. Research Program Report no. 59. L.S.U. Industrial Associates, Baton Rouge, LA.

DUBOIS, R.N. (1977) Predicting beach erosion as a function of rising sea level. *J. Geol.* **85**, 470.

EDELMAN, T. (1968) Dune erosion during storm conditions. *Proceedings, 11th International Conference on Coastal Engineering*, pp. 719–723. American Society of Civil Engineers, New York.

EDELMAN, T. (1970) Dune erosion during storm conditions. *Proceedings, 12th International Conference on Coastal Engineering*, pp. 1305–1307. American Society of Civil Engineers, New York.

EVERTS, C.H. (1987) Continental shelf evolution in response to a rise in sea level. In: *Sea-level Fluctuation* and *Coastal Evolution* (Eds Nummedal, D., Pilkey, O.H. & Howard, J.D.). Soc. Econ. Paleontol. Mineral. Spec. Publ. 41, 49–58.

FISK, H.N. & MCFARLAN, E., JR. (1955) Late Quaternary deltaic deposits of the Mississippi River. *Geol. Soc. Am. Spec. Pap.* **62**, 279–302.

FISK, H.N., MCFARLAN, E., JR., KOLB, C. & WILBERT, L. JR. (1954) Sedimentary framework of the modern Mississippi Delta. *J. Sediment. Petrol.*, **24**, 76–99.

FRAZIER, D.E. (1974) Depositional. episodes: their relationship to the Quaternary stratigraphic framework in the northwestern portion of the Gulf Basin. *Tex. Bur. Econ. Geol. Geol. Circ.* **44–1**, 28 pp.

GALLOWAY, W.E. & HOBDAY, D.K. (1983). *Terrigenous Clastic Depositional Systems*. Springer Verlag, New York, 423 pp.

GILBERT, G.K. (1877) Report on the geology of the Henry Mountains. *U.S. Geogr. Geol. Surv. Rocky Mtn. Region* 160 pp.

GLEICK, J. (1987) *Chaos: Making a New Science*. Penguin, New York, 352 pp.

GORDON, R.B., FLATHER, R.A., WOLF, J., KANTHA, L.H. & HERRING, H.J. (1987) Modeling the current response of continental shelf waters to winter storms: comparisons with data. *Offshore Tech. Conf.* **1**, 503–512.

GUMBEL, E.J. (1958) *Statistics of Extremes*. Columbia University Press, New York, 375 pp.

GUNN, R. (1944) A quantitative study of the lithosphere and gravity anomalies along the Atlantic Coast. *Franklin Inst. J.* **237**, 139–154.

HACK, J.T. (1960) Interpretation of erosional topography in humid temperate regions. *Am. J. Sci.* **258a**, 89–97.

HAQ, B.U., HARDENBOL, J. & VAIL, P.R. (1987) Chronology of fluctuating sea levels since the Triassic. *Science* **235**, 1156–1167.

HELLAND-HANSEN, W., KENDALL, C.G. ST. C., LERCHE, I. & NAKAYAMA, K. (1988) A simulation of continental basin margin sedimentation in response to crustal movements, eustatic sea level change, and sediment accumulation rates. *Math. Geol.* **20**, 777–802.

HERBICH, J.B. (1977) Wave-induced scour around offshore pipelines. *Preprints, Offshore Technology Conference*, **4**, 79–84.

HIRANO, M. (1968) A mathematical model of slope development: an approach to the analytical theory of erosional topography. *J. Geosci. Osaka City Univ.* **11**, 13–52.

HOROWITZ, D.H. (1976) Mathematical modeling of sediment accumulation in prograding deltaic systems. In: *Quantitative Techniques for the Analysis of Sediments. Computers in Geoscience Series*, pp. 105–120. Pergamon Press.

IWAGAKI, Y. & NODA, E. (1962) Laboratory study of scale effects in two-dimensional beach processes. *Proceedings of 8th International Conference on Coastal Engineering*, pp. 194–210. American Society of Civil Engineers, New York.

Jervey, M.T. (1988) Quantitative geological modelling of siliciclastic rock sequences and their seismic expression. In: *Sea-level Changes—an Integrated Approach* (Eds Wilgus, C., Hastings, B., Posamentier, H., van Wagoner, J., Ross, C. & Kendall, C.). Soc. Econ. Paleontol. Mineral. Spec. Publ. 42, 47–70.

JOHNSON, D.W. (1919) *Shore Processes and Shoreline Development* (Hafner Facsimile edn, 1952). Columbia University Press, New York, 584 pp.

JOHNSON, J.W. (1949) Scale effects in hydraulic models involving wave motion. *Trans. Am. Geophys. Union*, **30**, 517–525.

JOHNSON, J.W. & EAGLESON, P.S. (1966) Coastal processes. In: *Estuary and Coastline Hydrodynamics* (Ed. Ippen, A.T.) pp. 404–492. McGraw Hill, New York.

JONSSON, I.G. (1966) Wave boundary layers and friction factors. *Proceedings, 10th Conference on Coastal Engineering*, Vol. 1, pp. 127–148. American Society of Civil Engineers, New York.

KENYON, P.M. & TURCOTTE, D.L. (1985) Morphology of a delta prograding by bulk sediment transport. *Geol. Soc. Am. Bull.* **96**, 1457–2465.

Knox, J.C. (1975) The concept of the graded stream. In: *Theories of Landform Development* (Eds, Melhorn, W.N. & Flemal, R.C. pp. 169–198. George Allen & Unwin, London.

LINDSAY, J.F., PRIOR, D.B. & COLEMAN, J.M. (1984) Distributary mouth bar development and the role of landslides in delta growth, South Pass, Mississippi Delta. *Bull., Am. Assoc Petrol. Geol.* **68**, 1732–1743.

LEOPOLD, L.B. & LANGBEIN, W.B. (1962) The concept of entropy in landscape evolution. *U.S. Geol. Surv. Prof. Pap.* **500-A**, 20 pp.

LEOPOLD, L.B. & MADDOCK, T. (1953) The hydraulic geometry of stream channels and some physiographic implications. *U.S. Geol. Surv. Prof. Pap.* **252**, 57 pp.

LEOPOLD, L.B., WOLMAN, M.G. & MILLER, J.P. (1964) *Fluvial Processes in Geomorphology*. W.H. Freeman, San Francisco, 522 pp.

LONGUET-HIGGINS, M.S. & STEWART, R.W. (1963) A note on wave setup. *J. Mar. Res.* **21**, 4–10.

MACKIN, J.H. (1948) Concept of the graded river. *Bull. Geol. Soc. Am.* **59**, 463–512.

MCCOWAN, J. (1894) On the highest wave of permanent type. *London, Edinburgh, Dublin Philos. Mag. J. Sci.* **38**.

Middleton, G.V. (1977) Hydraulic interpretation of sand size distributions. *J. Geol.* **84**, 405–26.

MOORE, D.G. & CURRAY, J.R. (1964) Wave-base, marine profile of equilibrium and wave built terraces, discussion. *Geol. Soc. Am. Bull.* **47**, 1267–1274.

NODA, E.K. (1972) Equilibrium profiles of model beaches. *J. Waterways Harbors Div., Am. Soc. Civ. Eng.* **98**, 511–527.

PILKEY, O.H. & DAVIS, T.W. (1987) An analysis of coastal recession models: North Carolina Coast: In: *Sea-level Fluctuation and Coastal Evolution* (Eds Nummedal, D., Pilkey, O.H. & Howard, J.D.). Soc. Econ. Paleontol. Mineral. Spec. Publ. 41, 59–68.

PITMAN, III, W.C. (1978) Relationship between eustasy and stratigraphic sequences of passive margins. *Geol. Soc. Am. Bull.* **89**, 1389–1403.

POSAMENTIER, H.W., JERVEY, M.T. & VAIL, P.R. (1988) Eustatic controls on clastic deposition, I—conceptual framework. In: *Sea-level Changes: an Integrated Approach* (Eds Wilgus, C.K., Hastings, B., Posamentier, H., van Wagoner, J., Ross, C. & Kendall, C.). Soc. Econ. Paleontol. Mineral. Spec. Publ. 42, pp. 109–124.

PYNE, S. (1975) The mind of Grove Karl Gilbert. In:

*Theories of Landform Development* (Eds Melhorn, W.N. & Flemal, R.C.) pp. 277–298. George Allen & Unwin, London.

QUAYLE, R.G. & FULBRIGHT, D.C. (1975) Extreme wind and wave return periods for the U.S. coast. *Mariners Weath. Log*, **190**, 67–70.

ROSEN, P.S. (1978) A regional test of the Bruun Rule on shoreline erosion. *Mar. Geol.* **26**, M7–M16.

SCHEIDEGGER, A.E. (1961) *Theoretical Geomorphology.* Springer-Verlag, Berlin, 334 pp.

SCHROEDER, W.W., HUH, O.K. ROUSE, L.J., JR & WISEMAN, W.J., JR. (1986) Satellite observation of the wind-driven circulation east of the Mississippi Delta. Research Program Report 53, L.S.U. Industrial Associates, Baton Rouge, LA. 14 pp.

SCHUMM, S.A. (1975) Episodic erosion; a modification of the geomorphic cycle. In: *Theories of Landform Development* (Eds Melhorn, W.N. & Flemal, R.C.) pp. 69–86. George Allen & Unwin, London.

SCHWARTZ, M.L. (1965) Laboratory study of sea-level rise as a cause of shore erosion. *J. Geol.* **73**, 528.

SCHWARTZ, M.L. (1967) The Bruun theory of sea-level rise as a cause of shore erosion. *J. Geol.* **75**, 76.

SCHWARZACHER, W. (1975) *Sedimentation Models in Quantitative Stratigraphy*, Vol. 19, *Developments in Sedimentology*. Elsevier, New York, 382 pp.

SCRUTON, P.C. (1956) Oceanography of Mississippi Delta sedimentary environments. *Bull., Am. Assoc. Petrol. Geol.* **40**, 2864–2952.

SILVERIA, W.A. & BRILLINGER, D.R. (1978) On maximum wave heights of severe seas. *Offshore Tech. Conf.* pp. 1563–1567. Soc. Expl. Geop., Richardson, TX.

SLOSS, L.L. (1962) Stratigraphical models in exploration. *J. Sediment. Petrol.* **32**, 415–422.

STRAHLER, A.N. (1952) Dynamic basis of geomorphology. *Geol. Soc. Am. Bull.* **63**, 923–928.

SUNAMARA, T. & HORIKAWA, K. (1974) Two-dimensional beach transformation due to waves. *Proceedings, 14th International Conference on Coastal Engeering*, pp. 920–937. American Society of Civil Engineers, New York.

VAIL, P.R. (1987) Seismic stratigraphy interpretation using sequence stratigraphy. Part 1: seismic stratigraphy interpretation procedure. In: *Atlas of Seismic Stratigraphy*, Vol. 1 (Ed. Bally, A.W.). Am. Assoc. Petrol. Geol. Stud. Geol. 27, 1–10.

VAIL, P.R., MITCHUM, R.M. JR., THOMPSON, S. III, SANGREE, J.R. DUBB, J.N. & HASLID, W.G. (1977) Sequence stratigraphy and global changes of sea level. In: *Seismic Stratigraphy—Applications to Hydrocarbon Exploration* (Ed. Payton, C.E.). Mem., Am. Assoc. Petrol. Geol. 26, 49–205.

VAN WAGONER, J.C., POSAMENTIER, H.W., MITCHUM, R.M., JR., VAIL, P.R., SARG, J.F., LOUTIT, T.S. & HARDENBOL, J. (1988) An overview of the fundamentals of sequence stratigraphy and key definitions of sequence stratigraphy. In: *Sea-level Changes: an Integrated Approach* (Eds C.K. Wilgus, B.S. Hastings, C.G. St. C. Kendall, H.W. Posamentier, C.A. Ross, & J.C. Van Wagoner). Spec. Publ. Soc. Econ. Paleontol. Mineral. 42., 407 pp.

VENING-MEINESZ, F.A. (1931) Une nouvelle methode pour la reduction isostatique regionale de l'intensite de la pesanteur. *Bull. Geod.* **29**.

VINCENT, C.E., YOUNG, R.A. & SWIFT, D.J.P. (1981) Bed-load transport under waves and currents. *Mar. Geol.* **39**, 71–80.

WALKER, R. (Ed.) (1981) *Facies Models*. Geoscience Canada, reprint series I. Geological Association of Canada, Waterloo, Ontario, 211 pp.

WARD, E.G., EVANS, D.J. & POMPA, J.A. (1978) Extreme wave heights along the Atlantic coast of the United States. *Offshore Tech. Conf.* **OTC 2846**, 315–324.

WATERS, C.H. (1939) *Equilibrium slopes of sea beaches*. Unpublished MS thesis, University of California, 129 pp.

WATTS, A.B. & THORNE, J.A. (1984) Tectonics, global changes in sea level and their relationship to stratigraphical sequences at the U.S. Atlantic continental margin. *Mar. Petrol. Geol.* **1**, 319–339.

WELLS, J.T. (1983) Dynamics of coastal fluid muds in low, moderate, and high tide-range environments. *Can. J. Fish. Aquat. Sci.* **40** (Suppl. 1), 130–142.

WHEELER, H.E (1964) Base level, lithosphere surface and time-stratigraphy. *Geol. Soc. Am. Bull.* **75**, 599–610.

WOLMAN, M.G. (1955) The natural channel of Brandywine Creek, Pennsylvania. *U.S. Geol. Surv. Prof. Pap.* **271**, 128 pp.

WRIGHT, L.D. & COLEMAN, J.M. (1972) River delta morphology: wave climate and the role of the subaqueous profile. *Science* **172**, 282–284.

# Sedimentation on continental margins, III: the depositional fabric—an analytical approach to stratification and facies identification

J.A. THORNE,[*] E. GRACE,[†] D.J.P. SWIFT[‡] and A. NIEDORODA[§]

[*] Research and Technical Services, ARCO, Oil and Gas Company, Plano, TX 75075, USA; [†] Chevron Oilfield Research Company, La Habra, CA 90633-6446, USA; [‡] Department of Oceanography, Old Dominion University, Norfolk, VA 23529-0276, USA; and [§] Environmental Science Services, Box 1703, Gainsville, FL 32602-1703, USA

## ABSTRACT

This paper presents an analytical approach to the problem of storm bed formation and preservation in the shallow marine environment. A random series of storms are assumed to rework and aggrade the depositional surface. Each storm reworks a thickness, $\Delta\eta$, and deposits a small amount of additional sediment such that the accumulation rate remains a constant, $\mathring{a}$. The storm bed thickness, $\Delta\eta$, is calculated as a function of the size class of each storm. In general, larger storms, which rework a greater amount of sediment $\Delta\eta$, occur less frequently than smaller storms. Storms can therefore be classed by their return period, $T$. An earlier formulation for computing storm bed thickness has been modified to take into account the role of *bedload armouring* as a control on storm bed thickness. In the formulation, a sediment transport computation is used to predict: (1) the mean, preserved, event bed thickness; and (2) the percentage of the sedimentary column consisting of identifiable single-event beds. Parameters for the sediment transport computation include water depth, grain-size distribution, accumulation rate and wave climate parameters $\alpha$ and $\beta$, which determine the return period of storms of various sizes.

The results of the computations are used to develop an analytical model for storm bed thickness and preservation. The model is built around two basic relationships. In the *power law approximation*, $\Delta\eta = a'T^{p}$, where $\Delta\eta$ is bed thickness for a storm of a given return period $T$, $p$ is return period power and $a'$ is the return period intercept, or the minimum bed thickness. A second basic relationship describes the extent of preservation of the deposited bed in terms of a reworking ratio, $r$, the ratio of the minimum bed thickness $a'$ to the accumulation rate, $\mathring{a}$. As $r$ increases, more of each deposited bed is resuspended by the next event and incorporated into the next bed. Knowing, $r$, it is possible to estimate the cut-off percentage, $c\%$, which is percentage of the bed that must remain for successive deposits to be identifiable as single-event beds.

The results of the model are applied to measured sections of bed thickness in the Upper Cretaceous Mesaverde Formation of northwestern Wyoming. Model parameters are estimated from the palaeogeography of the Western Interior Seaway. The computations show that water depth and accumulation rate vary characteristically with stratigraphic sequence position. The preferred interpretation shows that fifth-order cycles within the Mesaverde Formation are created by sea-level cycles of approximately 100 000 years duration and 2 m amplitude.

More generally, our study shows that the reworking ratio, $r$, is an important indicator of the stratification pattern in shallow marine settings. It shows that shifts in regime scale parameters (accumulation rate; $\mathring{a}$, sea-level change rate, **R**; sediment input rate, **Q**) are expressed at subequilibrium (event) scales by changes in the event parameters $a'$, $r$ and $c\%$. These changes in turn modify the stratification and grain-size patterns, which taken together constitute sedimentary facies.

## INTRODUCTION

The episodic and stochastic nature of shallow marine flow creates stratified deposits. The scale of stratification ranges from laminae (strata less than a centimetre thick; Ingram, 1954) to beds of a metre in thickness. Such beds, created by brief, high-frequency events, such as storms and tidal flows, have been described as *event strata* (Seilacher, 1982; Snedden & Nummedal, this volume). Event strata thus constitute a record of depositional history. Swift & Thorne (this volume, 3-31) and Thorne & Swift (this volume, 33-58) have argued that at large time and space-scales, shelf sedimentation can be treated as an equilibrium process, and the physical processes that underlie this equilibrium can be approached through regime theory. In this part, we examine shallow marine depositional processes at the time and space-scales of the depositional events themselves, and develop an analytical approach with which to relate these subequilibrium-scale processes to the larger scale processes of the depositional regime.

Event strata may be distinguished from the thicker lithostratigraphic units that are the result of the slow shift of depositional environments caused by changes in sea-level, and patterns of subsidence and sediment distribution. These larger scale units are facies, or more precisely, the lithosomes of Wheeler & Mallory (1956). They are in turn the building blocks of *depositional systems* (Fisher & McGowen, 1967), and the *depositional system tracts* of sequence stratigraphy (van Wagoner *et al.*, 1987). The basic premise of sequence stratigraphy states that the arrangement and architecture of depositional sequence tracts can be understood as a response to the geohistorical variables of sea-level, subsidence, oceanographic conditions, and sediment supply, which taken together define the sedimentary regime (see discussion by Swift *et al.*, this volume, 153-187). Event strata are the fundamental units from which these larger scale sedimentary units are formed, and they should in turn reflect the geological time-history of these stratigraphic variables.

There are, however, serious problems in interpretation of event strata, as a consequence of the probabilistic nature of the high-frequency oceanographic processes that form them (Dott, 1988). Numerous authors have derived theoretical expressions for preserved bedding thickness distributions as a result of stochastic variation between erosion and deposition (Kolmogorov, 1951; Schwarzacher, 1975; Dacey, 1979; Nishiwaki, 1979; Thompson, 1984; Zydorowicz & Matysiak, 1984). These studies did not model explicitly the physical processes responsible for erosion or deposition. They can be used to interpret the stochastic nature of bedding formation, but not to infer the geological history of stratigraphic processes. Stratification preserves a stratigraphic record of both the stochastic depositional event history and the secular changes in stratigraphic geohistorical variables. In order to interpret stratification patterns, therefore, it is necessary to understand how geohistorical variables affect statistical bedding parameters.

This paper presents a largely theoretical approach to the problem. The focus of this study is on the development of storm-generated stratification, though the methods presented may be extended to other environments. Storm stratification is the dominant class of event stratification on continental shelves. Other types of stratification are not considered in this study. Fair-weather flow, for example, generally is not capable of moving significant quantities of sediment on the continental shelf (e.g. Scruton, 1956). Tidal regimes sufficiently intense to create stratification occur on approximately 15% of the worlds shelves (Swift *et al.*, 1981). Stratification created by the action of tides is a more nearly deterministic process, although it is commonly storm modified. The distinctively astronomical imprint in tidal stratification is a subtle one (Visser, 1980). For most purposes, tidal strata can be treated as though deposited by semi-diurnal storms of extremely regular period.

In this paper, the analytical theory of storm bed stratification is presented in two parts. The theory of *storm bed generation* is described and the concept of dynamic bedload armouring is introduced. However, the pattern of event bed sequences reflects not only the processes of storm bed formation but the subsequent modification of each bed by later storm events. A theory of *storm bed preservation* is therefore presented. Finally, the theoretical results of these first sections are applied to measured sections of bed thickness from the Mesaverde Formation of northwestern Wyoming.

## GENERATION OF STORM BEDS

### General

In an earlier paper (Niedoroda *et al.*, 1989), we presented an initial attempt to quantify the development of storm-generated stratification. The manner in which the model views storm bed formation is indicated in Swift & Thorne (this volume, 3–31, Fig. 12). In this schematic figure, the curve represents the waxing and waning of a storm-generated bottom current over a period of several days. As boundary shear stress intensifies owing to increases in wave height and current velocity, the near-bottom water column can hold at each successive moment more than it did during the previous moment (has greater capacity). As soon as the boundary layer adjusts to the velocity increase, erosion ceases. The sediment load is in equilibrium with the capacity of the flow to hold sediment, and as a result the bottom is dynamically armoured. For every grain that is swept up into the current, another one must fall out of the boundary layer on to the bottom. As the current wanes, however, capacity decreases and all of the sediment is eventually deposited. The resulting bed is a record of the later, depositional portion of the flow event, while the basal unconformity is the record of the earlier erosional period. In event beds, coarser grains are stored in the basal portions, because they are the first to be deposited as the current wanes. During the rising phase of the next flow event, much of the finer, upper portion of the original bed may be re-entrained, but the coarse basal lag is more likely to remain buried. As noted by Swift & Thorne (this volume, 3–31), the greater probability of permanent burial over successive events that is associated with coarser grain sizes gives rise to *progressive sorting* as the sediment load passes through the dispersal system.

In the model (Niedoroda *et al.*, 1989), storm deposits on open continental shelves are be modelled by applying deterministic marine sediment dynamical relationships to a statistical characterization of storm-driven oceanographic conditions. In general, larger storms, which rework a greater thickness of sediment, $\Delta\eta$, occur less frequently than smaller storms. Storms can be classed, therefore, by their *return period, T*. A numerical model of graded storm bed formation requires quantitative expressions for: (1) storm event recurrence intervals on coasts of low to high wave energy; and (2) the thickness of sediment that will be resuspended by a storm event of a given size as a function of substrate grain-size distribution and water depth.

## Wave forecasting

Coastal engineering methods are be used to estimate the return period of extreme wave heights. Storm return period, $T$, and deep-water significant wave height, $H$, are related through the Gumbel distribution (e.g., Gumbel, 1958; Augustine & Maxwell, 1978; Silveria & Brillinger, 1978; Ward *et al.*, 1978)

$$T^{-1} = 1 - \exp\left\{-\left[\exp\left(\frac{-(H-\alpha)}{\beta}\right)\right]\right\} \quad (1)$$

Where $\alpha$ and $\beta$ are empirical wave climate coefficients. The variation of coefficients $\alpha$ and $\beta$ is well studied by coastal engineers (e.g. Quayle & Fulbright, 1975). High values of these coefficients indicate harsh wave-climate conditions, while low values indicate mild conditions. The major control on deep-water wave climate appears to be open ocean fetch. Limits on wave size owing to fetch are well known (Bretshneider, 1966a). Hsu (1986), in an empirical study of measured wave heights, found that significant wave height is proportional to the square root of fetch. The data of Quayle & Fulbright (1975) on the variation of coefficients $\alpha$ and $\beta$ along the coastline of the USA indicate that coefficient $\alpha$ is far more variable then coefficient $\beta$.

Equation (1), along with coefficients $\alpha$ and $\beta$, can be used to forecast the storm event recurrence intervals on coasts of low to high wave energy. An estimate of the thickness of sediment that will be resuspended by a storm event of a given size can be based on empirical and semi-empirical formulations of marine sediment dynamics. Many such formulations exist and no one set of empirical or semi-empirical expressions has been adopted universally by coastal engineers.

## Sediment transport formulations

For the purpose of this study two representative formulations have been chosen that differ critically in how they choose to dynamically armour the sediment surface to further erosion by large storm events. The first formulation is described fully in our earlier paper (Niedoroda *et al.*, 1989) and will not be discussed in detail in this paper. In Niedoroda *et al.* (1989) sediment entrained by large waves and currents was divided into three types on the basis of size class: (1) large immobile grains; (2) grains moving as bedload; and (3) grains moving as suspended load. In this theory, bedload and suspended load respond to the combined wave–current shear stress, which is computed by the method of Madsen & Grant (1976), assuming that the angle between wave and current direction is $c. 0°$. This combined shear stress is then used to compute bedload using the formulation of Yalin (1963), the reference concentration of suspended load using the formulation of Smith (1977), and a ratio of total suspended load to this reference concentration from an empirical fit to the data of Kachel (1980). In the Niedoroda et al. (1989) model, the total suspended load for each fraction is not affected by the formation of a bedload layer. In Niedroda et al. (1989) we assumed that such a layer does not form an impenetrable boundary to the layer beneath it.

The formulation presented in this paper is based on the concept of *bedload armouring* and is described fully in Appendix I. In this formulation, sediment is partitioned into two states: (1) an armouring bedload layer; and (2) a transitional and suspended load layer. The portion of each size class existing in each state depends on the ratio of settling velocity to combined wave–current shear stress for each grain size. By determining the total fraction of grains in the bedload layer, $BL\%$, and the thickness of the bedload layer, $BL_{thick}$ (using the formulation of Shi et al. (1985) and Sleath (1978)), the event bed thickness is determined from:

$$\Delta\eta = \frac{BL_{thick}}{BL_\%} \quad (2)$$

A second major difference between the two formulations is that the earlier approach (Niedoroda et al., 1989) is applicable only for low suspended sediment concentrations. The approach outlined in Appendix I, on the other hand, is applicable only for high suspended sediment concentrations. For the extreme storm events under consideration here, high suspended sediment concentrations are generally the rule. Formulations capable of handling intermediate cases have been developed (e.g. Smith & McLean, 1977; Kachel, 1980; Sheng & Butler, 1982). Unfortunately these codes are computationally prohibitive (N. Kachel, pers. comm., 1985).

### Model parameters

Model parameters, for either formulation, include water depth, grain-size distribution and wave climate parameters α and β, which determine the return period of storms of various sizes. Grain-size distribution, for the purpose of this modelling study, is given by the admixture of coarse and fine populations. The relative volume of the coarse fraction is given by parameter $f$. Framework grains are assumed to be log-normally distributed with a modal grain size, $d$, and a standard deviation in phi units of $s$. Additional coefficients, $c_1$ and $c_2$, relate wave height, $H$, to the component of the current velocity in the same direction as wave motion, $U_c$, and to wave period, $p$, through the relationships:

$$p = c_1 \sqrt{H} \quad (3)$$

$$U_c = c_2 H \quad (4)$$

where $H$ is in metres and $p$ is in seconds. Bretshneider (1966b) found that the coefficient $c_1 = 3.893$. The proportionality constants are from the empirical observations under conditions of short fetch and high wind speed. The linear relationship of equation (4) is based on the fact that geostrophically balanced shelf currents and wave heights both respond linearly to surface wind shear stress (Allen, 1985). Coefficient $c_2$ is generally very close to 0 on the shoreface because the angle between waves and current motion is nearly perpendicular in this area. This small angle results from the fact that: (1) geostrophically balanced shelf currents are topographically steered and flow along the sea-floor parallel to the contours (see discussion by Swift & Thorne, this volume, pp. 3–31); and (2) over the shoreface, where geostrophic currents are damped out by the friction of the shoaling bottom, the waves are refracted into parallelism with bottom contours (Komar, 1976).

Maximum wave height, $H_{max}$, is limited in shallow water by wave breaking. As a function of water depth, $h$, this criteria is given by:

$$H_{max} = b_{coef} \cdot h \quad (5)$$

where the breaking coefficient, $b_{coef}$, is generally taken to be 0.78 (Sunamura & Kraus, 1985). A range for the value of this coefficient between a minimum of 0.73 and a maximum of 0.87 has been reported in the literature (Dean & Eagleson, 1966).

### Comparison of model results for bed formation

Figure 1, based on our earlier approach (Niedoroda et al., 1989), shows the expected bed thickness for different water depths created by resuspension under the action of large storm events of different return periods (model parameters are given in the figure caption). For example, reading the maximum condition from the upper right-hand corner of the figure, the once in 100 000 year storm can resuspend 350 cm at 25 m or 30 m water depth. The initial model shows little dependence of bed thickness on water depth in the range between 20 m and 30 m.

Continental margin sedimentation—depositional fabric

**Fig. 1.** Expected resuspended bed thickness, $\Delta\eta$, as a function of storm return interval predicted by the formulation of Niedoroda *et al.* (1989). The curves shown are computed for 20 m, 25 m and 30 m of water. Model parameters are: (1) grain-size distribution: $d = 2.5$, $s = 1$, $f = 0.5$; (2) wave and current climate parameters: $\alpha = 2.3$, $\beta = 1.6$, $b_{coef} = 0.78$, $c_1 = 3.893$, $c_2 = 0.005$.

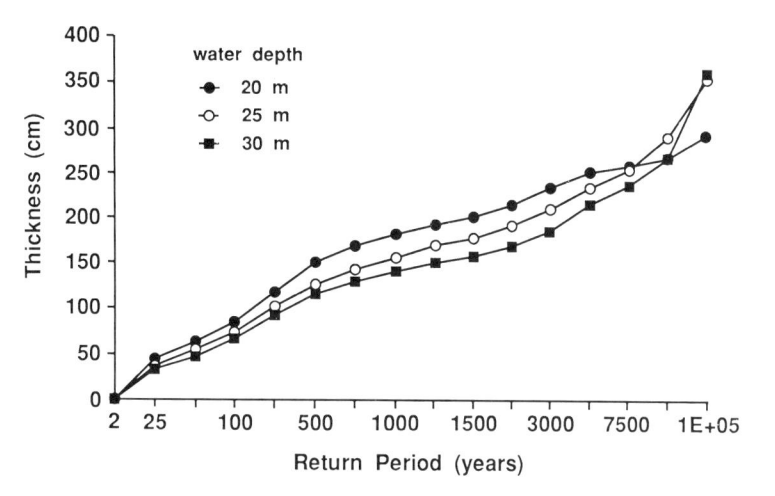

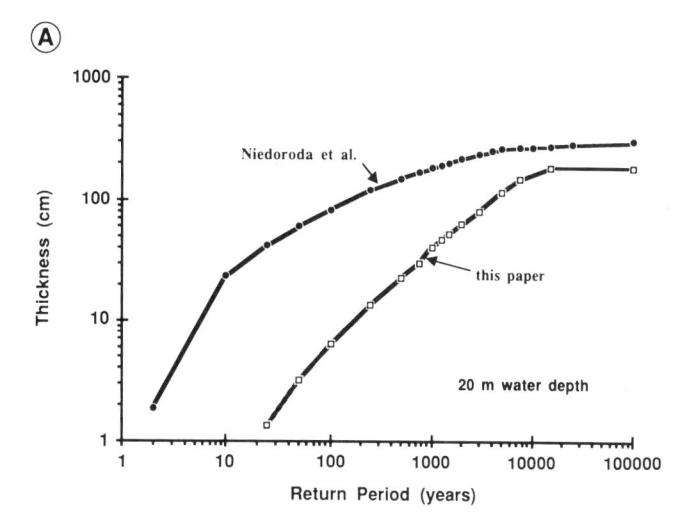

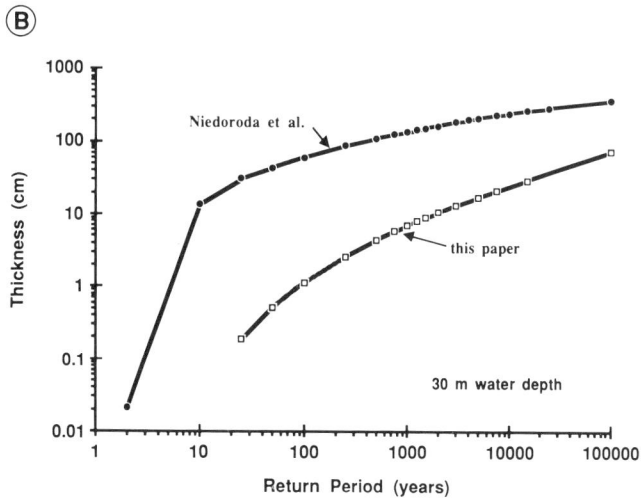

**Fig. 2.** Expected bed thickness, $\Delta\eta$, as a function of storm return interval predicted by the formulation of Niedoroda et al. (1989) versus the bed armouring approach described in Appendix I. The curves shown are computed for 20 m (A) and 30 m (B) of water. Model parameters are set to correspond to the Campanian Mesaverde Formation of the Western Interior Seaway as follows: (1) grain-size distribution: $d = 3.0$, $s = 0.75$, $f = 0.5$; (2) wave and current climate parameters: $\alpha = 2.0$, $\beta = 1.6$, $b_{coef} = 0.86$, $c_1 = 3.893$, $c_2 = 0.005$.

Figure 2A shows a comparison between the initial model and the bed-load armouring approach described in Appendix I for 20 m of water. Parameters for the curves shown have been chosen based on the Mesaverde Formation from the northern portion (Wyoming) of the Campanian Western Interior Seaway (see below). The initial approach predicts that large, 1 m size beds are created by storm return periods larger than 200 years. The bedload armouring approach, however, predicts that beds of over a metre thickness occur only for events of return period larger than 5000 years.

Differences between the two approaches are even more profound in 30 m water depth (Fig. 2B). Bed thicknesses estimated by the initial approach are consistently 10–20 times greater than those predicted by the bedload armouring approach.

### Power law approximation to bed thickness

In general, bed thickness is a function of storm return period, wave climate, water depth and grain-size distribution. We can approximate this relationship by a power function:

$$\Delta\eta = a'T^p \qquad (6)$$

where coefficients $a'$ and $p$ are functions of wave climate, water depth and grain size. Coefficients $a'$ and $p$ are referred to as the *return period intercept* and the *return period power*, respectively. If the return period is given in years, then the coefficient $a'$ can be thought of as a minimum estimate of the bed thickness (cm) that will be resuspended over the length of a year of observation. Coefficient $p$ is an indication of the importance of return period on bed thickness. If this coefficient is very low, resuspension events are all roughly the same size. If this coefficient is high, rare, intense storms create significantly greater bed thicknesses than do frequent, mild storms.

Coefficients $a'$ and $p$ can be found by best-fitting a straight line to a log–log plot of bed thickness versus return period. As can be seen in Fig. 2, a straight-line fit is a good approximation to the results from the bed armour model but not to the initial model. A straight line, however, can be fit to a segment of the initial model curves between a limited range of return periods. This range is set by choosing those return periods that dominate the preserved stratigraphic record of event beds.

## PRESERVATION OF STORM BEDS

Though equation (6) governs the initial thickness distribution of storm beds, it is not a measure of preserved thicknesses. As has been pointed out previously by Crowley (1984), the preservation of time lines in sedimentary sequences, i.e. temporal completeness, depends critically on sedimentation rate. As sedimentation rate decreases, low magnitude events are progressively eliminated from the record. Bed statistics, such as the bed thickness distribution, therefore, depend strongly on the processes of overprinting and partial preservation implicit in a sequence of independent storm events.

In our earlier model (Niedoroda *et al.*, 1989), we describe a numerical model for the creation of synthetic stratigraphic columns that examines the problem of bed preservation. In this preliminary numerical model no attempt is made to describe the internal grading of the resulting beds. In any one step, the model simulation undertakes the following procedures: (1) it randomly chooses a storm size, $T$, from the extreme event distribution (equation (6))—here $T$ is the return period, or average recurrence interval of a storm of that intensity; (2) it resuspends the top upper $\Delta\eta(T)$ centimetres of sediment, where $\Delta\eta(T)$ is a function of $T$; and (3) it deposits $\Delta\eta(T) + å$ centimetres of sediment, where $å$ is the net accumulation rate (cm) per storm event. In the following section we present an analytical rather than a numerical approach to the problem. This alternative approach allows appreciation of the role of each of the governing model parameters

### Parametric approach to bed preservation

The ratio of return period intercept, $a'$, from equation (6), divided by the accumulation rate, $å$, is a measure of depositional reworking:

$$r = a'/å \qquad (7)$$

If this ratio is large, then the resuspended thickness, $\Delta\eta$, is large compared with the accumulation of new sediment $å$ during any storm event. The ratio $r$ is equivalent to the physical mixing parameter, $h$, of Nittrouer & Sternberg, 1981;

$$h = \frac{l_p f_p}{å} \qquad (8)$$

where $l_p$ is the depth (cm) to which the sea-bed is eroded by storms, $f_p$ is the storm frequency and $å$ is the accumulation rate (cm yr$^{-1}$) (see discussion by

Swift & Thorne, this volume, pp. 3–31). The equivalency between the two equations holds despite the extra term in equation (8) because the accumulation rate, $å$, is the rate per storm event (cm $T^{-1}$, where $T$ is the return period), whose reciprocal is frequency.

Preserved thicknesses are generally distinguished by changes in grain size or sedimentary structure from bottom to top of a graded bed. When a large percentage of the initial thickness of the bed is preserved these changes can be distinguished readily. As the percentage of preserved beds decreases, the bed becomes harder to distinguish, until at some cut-off percentage the bed cannot be distinguished at all (Nittrouer & Sternberg, 1981). It is difficult, for example, to distinguish a thick series of storm beds from a single-event bed if the series is amalgamated (sandstone beds have no intervening shales). The cut-off percentage, $c\%$ ($\Delta\eta$), in general, may be a function of $\Delta\eta$ and the initial bed grading. In this study we assume that the cut-off percentage is a constant, $c\%$.

Appendix II gives a derivation of analytical expressions for mean distinguishable preserved bed thickness, $\Delta\bar{n}$, and the percentage of a sedimentary column that is composed of distinguishable event beds, $eb\%$. This analytical approach is based on the power law approximation of equation (6). Governing parameters of the analytical model are accumulation rate, $å$, reworking ratio, $r$, cut-off percentage, $c\%$, and return period power, $p$.

The effect of the governing parameters on bed thickness is best illustrated by considering variations in these parameters one at a time. We take as a

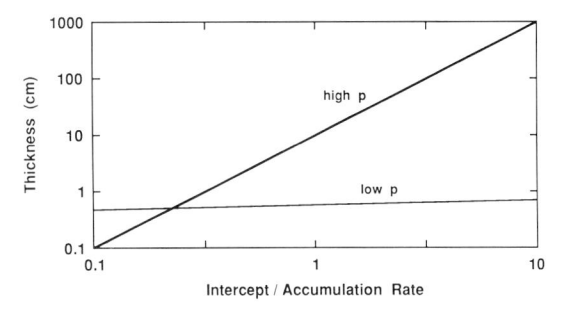

**Fig. 4.** Mean preserved bed thickness increases with reworking ratio while $c\%$ and $p$ remain constant. The high value of return power is $p = 0.5$. The low value of return period power is $p = 0.17$.

reference case an accumulation rate of 1 cm per event, a reworking ratio of 1, a cut-off percentage of 50% and a return period power of 0.5.

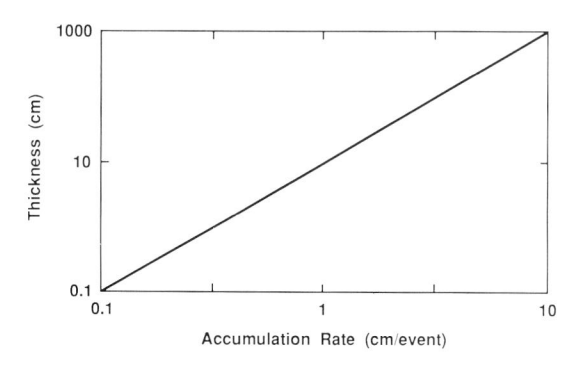

Figure 3 shows that mean preserved bed thickness increases proportionally with accumulation rate, $å$, while, $r$, $c\%$, and $p$ remain constant. In this experiment the intercept $a'$ changes so as to maintain a constant reworking ratio with variations in $å$. This result can be explained as follows. The number of preserved events does not change with $å$, as long as $r$, $c\%$ and $p$ remain constant. The total thickness of accumulation increases proportionally with $å$ and, thus the mean thickness per preserved event increases proportionally as well.

Figure 4 shows that mean preserved bed thickness increases with the reworking ratio, $r$, while $c\%$ and $p$ remain constant. Two cases are shown. The first, with a high value of return power, $p = 0.5$, shows an extreme dependence of thickness on reworking ratio. The second, with a low value of return period power, $p = 0.17$, shows only a slight increase of thickness with reworking ratio. This result can be explained as follows. A low reworking ratio results in preservation of a large number of small-magnitude event beds as sedimentary particles are buried after relatively few erosion–deposition events. The mean thickness is, thus, low. As the reworking ratio increases, the thin beds of frequent, small-magnitude storms are progressively removed, and the mean bed thickness preferentially reflects the largest events. When the return period power, $p$, is high, these infrequent events are significantly larger than more frequent events and, thus, this increase in mean thickness is extreme.

Figure 5 shows that mean preserved bed thickness increases with the minimum preservation cut-off

**Fig. 3.** Mean preserved bed thickness increases proportionally with accumulation rate, while $r$, $c\%$, and $p$ remain constant. Constant values are a working ratio of 1, a cut-off percentage of 50% and a return period power of 0.5.

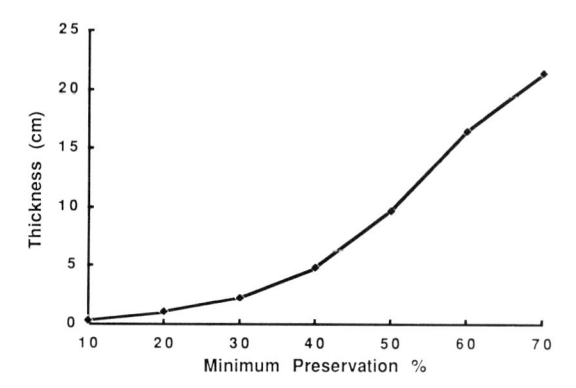

**Fig. 5.** Mean preserved bed thickness increases with the minimum preservation cut-off percentage while $å$, $r$ and $p$ remain constant.

percentage while $å$, $r$, and $p$ remain constant. This result can be explained as follows. Only large-event beds are likely to be well preserved. If a high preservation percentage is required to identify an event bed, only thick beds will enter into the calculation of mean thickness.

Figure 6 shows that the percentage of the total sediment column that is composed of distinct, single-event beds increases with the return period power for various values of the cut-off percentage, $c\%$. The curves shown are independent (within a few per cent) of variations in $å$ or $r$. The shape of the curves can be explained as follows. If the return period power is very low, then the thickness of the resuspension events, $\Delta\eta$, does not significantly increase with return period. In this case, the deposits of each event are invariably cannibalized by subsequent events. The net effect of many such events, mostly of roughly the same size, is to produce a large proportion of multiple-event beds rather than single event beds. These beds will appear to be single because they are thinner than the cut-off percentage, but in fact they will be multiple. For a high accumulation rate, for example, the resultant stratigraphic column will consist largely of poorly preserved, small return period beds. For high return period powers, the greater variation in size between low and high-frequency event beds allows preservation of the long return period events as discrete beds.

Figure 6 suggests that variations of return period power affect the percentage of single-event beds versus multiple-event beds. One cause of variation in the return period power is water depth. Extreme wave heights are not possible in shallow water owing to wave breaking. In this region large storms of greatly different magnitudes produce initial bed thicknesses that are similar because maximum wave height is limited by the water depth. The return period power $p$ in equation (6), therefore, approaches 0 in this region. Stratification within this zone should be characterized by multiple-event beds only.

The return period of preserved event beds in any stratigraphic column is a function of the process parameters, $r$, $c\%$ and $\dot{p}$. These return periods can be characterized by the return period of the smallest storm that on average will leave a preserved event bed, i.e the storm return period $T = T(m)$, where $m$ is the initial thickness of such a bed. The wave height of return period $T(m)$ can be found using equation (1), given coefficients $\alpha$ and $\beta$. The depth

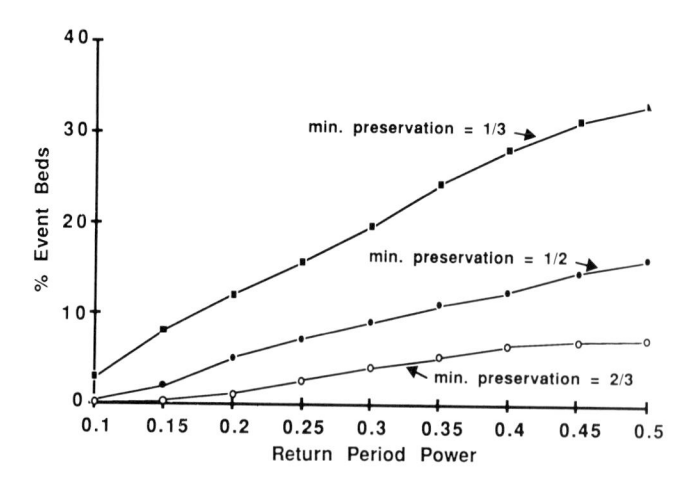

**Fig. 6.** The percentage of the total sediment column that is composed of distinct event beds increases with the return period power for various values of the cut-off percentage $c\%$. The curves shown are independent of variations in $a$ or $r$.

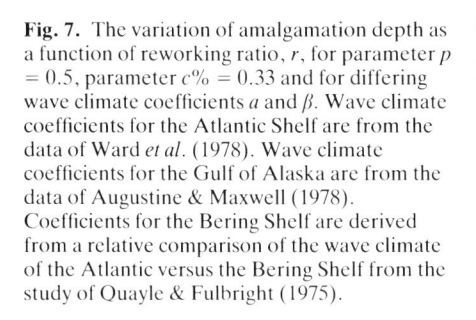

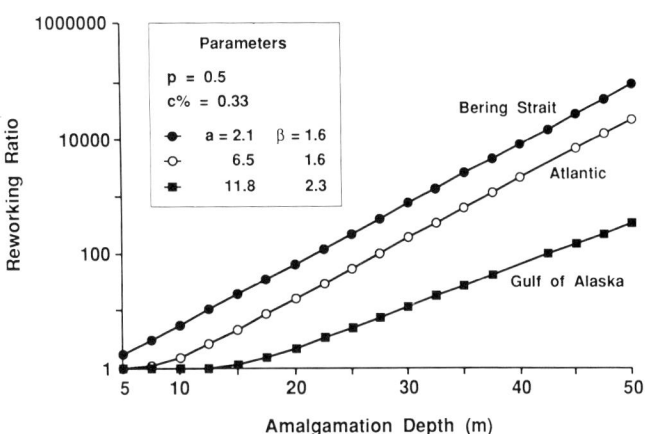

**Fig. 7.** The variation of amalgamation depth as a function of reworking ratio, $r$, for parameter $p = 0.5$, parameter $c\% = 0.33$ and for differing wave climate coefficients $a$ and $\beta$. Wave climate coefficients for the Atlantic Shelf are from the data of Ward *et al.* (1978). Wave climate coefficients for the Gulf of Alaska are from the data of Augustine & Maxwell (1978). Coefficients for the Bering Shelf are derived from a relative comparison of the wave climate of the Atlantic versus the Bering Shelf from the study of Quayle & Fulbright (1975).

at which such a wave breaks can be found using equation (5). This critical amalgamation depth separates deeper water sedimentation regimes with a full spectrum of wave heights characterized by a high value of parameter $p$ from shallower depths characterized by low values of parameter $p$.

Figure 7 shows the variation of amalgamation depth as a function of reworking ratio, $r$, for parameter $p = 0.5$, parameter $c\% = 0.33$ and for differing wave climate coefficients $\alpha$ and $\beta$. For example, for a reworking ratio of 10, for $\alpha = 6.5$ and $\beta = 1.6$, the critical amalgamation depth is 17.5 m. These wave climate coefficients correspond to the wave climate of the Baltimore Canyon area of the Atlantic shelf (Ward *et al.*, 1978). The variation of amalgamation depth as a function of reworking ratio is also shown for two areas in Alaska. The wave climate of the Gulf of Alaska is extremely harsh. The wave climate of the Bering Sea, however, is quite mild owing to the limited fetch of the seaway, which is protected by the Aleutian Islands (Bird & Schwartz, 1985).

## LIMITATIONS OF THE THEORY

The major limitation of the approach described here is the assumption that the bed response to an extreme storm event is independent of the history of preceding storm events. A large event, by this theory, always erases the record of preceding smaller events if they have not been buried sufficiently. More advanced numerical simulations of the bed preservation problem (not described here) indicate that this assumption may fail at high sedimentation rates. In particular, preliminary simulations indicate that a growing sediment pile can be armoured dynamically by stratal amalgamation. In this process, deposition of a storm bed is followed by one or more amalgamation episodes, in which additional sand layers are added without intervening mud beds. The process creates a multiple-event sand bed that armours the bottom, and until buried, suppresses the creation of thicker beds. Amalgamation occurs because if one event follows another with nearly the same intensity, the second one strips off the fine-grained top of the first and then deposits its own sand on top of the truncated base of the first bed. The resulting expanded sand bed limits the possible depth of erosion for any successive strata-forming event until it is sufficiently buried.

## STRATA-FORMING PROCESSES AND REGIME SEDIMENTATION: EXAMPLE FROM THE MESAVERDE FORMATION

### General

The preceding section has presented an analysis of strata formation in the shallow marine environment. How do such event-scale processes modify long-term and large-scale regime processes, and how are event-scale processes modified by them? How does event stratigraphy, caused by strata-forming processes, relate to sequence stratigraphy? In order to explore this issue, we have examined event stratigraphic sequences in the lower Mesaverde Formation (Campanian) of the Bighorn Basin, Wyoming.

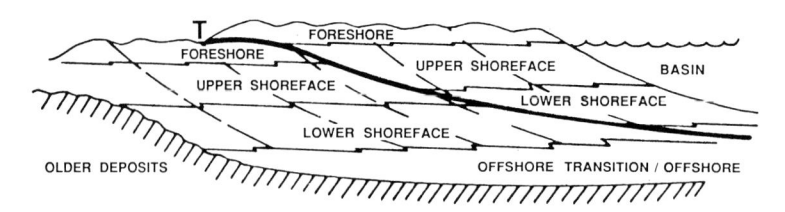

Fig. 8. Schematic diagram illustrating the relationship of lithostratigraphic to chronostratigraphic units in the Mesaverde Formation. Lithologic units are labelled with environmental names. Dashed, more steeply inclined lines are successive shoreface positions (time surfaces). The transgressive surface is labelled 'T'. (From Wright, 1986.)

### Character of the Mesaverde Formation

In the Mesaverde escarpment of the southern Bighorn Basin, the lower Mesaverde Formation progrades stepwise from east to west as a series of en echelon, overlapping, strandplain deposits that climb in the section to the east (Keefer, 1972). The lower contact of the Mesaverde Formation consists of 20–30 m of very thin bedded to medium bedded sandstone, alternating with similarly stratified shale. The sandstones are storm beds deposited on the lower shoreface and inner shelf. Each bed is an event bed that pinches out into the shale to the east, and merges with one of the many tongues of the lower Mesaverde to the west. In the main body of a sandstone tongue, the facies succession (lower shoreface–upper shoreface–beach) dips gently to the west while event beds dip gently to the east. This pattern of intersecting lithologic units and time surfaces is typical of Mesaverde and equivalent sandstones (Wright, 1986; Fig. 8).

The lower Mesaverde Formation in the Bighorn Basin is cyclic at several spatial scales. The upper surfaces of the strandplains locally retain vestiges of beach ridge morphology, with ridges at 50–70 m spacings, although more commonly, the subsequent transgression has erased this feature. Stratal bundling, of several metres vertical amplitude, can be traced through sandstone tongues. The bundles descend gently to the east as the tongue climbs to the west, but do not become well defined until near the bottom of the tongue where they become discrete subtongues splaying out into the shale. We infer that the bundles are successive progradational prisms, which correspond to successive beach ridges.

In addition, the lower Mesaverde is cyclic at the larger scale of the sandstone tongues themselves (Fig. 9B). Individual tongues can be traced for 5–10 km, from their landward pinchout in lagoonal deposits, to their seaward (eastward) pinchout in marine shale. Each sandstone tongue represents a separate episode of strandplain progradation into the subsiding western interior seaway, followed by transgression and shoreline retreat. The lower Mesaverde in our study area lies between the approximate mid-points of the *Scaphites hippocrepis* and *Bacculites* sp. (smooth) biozones (Gill & Cobban, 1973). Since the unit is approximately 100 m thick, and represents 1–1.5 my (Gill & Cobban, 1973), the 10–12 tongues that appear in most sections may represent durations of *c.* 100 000 my.

The cyclic nature of the Mesaverde Formation has been described by Hollenshead & Pritchard (1961), Ryer (1977, 1983), and Wright (1986). Many explanations of this characteristic have been offered. Wright (1986), following Ryer (1977, 1983), estimates the duration of the cycles at >100 000 years, and tentatively identifies the cycles as the fourth or fifth-order cycles of Vail *et al.* (1977). Recently, Barron *et al.* (1985), have identified similar cycles in the limestones of the basin axis, and have attributed them to climatic cycles at Milankovitch frequencies. However, Wright (1986) noted that, with regard to the Mesaverde cycles, '[the dilemma of] origin..., as well as the [determination] of duration values for short term cycles, remain as problems for studies of Cretaceous strand lines.'

### Procedures

The inner shelf and lower shoreface beds of the lower Mesaverde tongues contain well-preserved storm beds ranging in thickness from 1 cm to over a metre and are ideal subjects for a statistical analysis of the relationship of event stratigraphic to sequence

**Fig. 9.** Stratification characteristics of the lower Mesaverde Formation in the study area. (A) Low oblique aerial photograph of the Mesaverde escarpment at South Sand Draw. (B) The top of interval A (maximum flooding surface) at Cottonwood Creek. (C) Pinchout of a shale bed (at arrow) beneath a downlapping lowstand bed and the basal sequence boundary, Sand Draw.

stratigraphic successions. In order to undertake such an analysis, we collected data for bed thickness variations from five measured sections on the Mesaverde escarpment near Thermopolis, in the Big Horn Basin (Fig. 10). Measurements were made of the bedding thickness of sandstone, silt and shale units. The original measured thickness of individual lithologies were grouped into fining upward storm bed sequences. Thick shaley intervals were assumed to consist of multiple thin event beds, which could not be resolved by field observation.

Several prominent beds can be correlated visually between the four sections by use of photograph mosaics. Figure 11 shows a bed-thickness section in which the top of one such bed is used as datum. Except for the top horizon, the correlation between the four sections is inferred from the pattern of bed thicknesses. We infer that the section is the product of cyclic sedimentation. However, we do not use the Exxon terminology for sequence architecture (highstand systems tract, etc.) because the inferred sequences are of small scale and simple construction. We can recognize neither depositional systems tracts or depositional systems, only facies, for which we use the neutral term 'deposit.' See Swift *et al.* (this volume, pp. 153–187) for a discussion of stratigraphic nomenclature.

The most distinct aspect of the section shown is the presence of correlatable thick-bedded intervals, which we have identified as interval A and interval B in Fig. 11. Both these intervals contain progressively less net sand from west to east across the section. Interval A can be correlated across the entire section and is overlain by a thick shaley section. Intervals A and B cap parts of the section that are characteristically thinner bedded.

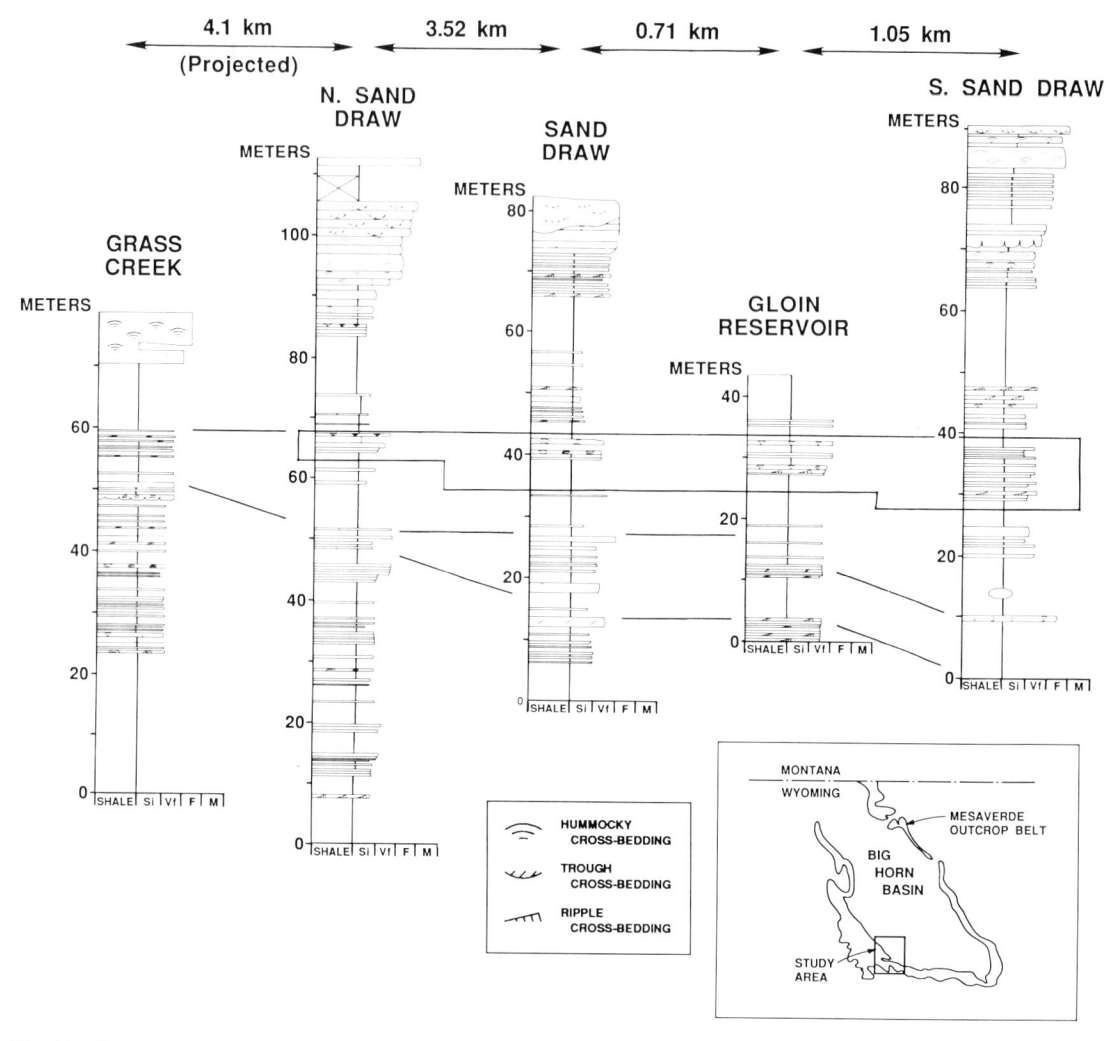

**Fig. 10.** Cross-section through a portion of the lower Mesaverde Formation near Thermopolis, Wyoming. Box lying across columnar sections indicates portion of section subjected to detailed study. Inset map shows location of study area.

## ESTIMATING MODEL PARAMETERS FOR THE MESAVERDE FORMATION

### General

The theoretical results illustrated in Figs. 1–7 suggest that the observed stratigraphy of thin and thick-bedded storm beds can be explained by variations with process parameters $r$, $p$ and $å$, in response to variations in regime parameters through one of the small-scale sea-level cycles. In order to quantitatively test this hypothesis, it is necessary to estimate grain-size distribution parameters, wave and current climate parameters, and bed preservation parameters for the Mesaverde Formation in the area of study.

As suggested in the previous section, wave climate is largely a function of fetch. An estimate of the fetch of the Cretaceous interior seaway can be based on reconstructions of the Campanian palaeogeography of Wyoming. Computer simulations using a global climate model for the Campanian (Parrish et al., 1984) indicate that peak storm winds were from the northwest and northeast. Palaeogeography of the seaway at this time (Kauffman, 1984) shows that the fetch of such winds is c.

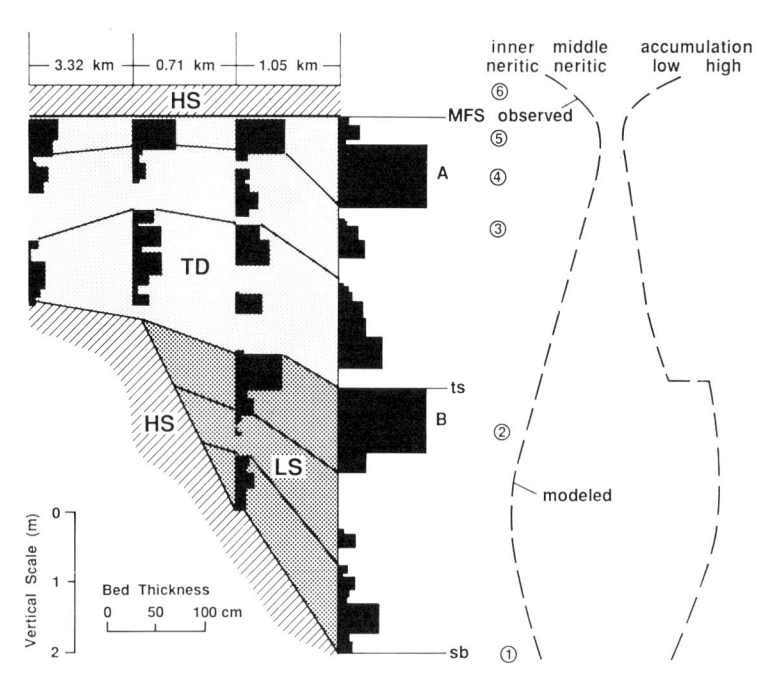

Fig. 11. Bed thickness cross-section of the area of detailed study, showing the relationship between the time history of water depth and accumulation and the observed stratal geometries. HS = Highstand deposits; TD = transgressive deposits; LS = lowstand deposits; MFS = maximum flooding surface; ts = transgressive surface; sb = sequence boundary.

1000 km. A fetch of this size is roughly equivalent to the fetch of the modern Bering Sea. We assume, therefore, that the wave climate parameters, α and β, of the Cretaceous seaway are similar to the modern Bering Sea (see Table 1). We do not consider the effects of tropical hurricanes originating in the Tethys ocean and being steered northward into the Western Interior Seaway. Recent numerical simulations (Barron, 1988) indicate that the extent of steering of hurricanes into the seaway may vary from epoch to epoch depending on the palaeogeography of the seaway.

The current climate of the seaway enters the model through parameter $c_2$ in equation (4). To estimate this parameter the diagnostic shelf-current model of Galt (1980) and Han et al. (1980) has been used to estimate currents owing to wind forcing, from reconstructions of the Campanian palaeogeography of the Western Interior Seaway. Figure 12 shows the computed bottom current velocity for an exerted wind shear stress of 1 dyne cm$^{-2}$ from the northwest (Fig. 12A) and northeast (Fig. 12B). Using an aerodynamic roughness height of 0.01 cm (Kraus, 1972) and the density of air at 10°C, this imparted stress corresponds to wind speed at 10 m above the ground of 8.17 m s$^{-1}$. The expected significant wave height for this wind velocity is 1.29 m (Frost, 1966). As can be seen in Fig.

12, the cross-shelf current created by these conditions ranges from less than 1 cm s$^{-1}$ to c.3 cm s$^{-1}$. The corresponding variation in parameter $c_2$ is from 0.005 to 0.02. The low value of this coefficient and the low sensitivity of model results to variations in this coefficient (see below) indicate that numerical modelling of currents is not, in general, a necessary prerequisite to applying the theory as presented in this paper.

Grain-size data for sandstones within the Mesaverde Group have been reported previously for the Kenilworth Member in central Utah (Swift et al., 1987). This study indicates that the modal grain size of the coarse fraction depends on sandstone facies. Sandstones exhibiting hummocky cross-stratification have a modal grain size of 2.75–400 phi and a phi standard deviation of c.0.25. The thin-bedded, graded facies shows similar characteristics. A modal grain size and phi standard deviation is taken as $d = 2.75$ to 4.00 phi, and $s = 0.5$ to 0.75 phi.

The percentage coarse fraction, $f$, is a poorly known parameter (Table 1). What is required is not the percentage of fines input into the system, because most of the fine component will be bypassed into deeper water, but the distribution of coarse to fine material that on average make up the original graded deposit of a large storm event bed. A quali-

**Table 1.** Input parameter values for Mesaverde study

| Variable | Range | Description |
|---|---|---|
| $a$ | 1.7–2.3 m | Most-probable, largest, significant wave height in a given year |
| $\beta$ | 1.5–1.7 | Scales increase of wave heights with return period |
| $c_1$ | 3.893 s m$^{-2}$ | Relates wave height to wave period |
| $c_2$ | 0.005–0.02 s$^{-1}$ | Relates significant wave height to current velocity |
| $b_{coef}$ | 0.73–0.87 | Relates wave height to breaking depth |
| $d$ | 2.75–3.25 phi | Modal grain size of coarse population |
| $\sigma$ | 0.5–0.75 phi | Standard deviation of log-normal distribution |
| $\nu$ | $8.2 \times 10^{-6}$ m$^2$ s$^{-1}$ | Kinematic viscosity of water at 0% sediment concentration |
| $f$ | 0.25–0.50 | Volume fraction of coarse population |
| net $\mathring{a}$ | c.125 m my$^{-1}$ | Net accumulation rate |
| $h$ | Inner–middle neritic | Qualitative estimate of water depths |
| $c\%$ | 0.33 | Minimum preservation cut-off percentage |

tative argument for this fraction can be made as follows. The model results below predict that on average event beds are 50% preserved; i.e. on average their top halves have been stripped by subsequent erosion. In approximately half the graded event beds studied, clay or silt drapes are thin or missing. These observations can be explained if, on average, a graded event bed is deposited with a bottom half of well to moderately sorted sand and coarse silt and a top half of draping fines.

Maps of accumulation rate, $\mathring{a}$, based on ammonite zonations (Gill & Cobban, 1973) show that the net accumulation rate of the Mesaverde Formation in the area of Thermopolis, Wyoming is c.125 m my$^{-1}$. Accumulation rate is expected to vary with sequence position, reaching a minimum during maximum flooding owing to sediment starvation (see also Swift & Thorne, pp. 3–31).

A qualitative, independent estimate of water depths within the measured sections can be obtained for both sandy and shaley parts of the section. The sand beds of intervals A and B show large-scale, hummocky cross-stratification, which we interpret to indicate middle or lower shoreface (inner neritic) conditions. Analysis of the Foraminifera of thick shaley intervals shows a shoaling upward trend ranging from middle to inner neritic (upper right of Fig. 11). Middle neritic depths are characterized by an assemblage of planktonic, calcareous and arenaceous forms, transitional depths by calcareous

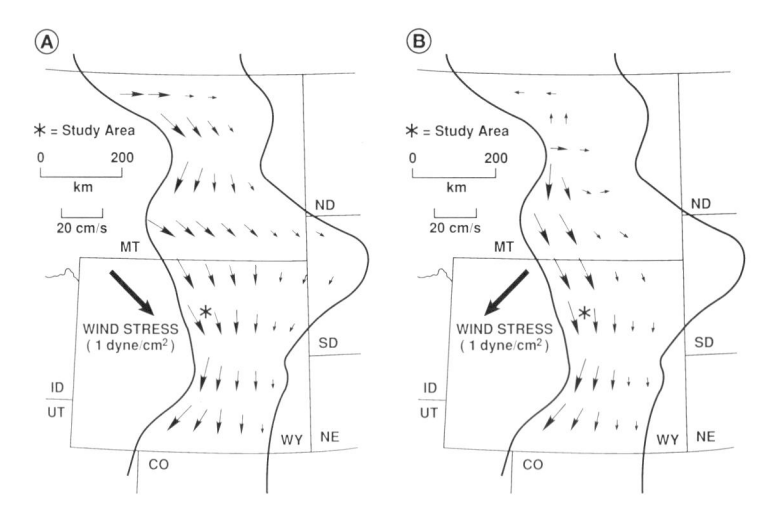

**Fig. 12.** Steady-state current created by a wind shear stress of 1 dyne cm$^{-2}$ from (A) the northwest and (B) the northeast. The cross-topographic component of bottom velocity ranges from less than 1 cm s$^{-1}$ to c.3 cm s$^{-1}$. Reconstruction of palaeogeography is based on palaeoshorelines of Gill & Cobban (1973). (From Parrish et al., 1984).

and arenaceous forms, and inner neritic depths by arenaceous forms only.

The minimum preservation cut-off percentage, $c\%$, is perhaps the most difficult parameter to assess independently. A value of $c\% = 0.33$ is chosen on the basis of Fig. 6. This figure shows that if $c\%$ is greater than 0.33, only a small percentage of the sediment column is predicted to consist of distinguishable event beds. We consider values significantly less than 0.33% to be intuitively unreasonable.

## ESTIMATING RETURN PERIOD POWER AND INTERCEPT

The model parameters shown in Table 1 can be used to predict the bed thickness versus storm return period relationship using either our earlier approach (Niedoroda et al., 1989) or the bedload armouring approach described in this paper. This relationship has been computed for water depths ranging between 16 and 30 m. Figure 2 shows the results of this calculation for 20 and 30 m water depth using preferred values for the model parameters.

The results of the analysis are simplified by using the power law approximation of equation (6). Coefficients $a'$ and $p$ from equation (6) are found by best-fitting a straight line to a log–log plot of bed thickness versus return period by one of two procedures. In the first method the fit is done at 12 different return periods between 100 and 15 000 years. This method is referred to as a 'curve' fit. In the second method, coefficients $a'$ and $p$ are found from the bed thickness at two end-points only: $T_{90}$ and $T_{40}$. The value $T_{90}$ is the return period that creates a bed whose base has a 90% chance of preservation. The value $T_{40}$ likewise has a 40% chance of preservation. The statistics of preserved event beds should reflect mainly storm events of return periods between these two extremes

Figure 13 shows the variation with water depth of power law coefficients $a'$ and $p$ for the bedload armouring approach. The end-point and curve-fit procedures give equivalent results in water depths of 20 m or greater. In water depths shallower than 20 m, the end-point fit is preferable to the straight line curve-fit.

The calculated return period power from the bedload armour method varies from about 0.45 to 0.65. A maximum is reached in water depths of 20 m. In water depths less than 20 m, the effect of wave breaking decreases return period power.

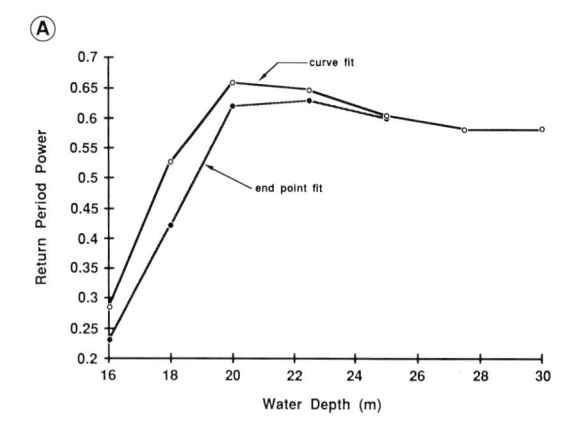

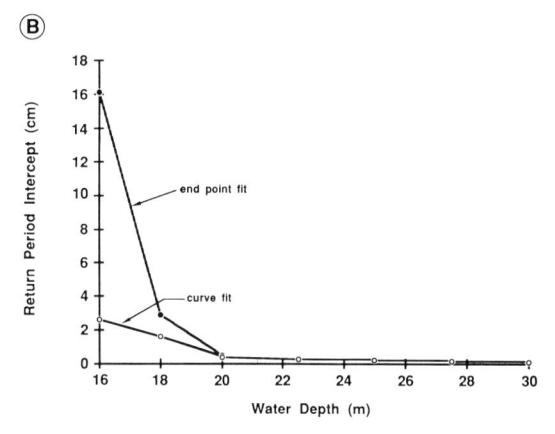

**Fig. 13.** Variation of return period power and intercept with water depth computed using the bedload armouring model. 'Curve fit' and 'end-point' procedures are described in the text. Model parameters are set to correspond to the Campanian Mesaverde Formation of the Western Interior Seaway as given in the caption to Fig. 2 and described in the text.

In greater water depths return period power also decreases owing to diminishing wave shear stresses.

Low return period power ($p < 0.35$) in water depths less than 16 m indicates that preserved deposition in these depths is characterized by the amalgamation of beds. Taking, for example, a minimum preservation cut-off, $c\%$, of 33%, Fig. 6 shows that if the return period power is less than 0.35 then the percentage of the total sediment column composed of distinct event beds is less than 25%.

Our earlier approach (Niedoroda et al., 1989) predicts different results for the return period power in water depths between 16 and 30 m (Fig. 14). Coefficients $a'$ and $p$ have been computed, in this

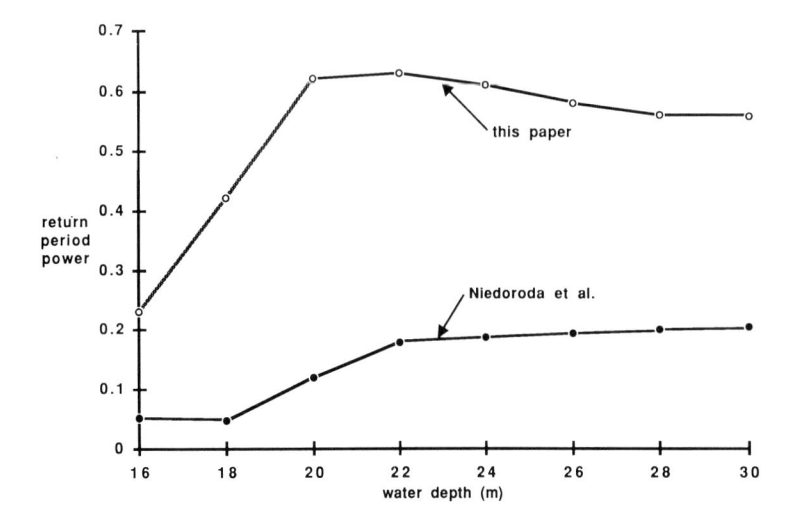

**Fig. 14.** Predicted variation of return period power with water depth based on Niedoroda *et al.* (1989) and bedload armouring approaches. Model parameters are set to correspond to the Campanian Mesaverde Formation of the Western Interior Seaway as given in the caption of Fig. 3 and described in the text. The bedload armour approach is preferred since observed stratigraphic sections indicate that the return period power is approximately 0.5.

case, using the end-point method only, for a sedimentation rate of $250 \text{ m my}^{-1}$. Most importantly, return period power is less than $c.0.2$ for the entire range of water depths. Again using Fig. 6, and a $c\%$ of 33%, this predicts that the observed stratigraphic columns of the Mesaverde Formation should consist of less than $c.12\%$ of single-event beds. This percentage is far less than observed (Fig. 11). We conclude, therefore, that the theory of bed formation presented in our earlier analysis may tend to overestimate bed thickness.

## SENSITIVITY ANALYSIS

It is important to determine whether model predictions are highly sensitive to the degree of uncertainty in assessing the values of model parameters. We refer to parameter values used in calculating the variation of return period power and intercept with depth shown in Figs 13, 14 as *standard* values. A suggested range for model parameters is shown in Table 1. Figure 15 shows the change in the predicted thickness versus return period curve resulting from changing one parameter at a time away from the standard. All five curves, as shown, nearly overlap, indicating that derived values of return period power and intercept are not sensitive to these parameter variations.

The upper end of the curve shows more sensitivity to changes in model parameters. This is particularly so for variations in the breaking coefficient $b_{\text{coef}}$ (Fig. 16). For the case where $b_{\text{coef}} = 0.78$, the wave height of any storm with return period greater than 5000 years is affected by wave breaking in water depths of 20 m. If $b_{\text{coef}} = 0.86$, however, only storms of greater than a 15 000-year return period are affected by breaking at this depth.

We do not illustrate the sensitivity to model parameters of our earlier approach; refer to Niedoroda *et al.* (1989). Derived values of return period power, for all tested conditions, remain significantly lower than 0.5. The remainder of this paper discusses the implications of the bedload armour model for quantitative analysis of the bed thickness relationships as shown in Fig. 11.

## QUANTITATIVE MODEL OF BED-THICKNESS CROSS-SECTION

Quantitative application of the theory described in the preceding sections to the beds measured in the Mesaverde Formation is shown in Fig. 17. In this figure the observed bed thicknesses and the modelled variation of return period power and intercept with water depth (Fig. 13) are used to predict the time evolution of water depth and accumulation rate during the accumulation of the deposits.

In Fig. 11, the interpreted geological history begins (point 1) on the inner shelf, before a prograding strandplain. The concurrent geohistory of water depth and accumulation is indicated in Fig. 17 by bold arrows, which trace the evolution of the section. In Fig. 16, point 1 occurs in a water depth of

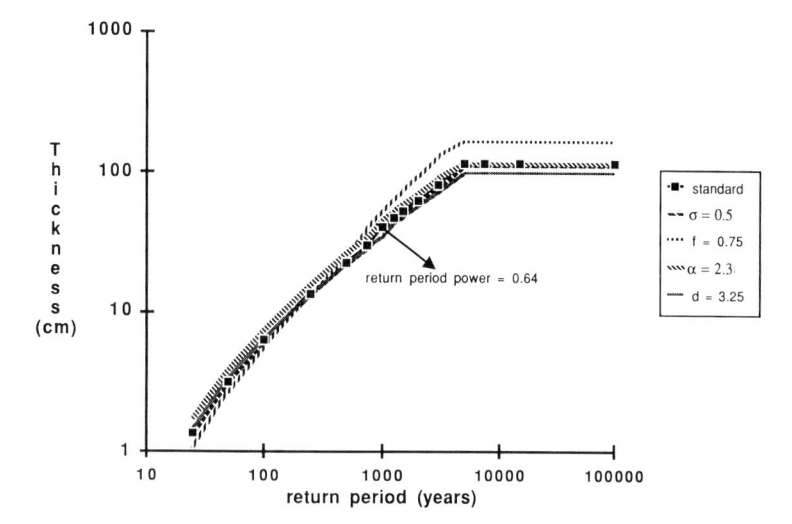

**Fig. 15.** Sensitivity of bedload armour model to variations in model parameters from the standard case. Four alternative curves are shown each differing from the standard case by one parameter only, as shown in the legend.

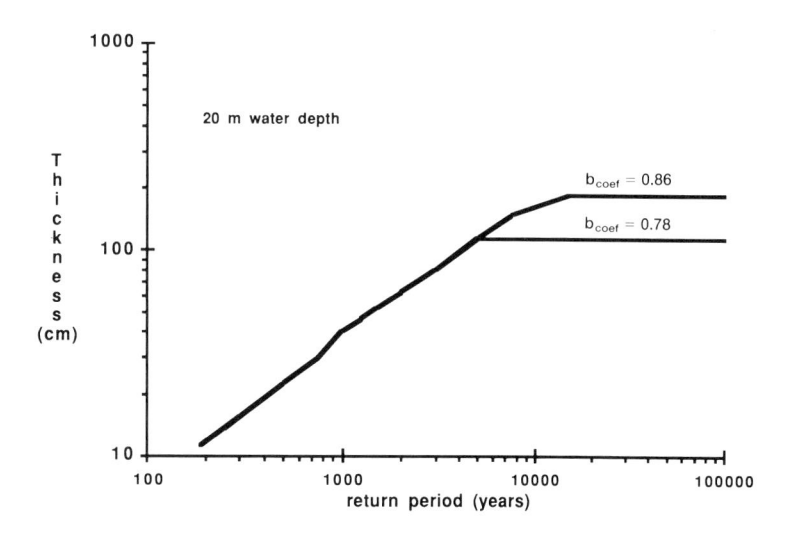

**Fig. 16.** Sensitivity of the bedload armour model to variations in model parameter $b_{coef}$.

25 m, with an accumulation rate of $c.\ 200\,\text{m my}^{-1}$. Progressing up-section, from point 1 to point 2, the thin beds below interval B grade up into the thick-bedded storm beds of interval B. In Fig. 16 the change is interpreted as the result of shallowing of the order of 3 m.

Interval B (Fig. 11) is capped by a rippled, burrowed, slightly coarser surface, which we infer to be a transgressive surface (ts) between points 2 and 3 (Fig. 9B). Deeper water conditions are re-established above the transgressive surface and bed thickness progressively decreases. From point 3 to 4, however, bedding thickness increases once more, and so does grain size, from an estimated 62–125 μ (very fine sand) to 125–250 μ (fine sand). We infer that both the thick beds and the coarser sand is a consequence of a decrease in the sediment accumulation rate, from 250 to 125 m my$^{-1}$. By analogy with the modern Texas coast (Sahl et al., 1987), the highly turbid, low salinity, coast-wise flowing water of the shoreface and adjacent inner shelf would have been separated from shelf water by a coastal front. The continuing transgression would have shifted the front from offshore to inshore of the depositional site, and the accumulation rate would have decreased in response. With a decreased depositional rate, the reworking ratio, $r$, would have increased. The preservation bias of beds would have changed accordingly to favour long return period (thicker) beds. The increased reworking ratio also would have increased the efficiency of progressive sorting, so that deposited grain size would have increased.

Transgression continues between points 4 and 5 but the decreasing accumulation rate has increased $c\%$, the minimum preservation percentage cut-off. The thicker storm beds no longer pass intact through the zone of shallow burial, but undergo serial restructuring before they are finally buried deeply enough to be immune to change. The zone is capped by a well defined, burrowed and rippled surface, which we interpret as a surface of maximum flooding. The surface is overlain by highstand shales (HS), which show shallowing upward water depths from point 5 to point 6.

## COMPATIBILITY WITH FIELD RELATIONSHIPS

The geohistory of water depth in Fig. 17 has an amplitude of 3 m. The total duration of this cycle is approximately 100 000 years. Assuming Airy isostasy, eustatic sea-level changes are $c.70\%$ of water-depth changes (e.g. Thorne & Watts, 1989). The model results suggest, therefore, that high-order stratigraphic cycles within the Mesaverde Formation are created by $c.100 000$-year sedimentation cycles of 2 m amplitude. We must use the neutral term 'sedimentation cycles' because we can establish only shoaling or deepening. We cannot distinguish between eustatic sea-level cycles, relative sea-level cycles owing to local crustal movements, or cycles of shallowing and deepening simply owing to fluctuating sediment input, perhaps associated with Milankovitch type rainfall cycles in the source terrain.

The latter model, in which the cycles result from relative rather than absolute sea-level variations, is consistent with regional field relationships. Here on the western flank of the seaway, subsidence rates were as high as 20 cm per 1000 years (Cross & Pilger, 1978; Kominz & Bond, 1986). It is not clear that these rates were ever exceeded by eustatic rates, so that it is possible to propose a sequence-forming model in which sea-level rarely could fall faster than the subsiding crust. In this scheme, transgressions and regressions are not controlled by eustasy, but rather by the interaction between sediment input and subsidence. Transgressions and regressions occur as sediment input lags behind or exceeds subsidence and the creation of accommodation space (see discussion of sequence formation in the Cretaceous Western Interior Basin (Swift et al., this volume, pp. 89–152)).

In this sedimentation-dependent model ($Q$-dependent model; see Thorne & Swift, this volume, pp. 33–58), cyclic variations in eustatic sea-level would result in a step-like modulation of relative sea-level rise. However, sea-level falls would not occur. The local depositional style of the Mesaverde is consistent also with this aspect of the model. The main transgressive surfaces in the Mesaverde have little relief. Such sea-level falls as occurred apparently did not significantly incise the subaerial suface. Estuaries did not form during transgressions. Instead, floodplains continued to aggrade, and micro-birdfoot deltas built out into the protected waters of the back-barrier lagoons, as the lagoons shifted landward (Kamola, 1984).

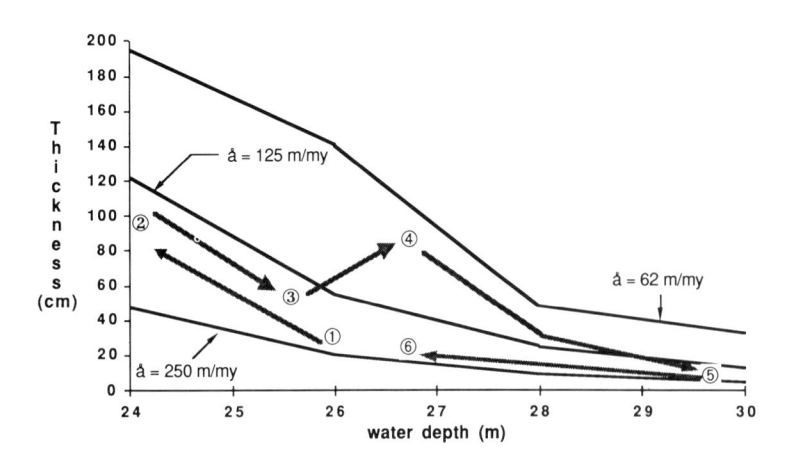

**Fig. 17.** Quantitative sequence stratigraphic interpretation of observed bed thickness cross-section based on prediction of mean bed thickness as a function of water depth and accumulation rate. The geohistory of water depth and accumulation is indicated by bold arrows which point in the direction of time and trace a path between successive points labelled 1 to 6.

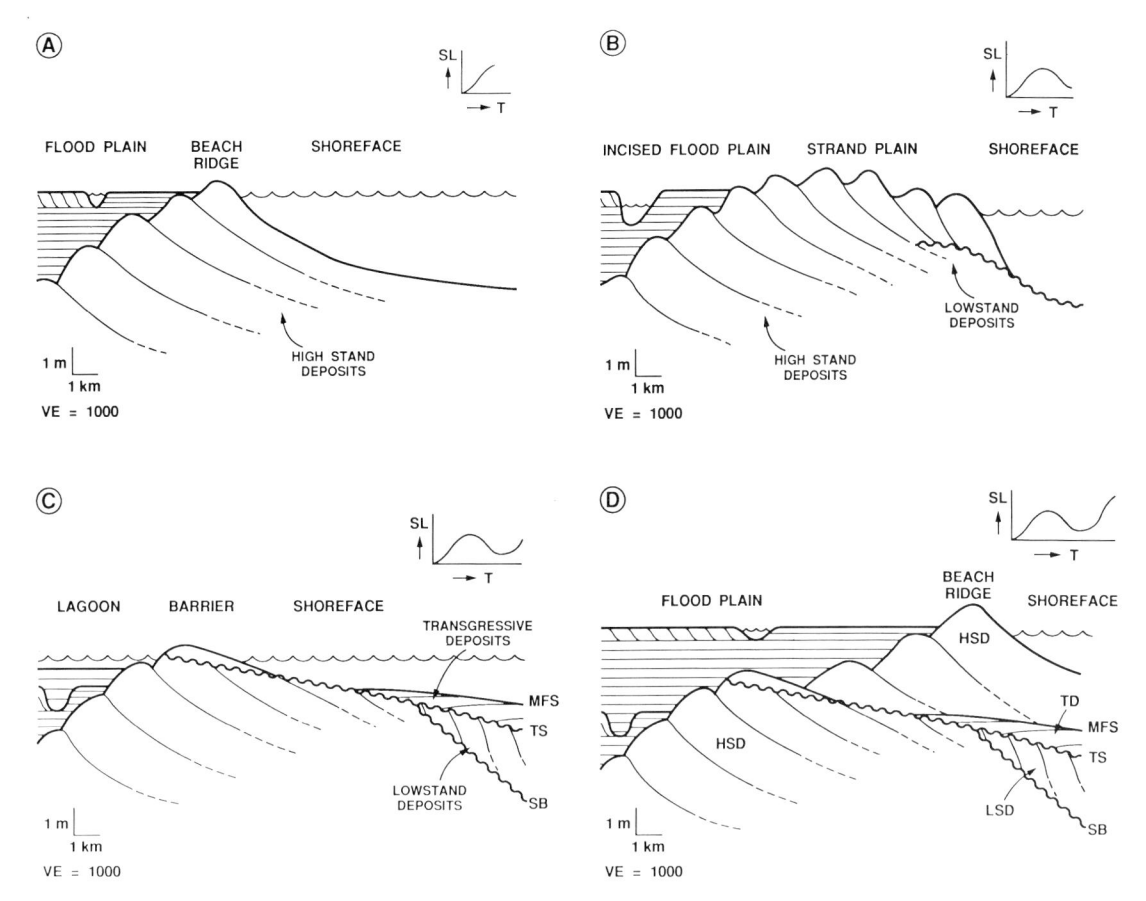

Fig. 18. Enlargement of the shallow marine portion of the sequence, showing the downlapping character of the highstand and lowstand deposits, versus the onlapping character of the transgressive deposits. A, late sea-level rise 1; B, sea-level fall; C, early sea-level rise; D, late sea-level rise.

However, other aspects of Mesaverde field relationships suggest that small-scale sea-level falls *did* occur during at least some of the cycles, as indicated in the analysis of strata thickness. The interaction that we have inferred between sedimentation at the event scale and at the regime scale carries implications concerning the architecture of the small-scale sequences. It suggests that the internal surfaces that characterize third-order sequences are present (Van Wagoner et al., 1988), and that consequently, deposits analogous to third-order systems tracts are present. This interpretation is at odds with that of previous workers. Ryer (1977, 1983) and Wright (1986) have described these small-scale sequences as highly asymmetric in nature, in that only regressive deposits are preserved (Fig. 8). Transgression is an erosive process, while regression is a depositional one; transgression cuts a transgressive surface.

Regression shifts inner shelf, lower shoreface, upper shoreface, and subaerial facies tracts back over the transgressive surface and buries it.

In the Mesaverde escarpment of the Southern Big Horn basin, however, the stratigraphy is more complex than this simple, asymmetrical model permits. Overflights of the Mesaverde escarpment indicate that between 20 and 30% of the sandstone units beneath the main body of the lower Mesaverde pinch out landward as well as seaward. Some of these lenticular bodies may be prodelta plume deposits, extending down-shelf from the tips of cuspate deltas, into the plane of the cross-section, as described by Palmer & Scott, 1984.

In most cases, however, they are explained more easily as lowstand or transgressive deposits. Figure 18 presents a conceptual model for the evolution of these deposits during a sea-level cycle in which the

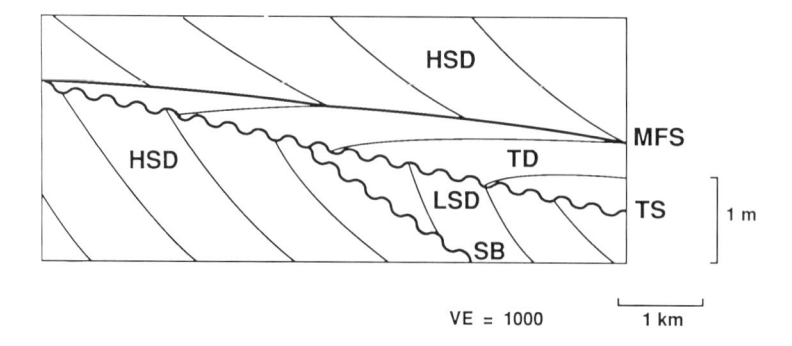

Fig. 19. Model for the deposition of a fifth-order sequence. HSD = highstand deposit; LSD = lowstand deposit; TD = transgressive deposit, SB = sequence boundary, TS = transgressive surface, MFS = maximum flooding surface.

eustatic component of sea-level variation is sufficiently strong to induce periodic sea-level falls (*R*-dependent model). The cycle illustrated in the figure begins during late sea-level rise, as a strandplain progrades into the subsiding basin, climbing in the section as it does so (panel A). However, the trajectory of climb flattens out as highstand is attained (panel B). As sea-level begins to fall, the coastal profile is no longer a constructional profile, but instead is incised into the sea-floor by submarine erosion, as the strandplain builds seaward over it (see Dominguez et al., this volume).

In panel C, the subsequent sea-level rise becomes rapid enough to overcome the sediment influx. The coastal profile flattens into its characteristic transgressive shape (see Swift et al., this volume pp. 89–152), and is driven back through erosional retreat across the subsiding strandplain, which is truncated by the retreat process. The inner profile is erosional. However, bottom wave power decreases more rapidly seaward than does sediment input, so that at some distance seaward, the dynamic accumulation rate ($\mathring{A} = Q/P$; see Swift & Thorne, this volume, 3–31) is sufficiently high to allow accumulation of onlapping transgressive beds. In panel D, regression resumes, burying lowstand and transgressive lenses beneath the main body of the strandplain.

The *R*-dependent model places constraints on stratal geometry; onlapping transgressive beds contrast with downlapping lowstand and highstand beds (Fig. 19). The model does not necessarily contradict the model of Fig. 8, but it does add a more complex stratigraphy to the inner shelf zone seaward of the area figured.

The analysis of bedding thickness in the area of detailed study (Fig. 11) indicates a stratigraphy compatible with the *R*-dependent sea-level cycle model described by Fig. 18. The pattern of onlap against the surface marked SB in Fig. 11 suggests that it is a high-order sequence boundary. There is no evidence for subaerial exposure at this horizon. If relative sea-level changes are in fact involved in the formation of these sequences, then this surface can be explained as a marine bypass surface, formed as the shoreface profile was incised into the inner shelf floor during the time of maximum sea-level fall rate (Fig. 18B). In this case, interval B becomes a downlapping, lowstand deposit, nourished during the sea-level fall and lowstand by sea-floor erosion occurring further landward.

In this interpretation, the upper bounding surface of interval B corresponds to a transgressive surface, cut during the subsequent period of sea-level rise as the steep progradational profile of the preceding transgression became a flattened, erosion-dominated transgressive profile (Fig. 18C). Interval A thus consists of onlapping transgressive beds nourished by erosional shoreface retreat (see Swift & Thorne, this volume, pp. 3–31). Similarly, the top of interval A, the most strongly developed surface in the section, becomes the maximum flooding surface. The highstand shales and sandstones above it would downlap on to the maximum flooding surface, in contrast to the onlapping relationships of the lowstand and transgressive sandstones below (Fig. 19). Downlapping and onlapping relationships occur at such very small angles of convergence that generally they are not visible at the scale of the outcrop. In some cases, however, disconformable surfaces within sandstone beds can be seen to pass laterally into shale seams, which in turn, thicken into shale interbeds (Fig. 9C).

If the Mesaverde cycles are only cycles of shoaling and deepening as sediment input into the subsiding basin fluctuates, then the simple sequence architecture of Fig. 8 would prevail. However, our analysis leads us to conclude that full sequence architecture, with onlapping lowstand and transgressive deposits, is present in at least some of the small-scale sequences. We therefore suggest that sea-level fluctuations, including at least occasional sea-level falls contributed to the regime governing the high-order depositional sequences.

## CONCLUSIONS

This paper presents an analytical model for storm bed formation, and applies it to the problem of relating strata thickness patterns to the sedimentary regime as inferred from depositional sequence architecture. It is based on a modification of our original sediment transport formulation that is more sensitive to bedload armouring.

The model is built around two basic relationships. In the power law equation, bed thickness is seen as a function of the storm climate; $\Delta\eta = a'T^p$, where $\Delta\eta$ is bed thickness, $T$ is the return period, $p$ is the return period power, $a'$ is the return period intercept (thickness for $T=1$ year) and $\dot{a}$, the accumulation rate.

A second basic relationship describes the extent of preservation of the deposited bed in terms of a reworking ratio, $r = a'/\dot{a}$. As $r$ increases, more of each deposited bed is resuspended by the next event and incorporated into the next bed. Knowing, $r$, it is possible to estimate the cut-off percentage, $c\%$, which is the percentage of the bed that must remain for successive deposits to be identifiable as single-event beds. The cut-off percentage may, in general, be a function of bed thickness and the vertical grain-size gradient. If the wave climate can be parameterized as a Gumbel distribution, then the critical amalgamation depth can be calculated.

The study has shown that variations of event bed thickness are diagnostic of changes in water depth and accumulation rate, which are parameters of the depositional regime. Further, the presence of preserved single-event beds, as opposed to amalgamated storm beds, can be used to put quantitative constraints on the palaeoenvironment of deposition. An increase in the reworking ratio, leading to an increase in mean event bed thickness, can arise from either a decrease in water depth or a decrease in accumulation rate. Water depth and accumulation rate vary characteristically with position within a depositional sequence, and variation in bed thickness therefore can be used to help distinguish between alternative sequence stratigraphic interpretations. In order to illustrate this application, the bed thickness model to palaeoceanographic conditions of the Campanian Western Interior Seaway has been calibrated. The results suggest that high-order stratigraphic sequences within the Mesaverde Formation are created by $c.100\,000$-year shoaling cycles of 2 m amplitude. Stratigraphic constraints indicate that relative sea-level changes were involved.

More generally, our study of strata thickness in the Mesaverde Formation shows that the reworking ratio, $r$, is often the dominant control on both bed thickness and grain size in shallow marine settings. Swift & Thorne (this volume, pp. 3–31) and Thorne & Swift (this volume, pp. 33–58) have presented a regime model of shelf sedimentation, in which the equilibrium shelf surface is governed by mutual adjustments in the geohistorical parameters of sediment input rate and character, relative sea-level fluctuations and the fluid power expenditure. The analysis of subequilibrium processes presented here shows how regime-scale parameters affect the event-scale parameters, and how these in turn control the stratification and grain-size patterns, which taken together constitute sedimentary facies. Swift *et al.* (this volume, pp. 89–152) examine the implication of these relationships for shelf depositional systems.

## ACKNOWLEDGEMENTS

We are grateful to Dr Peter Thompson for the foraminiferal analysis of the Mesaverde beds, and to Dr Charles Adams for taking part in the field work.

## APPENDIX I: BEDLOAD ARMOURING MODEL

Marine and fluvial sediment transport has been studied quantitatively since the days of Sorby (1853, 1855). Though much progress has been made, sediment transport science still relies largely on empirical or semi-empirical formulations that are calibrated on a limited range of field or laboratory experiments (Kennedy, 1983; Vanoni, 1984). The geologically important conditions of extreme storm events, unfortunately, are not, to date, sufficiently calibrated by this empirical approach. The model presented in this appendix, therefore, is based on a rather preliminary semi-theoretical approach

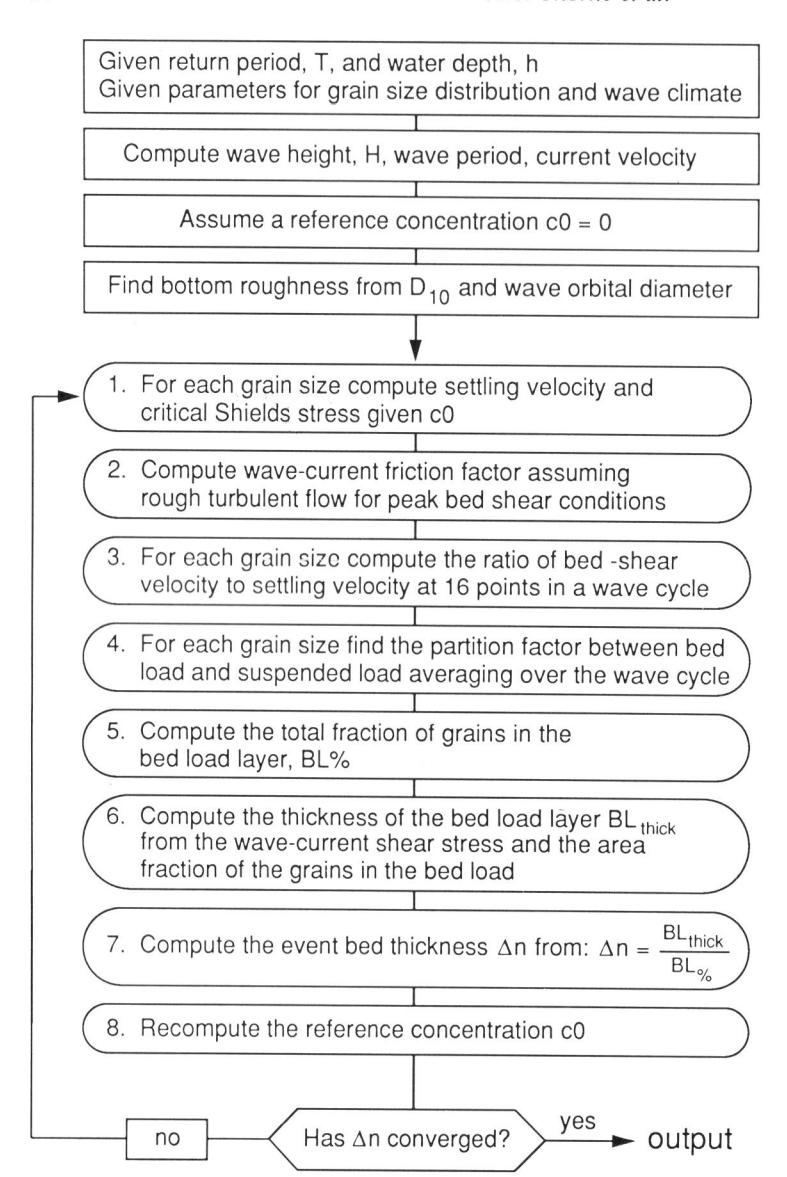

```
┌─────────────────────────────────────────────┐
│ Given return period, T, and water depth, h  │
│ Given parameters for grain size distribution and wave climate │
└─────────────────────────────────────────────┘
                    │
┌─────────────────────────────────────────────┐
│ Compute wave height, H, wave period, current velocity │
└─────────────────────────────────────────────┘
                    │
┌─────────────────────────────────────────────┐
│ Assume a reference concentration c0 = 0     │
└─────────────────────────────────────────────┘
                    │
┌─────────────────────────────────────────────┐
│ Find bottom roughness from D₁₀ and wave orbital diameter │
└─────────────────────────────────────────────┘
                    │
  ┌─→ 1. For each grain size compute settling velocity and
  │      critical Shields stress given c0
  │   2. Compute wave-current friction factor assuming
  │      rough turbulent flow for peak bed shear conditions
  │   3. For each grain size compute the ratio of bed-shear
  │      velocity to settling velocity at 16 points in a wave cycle
  │   4. For each grain size find the partition factor between bed
  │      load and suspended load averaging over the wave cycle
  │   5. Compute the total fraction of grains in the
  │      bed load layer, BL%
  │   6. Compute the thickness of the bed load layer BL_thick
  │      from the wave-current shear stress and the area
  │      fraction of the grains in the bed load
  │   7. Compute the event bed thickness Δn from: Δn = BL_thick / BL_%
  │   8. Recompute the reference concentration c0
  │           │
  │      ⟨Has Δn converged?⟩ --yes--> output
  └──no──┘
```

**Fig. AI.1.** Flowchart of algorithm for computing bed thickness based on the bed armouring approach. Steps labelled 1 to 8 are described in the text.

that attempts to describe sediment transport under extreme storm conditions. We recognize that our approach is somewhat heuristic and will require theoretical modification and empirical calibration at a later date.

Any formulation of sediment transport depends critically on whether the cohesive properties of sediments are considered. The Mesaverde Formation is a rather fine-grained clastic system and some question exists as to whether sediment transport is affected by cohesion. Studies of an analogous modern shelf, Norton Sound (see discussion in text), indicate that cohesion of surficial sediments is minor owing to the high water content of the sediments (Drake & Cacchione, 1986). In Norton Sound, water content is maintained by frequent wave stirring and bioturbation. The presence of surficial cohesive sediments on modern shelves appears to result from subaerial exposure, which reduces the water content of sediments (Amos & Mosher,

1985). Our field studies of the Mesaverde Formation indicate no evidence for subaerial exposure. We, therefore, have modelled sediment transport assuming cohesionless sediments.

Extreme storm events produce intense wave-dominated oscillatory flow at the sediment surface. One of the few studies of sediment motion as a result of intense oscillatory flow is that of Ahilan & Sleath (1987). This study showed that the bedload formulation of Sleath (1978) was in closest agreement to experimental sediment transport data compared with other formulations for bedload transport. Ahilan & Sleath (1987) calculated bed roughness according to the method of Grant & Madsen (1982), in which bed roughness is created by the combination of a sediment transport component and a form drag component. For large bedforms, such as ripples or hummocks, the form drag component to roughness dominates. It is thus critical to account for form drag in any model of storm stratification.

The program flow of the algorithm to compute bed thickness based on bedload armouring is shown in Fig. AI.1. Many of the steps shown are standard computations of sediment transport theory. The interested reader is referred to the work of van Rijn (1984a,b) and Madsen & Grant (1976). Several steps, though, bear further explanation as they are introduced for the first time in this paper (Fig. AI.1).

## Step 1

Sediment settling velocity and the Shield's parameter are from van Rijn (1984b) and van Rijn (1984a) respectively. Both quantities are functions of fluid density and viscosity, which depend on sediment concentration. Solution for concentration effects on these parameters must be done iteratively since sediment concentration is unknown until settling velocity and the Shield's parameter are specified, and the settling velocity and Shield's parameter, in turn, depend on sediment concentration. This iteration is shown in Fig. AI.1 as a large loop repeating until the computed bed thickness converges. The effect of sediment concentration on fluid viscosity is taken from van Rijn (1984b). The effect of sediment concentration on fluid density is computed assuming a sediment specific gravity of 2.65.

## Step 2

*Rough turbulent friction factor*

The wave-current friction factor, $f_{cw}$, for rough turbulent conditions, is given by Madsen & Grant (1976) as a function of six parameters as follows:

$$f_{cw} = f(A_{bm}, u_{bm}, U_c, k_b, h, \nu) \quad (AI.1)$$

where $A_{bm}$ is the maximum water particle excursion amplitude associated with wave motion, $u_{bm}$ is the maximum magnitude of the near-bed horizontal velocity, $U_c$ is the magnitude of the steady current, $k_b$ is the bottom roughness, $h$ is the water depth and $\nu$ is the kinematic viscosity.

The form drag component of bed roughness, $k_b$ is, in general, a function of bedform height, $\Delta$, and bedform wavelength $\lambda$ (Yalin, 1972). Various functional forms have been suggested for this relationship including those of Grant & Madsen (1982) given by:

$$k_b = \frac{28\Delta^2}{\lambda} \quad (AI.2)$$

and van Rijn (1984b) given by:

$$k_b = 1.1\Delta[1 - \exp(-25(\Delta/\lambda))] + 4D_{90} \quad (AI.3)$$

Equations (AI.2) and (AI.3) predict magnitudes of the roughness, $k_b$, which can be an order of magnitude different. This discrepancy may be explained by the argument of Grant & Madsen (1982) that many forms of boundary roughness occur depending on the scale of the roughness elements and the nature of eddies formed in the fluid. Equation (AI.2) is calibrated using flume observations of short-period oscillatory flow made by Carstens *et al.* (1969). Equation (AI.3) is calibrated by observations of unidirectional flow. We have chosen to use equation (AI.3) as a better approximation to roughness conditions that occur during very long period (>15 s) oscillatory flows characteristic of extreme storm events (return period > 1000 years). Our model, therefore, should not predict accurately the sediment resuspension occurring in smaller storms (return period ≪ 1000 years). Further theoretical and laboratory work is necessary to establish the functional behaviour of form drag under a variety of oscillatory flow conditions.

Recent flume observations made by Southard *et al.* (1990) indicate that bedform parameters $\Delta$

and $\lambda$ are linked strongly to wave orbital diameter under conditions that create hummocky cross-stratification. Forty-nine observations of bedform height and spacing were made for various conditions of sediment size, orbital velocity and orbital diameter. Discounting four cases in which two-dimensional ripples showed prominent small-scale superimposed ripples, the observations show that an excellent fit can be made to the data using the linear relationship:

$$\Delta = 0.156\lambda = 0.0977 A_{bm} \quad \text{(AI.4)}$$

The relationship given in equation (AI.4) should hold for storm-generated stratification when bedforms are present. We have chosen to use equation (AI.4) as a general relationship between bed roughness and maximum wave orbital diameter by arguing in the following way. It is well known that high flow regime conditions produce plane bed conditions in which equation (AI.4) will not be applicable. During the intensification of a storm, however, the flow regime must pass through a stage in which equation (AI.4) *is* applicable. It is at this stage that the maximum bed roughness and the maximum bed-load transport is present. The modelling approach used here attempts to evaluate these maximum conditions that generate the deepest depth in the sediment column to which remobilization occurs.

*Laminar friction factor*

Many authors have considered the effects of large sediment concentrations on damping turbulence (e.g. Smith & McLean, 1977; Sheng & Butler, 1982; Wiberg & Smith, 1983). Damping of turbulence is governed by the Richardson number of the flow, which is given by:

$$Ri = \frac{-g(s-1)\frac{\partial C}{\partial z}}{\left(\frac{\partial u}{\partial z}\right)^2} \quad \text{(AI.5)}$$

where $s$ is the specific gravity of the sediment, $C$ is sediment volume concentration, $u$ is fluid velocity and $z$ is depth in the water column from the sea-bed. At a critical value for the Richardson number, turbulence is damped, and the flow becomes laminar. Though estimates of the critical Richardson number differ (Sheng & Butler, 1982), the Richardson number computed from concentration and velocity gradients in the vicinity of the bed for storm events resuspending over 10 cm of sediment will generally exceed these values.

For laminar flow, sediment diffusivity is governed by the kinematic viscosity of the fluid, $\nu$. The wave-current friction factor, $f_{cw}$, is then given by:

$$f_{cw} = \sqrt{\frac{4\nu}{u_b A_b}} \quad \text{(AI.6)}$$

where $u_b$ is the magnitude of the near-bed horizontal velocity and $A_b$ is the excursion amplitude of water particles relative to the bed. The near-bed velocity will change throughout a wave cycle and, in general, is a vectorial sum of wave and current components.

*Maximum bed-shear approximation*

The rough turbulent friction factor from equation (AI.1) using expression (AI.4) is, for extreme storm events, two to three orders of magnitude larger than the laminar friction factor in equation (AI.6). This extreme difference in friction depending on flow regime can lead to a highly time-dependent process feedback which evolves over many wave cycles as follows: (1) bedforms are generated under rough turbulent conditions; (2) peak bottom shear stresses are generated and the sediment column is remobilized to its deepest depth; (3) continued resuspension damps turbulence sufficiently so that laminar flow is established, thereby dropping bottom shear stresses precipitously; (4) owing to low bottom shear stress, sediment falls out of suspension without being resuspended; and (5) when sediment concentrations are low enough rough turbulent conditions are re-established. A true time-dependent model of this process feedback is beyond the scope of this study. We have instead used equation (AI.1) to estimate friction factors under the assumption that *peak* bottom shear stresses can be estimated using this rough turbulent formulation.

**Step 3**

Bed-shear velocity is computed as per Madsen & Grant (1976) using first-order wave theory and the friction factor from equation (AI.1).

**Step 4**

Many authors have made use of the suspension criteria of Bagnold (1966) given by:

$$\frac{U*_{\mathrm{crs}}}{W_s} = 1 \qquad (AI.7)$$

where $U*_{\mathrm{crs}}$ is the critical bed-shear velocity for the initiation of suspension and $W_s$ is the settling velocity. Van Rijn (1984b) has shown that the suspension criteria of Bagnold (1966) must be modified for particles sizes smaller than fine sand. Experimental results indicate that this ratio depends on grain size $D$ through a particle parameter $D*$ given by:

$$D* = D \left[ \frac{(s-1)g}{v^2} \right]^{1/3} \qquad (AI.8)$$

A partition function between bedload and suspended load is therefore of the form:

$$P = f\left[ \frac{U*}{W_s}, D* \right] \qquad (AI.9)$$

Studies of entrainment of suspended load have shown an explosive dependence on the shear stress to settling velocity ratio. Akiyama & Stefan (1985), summarizing seven sets of laboratory and field data, showed that a doubling of this ratio can increase sediment entrainment by four orders of magnitude.

We have used the theoretical work of Murphy (1985) to derive an explicit form of the partition function in equation (AI.9). Murphy (1985) studied the equilibrium between erosion and deposition at the interface between the bedload and suspended load layers. He found that the equilibrium concentration of suspended sediment at the bottom of the suspended load zone, $C_{\mathrm{eq}}$, depends only on the shear stress to settling velocity ratio for sand size particles. A suitable approximation to his theoretical results is given by:

$$C_{\mathrm{eq}} = 10 - \left[ \left( \frac{\log_{10}(U*/W_s) + 2.45}{4.067} \right)^{-2.86} \right] \qquad (AI.10)$$

To extend the range of this function to particles smaller than fine sand, equation (AI.10) is modified to:

$$C_{\mathrm{eq}} = 10 - \left[ \left( \frac{\log_{10}(U*/W_s \cdot f(D*)) + 2.45}{4.067} \right)^{-2.86} \right] \qquad (AI.11)$$

where the function $f(D*)$ is based on van Rijn (1984b) and is given by:

$$f(D*) = \begin{cases} \dfrac{\varsigma w_s}{\sqrt{\theta_{\mathrm{cr}}(s-1)gD}} & \text{for } D* < 1 \\ D*/10 & \text{for } 1 < D* < 10 \\ 1 & \text{for } D* > 10 \end{cases} \qquad (AI.12)$$

where $\varsigma$ is a constant chosen to ensure continuity of the function at $D* = 1$ and $\theta_{\mathrm{cr}}$ is the critical Shields parameter. The function given by equations (AI.11) and (AI.12) is a good fit to the data given in Akiyama & Stefan (1985). The fraction of given grain size in bedload, $P$, is then given by:

$$P(U*, W_s, D*) = \frac{C_{\mathrm{eq}}(1)}{C_{\mathrm{eq}}(1) + C_{\mathrm{eq}}\left(\dfrac{U*}{W_s}, D*\right)} \qquad (AI.13)$$

The bed-shear velocity $U*$ varies throughout the wave cycle (e.g. Madsen & Grant, 1976). The partition factor in equation (AI.13) is, therefore, averaged over 16 parts of a wave cycle.

### Step 5

The total fraction of grains in the bedload, $BL_\%$ is computed by summing the components for each grain size from equation (AI.13) averaged over a wave cycle. For this calculation the sediment grain-size distribution has been split into 17 fractions.

### Step 6

The thickness of the bedload layer, $BL_{\mathrm{thick}}$, is computed using the formulation of Shi et al. (1985) based on the transport formulation of Sleath (1978).

### Step 7

The total integrated suspended load in the water column, $\Sigma C$, in volume per unit area is given by:

$$\Sigma C = \frac{(\Delta \eta - BL_{\mathrm{thick}})}{(1 - \phi)} \qquad (AI.14)$$

where $\phi$ is the sediment porosity. The near-bed concentration $c_0$ has been related empirically to the integrated concentration by Niedoroda et al. (1989) based on the data of Kachel (1980) on the mid- to outer Washington Shelf. The derived expression is given by:

$$\Sigma C = 0.308 \cdot c_0 \cdot \frac{h}{2} \quad (\text{AI.15})$$

where $h$ is the water depth.

The question becomes whether the coefficient 0.308 in equation (AI.15) derived empirically under relatively low concentration conditions also will be appropriate for the extreme high concentration events under consideration in this study. In the high concentration limit the flow becomes stratified throughout the boundary layer between the sea-bed and $h/2$. The momentum diffusivity of such a flow is simply the fluid kinematic viscosity. The horizontal velocity profile is thus a simple parabolic form of a laminar boundary layer (e.g. Kachel, 1980). Assuming that the Richardson number controlling stratification is approximately constant throughout the layer the shape of the concentration profile can be determined from (equation AI.5) as proportional to the velocity gradient squared. After integrating with depth, the derived relation between $\Sigma C$ and $c_0$ for a high concentration is:

$$\Sigma C = 0.250 \cdot c_0, \frac{h}{2} \quad (\text{AI.16})$$

The coefficient 0.250 in equation (AI.16) does not differ significantly from the coefficient in equation (AI.15). Thus we do not introduce a significant error by using equation (AI.16) to compute the near-bed concentration, $c_0$.

## APPENDIX II: THEORY OF STORM BED PRESERVATION

In general, the relationship between initial storm bed thickness and storm return period can be approximated by a power function:

$$\Delta\eta = aT^{1/q} \quad (\text{AII.1})$$

where coefficients $a$ and $q$ are functions of wave climate, water depth and grain size. In the formulation that follows, return period is a dimensionless number representing the ratio of the average time between storms of a given size class and the average time between storms of any size class. Accumulation rate is then given as a net deposition per storm event. The probability distribution $p(\Delta\eta)$ can be derived from equation (AII.1) from the definition of return period (see Gumbel, 1958). The probability $P(\Delta\eta)$, that an event exceeds magnitude $\Delta\eta$ is given by:

$$P(\Delta\eta) = \frac{1}{1+T} = \frac{1}{1+(\Delta\eta/a)^q} \quad (\text{AII.2})$$

The probability function $p(\Delta\eta)$ is then given by:

$$p(\Delta\eta) = \frac{\partial P(\Delta\eta)}{\partial \Delta\eta} = [1+(\Delta\eta/a)^q]^{-2} \frac{q}{a}(\Delta\eta/a)^{q-1} \quad (\text{AII.3})$$

For reasonable limits on sedimentation time, the numerical model of Niedoroda et al. (1989) shows that the probability of preserving events of return period not much larger than $T=1$ is vanishingly small. Equation (AII.3) can be simplified for return periods significantly larger than 1 as:

$$p(\Delta\eta) = \frac{q}{a}(\Delta\eta/a)^{-q-1} \quad T \gg 1 \quad (\text{AII.4})$$

Once a bed of thickness $\Delta\eta$ is deposited, it is buried at a rate $\mathring{a}$. The preservation potential of a bed of return period significantly greater than 1, therefore, is given by the joint probability that any storm event at a time $t$ reworks a depth greater than $\Delta\eta + \mathring{a}t$. We first examine the case for a bed that has a high probability of preservation. In this case the probability of erasure by any succeeding event is much less than 1 and the joint probability for all events is given by the sum of succeeding events. Expressing this summation in the limit as an integral we have for the probability of preservation $pr(\Delta\eta)$:

$$pr(\Delta\eta) = 1 - \int_0^\infty a^q(\Delta\eta + \mathring{a}t)^{-q}\partial t \quad (\text{AII.5})$$

where $t$ is a 'time' variable representing a continuous form of the discrete variable: number of events since deposition. Performing the integration we have:

$$pr(\Delta\eta) = 1 - \frac{a^q}{\mathring{a}(q-1)\Delta\eta^{q-1}} \quad pr(\Delta\eta) \gg 0 \quad (\text{AII.6})$$

For the case where the probability of preservation is small the joint probability is not a simple addition as in equation (AII.6) but is instead given as a function of the error function, erfc (see Cox & Miller, 1965):

$$pr(\Delta\eta) = \text{erfc}\left(\frac{a^q}{\mathring{a}(q-1)\Delta\eta^{q-1}}\right) \quad (\text{AII.7})$$

The expected mean bed thickness, $\Delta\bar{\eta}$, is an average of each thickness class weighted by its expected probability of occurrence and mean preservation. However, when only a small fraction of an original depositional or reworking event is preserved, a distinct bed cannot be measured from core or outcrop data. What generally distinguishes a bed are changes in grain size or sedimentary structure from bottom to top. When a large percentage of the initial thickness of the bed is preserved these changes can be distinguished readily. As the percentage decreases the bed becomes harder to distinguish until at some cut-off percentage $c\%$ ($\Delta\eta$) the bed cannot be distinguished at all. The cut-off percentage $c\%$ ($\Delta\eta$) may, in general, be a function of $\Delta\eta$, and the initial bed grading. In this study we assume that the cut-off percentage is a constant.

The mean observable preserved thickness, $\Delta\bar{\eta}_0$, is an average of each thickness class whose mean preservation is greater than $c\%$, weighted by its expected probability of occurrence and probability of preservation. The minimum observable size class, $\Delta\eta = m$, is determined by the condition:

$$c\% = \frac{\int_0^1 pr(xm)[1 - pr(xm)](1-x)dx}{\int_0^1 pr(xm)[1 - pr(xm)]dx}$$

(AII.8)

Solution of equation (AII.8) for $m$ can be found using *Regula Falsi* iteration. The mean distinct bed thickness $\Delta\bar{\eta}_m$ is now found by integrating over all size classes larger than $m$:

$$\Delta\bar{\eta}_m = \frac{\int_m^\infty \frac{\int_0^1 pr(x\Delta\eta)[1 - pr(x\Delta\eta)](1-x)dx}{\int_0^1 pr(x\Delta\eta)[1 - pr(x\Delta\eta)]dx} p(\Delta\eta)\Delta\eta\partial\Delta\eta}{\int_m^\infty p(\Delta\eta)\partial\Delta\eta}$$

(AII.9)

Equation (AII.9) can be integrated numerically by *Gauss–Legendre* quadrature.

The return time of the minimum observable size class, $T_m$ from equations (AII.3) and (AII.5) is:

$$T_m = (a \cdot m)^{-q} \qquad \text{(AII.10)}$$

The percentage of the sediment column that is made of distinguishable event beds, $eb\%$, is then given by:

$$eb\% = \frac{\Delta\bar{\eta}_m}{T_m \bar{a}} \qquad \text{(AII.11)}$$

# REFERENCES

AKIYAMA, J. & STEFAN, H. (1985) Turbidity current with erosion and deposition. *J. Hydraul. Eng.*, **111**, 1473–1495.

AHILAN, R.V. & SLEATH, J.F.A. (1987) Sediment transport in oscillatory flow over flat beds. *J. Hydraul. Eng.* **113**, 308–321.

ALLEN, J.R.L. (1985) *Principles of Physical Sedimentation*. George Allen & Unwin, London, 272 pp.

AMOS, C.L. & MOSHER, D.C. (1985) Erosion and deposition of fine-grained sediments from the Bay of Fundy. *Sedimentology*, **32**, 815–832.

AUGUSTINE, F.E. & MAXWELL, F.D. (1978) Extreme wave heights in the Gulf of Alaska. *Offshore Technical Conference*, pp. 1551–1557. Soc. Exp. Geop., Richardson, TX.

BAGNOLD, R.A. (1966) An approach to the sediment transport problem from general physics. *U.S. Geol. Surv. Prof. Pap.* **422-1**.

BARRON, E. (1988) The capabilities and limitations of climate models with reference to the sedimentary system. Paper presented at the *Quantitative Dynamic Stratigraphy Workshop*, Lost Valley Ranch, CO, February 1988.

BARRON, E.J., ARTHUR M.A. & KAUFFMAN, E.G. (1985). Cretaceous rhythmic bedding sequences: a plausible link between orbital variations and climate. *Earth Planet. Sci. Lett.* **72**, 327–340.

BIRD, E.C.F. & SCHWARTZ, M.L. (1985) *The World's Coastline*. Van Nostrand Reinhold, New York, 1071 pp.

BRETSHNEIDER, C.L. (1966a) Wave generation by wind, deep and shallow water. In: *Estuary and Coastline Hydrodynamics* (Ed. Ippen. A.T.) pp. 133–196. McGraw Hill, New York.

BRETSHNEIDER, C.L. (1966b) Engineering aspects of hurricane surge. In: *Estuary and Coastline Hydrodynamics* (Ed. Ippen, A.T.) pp. 231–256. McGraw Hill, New York.

GARSTENS, M.R., NEILSON, R.M. & ALTINBILEK, H.D. (1969) *Bed Forms Generated in the Laboratory Under Oscillatory Flow; Analytical and Experimental Study*. Technical Memorandum 28, U.S. Army Corps of Engineering, Coastal Engineering Research Center.

Cox, D.R. & MILLER, H.D. (1965) *The Theory of Stochastic Processes*. Wiley, New York, 398 pp.

CROSS, T.A., & PILGER, R.H. (1978) Tectonic controls of late Cretaceous sedimentation, western interior, USA. *Nature* **274**, 653–657.

CROWLEY, K.D. (1984) Filtering of depositional events and the completeness of the sedimentary record. *J. Sediment. Petrol.* **54**, 127–136.

DACEY, M.F. (1979) Models of bed formation. *Math. Geol.* **11**, 655–668.

DEAN, R.G. & EAGLESON, P.S. (1966) Finite amplitude waves. In: *Estuary and Coastline Hydrodynamics* (Ed. Ippen, A.T.) pp. 93–132. McGraw Hill, New York.

DOTT, R. (1988) An episodic view of shallow marine clastic sedimentation. In: *Tide-influenced Sedimentary Environments and Facies* (Eds de Boer, P.L., van Gelder, A. & Nio, S.D.) pp. 3–12. Reidel, Dordrecht.

DRAKE, D.E. & CACCHIONE, D.A. (1986) Field observations of bed shear stress and sediment resuspension on continental shelves, Alaska and California. *Continental Shelf Res.* **6**, 415-429.

FISHER, W.L. & MCGOWEN, J. (1976) Depositional systems in the Wilcox Group of Texas and their relationship to the occurrence of oil and gas. *Trans. Gulf Coast Assoc. Geol. Soc.* **17**, 105-125.

FROST, B.A. (1966) The relation between Beaufort Force, wind speed and wave height. *Sci. Pap. Meteorol. Office*, 25, 22 pp.

GALT, J.A. (1980) A finite element selection procedure for the interpolation of current data in complex regions. *J. Phys. Ocean.* **10**, 1984-1997.

GAYNOR, G.C. & SWIFT, D.J.P. (1988) Shannon sandstone depositional model: sand ridge formation on the Campanian western interior shelf. *J. Sediment. Petrol.* **58**, 868-880.

GILL, J.R. & COBBAN, W.A. (1973) Stratigraphy and geological history of the Montana Group and equivalent rocks, Montana, Wyoming and North and South Dakota. *U.S. Geol. Surv. Prof. Pap.* **776**, 37 pp.

GRANT, W.D. & MADSEN, O.S. (1982) Moveable bed roughness in unsteady oscillatory flow. *J. Geophys. Res.* **87**, 469-481.

GUMBEL, E.J. (1958) *Statistics of Extremes.* Columbia University Press, New York, 375 pp.

HAN, G.C., HANSEN, D.V. & GALT, J.A. (1980) Steady state diagnostic model of New York Bight. *J. Phys. Ocean.* **10**, 1998-2020.

HOLLENSHEAD, C.T. & PRITCHARD, R.L. (1961) Geometry of producing Mesaverde Shandstones, San Juan Basin. In: *Geometry of Sandstone Bodies* (Eds Peterson, J.A. & Osmond, J.C.) pp. 98-118. Am. Assoc. Petrol. Geol., Tulsa.

HSU, S.A. (1986) Meteorological effects on the determination of significant wave height under fetch-limited conditions. Research Program Report no. 19, L.S.U. Industrial Associates. 22 pp.

INGRAM, R.L. (1954) Terminology for the thickness of stratification and parting units in sedimentary rocks. *Geol. Soc. Am. Bull.* **65**, 937-938.

KACHEL, N.B. (1980) A time dependent model of sediment transport and strata formation on a continental shelf. PhD thesis, University of Washington, Seattle, WA, 123 pp.

KAMOLA, D.L. (1984) Trace fossils from marginal marine facies of The Spring Canyon member, Blackhawk Formation (Upper Cretaceous), east central Utah. *J. Paleontol.* **58**, 521-549.

KAUFFMAN, E. (1984) Paleobiogeography and evolutionary response dynamics in the Cretaceous western interior seaway of North America. In: *Jurassic-Cretaceous Biochronology and Paleogeography of North America*, (Ed. Westermann, G.E.G.). Geol. Soc. Can. Spec. Pap. 27, 273-306.

KEEFER, W.R. (1972) Frontier, Cody, and Mesaverde Formations in the Wind River and southern Big Horn Basins, Wyoming. *U.S. Geol. Surv. Prof. Pap.* **495-E**, 23 pp.

KENNEDY, J.F. (1983) Reflections on rivers, research, and Rouse. *J. Hydrol. Eng.* **109**, 1254-1271.

KOLMOGOROV, A.N. (1951) Solution of a problem in probability theory connected with the problem of the mechanism of stratification. *Ann. Math. Soc. Trans.* **53**, 171-177.

KOMAR, P. (1976) *Beach Processes and Sedimentation.* Prentice Hall, Englewood Cliffs, NJ, 423 pp.

KOMINZ, M.A. & BOND, G.C. (1986) Geophysical modeling of the thermal history of foreland basins. *Nature* **320**, 2520-2526.

KRAUS, E.B. (1972) *Atmospheric-Ocean Interaction.* Clarendon Press, Oxford, 275 pp.

MADSEN, O.S. & GRANT, W.D. (1976) *Sediment Transport in the Coastal Environment.* MIT Department of Civil Engineering Report 209, Michigan Institute of Technology, 105 pp.

MURPHY, P.J. (1985) Equilibrium boundary condition for suspension *J. Hydrol. Eng.* **111**, 108-115.

MURPHY, P.J. & AGUIRRE, E.J. (1985), Bed load or suspended load. *J. Hydrol. Eng.* **111**, 93-107.

NIEDORODA, A.W., SWIFT, D.J.P. & THORNE, J.A. (1989) Modeling shelf storm beds: controls of bed thickness and bedding sequence. In: *Shelf Sedimentation, Shelf Sequences and Related Hydrocarbon Accumulation.* (Eds Morton R.A. & Nummedal, D.) Gulf Coast Section, Society of Economic Paleontologists and Mineralogists, Seventh Annual Research Conference Proceedings, pp. 15-39.

NISHIWAKI, N. (1979) Simulation of bed-thickness distribution based on waiting time in the Poisson process. In: *Geomathematical and Petrophysical Studies in Sedimentology* (Eds Gill, D. & Merriam, D.F.) pp. 17-31. Pergamon Press, Oxford.

NITTROUER, C.A. & STERNBERG, R.W. (1981) The formation of sedimentary strata in an allochthonous shelf environment: the Washington Continental Shelf. *Mar. Geol.* **42**, 201-232.

PALMER, J.J. & SCOTT, A.J. (1984) Stacked shoreline and shelf sandstone of La Ventana Tongue (Campanian), Northwestern New Mexico. *Bull., Am. Assoc. Petrol. Geol.* **68**, 74-91.

PARRISH, J.T., GAYNOR, G.C. & SWIFT, D.J.P. (1984) Circulation in the Cretaceous western interior seaway of North America, a review. In: *Mesozoic of Middle North America* (Ed. Scott, D.F.). Mem. Can Soc. Petrol. Geol. 9, 221-231.

QUAYLE, R.G. & FULBRIGHT, D.C. (1975) Extreme wind and wave return periods for the U.S. coast. *Mariners Weath. Log* **190**, 67-70.

RYER, T.A. (1977) The pattern of Cretaceous shallow marine sedimentation, Coalville and Rockport areas, Utah. *Geol. Soc. Am. Bull.* **88**, 177-188.

RYER, T.A. (1983) Transgressive-regressive cycles and the occurrence of coal in some Upper Cretaceous strata of Utah. *Geology* **11**, 201-210.

SAHL, L.E., MERRILL, W.J., MCGRAIL, D.W. & WEBB, J.A. (1987) Transport of mud on continental shelves: evidence from the Texas Shelf. *Mar. Geol.* **76**, 33-43.

SCHWARZACHER, W. (1975) *Sedimentation Models in Quantitative Stratigraphy*, Vol. 19, *Developments in Sedimentology.* Elsevier, New York, 382 pp.

SCRUTON, P.C. (1956) Oceanography of Mississippi Delta sedimentary environments. *Bull., Am. Assoc. Petrol. Geol.* **40**, 2864-2952.

SEILACHER, A. (1982) General remarks about event deposits. In: *Cyclic and Event Stratification* (Eds Einsele, G. & Seilacher, A.) pp. 161-174. Springer Verlag, New York.

SHENG, Y.P. & BUTLER, H.L. (1982) Modeling coastal

currents and sediment transport. *Coastal Engineering Conference, 1982*, pp. 1127–1148.

SHI, N.C., LARSEN, L.H. & DOWNING, J.P. (1985) Predicting suspended sediment concentration on continental shelves. *Mar. Geol.* **62**, 255–275.

SILVERIA, W.A. & BRILLINGER, D.R. (1978) On maximum wave heights of severe seas. *Offshore Tech. Conf.*, pp. 1563–1567. Soc. Expl. Geop., Richardson, TX.

SLEATH, J.F.A. (1978) Measurements of bedload in oscillatory flow. *Proc. Am. Soc. Civ. Eng. J. Waterway Port Coastal Ocean Div.* **104**, 291–307.

SMITH, J.D. (1977) Modelling of sediment transport on continental shelves. In: *The Sea* (Eds Goldberg, E.D., McCave, I.N., O'Brien, J.J. & Steele, J.H.) Vol. 6, pp. 539–577. Wiley, New York.

SMITH, J.D. & MCLEAN, S.R. (1977) Boundary layer adjustments to bottom topography and suspended sediment. In: *Bottom Turbulence* (Ed. Nihoul, J.C.J.) pp. 1213–1250. Elsevier, Amsterdam.

SORBY, H.C. (1853) On the oscillation of currents drifting the sandstone beds of the south-east of Northumberland, and on their general direction in the coal field in the neighbourhood of Edinburgh. *Geol. Polytechnic Soc. West Riding Yorkshire, Rep. Proc.* **3** (1852), 232–240.

SORBY, H.C. (1855) On the motion of waves, as illustrating the structure and formation of stratified rocks. *Proc., Yorkshire Geol. Soc.* **3**, 372–378.

SOUTHARD, J.B., LAMBIE, J.M., FEDERICO, D.C., PILE, H.T. & WEIDMAN, C.R. (1990) Experiments on bed configurations in fine sands under bidirectional purely oscillatory flow, and the origin of hummocky cross-stratification. *J. Sediment. Petrol.* **60**, 1–17.

SUNAMURA, T. & KRAUS, V.C. (1985) Prediction of average mixing depth of sediment in the surf zone. *Mar. Geol.* **62**, 1–12.

SWIFT, D.J.P., CLARKE, T., NEIDORODA, A.W., YOUNG, R.A. & VINCENT, C.E. (1981) Sediment transport in the Middle Atlantic Bight of North America, synopsis of recent observations. In: *Holocene Marine Sedimenttation in the North Sea Basin*. (Eds Nio, S.D., Schutenholm, R.T.E. & Van Weering, T.C.E). Spec. Publ. Int. Assoc. Sediment. **5**, 361–383.

SWIFT, D.J.P., HUDLESON, P.M., BRENNER, R.L. & THOMPSON, P. (1987) Shelf construction in a foreland basin: storm beds, shelf sand bodies, and shelf-slope depositional sequences in the Upper Cretaceous Mesaverde Group, Book Cliffs, Utah. *Sedimentology*, **34**, 423–457.

THOMPSON, R. (1984) A stochastic model of sedimentation. *Math. Geol.* **16**, 753–778.

THORNE, J.A. & WATTS, A.B. (1989) Quantitative analysis of North Sea subsidence. *Bull., Am, Assoc. Petrol. Geol.* **73**, 88–116.

VAIL, P.R., MITCHUM, R.M. JR., THOMPSON III, S., SANGREE, J.R., DUBB, J.N. & HASLID, W.G. (1977) Sequence stratigraphy and global changes of sea level. In: *Seismic Stratigraphy–Applications to Hydrocarbon Exploration* (Ed. Payton, C.E.). Mem., Am. Assoc. Petrol. Geol. **26**, 49–205.

VAN RIJN, L.C. (1984a) Sediment transport, part 1: bed load transport. *J. Hydrol. Eng.* **110**, 1431–1456.

VAN RIJN, L.C. (1984b) Sediment transport, part 2: suspended load transport. *J. Hydrol. Eng.* **110**, 1613–1640.

VAN WAGONER, J., MITCHUM, R., JR., POSAMENTIER, H. & VAIL, P. (1987) Seismic stratigraphy interpretation using sequence stratigraphy. Part 2: key definitions of sequence stratigraphy. In: *Atlas of Seismic Stratigraphy*, (Ed. Bally, A.W.) Vol. 1. Am. Assoc. Petrol. Geol. Stud. Geol. **27**, 11–14.

VAN WAGONER, J.C., MITCHUM, R. JR., POSAMENTIER, H.W., VAIL, P.R. SARG, J.F., LOUTIT, T.S. & HARDENBOL, J. (1988) An overview of the fundamentals of sequence stratigraphy and key definitions of sequence stratigraphy. In: *Sea-level Changes: an Integrated Approach* (Eds Wilgus, C.K., Hastings, B.S., Posamentier, H., van Wagoner, J., Ross, C.A. & Kendall, C.G.St.C.). Soc. Econ. Paleontol. Mineral. Spec. Publ. **42**, 39–47.

VANONI, V.A. (1984) Fifty years of sedimentation. *J. Hydrol. Eng.* **110**, 1022–1057.

VISSER, M.J. (1980) Neap-spring cycles reflected in Holocene subtidal large-scale bedform deposits: a preliminary note. *Geology* 8, 5453–5456.

WARD, E.G., EVANS, D.J. & POMPA, J.A. (1978) Extreme wave heights along the Atlantic coast of the United States. *Offshore Tech. Conf.* **2846**, 315–324.

WHEELER, H.E. & MALLORY, S.V. (1956) Factors in lithostratigraphy. *Bull., Am. Assoc. Petrol. Geol.* **40**, 2711–2723.

WIBERG, P. & SMITH, J.D. (1983) A comparison of field data and theoretical models for wave–current interactions at the bed on the continental shelf. *Continental Shelf Res.* **2**, 147–162.

WRIGHT, R. (1986) Cycle stratigraphy as a paleogeographic tool: Point Lookout Sandstone, southeastern San Juan Basin, New Mexico. *Geol. Soc. Am. Bull.* **96**, 661–673.

YALIN, S.J., (1963) An expression for bed-load transportation. *Proc. Am. Soc. Civil Eng. J. Hydraulics Div.*, **89** (HY3), 221–250.

YALIN, M.S. (1972) *Mechanics of Sediment Transport*. Pergamon Press, Oxford, 332 pp.

ZYDOROWICZ, T. & MATYSIAK, S.J. (1984) Stochastic analysis of lithologic sequences: bed thickness distribution. *Math. Geol.* **16**, 315–326.

# Sedimentation on continental margins, IV: lithofacies and depositional systems

D.J.P. SWIFT*, S. PHILLIPS[†] and J.A. THORNE[‡]

*Department of Oceanography, Old Dominion University, Norfolk, VA 23529, USA; [†] ARCO Alaska Inc., 700 G St., Anchorage, AK 99501, USA; and [‡] Research and Technical Services ARCO Oil and Gas Co., 2300 W. Plano Parkway, Plano, TX 23575, USA

## ABSTRACT

Lithologic variation in shallow-marine sediments is best understood in terms of *depositional systems* (assemblages of process-related lithofacies). Each depositional system is the product of a *dispersal system*, or assemblage of flow-linked depositional environments. Two useful criteria for classifying the small-scale depositional systems are: (1) the grain size bias of the regional dispersal system; and (2) the type of local dispersal system that created the deposit.

Dispersal systems on tectonically youthful continental margins (rift margins) tend to be regionally biased towards coarse-grained sediment. In most tectonic settings, however, the primary regional controls of grain size are the base level shifts and fluctuations in sediment input that result in transgressive or regressive situations. Transgressive settings have higher bypassing and reworking ratios, and accumulated coarser lithofacies than do regressive settings. Therefore, a primary criterion for classifying depositional systems is whether they occur in transgressive or regressive settings.

Dispersal systems have a common pattern, which consists of an eroding *source environment* attached to a series of *depositional environments* whose linking current regime is characterized by a downstream decrease in competence. The most important fluid dynamical parameter in such systems is the *reworking ratio*, $r = a'/å$, where $a'$ is the minimum depth of erosion, and $å$ is the accumulation per event. Both the fluid power and the accumulation rate decrease downstream, but the fluid power gradient is steeper, so that the reworking ratio also decreases. These gradients may reverse in the distal environment, so that the reworking ratio increases once more.

The resulting depositional systems likewise have a common pattern, which can be deduced from the strata-forming principles outlined by Thorne *et al.* (this volume, pp. 59–87). They consist of a *source diastem* that overlies, or is overlain by, a lithofacies succession that becomes increasingly finer grained with distance above or below the diastem. The proximal (near-source) portion of the depositional system is characterized by a condensed section as a consequence of the high reworking ratios that prevailed in the depositional environment (*amalgamated sand lithofacies*), while the distal environment experiences lower reworking ratios and is characterized by an expanded section (*Interbedded sand and mud lithofacies; laminated or bioturbated mud lithofacies*). However, in some settings, high reworking ratios also may occur in the the far distal environment, when the decrease in fluid power is more than compensated for by the decrease in sediment input. These distal condensed sections are the classical condensed section of stratigraphers.

Depositional systems are characterized by a variety of spatial arrangements. *Conjugate systems* are stacked together and are separated by a shared source diastem. *Serial systems* also share a source diastem and are arranged in tandem, with both systems above or below the source diastem. *Interfingering systems* share a common distal facies. *Superimposed* systems have been overprinted by several depositional environments. If migration of the source environment has destroyed one of the depositional systems originally in contact with it, the resulting contact is an *occluded contact*.

Regressive depositional systems occur as *delta mouth bars*, as *prodelta plumes* in response to detachment of the coastal boundary flow, or as *regressive shoreface-shelf systems*. The distal portions of shoreface-shelf systems may develop seaward facing, constructional, *terrace-scarp complexes*, with clinoform internal structure. Where the shelf widens or deepens so that the along-shelf flow must decelerate, an along-shelf fining *deceleration sheet* may result. Relative proportions of lithofacies vary from subsystem to subsystem on regressive shelves, but always with the same spatial relationships (i.e. undergo allometric variation).

Transgressive depositional systems consist of the same lithofacies as regressive depositional systems, but the proportions are often very different. Transgressive sands occur as erosion-truncated, discontinuous, *barrier and inlet sand systems* below a ravinement surface, cut by erosional shoreface retreat. Several back-barrier subsystems may be present, organized around barrier washover, inlet and tidal channel sources. Transgressive back-barrier systems prograde landward, and are capped by their source diastems. Like regressive systems, they consist of both condensed and expanded sections, but their stratification parameters decrease landward, from the barrier to the lagoonal shoreline, not seaward. The shoreface itself is erosional.

Further seaward on the shelf, transgressive facies occur as thin, discontinuous, *coarse-grained sand sheets*. These sand sheets overlie the source diastem, which in this case is the ravinement surface, or the associated flooding surface that develops seaward of the lowstand shoreline as the transgression proceeds. The proximal condensed lithofacies (amalgamated sand facies) of coarse-grained sheet sands may extend for hundreds of kilometres across and along shelf; expanded lithofacies are displaced towards or over the shelf edge, or along shelf, towards zones of shelf widening and flow deceleration. Coarse-grained sand sheets may be localized as cape-associated and estuarine *shoal retreat massifs*, which may be dissected by cross-massif tidal or storm flows into comb-like *sand ridge arrays*. *Sand ridge fields* also may occur on the submarine shelf plains between massifs, if the sand supply is adequate.

## INTRODUCTION

Swift & Thorne (this volume, pp. 3–31) synthesized recent studies of shelf sedimentation into a general descriptive model. The model is built around the central concept of a *depositional regime*. At small time and space-scales, the scale of *depositional events*, depositional variables fluctuate markedly. However, at large time and space-scales, depositional regimes are considered to exist in a state of equilibrium, governed by mutual adjustments of regime variables (the rate and character of sediment input, $Q$, the rate of relative sea-level fluctuation, $R$, and the rate of sediment transport, $D$). These variables are related by the regime ratio, $R \cdot D/Q \cdot M$ (see Swift & Thorne, this volume, p. 13). Event-scale processes are related to regime-scale processes, in that repeated depositional events lead to the progressive sorting process that drives lithofacies differentiation.

Thorne & Swift (this volume, pp. 33–58) and Thorne *et al.* (this volume, pp. 59–87) explore the implications of the general model: Thorne & Swift approach the regime concept from an analytical point of view and present numerical simulations of shelf regimes; Thorne *et al.* analyse subequilibrium strata-forming processes (event-scale processes) in terms of the parameters of strata formation, and examine the manner in which strata-forming processes affect regime variables as the deposits accumulate.

In this section we continue to examine the two basic themes of event and regime sedimentation. Here, we are concerned with the effect of event and regime processes at intermediate spatial scales; the scales of dispersal systems and their corresponding deposits (depositional systems). Concepts of regime sedimentation deduced by Thorne & Swift (this volume, pp. 33–58) are used to analyse the behaviour of dispersal systems. Concepts of strata formation presented by Thorne *et al.* (this volume, pp. 59–87) are applied to the problem of lithofacies differentiation. These approaches are used to develop a systematic description of shallow-marine depositional systems.

### Dispersal systems and depositional systems: statement of the problem

Swift & Thorne (this volume, p. 23) have argued that *depositional systems* (process-related facies assemblages; Fisher & McGowen, 1967) are the 'fossilized hard parts' of *dispersal systems* (flow-linked assemblages of depositional environments; Brown & Fisher, 1977). In these 'fossil dispersal systems', strata are the record of successive sorting events that cause facies differentiation.

Dispersal systems have been described by Swift & Thorne (this volume, pp. 23–25). They consist of several characteristic elements; an (erosional) source environment, and a series of down-current depositional environments that serve as successive partial sinks (Fig. 1A). These elements have been modelled as a Markov process in Fig. 11B, Swift & Thorne, this volume, p. 18.

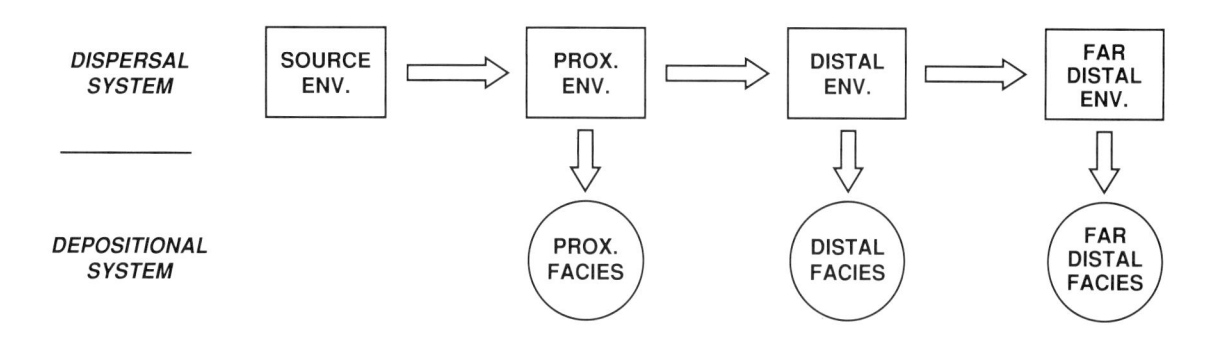

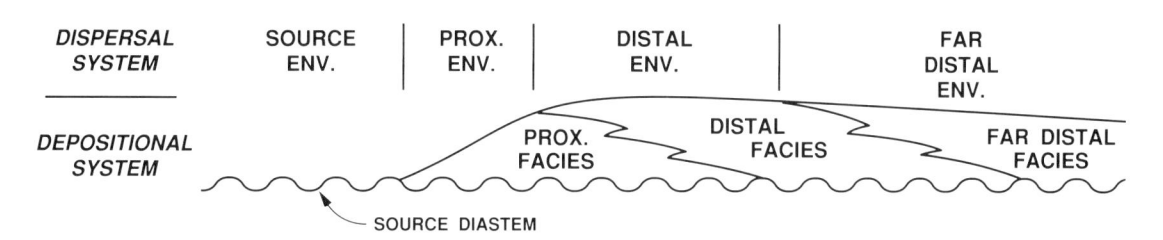

Fig. 1. Dispersal systems and depositional systems. (A) Schematic diagram illustrating a dispersal system and its corresponding depositional system in terms of the sediment budget. (B) Geometric relationships between a dispersal system and its corresponding depositional system.

In any continental margin section, there is usually only one such system (for instance, the eroding Canadian Shield, and the sediment accumulating environments of the Mississippi river, delta and deep-sea fan. The most distal environment in this system is the Gulf of Mexico abyssal plain, where muds from the Mississippi dispersal system interfinger with muds from Cuba and the Yucatan platform. It is convenient, however, to view continental margin dispersal systems as consisting of a hierarchy of subsystems. The Mississippi River is such a subsystem; the cut-bank side of the river channel is its eroding source environment, while the point bar is its proximal depositional environment and the floodplain its distal depositional environment. Similarly, the west Louisiana shelf is another subsystem; the surf zone is its eroding source environment, the shoreface is its proximal depositional environment and the outer shelf its distal depositional environment.

In order to make sense of shallow-marine depositional systems (subsystems in the context of the preceeding discussion), it is necessary to consider two aspects. On one hand, the large scale, continental margin systems of which they are a part vary widely. We know this best from experience of the rock record. The coarse grained, shale-free fill of the Jurassic East Greenland basin (Surlyk & Noe-Nygard, 1991) is strikingly dissimilar from, say, the Cenozoic section of the Atlantic continental margin section (Poag, 1985). Can such interregional variation be rationalized? On the other hand, there is the problem of lithofacies differentiation. If depositional systems are lithofacies assemblages, then do their component lithofacies follow any kind of basic plan? As rock volumes do they have characteristic shapes or characteristic types of contact?

These issues are considered in this paper. Since lithofacies differentiation is largely a matter of the segregation of sediment into bodies of differing grain size, it is necessary to explore the relative roles of tectonic evolution of the continental margin, versus changes in base level owing to relative sea-level movements, as regional controls of grain size in continental margin dispersal systems.

We also examine the anatomy of the resulting depositional systems. The principles of strata formation outlined by Thorne *et al.* (this volume, pp. 59–87) allow us to deduce lithofacies patterns in depositional systems as responses to progressive sorting, and to define them in terms of stratification parameters. We conclude that depositional systems are conveniently divided into transgressive and regressive families, and that within each family there are classes deposited in intra-coastal, inner shelf, and outer shelf systems, respectively. These classes are systematically described.

## ARCHITECTURE OF DEPOSITIONAL SYSTEMS

### Regional controls of grain size: tectonism versus base level changes

*Textural versus compositional maturity*

The petrographic attributes of shallow-marine sandstones have long been known to be sensitive indicators of the tectonic history of the source terrain, as well as keys to the depositional regime (Krynine, 1942). The compositional maturity of marine sandstones is a measure of the extent to which labile minerals (primarily ferromagnesian and alkalic aluminosilicates) and labile rock fragments have been removed through chemical weathering or abrasion, and is a function of climate, and the relief and composition of the source terrain (Klein, 1982). The presence of carbonate sediment also carries information concerning sediment supply and climate. The stage of evolution of the basin is very important in that in mature basins, clastic sediments tend to be polycyclic, and to have had most of their labile constituents removed during repeated transgressive–regressive episodes (Folk, 1960)

*Textural maturity and the dispersal system*

The textural maturity of marine sandstones, however, provides a record of a very different kind. Textural maturity is a response to the pattern of fluid power expenditure during transportation and deposition, rather than the character of the source terrain. In the case of grain size, the prime textural attribute, maturity occurs over periods that are shorter by several orders of magnitude than the time required for compositional maturity. Sediment eroded from the terminal moraine of western Long Island, for instance, is deposited as sediment with high sorting values within hundreds of metres of the beach, and in some cases within days of erosion (Liu & Zarillo, 1990)

Textural maturation in the marine environment has been described by Swift & Thorne (this volume, p. 19) as a progressive sorting process, in which the sorting system consists of a source (e.g. the beach), and a series of partial sinks (e.g. successive stations on the shoreface and shelf). As sediment moves intermittently, seaward, the probability of re-entrainment during each successive transport event is greatest for the finest grains. The coarsest particles are thus deposited mainly in the proximal stations, and the finest in the distal stations, so that the resulting deposits become finer down the dispersal pathway.

In this kind of system, grain size at a station is a function of: (1) the initial input distribution; (2) the sorting history, reflecting the mean power gradient of the dispersal system; and (3) the mean fluid power of the depositional site (Swift & Ludwick, 1976). Because factors (1) and (2) are present as well as (3), a depositional environment cannot be identified by its absolute grain size. Instead, its grain size must be normalized relative to that of the rest of the dispersal system. Figure 2A presents a progressive sorting curve, in which phi mean diameter has been plotted against travel distance. The initial grain size is indicated by $\phi_0$, while sorting efficiency is described as the down-path change in grain size, $\Delta\phi$, divided by the path length, $x$. Exponential sorting curves, such as the one shown in Fig. 2A, can be described by a variant of Sternberg's law:

$$\phi_x = \phi_0 \exp(-\alpha x)$$

Where $\alpha$ is a constant determining the shape of the exponential curve, and therefore a measure of sorting efficiency (Sternberg, 1875, in Blatt *et al.*, 1980). Sorting efficiency is a response to the intensity and time pattern of energy expenditure (hydraulic climate), and its relationship to the depositional rate. The hydraulic climate can be parameterized as the Gumbel distribution, and its relationship to the depositional rate can be parameterized as the reworking ratio. Gumbel distributions and reworking ratios are described by Thorne *et al.* (this volume, pp. 59–87).

*Textural maturity and basin evolution*

The characteristics of the dispersal system change rapidly during the earliest stages of basin evolution, and become relatively stable in later stages. The elevated source terrains of rift basins provide an abundant supply of coarse sediment. Sedimentation rates are high, and the stream and shelf profiles are so short and steep that gravel and sand fractions are bypassed in significant quantities throughout most or all of the dispersal system (Veevers, 1981). As the rift floods, alluvial fans become fan-deltas, but continue to bypass coarse sediment (Holmes, 1965; Fraser & Suttner, 1986). During the ensuing transgression, coarse sediment is reworked and redistributed (Manspieser, 1985).

However, during successive highstands, the continental margin progrades. As its slopes are reduced, the sorting efficiency of the subaerial portion of the dispersal system rises rapidly. Fan-deltas become deltas. In this transition, they lose the ability to deliver gravel to the marine environment, and produce coarse-truncated distributions of fine and very fine sand, as well as a voluminous wash load of mud (Holmes, 1965; Fraser & Suttner, 1986). The fine and very fine sand classes travel in short-term suspension, and like the wash load, are largely bypassed by the shoreline, no matter what the characteristics of the dispersal system. As the basin continues to evolve, marginal sediment prisms thicken and broaden through subsequent sea-level cycles, but for long periods there is relatively little change in the character of sediment delivery. For example, in the case of the Atlantic continental margin, anomalously coarse marine deposits persist into the Jurassic (e.g. Surlyk & Noe-Nygard, 1991), but dispersal systems were stabilized by Cretaceous time (Swift & Heron, 1969). For stabilized systems, beach grain sizes are rarely finer than 2.0–2.5 phi (fine sand), so that sediment bypassed to the shelf is generally coarse-truncated at this value. Finer beach deposits occur mainly in the humid tropics, where deep lateritic weathering greatly enhances the ratio of mud to sand (Rhodes, 1982).

The clock of this evolutionary sequence for dispersal systems can be partially and briefly reset by eustatic sea-level changes. Type 1 sea-level falls (over the shelf edge; Posamentier *et al.*, 1988) can rejuvenate the continental margin to the point when fan-deltas once more leak gravels on to the shelf (see discussion by Thorne & Swift, pp. 229–230), although the sorting efficiency of the dispersal system is restored at highstand.

Likewise, if a divergent margin should become convergent, episodic rejuvenation will occur in response to changes in relative sea-level and sediment input induced by tectonism (Miall, 1984). However, such rejuvenations of the sediment dispersal system are usually brief relative to those of the rifting state, and their deposits relatively minor. Most mature convergent margins, and also the fore-arc and back-arc basins of divergent margins, deposit compositionally immature rocks (wackes and arkoses), however, they maintain mature dispersal systems and deposit texturally mature sediments most of the time (e.g. Chan & Dott, 1983). Several papers in this volume (Keith, pp. 457–488; Krause & Nelson, pp. 427–456; Riemersma & Chan, pp. 489–510) present an example of such behaviour in the Cretaceous western interior basin, a back-arc basin of the convergent western margin of the USA.

In the Cretaceous western interior basin, complex interactions of sea-level change and thrust plate loading repeatedly caused flooding of the basin, right up to the tectonic front along its western margin (Lawton, 1982). At maximum flooding, fan-deltas interfingered with marine shales. Then as the rate of subsidence slowed, braid plain, meander plain, strandplain and highstand shelf deposits prograded seaward, underwent transgression, and prograded again (Fouch *et al.*, 1983). Gravels were bypassed seaward in quantity several times, notably in the Turonian and Coniacian Frontier and Cardium Formations (Merewether, 1980; Keith, this volume; Krause & Nelson, this volume), but the great bulk of these formations consists of relatively fine-grained, highly differentiated sands and shales deposited by mature dispersal systems. In the marine lithofacies of these deposits, the major textural distinctions reflect not the degree of tectonic stability, but instead the difference between transgressive and regressive regimes.

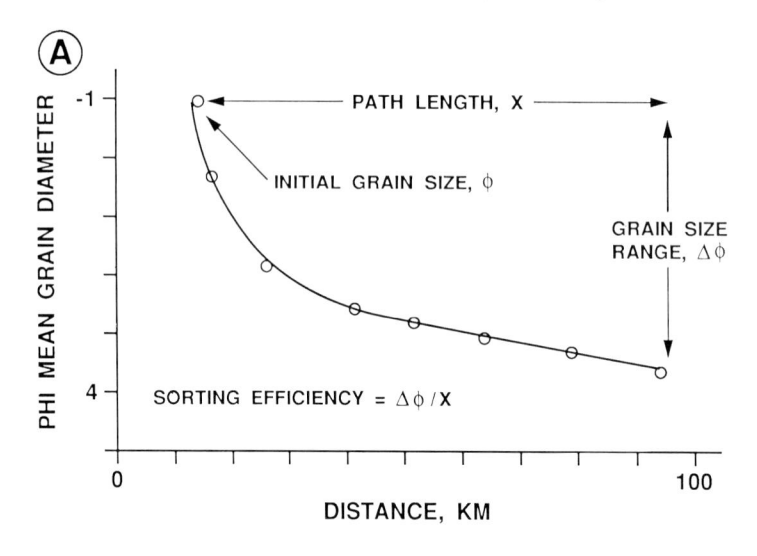

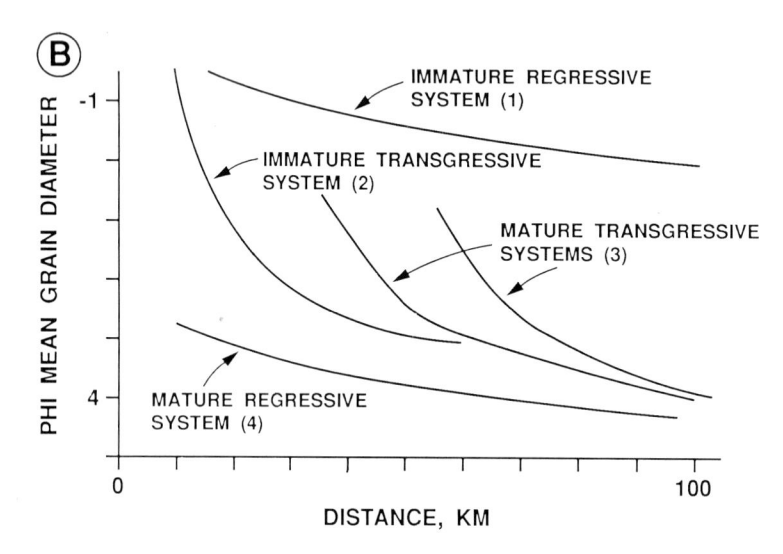

**Fig. 2.** Grain-size profiles of marine dispersal systems. (A) Grain size parameters of a dispersal system. (B) Comparison of grain-size characteristics of an immature dispersal system with a mature transgressive system and a mature regressive system.

*Textural maturity and grain-size profiles*

The sorting profiles of marine dispersal systems in evolving basins tend to follow three basic patterns (Fig. 2B). Prograding (regressive) immature margins are nourished by alluvial fans or fan-deltas (Tag *et al.*, 1991). The initial sediment input is coarse. Because of the relief, initial grain size is large, but because of the very high rate of production of sediment the accumulation rate is high, and the reworking ratio is low. The grain-size curve (type 1) in Fig. 2B is a coarse one (is shifted toward the top of the plot) but it is relatively flat (sorting efficiency is low).

In contrast, mature regressive settings contain broad subaerial depositional environments between the fall line and the shoreline. Extensive progressive sorting of the transported load occurs in these environments, and sediment delivered to the shoreline has been coarse-truncated. Allen's (1964) analysis of the Niger Delta and shelf provides a classic case of a mature, regressive deposit, in which successive fluvial and deltaic environments deplete the transport system of coarse components, until only fine, suspendable sediment is delivered to the shoreline. In such systems marine sandstones are commonly fine to very fine grained and very well sorted.

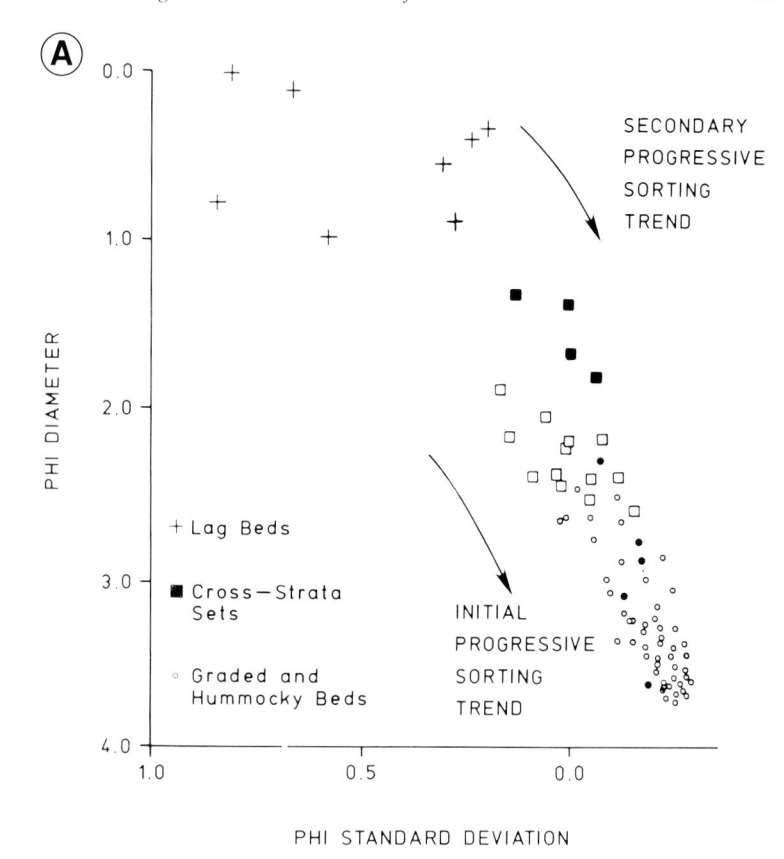

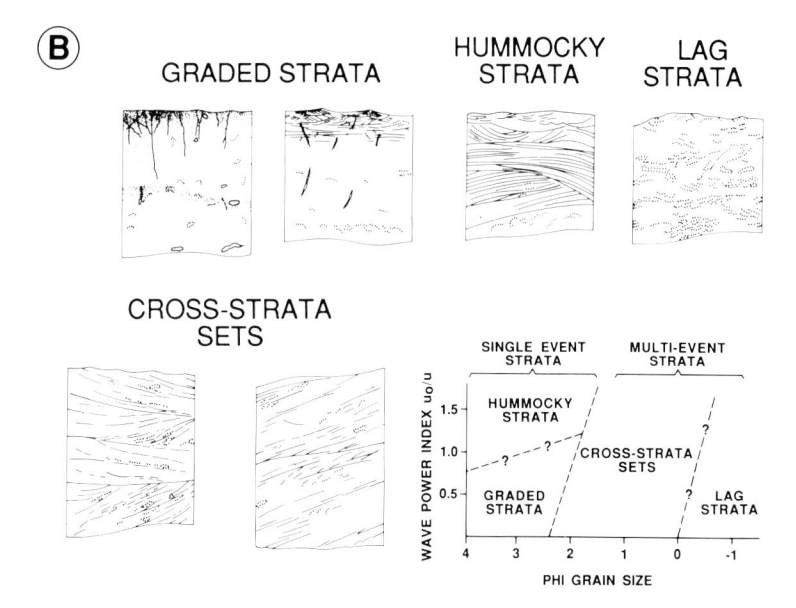

Fig. 3. Textures and structures of depositional systems that have undergone progressive sorting. (A) Phi mean diameter and standard deviation of samples from the Mesaverde Sandstone, Utah (Swift *et al.*, 1987). (B) Drawings of box cores of characteristic stratal types; diagram of relationship of stratal types to velocity structure (ratio of maximum wave orbital velocity component, $u_0$, to the mean velocity component, $\bar{u}$ (Swift *et al.*, 1986b).

Mature regressive systems are nourished by deltas and accumulation rates are high. The resulting grain-size profiles (type 4) are flat but are positioned low in the diagram (Fig. 2B). Sorting efficiency is poor for the fine sediment of the marine portion of the system because the accumulation per event ($\mathring{a}$) is high relative to the resuspension depth ($\mathring{a}$), and the reworking ratio ($r = a'/\mathring{a}$; see Thorne *et al.*, this volume, p. 64) is therefore low.

Transgressive dispersal systems are nourished by erosional shoreface retreat (see discussion by Swift & Thorne, this volume, p. 13). As a consequence, accumulation rates are low. The reworking ratio is high, and sediment is subjected to a very efficient progressive sorting process, in which it is deposited and re-entrained many times. The grain-size profiles of the dispersal system may start low or high in the diagram, but extend over a considerable grain-size range (Fig. 2B). Older sediments are penetrated by the shoreface retreat process, so coarse lags that were sequestered during the previous sedimentary cycle may be returned to the system. Transgressive dispersal systems, unlike regressives ones are not usually well integrated, and several dispersal systems may coexist on a transgressive shelf, centred on localized areas of erosion.

Efficient progressive sorting systems are also efficient concentrators of authigenic constituents, and transgressive, autochthonous sandstones are characterized by high concentrations of skeletal or oolitic carbonate and phosphorite grains, glauconite, and similar minerals (Emery, 1968; Loutit *et al.*, 1988).

*Textural maturity and primary structures*

An example of the grain-size fractionation of sediment into lithofacies in transgressive and regressive depositional systems is presented in Fig. 3A. The data is drawn from sandstones of the Blackhawk Formation, Mesaverde Group, of the Book Cliffs, Utah (Swift *et al.*, 1987). Samples from a prograding sandstone tongue form a continuum when phi mean diameter is plotted against phi standard deviation (open symbols, labelled 'initial progressive sorting trend'). In this strandplain sandstone, the sand of the graded and hummocky beds of the lower shoreface have been winnowed from cross-stratified beds of the upper shoreface. The top of the strandplain sandstone is capped by a coarse, transgressive lag deposit. The lag deposits form part of a secondary progressive sorting trend (crosses and closed symbols), which has been generated by the reworking of the deposits of the first one.

The primary structures of these beds are determined in part by petrographic attributes. Drawings from box core of lag and cross-stratified beds from the Atlantic shelf are shown in Fig. 3B. Their coarse grains travelled as bed load, and their internal structure reflects bedform migration. Lag beds are also enriched in shell material (dotted lines). Hummocky and graded beds are composed of finer sand that was transported mainly in suspension, and the internal structure of such beds is usually a Bouma sequence (Bouma, 1962). If, however, the ratio of the wave orbital velocity component to the mean flow component was high during deposition, the vertical sequence of primary structures in the bed may be dominated by a hummocky division (Dott & Bourgeois, 1982).

*Textural maturity and topographic profiles*

Petrographic differences in the proximal lithofacies of transgressive and regressive depositional systems lead to differences in geomorphology during deposition, and in the geometry of the resulting deposit. The fine to very fine-grained regressive sands are so easily resuspended by wave orbital currents, and settle so slowly once suspended, that steep slopes cannot be sustained on their depositional surfaces. Seaward of the shoreface, slopes on regressive shelves are measured in hundredths of a degree. In contrast, the coarser transgressive shelf sands can often sustain slopes on constructional marine landforms up to 5°, as on the down-current flanks of tide-built sand ridges on Georges Bank (Mann *et al.*, 1981).

**Lithofacies arrays as responses to stratification processes and progressive sorting**

*Towards a genetic classification of lithofacies*

Swift & Thorne (this volume, p. 23) argued that depositional lithofacies are defined by horizontal grain-size gradients (lithofacies changes), and cyclic vertical grain-size gradients (stratification patterns). Thorne *et al.* (this volume, pp. 59–87) have approached strata-forming processes from an

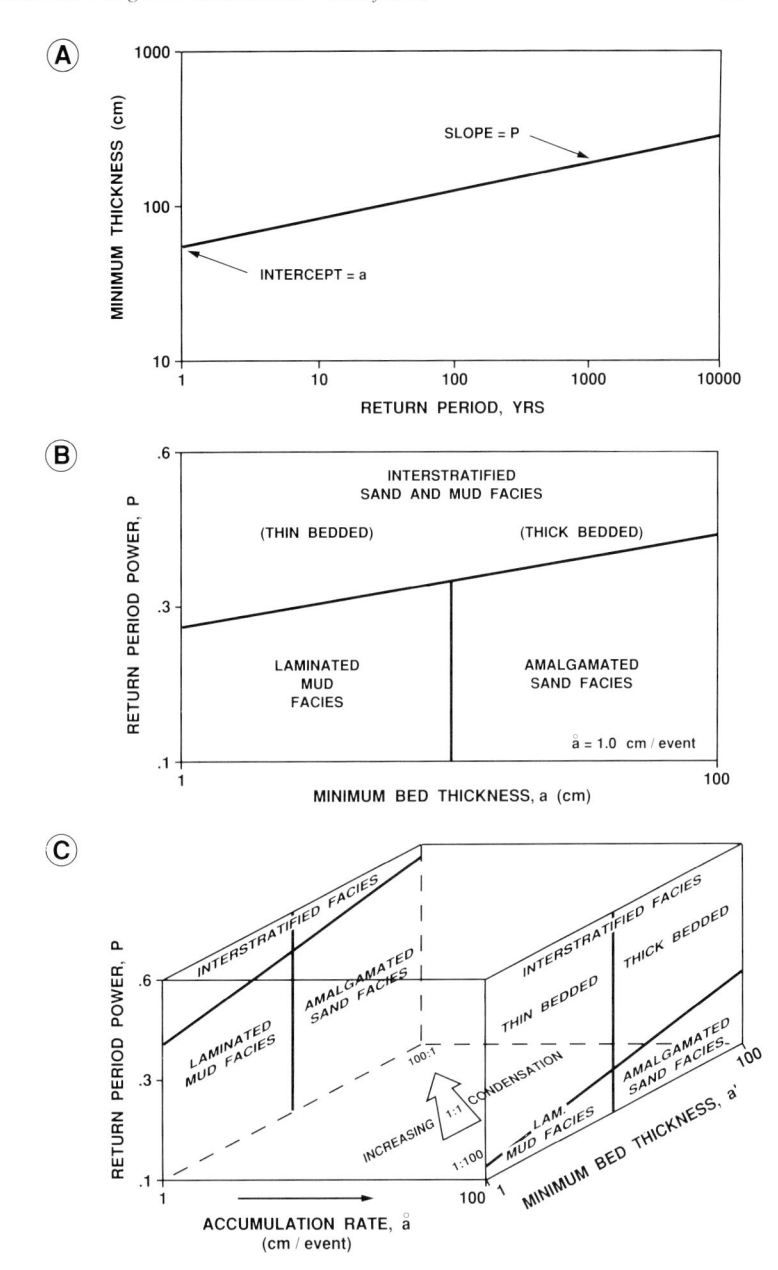

Fig. 4. The relationship between strata-forming processess and depositional facies. (A) Plot of minimum grain size, $a'$, against return period, $T$, illustrating the power law of strata thickness. (B) Depositional facies as a function of $a'$ and $p$. (C) Depositional facies as a function of $a'$, $p$, and $\mathring{a}$.

analytical point of view. In this section, we use the principles formulated in the previous sections to derive the anatomy of a shallow-marine lithofacies assemblage. In subsequent sections we show that this genetic scheme of lithofacies classification is applicable to all shallow-marine depositional systems.

Thorne et al. (this volume, p. 64) present a power law for event bed thickness, $\Delta\eta$, as

$$\Delta\eta = a'T^p$$

where $a'$ is return period intercept or 'minimum bed thickness', $T$ is the return period, and $p$ is the return period power. This relationship of stratification variables is presented graphically in Fig. 4A. The relationship is an important one for understanding the nature of shallow-marine facies. A unique stratification curve must characterize each depositional environment along the sediment transport pathway of a dispersal system.

The return period power, $p$, appears in Fig. 4A as the slope of the curve. It specifies the dynamic range of the depositional agent; the relative abundance of mild (short return period) versus intense (long return period) storms, for instance, or mild versus intense tidal flows. The higher the return period power, the greater the variation in event intensity, and therefore the greater the range in thickness and contrast in maximum grain size for beds of the lithofacies.

The minimum bed thickness, $a'$, appears in Fig. 4A as the $y$ intercept. It specifies the intensity of the threshold event. For small values of minimum bed thickness, all beds of the lithofacies will be relatively thin and fine. For large values, all beds will be relatively thick and coarse. Thus by varying $p$ and $a'$, a variety of distinctive lithofacies can be described. As the successive lithofacies of a depositional system are measured, the curve of Fig. 4A must steepen, flatten and shift up or down.

A third parameter must be allowed to vary if the complete range of depositional lithofacies is to be reproduced. The accumulation rate, $å$, together with the minimum bed thickness, $a'$, determines the working ratio, $r = a'/å$, and thus specifies the degree of condensation (see Thorne & Swift, this volume, pp. 33–58). As condensation increases ($r$ increases), thin beds are selectively destroyed, and thicker beds lose their fine grained tops. Grain size and thickness variation are minimized.

Figure 4B presents the nature of lithofacies variation in response to variation of $p$ and $a'$ for an intermediate value of $å$. For low values of the return period power ($p$), a *laminated mud lithofacies* or an *amalgamated sand lithofacies* may form, depending on $a'$, a measure of minimum event intensity. While custom has dictated different modifying terms for these two facies, the implication of the modifiers is similar; the deposits are stratified, but exhibit little grain-size variation.

For high values of $p$, there is greater variation in the maximum grain size per bed, as well as in bed thickness. An *interstratified sand and mud lithofacies* results. However, this lithofacies tends to be suppressed at high values of minimum bed thickness, $a'$.

As $a'$ increases relative to $å$, the reworking ratio, $r$, increases. Condensation results, and vertical grain-size variation is suppressed. Consequently the boundary between the interstratified lithofacies and the two homogeneous facies occurs at higher values of $p$ as $a'$ increases.

The relationships between all three variables are presented graphically in Fig. 4C as a three-dimensional solid. Figure 4B constitutes a vertical slice through the centre of the solid shown in Fig. 4C. The critical gradient of reworking ratios can be seen on the floor of the solid. As $r$ increases across the floor of the solid, the strata undergo condensation. Coarse-grained bed bases increase in volume relative to thinner, finer beds and bed tops. Vertical grain-size variation decreases. The sand content increases.

It should be noted that Fig. 4C is not analogous to a phase diagram for a chemical system in the same manner that a bedform diagram is analogous to a phase diagram. In a bedform diagram (e.g. Harms et al., 1982), stability fields for bedforms are described in terms of boundary shear stress, grain size and depth. Field boundaries represent discontinuities where one class of bedform is transformed rapidly into another as variables changes. In Figs 4B, 4C, lithofacies variation is continuous, and the exact location of boundaries is arbitrary.

The lithofacies boundaries shown in Figs 4B, 4C are abitrary for a further reason; an important variable has not been specified. Grain-size variation is treated in Fig. 4(B,C) as a dependent variable; a product of the mean fluid power at the station of observation, parameterized as the return period power. However, as noted by Swift & Thorne (this volume, p. 19), the grain size deposited at a station in a dispersal system is only partly a response to the local fluid power; it also depends on the grain size supplied from upstream (Fig. 3B). There is therefore an inherited bias in a given system towards sand-rich or mud-rich lithofacies. In addition, the inherited aspect of grain size contributes to the value of $c\%$, the cut-off percentage that controls the transition from amalgamated to interstratified lithofacies (see Thorne et al., this volume, p. 65). Figure 4C is thus a single example from a family of diagrams illustrating lithofacies response to differing mixes of inherited grain sizes in dispersal systems.

Figure 4C suggests that stratal parameters are the dominant controls of lithofacies variation, and that dispersal systems are characterized by three basic lithofacies; an amalgamated sand lithofacies, an

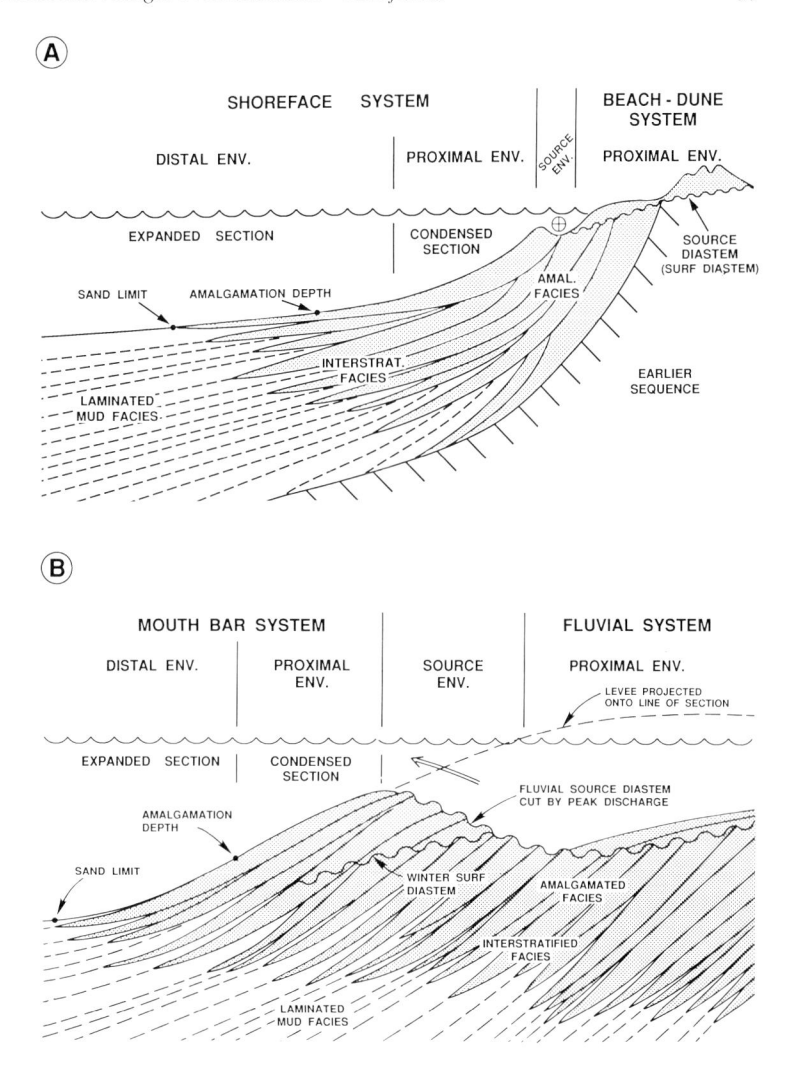

**Fig. 5.** Lithofacies of regressive depositional systems. (A) Lithofacies of strandplain and regressive shoreface shelf systems. Cross-circle is current arrow, seen tail-on. (B) Lithofacies of delta mouth bar systems. See text for explanation.

interstratified sand and mud lithofacies, and a laminated mud lithofacies. It should be clear, however, from the preceding discussions, that variation in stratal and grain-size parameters can result in numerous modications of this basic scheme. When varietal lithofacies traits are also considered, such as the degree of bioturbation, or petrographic or faunal composition, lithofacies complexity can become extreme.

*Example: lithofacies differentiation on a prograding shelf*

Since a depositional system has been defined as an assemblage of process-related lithofacies (Fisher & McGowen, 1967), a depositional system should appear as a discrete volume within the three-dimensional solid of Fig. 4B. However, the shape of the system within the parameter space of Fig. 4B is not the same as the shape of the system in the rock record. In this section we examine the lithofacies architecture and its relationship to strata-forming processes in real depositional systems. We take as our example a well-studied system; the prograding shelf system (Fig. 5A).

The lithofacies of prograding shelf systems are characterized by down-column and seaward-fining grain-size gradients (Fig. 5A). The stratification pattern also changes systematically in a down-column and seaward direction as more distal facies are encountered (Galloway & Hobday, 1983).

These characteristics may be understood by reference to the dispersal system that creates them. In the model presented by Swift & Thorne (this volume, pp. 13–15), the wave driven, sand-charged, littoral current serves as the local source. Episodic storm flows trend along coast and undergo downwelling (see Fig. 8 and accompanying text, Swift & Thorne, this volume, p. 15). They sweep sand obliquely down the shoreface through proximal and distal shoreface environments. Each grain takes part in a series of transport events before it finally undergoes permanent burial, or over a period of years or decades, escapes the shoreface altogether. As the sand is transported into deeper water, where wave orbital currents are weaker, it undergoes progressive sorting (see definition by Swift & Thorne, this volume, p. 19). In this process, the upper, finer, portion deposited by the last event is most likely to be entrained by the next event and bypassed seaward. Thus, successively finer fractions are preferentially deposited at each successive seaward location on the dispersal pathway, so that the proximal and distal shoreface lithofacies are distinguished by grain size.

The stratification pattern also changes systematically in a seaward direction (Fig. 5A), as a consequence of the weakening of near-bottom wave orbital motions with depth, and as a consequence of the increasingly finer character of the sand supplied. The changes in pattern are complex. They are best understood in terms of stratification parameters (return period power $p$, the minimum bed thickness $a'$ (or minimum resuspension depth), the accumulation per event $å$, the reworking ratio $r$, and the cut-off percentage $c\%$). These parameters, which form the basis of the following discussion, are described by Thorne et al. (this volume, pp. 64–67).

On the upper shoreface, the return period power, $p$, is relatively low because the shallow water depth limits wave height. Extreme wave heights are not possible in shallow water owing to wave breaking. In this region, storms of different magnitudes produce bed thicknesses which are similar owing to the fact that maximum wave height is limited by wave breaking. However, the return period curve is set high (there is a very large value for the minimum bed thickness, $a'$), also by virtue of the shallow depth. That is, the depth of reworking of the 'annual' scale storm is high. It is in fact markedly higher than the mean accumulation per storm ($å$), so that the reworking ratio, $r = a'/å$, is also high. As a consequence, the preservation potential of beds is low. The section is significantly 'cannibalized'. Surviving beds are the basal sections of storm beds with the longest return period possible in such shallow water.

Because of the high minimum bed thickness, beds are thick on the upper shoreface (20–100 cm thick). Because the return period power is low (peak storm waves are all about the same), the beds in this proximal lithofacies are all about the same size. Because the reworking ratio is high, much of each bed has been lost. The preservation percentage therefore tends to fall below the cut-off percentage, $c\%$. As a result, the beds are amalgamated (have lost their fine-grained tops). Beds tend to be multiple-event beds (composites of single-event beds whose grain sizes do not differ sufficiently to be distinguished). The amalgamation of multiple-event beds results in a characteristic *condensed section*. It is condensed because the minimum bed thickness $a'$ (resuspension depth of the storm of smallest return period) is large relative to the accumulation rate $å$; thus much of each new bed is resuspended, so that the net accumulation thickness is significantly less than the gross value (see discussion of reworking ratio, Thorne et al., this volume, p. 64). The term 'condensed section' is here used in the literal sense defined by stratification parameters, but the term is also applicable in the conventional stratigraphic sense. In a Wheeler diagram, where horizontal distance is plotted against time instead of against vertical distance, upper shoreface beds are widely separated and their poor preservation potential is obvious (see Fig. 16, Swift & Thorne, this volume, p. 23). The relationship between the condensed section of Fig. 5A and the classic condensed section of stratigraphers is analysed in a following section entitled 'lithofacies patterns and depositional systems'.

The preceding discussion indicates that although the upper shoreface beds are formed above 'fair-weather wave base,' they are storm deposits. Because of the high reworking ratio (high degree of cannibalization of beds), the *surviving* beds are the truncated bases of extreme-event deposits. Fair-weather deposits are thin. They are deposited but are not preserved because the episodic storm currents scour down through them to the last thick storm deposit. This conclusion is consonant with engineering studies, which show that abrupt changes in beach profiles occur during storms, not between them (Hands, 1983).

On the lower shoreface and inner shelf, the return period power, $p$, increases in response to increasing water depth; the water is too deep to filter out the waves of extreme events. However, the return period curve is 'set' lower; that is, the return period intercept, $a'$ (thickness of reworking by the waves of the 'annual' scale storm) is smaller. The sedimentation rate (accumulation per storm), $å$, decreases seaward with distance from the surf zone, but not as fast as $a'$, so that the reworking ratio, $r = a'/å$, decreases. Preservation potential is higher, and the section is more nearly complete. Because of the high minimum bed thickness, beds in this distal facies are thin on the upper shoreface (1–20 cm thick). Because the return period power is high (peak storm wave heights vary markedly), the bed thickness is variable. Because the reworking ratio is low, the mud tops of storm beds tend to be preserved. The preservation percentage therefore rises above the cut-off percentage, $c\%$. As a result, sand beds are discrete (separated from each other by mud intercalations) and are single-event beds. The distal facies thus constitutes an *expanded section*.

The transition between the proximal and distal lithofacies is characterized by an *amalgamation depth* (Fig. 5A). Seaward of this limit, the mud caps deposited in the last stages of waning storm flow survive through the accumulation process. In the inner part of the distal facies, the number of sand beds per metre of section increases seaward, as the reworking ratio drops; here more of the thinner sand beds are surviving. Sand bed frequency rises to a maximum, then begins to decrease (Aigner & Reineck, 1982). The decrease is a response to progressive sorting. It occurs because the transported load is constantly exchanging particles with the substrate. However, the exchange is biased; the flow deposits coarser particles and in return acquires finer ones. As a result, the flow becomes sand depleted. It is still competent to carry sand (has enough fluid power) but only has mud available to it. For a given transport event, sand depletion generally occurs before the flow reaches depths at which there is no longer sufficient wave orbital motion to suspend the sand. The problem is supply, not energy. This depletion process results in a second important boundary; the *sand limit* (Fig. 5A).

On the outer shelf, the grain-size range is significantly limited by progressive sorting, and the cut-off percentage ($c\%$) rises. Mud interbeds are increasingly multiple-event beds; that is, they are no longer simple end-of-storm mud caps on sand tempestites, but become successions of *mud tempestites* (Pilkey *et al.*, 1978; Aigner & Reineck, 1982; Nelson, 1982; Snedden & Nummedal, this volume). In this manner, distal shoreface lithofacies pass seaward into storm-stratified shelf mudstones.

## THEORY OF DEPOSITIONAL SYSTEMS

### Depositional systems and subsystems

It is possible to generalize from the proceeding example of lithofacies differentiation to a generic depositional system. To begin with, it is necessary to look more closely at the related concepts of *dispersal system* (assemblage of flow-linked depositional environments; Swift & Thorne, this volume, p. 23) and the *depositional system* that the dispersal system creates (assemblage of process-related facies; Fisher & McGowen, 1967).

### The source diastem: geometry of depositional systems

Walther's law says that as a depositional system aggrades, its parent depositional environments tend to migrate laterally, so that the deposits of one environment overlie the deposits of another (Walther, 1874; in Blatt *et al.*, 1982). As the environments migrate, their source environment migrates with them. Consequently, a depositional system will underlie or overlie its *source diastem* (Fig. 1B). The relationship of the depositional system to its source diastem depends on the relative values of the terms in the regime ratio, $R \cdot D/Q \cdot M$ (see Swift & Thorne, this volume, p. 13). If the supply terms on the right side (sediment input, grain size) greatly exceed the accommodation terms on the left side (sea-level, transport intensity), then the dispersal system is accepting significantly more sediment than it can accommodate, and supply-dependent morphodynamics prevail. The source environment migrates over its own deposits and the resulting depositional system is capped by its source diastem. This situation holds for prograding shoreface systems as described in the previous section, and in Fig. 6. In this case, the source diastem is cut by the along-shore trough, which is scoured by the wave-driven along-shore current, as it migrates seaward (Fig. 6D). It is also true for washover depositional sys-

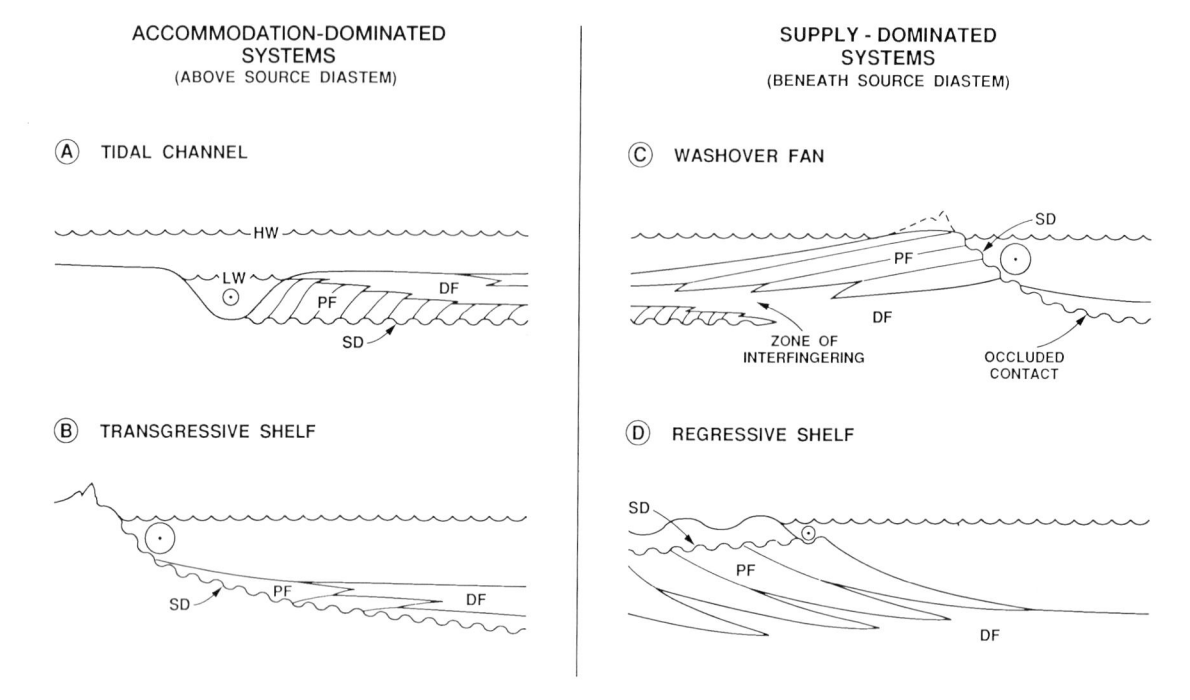

**Fig. 6.** Classes of depositional systems, and types of contact between systems. Accommodation-dominated depositional systems are formed from dispersal systems that have a sediment deficit. The depositional system therefore accumulated over its source diastem. Supply-dominated systems overlie their source diastems. HW = high water; LW = low water; PF = proximal facies; DF = distal facies; SD = source diastem. Panel C is a continuation of Panel A. Panel B overlaps with panel C. Circle with central dot in each panel is axis of current responsible for source diastem. Area of circle is proportional to cross-sectional area of current but not to its intensity.

tems in a transgressive back-barrier environment. In this case the ravinement diastem (Fig. 6C) is cut by erosional shoreface retreat (see discussion by Swift & Thorne, this volume, p. 17).

In contrast, in transgressive lagoons, tidal channel dispersal systems experience supply terms that only slightly exceed the accommodation terms, and accommodation-dependent morphodynamics prevail. The migrating thalwegs of the meanders scour out surfaces over which the tidal channel lithofacies accumulate through a process of lateral accretion. Since little material is being added, sediment simply can be transferred from one side of the channel to the other. Here the depositional system overlies its source diastem (Fig. 6A). For transgressive shoreface dispersal systems, the accommodation terms dominate the regime relationship. Accommodation-dependent morphodynamics prevail here also, and the resulting depositional system also rests on its source diastem; in this case, the ravinement surface (Fig. 6B).

## Classes of contact between depositional systems

Depositional systems can be delineated so completely by analysis of the stratal architecture of their component lithofacies, as in Fig. 5, that it becomes possible to account fully for the volume of a deposit by assigning its component strata to one system or another. In such a scheme of total accountability, several classes of contact between adjacent depositional systems can be observed. Two dispersal systems can be nourished by the same source environment, so that the resulting depositional systems lie back to back, with their proximal faces in contact across the source diastem. In Fig. 6D, the coupled strandplain and regressive shelf systems bear this relationship to each other. The shelf system progrades seaward, the surf-cut source diastem truncates its upper beds, and the beach-dune system progrades seaward across the source diastem. In this paper, such back-to-back depositional systems are designated *conjugate systems*.

In another fundamental relationship, the distal lithofacies of two depositional systems may interfinger. In Fig. 6C, the tidal channel system and washover fan system are *interfingering systems*. In this example, two dissimilar systems interfinger. However, similar systems may lie side by side, as in the case of an alluvial plain, in which the distal overbank lithofacies of one river interfingers with the overbank lithofacies of another. Less obviously, a regressive continental shelf may bear side-by-side shelf dispersal systems, fed by different rivers. In both the similar and dissimilar interfingering cases, it is the distal lithofacies that are in contact.

In a third relationship the source diastem of a migrating dispersal system, may completely destroy the depositional system over which it is passing, so that the depositional system that it deposits rests not on the second system, but on a third system. In Fig. 6C, the advancing ravinement diastem completely destroys the beach-dune depositional system, so that on the inner shelf, the proximal lithofacies of the transgressive shelf depositional system rests on the truncated distal lithofacies of the washover fan system. The contact of the shelf and washover systems is an occluded contact.

In a fourth relationship (not figured), two dispersal systems may be superimposed to form a third, which has some of the characteristics of both. The resulting depositional systems are such *superimposed systems*. For instance, delta mouth bars are formed by the overprinting of shoreface on fluvial channel environments. They tend to contain both surf base and thalweg diastems, the respective source diastems of the parent dispersal systems (see following discussion of delta mouth bar systems).

Finally, depositional systems may be *serial systems* (also not figured). In this case their parent dispersal systems are components of a large-scale dispersal system, as in the case of the Mississippi Valley-Gulf of Mexico continental margin system described above. In these cases, some portion of the upstream system serves as the source environment for the downstream system. The source environment of the upstream system may feed directly into the source environment of the downstream environment, as in the case of an alluvial dispersal system bearing a serial relationship to a deltaic dispersal system. In another kind of serial relationship, a source environment for a downstream system may be created by the rejuvenation of the distal environment of an upstream system, as when the backshore beds of the berm surface of a beach become the source for a dune-aeolian-flat system.

## Lithofacies patterns in depositional systems

The depositional environments of a dispersal system, distinguished primarily by their hydraulic climates, are arranged in systematic fashion downcurrent from the source environment (Fig. 1). The lithofacies of a depositional system are distinguished primarily by their vertical grain-size patterns (stratal patterns) and horizontal grain-size ranges (lithologies). They are stacked above or below the source diastem so that grain size decreases away from the diastem. The transition from the condensed lithofacies to the expanded lithofacies reflects the hydraulic gradients of the original dispersal system, from proximal environments with high reworking ratios to distal environments with low reworking ratios. Lithofacies patterns tend to be allometric in nature (Thompson, 1961); proportions of different facies vary from system to system, but not the sequence, or relative position within the system.

The use of the term 'condensed section' to describe the near-source (proximal) lithofacies in the proceeding paragraph *does not* compete with the way in which the term is used by classical stratigraphers. The proximal condensed section of a depositional system is condensed because in it, gross deposition has exceeded or has greatly exceeded net deposition; much or most of the section is in fact missing owing to intermittent erosion. The section is 'apparently' condensed because earlier deposited material had been reused by later events. It is 'truly' condensed because much of the material deposited originally has been by-passed, leaving a lag residue. It is also condensed in the sense that it is physically thinner than the adjacent expanded section; for a given time interval, it thins to a feather edge as the beds of that interval approach the source diastem.

The relationship between the term 'condensed section' as used in this paper and as used previously is best understood by recourse to a transgressive setting. In such settings (Fig. 7A), the same lithofacies occur as in the case of a regressive shelf (Fig. 5A), although the relative volumes differ, and the vertical sequence is reversed (to go up column is to encounter more basinward facies). In the transgressive case, the shelf system is prograding land-

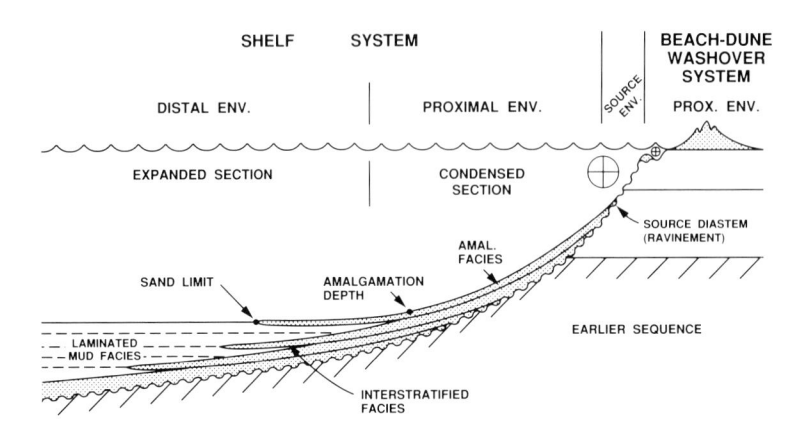

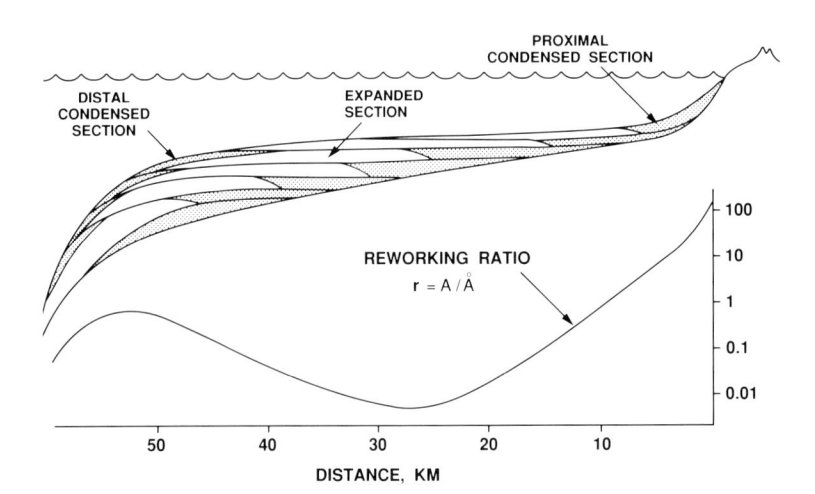

**Fig. 7.** Lithofacies on a transgressive shelf. (A) Inner shelf detail; (B) general view, showing relation of condensed sections to reworking ratio. Cross-circle is current arrow, seen tail-on.

ward over the back-barrier system. The source diastem separating them is no longer cut by a prograding surf zone, but is instead the ravinement surface (see Swift & Thorne, this volume, p. 17 for a discussion of ravinement surface).

In the transgressive case, a *distal condensed section* tends to occur in addition to the proximal one (Fig. 7B). This is the condensed section of taphonomists and classical stratigraphers (e.g. Kidwell and Jablonski, 1988; Roberts & Coleman, 1988) and of seismic stratigraphers (Loutit *et al.*, 1988). Both the distal and the proximal condensed sections are the result of elevated reworking ratios. The situations are not completely parallel, however. In the proximal environment, both the minimum bed thickness $a'$ (measure of hydraulic climate), and the accumulation rate $å$ are high. The increase in $a'$ is larger than the increase in $å$, however, so that $r$ is large. In the distal environment, both $a'$ and $å$ are small. The reworking ratio begins to rise toward the outer sector because the drop in the accumulation rate $å$ eventually becomes more important than the drop in the fluid power parameter $a'$. Coarse particles have been filtered out by progressive sorting in their passage through the dispersal system, but distal condensed sections may become coarser through the addition of allochemical particles; skeletal or microskeletal calcium carbonate, phosphate concretions or glauconite. Distal condensed sections are typically capped by an erosional surface. These surfaces have been called surfaces of omission or non-deposition (Kennedy & Garrison, 1975). However, the term 'non-deposition' is incorrect. While sediment accumulation has ceased, fluid power expenditure has not, so that net erosion occurs. As a consequence of such rejuvenation, these surfaces may serve as the source diastem for thin, slowly accumulating depositional systems that are forming further seaward, over the slope. The condensed section of classical stratigraphy and lithostratigraphy (e.g. Loutit et al., 1988) is thus a special case of the more general situation described in this analysis.

### Depositional systems versus sand and sandstone bodies

Depositional systems consist of a lithofacies assemblage in which lithofacies are arranged above or below a source diastem in a vertical and horizontal succession of decreasing grain size. A depositional system tends to have a core sand or sandstone body (proximal facies) embedded in a mantle of distal facies. There is therefore a general correspondence between types of depositional subsystems and types of sand bodies. It is possible to talk of a fine-grained (regressive) shelf sandstone sheet as the critical reservoir element in a regressive shelf depositional system, which may also contain an extensive distal (shale) facies.

Complications arise, however, because in many cases, the proximal (sand depositing) depositional environments of the different dispersal systems may have been nourished by a single source environment that lay between them, so that their respective depositional systems are conjugate. As these environments shift in response to regime variables, the amalgamated sand lithofacies of the two systems will be deposited back to back across a source diastem, and the resulting sand body will be a composite one. For instance, in Fig. 5A a strandplain sand body is being created as shoreface, beach and dune environments migrate seaward. The corresponding sand lithofacies are being deposited on their respective sides of the source diastem, resulting in a composite strandplain–shelf sand body. Further stacking of proximal lithofacies may occur when a strandplain is deposited during a sea-level fall, then transgressed. The strandplain is incised, and its valleys accumulate a proximal fluvio-estuarine facies. During the transgression, the sequence may be capped by a proximal transgressive facies. This sequence of events leads to the so-called 'lowstand sandstones' (e.g. Riemersma & Chan, this volume, pp. 489–510).

### Depositional systems and morphodynamics

Thorne & Swift (this volume, pp. 33–44) have stressed the dynamics of the depositional surface as the key to the geometry and internal architecture of sedimentary bodies. In the following review of classes of shallow-marine depositional systems, we will begin each section with a description of the morphodynamics leading to formation of the deposit.

## CLASSES OF SHALLOW-MARINE DEPOSITIONAL SYSTEMS

### Transgressive and regressive families

An analysis of continental margin dispersal systems, and the resulting depositional systems has been presented above. It provides a rational basis for the classification of shallow-marine depositional systems. The lithologic contrast between systems deposited in transgressive, autochthonous regimes and those deposited in regressive, allochthonous regimes is thus so marked that it can serve as a primary criterion for the classification of depositional systems into transgressive and regressive families. The families can be differentiated further on the basis of the dynamics of their dispersal systems into an intra-coastal group (landward of the main shoreline), an inner shelf group (Table 1). Such groupings of depositional systems have been called *depositional systems tracts* (a linkage of contemporaneous depositional systems; Brown & Fisher, 1977). In Table 1, an informal intermediate designation of *complex* is also used. The term is required because intermediate scales of organization occur. For instance, we will show in the following discussion that tidal-delta–tidal-channel complexes consist of two systems, each with its source diastem,

**Table 1.** Shallow marine depositional systems, complexes, and systems tracts

|  | Regressive settings | Transgressive settings |
|---|---|---|
| Intra-coastal settings | Regressive intra-coastal systems tract | Back-barrier systems tract |
|  | Standplain or chenier plain systems | Beach-dune–washover-fan complexes |
|  | Deltaic-channel–mouth-bar complexes | Tidal-delta–tidal-channel complexes |
| Shelf settings | Regressive shelf systems tract | Transgressive shelf systems tract |
|  | Regressive shoreface-shelf systems | Transgressive shoreface-shelf systems |
|  | Prodelta plume systems | Sand ridge complexes |
|  | Fine-grained deceleration sheets | Coarse-grained deceleration sheets |

and proximal and distal facies, and that within the tidal delta system several subsystems may be discerned.

Following Brown & Fisher (1977), no attempt is made here to restrict systems tracts to specific positions within depositional sequences, such as 'highstand' or 'lowstand' systems tract. Such names are based on characteristics extrinsic to the systems tract. In contrast, the distinction between regressive shelf systems tracts and transgressive shelf systems tracts is made for intrinsic reasons; the component systems are different in each case.

The depositional systems tracts of Table 1 are not the same as the depositional systems of seismic stratigraphers (e.g. Vail *et al.*, 1977; Posamentier *et al.*, 1988). As originally defined, these bodies are contemporaneous in age, and are so thin relative to their extent as to be reasonably designated by the two-dimensional term 'tract'. However, seismic stratigraphers have used the term 'systems tract' to describe major subdivisions of third-order depositional sequences (lowstand systems tract, trangressive systems tract, etc.). Such thick deposits represent accumulations of considerable time duration. They commonly consist of stacked, high-frequency depositional sequences, sometimes known as parasequences (Vail *et al.*, 1977), that are deposited in response to short-period oscillations of regime variables. High-frequency sequences frequently consist of a diastem or thin transgressive basal deposit, followed by a thicker, upward fining, regressive deposit. A 'transgressive systems tract' is said to consist of 'back-stepping parasequences' (Posamentier *et al.*, 1988), and therefore consists predominantly of *regressive* depositional systems, as defined in this paper. We therefore restrict the term 'depositional systems tract' to deposits that have accumulated under conditions of essentially constant regime, as defined by the regime variables $Q$, $R$, $M$ and $D$. The observed partitioning of third-order sequences is both real and significant, and we use the provisional term 'geometric systems tracts' to designate them.

The relationship between depositional systems and depositional sequences is discussed in greater detail by Swift *et al.* (this volume, pp. 153–187) and Thorne & Swift (this volume, pp. 189–255).

## REGRESSIVE DEPOSITIONAL SYSTEMS

### General

The regressive family of shelf depositional systems are those created by regressive, supply-dominated regimes. Supply-dominated regimes have been characterized by Swift & Thorne (this volume, p. 13) as occurring whenever the supply terms (sediment input, $Q$, and grain size, $M$, dominate over the accommodation terms, sea-level change, $R$, and sediment transport, $D$, in the regime ratio $R \cdot D/Q \cdot M$. In such a setting, shorelines prograde, river mouths become equilibrium estuaries or deltas, and river mouth bypassing is the dominant mode of sediment supply to the shelf. Coastal profile translation can be described by the $Q$-dependent model (Thorne & Swift, this volume, pp. 39–41) when the regression is the result of a high sediment supply (depositional regression), rather than to a falling sea-level (erosional regression; Curray, 1964). The ratio between the fluvial sediment discharge per unit length of shoreline and the mean annual wave power is a major parameter determining the character of supply-dominated regimes (Wright & Coleman, 1977). This value tends to be highest on passive margins near river mouths.

The ratio of fluvial to tidal discharge over the tidal cycle affects the depositional regime. Increased tidal flushing enhances river mouth bypassing and retards delta formation (Rhine, Ganges Rivers; Coleman & Prior, 1982) or inhibits it altogether. Thus, as the tidal range increases relative to sediment input, tidal deltas pass into equilibrium estuaries (e.g. Amazon River mouth; Coleman & Prior, 1982). In the Amazon case, the mouth is landward of the salt wedge, throughout much of the year, but still has a large tidal range.

Sediment transport in allochthonous regressive shelf settings has been described by Swift & Thorne (this volume, p. 25) as diffusive rather than advective in nature; that is, the net sediment flux tends to be offshore as well as along shore, in response to the offshore gradient of decreasing wave energy flux into the bottom. As a consequence, lithofacies boundaries tend to be shelf parallel, and gradational in nature. Seaward of the sandy shoreface, most of the shelf surface is covered with mud. In the following sections we describe in detail the regressive depositional systems that form in response to supply-dominated regimes.

## Regressive coasts and deltas

Considerable confusion exists concerning the nature of *deltas* and *deltaic deposits* in the rock record. The confusion stems in part from the relationship between geomorphological and stratigraphic concepts (Galloway, 1975). A delta is a constructional coastal landform; it is a land *surface* distributed about a river mouth that protrudes into the sea. A deltaic deposit, in contrast, is a sediment or rock *volume* of deltaic origin, deposited over a single sedimentary cycle. Many such deposits may be stacked within a basin, but in such cases, the depositional relief at any given time is generally much less than the aggregate thickness.

Further confusion stems from the fact that a deltaic dispersal system is a supersystem. It consists of subsystems, such as a delta-channel–delta-plain system, a delta mouth bar system, and flanking strandplain–shoreface systems (Fisher *et al.*, 1969). In the broadest sense all prograding coasts are deltas, and the term is frequently so used. However, a classification of regressive coasts in which there is a single class is of little use, hence in this paper the term 'deltaic deposit' is restricted to channel and overbank deposits that protrude into a marine setting, and the immediately associated marine facies. Since this paper examines marine systems, its concern lies with the three of these deltaic systems that are marine; mouth bar systems, strandplain–shelf complexes, and prodelta plume systems; and the discussion will be limited to them.

## Delta mouth bar systems

### Morphodynamics of delta mouth bars

The term delta mouth bar is used here in preference to 'distributary mouth bar,' since distributaries are common only in river-dominated (crescentic, lobate or elongate) deltas (Fisher *et al.*, 1969). 'Mouth bars' are not unique to delta mouths. In fact, they occur off any kind of coastal inlet, including tidal inlets and estuaries. Delta mouth bars are planoconvex to biconvex lenses in cross-section, are ellipsoidal to kidney-shaped in plan view, and are 2–20 m thick and 0.2–20 km in longest extent (Coleman & Prior, 1982).

L.D. Wright (1977) has shown that mouth bar morphology depends on the relationship among three hydrodynamical forces; outflow inertia, turbulence induced by bed friction seaward of the mouth, and outflow buoyancy. In a delta mouth dominated by inertial flow, the bar does not shoal enough to seriously constrict the flow. This type of effluent is associated with steep-gradient streams entering relatively deep water (Fig. 8A). In the marine environment, the bars of fan-deltas may approach this configuration (Fraser & Suttner, 1986).

More commonly, deltaic effluents deposit sufficient sediment seaward of the mouth to cause sea-floor shoaling and frictional retardation of the effluent jet (L.D. Wright, 1977). A positive feedback results; frictional retardation of the flow with attendant spreading causes more spreading and more retardation (Oertel, 1988). As the delta mouth evolves, stability is ultimately re-established by the creation of bifurcating channels, separated by 'middle ground' shoals (Fig. 8B).

L.D. Wright (1977) shows that in most marine deltas, the outflow spreads as a buoyant plume above the denser underlying salt-water, and a buoyancy-dominated depositional pattern results. In a buoyant jet that is not in contact with the bottom, the surface water diverges as it flows seaward, while the water in the bottom part of the jet converges, so that in cross-section, two helical flow cells are seen.

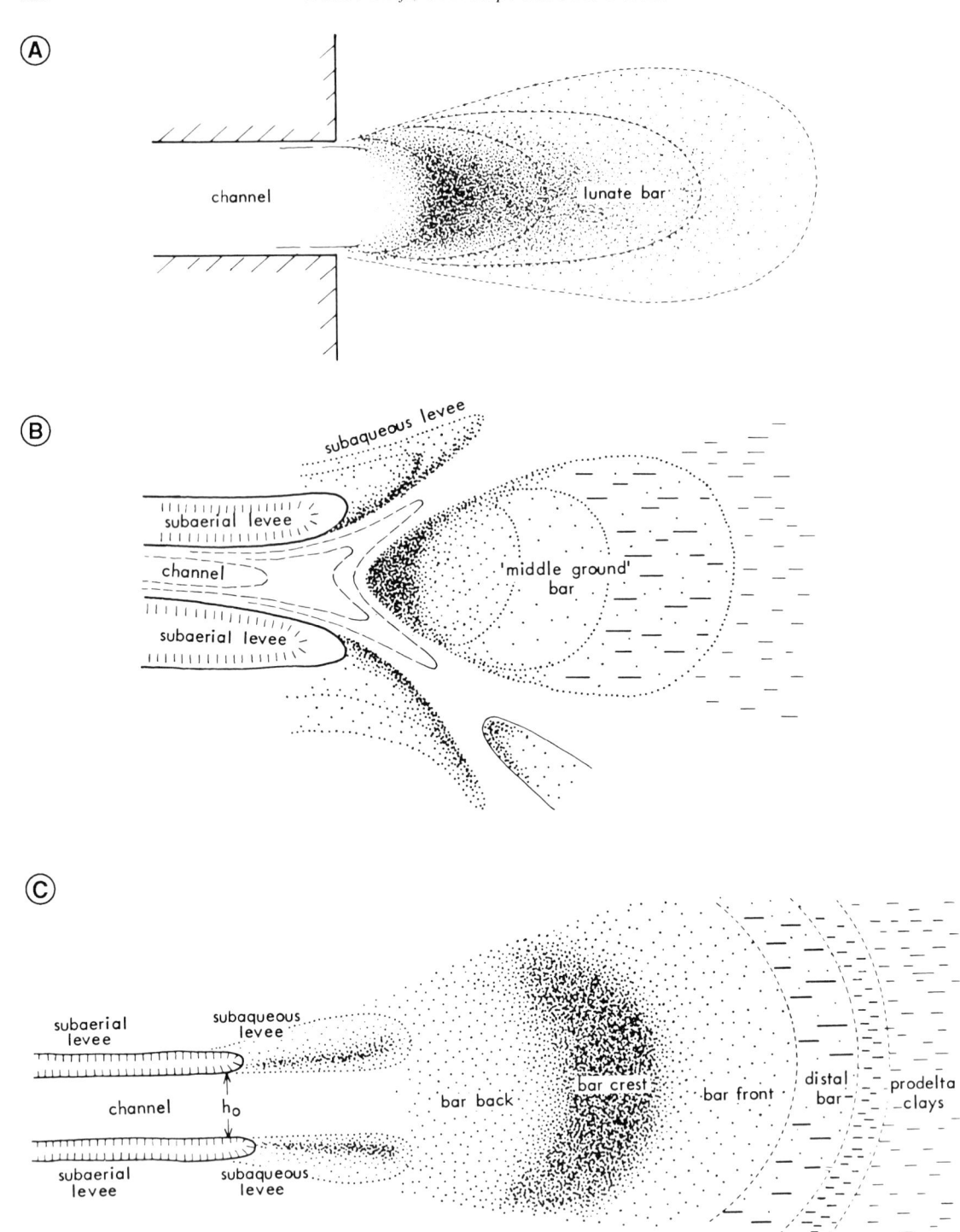

**Fig. 8.** Dynamics of delta mouth bars. (A) Inertia dominated effluent; (B) friction dominated effluent; (C) buoyant effluents. (From L.D. Wright, 1977.)

This flow pattern leads to the formation of long, straight, subaqueous levees, and as a consequence, to straight, digitate distributaries with comparatively high depth:width ratios (Fig. 8C). The major distributaries of the Mississippi delta exhibit this pattern, in contrast to the friction-influenced crevasse splays that have occurred upstream in the recent past (L.D. Wright, 1977).

Many rivers empty into macrotidal environments. Tides have three basic effects on river mouths: (1) bidirectional transport prevails (although the signature of flood transport may be erased by the ebb or vice versa); (2) tidal mixing minimizes vertical density gradients; and (3) the range of position of the land–sea interface and of the zone of river–sea interaction is greatly extended, both vertically and horizontally. In tidally dominated river mouths, the tidal current cycle tends to be asymmetrical so that at a given point, the sea-floor is ebb-dominated or flood-dominated (Ludwick, 1970; L.D. Wright, 1977). Ebb and flood-dominated sediment transport tends to occur in interfingering, mutually evasive channels, separated by partitioning ridges. In such settings, the continuous delta-mouth bars of micro- and mesotidal river mouths become 'zig-zag submarine spits' (Ludwick, 1974).

River mouths in rigorous wave climates experience direct modification of the expansion and deceleration patterns of outflow (L.D. Wright, 1977). The outflow refracts and steepens waves in such a way as to concentrate power on the effluent and to cause breaking in abnormally deep water. The resultant wave-induced set-up opposes outflow, while wave breaking enhances the mixing and momentum exchange between effluent and ambient waters. The effect is to cause very rapid deceleration and loss of sediment transporting ability within short distances from the outlet.

The delta mouth bar dispersal system is a superimposed system (see previous discussion of dispersal system geometry). In delta mouth bar environments, shoreface and alluvial channel processes operate alternately, and sometimes simultaneously, so that the lithofacies architecture of a shoreface is overprinted by the lithofacies architecture of an alluvial channel (Fig. 5B). The forward face of the delta mouth bar can be viewed as a shoreface wrapped around the river mouth. Shoreface processes dominate on the flanks of the delta mouth bar, while channel processes dominate along the axis. The following discussion of mouth bar lithofacies will begin with a synthesis of shoreface lithofacies architecture, before proceeding to the more complex situation in the axial zone.

*Mouth bar lithofacies: shoreface deposits on mouth bar flanks*

The lithofacies successions of mouth bar flanks resemble those of prograding shorefaces (Fig. 4A and accompanying text) formed by storm wave and current processes. In these successions, a *laminated shale lithofacies* is overlain by an *interstratified sand and shale lithofacies*, which in turn is overlain by an *amalgamated sandstone lithofacies*, capped by a surf-cut source diastem (Swift *et al.*, 1987; Fig. 5B).

The laminated mud lithofacies consists of silt and silty clay lamina and thin beds (bedding thickness terminology from Ingram, 1954), which formed as mud tempestites. Thicker units may have rippled tops and sharply defined bases. Stratification may be locally obliterated by bioturbation. The interstratified lithofacies consists of fine to very fine grained, thin-bedded to very thin-bedded sandstones interbedded with laminated siltstone and shale. Sandstone beds have erosional bases, fine upwards, and have bioturbated or rippled tops. The thicker sandstone beds may exhibit the fining upward texture characteristic of suspension deposits, or may be hummocky. However, as bed thickness and grain size decrease, ripple cross-laminations become the dominant structure. Flaser, wavy and lenticular-bedded variants occur as beds become finer and thinner (Ruby, 1981). The interstratified lithofacies is deposited on the lower shoreface and inner shelf.

The interstratified lithofacies is overlain by an amalgamated sandstone lithofacies, in which no silt or shale interbeds occur (condensed section of Fig. 5B). Such successions begin with hummocky fine to very fine-grained sandstone beds with rippled tops, deposited in the middle shoreface (Swift *et al.*, 1987). The horizontal spacing of the hummocks and swales tends to increase up-section, as does bed thickness (Leckie & Walker, 1982). The hummocky beds are commonly overlain by fine to medium-grained trough cross-strata sets deposited by waning storm flows in the surf zone. These cross-strata sets comprise the upper shoreface deposits, and have been described from modern environments by Howard & Reineck (1981) and Short (1984). They may be capped by fine to medium-grained, low angle, wedge-shaped cross-strata sets deposited in beach and dune environments (Campbell, 1979).

Bounding surfaces of lithofacies in shoreface depositional systems are created as depositional environments shift seawards (Fig. 5A). The most sharply defined bounding surface is the source diastem cut by the seaward migration of the surf zone. It separates the amalgamated sandstone lithofacies of the shoreface dispersal system from the amalgamated sandstone lithofacies of the beach-dune dispersal system. The amalgamated sandstone lithofacies is divided from the interbedded sandstone and shale lithofacies by an amalgamation surface, defined by the landward tips of shale interbeds, and is separated from the underlying laminated shale lithofacies by a sand limit surface, defined by the seaward tips of sand beds.

*Mouth bar lithofacies: flood deposits in the central bar mouth*

In the axial zone of delta mouth bars, fluvial processes dominate. Studies of the distributary mouth bars of the Mississippi delta (Wright & Coleman, 1974) indicate that the bar aggrades rapidly during flood stage. Consequently, stratification in the bar must be primarily of flood origin, and the beds there are primarily 'inundites' (flood beds; Seilacher, 1982) rather than 'tempestites' (storm beds; Ager, 1974). The inundite subfacies can be described by the same parameters as the tempestite subfacies; $T_i$ for an inundite's flood return period, versus $T_t$, the storm return period for a tempestite. The inundite subfacies constitutes a condensed section with high values of $r$ and $a'$, much as does the amalgamated tempestite subfacies. Inundites must interfinger with tempestites in a seaward direction and on either flank, but the relationships between the two stratification types are poorly understood.

As a consequence of the superposition of dispersal systems, two kinds of source diastem can be seen in Fig. 5B; storm-cut shoreface diastems (surf diastems) develop when major storm events entrain flood deposits and move them laterally along the coast to flanking strandplains. In addition, the prograding river channel cuts a source diastem (thalweg diastem) along its thalweg during flood events. Because of the central intrusion of the flood beds, and the associated high depositional rate per event, å, the amalgamated lithofacies tends to be thickened in the delta mouth bar relative to the equivalent lithofacies in the flanking strandplain deposits.

On shelves seaward of large deltas, such as the Louisiana prodelta shelf, flood stratification extends further offshore, in part because the freshwater discharge of the river creates a strong vertical salinity gradient. This freshwater shield decouples the deeper bottom waters from surface wind stress, thus inhibiting the downward flux of momentum, and preventing the formation of storm strata. Here the reworking ratio falls to zero and stratification is largely owing to the snaking back and forth of the turbid surface plume as it is pushed around by wind stress (Adams *et al.*, 1988). Since storm diastems are largely lacking, the deposits appear massive or vaguely banded.

Because delta mouth bars occur in regressive settings where all depositional environments are shifting seaward, delta mouth bars often have been cut through by, and partially destroyed by, the main channel. In the extreme case (finger bars of the Mississippi delta; Fiske *et al.*, 1955), delta mouth bars are distributary mouth bars which lay a continuous track of bar facies. As this lithofacies subsides into the underlying shales, the channel cuts through and advances over it, so that the basal bar lithofacies is disconformably overlain by a fluvial channel facies, separated by the channel-cut source diastem (Fig. 5B).

**Prodelta plume systems**

*Prodelta plume morphodynamics*

Delta mouth bar and strandplain–shoreface depositional systems are often associated with prodelta plume systems in regressive settings (Barratt, 1982; Patterson, 1983; Palmer & Scott, 1984). Prodelta plume systems are typically 2–20 m thick, 2–10 km wide and 5–50 km long. They are crescentic or plume-shaped deposits in plan view. Where sufficient control is available, their proximal ends can be seen in plan view to be attached to the cuspate deltaic projections along the seaward margin of strandplain deposits. In cross-section, they appear as lenticular sandstones embedded in shale, beneath strandplain sandstones or basinward of their pinch outs.

It has been proposed that the sand body on the Nile Prodelta Shelf in the vicinity of the Damietta distributary mouth is a modern analogue of such

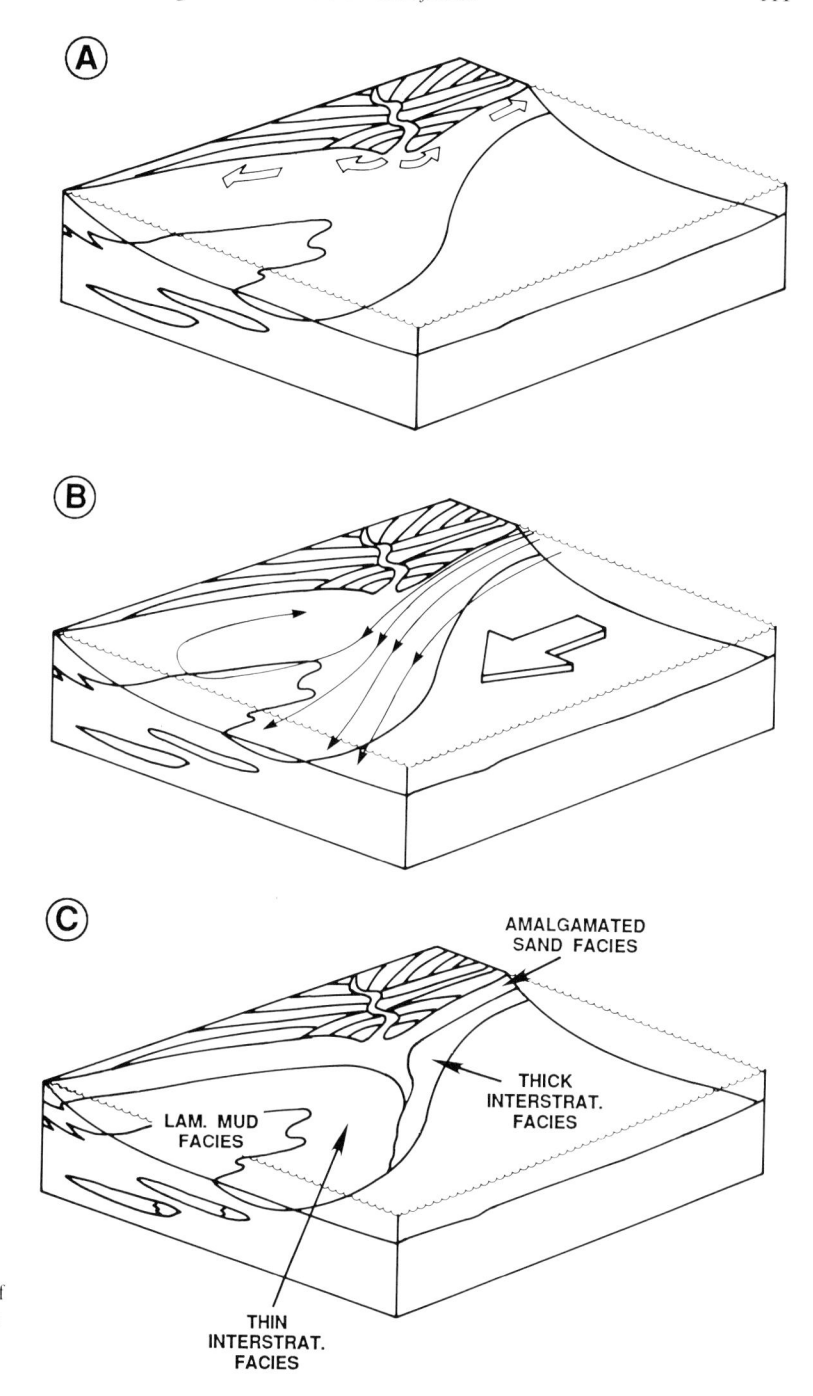

**Fig. 9.** Dynamics and lithofacies of a prodelta plume. (A) Flood stage; (B) storm flow; (C) lithofacies. (Modified from Swift *et al.*, 1987.)

ancient prodelta plume deposits as those of the Cretaceous Navarro field of Texas (Patterson, 1983). The dynamics of formation of the Damietta prodelta plume deposit have been described by Coleman *et al.* (1981). Prior to the erection of the Aswan Dam, the annual Nile flood brought large volumes of sand to the delta front (Sharaf El Din, 1977). The sand was deposited in the Damietta mouth bar, and in other distributary mouth bars. During the rest of the year, it was resuspended, entrained into the wave-driven littoral current, and redeposited on the flanking beaches of the distributary mouths.

However, during peak transport events, sand appears to have leaked out on to the shelf downcurrent from distributary mouths. The probable hydraulic machinery controlling this transport path has been observed at the Damietta mouth. Here, during periods of strong winds out of the west, a shelf-wide eastward flow develops (Murray et al., 1981). The flow hugs the shoreface, but cannot 'make the turn' at the Damietta mouth. A reverse pressure gradient forms. The coastal boundary layer detaches, and in plan view, a back eddy develops in the zone of separation (Fig. 9B).

Grain-size data indicates a complex history for the Damietta prodelta plume. The mean diameter is relatively course (about 1.8 phi), coarser than that of the adjacent shore face. The relatively coarse nature of the sand suggests that the formative plase of the plume may have ended, and that its surface is now undergoing winnowing, possibly in response to sediment starvation on the shelf since the advent of the Aswan Dam (Stanley, 1988). The anomalous grain size has been taken as evidence that Damietta mouth is not a good modern analogue (Schieing & Gaynor, 1991). However, the necessary hydraulic machinery is clearly in place, and examples capped by coarser transgressive sands are in fact common in the rock record (Barratt, 1982). Prodelta plume depositional systems are formed frequently during fourth or fifth-order sea-level fluctuations. Many are thus parasequences as well as depositional systems, and as such are topped by transgressive surfaces (see discussion of parasequences by Swift et al., this volume, pp. 153–187).

*Prodelta plume lithofacies*

Prodelta plume sandstones have been observed in outcrop in the Cretaceous Blackhawk Formation (Mesaverde Group) of Utah (Swift et al., 1987). They appear as lenses in the Mesaverde escarpment that pinch out both seaward and landward, but their plume-like plan geometry can be resolved only in the subsurface. Examples from hydrocarbon reservoirs, in which there is sufficient density of wells to establish plan geometry, include the Fales Member, Mesaverde Formation, West Poison Spider Field, Wind River Basin, Wyoming (Barratt, 1982);  Navarro Group, East Central Texas (Patterson, 1983); the La Ventana Tongue, Point Lookout Formation, San Juan Basin, New Mexico (Palmer & Scott, 1984); and the Woodbine Formation of the east Texas Basin (Phillips & Swift, 1985; Phillips, 1987).

Prodelta plume deposits resemble truncated versions of the shoreface deposits seen in regressive shoreface systems (Fig. 9C). The basal plume deposit is a *laminated mud lithofacies*. It is overlain by very thin-bedded to thin-bedded sandstones intercalated with shale beds of equal thickness (*interstratified sand and mud lithofacies*). As the plume deposits are traced up-section and along the shelf in the sourceward direction, sandstone beds thicken and lose their mud interbeds (*amalgamated sandstone lithofacies*). They develop laminated to massive size-graded lower divisions, overlain by ripple cross-laminated upper divisions (Patterson, 1983). Plume deposits may be overlain by amalgamated hummocky beds at their nearshore end, and in general there is a proximal to distal lithofacies trend that parallels the top to bottom trend (Phillips & Swift, 1985). If the sequence has been transgressed, it may be capped by a medium-grained, trough cross-stratified sandstone (Barratt, 1982).

**Strandplain and regressive shoreface-shelf systems**

*Strandplain and shoreface-shelf morphodynamics*

Strandplain and regressive shoreface-shelf systems are conjugate systems that lie on either side of a source diastem cut by the seaward migrating wave-driven littoral current (see Fig. 6D and accompanying text). The morphodynamics of their respective dispersal systems are likewise closely linked, and they will be described together. Strandplain sandstones are sheet-like deposits 10–20 m thick, extending 1–30 km seaward, and 5–100 km along shore (McCubbin, 1982). Strandplain systems develop on the flanks of wave-dominated deltas, and have been considered to be deltaic subenviroments (Brown & Fisher, 1980). However, this approach lumps all regressive coasts into a single littoral environment, which is counter to the purposes of classification; see previous discussion.

Strandplain–shelf systems are characteristically associated with deltas supplied with a relatively coarse (sandy) load, undergoing relatively intense wave attack, so that lobate to cuspate morphologies

$P = 3 \times 10^8$ ergs/cm·s

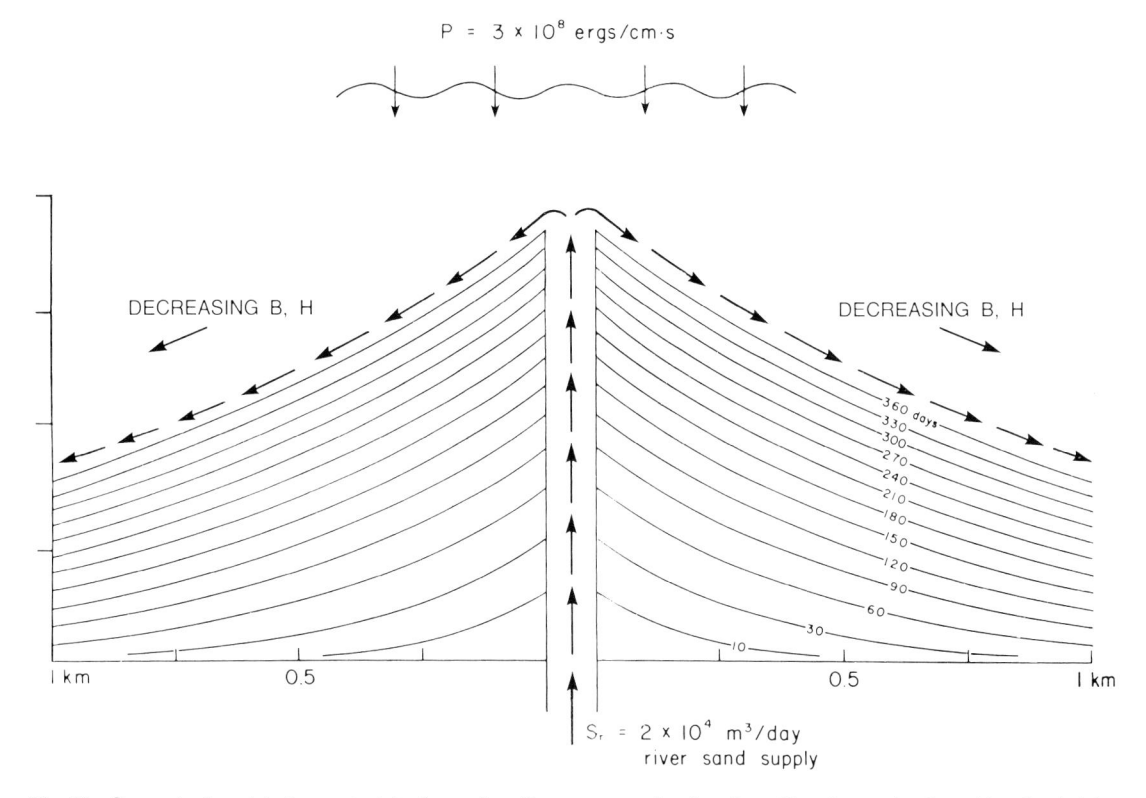

DECREASING B, H

DECREASING B, H

360 days
330
300
270
240
210
180
150
120
90
60
30
10

1 km — 0.5 — 0.5 — 1 km

$S_r = 2 \times 10^4$ m³/day
river sand supply

**Fig. 10.** Conceptual model of cuspate delta formation. Transport rate is a function of breaker angle, $B$, and breaker height, $H$. (From Komar, 1973.)

result (Coleman & Prior, 1982). Two classes of strandplain–shelf systems may be distinguished, depending on whether the coast is in regression because sediment input is greater than the accommodation created by sea-level rise (rising sea-level strandplains), or whether it is in regression because sea-level is falling (falling sea-level strandplains). Rising sea-level strandplains are formed by $Q$-dependent profile translation (see discussion by Thorne & Swift, this volume, pp. 39–41). The source diastem is cut by surf erosion. A trivial portion of the section is missing in the diastem, which is generally poorly developed. In a falling sea-level strandplain, however, the coastal profile is $R$-dependent (see discussion by Thorne & Swift, this volume, pp. 37–39). The whole coastal profile is involved in cutting the source diastem in a reversal of the ravinement process: sediment is eroded from the inner shelf floor and transferred to the shoreface (see Dominguez & Wanless, this volume, pp. 259–282).

The dynamics of coastwise dispersal from delta mouths to flanking strandplains has been modelled by Komar (1973). Komar shows that in such deltas, the plan view and cross-sectional shape of the shoreface interact with the pattern of refracting waves, until an equilibrium configuration is achieved. Komar's idealized delta is cuspate in plan view, and the cusp is oriented normal to the direction of wave approach (Fig. 10).

The critical depositional surfaces in such a cuspate delta are the two shorefaces that curve away from the apex rather like blades of a V-shaped ploughshare (Fig. 10). These shoreface surfaces both cause and are caused by the wave refraction pattern. As the shoaling waves move towards the beach, the wave rays bend towards the submarine projection of the delta apex, and the wave front rotates toward an orientation parallel to the shore. Adjustment of the wave to the curve of the shoreline is least complete at the delta apex, where the greatest rotation is required, so that the resulting breaker angle is greatest nearest the apex, and least near the base on either side of the delta. Because the wave rays converge towards the apex, wave energy is concentrated there, and wave height is greatest. Thus

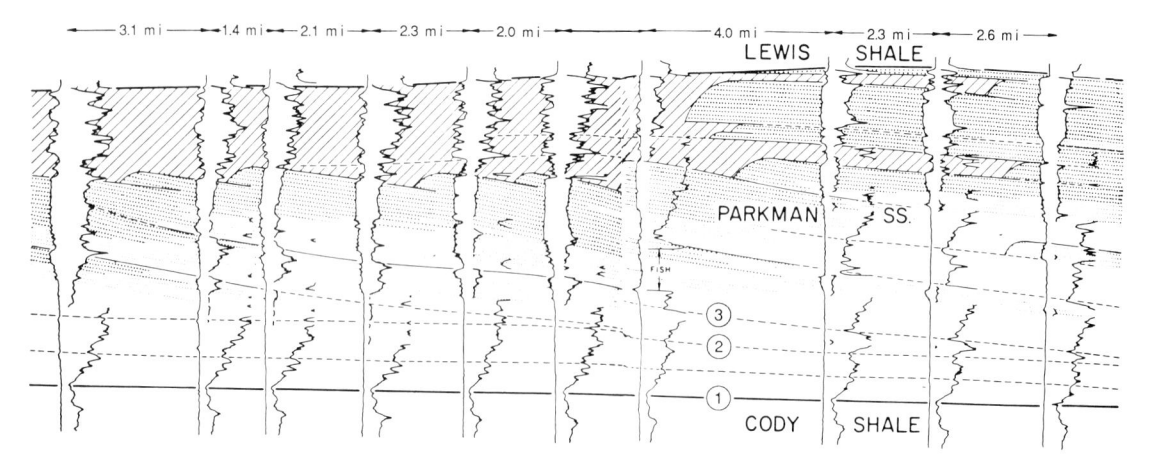

**Fig. 11.** The strandplain depositional system. Mesaverde sandstone in the Powder River Basin, Wyoming, as revealed by well logs. Diagonal hatching: fluvial sandstone and shale. Dotted lines are shoreface sandstone beds. Open is inner shelf shale. (From Asquith, 1970.)

each sector of the beach experiences a down-current decrease in driving force (determined by wave height and breaker angle), and therefore in its capacity for sediment transport. Sediment is deposited and the shoreface progrades.

Progradation on the flanks of cuspate deltas is not continuous, but episodic. The fundamental scale of periodicity is controlled by the weather and the passage of seasons, which control wave power, and by the hydrological cycle of the river, which controls sediment supply. In mid-latitudes, for instance, peak fluid and sediment discharge of the river tends to occur in the spring, and the delta mouth bar aggrades rapidly at this time (Wright & Coleman, 1977). Storm-generated wave trains, however, are most frequent and most intense in the autumn and winter. During storms, breaker height increases. The delta mouth bar and the flanking beaches are eroded. Sand is transported from the delta apex towards its flanks, and accumulates beneath the breaker as a break-point bar. Sand transport during this period is very intense in a narrow zone less than 100 m wide, in the longshore trough between the bar and the beach.

At peak flow, the storm-amplified, wave-driven littoral current may merge with the shelf flow field in the form of a downwelling coastal jet that may be several kilometres wide (Swift et al., 1985a,b; see discussion by Swift & Thorne, this volume, p. 10). During this period, much of the sand in transport in the surf zone may be swept seaward in storm-amplified rip currents, and may rain out into the downwelling coastal flow. Sand so transported is swept obliquely down the shoreface on to the inner shelf floor, where it is permanently lost to the littoral system. As wave heights decrease and the littoral current wanes, the sand remaining on the upper shoreface migrates landward over the source diastem as a break-point bar and eventually welds to the beach (Carter, 1986). Some of the segments added to the beach in this fashion contribute to the widening beach, resulting in a new dune line. The strandplain thus progrades by adding successive *beach ridges*. In three dimensions, the strandplain deposit grows by adding successive beach ridge prisms, consisting of a beach ridge and shoreface apron, separated from each other by erosional surfaces (Fig. 11). The surface of the strandplain is corrugated by these ridges. Beach ridges are typically organized into beach ridge sets, which reflect longer scale, climatic periodicity (Curray et al., 1969).

Deltas servicing elevated source terrains may have significant quantities of gravel sized material delivered to them, and become *fan deltas* whose strandplains consist of successive *storm beaches* (gravel ridges; Fraser & Suttner, 1986). In deltas whose river loads contain a higher proportion of mud, strandplains are replaced by *chenier plains*, in which the individual beach ridges rise from a muddy substrate (Gould & McFarlan, 1959; Augustinus, 1989). As an end member case, *mud-flat coasts* (Rhodes, 1982) form around sand-free river mouths.

*Proximal lithofacies of regressive shelves*

The stratification pattern of strandplain and regressive shoreface-shelf depositional systems is presented in Fig. 5A and accompanying text, and the lithofacies of prograding shorefaces has been described in the discussion of mouth bar flanking deposits. In strandplains, a distal *laminated shale lithofacies* is overlain by an *interstratified sandstone and shale lithofacies* and an *amalgamated shale lithofacies*, as on mouth bar flanks. In vertical successions, strandplain sandstones coarsen upwards through their several facies, from very fined-grained sandstone to fine or medium-grained sandstone. In general, grain size and primary reservoir properties reach a maximum in the upper beds, where, however, postdepositional calcite or silica cementation may significantly diminish reservoir quality (McCubbin, 1982)

*Distal lithofacies of regressive shelves: fine sediment dispersal*

Like sand, fine sediment tends to move along shelf in the direction of prevailing direction of transport, but the two dispersal systems are only loosely coupled. Much of the sand moves in the littoral drift, a wave-driven sand stream 100 m or less wide (Komar, 1983). However, mud moves in a 'nearshore advective mud stream' that may be 3 km or more wide (McCave, 1972, Oertel, 1976; Oertel & Dunstan, 1984). The term is misleading, because the mud is passively carried by a coastal band of brackish water that contains most of the fresh water released by upstream deltas. Prior to the construction of the Aswan dam, the Nile floods generated such a mud stream. Flow took the form of a coastal jet, which left the Damietta mouth of the Nile with velocities of over $100 \, \text{cm s}^{-1}$, flowed north along the Sinai coast, and retained velocities of $25 \, \text{cm s}^{-1}$ as far north as Mount Carmel, almost to Lebanon (Sharaf el Din, 1977). An advective mud stream also has been reported flowing north from the Amazon River mouth along the Brazilian and Guyana coasts (Rine & Ginsburg, 1985). Such inner shelf advective mud streams may in their proximal regions have suspensions dense enough to constitute fluid muds (Wells & Roberts, 1980).

The relationship between fine sediment transport on the inner shelf, and its dispersal across the shelf is best documented for the Mississippi delta and northwest Gulf of Mexico shelf (Fig. 12). During the spring flood, a metre or more of sediment accumulates off the active distributary mouths (Adams *et al.*, 1988). The sediment slides away, compacts and is buried, or is resuspended by wind driven currents during frontal passages of the succeeding winter, and is transported westward along the shelf. A 'nearshore advective mud stream' extends west from the Mississippi delta along the Louisiana and Texas Coasts as far as Brownsville (Sahl *et al.*, 1987). The flow is pressed against the coast by Coriolis veering (to the right in the Northern Hemisphere), and is impelled by the forces of buoyant spreading. It is separated from the main shelf-water mass by a tilted halocline or inner shelf front. The brackish, turbid coastal water overrides more saline shelf water along the halocline which is forced down towards the sea floor. Fine sediment and organic debris rain out of the nutrient-rich flow above the halocline into the oxygen deficient waters beneath (Rabalais, 1988). Muddy sediment escapes across the shelf from the coastal mud streams primarily during storms, when the frontal surface is disrupted by coastal storm currents (Holmes, 1982).

*Distal lithofacies of regressive shelves: constructional scarps*

The better-studied allochthonous shelves on modern continental margins are generally mud-floored (*laminated or bioturbated mud facies*), and fine regressive sands are limited to a shoreface and inner shelf lithofacies a few kilometres wide. However, the rock record shows that sheet-like, fine-grained regressive sandstones, extending for many tens of kilometres, may also occur. The explanation for these extended proximal deposits must be sought in the stage of shelf evolution, and also in the geometry of the shelf surface. As the depositional cycle passes into highstand, and the rate of relative sea-level rise decreases, the ratio between shelf aggradation and shelf progradation shifts towards progradation (Posamentier & Vail, 1988). The rate of bypassing increases, and is reflected in a higher reworking ratio (Thorne *et al.*, this volume, p. 64), and selective removal of fine sediment. At this late stage, the inner shelf sand lithofacies can expand across the shelf. In such regressive, fine-grained sheet sands, the distal *interstratified sandstone and shale facies* is dominant. A small volume of hummocky subfacies of the *amalgamated sandstone facies* also may occur.

Regressive shelf systems tend to develop con-

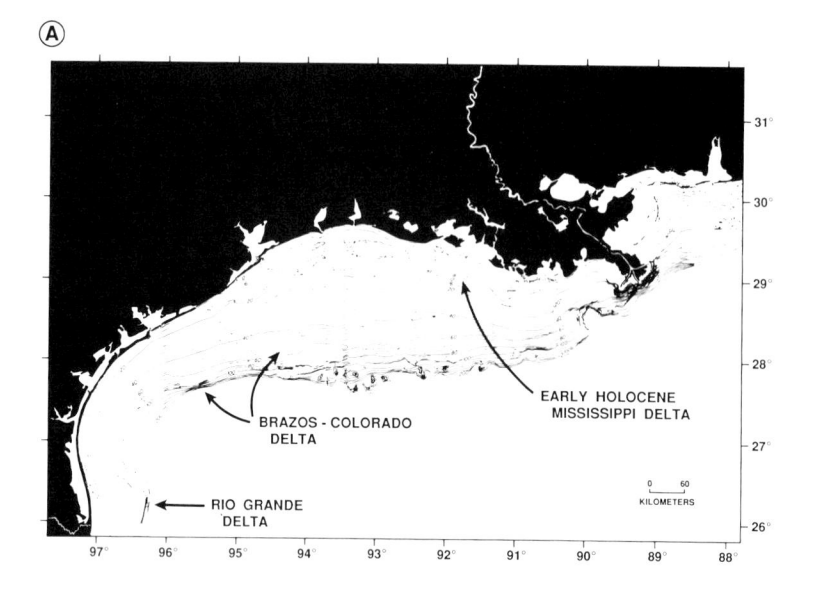

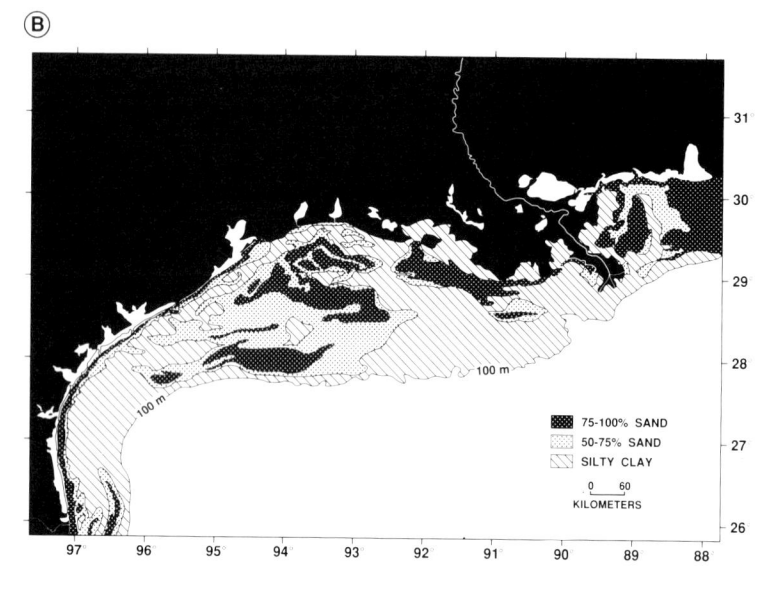

Fig. 12. Depositional framework of the northwest Gulf of Mexico shelf. (A) topography. (From Ballard & Uchupi, 1970.) (B) Sediment distribution. (From Frazier, 1974.)

structional scarps as they prograde across the shelf. This morphodynamical behaviour is paralleled by the growth of clinoform internal structure. Kuehl et al. (1982) have reported such behaviour from the mud sheet on the Amazon shelf. Clinoform prograding mud deposits may be capped by regressive fine-grained sheet sands. The stacked clinoform sequences of the lower Kuparuk Formation of the Alaskan North Slope are each capped by thin fine-grained sheet sands (Masterson & Paris, 1976). The sheet sands extend from the flat-lying (ondaform) part of the sequences into the inclined (clinoform) portions, clinoform structure and thicken there, indicating that they not only transported downshelf, but also were bypassed seaward on to the clinoform face. Constructional scarps also have been reported from the East China shelf (Demaster et al., 1985), the Bengal shelf (Nittrouer et al., 1986), and a constructional scarp also probably forms the south boundary of the Adriatic shelf (van Straaten, 1965a; see discussion by Thorne & Swift, this volume, pp. 218–219). The Northwest Gulf of Mexico may

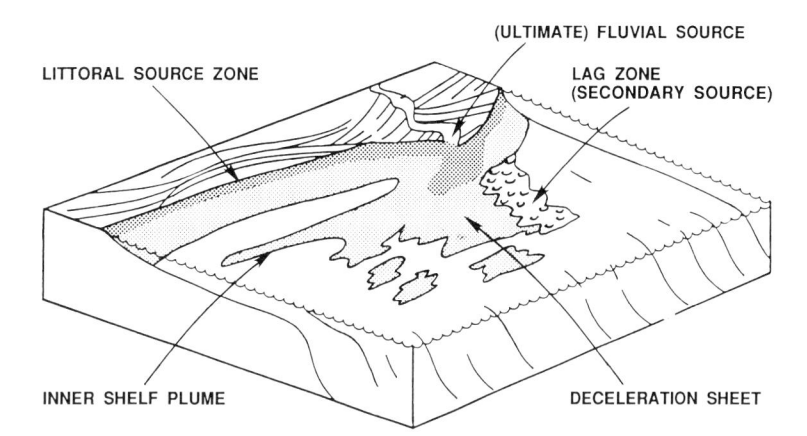

**Fig. 13.** Deceleration sheet on a regressive shelf. (From Phillips, 1987.)

exhibit an early stage of scarp formation; an oversteepened submarine slope exists just landward of the shelf edge in the vicinity of the Mississippi delta (see Fig. 12 and analysis of Louisiana shelf sedimentation by Thorne & Swift, this volume, pp.189–255).

*Distal lithofacies of regressive shelves: deceleration sheets*

However, deposition on regressive shelves is not a simple two-dimensional process. That is, the resulting lithofacies pattern is not constant in the alongshelf direction. Instead it tends to be controlled by the point-source nature of riverine sediment input, and the direction of prevailing flow. Shoaling of the shelf before and immediately down-current of prograding deltas results in acceleration of the alongshelf flow and down-shelf bypassing of sediment (see Fig. 17 of Swift & Thorne, this volume, p. 24). Further down-shelf, where the flow expands and decelerates, sediment is deposited as a succession of down-shelf fining facies. These deposits are in a sense plume-like, but are much larger in scale and are more nearly sheet-like in geometry (thickness to width ratios $>1000:1$) than the depositional systems described as prodelta shelf plumes in the preceding section. The inner margin of such lobate deposits may in fact be a prodelta plume.

Such regressive plume-like deposits that occupy much of the shelf are *deceleration sheets* (Fig. 13). Their plan-view geometry is determined by interaction of wave power and sediment concentration gradients on the sea-floor (McCave, 1972, fig. 94). In the nearshore zone, near-bottom sediment concentrations are high (nearshore advective mudstream). Immediately down-shelf from major deltas, the sediment accumulation rate $\dot{a}$ is high relative to the minimum resuspension depth $a'$. The resulting reworking ratio, $r = a'/\dot{a}$, may be so high that sand-free coasts (mudflat coasts; Rine & Ginsburg, 1985) may occur. Further down-shelf, suspended sediment concentration decreases, the sediment accumulation rate $\dot{a}$ decreases relative to $a'$ and the higher reworking ratio prevents inshore mud deposition. As a consequence the muddy lithofacies begins to separate from the shoreline, and an intervening proximal sandy lithofacies appears.

Further seaward, bottom wave power is reduced by the greater depth, so mud is deposited. In the vicinity of the parent delta, the plume may continue seaward over the shelf edge. Further down-shelf, however, the plume may have a distinct outer boundary. Wave power is reduced in this outer shelf zone, but at this greater distance from the source delta, the sediment supply is reduced even further. As a result, the older, transgressive lithofacies may be exposed on the outer shelf floor as a distal condensed section (Fig. 7A).

A muddy bottom plume carpets the shelf floor to the west of the Mississippi delta, constituting a time-averaged record of westward and offshore fine sediment flux. It extends from the delta along the length of the Texas shelf as far as Brownsville, near the Mexican border, a distance of some 900 km (Fig. 12B). Other Texas rivers, notably the Brazos and the Trinity, make contributions. Such local contributions are apparent in the sand and silt fractions (Mazzullo & Withers, 1984; Mazzullo & Crisp, 1988), but the distinctive clay mineral suite of the Mississippi river sediment load is detectable to the end of the plume (Holmes, 1982). The plume is not continuous, but contains 'windows', similar to those on the Narayit shelf plume (see Swift & Thorne, this volume, p. 26, Fig. 19). The windows expose the lowstand Brazos–Colorado and Rio Grande deltas and an early Holocene phase of the Mississippi delta. The deltas have little relief, but enough to increase the reworking ratio and create a condensed facies. The plume thickens to over 15 m in the recess between the two southern deltaic bulges (Berryhill, 1986).

The plume can be considered as the product of a regressive regime only as far west as the Texas–Louisiana border. In the near-delta, regressive portion of the plume, mud beds downlap on to the outer shelf where the transgressive surface is still exposed (Suter & Berryhill, 1985). Beyond the Texas–Louisiana border, however, the regressive chenier plain coast changes to a barrier island coast, and a proximal (sandy) inner shelf lithofacies appears. At this point, the muddy beds of the plume presumably begin to onlap the inner shelf sand lithofacies (begin to exhibit the behaviour of a transgressive regime). On the central Texas shelf between the ancestral Brazos–Colorado and Rio Grande delta, the stratal geometry of the outer shelf has not been resolved. However, if this sector is indeed transgressive, as its barrier coast suggests, then the stratal geometry would be that of a *paratruncation*, where the time surfaces (which onlap landward) converge asymptotically seaward with the depositional surface (Vail *et al.*, 1977). Such a seaward-aging surface is an evolving *flooding surface* (Posamentier *et al.*, 1988; see Fig. 15A, Thorne & Swift, this volume, p. 216).

Deceleration sheets appear to be common in the rock record. Fine-grained sand sheets in the Dakota Sandstone of New Mexico (Two Wells and Paguate Members; Noon, 1980) appear to bear a plume-like relationship to the Four Corners delta and are probably deceleration sheets.

Sedimentation rates are high on allochthonous shelves, and the accumulating sediment is fine grained and water-rich (Holmes, 1982). Under such conditions listric faulting is often widespread, resulting in active fault scarps on the shelf surface (Coleman & Prior, 1982). Shelf sand sheets tend to thicken on the down-shelf side of such scarps.

# TRANSGRESSIVE DEPOSITIONAL SYSTEMS, I: BARRIER–LAGOON–ESTUARY SYSTEMS

## General

The transgressive family of shallow marine dispersal is created by transgressive accommodation-dominated regimes. Accommodation-dominated regimes have been characterized by Swift & Thorne (this volume, p. 17) as occurring whenever sediment is entering the shelf in lesser quantities than can be accommodated by compaction and sea-level rise, or can be swept away by tidal or storm currents. These relationships can be described by a configuration of the regime ratio $R \cdot D/Q \cdot M$ in which the accommodation terms (sea-level change, $R$, and transport intensity, $D$) are more important than the supply terms (sediment input, $Q$, grain size, $M$). In such a setting, shorelines retrograde, and river mouths become estuaries. Shoreface bypassing by means of erosional shoreface retreat is the dominant mode of sediment supply, and can be described by an $R$-dependent model of coastal profile translation (see Thorne & Swift, this volume, pp. 37–39). The debris of erosional shoreface retreat is incorporated into the leading edge of an onlapping shelf sand sheet. Under suitable conditions, the sand sheet develops sand ridges. If the transgression is sufficiently 'leaky', the sand sheet may in turn be onlapped by a transgressive mud deposit.

## Classes of transgressive coast

### General

The morphodynamics of transgressive shorelines are controlled by the wave climate, the tide range, and the gradient and relief of the surface being transgressed (physiographic maturity). The formation of barrier–lagoon–estuary systems is a common response to transgression on shelves flanked by coastal plains incised by their drainage systems during Pleistocene sea-level falls. Variations in pattern are caused by variations in the tidal range and in the wave climate.

### Microtidal transgressive coasts

On microtidal coasts, barrier systems are long and continuous. Tidal inlets are characterized primarily by flood tidal deltas that build from the tidal inlet into the lagoon behind the inlet, using sand extracted from the littoral drift system. Microtidal and mesotidal inlets tend not to build large ebb tidal deltas on the shelf before the inlet, because the oceanic wave regime tends to destroy such features faster than they can be built.

Estuaries on microtidal coasts tend to have dendritic outlines as a consequence of having flooded faster than they can be filled by sediment (e.g. Gulf coast of USA; Bernard & LeBlanc, 1965). As the tide range increases, so does the ease of sediment transport (regime variable $D$; Swift & Thorne, this volume, p. 13), and the lower reaches take on a trumpet-shaped pattern, with flaring mouths and straight stems, while the upper reaches may meander. The pattern is characteristic of equilibrium estuaries, whose channels have closed down to the equilibrium configuration. In such a configuration, the cross-sectional area decreases upstream as a function of the decreasing intertidal volume serviced through the cross-section (O'Brien, 1969; e.g. Georgia Bight of the USA Atlantic coast; Oertel, 1988).

*Mesotidal and macrotidal transgressive coasts*

On mesotidal coasts with moderate to mild wave climates, inlet width varies inversely with the tide range. In mesotidal inlets, the rate of sand transport is so high that large ebb tidal deltas can be maintained despite wave activity (Hayes, 1979). Inlet width increases and barrier length decreases as tide range increases, until the coast passes into the macrotidal category. At this point the barriers vanish, and the shoreline is dominated by shore-normal tidal channels and their ebb tidal deltas. The channels may be either 'blind' channels or true estuaries connected to rivers (German Bight of the North Sea; Ehlers, 1988). However, high mesotidal and macrotidal coasts with rigorous wave climates cannot store their sand in barriers or in ebb deltas. Instead, sand storage is subtidal, in tidal sand ridge complexes, equivalent to the storm-built sand ridges of tide-built coasts, which continue to evolve on the shelf floor (Anglian coast of the North Sea; Swift, 1975a).

*Rocky transgressive coasts*

Where the modern transgression overlaps a lithified basement of significant relief, the submerging surface is less easily remodelled by the advancing sea and sediment is in relatively limited supply. The shoreface cannot maintain continuity, and breaks up into a complex series of pocket beaches, tombolos, and other rock-defended sand bodies. Sand body geometry attains maximum complexity on such coasts (Zenkovich, 1967). The ratio of freshwater input to contained salt-water in the intra-coastal water bodies of such coasts is frequently so low, and their shapes are so irregular, that they are commonly called bays rather than estuaries or lagoons.

## Morphodynamics of barrier–lagoon–estuary systems: the microtidal case

*Definitions*

Barrier–lagoon–estuary depositional systems are characteristic of transgressive coasts. The term 'barrier' is used here in a restrictive sense for a coastal landform with both lagoonal and oceanic shorelines, both of which (at time-scales of several thousand years) are moving landward. 'Regressive' barrier systems are, by this definition a null class, since in this case, the lagoonal shoreline is fixed, while the oceanic shoreline is moving seaward. Such 'barriers' are no longer barriers as defined, but are turning into strandplains as the transgression turns into regression.

*Genesis of barrier–lagoon–estuary coasts*

The origin of barrier coasts has been debated for many years (summary in Schwartz, 1971). The basic causes of barrier formation now seem reasonably clear. (Swift, 1975b; Niedoroda *et al.*, 1985). Barriers originate primarily by a process of mainland beach detachment (Hoyt, 1967). The process occurs because shoreline retreat during a transgression is not simply a matter of passive drowning. Rather, the shoreface undergoes erosional retreat so that the land surface is bevelled. The debris of erosional shoreface retreat may be transferred in part to the

**Table 2.** Examples of transgressive shoreline classes

|  | Mild wave climate | Rigorous wave climate |
| --- | --- | --- |
| Low-relief microtidal | Middle Atlantic Bight | USA Atlantic coast (Swift, 1975b) |
| High meso to macrotidal | USA Georgia coast (Howard & Reineck, 1972) German Bight, North Sea (Ehlers, 1988) | Anglian coast, UK (Robinson, 1980) |
| High relief | New England coast (Shepard & Wanless, 1971) | Norwegian coast (Holtedahl & Bjerkli, 1975) |

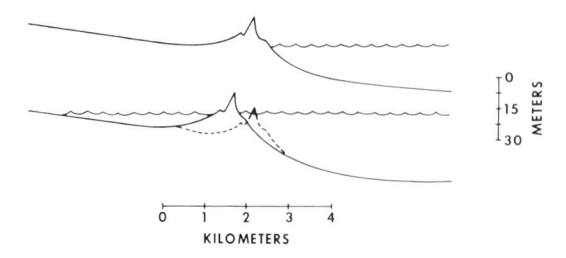

**Fig. 14.** Mainland beach detachment. A mainland beach is nourished by littoral drift and keeps pace with sea-level rise. Swale behind beach is not nourished, and floods. During sea-level rise, storm washovers drive landward barrier migration. (Modified from Hoyt, 1967.)

beach and dune (Swift, 1968; Niedoroda *et al.*, 1985; Fig. 7 of Swift & Thorne, this volume, pp. 16–17). The beach and dune can thus grow upwards as sea-level rises, and can respond to patterns of wave refraction by along-shore sediment transfer and coastal straightening (Swift, 1975b). The swale behind the dune, however, is sediment starved and tends to flood. As a result the mainland beach tends to detach (Hoyt, 1967), so that the transgression proceeds with two shorelines, a relatively straight oceanic shoreline of barrier islands and spits, and an irregular intra-coastal shoreline of lagoons and estuaries. In the detachment process, back-barrier flooding starts where a beach abuts an estuary. The beach is progressively detached from the mainland, as a lagoon creeps in behind it (Fig. 14), turning the mainland beach into a barrier system.

*Barrier migration*

Barriers do not remain in place, but migrate landward as sea-level rises. Migration is accomplished in part by a tank-tread movement of sand described as *barrier roll-over* (Leatherman, 1983). Sand eroded from the shoreface moves on to the beach under the impetus of landward-asymmetric fair-weather wave surge (Short, 1984); on dissipative beaches, the sand may migrate as a bar that eventually welds to the berm (Wright *et al.*, 1979). Sand is blown from the beach into dunes, and across them on to the back-barrier environment. Storms also transport sand landward as washover fans. Much of the washed over sand remains on the barrier surface, and relatively minor amounts reach the bay shore, so that the net effect is to build the barrier upwards (Leatherman, 1983).

Leatherman (1983) has pointed out that the cycle has an important horizontal component of movement, in which sand moves along the surf zone of the barrier and into an inlet, where it is stored in an ebb tidal delta. Inlets may migrate down-drift, and new inlets form as others close. Eventually, much of the back barrier area consists of coalescing ebb tidal deltas of varying ages, which extend the barrier landward. Movement of the barrier's oceanic shoreline tends to be discontinuous over time-scales of several thousand years. When it eventually shifts, dune and washover processes shift landward with it, so that the barrier 'rolls over' on to the back barrier inlet platform. The cycle of platform building and inlet formation then begins anew (Fig. 15). Barrier roll-over is a probabilistic process in that two complementary subprocesses of back-barrier platform construction and shoreline shift are not coupled closely, but instead proceed independently.

*Barrier retreat modes*

Barriers on the Middle Atlantic shelf of North America appear to have undergone discontinuous retreat by a process of intermittent roll-over at time-scales of several thousand years or less (*barrier roll-over*, Fig. 15). The process is spatially continous, but the migration rate has varied. At larger time-scales, however, the retreat process also appears temporally continuous (*continuous barrier retreat*, Fig. 16). The evidence for this conclusion comes from the shelf in front of the modern barrier chain (Niedoroda *et al.*, 1985). Vibracores show that at most places, the barrier is missing in the stratigraphic sequence, its place being taken by an erosional surface cut by the shoreface retreat process, the *ravinement surface* (Swift, 1968, after Stamp, 1921; Niedoroda *et al.*, 1985).

Until recently, there has been some doubt concerning the continuity of the retreat process. The modern barriers at the leading edge of the ravinement surface contain wood fragments and molluscan fossils (Fischer, 1961) that are locally as old as several thousand years, and for this reason, the barriers have been said to have formed within the last several thousand years.

We know that the modern Atlantic shoreline has undergone considerable change in the last several thousand years (e.g. Leatherman & Joneja, 1980; Swift & Moslow, 1982). The change may be the result, in part, of the major reduction in the rate of sea-level rise 4000–7000 years ago (Dillon & Oldale,

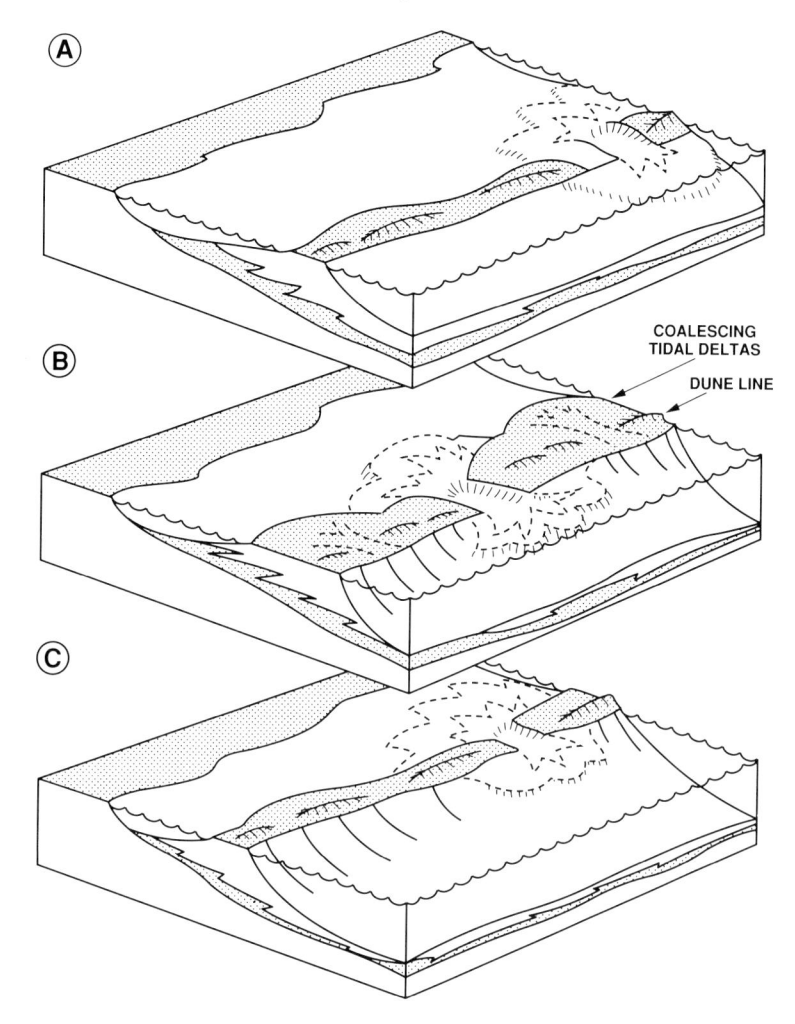

**Fig. 15.** Barrier roll-over: (A) initial narrow barrier; (B) construction of tidal delta platform; (C) barrier roll-over shoreface retreat. Repeated breaching of barrier by ephemeral inlets builds back-barrier platform. Eventually a major event erodes back the shoreface and dumps much of the resulting sand on the barrier top and back-barrier platform. (Based on Leatherman, 1985.)

1978). However, the evidence indicates modification of pre-existing migrating barriers, not barrier initiation. The marine evidence indicates that the barriers were not 'formed' in place, but are descended instead from predecessors that lay immediately seaward. More precisely, simultaneous barrier-front destruction and back-barrier formation occurred continuously as the Middle Atlantic barrier chain swept across the subsiding coastal plain during the course of the Holocene transgression (Niedoroda et al., 1985). Individual barrier sections may have detached, migrated, and vanished as part of the process of inlet formation and closure. There appears to be at least one, and probably several, periods when the whole process came to a halt, so that for a while, strandplains grew seaward; see discussion of mid-Holocene regression on the Atlantic shelf by Swift et al. (this volume, pp. 161–164). At timescales of several thousand years, however, the chain as whole moved continuously landward most of the time.

Evidence is beginning to accumulate to show that, on the Atlantic shelf, the continuous barrier retreat mode is a general category that includes several distinct types. For instance, the New Jersey, Delmarva (Delaware–Maryland–Virginia), and Virginia–North Carolina coastal compartments constitute a repeating coastal pattern, in which a mainland beach at the northeast end is followed by an open lagoon sector, followed by a marsh-filled lagoon sector at the southwestern end (Fisher, 1968). The drift-divergence nodes in the respective littoral drift cells tend to lie on the barrier spit that extends south from each headland. In other words, the mean

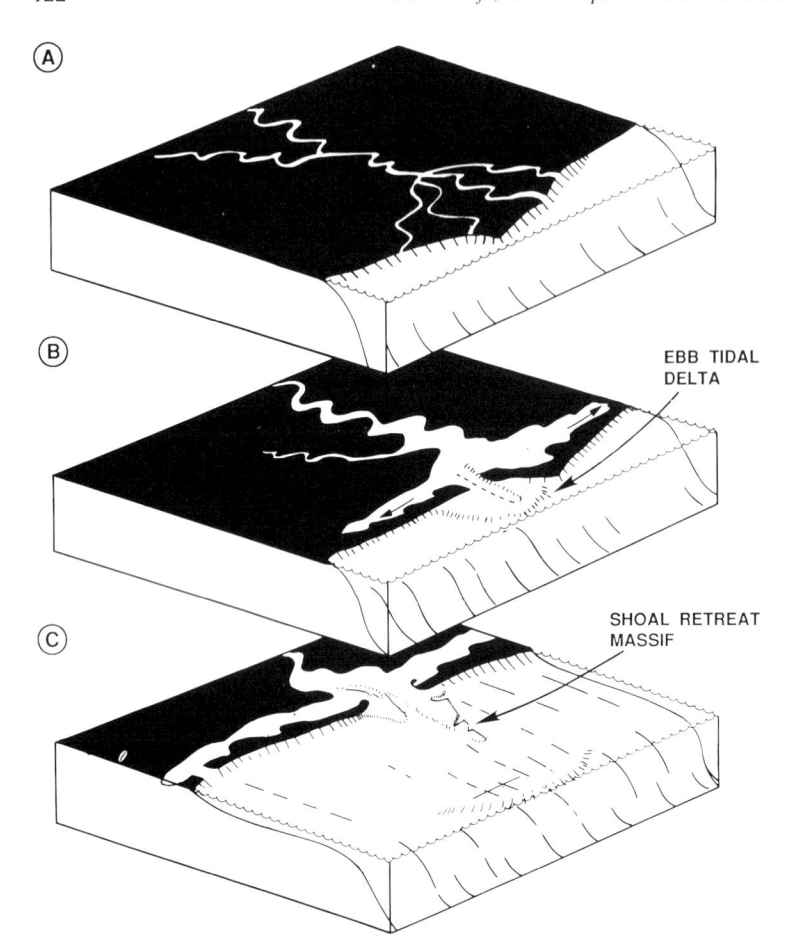

**Fig. 16.** Continuous barrier retreat. (A) A shelf-edge delta has formed at lowstand. (B) Transgression has begun. The delta plain has submerged, and its beach has detached to form a cuspate barrier. Arrows indicate lateral extension of the lagoons as submergence continues. (C) Continued transgression and retreat. The barrier system has retreated back across the shelf. The retreat path of the estuary mouth is marked by a shoal retreat massif. Compare with Fig. 19.

annual direction of littoral drift is *from* the node on the spit, *to* the eroding headland. These open-lagoon barriers are therefore probably sourced primarily by inlet throat erosion rather than headland erosion (Swift, 1975b).

Oertel (1987) has pointed out that the open-lagoon barriers and the marsh-lagoon barriers in these coastal compartments must have rather different retreat modes (Fig. 17). The marsh-filled lagoons have developed this configuration because their retreat has brought them up against the subaerial shorefaces of fossil Pleistocene strandlines. Because of the steeper gradient, the lagoons are narrow enough to fill with marsh. Because they fill with marsh, they develop well-organized systems of drainage channels, whose cyclic tidal discharge cuts deep (10–30 m) inlets. Because the inlets are so deep, and because their service channels extend back to the subaerial stream channels, they tend to lock on to the marsh-buried subaerial drainage system. Thus the *marsh-lagoon mode* of barrier retreat is controlled by antecedent topography (Fig. 17A).

In contrast, the inlets of the open lagoon barriers are shallow and unstable. They form more or less at random, migrate laterally, and close after a few centuries. Since the ebb tidal effluent is not preconstricted by a deep lagoonal channel, the inlet does not become deeply incised. In the *open lagoon mode* (Fig. 17B), the shelf before the barrier is peppered with short inlet-fill trenches that show much less relationship to the antecedent topography (Field, 1980, fig. 4).

In addition to the modes of continuous retreat described above, discontinuous modes exist. On the Atlantic shelf, evidence for interruptions of con-

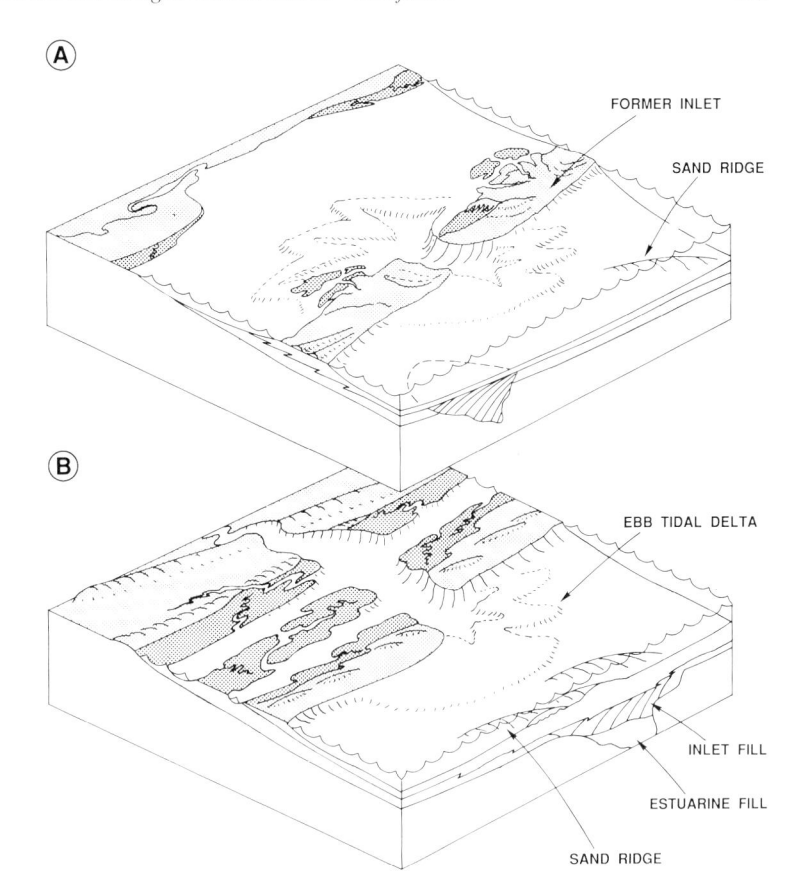

**Fig. 17.** Classes of continuous barrier retreat. (A) Marsh-lagoon mode of barrier retreat. (B) Open lagoon mode of barrier retreat. (Based on Oertel, 1987.)

tinuous retreat exist in the form of shelf scarps. They represent fossil lower shorefaces, deposited when the rate of sea-level rise slowed or reversed, and progradation briefly interrupted barrier retreat. When retrogradation resumed, ravinement resumed, and the top portion of the shoreface was clipped off (Fig. 18B). The stratigraphic consequences of this process are described by Swift *et al.* (this volume, pp. 161–164).

*Cyclic barrier retreat*, is a discontinuous mode described by Penland *et al.* (1985) from the Mississippi delta. These authors show that as a subdelta is abandoned, and its mouth bars are eroded, it becomes flanked by strandplains, which as transgression proceeds become retreating spits. Eventually, the mainland beach detaches, creating a barrier lagoon system. In the mild wave climate of the Gulf of Mexico, the barriers have shallow shoreface bases (*c*.5 m) and retreat slowly under wave attack. However, the retreat of the lagoonal shoreline is a passive flooding process, and, because of the high compaction and subsidence rates in the delta region, it is a rapid one. As a consequence, wide lagoons open up, which remain shallow owing to the extremely high input of suspended sediment. As the lagoonal mud deposit thickens, the retrograding inlets become less able to cut through it to mine sand from the substrate. Because of this, the retreating barrier system cannot replenish itself easily. The retreating barriers steadily lose sand to a shelf sand sheet that forms behind them as they retreat. They eventually starve and undergo *barrier overstep*. The overstepped barriers may exist for some time as submarine shoals before becoming buried by shelf mud (Figs. 18A, 19). In this cyclic retreat mode the oceanic shoreline retreats more slowly than the mainland shoreline, but for different reasons than in the case of open lagoon retreat mode.

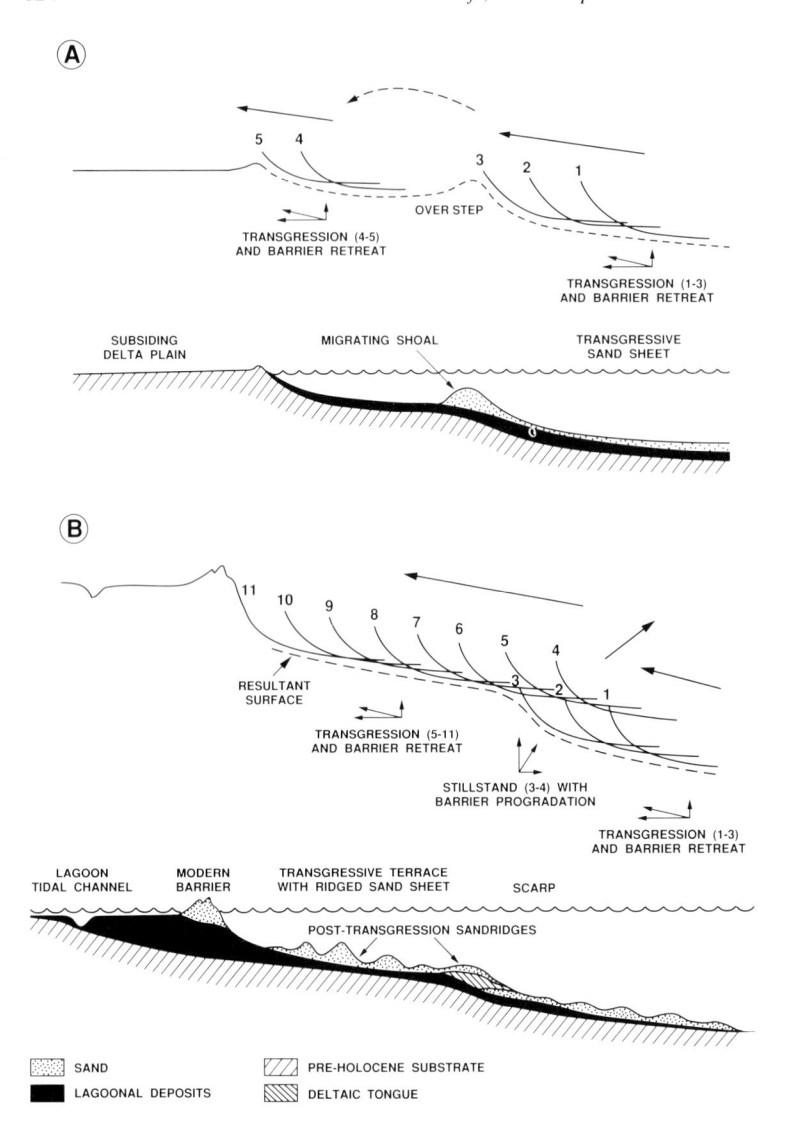

Fig. 18. Barrier overstep (A) versus barrier step-up (B).

## Morphodynamics of barrier–lagoon–estuary systems: the mesotidal case

Our understanding of the coastal morphodynamics in transgressive, high-tidal settings comes mainly from studies of the Georgia Bight of the USA southern Atlantic shelf (Hails & Hoyt, 1968; Howard & Reineck, 1972; Oertel, 1977; Howard & Frey, 1985), and of the littoral setting of the North Sea (van Straaten, 1965b; Reineck & Singh, 1980; Robinson, 1980; Ehlers, 1988). In particular, the Georgia Bight of the USA southern Atlantic shelf, and the German Bight of the North Sea, exhibit similar coastal configurations. Hayes (1979) has noted that the barrier islands in these mesotidal settings tend to be short and 'drumstick-shaped', with wide tidal inlets. Inlet width increases and inlet spacing decreases towards the centre of each bight. These geomorphological trends appear to be a response to the increase in tidal range toward the head of each bight, as the tidal wave is amplified by the decreasing cross-sectional area. In the German Bight, barrier islands shrink and nearly vanish at the bight apex, where the coastal morphology consists of large, side-by-side inlets. This effect is enhanced by the presence of several large rivers (Elbe, Weser,

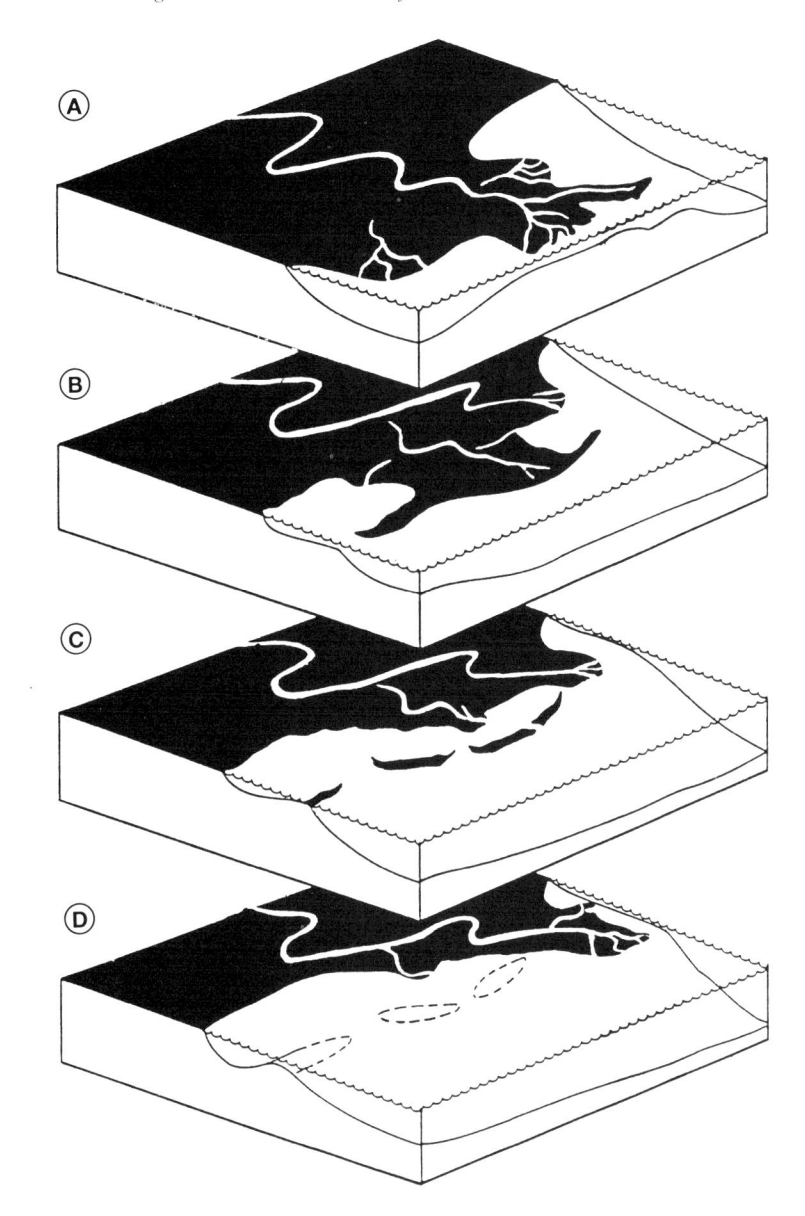

**Fig. 19.** Cyclic barrier retreat.
(Based on Penland et al., 1985.)

Eider), whose mouths lie among 'blind inlets' not backed by rivers (Ehlers, 1988).

### Sediment dispersal on transgressive coasts

*Sand dispersal on a migrating barrier*

Migrating barriers obtain their sand from the substrate over which they migrate (Fig. 20). This substrate comes intermittently to the surface in the form of headlands (mainland beaches), to which the barriers attach as spits. The headland–barrier system can be viewed as an attempt by oceanic processes to maintain a continuous shoreface. The shoreface is incised into the eroding headland (erosional shoreface), and the resulting debris is transported down-drift, to build a barrier spit with a complementary, constructional shoreface (Swift, 1975b). The shorefaces of eroding headlands are thus a primary sediment source. As they retreat they release sand deposited during a previous sedimentary cycle.

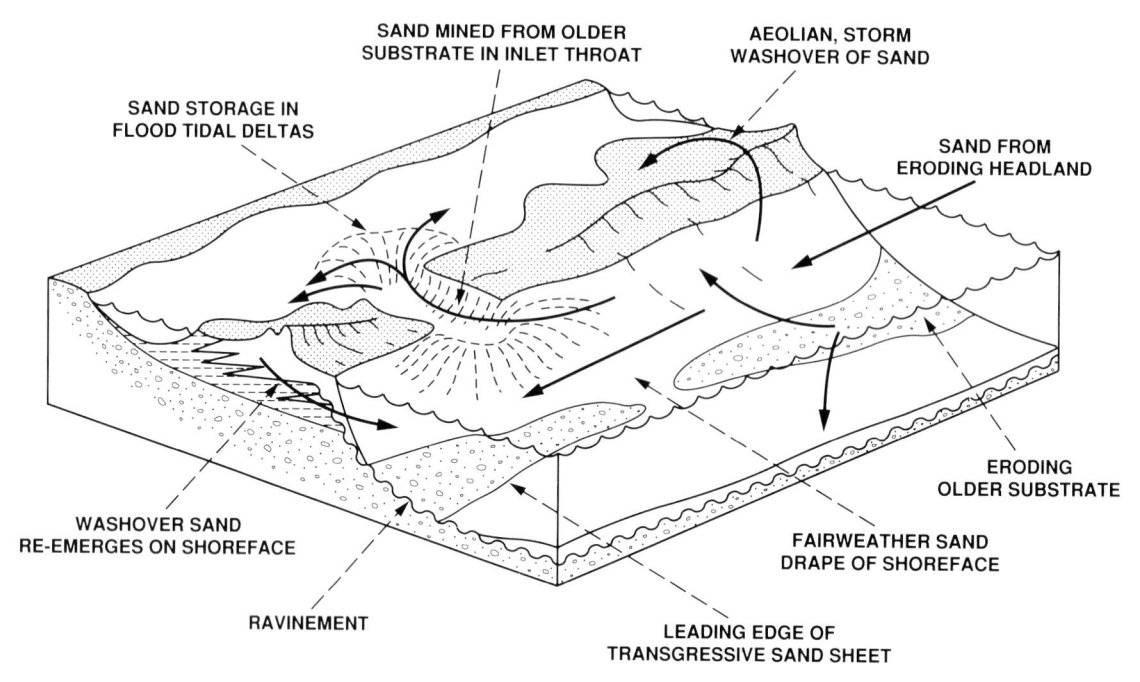

**Fig. 20.** Sand budget of a migrating barrier. See text for explanation.

In contrast, barrier shorefaces are secondary sources. Their erosional retreat mainly recycles washover or inlet sand originally deposited at the back of the barrier. Barriers form over buried valleys. The lower portion of the barrier shoreface may erode into lagoonal muds that have been over-ridden in the retreat process, but it generally cannot reach the antecedent substrate. Tidal inlets, however, commonly *are* primary sources of sand. They may scour 15 or 30 m down, make contact with antecedent substrate, and release large amounts of sand from the previous depositional cycle.

Updrift and offshore contributions to barriers are balanced by washover, offshore and down-drift losses. The along-shore component of the budget is controlled by littoral drift, which may exceed $10^6 \, m^3 \, yr^{-1}$ (Komar, 1983). Headlands retreat more slowly than barriers, and tend to form salients, so that wave refraction causes a gradient of decreasing wave power down the flanking barriers, much as in the case of strandplains (Fig. 10). The littoral current loses fluid power and the capacity to transport sediment in a down-drift direction, and sand is deposited (May & Tanner, 1973). The fraction of sand thus extracted from the littoral drift is typically 1–10% of the net annual along-shore flux (Komar, 1983).

In the onshore–offshore component of the budget, sand losses to a barrier segment occur at varying time-scales. Washover sand is lost temporarily, but at longer time-scales must be considered as banked, since it re-emerges at the shoreface. Sand is moved off the beach into storage on the break-point bar during every storm, and tends to return during fair-weather periods, carrying contributions from the along-shore flux with it. However, during peak storm flow, sand may be swept down the shoreface to be incorporated into the leading edge of the transgressive sand sheet (Swift *et al.*, 1985), where it is permanently lost from the littoral system. The loss rate is approximately 1–10% of the rate of recycling of sand in the beach cycle; of the same order as contribution from the along-shore drift (Wright *et al.*, 1979).

Thus a migrating barrier is supplied with sediment from relatively short-lived point sources (headlands, inlets). When sediment from these sources is in abundant supply, the retreat rate is slowed. In extreme cases, when an abandoned delta distributary mouth is being transgressed, the supply of sand may be so great that retreat may be reversed, so that flanking strandplains may grow forward as the mouth bar erodes back (Caminada strandplain, LaFourche subdelta, Mississippi delta; Penland *et* al., 1981; Penland & Boyd, 1982; Penland & Suter, 1989). However, when local sand sources are not available, the barrier must rely on its banked sand supply to balance offshore losses. As it withdraws this capital, the barrier becomes lower and thinner. At some point, the barrier is thin enough for storm washovers to readily cross to the landward side of the barrier, and the retreat process becomes overwash-dominated (Pilkey & Davis, 1987). The overwash frequency rate, and therefore the barrier migration rate, increase abruptly. They remain high until a new sand supply is reached. If it is not reached in time, the stored sand of the barrier is exhausted, and the barrier is overstepped.

*Fine-sediment dispersal in lagoons and estuaries*

Fine-sediment dispersal in the back barrier environment is coupled only loosely to the coastal sand budget. On transgressive coasts, the immediate source of fine sediment is the nearshore turbid zone on the adjacent inner shelf (McCave, 1972; Swift *et al.*, 1985a, b) where it is kept in suspension by wave stirring, and is trapped against the coast by baroclinic coastal currents and by the weak estuarine circulation characteristic of the inner shelf (Sahl *et al.*, 1987). Local rivers also contribute fine sediment, but within the back-barrier region mud deposits unusually contain open marine microfossils, suggesting that most of the material has been derived recently from the shelf (Nichols, 1989). Within the back-barrier environment, the primary fine sediment sinks are the fringing marshes (Frey & Basan, 1985). The ultimate source of this material is the ebb tidal effluent of upstream estuaries and inlets, and shoreface erosion of older Holocene lagoonal deposits overrun by the barrier (Swift *et al.*, 1985b). On the Atlantic continental shelf, the annual discharge of fine sediment by rivers is approximately equal to the volume deposited on back-barrier marshes (Meade, 1972). Much of this material is exhumed on the shoreface by the ravinement process and returned to the back-barrier environment through tidal inlets.

The intra-coastal system of lagoonal and estuarine water bodies is thus serviced by the nearshore turbid zone through tidal inlets and estuary mouths. It is able to build up suspended sediment concentrations several orders of magnitude higher than those of its parent water mass through a series of fluid dynamic trapping mechanisms (settling lag, time velocity asymmetry of the tidal cycle; Postma, 1967). The turbid intra-coastal water mass, in turn, exchanges sediment with the floors of lagoons and estuaries, and with their fringing mud flats and marshes during every tidal cycle. Owing to their high suspended sediment concentrations, lagoons and estuaries tend to shoal until their depth can be maintained by wave and tidal scour. These current-graded surfaces rise with rising sea-level until they are overrun by the advancing barrier system (Nichols, 1989).

Estuaries are characterized by zones of maximum stratification where fresh river water overrides denser salt-water. Salt-water is mixed by the tide upwards into the seaward flowing fresh water, and more salt-water must flow in along the estuary floor to replace it (estuarine circulation; Pritchard, 1967). Suspended river sediment settling into the saline underflow is transported back up-estuary, where it rises into the seaward flowing upper layer and flows seaward once more. This zone of sediment recirculation therefore becomes a *turbidity maximum* (Postma, 1967). As the turbidity maximum migrates up-estuary during the course of the transgression, so does the muddy shoal associated with it, and the fluvial sand deposits of the estuary head delta are buried.

## Lithofacies of barrier–lagoon–estuary depositional systems

### General

Back barrier dispersal systems consist of a series of interacting subsystems of great compexity (washover fan, ebb tidal delta, flood tidal delta, estuary head delta, tidal channel systems). Each deposits its own lithofacies assemblage arranged systematically below or above the source diastem (see preceding discussion of depositional systems architecture). The distal facies of each assemblage interfingers with distal lithofacies of others. Forcing mechanisms are complex; the beds are essentially inundites (flood beds) in the washover fans, tidalites (tidal beds) on ebb tidal deltas and in the tidal channels, and may be tempestites (storm beds) on the lagoonal interfluves (genetic rock names from Ager, 1974; Seilacher, 1982).

### Barrier–lagoon–estuary depositional systems

Three dispersal systems occur on barrier islands; the beach, dune–aeolian-flat, and washover fan systems. They are lumped as system 1 in Fig. 21B. The beach

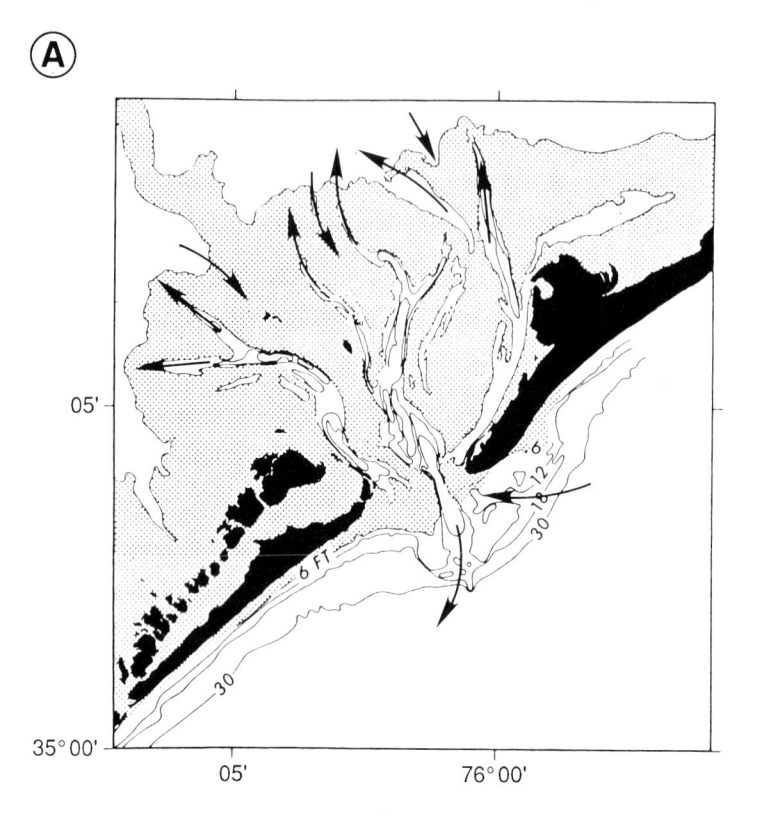

**Fig. 21.** Back-barrier dispersal systems. (A) Physiography of a tidal inlet; Okracoke inlet, North Carolina coast. Ebb-flood channel pairs are marked with arrows.

and washover fan systems share the eroding upper shoreface as a source environment, and all three systems are conjugate with the transgressive shelf system (Figs. 21, 22; see pp. 102–103 for terminology). The equivalent depositional systems, to the extent that they are preserved, are capped by the ravinement diastem.

Beach systems are built by swash aggradation during periods of fair-weather swell (Davis, 1985). Beach depositional systems are mud-free, and build a distal lithofacies of backshore beds. The dune–aeolian-flat dispersal system bears a serial relationship to the beach system. These systems are nourished by sand carried along the surf zone of the barrier face by the wave-driven littoral current. Transport occurs mainly during storms, alternating with periods of storage in the bar and beach prism.

Some of the same storms drive sand over the barrier crest and into the lagoon, through washover chutes to form the washover fan system (Leatherman, 1979; Hennessy & Zarillo, 1987). The erosion surface that underlies the break-point bar and beach prism extends into the dunes as the floors of washover chutes (Boothroyd et al., 1985), and constitutes the leading edge of the source diastem (ravinement diastem).

The proximal deposits of the beach, dune–aeolian-flat and washover fan systems are largely destroyed by the ravinement process. Except under exceptional circumstances, beach and dune deposits are not preserved, but the cross-stratified proximal *amalgamated lithofacies* of the washover deposits may be (Boothroyd, 1985). As amalgamated washover beds pass into a more distal *interstratified facies*, mud beds containing *Spartina* rhizomes may appear (Oertel et al., 1989), or in the tropics, the root casts of mangroves (Allen, 1964). Because of the tide-dominated nature of the depositional environment and the limited wave activity, *Scolithos*-like vertical burrows tend to be abundant (Frey & Basan, 1985) in intertidal and subtidal sand beds, while the more distal lithofacies may be heavily bioturbated (*bioturbated mud facies*) rather than laminated. In cross-section, washover fan deposits coarsen upwards from distal to proximal facies, and are capped by the ravinement unconformity.

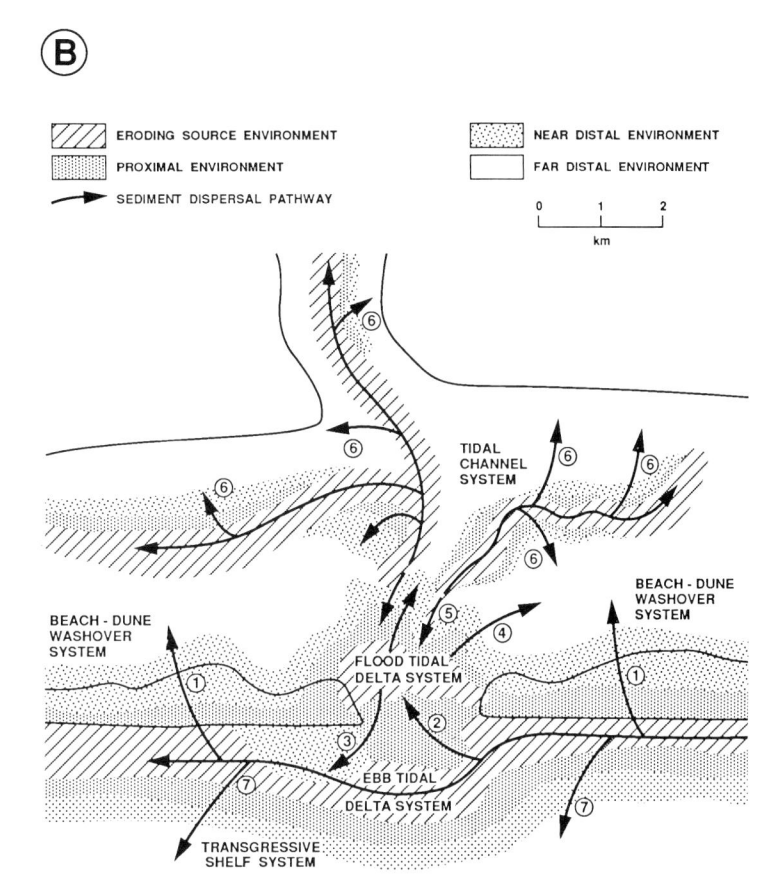

**Fig. 21.** (B) Idealized back-barrier dispersal systems. Arrows define sediment dispersal paths that begin in eroding source environments and end in distal depositional environments. 1 = beach-dune-washover system; 2 = seaward face, ebb tidal delta system; 3 = landward face, ebb tidal delta system; 4 = seaward face, flood tidal delta system; 5 = landward face, flood tidal delta system; 6 = tidal channel system; 7 = transgressive shelf system.

*Ebb-delta–flood-delta–tidal-channel depositional systems*

At sections through tidal inlets, a complex sequence of depositional systems may be discerned (Ward & Ashley, 1989; Figs 21, 22). Terms used in the following discussion of back-barrier depositional systems is based on the previous analysis of depositional systems architecture; see section entitled 'classes of contact between depositional systems'.

Ebb tidal deltas form on the seaward side of inlets. On the updrift side of the ebb tidal delta, flood-dominated tidal channels extend up the face of the inlet and penetrate the delta-top platform, where they alternate with ebb-dominated channels from the inlet throat (Oertel, 1988; Figs 21, 22).

Ebb tidal deltas are the product of the interfingering of two dispersal systems, a seaward-side system and an inlet-side system (systems 2 and 3, Fig. 21B). On the seaward face, the ravinement surface of the upper shoreface extends up into the floors of the flood-dominated tidal channels that penetrate the delta-top platform (Moslow & Tye, 1985; Sha, 1989), and constitutes the source environment for the seaward-side system. Sand moves up into the channels from the eroding upper shoreface (dispersal system 2, Fig. 21B). It also moves laterally along the crest of the ebb delta platform, from the up-drift side to the down-drift side of the inlet (Oertel, 1988; Oertel *et al.*, 1989), and the dispersal system of the delta's seaward face can be regarded as two superimposed systems; a tide-driven system (system 1, Fig. 21B), and a surf-driven system (unlabelled).

As these two seaward-side systems interact, tidal channels and their intervening shoals migrate along the crest of the ebb delta platform toward the down-drift side (Ehlers, 1988). Lateral accretion in the channels forms a proximal *amalgamated facies*. As the channels become inactive, they may accumulate mud and develop an *interstratified lithofacies* (Howard & Reineck, 1972).

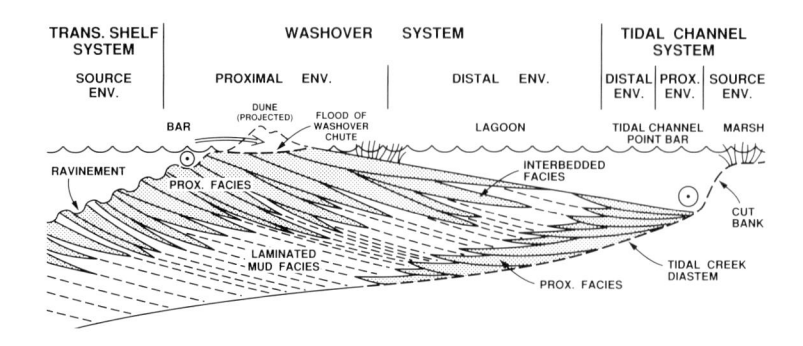

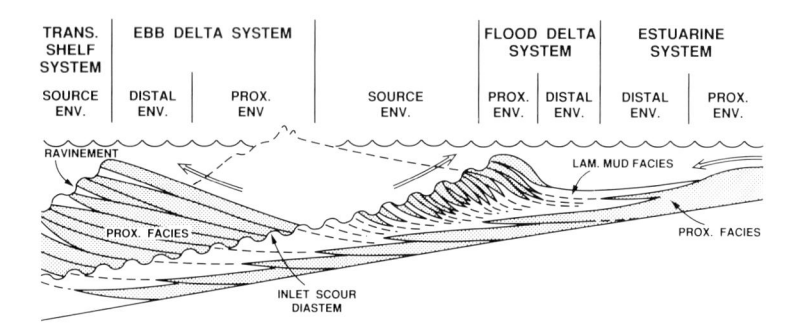

Fig. 22. Barrier-lagoon lithofacies. (A) The barrier case; (B) the inlet case. Dot-circle is current arrow, seen head-on.

The seaward-side dispersal system of the ebb tidal delta interfingers with a system on the inlet side of the delta (system 3, Fig. 21B). The inlet-side ebb delta system is sourced from the eroding inlet floor (Sha, 1989). This inlet-side system also receives sand from the wave-driven littoral current on the up-drift side of the inlet during flood tide (Oertel, 1988). A proximal *amalgamated facies* is built by tidal megaripples migrating along dispersal system 3, up the 'ramp to the sea'; the slope from the inlet floor up to the crest of the ebb tidal delta (Oertel & Howard, 1972). This lithofacies is enriched in shell debris by the condensation process (Shepard & Moore, 1954, 1955). The lithofacies passes seaward into a more distal (but still amalgamated) lithofacies deposited by migrating swash bars on the delta top, driven by the surf zone as well as tidal processes.

As the transgression continues, the ebb delta front undergoes erosional shoreface retreat, while its aggrading rear surface (ramp to the sea) backfills the inlet's scour trench. The depositional system of the delta's seaward face is largely destroyed in this process, and in the end, landward-dipping beds from the inlet-side system are sandwiched between the inlet-floor diastem and the ravinement diastem (Fig. 22). The surviving proximal lithofacies is a cross-stratified, amalgamated sand lithofacies that coarsens upwards.

The flood tidal delta is likewise built by the interfingering of the distal ends of two tidally driven dispersal systems across the delta crest. The situation here is a little simpler, since no surf zone is superimposed. Inlet scour encroaches on the inlet-side face of the flood tidal delta, while its landward face aggrades into the lagoon. In this process, the inlet-side deposits and most of the lagoon-side deposits of the flood tidal delta are destroyed. However, tidal inlets are typically short-lived (life spans of decades or centuries), and when the inlet closes, beds of the ebb tidal delta's lagoon-side deposition may survive. They may also survive on the flanks of the inlet. Where they do survive, they lie between the capping ravinement diastem, and locally, a basal source diastem cut by migration of channels on the lagoon face of the deltas (Sha, 1989; Fig. 22). The sequence fines upwards, but may be capped by coarser beds, if deposits from the inlet-side depositional system are preserved. The *amalgamated sand lithofacies* of the flood tidal delta tends to be cross-stratified or micro-cross-laminated (Boothroyd *et al.*, 1985). Tidal bundling may be apparent (Visser, 1980). The *interstratified lithofacies* exhibits complex flaser bedded, wavy bedded and lenticular sand beds (Reineck & Singh, 1980) and abundant *Scolithos*-like burrows. Horizons of spartina or mangrove rhizomes are frequent (Allen, 1964; Oertel *et al.*, 1989). The *distal mud lithofacies* may be highly bioturbated.

The dispersal system of ebb-dominated tidal channels that builds the lagoonal side of the flood delta bears a serial relationship to a tidal channel dispersal system in the lagoon (see section on depositional system contact relationships in discussion of depositional systems architecture). Sand is eroded from, or bypassed across, the floors of tidal channels cut into the lagoonal side of the flood delta. It is transported along tidal channels into the lagoon, where it accretes on point bars (Reinson, 1979; Fitzgerald & Penland, 1987). During spring high tides and storm surges, sand is carried overbank on to the interchannel flat (Boothroyd, 1985; depositional system 7, Fig. 21B). The *amalgamated sand lithofacies* of the lagoonal channel depositional system (Fig. 22A) is cross-stratified, and is characterized by shell lags and clay pebble gravels eroded from cutbanks (van Straaten, 1956; Reineck & Singh, 1980). The *interstratified lithofacies* grades laterally into a laminated or bioturbated mud lithofacies (Nichols, 1989), which interfinger with the muds of the washover fan system (is a common distal *laminated mud lithofacies* for both the tidal channel system and the washover fan system).

During transgression, inlets may retreat landward over antecedent valleys (Oertel, 1987). Inlet systems may thus prograde over the deposits of the dispersal system of the estuary head delta, created by the sudden loss of competence of the river as it enters the estuary (Fig. 22B).

*Lithofacies variations in barrier–lagoon–estuary systems*

As the barrier–lagoon–estuary depositional system is traced along strike, the relative thicknesses of the distal and overlying proximal lithofacies vary (Reinson, 1979; Boothroyd *et al.*, 1985). In the vicinity of an inlet, the basal lagoonal mud and interstratified sandstone and shale lithofacies are reduced relative to the overlying amalgamated sandstone lithofacies associated with the ebb tidal delta (Fig. 22B). As noted above, the landward migrating inlet may have trenched through the entire section, and the trench may have filled subsequently with inlet margin and ebb tidal delta deposits. Inlets may have been migrating landward along antecedant fluvial channels, so that the inlet sandstones overlie estuarine shales (Stone, 1972). Very large antecedent channels do not become closed off by barriers, but tend to persist as open estuaries, whose deposits are capped by estuary mouth shoals that resemble enlarged versions of inlet ebb and flood delta complexes (Ludwick, 1970; Coleman *et al.*, 1989).

As the tide range increases from microtidal to mesotidal, sandy estuarine and lagoonal tidal channel deposits increase in the lower part of the section at the expense of the muddy lagoonal deposits, while flood tidal delta deposits expand in the upper part of the section at the expense of washover fan and ebb tidal delta deposits (Boersma & Terwindt, 1981; Ehlers, 1988). Wide mesotidal estuaries develop large-scale estuary mouth shoals (the 'tidal delta' of L.D. Wright (1977) is equivalent to the estuary mouth shoal of an equilibrium estuary). In these depositional environments, primary structures indicative of estuarine tidal sedimentation are often well developed. These include tidal bundling with reactivation, full vortex and slackening structures, (Boersma & Terwindt, 1981), and double clay drapes (Visser, 1980).

# TRANSGRESSIVE DEPOSITIONAL SYSTEMS, II: SHOREFACE-SHELF SYSTEMS

### Transgressive shelf regimes

Transgressive shelf regimes can be subdivided into storm-dominated regimes, tide-dominated regimes and regimes characterized by intruding oceanic currents (Curray, 1960; Swift *et al.*, 1981). At mid-latitudes, storm-dominated autochthonous shelves typically experience three to five major transport events per year (Swift *et al.*, 1986a). Tide-dominated regimes are characterized by mesotidal or macrotidal

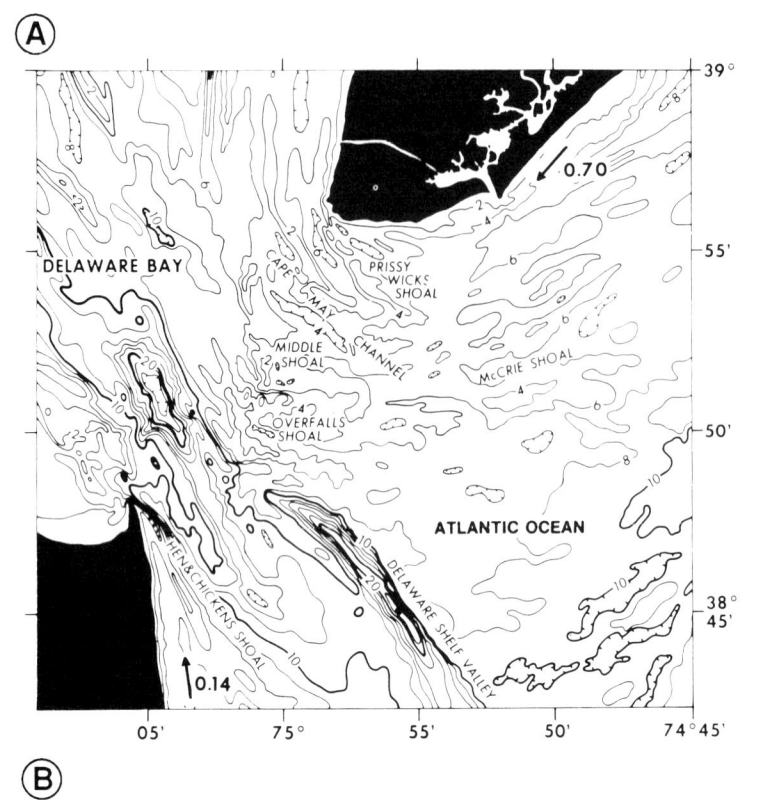

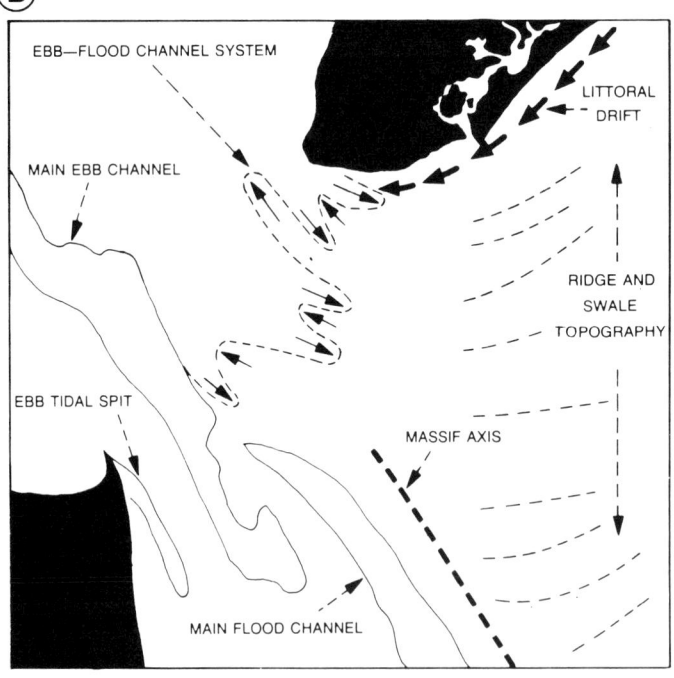

**Fig. 23.** Delaware estuary mouth shoal. (A) Topography; (B) geomorphological elements. (Modified from Swift, 1973.)

ranges (Davies, 1964), or by mid-tide surface velocities in excess of 25 cm s$^{-1}$ (Swift et al., 1981), and where the regime is semi-diurnal, experience two transport events per day. Regimes characterized by intruding oceanic currents, such as the North American Atlantic shelf at Cape Hatteras (Hunt et al., 1977), or the southeastern African shelf (Flemming, 1980, 1988) experience more nearly continuous transport. The proximal sand lithofacies of these three classes of hydraulic regime are distinguished by increasing depositional relief in the form of large-scale bedforms (sand ridges and sand waves) and sand shoals, as a consequence of the corresponding increase in mean annual fluid power expenditure (a measure of the regime variable $D$, the sediment transport function).

Transgressive shelf regimes are accommodation-dominated (the regime ratio $Q \cdot M/R \cdot D$ is small). As a consequence, the sediment supply, $Q$, is small relative to the rate of sea-level rise and to the fluid power available to disperse it. Fine sediment tends to be largely or completely bypassed over the shelf edge. A transgressive shelf system therefore tends to be dominated by its coarser proximal lithofacies.

**Transgressive morphodynamics: storm-dominated example**

At regional scales, coarse-grained sheet sandstones are best explained as autochthonous responses to transgression. They have been eroded out of the underlying substrate during the passage of the shoreline as the shoreface underwent erosional retreat in response to rising sea-level (see discussion by Swift & Thorne, this volume, pp. 15–17). In this manner, a mantle of relatively coarse, clean sand is left on the ravinement surface cut by the retreat process (see preceding discussion of barrier retreat modes). On modern continental shelves, the sheet is discontinuous, ranging from 0 to 10 m thick (Stride, 1982; Swift et al., 1986b). Basal gravels, Holocene back-barrier deposits or older Pleistocene or Tertiary deposits are exposed in windows through the sand sheet (Belderson & Stride, 1966; Swift, 1976b; Swift et al., 1986b).

At scales of tens of kilometres, the structure of the sand sheet of the storm-dominated middle Atlantic Bight reflects variations in the littoral sediment dispersal system at the leading edge of the transgression. Major variations occur at the mouths of major estuaries; Chesapeake Bay, Delaware Bay, and the Hudson River estuary. Zones of rapid loss of littoral transport capacity at the up-drift sides of these estuary mouths result in the aggradation of estuary mouth shoals on the up-drift side (Swift, 1973). The surfaces of these shoals are stabilized by interfingering ebb-dominated and flood-dominated channel systems (Fig. 23). The shelf tidal wave, propagating into the estuary, is abruptly slowed by the frictional drag of the shallow bottom, so that the gradient of tidal phase is steep at the estuary mouth (Swift & Ludwick, 1976). Shallow sand bottoms are unstable under such circumstances, and tend to be modified into such ebb-flood channel systems. Most of the tidal exchange is accomplished by a large-scale, ebb-channel–flood-channel pair on the down-drift side of the estuary mouth (Ludwick, 1970; Fig. 23). As the estuary mouth shifts landward in response to sea-level rise, the estuary mouth shoal and the main flood channel shift landward with it, giving rise to a *shelf valley complex*, a morphological and stratigraphic entity consisting of a levee-like high (*shoal retreat massif*) paired with a *shelf valley* (Fig. 24). These morphological features together constitute the retreat path of the estuary mouth during the Holocene transgression (Swift et al., 1972; Swift, 1973). As they are traced seaward, they are seen to terminate in a mid-shelf or shelf-edge delta, the morphological equivalent of a lowstand systems tract. Similarly, the littoral drift convergence at the large cuspate forelands along the southern margin of the Middle Atlantic Bight has created cape-associated shoals, and *cape shoal retreat massifs* (Swift et al., 1972, 1973). Cape-associated features probably started as lowstand cuspate deltas. When the transgression began, they became retreating cuspate forelands, locked into a pattern of positive feedback, whereby the foreland morphology focuses wave energy on its apex, creating a littoral drift convergence there, and so generating the massif. The massif in turn perpetuates the wave refraction pattern that drives the littoral drift. Powerful storm flows sweep southward across cape retreat massifs (Hunt et al., 1977), which are consequently re-moulded into ridge and swale systems (Fig. 25).

The interfluves between shelf valley complexes in the middle Atlantic Bight are characterized by sand-ridge fields. Sand ridges are 0–10 m high, 2–4 km apart, and make angles of 25–45° with the coastline and the regional trend of the shoreline. The angles open into the prevailing southwesterly flow (Duane et al., 1972; Swift & Field, 1981). The ridges form on the shoreface in response to storm flows, and are let down on the shelf floor as the shoreface erodes out

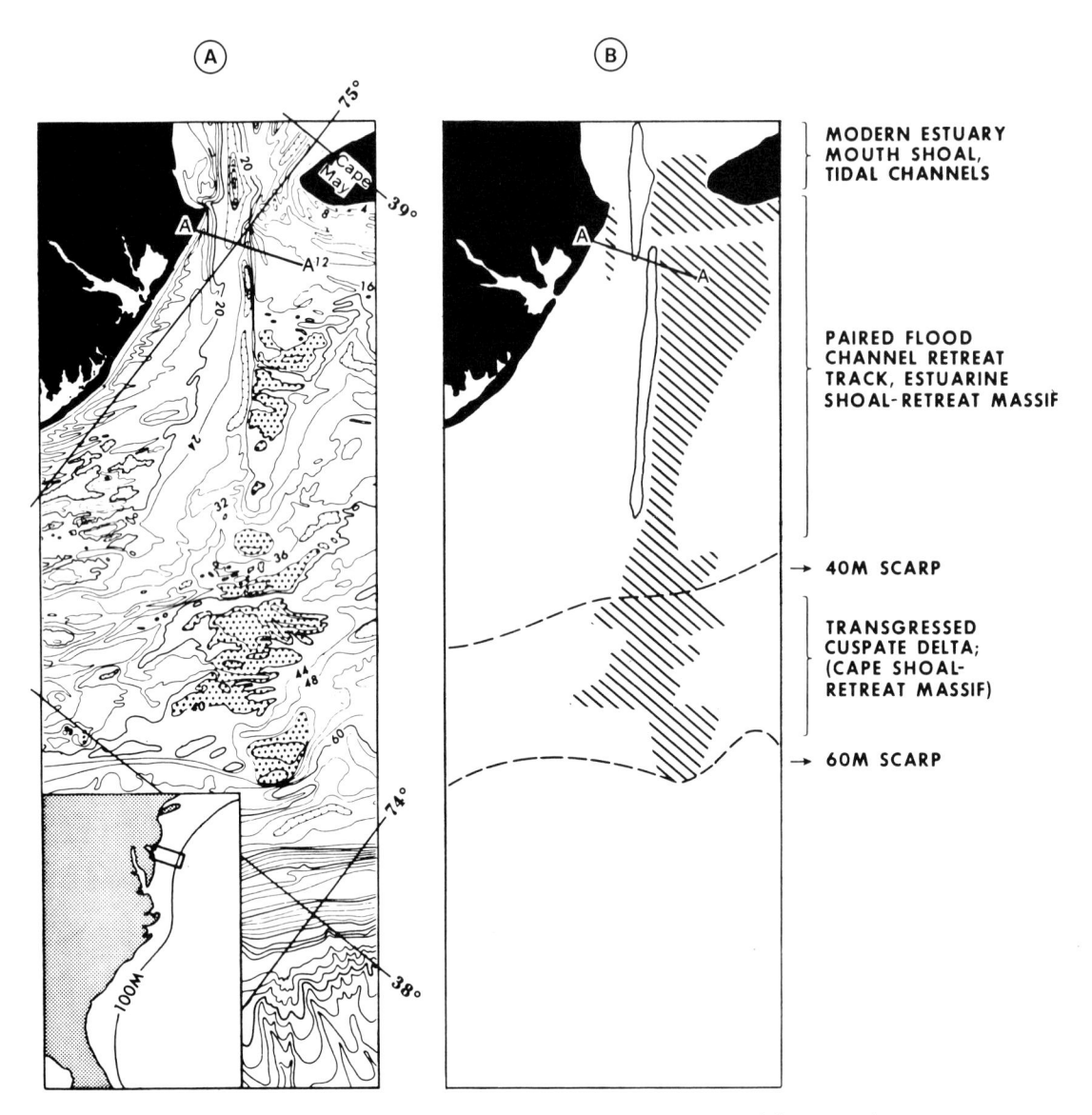

**Fig. 24.** The Delaware shelf valley complex, Middle Atlantic shelf of North America. (A) Topography; (B) geomorphological elements. (Modified from Swift, 1973.)

from under them (Swift *et al.*, 1973; Fig. 26). Boscar-Karakiewicz & Bona (1986) have proposed that the ridges are responses to infra-gravity waves concurrent with periods of storm resuspension. However, the infra-gravity wave model does not seem to explain the orientation of the ridges with respect to the shoreline and to the prevailing shelf flow direction as well as does the stability model proposed by Huthnance (1982); see also Figueiredo *et al.* (1981) and Parker *et al.* (1982). The sand ridge topography extends without interruption from the plateau-like interfluves between shelf valley complexes into estuarine and cape-associated massifs, where the ridges attain their maximum height. As the water column deepens over shelf sand ridges and the shoreline recedes, ridges undergo a continuing evolution (Fig. 26). Ridge height and spacing increases, while side slopes decrease (Swift

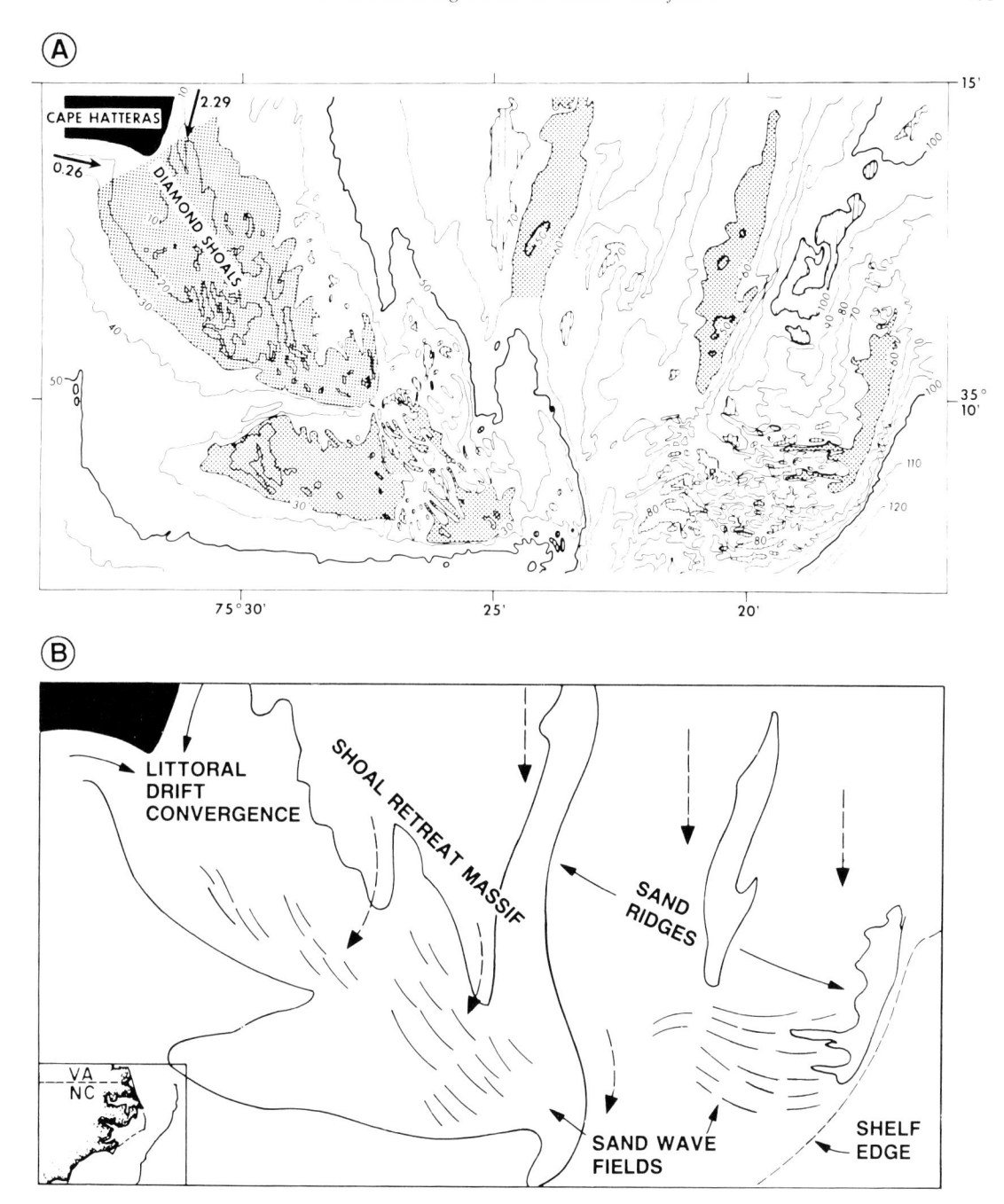

Fig. 25. Diamond Shoals, a cape shoal-retreat massif. (A) Topography; (B) geomorphological elements. Dashed arrows indicate prevailing direction of storm flow. (Modified from Hunt et al., 1977.)

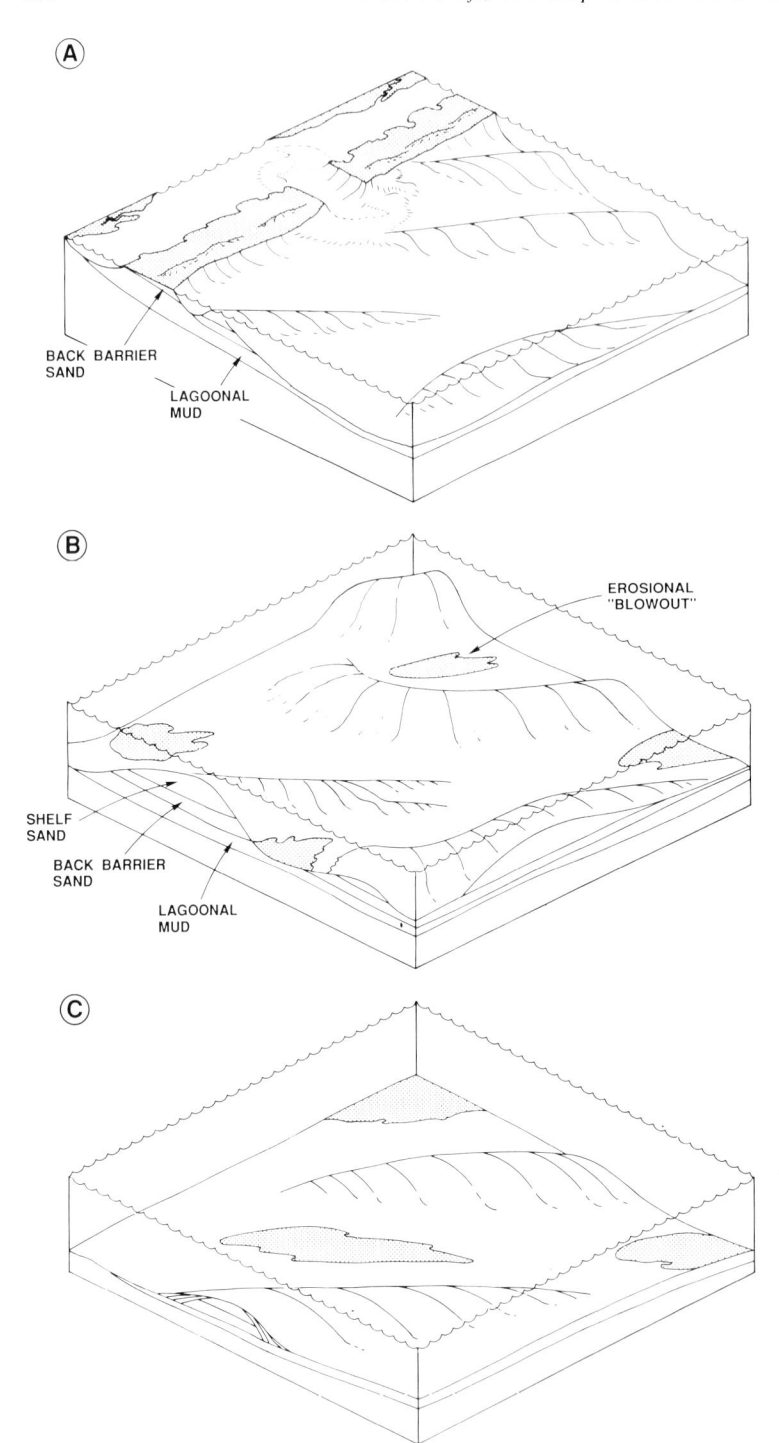

**Fig. 26.** Evolution of a sand-ridge field. (A) Formation on the shore face. (B) Inner shelf stage; barrier has migrated landward. Some ridges still have curved nearshore ends. (C) Outer shelf stage. Some ridges have moved, others have combined. Activity wanes. (Based on Swift & Field, 1985.)

*et al.*, 1972, 1973; Swift & Field, 1981). The rate of change decreases with time, as the ridges become senescent or *moribund* (term from Kenyon *et al.*, 1981).

**Transgressive morphodynamics: tidal examples**

On Nantucket shoals and Georges bank on the North American Atlantic shelf, and again in the Southern Bight of the North Sea, and in the outer Celtic Sea, similar arrays of sand ridges have formed in response to a tidal, rather than a storm-dominated regime (Twichell, 1983; Kenyon *et al.*, 1981). Tidal sand ridges may be longer and higher than storm-built sand ridges but otherwise resemble them closely in lithofacies and geometry (Swift, 1975a), and are probably formed by the same basic hydrodynamic mechanism. Like storm-built ridges, tidal ridges are typically oriented obliquely with respect to the prevailing flow, as required by the Huthnance stability model (Huthnance, 1982; see also discussion in Parker *et al.*, 1982).

On these tidal shelves, retreating coastal salients have given rise to ridged shoal retreat massifs, analogous in many respects to those of the storm-dominated Middle Atlantic shelf (Swift, 1975a; Kenyon *et al.*, 1981; Fig. 27). Shoreface-connected tidal sand ridges form as complex ebb-flood channel systems on the retreating shoreface in regions characterized by steep coastwise gradients in tidal phase. After ridges are detached from the retreating shoreface and stranded on the shelf floor, they retain for some time the parabolic and double-parabolic (sigmoidal) patterns characteristic of paired ebb-flood channel systems. However, when a critical depth limit is passed the flow pattern and corresponding morphology shifts. Sigmoidal ridges defining paired ebb and flood channels are re-moulded and become linear, parallel ridges, whose intervening channels each have an ebb- and flood-dominated side (Swift & Ludwick, 1976).

**Sediment dispersal systems on transgressive shelves**

Transgressive shelves experience autochthonous, supply-dominated depositional regimes, in which the sediment input is derived largely from shoreface erosion (see discussion of supply-dominated regimes by Swift & Thorne, this volume, p. 17). Their dispersal systems are primarily advective, because diffusion constants associated with their relatively coarse sediments are low. Along-shelf advection is a more important aspect of transport than diffusion, and lithofacies boundaries tend to trend across-shelf, normal to the direction of transport (see discussion of advective dispersal systems by Swift & Thorne, this volume, p. 25).

Just as transgressive shelf morphological elements occur in a hierarchical array of spatial scales, so do transgressive shelf dispersal systems. At scales of tens to hundreds of kilometres, regional variations in shelf width and depth result in zones of flow acceleration and shelf-floor erosion, followed by down-shelf zones of flow expansion and deceleration, and shelf-floor deposition. In the Middle Atlantic Bight of the North American Atlantic shelf for instance, the orientation of successive coastal compartments rotates from southwest through south to southeast, as the compartments are observed from north to south (see Fig. 18 and associated discussion of Swift & Thorne (this volume, pp. 24–25)). During periods of peak storm flow the currents leaving one coastal segment accelerate and reorient as they impinge on the up-current end of the next coastal segment. The sea-floor is scoured and the eroded sediment is redeposited in a series of down-current lithofacies of decreasing mean grain size (Swift *et al.*, 1986b). The energetic, high meso-tidal to macrotidal regime of the shelf around the British Isles has developed an even more complex pattern of large-scale (>200 km) sediment dispersal paths, between zones of divergence ('bed load partings', Kenyon & Stride, 1970), and zones of convergence (Fig. 28). Sediment transport is oriented down the gradient of mid-tide surface velocities (Kenyon & Stride, 1970).

In many cases, smaller dispersal subsystems are superimposed on the larger scale systems, with scales of tens of kilometres. On the Middle Atlantic Bight of North America, such subsystems are generated by along-shelf flow across the shore-normal constructional topography created at the foot of the retreating shoreface (Fig. 29). Shoal retreat massifs are dissected by shelf flow into comb-like arrays of sand ridges and swales whose 'teeth' point into the prevailing direction of storm current approach (compare Figs 23, 24, 25; other cases are described in Swift *et al.*, 1977, 1978). Downcurrent-fining lithofacies accumulate on the lee flanks of the massifs (Swift *et al.*, 1977, 1978; Fig. 30). Finally, the sand ridges themselves constitute small-scale sediment dispersal subsystems, in which up-current sides are winnowed, creating coarse lags. Fines are bypassed over the crest to the down-current side,

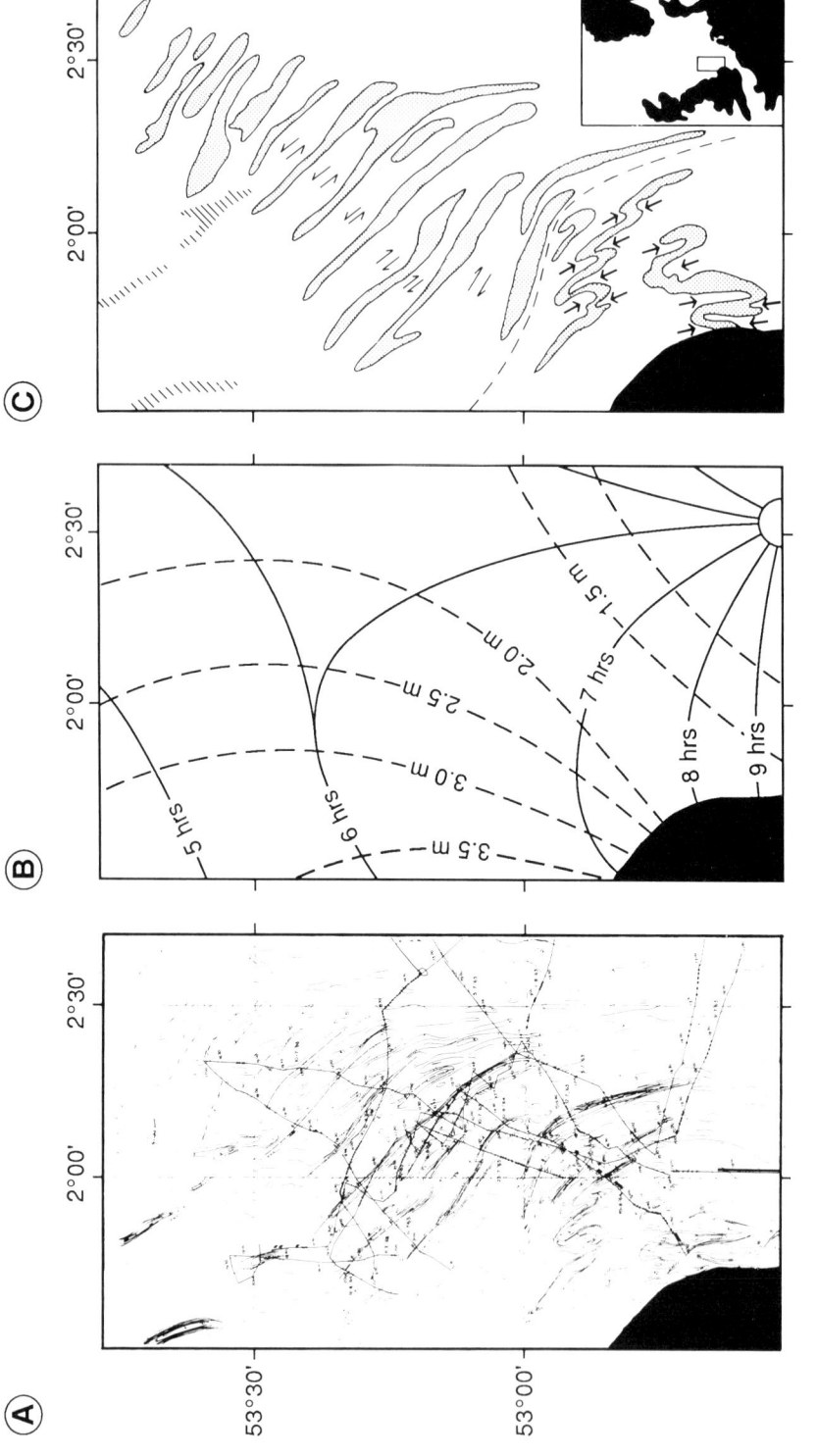

Fig. 27. The Norfolk banks, a tidal shoal retreat massif. (Modified from Houbolt, 1968.) (A) Topography. (B) Tidal phase and co-range lines. (C) Geomorphological elements. Paired arrows indicate sense of ebb- and flood-dominant flow. Dashed line separates parabolic channel province from linear channel province. Diagonal hatching designates *Rinnentalers* (subglacial stream valleys). (From Houboult, 1968.)

**Fig. 28.** Pattern of sediment dispersal on the shelf around the British Isles. (From Stride, 1982.)

accumulating as successively more fine-grained lithofacies. See Stubblefield & Swift (1981) for a modern example, and Gaynor & Swift (1988) for an ancient example.

**Proximal lithofacies of transgressive shelf systems**

The initial lithofacies of the shelf system on most modern shelves is a coarse, discontinuous lag, which may occur as a ridged sheet (Belderson & Stride, 1966; Emery, 1968; Swift et al., 1986b). The leading edge of this sheet lies at the foot of the shoreface, and it forms from the debris of the erosional retreat process (Swift et al., 1972). Quaternary lowstands rejuvenated rivers as they crossed the exposed shelf, and in many cases, the lowstand shelf valleys contain gravel. The retreating shoreface has breached these gravelly valley fills and has redistributed them as a marine basal transgressive gravel (Belderson & Stride, 1966; Figueiredo et al., 1981). Where they occur, such gravels typically form in thin, discontinuous, shore-parallel bands, marking successive phases of shoreface retreat. The gravel is overlain by several metres of sand. The sand is discontinuous and in areas of strong storm or tidal currents may be lacking altogether, so that the gravel is exposed at the sea-floor. As noted, the basal gravel here continues to be reworked by shelf currents. The sands and gravels are cross-stratified to laminated and upward fining, and constitute a proximal *amalgamated sand and gravel facies*. This facies constitutes the 'relict' sediment of earlier workers (Emery, 1968). The sand and gravel is 'relict' in the sense that it is of autochthonous origin. However, it is not correct to state as has been done (Curray, 1965) that the texture (grain size) is 'relict', and 'out of equilibrium' with the prevailing hydraulic regime. Bottom grain size on the shelf is not simply a response to the sediment transport function $D$, but to the regime ratio $R \cdot D/Q \cdot M$, which in turn controls $r$, the

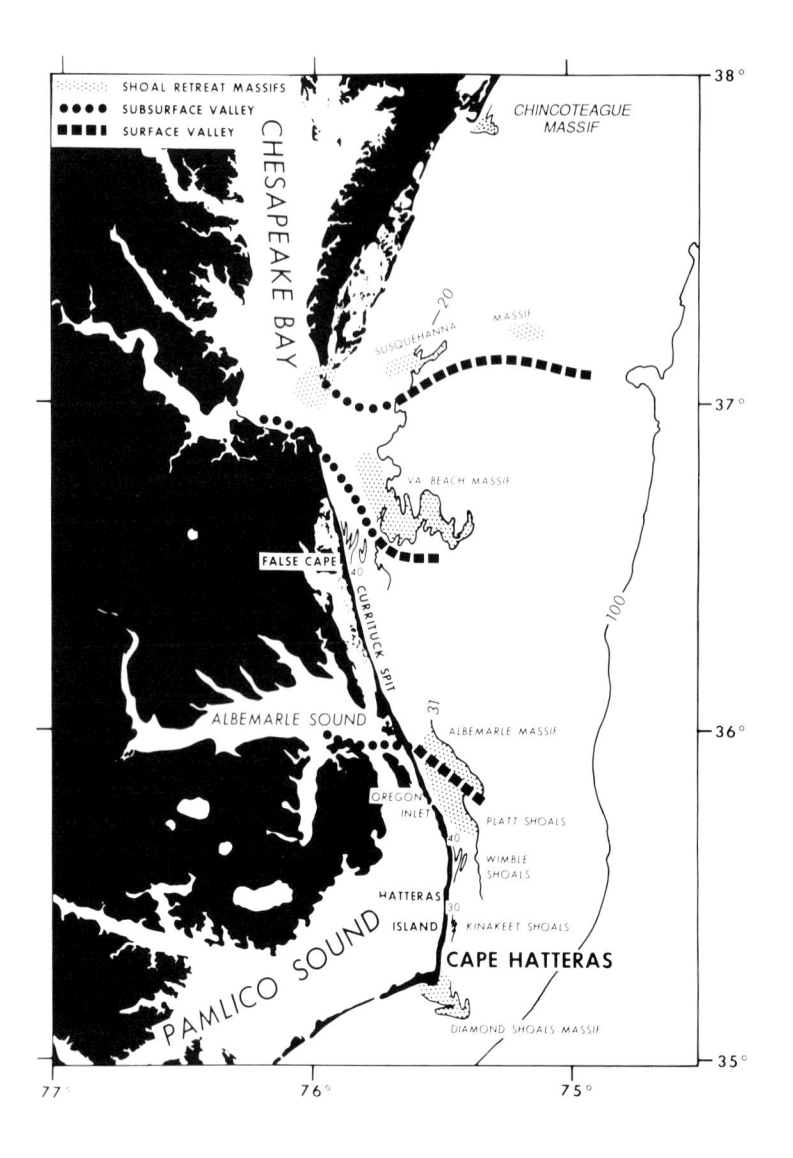

**Fig. 29.** Shoal retreat massifs of the Middle Atlantic shelf, North America. (Modified from Swift *et al.*, 1973.)

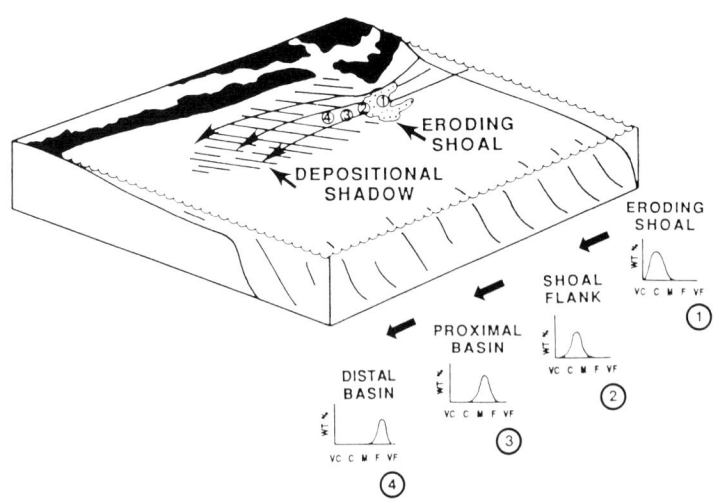

**Fig. 30.** Sediment dispersal in the vicinity of Chincoteague shoal, Middle Atlantic shelf, North America.

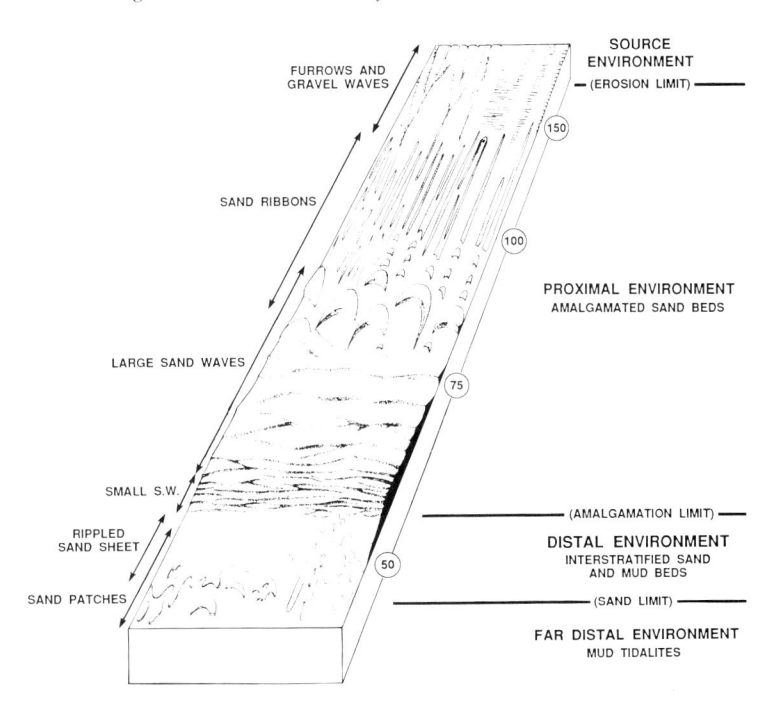

**Fig. 31.** Lithofacies of a sediment dispersal pathway on the shelf around the British Isles. (From Stride, 1982.)

reworking ratio. 'Relict' sediment is an equilibrium response to high values of the accommodation/supply ratio, which in turn induces a high reworking ratio. Relict sediments form at the high end of the condensation gradient on the floor of the facies solid (Fig. 4C).

Lithofacies of coarse-grained sheet sands are created by the progressive sorting of the transported sediment into zones of successively finer material down sediment dispersal pathways. These relationships are most strikingly developed on the tide-swept shelf around the British Isles, whose sediment dispersal pathways are described above. Sediment moving down these pathways is derived from the retreating shoreface, and also from 'bedload partings' on the shelf floor (Kenyon & Stride, 1970), whose locations are determined by the tidal dynamics of the shelf water body. Along parting zones, the transgressive section is locally scoured off down to bedrock. Further down the sediment transport path, gravel, then coarse sand accumulates. Gravel lithofacies may be moulded into sedimentary furrows (Flood, 1983) or gravel waves. Yet further down-current, sand ribbons appear (Mclean, 1981), sometimes emerging from sedimentary furrows (Flood, 1983). As the sand sheet thickens, barchan sand waves appear, and eventually sand wave fields (Fig. 31). As the transported load continues to become finer grained down the sediment transport path, the bedload to suspended load ratio decreases. The sand sheet passes through a thickness maximum. Bedforms become subdued and eventually disappear. Similar sediment dispersal pathways have been identified on the Atlantic continental shelf, across the Virginia Beach Massif (Swift et al., 1977), and on Sable Island Bank (Amos & Nadeau, 1988).

The down-current changes in grain size, and the resulting shift from bedload to suspended load transport leads to a concomitant change in stratification pattern. The lithofacies assemblage of a transgressive shelf depositional system consists of the same basic elements as does the fine-grained sand sheet, but in most cases, the volumetric proportions of fine and coarse lithofacies are reversed, so that the coarse lithofacies dominate (Fig. 31). As a consequence, the proximal lithofacies (amalgamated sandstone facies) is coarse-grained, highly condensed, and characterized by primary structures resulting from repeated episodes of erosion and bedload transport, so that cross-strata sets tend to alternate with allochem-enriched lags. In regressive systems, this lithofacies is limited to the beach and surf-zone deposits of a prograding shoreface. In the transgressive case, the proximal lithofacies is deposited at the foot of the eroding shoreface, or immediately down-current from a bedload parting, and may extend for tens or hundreds of kilometres down the dispersal path.

On regressive shelves, the bulk of the depositional system is taken up by the discretely bedded, expanded, fine-grained distal facies, where sand and mud are intercalated, and sand beds fine down-current into mud tempestites (or tidalites as the case may be). In the coarse-grained dispersal systems around the British Isles, the discretely bedded distal lithofacies is not always realized. The zone of maximum accumulation and sand wave development may pass down-current through a zone of sand patches (*interstratified sand and mud facies*) into a *laminated mud facies* (Stride, 1982; Fig. 31) or transport of coarse and fine sediment may decouple, with the bedload dispersal path terminating in a bedload convergence, while the suspended sediment bypasses over the shelf edge (Kenyon & Stride, 1970).

### Distal lithofacies of transgressive shelf systems

Some transgressive shelves, however, are noticeably more 'leaky.' Shoreface erosion releases fines in greater quantities. The fines are bypassed across the inner shelf, but further seaward where the wave regime expends less fluid power, they may come to rest. McCave (1972) has shown that deposition depends not only on the hydraulic climate of waves and currents in the benthic boundary layer, but also on the near-bottom sediment concentration. If the fluvial sediment discharge per kilometre of coastline is sufficiently high, and the fluvial load sufficiently fine, a mixed autochthonous–allochthonous regime occurs, in which fluvial fines accumulate together with the autochthonous fines winnowed out of the substrate by erosional shoreface retreat. This situation, like that of sandy transgressive shelves, requires that the regime ratio, $R \cdot D/Q \cdot M$ be accommodation-dominant (right-hand side be greater). For muddy shelves, however, the ratio is closer to unity than it is for the fully autochthonous case.

In both the autochthonous and allochthonous cases, an offshore shelf mud lithofacies onlaps landward on to the inner shelf sand facies, burying sand ridges that formed when the sand was still exposed at the surface (Twichell *et al.*, 1980; Maldonado *et al.*, 1983). Time-planes pass from the onlapping mud into the underlying sand and gravel facies. Storms sweep inner shelf sand on to the outer shelf mud, then the mudline advances landward over the sand, so that the two lithofacies intertongue, and sand beds as well as mud beds onlap the inner shelf (Maldonado *et al.*, 1983; see Fig. 13 of Thorne & Swift, this volume, p. 214). The onlapping character is diagnostic of *transgressive* mud deposits, and distinguishes them from downlapping highstand mud deposits.

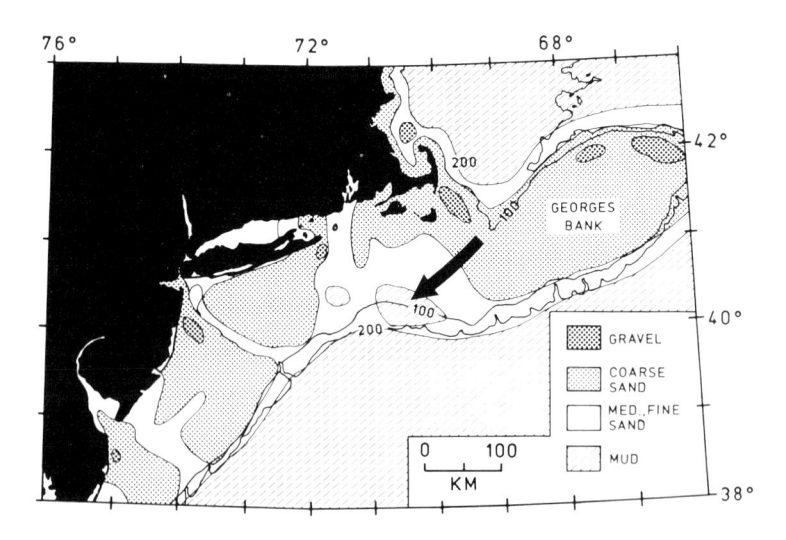

**Fig. 32.** Sediment-distribution on the southern New England shelf, showing the 'mud patch' and its showing the 'mud patch' and its possible relationship to the drowned Georges Peninsula. Arrow indicates probable progressive sorting pathway. Compare with Fig. 16 of Swift & Thorne, this volume, pp. 3–31. Grain size information from Schlee (1968).

The transgressive mud deposits of the southern New England shelf (Twichell *et al.*, 1980) and of the Irish Sea (Belderson, 1964) may be dominantly autochthonous mud deposits. Each constitutes an accumulation zone at the end of a well-defined, along-shelf, sediment transport pathway that begins at a zone of transport divergence and erosion on the shelf. On the southern New England shelf, the sediment transport path begins on the eroding surface of Nantucket shoals, and extends for approximately 280 km into the 'mud patch' of the southern New England shelf (Twichell *et al.*, 1980; Bothner *et al.*, 1981; Fig. 32). Although the lithofacies of this system have been little studied, the full range of distal lithofacies presumably occurs (interstratified sand and mud, amalgamated mud). In the case of the Southern New England mud patch, the deposit may be the by-product of erosional shoreface retreat across the ancestral 'Georges Peninsula', whose present expression is Nantucket shoals and Georges Bank shoals (Fig. 32).

Transgressive allochthonous mud deposits are common on continental shelves. They occur as great plume-shaped deposits of fine sediment that stretch for hundreds of kilometres along the shelf floor from the mouths of major rivers in the prevailing direction of sediment transport. These plumes are the equivalent of the deceleration sheets of regressive shelves. A continous sequence exists between fully transgressive deceleration sheets bordering retrograding shorelines, and regressive examples, in which the shoreline is prograding for a long distance down-drift from the river mouth. One of the best-documented transgressive examples is the plume associated with the Columbia River (Fig. 33). It consists of a mid-shelf band of allochthonous clayey silt resting on basal autochonous sands (Nittrouer & Sternberg, 1981). On a map of accumulation rates derived from radioisotopic analysis, it appears as an anomaly, with rates decreasing longitudinally down the plume away from the source, and decreasing laterally away from the axis of the plume (Fig. 10B). The autochthonous sands are exposed on the shelf floor on the sediment-starved outer shelf seaward of the plume, and are being created on the wave-agitated inner shelf landward of the plume. Recent studies of the movement of Mount St Helens ash down the plume show that the rates of incremental, storm-driven, along-shelf transport are grain-size dependent, with the finest sediment travelling the most rapidly (Healy-Ridge & Carson, 1987; Fig. 33C).

Plume-like, allochthonous mud deposits of probable transgressive (onlapping) character also occur on the Niger shelf west of the Niger delta (Allen, 1965), on the Spanish Mediterranean shelf south of the Ebro Delta (Maldonado *et al.*, 1983), and in the Frisian and German Bight sectors of the North Sea, where they are, at least in part, derived from the suspended load of the Rhine (McCave, 1973).

## CONCLUSIONS

The concept of facies models has been an outgrowth of the heightened petroleum exploration of the last 20 years. The descriptive and conceptual nature of the modelling process has been adeptly summarized in Roger Walker's (1979) cartoon, in which local examples are examined, and 'local variability [is] distilled away', leaving 'the pure essence of environmental summary'. Facies models have been recipes for evaluation, akin to the protocols of biologists, or the decision trees of artificial intelligence schemes. Despite this (or because of it) they have been very useful in understanding the organization of sedimentary rocks and in finding oil.

Meanwhile geological science as a whole has been growing more complex and, in response to the computer revolution, more quantitative. Our purpose in this paper has been to show that we have reached a stage where quantitative analytical techniques can be used in facies models also. Thorne *et al.* (this volume, pp. 59–87) have taken the results of a one-dimensional numerical model of strata formation (Niedoroda *et al.*, 1987), and have used it to develop an analytical model; that is, they have examined strata parameters one at a time, and quantified their relationship to other parameters. The result is a set of rules for strata formation. Figure 4 shows how these rules can be used to derive depositional facies. The rules indicate that the various stratification patterns that we view as depositional facies are created by variations in stratal condensation as defined by the reworking ratio; the ratio of the minimum reworking depth to the rate of accumulation ($r = a'/\mathring{a}$). The rules show that facies can be assigned to three basic classes on the basis of their degree of condensation; an amalgamated sand lithofacies, an interstratified sand and mud lithofacies, and a laminated or bioturbated mud litho-

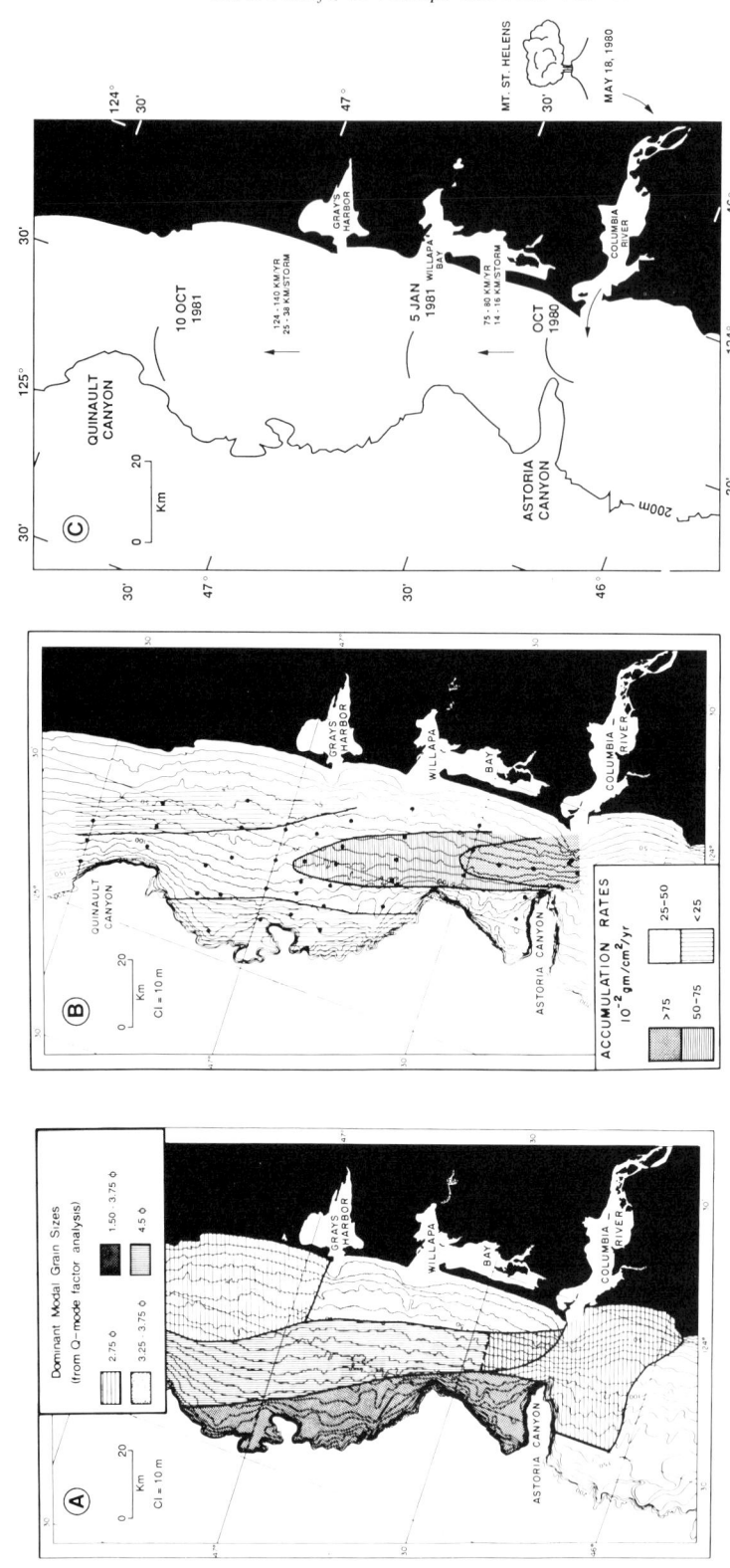

**Fig. 33.** Allochthonous transgressive sedimentation on the Washington Shelf. (A) Distribution of mean grain sizes on the Washington Shelf. (A) Distribution of mean grain sizes, revealing 'diffusive' pattern (see Swift & Thorne, this volume, pp. 3–31). (From Nittrouer et al., 1979.) (B) Accumulation rates, revealing lateral transport pattern. (From Nittrouer & Sternberg, 1981.) (C) Northward movement of Mount St Helens ash on the Washington Shelf. (From Healy-Ridge & Carson, 1987.)

facies. While there may be hundreds of superficial variations, these are the three basic responses to the facies-making processes of differential transport and strata formation.

These insights allow much more specific descriptions of depositional systems than has hitherto been possible (depositional systems are assemblages of process-related facies; Fisher & McGowen, 1967). The descriptions are based on the related concept of dispersal systems. Dispersal systems are flow-linked assemblages of depositional environments (this paper). A dispersal system is organized along a hydrodynamic gradient resulting in a downstream decrease in competence, so that it consists of a source environment and a set of depositional environments.

Sediment movement through a dispersal system builds a depositional system. Sediment movement is episodic, so that the resulting depositional system is event-stratified (has cyclic vertical grain-size gradients). The intermittent nature of the dispersal induces progressive sorting, so that there also is a horizontal gradient of downstream decrease in grain size. The decrease is the result of progressive trapping of the coarsest particles in bed soles, and it feeds back into bed structure; as downstream beds become finer, they are increasingly deposited from suspension rather than traction, and cross-stratification gives way to Bouma structures (Fig. 3).

Because the constituent lithofacies can be described fully in terms of their stratal parameters, it becomes possible to map out the extent of depositional systems composed of these facies (Figs 5, 7, 22) and to determine the nature of their contacts (Fig. 6). Depositional systems are characterized by a variety of spatial arrangements. Conjugate systems are stacked one above the other, and are separated by a shared source diastem. Serial systems also share a source diastem but are arranged in tandem, with both systems above or below the source diastem. Interfingering systems share a common distal facies. Superimposed systems have been overprinted by several depositional environments. If migration of the source environment has destroyed one of the depositional systems originally in contact with it, the resulting contact is an occluded contact.

Depositional systems can be divided into regressive and transgressive families on the basis of lithology, stratification pattern and geometry. Regressive settings tend to form large, integrated sediment dispersal systems supplied with allochthonous (river-derived) sediment. Depositional systems on regressive shelves are fine to very fine-grained because they are the end product of extensive fluvial and littoral progressive sorting systems. This textural characteristic is preserved in the regressive shelf environment because rapid sediment input and accumulation reduces the reworking ratio. The proximal condensed lithofacies is generally limited to the beach and upper shoreface environments. The bulk of the stratification pattern is an expanded one, in which much of the depositional record is preserved, and in much of the section sand beds are separated from each other by intercalated clay beds. Because the suspended load processes are dominant during the deposition of the sand beds, they frequently contain Bouma sequences (Bouma, 1962).

Dispersal systems on transgressive shelves are autochthonous (self-sourced); accumulation rates are low. Transgressive dispersal systems are relatively small in scale. Back-barrier dispersal systems are especially diverse. The associated depositional systems form a complexly interlocking pattern in three dimensions, with conjugate, serial and interfingering contacts. Proximal lithofacies of amalgamated sand beds on transgressive shelves make up a much greater proportion of the sedimentary volume relative to mud-rich distal lithofacies than they do on regressive shelves. On many transgressive shelves, the muddy distal lithofacies has been pushed over the shelf edge, and most of the shelf surface is taken up by the proximal lithofacies (amalgamated sand lithofacies).

Because the analytical approach to facies outlined in this paper allows depositional systems to be mapped out, depositional systems tracts also can be resolved explicitly (linkages of contemporaneous

**Table 3.** Lithology defined versus geometry defined lithosomes

| Spatial scale | Lithology defined | | Geometry defined |
|---|---|---|---|
| Event stratigraphy | | | Lamina<br>Lamina set<br>Bed |
| Systems stratigraphy | Lithofacies depositional system | = | Bed set |
| Sequence stratigraphy | Depositional systems tract | = | Parasequence<br>Geometric systems tract<br>Depositional sequence |

depositional systems; Brown & Fisher, 1977; see Table 1). However, the depositional systems tracts determined by facies analysis are not those of seismic stratigraphers (e.g. Vail et al., 1977; Posamentier et al., 1988). Facies-defined depositional systems assemblages are so thin relative to their extent as to be reasonably designated by the two-dimensional term 'tract'. In contrast, seismic stratigraphers have used the term 'systems tract' to describe thicker bodies that are the major subdivisions of third-order depositional sequences (lowstand systems tract, trangressive systems tract, etc; Posamentier & Vail, 1988). Depositional systems in these subdivisions are multiplied by parasequence repetition, hence they are more accurately styled 'parasequence sets' (term of Van Wagoner et al., 1988). Because the observations have been made by seismic means, these lithosomes are defined not by their constituent facies, but by their geometry, by parasequence architecture (progradational or retrogradational), and by reflector geometry (onlapping or downlapping). We restrict the term 'depositional systems tract' to deposits that have accumulated under conditions of essentially constant regime, as defined by the regime variables $Q$, $R$, $M$ and $D$. However, the observed partitioning of third-order sequences into geometrically defined subunits is both real and significant, and we use the provisional term 'geometric tracts' to designate them.

Facies-defined and geometric schemes of lithosome classification are compared in Table 3. Lithosomes are ranked in descending order of increasing size. We have included the stratal terms of Campbell (1967) so that the relationship between event stratigraphy, systems stratigraphy and sequence stratigraphy can be seen. The relationship between depositional systems and depositional sequences is discussed in greater detail by Swift et al. (this volume, pp. 153–187) and Thorne & Swift (this volume, 189–255).

## REFERENCES

ADAMS, C.E., JR., SWIFT, D.J.P. & COLEMAN, J.M. (1988) Bottom currents and fluviomarine sedimentation on the Mississippi Prodelta Shelf, February–May. *J. Geophys. Res.* **92**, 14585–14609.

AGER, D.V. (1974) Storm deposits in the Jurassic of the Morrocan high Atlas. *Palaeogeogr. Palaeoclimatol. Palaeoecol.* **15**, 83–93.

AIGNER, T. & REINECK, H.E. (1982) Proximity trends in modern storm sands from the Heligoland Bight (North Sea) and their implications for basin analysis. *Senckenbergiana Maritima* **14**, 183–215.

ALLEN, J.R.L. (1964) The Quaternary Niger Delta and adjacent areas: Sedimentary environments and facies. *Bull. Am. Assoc. Petrol. Geol.* **49**, 547–600.

AMOS, C.L. & NADEAU, O.C. (1988) Surficial sediments of the outer banks, Nova Scotian Shelf, Canada. *Can. J. Earth Sci.* **25**, 1923–1944.

ASQUITH, D.O. (1970) Depositional topography and major marine environments, Late Cretaceous, Wyoming. *Bull. Am. Assoc Petrol. Geol.* **54**, 185–202.

AUGUSTINUS, P.G.E.F. (1989) Cheniers and chenier plains: a general introduction. *Mar. Geol.* **90**, 219–230.

BALLARD, R.D. & UCHUPI, E. (1970) Morphology and Quaternary history of the continental shelf of the Gulf Coast of the United States. *Bull. Marine Sci.* **20**, 547–559.

BARRATT, J.C. (1982) *Fales member (Upper Cretaceous) deltaic and shelf bar complex, Central Wyoming.* MS thesis, University of Texas, Austin, TX, 120 pp.

BELDERSON, R.H. (1964) Sedimentation in the Western half of the Irish Sea. *Mar. Geol.* **2**, 147–163.

BELDERSON, R.H. & STRIDE, H.R. (1966) Tidal current fashioning of a basal bed. *Mar. Geol.* **4**, 237–257.

BERNARD, H.A. & LEBLANC, R.J. (1965) Resume of the Quaternary Geology of the Northwestern Gulf of Mexico Province. In: *The Quaternary of the United States* (Eds Wright, H.E., Jr & Freg, D.G.) pp. 137–185. Princeton University press, Princeton, NJ.

BERRYHILL, H.L. (1986) Late Quaternary lithofacies and structure, northwest Gulf of Mexico: interpretations from Seismic Data. *Am. Assoc. Petrol. Geol. Stud. Geol.* **23**, 289 pp.

BLATT, H., MIDDLETON, G.V. & MURRAY, R. (1980) *Origin of Sedimentary Rocks*, pp. 90–124. Prentice Hall, Englewood Cliffs, NJ.

BOERSMA, J.R. & TERWINDT, J.H.J. (1981) Neap-spring tide sequences of intertidal shoal deposits in a mesotidal estuary. *Sedimentology* **28**, 151–170.

BOOTHROYD, J.C. (1985) Tidal inlets and tidal deltas. In: *Coastal Sedimentary Environments* (Ed. Davis, R.A., Jr.) pp. 445–532. Springer Verlag, New York.

BOOTHROYD, J.C., FRIEDRICH, N.E. & McGINN, S.R. (1985) Geology of microtidal coastal lagoons: the Rhode Island example. *Mar. Geol.* **54**, 35–76.

BOSCAR-KARAKIEWICZ, B. & BONA, J.L. (1986) Wave dominated shelves: a model of sand wave formation by progressive infragravity waves. In: *Shelf Sands and Sandstone Reservoirs* (Eds Knight, R.J. & McLean, J.R.). Can. Soc. Petrol. Geol. Mem. **11**, 163–179.

BOTHNER, M.H., SPIKER, E.C., JOHNSON, P.P., RENDIGS, R.R. & ARUSCAVAGE, P.J. (1981) Geochemical evidence for modern sediment accumulation on the continental shelf off New England. *J. Sediment Petrol.* **51**, 281–292.

BOUMA, A.H. (1962) *Sedimentology of Some Flysch Deposits.* Elsevier, Amsterdam, 168 pp.

BROWN, L.F., JR. & FISHER, W.L. (1977) Seismic stratigraphic interpretation of depositional systems: examples from Brazilian rift and pull-apart basins. In: *Seismic Stratigraphy—Applications to Hydrocarbon Exploration* (Ed. Clayton, C.E.). Mem. Am. Assoc. Petrol. Geol. **26**, 218–248.

BROWN, L.F., JR. & FISHER, W.L. (1980) Seismic Stratigraphic Interpretations and Petroleum Exploration. *Am. Assoc. Petrol. Geol. Educ. Course Note Ser.* **16**, 181 pp.

CAMPBELL, C.V. (1967) Lamina, laminaset, bed, and bedset. *Sedimentology* **8**, 7–26.

CAMPBELL, C.W. (1979) Model for beach shoreline in Gallup Sandstone (Upper Cretaceous) of Northwestern New Mexico. *N. M. Bur. Mines Miner. Res. Circ.* **164**, 32 pp.

CARTER, R.W.G. (1986) The morphodynamics of beach ridge formation, Magilligan, Northern Ireland, *Mar. Geol.* **73**, 191–214.

CHAN, M.A. & DOTT, R.H. JR. (1983) Shelf and deep sea sedimentation in an Eocene forearc basin, Western Oregon—fan or non fan? *Bull. Am. Assoc. Petrol. Geol.* **67**, 2100, 2116.

COLEMAN, J.M. & PRIOR, D.B. (1982) Deltaic environments of deposition. In: *Sandstone Depositional Environments* (Eds Scholle, P.A. & Spearing, D.). Mem., Am. Assoc. Petrol. Geol. **31**, 139–178.

COLEMAN, J.M., ROBERTS, H.H., MURRAY, S.P. & SALAMA, M. (1981) Morphology and dynamic sedimentology of the eastern Nile Delta shelf. *Mar. Geol.* **42**, 301–326.

COLEMAN, S.M., BERGQUIST, C.R., JR. & HOBBS, C.H., III (1989) Structure, age and origin of the Bay-mouth shoal deposits, Chesapeake Bay, Virginia. *Mar. Geol.* **83**, 95–113.

CURRAY, J.R. (1960) Sediments and history of Holocene transgression, continental shelf, northwest Gulf of Mexico. In: *Recent Sediments, Northwestern Gulf of Mexico* (Eds Shepard, F.P., Phleger, F. & Van Andel, Tj.) Am. Assoc. Petrol. Geol. 221–226.

CURRAY, J.R. (1964) Transgressions and regressions. In: *Papers in Marine Geology* (Ed. Miller, R.C.) pp. 175–203. McMillan, New York.

CURRAY, J.R. (1965) Late Quaternary history, continental shelves of the United States. In: *The Quaternary of the United States* (Eds Wright, H.E. & Frey, D.G.) pp. 723–735. Princeton University Press, Princeton, NJ.

CURRAY, J.R., EMMEL, F.J. & CRAMPTON, P.J.S. (1969) Holocene history of a strandplain. In: *Simposio Internacional de Lagunas Costeras* (Eds Castañares, A.A. & Phleger, F.B.) pp. 63–100. Univ. Nac. Autonoma De Mexico, Mexico City.

DAVIS, R.A., JR. (1985) Beach and nearshore zone. In: *Coastal Sedimentary Environments* (Ed. Davis, R.A., Jr.) Springer-Verlag, New York.

DAVIES, J.L. (1964) A morphogenic approach to world shorelines. *Z. Geomorphol.* **8**, 127–142.

DEMASTER, D.J., MCKEE, B.A., NITTROUER, C.A., QIAN, J.C. & CHENG, G.D. (1985) Rates of sediment accumulation and particle reworking based on radiochemical measurements from continental shelf deposits in the East China Sea. *Continental Shelf Res.* **4**, 143–158.

DILLON, W.P. & OLDALE, R.N. (1978) Late Quaternary sea level curve: reinterpretation based on sealevel glacioeustatic influence. *Geology*, **6**, 56–60.

DOTT, R.H., JR. & BOURGEOIS, J. (1982) Hummocky cross-stratification: Significance of its variable bedding sequences. *Geol. Soc. Am. Bull.* **93**, 663–680.

DUANE, D.B., FIELD, M.E., MEISBERGER, E.P., SWIFT, D.J.P. & WILLIAMS, S.J. (1972) Linear shoals on the Atlantic continental shelf, Florida to Long Island. In: *Shelf Sediment Transport: Process and Pattern* (Eds Swift, D.J.P., Duane D.B. & Pilkey, O.H.) pp. 447–498. Dowden Hutchinson & Ross, Stroudsburg, PA.

EMERY, K.O. (1968) Relict sediments on continental shelves of the world. *Bull. Am. Assoc. Petrol. Geol.* **52**, 445–464.

EHLERS, J. (1988) *The Morphodynamics of the Wadden Sea*. Balkema, Rotterdam, 397 pp.

FIELD, M.E. (1980) Sand bodies on coastal plain shelves: Holocene record of the U.S. inner shelf off Maryland. *J. Sediment. Petrol.* **50**, 505–528.

FIGUEIREDO, A.G., JR., SWIFT, D.J.P. & STUBBLEFIELD, W.L. (1981) Submarine sand ridges on the inner Atlantic Shelf of North America: morphometric comparisons with Huthnance stability model. *Geomarine Lett.* **1**, 1687–1691.

FISCHER, A.G. (1961) Stratigraphic record of transgressing seas in the light of sedimentation on the Atlantic Coast of New Jersey. *Bull., Am. Assoc. Petrol. Geol.* 1646–1667.

FISHER, J.J. (1968) Barrier island formation: discussion. *Geol. Soc. Am. Bull.* **79**, 1421–1426.

FISHER, W.L. & MCGOWEN, J.H. (1967) Depositional systems in the Wilcox Group of Texas and their relationship to the occurrence of oil and gas. *Trans. Gulf Coast Assoc. Geol. Soc.* **17**, 105–125.

FISHER, W.L., BROWN, L.F., JR., SCOTT, A.J. & MCGOWEN, J.H. (1969) *Delta Systems in the Exploration for Oil and Gas*. Texas Bureau of Economic Geology, Austin, not consecutively paged.

FISKE, H.N., MCFARLAN, E. JR., KOLB, C.R. & WILBERT, L.J. JR. (1955) Sedimentary framework of the modern Mississippi Delta. *J. Sediment. Petrol.* **24**, 76–99.

FITZGERALD, D.M. & PENLAND, S. (1987) Backbarrier dynamics of the east Frisian Islands. *J. Sediment. Petrol.* **57**, 746–754.

FLEMMING, B.W. (1980) Sand transport patterns on the continental shelf between Durban and Port Elizabeth. *Sediment. Geol.* **26**, 179–206.

FLEMMING, B.W. (1988) Pseudo-tidal sedimentation in a non tidal shelf environment (southeastern African margin). *Tide-influenced Sedimentary Environments and Facies* (Eds Deboer, P.L., van der Gelder, P.L. & Nio, S.D.) pp. 167–180. D. Reidel, Dordrecht.

FLOOD, R.D. (1983) Classification of sedimentary furrows and a model for furrow initiation and evolution. *Geol. Soc. Am. Bull.* **94**, 630–639.

FOLK, R.L. (1960) Petrography and origin of the Tuscarora, Rose Hill and Keefer Formations, Lower and Middle Silurian of eastern West Virginia. *J. Sediment Petrol.* **30**, 1–58.

FOUCH, T.D., LAWTON, T.F., NICHOLS, D.J., CASHION, W.B. & COBBAN, W.A. (1983) Patterns and timing of synorogenic sedimentation in Upper Cretaceous rocks of central and northeast Utah. In: *Mesozoic Paleogeography of West Central United States* (Eds Reynolds, M.E. & Dolly, E.D.) pp. 305–336. Rocky Mountain Section, Soc. Econ. Paleontol. Mineral., Denver, CO.

FRASER, G.S. & SUTTNER, L. (1986) *Alluvial Fans and Fan Deltas*. Human Resources Development Corporation, Boston, 199 pp.

FRAZIER, D.E. (1974) Depositional-episodes: their relationship to the Quaternary stratigraphic framework in the northwestern portion of the Gulf Basin. *Tex. Univ. Bur. Econ. Geol. Geol. Circ.* **44-1**, 28 pp.

FREY, R.W. & BASAN, P.B. (1985) Coastal salt marshes. *Coastal Sedimentary Environments* (Eds Davis, R.A.) pp. 225–301. Springer-Verlag, New York.

GALLOWAY, W.E. (1975) Process framework for describing the morphologic and stratigraphic evolution of deltaic depositional systems. In: *Deltas: Models for Exploration* (Ed. Broussard, M.I.) pp. 87–98. Houston Geological Society, Houston.

GALLOWAY, W.E. & HOBDAY, D.K. (1983) *Terrigenous Clastic Depositional Systems*. Springer-Verlag, New York, 423 pp.

GAYNOR, G.C. & SWIFT, D.J.P. (1988) Shannon sandstone depositional model: sand ridge dynamics on the Campanian western interior shelf. *J. Sediment Petrol.* **58**, 868–880.

GOULD, H.R. & MCFARLAN, E., JR. (1959) Geologic history of the chenier plain, southwestern Louisiana. *Trans. Gulf Coast Assoc. Geol. Soc.* **9**, 1959.

HAILS, J.R. & HOYT, J.H. (1968) Barrier development on submerged coasts: problem of sea-level changes from a study of the Atlantic coastal plain of Georgia, and part of the east Australian coast. *Z. Geomorphol.* **7**, 24–55.

HANDS, E.B. (1983) The great lakes as a text model for profile responses to sea level changes. In: *CRC Handbook of Coastal Processes and Erosion* (Ed. Komar, P.D.) pp. 167–190. CRC Press, Boca Raton, FL.

HARMS, J.C., SOUTHARD, J.B. & WALKER, R.G. (1982) *Structures and Sequence in Clastic Rocks*. Tulsa, OK, Soc. Econ. Paleon. Mineral. Short Course 9.

HAYES, M.O. (1979) Barrier island morphology as a function of wave and tidal regime. In: *Barrier Islands* (Ed. Leatherman, S.P.) pp. 1–27. Academic Press, New York.

HEALY-RIDGE, M.J. & CARSON, B. (1987) Sediment transport on the Washington Shelf: estimates of dispersal rates from the Mt. St. Helens Ash. *Continental Shelf Res.* **7**, 759–772.

HENNESSY, J.T. & ZARILLO, G.A. (1987) The interrelation and distinction between flood-tidal delta and washover deposits in a transgressive Barrier Island. *Mar. Geol.* **78**, 35–56.

HOLMES, A. (1965) *Principles of Physical Geology*. Thomas Nelson Sons, London, 1288 pp.

HOLMES, C.W. (1982) Geochemical indices of fine sediment transport, Northwest Gulf of Mexico. *J. Sediment. Petrol.* **52**, 307–321.

HOLTEDAHL, H. & BJERKLI, K. (1975) Pleistocene and recent sediments of the Norwegian Continental Shelf (62°N–71°N) and the Norwegian Channel Area. *Norges Geol. Unders.* **316**, 241–252.

HOUBOLT, J.H.C. (1968) Recent sediment in the Southern Bight of the North Sea. *Geol. Mijnbouw* **47**, 245–273.

HOWARD, J.D. & FREY, R.W. (1985) Physical and biogenic aspects of back-barrier sedimentary sequences, Georgia Coast, U.S.A. *Mar. Geol.* **63**, 77–127.

HOWARD, J.D. & REINECK, H.E. (1972) Georgia coastal region, Sapelo Island, USA: sedimentology and biology. IV: physical and biogenic sedimentary structures of the nearshore Gulf. *Senckenbergiana Maritima* **4**, 47–79.

HOWARD, J.D. & REINECK, H.E. (1981) Depositional facies of high energy beach-to-offshore sequence: comparison with low energy sequence. *Bull. Am. Assoc. Petrol. Geol.* **65**, 807–827.

HOYT, J.H. (1967) Barrier island formation. *Geol. Soc. Am. Bull.* **78**, 1123–1136.

HUNT, R.E., SWIFT, D.J.P. & PALMER, H. (1977) Constructional shelf topography, Diamond Shoals, North Carolina. *Geol. Soc. Am. Bull.* **88**, 299–311.

HUTHNANCE, J.M. (1982) On one mechanism forming linear sand banks. *Estuar. Coast. Shelf Sci.* **14**, 79–99.

INGRAM, R.L. (1954) Terminology for the thickness of stratification and parting units in sedimentary rocks. *Geol. Soc. Am. Bull.* **65**, 937–938.

KENNEDY, W.J. & GARRISON, R.E. (1975) Morphology and genesis of nodular hardgrounds in the Upper Cretaceous of southern England. *Sedimentology* **22**, 311–386.

KENYON, N.H. & STRIDE, A.H. (1970) The tide-swept continental shelf sediments between the Shetland Isles and France. *Sedimentology* **14**, 159–175.

KENYON, N.H., BELDERSON, R.H., STRIDE, A.H. & JOHNSON, M.A. (1981) Offshore tidal sand banks as indicators of net sand transport and as potential deposits. In: *Holocene Marine Sedimentation in the North Sea Basin* (Eds Nio, S.D., Schuttenholm, R.T.E. & van Weering, T.C.E.). Spec. Publ. Int. Assoc. Sediment. **5**, 361–383.

KIDWELL, S. & JABLONSKI, D. (1983) Taphonomic feedback: ecological consequences of shell accumulation. In: *Biotic Interactions in Recent and Fossil Benthic Communities* (Eds Tevesz, M.J. & McCall, P.L.) pp. 195–248. Plenum Press, New York.

KLEIN, G. DEV. (1982) Probable sequential arrangement of depositional systems on cratons. *Geology* **10**, 17–22.

KOMAR, P.D. (1973) Computer models of delta growth due to sediment input from rivers and longshore transport. *Geol. Soc. Am. Bull.* **84**, 2217–2219.

KOMAR, P.D. (1983) Nearshore currents and sand transport on beaches. In: *Physical Oceanography of Coastal and Shelf Seas* (Ed. Johns, B.) pp. 76–109. Elsevier, Amsterdam.

KRYNINE, P.D. (1942) Differential sedimentation and its products during one complete geosynclinal cycle. *Ann. Prim. Congr. Panamericano de Ingen de Minas Geolog. (Chile), Geologica*, Part 1, Vol. 2, pp. 537–561.

KUEHL, S.A., NITTROUER, C.A. & DEMASTER, D.J. (1982) Modern sediment accumulation and strata formation on the Amazon Continental Shelf. *Mar. Geol.* **49**, 279–300.

LAWTON, T.F. (1982) Lithofacies correlations within the upper Cretaceous Indianola group, Central Utah. In: *Overthrust Belt of Utah*. Utah Geol. Assoc. Publ. 10, 332, pp.

LEATHERMAN, S.P. (1979) Migration of Assateague Island, Maryland, by inlet and overwash processes. *Geology* **7**, 104–107.

LEATHERMAN, S.P. (1983) Barrier dynamics and landward migration with Holocene sea-level rise. *Nature* **301**, 415–418.

LEATHERMAN, S.P. (1985) Geomorphic and stratigraphic analysis of Fire Island, New York, *Mar. Geol.* **63**, 153–172.

LEATHERMAN, S.P. & JONEJA, (1980) *Geomorphic Analysis of South Shore of Long Island Barriers*. National Park

Service Coastal Research Unit, University of Massachusetts, Report 47, 161 pp.

LECKIE, D.A. & WALKER R.G. (1982) Storm and tide dominated shorelines in Cretaceous Moosebar–Lower Gates interval—outcrop equivalents of deep basin gas trap in Western Canada. *Bull. Am. Assoc. Petrol. Geol.* **66**, 138–157.

LIU, J.T. & ZARILLO, G.A. (1990) Shoreface dynamics: evidence from bathymetry and surficial sediments. *Mar. Geol.* **94**, 37–54.

LOUTIT, T.S., HARDENBOL, J., VAIL, P.R. & BAUM, J.R. (1988) Condensed sections: the key to age dating and correlation of continental margin sequences. In: *Sea-level Changes: an Integrated Approach* (Eds Wilgus, C.K., Hastings, B.S., Kendall, C.G. St. C., Posamentier, H.W., Ross, C.A. & van Wagoner, J.C.). Soc. Econ. Paleontol. Mineral. Spec. Publ. **42**, 183–216.

LUDWICK, J.C. (1970) Sand waves and tidal channels in the entrance to Chesapeake Bay. *Virginia J. Sci.* **21**, 178–184.

LUDWICK, J.C. (1974) Tidal currents and zig-zag sand shoals in a wide estuary entrance. *Geol. Soc. Am. Bull.* **85**, 717–726.

MALDONADO, A., SWIFT, D.J.P., YOUNG, R.A., HAN, G., NITTROUER, C.E., DEMASTER, D., REY, J., PALOMO, C., ACOSTA, J., BALLESTER, C.E. & CASTELLVI, I. (1983) Sedimentation on the Valencia continental shelf: preliminary results. *Continental Shelf Res.* **3**, 195–211.

MANN, R.G., SWIFT, D.J.P. & PERRY, R. (1981) Size classes of low transverse bedforms in a subtidal environment, Nantucket Shoals, North American Atlantic Shelf. *Geomarine Lett.* **1**, 39–43.

MANSPIESER, W. (1985) Early Mesozoic history of the Atlantic Continental Margin. In: *Geologic Evolution of the United States Atlantic Margin* (Ed. Poag, C.W.) pp. 1–23. Hutchinson Ross, New York.

MASTERSON, W.D. & PARIS, C.E. (1987) Depositional history and reservoir description of the Kuparuk River Formation, North Slope Alaska. In: *Alaskan North Slope Geology* (Eds Tailleur, I. & Wiemar, P.). Pacific Section, Soc. Econ. Paleontol. Mineral. 1, 298 pp.

MAY, J.P., & TANNER, W.F. (1973) The littoral power gradient and shoreline changes. In: *Coastal Geomorphology* (Ed. Coates, D.R.) pp. 43–60. Publications in Geomorphology, State University of New York, Binghamton, NY.

MAZZULLO, J. & CRISP, J. (1988) Source and dispersal of coarse silt on the South Texas Continental shelf. *Mar. Geol.* **69**, 131–148.

MUZZULLO, J. & WITHERS, K.D. (1984) Sources, distribution and mixing of late Pleistocene and Holocene sands on the South Texas Continental shelf. *J. Sediment. Petrol.* **54**, 113–131.

McCAVE, I.N. (1972) Sediment transport and escape of fine-grained sediment from shelf areas. In: *Shelf Sediment Transport, Process and Pattern* (Eds Swift, D.J.P., Pilkey, O.H. & Duane, D.B.) pp. 215–248. Dowden Hutchinson & Ross, Stroudsburg, PA.

McCAVE, I.N. (1973) Mud in the North Sea. In: *North Sea Science* (Ed. Goldborg, E.) pp. 75–100.

McCUBBIN, D.G. (1982) Barrier island and strandplain facies. In: *Sandstone Depositional Environments* (Eds Scholle, P.A. & Spearing, D.). Mem., Am. Assoc. Petrol. Geol. **31**, 247–280.

MCLEAN, S.R. (1981) The role of non uniform roughness in the formation of sand ribbons. *Mar. Geol.* **42**, 49–74.

MEADE, R.H. (1972) Sources and sinks of suspended matter on continental shelves. In: *Shelf Sediment Transport, Process and Pattern* (Eds Swift, D.J.P., Pilkey, O.H. & Duane, D.B.) pp. 249–262. Dowden Hutchinson & Ross, Stroudsburg, PA.

MEREWETHER, E.A. (1983) The Frontier Formation and Mid-Cretaceous Orogeny in the foreland of southwestern Wyoming. *Mountain Geol.* **20**, 121–138.

MIALL, A.D. (1984) *Principles of Sedimentary Basin Analysis*. Springer-Verlag, New York, 490 pp.

MOSLOW, T.F. & TYE, R.S. (1985) Recognition and characterization of Holocene tidal inlet sequences. *Mar. Geol.* **63**, 129–152.

MURRAY, S.P., COLEMAN, J.M., ROBERTS, H.H. & SALAMA, M. (1981) Accelerated currents and sediment transport of the Damietta Nile Promontory. *Nature* **293**, 51–54.

NELSON, L.H. (1982) Modern shallow water graded sand layers from storm surges, Bering Shelf: a mimic of Bouma sequences and turbidite systems. *J. Sediment. Petrol.* **52**, 537–545.

NICHOLS, M.M. (1989) Sediment accumulation rates and relative sea level rise in lagoons. *Mar. Geol.* **88**, 201–220.

NIEDORODA, A.W., SWIFT, D.J.P., FIGUEIREDO, A.G. & FREELAND, G.L. (1985) Barrier island evolution, Middle Atlantic Shelf, U.S.A. Part II: evidence from the shelf floor. *Mar. Geol.* **63** 363–396.

NITTROUER, C.A. & STERNBERG, R.W. (1981) The formation of sedimentary strata in an allochthonous shelf environment: the Washington Continental Shelf. *Mar. Geol.* **42**, 201–232.

NITTROUER, C.A., STERNBERG, R.W., CARPENTER R. & BENNETT, J.J. (1979) The use of $^{210}$Pb geochronology as a sedimentological tool: application to the Washington Continental Shelf. *Mar. Geol.* **21**, 297–316.

NITTROUER, C.A., DEMASTER, D.J., KUEHL, S.A. & MCKEE, B.A. (1986) Association of sand with mud deposits accumulating on continental shelves. In: *Shelf Sands and Sandstones* (Eds Knight, R.J. & McLean, J.R.). Can. Soc. Petrol. Geol. Mem. **11**, 17–25.

NOON, P.L. (1980) *Surface to subsurface stratigraphy of the Dakota Sandstone (Cretaceous) and adjacent units along the eastern flank of the San Juan Basin, New Mexico and Colorado.* MS thesis, Bowling Green State University, 133 pp.

O'BRIEN, M.P. (1969) Dynamics of tidal inlets. In: *Simposio Internacional de Lagunas Costeras* (Eds Castañares, A. & Phleger, F.) pp. 397–406. Univ. Nac. Autonoma De Mexico, Mexico City.

OERTEL, G.F. (1976) Characteristics of suspended sediments in estuaries and nearshore waters of Georgia. *Southeast. Geol.* **18**, 107–118.

OERTEL, G.F. (1977) Geomorphic cycles in ebb deltas and related patterns of shore erosion and accretion. *J. Sediment. Petrol.* **47**, 1121–1131.

OERTEL, G.F. (1987) Backbarrier and shoreface controls on inlet channel orientation. Coastal Sediments '87, Waterways Division, American Society of Civil Engineers, New Orleans, 12–14 May 1987, pp. 2022–2029.

OERTEL, G.F. (1988) Process of sediment exchange between tidal inlets, ebb tidal deltas and barrier islands. In: *Hydrodynamics and Sediment Dynamics of Tidal Inlets* (Eds Aubrey, D.G. & Weishar, L.) pp. 297–318. Springer-Verlag, New York.

OERTEL, G.F. & DUNSTAN, W.M. (1981) Suspended sediment distribution and certain aspects of phytoplankton production off Georgia, U.S.A. *Mar. Geol.* **40**, 171–197.

OERTEL, G.F. & HOWARD, J.D. (1972) Wave circulation and sedimentation at estuary entrances on the Georgia Coast. In: *Shelf Sediment Transport, Process and Pattern* (Eds Swift, D.J.P., Pilkey, O.H. & Duane, D.B.) pp. 411–428. Dowden Hutchinson & Ross, Stroudsburg, PA.

OERTEL, G.F., KEARNEY, M.S., LEATHERMAN, S.P. & WOO, H.J. (1989) Anatomy of a barrier platform, outer barrier lagoon Southern Delmarva Peninsula, Virginia. *Mar. Geol.* **88**, 303–318.

PALMER, J.J. & SCOTT, A.J. (1984) Stacked shoreline and shelf sandstone of La Ventana Tongue (Campanian), Northwestern New Mexico. *Bull., Am. Assoc. Petrol. Geol.* **68**, 74–91.

PARKER, G., LANFREDI, N.W. & SWIFT, D.J.P. (1982) Substrate response to flow in a southern hemisphere ridgefield. *Sediment. Geol.* **33**, 195–216.

PATTERSON, J.E., JR. (1983) *Exploration potential and variations in shelf plume sandstones, Navarro Group (Mastrichtian), east central Texas.* MS thesis, University of Texas, Austin, TX, 91 pp.

PENLAND, S. & BOYD, R. (1982) Mississippi delta barrier systems: summary and overview. In: *Deltaic Sedimentation of the Louisiana Coast* (Ed. Nummedal, D.) p. 71. Society of Economic Paleontologist and Minerologists, Tulsa, OK.

PENLAND, S. & SUTER, J.R. (1989) The geomorphology of the Mississippi River chenier plain. *Mar. Geol.* **90**, 231–258.

PENLAND, S., BOYD, R., NUMMEDAL, D. & ROBERTS, H.H. (1981) Deltaic barrier development on the Louisiana coast. *Trans. Gulf Coast Assoc Geol. Soc.* **XXXI**, 471–476.

PENLAND, S., SUTER, J.R. & BOYD., R. (1985) Barrier island arcs along abandoned Mississippi river deltas. *Mar. Geol.* **63**, 197–233.

PHILLIPS, S. (1987) *Shelf sedimentation and depositional sequence stratigraphy of the upper Cretaceous Woodbine–Eagleford groups, east Texas.* PhD dissertation, Cornell University, Ithaca, NY 232 pp.

PHILLIPS, S. & SWIFT, D.J.P. (1985) Shelf sandstones in the Woodbine–Eagleford Interval, east Texas: a review of depositional models. In: *Shelf Sands and Sandstone Reservoirs* (Eds Tillman, R.W., Swift, D.J.P. & Walker, R.G.) Soc. Econ. Paleontol. Mineral. Short Course Notes 13, 503–588.

PILKEY, O.H. & DAVIS, T.W. (1987) An analysis of coastal recession models: North Carolina Coast. In: *Sea-level Fluctuation and Coastal Evolution* (Eds Nummedal, D., Pilkey, O.H. & Howard, J.D.). Soc. Econ. Paleontol. Mineral. Spec. Publ. 41, 59–68.

PILKEY, O.H., TRUMBULL, J.V.A. & BUSH, D.M. (1978) Equilibrium shelf sedimentation, Rio De La Plata Shelf, Puerto Rico. *J. Sediment. Petrol.* **48**, 389–400.

POAG, W.C. (1985) Depositional history and stratigraphic reference section for central Baltimore Canyon Trough. In: *Geologic Evolution of the United States Atlantic Margin* (Ed. Poag, C.W.) pp. 217–264. Hutchinson Ross, New York.

POSAMENTIER, H.W. & VAIL, P.R. (1988) Eustatic controls on clastic deposition, II—sequence and systems tract models. In: *Sea-level Changes: an Integrated Approach* (Eds Wilgus, C.K., Hastings, B.S., Kendall, C.G.St.C., Posamentier, H.W., Ross, C.A. & van Wagoner, J.C.). Soc. Econ. Paleontol. Mineral. Spec. Publ. 42, 124–154.

POSAMENTIER, H.W., JERVEY, M.T. & VAIL, P.R. (1988) Eustatic controls on clastic deposition, I—conceptual framework. In: *Sea-level Changes: an Integrated Approach* (Eds Wilgus, C.K., Hastings, B.S., Kendall, C.G.St.C., Posamentier, H.W., Ross, C.A. & van Wagoner, J.C.). Soc. Econ. Paleontol. Mineral. Spec. Publ. 42, 109–124.

POSTMA, H. (1967) Sediment transport and sedimentation in the marine environment. In: *Estuaries* (Ed. Lauff, G.H.) pp. 150–180. American Association for Advancement of Science, Washington, DC.

PRITCHARD, D.W. (1967) Observations of circulation in coastal plain estuaries. In: *Estuaries* (Ed. Lauff, G.H.) pp. 37–44. American Association for Advancement of Science, Washington, DC.

RABALAIS, N.N. (1988) Hypoxia on the continental shelf of the Northwestern Gulf of Mexico. In: *Proceedings, Symposium on the Physical Oceanography of the Louisiana–Texas Shelf* (Ed. Mitchell, T.) pp. 81–87. US Minerals Management Service, Metairie, LA.

REINECK, H.E. & SINGH, I.B. (1980) *Depositional Sedimentary Environments.* Springer-Verlag, New York, 549 pp.

REINSON, G.E. (1979) Facies models 14, barrier islands systems. *Geosci. Can.* **6**, 51–58.

RHODES, E.G. (1982) Depositional model for a chenier plain, Gulf of Carpenteria, Australia. *Sedimentology* **29**, 201–221.

ROBERTS, H.H. & COLEMAN, J.M. (1988) Lithofacies characteristics of shallow expanded and condensed sections of the Louisiana distal shelf and upper slope. *Trans. Gulf Coast Assoc. Geol. Soc.* **37**, 272–301.

RINE, J.M. & GINSBURG, R.N. (1985) Depositional facies of a mud shoreface in Suriname, South America—a mud analog to sandy shallow marine deposits. *J. Sediment. Petrol.* **55**, 633–652.

ROBINSON, A.H.W. (1980) Erosion and accretion along part of the Suffolk coast of east Anglia, England. *Mar. Geol.* **37**, 133–145.

RUBY, C.H. (1981) *Cretaceous Rocks of Western North America: a guide to Terrigenous Clastic Rock Identification.* Research Planning Institute, Columbia, SC, 100 pp.

SAHL, L.E., MERILL, W.J., MCGRAIL, D.W. & WEBB, J.A. (1987) Transport of mud on continental shelves: evidence from the Texas Shelf. *Mar. Geol.* **76**, 33–43.

SCHIEING, M.H. & GAYNOR, G.C. (1991) Shelf sand plume model: a critique. *Sedimentology*, in press.

SCHLEE, J. (1968) Atlantic continental shelf and slope of the United States—sediment texture of the northeastern part. *U.S. Geol. Surv. Prof. Pap.* **529**.

SCHWARTZ, M.L. (1971) The multiple causality of barrier islands. *J. Geol.* **79**, 91–94.

SEILACHER, A. (1982) General remarks about event deposits. In: *Cyclic and Event Stratification* (Eds Einsele, G. & Seilacher, A.) pp. 161–174. Springer-Verlag, New York.

SHA, L.P. (1989) Holocene–Pleistocene interface and three dimensional geometry of the ebb delta complex, Texel Inlet, The Netherlands. *Mar. Geol.* **89**, 207–228.

SHARAF EL DIN, S.H. (1977) Effect of the Aswan High Dam on the Nile flood and the estuarine and coastal circulation pattern along the Mediterranean Egyptian coast. *Limnol. Oceanogr.* **22**, 207–230.

SHEPARD, F.P. & MOORE, D.G. (1954) Sedimentary environments differentiated by course fraction studies. *Bull., Am. Assoc. Petrol. Geol.* **38**, 1792–1802.

SHEPARD, F.P. & MOORE, D.G. (1955) Central Coast sedimentation: characteristics of sedimentary environment, recent history and diagenesis. *Bull., Am. Assoc. Petrol. Geol.* **39**, 1463–1593.

SHEPARD, F.P. & WANLESS, H.R. (1971) *Our Changing Coastlines*. McGraw Hill, New York, 233 pp.

SHORT, A.D. (1984) Beach and nearshore facies: southeast Australia. *Mar. Geol.* **60**, 261–282.

STAMP, L.D. (1921) On cycles of sedimentation in the Eocene strata of the Anglo-Franco-Belgian basin. **58**, 108–114, 194–200.

STANLEY, D.J. (1988) Low sediment accumulation rates and erosion on the middle and outer Nile Delta shelf of Egypt. *Mar. Geol.* **84**, 111–117.

STONE, W.D. (1972) Stratigraphy and exploration of the Lower Cretaceous Muddy Formation, Northern Powder River Basin, Montana and Alaska. *Mountain Geol.* **9**, 355–378.

STRIDE, A.H. (1982) *Offshore Tidal Sands*. Chapman & Hall, New York, 222 pp.

STUBBLEFIELD, W.L. & SWIFT, D.J.P. (1981) Grain size variation across sand ridges, New Jersey Continental shelf, USA. *Geomarine Lett.* **1**, 28–32.

SURLYK, F. & NOE-NYGARD, N. (1991) Sand bank and dune facies associated with a wide intracratonic sea. In: *Three Dimensional Facies Architecture of Terrigenous Clastic Sediments and its Implications for Hydrocarbon Discovery. Concepts and Models Series* (Eds Miall, A.D. and Tyler, N.). Society of Economic Paleontologists and Mineralogists, Tulsa, OK, in press.

SUTER, J.H. & BERRYHILL, H.L., Jr. (1985) Late Quaternary Shelf margin deltas, Northwest Gulf of Mexico. *Bull. Am. Assoc. Petrol. Geol.* **69**, 77–91.

SWIFT, D.J.P. (1968) Coastal erosion and transgressive stratigraphy. *J. Geol.* **76**, 444–456.

SWIFT, D.J.P. (1973) Delaware Shelf Valley: estuary retreat path, not drowned river valley. *Geol. Soc. Am. Bull.* **84**, 2743–2748.

SWIFT, D.J.P. (1975a) Tidal sand ridges and shoal retreat massif. *Mar. Geol.* **18**, 105–133.

SWIFT, D.J.P. (1975b) Barrier island genesis: evidence from the Middle Atlantic Shelf, Eastern USA. *Sediment. Geol.* **14**, 1–43.

SWIFT, D.J.P. (1976a) Coastal sedimentation. In: *Marine Sediment Transport and Environmental Management* (Eds Stanley, D.J. & Swift, D.J.P.) pp. 245–310. Wiley, New York.

SWIFT, D.J.P. (1976b) Continental Shelf Sedimentation. In: *Marine Sediment Transport and Environmental Management* (Eds Stanley, D.J. & Swift, D.J.P.) pp. 311–410. Wiley, New York.

SWIFT, D.J.P. & FIELD, M.F. (1981) Evolution of a classic ridge field, Maryland Sector, North American Inner Shelf. *Sedimentology* **28**, 461–482.

SWIFT, D.J.P. & HERON, S.D. (1969) Stratigraphy of the Carolina Cretaceous. *Southeast. Geol.* **10**, 201–245.

SWIFT, D.J.P. & LUDWICK, J.C. (1976) Substrate response to hydraulic process: grain size frequency distributions and bedforms. In: *Marine Sediment Transport and Environmental Management* (Eds Stanley, D.J. & Swift, D.J.P.) pp. 159–196. Wiley, New York.

SWIFT, D.J.P. & MOSLOW, T.F. (1982) Holocene transgression in south central Long Island, New York, Discussion. *J. Sediment. Petrol.* **47**, 1454–1474.

SWIFT, D.J.P., KOFOED, J.W., SAULSBURY, F.P. & SEARS, P. (1972) Holocene evolution of the shelf surface, central and southern Atlantic shelf of North America. In: *Shelf Sediment Transport, Process and Pattern* (Eds Swift, D.J.P., Duane, D.B. & Pilkey, O.H.) pp. 100–148. Dowden Hutchinson & Ross, Stroudsburg, PA.

SWIFT, D.J.P., DUANE, D.B. & MCKENNY, T.F. (1973) Ridge and swale topography of the middle Atlantic Bight: secular response to the Holocene hydraulic regime. *Mar. Geol.* **15**, 227–247.

SWIFT, D.J.P., NELSEN, T.A., MCHONE, J., HOLLIDAY, B., PALMER, H. & SHIDELER, G. (1977) Holocene evolution of the inner shelf of southern Virginia. *J. Sediment. Petrol.* **47**, 1454–1474.

SWIFT, D.J.P., SEARS, P.C., BOHLKE, B. & HUNT, R. (1978) Evolution of a shoal retreat massif, North Carolina Shelf: evidence from areal geology. *Mar. Geol.* **27**, 19–423.

SWIFT, D.J.P., CLARKE, T., YOUNG, R.A. & VINCENT, C.E. (1981) Sediment transport in the Middle Atlantic Bight of North America, synopsis of recent observations. In: *Holocene Marine Sedimentation in the North Sea Basin* (Eds Nio, S.D., Schuttenholm, R.T.E. & van Weering, T.C.E.). Spec. Publ. Int. Assoc. Sediment. **5**, 361–383.

SWIFT, D.J.P., NIEDORODA, A.W., VINCENT, C.E. & HOPKINS, T.S. (1985a) Barrier island evolution Middle Atlantic Shelf, U.S.A. Part I: shoreface dynamics. *Mar. Geol.* **63**, 331–361.

SWIFT, D.J.P., STUBBLEFIELD, W.L., CLARKE, T.L., YOUNG, R.A., FREELAND, G.L., HARVEY, G. & HILLARD, B. (1985b) Sediment budget near the New York Dumpsites: implications for pollutant dispersal. In: *Wastes in the Ocean*, Vol. 6, *Nearshore Waste Disposal* (Ed. Duedall, I.) Wiley, New York, 835 pp.

SWIFT, D.J.P., HAN, G. & VINCENT, C.E. (1986a) Fluid process and sea floor response on a modern storm-dominated shelf: middle Atlantic Shelf of North America. Part 1, the storm current regime. In: *Shelf Sands and Sandstone Reservoirs* (Eds Knight, R.J. & McLean, J.). Can. Soc. Petrol. Geol. Mem. 11, 99–119.

SWIFT, D.J.P., THORNE, J.A. & OERTEL, G.F. (1986b) Fluid process and sea floor response on a storm dominated shelf: Middle Atlantic shelf of North America. Part II: response of the shelf floor. In: *Shelf Sands and Sandstone Reservoirs* (Eds Knight, R.J. & McLean, J.R.). Can. Soc. Petrol. Geol. Mem. 11, 191–211.

SWIFT, D.J.P., HUDELSON, P.M., BRENNER, R.L. &

THOMPSON, P. (1987). Shelf construction in a foreland basin: storm beds, shelf sand bodies and shelf-slope sequences in the Upper Cretaceous Mesaverde group, Book Cliffs, Utah. *Sedimentology* **34**, 423–457.

TAG, R.J., ABU-OUF, M. & EL SHATER, A. (1991) Characteristics of coastal sediments between Wadi Al-Fagh and Wadi Al Qunfidah, southeastern Red Sea. *Mar. Geol.*, in press.

THOMPSON, D.W. (1961) *On Growth and Form* (1969 abridged edn). Cambridge University Press, Cambridge, 346 pp.

TWICHELL, D.C. (1983) Bedform distribution and inferred sand transport on Georges Bank, United States Atlantic Continental Shelf. *Sedimentology* **30**, 695–710.

TWICHELL, W.C., MCCLENNEN, C.E. & BUTMAN, B. (1980) Morphology and processes associated with the accumulation of the fine grained sediment deposit on the Southern New England Shelf. *J. Sediment. Petrol.* **51**, 269–280.

VAIL, P.R., MITCHUM, R.M., JR., THOMPSON, S, III, SANGREE, J.R., DUBB, J.N. & HASLID, W.G. (1977) Sequence stratigraphy and global changes of sea level. In: *Seismic Stratigraphy—Applications to Hydrocarbon Exploration* (Ed. Payton, C.E.). Mem., Am. Assoc. Petrol. Geol. 26, 49–205.

VAN STRAATEN, L.M.J.U. (1956) Composition of shell beds formed in tidal flat environment in The Netherlands and in the Bay of Arcachon (France). *Geol. Mijnbouw, Niew Ser.* **18**, 211–226.

VAN STRAATEN, L.M.J.U. (1965a) Coastal barrier deposits in south and north Holland, in particular in the areas around Scheveningen and Ijmuiden. *Meded. Geol. Stichting Niewe Ser.* **17**, 41–75.

VAN STRAATEN, L.M.J.U. (1965b) Sedimentation in the northeastern part of the Adriatic Sea. In: *Submarine Geology and Geophysics* (Eds Whittard, W.H. & Bradshaw, W.) pp. 143–162. Butterworths, London.

VAN WAGONER, J.C., MITCHUM, R., POSAMENTIER, H.W., MITCHUM, R., JR., VAIL, P.R., SARG, J.F., LOUTIT, T.S. & HARDENBOL, J. (1988) An overview of the fundamentals of sequence stratigraphy and key definitions of sequence stratigraphy. In: *Sea-level Changes: an Integrated Approach* (Eds Wilgus, C.K., Hastings, B.S., Kendall, C.G. St. C., Posamentier, H.W., Ross, C.A. & Van Wagoner, J.C.). Soc. Econ. Paleontol. Mineral. Spec. Publ. 42, 39–47.

VEEVERS, J.J. (1981) Morphotectonics of rifted continental margins: in embryo (East Africa), youth (Africa–Arabia), and maturity (Australia). *J. Geol.* **89**, 57–82.

VISSER, M.J. (1980) Neap-spring cycles reflected in Holocene subtidal large-scale bedform deposits: a preliminary note. *Geology* **8**, 5453–5456.

WALKER, R.G. (1979) Facies models (2nd edn 1984). *Geosci. Can. Reprint Ser. Geol. Assoc. Can.* **1**, 317 pp.

WALTHER, J. (1874) *Einleitung in die Geologie als Historische*, Vol. 3. Fischer-Verlag, Wissenschaft. Jena, 229 pp.

WARD, L.G. & ASHLEY, G.M. (1989) Introduction: coastal lagoonal systems. *Mar. Geol.* **88**, 181–185.

WELLS, J.T. & ROBERTS, H.H. (1980) Fluid mud dynamics and shoreline stabilization Louisiana chenier plain. *Proceedings, 17th International Coastal Engineering Conference*, American Association of Civil Engineers, pp. 1382–1400.

WRIGHT, L.D. (1977) Sediment transport and deposition at river mouths: a synthesis. *Geol. Soc. Am. Bull.* **55**, 857–868.

WRIGHT, L.D. & COLEMAN, J.M. (1974) Mississippi River mouth processes: effluent dynamics and morphologic development. *J. Geol.* **82**, 751–778.

WRIGHT, L.D. & COLEMAN, J.M. (1977) Variations in morphology in major river deltas as functions of ocean wave and river discharge regimes. *Bull., Am. Assoc. Petrol. Geol.* **57**, 370–398.

WRIGHT, L.D., CHAPPELL, J., THOM, B.G., BRADSHAW, M.P. & COWELL, P. (1979) Morphodynamics of reflective and dissipative beach and inshore systems: southeastern Australia. *Mar. Geol.* **32**, 105–140.

WRIGHT, R. (1986) Cycle stratigraphy as a paleogeographic tool: Point Lookout Sandstone, southeastern San Juan Basin, New Mexico. *Geol. Soc. Am. Bull.* **96**, 661–673.

ZENKOVITCH, V.P. (1967) *Processes of Coastal Development*. Wiley Interscience, New York, 738 pp.

# Sedimentation on continental margins, V: parasequences

D.J.P. SWIFT[*], S. PHILLIPS[†] and J.A. THORNE[‡]

[*] *Department of Oceanography, Old Dominion University, Norfolk, VA 23529, USA;* [†] *ARCO Alaska Inc., 700 G St., Anchorage, AK 99501, USA; and* [‡] *Research and Technical Services, ARCO Oil and Gas Co., 2300 W. Plano Parkway, Plano, TX 23575, USA*

## ABSTRACT

Depositional facies, and arrays of depositional facies (systems and systems tracts) are organized into repeating vertical arrays called *depositional sequences*. The smallest, simplest sequences are a few metres thick, and may consist of upward-coarsening bed sets separated by flooding surfaces. Such high-frequency sequences are termed *parasequences*. Facies, systems and systems tracts are defined primarily on the basis of lithology, while sequences are defined primarily on the basis of their geometry. The two types of sediment body classification overlap at the scale of a parasequence; a parasequence is essentially a single depositional system. Since a systems tract is defined as 'a linkage of contemporaneous depositional systems' then a parasequence, if traced through more than one system, becomes a depositional systems tract.

Parasequences are believed to be responses to regime oscillations of thousands to hundreds of thousands of years duration. They tend to be monolithologic (to consist of simple, upward-coarsening successions of beds). However, such successions, traced towards the basin margin, tend to develop internal facies contrasts and evidence of significant erosion at bounding surfaces. We therefore use the term 'parasequence' for any scale of depositional sequence of higher frequency than the dominant frequency (usually third order).

At somewhat longer time-scales (hundreds of thousands or millions of years) parasequences become grouped into larger scale sequences. Such larger scale sequences exhibit a more complex structure, in which subunits have contrasting parasequence stacking patterns. Since the sequence subunits consist of repetitive successions of parasequences, they are properly styled *parasequence sets*. However, the designation 'systems tracts' for sequence subunits has entered the literature. Problems are minimized if the subunits are recognized as *geometric systems tracts*, defined on the basis of their parasequence stacking geometries, rather than as the much smaller scale depositional systems tracts, defined on the basis of facies architecture.

Comparison of the sequence structure of recent and subrecent deposits on modern continental shelves with that observed in older sections suggests that parasequence architecture varies with the regime parameter and the preservation parameter, described by Thorne & Swift (this volume, pp. 196–197). At large (accommodation-dominated) values of the regime ratio, and low values of the preservation parameter, regressive deposits generated by high-frequency regime oscillations are destroyed by the subsequent transgression, resulting in 'collapsed parasequences'. As the regime ratio decreases, asymmetrical back-stepping parasequences are created, in which the transgressive half-sequence is dominant. As the regime passes into the supply-dominated mode, parasequences develop downlapping geometry and the regressive half-sequence becomes dominant.

Parasequences in coastal settings may stem from complex origins, in which both autocyclic forcing (river avulsion) and allocyclic forcing (external to the basin) may play a role. Farther offshore, cyclicity may be in large measure allocyclic, a response to orbital forcing (the Milankovich mechanism). It is likely, however, that throughout much of Earth history, orbital forcing has modulated the depositional regime primarily by influencing rainfall patterns, and therefore sediment input, rather than through the medium of glacio-eustacy. It is also possible that non-linear interactions of regime variables (chaotic behaviour) can generate parasequence successions. This form of autocyclicity has been little studied.

# INTRODUCTION

Other papers in this volume have presented a regime model for continental margin sedimentation. Swift & Thorne (this volume, pp. 3–31) describe the depositional regime as governed at geological scales by interdependent variables, whose mutual adjustments maintain a state of dynamic equilibrium. The regime may be described in terms of equilibrium surfaces controlled by homeostatic responses of erosion or deposition. Thorne & Swift (this volume, pp. 33–58) develop an analytical foundation for this approach.

Thorne et al. (this volume, pp. 59–87) describe the high-frequency (flood, storm and tidal) kinetics of erosion, transportation and deposition on shallow continental margins, leading to the formation of event beds. The analysis of the dynamics of strata formation by Thorne et al. leads to rules for facies differentiation and the geometry of facies arrays (depositional systems and depositional systems tracts), as presented by Swift et al. (this volume, pp. 89–152). Depositional systems tracts are the largest scale of sedimentary body that forms under constant regime conditions, and are so defined by Swift et al. (this volume, pp. 89–152). However, Swift et al. note that as regime conditions vary cyclically or quasi-cyclically, yet larger scale patterns of stratigraphic organization appear, in which repetitive successions of depositional systems and system tracts are repeated as *depositional sequences* at a hierarchy of spatial scales (Vail et al., 1977b).

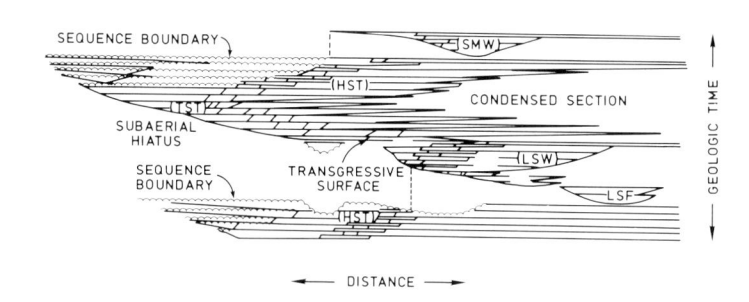

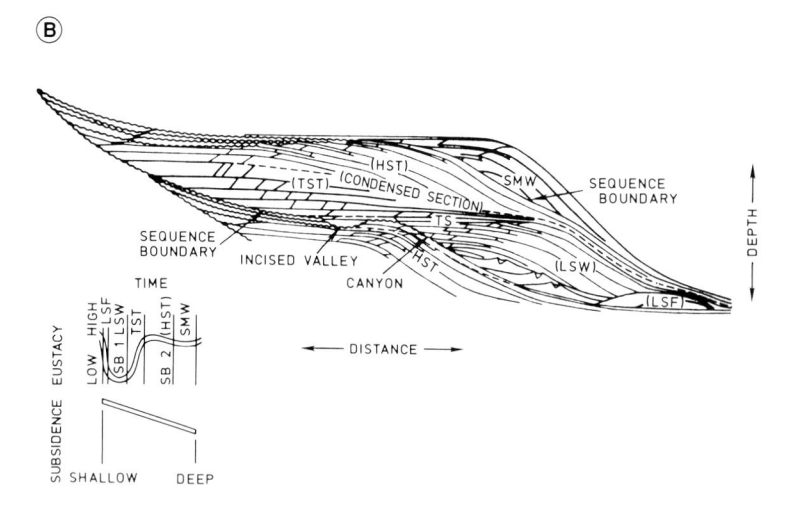

**Fig. 1.** The Vail (1987) model for a depositional sequence. (A) Chronostratigraphic diagram; (B) lithostratigraphic diagram. HST = highstand systems tract; TST = transgressive systems tract. Lowstand systems tract elements include: SMW = shelf-margin wedge; LSW = lowstand wedge; LSF = lowstand fan. (After Haq et al., 1987.)

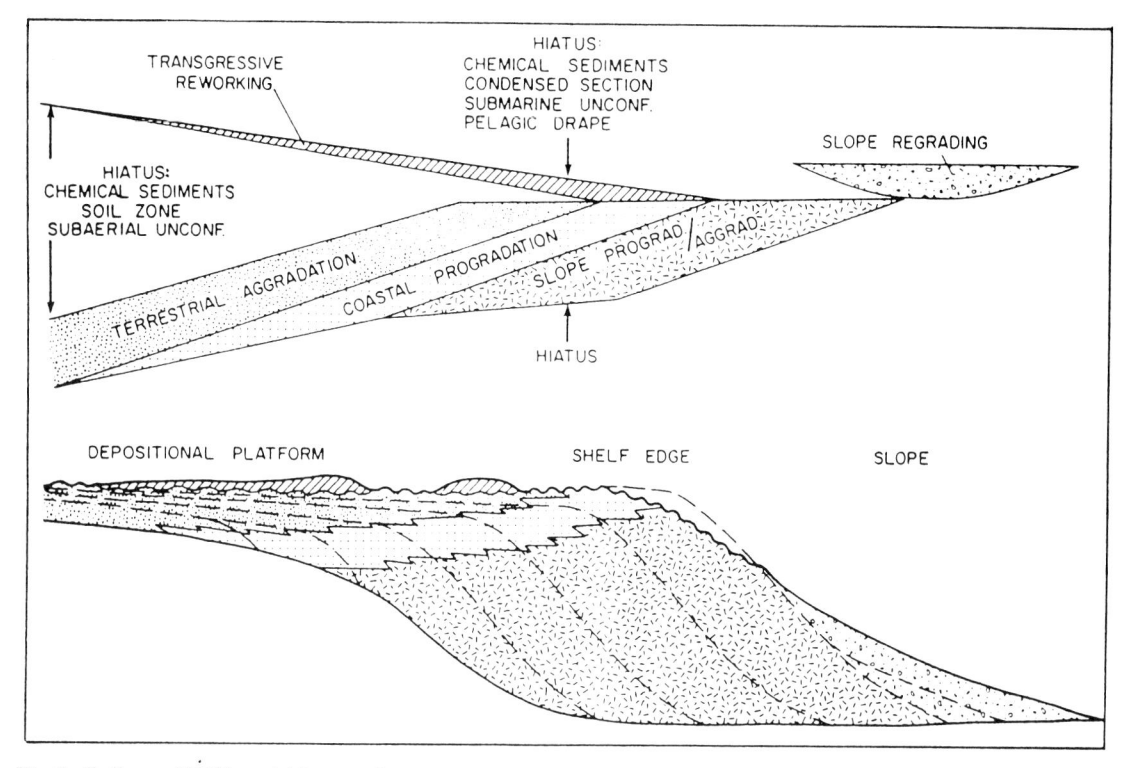

**Fig. 2.** Galloway (1989) model for genetic stratigraphic sequences. (From Galloway, 1989a.)

In this paper we focus on the zone of overlap between the two classifications. How do depositional systems and systems tracts fit into the larger depositional sequences? What mechanisms create these several scales of basin architecture? In order to investigate these problems we review existing sequence models. We review depositional systems and depositional sequences in the Quaternary record of modern continental margins. We then examine depositional sequences in the Neogene of the Atlantic continental margin, and depositional sequences of the Cretaceous Western Interior Basin. On the basis of these examples we draw conclusions concerning the relationship between systems stratigraphy and sequence stratigraphy.

## STATEMENT OF THE PROBLEM

### Vail versus Galloway models

We need to briefly compare the most recent models of depositional sequence formation (Vail, in Van Wagoner *et al.*, 1987; Galloway, 1989a,b; see Figs 1, 2 and Table 1) before we investigate sequence stratigraphic concepts. The Vail (1987) model consists of a hierarchy of lithostratigraphic units defined in terms of their stratal geometry. The units are more specifically *allostratigraphic* (Salvador, 1987), in the sense that they are bounded by unconformities or their correlative conformities, or by hiatuses,

**Table 1.** Sequence stratigraphic terminology

| Van Wagoner *et al.* (1988) | Galloway (1989a) |
|---|---|
| Facies | Facies |
| Depositional system | Depositional system |
| (Depositional) Systems tract[a] | Component (set) |
| Depositional sequence | Genetic stratigraphic sequence |

[a] Modifiers and substantives in parentheses are not always used.

such as flooding surfaces. This system stands outside of, and parallel to, formal stratigraphic taxonomy.

The basic unit in the Vail (1987) model is the depositional sequence, defined as 'a conformable succession of genetically related strata bounded above or below by unconformities or their correlative conformities' (Mitchum, 1977). Vailian sequences consist of a hierarchy of subunits. In order of increasing size, these are facies, depositional systems, and systems tracts. Facies, never clearly defined in the Vailian scheme, are implicitly rock volumes of uniform lithologic character. Depositional systems are 'three-dimensional assemblages of facies' (Van Wagoner *et al.*, 1987, 1988, 1990; versus 'assemblages of process-related facies', in the Fisher & McGowen paper, 1967). A systems tract is 'a linkage of contemporaneous depositional systems' (Van Wagoner *et al.*, 1987, 1988; after Brown & Fisher, 1977). Thus a systems tract is a three-dimensional rock body, as are its constituent facies masses and depositional systems, despite the strongly two-dimensional implication of the word 'tract'.

These systems stratigraphic units have been borrowed for the Vail (1987) model from papers by workers in the Texas Bureau of Economic Geology primarily in order to describe the structure within depositional sequences. In the Vail (1987) model, this structure is described in terms of 'systems tracts'. There are four such systems tracts in the Vailian terminology; *lowstand*, *highstand*, *transgressive*, and *shelf-margin* tracts.

Vailian systems tracts are distinguishable not by their component facies or depositional systems, but by their characteristic parasequence stacking patterns. In the Vail (1987) model, parasequences comprise a second category of substructure within a depositional sequence. A parasequence is a small-scale sequence bounded by flooding surfaces rather than unconformities (van Wagoner *et al.*, 1988). Thus it is a sequence stratigraphic term rather than a systems stratigraphic term (Table 3 of Swift *et al.*, this volume, p. 145). Parasequences may occur in groups of similar configuration known as *parasequence sets*.

In the Vailian scheme, the parasequences of transgressive systems tracts exhibit back-stepping patterns, while the parasequences of highstand, lowstand, and shelf-margin systems tracts exhibit 'prograding' (offlapping) patterns. The terms refer to the manner in which the locus of deposition shifts with time, as indicated by the position of the thickness maximum of the parasequence. Successive transgressive parasequences are deposited by a marginward shifting depocentre, while successive highstand parasequences are deposited by a basinward shifting depocentre (see Fig. 1). These parasequence sets constitute the systems tracts of the Vail (1987) model.

The boundaries of these systems tracts may be the sequence boundaries, or may be one of the two important flooding surfaces within the sequence. The *transgressive surface*, separates the lowstand and transgressive systems tracts, while the *maximum flooding surface*, separates the transgressive and highstand systems tracts (Fig. 1). The systems tracts also can be distinguished by the manner in which parasequence boundaries (or seismic reflectors) terminate against parasequence set boundaries. The parasequences of both transgressive and regressive sequence tracts onlap marginwards against their lower set boundaries. However, highstand parasequence boundaries downlap basinward against the lower set boundary (maximum flooding surface), while transgressive parasequence boundaries converge asymptotically against their upper boundary (also a maximum flooding surface). Such an asymptotic upper convergence is called a *paratruncation* (see discussion by Swift *et al.*, this volume, p. 118). These various classes of parasequence boundary truncations are illustrated in Fig. 1. The paratruncation type is better illustrated in Figs 13 and 15 of Thorne & Swift (this volume, pp. 214–216), who discuss parasequences and their bounding surfaces in greater detail.

In Galloway (1989a,b) terminology, 'genetic stratigraphic sequences' form the basic unit. Facies and depositional systems come in sets, referred to as *offlap components* and *onlap components*, that parallel the highstand and transgressive systems tracts of the Vail (1987) model. In both terminologies, the transgressive elements are capped by a marine flooding surface, while the subsequent regressive elements tend to be capped (at least near the basin margin) by a subaerial erosional surface. The models differ in that the Vailian sequence is measured from subaerial erosion surface to subaerial erosion surface, while the Galloway sequence, following the original Frazier (1974) model, extends from maximum flooding surface to maximum flooding surface.

Systems tracts and the related offlap and onlap

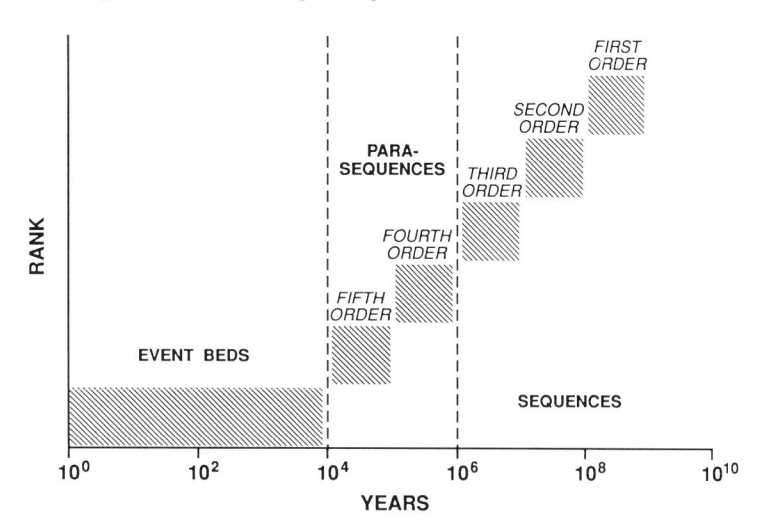

**Fig. 3.** Time-scales of transport events and sea-level cycles, and of corresponding depositional units. (After Vail *et al.* (1977a,) and Van Wagoner *et al.* (1990).)

components are thus well-defined entities in these schemes. They occupy specific positions within a sequence, and they are bounded by specific surfaces, such as subaerial unconformities and correlative conformities, or such hiatuses as transgressive surfaces or flooding surfaces. However, the nomenclatural system tends to break down for smaller scale units, especially in the Vail (1987) system. In illustrative diagrams, coloured portions of systems tracts are keyed as 'sandstones', 'mudstones' or simply 'deposits' instead of depositional systems (Van Wagoner *et al.*, 1988); or are keyed as 'facies', when scale considerations suggest that in fact depositional systems are meant (e.g. Posamentier *et al.*, 1988)

**Sequence hierarchies**

There is an additional complication, in that both the Vail (1987) and Galloway (1989a,b) models stress that sequences occur at several spatial scales (Table 2). The features tend to occur simultaneously, with lower frequency sequences overprinted by the higher frequency ones (Van Wagoner *et al.*, 1988, 1990; see Fig. 3). In the Vailian scheme fifth-order sequences are defined as being deposited by depositional cycles with periods of tens of thousands of years. Lower order sequences are defined by cycles of longer duration. Fourth-order cycles have periods of hundreds of thousands of years; while third-order cycles have periods of 1–5 million years. Second-order cycles are tens of millions of years long, while first-order cycles are hundreds of millions of years long (Fig. 3). Fourth and fifth-order sequences are commonly parasequences, bounded by flooding surfaces rather than by unconformities. The consequences of this are considered below.

The Galloway model has a similar hierarchy of small-scale *facies sequences* and large-scale *depositional complexes*, or *genetic stratigraphic sequences* (Galloway, 1989a; see Table 2). Facies sequences are the product of depositional events, while depositional complexes form during depositional episodes. The terms are Frazier's (1974), whose stratigraphic laboratory was the Quaternary section of the Louisiana shelf. Perhaps as a result, Frazier (1974) and Galloway (1989a) see the small-scale facies sequences as autocyclic in origin (term of Beerbower, 1964), resulting in the shifting of deltaic

**Table 2.** Scales of depositional patterns

|  | Vail (1987) model | Galloway (1989a,b) model |
|---|---|---|
| *Small-scale pattern* | Parasequence | Facies sequence |
| Equivalent duration | Paracycle | Depositional event |
| Origin | Eustacy | Shift of depocentre |
| *Large-scale pattern* | Depositional sequence | Depositional complex (genetic stratigraphic sequence) |
| Equivalent duration | Sea-level cycle | Depositional episode |
| Origin | Eustacy | Varied |

depocentres in response to channel avulsion (lobe switching). Frazier, however, saw his larger scale depositional complexes as responses to (allocyclic) eustatic oscillations of sea-level. His depositional episodes were the glacio-eustatic fluctuations of the Quaternary. In Galloway's (1989a) generalization of the model, however, the large-scale stratigraphic sequences may be of compound origin. In his analysis of the northwest Gulf of Mexico Cenozoic basin (Galloway, 1989b), Paleogene sequences of 4–5 my duration mainly result from large-scale autocyclicity, while eustacy becomes dominant as glacial sea-level oscillations increase in amplitude through the Neogene.

### Parasequences

In recent expansions of the Vail (1987) model, the concept of a parasequence as a special case of a depositional sequence is made more explicit (Van Wagoner, 1985; Van Wagoner et al., 1987, 1988, 1990; Posamentier & Vail, 1988; Mitchum & Van Wagoner, 1991). The requirement that parasequences be bounded by marine flooding surfaces rather than by unconformities (see above) has consequences that are contingent on the Vailian definitions of these terms. An unconformity in Vailian theory is 'a surface separating older from younger strata along which there is evidence of subaerial erosion or subaerial exposure with a significant hiatus indicated' (Van Wagoner et al., 1988). Correlative marine erosion may also occur. In contrast, a flooding surface is marked by lithologic or other criteria indicative of an abrupt increase in depth. A flooding surface may be accompanied by 'minor marine erosion but no subaerial erosion or basinward shift in facies' (Van Wagoner et al., 1988).

In Vailian theory, sea-level is seen as the operative regime variable and sea-level falls as the decisive event. Sequence-bounding unconformities are cut by sea-level falls. The sea-level fall results in a 'downward shift in coastal onlap' that juxtaposes disparate facies across sequence boundaries. The restriction of parasequence boundaries to flooding surfaces, however, means that a sea-level fall is not involved in their formation. They are not just scaled-down sequences but are different in kind. Parasequences coarsen upwards, as the deepening event is followed by sedimentation and shoaling, but there can be no basinward shift in facies resulting from a 'downward shift in coastal onlap', of the sort that creates substructure ('systems tracts') in lower frequency sequences. Consequently, Mitchum & Van Wagoner (1991) distinguish between parasequences as 'small, upward-coarsening stratigraphic units' and 'high-frequency sequences', which have all the elements of sequences, but occur at the high frequencies common to parasequences. In the following review of data from the rock record, we present reasons for using the terms interchangeably, for any scale of depositional sequence of higher frequency than the dominant frequency (usually third-order).

### Problems with existing models

Facies, depositional systems and depositional sequences are defined on the basis of lithology as determined in outcrop, or by inspection of cores. A depositional system is 'an assemblage of process-related facies' (Fisher & McGowen, 1967), while a systems tract is 'a linkage of contemporaneous depositional systems' (Brown & Fisher, 1977). In contrast, depositional sequences are defined on the basis of the geometry of reflectors and well-log correlations.

The two schemes of classification (lithology defined and geometry defined) do not mesh smoothly. Depositional systems and systems tracts, the two larger scales of lithological variation are related most closely to the smallest scale of depositional sequence or *parasequence* (a relatively conformable succession of beds or bedsets bounded by marine flooding surfaces or their correlative unconformities; Van Wagoner et al., 1988). A body that in plan view constitutes a depositional system or systems tract (classes of facies assemblage) tends in cross-section to be a parasequence. A depositional system or systems tract is the largest sediment body that can be deposited under uniform regime conditions (see Swift et al., this volume, p. 106). Such constant values for regime variables should be manifest as a monotonic vertical change in lithology at scales greater than mean bed thickness (depositional system should coarsen or fine upwards). Shifts in the ratio of regime variables ($R \cdot D/Q \cdot M$) in order to maintain equilibrium will terminate the depositional system with a flooding surface (will initiate a new parasequence).

Since a depositional systems tract is a 'linkage of *contemporaneous* systems' (Brown & Fisher, 1977), one should be able to track a parasequence from one depositional system (say a strandplain depositional system) into another (perhaps the equivalent shoreface–shelf depositional system. Thus both depositional systems and depositional systems tracts are facies concepts that are analogous to the geometric concept of parasequence.

Parasequences commonly have no other internal organization beyond their constituent depositional systems. However, larger scale sequences (third-order sequences; Vail et al., 1977a) tend to have a well-defined internal structure, expressed as internal regions of differing organization of parasequences (retrogradational versus progradational stacking of parasequences; Posamentier et al., 1988). As noted above, these regions are separated by well-defined internal bounding surfaces. Unfortunately, these depositional sequence subunits also have been designated 'systems tracts', even though they in fact consist of repetitive, cyclic *alternations* of depositional systems or systems tracts. Contradictions result. For instance, the 'transgressive systems tract' of Posamentier & Vail (1988) consists of 'retrograding parasequences'. The constituent parasequences tend to be upward-coarsening units. Thus, in facies terms, the 'transgressive systems tract' consists of regressive deposits.

Seismic stratigraphers have come by the term 'depositional systems tract' legitimately. It was proposed by members of the Texas Bureau of Economic Geology (Brown & Fisher, 1977); the group that had done the most thinking about scales of facies organization, and it was coined specifically to describe the basic building blocks of depositional sequences. Unfortunately, Brown & Fisher wrote before the distinction between high-order and third-order sequences was fully appreciated. In this paper we will distinguish between *depositional systems tracts*, bounded by flooding surfaces and identified primarily by their component lithology, and the larger scale *geometric systems tracts*, separated by more complex bounding surfaces, and identified primarily by the geometry of their reflectors or well-log correlations. In the following sections we test this model against data collected from Mesozoic and Cenozoic basin margins.

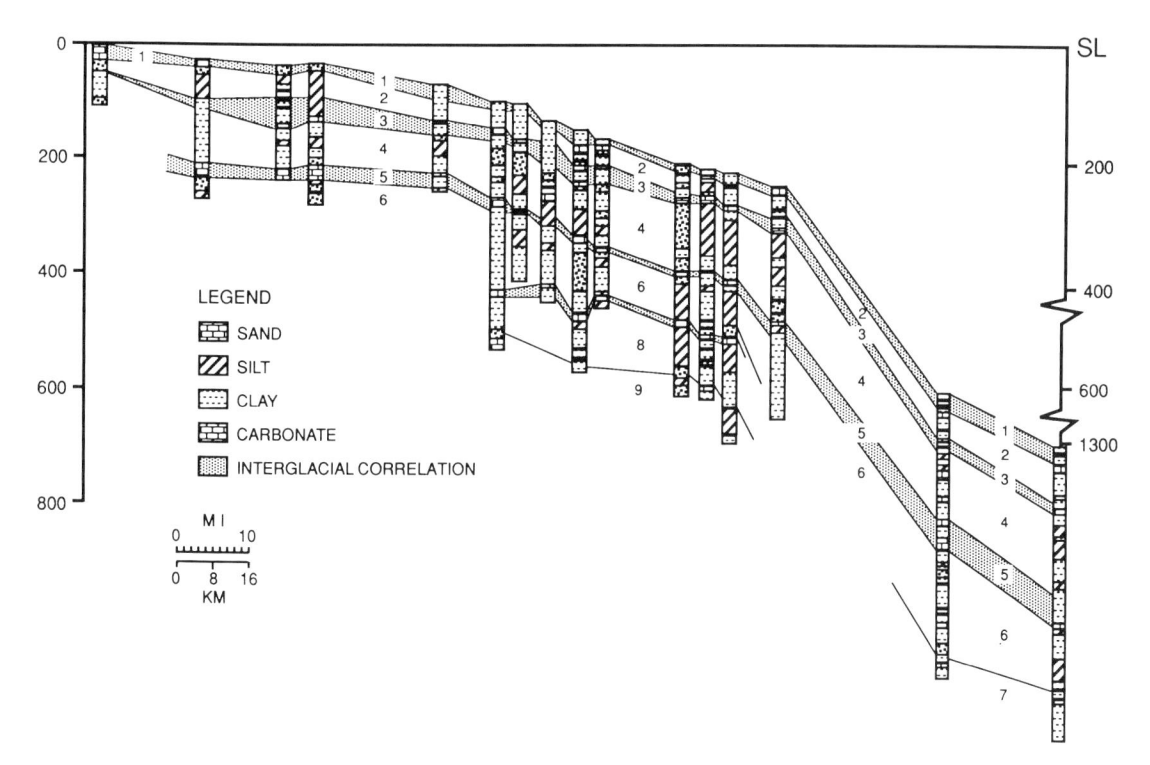

**Fig. 4.** Cross-section of the Louisiana shelf based on foundation borings and oxygen isotope ratios. Numbers refer to oxygen isotope stages. (From Coleman & Roberts, 1988.)

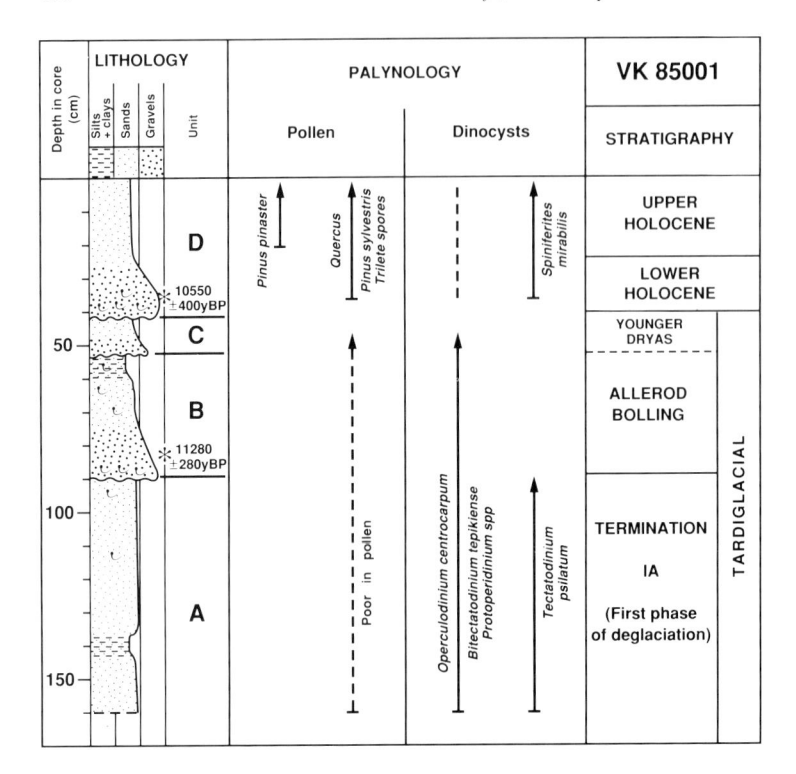

Fig. 5. Late Quaternary stratigraphy of the Aquitaine Shelf. (From Arbouille et al., 1991.)

## PARASEQUENCES ON QUATERNARY SHELVES

### Parasequences in the Quaternary of the Louisiana shelf

Coleman & Roberts (1988) and Roberts & Coleman (1988) have used data from platform foundation borings and seismic records to examine the cyclic Quaternary stratigraphy of the Louisiana shelf (Fig. 4). They used $\delta^{18}O$ values for stratigraphic control. The cores penetrated deposits as old as 240 000 years BP and contained a record of three Quaternary sea-level cycles. The $\delta^{18}O$ isotope record suggests a 100 000 year periodicity, i.e. fourth-order sea-level cycles as defined by Fig. 3. Each cycle has created a high-frequency depositional sequence consisting of a thin, back-stepping basal deposit and a thicker overlying downlapping deposit.

Sedimentation during periods of rising to high sea-level, deposited transgressive carbonate-rich muds and shell hashes, which occur as a thin, back-stepping drape across the shelf surface and down the slope. They are characterized by thin, slowly accumulated deposits referred to as 'condensed sections'. They pass, by facies change, into thicker back-stepping sand and gravel valley fills in the incised valleys of the inner shelf. These transgressive shelf deposits have wide lateral continuity, and high amplitude seismic response. They are the transgressive shelf depositional systems described by Swift et al. (this volume, pp. 105–143). The calcareous units locally occur as doublets, suggesting repetition by depositional cycles of yet higher frequency.

Lowstand sedimentation is characterized by thick, rapidly accumulated shelf-edge sequences, which contain some recognizably deltaic lithofacies similar to those of the modern Balize subdelta. The deposits exhibit well-defined depositional trends and a wide variety of seismic responses. Sand-rich and gravelly lithofacies are also present. The deposits are referred to as 'expanded sections'. They are regressive shelf depositional systems as defined by Swift et al. (this volume, pp. 106–107). Repetition of lithologies suggests depositional cycles of yet higher frequency. Accumulation rates for low sea-levels are three to 15 times greater than those deposited during high sea-levels.

A very similar picture has been reported by Suter

& Berryhill (1985) from the Louisiana–Texas margin, by Kindinger (1988) from the Alabama margin, and by Tesson et al. (1990) from the Rhône (French Mediterranean) margin. These late Quaternary sequences were deposited by sea-level cycles of sufficient amplitude (c.100 m) to result in good lithologic differentiation. However, because of the high-frequency nature of the cycles (c.100 000 years) relative to the rate of subsidence, highstand deposits (in the sense of Vail, 1987) are not preserved. The problem of highstand preservation potential is discussed in greater detail by Thorne & Swift (this volume, pp. 189–255).

**Early Holocene parasequences on the outer Aquitaine shelf**

Cyclic sedimentation in the early Holocene deposits of the outer Aquitaine shelf (French Atlantic shelf) has been reported by Arbouille et al. (1991). These authors describe a late Quaternary section 1.6 m thick (Fig. 5). It consist of four units separated by erosional bases. The units begin with a lag of shell debris and coarse sand, and fine upward into fine muddy sand or sandy mud. The oldest unit is older than 13 000 years BP; the others start at 13 000, 11 000 and 10 000 years BP, respectively. The units thus represent depositional episodes of a few thousand years duration at most, or fifth-order sequences as defined in Fig. 3.

The beds are interpreted as responses to stepwise increases in glacio-eustatic sea-level. During each surge, the rate of creation of accommodation increased abruptly, shifting the depositional regime to an accommodation-dominated mode ($R \cdot D > Q \cdot M$, see Swift & Thorne, this volume, p. 13). River sediment was trapped in estuaries, and autochthonous sedimentation prevailed. During each subsequent period of sea-level stabilization, the regime became more nearly supply-dominated ($R \cdot D > Q \cdot M$) and some fluvial mud was bypassed. Thus these parasequences are upward-fining rather than upward coarsening. They qualify as parasequences in the restricted sense of Van Wagoner et al. (1988) and Mitchum & Van Wagoner (1991), as well as in the sense of this section.

**Parasequences on modern transgressive coasts**

Parasequences are well documented from modern transgressive coasts. Van Straaten (1965) was able to show by means of cores and excavations that the Holocene transgression of The Netherlands coast attained its most landward position at about 4800 years BP, north of the (then) estuarine mouth of the Rhine. At that time the rate of sea-level rise slowed relative to the rate of sediment input, and the coast prograded as a strandplain from then to about 2850 years BP. Both the early transgressive system and the later regressive deposits are present. At the present time, the coast is retrograding once more. The regressive shoreface prograded across a sea-floor 4 m deep, but the present, retrograding shoreface cuts down 16 m, so the regressive deposits are being destroyed, and are being replaced seaward by a transgressive sand sheet whose surface is moulded into sand ridges.

A mid-Holocene progradation likewise characterizes the Brazilian and Argentine coasts (Dominguez & Wanless, this volume), between 4000 and 2000 years BP. Parker et al. (1982) have shown that on the coast of Buenos Aires Province, Argentina, a mid-Holocene oscillation of relative sea-level produced a facies sequence similar to that of The Netherlands coast. The regression was expressed as a narrowing of the adjacent Rio de la Plata estuary, as well as a seaward shift of the oceanic shoreline. As in the case of The Netherlands coast, the present transgression is destroying the mid-Holocene record and is replacing it with a ridged, transgressive sand sheet.

**Parasequences on modern transgressive shelves**

A parasequence in a better state of preservation appears to exist on the floor of the Middle Atlantic Bight, North American continental shelf (Fig. 6). Its geomorphological expression is a discontinuous scarp, described as the Midshelf Shoreline by Swift et al. (1972), and as the Block Island Shoreline by Dillon & Oldale (1978). Well-developed shelf valleys extend across the shelf, from the mouths of present estuaries out to the scarp, but in most cases these valleys lose much or all of their geomorphological expression beyond the scarp (Swift et al., 1972, 1977, 1978; Swift, 1973; McKinney et al., 1974). The shelf valleys are not simply drowned river valleys, but were shaped primarily by retreating estuary mouths, whose retreat paths they constitute (see Swift et al., this volume, pp. 133–135).

The New Jersey shelf has been cored above the scarp (Niedoroda et al., 1985; Rine et al., this volume, pp. 395–423), as has the Virginia shelf (Swift et al., 1977) and the North Carolina shelf

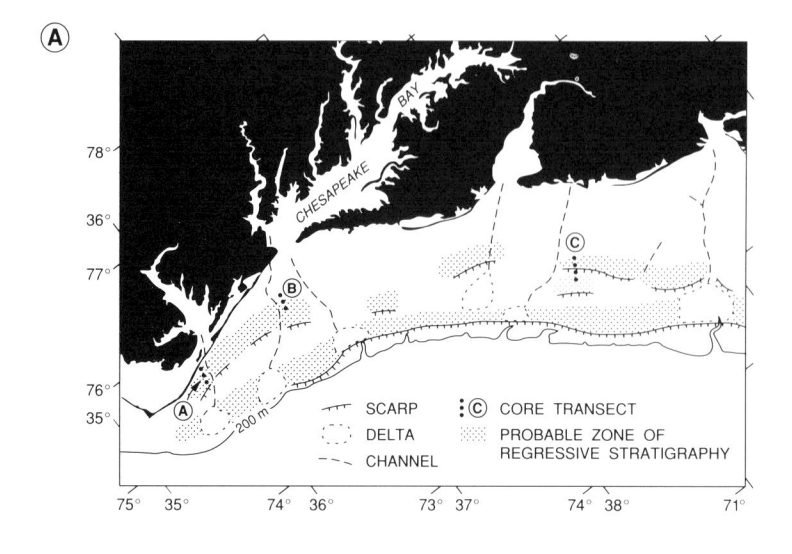

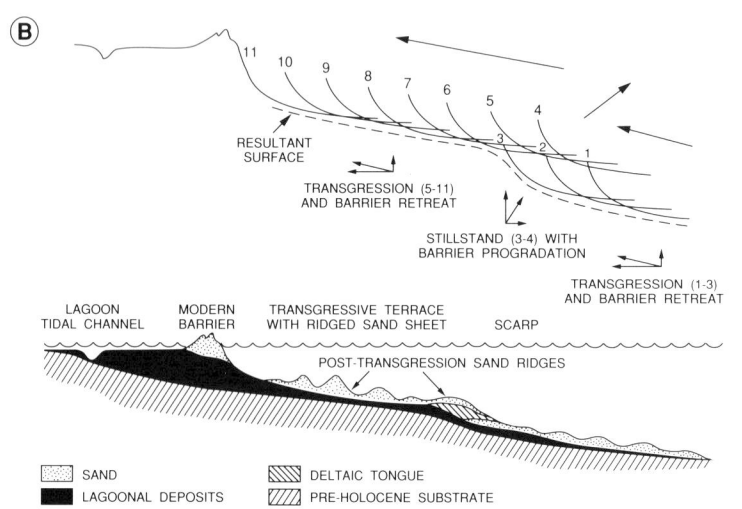

**Fig. 6.** Zone of early Holocene progradation on the middle Atlantic shelf. (Modified from Swift *et al.*, 1977, 1978.) (A) Regional distribution; (B) model for formation.

(Swift *et al.*, 1978). A similar twofold Holocene stratigraphy emerges in these areas, in which a lower unit consists of sand with intercalated mud beds, and an upper unit consists of mud-free sand (Fig. 7). The lower unit is interpreted as a delta–strandplain unit formed during the period of shoreline progradation that created the scarp, while the upper unit is attributed to reworking of the delta–strandplain surface as erosional shoreface retreat resumed. In contrast to The Netherlands and Argentine cases, the envelope of shoreface positions occupied during the resumed transgression lies *above* the lower envelope of progradational shoreface positions (Fig. 6B). As a consequence, the regressive deposits were only partially destroyed. Their leading edge is a scarp that constitutes a relict lower shoreface.

A second, more fragmentary system appears to lie seaward of the main scarp (Fig. 7a). Two additional scarps occur near the shelf edge: the Nichols and Franklin Shorelines (Veatch & Smith, 1939). These shelf-edge scarps are associated with shelf-edge delta systems (Swift *et al.*, 1972), and may represent two earlier sequences formed during the slow initial phase of the transgression.

The main parasequence of the Middle Atlantic shelf appears to be of compound origin. Its regional extent suggests that relative sea-level change may have been involved in its formation, as in the Vail

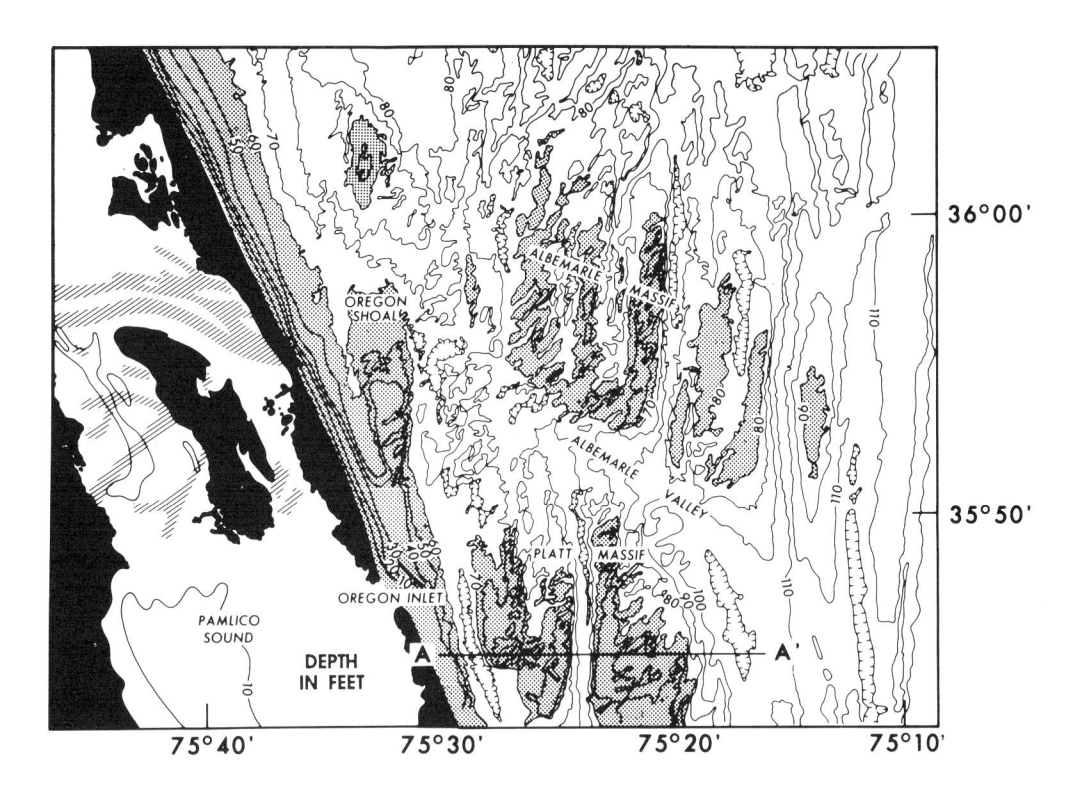

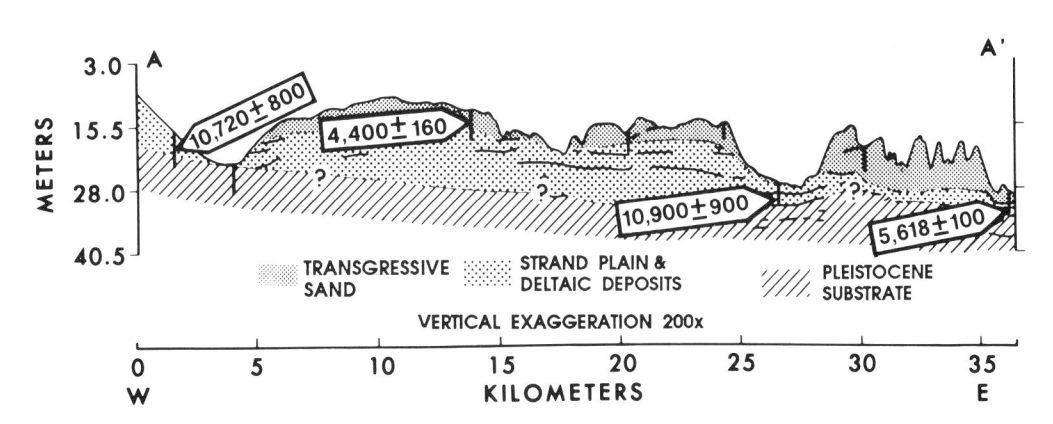

Continental margin sedimentation—parasequences

Fig. 7. Physiography of the mid-shelf shoreline in the vicinity of the Albemarle shelf valley. (A) Plan view. Diagonal ruling in lagoon indicates buried stream channel. Diagonal trend in contours on Albemarle massif is the result of sand waves superimposed on sand ridges. (B) Cross-section with radiocarbon ages. (From Swift *et al.*, 1978.)

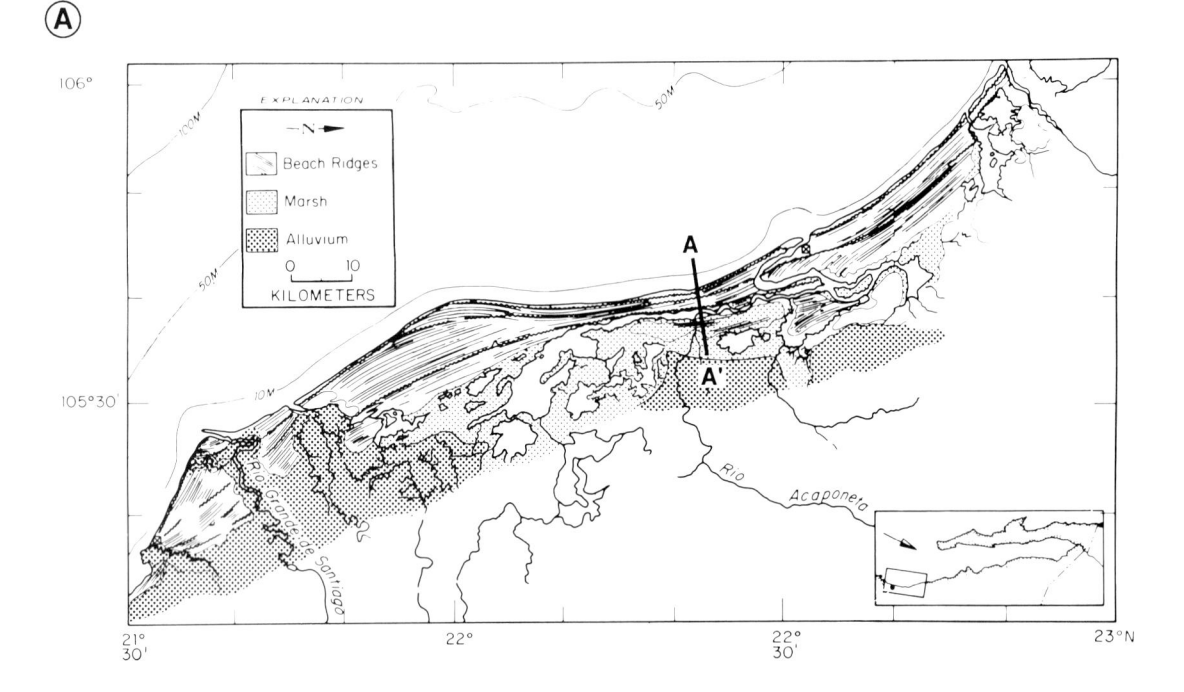

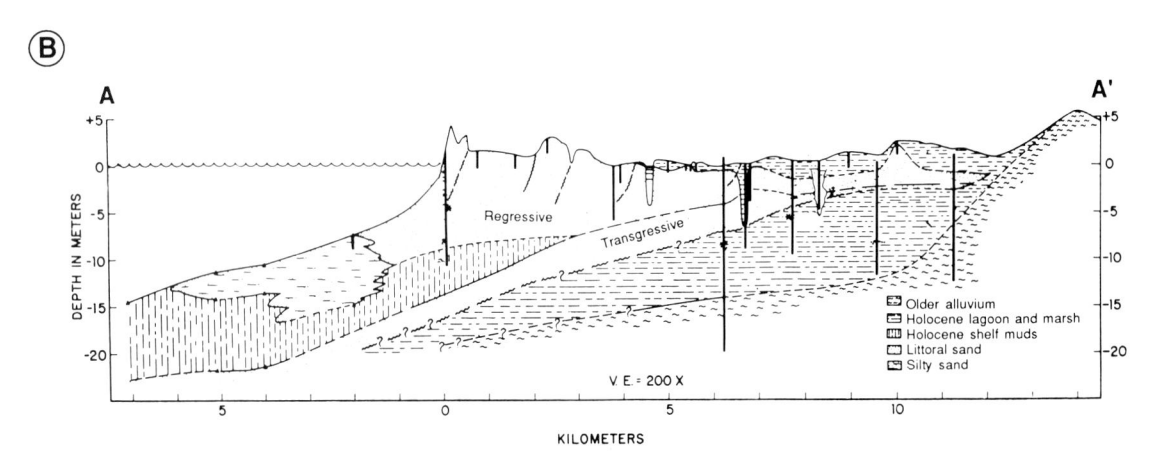

**Fig. 8.** Nayarit strandplain. (A) Plan view, showing through-going rivers, and relationship of strandplain, lagoonal system, and alluvial plain. (B) Cross-section through the Nayarit strandplain. (Modified from Curray *et al.*, 1969.)

(1987) model. Development was strongly influenced, however, by the proximity to river systems, as in the case of The Netherlands and Argentine examples.

Neither The Netherlands, Argentine nor USA Atlantic shelf examples qualify as parasequences in the restricted sense of the Vail (1987) model (Van Wagoner, 1988), because they exhibit facies differentiation, and are bounded by ravinements, rather than by flooding surfaces.

### Parasequences on modern highstand shelves

Parasequences probably exist on most shelves undergoing 'highstand' type progradation and aggradation (highstand in the sense of Vail, 1987). However, they have been well studied in only two localities; the Nayarit strandplain of Pacific Mexico (Curray *et al.*, 1969; Fig. 8), and the Mississippi delta (Frazier, 1974). The deposits of these environments have contrasting geometries. The Nayarit strandplain sands can be viewed as a sheet deposit formed by a coalescing series of wave-dominated deltas prograding seaward along a straight front, and serviced by a set of sinuous, but subparallel rivers. In the terminology introduced in this paper (Swift *et al.*, this volume, p. 105), the strandplain deposits constitute a depositional systems complex. However, the study of Curray *et al.* (1969) showed that the deposit consists primarily of strandplain systems. Deltaic systems, with recognizable channel deposits, constitute less than 15% of the section. Curray *et al.* (1969) have cored an interstratified sand and clay facies seaward of and beneath the modern Nayarit shoreline.

The Nayarit strandplain began to advance at about 4500 years BP (Curray *et al.*, 1969), but the regression was punctuated by episodic realignments of the coast with local truncation and erosion of older beach ridges, so that the beach ridges are grouped into sets oriented at slight angles to each other. These short-term transgressions appear to be associated with periods of climatic warming and cooling that resulted in reversals of littoral drift, causing local erosion. At about AD 500 years, however, the Rio de Santiago, the major river of the area, changed its course, resulting in a major coastal realignment.

In contrast to the Nayarit strandplain, the Mississippi delta is serviced by a single great trunk river, whose repeated avulsions over the past 6000 years have built a complex compound delta deposit of many subdelta deposits, with a radial structural plan (Frazier, 1967, 1974; Fig. 9). Frazier (1967) counted 16 separate delta lobes formed by

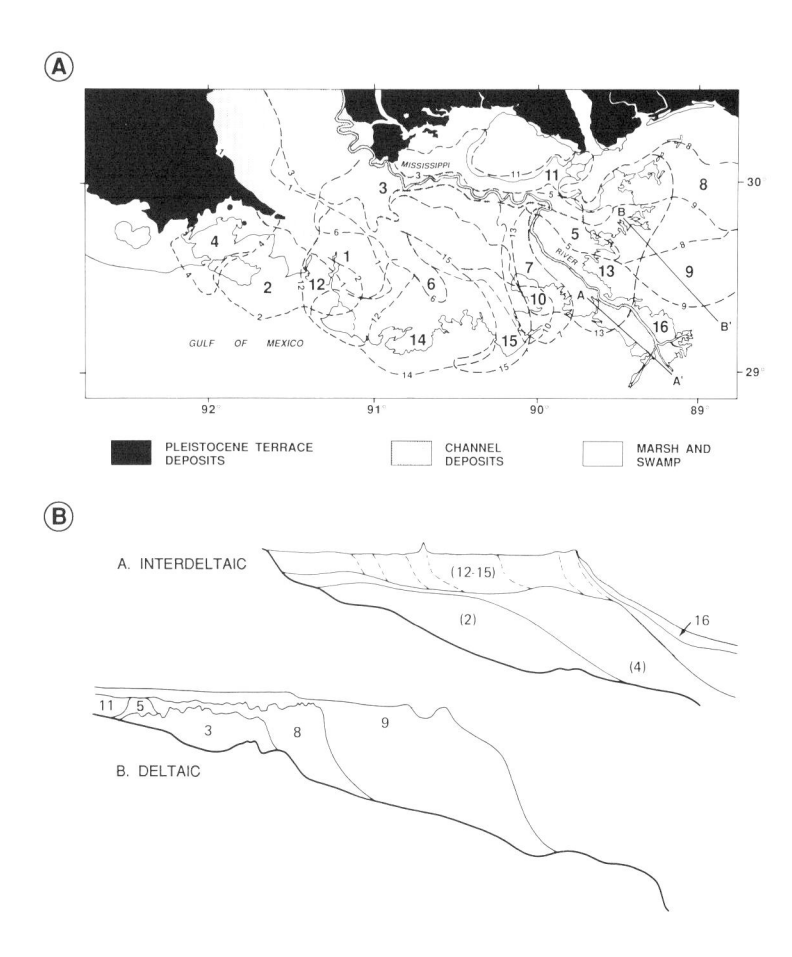

**Fig. 9.** Architecture of the Mississippi delta. (A) Plan view of delta lobes formed by the Mississippi River during the last 6000 years. (After Frazier, 1974.) (B) Cross-sections through the Mississippi delta. Numbers refer to delta lobes. (From Frazier, 1974.)

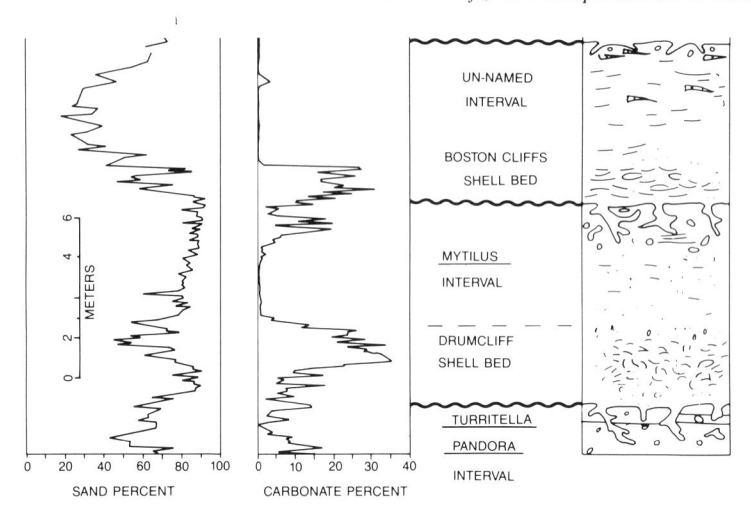

**Fig. 10.** Section through the middle Miocene, Calvert Cliffs, Chesapeake Bay. Irregular contacts are burrowed by *Thalassinoides*. (From G. Shideler, US Geological Survey, Denver, unpublished data.)

the Mississippi River in the past 6000 years (Fig. 9). The lobes are grouped into the Teche, St Bernard, Lafourche and Plaquemines (modern) delta complexes. Each lobe consists of a basal fine-grained prodelta shelf facies, an overlying sandy delta-front facies, and an uppermost fine-grained delta-plain facies. The development of each delta complex was not a continual process. River avulsion caused the cessation of growth in one lobe as progradation occurred in another. Avulsion is seen as a basically random process (Coleman & Gagliano, 1964; Frazier, 1967), but it may be superimposed on a deterministic processes, such as climatic change or glacio-eustatic sea-level fluctuations; Nummedal (pers. comm.) has suggested that the development of the modern (Balize) subdelta may be related to a small-amplitude sea-level fall associated with the 'Little Ice Age' of AD 1600–1800.

The basic sequence of the formation and subsequent foundering and destruction of the delta complexes has been described by Scruton (1960). Wells (1987) has described the initiation of a new lobe associated with the Atchafalaya River, an avulsion of the main Mississippi channel whose growth has been arrested partially by the Corps of Engineers. Penland *et al.* (1985) have described the cycle as one of barrier formation (see Swift *et al.*, this volume, p. 123). The LaFourche subdelta, the Chandeleur Islands, Ship Shoal and Trinity Shoal are seen as the equivalent of a time series in the transformation of a transgressed lobe, from subaerial delta plain through barrier island and submarine shoal to submarine sand plain (Suter *et al.*, 1987).

These examples of high-frequency (fifth-order) sequences on modern shelves with 'highstand' characteristics are, in Vailian terms, high-frequency sequences (Mitchum & Van Wagoner, 1991), rather than parasequences, since they are bounded by ravinements rather than by flooding surfaces, and exhibit facies differentiation. They are of mixed origin, being more strongly autocyclic in the Mississippi delta than in the Nayarit strandplain. The implications of the sequences of the Mississippi delta for the global cycle correlations of Haq *et al.* (1987) are discussed by Boyd *et al.* (1989).

## PARASEQUENCES IN THE MARYLAND MIOCENE

### Depositional setting

Middle Miocene deposits of the Salisbury embayment of the Atlantic continental margin are well exposed in the Calvert Cliffs section of Maryland on the western shore of Chesapeake Bay. Kidwell (1984) has shown that the Calvert Formation (Plum Point Member) and the Choptank Formations are divided by a series of erosional surfaces into six depositional sequences. The sequences are of the order of 840 000 years in duration and are therefore fourth-order sequences. Higher frequency variations are present in some of the shell beds (Kidwell, 1988). The diastems that separate them are less than one diatom biozone in duration. They take the form of burrowed firmgrounds; *Thalassinoides*-type galleries extend downward from each surface to depths of a metre or more (Fig. 10).

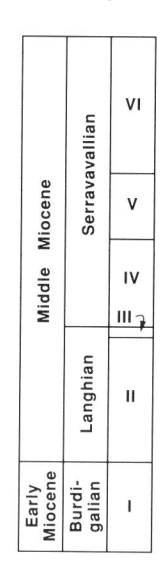

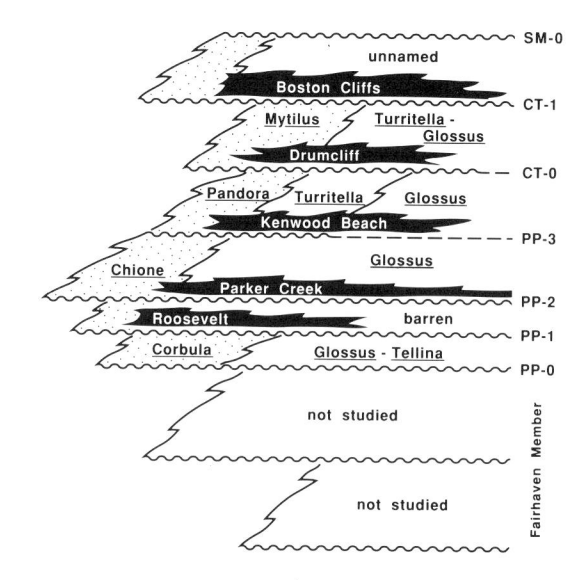

Fig. 11. Schematic representation of the middle Miocene of the Chesapeake Bay. (Modified from Kidwell, 1984.)

## Parasequences in the Maryland Miocene

Five of the six surfaces are overlain by beds of shelly sand or framework-intact shell beds, generally 60–150 cm thick. The shell beds are overlain in turn by fine- to very fine-grained bioturbated muddy sand or sandy mud that coarsens and becomes sandier up column in a very uniform manner (Fig. 10). Kidwell's (1984, 1988) faunal analysis shows that the basal shell beds were deposited in open to sheltered, shallow to very shallow sublittoral environments. These beds constitute the proximal amalgamated facies of transgressive depositional systems whose distal facies occur farther downdip. The shell beds rest on a source diastem of probable ravinement origin, and are capped by a secondary diastem (flooding surface) as a result of reduced sedimentation (source diastem concept is described by Swift et al., this volume, pp. 101–102). The upward-coarsening upper units exhibit systematic biofacies changes, grading upwards and landwards from deep sublittoral to very shallow sublittoral or intertidal biofacies. They constitute the distal deposits of a regressive shelf depositional system (see Swift et al., this volume, pp. 115–118). In most cases they are the bioturbated equivalent of an interbedded lithofacies.

The Calvert and Choptank sequences are entirely marine, as far inland as their updip pinch-out beneath the modern erosion surface (Fig. 11). The diastems separating them are at least flooding surfaces. They are called ravinements by Kidwell (1984), who used the term broadly. Despite the lack of confirming subaerial deposits, the diastems may be ravinements, in the restricted sense used in this paper, of a diastem cut into a subaerial surface by erosional shoreface retreat. The ravinement surface presently being incised into the inner Atlantic shelf is locally mantled with a shell bed (Figueiredo et al., 1980). Since it is created by erosional shoreface retreat, with ensuing condensation of the back barrier section, it consists of a mixture of lagoonal and open shelf species (Fischer, 1982). Similarly, Kidwell (1988) reports a mixture of 'fully marine' and 'polyhaline' species in several of the Miocene shell beds. Thus each small-scale sequence consists of a trangressive depositional system resting on its source diastem and overlain by a regressive depositional system, whose proximal facies and source diastem have been destroyed by the next ravinement (terms from Swift et al., this volume, pp. 101–103).

## Third-order sequences in the Maryland Miocene

The middle Miocene section of the Calvert Cliffs has a larger pattern of organization in addition to that of the six high-frequency sequences. The second diastem above the base of the Plum Point Member of the Calvert Formation (PP-2 in Fig. 11) is the Parker Creek Bone Bed. This sand bed contains disseminated bones and teeth of marine mammals and sharks, and local concentrations of poorly preserved shells. Myrick (1979) concluded that the marine mammal assemblage is the condensed record of at least two ecologically distinct faunas resulting from reduced sedimentation over a period of a thousand to tens of thousands of years. The Parker Creek Bone Bed was thus enriched in phosphatic fossils by selective dissolution during sea-floor exposure that was significantly prolonged relative to the duration of the diastems above and below. Kidwell (1984) concludes that the lower Plum Point Member records a transgressive phase of deposition, with maximum open shelf water depths attained in the massive barren interval of the PP-1 sequence. Above the Parker Creek Bone Bed, the Plum Point strata record an overall regressive phase, punctuated by transgressive pulses. The Parker Creek Bone Bed thus corresponds to the 'maximum flooding surface' of a third-order sequence, which, in the Vailian terminology (Vail et al., 1977b; Posamentier et al., 1988) separates 'transgressive' from 'highstand' systems tracts. Kidwell (1984) has mapped a disconformity (Ct-0 surface) at the top of the Calvert Formation with 7 m of relief, apparently formed by the incision of a subaerial drainage surface, between Calvert and Choptank time. The Drumcliff Shell Bed cuts across the top of the most pronounced low in the CT-0 surface, isolating a lagoonal deposit (Governor's Run Sand) between a ravinement surface at the base of the shell bed and the CT-0 surface. The CT-0 surface thus exhibits an incised valley, in Vail's (1987) terminology the identifying characteristic of a type 1 sequence boundary, cut by a rejuvenated drainage system after sea-level falls over the shelf edge (see discussion of the Vail (1987) model by Thorne & Swift, this volume, pp. 224–231).

## DEPOSITIONAL SYSTEMS AND PARASEQUENCES IN THE CRETACEOUS WESTERN INTERIOR BASIN

### General

Parasequences are well developed along the western margin of the Cretaceous Western Interior Basin, in the Rocky Mountains–Great Plains sector of the North American continent (Van Wagoner et al., 1990). They are well exposed in outcrop and well defined in the subsurface by wireline logs and occur in transgressive, regressive, coastal and basinal settings (Nummedal & Swift., 1987). Well-studied examples occur in the San Juan Basin and Powder River Basin of New Mexico (Asquith, 1970; Nummedal et al., 1987). These are successor basins, isolated from other portions of the Western Interior Basin by the Late Cretaceous Laramide deformation. Our review also will refer to the Bighorn, Alberta and East Texas Basins.

The marine Upper Cretaceous section consists of thin, basin-centre limestones (Greenhorn, Niobrara Formations, thick shallow-marine shales (Mancos Shale, Pierre Shale, and correlative shales) on the basin's western flank, and the littoral and fluvial sandstones and shales of the Mesaverde Group on the basin's edge. These units exhibit a 'clastic wedge' filling pattern (King, 1959), sourced from the overthrusted western margin of the basin. The clastic wedge is in fact a multiple wedge, in that it is organized into four second-order sequences, designated the Greenhorn, Niobrara, Clagget and Bearpaw 'cyclothems' (Hancock & Kauffman, 1979). These typically consist of a transgressive shale, a calcareous condensed section representing peak transgressive conditions (e.g. Greenhorn and Niobrara Limestones), and a regressive sequence of shale, littoral sandstone, and fluvial sandstone and shale.

The geometry of this foreland basin was the reverse of that seen on a divergent margin (Fig. 1 of Swift & Thorne, this volume, p. 5). Subsidence was most rapid beneath the fluvial, littoral and shallow-marine facies of the western margin, and decreased basinward, and marginal deposits consequently constitute a subsidence-expanded record. As a consequence, the basin-margin stratigraphy is seen to be highly hierarchical, with third, fourth and fifth-order as well as second-order sequences. Subsidence was a response to more or less continuous tectonic loading by advancing thrust plates (Jordan, 1982), and the architecture of the fill was controlled by the relationship between the regime variables of sediment imput, $M$ and the subsidence-controlled relative sea-level, $R$. Eustacy appears to have been an important but subordinate control, and became increasingly subordinate along the western margin during the later Cretaceous (Santonian and Campanian). Sediment input and subsidence tended to be coupled but lagged. In this kind of interaction, a major loading event leads to basin deepening, which at first outpaces sedimentation, followed by a heightened input

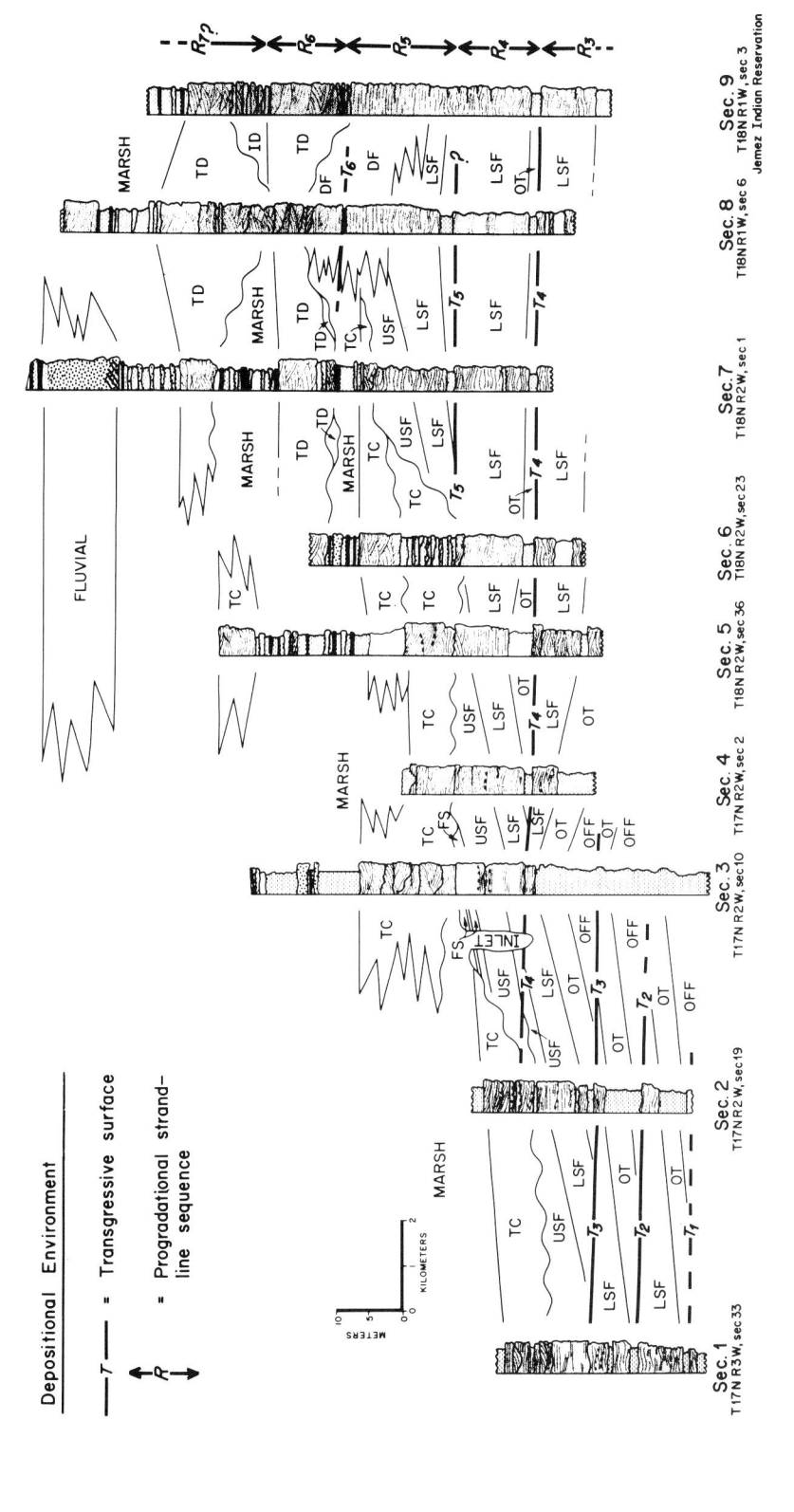

Fig. 12. Section through the Point Lookout Sandstone in the San Luis Valley, New Mexico. (From Wright, 1986.) OFF = offshore marine shale; OT = offshore transition: shale/siltstone/sandstone; LSF = lower shoreface sandstone; USF = upper shoreface sandstone; FS = foreshore sandstone; DF = delta front sandstone; TC = tidal creek channel sandstones/intertidal flat; TD = tidal distributary channel sandstones; ID = interdistributary sandstones/shales; MARSH = marginal to non-marine shales/sandstones; and FLUVIAL = non-marine channel sandstones.

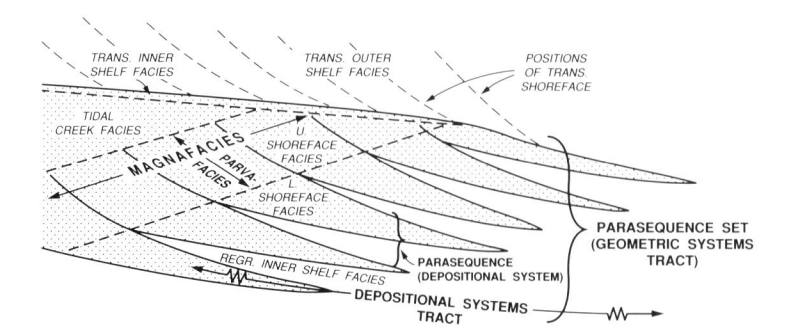

Fig. 13. Cartoon showing relationships between parvafacies, magnafacies, depositional systems and parasequence in a littoral sandstone.

of sediment from the eroding tectonic front, and progradation (see analysis of regime sedimentation in the Cretaceous Western Interior Basin presented by Thorne & Swift, this volume, pp. 241–244). Thus as sedimentation increases relative to subsidence ($Q \cdot M > R \cdot D$; see Swift & Thorne, this volume, p. 13), the shoreline overruns the basin; when sedimentation diminishes relative to subsidence ($R \cdot D > Q \cdot M$) the shoreline again retreats.

## Parasequences in regressive settings

### Regressive parasequences, Point Lookout Sandstone, San Juan Basin

The most comprehensive recent study of high-frequency sequences in the Cretaceous Western Interior Basin is that of Wright (1986), who examined high-frequency (>100 000 year) depositional cycles in the Point Lookout Sandstone (Fig. 12). The Point Lookout Sandstone was deposited by cyclic repetition of the proximal facies of a strandplain–shelf depositional complex (terminology from Swift et al., this volume, pp. 112–113). The depositional regime apparently was mesotidal in character. As a consequence, beach and dune depositional subsystems have been replaced largely by sheet sandstones deposited by laterally migrating tidal channels, overlain in turn by fluvial channel sandstones, overbank sandstones and claystones (Wright's lower delta and upper delta lithofacies).

Both river-connected and blind channels existed (Fig. 12). As a consequence, the surf-cut source diastem characteristic of microtidal strandplain–shelf complexes has been replaced by the thalweg-cut source diastem of the tidal channel system. The two main depositional systems of the complex (tidal channel and shelf systems) thus bear an occluded rather than a conjugate relationship (see discussion of source diastem types by Swift et al., this volume, p. 103). The shelf system consists of a well-developed assemblage of amalgamated sandstone (shoreface) lithofacies and interbedded sandstone and shale lithofacies (Wright's (1986) offshore transition lithofacies). Locally, a distributary mouth bar system has been superimposed on it (see discussion of superimposed systems in section on depositional system relationships by Swift et al., this volume, p. 103).

Wright's (1986) section through the Point Lookout Sandstone in the San Luis Valley reveals a complex relationship between the Point Lookout Sandstone, and its high-frequency depositional sequences, its component facies and individual strata within the facies (Fig. 12). The sandstone is 30–60 m thick and climbs in the section (becomes younger and higher basinward). It is composed of imbricated sequences 5–17 m thick that dip gently basinward. The diastemic contacts between successive systems are easily identifiable in the downdip region, where distal mudstones and shales of a younger sequence overlie amalgamated or interstratified sandstones of an older sequence.

Such surfaces are consistently traceable into the updip region, as thin mud partings separating massive amalgamated sandstones. Each sequence consists in turn of a proximal to distal array of regressive shelf facies. Facies contacts dip landward, so that each facies mass (lithosome) is replaced upward and landward by a more proximal shelf facies (Fig. 13). Finally, the diastems that form the erosional bases of individual storm strata dip basinward through facies boundaries, and can be traced through several lithosomes. These bed soles dip more steeply than do the system boundary diastems. They rested on a prograding shoreface, while the system boundary diastems represent transgressive shoreface surfaces, which are typically of gentler slope.

The sequences resemble Vailian parasequences, since they coarsen upwards and rest on erosional

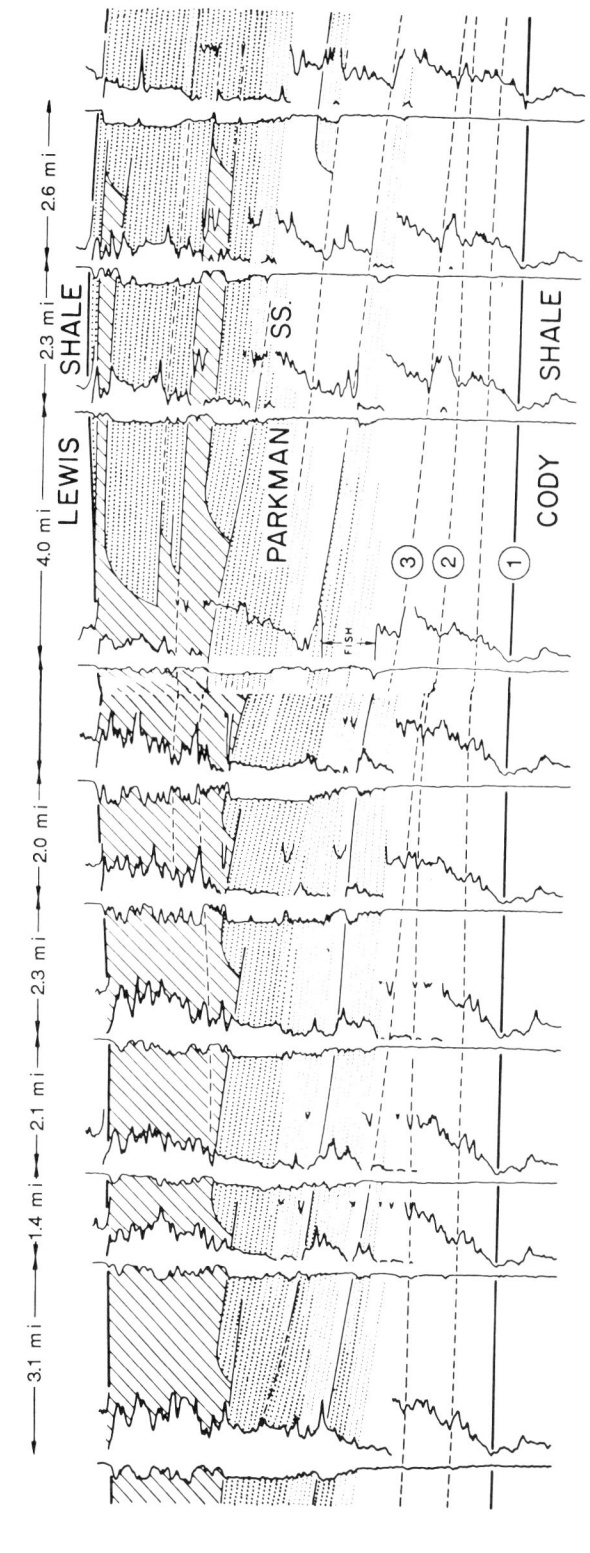

Continental margin sedimentation—parasequences

**Fig. 14.** Progradational parasequences in an inner shelf setting. Mesaverde sandstone in the Powder River Basin, Wyoming, as revealed by well-logs. Diagonal ruling: fluvial sandstone and shale. Dotted lines are shoreface sandstone beds. Clear is inner shelf shale. (From Asquith, 1970.)

surfaces whose time value is below biochronologic resolution. However, many of the characteristics of small-scale depositional sequences are present. The bounding surfaces are not simple flooding surfaces. They are ravinements in which 5–10 m of section and a subaerial surface have been destroyed. In the proximal portion of the sequences, the depositional cycle is highly asymmetric; basal transgressive deposits are limited to a thin (centimetres) mud parting. Detailed analysis of time-equivalent rocks further north (Thorne *et al.*, this volume, pp. 67–79) indicate that the transgressive system is lacking mainly because of the proximal setting. The only transgressive environment to be expected in less than 10 m of water is the erosional source environment. In the Lower Mesaverde Formation of the Big Horn Basin, depositional systems architecture becomes more apparent in the downdip section, within a few kilometres of the shoreface. The transgressive mudstone partings thicken to a metre or more, and the transgressive system becomes capped by a flooding surface. Transgressive beds toplap basinwards, while the regressive beds above it downlap on the flooding surface. In the far downdip section, however, the transgressive sections wedge out, and the section simplifies to regressive shale intervals, separated by indistinct flooding surfaces.

The basinward wedge-out of the trangressive beds is probably a consequence of transgressive sedimentation patterns. On modern mud-depositing shelves undergoing transgression, sedimentation rates drop markedly seaward of the 'nearshore advective mudstream' of the inner shelf (see Swift *et al.*, this volume, p. 115).

Thus in the cliffs of the San Louis escarpment, a high-frequency sequence architecture has been overprinted on the depositional systems pattern described by Swift *et al.* (this volume, pp. 90–105). The repetitive basinward shifts of depositional systems during successive depositional cycles results in large-scale lithofacies bodies consisting of stacked repetitions of a given lithofacies (Fig. 13). Such large-scale lithofacies bodies have been called *magnafacies*, and their component deposits *parvafacies* (Caster, 1934).

*Regressive parasequences in the Mesaverde Formation and Cody Shale, Wyoming*

Asquith (1970) and Curry (1973, 1976) have examined parasequences at the equivalent horizon in the subsurface of Wyoming (Fig. 14). Here the Parkman Sandstone Member of the Mesaverde Formation has prograded southwestward over the Cody Shale of the Powder River Basin. The Parkman Sandstone passes by facies change into the Cody Shale in much the same way that the Point Lookout Sandstone passes into the Mancos Shale. Specific facies are less easily resolved in the subsurface of the Powder River Basin than in the San Luis Cliffs, but the regional relationships are more readily apparent.

In the well-log cross-section, the Parkman Sandstone appears as a series of basinward-stepping small-scale depositional sequences. Each cycle is highly asymmetrical, coarsening upward to a transgressive surface, over which the cycle begins again. By inference, each is a conjugate pair of depositional systems, whose surf-cut source diastem separates an upper, regressive, beach–dune–washover fan depositional complex from a prograding shelf–shoreface depositional system (terms from Swift *et al.*, this volume, pp. 112–113, and see their Figs 5A, 6D). These relationships cannot be resolved on the Parkman well-log section of Fig. 14, but they can be seen in the equivalent outcrop sections measured in the San Louis cliffs (Fig. 13). Each successive shelf–shoreface depositional system passes from its source diastem through successive amalgamated sandstone facies and interstratified sandstone and shale facies to a laminated shale facies.

Upward coarsening, shale–silt sequences can be traced for many tens of kilometres from the Mesaverde and equivalent sandstones of the basin's western margin into the Mancos, Cody and Pierre Shales, both in the subsurface and, under favourable conditions, in outcrop. As the pattern becomes more subtle in downdip regions, a second pattern emerges, of episodic bentonite (volcanic ash)—contaminated beds of higher resistivity. As they are traced basinward by these markers, the parasequences pass through an expanded clinoform structure (Fig. 15), thin to fondoform (bottomset) beds, and pass by facies change into the limestone of the underlying Niobrara Formation (central basin lithofacies).

In the Lower Cretaceous Kuparuk River Formation of the Alaskan north slope, shallow-marine parasequences form a similar pattern of basinward-stepping clinoform lenses. Here each clinoform parasequence passes upwards into an interstratified sand and shale facies, capped by a marine erosion surface (Masterson & Paris, 1987)

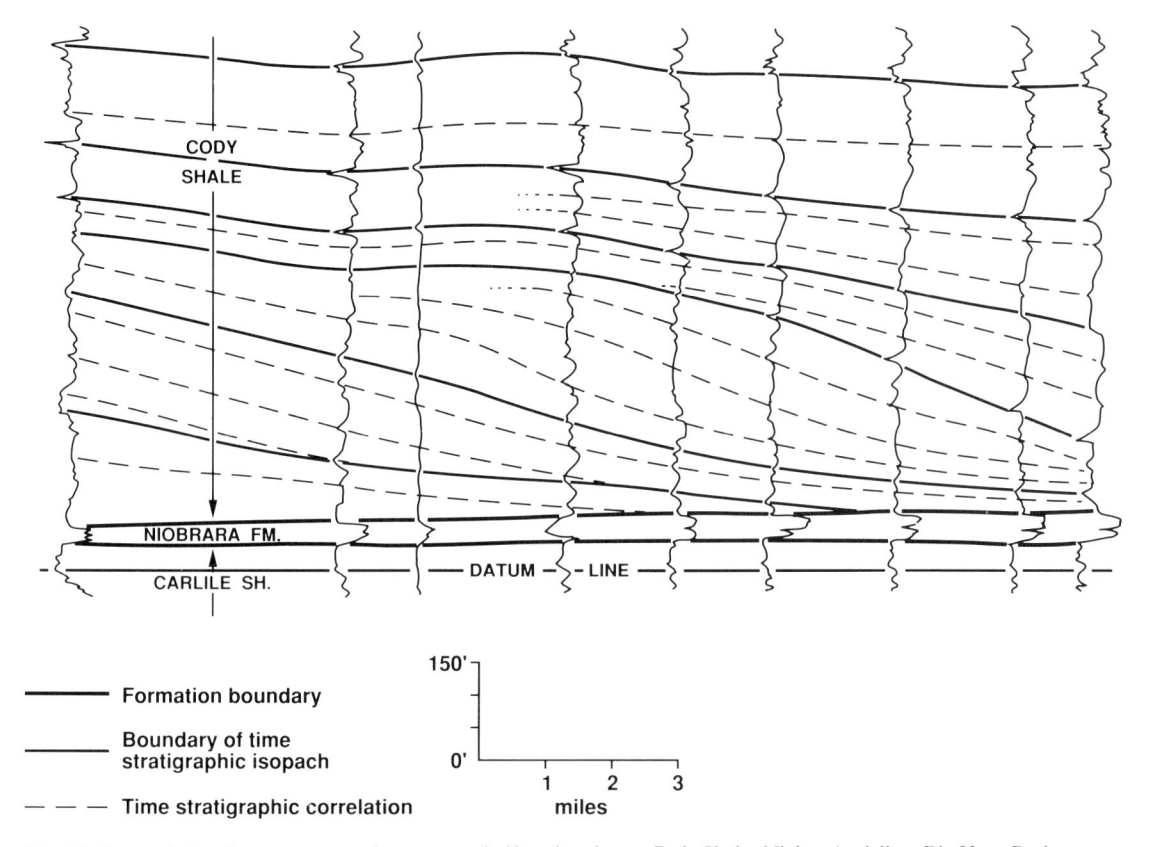

**Fig. 15.** Progradational parasequences in an outer shelf setting, lower Cody Shale, Nieber Anticline, Big Horn Basin, Wyoming. (From Asquith, 1970.)

## High-frequency condensed sequences in the Bridge Creek Limestone

The Niobrara Formation, like the underlying Greenhorn Formation is a marginward extension of the central basin limestone lithofacies, deposited during a time of maximum flooding. In the Vail (1987) model (Fig. 1), such limestones are condensed sections, which contain the maximum flooding surface, separating the transgressive and highstand systems tracts (Vail's (1987) terms). The Cody–Niobrara and Mancos–Greenhorn contacts, like the Parkman–Cody and Point Lookout contacts, are good places to study high-frequency sequence architecture. In each of these cases, cyclic, high-frequency changes in depositional regime have caused an intertonguing of lithologies across the facies boundary. The Bridge Creek Member of the Greenhorn Formation (San Juan Basin) is the best studied of these condensed sections, and its characteristics will be reviewed here.

Parts of the Bridge Creek Limestone (latest Cenomanian to early Turonian) Member of the Greenhorn Formation are markedly rhythmic, with alternations of light-coloured, bioturbated, pelagic limestone and less bioturbated to evenly laminated dark shaley limestone and calcareous shale (Barron et al., 1985). Hook & Cobban (1981) report three erosional surfaces within the upper metre of the Bridge Creek Limestone, which indicate a complex history of deposition and erosion (Fig. 16). All three surfaces show hardground development; the upper two have been partially phosphatized. They are characterized by coquina lenses composed of the encrusting oyster *Pycnodonte kansasense*.

Scattered through the unit are oyster-encrusted, partly phosphatized, internal moulds of the coiled Cretaceous ammonite *Pseudaspidoceras* sp. Most

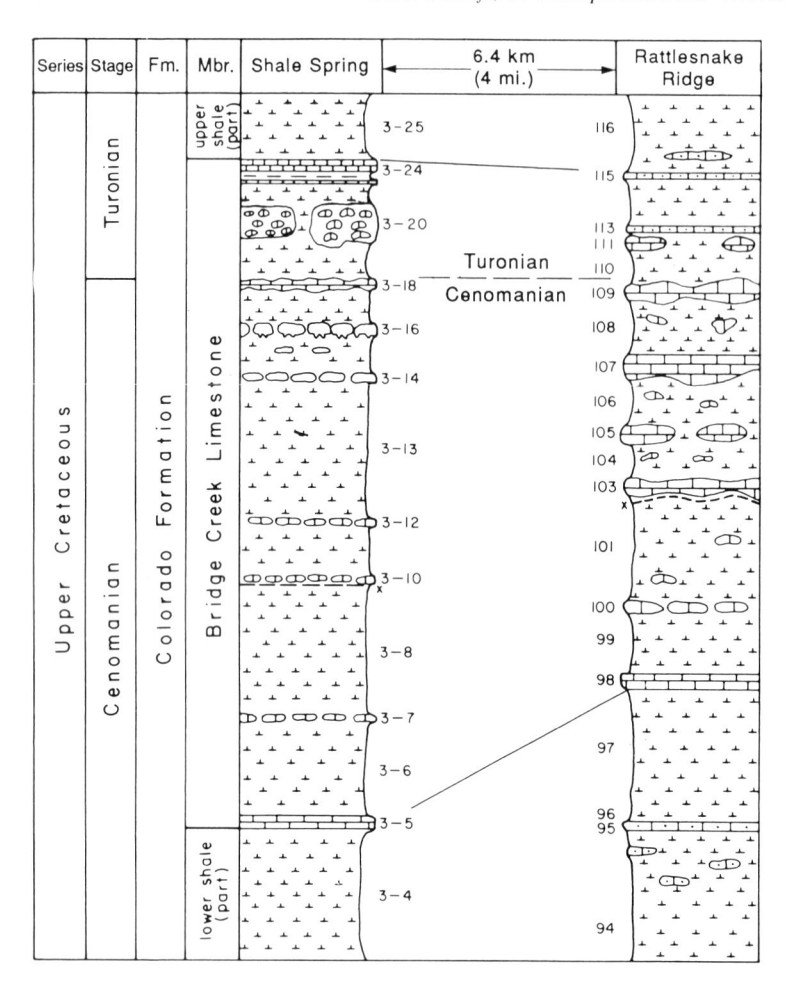

**Fig. 16.** Measured sections of the Bridge Creek Limestone member of the Greenhorn Formation, New Mexico (Cenomanian–Turonian condensed section). Units 109, 111 and 115 are capped by flooding surfaces. (From Hook & Cobban, 1981.)

are abraded on one side. All moulds are in horizontal position and most are corroded and encrusted or phosphatized on one or both sides. The moulds were formed within the substrate, cemented, exhumed by submarine erosion, stripped of their shells, abraded, corroded, mineralized and encrusted, and, in some cases, flipped over and encrusted, mineralized or corroded again. One of these hardgrounds is the Turonian maximum flooding surface.

A similar condensed section, highlighted by rhythmic alternations of shale and limestone and containing a mineralized erosional surface, has been described from the Cretaceous–Tertiary boundary of Alabama (Baum & Vail, 1988). An analysis of the stable isotope geochemistry of the Greenhorn example has led Barron *et al.* (1985) to suggest periodic variations in rainfall and runoff as a result of orbital forcing, leading to variations in water mass temperature, salinity, and degree of stratification.

Radiometric dates suggest that the periodicity is in the 20 000–100 000 year range.

### Parasequences in transgressive settings

*Transgressive parasequences in the Muskiki Formation, Alberta Basin*

The Coniacian Muskiki Formation of the Alberta Basin consists of 25–95 m of shale and siltsone (Plint, 1990). It rests disconformably on the eroded surface of the Cardium Formation, and grades upwards into the coarse siltstones and sandstones of the Coniacian–Turonian Marshybank Formation. Plint (1990) noted that although the Muskiki and Marshybank can be mapped easily in outcrop, the individual, upward-coarsening facies successions that comprise the fundamental 'building blocks' of the formations are less easily traced as a result of lateral facies and thickness changes. Only by correlating outcrop sections to a regional subsurface framework is it possible to make reliable correlations between outcrops (Fig. 17).

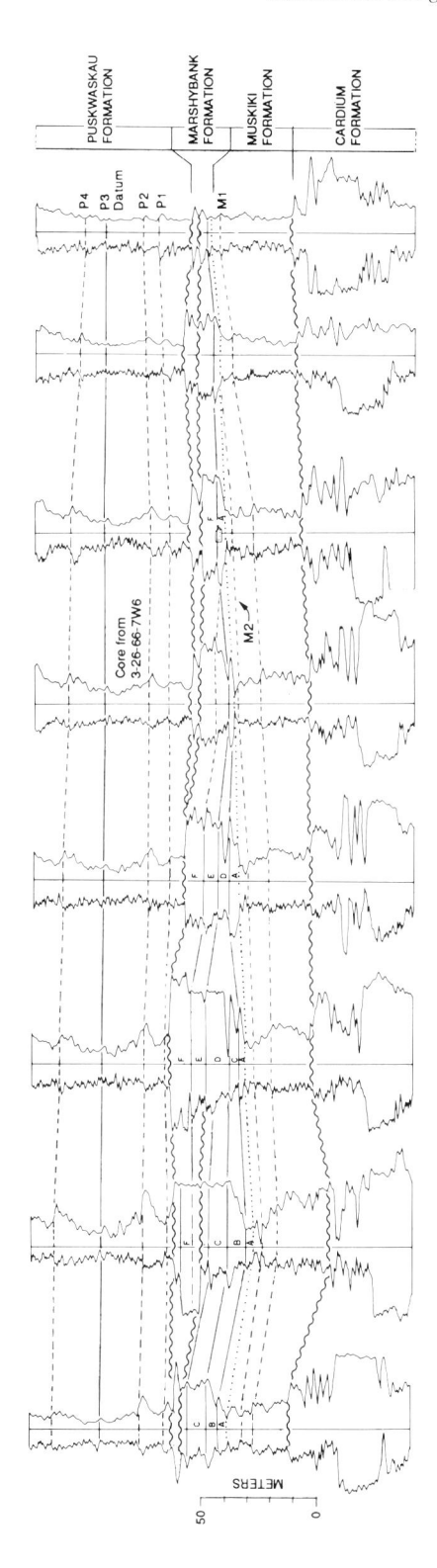

In a discussion of depositional cyclicity in the Muskiki and Marshybank Formations, Plint (1990) showed that, as a whole, these two formations form a depositional sequence in the sense of Vail (1987). The surface separating the two formations is identified as a maximum flooding surface, across which the Marshybank Formation downlaps. We are concerned here with the lower Muskiki Formation, designated by Plint (1990) as a transgressive systems tract (sense of Vail, 1987). Both outcrops and subsurface sections can be resolved into at least three upward-coarsening parasequences. Backstepping geometry is apparent; the parasequences thin basinward and their bounding surfaces converge basinward with the maximum flooding surface. The parasequences consist of shale that becomes silty near the top. They are locally capped with thin layers of chert pebbles, or in one case, several metres of bioturbated sandy siltstone. Each upward-coarsening parasequence can be viewed as a regressive shelf–shoreface depositional system (see Swift et al., this volume, pp. 112–118), capped by a transgressive source diastem. The sporadic gravel deposits on the source diastem suggests that the regressive episodes were followed by base level falls sufficient to bypass gravel through the shoreline and on to the shelf (see discussion of textural maturity and grain-size profiles; Swift et al., this volume, pp. 94–96).

*Transgressive parasequences and sand ridge deposits in the Woodbine Formation, East Texas Basin*

Our study of the East Texas Basin has examined the Cenomanian and Turonian stages (Early–Late Cretaceous, and has led to the recognition of two depositional sequences in the interval between the Buda Limestone and the Austin Chalk (Fig. 18), equivalent to the Woodbine and Eagleford Formations (Phillips & Swift, 1986; Phillips, 1987). In the subsurface, the Woodbinean sequence consists of a thick highstand systems tract (Vail (1987) terminology), capped by a deltaic sandstone (lower Harris Delta, Fig. 18). The Eaglefordian sequence, however, can be resolved into a lowstands systems tract (upper Harris Delta) overlain by a transgres-

**Fig. 17.** Well-log section through the Coniacian Muskiki Formation, Alberta Basin, showing upward-coarsening, back-stepping parasequences. (From Plint, 1990.)

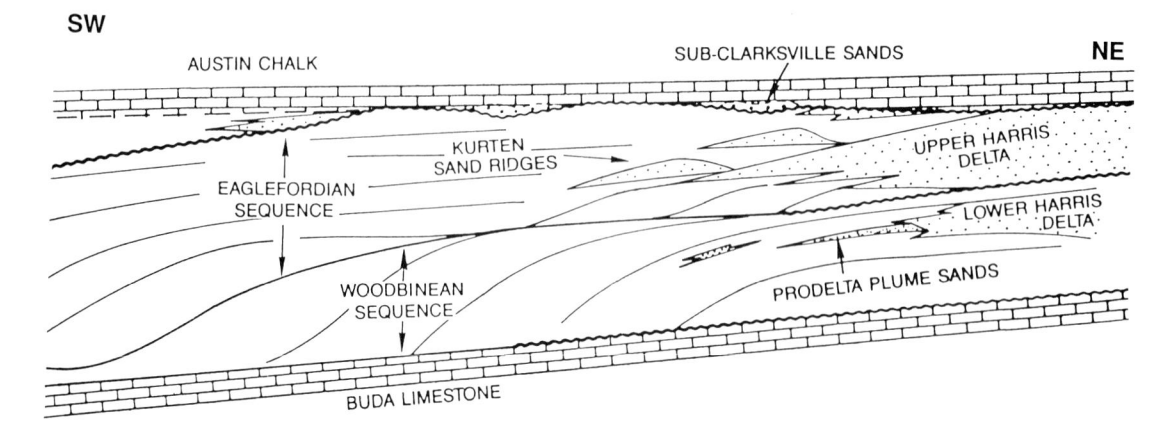

Fig. 18. Cross-section through the Cenomanian–Turonian section in the east East Texas Basin. (From Phillips, 1987.)

sive systems tract (Fig. 18; Vail (1987) terminology).

Parasequence architecture becomes apparent in the transgressive Eaglefordian at Kurten Field in Madison County, Texas. Here the producing sandstones form a back-stepping pattern against the ravinement surface cut across the top of the upper Harris Delta (Figs 18,19). Isopach maps of the Kurten sandstones reveal asymmetrical, lenticular bodies, whose long axes are oblique with respect to the trend of the palaeoshoreline (Fig. 20). Facies distribution is also asymmetrical; the coarsest facies (flaser bedded to cross-bedded sandstone; bioturbated flaser bedded to cross-bedded sandstone) occur on the steep southwestern flanks of the bodies, while the finer facies (laminated, bioturbated sandstone; intensely bioturbated sandstone) occur on the gentler northwestern flank.

The oblique orientation of the sand bodies, thickness and facies asymmetry, and association with a transgressive surface is typical of transgressive sand ridge deposits in both modern and ancient settings (Swift & Field, 1981; Gaynor & Swift, 1988; Swift et al., this volume, pp. 89–152). The back-stepping sand ridges are most economically interpreted as the product of paracycles of sedimentation similar to that described from the Atlantic shelf of North America. In this model, shorefaces prograde, creating strandplains. The strandplains are then cut back by erosional shoreface retreat, and the debris of the retreat process swept up as sand ridges on the resulting ravinement surface.

*Transgressive sequences and sand ridge deposits in the Tocito Sandstone, San Juan Basin*

Similar transgressive, upward-coarsening sandstone bodies occur somewhat higher in the section in the San Juan Basin of New Mexico (Fig. 21), also in association with deposits believed to be lowstand (Vail's (1987) definition). Here a Coniacian progradation culminated in the multiple tongues of the Gallup Sandstone (Molenaar, 1983). The Gallup is the progradational littoral magnafacies of the Coniacian stage, and is the equivalent of the Campanian Point Lookout magnafacies described above. The transgressive sandstone lenses are those of the Tocito Sandstone, a major petroleum reservoir in the San Juan Basin (Tillman, 1985).

Recent studies (Nummedal et al., 1989) suggest that the lower tongues of the Gallup Sandstone (F through to D, Fig. 21) are highstand deposits (Vail's (1987) terminology). Nummedal et al. (1989) noted that:

> The vertical aggradational pattern of Gallup tongues F through D, and the contrasting progradational architecture of the Gallup Sandstone tongues C through A are strongly suggestive of relative sea level rise during deposition of the lower Gallup Sandstone, followed by rapid fall and downward shift in facies for the upper tongues.

As the last of the Gallup tongues, the A through to C tongues are in fact in the right place at the right time to be the lowstand deposits called for by the Vail (1987) model. However, it could be argued from Fig. 21 that it is the F through to C tongues that appear to have the progradational stacking pattern, and the A and B tongues that appear to have the aggradational pattern. While inappropriate for Vail's (1987) eustacy dominated passive margin model, such a progradational to aggradational arrangement conforms to the $Q$-dominated clastic wedge model proposed for foreland basins (Swift

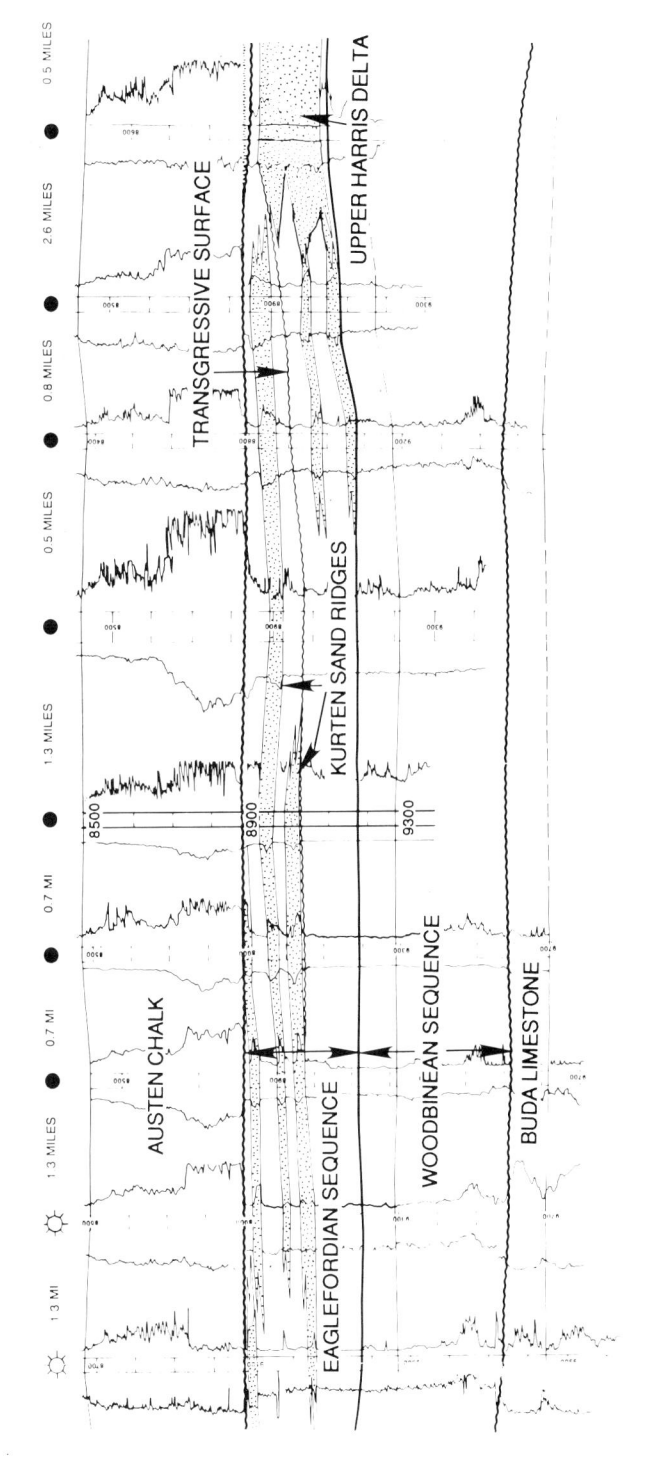

Continental margin sedimentation—parasequences 177

**Fig. 19.** Well-log section through the Kurten Field, East Texas Basin, showing sand-ridge deposits above the ravinement surface.

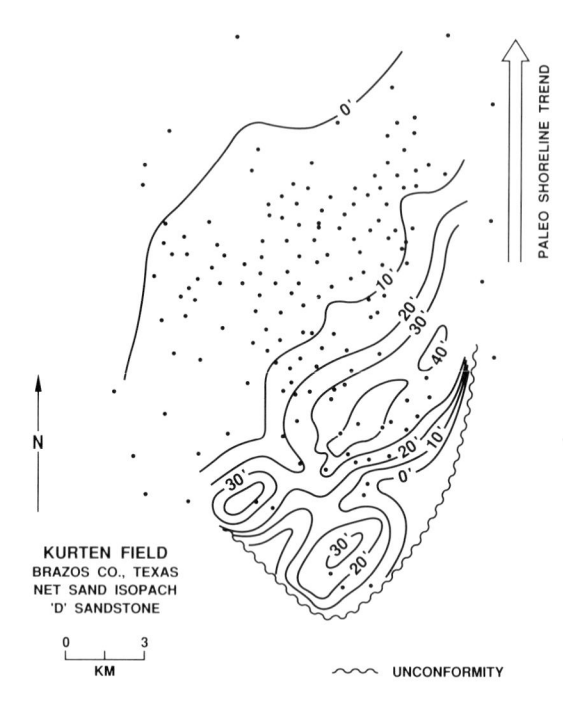

**Fig. 20.** Net isopach map of the 'D' Sandstone. Kurten Field, East Texas Basin. (From Turner & Conger, 1984.) Compare with Oregon Shoal, Fig. 7.

et al., 1987; see Thorne & Swift, this volume, pp. 241–244). In the clastic wedge model the basinward turnaround occurs because sediment supply decreases relative to subsidence. Under these conditions, and in the absence of significant variation in relative sea-level, the shift at turnaround, from a progradational parasequence stacking pattern to an aggradational one, is liable to be marked.

However, there is evidence that a relative sea-level fall was in fact involved in the construction of the Coniacian sequence. The uppermost Torrivio Member of the Gallup Sandstone is a medium to very coarse-grained and occasionally pebbly fluvial sandstone (Nummedal et al., 1989). It appears to have supplied the feeder channels to the A, B and C tongues. It rests on 10 m of finer grained 'non-marine' Gallup Sandstone of fluvial origin, and in a more basinward position, the lower Torrivio rests directly on the Gallup C tongue. Its base is an erosional contact with up to 15 m of relief. The Torrivio Sandstone is most economically explained as having been deposited as sea-level began to rise, after an episode of relative lowstand.

Basinward of the Gallup Sandstone, and on its upper surface, lie elongate, nearly shore-parallel lenses of marine Tocito Sandstone (Nummedal et al., 1987). A cross-stratified sandstone facies tends to occur on the shoreward sides of the ridges, with obliquely cross-ridge (southeasterly) palaeocurrent orientation. Interbedded, muddy bioturbated and ripple bedded sandstone facies occur on the down-current side. The oblique orientation and facies asymmetry of these deposits indicates that they formed as shelf sand ridges. Gravel beds on the upcurrent flanks of the Tocito lenses suggest a relative sea-level fall sufficient to bypass Torrivio gravel out on to the shelf. In several cases, a flooding surface marked by thin bioclastic calcarenite occurs in the shale several metres above the crest of the Tocito lens.

The more basinward Tocito lenses rest on erosional surfaces incised into the Mancos Shale. Towards the northwest and the basin margin they rest on the truncated surface of the Gallup Sandstone (Fig. 21). Yet farther to the northwest, they rest disconformably on back-stepping estuarine and lagoonal deposits (Dilco Coal Member), or on flood tidal delta deposits (Borrego Pass Sandstone). There is reason to believe, however, that the Tocito sand ridge deposits do not lie on a single continuous transgressive surface. Nummedal et al. (1989) show that bentonitic markers on well-logs, originally identified by McCubbin (1969), pass above the lower Tocito lenses, but below the B tongue of the Gallup. Here at the basinward limit of the Gallup, the more nearly aggradational parasequence architecture has inhibited the formation of a continuous transgressive surface. Instead, the successive Tocito lenses appear to be coupled with successive tongues of the Gallup Sandstone so that the marine erosion surface beneath each basinward Tocito lens passes landward over the top of its equivalent Gallup tongue (Fig. 21). The markers that occur above some of the Tocito lenses appear to be maximum flooding surfaces of parasequence scale. If this is the case, then the Gallup–Tocito parasequences express a relatively complete sequence structure. They appear to be of fourth-order duration (100 thousand to 1 million years).

## DISCUSSION

### General

The preceeding review of high-frequency sequences raises a series of questions concerning these patterns

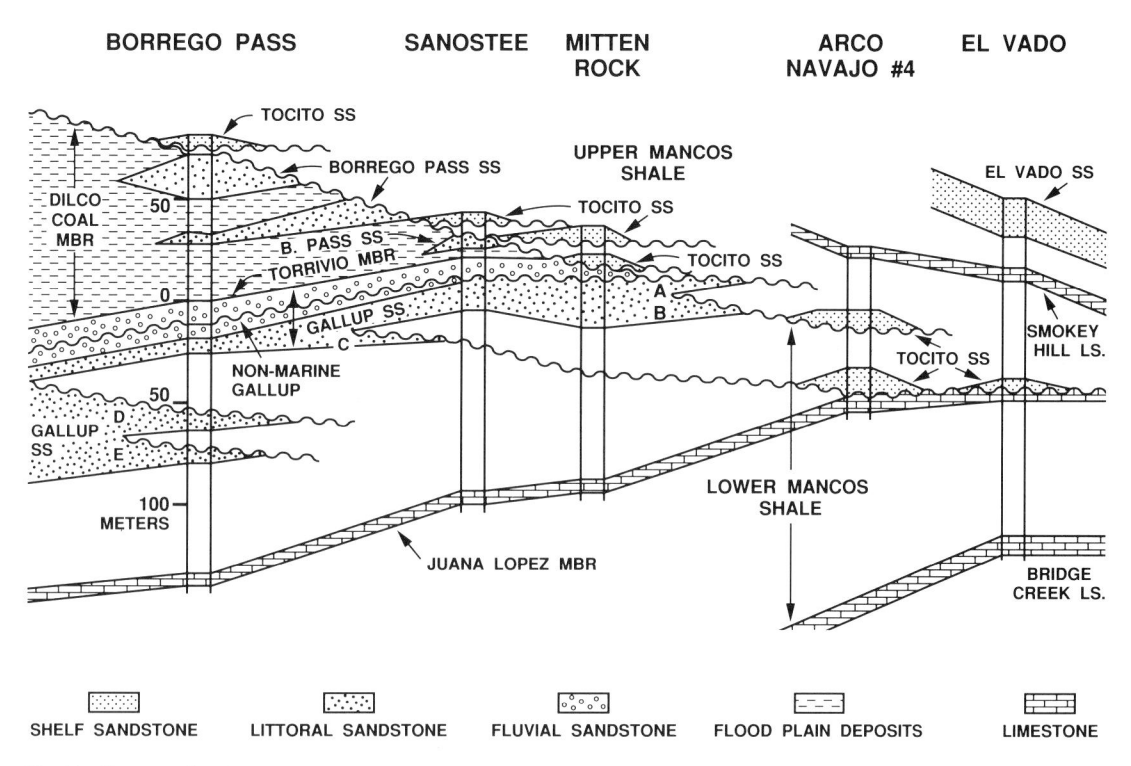

Fig. 21. Cross-section through the Coniacian stage in the northwest San Juan Basin, New Mexico, based on measured sections and wire-line logs. Much of the data are presented in greater detail in Nummedal et al. (1989).

of stratigraphic origin. Do they have a characteristic internal structure? How are they related to smaller and larger scales of depositional organization? What kinds of time-scales do they typically represent? What is their forcing? We will address these issues in turn.

## Internal structure of parasequences

As noted above, parasequences, in Vailian theory, are formed without sea-level falls. They do not experience basinward shifts of depositional environments, hence that consist of simple, upward-coarsening bed sets bounded by flooding surfaces. In contrast, high-frequency sequences are terminated by sea-level falls, with the resulting formation of discrete 'systems tracts'.

The examples of parasequences presented in this review generally confirm the observations of Van Wagoner et al. (1990) and Mitchum & Van Wagoner (1991). Shallow-marine parasequences are commonly simple, upward-coarsening bed sets separated by flooding surfaces. In regime terms, such a parasequence boundary is the product of the cyclic oscillation of the regime ratio $R \cdot D/Q \cdot M$ from an accommodation-dominated value $R \cdot D > Q \cdot M$ to a supply-dominated value $R \cdot D < Q \cdot M$ (see discussion by Swift & Thorne, this volume, p. 13).

Note that while the cycle may be symmetrical, the sequence is not. As the regime ratio $R \cdot D/Q \cdot M$ rises above unity, the reworking ratio ($r = a'/å$, see Thorne et al., this volume, p. 64) also rises above unity. The new equilibrium surface is a deeper one, and is obtained through scour of the sea-bed (see discussion of the equilibrium index by Swift & Thorne, this volume, pp. 10–13). When the regime ratio reverts to its supply-dominated original value, however, the equilibrium surface begins a shoaling, basinward shift, and the upward-coarsening half-sequence is deposited.

However, we have described a number of cases that are intermediate in nature between parasequences as upward-coarsening bed sets, and high-frequency sequences as systems tract assemblages. For instance, the tongues of the Gallup Sandstone, with their prograding strandplains and associated transgressive sand-ridge systems, are sufficiently differentiated to be considered high-

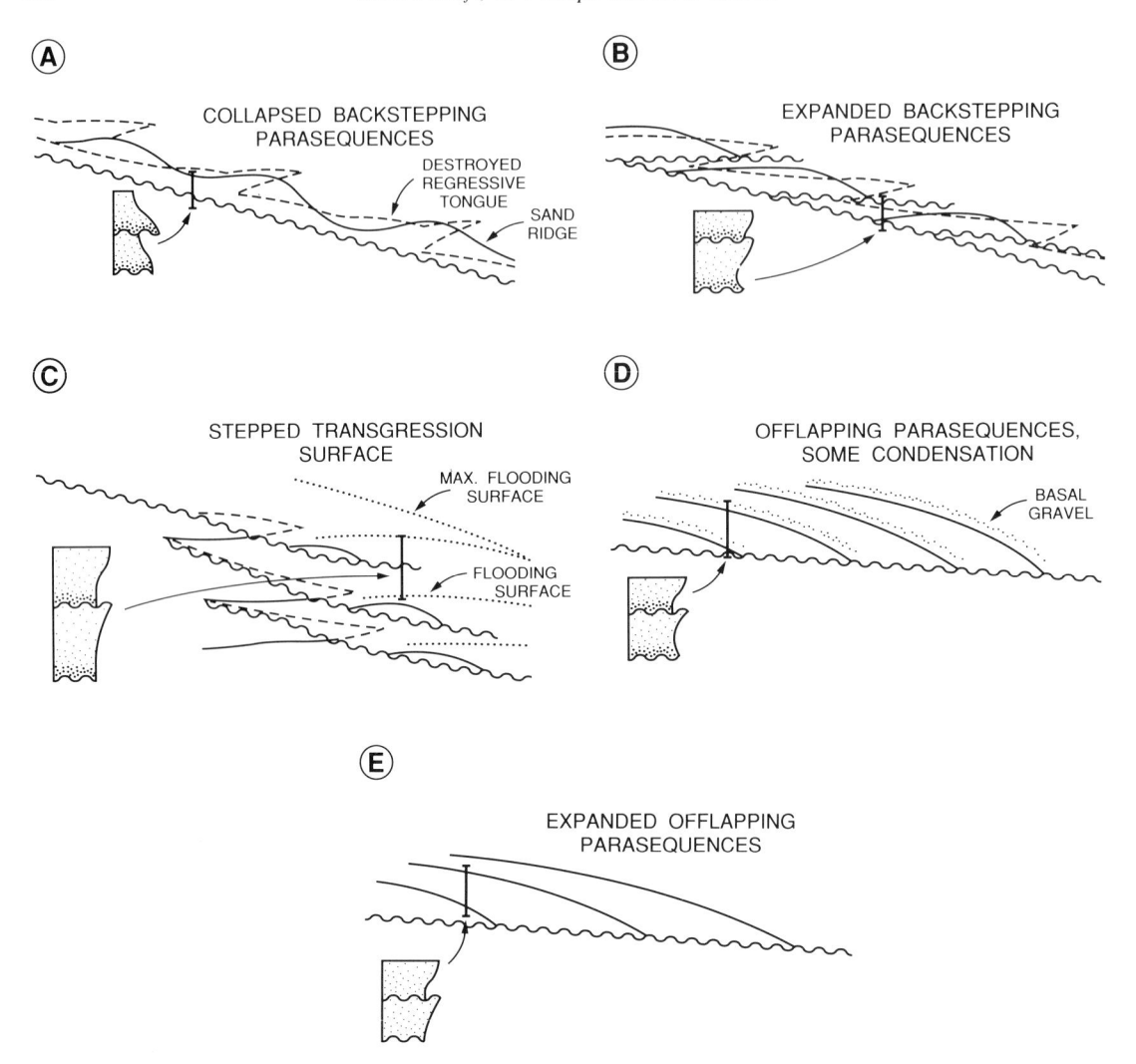

**Fig. 22.** Varieties of parasequence architecture. (A) Transgressive, high accommodation, ¥ ≫ 1. (B) Transgressive, medium accommodation, ¥ > 1. (C) Stillstand, low accommodation, ¥ = 1. (D) Regressive, low supply, ¥ < 1. (E) Regression, high supply, ¥ ≪ 1.

frequency (fourth-order) sequences in the Mitchum & Van Wagoner (1991) scheme. Yet it is very likely, by analogy with our review of the Bridge Creek Limestone beds, that by the time the Gallup sequences are traced into the appropriate maximum flooding surface (in this case the Smokey Hill Limestone Member of the Niobrara Formation; Fig. 21), they will be monolithologic beds of 30 cm or less thickness, and simple enough to be called parasequences in any definition.

The problem is placed in sharper focus by the case of the Maryland Miocene. The sequence boundaries are of no diagnostic use. They may well be ravinements rather than simple flooding surfaces. If they are ravinements, then there may well have been subaerial erosion as well as submarine erosion (see discussion of ravinement surfaces in Thorne & Swift, this volume, p. 17). The basal shell beds are suspiciously thick and complex. Thus it does not seem prudent to present parasequences and high-frequency sequences as mutually exclusive geometries.

## Parasequences: control by depositional regime

The examples of small-scale sequences that we have cited in this paper exhibit a systematic variation in response to the depositional regime (Fig. 5). In Fig. 5, we replace the accommodation/supply ratio, $R \cdot D/Q \cdot M$, with the symbol ¥ (see discussion of the accommodation/supply ratio and ¥ by Thorne & Swift, this volume, 195–197).

On The Netherlands shelf (van Straaten, 1965), the Argentine shelf (Parker *et al.*, 1982) and the Aquitaine Shelf (Fig. 5), high-frequency sequences have been deposited under conditions that, at large time-scales (3000 years) were strongly accommodation-dominated $(R \cdot D \gg Q \cdot M)$. At shorter time-scales, the regime oscillated, with each oscillation generating a shoreline advance followed by a shoreline retreat. In inner shelf settings, however, the preservation ratio (see Thorne & Swift, this volume, 196–197) was small, and the depth of incision of each transgressive shore face was greater than the thickness of the preceding regressive deposit, so that the regressive depositional systems were remoulded into back-stepping transgressive depositional systems (collapsed parasequences, ¥ ≫ 1; Fig. 22A). In outer shelf settings (Aquitaine shelf), the periods of more rapid sea-level rise appear as episodes of shoreface bypassing and sand accumulation, while the periods of stillstand or sea-level fall appear as episodes of rivermouth bypassing and outer shelf mud accumulation. The result is an upward *fining* parasequence. The reverse grain-size profile can be seen as a plastic distortion of the normal sequence rather than an unrelated phenomenon; the 'basal lag', which always fines upwards, has been thickened greatly relative to the capping muds, which, if examined closely enough, are likely to coarsen slightly upwards.

On the Middle Atlantic shelf of North America (Fig. 6), the long-term depositional regime was less accommodation-dominated (¥ > 1, Fig. 22B). The preservation ratio was larger, so that in each regime oscillation, some regressive deposits were preserved. Upward fining of the back-stepping transgressive parasequences is less developed, in part because the transgressive cap has been remoulded into a sand-ridge topography. Sand ridges commonly rise from a coarse basal deposit, but their upper sections tend to coarsen upwards, in response to migration of the coarse upcurrent flank over the downcurrent flank (Figueiredo *et al.*, 1980).

On the Coniacian shelf of New Mexico, the long-term (>100 000 year) depositional regime was nearly balanced between accommodation and supply (¥ ≈ 1) during the seaward turnaround of the Gallup Sandstone. The resulting parasequence architecture is aggradational rather than progradational, and the transgressive surface is not continuous but stepped (Fig. 22C). The preservation potential varied across the shelf, from low outer shelf values to higher inner shelf values. The variation appears to have been the result, in part, of a basinward decrease in sediment input, and in part of a marginward increase in the subsidence rate. As a result of the cross-shelf gradient in preservation potential *alone*, the parasequences are 'collapsed' transgressive deposits on the outer shelf, and 'expanded' regressive deposits on the inner shelf.

The Maryland Miocene section (Fig. 11) offers an instructive comparison to the New Mexico Coniacian section (Fig. 22), because it too tends to preserve both transgressive and regressive deposits. Cross-shelf gradients in preservation potential must have been gentle, because regressive and transgressive deposits are stacked in layer-cake fashion, rather than passing from one into the other by lateral facies change. Because these parasequences are known mostly from outcrop, their stacking geometry is poorly understood. However, the regressive half-sequences are usually thicker, and we assume that above the maximum flooding surface, the Maryland Miocene represents a more nearly supply-dominated section (¥ < 1, Fig. 22D).

Finally, the Mesaverde sandstones of the Campanian in the western interior basin (e.g., the Point Lookout Sandstone), and its correlative shales (Mancos and Cody Shales) must represent a supply-dominated end-member in this sequence (¥ ≪ 1, Fig. 22E). Transgressive episodes are present as hiatuses or a very thin regressive shale horizon, but condensation is generally not sufficient to create a basal gravel or coquina. Where the maximum flooding surface is observed in such supply-dominated settings (e.g. the Turonian Marshybank Formation, Fig. 17), the geometry is clearly offlapping.

## Origin of parasequences

Recent studies of parasequences and high-frequency sequences (Van Wagoner *et al.*, 1990; Mitchum & Van Wagoner, 1991) acknowledge the potential role of autocyclicity in the formation of nearshore parasequences, but stress eustacy as the formative mechanism for both fourth and fifth-order sequences. Glacio-eustatic sea-level cycles have been well documented from the Quaternary (see reviews of Coleman & Roberts (1988) and Tesson *et al.*, (1989) earlier in this paper), and inferred from the Triassic (Goldhammer *et al.*, 1987) and Pennsylvanian (Busch *et al.*, 1985). Glacio-eustacy has been related to cyclic variations in insolation, owing in turn to variations in the Earth's orbit (Milankovich mechanism, Hays *et al.*, 1976; Imbrie, 1982). A number of periodicities occur, including that owing to precession of the equinoxes (19 000 and 23 000 years), obliquity (41 000 years) and ellipticity (100 000 years).

Glacio-eustacy can be used to explain parasequence and sequence formation in the Tertiary (Matthews & Frohlich, 1991), but is unsatisfactory as a universal mechanism because the planet appears to have been ice-free for much of its history; glaciation has been an episodic phenomenon (Pitman & Golovchenko, 1983; Thorne & Watts, 1984). However, Milankovitch periodicity is a much more general determinant of climate, and may have driven the formation of parasequences through both 'icehouse' and 'greenhouse' periods (Fischer, 1982). Numerical simulations of Quaternary climate (Kutzbach & Guetter, 1984) indicate that Milankovitch cyclicity results in major oscillations in the extent and intensity of monsoonal climates at low and middle latitudes, which would result in significant variation in precipitation and in sediment yield throughout ice-free as well as glacial periods (Barron *et al.*, 1985; Grotzinger, 1986).

The Milankovitch mechanism thus appears to be an extremely important, if not dominant, forcing for high-frequency depositional cycles. However, autocyclicity is also important. Autocyclicity clearly plays an important role in determining depositional cyclicity in such major coastal depocentres as the Mississippi delta (see discussion on previous pages) or in the Carboniferous of the North American mid-continent (Busch *et al.*, 1985), through the mechanism of river avulsion.

There is reason to believe, however, that autocyclicity may have been embedded in the depositional regimes of the Cenozoic Gulf Coast Basin of North America, and in the Carboniferous depositional regime of the North American mid-continent, in a much more fundamental way. In these and in many other basins, the section consists of thousands of metres of alternating thin, shallow-water sandstones and shales. Mitchum & Van Wagoner (1991) note that in the thick, rapidly deposited depocentre of the Tertiary Gulf Coast Basin, rapid subsidence rates were counterbalanced by equally rapid sedimentation rates. 'This very closely balanced [class of] equilibrium produces a high sensitivity to very small, rapid oscillations in sea-level that is not present in areas of more moderate depositional rates'. Shallow-marine sea-floors behaving in this manner have been described by Swift & Thorne (this volume, pp. 9–12) as equilibrium surfaces, whose homeostatic behaviour can be defined in terms of an equilibrium index. More generally, we have described the equilibria in terms of an accommodation/supply ratio, $R \cdot D/Q \cdot M$. Because the interrelationships of these variables are poorly understood, we have generally presented the ratio semi-quantitatively, as an inequality; for instance $R \cdot D > Q \cdot M$. Thorne & Swift (this volume, pp. 195–197), however, define a regime parameter, $¥ = R \cdot D/Q \cdot M$, and expand the expression to include coefficients that establish a power law relationship among the variables. In this model, the depositional regime of the continental margin is seen as a non-linear system.

There has been considerable recent interest in a class of behaviour in complex non-linear systems that has come to be called 'chaotic' (Gleick, 1987). Such systems are said to exhibit 'extreme sensitivity to initial conditions'; that is, initially minor instabilities are greatly magnified, a condition known as the 'butterfly effect'. Fluids typically exhibit chaotic behaviour. The butterfly effect is so named because of the extreme sensitivity of weather systems to initial effects; the movement of a butterfly's wings in Hawaii might well result in a major weather system (storm) embedded in the westerly wind flow over Boston, thousands of kilometres and several days downstream. Similarly, in the turbulent flow characteristic of most fluids, energy cascades through successively smaller spatial scales (eddy to smaller eddy to thermal motion), while *information* moves upscale. That is, the eddies themselves are in a sense gigantic amplifications of thermal motion. It is important to note that such behaviour, while unpredictable, is none the less deterministic, and often can be described by relatively simple equations. It is not intrinsically probabilistic in the way that quantum behaviour is probabilistic.

If high-frequency variation in depositional regimes is in fact chaotic, then the periodicity expressed need not bear any simple relationship to forcing periodicity. For instance, it turns out that a dripping tap frequently exhibits chaotic behaviour. In such cases, the time interval between drips is not constant but varies irregularly within a relatively restricted range. A mean period can be computed for a period of observation and the successive deviations can be attributed to 'noise'. However, in fully chaotic behaviour, such statistics do not reflect the physical reality. There is no stable mean for the dripping tap. There is no time interval of any size that will yield the same mean on successive trials.

If, for instance, high-frequency fluctuations (perhaps $c.20\,000$ years or $c.100\,000$ years) of the Miocene shoreline of Texas (Galloway, 1989c) were in fact chaotic, then perturbations of *any* of the regime variables at amplitudes very possibly too small to measure would trigger transgressions and regressions of many tens or hundreds of kilometres. The resulting polyharmonic curve of shoreline position against time would superficially resemble one created by complex orbital forcing, but the basic harmonics would not be stable.

The implications of non-linear systems theory for complex, shallow-marine stratigraphic sections are clear (W. Galloway, University of Texas, pers. comm.). While the ultimate control may in fact be the relatively simple, deterministic, orbital forcing, the response to that control may be complex, unpredictable and technically chaotic—an unexpectedly large-scale form of autocyclic response.

## CONCLUSIONS

At large spatial scales, depositional facies, systems and systems tracts are organized into repeating vertical arrays called *depositional sequences*. Depositional sequences are the product of cyclic variation of regime variables. They are hierarchical in nature; depositional sequences may occur at several spatial scales in the same section, in response to regime oscillations occurring simultaneously at multiple frequencies. The smallest, simplest sequences are a few metres thick, and may consist of upward-fining bed sets separated by flooding surfaces. Such small-scale sequences, termed *parasequences*, are believed to be responses to regime oscillations of $10^4$ years duration (fifth-order cycles) to $10^5$ years duration (fourth-order cycles). At somewhat longer timescales ($10^5$ to $10^7$ years) sequences develop a more complex structure, marked by stronger facies contrasts, and can be referred to as high-frequency sequences. However, the distinction cannot be pressed too far. Parasequences, as strictly defined (upward-coarsening bedsets separated by flooding surfaces), tend to pass landward into more complex sequences exhibiting systems differentiation, and we extend the term 'parasequence' to describe high-frequency sequences of higher order than the dominant order (usually third-order).

Third-order sequences (millions of years duration) contain parasequences, and larger subunits distinguished by contrasting parasequence stacking patterns. The subunits have been referred to as systems tracts. The term is part of a system devised by workers at the Texas Bureau of Economic Geology, which includes facies, depositional systems (assemblage of process-related facies), and depositional systems tracts (linkage of contemporaneous facies). In fact, the subunits of depositional sequences are much larger features than the systems tracts so defined. Parasequences, not sequence subunits, correspond in scale to depositional systems and systems tracts, and a parasequence can be described fairly as a depositional system seen in side-view. Sequence subunits consist of repetitive successions of parasequences, and are properly styled *parasequence sets*. However, the 'systems tract' terminology has entered the literature. Problems are minimized if sequence subunits are recognized as *geometric systems tracts*, defined on the basis of their parasequence stacking geometries, rather than *depositional systems tracts*, defined on the basis of their facies architecture.

Comparison of the sequence structure of recent and subrecent deposits on modern continental shelves with that observed in older sections suggests that parasequence architecture varies with the accommodation/supply ratio ($\boldsymbol{R} \cdot \boldsymbol{D}/\boldsymbol{Q} \cdot \boldsymbol{M}$) and the preservation index; see Thorne & Swift (this volume, pp. 196–197) for definitions. At large (accommodation-dominated) values of the accommodation/supply ratio and low values of the preservation index, regressive deposits generated by high-frequency regime oscillations are destroyed by the subsequent transgression, resulting in 'collapsed parasequences'. As the accommodation/supply ratio decreases, asymmetrical back-stepping parasequences result, in which the transgressive half sequence is dominant. As $\boldsymbol{R} \cdot \boldsymbol{D}/\boldsymbol{Q} \cdot \boldsymbol{M}$ approaches 1, the stacking geometry becomes aggradational, and if the preservation index is still large, the transgressive surface breaks up into a stepped surface. As the regime passes into the supply dominated field ($R \cdot D/Q \cdot M < 1$), parasequences develop downlapping geometry and the regressive half-sequence becomes dominant.

Parasequences, as strictly defined (upward-coarsening bed sets separated by flooding surfaces), tend to pass landward into more complex sequences exhibiting systems differentiation, and we extend the term to describe high-frequency sequences of higher order than the dominant order (usually third-order). Parasequences in coastal settings may stem from complex origins, in which both autocyclic forcing (local river avulsion) and allocyclic forcing (external to the basin) may play a role. Further offshore, cyclicity may be in large measure allocyclic; a response to orbital forcing (the Milankovich mechanism). However, further study is needed in order to establish the extent to which non-linear interactions of regime variables (chaotic behaviour) can generate apparently periodic parasequence successions.

## REFERENCES

ARBOUILLE, D., WEBER, O., TURON, J.L. & CARBONEL, P. (1991) Holocene transgression of the continental shelf of northern Biscay Bay: biosedimentary record of glacioeustatic changes. *Mar. Geol.*, in press.

ASQUITH D.O. (1970) Depositional topography and major marine environments, late Cretaceous Wyoming. *Bull., Am Assoc. Petrol. Geol.* **54**, 1184–1224.

BARRON, E.J., ARTHUR, M.A. & KAUFFMAN, E.G. (1985) Cretaceous ryhthmic bedding sequences: a plausible link between orbital variations and climate. *Earth Planet. Sci. Lett.* **72**, 327–340.

BAUM, G.R. & VAIL, P.R. (1988) Sequence stratigraphic concepts applied to Paleogene outcrops, Gulf and Atlantic Basins. In: *Sea-level Changes: An Integrated Approach* (Eds Wilgus, C.K., Hastings, B.S., Kendall, C.G.St.C., Posamentier, H.W., Ross, C.A. and Van Wagoner, J.C.). Soc. Econ. Paleontol. Mineral. Spec. Publ. 42, 309–328.

BEERBOWER, J.R. (1964) Cyclothems and cyclic depositional mechanisms in alluvial plain sedimentation. *Kans. State Geol. Surv. Bull.* **169**(1), 31–42.

BOYD, R., SUTER, J. & PENLAND, S. (1989) Relation of sequence stratigraphy to modern depositional environments. *Geology* **17**, 926–929.

BROWN, L.F. & FISHER, W.L. (1977) Seismic stratigraphic interpretation of depositional systems: examples from Brazilian rift and pull-apart basins. In: *Seismic Stratigraphy—Applications to Hydrocarbon Exploration* (Ed. Payton, C.E.). Mem., Am. Assoc. Petrol. Geol. **26**, 218–248.

BUSCH, R.M., WEST, R.R., BARRETT, F.J. & BARRETT, T.R. (1985) Cyclothems versus hierarchy of transgressive–regressive units. In: *Recent Interpretations of Paleozoic Cyclothems: Guidebook*, pp. 141–153. Society of Economic Paleontologists and Mineralogists, Midcontinental Section, Tulsa, OK.

CASTER, K.E. (1934) The stratigraphy and paleontology of Northwestern Pennsylvania, Part 1. *Bull. Paleontol.* **21**, 185 pp.

COLEMAN, J.M. & GAGLIANO, S.M. (1964) Cyclic sedimentation in the Mississippi River deltaic plain. *Trans. Gulf Coast Assoc. Geol. Soc.* **14**, 67–80.

COLEMAN, J.M. & ROBERTS, H.H. (1988) Sedimentary development of the Louisiana Continental Shelf related to sea-level cycles, part II: seismic response. *Geomarine Lett.* **8**, 109–119.

CURRAY, J.R., EMMEL, F.J. & CRAMPTON, P.J.S. (1969) Holocene history of a strandplain. In: *Simposio Internacional de Lagunas Costeras* (Eds Castañares, A.A. & Phleger, F.B.) pp. 63–100. Univ. Nac. Autonoma De Mexico, Mexico City.

CURRY, W.H., III (1973) Parkman Delta in Central Wyoming. *Wyoming Geol. Assoc. Earth Sci. Bull.* **December** 5–18.

CURRY, W.H., III (1976) Late Cretaceous Teapot Delta of southern Powder River Basin, Wyoming. *Twenty-eighth Annual Field Conference—1976, Guidebook*, pp. 21–28. Wyoming Geological Association, Casper, WY.

DILLON, W.P. & OLDALE, R.N. (1978) Late Quaternary sea level curve: reinterpretation based on sea level glacio-eustatic influence. *Geology* **6**, 56–60.

FIGUEIREDO, A.G., SWIFT, D.J.P. & SANDERS, J. (1980) Storm-graded layers on inner continental shelves: examples from southern Brazil and the Atlantic coast of the central United States. *Sediment. Geol.* **31**, 171–180.

FISCHER, A.G. (1982) Long term climatic oscillations recorded in stratigraphy. In: *Climate in Earth History* (Eds Berger, W.H. & Crowell, J.C.) pp. 97–104. National Academy Press, Washington, DC.

FISHER, W.L. & MCGOWEN, J. (1967) Depositional systems in the Wilcox Group of Texas and their relationship to the occurrence of oil and gas. *Trans. Gulf Coast Assoc. Geol. Soc.* **17**, 105–125.

FRAZIER, D.E. (1967) Recent deltaic deposits of the Mississippi River; their development and chronology. *Trans. Gulf Coast Assoc. Geol. Soc.* **17**, 287–315.

FRAZIER, D.E. (1974) Depositional-episodes: their relationship to the Quaternary stratigraphic framework in the northwestern portion of the Gulf Basin. *Tex. Univ. Bur. Econ. Geol. Geol. Circ.* **44-1**, 28 pp.

GALLOWAY, W.E. (1989a) Genetic stratigraphic sequences in basin analysis—part I: architecture and genesis of flooding surface bounded depositional units. *Bull., Am. Assoc. Petrol. Geol.* **73**, 125–142.

GALLOWAY, W.E. (1989b) Genetic stratigraphic sequences in basin analysis—part II: application to Northwest Gulf of Mexico Cenozoic basin. *Bull. Am. Assoc. Petrol. Geol.* **73**, 143–154.

GALLOWAY, W.E. (1989c) Clastic facies models, depositional systems, sequences and correlation: a sedimentologist's view of the dimensional and temporal resolution of lithostratigraphy. In: *Quantitative Dynamic Stratigraphy* (Ed. Cross, T.A.) pp. 459–477. Prentice Hall, New York.

GAYNOR, G.C. & SWIFT, D.J.P. (1988) Shannon sandstone

depositional model: sand ridge formation on the Campanian western interior shelf. *J. Sed. Petrol.* **58**, 868–880.

GLEICK, J. (1987) *Chaos: Making a New Science.* Penguin Books, New York, 352 pp.

GOLDHAMMER, R.K., DUNN, P.A. & HARDIE, L.A. (1987) High frequency glacioeustatic sea level oscillations with Milankovitch characteristics recorded in middle Triassic platform Carbonates in Northern Italy. *Am. J. Sci.* **287**, 853–892.

GROTZINGER, J.P. (1986) Upward-shallowing platform cycles: a response to 2.2 billion years of low-amplitude, high-frequency (Milankovitch band) sea level oscillations. *Paleoceanography* **1**, 403–416.

HANCOCK, J.M. & KAUFFMAN, E.G. (1979) The great transgression of the Late Cretaceous. *J. Geol. Soc. London* **136**, 175–186.

HAQ, B.U., HARDENBOL, J. & VAIL, P.R. (1987) Chronology of fluctuating sea levels since the Triassic. *Science* **235**, 1156–1167.

HAYS, J.D., IMBRIE, J. & SHACKLETON, N.J. (1976) Variations in the Earth's orbit: pacemaker of the ice ages. *Science* **194**, 2212–2232.

HOOK, S.C. & COBBAN, W. (1981) Late Greenhorn (Mid-Cretaceous) discontinuity surfaces, southwestern New Mexico. *N. M. Bur. Mines Miner. Resour. Circ.* **181**, 5–21.

IMBRIE, J. (1982) Astronomical theory of the Pleistocene ice ages: a brief historical review. *Icarus* **50**, 408–422.

JORDAN, T.E. (1982) Thrust loads and foreland basin evolution, Cretaceous Western United States. *Am. Assoc. Petrol. Bull.* **65**, 2506–2520.

KIDWELL, S. (1984) Outcrop features and origin of basin margin unconformities in the lower Chesapeake Group (Miocene, Atlantic coastal plain). In: *Interregional Unconformities and Hydrocarbon Accumulation* (Ed. Schlee, J.S.). Bull., Am. Assoc. Petrol. Geol. 36, 37–58.

KIDWELL, S. (1988) Stratigraphic condensation of marine transgressive records: origin of major shell deposits in the Miocene of Maryland. *J. Geol.* **97**, 1–24.

KINDINGER, J.L. (1988) Seismic stratigraphy of the Mississippi–Alabama shelf and upper continental slope. *Mar. Geol.* **83**, 73–94.

KING, L.C. (1959) Denudational and tectonic relief in southeastern Australia. *Trans. Geol. Soc. S. Afr.* **62**, 113–138.

KUTZBACH, J.E. & GUETTER, P.J. (1984) The sensitivity of monsoon climates to orbital parameter changes for 9,000 years B.P.: experiments with the NCAR general circulation model. In: *Milankovich and Climate* (Eds Berger, A., Imbrie, J., Hays, J., Kukla, G. & Salzman, B.), pp. 124–136. D. Reidel, Dordrecht.

MASTERSON, W.D. & PARIS, C.E. (1987) Depositional history and reservoir description of the Kuparuk River Formation, North Slope, Alaska. In: *Alaskan North Slope Geology*, Vol. 1 (Eds Tailleur, I. & Wiemar, P.) pp. 95–107. Society of Economic Paleontologists and Mineralagists Pacific Section, Bakersfield, CA.

MATTHEWS, R.K. & FROHLICH, C. (1991) Toward convergence of dynamic stratigraphy and seismic sequence stratigraphy: orbital forcing of low frequency Glacioeustacy. *J. Geophys. Res.*, in press.

MCCUBBIN, D.G. (1969) Cretaceous strike-valley sandstone reservoirs, Northwestern New Mexico. *Bull., Am. Assoc. Petrol. Geol.* **53**, 2114–2140.

MCKINNEY, T.F., STUBBLEFIELD, W.L. & SWIFT, D.J.P. (1974) Large scale current lineations on the central New Jersey Shelf: investigations by side scan sonar. *Mar. Geol.* **17**, 79–102.

MITCHUM, R.M. (1977) Seismic stratigraphy and global changes in sea level, Part II: glossary of terms used in seismic stratigraphy. In: *Seismic Stratigraphy—Applications to Hydrocarbon Exploration* (Ed. Payton, C.E.). Mem., Am. Assoc. Petrol. Geol. 26, 205–212.

MITCHUM, R.M. & VAN WAGONER, J.C. (1991) Parasequences and high frequency sequences in well logs, core, and outcrops. In: *Sequence Stratigraphy as an Exploration Tool* (Eds Armentrout, J., Coleman, J., Galloway, W. & Vail, P.R.). Trans. Gulf Coast Assoc. Geol. Soc., in press.

MOLINAAR, C.M. (1893) Major depositional cycles of rocks and regional correlations of Cretaceous rocks, Southern Colorado Plateau. In: *Mesozoic Paleography of West Central United States* (Eds Reynolds, M.W. & Dolly, E.D.) pp. 204–224. Rocky Mountain Section, Soc. Econ. Paleontol. Mineral., Denver, CO.

MYRICK, A.C., JR. (1980) *Variation, taphonomy, and adoptation of the Rhabdosteidae (= Eurhinodelphidae; Odontoceti, Mammalia) from the Calvert formation of Maryland and Virginia.* Unpublished PhD thesis, University of California, 411 pp.

NIEDORODA, A.W., SWIFT, D.J.P., FIGUEIREDO, A.G. & FREELAND, G.L. (1985) Barrier island evolution, Middle Atlantic Shelf, U.S.A. Part II: evidence from the Shelf Floor. *Mar. Geol.* **63** 363–396.

NUMMEDAL, D. & SWIFT, D.J.P. (1987) Transgressive stratigraphy at sequence bounding unconformities—Some principles derived from Holocene and Cretaceous examples. In: *Sea-level Fluctuation and Coastal Evolution* (Ed. Nummedal, D.). Soc. Econ. Paleontol. Mineral. Spec. Publ. 41, 241–260.

NUMMEDAL, D., SWIFT, D.J.P. & KOFRON, B.M. (1989) Sequence stratigraphic interpretation of Coniacian strata in the San Juan Basin, New Mexico. In: *Shelf Sedimentation, Shelf Sequences and Related Hydrocarbon Accumulation* (Eds Morton, R.A. & Nummedal, D.). Seventh Annual Research Conference Proceedings, Society of Economic Paleontologists and Mineralogists, Gulf Coast Section.

PARKER, G., LANFREDI, N.W. & SWIFT, D.J.P. (1982) Substrate response to flow in a southern hemisphere ridge field. *Sediment. Geol.* **33**, 195–216.

PENLAND, S., SUTER, J.R. & BOYD, R. (1985) Barrier island arcs along abandoned Mississippi River Deltas. *Mar. Geol.* **63**, 197–233.

PHILLIPS, S. (1987) *Shelf sedimentation and depositional sequence stratigraphy of the Upper Cretaceous Woodbine–Eagle Ford Groups, East Texas.* PhD dissertation, Cornell University, Ithaca, NY, 232 pp.

PHILLIPS, S. & SWIFT, D.J.P. (1985) Shelf sandstones in the Woodbine–Eagle Ford Interval: a review of depositional models. In: *Shelf Sands and Sandstone Reservoirs* (Eds Tillman, R.W., Swift, D.J.P. & Walker, R.G.). Soc. Econ. Paleontol. Mineral. Short Course Notes 13, 503–588.

PITMAN, W.C. & XENIA GOLOVCHENKO (1983) The effect of sea level change on the shelfedge and slope of passive margins. In: *The Shelfbreak: Critical Interface on Continental Margins* (Eds Stanley, D.J. & Moore, G.T.). Soc. Econ. Mineral. Paleontol. Spec. Publ. 33, 41–58.

PLINT, A.G. (1990) Allostratigraphic correlation of the Muskiki and Marshybank formations (Coniacian-Santonian) in the Foothills and subsurface of the Alberta Basin. *Bull. Can. Petrol Geol.* 38, 288–306.

POSAMENTIER, H.W. & VAIL, P.R. (1988) Eustatic controls on clastic deposition II—sequence and systems tract models. In: *Sea-level Changes: an Integrated Approach* (Eds Wilgus, C.K., Hastings, B.S., Kendall, C.G.St.C., Posamentier, H.W., Ross, C.A. & Van Wagoner, J.C.). Soc. Econ. Paleontol. Mineral. Spec. Publ. 42, 124–154.

POSAMENTIER, H.W., JERVEY, M.T. & VAIL, P.R. (1988) Eustatic controls on clastic deposition, I—conceptual framework. In: *Sea-level Changes: an Integrated Approach* (Eds Wilgus, C.K., Hastings, B.S., Kendall, C.G.St.C., Posamentier, H.W., Ross, C.A. & Van Wagoner, J.C.). Soc. Econ. Paleontol. Mineral. Spec. Publ. 42, 109–124.

ROBERTS, H.H. & COLEMAN, J.M. (1988) Sedimentation styles and accumulation rates on the Louisiana shelf and slope: stacked condensed and expanded sections. *Trans. Gulf Coast Assoc. Geol. Soc.* 38, 584–591.

SALVADOR, A. (1987) Unconformity-bounded stratigraphic units. *Geol. Soc. Am. Bull.* 98, 232–237.

SCRUTON, P.C. (1960) Delta building and the deltaic sequence. In: *Recent Sediments, Northwest Gulf Of Mexico* (Eds Shepard, F.P., Phleger, F.B. & van Andel, Tj.H.) pp. 82–102. Am. Assoc. Petrol. Geol., Tulsa, OK.

SUTER, J.R. & BERRYHILL, H.L. JR. (1985) Late Quaternary shelf margin deltas, Northwest Gulf of Mexico. *Bull., Am. Assoc. Petrol. Geol.* 69, 77–91.

SUTER, J.R., BERRYHILL, H.L., JR. & PENLAND, S. (1987) Late Quaternary sea level fluctuations and depositional sequences, southwest Louisiana Continental Shelf. In: *Sea-level Fluctuation and Coastal Evolution* (Ed. Nummedal, D.). Soc. Econ Paleontol. Mineral. Spec. Publ. 41, 199–219.

SWIFT, D.J.P. (1973) Delaware Shelf Valley: estuary retreat path, not drowned river valley. *Geol. Soc. Am. Bull.* 84, 2743–2748.

SWIFT, D.J.P. & FIELD, M.F. (1981) Evolution of a classic ridge field, Maryland Sector, North American Inner Shelf. *Sedimentology* 28, 461–482.

SWIFT, D.J.P., KOFOED, J.W., SAULSBURY, F.P. & SEARS, P.C. (1972) Holocene evolution of the shelf surface, central and southern Atlantic shelf of North America. In: *Shelf Sediment Transport: Process and Pattern* (Eds Swift, D.J.P., Duane, D.B. & Pilkey, O.H.) pp. 499–574. Dowden Hutchinson & Ross, Stroudsburg, PA.

SWIFT, D.J.P., NELSEN, T.A., MCHONE, J., HOLLIDAY, B. PALMER, H. & SHIDELER, G. (1977) Holocene evolution of the inner shelf of southern Virginia. *J. Sediment. Petrol.* 47, 1454–1474.

SWIFT, D.J.P., SEARS, P.C., BOHLKE, B. & HUNT, R. (1978) Evolution of a shoal retreat massif, North Carolina Shelf: evidence from areal geology. *Mar. Geol.* 27, 19–423.

TESSON, M., GENSOUS, B., ALLEN, G.P. & RAVENNE, C. (1990) Late Quaternary deltaic lowstand wedges on the Rhône continental shelf, France. *Mar. Geol.* 91, 325–332.

THORNE, J.A. & WATTS, A.B. (1984) Seismic reflectors and unconformities at passive continental margins. *Nature* 311, 365–368.

TILLMAN, R.W. (1985) The Taocito and Gallup Sandstones, New Mexico. A comparison. In: *Shelf Sands and Sandstone Reservoirs* (Eds Tillman, R.W., Swift, D.J.P. & Walker, R.G.) Short Course 13, pp. 403–464. Soc. Econ. Paleontol. Mineral., Tulsa, OK.

TURNER, J.R. & CONGER, S.J. (1984) Environment of deposition and reservoir properties of the Woodbine Sandstone at Kurten Field, Brazos County, Texas. In: *Siliciclastic Shelf Sediments* (Eds Tillman, R.W. & Siemers, C.T.). Soc. Econ. Paleontol. Mineral. Spec. Publ. 34, 215–250.

VAIL, P.R., MITCHUM, R.M., JR. & THOMPSON, S., III (1977a) Seismic stratigraphy and global changes of sea level, part 4: global cycles of relative cycles of sea level. In: *Seismic Stratigraphy—Applications to Hydrocarbon Exploration* (Ed. Payton, C.E.). Mem. Am. Assoc. Petrol. Geol. 26, 83–97.

VAIL, P.R., MITCHUM, R.M., JR., TODD, R.G., WIDMIER, J.M., THOMPSON, S., III, SANGREE, J.R., BUBB, J.N. & HATLID, W.G. (1977b) Sequence stratigraphy and global changes of sea level. In: *Seismic Stratigraphy—Applications to Hydrocarbon Exploration* (Ed. Payton, C.E.) Mem., Am. Assoc. Petrol. Geol. 26, 49–205.

VAN STRAATEN, L.M.J.U. (1965) Coastal barrier deposits in south and north Holland, in particular in the areas around Scheveningen and Ijmuiden. *Meded. Geol. Stichting Niewe Ser.* 17, 41–75.

VAN WAGONER, J.C. (1985) Reservoir facies distribution as controlled by sea-level change. *Soc. Econ. Paleontol. Mineral. Ann. Midyear Meet. Abstr.* Golden, Colorado, pp. 91–92.

VAN WAGONER, J.C., MITCHUM, R.M., JR., POSAMENTIER, H.W. & VAIL, P.R. (1987) Part 2, key definitions of sequence stratigraphy. In: *Atlas of Seismic Stratigraphy* (Ed. Bally, A.W.). Am. Assoc. Petrol. Geol. Stud. Geol. 27, 11–12.

VAN WAGONER, J.C., MITCHUM, R.M., JR., POSAMENTIER, H.W., VAIL, P.R., SARG, J.F., LOUTIT, T.S. & HARDENBOL, J. (1988) An overview of the fundamentals of sequence stratigraphy and key definitions of sequence stratigraphy. In: *Sea-level Changes: an Integrated Approach* (Eds Wilgus, C.K., Hastings, B.S., Kendall, C.G.St.C., Posamentier, H.W., Ross, C.A. & van Wagoner, J.C.). Soc. Econ. Paleontol. Mineral. Spec. Publ. 42, 39–47.

VAN WAGONER, J.C., MITCHUM, R.M., JR., CAMPION, K.M. & RAHMANIAN, V.D. (1990) Siliciclastic sequence stratigraphy in well logs, cores and outcrops: techniques for high resolution correlation of time and facies. *Am.*

*Assoc Petrol. Geol. Meth. Expl. Ser.* **7**, 55 pp.

VEATCH, A.G. & SMITH, P.A. (1939) Atlantic submarine valleys of the United States and the Congo Submarine Valley, *Geol. Soc. Am. Spec. Pap.*

WELLS, J.T. (1987) Effects of sea level rise on deltaic sedimentation in South Central Louisiana. In: *Sea-level Fluctuations and Coastal Evolution* (Eds Nummedal, D., Pilkey, O.H. & Howard, J.D.). Soc. Econ. Paleontol. Mineral. Spec. Paper 41, 157–166.

WRIGHT, R. (1986) Cycle stratigraphy as a paleogeographic tool: Point Lookout Sandstone, southeastern San Juan Basin, New Mexico. *Geol. Soc. Am. Bull.* **96**, 661–673.

# Sedimentation on continental margins, VI: a regime model for depositional sequences, their component systems tracts, and bounding surfaces

J.A. THORNE* and D.J.P. SWIFT[†]

*Research and Technical Services, ARCO Oil and Gas Company, 2300 Plano Parkway, Plano, TX 75075, USA; and [†] Department of Oceanography, Old Dominion University, Norfolk, VA 23529-0276, USA

## ABSTRACT

This study evaluates the sequence architecture models of Vail (1987) and Galloway (1989) in terms of the regime concepts presented in the first five papers of this volume. Comparison of these models with modern and submodern shelf deposits, in which sediment dispersal systems are presently functioning, leads to the development of a new, regime-based model for their system tracts and bounding surfaces. To compare effectively the Vail, Galloway and regime-based models, four *geometric systems tracts* are defined that make no explicit reference to eustatic sea-level. Rough Vail or Galloway equivalents can be found for the (1) *basin fan*, (2) *back-step wedge*, and (3) *offlap wedge* geometric systems tracts. The fourth *alluvial fan* geometric systems tract has no equivalent. Geometric systems tracts are defined in terms of stratal geometry, primarily as observed through seismic records or well-log correlations. They are generally larger in scale and more inclusive in character than the facies-defined depositional systems tracts discussed by Swift *et al.* (this volume, pp. 89–152.)

Our survey of modern sedimentation, based mostly on high-resolution seismic profiles, largely confirms the descriptive aspects of the sequence architecture model of Vail. The model of Vail, however, may need to be modified to include several additional architectural elements that can be present under certain regime conditions. One modification concerns the transgressive surface. The transgressive surface is a surface whose time range encompasses the turnaround event, when paracycles of progradational sedimentation give way to paracycles of back-stepping sedimentation. A single transgressive surface may include portions of: (1) a ravinement surface formed by erosional shoreface retreat landward of the maximum basinward shoreline position; (2) a marine erosion surface that may form seaward of the maximum basinward shoreline position; and (3) a conformity between beds of the back-step wedge and beds of the preceding offlap wedge systems tract. Other elements recognized in this study are; (4) a time-transgressive coastal plain unconformity formed by fluvial entrenchment during offlap wedge progradation, which is not related in origin to the slope unconformity formed during type 1 sequence boundaries; (5) a wedge of back-barrier sediments between the transgressive surface and the ravinement surface; (6) a sedimentary wedge that onlaps the relict slope and downlaps the deep basin formed during transgression; and (6) the formation of extensive coastal-plain–slope unconformities that mimic sequence boundaries at times other than during falling sea-level. A few of these modifications have been suggested by Galloway.

The Galloway model, although giving a less detailed picture of architectural sequence elements, appears essentially realistic in its recognition that other regime variables besides relative sea-level change can largely control sequence architecture. The regime concepts of Galloway, however, are modified to include more explicitly the effects of regime variables, $R$, related to the rate of relative sea-level change, $D$, the rate of sediment transport, $Q$, the rate of sediment supply and, $M$, the sediment grain size.

Specifically we find that the product $Q \cdot M$ is a regime expression for effective sediment supply while the product $R \cdot D$ is a regime expression for effective accommodation potential. It is suggested that the ratio of accommodation to supply, defined as $¥ = R' \cdot D'/Q' \cdot M'$, is fundamental to the architectural response of sedimentation to changing regime conditions. A semi-quantitative model for each systems tract, the subsystems within it, and its bounding surfaces is proposed based on the theoretical response of *regime topographic profiles* to changes in $¥$, the accommodation/supply ratio. Hypothetical examples of sequence development in response to changes in: (1) sediment supply without changes in sea-level; and (2) sea-level without changes in sediment supply serve to illustrate the differences between the new regime model and the eustatic model of Vail.

The regime-based, ¥-dependent model is used to interpret observed sequence stratigraphic architecture of the Campanian–Maastrichtian deposits of the USA Western Interior. This analysis suggests that second-order (c.10 Ma) sequence cycles may be controlled by episodic subsidence caused by *in-sequence* thrusts, while third or higher order (<1 Ma) sequence cycles may be controlled by sediment supply changes caused by *out-of-sequence* thrusts.

A comparison of regime conditions for various time periods on the Northwestern Gulf of Mexico shelf and Middle Atlantic shelf sections indicates that the preservation potential of sequence architecture is dependent on the relative roles of sediment input ($Q$) and sea-level change ($R$). As sediment input and subsidence increase relative to the frequency and intensity of sea-level falls, the preservation potential of highstand deposits increases.

This study is the last of six papers in this volume that apply the concept of regime sedimentation to continental-margin deposition. These six papers have shown, at various spatial and temporal scales, that sedimentation systems are governed by interdependent variables, whose mutual adjustments tend towards a state of dynamic equilibrium. As a first approximation, semi-quantitative regime models can be developed in terms of equilibrium surfaces controlled by homeostatic responses of erosion or deposition.

# INTRODUCTION

## Goal

The concepts of sequence stratigraphy are generally acknowledged to constitute a potentially unifying idea, the testing of which is pressing business for the present generation of stratigraphers. Sequence concepts are well presented in the literature (e.g Vail *et al.*, 1977, 1984; Vail & Todd, 1981; Haq *et al.*, 1987; Vail, 1987; Jervey, 1988; Posamentier *et al.*, 1988; Galloway, 1989a; Lawrence *et al.*, 1990) and have been the subject of critical reviews (e.g. most recently, Miall, 1986; Cross & Lessenger, 1988; Hubbard, 1988; Kendall & Lerche, 1988; Fulthorpe & Carter, 1989; Galloway, 1989a; Christie-Blick *et al.*, 1991; Pitman, 1991; Thorne, 1991). The focus of this study is to explore the relationship between sequence stratigraphic concepts and the dynamic concepts presented in earlier papers in this volume.

## Approach to the problem

The first five papers in this volume have presented a regime model for shelf sedimentation, in which the shelf surface is seen as being in a state of long-term equilibrium with the regime variables of $Q$ and $M$, sediment input rate and grain size, $R$, the rate and sense of relative sea-level change, and $D$, the sediment transport rate. Swift & Thorne (this volume, pp. 3–31) present an overview of the model, and define two basic variants of the depositional regime. In an allochthonous mode, far-travelled sediment is provided from deltaic river mouths (river mouth bypassing). Allochthonous sedimentation is characterized by rapid, diffusive, sediment dispersal and accumulation. Thorne & Swift (this volume, pp. 33–58) undertake a systematic exploration of the regime concept, and link autochthonous regimes to $R$-dependent coastal equilibria and allochthonous regimes to $Q$-dependent coastal equilibria. Thorne *et al.* (this volume, pp. 59–87) consider processes of sedimentation at the subequilibrium (event stratigraphic) scale. They examine strata-forming processes and describe parameters that control stratification patterns. Swift *et al.* (this volume, pp. 89–152) apply stratification concepts to the problem of facies differentiation in the depositional systems of both allochthonous and autochthonous regimes. Swift *et al.* (this volume, pp. 153–187) consider the relationship between the regime model for sedimentation on shelf surfaces described in the first four papers of this volume, and parasequence models. This paper adopts a similar approach but, rather than looking at parasequences, considers the depositional architecture of *seismic-scale* sequence models proposed by stratigraphers for the sediment prisms of continental margins.

We begin with a brief definition of sequence stratigraphic and regime sedimentation terminology, which puts the Vail (1987) model, the Galloway

(1989a) model and the regime model of this study on a common ground for comparison. To explore the relationship between depositional sequences and the regime concepts presented in earlier papers of this volume, a comparison is made of sequence architecture, as set forth in the Vail (1987) and Galloway (1989a) models, against observations from modern (and submodern) shelves. This comparison is made first for descriptive observations of system tract geometry. In this section we attempt to avoid genetic interpretations of the role of eustacy and other regime variables in controlling sequence architecture. The role of these variables is treated in the next section in which the genetic depositional models of Vail (1987) and Galloway (1989a) are compared with observations of dynamic processes on modern shelves. The results of the comparison are encoded in tabular form (see Table 2). The comparison suggests a dominant control on sequence architecture by regional and temporal variations in regime variables. These variables can be explained in terms of a regime-based model for the general sequence architectural response to each of the geohistorical variables $Q$, $M$, $R$ or $D$, independently or in combination.

## SEQUENCE MODELS: A COMMON GROUND TERMINOLOGY

### Sequence stratigraphic terminology

*Depositional versus geometric systems tracts*

The Vail (1987) and Galloway (1989a) models use different sequence stratigraphic terminology. In order to avoid confusion, Swift *et al.* (this volume, pp. 153–187) briefly described (in their Tables 1 and 2) our understanding of several sequence stratigraphic terms used by Vail and Galloway. We extend this discussion in this section to terms involving depositional systems tracts.

Depositional systems tracts were discussed by Swift *et al.* (this volume, pp. 89–152 and 153–187). The Vail model has borrowed the terms 'depositional system' and 'depositional systems tract' from papers published by Texas Bureau of Economic Geology. A depositional system is 'an assemblage of process-related facies' (Fisher & McGowen, 1967), and a systems tract is 'a linkage of contemporaneous depositional systems' (van Wagoner *et al.*, 1987, 1988; after Brown & Fisher, 1977).

The Texas workers developed these terms in the course of extensive studies of subsurface stratigraphy, using wire-line log and core data. They used the terms to describe two successive scales of lateral lithologic differentiation beyond the basic (lithofacies) scale. In other words, a distal plume lithofacies may be part of a prodelta plume depositional system, which may in turn be part of a regressive shelf depositional systems tract (Table 1, Swift *et al.*, this volume, pp. 89–152). As originally defined, depositional systems and systems tracts are implicitly of uniform, or uniformly changing vertical composition; that is, they are deposited under conditions of essentially constant values of regime variables. When regime values change, the systems tract becomes overlaid by a second systems tract, or by the equivalent erosional surface. In shallow-marine settings, systems tracts commonly alternate in response to cyclic variations in regime variables, from accommodation-dominated to supply-dominated, at frequencies of tens or hundreds of thousand of years. The cyclic variation may create trangressive–regressive pairs of systems tracts, or more commonly repetitions of regressive systems tracts separated by flooding surfaces. These cycles and repetitions have been called *parasequences* (Vail *et al.*, 1977).

Meanwhile, seismic stratigraphers were examining the deposits of third-order sea-level cycles (third-order depositional sequences) on seismic records, and were observing a well-defined internal structure. At this larger spatial scale, facies relationships are less apparent, and geometric relationships come to the fore. Third-order depositional sequences are seen to consist of component parts, which are referred to as systems tracts in the Vail (1987) terminology; *lowstand*, *highstand*, *transgressive*, and *shelf-margin* tracts have been defined (Fig. 1). At these scales however, the 'systems tracts' consist of stacked parasequences, and contradictions occur. 'Transgressive' systems tracts may in fact consist of 'back-stepping parasequences', which are composed in turn of mainly *regressive* lithofacies. The problem is avoided by Galloway (1989a) who refers less formally to *offlap components* and *onlap components*, that roughly parallel the highstand and transgressive systems tracts of the Exxon workers (Fig. 2 versus Fig. 1).

Much of the discussion in this study deals with the differing approaches used by Vail (1987) and Galloway (1989a) in distinguishing stratigraphic sequences on the basis of cyclic repetition of

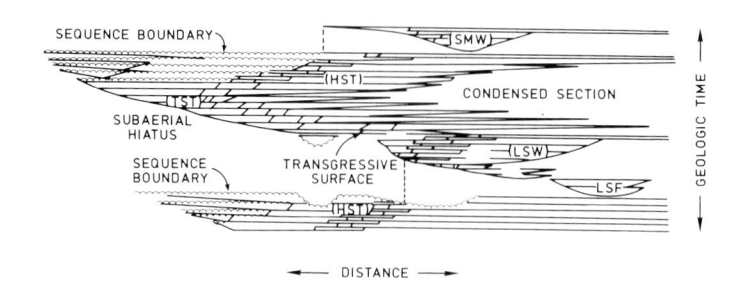

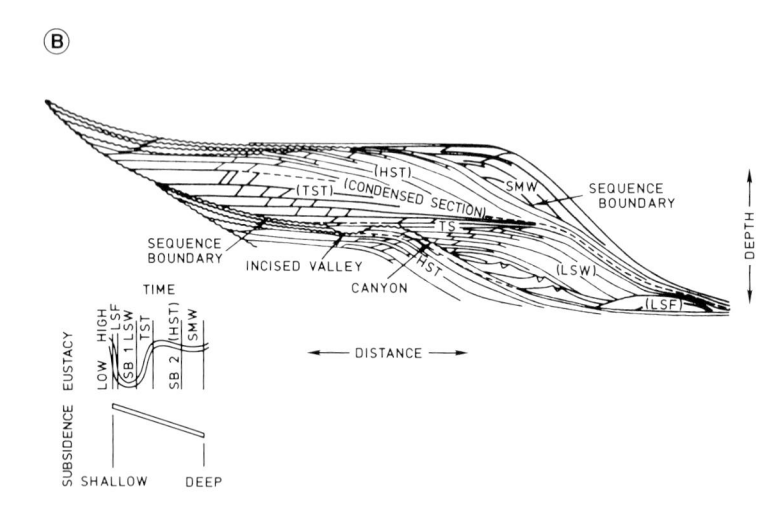

**Fig. 1.** The Vail (1987) model for a depositional sequence. (A) Chronostratigraphic diagram. (B) Lithostratigraphic diagram. HST = highstand systems tract; TST = transgressive systems tract. Lowstand systems tract elements include: SMW = shelf margin wedge; LSW = lowstand wedge; lsf = lowstand fan. (After Haq et al., 1987.)

sequence components. We wish to examine the Vail as well as the Galloway approaches in detail. We note that Vailian 'depositional systems tracts' are defined *not* by their facies or by their component depositional systems, but instead by the geometry of the constituent parasequence set (progradational and offlapping or retrogradational and backstepping) and the relationship of internal (parasequence) boundaries to systems tracts boundaries (onlapping; downlapping; Vail et al., 1977; Vail, 1987). We therefore use the provisional term *geometric systems tracts* to distinguish these sequence building blocks from the facies-defined depositional systems tracts of Swift et al. (this volume, pp. 89–152). The two terminologies mesh relatively smoothly. In some cases we will demonstrate a one-for-one congruence of depositional systems tracts and geometric systems tracts. More commonly a geometric systems tract consists of several facies-defined depositional systems tracts, multiplied by parasequence repetition.

*Sequence versus cycle boundaries*

The bounding surfaces of the depositional sequences themselves generally may be called sequence boundaries. Van Wagoner et al. (1987, 1988), however, have defined *sequence boundaries* as surfaces associated with subaerial erosion, and formed by falling sea-level. In this paper, therefore, we will use the more general term *cycle boundary* to refer to the bounding surfaces of either Vail (1987) or Galloway (1989a) and reserve the term sequence boundary for the Vail (1987) model.

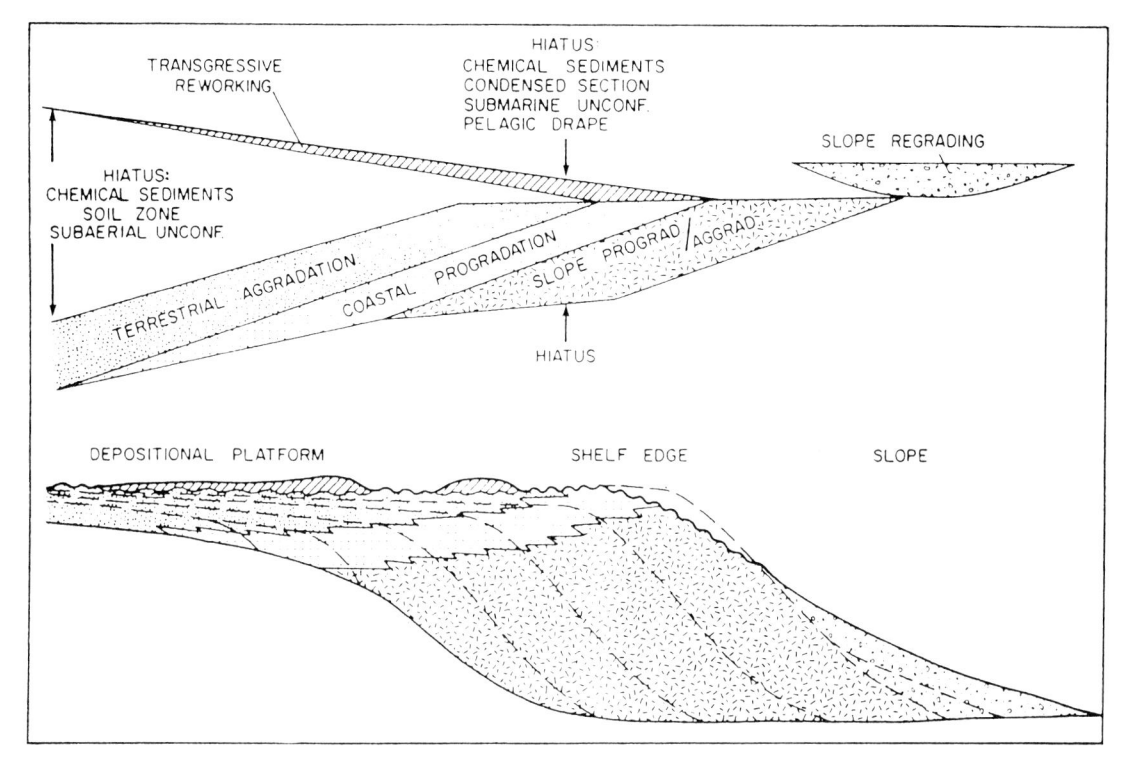

Fig. 2. Galloway model for genetic stratigraphic sequences. (From Galloway, 1989a.)

In the Vail (1987) model, the sequence boundary (type 1 or type 2) is an unconformity of interregional extent cut by a sea-level fall with a significant eustatic component, and it defines the cycle boundary. This concept is used to justify global correlation on the basis of type 1 sequence-bounding unconformities cut by high rates of eustatic sea-level fall, and considerable ingenuity has been expended in demonstrating synchronous mechanisms for the timing of the subaerial erosion and concurrent erosion on the continental slope by submarine canyons. The end of fluvial sedimentation and the beginning of subaerial and submarine erosion is seen as a critical event, since it creates the 'downward shift in coastal onlap' that is central to the Vail model (Vail et al., 1977, 1984; Posamentier & Vail, 1988, p. 129).

In the Galloway (1989a) model, subaerial erosion and continental slope erosion formed by submarine canyons are not seen as necessarily synchronous events. In fact, Galloway (1989a) describes them not as 'events' at all, occurring over a short period of time, but rather as ongoing processes that may occur throughout a depositional cycle. In the Galloway (1989a) model the sequence boundary of Vail (1987) is a coastal-plain–slope unconformity without any necessary eustatic, global, or precise time-stratigraphic significance.

The maximum flooding surface is the important bounding surface in the Galloway (1989a) model. Galloway (1989a), following Frazier (1974), argues that this surface is characteristically: (1) easily recognizable in seismic and well-logs; (2) fossil-rich; (3) of great regional extent; (4) more ubiquitous than subaerial erosion surfaces; (5) more apt to form in rapidly subsiding basins than subaerial erosion surfaces; and (6) define sequences of greater palaeogeographic stability then sequences bounded by subaerial unconformities.

A critical question raised by Galloway (1989a) is whether the Vailian sequence boundary can be identified dependably. If this boundary cannot be identified dependably, then separation of the highstand and lowstand systems tracts becomes very difficult when both of these systems tracts are dominated by prograding clinoforms in dip view.

## Geometric systems tract terminology

It is often difficult to analyse or critically assess the Vail (1987) model because many sequence stratigraphic terms are, by definition, based on the concept of a eustatic cycle, rather than on direct observation. For example, several terms used in this model, such as *lowstand systems tract*, include the term lowstand and thus, by definition, appear to imply that these terms refer to deposition that must occur during lowstands of eustatic sea-level. We are faced with a problem. What if we observe in the ancient a deposit that in all regards (architectural anatomy, depositional systems, depositional facies) is analogous to the description of a lowstand systems tract? Can we use this term to describe this deposit if we do not know if eustatic sea-level was at a lowstand or, in fact, have evidence to the contrary?

The terms used by Galloway (1989a) have no such built-in genetic implications. Unfortunately, this model does not provide the descriptive detail of the Vail (1987) model. There are, thus, several useful terms in the Vail (1987) model that have no equivalent in the Galloway (1989a) model.

In order to avoid confusion, in this study, the terms *lowstand deposits*, *transgressive deposits*, and *highstand deposits* refer to deposition that has been stratigraphically correlated to low, rapidly rising, or high periods of glacial eustatic sea-level. These terms are distinguished from their use in the Vail (1987) model. For example lowstand deposits of the Vail (1987) model are deposits that are interpreted to have the depositional architecture of the lowstand systems tract. It is possible, therefore, for modern lowstand deposits not to be part of the lowstand systems tract of Vail (1987) if their architecture does not resemble the Vailian model.

Most importantly, we introduce the following *geometric systems tracts: basin fan (fill)*, *back-step wedge (fill)*, *offlap wedge (fill)*, and *alluvial fan (fill)*. (To make the text more readable, the full geometric systems tract names are shortened in many cases.) Each of these systems tracts is defined by its dominant depositional architecture *without* reference to sea-level position, though a rough equivalent can be made to the Vail (1987) and Galloway (1989a) systems tracts (Table 1).

The modifiers *fan* and *wedge* are used to indicate the dominant three-dimensional geometry of depositional units within a systems tract. A depositional lens is fan-shaped if strike sections taken through the proximal end of the deposit show markedly less lateral extent of deposition than sections taken at other locations. A depositional lens is wedge-shaped if strike sections taken at various places through the deposit show a similar lateral extent of deposition. Wedges are typically elongate in the strike direction, while fans are typically elongate in the dip direction. The term fan is used, here, to describe the three-dimensional geometric shape of deposited sediments. The use of the term fan has a different connotation when referring to a depositional fan dispersal *system*. In this case it is implied that sediments are eroded, transported and deposited along a transport system that is fan-shaped in map view. A fan system, however, does not necessarily *deposit* a fan-shaped body of sediment.

The modifiers *back-step* and *offlap* are used to indicate the dominant stacking pattern of depositional units within a geometric systems tract. If the depocentres of successive depositional units generally move in a proximal direction, a back-stepping pattern is formed. If depocentres move in a distal direction an offlapping pattern is formed. The terms *back-step* and *offlap* are distinct from *onlap* and *downlap*. The aggradational fill of a basin, for example, can onlap up against the sides of the basin without forming a back-stepping stacking pattern.

The modifiers *alluvial* and *basin* are used to distinguish between fan-shaped deposits adjacent to the source terrain and fan-shaped geometries of the

Table 1. Approximate systems tract equivalents between Vail(1987), Galloway (1989a) and this paper

| Vail systems tracts | Galloway components | Geometric systems tracts |
|---|---|---|
| Lowstand | Offlap with shelf-edge delta | Basin fan/fill |
| Transgressive | Onlap | Back-step wedge/fill |
| Highstand | Offlap with platform delta | Offlap wedge/fill |
| Shelf margin | Mixed | Usually back-step wedge/fill |
| — | — | Alluvial fan/fill |

more distal subsiding basin. The alluvial fan (fill) geometric systems tract does not seem to have a direct equivalent in the terminology of Vail (1987) or Galloway (1989a). This study emphasizes shelf sedimentation, hence no further discussion is made of the alluvial fan (fill) geometric systems tract.

**Topographic fill sedimentation terminology**

The modifier *fill* is appended to the names of all geometric systems tracts. This modifier is used to distinguish between depositional geometries that are a direct reflection of current regime conditions (wedge or fan) and depositional fill of previous palaeotopography. In this latter case, much of the geometric depositional form is inherited from *previous* regime conditions. Thus, for example, we refer generically to a back-step wedge (fill), meaning that this geometric systems tract may be expressed *either* as a wedge or as a fill. There are three types of topographic fill deposits, which we discuss in turn.

*Gradient reduction deposits*

For many types of sediment transport processes, transport capacity increases with topographic slope. Sediment transport by rivers and by sediment gravity flows are clear examples of this relationship. Both rivers and sediment gravity flows can move greater amounts of sediment on higher topographic gradients. Slope control of transport capacity also applies to transport by sediment slumping, downslope creep, and in any area of relatively uniform fluid power where the net transport can be approximated as a diffusive process (Scheidegger, 1961).

It follows, that in any area of decreasing topographic gradient, there is a decrease of sediment transport capacity in the absence of other controlling factors. *Gradient reduction deposits* are formed by this decrease in transport capacity. These deposits are, most often, equivalent to the proximal and near-distal facies of the amalgamated sand lithofacies and interstratified sand and mud lithofacies that have been described by Swift *et al.* (this volume, pp. 89–152).

*Ponding deposits*

As a general rule, sediments tend to collect in topographic lows. Topographic lows, characteristically, are deeper and have lower topographic slopes than surrounding areas. These characteristics generally lead to a relatively low sediment transport capacity within the low compared with surrounding areas. *Ponding deposits* are formed by this decrease in sediment transport capacity. Geometrically, deposits of this type generally display an onlapping character. Ponded deposits are equivalent to the distal lithofacies (usually laminated or bioturbated mud facies) of depositional systems that have been described by Swift *et al.* (this volume, pp. 89–152).

*Draping deposits*

The last type of topographic fill deposit has a draping form. In this case the depositional geometry of the deposit is entirely controlled by pre-existing topography. Draping deposits are also a distal lithofacies.

# REGIME SEDIMENTATION TERMINOLOGY

**Accommodation to supply ratio**

Swift & Thorne (this volume, pp. 3–31) and Thorne & Swift (this volume, pp. 33–58) argued that at large time and space-scales, shelf sedimentation achieves a predictable steady-state regime parameterized by four *geohistorical regime variables*; $Q$, $M$, $R$ and $D$. These variables were used to show that when $Q \cdot M > R \cdot D$, the shelf is characterized by allochthonous sedimentation, and when $Q \cdot M < R \cdot D$, the shelf is characterized by autochthonous sedimentation. Swift *et al.* (this volume, pp. 89–152) applied these inequalities to an analysis of depositional systems and facies. In this part of the study we will attempt to extend this type of analysis to the larger spatial and temporal scales of depositional systems tracts that control the architecture of depositional sequences (see Table 2 and related text of Swift *et al.* (this volume, pp. 153–187) for a discussion of scale).

The product of regime variables $Q$ and $M$ is a measure of the *effective* sediment supply rate. Variable $Q$ is the clastic sediment supply rate to the marginal marine or marine environment of one section of a continental margin. Not all the sediment supplied to this section is deposited within this area; some sediment is transported to other sections of the margin. Fine sediments, particularly, are more easily transported out of the area than coarse sediments and, thus, the *effective* sediment supply of fines is relatively reduced compared with the coarse component. The product of regime variables $Q$ and $M$ accounts for this relative effect.

The product of regime variables $R$ and $D$ is a measure of the *effective* sediment accommodation potential. There is the least potential for sediment accommodation when relative sea-level, $R$, falls quickly and, conversely, there is the greatest potential when $R$ increases quickly. *Effective* sediment accommodation requires the additional factor of sediment transport rate, $D$. For example, a deep ocean basin provides a large amount of potential accommodation. If, however, there is no means of transporting sediment into the basin then this potential is not realized. The product $R \cdot D$, therefore, measures the *effective* sediment accommodation potential.

Many authors have pointed out that the ratio of sediment supply to sediment accommodation controls the architectural stacking patterns of retrogradation, aggradation and progradation (e.g. Curray, 1964; Galloway & Hobday, 1983; Cross & Lessenger, 1988; Jervey, 1988; Posamentier *et al.*, 1988). The three cases, therefore, $Q \cdot M < R \cdot D$, $Q \cdot M = R \cdot D$ and $Q \cdot M > R \cdot D$, correspond, respectively, to retrogradation, aggradation and progradation.

The importance of this relationship between sediment supply factors, $Q \cdot M$, and accommodation factors, $R \cdot D$, indicates that the ratio of accommodation to supply is a critical parameter in determining sequence architecture. In this paper we define the *regime parameter*, ¥, by the following expression:

$$¥ \equiv C_1 \frac{D^{C_2} \cdot C_3^{C_4 R}}{Q^{C_5} \cdot M^{C_6}} \quad (1)$$

where $C_1$ is a positive scaling coefficient, $C_2$, $C_5$ and $C_6$ are positive exponents that allow for a power law relationship between $D$, $Q$ and $M$ with ¥, and $C_3$ and $C_4$ are positive coefficients for a general function of $R$ that ranges from 0 to $\infty$. The coefficients $C_1$ through to $C_6$ are not known (though similar coefficients have been determined for regime relationships in rivers, Leopold *et al.*, 1964). A simpler notation for the regime parameter is therefore:

$$¥ \equiv \frac{D'R'}{Q'M'} \quad (2)$$

where a prime indicates that the geohistorical variables are modified. Note in particular that the range of $R'$ is from 0 to $\infty$ rather than from $-\infty$ to $\infty$. The regime variable ¥ is used in the section on sequence models to explain variations in sequence architecture (see Table 1). In much of this paper, as in previous papers in this volume, the primes have been dropped for simplicity. The symbol ¥ is normally used for the Japanese unit of currency, the yen. It may be referred to as y-bar. A related symbol, ¥̇, is the derivative with respect to time. It describes the rate and sense of change of ¥.

## Index of highstand preservation potential

We propose an *index of stratigraphic preservation potential*, $S_p$, as a measure of the probability of stratigraphic completeness. The preservation potential index indicates the likelihood of long-term preservation of complete depositional sequences including their highstand components within multiple stacked depositional sequences.

The importance of preservation potential has long been recognized. Barrell (1917) modelled the deposition of a stratigraphic succession as a consequence of base level change. He showed that stratigraphic preservation requires a long-term rise in base level that is large with respect to any oscillating components of rise and fall. Mathematically this can be written:

$$S_P = \frac{s}{hf} \quad (3)$$

where the change in base level, $L$, with time is given by:

$$L(t) = st + h \cdot \sin(ft) \quad (4)$$

where $s$ is the steady rate of rise of $L(t)$, and $h$ and $f$ are the amplitude and frequency of the sinusoidal component of $L(t)$.

If one applies equations (3) and (4) to periods of geological time lasting for many sinusoidal cycles, then the steady rate of rise of base level, $s$, is given to a first approximation by the long-term average rate of accumulation, $å$, giving:

$$S_P = \frac{å}{hf} \quad (5)$$

Equation (5) is a general expression for preservation potential. For highstand deposits, two additional factors enter. The first of these factors relates to the basin's flexural strength. When a basin is flexurally strong, deposition at any one point in the basin induces flexural subsidence in a broad area around that point and flexural uplift at large distances (Turcotte & Schubert, 1982). The area of subsidence is called the *flexural moat* and the area of uplift the *peripheral bulge*. When a basin is flexurally weak, on the other hand, flexural moats are narrow and peripheral bulges are relatively close. During the deposition of the lowstand systems tract, flexural subsidence can cause the preservation of previously deposited highstand deposits. If the basin is flexurally strong, highstand deposits are likely to fall in the flexural moat of the lowstand deposits. In this case flexural subsidence will act to preserve the deposits. If the basin is flexurally weak, then highstand deposits are likely to fall within the peripheral bulge of the lowstand deposits. In this case flexural uplift lowers the chance of highstand preservation.

A second factor that increases highstand preservation potential is the flux into the basin of terrigenous source material containing a high fraction of clay. Increased preservation is caused by a combination of two factors: (1) increased subsidence owing to compaction of underlying clay-rich deposits; and (2) increased resistance to erosion owing to cohesion of subaerially exposed clay-rich deposits.

In summary, highstand stratigraphic preservation potential is favoured by high accumulation rates, low magnitude, low-frequency sea-level changes, high basin flexural strength and clay-rich sediment input. If geohistorical variable $F$ represents basin flexural strength as a function of time and we substitute sediment input rate $Q$ for long-term accumulation rate $å$ we can write:

$$S_p = \frac{QF}{hfM} \quad (6)$$

### $Q$-dominated and $R$-dominated regimes

We have described basin sedimentation as, in general, a function of four geohistorical variables, $Q$, $M$, $R$ and $D$. In addition we have described stratigraphic preservation as a function of three additional variables: $h$, $f$ and $F$. Although, in general, these variables vary independently, it is useful to identify two end-member cases, which will be referred to as $Q$-dominated and $R$-dominated modes of sedimentation. In a $Q$-dominated mode of sedimentation, variations of sediment supply are far more dynamic than variations of any other regime variable. In an $R$-dominated mode of sedimentation, variations of relative sea-level are more dynamic than variations in any other regime variable.

These two end-member scenarios can greatly simplify genetic interpretations of sequence geometry. In an $R$-dominated mode, variations in the accommodation to supply ratio ¥ are proportional to and synchronous with variations in $R$. Variations in highstand preservation potential are inversely proportional to and synchronous with variations in $h$ and $f$. In a $Q$-dominated mode, variations in the accommodation to supply ratio ¥ are inversely proportional to, and synchronous with, variations in $Q$. Variations in highstand preservation potential are proportional to, and synchronous with, variations in $Q$.

## SEQUENCE MODELS VERSUS OBSERVATIONS ON MODERN SHELVES: ARCHITECTURAL DESCRIPTION OF SYSTEMS TRACTS

The sequence models of Vail (1987) and Galloway (1989a) are based largely on a compilation of observations from multichannel seismic and well-log information gathered for exploratory purposes. Data gathered from recent shallowly buried sediments, such as high-resolution seismic and cores, can serve as a useful calibration for these models.

Within the last several years, sequence stratigraphic concepts have been used with increased frequency to interpret seismic surveys of modern sediments. These studies, in general, show the power of this approach to stratigraphy. The geometric system tracts and architectural elements of the Vail (1987) and Galloway (1989a) models have been identified in such varying localities as the Gulf Coast of Louisiana (Boyd et al., 1989) and the Beaufort Sea of Alaska (Hequette & Hill, 1989). Depositional units imaged by high-resolution seismic have, to a great extent, shown readily interpretable patterns of toplap, truncation, downlap and onlap on their bounding surfaces.

In the following sections we examine each geometric systems tract in turn, comparing observations from surveys of modern sediments to the sequence models of Vail (1987) and Galloway (1989a). Our emphasis, perhaps unduly, is on exceptions to the Vail (1987) and Galloway (1989a) models.

## Basin fan geometric systems tract: models and observations

### Lowstand: Vail (1987) model

The Vail (1987) model, unlike the Galloway (1989a) model, provides a geological model for several subsystems within the basin fan systems tract (his lowstand systems tract). The terms referring to these subsystems have changed with successive versions of the Vailian concepts. Using the terms from Vail (1987), these subsystems are: (1) basin-floor fan, a penecontemporaneous set of mounded fan-lobe deposits lying above the type 1 sequence boundary, which in preserved depositional form appears seismically isolated from updip feeder systems; (2) slope fan, a penecontemporaneous set of levee, channel and overbank deposits that onlaps the type 1 sequence boundary on the slope and downlaps in the basin plain; and (3) a prograding complex, a 'regressive stratigraphic unit characterized by a progradational parasequence stacking' (Posamentier & Vail, 1988), during which time shelf-edge deltas build out from submarine canyon heads, while fluvial deposits onlap back up the incised valleys.

A basin-floor fan (Vail, 1987) is ideally recognized by: (1) bidirectional downlap on bottomsets of the deep basin; and (2) stratigraphic position immediately above the type 1 sequence boundary. The slope fan is ideally recognized by fan-shaped slope deposits that onlap a slope unconformity and downlap in the deep basin. The lowstand prograding complex is ideally recognized by wedge or fan-shaped prograding deposits that onlap a slope unconformity, downlap the basin floor or slope fan, and display toplap on their upper surface.

### Lowstand: Galloway (1989a) model

Galloway (1989a) equivalents of the Vail (1987) lowstand systems tract, or the basin fan geometric systems tract (this paper), are shelf-edge deltas and the associated slope deposits occurring during the final stage of offlap. In the Galloway (1989a) model, initial *platform delta* progradation at first reclaims the flooded platform of the previous episode. Galloway (1989a) describes the development of a basin fan systems tract as being the result of continued progradation. As progradation extends to the shelf edge, the thickness of shelf edge, delta front and prodelta facies increases, gravity remobilization assumes a more important role, and the position of major river systems becomes relatively fixed owing to incision of the shelf.

## Basin fan geometric systems tract: comparison with modern lowstand deposits

### General

The deposits of glacial lowstands have been recognized in abundance on modern continental shelves. Stratigraphic methods, including radiocarbon dating, correlation with the $\delta^{18}O$ isotope record, palynology and faunal analysis have served to identify deposits of glacial lowstands and interglacial times. This evidence, combined with high-resolution seismic and shallow core surveys, can be used to address whether recent lowstand deposits: (1) occur in a mappable basin-fan systems tract geometry; (2) are well described by the subsystems of basin-floor fan, slope fan, and prograding complex; and (3) display a chronostratigraphic evolution of succeeding subsystems as described by the Vail (1987) model.

### Seismic recognition of a basin fan (fill) geometry in recent lowstand deposits

The lowstand systems tract of Vail (1987) is identified by the presence of a sudden basinward shift in coastal onlap. Two scenarios create this identifying characteristic: (1) slope bypass and basin deposition in a fan-like form occurring immediately following the formation of a regional or globally correlatable coastal-plain–slope unconformity (type 1 sequence boundary of Vail, 1987); or (2) a prograding depositional system that changes depositional form from a wedge-shaped geometry to a fan-shaped geometry without undergoing a period of basin-wide erosion on the continental slope. The first scenario has been described extensively by Vail (1987). The second scenario, using the methodology of Vail (1987), is not generally called upon to explain downward shifts in coastal onlap.

To understand how and where a sudden downward shift in onlap is created by a change from wedge to fan-shaped deposition we use a hypothetical example of a series of three fans arranged along a continental margin, which for convenience are called fans A, B and C. All fan-shaped deposits have feeder systems at their updip end. Fans A, B and C, likewise have three feeder systems, which are separated geographically. A dip section of this continental margin, chosen at random, is not likely to pass directly down the axis of any of the three fans. It is more likely, that this section will pass between the feeders of the fans until at some down-dip location it intersects the deposits of one of the three fans. For example, one profile on the upper slope may pass between the feeders of fans A and B but, on the lower slope, intersect the deposits of fan B. Seismically this will create onlap of fan B on to an apparent slope surface of bypass. It is only along the fan axis, where axial deposits are thick enough, that the fan deposit will display downlap rather than onlap of the surface beneath it.

A sudden basinward shift in coastal onlap, as in the Vail (1987) lowstand model, indicates that succeeding deposition is part of the basin fan (fill) geometric systems tract. One line interpretations can be misleading, however, because this sudden shift may not be apparent on sections that go directly down the fan axis. Sequence interpretations based on seismic grids can better identify the characteristic fan shape of the fan (fill) geometric systems tract.

As we have just seen, the basin fan systems tract can be identified seismically when: (1) it stratigraphically sits above a regional or globally correlatable coastal-plain–slope unconformity (type 1 sequence boundary of Vail, 1987); or (2) it is that part of a prograding depositional system with a fan-shaped geometry. We discuss these two criteria in turn.

*Presence of an erosional type 1 sequence boundary.* Though evidence for Quaternary erosion of the coastal plain and slope can be found in numerous localities, there are only a few possible cases where these features have been interpreted as a genetically linked, synchronous, basin-wide coastal-plain–slope unconformity.

The Quaternary evolution of the northern Gulf coastal plain and Mississippi fan are often cited as providing strong support for the concept of a synchronous, basin-wide coastal-plain–slope unconformity (Beard *et al.*, 1982; Coleman & Roberts, 1988a,b; Weimer, 1989,1990; Feeley *et al.*, 1990). Weimer (1989) was able to identify significant erosion of underlying deposits by five of 17 Quaternary sequences mapped on the deep Mississippi fan. Feeley *et al.* (1990) presented evidence for slope erosion by several of the 13 mapped Quaternary sequences also mapped on the Mississippi fan. These studies indicate that slope erosion is present beneath some, but not the majority, of lowstand units of the Mississippi fan.

It is unclear, however, whether these documented cases of slope erosion can be related necessarily to a regional, synchronous unconformity created by canyon formation, sediment bypass and erosion on the upper slope, and fluvial entrenchment on the coastal plain. These processes are linked in the Vail (1987) model by assuming that canyon formation and upper slope bypass are a response to river incisement of an exposed shelf. Goodwin & Prior (1989), however, examined the basal fill of the upper Mississippi canyon and concluded that the canyon was initiated when it was located in a distal prodelta position geographically removed from any river source.

Fluvial entrenchment of the northern Gulf coastal plain is not clearly related to slope erosion features at the base of lowstand deposits. Suter & Berryhill (1985) have mapped a surface of fluvial incision of late Wisconsin age (Fig. 3). This coastal plain unconformity surface, unlike a Vail (1987) type 1 sequence boundary does not dip *below* glacial lowstand deposits but continues over the *top* of these deposits. The mapped extent of incision, in fact, continues to within a few kilometres of the latest lowstand shelf edge.

Fluvial entrenchment of the northern Gulf coastal plain appears, moreover, not to have created incised valleys, in the sense of Vail (1987) that, in the Vail (1987) model, act as point-sources for lowstand fan development. Suter & Berryhill (1985) and Kindinger (1988) have pointed out that late Wisconsin fluvial channels form an anastomosing pattern that covers the lowstand exposed shelf surface of Texas, Mississippi and Louisiana. This pattern suggests that the fluvial system at this time was able to avulse across the entire exposed surface. The anastomosing pattern of the preserved fluvial system represents, in this interpretation, the superimposed patterns of several generations of channel belts formed during a progressive period of little or no sediment accumulation. This picture of lowstand coastal-plain river system development is quite different than the conception of lowstand river incision of Vail (1987) in which lowstand rivers are trapped within narrow incised valleys.

There is evidence to indicate, moreover, that the upper continental slope of the Gulf coast was not generally an area of erosion and bypass during the last glacial lowstand. The early seismic studies of Sangree & Widmier (1977), Sidener *et al.* (1978),

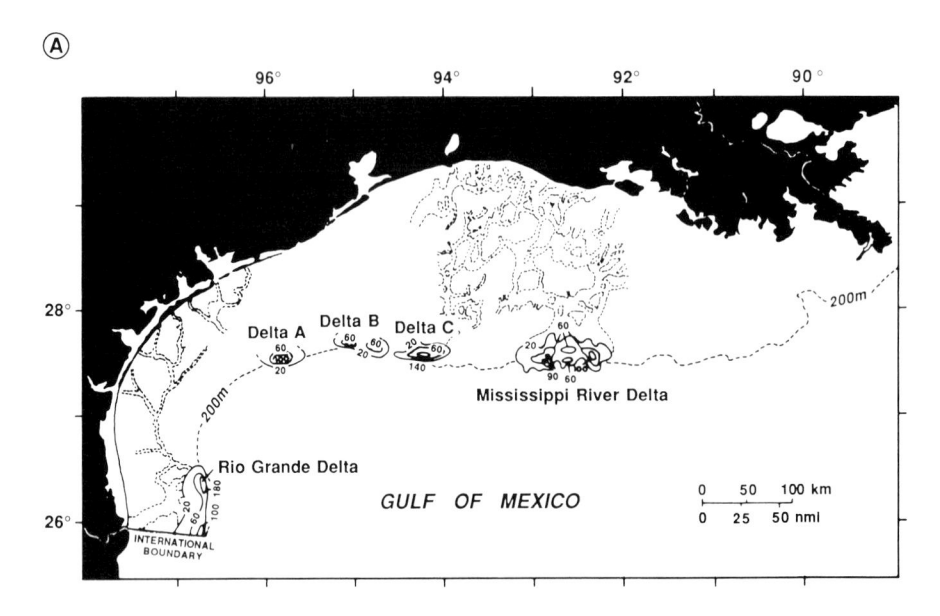

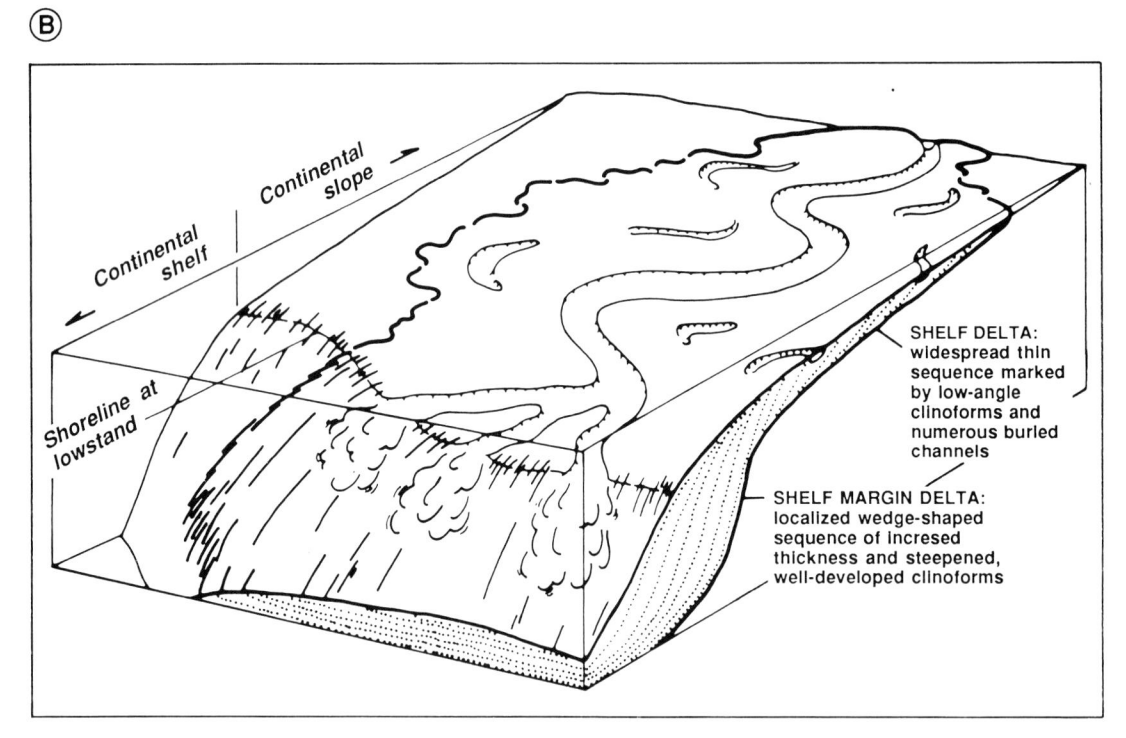

**Fig. 3.** Lowstand facies tract on the Louisiana Shelf. (A) Pre-Wisconsinan drainage pattern in the subsurface and associated lowstand deltas. (B) Conceptual model of a lowstand delta. (From Suter & Berryhill, 1985.)

and Woodbury et al. (1978) have shown that the late Pleistocene deposits of the shelf-break and upper continental slope of the Gulf coast have an up- and outbuilding progradational form. Van den Bold et al. (1987) documented erosive stripping of sediments on the Louisiana upper continental slope. These erosive features, however, were observed only on areas of topographic relief, such as fault scarps, and salt or shale diapirs. Biostratigraphic age dating of the timing of sea-floor erosion led the authors to suggest that erosion mainly occurred during Holocene sea-level rise owing to glacial melt-water discharge.

The evidence is generally lacking, therefore, to strongly link continental slope erosion to the formation of an interregional upper-slope–coastal-plain unconformity (type 1 sequence boundary) in the northern Gulf of Mexico. Because of the importance of the sequence boundary concept to the Vail (1987) model, we discuss this issue in a more general and theoretical sense in subsequent sections.

*Transition from fan to wedge-shaped deposits.* A second criterion for identifying the basin fan systems tract is an observed transition in prograding deposits from wedge-shaped units (the offlapping wedge systems tract) to fan-shaped units. We apply this criteria to the lowstand deposits of three areas: the Mississippi fan, the Rio Ebro delta of northeastern Spain, and the Rhône delta of France.

Pleistocene lowstand deposits of the northwest Gulf of Mexico shelf form a retrogressive system of coastal plain, continental shelf edge, continental slope and submarine fan deposition (Suter & Berryhill, 1985; Suter et al., 1987; Coleman & Roberts, 1988a,b). Importantly, both the offlapping wedge and basin fan systems tracts can be identified within the lowstand phase of sedimentation. Suter et al. (1987) using high-resolution seismic data, mapped the distribution of *shelf-phase* deltas (platform deltas of Galloway, 1989a) and subsequent late Wisconsin *shelf-margin* deltaic systems (shelf-edge deltas of Galloway, 1989a). Deltaic bodies are recognized by geomorphological pattern, high-angle clinoform seismic reflection patterns and association with buried river systems. In map view, platform deltas form a wedge-shaped deposit that is cut by later fluvial incision. Shelf-edge deltas, however, form fan-shaped deposits. Shelf-edge deltaic deposits attain much greater thicknesses (>180 m) than platform deltas and are elongate parallel to the shelf edge.

The high-resolution seismic profiles in the Berryhill (1987) seismic atlas provide a host of varying examples of downlap and onlap created by the shelf-edge deltas of the late Wisconsin glacial lowstand. Among the stratal geometries observed are: (1) downlap on to the outer shelf and on to the slope at depths of greater then 300 m; (2) several onlap surfaces within the late Wisconsin deposits separating succeeding lobes of shelf-edge deltaic deposits; (3) downlap on to the outer shelf and uppermost slope with onlap in deeper water depths caused by ponded turbidites in intra-slope basins. Similar features can be found in high-resolution seismic profiles of Morton & Price (1987) and Bouma et al. (1987).

It is difficult to fit these features to the Vail (1987) model for a type 1 sequence boundary. It is simpler to explain them as manifestations of the gradual, episodic shift from the offlapping wedge systems tract with its platform deltas to a basin fan systems tract with its shelf-edge deltas. This type of interpretation of lowstand architecture resembles more closely the model of Frazier (1974) and Galloway (1989a) than it does that of Vail (1987) or Posamentier & Vail (1988).

Though the Quaternary Mississippi fan is fan-shaped as a whole, individual units show significant variations. Isopach maps of four successive intervals of the modern Mississippi fan lobe (Bouma et al., 1985) show that the earliest mapped interval (horizons 40–50, age equivalent c.40 000 to c.50 000 years BP) contains broad sheet-like deposits that do not narrow or thicken towards the upper fan. Depositional units of the two latest intervals (horizons 0–30), however, narrow and thicken towards the upper fan. The mapped patterns suggest a transition from the offlap wedge geometric systems tract to the basin fan geometric systems tract occurring between horizons 30 and 40 (approximately 30 000–40 000 years BP). Further documentation of this interpretation requires a study correlating the evolution of shelf-edge and platform deltas to the changes in Mississippi fan-lobe geometry.

*Lowstand deposits within the offlap wedge systems tract.* The preceding example of Gulf coast Pleistocene deposition illustrates that lowstand deposits can fall into both the offlap wedge and the basin fan geometric systems tracts. We have chosen two

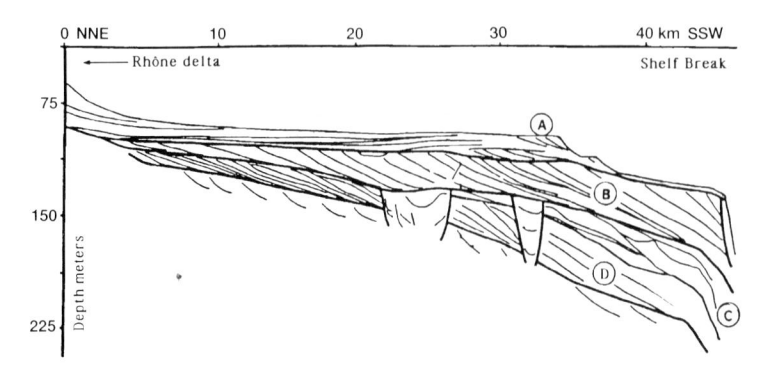

**Fig. 4.** High-resolution seismic profile of the Rhône delta shows deltaic and shelf sediments in series of stacked offlap wedges and backstep wedge/fill systems tracts. (From Tesson *et al.*, 1990.)

examples, the Ebro delta and the Rhône delta, to illustrate that lowstand deposits may fall entirely within the offlap wedge. Both areas are Mediterranean deltas and are within 400 km of each other. These two studies complement each other because the Ebro delta study focuses on seismic profiling of the continental slope, while the Rhône study focuses on seismic profiling of the shelf.

On the basis of a high-resolution seismic sequence stratigraphy, Field & Gardner (1990) propose a *prograding slope* model for the Ebro margin during Plio-Pleistocene time. This term is used by the authors to describe the up- and outbuilding wedge-like form of succeeding depositional units. The authors, moreover, recognize only one major discontinuity surface separating early Pleistocene from later deposits. Our interpretation of these observations is that neither criteria for identification of a basin fan systems tract is present on this margin.

At the present time, fluvial deposits of the Ebro River are trapped in deltaic and nearshore environments and a flooding surface is forming on the outer shelf (Maldonado *et al.*, 1983). The high-resolution seismic profiles of Field & Gardner (1990) do not extend much landward of the shelf break. What can be seen on the sections presented can be interpreted as downlap on Pleistocene flooding surfaces. Without more detailed study, the Ebro margin appears best described as a series of stacked offlapping wedge systems tracts. It is of interest to note that several of the seismic facies commonly associated with the basin fan systems tract are here associated with the offlap wedge systems tract. Field & Gardner (1990) note the presence of levee-channel deposits, and slide blocks on the continental slope and unconfined, sediment-gravity flows on the basin plain.

The high-resolution seismic profiles of the Rhône delta described by Tesson *et al.* (1990) give a good picture of deltaic and shelf sediments in a series of stacked offlap wedges and back-step wedge systems tracts (Fig. 4). Units B and D exhibit oblique clinoforms that have prograded out to the shelf break, forming a sedimentary wedge attaining a thickness of *c*.50 m. These units are referred to as *lowstand wedge* deposits by the authors. Their oblique progradational style is indicative of deltaic or strandline progradation (Sangree & Widmier, 1977). Tesson *et al.* (1990) argue that a relative sea-level rise of 90 m must have occurred since the deposition of Unit B to explain the current water depth of these originally shallow-water deposits. We interpret both units A and C (Fig. 4) as back-step wedge units. The geometry of these units will be discussed in a following section.

The lowstand wedge deposits of the Rhône delta are, in the geometric classification adopted here, part of the offlap wedge systems tract. Without independent evidence of the position of sea-level during their formation it would be hard to distinguish these deposits from similar highstand offlap wedge deposits.

*Seismic recognition of basin fan (fill) subsystems in recent lowstand deposits*

As we have reported earlier, the Vail (1987) lowstand systems tract (our basin fan (fill) geometric systems tract) is divided into three subsystems: the basin-floor fan, the slope fan and the prograding complex. We evaluate evidence from high-resolution seismic studies for each of these subsystems in turn.

*Recognition of basin-floor fan and the bidirectional downlap criterion.* One of the key criteria that has been suggested for recognition of a basin-floor fan is bidirectional downlap on the basin plain (Vail, 1987). Two questions immediately arise. What is the origin of observed bidirectional downlap features of Quaternary age? Secondly, can we find evidence that modern fan-lobe deposits of the basin plain would appear characteristically as isolated, mounded features on both dip and strike sections?

It is beyond the scope of this study to compile a complete catalogue of observed Pleistocene and modern bidirectional downlap features. A brief survey of these features, however, indicates that their origin can be quite diverse. Feeley et al. (1990) give several examples of bidirectional downlap of basin-floor fans of the Pleistocene Mississippi fan, which they suggest are the preserved depositional geometry of sediment-gravity flows on the basin plain. We note, however, that the interpreted downlap features are quite subtle, even on the highly laterally compressed seismic sections shown in their study (see their Fig. 12).

Other possible origins of seismic bidirectional downlap include: (1) levee-channel, overbank deposits viewed oblique to the channel axis (e.g. Piper & Normark, 1989, figs 2, 4, 5, 10); (2) apparent downlap caused by the cross-cutting relationships of stacked fan-lobes from the same or different sea-level cycles (e.g. Weimer, 1989); (3) mound-like features formed by differential compaction of originally flat-lying basin plain sediments, caused either by different amounts of overburden or lateral lithological changes (e.g. Droz et al., 1985, our interpretation of their fig. 5); (4) internally stratified, elongate, slide deposits viewed oblique to the slide axis (e.g., Savoye et al., 1990, our interpretation of their fig. 7; Trincardi & Normark, 1989, figs 4, 5); (5) base of slope slump blocks (e.g. McGregor, 1985, fig. 2; Bouma et al., 1987, fig. 13); (6) glacial debris lobes (Vorren et al., 1984, fig. 9); (7) internal folding caused by downslope creep (Coumes, 1987, fig. 4); and (8) contour current ridge deposits (Mougenot, 1985, fig. 16.)

Though the presence of bidirectional downlap may not be a unique identifier of a basin-floor fan, some workers suggest that the various origins for this geometric feature can be distinguished on the basis of seismic facies. Pacht et al. (1990) in a combined well-log and seismic study of the intraslope basins of offshore Louisiana, showed that basin-floor fans were seismically transparent and on well-logs contained thick blocky sands. In a contrasting example, Weimer (1989), working further offshore, found a chaotic seismic facies lying stratigraphically above slope unconformities of the deep fan. He interpreted them as shale-prone, massflow complexes.

No simple answer has emerged regarding the validity of the basin-floor fan concept. Owing to the importance of this concept to oil exploration, researchers in the hydrocarbon industry are evolving second-generation geological models of basin-floor fan deposition that account for different scales of canyon and fan morphology, different structural regimes, and other geological variables.

*Recognition of slope fans.* The slope–fan subsystem, as described by Vail (1987) and Posamentier et al. (1988), has two key elements: (1) it onlaps the type 1 sequence boundary on the slope and downlaps the basin-floor fan or the basin plain; and (2) it contains a set of levee-channel and overbank deposits with a retrogradational stacking pattern.

Few studies of modern fan systems have documented onlap–downlap relationships of deep-water sediments. As we have indicated earlier, a fan-shaped unit will appear, on dip lines that are not directly down the canyon axis, to onlap the slope beneath it simply as a consequence of the systems fan-shaped geometry. A down-the-axis view, however, of the modern Mississippi canyon (Goodwin & Prior, 1989) shows that canyon fill units do not pinch out against the canyon floor but are continuous with shelf deposition.

Studies of modern fan systems have, generally, documented the importance of levee-channel and overbank deposition on the upper and middle parts of large continental shelf fans (e.g. Kolla & Coumes, 1987; Damuth et al., 1988). Weimer (1990) estimated, using an extensive grid of conventional seismic data, that slope fans of the Mississippi fan consist of 80–100% levee-channel, overbank deposits and 0–20% chaotic mass-flow deposits.

Studies of modern fan systems indicate, however, that complete seismic sequence stratigraphic models of the slope–fan subsystem must recognize: (1) the extreme variability of size of levee-channel deposits; (2) the inclusion of diverse types of deep-water deposits within the slope fan; and (3) systematic changes in seismic facies, from upper, high-gradient regions of the fan, to lower, low-gradient regions.

It is beyond the scope of this study to summarize observations from fans around the world. The importance of morphological diversity should be emphasized, however. The size of levee-channel

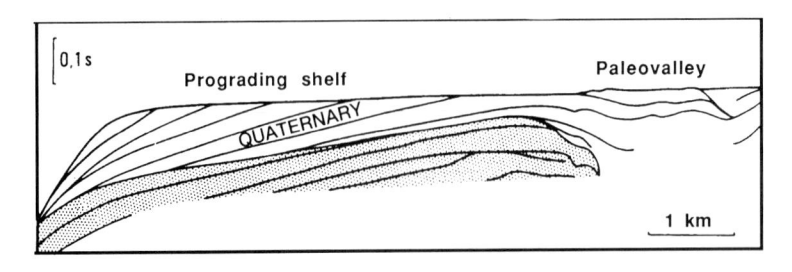

**Fig. 5.** High-resolution seismic profile of the Portuguese continental margin. (From Mougenot, 1985.) The Quaternary units on this seismic profile are interpreted as a prograding complex because they show outbuilding of a shelf-edge delta and contemporaneous filling of a palaeovalley in close proximity to the shelf edge.

deposits, for example, ranges from larger than 100 km across on the Indus fan (Kolla & Coumes, 1987) to smaller than 1 km for intra-slope basin channels on the Mississippi fan (Satterfield & Behrens, 1990). Slope depositional processes are highly diverse as well, including: channelized low-density sediment gravity flows, rotational slumps, slides, downslope creep, high-density sediment gravity flows, debris flows, pelagic sedimentation, and unchannelized turbidites (Stow, 1986). In view of this complexity, the slope–fan geological model of Vail (1987) seems a good first-order approximation to reality. The development of more advanced depositional models, requires a better understanding of how the observed morphological variations, from fan to fan or within one fan, are related to depositional regime.

*Recognition of prograding slope complexes.* The prograding slope subsystem of Vail (1987) has been well defined in a theoretical sense by Posamentier *et al.* (1988). Unfortunately, the identification of such a subsystem is particularly problematic because several prograding slope characteristics are shared by other systems tracts. For example, as noted earlier, Wisconsin lowstand deposits often have a prograding wedge form. It is tempting, but misleading, to interpret these deposits as forming the prograding complex rather than as part of the offlap wedge geometric systems tract. A prograding complex, as described by Posamentier *et al.* (1988), differs in three respects from an offlap wedge: (1) progradation of the unit occurs by infilling of the canyon head by a shelf-edge delta rather than by progradation of platform deltas; (2) coastal-plain river entrenchment has stopped; and (3) the unit downlaps a fan-shaped unit that, ideally, has a Vailian type 1 sequence boundary at its base. To add further to the possibilities for misinterpretation, these last two characteristics are shared by the back-step wedge systems tract.

Given these possibilities for misinterpretation, it is hard to evaluate whether the modern stratigraphic record contains a good example of a prograding slope complex. The architectural systems tract diagram of the Vail (1987) model (Fig. 1) appears to indicate that the prograding complex should make up the major component of deep-water fan system deposition. Surveys of modern continental fans (e.g. Weimer, 1990) indicate, however, that the majority of the sediment fill is not associated with deltaic progradation. Canyon fill in many of these systems appears to be caused by channel abandonment and subsequent deposition by overbank, debris flows and slides.

Though modern deep-water examples of the prograding complex may not exist, many deposits of the last glacial lowstand resemble the updip portions of the prograding slope complex. Stratigraphic sections through late Pleistocene deposits of the Texas shelf (Morton & Price, 1987) show contemporaneous progradation of shelf-edge deltas and aggradation of fluvial and deltaic floodplain deposits within shelf-edge incised valleys. This association also can be seen on the Portuguese continental margin (Fig. 5). The Quaternary units on this seismic profile are interpreted as a prograding slope complex because they show outbuilding of a shelf-edge delta and contemporaneous filling of a palaeovalley in close proximity to the shelf edge.

## Back-step wedge geometric systems tract: models and observations

*Transgression: Vail (1987) model*

The following synopsis should be compared with Fig. 1(A,B). The transgressive systems tract is described as a 'succession of back-stepping or retrogradational parasequences'. It is initiated by the first significant flooding event (whose record is the trans-

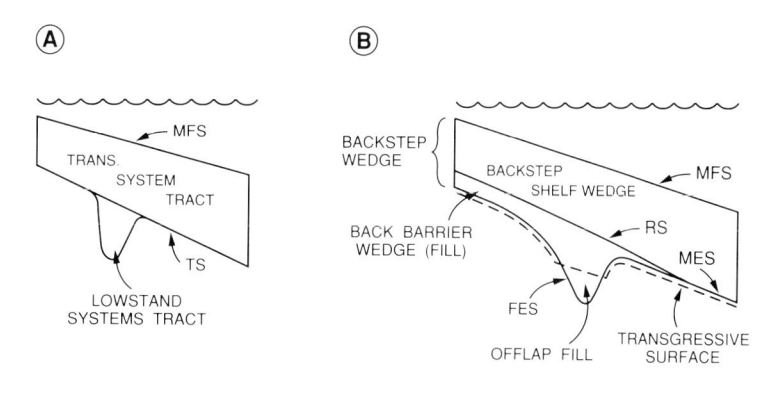

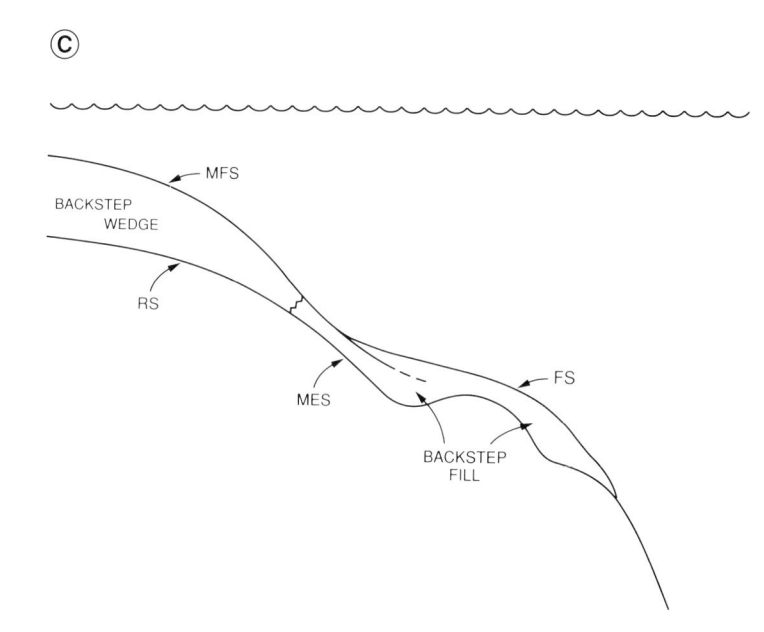

**Fig. 6.** Schematic illustration of surfaces bounding and within the back-step wedge (fill) systems tract. (A) Vail (1987) model. TS = transgressive surface; MFS = marine flooding surface. (B) Model presented in this paper, showing distinct transgressive and ravinement surfaces. RS = ravinement surface; MES = marine erosion surface. Transgressive surface is dashed. (C) Two generations of offshore back-step fill. Second generation accumulates as maximum erosion surface begins to form on back-step wedge.

gressive surface) after the time of maximum regressions of the lowstand systems tract (Posamentier & Vail, 1988). The parasequences onlap onto the transgressive surface in a landward direction, and downlap onto it in a basinward direction. The top of the transgressive systems tract is the 'surface of maximum flooding' and the several metres of section on either side of this surface may be sufficiently distinctive in lithologic terms to be identifiable as a condensed section. The surface of maximum flooding marks the change from a retrogradational to a progradational parasequence set.

Figure 6A shows the relationship of surfaces found in the Vail (1987) model. Incised valleys are filled with the deposits of the prograding complex capped by a back-stepping set of deposits bounded by the transgressive surface below and the maximum flooding surface above.

*Transgression: Galloway (1989a) model*

The Galloway (1989a) model offers a similar architecture. It recognizes a shelf systems tract and also a transgressive, barrier–lagoon systems tract, which is preserved by virtue of base level rise (Galloway, 1989a). These tracts are 'onlap components'.

In the Galloway (1989a) model, slope sedimentation is not terminated by relative sea-level rise. The transgressive period, following active outbuilding of the continental margin, is seen as a good time for extensive mass wasting and retrogradation of the upper slope and continental margin. As a result, a distinct apron of resedimented material onlaps the toe of the slope clinoform during the transgression, and thus is an 'onlap component.'

*Back-step wedge systems tract: comparison with modern transgressive deposits*

*General.* The depositional systems of the Holocene transgression have been well studied on modern continental shelves (Belknap & Fraft, 1985; Swift *et al.*, 1986; Kraft *et al.*, 1987), and have been described systematically by Swift *et al.* (this volume, pp. 89–152). The architecture of these systems appears, in general, to be more complicated than the proposed models of Vail (1987) and Galloway (1989a).

Back-step wedge (fill) deposits do not constitute a simple sheet, but rather a ribbed sheet (see Swift *et al.*, this volume, pp. 89–152). While the upper surface (maximum flooding surface) is relatively flat, its lower surface has 'ribs' or flange-like projections (fill deposits) that fit into the incised drainage pattern of the underlying subaerial erosion surface. The back-step wedge (fill) thus expresses a hybrid geometry, justifying its compound name. The structure within and directly above the flange-like projections is liable to be complex, and the term *shelf-valley complexes* has been adopted to describe the interrelated depositional systems and the geometry of these deposits (Swift *et al.*, 1972, 1973, 1977; see Swift *et al.*, this volume, pp. 89–152). As a consequence of this complexity, the back-step wedge (fill) geometric systems tract can be divided generally into several geometric subunits, bounded below by the transgressive surface and above by the maximum flooding surface (Fig. 6).

The transgressive surface is a surface whose time range encompasses the turnaround event, when paracycles of progradational sedimentation give way to paracycles of back-stepping sedimentation. Thus it caps parasequences with progradational architecture, and is overlain by parasequences with back-stepping architecture. The term 'transgressive surface' constitutes a relatively high level of conceptualization, and may be applied to surfaces already having names reflecting their formative mechanism. Thus, at any given point, the designation 'transgressive surface' may fall to a marine erosion surface, its onshore equivalent, the ravinement surface, or a fluvial entrenchment surface. The transgressive surface may also be a conformity. These classes of surface may each become the transgressive surface in turn, as the section is traced updip. In valley fills, the surfaces may occur in parallel in a vertical sequence, separated by depositional systems. However, at each point along the valley, only one such surface may be the trangressive surface. On the interfluves of a subaerial surface undergoing transgression, the multiple surfaces collapse into a single flooding surface (ravinement surface), which thus becomes the transgressive surface. The transgressive surface may be the marine erosion surface in a downdip setting, shift to a conformity within the valley fill further up dip and, finally, be expressed as the fluvial entrenchment surface. In a strict sense, the transgressive surface is simply the surface above which parasequences possess a back-stepping geometry.

*Valley fill deposits and the basal conformity.* Stratigraphic studies of valley fill deposits show that, with few exceptions (e.g. Fig. 5), these sediments are largely of Holocene transgressive age. On the modern Atlantic shelf only the thalweg deposits of the valley fills are of alluvial origin. Most cores that penetrate shelf valleys pass through estuarine deposits overlain by lagoonal silts and clays (Oertel *et al.*, 1989). Only a small portion include basal alluvial deposits (Belknap & Kraft, 1985; Figs 7, 8). The alluvial deposits in the thalwegs of these channels are diachronous, and are correlative with the overlying transgressive shelf sands of more seaward stations. Only the most seaward of the basal valley fills are correlative with the lowstand shelf edge and upper slope deposits (Belknap & Kraft, 1985; Coleman & Roberts, 1988a).

Although the valley fills of the USA Atlantic margin have been better studied, several stratigraphic studies indicate that the fill of lowstand incised channels on the Gulf coast and California margins is also largely of Holocene age. Morton & Price (1987) summarized evidence that incised valley fill of the Trinity River near Galveston Island is more than 50% of Holocene age. Darigo & Osborne (1986) found that channels near San Diego are floored by Pleistocene gravel and stream terrace deposits, but are largely filled by Holocene sediments.

It has been suggested that river valleys undergoing transgressive flooding cannot undergo fluvial back-step because they have equilibrium profiles and are at grade (Posamentier & Vail, 1988). However, this can be true only landward of the

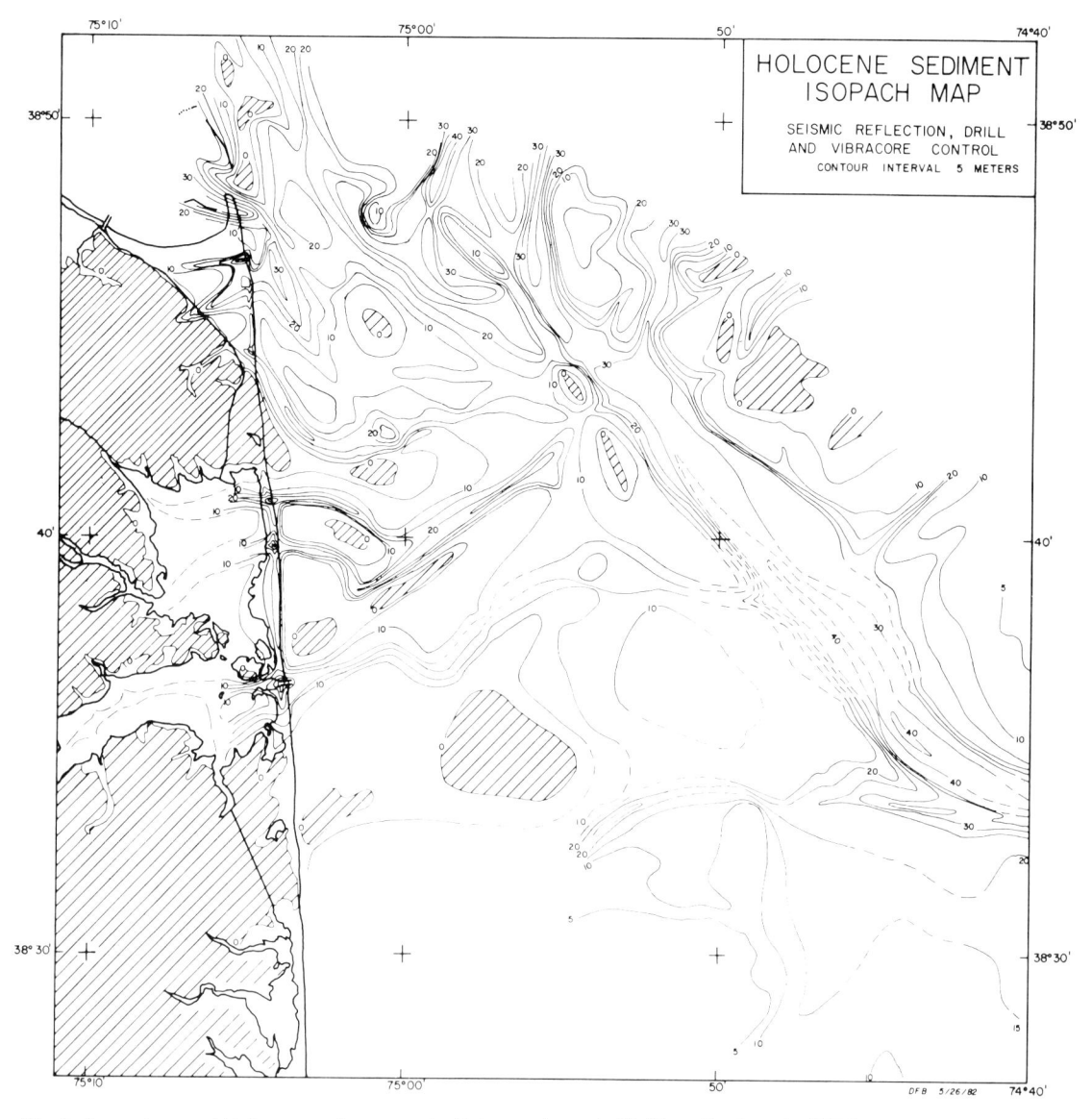

Fig. 7. Isopach map of Holocene sediment on the Delaware inner shelf. (From Kraft et al., 1987.)

hinge point. Seaward of the hinge, they must compensate subsidence with deposition in order to maintain Davisian grade (see Fig. 1 of Thorne & Swift, this volume, pp. 33–58). The back-stepping nature of valley fill deposition is generally hard to document on seismic profiles. This is particularly true where the ravinement surface cuts directly across the top of valleys (Fig. 6A), causing the fill to be separated from the overlying deposits. Knebel & Circe (1988), using an extensive grid of seismic reflection profiles and core data, mapped the back-stepping fill of Delaware Bay in a series of palaeogeographic maps showing the distribution of fluvial deposits within channel thalwegs, tidal wetlands, barrier–headland beach, and open estuarine deposits. Figure 9 shows a seismic example of this fill towards the mouth of Delaware Bay.

At some point downdip, back-stepping fluvial deposits may overlie fluvial deposits of the offlapping wedge (fill) geometric systems tract. Here, as

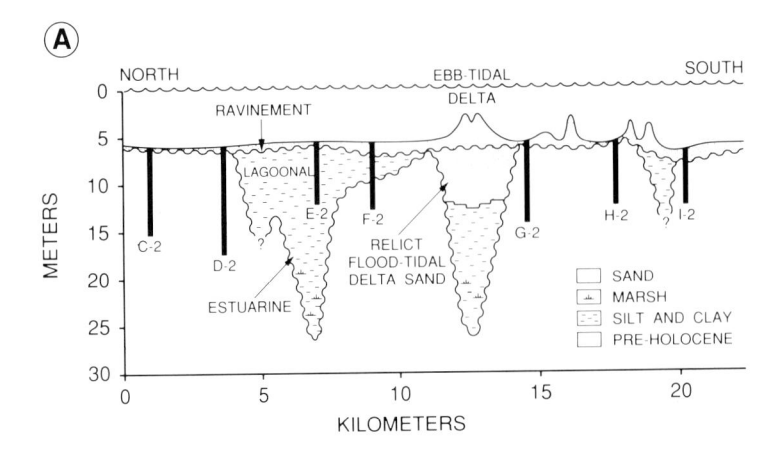

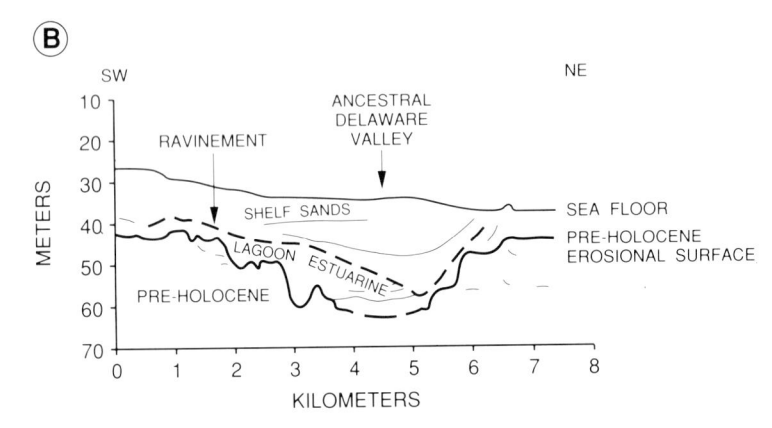

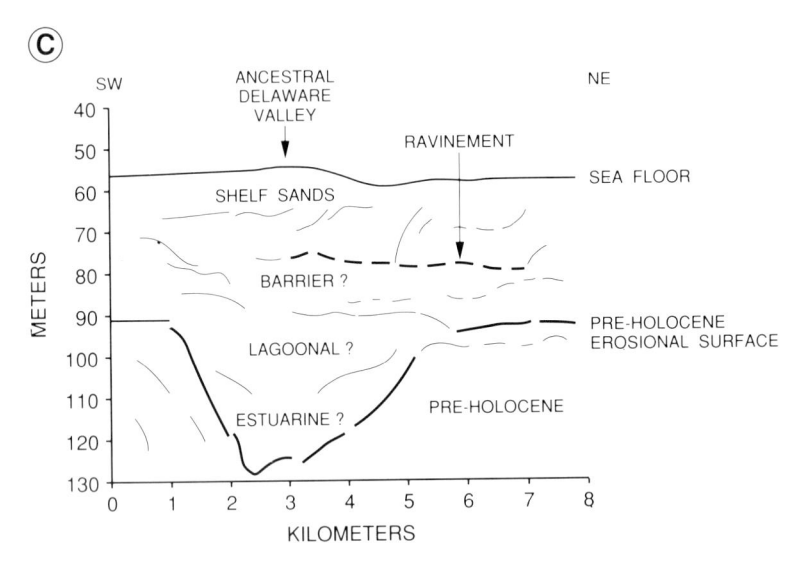

**Fig. 8.** Cross-sections of the Delaware–Maryland shelf, indicating twofold stratigraphy (shelf, back-barrier). Back-barrier deposits are more extensive on the outer shelf where the transgression was more rapid, and the depth of shoreface incision less. (Modified from Belknap & Kraft (1985) and Kraft *et al.* (1987).) A, inner shelf; B, mid-shelf; C, outer shelf.

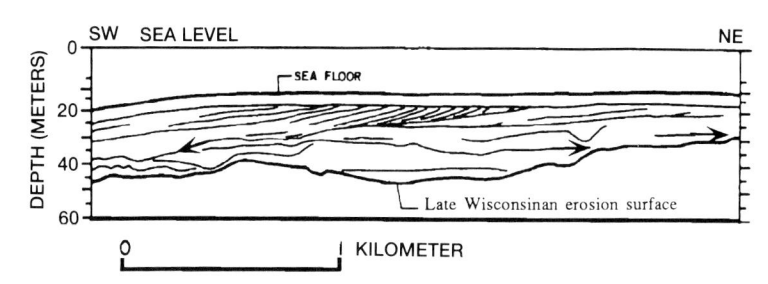

**Fig. 9.** High-resolution seismic profile showing back-stepping fill of palaeovalleys of the Delaware Bay estuary. (From Knebel & Circe, 1988.)

fluvial parasequences pass from offlapping to back-stepping geometry, the transgressive surface is a conformity.

*Ravinement and marine erosion surfaces.* The geometric systems tract described here as a back-step wedge (fill) is deposited on both sides of the transgressing shoreline. As it forms, the process of erosional shoreface retreat cuts a major erosional surface, the *ravinement surface*, which becomes incorporated into the wedge (see discussion by Swift & Thorne (this volume, pp. 3–31); Thorne & Swift (this volume, pp. 33–58); and Swift *et al.* (this volume, pp. 89–152). Swift *et al.* have shown that strict application of the source–sink model of a depositional system leads to the resolution of two distinct depositional systems tracts that are conjugate across a ravinement diastem; a back-barrier systems tract below the ravinement, and a transgressive shelf systems tract above it. However, since both units consist of back-stepping parasequences, they constitute a single geometric systems tract; the back-step wedge (fill); see Fig. 6B. If the ravinement surface climbs at a higher angle than the fluvial entrenchment surface then the back barrier depositional systems tract is preserved in a wedge-shaped form. If the ravinement surface climbs at a relatively low angle, these deposits are preserved only in incised valleys.

Unlike the systems tract bounding surfaces of the Vail (1987) model, the ravinement surface does not have any time-stratigraphic significance. The minimum criterion for a surface to have time-stratigraphic significance is that there is no unit above the surface that is stratigraphically younger than any of the units below the surface. Figure 10 shows a schematic illustration of the formation of a ravinement surface by shoreface retreat and barrier island migration. The resulting erosional surface has depositional units above the surface that are *older* than units below the surface.

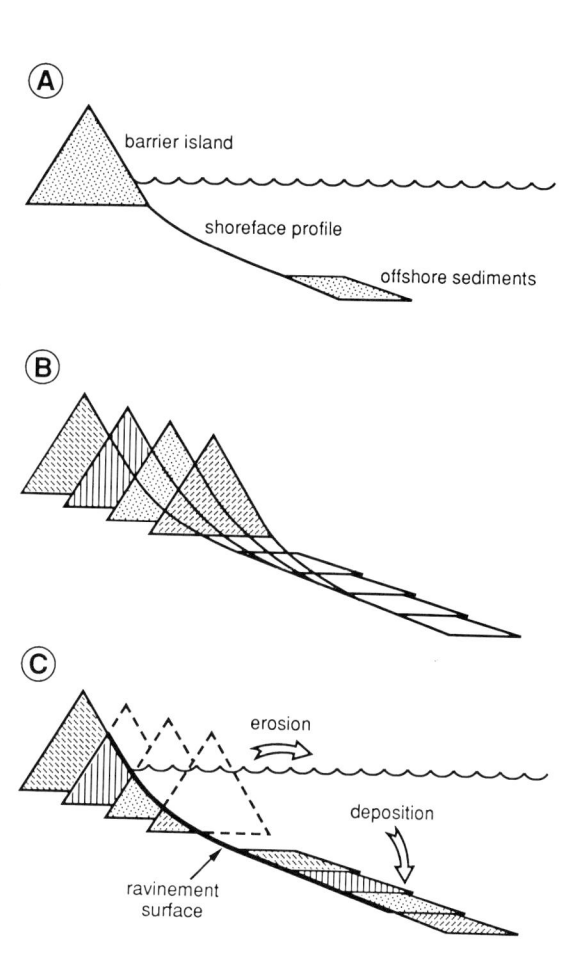

**Fig. 10.** Schematic illustration that deposits above the ravinement surface can be older than deposits below the ravinement surface. (A) During ravinement contemporaneous sedimentation occurs on the barrier island and seaward of the erosional shoreface. (B) The barrier island system is shown at four successive stages (four different shading patterns are used). (C) Stratigraphic section resulting from four stages in panel B. Stippled shelf deposition above the ravinement surface is older than vertically hatched barrier island deposition below ravinement surface.

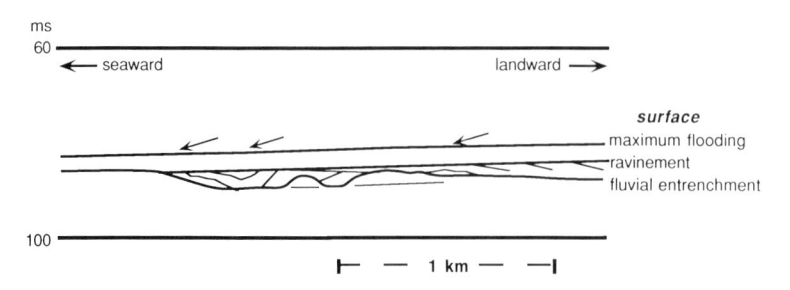

**Fig. 11.** Line-drawing of high-resolution seismic profile across the Korea Strait. (From Park & Yoo, 1988.) A landward thickening wedge occurs between the ravinement surface and the Pleistocene surface of fluvial entrenchment.

If the shoreline did not fall over the shelf edge during deposition of the preceding geometric systems tract (type 2 sea-level fall of Vail, 1987), then ravinement must pass seaward into a *marine erosion surface* (Nummedal & Swift, 1987). A marine erosion surface forms seaward of the shoreline as transgression begins. The rate of sediment input falls abruptly as lagoons and estuary begin to trap sediment, but wave tidal and wind current stirring of the sea-floor continue. The bypassing ratio $B$ increases until the water column deepens appreciably and fluid power decreases (see Swift & Thorne, this volume, pp. 3–31), and the sea-floor erodes. The marine erosion surface is a variant of the more general class of surfaces called transgressive surfaces.

*Back-barrier wedge (fill).* Where the ravinement surface climbs stratigraphically above the late Wisconsin coastal plain unconformity, estuarine and, particularly, lagoonal deposits tend to bridge over drainage divides (Figs 9, 10). In facies terms, these deposits constitute the back-barrier depositional systems tract (Table 1 of Swift *et al.*, this volume, pp. 89–152). They are presented here as a subunit within the backstep wedge (fill) geometric systems tract.

These deposits have a generally wedge-like form that thickens in the landward direction (e.g. Morton & Price, 1987, fig. 9). This subunit is a hybrid wedge-fill unit, with rib-like projections (fills) extending downwards from the wedge into the valleys of the fluvial entrenchment surface. There is no surface of discontinuity separating the upper wedge-like portion from the underlying fluvial and estuarine valley fills. Within valleys, Holocene back-stepping fluvial fill deposits conformably overlie prograding fluvial fills of the preceding downlap wedge geometric systems tract, if this latter systems tract is present. If not, the fluvial back-step fill rests on a fluvial entrenchment surface. The back-stepping fluvial fill passes upwards into a back-stepping estuarine fill by means of a time-transgressive facies change. The estuarine fill in turn passes upwards into the lagoonal deposits of the back-barrier wedge by time-transgressive facies change (Fig. 8). In depositional systems terms, the contact between the fluvial and estuarine systems is a serial contact; their proximal facies are contiguous. However, the contact between the estuarine and lagoonal systems is an interfingering contact, between their respective distal facies (see Fig. 22 of Swift *et al.*, this volume, pp. 89–152).

The wedge-like form of this type of depositional unit is well expressed in the early Holocene sediments of the Korea Strait (Park & Yoo, 1988). Figure 11 is a line-drawing of one high-resolution seismic profile across the strait. Three surfaces are identified on the basis of reflection termination patterns. A landward thickening wedge occurs between the ravinement surface and the Pleistocene surface of fluvial entrenchment.

*Back-step shelf wedge (fill).* Although maximum flooding or near-maximum flooding conditions exist on much of the world's margins, surprisingly few high-resolution seismic profiles are available that resolve the depositional geometry of Holocene transgressive deposits on the continental shelf. Where these deposits are thick, they are typically covered by subsequent offlap wedge progradation. Where these deposits are thin, they are hard to image seismically.

The Holocene deposits of the Rhône delta (unit A on Fig. 4) display an overall back-stepping geometry and wedge-like form. Seismic resolution is lost towards the southwest, where reflectors seem to converge asymptotically with the depositional surface. Tesson *et al.* (1990) interpret a back-stepping pattern for the step-like bathymetric profile on the south-southwest end of the unit.

A wedge of onlapping marine Holocene deposits

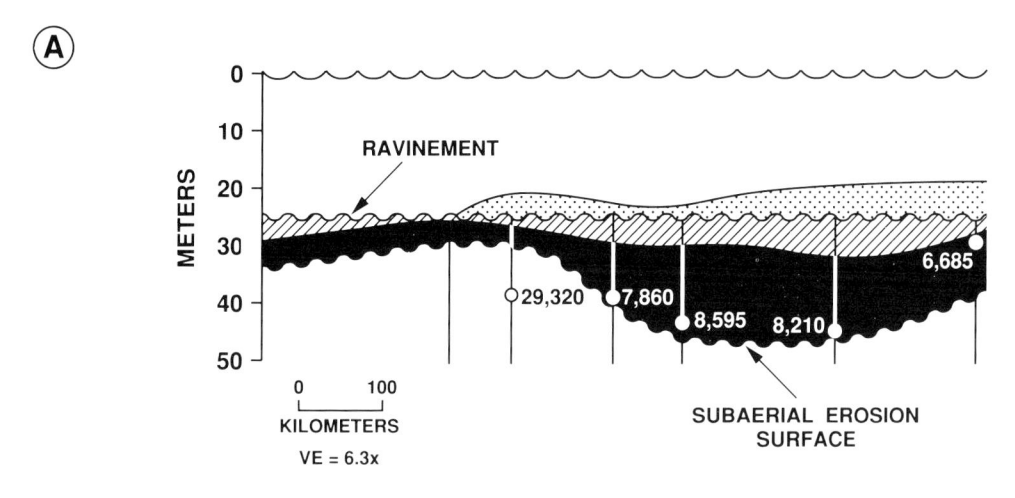

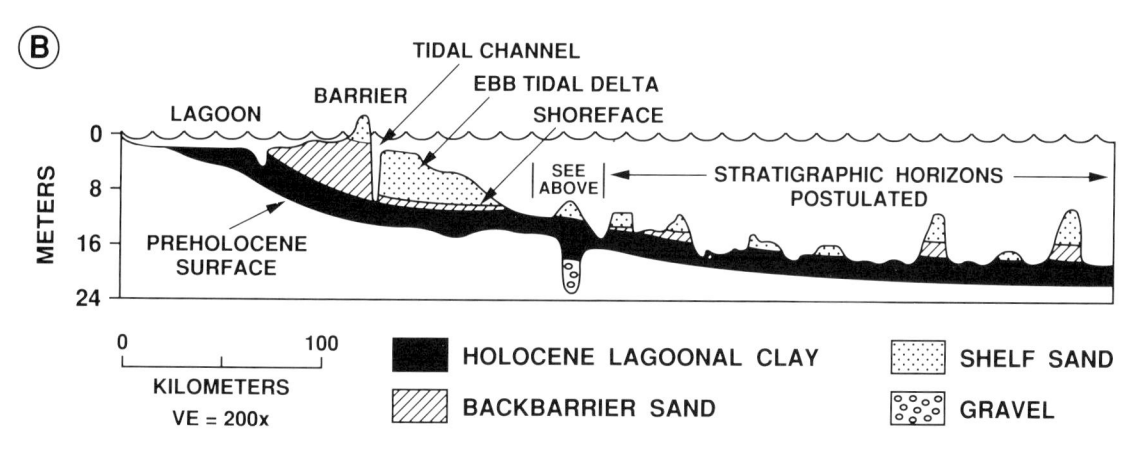

**Fig. 12.** Cross-section through the transgressive facies tract of the New Jersey Shelf. (A) Detail of the inner shelf, with radiocarbon dates showing cross-section through buried lagoonal channel, overlain by ridged sand sheet. (From Stahl et al., 1974.) (B) Shelf cross-section. Troughs between ridges have been eroded into pre-recent substrate. (From McKinney et al., 1974.)

lying above a ravinement or marine erosion surface can be identified readily on seismic profiles of the narrow continental shelf of San Diego County, California (Darigo & Osborne, 1986). The ravinement surface in this area is characterized by a series of marine terraces that are correlated to relative stillstands within the Holocene transgression previous to 11 000 years BP. The severe wave climate of this area accentuates marine erosion of the shelf.

The architectural stacking pattern of the Holocene wedge described by Darigo & Osborne (1986) is aggradational rather than retrogradational. The accumulation rate of Holocene deposition is higher in proximal areas of the shelf than on the upper slope of the continental shelf edge or on the canyon walls of the shelfward extension of the Carlsbad submarine canyon. No shift in accumulation patterns appears to have occurred from early to late Holocene. The resulting stratal pattern is a series of stacked, aggradational, shoreward-thickening wedges.

In many areas of the world, sediments between the ravinement surface and the maximum flooding surface are too thin to image seismically. Core studies of both the USA Atlantic continental shelf (Swift et al., 1972; Belknap & Kraft, 1985) and the North Sea (Belderson et al., 1966), demonstrate, however, that transgressive shelf deposits overlie a ravinement surface, which in turn overlies backbarrier deposits (Figs 8, 9, 12).

On many modern transgressive shelves, the transgressive sand sheet advancing over the ravinement surface is very patchy. The surface between the patches continues to undergo erosion, much as deflation depressions (blowouts) do on desert surfaces. The patches tend to migrate, exposing further sectors of the ravinement to renewed erosion, so that erosion may be said to be going on *through the sand sheet* (Stubblefield & Swift, 1976; Fig. 12B; see also Fig. 26 of Swift *et al.*, this volume, pp. 89–152).

On modern shelves near the mouths of major rivers, Holocene muds overlie, and onlap landward on to the inner shelf sand facies, burying earlier Holocene sand ridges (Twichell *et al.*, 1981; Maldonado *et al.*, 1983). The contact between the two classes of deposit is a gradual facies change. In facies terms, the inshore sands comprise the amalgamated sand lithofacies of the trangressive shoreface–shelf depositional system (see Swift *et al.*, this volume, pp. 89–152), while the Holocene muds constitute the distal interstratified sand and mud lithofacies, and the far distal laminated or bioturbated mud lithofacies of the same depositional system. High-resolution seismic profiles from the USA Atlantic shelf (Twichell *et al.*, 1981), USA Gulf shelf (Suter *et al.*, 1987; Kindinger, 1988) and the Amazon shelf (Alexander *et al.*, 1986) show that time lines within onshore Holocene muds pass into the basal sand (proximal amalgamated sand facies) and onlap the underlying ravinement surface. The onlapping character of these deposits is diagnostic of back-step wedge systems tract, and distinct from deposits of the downlapping offlap wedge systems tract.

*Offshore back-step fills.* Many continental shelves that are today characterized by wedge-form sandy deposits on the inner shelf, bear offshore muds with fill geometries.

On the outer Texas shelf, the buried Pleistocene surface has considerable relief in the form of protruding shelf-edge deltas, separated by embayments (Berryhill, 1987). These features constitute a prograding complex of the late Pleistocene basin fan systems tract. As the deltas flooded, thin, reworked basal sands were buried by mud that filled the embayments and bridged over the deltas (Swift *et al.*, this volume, pp. 89–152; Fig. 12). The Holocene sequence is a back-step wedge (fill) systems tract. On the outer shelf, it is primarily a fill; its only wedge trait is its nearly level upper surface. As traced on to the flatter inner shelf surface, however, it rapidly develops a full wedge geometry.

The 'mud patch' of the southern New England shelf (Twichell *et al.*, 1981) is a back-step fill that occupies the head of the Connecticut shelf valley (Swift *et al.*, 1972). It constitutes the distal facies of a regressive deceleration sheet (Fig. 13 of Swift *et al.*, this volume, pp. 89–152) that extends southwestwards along the Southern New England shelf from Nantucket shoals (their fig. 32).

Holocene topographic fill deposits also can be found further out on the shelf edge and upper slope. Here they may be separated physically from the back-step wedge of the shelf surface, occurring as ponded topographic fill geometries in intra-canyon areas. Petrologic, foraminiferal, and oceanographic evidence indicates that the slope and rise segments of modern canyons on autochthonous shelves are actively transporting significant amounts of sediment (e.g. Stanley *et al.*, 1986). High-resolution seismic studies document that these deposits are commonly large slide blocks and small-scale slumps (Knebel & Carson, 1979; Savoye *et al.*, 1991). Where submarine canyons intersect the shelf, canyon heads can also collect sediments from longshore currents, storm-created nepheloid layers, or tidal transport (Twichell, 1983; Droz *et al.*, 1985; Carson *et al.*, 1986; Dingler & Anima, 1989).

Holocene back-step fill deposits are particularly voluminous in areas influenced by glacial meltwater discharge and sediment-laden icebergs. For example, sedimentation in the northeastern North Sea, formed during glacial retreat from 15 000 years BP to 8000 years BP, shows a back-stepping pattern on the shelf (the back-step wedge systems tract) and a thick base-of-slope onlapping deposit that infills the deep-water Norwegian Channel (Salge & Wong, 1988, fig. 4). Dinter (1985) suggests a similar origin for a Holocene wedge of glaciomarine sediments on the Alaskan Beaufort shelf, which is of negligible thickness on the middle shelf but thickens to greater than 40 m on the outer shelf.

## Offlap wedge geometric systems tract: models and observations

### Highstand: Vail (1987) model

In the Vail (1987) model (Fig. 1), eustatic sea-level fluctuations dominate the stratigraphic record (Jervey, 1988). As eustatic sea-level rise passes the inflection point and begins to rise more slowly, a point is reached when accommodation is no longer created rapidly enough to satisfy the sediment supply. The shoreline passes through stillstand and begins to prograde. Since the rate of creation of accommodation is decreasing, the highstand systems tract is characterized by an increasingly progradational parasequence stacking pattern (Posamentier et al., 1988). Successive strata downlap seaward, burying the maximum flooding surface. In the Vail (1987) model, fluvial deposits are an important part of the highstand systems tract. Depositional sequences are seen as episodes of coastal onlap, and the fluvial deposits are part of the onlap record. Fluvial sediments continue to onlap the coastal plain as long as relative sea-level rises, even after the rise rate has become so slow that the shoreline begins to prograde. Thus, in the Vail (1987) model, the end of fluvial onlap is the end of the sequence.

*Highstand: Galloway (1989a) model*

The offlap components of the Galloway (1989a) model (Fig. 2) include: (1) sandy fluvial, delta plain and bay–lagoon facies; (2) progradational deposits of the shore zone; and (3) mixed aggradational upper slope and progradational lower slope facies (Galloway, 1989a). Initial delta progradation at first reclaims the flooded platform of the previous episode. These deposits are equivalents of the Vail (1987) highstand systems tract or the offlapping wedge geometric systems tract. Galloway (1989a) argues, however, that base-of-slope lobes are not necessarily confined to the early period of relative sea-level fall and shelf incision, but may be interspersed within the muddier offlap apron.

*Offlap wedge geometric systems tract: comparison with modern highstand deposits*

*General.* Active sedimentation in an offlap wedge geometry is found on many continental shelves at the present time, particularly in areas of high sedimentation rate near the mouths of major rivers (Nittrouer et al., 1986). High-resolution seismic profiles of these deposits typically show a sigmoid oblique progradational pattern that downlaps a well-defined condensed interval in deeper water (e.g Kindinger, 1988; Lykousis & Chronis, 1989). The wedge-shape of these units is typified by the late Holocene deposits of the Thermaikos Plateau in the Aegean Sea (Lykousis & Chronis, 1989). The thickness of these deposits decreases systematically with distance from shore along a mapped stretch of coastline c. 50 km long.

As a first-order description, the portrayal of the marine portions of the highstand system tract by the Vail (1987) and Galloway (1989a) models is largely adequate. Several important issues are raised, however, by modern studies concerning: (1) the regional extent of a synchronous maximum flooding surface; and (2) the origin of the *depositional shoreline break*, which plays an important role in the Vail (1987) model.

*Condensed intervals and the maximum flooding surface.* Maximum flooding surfaces are important internal boundaries in the Vail (1987) model, where they separate the transgressive systems tract from the highstand tract. In the Galloway (1989a) model, they form the cycle boundaries.

The maximum flooding surface is an isochronous boundary that occurs at that point in time when the shoreline has reached its maximum landward excursion. In the rock record, stratigraphic sections that show evidence of condensed sedimentation, characteristic of the maximum flooding period, may span appreciable lengths of time. The time of maximum flooding itself is marked by an erosional surface within the condensed section that forms as sediment input falls to zero. The erosion is limited in time as well as space. It begins on the outer shelf and expands landward until much of the shelf surface may be affected. The landward edge of the zone of erosion then shifts back toward the shelf edge as deposition of the offlap wedge begins. Thus erosion on the outer shelf may last longer than erosion on the inner shelf by one or more orders of magnitude. The entire palaeontological record of the hiatal period is commonly retained in the condensed section; much of it on the flooding surface itself. Thus biozones tend to begin and end on condensed sections (Baum & Vail, 1988). Galloway (1989a) stresses the utility of the maximum flooding surface as a cycle boundary for this reason.

In detail, the history of condensed sedimentation may be complicated. For example, high-resolution seismic profiles of the Amazon continental shelf (Alexander et al., 1986) document condensed intervals, overlain by Holocene muds, followed by renewed condensed sedimentation.

The evolution of a modern surface of non-deposition has been examined on the Spanish

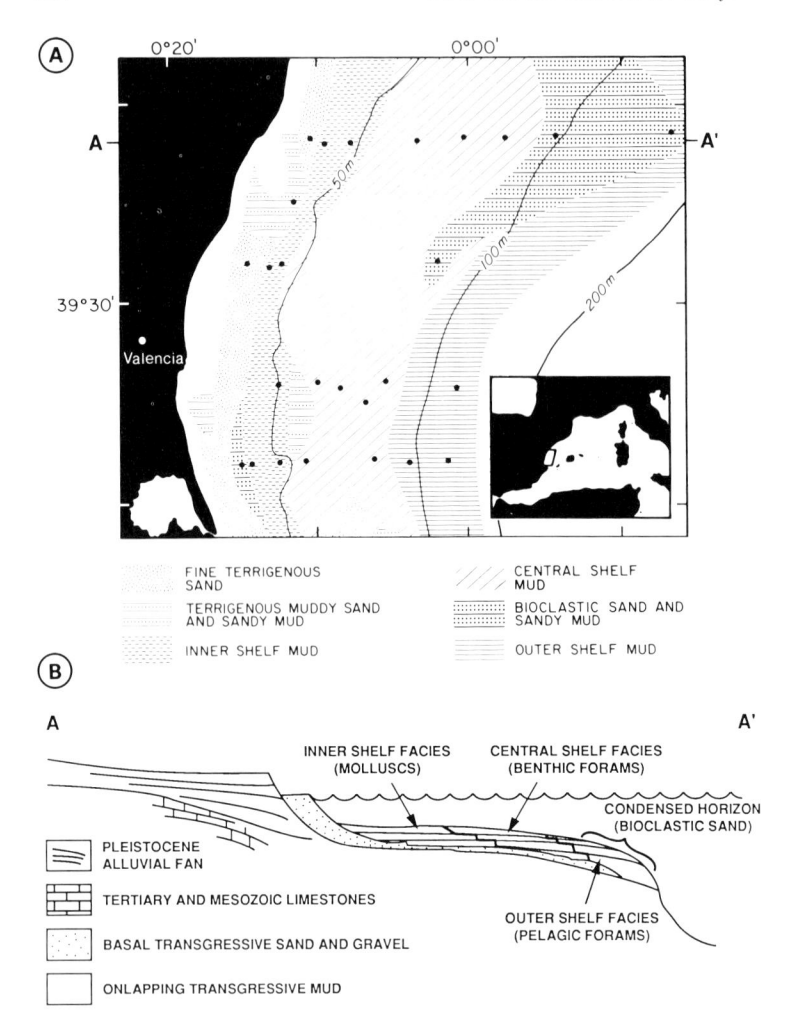

**Fig. 13.** Transgressive facies tract and maximum flooding surface on the Spanish Mediterranean shelf. (A) Facies distribution; (B) cross-section. (After Maldonado et al., 1983.)

Mediterranean shelf by Maldonado et al. (1983). A nearshore mud facies with a primarily molluscan fauna begins seaward of the sandy shoreface. It gives way seaward to a mud facies with a Foraminifera-rich, more pelagic fauna and lower sedimentation rates (Fig. 13). On the outer shelf, mud is mantled by a thin skin of skeletal sand. The sand defines an area of non-deposition. It corresponds to a *firm ground*, otherwise described from the rock record as a *burrowed omission surface* by Bromley (1975), and by Kennedy & Garrison (1975).

The surface of erosion on the outer Spanish shelf constitutes the distal extent of condensed sedimentation. With continued Holocene transgression, the shoreface profile moves landward, and time-transgressive condensed sedimentation can be expected to propagate shoreward. Time surfaces in the transgressive shoreface–shelf system, therefore, onlap the shelf in a landward direction and merge asymptotically with the flooding surface in a basinward direction (Fig. 13).

Beard et al. (1982), Weimer (1989), and Pacht et al. (1990), among others, have used continuous reflectors arising from condensed sections as the basis for sequence stratigraphic correlations of slope and deep-basin Quaternary sediments of the northwest Gulf of Mexico. Coleman & Roberts (1988b) found that continuous reflectors, on the outer continental shelf, are also associated with condensed sedimentation characterized by shell beds and massive clay deposits. It is important,

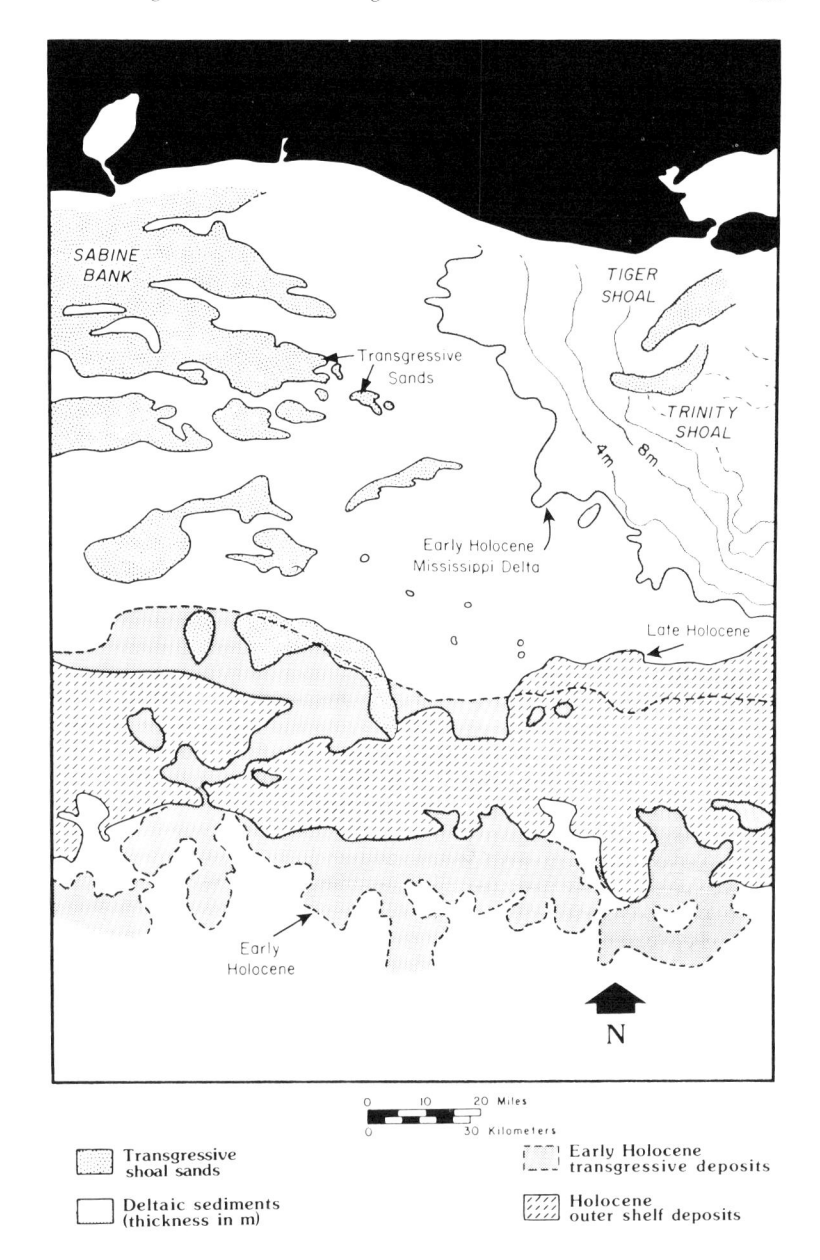

**Fig. 14.** Holocene deposits of the Western Louisiana Shelf. (From Suter *et al.*, 1987.)

therefore, to understand the regional variations of facies such as shell beds and clay deposits within the condensed interval on the northwestern Gulf shelf.

The study of Frazier (1974) describes the Louisiana shelf seaward of the Mississippi delta as a 'hiatal surface' in which coarse-grained, sandy, shelly deposits are seen as 'relict shoreline sands', now exposed on the sea-floor or thinly veneered with mud. This characterization is essentially correct, though Frazier's 1974 map (his fig. 17) probably exaggerates the proportion of the Louisiana–Texas shelf-floor that is 'relict' (i.e. consists of proximal amalgamated sand facies; see discussion of 'relict' sediments by Swift *et al.* (this volume, pp. 89–152)). More recent evidence indicates that muds are actively accumulating on much of the Texas shelf (Holmes, 1982) and that some of the sand shoals west of the Mississippi delta are not early trans-

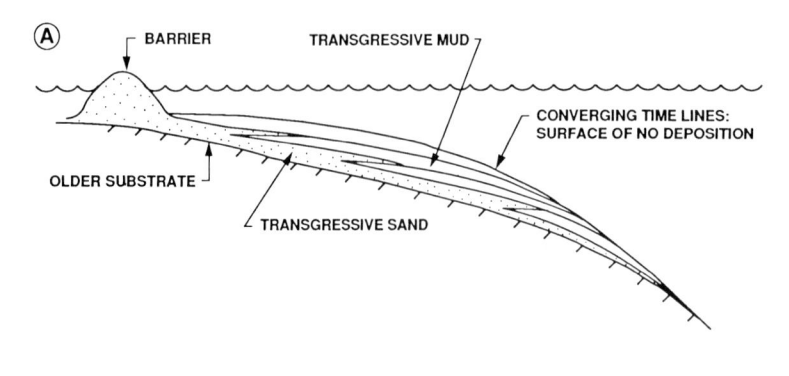

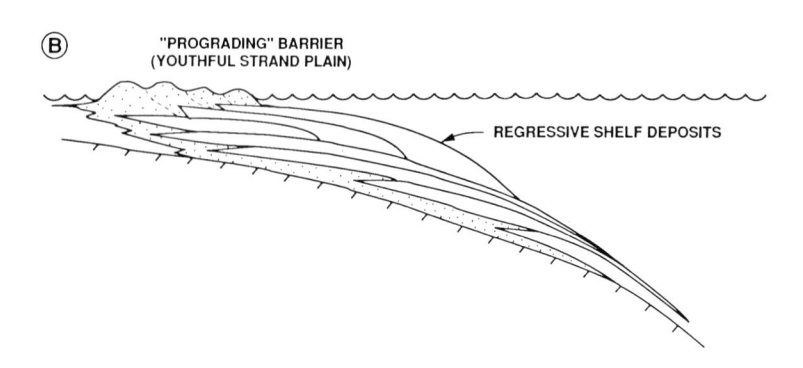

**Fig. 15.** Interpretations of the facies architecture of the central Texas shelf mud sheet. (A) As a transgressive deposit; (B) as a regressive deposit.

gressive deposits, but instead are recently transgressed deltaic lobes (Suter *et al.*, 1987; Tiger and Trinity shoals in Fig. 14).

Roberts & Coleman (1988) have collected abundant cores on the Louisiana upper slope and outer shelf, and are able to show that the diagnostic characteristics of a condensed section are present. The sediment is an unstratified, highly burrowed and pelleted mud. It is characterized by a high carbonate content (owing both to a Foraminifera-rich skeletal fraction and to diagenetic carbonate), cementing burrows, microfossil tests and diagenetic pyrite.

The Louisiana shelf has been swept by the transgressing surf zone, and mantled in the debris of erosional shoreface retreat (proximal amalgamated sand facies; Swift *et al.*, this volume, pp. 89–152). Transgressive deposits are particularly thick in the Sabine Bank area (Suter *et al.*, 1987; Fig. 14). Back-barrier Holocene and Pleistocene deposits locally outcrop on the shelf floor between Sabine Bank and the Trinity shoal region (white area, Fig. 14). However, most cores show that they are overlain by at least a decimetre and usually several metres of mud or muddy sand, frequently grading down into a basal shell lag (Roberts & Coleman, 1988). Two lithofacies of the transgressive shoreface–shelf depositional system are present in this deposit: (1) a basal shelly sand left as a debris sheet by the process of erosional shoreface retreat (proximal amalgamated sand lithofacies; see definition by Swift *et al.* (this volume, pp. 89–152)); and (2) an overlying distal bioturbated mud facies. Where the transgressive system is thin, the mud appears primarily as a muddy fraction within the upper part of the basal sand, into which it has been reworked by burrowing organisms. Such a mixed transgressive facies has been reported from the adjacent Gulf shelf (Curray, 1960; Swift *et al.*, 1971), from the USA Atlantic continental shelf (Figueiredo *et al.*, 1980), from the East China Sea (Nittrouer *et al.*, 1984), and from the North Sea (McCave, 1973). While thin, this transgressive deposit nevertheless separates the ravinement surface and the maximum flooding surface.

The stratigraphic architecture of the central Texas

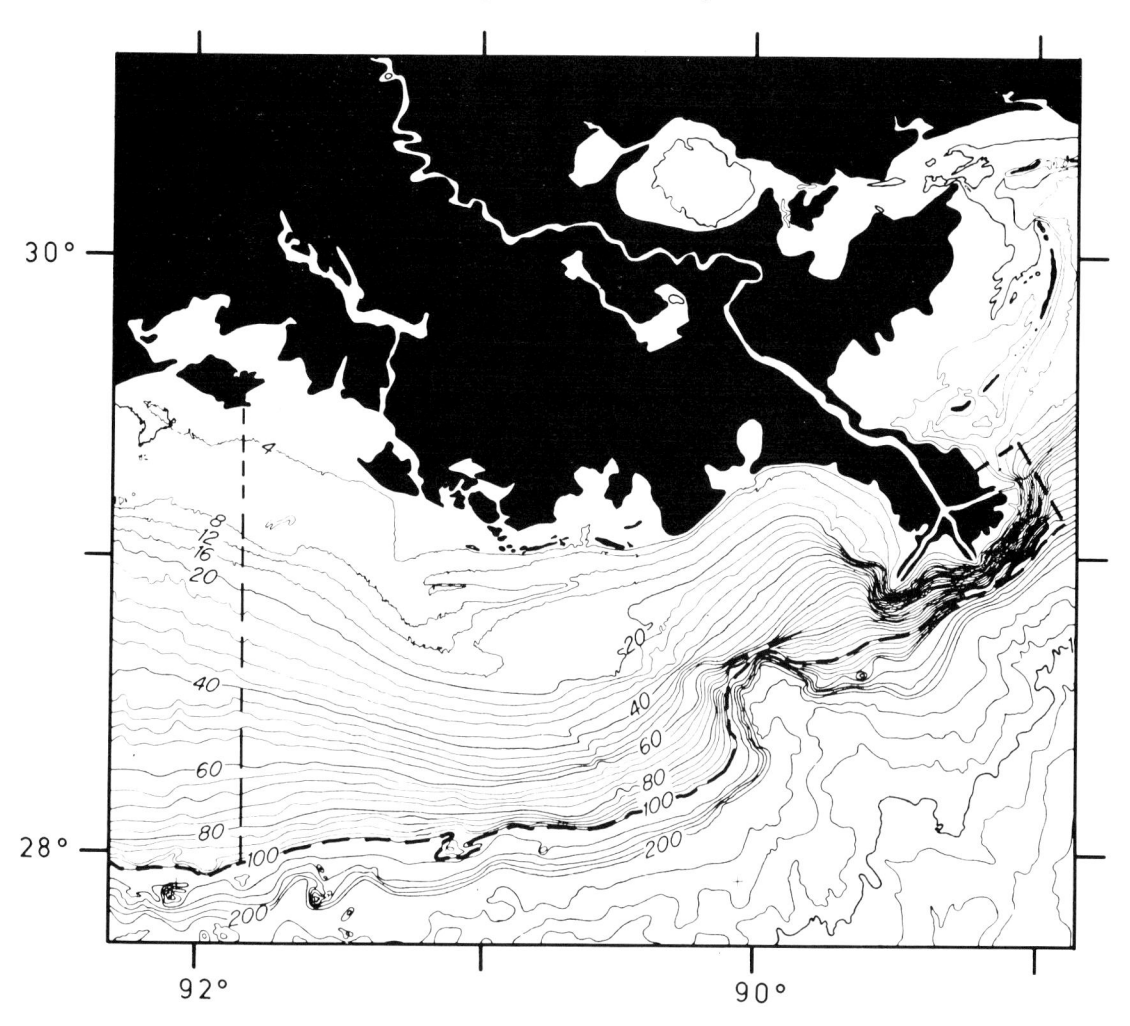

**Fig. 16.** Physiography of the Mississippi delta. The delta has prograded to the shelf edge. As a shelf-edge delta, the Mississippi delta has a much higher 'delta front' than the typical shoreface (c. 5–10 m) of a platform delta.

shelf is not as well understood as the Louisiana shelf. Two alternative interpretations of the central Texas shelf mud sheet are possible depending on whether or not its upper beds are prograding (Fig. 15). If the mud sheet is transgressive, then it intertongues with onlapping storm beds that extend into it from the basal transgressive sand (Snedden & Nummedal, this volume), while if it is regressive, it intertongues with downlapping storm beds that extend into it from the upper, regressive littoral sand. In the latter case, these beds would downlap on to a maximum flooding surface.

In Swift *et al.* (this volume, pp. 89–152) the distal laminated mud lithofacies of the Louisiana–Texas shelf is described as a bottom mud plume (deceleration sheet) trending westward along the Louisiana–Texas shelf from its source at the Mississippi delta. They suggest that the bottom plume is regressive as far west as the Louisiana–Texas border, but is transgressive from there to Mexico. Thus Fig. 15B may apply to the Louisiana shelf, while Fig. 15A would apply to the Texas shelf. The nearfield plume would, thus, be 'folded over on itself' as a back-step–offlap couplet, while the farfield plume would be a simple back-step wedge, capped by a maximum flooding surface. Holmes (1982) reports that accumulation rates, as revealed by $^{210}$Pb activity profiles, fall to zero on the southern Texas shelf.

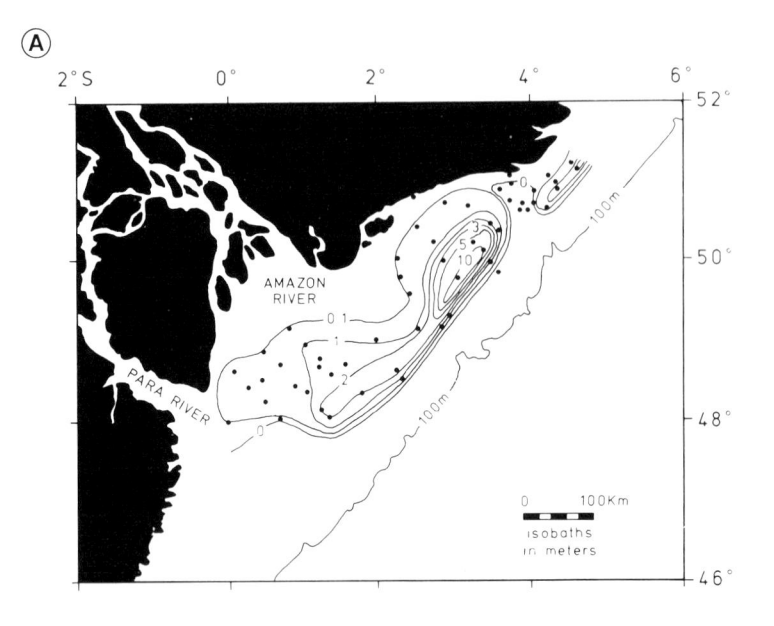

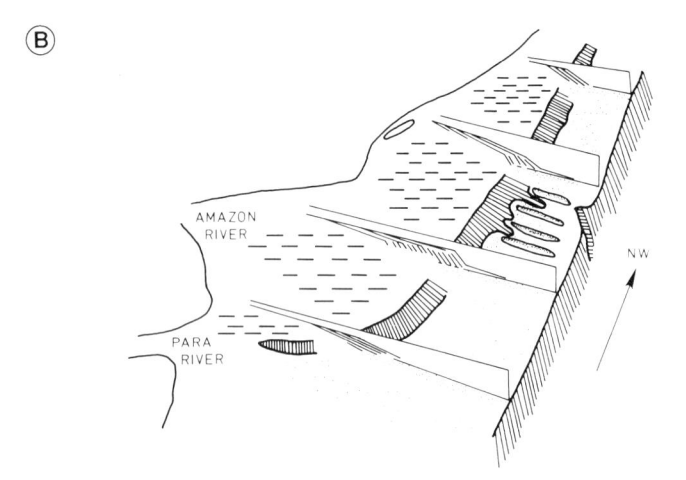

Fig. 17. Highstand sedimentation on the Amazon shelf. (A) Accumulation rates; (B) Cartoon of facies architecture, showing prograding scarp. (From Kuehl et al., 1982.)

*Prograding scarps.* The *depositional shoreline break* plays an important role in the Vail model where it is defined as 'the position on the shelf landward of which the depositional surface is at or near base level, usually sea-level, and seaward of which the depositional surface is below sea-level' (van Wagoner et al., 1988). The depositional shoreline break is described as coinciding 'with the seaward end of the stream mouth bar in a delta, or the upper shoreface in a beach environment'. The depositional shoreline break thus appears on seismic records as the base of a series of updip (shoreface) clinoform beds.

A possible alternative origin for this seismically observed change in bedding inclination may be as deeper water prograding scarps of the offlap wedge shelf systems. As highstand shelf systems mature, they tend to develop depositional scarps (see discussion of depositional scarps by Swift et al. (this volume, pp. 89–152)). One class of depositional scarp is associated with offlap wedge deltas, such as the Mississippi delta (Fig. 16). As highstand deltas prograde to the shelf edge, they shift from platform delta configurations to shelf-edge delta configurations. Platform deltas have shallow (6–15 m) shoreface profiles (Galloway & Hobday, 1983). Shelf-edge deltas build marine terrace-scarps with 60 m or more of relief as platforms for their progradation. Their profiles are therefore sigmoid, with shallow, concave shorefaces, and convex terrace-scarp systems (Fig. 17).

Constructional terrace-scarp complexes may form early in the offlap sequence. When this occurs, their development is not linked as closely to delta development. As the surface of the main deposit builds up to the wave-determined equilibrium depth (see discussion by Swift & Thorne (this volume, pp. 3–31)), the leading edge pushes seaward into progressively deeper shelf water, and develops a seaward-inclined face. Studies of depositional rates on the prograding mud sheet of the Amazon shelf (Kuehl *et al.*, 1982) show that as the forward face of the deposit moves seaward, its angle of inclination increases. Thus, a feedback transpires: the incipient prograding scarp becomes an increasingly efficient surface of deposition relative to the surface of bypassing behind it. It captures an increasingly larger share of regional deposition, so that its height and inclination increase yet further (Fig. 17). As a result, prograding shelf mud sheets tend to develop a constructional terrace-scarp morphology and a clinoform internal structure. Examples associated with major rivers can be seen on the Amazon, east China, and Bengal shelves (Nittrouer *et al.*, 1986; Fig. 16), and in the upper Adriatic Sea (Van Straaten, 1965).

*Offlap fill deposits.* Secondary offshore depocentres caused by topographic fill sedimentation occur during offlap wedge deposition, as well as back-step wedge deposition. A particularly well-documented example occurs on the continental margin of the northwestern Aegean Sea (Lykousis & Chronis, 1989). In this area the primary depocentre is an offlapping Holocene delta showing an oblique-sigmoid progradational pattern. A secondary depocentre occurs on the continental slope as a thin (1–2 ms) draping fine-grained deposit. Separating the two depocentres is an area of relict Pleistocene sediment on the outer shelf. Lykousis & Chronis (1989) estimate that only minor amounts (10–20%) of river suspended load reaches the secondary offshore depocentre.

### Summary chart of sequence architecture elements

Our review of modern sedimentation largely confirms the descriptive aspects of the sequence architecture model of Vail (1987). The Galloway (1989a) model, although giving a less detailed picture of architectural sequence elements, is more realistic in its recognition that other regime variables besides relative sea-level change can control sequence architecture.

The model of Vail (1987), however, may need to be modified to include several additional architectural elements that can be present under certain regime conditions. The additional elements recognized in this study are: (1) a time-transgressive coastal plain unconformity formed by fluvial entrenchment during offlap wedge progradation, which is not of the same origin as the type 1 sequence boundary, coastal-plain—continental-slope, unconformity of Vail (1987); (2) a significant marine erosion surface formed on the shelf during transgression; (3) the importance of ravinement in forming an erosional surface diastemic time value within the transgressive systems tract; (4) a wedge of back-barrier sediments between the transgressive surface and the ravinement surface; (5) a wedge that onlaps the relict slope and downlaps the deep basin formed during transgression; and (6) the formation of extensive coastal-plain—continental-slope unconformities that mimic sequence boundaries at times other than the maximum rate of fall of eustatic sea-level. Several of these modifications have been suggested by Galloway (1989a).

Table 2 provides a summary chart of the various surfaces, and geometric systems tracts and subtracts, that make up a complete sequence stratigraphic description of clastic sediments. With the addition of the three types of fill deposits (reduction in downslope gradient, ponding and draping), the architectural elements in this chart provide a systematic classification for interpreting the geometric relationships in the stratigraphic record.

## SEQUENCE MODELS VERSUS OBSERVATIONS ON MODERN SHELVES: CONTROLLING SEDIMENTATION PROCESSES

### General

Three columns are provided in Table 2 to indicate the controlling regime variables for each architectural element as suggested by the geological models of Vail (1987), Galloway (1989a) and this paper. The relationships suggested by these three

**Table 2.** Summary chart of sequence stratigraphic elements[a]

| Architectural element | Regime variables (Vail, 1987) | Regime variables (Galloway, 1989a) | Regime variables (this paper) |
|---|---|---|---|
| *Basin fan (fill)* | | | |
| Basin fan (canyon cut) onlap surface (Vailian type 1 sequence boundary) | $\dot{R} = 0, R < 0$ | $D \gg Q$ | $\dot{¥} \gg 0$ |
| Fluvial entrenchment surface (Vailian type 2 sequence boundary) | — | $Q > R$ | $\dot{¥} < 0, ¥ < 1$ |
| Fan showing bidirectional downlap on bottomsets of the deep basin (basin-floor fan) | $R < 0, SL < 0$ | — | $\dot{¥} \gg 0$ |
| Fan onlaps slope and downlaps deep basin (slope fan) | $R > 0, SL < 0$ | $Q, M, R, D$ | $¥ < 1, |\partial ¥/\partial y| \gg 0$ |
| Fan or wedge onlaps sequence boundary and downlaps slope or basin-floor fan (prograding complex) | $R > 0, SL < 0$ | — | $\dot{¥} \ll 0, |\partial ¥/\partial y| \gg 0$ |
| *Backstep wedge (fill)* | | | |
| Transgressive surface | $R > 0, SL \cong 0$ | $Q < R$ | $\dot{¥} > 0, ¥ = 1$ |
| Ravinement surface | — | — | $¥ > 1$ |
| Marine erosion surface | — | $R > 0$ | $\dot{¥} > 0, ¥ > 1, D \gg 0$ Low rate of down-to-basin tectonic tilt |
| Back-barrier wedge (fill) | — | — | $¥$ high, $¥ > 1$ |
| Shelf wedge (fill) | $R > 0, SL \cong 0$ | $Q < R$ | $¥ > 1$ |
| Offshore back-step (fill) | — | $R > 0$ | $¥ > 1, D \gg 0$ |
| *Offlap wedge (fill)* | | | |
| Maximum flooding surface | $\dot{R} = 0, R > 0$ | $\dot{R} = 0, R > 0$ | $\dot{¥} < 0, ¥ = 1$ |
| Wedge downlaps on maximum flooding surface | $SL > 0$ | $Q > R$ | $¥ < 1, |\partial ¥/\partial y| \cong 0$ |
| Wedge onlaps coastal plain unconformity and downlaps relict slope (shelf-margin systems tract) | $R > 0, SL < 0$ | — | $\dot{¥} \ll 0, |\partial ¥/\partial y| \cong 0$ |

[a] $¥ \equiv D'R'/Q'M'$; a dot indicates derivative with respect to time; $|\partial ¥/\partial y|$ is the absolute magnitude of the change of $¥$ in the strike direction, $y$. $SL > 0$, sea-level above the depositional shoreline break; $SL < 0$, sea-level below the depositional shoreline break; $Q$ = clastic sediment supply rate to marginal marine or marine environment; $M$ = sediment grain size; $R$ = rate of relative sea-level change; and $D$ = rate of marine sediment transport.

models are largely complementary and represent a progressive evolution of ideas from the eustatic ($R$-dependent) model of Vail (1987) and Posamentier et al. (1988), to the eustatic plus sediment supply model of Galloway (1989a), to the more quantitative regime model of this paper.

In the following sections we first review several theoretical suggestions made by Posamentier et al. (1988) in an attempt to explain sequence geometries of the Vail (1987) model. This review indicates that the proposed theory is inconsistent with important observations of modern processes. To address this problem we then present a new, theoretical model for the control of regime variables on each of the architectural elements in Table 2. This model is, in many respects, highly conjectural and oversimplified. We hope that it will serve as a basis for hypothesis and interpretation, much in the same valuable way that the Posamentier et al. (1988) model has. To illustrate the predictive power of the proposed theory, we provide examples of sequence development controlled primarily by changes in sediment supply, versus examples controlled primarily by eustatic sea-level change.

**Evaluation of the Posamentier *et al.* (1988) *R*-dependent regime model**

*Timing of fluvial sedimentation (Posamentier et al., 1988)*

In the Vail (1987) model, episodes of systems tract formation are used as chronometers for the global correlation. Sequence-bounding unconformities are seen as responses to eustatic sea-level fall, and considerable ingenuity has been expended in demonstrating synchronous mechanisms for the timing of the subaerial erosion. The end of fluvial sedimentation and the beginning of subaerial erosion is seen as a critical event, since it creates the 'downward shift in coastal onlap' that is central to the model (Vail *et al.*, 1977,1984; Posamentier & Vail, 1988, p. 129).

In the Posamentier *et al.* (1988) model, coastal evolution during highstand and sea-level fall is seen as a response of two significant positions in the coastal profile, the *bayline* and the shoreline, to movements of an equilibrium point. The bayline was defined by Lowman (1949) as the line landward of the shoreline, connecting heads of estuaries and bays. The significance of the equilibrium point was originally analysed by Pitman (1978) and Pitman & Golovchenko (1983), who pointed out that when eustatic sea-level falls, the hinged, seaward-increasing subsidence characteristic of divergent margins must limit the seaward excursion of the shoreline. The shoreline can only migrate out to the point where the rate of sea-level fall is equal to the rate of subsidence. Thus the equilibrium point divides the continental margin profile into offshore and onshore sections with contrasting directions of relative sea-level change.

In the analysis of Posamentier & Vail (1988, fig. 16), coastal progradation begins as eustatic sea-level passes the rising sea-level inflection point. Posamentier & Vail (1988) attribute the progradation of the shoreline to a 'sufficient sediment supply... brought in by longshore drift...'. In this model, the bayline continues to retrograde as eustatic sea-level continues to rise. The bayline and shoreline thus diverge, so that the back-barrier lagoon expands. Rivers continue to be graded to their mouths at the retreating bayline. As the sea creeps up the river thalweg, the subaerial remnant of the river continues to experience grade (see Fig. 1 of Thorne & Swift (this volume, pp. 33–58)) so that no fluvial deposition can occur.

However, when the eustatic fall begins, the equilibrium point begins to move seaward. When it reaches the bayline, some time after eustatic fall has begun, the sense of bayline migration reverses, and the bayline and equilibrium point move seaward together (*relative* sea-level fall now begins). At this point, fluvial deposition is initiated. The rivers are graded to their mouths, and the mouths are shifting seaward. The profiles must also shift seaward, so that at a given position on the river profile, the equilibrium surface is rising and accommodation is being created (Posamentier & Vail, 1988, fig. 11).

When the falling sea-level inflection point is reached on the eustatic sea-level curve, the equilibrium point begins to migrate landward once more. In the Vail (1987) model, during transgression, rivers continue to experience grade (see above), so that fluvial sedimentation abruptly ceases. Because the Vail (1987) model stresses 'sequence bounding unconformities' the subaerial landscape is said to undergo a 'general wearing down' and 'a slow widespread degradation' as the transgression progresses (Posamentier & Vail, 1988).

These dynamics are illustrated in Fig. 18, in terms of a coastal onlap curve and other graphic conventions of the Vail (1987) model. In case A, the equilibrium point moves out on to the flooded region of the subsiding shelf, forcing the inner portion to emerge and undergo fluvial sedimentation, before submerging again. In case B, however, sea-level fall has been to slow, or subsidence is too high, for such an emergence to occur. No fluvial deposits are formed. The equilibrium point moves seaward of the hinge, but does not reach the bayline before the F inflection point (falling sea-level inflection point) is reached and the equilibrium point driven back. The transgression is slowed but not reversed. No subaerial unconformity is cut, and no downward shift in coastal onlap occurs.

*Timing of fluvial sedimentation: fluvial morphodynamics*

It is difficult to reconcile the Posamentier *et al.* (1988) scheme with fluvial morphodynamics (e.g. Johnson, 1931). A first problem concerns regrading

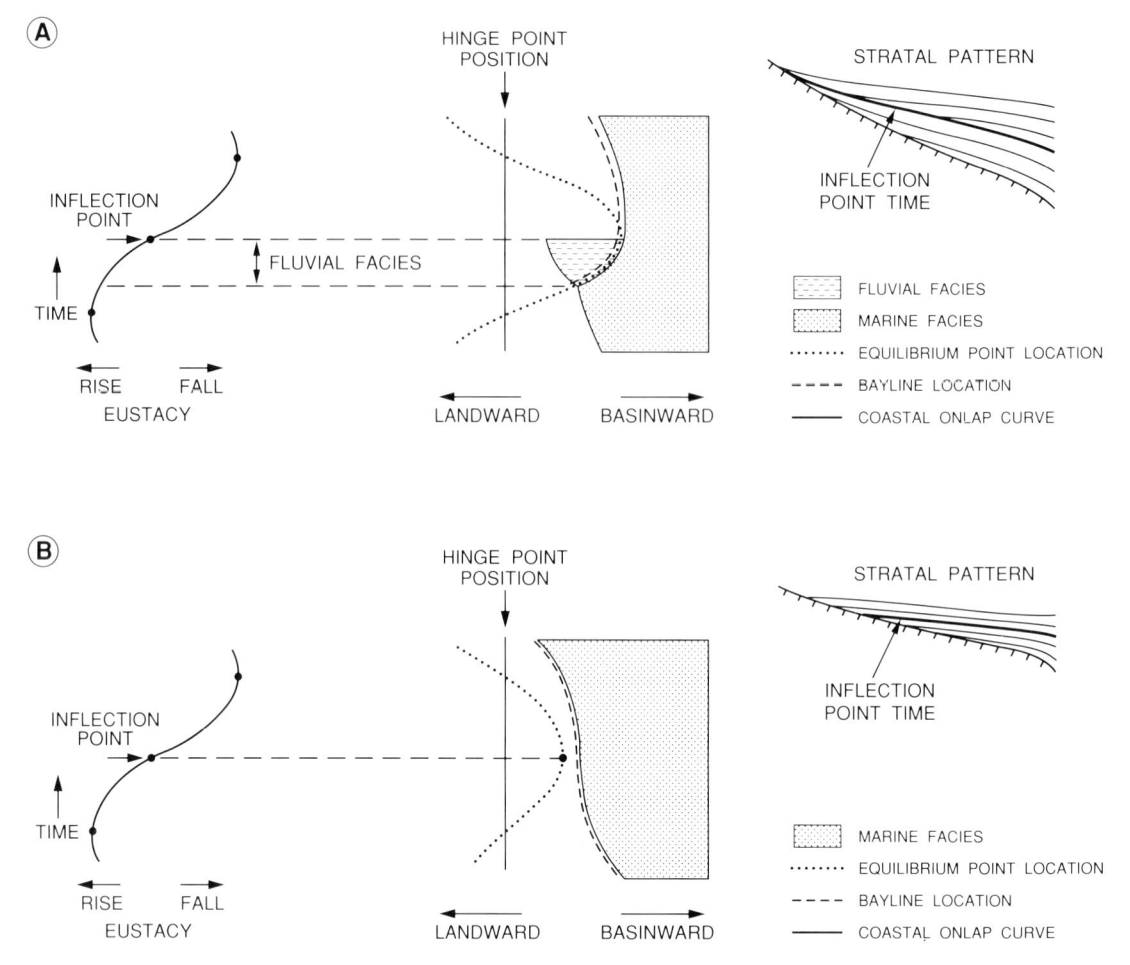

**Fig. 18.** Kinematics of a type 2 sea-level fall on a passive margin. (A) The case where eustatic sea-level fall is rapid relative to subsidence. (B) The case where eustatic sea-level fall is slow relative to subsidence.

of the fluvial profile as the shoreline migrates seawards. The Vail (1987) model assumes fixed river profiles during sea-level rise (Posamentier & Vail, 1988), but this assumption is valid only in the special case of 'elevator tectonics', where the rate of subsidence does not vary with distance across the line of section. On a hinging passive margin (Fig. 1A of Swift & Thorne, this volume, pp. 3–31), Gilbertian grade, with neither erosion nor deposition, cannot be maintained by rivers seaward of the hinge. Instead, a more generalized Davisian grade must prevail (see discussion of grade by Thorne & Swift, this volume, pp. 33–58), with thalwegs aggrading constantly to maintain their profiles against subsidence.

A second problem in fluvial morphodynamics concerns the rise of the graded profile during sea-level fall, resulting in subaerial accommodation. As relative sea-level falls, Posamentier & Vail (1988, fig. 11) show a curve-invariant seaward shift of the equilibrium river profile: the upstream end of the profile translates seaward with the downstream end. In reality the upstream end of a river profile must be fixed by its source, the drainage divide at its headwaters. Though drainage divides do move with changes in the geomorphological landscape, they do not move in a simple seaward fashion in response to base level fall. This problem is discussed in more detail in a subsequent section (formation of a fluvial entrenchment surface).

### Timing of fluvial sedimentation: *Q*-dependency of modern highstand coasts

A more serious problem concerns the relative movements of the bayline and shoreline as visualized in the Posamentier *et al.* (1988) model. Most of the world's coasts have experienced a reduction in the rate of sea-level rise between 4000 and 7000 thousand years ago (Milliman & Emery, 1968). It is not possible in practice to distinguish the eustatic and tectonic components of the recent rise, and it is perhaps not possible in principle. However, it is reasonable to assume that on most coasts, the last 4000 years corresponds to the beginning of highstand time in the Vail (1987) scenario, but in no case is shoreline progradation occurring at the same time as bayline retreat.

The reason for this is that before progradation can begin, supply of littoral drift must be so abundant that surf zone deposition exceeds losses to the offshore (see Fig. 20 of Swift *et al.*, this volume, pp. 89–152). Local erosion of headlands can supply enough sand to nourish a retreating barrier so that it can retain its integrity while retreating, but it cannot reverse the sense of shoreline movement from retreat to progradation, because the source would then be left behind. Coastal progradation occurs *only* when accommodation in back-barrier estuaries ceases to be adequate (see discussion by Swift & Thorne (this volume, pp. 3–31)). The estuaries then invert as bayhead deltas (Mobile Bay, Alabama coast; Schroeder, 1977), and the trunk streams cut through the barriers to form oceanic deltas (Brazos Delta, Texas coast, Morton, 1977; Costa de Nayarit, Curray *et al.*, 1969; Fig. 3 of Swift *et al.*, this volume, pp. 153–187).

When the back-barrier zone is filled, sand is released in sufficient quantities to nourish shoreline progradation, but the point to which the rivers are graded is no longer the bayline, as required by the Vail (1987) model. Rivers now extend through the back-barrier zone to the oceanic shoreline, and maintain well-formed channels as they do so. Thalwegs continue to deepen until the river mouth bar is reached (Wright & Coleman, 1974). The thalweg is in fact *below* sea-level for some distance before the mouth is reached, but the river flow is largely insulated from the oceanic pressure field. During flood, the water is driven through the channel to the mouth by the pressure gradient associated with the slope of the river surface (energy grade line). The river surface at flood lies above sea-level and slopes smoothly down to sea-level across the strandplain to the river mouth, as does its levees. It is to this oceanic mouth that the profile of grade must be attached, not to the bayline. These relationships can be seen clearly in the physiography of the Nayarit coast, as portrayed in the classic study of Curray *et al.* 1969 (Fig. 8 of Swift *et al.*, this volume, pp. 153–187).

The nature of bayline advance has been documented for strandplains by Curray *et al.* (1969), and for deltas by Tye & Kosters (1986). Fossil lagoons can be seen along the landward margins of most modern strand and chenier plains. They were captured when the barriers locked into position at maximum transgression, and began to prograde seaward (Fig. 8 of Swift *et al.*, this volume, pp. 153–187). In some cases they may have expanded somewhat in response to continued sea-level rise. In most cases, they are being filled by mangrove swamps or salt marshes, which trap alluvial silt and clay during high water. As sea-level rise continues, the accreting, seaward-moving, leading edge of the strandplain rises with it. The rear of the strandplain undergoes relative subsidence, and is buried by lateral meandering, avulsion and crevassing, and by the general seaward advance of the fluvial depositional system over it (Curray *et al.*, 1969; Fig. 8 of Swift *et al.*, this volume, pp. 153–187).

The dynamics of a single advancing delta, such as the Mississippi delta, is similar to that of a strandplain complex, such as the Narayit coast, except that the geomorphological plan is radial rather than rectilinear in arrangement (Coleman & Prior, 1982; Tye *et al.*, 1989; Fig. 9 of Swift *et al.*, this volume, pp. 153–187). Successive subdeltas prograde radially seaward from an avulsion point in contrast to the parallel progradation along a straight front that is a strandplain.

Thus, there seems to be no reason to assume that the position of the fluvial plain front in prograding strandplain and delta provinces is controlled only by eustatic sea-level rise. To do so is to argue for an *R*-dependent coastal equilibrium model for the offlap wedge, when a *Q*-dependent model is called for (see discussion of coastal equilibria by Thorne & Swift (this volume, pp. 33–58)). When Pitman & Golovchenko (1983) defined the equilibrium point, they made it clear that in the rising sea-level case, subsidence is offset by sedimentation as well as by sea-level fall. The shoreline corresponds to an equilibrium point where the rate of subsidence *plus* the sedimentation rate is equal to the rate of sea-level rise. The stage of highstand at which fluvial and littoral marine facies prograde is, thus, determined not simply by the relationship of eustacy and subsidence, but by the regime relationship as a whole. That is, progradation begins when supply factors *as a whole* exceed accommodation factors; $Q \cdot M > R \cdot D$ (see preceding discussion of the regime relationship).

The problem of $Q$-dependency affects the timing of the termination of fluvial sedimentation as well as its beginning. Posamentier & Vail (1988) note that if the fluvial sedimentation rate is low, the streams may still be catching up with rise of their equilibrium profiles after the migration of an $R$-dependent equilibrium point has reversed and is pushing the bayline landward. It is more likely, however, that fluvial sedimentation is sufficient to define a $Q$-dependent equilibrium point, as described by Pitman & Golovchenko (1983), which reverses and migrates landward significantly later than would an $R$-dependent equilibrium point; that is, significantly later than the falling sea-level inflection point in the eustatic curve. In terms of the regime relationship, the equilibrium point would move landward and the shoreline retrograde only when accommodation factors *as a whole* exceed supply factors; $Q \cdot M > R \cdot D$.

**Regime topographic profiles: control by accommodation/supply ratio**

*Regime topographic profiles*

Thorne & Swift (this volume, pp. 33–58) have illustrated the role of regime topographic profiles in controlling the geometry of sedimentary deposits. A simple numerical model of sequence stratigraphic evolution was developed based on an $R$-dependent coastal equilibrium profile. While the shape of this equilibrium profile did not change with time, the location of the profile shifted with the position of the shoreline in response to changes in relative sea-level, $R$.

Their simple model was able to reproduce several salient features of sequence architecture. In this section, however, we develop the theory of regime topographic profiles a step further in order to investigate the control of all four regime variables, $Q$, $M$, $R$ and $D$, on sequence stratigraphic patterns.

A typical topographic profile of the non-marine and shelf sections of a continental margin can be divided into four segments; shelf, shoreface, coastal plain and fluvial plain. In its fundamental meaning, the term *coastal plain* refers to a subaerial coastal surface underlain by gently seaward-dipping strata (Bates & Jackson, 1987). In this sense, a coastal plain extends from shoreline to the fall line. Here the strata wedge out with (often) angular unconformity over older, more indurated rooks, and waterfalls occur. We have used the term *coastal plain* in this broad sense earlier in the paper, when we referred to a coastal-plain–continental-slope unconformity. In this section, however, we are concerned with distinguishing among coastal provinces with differing gradients, formed by differing processes (Fig. 19). For the purpose of the following discussion, however, we restrict the term *coastal plain* to the relatively flat portion of coastal surface nearer the shoreline, and will refer to the steeper marginward portion as the *fluvial plain*.

Each of the sections of the coastal surface are characterized by different gradients (Fig. 19). The shelf and coastal plain segments have relatively low gradients, while the shoreface and fluvial plain segments have relatively high gradients. The slope of the topographic gradient on each segment is controlled by the regime variables, $Q$, $M$, $R$ and $D$, but, as we shall see, the regime control on each segment is of differing origin.

*Regime response of the fluvial plain*

The subject of equilibrium profiles developed by fluvial systems has been dealt with extensively by geomorphologists, hydraulic engineers, and fluvial

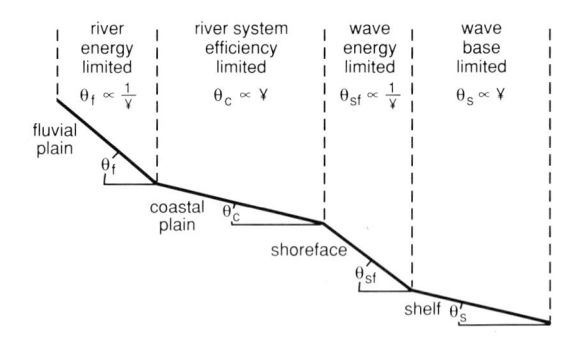

Fig. 19. Typical topographic profile of the non-marine and shelf sections of a continental margin divided into four segments with distinct topographic gradients. The control on each segment is either proportional to or inversely proportional to the accommodation/supply ratio, ¥.

sedimentologists (see Swift & Thorne, this volume, pp. 3–31). Although it is beyond the scope of this paper to give a thorough account of these studies, the general sense of the effect of each regime variable on river slope, $\theta_f$ can be based on the rather simple proportionality expression of Chorley et al. (1984) given by:

$$\text{River gradient} \propto \frac{\text{sediment load} \cdot \text{mean grain size}}{\text{mean water discharge}} \quad (7)$$

This proportionality expression is meant to give the sense of change, but not the magnitude of change, in river gradient caused by increasing or decreasing the other factors. Using regime notation, the sediment load is given by $Q$, and the mean grain size is given by $M$. In alluvial rivers the bedload sediment transport rate increases proportionally with mean water discharge while suspended transport increases as a power function of discharge (e.g. Leopold et al., 1964, fig. 7-11). Thus, we can rewrite equation (7) in regime notation as:

$$\theta_f \propto \frac{Q' \cdot M'}{D'} \quad (8)$$

where again we use primes to indicate modified functions of the original variables.

The response of river gradients to changes in relative sea-level follows from the regime concept of Davisian grade (see Thorne & Swift, this volume, pp. 33–58). High rates of relative sea-level fall ($R'$ low), whether caused by eustatic sea-level fall or by tectonic uplift, lead to landscape rejuvenation and steep rivers. Relative sea-level rise ($R'$ high) leads to landscape peneplanation and low gradient rivers. Thus $\theta_f$ is inversely proportional to regime variable $R'$. The full expression for $\theta_f$ is given by:

$$\theta_f \propto \frac{Q' \cdot M'}{D' \cdot R'} = ¥ \quad (9)$$

*Regime response of the coastal plain*

Topographic profiles of the coastal plain are controlled by the response of the integrated river, flood plain and delta plain system to long-term changes in regime variables. Studies by hydraulic engineers of regime relationships in river channels do not consider the interrelationships between the river system, the flood plain and the delta plain. Thus, these studies, though useful in the previous section on fluvial channels, cannot be applied directly to the regime behaviour of the coastal plain.

An alternative approach to coastal plain regime is based on the idea of coastal plain system sediment transport efficiency. The following discussion of efficiency is quite abbreviated and we have not given specific literature references to pertinent studies. The interested reader is referred to texts by Leopold et al. (1964), Chorley et al. (1984) and Dingman (1984).

All coastal plain systems transport sediment through the action of water running downslope (we neglect wind in this formulation). The amount of water that is available to carry sediment is fixed by the hydrological cycle of precipitation and evaporation. If two coastal plain systems receive the same amount of available water, the one that transports more sediment is the more efficient.

Sediment transport efficiency in controlling coastal plain morphology varies directly with the accommodation to supply ratio. If this ratio is low (supply is larger than accommodation within the coastal plain), an efficient coastal plain system will maintain itself by transporting large amounts of sediment out of the drainage basin into marine environments. If this ratio is high (supply lower than accommodation), on the other hand, an inefficient system, with relatively little sediment bypass to the marine environment, will operate instead. Algorithmically we can write this relationship as:

$$\text{Sediment transport efficiency} \propto \frac{D'R'}{Q'M'} = ¥ \quad (10)$$

The sediment transport efficiency of the coastal plain can be related to the gradient of its topographic regime profile. Large rivers, which collect water from large drainage areas, are inherently more efficient than small rivers. When water flows in small river channels much of its power to transport sediment is lost by frictional dissipation by the river banks and channel floor. Water flowing in larger rivers loses relatively less energy to frictional effects, and therefore reaches higher downstream velocities. The results of higher flow velocities are profound on sediment transport capacity: sediment is at rest for less time during the year, more of the load can be carried in suspension rather than in bedload, and both suspended load and bedload move downstream at a greater velocity.

Large trunk rivers have downstream gradients that are one to three orders of magnitude less than their small upstream tributaries (e.g. Leopold et al., 1964, fig. 7-23). The reasons for this relationship are complex. One important controlling process is a feedback between drainage basin size and drainage basin elevation above sea-level. Drainage basins at low elevations will tend to capture the drainage of topographically higher areas. To maintain low elevations throughout a drainage basin requires a low downstream gradient. Thus large rivers fed by large drainage basins must maintain their low downstream gradient or they will loose their water supply to other areas that are topographically lower. Algorithmically, the relationship between coastal plain sediment transport efficiency and downstream slope of the coastal plain can be written:

Coastal plain sediment transport efficiency $\propto \theta_c$ (11)

Substituting equation (9) into equation (8) results in:

$$\theta_c \propto \frac{D'R'}{Q'M'} = ¥ \quad (12)$$

*Regime response of the shoreface*

The control of the equilibrium topographic profile of the shoreface has been studied by many authors (see Thorne & Swift, this volume, p. 39, for references). Moore (1982) found that the steepness of shoreface profiles increased with sediment grain size. Kenyon & Turcotte (1985) studied equilibrium shoreface profiles of prograding deltas. They found that shoreface slope was linearly proportional to sediment supply, $Q$, inversely proportional to the sediment transport rate, $D$ (parameterized in their model by an offshore diffusion coefficient $K$), and inversely proportional to the elevation difference between delta topsets and delta bottomsets. The elevation difference between delta topsets and bottomsets is related to relative sea-level change. If sea-level is rising relatively quickly ($R'$ is high) than delta topsets will build up and out in order to maintain a shallow water depth. Delta bottomsets, however, will be sediment starved and, thus, as sea-level rises, the elevation difference between delta topsets and bottomsets will increase. The opposite case will occur for relatively quick sea-level falls ($R'$ low). Using these observed relationships between shoreface slope and regime variables from Moore (1982) and Kenyon & Turcotte (1985) we can write:

$$\theta_{sf} \propto \frac{Q' \cdot M'}{D' \cdot R'} = \frac{1}{¥} \quad (13)$$

*Regime response of the shelf surface*

The regime topographic profile of the shelf is essentially a graded profile that develops in response to the decay of marine transport energy with water depth (Johnson, 1919). In the absence of relative sea-level changes, grade is achieved at a position, $x$, along the topographic profile, when the sediment transport rate to and from that position is identical (Fig. 3 and accompanying text of Swift & Thorne (this volume, pp. 3–31)). In regime terms, if the clastic sediment supply rate to position $x$ is $Q$, and the grain size of the material is $M$, then the regime topographic profile adjusts so that the sediment transport rate, $D$, can totally bypass this material, providing that the rate of relative sea-level change, $R$, is 0.

Though the influence of currents dominates the net direction of sediment transport on the shelf, the severity of wave climate largely determines the net magnitude of offshore and onshore sediment transport (e.g. Vincent, 1986). The shear stress imparted to the sea-bed by wave orbital action is typically an order of magnitude or more greater than the shear stress due to currents (see Thorne et al., this volume, pp. 59–87).

Sea-bed, wave-orbital, velocities, and the consequent shear stresses that move sediments, decay rapidly with water depth. Hence the concept of a *wave base* (Gulliver, 1899), below which sediments are transported at greatly diminished rates.

For the regime sediment transport rate $D$ we can write:

$$D = f\left(\frac{1}{z(x)}\right) \quad (14)$$

where $f$ is a monotonically increasing function and $z(x)$ is water depth at position $x$.

To illustrate the concept of grade, for the sake of simplicity, assume that the function for $D$ is given by:

$$D = \frac{D_0}{z(x)} \quad (15)$$

where $D_0$ is a constant proportional to wave climate intensity.

It is also well known that fine sediments can be transported at significantly lower velocities than coarse sediments (e.g. Hjulstrom, 1939). The regime equation for grade is therefore:

$$Q = \frac{D_0}{z(x) \cdot M'} \quad (16)$$

The gradient of the shelf surface, $\theta_s$, is given by:

$$\theta_s = \tan^{-1}[z(x)/x] \cong z(x)/x \qquad (17)$$

Assuming, for simplicity, that the topographic profile $z(x)$ is linearly proportional to distance from shore, $x$, we have from equations (16) and (17):

$$\theta_s \propto \frac{D_0}{Q \cdot M'} \qquad (18)$$

Equation (18) was derived in the absence of relative sea-level changes. The gradients of modern shelf surfaces are highly influenced by the rapid Holocene eustatic sea-level rise. The depth of the shelf edge, under these conditions, may largely reflect the total rise in sea-level during the Holocene (Shepard, 1948). Modifying equation (18) to account for the effects of rapid changes in relative sea-level results in:

$$\theta_s \propto \frac{R' \cdot D'}{Q' \cdot M'} = ¥ \qquad (19)$$

where we have used the modified forms of the regime variables to account for the possible non-linear relationships between $R$ and $D$ and accommodation on one hand, and between $Q$ and $M$ and supply on the other.

## SEQUENCE ARCHITECTURE AND THE ACCOMMODATION/SUPPLY RATIO

### General

A regime model for sequence stratigraphy can be based on the geohistorical variation of the accommodation to supply ratio, $¥$. This model, which we will refer to as the $¥$-dependent model, can serve as the basis for interpretation of sequence stratigraphic patterns in terms of a sequence of different regimes, each characterized by: (1) a high or low accommodation/supply ratio; (2) an increasing or decreasing accommodation/supply ratio; and (3) a quickly or slowly changing accommodation/supply ratio. The fourth column of Table 2 relates these three characteristics to the formation of sequence stratigraphic bounding surfaces, and geometric systems tracts and subtracts.

The $¥$-dependent model is a conceptual model, not a quantitative or numerical model. The proposed correlation between changes in $¥$ and sequence architecture (Table 2) cannot be used as the basis for a forward or inverse numerical model. The modified geohistorical regime variables $Q'$, $M'$, $R'$ and $D'$, which make up the accommodation to supply ratio, are intended to indicate the sense of change in $¥$, but not the magnitude. Moreover, if two geohistorical variables are changing simultaneously the $¥$-dependent model does not predict the sense of change in $¥$ unless both variables act to change $¥$ in the same direction.

In the following section we apply regime principles in order to determine the origin and dynamics of each of the elements (surfaces and volumes) reviewed in the section entitled 'architectural description of systems tracts.' The successive dynamical analyses should be compared against the original descriptions and their summary in Table 2.

### Formation of the basin fan (canyon-cut) onlap surface

As discussed previously, an abrupt downward shift in coastal onlap occurs: (1) following the formation of a regional or globally correlatable coastal plain unconformity and its canyon-cut slope extension (Vailian type 1 sequence boundary); or (2) during any change within a progradational unit from a wedge-shaped geometry to a fan-shaped geometry (see section entitled 'seismic recognition of basin fan (fill) geometry . . .'). The term *canyon-cut* refers to a set of processes which include canyon incisement, sediment bypass and erosion on the upper slope, contemporaneous with fluvial entrenchment on the coastal plain.

In both of the cases listed above, basin fans onlap the continental slope. The onlap surface formed in either case can be referred to as the *basin fan (canyon-cut) onlap surface*.

Our review of modern lowstand sedimentation (see above) suggests that the formation of a basin fan onlap surface can be associated with the transition from platform delta to shelf-edge delta deposition. The formation of a major canyon-cut surface does not appear to be a necessary element of this transitional period.

The transition from platform to shelf-edge delta conditions is characterized by a sudden increase in the offshore sediment transport rate, $D$, caused by the proximity of the delta to a relatively steeply inclined slope. Transport can occur by various sediment gravity flow mechanisms or by combined wave–current shelf flows that have an increased downslope gravitational component.

The association of a sudden increase in $D$ and the initiation of basin fan deposition suggests the hypothesis that the formation of a basin fan (canyon-cut) surface can be triggered by any sufficiently rapid increase in the accommodation/supply ratio ¥. In Table 2, this condition is written $\dot{¥} \gg 0$. Modern processes of canyon initiation are too poorly known to critically evaluate this hypothesis. Several inferences, however, can be based on this hypothesis, which can be tested through stratigraphic studies.

*Basin fan initiation caused by rapid transgression*

A rapid increase in the accommodation/supply ratio can be caused by a combination of factors. One such scenario is: (1) a rapid increase in relative sea-level, $R$, which does not cause a consequent decrease in sedimentation transport rate, $D$; associated with (2) a rapid decrease in sediment supply variables, $Q$ and $M$. This geological scenario for basin fan initiation is identical with the *destructive slope* model of Brown & Fisher (1977, 1980). These authors view continental slope sedimentation as a continuing balance between destructive and constructive processes. When sediment supply for the upper slope diminishes, Brown & Fisher (1980) suggest that construction of the slope terminates and erosion of the slope and outer shelf predominates.

The numerous geological examples cited by Brown & Fisher (1977, 1980) for their destructive slope model do not bear repeating here. A particularly noteworthy example comes from the mixed clastic–carbonate, Late Cretaceous and Tertiary sediments of offshore Brazil. Brown & Fisher (1977, 1980) note that erosion of submarine canyons and deposition of deep-marine, onlapping, deposits (basin fan (fill) systems tract) occurs during formation of a drowning unconformity on the carbonate platform. Regime conditions are: a sudden rapid sea-level rise (high $R$), and a sudden decrease in sediment supply (low $Q \cdot M$) from shelf carbonate productivity. These factors combine to cause a sudden increase in the accommodation/supply ratio.

More recently, Dingus (1987), in a stratigraphic study of the large Middle Wilcox Yoakum Canyon of the Texas coastal plain, concluded that canyon formation was initiated in response to a sudden lowering of sediment supply followed by rapid sea-level rise. The Middle Wilcox fill of the Yoakum Canyon is composed of a thick section of silt and mud, indicating continued sedimentation by prodelta deposits.

It should be noted that a rapid increase in relative sea-level, $R$, in most circumstances leads to a rapid increase in water depth, and a consequent rapid decrease in sediment transport rate $D$ (see equation (14)). Thus, the change in accommodation, a product of these two geohistorical variables, $R \cdot D$, may be small. Initiation of basin fan systems tract deposition therefore does not necessarily follow a very rapid sea-level rise.

*Basin fan initiation caused by a rapid increase in wave or current energy*

A sudden increase in sediment transport rate, $D$, can be caused by a sudden increase in wind-driven or tidal currents or a sudden increase in the severity of the wave climate. It is likely that a sudden increase in $D$ will bring about a sudden increase in the accommodation/supply ratio ($\dot{¥} \gg 0$). This geological scenario is, thus, an additional model for the formation of a basin fan (canyon cut) onlap surface.

The Late Jurassic basin fans of the northern North Sea are a likely candidate for this type of scenario. The sudden downward shift in coastal onlap at the end of Kimmeridgian time figures predominantly in the coastal onlap chart of the North Sea constructed by Vail & Todd (1981). The end of Kimmeridgian time was also a period of rapid oceanographic change in the North Sea. The highly organic-rich Kimmeridge Clay Formation was deposited in a seasonally stratified, relatively shallow-water basin with little wave or current action (S. Brown, 1984). The termination of Kimmeridge Clay Formation deposition is believed to be a sudden, isochronous, oceanographic change, marked by a rapid increase in marine circulation within the North Sea (Rawson & Riley, 1982). This association of a sudden change in marine transport energy, with a correlative large, sudden downward shift in coastal onlap suggests that fan formation at this time was initiated by a sudden increase in the accommodation/supply ratio.

*Autocyclic basin-fan—prograding-complex sequences*

Autocyclic switching between periods of aggradation and erosion are familiar to students of arroyo formation (Schumm, 1976). Autocyclic processes involving submarine basin fan development are a predicted consequence of the proposed ¥-dependent model during high rates of sediment supply and low rates of subsidence.

The geological scenario for this proposed form of autocylicity is as follows: (1) sediments prograde in a wedge-shaped form; (2) sediments prograde to form a shelf-edge delta; (3) a sudden increase in $D$ owing to increased downslope transport causes the formation of basin fan (canyon-cut) onlap surface; (4) the greater amount of sediment bypass to the deep basin causes the rate of progradation to slow; (5) shelf-edge delta lobes stack in a fan-formed, aggradational fashion; (6) the aggradation of the shelf-edge delta causes the transport path between updip and downdip areas to be blocked, leading to a sudden decrease in $D$; (7) the sudden decrease in $D$ causes updip sediment progradation to resume in a wedge-like form; and (8) return to stage 1.

A possible example of this proposed form of autocylicity is found in the prograding units of the Cretaceous Nanushuk–Torok Group on the North Slope, Alaska. Within the Albian, several hundred basin fan (canyon-cut) onlap surfaces can be identified (Molenaar, 1985; McMillen, 1991). The geometry observed on dip lines of the progradational units is a repetitive alteration of slope fan and prograding complex subsystems.

*Continued offlap wedge deposition caused by a rapid decrease in relative sea-level*

The Vail (1987) model associates the formation of type 1 sequence boundaries with rapid rates of sea-level fall. The ¥-dependent model, on the other hand, proposes that periods of rapid sea-level fall, unless accompanied by changes in other regime variables, should be times of continued offlap wedge deposition.

A modern example of this phenomenon occurs in the Gulf of Corinth, Greece. Active faulting has caused at least 950 m of Pleistocene uplift of mountains of the southern side of the Gulf (Piper *et al.*, 1990). This rapid relative sea-level fall has caused an overall progradation of a sedimentary wedge consisting of a series of coalesced fan-deltas and a series of coalesced submarine fan-apron deposits in deeper water (Ferentinos *et al.*, 1988).

**Formation of the fluvial entrenchment surface**

Vail (1987) distinguished between type 1 sequence boundaries, which are interregional coastal plain and slope unconformities, and type 2 sequence boundaries, which are also coastal plain unconformities but are (seismically) conformable on the slope. The Vail (1987) model suggests that this difference, between type 1 and type 2 sequence boundaries, is essentially one of degree: the greater degree of sea-level fall in type 1 events leads to the greater extent of the unconformity.

Our review of Quaternary erosional surfaces (see section entitled 'presence of an erosional type 1 sequence boundary') suggests that the regime conditions leading to the development of an interregional coastal plain unconformity (type 2 sequence boundary) may be largely unrelated to the regime conditions leading to the formation of a slope–coastal-plain unconformity (type 1 sequence boundary). Thus the distinction between type 1 and type 2 sequence boundaries is not a matter of degree, but more fundamental in origin.

The formation of an interregional fluvial entrenchment surface can be understood in terms of the characteristic regime response of the topographic profiles of the fluvial and coastal plain to changes in the accommodation/supply ratio during falling sea-level (see sections entitled 'timing of fluvial sedimentation: fluvial morphodynamics' and 'timing of fluvial sedimentation: $Q$-dependency of modern highstand coasts'). Figure 20 illustrates, for two different cases, the position of regime topographic profiles during two stages of sea-level.

In the first case (Fig. 20A), sea-level falls while the accommodation/supply ratio remains constant. The river profiles, following Butcher (1989), are assumed to be tied at two points, the river mouth at one end and the source (drainage divide) at the upstream end. In this scheme, one point (drainage divide) is fixed, and the profile changes shape as the other point (river mouth) shifts. As the river mouth moves seaward, the profile must stretch between the two points and become flatter, so that at a given station the profile rises, thus creating subaerial accommodation (Fig. 20A). This first scenario creates localized areas of fluvial entrenchment located where the seaward shift of the more steeply dipping fluvial plain has cut into the previous shallow-dipping coastal plain. As shown in the figure, this form of fluvial entrenchment is not expected to link up as an interregional unconformity.

In the second case (Fig. 20B), sea-level falls while the accommodation/supply ratio decreases. The river profiles are again assumed to be tied at two points, the river mouth and the drainage divide, but

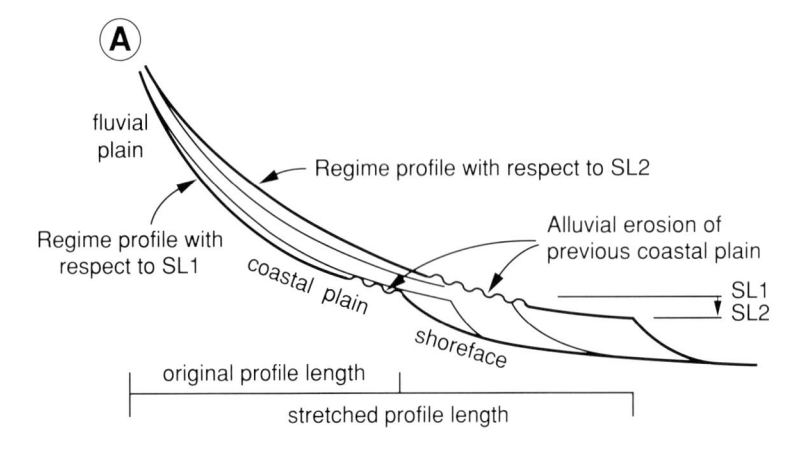

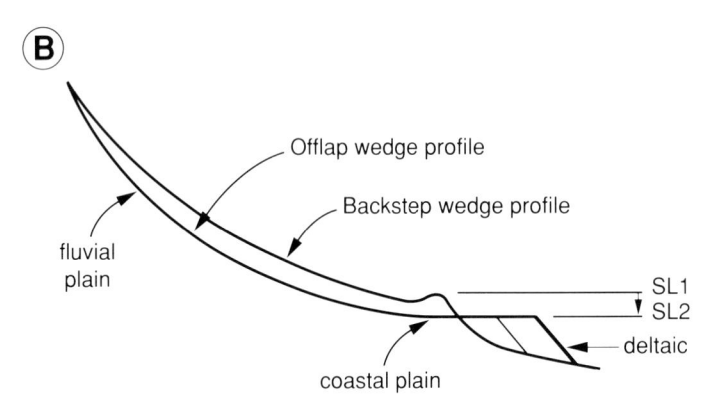

**Fig. 20.** Schematic illustration of the formation of a fluvial entrenchment surface. (A) Sea-level falls while the accommodation/supply ratio remains constant. The profile 'stretches', creating accommodation. (B) Sea-level falls while the accommodation/supply ratio, ¥, decreases. The profile changes from a back-step wedge shape (high ¥) to an offlap wedge shape (low ¥). Steepening of the inner coastal plain and flattening of the outer coastal plain causes a fluvial entrenchment surface to be cut.

in this case the form of the profile changes owing to the change in the accommodation/supply ratio. In response to the decrease in ¥, the gradient of the fluvial plain, $\theta_f$, steepens, while the gradient of the coastal plain, $\theta_c$, flattens. The net result of the movement of the regime topographic profile is interregional fluvial entrenchment of the coastal plain. An interregional fluvial entrenchment surface thus appears to be formed as a characteristic regime response of the coastal plain to a decreasing accommodation/supply ratio during offlap wedge progradation.

In a later section ('development of the offlap wedge systems tract') it will be demonstrated that offlap wedge deposition occurs when the accommodation/supply ratio is less than 1. In Table 2, the regime conditions for the formation of a fluvial entrenchment surface (type 2 sequence boundary) are given as $\dot{¥} < 0$ and $¥ < 1$.

## Development of the basin-floor fan

Our review of modern basin-floor fan deposits (section entitled 'basin-floor fan . . .') indicates that the identification of such deposits may be problematic on the basis of proposed seismic criteria, such as bidirectional downlap and internal seismic facies. It is likely, however, that the initial deposits associated with the formation of a basin fan (canyon-cut) onlap surface may be diagnostically different from subsequent units of the basin fan systems tract (e.g. Mutti, 1985). In view of the difficulties of correct interpretation of basin-floor fan features, the descriptive model of Vail (1987), for basin floor fans remains a good working hypothesis.

In keeping with the ideas presented in the section entitled 'formation of the basin fan (canyon-cut) onlap surface', a necessary, but not sufficient, regime condition for creation of a basin-floor fan is a sufficient rapid increase in the accommodation/supply ratio to form a basin fan (canyon-cut) onlap surface. In Table 2 this regime condition is written $\dot{¥} \gg 0$. It is beyond the scope of this study to discuss additional interactions with regime variables that affect the development of basin-floor fan sedimentation.

### Development of the slope fan

Our review of modern submarine fan deposition (section entitled 'seismic recognition of basin fan subsystems...') demonstrates that the slope fan subsystem is, volumetrically, the most important component of the basin fan systems tract. Two factors appear to be important in the development of slope fan deposition: (1) a point source for sedimentation, such as a canyon head, a shelf-edge delta, or a shelf-edge embayment; and (2) allochthonous deposition on the shelf in which the supply terms are larger than the accommodation terms ($Q \cdot M > R \cdot D$). Both of these factors appear to be necessary conditions for slope fan deposition. Given the first condition, but not the second, deep-water sedimentation can continue as various forms of topographic fill deposition (see section entitled 'offshore backstep fills'). However, these deposits can be distinguished from true slope fan deposition by their depositional geometry, internal seismic facies and slow accumulation rate. Given the second condition ($Q \cdot M > R \cdot D$), but not a point source for deposition, deep-water deposition takes the form of an offlap wedge (see section entitled 'lowstand deposits within the offlap wedge systems tract').

In Table 2, the first of these necessary conditions for slope fan deposition is given by the regime expression $|\partial ¥/\partial y| \gg 0$, where $|\partial ¥/\partial y|$ is the absolute magnitude of the change of $¥$ in the strike direction, $y$. For point-sourced deposition the magnitude of $¥$ changes rapidly along strike. The second condition ($Q \cdot M > R \cdot D$) is expressed by $¥ < 1$.

### Development of the prograding complex

Modern examples of the prograding complex subsystem (see section entitled 'prograding complex') indicate that this depositional unit results from contemporaneous fluvial aggradation within incised valleys and shelf-edge deltaic sedimentation. The rate of shoreline progradation during the formation of shelf-edge deltas is significantly less than during platform delta deposition. Shelf-edge delta lobes, therefore, tend to stack in a more aggradational fashion than platform delta lobes. This aggradation of the shelf-edge delta complex can cause the transport path between updip fluvial and downdip deep marine areas to be blocked. The sediment supplied by the alluvial system to the coastal plain incised valleys, in this scenario, can no longer be bypassed to the slope, and consequently infills the incised valleys.

In Table 2, we write the general condition for formation of the prograding complex subsystem as $\dot{¥} \ll 0, |\partial ¥/\partial y| \gg 0$. The first condition ($\dot{¥} \ll 0$) is a rapid decrease in the accommodation/supply ratio. The geometric response to a rapid decrease in this ratio is a unit that displays a combination of onlapping, aggradational and progradational geometries. Sediment that is allochthonous to the shelf system, owing to the sudden decrease in accommodation, is deposited in an aggradation or retrogradational style. Sediment that is autochthonous to the shelf system can, for a short period of time, continue to prograde at a diminished rate.

The second condition for formation of a prograding complex subsystem ($|\partial ¥/\partial y| \gg 0$) stipulates that along-strike variations in $¥$ are relatively large. This condition ensures that the prograding complex is a subsystem of the overall fan-shaped basin fan (fill) geometric systems tract. This condition is necessary to distinguish the prograding complex from the *shelf margin systems tract* of Vail (1987), which also displays a combination of aggradational and progradational geometries, but has a more wedge-like form (see later section entitled 'development of the shelf margin systems tract').

### Formation of the transgressive surface

The transgressive surface is an isochronous surface that separates geometric systems tracts. In the eustatic Vail (1987) model it separates lowstand and transgressive systems tracts. Our review of modern and submodern deposition suggests, more generally, that the transgressive surface separates two contrasting regimes: (1) $Q \cdot M < R \cdot D$; the back-step wedge systems tract in which sediment supply terms are smaller than potential accommodation; and (2) $Q \cdot M > R \cdot D$; the offlap wedge or basin fan systems tract in which sediment supply terms are larger than potential accommodation.

The transgressive surface can merge with other surfaces within the back-step wedge systems tract, such as ravinement or marine erosion surfaces. The transgressive surface, however, is fundamentally different from these secondary surfaces. They are formed by diachronous erosion, whereas the isochronous transgressive surface is created by a shift to a back-stepping stacking pattern of successive parasequences. This distinction has been lost in much of the work based on the Vail (1987) model. Van Wagoner *et al.* (1990), for example, place back-stepping, fluvial and estuarine, incised-valley-fill deposits within the lowstand systems tract rather than the transgressive systems tract. This confusing interpretation results from assuming that the ravinement surface above the estuarine deposits is the transgressive surface.

The transgressive surface is a turnaround surface created during increasing accommodation or decreasing supply ($\dot{¥} > 0$) by a shift in the regime inequality from $Q \cdot M > R \cdot D$ to $Q \cdot M < R \cdot D$. In Table 2, therefore, we write the general condition for formation of the transgressive surface as: $¥ = 1$, $\dot{¥} > 0$.

## Formation of the ravinement surface

Ravinement surfaces are cut by erosional shoreface retreat through storm current action that extends from the surf zone down to the lower shoreface (Stamp, 1921; Swift, 1968; Niedoroda *et al.*, 1985; see discussion by Swift & Thorne (this volume, pp. 3–31) and Swift *et al.* (this volume, pp. 89–152)). The ravinement process may result in the loss of 10 m or more of section (Swift, 1968; Niedoroda *et al.*, 1985; Belknap & Kraft, 1985). Shoreface retreat occurs whenever $Q \cdot M < R \cdot D$. In Table 2, therefore, the regime condition for generation of a ravinement surface is given as: $¥ > 1$.

## Formation of a marine erosion surface

Marine erosion surfaces develop as the seaward extensions of ravinement surfaces. When the transgression begins: (1) lagoons and estuaries form; (2) river sediment is trapped in these intra-coastal water bodies and no longer reaches the shelf; and (3) waves and currents continue to stir the shelf floor causing the sea-bed to erode.

Pitman (1991) has described a numerical model of marine erosion occurring during transgression based on an *R*-dependent regime topographic profile (see Thorne & Swift, this volume, pp. 33–58) for the coastal plain and shelf. In the Pitman model, shelves are seen as commonly possessing steeper gradients than the adjacent coastal plain. Consequently even a modest relative sea-level rise will cause extensive regrading of the shelf surface through submarine erosion, shifting the landward limit of sedimentation (landward limit of coastal onlap), out toward the shelf edge or even over it.

The problem of shelf regrading is amenable to an analysis in which the topographic profile includes four segments: a relatively steep fluvial plain (braid plain or meander plain), a relatively flat coastal plain (meander plain, marsh–lagoon or delta plain), a steep shoreface and a relatively flat shelf (Fig. 19). The formation of a marine erosion surface can be explained by the characteristic response of each segment of a regime topographic profile to changes in the accommodation/supply ratio during shoreface retreat (see section entitled 'regime topographic profiles . . .'). Figure 21 illustrates, for two different

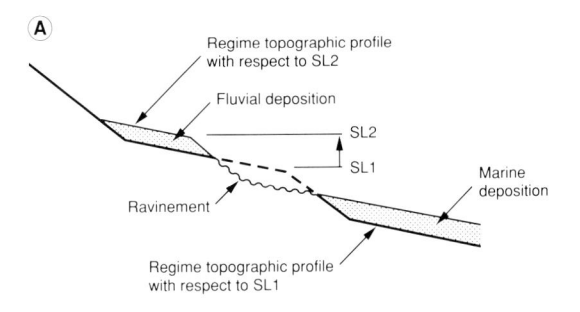

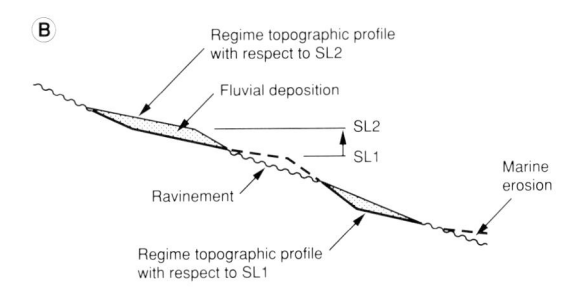

Fig. 21. Schematic illustration of the formation of a marine erosion surface. (A) Shoreface retreat occurs while the accommodation/supply ratio remains constant. The movement of the regime topographic profiles creates accommodation on the shelf. (B) Shoreface retreat occurs while the accommodation/supply ratio, $¥$, increases. The changing regime topographic profile causes marine erosion on the shelf.

cases, the position of regime topographic profiles during two stages of sea-level.

In the first case (Fig. 21A), shoreface retreat occurs while the accommodation/supply ratio remains constant. Therefore, the slopes on each segment of the regime topographic profile do not change during shoreface retreat. This first scenario creates a ravinement surface that has no seaward extension as a marine erosion surface.

In the second case (Fig. 21B), shoreface retreat occurs while the accommodation/supply ratio increases ($\dot{\yen} > 0$). As a consequence, the coastal plain and shoreface topographic segments flatten while the fluvial plain and shelf segments steepen. These slope changes lead to the formation of a marine erosion surface that is contemporaneous with shoreface retreat.

In Table 2 we give four regime conditions for formation of a marine erosion surface during shoreface retreat. The first condition ($\dot{\yen} > 0$) requires the accommodation/supply to increase. The second ($\yen > 1$) is the condition for shoreface retreat. The third condition ($D \gg 0$) is added to eliminate cases in which marine erosion does not occur owing to lack of sufficient fluid power to transport sediment on the shelf ($D$ near 0). This third condition is particularly important during a large, rapid deepening event on the shelf. Marine erosion, in this case, can be inhibited by development of water depths near or below storm wave base. A fourth condition is that the rate of down-to-basin tilting of the shelf owing to tectonic subsidence be small compared with the down-to-basin tilt of the new shelf surface. If this condition is not met the shelf regime topographic surface formed during shoreface retreat will not intersect the subsided, transgressed, shelf surface as in Fig. 21B. As a consequence, marine erosion is prevented. This last condition is particularly important on passive margins on which the rates of down-to-basin tectonic tilting are large compared with changes in regime topographic gradients. The tectonic tilt induced by foreland basin subsidence is of opposite sign to passive margin tilt, so that this subsidence contributes rather than detracts from the formation of marine erosional surfaces.

## Development of the back-step wedge systems tract

The stratigraphic architecture of the back-step wedge (fill) systems tract depends on the geometric relationships between the maximum flooding surface and the transgressive surface, in its various manifestations as fluvial entrenchment surface, the ravinement surface, or marine erosion surface (see Fig. 6 and accompanying text). Valley fill deposits form the basal deposits of this systems tract, except where they are part of the prograding complex (see section entitled 'prograding complexes'). Regime conditions favourable for prograding-complex development (see Table 2) lower the potential for subsequent back-stepping valley fill.

Above the interfluves, the valley fills expand into a wedge-shaped unit, bounded above by the ravinement surface (see section entitled 'back-barrier wedge (fill)'). This tract component is created if the ravinement surface climbs at a higher angle than the fluvial entrenchment surface (Fig. 22B). This angle of climb will be high if the longitudinal gradient of the coastal plain is high during shoreface retreat. Thus a high accommodation/supply ratio favours a steep angle of climb and preservation of a wedge-shaped unit below the ravinement surface (Table 2).

Back-stepping parasequences between the maximum flooding surface and the ravinement surface are an omnipresent feature of shoreface retreat (see section entitled 'back-step shelf wedge' and Swift et al. (this volume, pp. 153–187)). The regime condition for development of this subsystem is thus $\yen$ high, $\yen > 1$.

## Formation of offshore back-step fills

In the section entitled 'offshore back-step fills' we have described back-step topographic fill deposits composed of mud. Some are contiguous with the back-step wedge, and constitute its distal facies, while others form separate offshore depocentres. The formation of this type of topographic fill deposit is related to hydraulic conditions on the outer shelf. Fines are bypassed across the inner shelf, but further seaward where the wave regime expends less fluid power, they may come to rest. The muds may be river-derived within the present sedimentary cycle (allochthonous), derived from shelf-floor and coastal erosion (autochthonous), or of mixed origin. Swift et al. (this volume, pp. 89–152) have described the transgressive mud deposits of the southern New England shelf (Twichell et al., 1981) and of the Irish Sea (Belderson, 1964) as dominantly autochthonous mud deposits. Each constitutes an accumulation zone at the end of a well-defined, along-shelf, sediment transport pathway that begins at a zone of transport divergence and erosion on the shelf (see their Fig. 32).

The regime conditions for offshore deposition of either (1) mixed autochthonous–allochthonous sediments or (2) autochthonous sediments can be written: $¥ > 1$, $D \gg 0$. The first condition places these deposits within the back-step wedge (fill) systems tract. The second condition arises from the requirement that an offshore secondary depocentre requires a landward zone of sediment bypass with a relatively high rate of sediment transport.

### Formation of the maximum flooding surface

The maximum flooding surface is ideally an isochronous boundary that occurs at that point in time when the shoreline has reached its maximum landward excursion. At times prior to maximum flooding the shoreline is retreating and supply is less than accommodation ($Q \cdot M < R \cdot D$).

A simple model of this process is suggested by the *ramp surface* of Everts (1987); the envelope of positions of the aggradational lower limb of the equilibrium shoreface profile as it shifts landward. In the absence of an offshore transgressive mud facies (see section entitled 'back-step fill deposits'), the landward shift of the ramp surface will create a back-stepping wedge of coarse sediment whose deposition is characterized by a high reworking ratio. These time-transgressive deposits, described as the amalgamated sand lithofacies by Swift *et al.* (this volume, pp. 89–152), constitute a *proximal condensed section* of autochthonous origin, whose 'reworked' surface consists of the 'relict' sediments of earlier workers (Emery, 1968; see discussion of 'relict' sediments by Swift *et al.* (this volume, pp. 89–152)).

Under appropriate regime conditions, however, the ramp surface does not persist as a static entity. Instead, it can be considered a responsive equilibrium surface (see discussion of equilibrium index by Swift & Thorne (this volume, pp. 3–31)). As flooding continues, a distal laminated mud lithofacies may accumulate beneath the rising equilibrium surface (see discussion of shoreface–shelf depositional system by Swift *et al.* (this volume, pp. 89–152)). As maximum flooding approaches, the sediment supply drops to such low values that the reworking ratio rises once more and a *distal condensed section* develops (see discussion of proximal and distal condensed sections by Swift *et al.* (this volume, pp. 89–152)). After maximum flooding, the shoreline advances, supply is greater than accommodation ($Q \cdot M > R \cdot D$), and the seaward shift of the ramp surface creates an offlapping wedge of deposits that terminates condensed sedimentation.

The maximum flooding surface, therefore, is a turnaround surface created during decreasing accommodation or increasing supply ($\dot{¥} < 0$) by a shift in the regime inequality from $Q \cdot M < R \cdot D$ to $Q \cdot M > R \cdot D$. In Table 2, therefore, we write the general condition for formation of the transgressive surface as $¥ = 1$, $\dot{¥} < 0$.

### Development of the offlap wedge systems tract

The development of an offlap wedge systems tract is specified by two regime conditions: $¥ < 1$, $|\partial ¥/\partial y| \cong 0$. The first condition ($¥ < 1$), indicates that supply factors are larger than accommodation factors: $Q \cdot M > R \cdot D$ and, thus, the basinward movement of the depocentre creates an offlapping pattern. The second condition ($|\partial ¥/\partial y| \cong 0$), distinguishes offlap wedge sedimentation from the slope fan component of the basin fan systems tract. During offlap wedge deposition, along-shelf variations of the accommodation/supply ratio are relatively small compared with slope fan deposition.

### Development of the shelf margin systems tract

In this paper we have devoted little space to the *shelf margin systems tract* of the Vail (1987) model. The sequence architecture of this systems tract (Fig. 1) appears to be a hybrid form containing both back-step wedge and offlap wedge elements. We offer two hypotheses to account for this geometry:

1 The shelf margin systems tract of Vail (1987) is a composite of one or several sets of alternating offlap and back-step wedge systems tracts, which owing to inadequate seismic resolution cannot be distinguished.

2 The shelf margin systems tract is similar in origin to the prograding complex of the basin fan systems tract, which also displays penecontemporaneous onlapping, aggradational and prograding components (see section entitled 'development of prograding complex'). It is distinguished from a prograding complex by a more wedge-like form characterized by deposits of platform deltas rather than shelf-edge deltas. The regime conditions for this scenario are written: $\dot{¥} \ll 0$, $|\partial ¥/\partial y| \cong 0$.

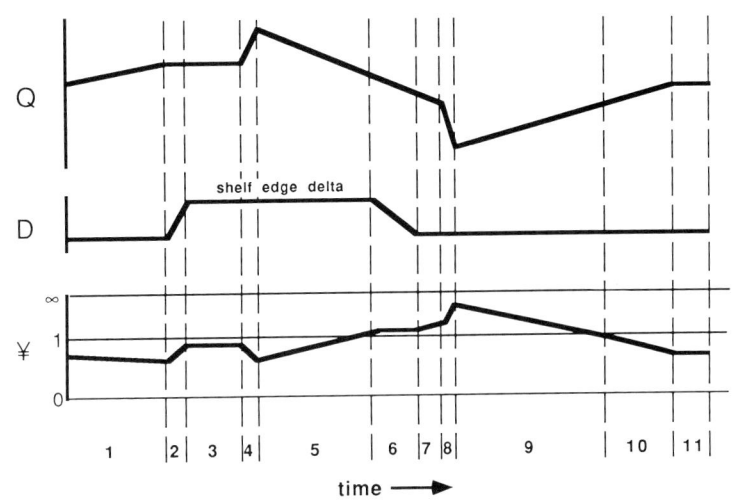

**Fig. 22.** The sequence development caused by hypothetical changes in the rate of clastic sediment supply, $Q$, of a constant grain-size distribution during a period of steadily rising relative sea-level without any fluctuations in the rate of rise. The formation of systems tracts and depositional surfaces during each of the 11 stages is conceptually derived from the geohistory of ¥ based on the regime conditions listed in Table 2.

systems tract
1. offlap wedge
2. basin floor fan
3. slope fan
4. prograding complex
5. slope fan
6. backstep wedge
7. backstep wedge
8. canyon cut/ basin floor fan
9. backstep canyon fill
10. offlap wedge
11. offlap wedge

surface
1. fluvial entrenchment
2. basin fan onlap

6. transgressive/ ravinement
7. ravinement/ marine erosion
8. canyon fill onlap

10. maximum flooding/fluvial entrenchment

## USE OF THE REGIME MODEL FOR SEQUENCE INTERPRETATION

### Hypothetical example of $Q$-dominated sedimentation

The regime model of continental margin sedimentation can be illustrated by scenarios in which the consequences of a given setting of regime variables can be examined through time. Figure 22 illustrates the sequence development caused by hypothetical changes in the rate of clastic sediment supply, $Q$, of a constant grain size distribution during a period of steadily rising relative sea-level without any fluctuations in the rate of rise. Changes in the sediment transport rate, $D$, occur depending on the proximity of deltaic sedimentation to the continental margin shelf edge. Changes in sediment transport rate do not occur as a result of any other external factor. The geohistory of variable ¥ is derived from the fluctuations of variables $Q$ and $D$ (Fig. 22). The formation of geometric systems tracts and bounding surfaces is conceptually derived from the regime conditions listed in Table 2. To illustrate the application of this table the 11 stages labelled in Fig. 22 are discussed in turn.

During stage 1 (Fig. 22) the rate of sediment supply increases from an initially moderately high value. Supply is larger than accommodation (¥ < 1) causing progradation of platform deltas of the offlap wedge systems tract. The decrease of the accommodation/supply ratio with time ($\dot{¥}<0$) causes a fluvial entrenchment surface to form as the coastal plain flattens and the fluvial plain steepens. By stage 2, the platform deltas reach the continental margin shelf edge causing a sudden increase in sediment transport by downslope gravitational processes and a consequent sudden increase in the accommodation/supply ratio ($\dot{¥}\gg 0$). A major canyon is not formed at this time, but the presence of shelf-edge deltas causes the formation of a basin-floor fan resting on a basin fan onlap surface. During stage 3, the sediment transport rate stabilizes with supply still greater than accommodation. The position of the shelf-edge deltas remains relatively stationary and thus rapid spatial variations in the accommodation/supply ratio occur up and down the shelf edge ($|\partial ¥/\partial y| \gg 0$). These conditions lead to ongoing slope fan deposition.

During stage 4, a sudden increase in sediment supply causes a sudden decrease in the accommodation/supply ratio ($\dot{¥} \ll 0$). As a consequence, incised valleys are unable to fully bypass the sediment to the shelf edge. These valleys aggrade as the shelf-edge delta progrades, forming a prograding complex. During stage 5, slope fan conditions are re-established as sediment supply begins to decrease. Aggradation within the incised valleys is no longer necessary during this stage. Sediment supply continues to decrease until, at the start of stage 6, accommodation and supply are roughly equal. As the balance shifts in favour of accommodation, a transgressive surface begins to form, and the depositional systems of the back-step wedge begin to accumulate over it. Both sediment supply and sediment transport rate decrease concurrently during this stage causing the accommodation/supply ratio to remain relatively constant. A ravinement surface is cut during this time but, since the accommodation/supply ratio is not increasing, it has no extension as a marine erosion surface. In stage 7, downslope sediment gravity flows that tap autochthonous sediments have reached a background low level. Sediment supply continues to fall, raising the accommodation/supply ratio, and steepening the regime topographic gradient of the shelf. Assuming that water depths on the outer shelf are not below storm wave base during this stage, regime conditions ($\dot{¥} > 0$, $¥ > 1$, and $D \gg 0$) can cause marine erosion.

During stage 8, a sudden drop in sediment supply and sudden increase in the accommodation/supply ratio precipitates the formation of a major canyon (e.g. resembling the Yoakum Canyon). The low sediment supply during this time eliminates any possibility of forming progradational parasequences within this interval. As a consequence, headward erosion of the canyon proceeds into shallow water and is not arrested by progradational infill into the canyon head. This first stage of canyon evolution deposits a basin-floor fan consisting of eroded shelf material on a canyon-cut onlap surface. The shoreface continues to retreat during stage 9 ($¥ > 1$).

A second depocentre forms in the canyon where various forms of fine-grained topographic fill accumulate, burying the shelfward extension of the canyon. Sediment supply continues to increase until, at the start of stage 10, accommodation and supply are roughly equal. As the balance shifts in favour of supply, a maximum flooding surface is formed beneath systems tracts of the offlap wedge. During stage 10 the decrease of the accommodation/supply ratio with time ($\dot{¥} < 0$) causes a fluvial entrenchment surface to form. However, during stage 11, sediment supply is constant and offlap wedge deposition proceeds without formation of a fluvial entrenchment surface.

## Hypothetical example of *R*-dominated sedimentation

Figure 23 illustrates the sequence development caused by two hypothetical cycles in eustatic sea-level, during a period of steady subsidence, sediment supply, sediment transport rate and input grain-size distribution. The rate of change of relative sea-level is computed from the combination of eustatic sea-level and subsidence (Fig. 24). Changes in the sediment transport rate, *D*, occur depending on the proximity of deltaic sedimentation to the continental margin shelf edge. The geohistory of variable ¥ is derived from the fluctuations of variables *R* and *D* (Fig. 23). The formation of systems tracts and depositional surfaces is conceptually derived from the regime conditions listed in Table 2.

One intention of this figure is to show that basin fan systems tract formation can be favoured by low amplitude, longer period, sea-level cycles as opposed to higher amplitude, short period, sea-level cycles. This conclusion is opposite to that of Jervey (1988) and Posamentier *et al.* (1988).

During stage 1 (Fig. 23) the rate of sea-level rise lessens from a moderately high value. Supply is less than accommodation ($¥ > 1$) causing ongoing back-step wedge deposition. The rate of sea-level rise continues to decrease until, at the start of stage 2, accommodation and supply are roughly equal. As the balance shifts in favour of supply, a maximum flooding surface is formed and is buried beneath systems tracts of the offlap wedge. The decrease of the accommodation/supply ratio with time ($\dot{¥} < 0$) causes a fluvial entrenchment surface to form as the coastal plain flattens and the fluvial plain steepens. During stage 3, supply is still larger than accommodation ($¥ < 1$). However, the rate of sea-level fall decreases during this period ($\dot{¥} > 0$) and, thus, fluvial entrenchment ceases. By stage 4, the rate of relative sea-level rise is sufficiently rapid to cause the balance between accommodation and supply to shift in favour of accommodation ($¥ > 1$). A trans-

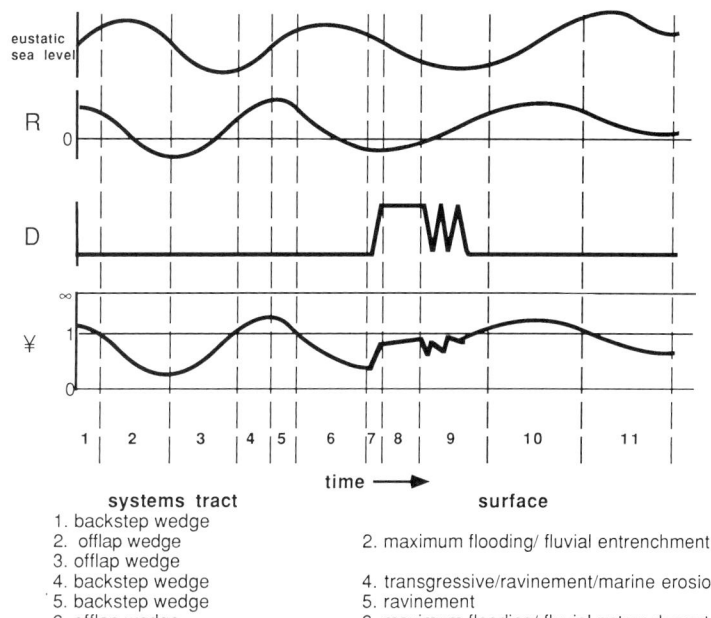

**Fig. 23.** The sequence development caused by two hypothetical cycles in eustatic sea-level, during a period of steady subsidence, sediment supply, sediment transport rate, and input grain-size distribution. The formation of systems tracts and depositional surfaces during each of the 11 stages is conceptually derived from the geohistory of ¥ based on the regime conditions listed in Table 2.

systems tract
1. backstep wedge
2. offlap wedge
3. offlap wedge
4. backstep wedge
5. backstep wedge
6. offlap wedge
7. basin floor fan
8. slope fan
9. prograding complex/ slope fan
10. backstep wedge
11. offlap wedge

surface
2. maximum flooding/ fluvial entrenchment
4. transgressive/ravinement/marine erosion
5. ravinement
6. maximum flooding/ fluvial entrenchment
7. basin fan/canyon cut onlap
10. transgressive
11. maximum flooding

gressive surface begins to form, and the depositional systems of the back-step wedge begin to accumulate over it. A ravinement surface forms within the wedge, and, possibly, a marine erosion surface beneath it (required conditions are $¥>1$, $\dot{¥}>0$). This latter surface is only formed if the water depths on the outer shelf have fallen to levels near storm wave base during this stage ($D \gg 0$) and if the rate of down-to-basin tectonic tilt of the transgressed shelf surface is small. Marine erosion is absent in all cases during stage 5, however, owing to the decrease in the accommodation/supply ratio ($¥>1$, $\dot{¥}<0$).

Regime conditions during stage 6 ($¥<1$) cause a prolonged period of offlap wedge sedimentation that downlaps a maximum flooding surface. The rate of sea-level fall is less during this cycle than the previous cycle and thus there is more time for an offlap wedge to prograde. By stage 7, prograding platform deltas reach the continental margin shelf edge causing a sudden increase in sediment transport by downslope gravitational processes and a consequent sudden increase in the accommodation/supply ratio ($\dot{¥} \gg 0$). A major canyon formed at this time builds a basin-floor fan resting on a canyon-cut onlap surface. The position of the shelf-edge deltas remains relatively stationary, and thus marked spatial variations in the accommodation/supply ratio occur up and down the shelf edge ($|\partial¥/\partial y| \gg 0$). Accommodation remains less than supply during stage 8 and, thus, regime conditions ($|\partial¥/\partial y| \gg 0$, $¥<1$) create slope fan deposition.

In stage 9 we indicate the possible formation of a series of prograding complexes by autocyclic processes described in the section entitled 'autocyclic basin-fan–prograding-complex sequences'. The sudden drops in the accommodation/supply ratio shown for this stage occur during intervals when incised valleys are unable to fully bypass sediment to the shelf edge. These valleys aggrade while the shelf-edge delta progrades, forming a series of prograding complexes. During stages 10 and 11 the accommodation/supply ratio rises above a value of unity, then falls below it once again. This episode creates a stacked sequence of back-step and offlap wedge deposits bounded by maximum flooding and transgressive surfaces. However, the rate of change of the accommodation/supply ratio during these stages is approximately 0. Regime conditions, therefore, do not create marine erosion or fluvial entrenchment surfaces.

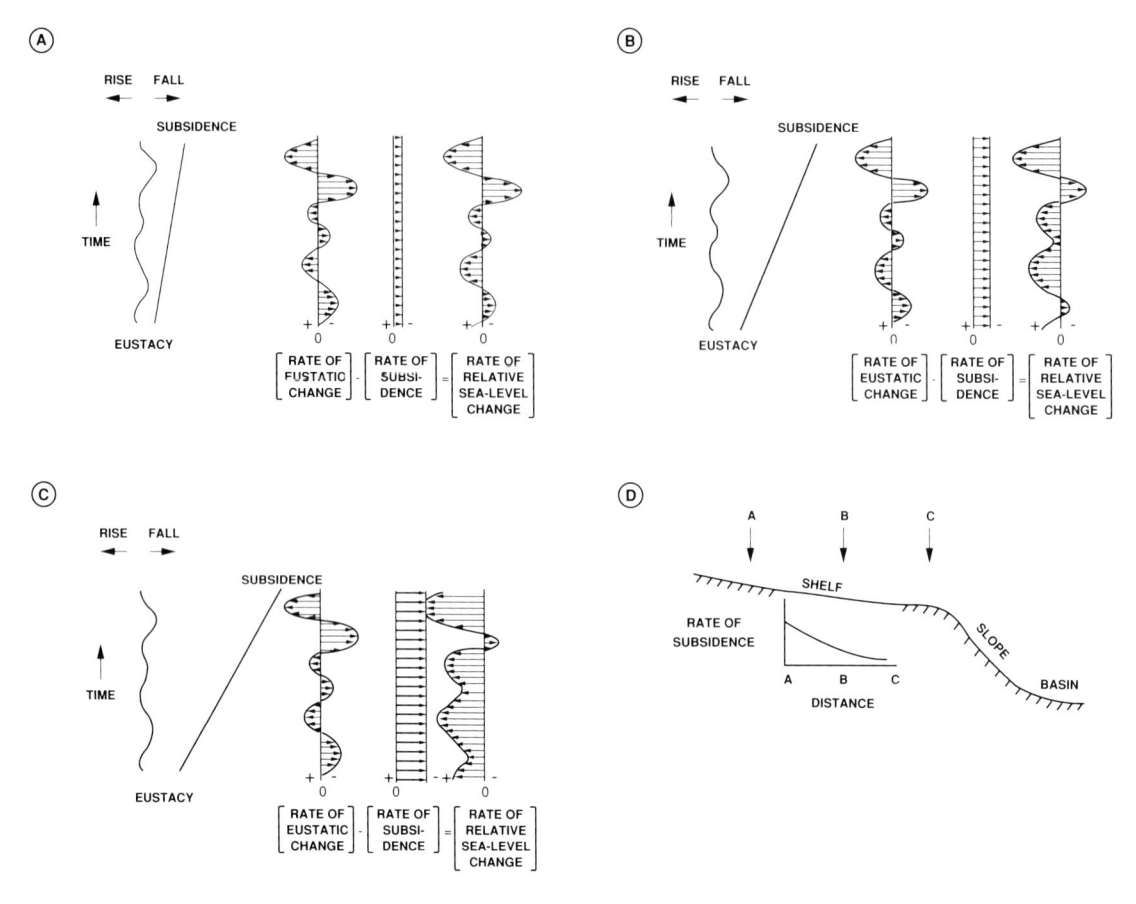

Fig. 24. The relationship between the stratigraphic process variables of the continental margin and relative sea-level (sediment accommodation potential). (Modified from Posamentier *et al.*, 1988.)

## PATTERNS OF *Q*- AND *R*-DOMINATED SEDIMENTATION: EXAMPLE OF THE CRETACEOUS WESTERN INTERIOR SEAWAY

*Q*- and *R*-dominated sedimentation patterns can be illustrated by recourse to the depositional history of the Cretaceous Western Interior Basin. A stratigraphic section of the Cretaceous Western Interior Basin of Wyoming (Fig. 25A) has the geometry of a clastic wedge (King, 1959), in which an asymmetrically subsiding basin is filled from its steeper side. The wedge is a compound one; the depositional regime varied so that sometimes the supply terms dominated ($Q \cdot M > R \cdot D$) and drove the shoreline basinward, while at other times the accommodation terms dominated ($Q \cdot M < R \cdot D$) so that the shoreline retreated back toward the tectonic front.

Two phases of subsidence are clearly visible in Fig. 25A as two contrasting stratigraphic patterns. The Albian–Santonian section, deposited during the early, low-subsidence epoch, is only several hundred metres thick throughout most of Wyoming, although it thickens to several times that value in the western foredeep. In this thinner Albian–Santonian section, two major progradational episodes are evident (Muddy Sandstone, Frontier Formation), but they are so attenuated that their wedge geometry is not apparent.

During Campanian–Maastrichtian sedimentation, the rate of accumulation increased by fivefold (and locally by an order of magnitude) compared with the earlier period. In Fig. 25A, the Adaville, Mesaverde and Lance Formations comprise well-defined clastic wedges. Throughout the period of heightened subsidence successive progradational

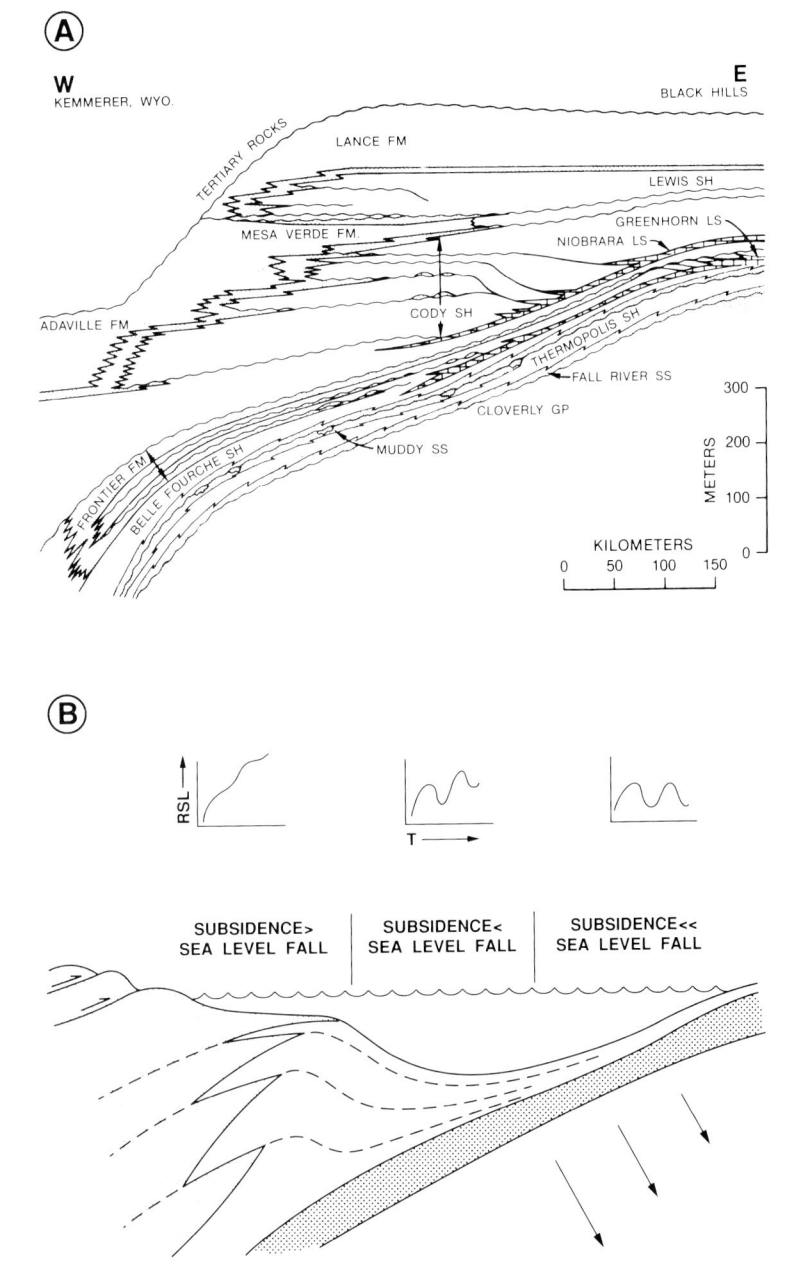

**Fig. 25.** Architecture of the Cretaceous Western Interior Basin. (A) Cross-section of the Cretaceous western interior basin of Wyoming. (After Weimer (1983) and Gill & Cobban (1973).) (B) Diagram illustrating depositional dynamics in the Cretaceous Western Interior Basin during periods of maximum rate of subsidence.

events built shale prisms out from the shoreline, with shelf–slope morphology (Cody Shale, Asquith, 1970; Lewis Shale, Krystinik, 1984). During Cody time, the successive shelfal prisms intertongued with the Niobrara Limestone of the central and eastern seaway.

Periodic changes in $Q$ as well as in $R$ can cause periodic changes in $¥$, leading to similar sequence stratigraphic patterns. We believe, however, that it is still possible to distinguish between $Q$-dominated and $R$-dominated sequence patterns in the western interior. Our favoured interpretation is that relative sea-level falls played an important role in the formation of the thinner Albian–Santonian section, while changes in sediment supply dominated the formation of third-order sequences within the thicker Campanian–Maastrichtian section (see Thorne et al. (this volume, pp. 59–87) for a relative sea-level interpretation of fourth-order Campanian cycles).

### Albian–Santonian R-dominated sedimentation

Incised valleys within the Albian of the Powder River and Denver Basins are classic examples of erosional features cut by falling sea-level (Stone, 1972; Weimer, 1983; Van Wagoner et al., 1990). The depth of incision of these features into the underlying Skull Creek Shale can reach 30 m. A large base level fall is implied by this large depth of incision. The Albian Muddy Sandstone makes up the incised valley fill (Nummedal & Swift, 1987); its basal portion consists of estuarine valley fills between the ravinement surface and a basal subaerial erosion surface, while its upper portion is a transgressive shelf deposit.

The Frontier Formation also contains multiple unconformities (Merewether, 1983). The unconformities are veneered with gravel, suggesting base level falls sufficient to bypass gravel on to the exposed shelf surface. While three main sandstones are shown in Fig. 25A, as many as 13 are present locally (Goodell, 1962). They are thin (1–10 m) strandplain sandstones capped by gravelly erosional surfaces. Some of the strandplain sandstones can be traced for 60 km or more. Such sheets are not the product of single progradations, however. The shoreline progrades 5–10 km during a single episode of sea-level fall, then undergoes transgression, then progrades again, at a stratigraphic level higher than the transgressed top of the last segment. The strandplain sandstones thus extend basinward in successively younger steps; that is, a gentle 'stratigraphic climb' is apparent.

### Campanian–Maastrichtian Q-dominated sedimentation

The sequence pattern changes abruptly in the Campanian section. The strandplain sandstone tongues of the Mesaverde Formation and equivalent units are markedly thicker than those of the Muddy Sandstone and the Frontier Formation. The Point Lookout Sandstone, for instance, is commonly 20–30 m thick. Like the Frontier units, it prograded basinward in short, progressively younger segments, but in the Point Lookout Sandstone, the segments overlap heavily, so that at any given section, two or three overlay each other (R. Wright, 1986).

Regional fluvial entrenchment surfaces appear only occasionally to have affected the Campanian section. Shallow fluvial channels are abundant within the fluvial deposits of the Campanian–Maastrichtian sections. Both small and large rivers within this fluvial system form deltas (Weimer, 1983), but there is little evidence of fluvial entrenchment and the development of high-efficiency, large drainage basin, river systems (see section entitled 'formation of the fluvial entrenchment surface'). Local ravinement surfaces are abundant in the Mesaverde and equivalent sandstones, but there appears to be little loss of section associated with them. These ravinements terminate landward in barrier-overwash deposits, which preserve the last position of the barrier whose retreat formed the ravinement. After formation, these 'turnaround' barriers became the anchoring units of strandplains that climb basinward above the subsiding ravinement (Swift et al., 1987, fig. 19). Lagoonal deposits preserved behind these turnaround barriers contain 'micro-birdfoot deltas' (Kamola, 1984) rather than estuaries, indicating that the subsiding fluvial plain continued to aggrade during submergence.

The clearest example of an unconformity occurs at the base of the Teapot Sandstone of Wyoming. Gill & Cobban (1973) report a missing section equivalent to the Judith River beds of Montana (four ammonite zones at about 500 000 years each). Ammonite zones below the unconformity are more steeply dipping; between Red Bird and Cottonwood Creek the unconformity cuts across 14 faunal zones. It increases in time value toward the Granite Mountains on the southern flank of the Wind River Basin, where the basal Teapot Sandstone incorporates gravel with cobble-sized clasts (Gill & Cobban, 1966). The unconformity appears to have resulted from uplift of the Granite Mountains during the mid-Campanian, rather than from eustacy. No entrenched valley fills similar to those of the Albian Muddy Sandstone have been observed on this surface.

The Castlegate Sandstone of the Book Cliffs (uppermost Campanian) likewise rests on an unconformity of sufficient duration to be resolved palaeontologically (Fouch et al., 1983). Van Wagoner et al. (1990) map Vailian lowstand systems tracts within this horizon on the basis of well-log and outcrop studies. They interpret 'lowstand systems' in this area to consist mainly of estuarine deposits. However, as discussed earlier, estuarine sedimentation falls entirely within the back-step wedge systems tract and not the basin fan systems tract (see section entitled 'back-step wedge systems tract...'). Like the Teapot Sandstone, the Castlegate Sandstone appears to have been formed in response to regional uplift, rather than in response to eustacy.

No other erosional surface has removed as much as an ammonite zone within the Campanian (Gill & Cobban, 1973). Several other diastems occur, however, the best-known one is a widespread diastem at the base of the Eagle and Shannon Sandstones (Gaynor & Swift, 1988). This erosional surface has been interpreted as a fluvial entrenchment surface on the basis of well-log correlations by Van Wagoner et al. (1990). However, detailed field studies in the Powder River Basin (Gaynor & Swift, 1988), and in the shoreline equivalent rocks of the Big Horn Basin have failed to reveal entrenchment features, as typified by the earlier Albian deposits.

The unconformity at the base of the Shannon is exposed along the west flank of the Powder River Basin and is well exposed in the Salt Creek anticline (Gaynor & Swift, 1988). In this region Shannon sands rest directly on mudstones of the Gammon Ferruginous Member of the Pierre Shale. Isolated gravel-sized phosphate clasts are present locally on this surface indicating marine erosion. When traced in the field, this unconformity surface has no marked relief, and is nowhere overlain by estuarine deposits. The same horizon can be traced to the Lower Mesaverde Formation of the Big Horn Basin at Nowater Creek, Zimmerman Butte and Lucerne (Keefer, 1972; Gill & Cobban, 1973). At these locations it appears as a flat ravinement surface, capping the lower Mesaverde Formation. Fluvial channels are numerous at this horizon of the Mesaverde (as well as above and below) but show no indication of entrenchment.

The shelf–slope topography of the western interior during the Campanian–Maastrichtian is very nearly a ramp setting (Asquith, 1970; Van Wagoner et al., 1990). Maximum gradients on the Wyoming slope reach only 1° during deposition of the Gammon and Mitten Formations (Thorne & Swift, 1985). The low slope gradients appear to inhibit the formation of basin fan systems tracts. Fan-shaped deposits with feeder channels are present locally at the Campanian shelf edge (Asquith, 1970), but the sequence stratigraphic geometry of these units, constructed through E-log sections, is better interpreted as back-step fill deposits than as basin fan systems tract deposits. Well-defined fans occur near the top of the Maastrichtian (Krystinik, 1984). Here, however, they are associated with accelerated uplift and basin formation at the onset of the Laramide Orogeny.

In summary, the overall architecture of Campanian and Maastrichtian deposition consists of stacked, highly asymmetric sequences, in which sedimentation is partitioned between a thickened offlap wedge systems tract and a much thinner back-stepping systems tract (Swift et al., 1987). The cycle-bounding surfaces are transgressive surfaces (ravinements and their correlative marine erosion surfaces), rather than unconformities cut by fluvial entrenchment. Because the surface being transgressed has little relief, the back-barrier system is generally destroyed in the ravinement process, and back-step wedge deposits are consequently thin.

## ¥-Dependent models for the Cretaceous Western Interior Basin

Figure 26(A–D) illustrates alternative regime histories for Campanian–Maastrichtian sequence development on the western margin of the Cretaceous foreland basin, based on the ¥-dependent model of the previous section. Each alternative history can be thought of as a different 'model run', though the models shown are conceptually derived from Table 2 rather than created by numerical computer simulation.

The regime histories shown in Fig. 26(A,B) are *R*-dependent. Predicted sequence patterns are created by cycles in relative sea-level without any variation in other regime variables. The relative sea-level curve shown at the top of each figure combines a short-term cyclic change of relative sea-level with a large component of long-term rise. This combination of short and long-term relative sea-level is appropriate for the western margin of the foreland basin during periods of maximum tectonic subsidence (Fig. 24B).

Though changes in the accommodation/supply ratio, ¥, are controlled by changes in *R*, the critical times at which ¥ = 1 are modulated by the constant values of the other regime variables. The regime history shown in Fig. 26A has, at all times, a higher value of ¥ than the regime history of Fig. 26B. In Fig. 26A, accommodation is less than supply (¥ < 1) only during periods of slow relative sea-level rise or fall. Accommodation is greater than supply (¥ > 1)

242 J.A. Thorne and D.J.P. Swift

**(A) HIGH-¥, R-DEPENDENT MODEL**

**(B) LOW-¥, R-DEPENDENT MODEL**

**(C) Q-DEPENDENT MODEL**

**(D) Q-DEPENDENT MODEL (against thickness)**

LEGEND FOR PANELS A, C, D

| systems tract | surface |
|---|---|
| 1. backstep wedge | |
| 2. offlap wedge | 2. maximum flooding / fluvial entrenchment |
| 3. offlap wedge | |
| 4. backstep wedge | 4. transgressive / ravinement / marine erosion |
| 5. backstep wedge | 5. ravinement |
| 6. offlap wedge | 6. maximum flooding |
| 7. backstep wedge | 7. transgressive / ravinement / marine erosion |
| 8. backstep wedge | 8. ravinement |
| 9. offlap wedge | 9. maximum flooding |

LEGEND FOR PANEL B

| systems tract | surface |
|---|---|
| 1. backstep wedge | |
| 2. offlap wedge | 2. maximum flooding / fluvial entrenchment |
| 3. offlap wedge | |
| 4. backstep wedge | 4. transgressive / ravinement |
| 5. offlap wedge | 5. maximum flooding / fluvial entrenchment |
| 6. offlap wedge | |
| 7. backstep wedge | 7. transgressive / ravinement |
| 8. offlap wedge | 8. maximum flooding / fluvial entrenchment |

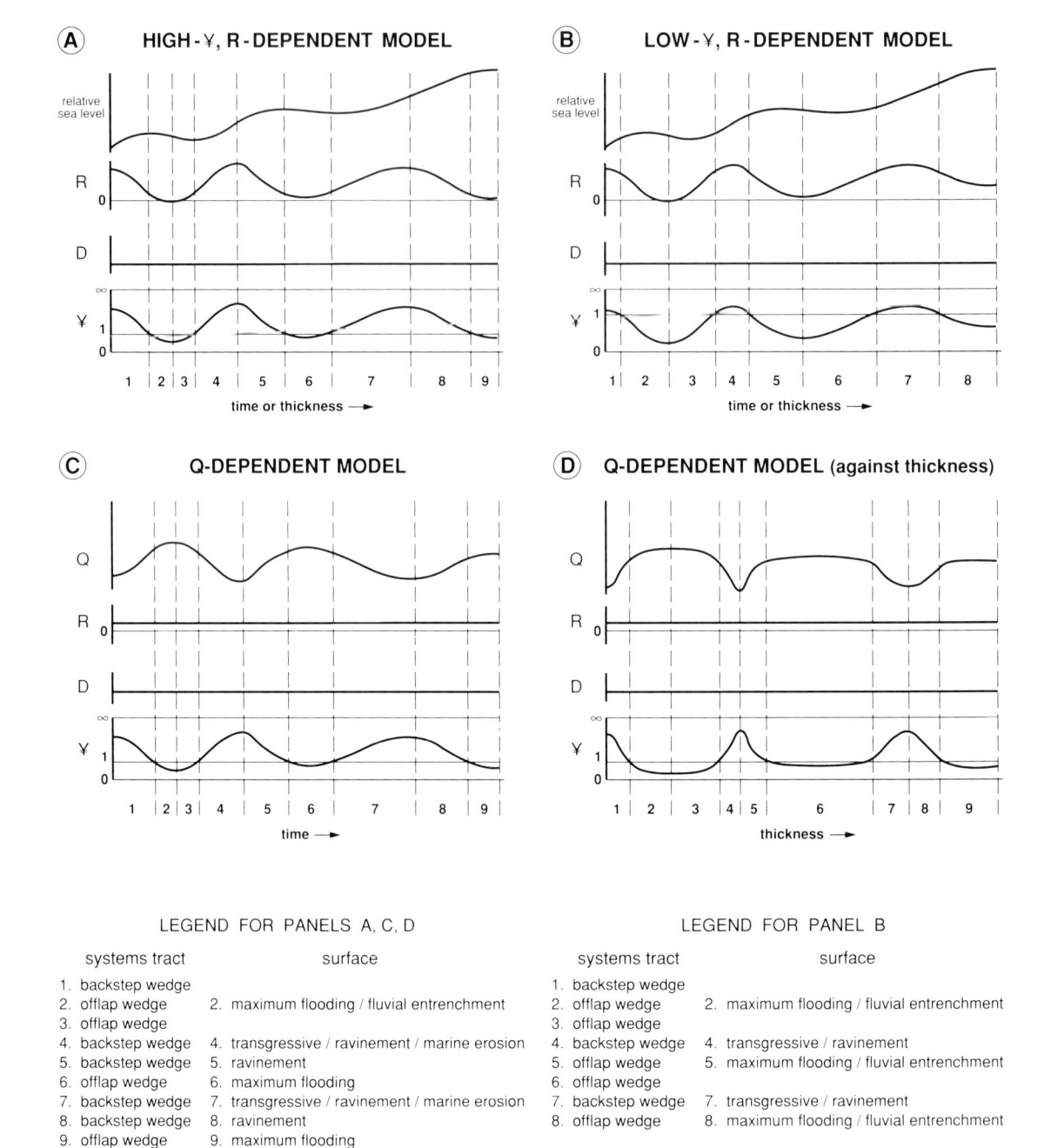

**Fig. 26.** Regime history for Campanian–Maastrichtian sequence development on the western margin of the western interior foreland basin. The formation of systems tracts and depositional surfaces is conceptually derived from the geohistory of ¥ based on the regime conditions listed in Table 2. (A) High ¥ value, $R$-dependent case: changes in the accommodation/supply ratio, ¥, are controlled by changes in relative sea-level, $R$. (B) Low ¥ value, $R$-dependent case: same as (A) except that the accommodation/supply ratio is lower. (C) $Q$-dependent case: changes in the accommodation/supply ratio, ¥, are controlled by changes in sediment supply. (D) Same as case (C) except that the horizontal ordinate is proportional to thickness instead of time.

during all other parts of the sea-level cycle. In Fig. 26B, conversely, accommodation is greater than supply (¥ > 1) only during periods of very fast relative sea-level rise. Accommodation is less than supply (¥ < 1) during all other parts of the sea-level cycle.

Both models (Figs 26A, 26B) show sequences consisting of offlap wedge systems tract and back-step wedge systems tract components. Progradation of the offlap wedge occurs not because sea-level falls, but because sea-level rise decreases to the point that sediment supply exceeds accommodation; $Q \cdot M < R \cdot D$. When relative sea-level begins to rise at a faster rate, accommodation once more begins to exceed sediment input; $Q \cdot M > R \cdot D$. The asymmetry of the cycles, however, varies from Fig. 26A to Fig. 26B. In the high ¥ value model (Fig. 26A) offlap components are thin compared with back-step components of the cycle. In the low ¥ value model (Fig. 26B), conversely, offlap components are thick compared with back-step components.

The low and high ¥ value, $R$-dependent models predict a different succession of sequence stratigraphic surfaces. In the low ¥ value model, fluvial erosion surfaces are formed during the first part of each offlap wedge (stages 2, 5 and 8, Fig. 26B) owing to the rapid decrease in ¥ during this stage. The high ¥ value model, however, creates a fluvial entrenchment surface on only one out of three offlap wedges (stage 2). The ¥ curve falls below 1 at the start of stages 6 and 9; during these stages, however, the accommodation:supply ratio does not significantly fall and thus a fluvial entrenchment surface is not formed. In the high ¥ value model, marine erosion surfaces are formed during the first part of each back-step wedge (stages 4 and 7, Fig. 26A) owing to the rapid increase in ¥ during this stage. The low ¥ value model, however, does not create marine erosion surfaces. The ¥ curve rises above 1 at the start of stages 4 and 7. During these stages, however, the accommodation to supply ratio does not continue to rise and thus a marine erosion surface is not formed.

Figures 26C, 26D illustrate a third alternative regime history for Campanian–Maastrichtian sequence development. This model shows $Q$-dependent variations in the accommodation:supply ratio. Predicted sequence patterns are created by cycles in sediment supply without any variation in other regime variables. The sediment supply curve shown at the top of each figure creates the same variation in ¥ and the same stratigraphic surfaces as the high ¥ value, $R$-dependent model of Fig. 26A.

In the $Q$-dependent model, marine erosion surfaces are formed during the first part of each back-step wedge (stages 4 and 7, Fig. 26C) owing to the rapid increase in ¥ during this stage. In this case, however, owing to changes in sediment supply, the time axis is not a good reflection of the thickness of each stage of deposition. Figure 26D is the same as Fig. 26C except that the succession of systems tracts and surfaces are arranged along a thickness axis instead of a time axis. This causes stages with a high sediment supply to be expanded relative to stages with low sediment supply. Figure 26D has a different cycle asymmetry than Fig. 26A. In the high ¥ value, $R$-dependent model (Fig. 26A), offlap components are thin compared with back-step components of the cycle. In the $Q$-dependent model (Fig. 26D), conversely, offlap components are thick compared with back-step components.

## Comparison of observed and model sequence architecture

Of the three models (Fig. 26A–D) the $Q$-dependent model is the best match to the observed sequence stratigraphic architecture of the Campanian–Maastrichtian deposits of the USA western interior. Points of comparison are as follows:

**1** Sequences consist of an alternation of back-step and offlap wedge systems tracts.
**2** Back-step wedge deposits are thinner than offlap wedge deposits.
**3** Fluvial entrenchment surfaces can occur but are rare.
**4** Ravinement and marine erosion are associated with transgression.

Neither of the $R$-dependent models fits all four points of comparison that favour the $Q$-dependent model. The high ¥ value, $R$-dependent model creates cycles with the wrong asymmetry: back-step components are thicker than offlap components. The low ¥ value, $R$-dependent model creates too many fluvial entrenchment surfaces and does not predict significant marine erosion.

All of the three models for the Cretaceous Western Interior Seaway are overly simplistic in the assumption of either uniform rates of either subsidence or sediment input. More realistic models that couple sediment supply and subsidence must consider the complexities of foreland basin tectonics, such as the tempo of loading events, lithospheric stress relaxation causing the migration of peripheral bulges, and the phase lag of sedimentation with respect to the time of maximum uplift (Jordan, 1982; Quinlan & Beaumont, 1984; Swift et al., 1985).

The success of the proposed $Q$-dependent model in explaining the salient features of Campanian–Maastrichtian sequence architecture is perhaps the more convincing because of its simplicity. Stratigraphic patterns unique to this area are explained as a simple succession of periods of slowly increasing, rapidly increasing, slowly falling, and rapidly falling sediment supply. The conceptual model presented in this paper, therefore, can be used to interpret sequence stratigraphic patterns using a succession of relatively few, characteristically different, regime conditions.

### $R$ versus $Q$ control of 10 Ma and 1 Ma cycles: in- versus out-of-sequence thrusts

The results presented in this study support a $Q$-dominated mode of sedimentation for third-order ($c.1$ Ma) cycles within the western interior foreland. Previous results (Swift et al., 1985), however, suggest an $R$-dominated mode of sedimentation for second-order ($c.10$ Ma) cycles within the western interior foreland. This difference may be characteristic of foreland basins. Recent studies of convergent orogens indicate that foreland thrusts can be separated into two types (Morley, 1988). The first *in-sequence* type of thrust breaks out well basinward of the previous thrust front. This type of thrust thickens the continental crust and moves the locus of thrust loading basinward, thus causing a significant pulse of subsidence in the foreland basin (Beaumont, 1981; Tankard, 1986). The second *out-of-sequence* type of thrust is internal to the foreland fold-and-thrust belt and serves to bring material at the base or the orogenic wedge closer to the surface. This type of thrust maintains the balance between the influx of material at the toe of the wedge and erosion off the top of the uplifted thrust belt (Chapple, 1978; Davis et al., 1983).

The episodic nature of these two types of thrusts within the Rocky Mountain foreland may be quite different. Major in-sequence thrusts occur during the Aptian through to Campanian of Idaho, Wyoming and Utah with an episodic spacing of $c.10$ Ma (Wiltchko & Dorr, 1983; Swift et al., 1985). Out-of-sequence thrusts, however, must occur significantly more often in order to allow the orogenic wedge to maintain a roughly steady-state topography (Covey, 1986).

This difference in the episodic nature of in- and out-of-sequence thrusts suggests that second-order sequence cycles may be controlled by episodic subsidence caused by in-sequence thrusts, while third or higher order sequence cycles may be controlled by sediment supply changes caused by out-of-sequence thrusts. Thus, foreland basins, in general, may show $R$-dominated sedimentation for second order ($c.10$ Ma) cycles and $Q$-dominated sedimentation for third or higher order ($<1$ Ma) cycles.

## VARIATIONS IN HIGHSTAND PRESERVATION POTENTIAL

The regime model of the previous sections is a *depositional* model. It can be used to interpret sequence stratigraphic patterns in areas where later erosion has not removed large parts of the stratigraphic record. In general, however, later erosion can remove the stratigraphic record of entire systems tracts making the interpretation of the resulting sequence pattern more difficult.

Systematic variations in systems tract completeness can be understood using the concept of an *index of highstand preservation potential* (see section so entitled). As indicated previously, this index is to a large extent controlled by the relative magnitude of the changes in sediment supply, $Q$, or of relative sea-level, $R$. To illustrate variations in highstand preservation potential we will consider two cases: (1) the northwestern Gulf shelf; and (2) the Atlantic shelf.

### Highstand preservation on the Gulf shelf

Highstand preservation is favoured by high flexural rigidity, clay-rich sediment supply, high rates of sediment input, and low rates of relative sea-level change (see equation (6)). The relative change in these factors can be used to predict systematic changes in highstand preservation in the Paleogene versus the Pleistocene of the northwestern Gulf (Fig. 27A). The first two of these factors are not expected to change systematically during the Cenozoic. The flexural rigidity throughout the Cenozoic is moderately low (Nunn, 1990). The clastic sediment input into the basin during the Paleogene derives largely from a mixture of sedimentary and igneous terrains uplifted during the

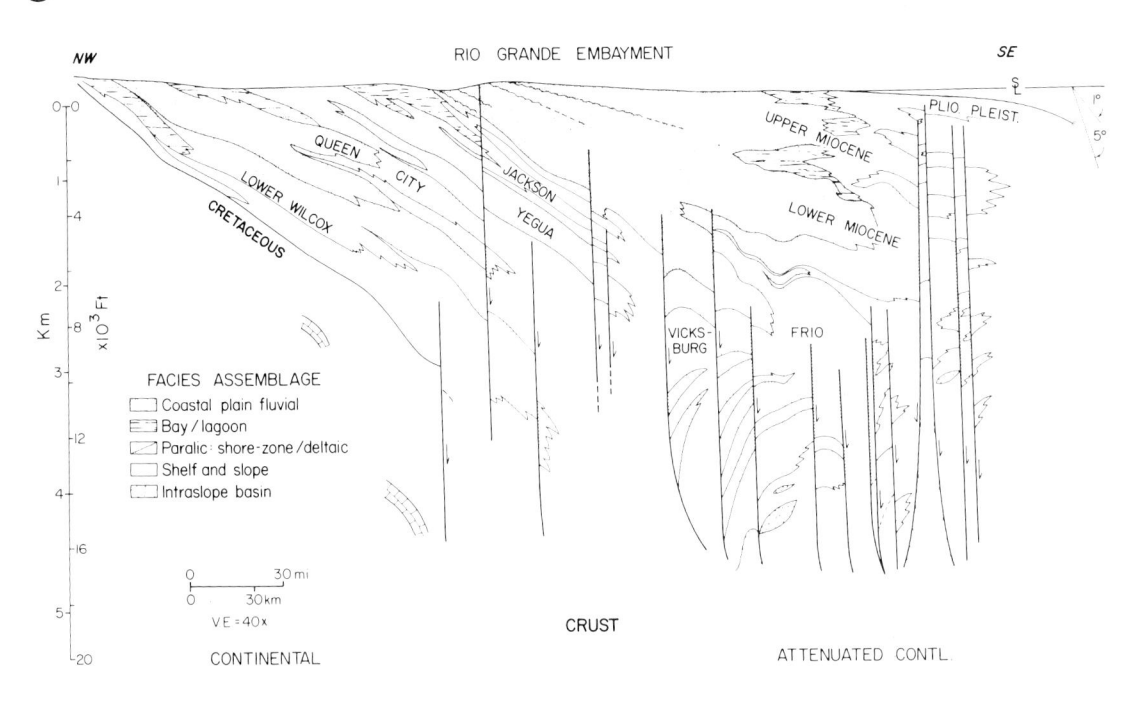

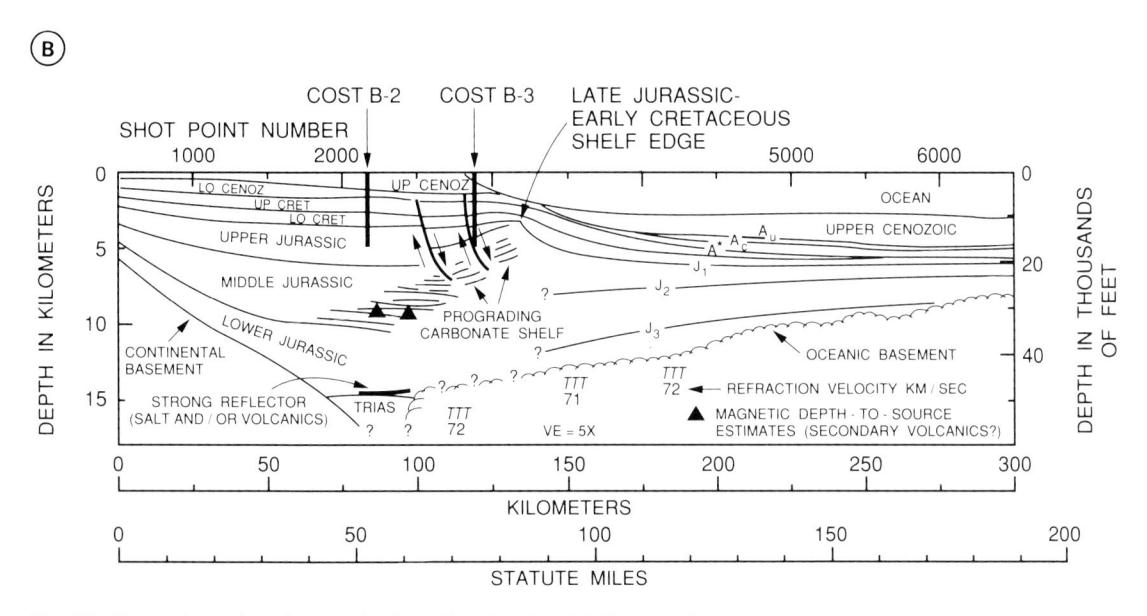

**Fig. 27.** Comparison of passive margin depositional styles. (A) Cross-section through the Rio Grande depocentre, northwest Gulf of Mexico shelf. (From Galloway, 1989b.) (B) Cross-section through the New Jersey Shelf. (From Poag, 1985, after Grow, 1980.) Note differing vertical exaggerations.

Laramide orogeny within the western interior of North America (Winker, 1982). The products of such a diverse source terrain can be expected to contain a wide distribution of grain sizes. The clastic source for Pleistocene sediments is similarly derived from epeirogenic uplift of the Rocky Mountains and Great Plains (Galloway, 1989b).

The sediment supply to the northwest Gulf of Mexico, is quite high during the Paleogene, and higher still in the Pleistocene. Hay *et al.* (1989) calculate the sediment input to the northwest Gulf as ranging between $1.2 \times 10^{20}$ g Ma$^{-1}$ and $2.1 \times 10^{20}$ g Ma$^{-1}$ during the Paleogene. Their calculated mass volume for Quaternary sediment input is approximately $8 \times 10^{20}$ g Ma$^{-1}$.

Paleogene sea-level cycles on the Haq *et al.* (1987) curve have approximately 50 m total amplitude and a frequency of $c.0.5$ cycles Ma$^{-1}$. It is a matter of some debate whether the Paleogene cycles of the Haq *et al.* (1987) curve are a true indication of eustatic changes (Miller & Kent, 1987; Matthews, 1988). In this study it is assumed that, although the specific timing and magnitude of any one cycle may be inaccurate, the approximate average magnitude and frequency of the cycles is correctly determined. At the other extreme, glacio-eustatic cycles of $c.100$ m total amplitude, and $c.10$ cycles Ma$^{-1}$, have dominated the Pleistocene record of sea-level change (e.g. Hays *et al.*, 1976).

The ratio of the highstand preservation index in the Paleogene versus the Pleistocene is found from substituting values for, $Q$, $h$, and $f$ into equation (6):

$$\frac{S_p(\text{Pleistocene})}{S_p(\text{Paleogene})} = \frac{8 \cdot 10^{20}}{100 \cdot 10} \cdot \frac{50 \cdot 0.5}{2 \cdot 10^{20}} = 0.1 \quad (19)$$

Equation (19) indicates that the highstand stratigraphic preservation potential index for the Pleistocene is only one-tenth of the highstand preservation potential index for the Paleogene. In the remainder of this section we compare briefly the stratigraphy of the Paleogene and Pleistocene to illustrate the importance of highstand preservation potential.

Galloway (1989b), drawing on his own earlier studies, the work of his students, and on the work of Winker (1982), has shown that the Paleogene stratigraphy of the northwest Gulf of Mexico is characterized by sequences showing systems tracts of both basin offlap and periods of transgression and marine flooding of the depositional platform (Fig. 27A). Based on these general observations, it appears that the Paleogene of the northwest Gulf was a period of high stratigraphic preservation potential of the back-step and offlap wedge geometric systems tracts.

The $Q$-dominated mode of sedimentation during the Paleogene appears to have controlled the overall architectural style of the period. Continental margin outbuilding was concentrated at one or more shelf-edge deltaic depocentres separated by interdeltaic bights. Depocentres remained fixed during a depositional episode but commonly relocated during transgression and flooding, often at a scale of hundreds of miles in large systems (Galloway, 1981). Such depocentres on the northwest Gulf of Mexico continental margin have typically exhibited sedimentation rates of tens of metres per thousand years during the Paleogene. Their episodic growth during Paleogene time, resulting in the formation of 'depositional stratigraphic sequences', was related closely to tectonic events of the source terrain, which in turn affected the rate and location of sediment supply, and the basin margin response to loading.

However, during the Pleistocene, the shelf depositional pattern changed to one where thin 'condensed sections' accumulated at high sea-level, while thick, rapidly deposited 'expanded sections' were deposited at low sea-level (Coleman & Roberts, 1988a). The change was a function of changing preservation potential; the thick highstand deposits no longer had the time to subside to safety as both the amplitude and frequency of sea-level oscillations increased. Deltas that formed as each rise slowed were cannibalized by their own incising trunk streams during each subsequent fall.

## Highstand preservation on the Atlantic shelf

In contrast to the large-scale, well-integrated interior drainage characteristic of the northwestern Gulf continental margin, the central Atlantic margin (Fig. 27B) is a strike coast, bordered by a marginal mountain system and drained largely by subsequent streams. As a consequence, the sedimentation rate diminished rapidly after the Mesozoic period of rifting, initial separation and high relief, and was an order of magnitude lower than that of the Gulf throughout most of the Cenozoic (Poag, 1985, Ward & Strickland, 1985). While much of the Paleogene Gulf section consists of deltaic depocentres, the thinner Atlantic deposits are almost entirely marine

(Ward & Strickland, 1985). The Paleogene outcropping on the coastal plain consists of a succession of thin (several metres), muddy, glauconitic sands separated by disconformities (Fig. 27B). Each unit appears to be open marine, onlapping and transgressive; marine marginal and highstand deposits are missing. This stratigraphic pattern is similar to the pattern reported by Coleman & Roberts (1988a) from the Quaternary of the Louisiana shelf. The low highstand preservation potential, however, results not from an increase in magnitude and frequency of eustatic cycle, as in the Louisiana shelf case, but from a reduction in $Q$.

The pattern reversed during the Miocene with a pulse of high sediment supply and of good highstand preservation. Poag (1985) reports a thick Miocene section from the offshore shelf of the Baltimore Canyon Trough that consists of a series of prograding platform delta wedges. The seismic stratigraphy of this area (Greenlee et al., 1988) displays a series of well-defined offlap wedge systems tracts alternating with basin fan systems tracts.

Preservation of Miocene highstand deposits is also documented in early to middle Miocene sediments of the Salisbury Embayment of Chesapeake Bay. Ten disconformity bounded units are present, four of which consist of well-defined transgressive–regressive cycles of approximately 750 000 years duration. In the cycles, transgressive shell beds are overlain by thicker regressive muddy sands (Kidwell, 1984). The sequences are separated by disconformities of ravinement origin (Kidwell, 1984; see discussion by Swift et al. (this volume, pp. 153–187)).

The Pleistocene of the middle Atlantic shelf is less well known than in the case of the Gulf shelf. On the inner shelf surface, stacked sequences, lacking in highstand deposits, consist of back-barrier depositional systems, separated by ravinement surfaces (Niedoroda et al., 1985). On the shelf edge, however, the Pleistocene thickens to more than 400 m and exhibits clinoform internal structure (Poag, 1985, fig. 6-3). Geomorphological interpretation indicates that these deposits are lowstand shelf-edge deltas (Swift et al., 1972), similar to those described from the Louisiana shelf by Coleman & Roberts (1988). Thus the Pleistocene section of the Atlantic continental shelf, deposited under conditions of a low ratio of sediment input to the frequency and intensity of sea-level falls, appears to have a low highstand preservation potential owing to cannibalization of earlier deposits during each successive sea-level cycle.

## CONCLUSIONS: DEPOSITIONAL SEQUENCES AND DEPOSITIONAL REGIMES

This study has evaluated the sequence architecture models of Vail (1987) and Galloway (1989a) in terms of the regime concepts presented in the five preceding papers of this volume. Comparison of these models with modern and submodern shelf deposits, in which sediment dispersal systems are presently functioning, is the basis for a new, regime-based model for depositional sequences, their component systems tracts and bounding surfaces. To effectively compare the Vail (1987), Galloway (1989a) and regime-based models, we introduce four *geometric systems tracts* that make no explicit reference to eustatic sea-level. Rough Vail (1987) or Galloway (1989a) equivalents can be found for the (1) basin fan, (2) back-step wedge and (3) offlap wedge geometric systems tracts. The fourth, the alluvial fan geometric systems tract, has no equivalent.

Our survey of modern sedimentation, based largely on high-resolution seismic profiles, largely confirms the descriptive aspects of the Vail (1987) systems tracts, and within them, the depositional systems of the more facies oriented Galloway (1989a) model. In our review, regime theory has provided a dynamic explanation for sequence geometry, and we have been able to interpret depositional sequences as the historical record of homeostatic responses of the depositional system to shifts in regime variables.

The model of Vail (1987), however, may need to be modified to include several additional architectural elements that can be present under certain regime conditions. The additional elements recognized in this study are: (1) a time-transgressive coastal plain unconformity formed by fluvial entrenchment during offlap wedge progradation, which is not related in origin to type 1 sequence boundary slope unconformities; (2) a significant marine erosion surface formed on the shelf during transgression; (3) the importance of ravinement in forming an erosional surface of small time value within the transgressive systems tract; (4) a wedge of back-barrier sediments between the transgressive surface and the ravinement surface; (5) a sedimentary wedge that onlaps the relict slope and downlaps the deep basin formed during transgression; and (6) the formation of extensive coastal-plain–slope unconformities that mimic sequence boundaries at times other than during falling sea-level. A few of these modifications have been suggested by Galloway (1989a).

The Galloway (1989a) model, although giving a less detailed picture of architectural sequence elements, appears essentially realistic in its recognition that other regime variables besides relative sea-level change can largely control sequence architecture. The regime concepts of Galloway (1989a), however, are modified to include more explicitly the effects of regime variables, $R$, related to the rate of relative sea-level change, $D$, the rate of sediment transport, $Q$, the rate of sediment supply and $M$, the sediment grain size.

Specifically we find that the product $Q \cdot M$ is a regime expression for effective sediment supply, while the product $R \cdot D$ is a regime expression for effective accommodation potential. It is suggested that the ratio of accommodation to supply, defined as $¥ = R' \cdot D'/Q' \cdot M'$, is fundamental to the architectural response of sedimentation to changing regime conditions. A semi-quantitative model for each systems tract, the subsystems within it, and its bounding surfaces is proposed based on the theoretical response of *regime topographic profiles* to change in $¥$, the accommodation/supply ratio. Hypothetical examples of sequence development in response to changes in (1) sediment supply without changes in sea-level and (2) sea-level without changes in sediment supply, serve to illustrate the differences between the new regime model and the eustatic model of Vail (1987).

The proposed regime-based model is only a first step in the development of quantitative stratigraphic models of sequence development. Present models do not adequately describe the morphodynamics of intra-sequence and cycle-bounding surfaces. In regime theory (Thorne & Swift, this volume, pp. 33–58), sediment masses and bounding surfaces are the volumes swept out by, and the final positions occupied by, the equilibrium sedimentation surface (Wheeler, 1964). Most existing numerical models, however, approximate bounding surfaces as non-depositional surfaces (Jervey, 1988; Helland-Hansen *et al.*, 1988; Thorne & Swift, this volume, pp. 33–58). Realistic models must also incorporate fluvial morphodynamics into modelling schemes. The problem is a complex one, although fluvial profiles and the regimes that they define have an extensive literature (Chorley *et al.*, 1984).

In conceptual fashion, we have applied the regime based, $¥$-dependent model to aid in interpretation of the observed sequence stratigraphic architecture of the Campanian–Maastrichtian deposits of the USA western interior. This analysis suggests that second-order (*c.* 10 Ma) sequence cycles may be controlled by episodic subsidence caused by in-sequence thrusts while third or higher order (<1 Ma) sequence cycles may be controlled by sediment supply changes caused by out-of-sequence thrusts.

A comparison of regime conditions for various time periods on the northwestern Gulf of Mexico shelf and middle Atlantic shelf sections indicates that the preservation potential of sequence architecture is dependent on the relative roles of sediment input ($Q$) and sea-level change ($R$). As sediment input and subsidence increase relative to the frequency and intensity of sea-level falls, the preservation potential of highstand deposits increases.

Proponents of the Vail (1987) model intermittently acknowledge the role of sediment supply (for instance Posamentier & Vail, 1988, p. 125). Elsewhere, however, sequence architecture is seen as a response to eustacy that can be used to construct cycle charts (Haq *et al.*, 1987). Posamentier & Vail (1988) note that each systems tract is associated with a segment of the eustatic curve, although *the exact timing of any given systems tract in any given basin will depend on local subsidence and sediment supply* (their italics). However, our review supports the greater stress placed on the role of sediment input by Winker (1982) and Galloway (1989b) in their studies of the Cenozoic basin of the Texas Continental margin. We conclude that the *general* timing of any given systems tract is liable to be controlled by the combined effects of sediment supply, sediment transport rate and relative sea-level, on the ratio of accommodation to supply.

This study is the last of six papers applying the concept of regime sedimentation to continental margin deposition. Variation in depositional style as a result of different 'settings' of the regime variables of sediment input, $Q$, sediment grain size, $M$, relative sea-level fluctuation, $R$, and the sediment transport function, $D$, have been touched on at many points in the successive papers. These six papers have shown, at various spatial and temporal scales, that sedimentation systems are governed by interdependent variables, whose mutual adjustments tend towards a state of dynamic equilibrium. As a first approximation, semi-quantitative regime models can be developed in terms of equilibrium surfaces controlled by homeostatic responses of erosion or deposition.

This paper, and those of Thorne et al. (pp. 59–87) and Swift et al. (pp. 89–152 and 153–187) have examined recent models for depositional sequence architecture in the light of the regime concepts expressed by Swift & Thorne (pp. 3–31) and Thorne & Swift (pp. 33–58). We conclude that regime concepts are central to the understanding of stratigraphy. Regime concepts serve to explain the formation of sedimentary facies, bounding surfaces, systems tracts and variations in preserved sequence architecture. These six papers are in a sense a tribute to Harry Wheeler, who realized 27 years ago that all stratigraphy is created by the movement of surfaces on which erosion and deposition is in equilibrium, and who first explored the resulting consequences (Wheeler, 1964). Wheeler (1964) showed that the stratigraphic problem is the inverse of the geomorphological problem. The regime concept has been developed largely by geomorphologists concerned with the denudation of an emergent landscape. Stratigraphers are concerned with the blanketing and burial of a subsiding one, and, thus, regime theory also can be applied usefully to this problem. Simply put, dynamic stratigraphy is a matter of morphodynamics.

Harry Wheeler was clearly a person before his time. He wrote before high-speed computers were widely available, when analytical and numerical techniques were restricted to simple, tractable problems. Stratigraphy is the science of a historical record; a discipline whose great experiments have been conducted only once. Observation and measurement will never lose their primacy in our discipline. However, important advances can be henceforth conducted with the aid of numerical techniques, which can bring geological time and geophysical spatial scales within the range of experiment.

## REFERENCES

ALEXANDER, C.R., NITTROUER, C.A. & DEMASTER, D.J. (1986) High-resolution seismic stratigraphy and its sedimentological interpretation on the Amazon continental shelf. *Continental Shelf Res.* **6**, 337–357.

ASQUITH, D.O. (1970) Depositional topography and major marine environments, late Cretaceous Wyoming. *Bull., Am. Assoc. Petrol. Geol.* **54**, 1184–1224.

BARRELL, J. (1917) Rhythms and measurement of geological time. *Geo. Soc. Am. Bull.* **28**, 754–904.

BATES, R.L. & JACKSON, J.A. (1987) *Glossary of Geology*, 3rd edn. American Geologic Institute, Alexandria, VA, 788 pp.

BAUM, G.R. & VAIL, P.R. (1988) Sequence stratigraphy concepts applied to Paleogene units, Gulf and Atlantic Tertiary Basins. In: *Sea-level Changes: an Integrated Approach* (Eds Wilgus, C.K., Hastings, B.S., Kendall, C.G.St.C., Posamentier, H.W., Ross, C.A. & van Wagoner, J.C.). Soc. Econ. Paleontal. Mineral. Spec. Publ. 42, 124–154.

BEARD, J.H., SANGREE, J.B. & SMITH, L.A. (1982) Quaternary chronology, paleoclimate, depositional sequences and eustatic cycles. *Bull., Am. Assoc. Petrol. Geol.* **66**, 158–169.

BEAUMONT, C. (1981) Foreland basins. *Geophys. J. R. Astron. Soc.* **65**, 291–329.

BELDERSON, R.H. (1964) Sedimentation in the western half of the Irish Sea. *Mar. Geol.* **2**, 147–163.

BELDERSON, R.H., KENYON, N.H. & STRIDE, A.H. (1966) Tidal current fashioning of a basal bed. *Mar. Geol.* **4**, 237–257.

BELKNAP, D.F. & KRAFT, J.C. (1985) Influence of antecedent geology on stratigraphic preservation potential and evolution of Delaware's barrier systems. *Mar. Geol.* **63**, 235–262.

BERRYHILL, H.L. (Ed.) (1987) Late Quaternary facies and structure, northern Gulf of Mexico: interpretations from seismic data. *Am. Assoc. Petrol. Geol. Stud. Geol.* **23**, 289 pp.

BOUMA, A.H., STELTING, C.E. & COLEMAN, J.M. (1985) Mississippi Fan, Gulf of Mexico. In: *Submarine Fans and Related Turbidite Systems* (Eds Bouma, A.H., Normark, W.R. & Barnes, N.E.) pp. 143–150. Springer-Verlag, New York.

BOUMA, A.H., STELTING, C.E. & FEELEY, C.E. (1987) High-resolution seismic reflection profiles. In: *Atlas of Seismic Stratigraphy*, Vol. 1 (Ed. Bally, A.W.). Am. Assoc. Petrol. Geol. Stud. Geol. **27**, 72–95.

BOYD, R., SUTER, J. & PENLAND, S. (1989) Relation of sequence stratigraphy to modern depositional environments. *Geology* **17**, 926–929.

BROMLEY, R.G. (1975) Trace fossils at omission surfaces. In: *The Study of Trace Fossils* (Ed. Frey, R.W.) pp. 339–428. Springer-Verlag, New York.

BROWN, L.F. & FISHER, W.L. (1977) Seismic stratigraphic interpretation of depositional systems: examples from Brazilian rift and pull-apart basins. In: *Seismic Stratigraphy—Applications to Hydrocarbon Exploration* (Ed. Payton, C.E.). Mem., Am. Assoc. Petrol. Geol. **26**, 218–248.

BROWN, L.F. & FISHER, W.L. (1980) Seismic stratigraphic interpretation and petroleum exploration. *Am. Assoc. Petrol. Geol. Educ. Course Notes* **16**, 181 pp.

BROWN, S. (1984) Jurassic. In: *Introduction to the Petroleum Geology of the North Sea* (Ed. Glennie, K.W.) pp. 103–132. Blackwell Scientific, Oxford.

BUTCHER, S.W. (1989) The nickpoint concept and its implications regarding onlap to the stratigraphic record. In: *Quantitative Dynamic Stratigraphy* (Ed. Cross, T.A.) pp. 375–385. Prentice Hall, Englewood Cliffs, NJ.

CARSON, B., BAKER, E.T., HICKEY, B.M., NITTROUER, C.A., DEMASTER, D.J., THORBJARNARSON, K.W. & SNYDERB, G.W. (1986) Modern sediment dispersal and accumulation in Quinault submarine canyon—a summary. *Mar. Geol.* **71**, 1–18.

CHAPPLE, W.M. (1978) Mechanics of thin-skinned fold and thrust belts. *Geol. Soc. Am. Bull.* **89**, 1189–1198.

CHORLEY, R.J., SCHUMM, S.A. & SUGDEN, D.E. (1984) *Geomorphology*. Methuen, New York, 498 pp.

CHRISTIE-BLICK, N., MOUNTAIN, G.S. & MILLER, K.G. (1991) Seismic record of sea level change. In: *Studies in Geophysics*, pp. 116–140. National Academy of Sciences, Washington, DC.

COLEMAN, J.M. & PRIOR, D.B. (1982) Deltaic environments of deposition. In: *Sandstone Depositional Environments* (Eds Scholle, P.A. & Spearing, D.). Mem., Am. Assoc. Petrol. Geol. 31, 139–178.

COLEMAN, J.M. & ROBERTS, H.H. (1988a) Sedimentary development of the Louisiana Continental Shelf related to sea level cycles, part I: sedimentary sequences. *Geomarine Lett.* **8**, 63–108.

COLEMAN, J.M. & ROBERTS, H.H. (1988b) Sedimentary development of the Louisiana Continental Shelf related to sea level cycles, part II: seismic response. *Geomarine Lett.* **8**, 109–119.

CONEY, P.J. (1971) Cordilleran tectonics and North American plate motion. *Am. J. Sci.* **2723**, 603–628.

COVEY, M. (1986) The evolution of foreland basins to steady state: evidence from the western Taiwan foreland basin. In: *Foreland Basins* (Eds Allen & Homewood, P.) Spec. Publ. Int. Assoc. Sediment. **8**, 77–90.

COUMES, F. (1987) Seismic stratigraphy of giant mass movements, including megaturbidites, on recent continental margins. *Geomarine Lett.* **7**, 113–117.

CROSS, T.A. & LESSENGER, M.A. (1988) Seismic stratigraphy. *Ann. Rev. Earth Planet. Sci.* **16**, 319–354.

CURRAY, J.R. (1960) Sediments and history of Holocene transgression, Continental Shelf, Northwest Gulf of Mexico. In: *Recent Sediments, Northwestern Gulf of Mexico* (Eds Shepard, F.P., Phleger, F. & van Andel, Tj.) pp. 221–226. Am. Assoc. Petrol. Geol., Tulsa, OK.

CURRAY, J.R. (1964) Transgressions and regression. In: *Papers in Marine Geology (Shepard Commemorative Volume)* (Ed. Miller, R.L.) pp. 175–203. Macmillan, New York.

CURRAY, J.R., EMMEL, F.J. & CRAMPTON, P.J.S. (1969) Holocene history of a strand plain. In: *Simposio Internacional de Lagunas Costeras* (Eds Castañares, A.A. & Phleger, F.B.) pp. 63–100. Univ. Nac. Autonoma De Mexico, Mexico City.

DAMUTH, J.E., FLOOD, R.D., KOWSMANN, R.O., BELDERSON, R.H. & GORINI, M.A. (1988) Anatomy and growth pattern of Amazon deep-sea fan as revealed by long-range side-scan sonar (GLORIA) and high-resolution seismic studies: *Bull., Am. Assoc. Petrol. Geol.* **72**, 885–911.

DARIGO, N.J. & OSBORNE, R.H. (1986) Quaternary stratigraphy and sedimentation of the inner continental shelf, San Diego County, California. In: *Shelf Sands and Sandstone Reservoirs* (Eds Knight, R.J. & McClean, J.R.). Can. Soc. Petrol. Geol. Mem. **11**, 73–98.

DAVIS, D., SUPPE, J. & DAHLEN, F. (1983) Mechanics of fold and thrust-belts and accretionary wedges. *J. Geophys. Res.* **88**, 1153–1172.

DINGLER, J.R. & ANIMA, R.J. (1989) Subaqueous grain flows at the head of Carmel submarine canyon, California. *J. Sediment. Petrol.* **59**, 280–286.

DINGMAN, S.L. (1984) *Fluvial Hydrology*. W.H. Freeman, New York, 383 pp.

DINGUS, W.F. (1987) *Morphology, paleogeographic setting, and origin of the Middle Wilcox Yoakum Canyon*. MS thesis, University of Texas, Austin, pp. 77.

DINTER, D.A. (1985) Quaternary sedimentation of the Alaskan Beaufort shelf: influence of regional tectonics, fluctuating sea levels, and glacial sediment sources. *Tectonophysics* **114**, 133–161.

DROZ, L., BLAREZ, E., MASCLE, J. & BOKO, S. (1985) The 'Trou Sans Fond' deep-sea fan (off Ivory Coast, Equatorial Atlantic). *Mar. Geol.* **67**, 1–11.

EMERY, K.O. (1968) Relict sediments on continental shelves of the world. *Bull., Am. Assoc. Petrol. Geol.* **52**, 445–464.

EVERTS, C.H. (1987) Continental shelf evolution in response to a rise in sea level. In: *Sea-level Fluctuation and Coastal Evolution*. (Eds Nummedal, D., Pilkey, O. & Howard, J.). Soc. Econ. Paleontol. Mineral. Spec. Publ. 41, 49–58.

FEELEY, M.H., MOORE, T.C., JR., LOUTIT, T.S. & BRYANT, W.R. (1990) Sequence stratigraphy of Mississippi fan relation to oxygen isotope sea level index. *Bull. Am. Assoc. Petrol. Geol.* **74**, 407–424.

FERENTINOS, G., PAPATHEODOROU, G. & COLLINS, M.B. (1988) Sediment transport processes on an active submarine fault escarpment: Gulf of Corinth, Greece. *Mar. Geol.* **83**, 43–61.

FIELD, M.E. & GARDNER, J.V. (1990) Pliocene–Pleistocene growth of the Rio Ebro margin, northeast Spain: a prograding-slope model. *Geol. Soc. Am. Bull.* **102**, 721–733.

FIGUEIREDO, A.G., SWIFT, D.J.P. & SANDERS, J. (1980) Storm-graded layers on inner continental shelves: examples from southern Brazil and the Atlantic coast of the central United States. *Sediment. Geol.* **31**, 171–180.

FISHER, W.L. & MCGOWEN, J. (1967) Depositional systems in the Wilcox Group of Texas and their relationship to the occurrence of oil and gas. *Trans. Gulf Coast Assoc. Geol. Soc.* **17**, 105–125.

FOUCH, T.D., LAWTON, T.F., NICHOLS, D.J., CASHION, W.B. & COBBAN, W.A. (1983) Patterns and timing of synorogenic sedimentation in Upper Cretaceous rocks of central and northeast Utah. In: *Mesozoic Paleogeography of West Central United States* (Eds Reynolds, M.E. & Dolly, E.D.) pp. 305–336. Soc. Econ. Paleontol. Mineral., Denver, CO.

FRAZIER, D.E. (1974) Depositional episodes: their relationship to the Quaternary stratigraphic framework in the northwestern portion of the Gulf Basin. *Tex. Univ. Bur. Econ. Geol. Geol. Circ.* **44-1**, 28 pp.

FULTHORPE, C.S. & CARTER, R.M. (1989) Test of seismic sequence methodology on a Southern Hemisphere passive margin: the Canterbury Basin, New Zealand. *Mar. Petrol. Geol.* **6**, 349–359.

GALLOWAY, W.E. (1981) Depositional architecture of Gulf coastal plain fluvial systems. In: *Recent and Ancient Non-marine Depositional Environments, Models for Exploration* (Eds Etheridge, F.G. & Flores, R.M.). Soc. Econ. Paleontol. Mineral. Spec. Publ. 31, 127–155.

GALLOWAY, W.E. (1989a) Genetic stratigraphic sequences in basin analysis—part I: architecture and genesis of flooding surface bounded depositional units. *Bull., Am. Assoc. Petrol. Geol.* **73**, 125–142.

GALLOWAY, W.E. (1989b) Genetic stratigraphic sequences in basin analysis—part II: application to Northwest Gulf of Mexico Cenozoic basin. *Bull., Am. Assoc. Petrol. Geol.* **73**, 143–154.

GALLOWAY, W.E. & HOBDAY, D.K. (1983) *Terrigenous Clastic Depositional Systems*. Springer-Verlag, New York, 423 pp.

GAYNOR, G.C. & SWIFT, D.J.P. (1988) Shannon Sandstone depositional model: sand ridge formation on the Campanian western interior shelf. *J. Sediment. Petrol.* **58**, 868–880.

GILL, J.R. & COBBAN, W.A. (1966) Regional unconformity in late Cretaceous, Wyoming. In: *Geological Survey Research, 1966*. U.S. Geol. Surv. Res. Pap. 550-B, B7–B20.

GILL, J.R. & COBBAN, W.A. (1973) Stratigraphy and geological history of the Montana Group and equivalent rocks, Montana, Wyoming and North and South Dakota. *U.S. Geol. Surv. Prof. Pap.* **776**, 37 pp.

GOODELL, H.G. (1962) The stratigraphy and petrology of the Frontier Formation of Wyoming. *Seventeenth Annual Field Conference, Guidebook*, pp. 173–210. Wyoming Geological Association, Laramie.

GOODWIN, R.H. & PRIOR, D.B. (1989) Geometry and depositional sequences of the Mississippi Canyon, Gulf of Mexico. *J. Sediment. Petrol.* **59**, 318–329.

GREENLEE, S.M., SCHROEDER, F.W. & VAIL, P.R. (1988) Seismic stratigraphic and geohistory analysis of Tertiary stratigraphy from the continental shelf off New Jersey; calculation of eustatic fluctuations from stratigraphic data. In: *The Atlantic Continental Margin: The Geology of North America*, Vol. I-2 (Eds Sheridan, R.E. & Grow, J.A.) pp. 437–444. Geol. Soc. Am., Boulder, CO.

GROW, J.A. (1980) Deep structure of the Baltimore Canyon Trough in the vicinity of the Cost B-3 well. In: *Geological Studies of the COST B-3 Well, United States Mid Atlantic Slope Area* (Ed. Scholle, P.E.) U.S. Geol. Survey Circ. 833, 117–125.

GULLIVER, F. (1899) Shoreline topography. *Am. Acad. Arts Sci. Proc.* **34**, 151–258.

HAQ, B.U., HARDENBOL, J. & VAIL, P.R. (1987) Mesozoic and Cenozoic chronostratigraphy and cycles of sea-level change. In: *Sea-level Changes: An Integrated Approach* (Eds Wilgus, C.K., Hastings, B.S., Kendall, C.G.St. C., Posamentier, H.W., Ross, C.A. & van Wagoner, J.C.). Soc. Econ. Paleontol. Mineral. Spec. Publ. 42, 72–108.

HAY, W.W., WOLD, C.N. & SHAW, C.A. (1990) Mass-balances paleogeographic maps: background and input requirements. In: *Quantitative Dynamic Stratigraphy* (Ed. Cross, T.) pp. 261–276. Prentice Hall, Englewood Cliffs, NJ.

HAYS, J.D., IMBRIE, J. & SHACKLETON, N.J. (1976) Variations in the Earth's orbit: pacemaker of the ice ages. *Science* **194**, 2212–2232.

HELLAND-HANSEN, W., KENDALL, C.G. ST. C., LERCHE, I. & NAKAYAMA, K. (1988) A simulation of continental basin margin sedimentation in response to crustal movements, eustatic sea level change, and sediment accumulation rates. *Math. Geol.* **20**, 777–802.

HEQUETTE, A. & HILL, P.R. (1989) Late Quaternary seismic stratigraphy of the inner shelf seaward of the Tuktoyaktuk Peninsula, Canadian Beaufort Sea. *Can. J. Earth Sci.* **26**, 1990–2002.

HJULSTROM, F. (1939) Transportation of detritus by moving water, part 1, transportation. In: *Recent Marine Sediments; a Symposium* (Ed. Trask, P.D.) pp. 5–31. Murby, London.

HOLMES, C.W. (1982) Geochemical indices of fine sediment transport, Northwest Gulf of Mexico. *J. Sediment. Petrol.* **52**, 307–321.

HUBBARD, R.J. (1988) Age and significance of sequence boundaries on Jurassic and early Cretaceous rifted continental Margins. *Bull., Am. Assoc. Petrol. Geol.* **72**, 49–72.

JERVEY, M.T. (1988) Quantitative geological modeling of siliciclastic rock sequences and their seismic expression. In: *Sea-level Changes: An Integrated Approach* (Eds Wilgus, C.K., Hastings, B.S., Kendall, C.G.St.C., Posamentier, H.W., Ross, C.A. & van Wagoner, J.C). Soc. Econ. Paleontol. Mineral. Spec. Publ. 42, 47–69.

JOHNSON, D.W. (1919) *Shore Processes and Shoreline Development*, facsimile edn 1965. Hafner, New York, 584 pp.

JOHNSON, D.W. (1931) *Stream Sculpture on the Atlantic Slope*. Columbia University Press, New York, 239 pp.

JORDAN, T.E. (1982) Thrust loads and foreland basin evolution, Cretaceous Western United States. *Bull., Am. Assoc. Petrol. Geol.* **65**, 2506–2520.

KAMOLA, D.L. (1984) Trace fossils from marginal marine facies of The Spring Canyon member, Blackhawk formation (Upper Cretaceous), east central Utah. *J. Paleontol.* **58**, 521–549.

KEEFER, W.R. (1972) Frontier, Cody, and Mesaverde Formations in the Wind River and southern Big Horn Basins, Wyoming, U.S. *U.S. Geol. Surv. Prof. Pap.* **495e**, 23 pp.

KENDALL, C.G.ST.C. & LERCHE, I. (1988) The rise and fall of eustacy. In: *Sea-level Changes: An Integrated Approach* (Eds Wilgus, C.K., Hastings, B.S., Kendall, C.G.St.C., Posamentier, H.W., Ross, C.A. & van Wagoner, J.C.). Soc. Econ. Paleontol. Mineral. Spec. Publ. 42, 3–18.

KENNEDY, W.J. & GARRISON, R.E. (1975) Morphology and genesis of nodular hardgrounds in the Upper Cretaceous of southern England. *Sedimentology* **22**, 311–386.

KENYON, N.H. & TURCOTTE, D.L. (1985) Morphology of a delta prograding by bulk sediment transport. *Geol. Soc. Am. Bull.* **96**, 1457–1465.

KIDWELL, S. (1984) Outcrop features and origin of basin margin unconformities in the lower Chesapeake Group (Miocene, Atlantic Coastal plain. In: *Interregional Unconformities and Hydrocarbon Accumulation* (Ed. Schlee, J.S.). Mem., Am. Assoc. Petrol. Geol. **36**, 37–58.

KINDINGER, J.L. (1988) Seismic stratigraphy of the Mississippi–Alabama shelf and upper continental slope: *Mar. Geol.* **83**, 73–94.

KING, P. (1959) *The Evolution of North America*. Princeton University Press, Princeton, NJ, 190 pp.

KNEBEL, H.J. & CARSON, B. (1979) Small-scale slump deposits, Middle Atlantic Continental Slope, off eastern United States. *Mar. Geol.* **29**, 221–236.

KNEBEL, H.J. & CIRCE, R.C. (1988) Late Pleistocene drainage systems beneath Delaware Bay. *Mar. Geol.* **78**, 285–302.

KOLLA, V. & COUMES, E. (1987) Morphology, internal structure, seismic stratigraphy, and sedimentation of

Indus Fan. *Am. Assoc. Petrol. Geol. Bull.* **71**, 650–677.

KRAFT, J.C., CHRZASTOWSKI, M.J., BELKNAP, D.F., TOSCANO, M.A. & FLETCHER, C.F. (1987) The transgressive barrier lagoon coast of Delaware: morphostratigraphy, sedimentary sequences and responses to relative rise in sea level. In: *Sea Level Fluctuations and Coastal Evolution* (Eds Nummedal, D., Pilkey, O.H. & Howard, J.D.). Soc. Econ. Mineral. Paleontol. Spec. Pap. 41, 129–145.

KRYSTINIK, LF. (1984) Turbidite fans in Upper Cretaceous Pierre Shale, Eagle Basin, Colorado: a new reservoir facies (abstr.). *Bull. Am. Assoc. Petrol. Geol.* **67**, 498.

KUEHL, S.A., NITTROUER, C.A. & DEMASTER, D.J. (1982) Modern sediment accumulation and strata formation on the Amazon Continental Shelf. *Mar. Geol.* **49**, 279–300.

LAWRENCE, D.T., DOYLE, M. & AIGNER, T. (1990) Stratigraphic simulation of sedimentary basins. *Mem., Am. Assoc. Petrol. Geol.* **74**, 273–295.

LEOPOLD, L.B., WOLMAN, M.G. & MILLER, J.P. (1964) *Fluvial Processes in Geomorphology*. Freeman, San Francisco, 522 pp.

LOWMAN, S.W. (1949) Sedimentary facies in the Gulf Coast of Texas. *Geol. Soc. Am. Bull.* **33**, 1939–1997.

LYKOUSIS, V. & CHRONIS, G. (1989) Mechanisms of sediment transport and deposition: sediment sequences and accumulation during the Holocene on the Thermaikos plateau, the continental slope, and basin (Sporadhes basin), northwestern Aegean Sea, Greece. In: *Geological Aspects and Tectonic Evolution of Mediterranean Seas* (Ed. Makris, J.). Mar. Geol. **87**, 15–26.

MALDONADO, A., SWIFT, D.J.P., YOUNG, R.A., HAN, G., NITTROUER, C.E., DEMASTER, D.J., REY, J., PALOMO, C., ACOSTA, J., BALLESTER, C.E. & CASTELLVI, I. (1983) Sedimentation on the Valencia continental shelf: preliminary results. *Continental Shelf Res.* **3**, 195–211.

MATTHEWS, R.K. (1988) Sea-level history. *Science* **241**, 597–599.

McCAVE, I.N. (1973) Mud in the North Sea. In: *North Sea Science* (Ed. Goldberg, E.) pp. 75–100. MIT Press, Cambridge, MA.

McGREGOR, B.A. (1985) Role of submarine canyons in shaping the rise between Lydonia and Oceanographer canyons, Georges Bank. *Mar. Geol.* **62**, 277–293.

McKINNEY, T.F., STUBBLEFIELD, W.L. & SWIFT, D.J.P. (1974) Large scale current lineations on the central New Jersey Shelf: investigations by side scan sonar. *Mar. Geol.* **17**, 79–102.

McMILLEN, K.J. (1991) Seismic stratigraphy of Lower Cretaceous foreland-basin turbidites of the North Slope, Alaska. In: *Seismic Facies and Sedimentary Processes of Submarine Fans and Turbidite Systems* (Eds Link, M. & Weimer, P.). Springer-Verlag, New York, in press.

MEREWETHER, A.E. (1983) The Frontier Formation and Mid-Cretaceous orogeny in the foreland of southwestern Wyoming. *Mountain Geol.* **20**, 121–138.

MIALL, A.D. (1986) Eustatic sea level changes interpreted from seismic stratigraphy: a critique of the methodology with particular reference to the Jurassic North Sea record. *Bull., Am. Assoc. Petrol. Geol.* **70**, 131–137.

MILLER, K.G. & KENT, D.V. (1987) Testing Cenozoic eustatic changes: the critical role of stratigraphic resolution. *Cushman Foundation Foraminiferal Res. Spec. Publ.* **24**, 51–56.

MILLIMAN, J.D. & EMERY, K.O. (1968) Sea levels during the past 35 000 years. *Science* **162**, 1121–1123.

MOLENAAR, C.M. (1985) Subsurface correlations and depositional history of the Nanushuk Group and related strata, North Slope, Alaska. In: *Geology of the Nanushuk Group and Related Rocks, North Slope, Alaska* (Ed. Huffman, A.C., Jr.). U.S. Geol. Surv. Bull. **1614**, 37–60.

MOORE, B. (1982) *Beach Profile Evolution in Response to Changes in Water Level and Wave Heigthts*, M. Sc. thesis. University of Delaware, Newark.

MORLEY, C.K. (1988) Out-of sequence thrusts. *Tectonics* **7**, 539–561.

MORTON, R.A. & PRICE, W.A. (1987) Late Quaternary sea-level fluctuations and sedimentary phases of the Texas coastal plain and shelf. *Soc. Econ. Paleontol. Mineral. Spec. Publ.* **41**, 181–198.

MORTON, R.A. (1977) Historical shoreline changes and their causes, Texas Gulf Coast. *Trans. Gulf Coast Assoc. Geol. Soc.* **27**, 352–364.

MOUGENOT, D. (1985) Progradation of the Portuguese continental margin: interpretation of seismic facies: *Mar. Geol.* **69**, 113–130.

MUTTI, E. (1985) Turbidite systems and their relations to depositional sequences. In: *Provenance of Arenites* (Ed. Zuffa, G.G.) pp. 65–93. D. Reidel, Dordrecht.

NIEDORODA, A.W., SWIFT, D.J.P., FIGUEIREDO, A.G. & FREELAND, G.L. (1985) Barrier island evolution, Middle Atlantic Shelf, U.S.A. Part II: evidence from the shelf floor. *Mar. Geol.* **63**, 363–396.

NITTROUER, C.A., DEMASTER, D.J. & McKEE, B.A. (1984) Fine scale stratigraphy in proximal and distal deposits of sediment in the East China Sea. *Mar. Geol.* **61**, 13–24.

NITTROUER, C.A., DEMASTER, D.J., KUEHL, S.A. & McKEE, B.A. (1986) Association of sand with mud deposits accumulating on continental shelves. In: *Shelf Sands and Sandstones* (Eds Knight, R.J. & McLean, J.R.). Can. Soc. Petrol. Geol. Mem. 11, 17–25.

NUMMEDAL, D. & SWIFT, D.J.P. (1987) Transgressive stratigraphy at sequence bounding unconformities— some principles derived from Holocene and Cretaceous examples. In: *Sea-level Fluctuation and Coastal Evolution* (Ed. Nummedal, D.). Soc. Econ. Paleontol. Mineral. Spec. Publ. 41, 241–260.

NUNN, J.A. (1990) Relaxation of continental lithosphere: an explanation for Late Cretaceous reactivation of the Sabine Uplift of Louisiana–Texas. *Tectonics* **9**, 341–359.

OERTEL, G.F., KEARNEY, M.S., LEATHERMAN, S.P. & WOO, H.J. (1989) Anatomy of a barrier platform, outer barrier lagoon, Southern Delmarva Peninsula, Virginia. *Mar. Geol.* **88**, 303–318.

PACHT, J.A., BOWIN, J.H. & SHAFFER, B.L. & POTTORF, B.R. (1990) Sequence-stratigraphy of Plio-Pleistocene depositional facies in the offshore Lousiana south additions. In: *Gulf Coast Sequence Stratigraphy: Concepts and Practices*, Eleventh Annual GCSSEPM Foundation Research Conference, 2–5 December 1990, Houston, TX, pp. 269–285. Earth Enterprises, Austin.

PARK, S.C. & YOO, D.G. (1988) Depositional history of Quaternary sediments on the continental shelf of the southeastern coast of Korea (Korea Strait). *Mar. Geol.* **79**, 65–75.

PIPER, D.J.W., KONTOPOULOS, N., ANAGNOSTOU, C.,

CHRONIS, G. & PANAGOS, A.G. (1990) Modern fan deltas in the western Gulf of Corinth, Greece. *Geomarine Lett.* **10**, 5–12.

PIPER, D.J.W. & NORMARK, W.R. (1989) Late Cenozoic sea-level changes and the onset of glaciation: impact on continental slope progradation off eastern Canada. *Mar. Petrol. Geol.* **6**, 236–347.

PITMAN, W.C., III (1978) Relationship of eustacy and stratigraphic sequences of passive margins. *Geol. Soc. Am. Bull.* **89**, 1389–1403.

PITMAN, W.C., III (1991) *Earth Planet. Sci. Lett.*, in press.

PITMAN, W.C. & GOLOVCHENKO, X. (1983) The effect of sea level change on the shelf edge and slope of passive margins. In: *The Shelfbreak: Critical Interface on Continental Margins* (Eds Stanley, D.J. & Moore, G.T.). Soc. Econ. Mineral. Paleontol. Spec. Publ. **33**, 41–58.

POAG, W.C. (1985) Depositional history and stratigraphic reference section for central Baltimore Canyon Trough. In: *Geologic Evolution of the United States Atlantic Margin* (Ed. Poag, C.W.) pp. 217–264. Hutchinson Ross, New York.

POSAMENTIER, H.W. & VAIL, P.R. (1988) Eustatic controls on clastic deposition, II—sequence and systems tract models. In: *Sea-level Changes: An Integrated Approach* (Eds Wilgus, C.K., Hastings, B.S., Kendall, C.G.St.C., Posamentier, H.W., Ross, C.A. & van Wagoner, J.C.). Soc. Econ. Paleontol. Mineral. Spec. Publ. **42**, 124–154.

POSAMENTIER, H.W., JERVEY, M.T & VAIL, P.R. (1988) Eustatic controls on clastic deposition, I—Conceptual framework. In: *Sea-level Changes: an Integrated Approach* (Eds Wilgus, C.K., Hastings, B.S., Kendall, C.G. St. C., Posamentier, H.W., Ross, C.A. & van Wagoner, J.C.). Soc. Econ. Paleontol. Mineral. Spec. Publ. **42**, 109–124.

QUINLAN, G.M. & BEAUMONT, C. (1984) Appalachian thrusting, lithospheric thrusting, flexure and the Paleozoic stratigraphy of the eastern interior of North America. *Can. J. Earth Sci.* **21**, 973–996.

RAWSON, P.F. & RILEY, L.A. (1982) Latest Jurassic–early Cretaceous events and the 'Late Cimmerian Unconformity' in the North Sea area. *Bull., Am. Assoc. Petrol. Geol.* **66**, 2628–2648.

ROBERTS, H.H. & COLEMAN, J.M. (1988) Lithofacies characteristics of shallow expanded and condensed sections of the Louisiana distal shelf and upper slope. *Trans. Gulf Coast Assoc. Geol. Soc.*, **37**, 272–301.

SALGE, U. & WONG, H.K. (1988) Seismic stratigraphy and Quaternary sedimentation in the Skagerrak (northeastern North Sea). *Mar. Geol.* **81**, 159–174.

SANGREE, J.B. & WIDMIER, J.M. (1977) Seismic interpretation of clastic depositional facies. In: *Seismic Stratigraphy—Applications to Hydrocarbon Exploration* (Ed. Payton, C.E.). Mem., Am. Assoc. Petrol. Geol. **26**, 165–184.

SATTERFIELD, W.M. & BEHRENS, E.W. (1990) A late Quaternary canyon/channel system, northwest Gulf of Mexico continental slope. *Mar. Geol.* **92**, 51–67.

SAVOYE, B., COCHONAT, P. & PIPER, D.J.W. (1991) Seismic evidence for a complex slide near the wreck of the *Titanic*: model of an instability corridor for non-channeled gravity events. *Mar. Geol.* **91**, 281–298.

SCHEIDEGGER, A.E. (1961) *Theoretical Geomorphology*. Springer-Verlag, Berlin, 334 pp.

SCHROEDER, W.W. (1977) Sea: truth and environmental characterization studies of Mobile Bay, Alabama, utilizing ERTS-1 data collection platforms. *Remote Sens. Environ.* **6**, 27–43.

SCHUMM, S.A. (1976) Episodic erosion: a modification of the geomorphic cycle. In: *Theories of Landform Development* (Eds Melhorn, W.N. & Flemal, R.C.) pp. 69–86. Publications in Geomorphology, State University of New York, Binghamton, New York.

SHEPARD, F.P. (1948) *Submarine Geology*. Harper & Bros, New York, 348 pp.

SIDENER, B.R., GARTNER, S. & BRYANT, W.R. (1978) Late Pleistocene geologic history of Texas outer continental shelf and upper continental slope. In: *Framework, Facies, and Oil-trapping Characteristics of the Upper Continental Margin* (Eds Bouma, A.H., Moore, G.T. & Coleman, J.M.). Am. Assoc. Petrol. Geol. Stud. Geol. **7**, 243–266.

STAHL, L., KOCZAN, J. & SWIFT, D.J.P. (1974) Anatomy of a shore-face-connected sand ridge on the New Jersey shelf: implications for the genesis of the Shelf Surficial sand sheet. *Geology* **2**, 117–120.

STAMP, L.D. (1921) On cycles of sedimentation in the Eocene strata of the Anglo-Franco Belgian basin. *Geol. Mag.* **58**, 108–114, 194–200.

STANLEY, D.J., CULVER, S.J. & STUBBLEFIELD, W.L. (1986) Petrologic and foraminiferal evidence for active downslope transport in Wilmington Canyon. *Mar. Geol.* **69**, 207–218.

STONE, W.D. (1972) Stratigraphy and exploration of the Lower Cretaceous Muddy Formation, Northern Powder River Basin, Montana and Alaska. *Mountain Geol.* **9**, 355–378.

STOW, D.A.V. (1986) Deep clastic seas. In: *Sedimentary Environments and Facies*, 2nd edn (Ed. Reading, H.G.) pp. 399–444. Blackwell Scientific, Oxford.

STUBBLEFIELD, W.L & SWIFT, D.J.P. (1976) Ridge development as revealed by sub-bottom profiles on the central New Jersey Shelf. *Mar. Geol.* **20**, 315–334.

SUTER, J.H. & BERRYHILL, H.L., JR. (1985) Late Quaternary shelf margin deltas, Northwest Gulf of Mexico. *Bull., Am. Assoc. Petrol. Geol.* **69**, 77–91.

SUTER, J.R., BERRYHILL, H.L. & PENLAND, S. (1987) Late Quaternary sea level fluctuations and depositional sequences, southwest Louisiana Continental Shelf. In: *Sea-level fluctuation and Coastal Evolution* (Ed. Nummedal, D.). Soc. Econ. Paleontol. Mineral. Spec. Publ. 41, 199–219.

SWIFT, D.J.P. (1968) Coastal erosion and transgressive stratigraphy. *J. Geol.* **76**, 444–456.

SWIFT, D.J.P., STANLEY, D.J. & CURRAY, J.R. (1971) Relict sediments: a reconsideration. *J. Geol.* **79**, 322–346.

SWIFT, D.J.P., KOFOED, J.W., SAULSBURY, F.P. & SEARS, P. (1972) Holocene evolution of the shelf surface, central and southern Atlantic shelf of North America. In: *Shelf Sediment Transport: Process and Pattern* (Eds Swift, D.J.P., Duane, D.B. & Pilkey, O.H.) pp. 499–574. Dowden Hutchinson & Ross, Stroudsberg, PA.

SWIFT, D.J.P., DUANE, D.B. & MCKINNEY, T.F. (1973) Ridge and swale topography of the Middle Atlantic Bight: secular response to the Holocene hydraulic regime. *Mar. Geol.* **15**, 227–247.

SWIFT, D.J.P., NELSEN, T.A., MCHONE, J., HOLLIDAY, B.,

PALMER, H. & SHIDELER, G. (1977) Holocene evolution of the inner shelf of southern Virginia. *J. Sediment. Petrol.* **47**, 1454–1474.

SWIFT, D.J.P., THORNE, J.A. & NUMMEDAL, D. (1985) Sequence stratigraphy in foreland basins: inferences from the Cretaceous Western Interior. Paper 4846, *17th Annual Offshore Technology Conference*, Houston, TX, 6–9 May 1985.

SWIFT, D.J.P., THORNE, J.A. & OERTEL, G.F. (1968) Fluid process and sea floor response on a storm dominated shelf: Middle Atlantic shelf of North America. Part II: response of the shelf floor. In: *Shelf Sands and Sandstone Reservoirs* (Eds Knight, R.J. & McLean, J.R.). Can. Soc. Petrol. Geol. Mem. **11**, 191–211.

SWIFT, D.J.P., HUDLESON, P.M., BRENNER, R.L. & THOMPSON, P. (1987) Shelf construction in a foreland basin: storm beds, shelf sandbodies, and shelf–slope depositional sequences in the Upper Cretaceous Mesaverde Group, Book Cliffs, Utah. *Sedimentology* **34**, 423–457.

TANKARD, A.J. (1986) On the depositional response to thrusting and lithospheric flexure: examples from the Appalachian and Rocky Mountain basins. In: *Foreland Basins* (Eds Allen, P. & Homewood, P.). Spec. Publ. Int. Assoc. Sediment. **8**, 369–392.

TESSON, M.B., GENSOUS, G., ALLEN, P. & RAVENNE, CH. (1990) Late Quaternary deltaic lowstand wedges on the Rhône continental shelf, France. *Mar. Geol.* **91**, 325–332.

THORNE, J.A., (1991) An analysis of the implicit assumptions of the methodology of seismic sequence stratigraphy. In: *Studies in Continental Margin Geology* (Ed. Watkins, J.S.). Am. Assoc. Petrol. Geol. Spec Publ., in press.

THORNE, J.A. & SWIFT, D.J.P. (1985) A theory of tidal amplification in coastal embayments and open shelves: application to ancient marine sedimentation. *Expanded Abstracts, Symposium on Tidal Sedimentation*, Utrecht, 28–30 August, pp. 85–89.

TRINCARDI, F. & NORMARK, W.R. (1989) Pleistocene Suvero slide, Paola basin, southern Italy. *Mar. Petrol. Geol.* **6**, 324–335.

TURCOTTE, D.L. & SCHUBERT, G. (1982) *Geodynamics: Applications of Continuum Physics to Geologic Problems*. Wiley, New York, 450 pp.

TWICHELL, W.C. (1983) Geology of the head of Lydonia Canyon, U.S. Atlantic outer continental shelf. *Mar. Geol.* **54**, 91–108.

TWICHELL, W.C., MCCLENNEN, C.E. & BUTMAN, B. (1981) Morphology and processes associated with the accumulation of the fine-grained sediment deposit of the southern New England Shelf. *J. Sediment. Petrol.* **51**, 269–280.

TYE, R.S. & KOSTERS, E.C. (1986) Styles of interdistributary basin sedimentation: Mississippi Delta Plain, Louisiana. *Trans. Gulf Coast Assoc. Geol. Soc.*, **36**, 575–587.

VAIL, P.R. (1987) Seismic stratigraphy interpretation using sequence stratigraphy. Part 1: seismic stratigraphy interpretation procedure. In: *Atlas of Seismic Stratigraphy*, Vol. 1 (Ed. Bally, A.W.). Am. Assoc. Petrol. Geol. Stud. Geol. **27**, 1–10.

VAIL, P.R. & TODD, R.G. (1981) Northern North Sea Jurassic unconformities, chronostratigraphy and sea-level changes from seismic stratigraphy. *Petroleum Geology of the Continental Shelf of North-West Europe* (Eds Illing, L.V. & Hobson, G.D.) pp. 216–235. Heydon, London.

VAIL, P.R., MITCHUM. R.M., JR., THOMPSON, S., III, SANGREE, J.R., DUBB, J.N. & HASLID, W.G. (1977) Sequence stratigraphy and global changes of sea level. In: *Seismic Stratigraphy — Applications to Hydrocarbon Exploration* (Ed. Payton, C.E.). Mem., Am. Assoc. Petrol. Geol. **26**, 49–205.

VAIL, P.R., HARDENBOL, J. & TODD, R.G. (1984) Jurassic unconformities, chronostratigraphy, and sea-level changes from seismic stratigraphy and biostratigraphy. In: *Interregional Unconformities and Hydrocarbon Accumulation* (Ed. Schlee, J.S.). Mem., Am. Assoc. Petrol. Geol. **36**, 1129–1144.

VAN DEN BOLD, M.C., MOSLOW, T. & COLEMAN, J. (1987) Origin and timing of seafloor erosion on the Louisiana continental slope. *Trans. Gulf Coast Assoc. Geol. Soc.* **37**, 487–498.

VAN STRAATEN, L.M.J.U. (1965) Sedimentation in the northeastern part of the Adriatic Sea. In: *Submarine Geology and Geophysics* (Eds Whittard, W.H. & Bradshaw, W.) pp. 143–162. Butterworths, London.

VAN WAGONER, J.C., MITCHUM, R.M., JR., POSAMENTIER, H.W. & VAIL, P.R. (1987) Part 2, key definitions of sequence stratigraphy. In: *Atlas of Seismic Stratigraphy*, Vol. 1 (Ed. Bally, A.W.). Am. Assoc. Petrol. Geol. Stud. Geol. **27**, 11–1.

VAN WAGONER, J.C., POSAMENTIER, H.W., MITCHUM, R.M, JR., VAIL, P.R., SARG, J.F., LOUTIT, T.S. & HARDENBOL, J. (1988) An overview of the fundamentals of sequence stratigraphy and key definitions of sequence stratigraphy. In: *Sea-level Changes: An Integrated Approach* (Eds Wilgus, C.K., Hastings, B.S., Kendall, C.G.St.C., Posamentier, H.W., Ross, C.A. & van Wagoner, J.C.). Soc. Econ. Paleontol. Mineral. Spec. Publ. **42**, 39–47.

VAN WAGONER, J.C., MITCHUM, R.M., JR., CAMPION, K.M. & Rahmanian, V.D. (1990) Siliciclastic sequence stratigraphy in well logs, cores, and outcrops. *Am. Assoc. Petrol. Geol. Meth. Expl. Ser.* **7**, 55 pp.

VINCENT, C. (1986) Processes affecting sand transport on a storm-dominated shelf. In: *Shelf Sands and Sandstone Reservoirs* (Eds Knight, R.J. & McLean, J.R.). Can. Soc. Petrol. Geol. Mem. **11**, 121–132.

VORREN, T.O., HALD, M. & THOMSEN, E. (1984) Quaternary sediments and environments on the continental shelf off northern Norway. *Mar. Geol.* **57**, 229–257.

WARD, L.W. & Strickland, G.L. (1985) Outline of Tertiary stratigraphy and depositional history of the U.S. Atlantic coastal plain. In: *Geologic Evolution of the United States Atlantic Margin* (Eds Poag, C.W.) pp. 87–123. Hutchinson Ross, New York.

WEIMER, R.J. (1983) Relationship of unconformities, tectonics and sea level changes, Cretaceous of Denver Basin. In: *Mesozoic Paleogeography of the Rocky Mountains* (Eds Reynolds, M.W. & Dolly, E.D.) pp. 359–376. Soc. Econ Paleontol. Mineral. Rocky Mountain Section.

WEIMER, P. (1989) Sequence stratigraphy of the Mississippi Fan (Plio-Pleistocene), Gulf of Mexico. *GeoMarine Lett.* **9**, 185-272.

WEIMER, P. (1990) Sequence stratigraphy, facies geometries, and depositional history of the Mississippi fan, Gulf of Mexico. *Bull., Am. Assoc. Petrol. Geol.* **74**, 425-453.

WHEELER, H.E. (1964) Base level, lithosphere surface and time-stratigraphy. *Geol. Soc. Am. Bull.* **75**, 599-610.

WILTCHKO, D.V. & DORR, J.A. JR. (1983) Timing of deformation in overthrust belt and foreland of Idaho, Wyoming, and Utah. *Bull., Am. Assoc. Petrol. Geol.* **67**, 1304-1322.

WINKER, C.D. (1982) Cenozoic shelf margins, Northwest Gulf of Mexico Basin. *Trans, Gulf Coast Assoc. Geol. Soc.* **32**, 427-448.

WOODBURY, H.G., SPOTTS, J.H. & AKERS, W.H. (1978) Gulf of Mexico continental-slope sediments and sedimentation. In: *Framework, Facies, and Oil-trapping Characteristics of the Upper Continental Margin* (Eds Bouma, A.H., Moore, G.T. & Coleman, J.M.). Am. Assoc. Petrol. Geol. Stud. Geol. 7, 117-137.

WRIGHT, L.D. & COLEMAN, J.M. (1974) Mississippi River mouth processes: effluent dynamics and morphologic development. *J. Geology* **82**, 751-778.

WRIGHT, R. (1986) Cycle stratigraphy as a paleogeographic tool: Point Lookout Sandstone, southeastern San Juan Basin, New Mexico. *Geol. Soc. Am. Bull.* **96**, 661-673.

# Regressive Shelf and Coastal Sandstones

# Facies architecture of a falling sea-level strandplain, Doce River coast, Brazil

J.M.L. DOMINGUEZ[1] and H.R. WANLESS

*Rosenstiel School of Marine and Atmospheric Sciences, Marine Geology and Geophysics Division, University of Miami, 4600 Rickenbacker Causeway, Miami, Florida 33149-1098, USA*

## ABSTRACT

The Doce River strandplain, located on the east coast of Brazil, comprises Holocene and Pleistocene beach-ridge terraces, fluvial, lagoonal, freshwater swamp and mangrove swamp deposits.

The Holocene evolutionary history of this strandplain was controlled by the interaction of sea-level changes, riverine sediment supply and wave reworking. The rise in sea-level that followed the last glacial period drowned the Pleistocene strandplain of the Doce River. A barrier-island–lagoonal system formed as a result of this drowning. After 7 ka, with decreasing rates of sea-level rise, the Doce River began emptying into the palaeolagoon and built an intra-lagoonal delta.

During deposition of the late Holocene portion of the strandplain, sea-level dropped 4–5 m from its maximum at about 5.1 ka. During this drop the shoreline downdrift of the river mouth advanced through a mechanism of coastwise extension of sand spits, generating in this process elongate lagoons that later became low-lying areas separated from each other by beach-ridge sets. From 3.8 to 3.5 ka, sea-level rapidly rose 2–3 m above present. The aforementioned low-lying areas were invaded by the sea and became interconnected bays. The updrift side, which prograded through the successive accretion of beach ridges, remained relatively unaffected during this rise. After 3.5 ka, progradation resumed in association with a drop in sea-level.

Theoretical considerations of the Bruun's rule of translation of the shoreface profile shows that during a rise in sea-level, lower shoreface deposits will be preserved preferentially. Conversely, during a drop, only those sediments deposited in the foreshore/upper shoreface will survive. Testing of this model in the study area did not produce conclusive results, probably because the large volume of riverborne sediments masked the effect of sea-level changes during deposition of the littoral zone sequences.

## INTRODUCTION

The purpose of this study is to document how changes in relative sea-level during the Holocene have affected the three-dimensional structure and facies evolution of the Quaternary deposits occurring on the Doce strandplain, located on the east coast of Brazil (Fig. 1).

Studies of coastal evolution of beach-ridge strandplains along the east coast of Brazil (Dominguez, 1984; Dominguez *et al.*, 1987), show that under conditions of sea-level rise, barrier-island–lagoonal systems are the dominant mode of sedimentation. Rivers do not reach the inner shelf during these periods, but tend to construct deltas within lagoons, bays and estuaries. Beach-ridge strandplains are virtually absent. By contrast, sea-level lowering poses highly unfavourable conditions for the genesis and the maintenance of barrier-island–lagoonal systems. Lagoons and bays become emergent and portions of the sandy shoreline rapidly prograde, giving origin to beach-ridge strandplains.

On the USA Atlantic and Gulf coasts, relative sea-level has been rising continuously during the

[1] Present Address: Instituto de Geociencias—UFBa, Rua Caetano Moura 123, Federação 40210, Salvador, Bahia, Brazil.

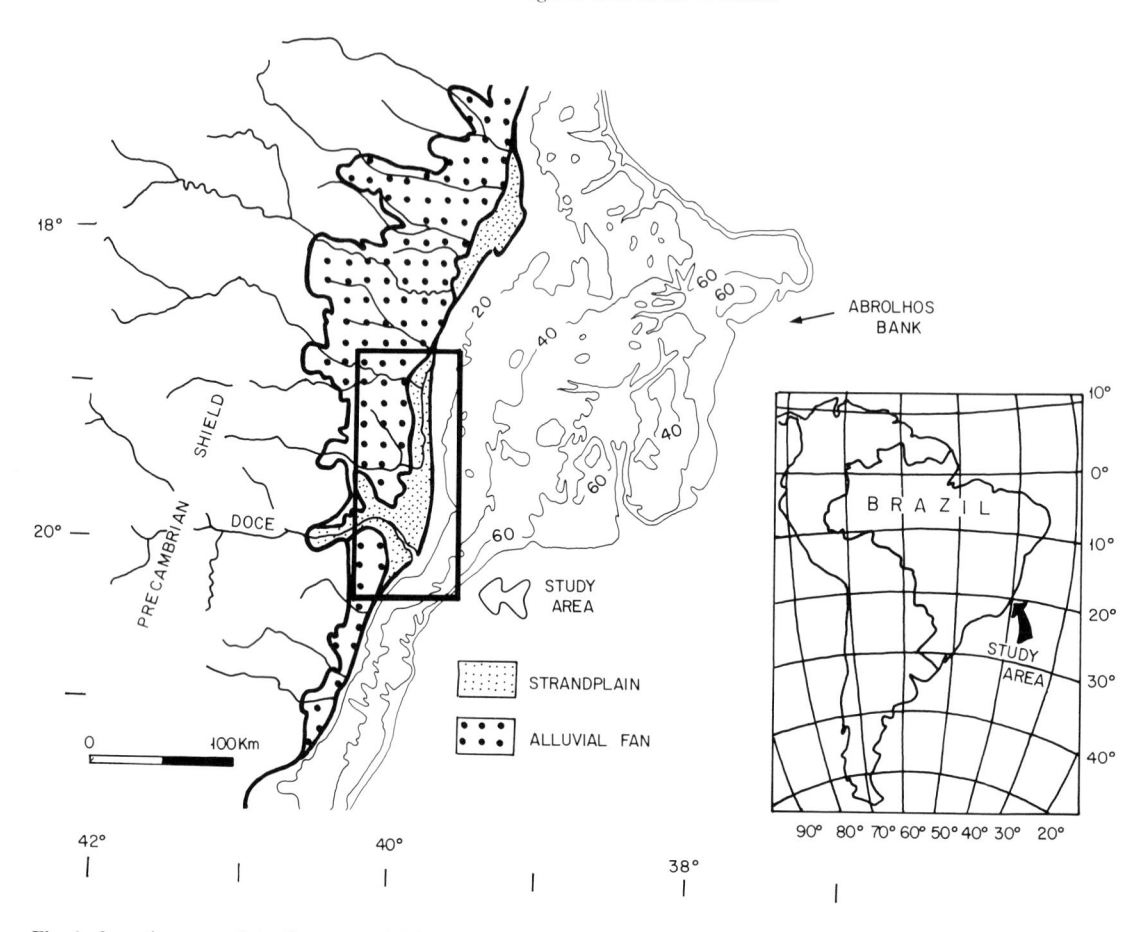

**Fig. 1.** Location map of the Doce strandplain. Also depicted are the three physiographic units of the emerged part of the east Brazil continental margin, and the physiography of the continental shelf. Isobaths are in metres.

Holocene, and as a result, strandplains are not common. Barrier-island–lagoonal systems are the dominant coastal features. By contrast, along the east coast of Brazil, where relative sea-level has fallen during the last 5000 years, barrier-island–lagoonal systems, although abundant in the past (during a period of rising sea-level), are absent today and have been replaced by extensive beach-ridge strandplains.

Beach-ridge strandplains occur associated with the major rivers emptying on to the east coast of Brazil. During deposition of these strandplains, sea-level has dropped 4–5 m from its maximum about 5.1 ka. This drop, however, was not regular but interrupted by high-frequency sea-level oscillations as large as 2–3 m acting on time-scales of 200–300 years (Martin *et al.*, 1979).

Prior to this study, the coast of Nayarit, Mexico (Curray *et al.*, 1967), has been the sole well-documented example of a modern beach-ridge strandplain, and is commonly cited in textbooks and scientific papers. Curray's classical paper, however, lacks both a documentation of the Quaternary sea-level changes and sufficient information for interpreting the rock record (e.g. structural facies sequence, sedimentary structures, etc.). This study of the Doce strandplain, therefore, provides: (1) a modern analogue for the accumulation of ancient strandplain sequences; and (2) a modern analogue for the response of strandplain systems to changes in relative sea-level.

The approach of this study was to use aerial photographs to map the various Quaternary deposits occurring on the Doce strandplain. This mapping was integrated with: (1) the knowledge of Quaternary sea-level history and coastal evolution along the east coast of Brazil; (2) previously reported radiocarbon dates for the region; and (3) wave and wind patterns, to derive a detailed two-dimensional evolutionary history for the Doce strandplain. The vibracoring data obtained during this study helped in improving this evolutionary history and to unravel the three-dimensional architecture of the strandplain.

The way this paper is organized closely follows the approach used in this study. We first describe the two-dimensional evolutionary history of the strandplain then document the sedimentary record of: (1) lagoonal sequences deposited during two episodes of sea-level rise (7.0–5.1 ka and 3.8–3.5 ka); and (2) littoral zone sequences deposited during the general drop in sea-level that followed the 5.1 ka highstand.

One might suggest that this approach sounds like circular reasoning and that we should start at a lower level of abstraction and generalization (i.e. sedimentary facies description) and then proceed to higher levels of abstraction (i.e. sea-level history then evolutionary history). We do not agree. Late Quaternary sea-level history has exerted a fundamental control in determining styles of sedimentation in the coastal zone (Dominguez et al., 1987). It is our opinion that without a good control of the sea-level history it is extremely difficult to interpret the evolution of coastal areas and their three-dimensional architecture. A reasonably good sea-level history control (to be discussed in the next sections) was already available for the east coast of Brazil before we started this study. Also a very simplified two-dimensional evolutionary model of the Doce strandplain had been published previously in Dominguez et al. (1987). These are the reasons we decided to start at a higher level of abstraction, first improving the two-dimensional evolutionary model of the strandplain. We then used this improved model to choose vibracoring sites.

## REGIONAL SETTING

The Doce River strandplain has a total area of 2500 km$^2$ stretching along almost 180 km of the shoreline (Fig. 1). A line of fossil sea-cliffs, carved into coalescing alluvial fans of Tertiary age, marks the landward limit of the strandplain.

The Doce River drains an area of 83 000 km$^2$, with average annual discharge of 693 m$^3$ s$^{-1}$. Peak discharges of as much as 2000 m$^3$ s$^{-1}$ are reported for summer months (Bandeira et al., 1975). The Doce River is estimated to carry an annual load of 4 × 10$^6$ tons of suspended solids (Bandeira et al., 1975).

The climate in the coastal zone is hot and humid, with a well-defined dry season (autumn–winter). During the summer, northeasterly and easterly trade winds originating in the South Atlantic anticyclone cell blow along the east coast of Brazil. During the winter, southeasterly trades dominate. As winter is also the season when the anticyclone cell tends to be more intense, the trade winds are strongest in the winter and weakest in the summer. During the winter, south-southeast winds associated with the advance of the polar air masses reinforce the southeasterly trades generated at the anticyclone cell. Gale force winds (25 m s$^{-1}$) have been reported associated with the advance of these polar air masses at the Doce strandplain area (Bandeira et al., 1975).

The east coast of Brazil is microtidal. Spring tide range at the Doce strandplain is 1.5 m (Diretoria de Hidrografia e Navegação, 1981). Tides are semi-diurnal.

No direct, long-term wave measurements are available for the study area. Thus, the wave climate affecting the Doce strandplain was deduced from published wave statistics (Hogben & Lumb, 1967; US Navy, 1978). These statistics show that north-northeast waves are present all year; however, beginning in April and extending through to August, east-southeast waves comprise a significant percentage of the wave-trains impinging on the study area. Thus, during the spring (September–November) and the summer (December–February) north-northeast waves with heights of 1.0 m and periods of 5 s or less dominate on the east coast of Brazil. During the autumn (April–May) and the winter (June–August); east-southeast waves with heights of 1.5 m and periods of 6–7 s are very distinctive.

Three important physiographic units characterize the east coast of Brazil at the study area (Fig. 1): (1) the Quaternary strandplain, object of attention in this study; (2) the Tertiary coalescing alluvial fans, which form an almost flat, seaward-dipping surface, 10–40 km wide, and 10–100 m in elevation; and (3) the hilly Precambrian shield, 200–1000 m in elevation.

The most important feature of the continental shelf seaward of the study area is the Abrolhos Bank (area of 48 000 km$^2$). This bank has a very flat surface with a declivity of 45 cm km$^{-1}$. The shelf edge occurs at a depth of only about 70 m. The inner

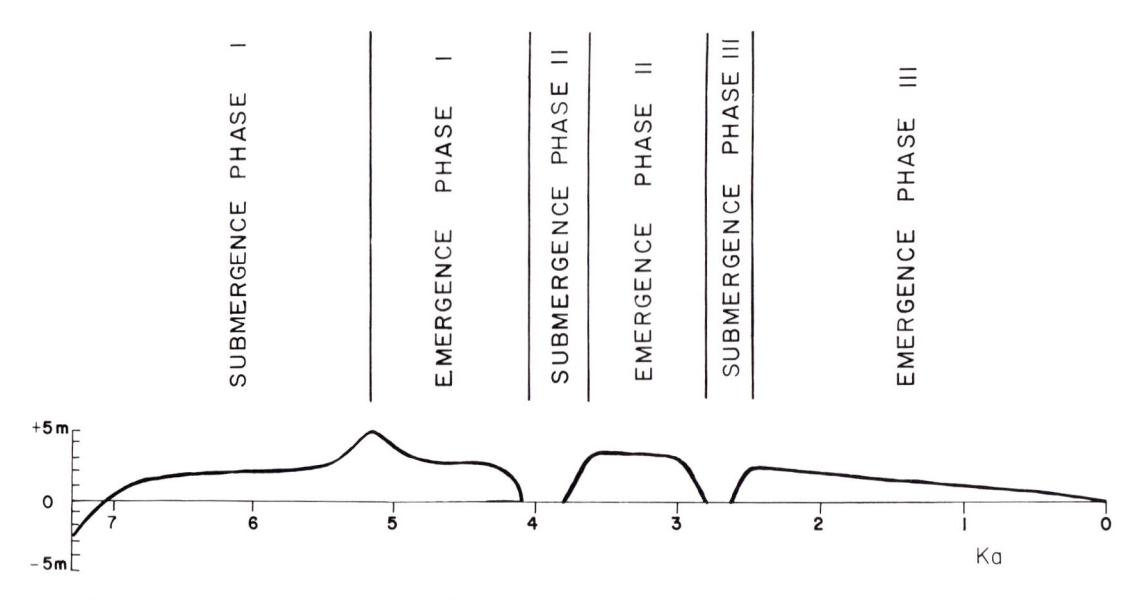

Fig. 2. Sea-level curve for the east coast of Brazil. (After Martin et al., 1979.)

shelf is floored mainly by quartz-feldspathic sands, with some muddy sand, largely derived from the Precambrian shield. This inner shelf sand sheet is locally covered by terrigenous muds, as off the Doce River mouth. At this region, the mud contains a substantial amount of rock fragments, mica (up to 25%) and plant fragments (Melo et al., 1975). The middle and outer shelves are covered by an almost pure carbonate sand, much of which is relict or palimpsest. Seaward of the Doce River, however, quartzose sands cover much of the middle shelf in addition to the inner shelf. Large parts of the Abrolhos Bank appear to be covered by a continuous, dense, reef-like growth of encrusting coralline algae (Melo et al., 1975).

During the last 10 years relative sea-level curves have been constructed for different sectors of the east-southeast coast of Brazil (Martin & Suguio, 1978; Martin et al., 1979, 1985; Suguio et al., 1979). Sea-level indicators used in constructing the curves included: (1) vermetid gastropod incrustations; (2) beach-ridge terraces; (3) sandstone reefs (cemented upper shoreface deposits); (4) beach rocks (cemented foreshore deposits); (5) coral; (6) calcareous algae; (7) shell middens; and (8) lagoonal sediments. These curves show an important sea-level highstand around 5.1 ka, followed by a general drop of relative sea-level since that time (Fig. 2). This drop was not regular, but interrupted by high-frequency sea-level oscillations. In each of these oscillations, sea-level first dropped to slightly below the present level to be followed later by a sharp rise to take sea-level approximately to its former position. These high-frequency oscillations thus define, for the east-southeast coast of Brazil, three phases of submergence or periods of relative sea-level rise (7.0–5.1 ka, 3.8–3.5 ka and 2.7–2.5 ka) alternating with three phases of emergence or periods of relative sea-level fall (5.1–3.8 ka, 3.5–2.7 ka and 2.5 ka to the present) (Fig. 2).

## METHODS

Aerial photographs were used for detailed mapping of the various sedimentary deposits occurring on the Doce strandplain. Careful mapping of beach-ridge sets, their relative orientation, and truncation patterns were emphasized during this phase of the research. Beach-ridges constitute lines of growth of the shoreline, thus their careful documentation enables the history of development of the strandplain to be reconstructed in detail. This mapping was integrated with (1) radiocarbon dates reported for the Doce strandplain by Suguio et al. (1981) and (2) previous knowledge of the Quaternary sea-level history for the region, to reconstruct the two-dimensional history of the strandplain. The

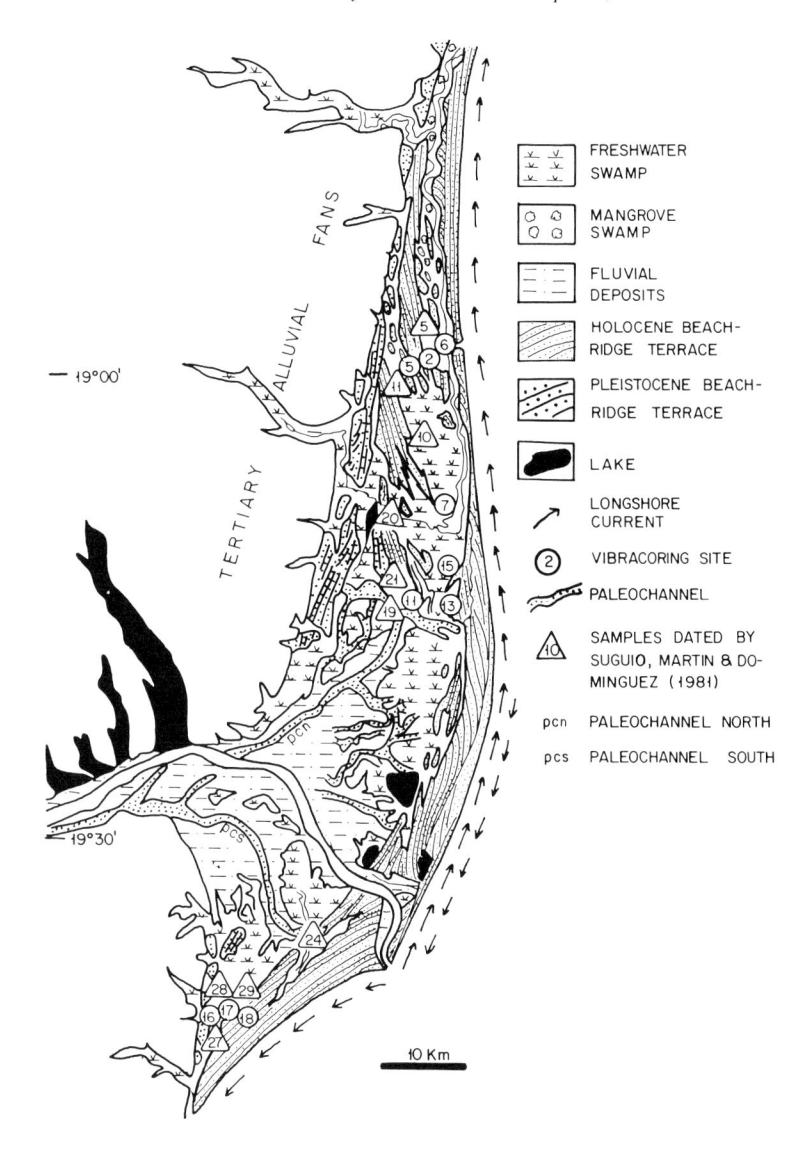

**Fig. 3.** Map showing major geomorphological units occurring on the Doce strandplain. Numbered circles show location of vibracoring sites. Numbered triangles show collection sites of samples dated by Suguio *et al.* (1981). These samples were collected in spoils of drainage ditches and drainage canals. Site 5, $3540 \pm 150$ years (shells); site 10, $3300 \pm 300$ (shells); site 11, $3940 \pm 150$ years (wood); site 19, $6020 \pm 200$ years (*Ostrea* sp.); site 20, $3550 \pm 150$ years (*Ostrea* sp.); site 21, $5550 \pm 200$ years (shells); site 24, $3430 \pm 150$ years (wood); site 27, $6280 \pm 200$ years (*Anomalocardia brasiliana*); site 28—sample RD-38 (*A. brasiliana*), $5600 \pm 200$ years, and sample RD-33 (wood), $6900 \pm 200$ years); site 29—sample RD-34 (*A. brasiliana*), $5800 \pm 200$ years, and sample RD-35 (wood) $7600 \pm 200$ years).

samples dated by Suguio *et al.* (1981) were collected in: (1) spoils of drainage canals; (2) shallow pits; and (3) outcrops (see Fig. 3 for location of these samples).

Fieldwork consisted almost exclusively of vibracoring of beach-ridge terraces and palaeo-lagoonal deposits occurring on the Doce River strandplain. The device used follows the design provided in Lanesky *et al.* (1979). Maximum penetration was 6 m. A total of 11 vibracores were recovered (see Fig. 3 for location).

Radiocarbon dating of samples collected in

this study (total of four) and those reported in Suguio et al. (1981) (total of 12) were performed at the Laboratorio de Fisica Nuclear Aplicada da Universidade Federal da Bahia, Brazil, under the supervision of J.M. Flexor and E.A. Azevedo. Activities of $^{14}$C were measured in a 1.2-1 proportional counter filled with $CO_2$ at 1 atm. Ages were calculated using schemes reported in Broecker & Olson (1961), Stuiver & Polach (1977) and Stuiver (1980). Ages are based on $^{14}$C half-life of 5568 years. Values reported are corrected for fractionation effects.

## GEOMORPHOLOGICAL UNITS OF THE DOCE STRANDPLAIN

Four major geomorphological units are present on the Doce strandplain (Fig. 3): (1) beach-ridge terraces; (2) alluvial ridges; (3) freshwater swamp; and (4) mangrove swamp.

The beach-ridge terraces are the most conspicuous geomorphological unit on the Quaternary strandplain. These terraces can be grouped into two sets using as criteria age, morphologic characteristics and elevation. Pleistocene beach-ridge terraces reach 6–10 m above the present high-tide level. They rest in direct contact with the fossil sea-cliffs of the Tertiary alluvial fans. Martin et al. (1982) assigned an age of 120 ka to these terraces based on the Io/U method.

The Holocene beach-ridge terraces are located seaward of the Pleistocene ones, from which they are separated by a low-lying swampy area. The height of the Holocene beach-ridge terraces varies from 4 m (most landward) to a few centimetres (most seaward) above present high-tide level. The surface of the Holocene terraces slopes gently seaward suggesting that their construction took place in a period of dropping sea-level. In contrast to the Pleistocene terraces, the surface of the Holocene terraces is characterized by perfectly preserved beach ridges. Radiocarbon dating of shells and wood collected in the Holocene beach-ridge terraces provided ages that are younger than 5 ka (Suguio et al., 1981).

The alluvial ridges comprise the natural levees, palaeochannels, and point bar environments where coarse-grained sediments were deposited (Fig. 3).

The freshwater swamps occupy the following settings: (1) the low-lying areas separating the alluvial ridges; (2) the swales separating individual beach ridges; and (3) the low-lying areas separating the Pleistocene from the Holocene beach-ridge terraces. In this last setting, excavation of pits and drainage canals has shown the subsurface existence of argillaceous lagoonal sediments. Radiocarbon dating from shell and wood in those sediments provided ages older then 5.0 ka (Suguio et al., 1981).

The mangrove swamps occur in association with tidal creeks and protected areas created by the coastwise extension of hooked spits (Fig. 3).

## TWO-DIMENSIONAL EVOLUTIONARY SCHEME OF THE DOCE STRANDPLAIN

The reconstruction of the two-dimensional evolutionary history of the Doce strandplain is based on the integration of various sources of information: (1) mapping of Quaternary deposits; (2) careful documentation of beach-ridge sets, their orientation and truncation patterns; (3) cross-cutting relationships between beach-ridge sets and palaeochannels; (4) known sea-level history for the east coast of Brazil; (5) published radiocarbon dates reported in Suguio et al. (1981); and (6) wave patterns. Vibracoring information was later used against this background to document the three-dimensional facies evolution and architecture of the strandplain. A more detailed account of this evolutionary history can be found in Dominguez (1987). This history is summarized below.

1 The rise in sea-level that followed the last Glacial period drowned the Pleistocene strandplain. Barrier-island–lagoonal systems were created as a result of this drowning (Fig. 4A). The palaeolagoons are well-preserved today, being represented by the low-lying mud-filled areas that separate the Pleistocene from the Holocene beach-ridge terraces. As discussed before, radiocarbon dates from shells and wood collected in shallow pits and drainage canals provided ages older then 5.0 ka (Suguio et al., 1981). Probably, after 7 ka, with decelerating rates of sea-level rise, the Doce River could then infill its estuary, and began emptying into the palaeolagoon. During this time the Doce River split in a multitude of small, narrow, meandering channels (Fig. 4A). Reconstruction of the barrier island system shows that by the time of the maximum of submergence phase I (5.1 ka), the shoreline could be subdivided into a convex and a concave seaward section. The boundary between the two sections was located

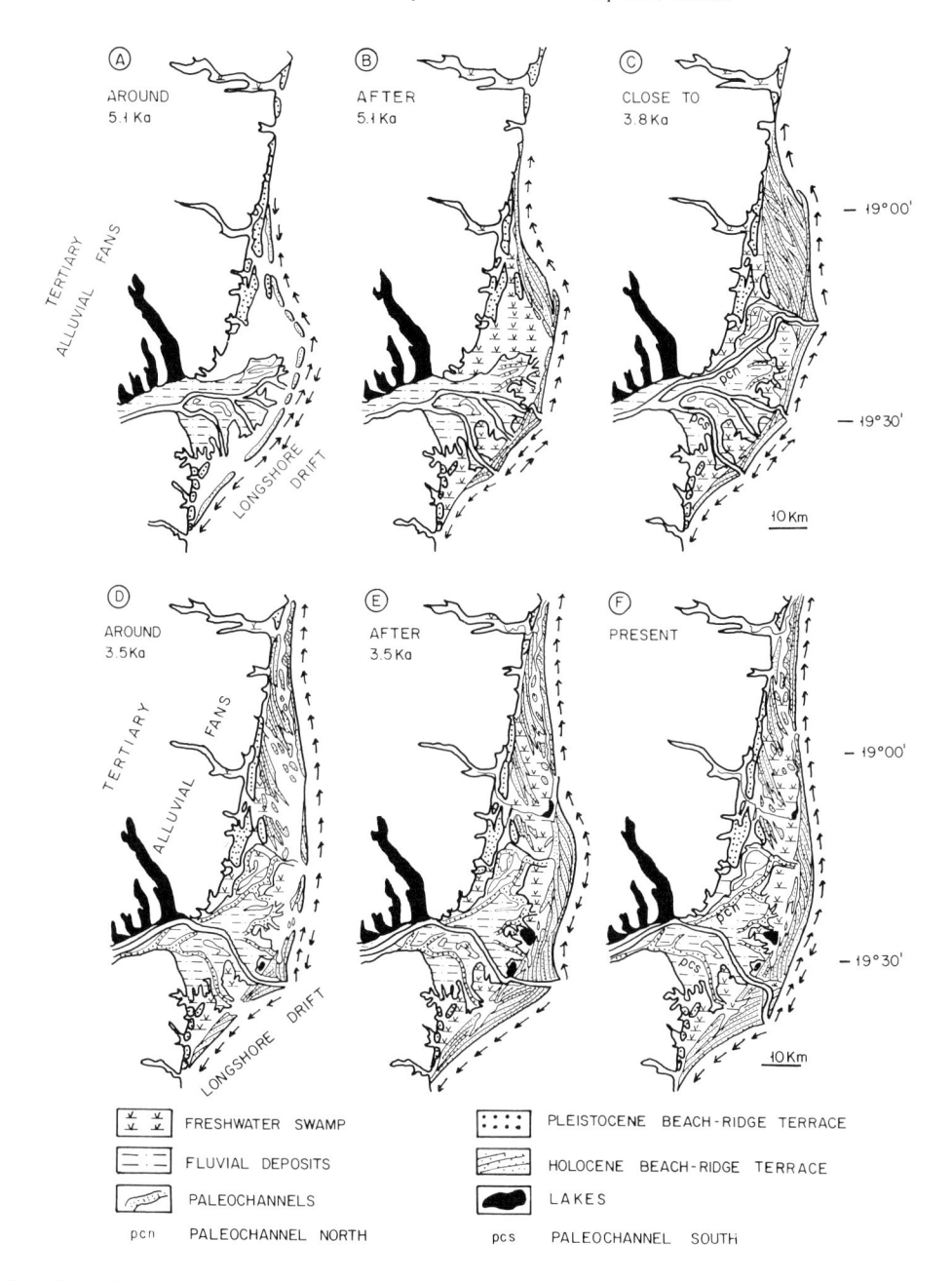

**Fig. 4.** Two-dimensional evolutionary model for the Doce strandplain. (A) Sea-level rise during the period 7.0–5.1 ka drowned the Pleistocene strandplain. The Doce River began building an intra-lagoonal delta. (B) When sea-level dropped after 5.1 ka the lagoonal environment was replaced by a freshwater swamp. Coastwise extension of sand spits in the northern part of the strandplain gave origin to elongate lagoons that later became low-lying areas separating beach-ridge sets. (C) Close to 3.8 ka an extensive beach-ridge set, including elongate low-lying areas, had been deposited in the northern part of the strandplain. (D) The rise in sea-level during the period 3.8–3.5 ka invaded the low-lying areas of the Holocene beach-ridge terrace and drowned the river mouths. Fluvial discharge was concentrated in one sole channel. (E) During the sea-level drop after 3.5 ka, most of the riverborne sediments were transported northward resulting in the development of an 'asymmetrical projection with lateral alimentation' (Zenkovitch, 1967, p. 385). (F) In a more recent time the Doce River changed its lower course to occupy the new river mouth.

about 19°20' latitude (Fig. 4A). In the concave section, the wave-trains approaching from east-southeast produced a dominant northward-directed longshore drift. This direction for the longshore drift is confirmed by the fact that during most of the Holocene, sediments were predominantly transported northwards (see next stages). Divergence of longshore drift characterized the convex section particularly around 19°30' latitude (Fig. 4A). It is assumed that a stronger drift will be generated by waves approaching from east-southeast because they are higher and therefore have a greater wave power. Northeast waves would be important only during those times when the east-southeast waves were subdued.

**2** After building the intra-lagoonal delta, the Doce River broke through the barrier island, reaching the open sea (Fig. 4B). Judging from cross-cutting relationships between beach-ridge sets and palaeo-channels, the flow of the Doce River by that time was probably concentrated into two or three major distributaries. Because the Doce River distributaries emptied into the sea in the convex section of the shoreline, which was characterized by divergence of the longshore drift, riverborne sands were rapidly distributed along the coast (Fig. 4B). Cusp-shaped beach-ridge sets were deposited associated with these river mouths. However, because of the dominant northward-directed longshore drift, riverborne sediments were preferentially transported north of the river mouths. Note that, because of the particular orientation of the shoreline with respect to the prevailing wave-trains, essentially no sediment from south or north of the study area could effectively enter the study area (Fig. 4B). Therefore, contrary to what has been observed for other strandplains along the east-southeast coast of Brazil, strike-feeding was almost non-existent at the Doce strandplain (Dominguez et al., 1983; Dominguez, 1986).

At 19°20' latitude the barrier island chain inflected abruptly landward, giving origin to a broad embayment (Fig. 4B). The shoreface, nourished by the northward-directed longshore drift and trying to maintain its lateral continuity, then began extending into this coastal void. As a result of this process, sand spits formed and continued extending northward into the open ocean far away from the shoreline in the concave section. As the spit extended coastwise, refraction of the south-southeast waves and waves impinging from east-northeast drove it landward until its northern extremity welded to the shoreline in the concave side. In several cases, this process gave origin to an extensive elongate lagoon separated from the open ocean by the spit itself (Fig. 4B,C). These lagoons later became the low-lying elongate areas separating the beach-ridge sets in the northern part of the strandplain.

**3** Cross-cutting relationships between the palaeo-channel north (PCN) and the beach-ridge sets indicate that sometime after 5.1 ka, channel switching occurred to originate that palaeochannel and divert part of the Doce River discharge (Fig. 4C). A reasonably large cusp-shaped beach-ridge set was deposited associated with this river mouth. Again, most of the riverborne sediments were transported northward to infill the re-entrant of the shoreline, following the same mechanisms described above. Cross-cutting relationships of beach-ridge sets also indicate that the other two distributaries of the Doce River were still active by this time. Lack of additional information, however, makes it impossible to establish an accurate chronology for the period of occupation of each palaeochannel.

**4** The rapid rise in sea-level associated with submergence phase II (3.8–3.5 ka) caused very dramatic changes in the Doce strandplain. The river mouths that were still active by that time were drowned (Fig. 4D). This drowning is evidenced by the fact that today, as visible in aerial photographs, the boundaries of the lower reaches of the palaeo-channels are somewhat diffuse and partially covered by freshwater swamps. The rise in sea-level also flooded the low-lying areas of the strandplain. Most of these low-lying areas were located in the northern part of the strandplain and comprised the elongate lagoons that formed in association with the coastwise extension of sand spits (Fig. 4D). Several radiocarbon dates of shell-middens, shells, and wood collected in these low-lying areas provided ages around 3.5 ka (Suguio et al., 1981). These dates coincide with the maximum of submergence phase II. No evidence of submergence phase II was found in the southern part of the strandplain, except for the drowned aspect of the lower reaches of the palaeochannel south (PCS, Fig. 4D). The reason for this is because in the southern part, no low-lying areas were available to be invaded by the sea.

Also, no evidence was found for re-occupation by the sea of the palaeolagoon formed during submergence phase I. All radiocarbon dates from this area are older than 5.1 ka (Suguio et al., 1981).

The rapid rise in sea-level during submergence phase II also caused disturbances in the fluvial equilibrium of the Doce River distributaries. As a result, these distributaries were abandoned and the flow was concentrated in a single channel, the present one (Fig. 4E). We prefer this interpretation as opposed to a simple autocyclic channel switching model for various reasons. In the Quaternary framework of the Brazilian coastal evolution, two main causes can be argued to explain fluvial imbalances: first, the loss of the river's transporting efficiency through an overextension of the lower river course as a result of progradation (autocyclic model) and second, a rapid positive oscillation of sea-level, which, by drowning the river's mouth, reduces the river's sediment transporting efficiency (allocyclic model). The second alternative is believed to be the more plausible one to explain the channel switching. From 5.1 to 3.8 ka, sea-level has dropped, thus enhancing the river's transporting efficiency. Furthermore, this channel switching occurred associated with an important marine invasion of the strandplain, and affected the active distributaries apparently at the same time. A major rearrangement of the shoreline occurred after this invasion, as indicated by cross-cutting relationships between beach-ridge sets and palaeochannels.

5 After 3.5 ka, progradation resumed, this time associated with the new river mouth (Fig. 4E). This progradation coincides with emergence phase II of the sea-level curve. Once more, because of the location of this river mouth with respect to the prevailing wave systems, riverborne sediments were distributed to both sides of the river mouth, and a cusp-shaped beach-ridge set was deposited. The stronger northward component of the longshore drift caused a larger volume of riverborne sediment to be transported northward (Fig. 4E). Submergence phase III left no identifiable record at the present surface of the Doce strandplain. Very probably there were no low-lying areas to be flooded by the sea during that rise. Evidence for drowning of the river mouth, if at one time present, was later eroded when the river mouth changed to its present position (Fig. 4F).

6 At a very recent point in time an abrupt change in

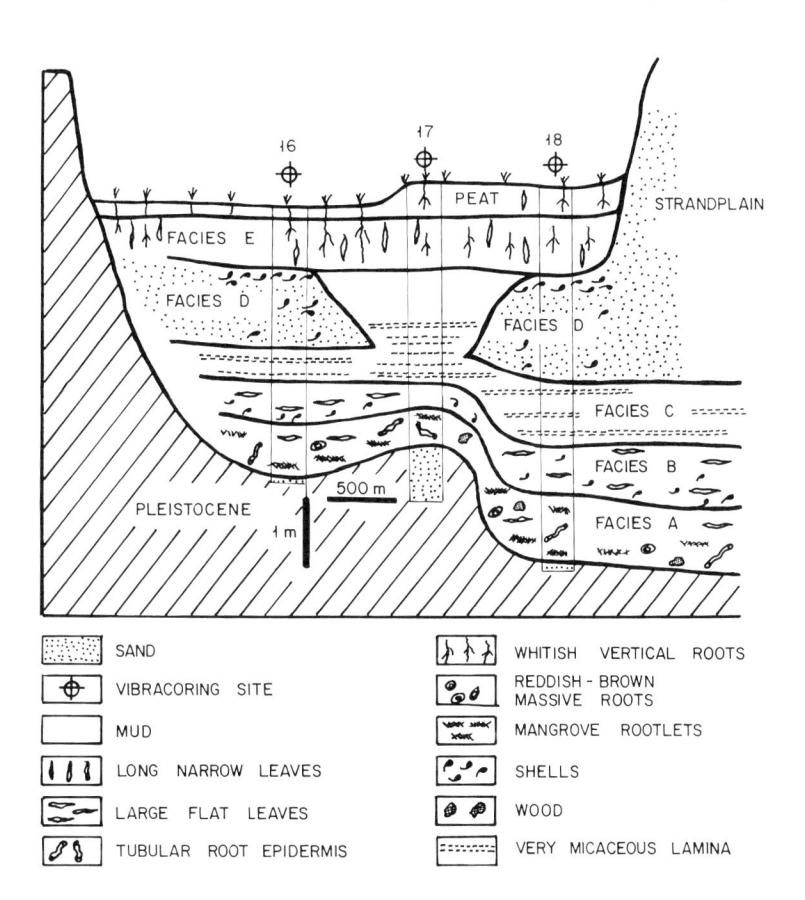

**Fig. 5.** Schematic cross-section across the palaeolagoonal fill associated with submergence phase I (7.0–5.1 ka), with core descriptions and correlations. Vibracoring sites are shown in Fig. 3. Cores were plotted using as a datum the contact between facies E and the freshwater peats. It was not possible to position the profile with respect to mean sea-level.

the low river course took place (Fig. 4F) resulting in the occupation of the present river mouth. Because of the position of this new mouth with respect to the wave-trains, riverborne sediments are transported today predominantly to the southern shoreline. Almost no deposition is taking place north of the river mouth today. This conclusion is further corroborated by the cross-cutting relationships between beach-ridge sets and alluvial channels. These relationships show that, whereas the shoreline north of the river mouth has shown almost no signs of seaward progradation, progradation of the southern shoreline has been extensive.

## LAGOONAL DEPOSITS ASSOCIATED WITH THE SUBMERGENCE PHASE I (7.0–5.1 ka)

The sedimentary fill of the palaeolagoonal depression formed during submergence phase I (7.0–5.1 ka) has been investigated in the southern section of the Doce strandplain (Fig. 5). Three vibracores were recovered from this palaeolagoonal infill (RD-16, RD-17 and RD-18). The substrate beneath the palaeolagoon deposits is dark brown (5YR 2/1, 5YR 2/2, 10YR 2/2, Geological Society of America Rock Color Chart) coarse to very coarse sands. These sands represent remnants of the Pleistocene beach-ridge terraces that border the palaeolagoon in its western side. The Pleistocene beach-ridge terraces are characterized by very clean white sands on the top. Towards the base, however, these sands become cemented by humic acids and iron oxides giving origin to a 'coffee-rock'. It is remnants of this 'coffee-rock' that have been penetrated by the vibracores.

### Sedimentary facies

Five sedimentary facies were identified in the palaeolagoonal fill (Table 1). Sediment types have been grouped into sedimentary facies based on: (1) sediment colour; (2) sediment texture; (3) types of plant remains; (4) molluscan assemblages; and (5) physical and biological sedimentary structures (Table 1). These sedimentary facies and their environmental interpretation are described as follows (see Fig. 5).

**Table 1.** Characteristics of facies infilling the palaeolagoon associated with submergence phase I (7.0–5.1 ka)

| Facies | Environment | Texture | Primary sedimentary structures | Plant remains | Molluscs |
|---|---|---|---|---|---|
| A | Mangrove swamp | Dark grey massive mud | None | Fibrous rootlets Massive roots Root epidermis Woodworm/teredinid borings | A. brasiliana B. exustus C. plana |
| B | Open lagoon | Dark to medium grey massive mud | None | Large flat leaves Bits of leaves and wood | A. brasiliana C. cancelata Tellina sp. C. swiftiana M. constricta Heleobia sp. C. caraboboensis C. plana |
| C | Intra-lagoonal prodelta | Light olive grey to yellowish grey micaceous mud | Planar parallel lamina, mica plates segregated into laminae | None | None |
| D | Lagoon margin | Olive grey muddy micaceous fine to medium sand | Planar parallel bedding, mica plates segregated into laminae | Small roots and narrow leaves | A. brasiliana P. pectinatus T. plebeius Heleobia sp. |
| E | Freshwater swamp | Yellowish grey to brown sandy mud | None | Long whitish roots Grass-like leaves | None |

*Facies A*

This facies is characterized by very plastic muds showing a very distinctive grey colour (N3). Five different types of plant remains are present: (1) small brownish fibrous rootlets; (2) reddish-brown massive roots (1–2 cm in diameter) oriented predominantly in a horizontal position; (3) wrinkled reddish-brown tubular root epidermis occurring in vertical or diagonal position; (4) dark-red wood with teredinid borings; and (5) large flat leaves. A molluscan fauna is almost completely absent, except for a narrow interval in core RD-18 where *Anomalocardia brasiliana*, *Brachidontes exustus* and *Crepidula plana* occur. The types of plant remains provide reasonable confidence in interpreting this facies as being deposited in a mangrove swamp environment.

*Facies B*

This facies is characterized by dark to medium grey (N3–N5) very plastic muds. The contact with facies A below is gradational. Two types of plant remains were found: (1) large flat leaves lying in a horizontal position; and (2) bits of leaves and wood scattered throughout the facies. The bivalve assemblage is dominated by *A. brasiliana*, followed in decreasing order of abundance by *Chione cancellata*, *Tellina* sp., *Corbula swiftiana* and *Macoma constricta*. Many of these bivalves occur in life position. The gastropod assemblage is dominated by *Heleobia* sp., followed by *Cyclostremiscus caraboboensis* and *C. plana*. This facies is interpreted as being deposited in a quiet subtidal environment where leaves and detrital organic matter could settle on the bottom. This environment is referred to here as a brackish-water pond/open lagoon.

*Facies C*

This facies is characterized by light olive grey (5Y 6/1) to yellowish grey (5Y 7/2) well-laminated muds. The contact with facies B below is gradational. Layering is expressed by variations in hues of grey and by the presence of laminae and beds (1–4 cm thick) of very micaceous, very fine sand. Rare small cylindrical burrows (1 cm in diameter) filled with micas are present. No molluscs were found in this facies except for a small shell fragment in core RD-18. The absence of molluscs and burrowing, and the sedimentary characteristics indicate rapid deposition in a calm subaqueous environment with periodic introduction of coarser sediment (the micaceous layers). The close proximity to the palaeochannel south (PCS) of the intra-lagoonal delta of the Doce River (Fig. 3) provides strong evidence for the interpretation of this facies as being deposited in an intra-lagoonal prodelta environment.

*Facies D*

This facies is characterized by olive grey (5Y 4/1, 5Y 6/1), muddy, micaceous, fine to medium-grained sand. The contact with facies C below varies from abrupt to gradational. Planar parallel bedding dominates, but some cross-bedded sets 6–10 cm in thickness have been observed. The mica plates are usually segregated into laminae. In core RD-16, mud beds 2–6 cm in thickness occur. The bivalve assemblage in this facies is dominated by *A. brasiliana*, followed by *Phacoides pectinatus* and *Tagelus plebeius*. The gastropod assemblage is almost completely dominated by *Heleobia* sp., which also is present in large numbers. A 20-cm-thick bed, in which *A. brasiliana* and *Heleobia* sp. are present in large numbers, characterizes the top of facies D. Facies D is interpreted as being deposited in a subtidal to intertidal environment close to the margins of the palaeolagoon. The basis for this interpretation is: (1) proximity of the vibracoring sites RD-16 and RD-18 to the borders of the palaeolagoon; and (2) the bivalves *P. pectinatus* and *T. plebeius* of the molluscan assemblage are common inhabitants of lagoon margins (Andrews, 1981). The presence of small roots and narrow leaves within the sediment indicate that the lagoon margin was somewhat vegetated. In core RD-17, facies D is not developed in the equivalent stratigraphic interval. Instead, burrowed micaceous sandy muds occur. The burrows, about 1 cm in diameter, are filled with very micaceous sandy mud. No molluscs were found in core RD-17.

*Facies E*

This facies has a characteristic yellowish grey (5Y 7/2) to yellowish brown (10YR 5/4, 10YR 4/2) colour. The contact with facies D below is abrupt. It is composed of muddy, fine to medium-grained sands and sandy muds. Two types of plant remains are present in this facies: (1) long whitish roots that penetrate the entire facies; and (2) long, narrow, grass-like leaves. No mollusc shells were found in this facies. Facies E is interpreted as being deposited in a freshwater swamp environment. The basis for this interpretation is: (1) similarity of the plant remains found in facies E with those thriving in the freshwater swamps today; and (2) the close proximity with the freshwater swamp that occurs on top of the palaeolagoonal fill. Facies E is capped by freshwater peats 10–30 cm in thickness.

## Depositional history of the lagoonal deposits associated with submergence phase I (7.0–5.1 ka)

In trying to reconstruct the depositional history of infilling of the palaeolagoon, radiocarbon dates of wood and shells published in Suguio *et al.* (1981) will be used because the amount of material found in the cores was not sufficient for radiocarbon dating. These samples were collected in spoils of drainage canals in the same area where the cores were taken. Although the exact stratigraphic position of the samples are not known, it is still possible, by comparing the stratigraphy of the cores and the type of the material dated, to draw some conclusions about the timing of the episodes involved in the infilling of the palaeolagoon. The location where these samples were collected are shown in Fig. 3. Two of the samples dated are pieces of wood and two are shells of *A. brasiliana*. Most of these shells still have both valves attached indicating little or no transport. In the palaeolagoonal fill the most probable interval to find pieces of wood is facies A (mangrove swamp). Accordingly, these samples provided ages of 7600 ± 200 years (sample RD-35) and 6900 ± 200 years (sample RD-33). The older age is for the sample collected in a more seaward location (Fig. 3), thus indicating an early invasion of the sea in this place. We believe these are good ages for the basal portion of the lagoonal infill (mangrove swamp facies). At the Jequitinhonha strandplain, radiocarbon dating of mangrove rootlets collected in the basal portion of the lagoonal infill associated with submergence phase I provided an age of 7140 ± 290 years (Dominguez, 1987). This age suggests that marine inundation on the east coast of Brazil began around 7.0 ka. The shells of *A. brasiliana* may have come from two stratigraphic intervals: (1) facies B (brackish-water pond/open lagoon); and (2) facies D (sandy lagoon margin). Facies B is not a good choice because the individuals in this facies are mostly juvenile forms. Also, because this facies replaces facies A (mangrove swamp) upcore, the dates obtained are too young (5.6–5.8 ka). Thus, it is suggested that the most probable stratigraphic interval from where these samples came is the shell-rich bed that occurs on top of facies D (sandy lagoon margin). *Anomalocardia brasiliana* occurs in large numbers in this bed and it is represented mostly by large individuals.

The sequence of events involved in the infilling of the palaeolagoonal depression at the Doce strandplain is described as follows.

1 The first signs of marine inundation in the palaeolagoon are represented by facies A (mangrove swamp), resting directly on top of the Pleistocene 'coffee-rock'. This event took place about 7.6 ka in the more distal (seaward) parts of the palaeolagoon and around 6.9 ka in the more proximal (landward) parts. The mangrove swamp environment had a very short duration and was rapidly drowned by the continued rising sea-level, being replaced by a brackish-water pond/open lagoon environment (facies B).

2 After infilling its estuary, the Doce River began emptying into the palaeolagoon. Deposition of the intra-lagoonal delta began. As a result of progradation of this delta, particularly of its southern distributaries, prodeltaic muds (facies C) were deposited in the distal (seaward) parts of the palaeolagoon.

3 Subsequently, the intra-lagoonal prodelta environment (facies C) was replaced by a sandy lagoon margin environment (facies D). In core RD-18 the contact between these two facies is very abrupt (Fig. 5). Because this core is located very close to the palaeobarrier-island, it can be assumed that the sharp contact and the source of sand for facies D may be associated with washover activity. In core RD-16 the contact is more gradational and characterized by an interval with interbedded mud and sand. Remnants of the Pleistocene beach-ridge terrace outcropping at the western margin of the palaeolagoon, were probably the source of sand in this case. The overwhelming numbers of *Heleobia* sp. in the gastropod assemblage of facies D indicate that low-salinity conditions prevailed in this part of the palaeolagoon. Most species of *Heleobia* are characteristic of low-salinity to freshwater environments (Silva & Davis, 1983). The low salinity in the palaeolagoon probably results from the presence of the intra-lagoonal delta of the Doce River. It also indicates that during deposition of facies D, the Doce River was still emptying into the palaeolagoon. The fact that the sandy lagoon margin environment (facies D) is abruptly replaced upcore

by a freshwater swamp environment (facies E) suggests that the former was deposited close to the maximum of submergence phase I (5.1 ka). Thus, the dates of the *A. brasiliana* shells mentioned before are probably anomalous ages. These anomalous ages may have resulted from two causes: (1) dates represent mixing of shells of different ages because the samples were collected without stratigraphic control; and/or (2) the ages are apparent ages because they do not include reservoir correction (E.A. Azevedo, pers. comm.), which in this case might have been an important factor because of high freshwater input from the intra-lagoonal delta.

**4** The drop in sea-level that followed the maximum of submergence phase I resulted in replacement of the palaeolagoon by a freshwater swamp environment (facies E). Muds originating from river bank overflow, particularly from palaeochannel south (PCS) (Fig. 3), were initially deposited in those swamps. Subsequently, with concentration of the river flow in a sole channel, the present one, the alluvial ridge of palaeochannel south (PCS) offered an obstacle to the mud-laden overflow waters. Peat deposits could then accumulate.

## LAGOONAL DEPOSITS ASSOCIATED WITH SUBMERGENCE PHASE II (3.8–3.5 ka)

Of all the strandplains along the east-southeast coast of Brazil, the Doce strandplain preserves the best record of submergence phase II (3.8–3.5 ka). The reason for the existence of such a well-preserved record is because in this strandplain extensive low-lying areas were available to be invaded by the sea.

Four vibracores were taken in these low-lying areas in order to document the sequence of events associated with the origin and evolution of these features, and the subsequent invasion by the sea during submergence phase II. Figure 3 shows the location of vibracoring sites (RD-7, RD-11, RD-13 and RD-15).

The substrate of these elongated lagoons is composed of very micaceous, light olive grey (5Y 6/1), coarse silt to very fine sand, with occasional mud interbeds. The bivalve assemblage is dominated by *Cyrtopleura costata*, *Mulinia cleryana*, *Lunarca ovalis*, *Strigilla carnaria* and *Tellina* sp. The gastropods are represented by *Turbonilla* sp., *Olivella* sp. and *Heleobia* sp. These sediments are interpreted as being deposited in a lower shoreface environment because of the similarity of fauna and sediment types between this facies and lower shoreface deposits found in the Holocene beach-ridge terraces. Although there is some variability in the sediment types and thicknesses among the cores, a consistent pattern of facies evolution has been observed (Fig. 6). The contacts between all sedimentary facies described below are gradational.

### Facies A

This facies is composed of well-laminated, well-bedded, very plastic, light olive grey muds resting on top of lower shoreface sands. The contact between the two units is gradational. Bedding and laminae are made apparent by the presence of: (1) very micaceous laminae (1–5 mm thick); (2) detrital organic matter laminae; and (3) beds (3–4 cm thick) showing different hues of grey. In core RD-13, yellowish grey mud laminae associated with rounded wood pebbles, similar to those found at the Doce River mouth today, are also present in this facies. Vegetal remains are sparse and are usually represented by long wrinkled roots in vertical position and small rootlets that occur scattered throughout the facies. The abundance of very micaceous laminae decreases towards the top of the facies. In some cores, bedding and laminae are almost absent in the upper part of the facies. Molluscs are restricted to the basal part. In the remainder of the facies they are absent or represented by one or two species. The bivalve assemblage is represented by *M. cleryana*, *L. ovalis*, *C. costata* and *C. cancellata*. The gastropod assemblage is dominated by *Heleobia* sp., followed by *Odostomia* sp. and *Solariorbis schumoi*. The number of mollusc species decrease towards the top of this facies. In core RD-11, radiocarbon dating of *M. cleryana* shells collected in sediments of this facies provided an age of 4959 ± 200 years. The shells dated are very well preserved and unbroken, suggesting little or no transport.

### Facies B

This facies is characterized by medium grey (N5) to light olive grey (5Y 6/1), well-laminated muds, and was found only in cores RD-13 and RD-15. Sediment characteristics in facies A and B are very similar. The subdivision of the sediments into two separate facies was based on the fact that the number of mollusc species increases upward in

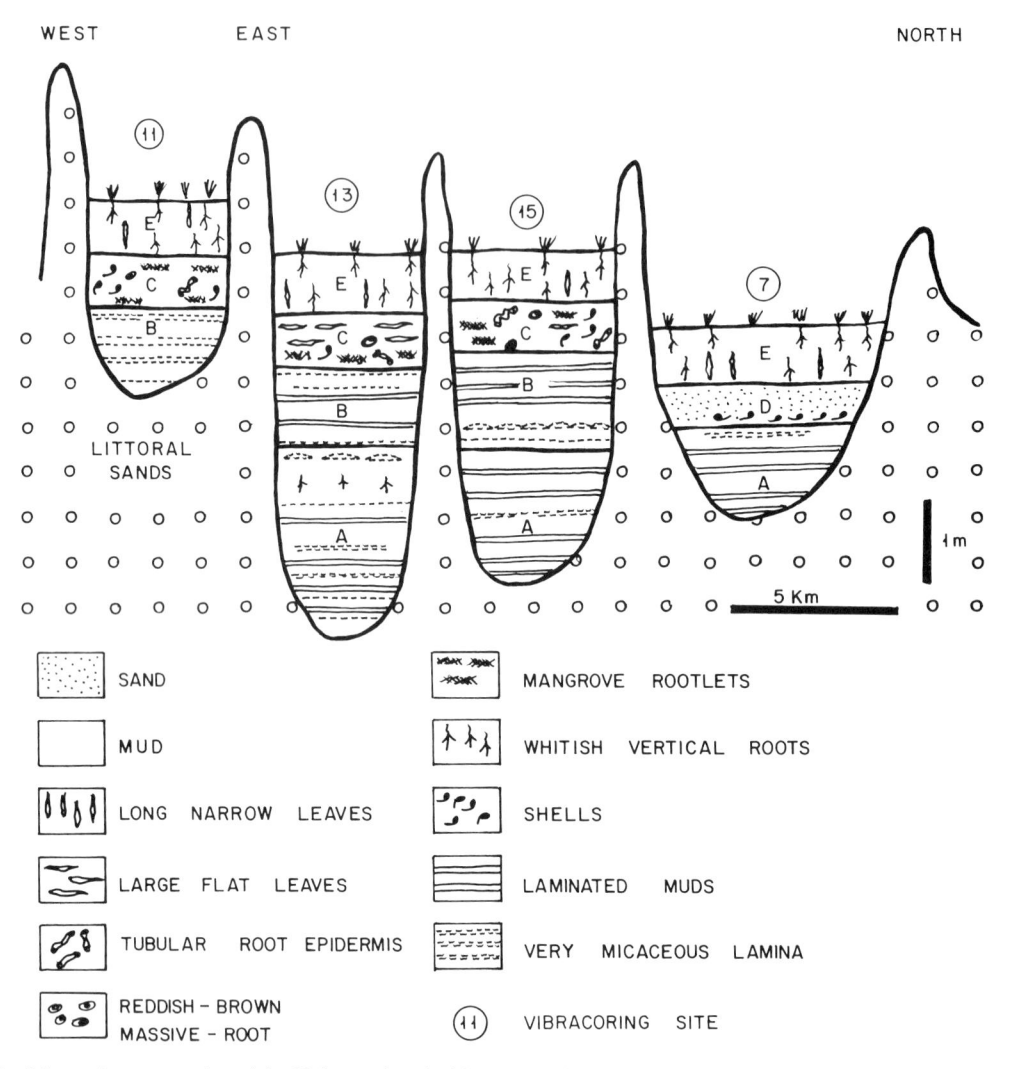

**Fig. 6.** Schematic cross-section of the Holocene beach-ridge terrace in the northern part of the Doce strandplain containing elongate low-lying areas between sand ridges, showing core descriptions and correlations. Vibracoring sites are shown in Fig. 3. Letters indicate the different facies recognized in the vibracores and described in detail in text. All contacts are gradational. Sand ridges are drawn with decreasing altitudes because sea-level had dropped during deposition. It was not possible to position the profile with respect to mean sea-level.

facies B as opposed to facies A. The contact between the two facies is placed exactly where very micaceous laminae reappear in the core. This reappearance coincides with the increase in the number of mollusc species. In core RD-13, yellowish (5Y 7/2) mud laminae were also observed in this facies. Bivalves are represented by *C. cancellata*, *M. cleryana*, *C. swiftiana*, and *Tellina* sp. Gastropods are represented mostly by *Heleobia* sp. Also present are *S. schumoi* and *C. caraboboensis*.

*Facies C*

This facies is characterized by dark grey (N3) plastic muds. It was found in cores RD-11, RD-13 and RD-15. The types of vegetal remains are the same as those present in facies A (mangrove swamp) of the palaeolagoonal fill sequence deposited during submergence phase I. The bivalves are represented by *C. swiftiana*, *C. cancellata* and *A. brasiliana*. The gastropod assemblage is dominated by *C.*

*caraboensis* and *Heleobia* sp. Radiocarbon dating of oyster shells collected in sediments of this facies at vibracoring sites RD-11 and RD-15 provided ages of 4790 ± 210 and 3560 ± 190, respectively. These shells were collected in shallow pits adjacent to vibracoring sites. The oyster shells were still attached to large pieces of dark-red wood with teredinid borings suggesting that they have not experienced transport.

*Facies D*

This facies was found only in core RD-7. It is composed of medium grey (N5) fine to medium-grained sand with some mud laminae and lenses. The lower 8 cm of the facies is shell-rich. Bivalves are represented by *L. ovalis*, *C. costata*, and *M. cleryana*. The gastropod assemblage is dominated by *Heleobia* sp. and *Olivella* sp. Except for this basal shell-rich layer, no other mollusc shells were found in this facies.

*Facies E*

This facies was found in cores RD-7, RD-13 and RD-15. It is characterized by light brownish grey (5YR 6/1) plastic muds totally penetrated by long whitish roots. Long grass-like leaves are also present. This facies is very similar to the facies capping the palaeolagoonal fill sequence deposited during submergence phase I (freshwater swamp). A freshwater peat (dusky yellowish brown, 10YR 2/2) occurs in the upper 20–25 cm of facies E.

**Depositional history of the lagoonal deposits associated with submergence phase II (3.8–3.5 ka)**

1 The coastwise progradation of sand spits in the northern part of the strandplain isolated portions of the shoreface and aborted its normal evolution (Fig. 7A). These portions of the shoreface then became the substrate of the elongate lagoons, which were separated from the open ocean by the sand spit. Water depth in each of the elongate lagoons was a function of the offshore distance at which the associated spits stabilized. If stabilization of the sand spit took place very close to the shoreline, the lagoon was shallow; deeper if stabilization took place well offshore. Once formed, the lagoons began collecting sediment from suspension (facies A). Some of this sediment had a riverine origin. This is very well illustrated in core RD-13 by the presence

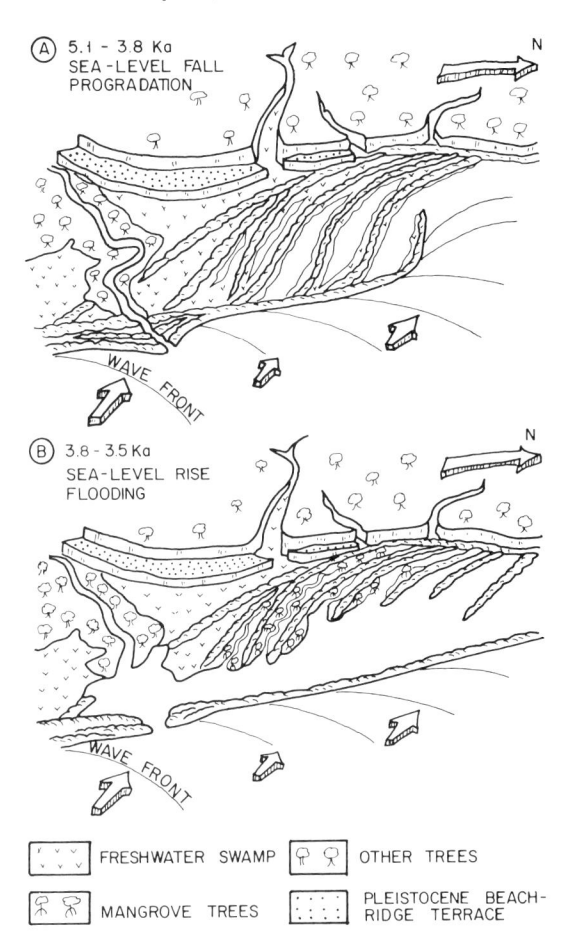

**Fig. 7.** Evolutionary model of the northern part of the Doce strandplain during: (1) emergence phase I (5.1–3.8 ka); and (2) submergence phase II (3.8–3.5 ka). During emergence phase I, lateral extension of sand spits gave origin to elongate lagoons, which after final closure evolved to freshwater lakes (Fig. 7A). During submergence phase II the sea drowned the river mouth of the palaeochannel north (PCN) and invaded the low-lying areas, which became interconnected bays (Fig. 7B).

of laminae of yellowish grey (5Y 7/2) mud and rounded wood pebbles. The decrease in abundance of these laminae toward the top of facies A suggests contined closure of the lagoon by northward extension of the sandy spit. The upward decrease in micaceous laminae in facies A, as observed in cores RD-7 and RD-15, probably records the same phenomenon. The decrease in the number of mollusc species observed, with almost complete disappearance of molluscs at the top, also reflects the final closure of the elongate lagoon with the onset of freshwater or very low-salinity conditions. After the final closure of the lagoon, elongate lakes occupied these depressions (Fig. 7A). No evidence of prolonged subaerial emergence was found, but these lakes were probably very shallow and supported some kind of vegetation, as indicated by the presence of rootlets. Muddy sediments were still occasionally supplied to these lakes during river bank overflow of palaeochannel north (Fig. 7A).

At vibracoring site RD-11, the sedimentation rate was sufficient to aggrade the lagoon floor to the intertidal zone before total closure by the sand spit. As a result, mangroves (facies C) rapidly colonized the low-lying area. Radiocarbon dating of facies C in core RD-11 (4.79 ka) and the underlying facies A (elongate lagoon) (4.95 ka) supports this interpretation.

**2** When sea-level rose during submergence phase II, the elongate lakes were flooded by the sea and became interconnected (Fig. 7B). This inundation of the sea is indicated by the increase in the number of mollusc species that occur right at the base of facies B in cores RD-13 and RD-15, and by facies D in core RD-7. No evidence was found that the low-lying area at vibracoring site RD-11 was invaded by the sea during this episode. Radiocarbon dating indicates that the lagoon floor at vibracoring site RD-11, aggraded to the intertidal zone about 4.9 ka, thus, during a time when sea-level was approximately 4 m above the present level. Therefore, during submergence phase II, when sea-level reached no more than 3.5 m above the present level, the sea could not invade this area.

Because of the drowning of the river mouth during submergence phase II, at least part of the discharge of palaeochannel north began emptying into the lagoon that resulted from the interconnection of the individual elongate lakes (Fig. 7B). Evidence for river discharge into the lagoon is indicated in facies B of core RD-13 by the presence of laminae of yellowish grey mud, and very micaceous coarse silt. These characteristics make facies B of core RD-13 very similar to facies C (intra-lagoonal prodelta) of the palaeolagoonal fill deposited during submergence phase I. In core RD-15, a similar increase in the number of the very micaceous laminae occurs in the same stratigraphic interval (lower half of facies B). The abundance of micaceous laminae, however, is greatly reduced in this core, as compared with core RD-13, thus, possibly reflecting the greater distance of vibracoring site RD-15 from the riverine source. The sandier character of facies D probably reflects the proximity of vibracoring site RD-7 to the beach ridges. The abundance of molluscs able to withstand low salinities, such as *M. cleryana* and *Heleobia* sp. in facies B of cores RD-13 and RD-15, indicates that brackish-water, low-salinity conditions prevailed in the palaeolagoon, either because of the existence of a nearby fluvial discharge emptying into the palaeolagoon, or because of rain-water.

**3** In cores RD-13 and RD-15, deposition of the micaceous laminated muds of facies B brought the sediment surface to the appropriate depth for mangrove colonization (facies D). The age of 3.5 ka of *Crassostrea rhizophorae* shells collected in facies D of core RD-15 coincides with the maximum of submergence phase II, thus, corroborating the interpretation above. Several other radiocarbon dates in Suguio *et al.* (1981), of samples of wood, shell, and shell middens collected at the low-lying areas of the northern part of the strandplain, have also provided ages around 3.5 ka, attesting to the invasion of these areas during submergence phase II. At vibracoring site RD-7, a mangrove swamp facies did not develop, probably because sediment supply was not sufficient to bring the lagoonal floor to the appropriate depth for mangrove colonization. Therefore, most of the lagoonal floor remained subtidal. This situation reflects the greater distance of vibracoring site RD-7 from the riverine source as compared with vibracoring sites RD-13 and RD-15.

**4** Finally, with the drop in sea-level after 3.5 ka, freshwater swamp deposition (facies E) ensued. Freshwater peats presently cap the palaeolagoonal sequence.

## LITTORAL ZONE DEPOSITS

The term littoral zone is used here in the sense proposed by Komar (1976) to describe that zone across the beach and into the water to a depth at which the sediment is less actively transported by surface waves.

Only 4 cores (RD-2, RD-5, RD-6 and RD-10) were acquired in the littoral zone deposits of the Doce strandplain (Fig. 3). The main reasons for this reduced number were: (1) the difficulty of penetration of the core barrel in these sandy sediments; and (2) lack of access routes, such as roads and streams, precluded vibracoring at many of the planned sites. Therefore, the results discussed here are not thoroughly representative of the littoral zone deposits in the study area. Proximity to the river mouths was avoided because of the presence of coarse sand. Also, no vibracores were taken updrift of the river mouth because of the absence of access routes. Thus, the facies sequences discussed in the next sections better represent conditions prevailing in the portion of the strandplains located downdrift of the river mouth. In order to increase the total length of the sequences sampled, vibracoring was performed preferentially along the eroding banks of rivers and streams. Thus, the upper part of the sequences described in this section were observed in outcrops of the beach-ridge terraces at the vibracoring sites.

Littoral zone deposits of the Doce strandplain are characterized by moderately well sorted, medium to very fine-grained sands. A vertical facies sequence model based on observations of vibracores and outcrops is shown in Fig. 8. The sedimentary facies that compose the littoral zone deposits at the Doce strandplain are, in descending order, L1, L2 and L3 (Fig. 8).

## Facies L1

Facies L1 is characterized by very pale orange (10YR 8/2) to yellowish grey (5Y 8/1), moderately well sorted, medium to fine-grained sands. The predominant sedimentary structure is planar, even laminae dipping gently seaward. Thicknesses of facies L1 observed at vibracoring sites RD-2 and RD-5 are 1.9 and 1.4 m respectively. Facies L1 is interpreted as being deposited in a foreshore environment, under the influence of swash action. This interpretation is based on the great similarity between the sedimentary structures observed in this facies and those present in modern foreshore deposits. The upper 0.5-1.0 m is structureless because of soil development and might include sediments deposited in a backshore environment, either by wave or aeolian activity. It was not possible to measure the thickness of facies L1 at vibracoring sites RD-10 and RD-6. The combined thicknesses of sediment above facies L3 (lower shoreface) in these two sites are 2 and 3.5 m, respectively, which compares well with the thicknesses for the equivalent intervals at vibracoring sites RD-2 and RD-5. No molluscs were found in facies L1.

## Facies L2

This facies is characterized by pale yellowish brown (10YR 6/2) to very pale orange (10YR 8/2), moderately well sorted, medium-grained sands. The sands are cross-bedded with intercalations of planar, even-laminated beds. Facies L2 is interpreted as being deposited in an upper shoreface environment (surf zone). The basis for this interpretation is: (1) the presence in facies L2 of cross-bedding, suggesting deposition under the influence of unidirectional currents, which are characteristic of the surf zone; and (2) facies L2 is located immediately below facies L1, which was interpreted as being deposited in a foreshore environment. The thickness of facies L2, as measured at vibracoring sites RD-2 and RD-5, is 0.7 m. No molluscan fauna was observed in facies L2.

## Facies L3

This facies is characterized by olive grey (5Y 4/1) to light olive grey (5Y 6/1) very fine-grained sands. The

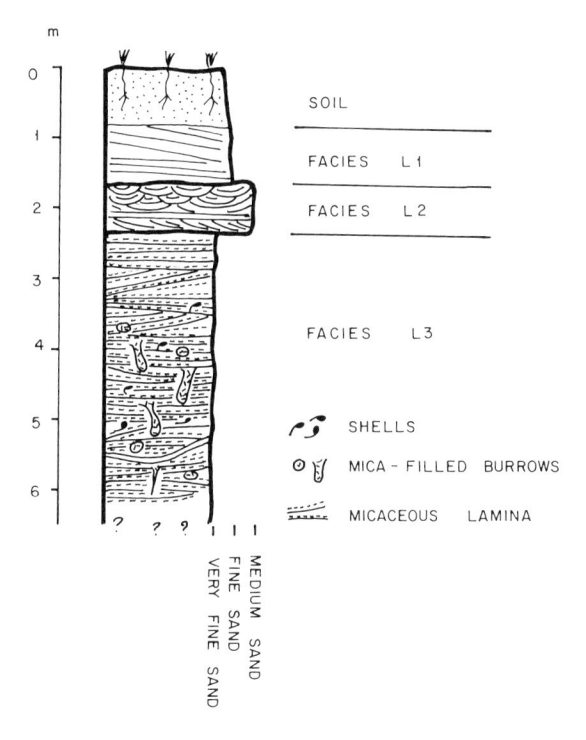

**Fig. 8.** Schematic vertical facies sequence of the littoral zone deposits at the Doce strandplain, based on observations made on: (1) cores RD-2, RD-5, RD-6 and RD-10; and (2) outcrops at vibracoring sites. Environmental interpretations of the sedimentary facies are: (1) facies L1 = foreshore; (2) facies L2 = upper shoreface; (3) facies L3 = lower shoreface. Total thickness of facies L3 (lower shoreface) is not known because vibracores have not penetrated the entire sequence.

sands are moderately well sorted and extremely micaceous. The dominant physical sedimentary structure is planar, even laminae, which occur in sets about 5–10 cm thick. These sets form a low-angle cross-bedding. The sets are separated by truncation surfaces which are lined in some cases with mud balls. The laminae are expressed by the segregation of micas. Cross-bedded sets 5 cm in thickness occur intercalated in the parallel laminae. Very micaceous beds 3–10 cm in thickness occur in core RD-5. Cross-bedded sets are usually coarser (fine to medium) grained sands than the planar-laminated beds (very fine-grained sands). Organic debris is normally associated with the very micaceous beds. Mud beds about 1 cm in thickness are common in core RD-2. Degree of burrowing is variable. Whereas scattered burrows are present in core RD-5, burrowing is extensive in core RD-2. Burrowing can be readily identified in the cores because the mica flakes infilling them show no preferential orientation as opposed to the parallel-laminated intervals, where the mica plates are all oriented. The most common burrow type is circular in cross-section with a diameter of about 1 cm. Micaceous, very fine sand showing planar laminae is also very characteristic of cores RD-10 and RD-6. Mud layers (0.5–1.0 cm in thickness) occur intercalated in the micaceous sand in core RD-10. Facies L3 is interpreted as being deposited in a lower shoreface environment (zone of shoaling waves). The basis for this interpretation is: (1) the stratigraphic position of the facies immediately below inferred upper shoreface deposits; (2) the dominance of planar, even laminae; and (3) the dominance of very fine sand. A well-defined molluscan faunal assemblage is present in facies L3. The bivalve assemblage is dominated by *C. costata*, *M. cleryana*, *L. ovalis* and *Sphenia antilhensis*. The gastropod assemblage is dominated by *Olivella* sp., *Anachis obesa* and *Polinices* sp. All of these genera, except for *C. costata*, are common in mollusc shell accumulations on Georgia beaches (Dorjes *et al.*, 1986). Echinoderm fragments are also very common in these sediments.

Radiocarbon dating of *A. brasiliana* shells from lower shoreface deposits in core RD-2 provided an age of 3420 ± 180 years. Suguio *et al.* (1981) reported an age of 4000 ± 150 years for *A. brasiliana* shells collected in the lower shoreface deposits in an outcrop located very close to vibracoring site RD-5. These dates indicate that the lower shoreface sands were deposited when the shoreline prograded during emergence phase I (5.1–3.8 ka).

The abundance of micas in the lower shoreface sediments probably indicates a fluvial origin for them. Melo *et al.* (1975) report that the inner

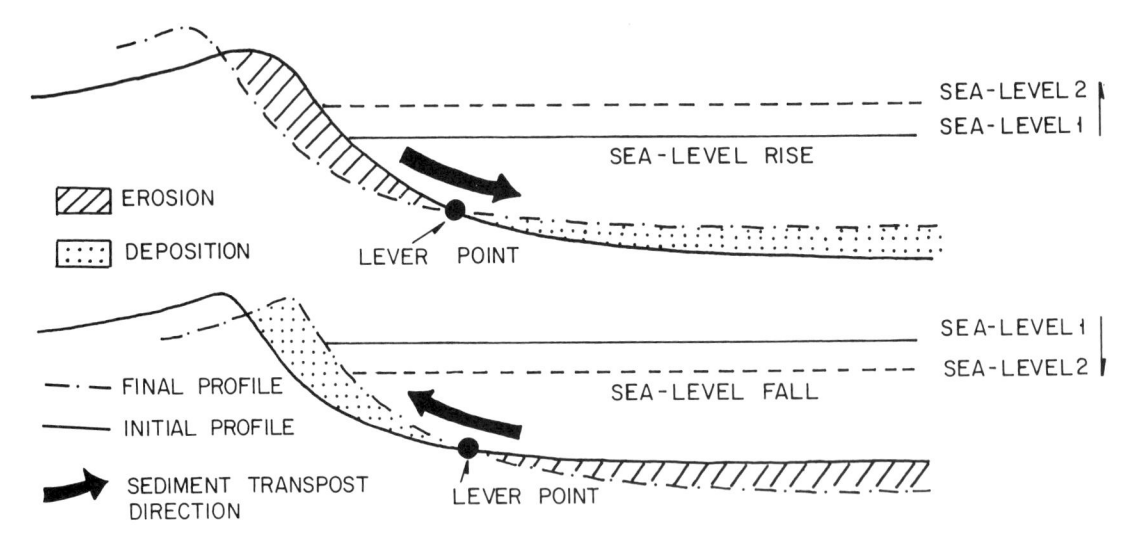

Fig. 9. Bruun's rule of translation of the shoreface profile. During a rise in sea-level, erosion prevails in the upper part of the profile causing the shoreline to recede. Conceptually this shore erosion supplies material to build upward the outer part of the responding profile. During a fall in sea-level the inverse is true. The seaward and downward translation of the shoreface profile causes erosion of the outer part of the profile. This erosion then supplies material to prograde the upper part of the profile seaward.

shelf, in front of the Doce River strandplain, is covered with muds and sandy muds and that these muds contain substantial amounts of mica and plant fragments. The same authors report a mollusc–echinoderm assemblage off the Doce River area, showing little evidence of reworking. Echninoderm fragments are also common in the lower shoreface sediments at the Doce strandplain. Thus, the great similarity between the sediments present on the inner shelf today, and the lower shoreface deposits (facies L3) recovered through vibracoring, suggests that the lower shoreface sediments were provided by the river during emergence phase I. In addition, the evolutionary history of the strandplain indicates that large amounts of riverborne sediments were actively transported northward during emergence phase I.

Information about the total thickness of the lower shoreface deposits is missing because a complete littoral zone sequence was not recovered during vibracoring operations.

## BRUUN'S RULE AND THE STRATIGRAPHIC RECORD OF LITTORAL ZONE SEQUENCES

During the initial stages of this research, the need for a model describing the response of the littoral zone to a fall in relative sea-level was immediately felt. Such a model was necessary to guide interpretation of the sedimentary sequences recovered through vibracoring. The approach was to use well-accepted models for the response of the littoral zone profile to a sea-level rise (Bruun's rule; Bruun, 1962), in order to generate a model describing the response of that profile to a drop in sea-level.

According to Bruun (1962), if the shoreface profile is accepted as an equilibrium response of the sea-floor to the coastal fluid power expenditure, then the effect of a sea-level rise could be deduced as a landward and upward translation of the profile (Fig. 9A). As the water level rises, erosion prevails in the upper part of the coastal profile causing the shoreline to recede. Conceptually this erosion supplies material to build upward the outer part of the responding profile.

Although Bruun's rule was originally developed for conditions of sea-level rise, several authors have proposed that a reversal of the rule also should be valid under falling sea-level conditions (Schofield, 1975; Swift, 1976; Tastet, 1979). A sea-level fall would result in a seaward and downward translation of the coastal profile, resulting in erosion of the outer part of the profile (Fig. 9B). Conceptually, this erosion will supply material to prograde the upper part of the profile seaward. Leontiev (1949) (in Zenkovitch, 1967, p. 528) has proposed this mechanism during his study of the Caspian Sea coast. Rikhter (1953) (in Zenkovitch, 1967, p. 528) has also invoked this mechanism to explain the great accumulation of sediments near the coast, and the simultaneous denudation of areas of the bottom in the Caspian Sea.

To understand which facies are to be preferentially preserved in the stratigraphic record in response to changes in relative sea-level, it is important to determine the position of the 'lever point' on the evolving profile. The lever point is here defined as the point of intersection between the initial profile and the profile adjusted to submergence or emergence (Fig. 9). Illustrations of Bruun's rule in several publications suggest that the lever point is expected to occur at a shallow subtidal level. Bruun (1962) calculated that, for the southeast coast of Florida, the intersection point between the old profile, without a rise in the water level, and a new profile corresponding to a 1-m rise in water level, is located at a depth of 2 m in the original profile (or at a depth of 3 m in the new profile). Also, the depth of erosion of back-barrier deposits resulting from shoreface retreat, as shown in numerous shore-normal cross-sections of barrier island coasts (e.g. Kraft, 1971; Belknap & Kraft, 1985), appears to confirm a shallow subtidal position for the lever point. This depth would correspond approximately to the boundary between the upper shoreface (surf zone) and the lower shoreface (zone of shoaling waves). As a result of the translation of the shoreface profile, the lever point will describe a line (or surface if the coastal zone is considered three-dimensionally), which, if preserved, will be represented in the stratigraphic record as an erosional surface (Fig. 10), also known as 'ravinement surface'. During a rise in relative sea-level, sediments deposited seaward of the lever point (i.e. lower shoreface) will be preserved favourably, resting on top of that erosional surface (Fig. 10). These lower shoreface sediments can be envisioned as the transgressive sands occurring on top of 'ravinement surfaces' and ubiquitously depicted in coastal zone profiles off transgressive shorelines. Sediments located landward of the lever point (i.e. foreshore/upper shoreface) will be eroded. During a fall in sea-level, the inverse is true. Sediments

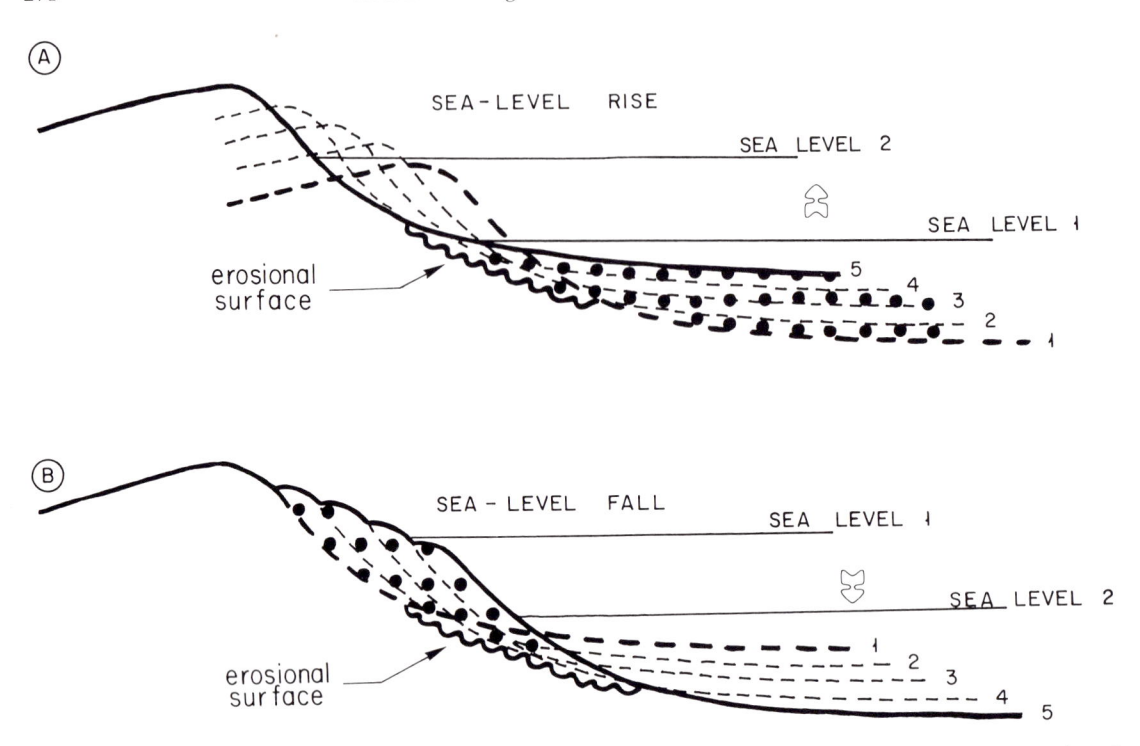

**Fig. 10.** Translation of the shoreface profile and preservation of the littoral zone deposits. (A) As a result of translation of the shoreface profile the lever point (see Fig. 9) will describe a line (or surface in three-dimensions), which if preserved will be represented in the stratigraphic record as an erosional surface. During a rise in sea-level, sediments deposited in the lower shoreface (black dots) will be preserved favourably, resting on top of this erosional surface. (B) During a fall in sea-level, sediments deposited in the foreshore/upper shoreface (black dots) will be preserved favourably, resting on top of the erosional surface.

deposited landward of the lever point (foreshore/upper shoreface) will be preserved favourably and will rest on top of the erosional surface, while those sediments located seaward of the lever point (lower shoreface) will be eroded (Fig. 10). It should be noted that a complex array of interdependent variables and their interactions, for example, rates of sea-level change and sediment supply, storminess, steepness and curvature of the shoreface profile, initial slope of the substrate, angle of translation of the shoreface, sediment texture, type of coast (mainland versus barrier island coast), wave regime, tidal range, etc., will exert an influence in determining the position of the lever point in the shoreface profile, therefore affecting the preservation of shoreline deposits in the stratigraphic record.

An important implication of Bruun's rule is that littoral zone sequences deposited during an episode of sea-level rise immediately followed by a drop in sea-level, such as the case for the east coast of Brazil, will be a composite sequence in which the lower shoreface sediments were deposited during a rise in sea-level and the upper shoreface/foreshore sediments were deposited during the subsequent drop in sea-level. This composite sequence mimics a sequence resulting from the simple progradation of the shoreline under conditions of excess sediment supply and stable or slowly rising sea-level.

One should always remember that the reasoning above is basically two-dimensional and assumes a closed material balance system. Longshore sediment supply will certainly affect the applicability of the Bruun's rule to three-dimensional situations. This is certainly the case at the Doce strandplain, where no positive evidence for composite littoral zone sequences was found. All vibracoring at the Doce strandplain was carried out in the beach-ridge system deposited during emergence phase I (5.1–3.8 ka). Radiocarbon dating of the lower shoreface sediments, as discussed previously, shows that they also were deposited during that period. Because of the very peculiar mode of advance of the shoreline, high rates of progradation characterized the northern part of the Doce strandplain during that time. These high rates of progradation have probably masked the effects of the drop in sea-level during the same period. Furthermore a complete littoral zone sequence was not recovered at the vibracoring sites. Thus, important information about the total thickness of the lower shoreface deposits and the nature of the sediments upon which they rest is missing.

## CONCLUDING REMARKS

This study has examined the influence of Quaternary sea-level changes on the depositional architecture of the Doce River strandplain located on the east coast of Brazil. This research has improved our understanding of the response of strandplains associated with river mouths to changes in relative sea-level in the following ways:

**1** During sea-level rises of the order of tens of metres the strandplain is extensively drowned and replaced by barrier-island–lagoonal systems. The sedimentary record deposited in these lagoons rest on top of Pleistocene beach-ridge deposits. Mangrove swamp deposits occur in the basal part of the sequence. These mangroves are rapidly drowned and replaced by subtidal environments, where prodeltaic muds of an intra-lagoonal delta are deposited. The sequence is capped by freshwater swamp deposits on top of which freshwater peats occur.

**2** High-frequency sea-level oscillations (2–3 m in amplitude over a time-scale of 200–300 years) are expected to cause flooding and changes in the coastal zone only in those sectors of the strandplain located downdrift of the river mouth. In this region, the shoreline advances through a mechanism of coastwise progradation of sand spits or hooked spits, generating in this process low-lying areas that later will serve, in conjunction with the river mouths, as pathways for the invasion by the sea. The updrift side, which progrades through the successive accretion of beach-ridges, is not affected by these high-frequency sea-level oscillations. Extensive erosion of the shoreline may occur, however, on the updrift side. A sea-level rise of the order of 2–3 m, although not sufficient to generate barrier-island–lagoonal systems, sets the initial stage for further flooding. If sea-level continues to rise, progression of flooding will isolate a barrier island from the mainland, and a well-established lagoonal system will form. Another important effect associated with high-frequency sea-level oscillations is the drowning of river mouths, reducing the transporting efficiency of the fluvial effluent. As a result, the river changes its lower course seeking a more efficient path to the sea.

**3** Theoretical considerations of Bruun's rule of translation of the shoreface profile shows that preservability of the different facies of the littoral zone will differ during rises and falls in sea-level. During a rise in sea-level, lower shoreface deposits will be preserved preferentially. Conversely, during a fall in sea-level, only those sediments deposited in the foreshore/upper shoreface will survive. Testing of this model in the study area did not produce conclusive results. Logistical problems restricted the vibracoring operations to those sectors of the strandplain located downdrift of the river mouth. Introduction of a large volume of riverborne sediments to those sectors, conceivably could mask the effects of the drop in sea-level during deposition of the littoral zone sequences

Broad testing of these concepts to both modern and ancient strandplain sequences will certainly prove useful for better understanding of the stratigraphy and genesis of ancient sandstone shoreline sequences. Detailed work in regions where good outcrops of ancient littoral zone sequences are present show that littoral zone sequences are composite in nature and are deposited under the influence of changing relative sea-levels (e.g. Miocene of the southeast Caliente Range, California (Clifton, 1981); Late Cretaceous Blackhawk Formation in the Book Cliffs, Utah (Balsley, 1983)).

Finally, one might ask why in the entire paper we avoided using the term delta to describe the study area. The very broad definitions of delta systems in use today have resulted in the grouping of a broad range of coastal accumulations whose only common feature is the geographical association with a river mouth (Dominguez, 1986). At the time the deltaic sedimentation models currently in use were developed, many of the areas were very poorly understood—particularly those needing modifiers, such as wave-dominated or tide dominated ones. Circularity of reasoning did the rest. Inevitably phenomena described for well-studied areas, such as the Mississippi River, were uncritically extrapolated and assumed to be valid for all other members of the so-called delta category. So far the application of the term delta to wave-dominated settings, such as the Doce River strandplain, has resulted in a series of misconceptions, and represented a major obstacle to the understanding of the evolution of those areas (Dominguez et al., 1987).

Finally, in connection with this aspect we would like to quote the following paragraph from Curray et al. (1967)—which predates the formulation of the sedimentation models mentioned above—about their study of the Nayarit coastal plain (Mexico), which is very similar to the Doce strandplain. 'The Nayarit coastal plain is a many faceted problem. It can be considered a type of delta, or a type of barrier, lagoon, strandplain coast. It can be studied as an example of the turning point from transgression to regression. It can be used as a modern model for deposition of regressive sheet sands so important in the geological record.'

In the geological literature the Nayarit coastal plain is usually referred to as a classical example of regressive sheet sands (e.g. Reineck & Singh, 1980). If it were studied today it would be termed a typical example of a wave-dominated delta. Circularity of reasoning would rapidly allow identification of a delta-front, delta plain, pro-delta, distributary mouth bars and delta cycles. As a result the many faceted aspect of the history of this plain, as observed by Curray, would have been easily overlooked.

## ACKNOWLEDGEMENTS

Financial support for this study was provided by the Programa de Pesquisa e Pos-Graduação em Geofisica of the Universidade Federal da Bahia (PPPG–UFBa–Brazil), the Geological Society of America (Grants-in-aid no. 3564-86), and the Coordenação de Aperfeicoamento de Pessoal de Nivel Superior (CAPES–Ministry of Education–Brazil).

Thanks are due to the Laboratorio de Geologia Marinha e Geofisica of the Universidade Federal Fluminense (LAGEMAR–UFF–Brazil), and Observatorio Nacional (ON–CNPq–Brazil) for logistical support.

The senior author is particularly grateful to D. J. P. Swift for his encouragement.

This paper forms part of a PhD dissertation written at the University of Miami, under the supervision of Dr Harold R. Wanless.

## REFERENCES

ANDREWS, J. (1981) *Texas Shells, a Field Guide*. University of Texas Press, Austin, 175 pp.

BALSLEY, J.K. (1983) *Cretaceous Wave-dominated Delta Systems—Book Cliffs, East Central Utah, Fieldguide*. Oklahoma City Geological Society, 219 pp.

BANDEIRA, JR., A.N., PETRI, S. & SUGUIO, K. (1975) *Projeto Rio Doce*. Internal Report, Petroleo Brasileiro S.A., Brazil 203 pp.

BELKNAP, D.F. & KRAFT, J.C. (1985) Influence of antecedent geology on stratigraphic preservation potential and evolution of Delaware's barrier systems. *Mar. Geol.* **63**, 235–262.

BROECKER, W.S. & OLSON, E.A. (1961) Lamont radiocarbon measurements VIII. *Radiocarbon* **3**, 176–204.

BRUUN, P. (1962) Sea-Level rise as a cause of shore erosion. *Am. Soc. Civ. Eng. Proc. J. Water. Harbors Div.* **88**, 117–130.

CLIFTON, H.E. (1981) Progradational sequences in Miocene shoreline deposits, southeastern Calient Range, California. *J. Sediment. Petrol.* **51**, 165–184.

CURRAY, J.R., EMMEL, F.J. & CRAMPTON, P.J.S. (1967) Holocene history of a strandplain, lagoonal coast, Nayarit, Mexico. In: *Lagunas Costeras, Un Simposio* (Eds Casteñares, A.A. & Phelger, F.B.) pp. 63–100. Universidad Nacional Autonoma de Mexico, Mexico.

DIRETORIA DE HIDROGRAFIA E NAVEGAÇÃO (1981) Tabuas de Mares, Costa do Brasil e Portos Estrangeiros. *Marinha do Brasil*. Diretoria de Hidrografia e Naregação, Brazil 127 pp.

DORJES, J., FREY, R.W. & HOWARD, J.D. (1986) Origins of, mechanisms for, mollusc shell accumulations on Georgia beaches. *Senckenbergiana Maritima* **18**, 1–43.

DOMINGUEZ, J.M.L. (1984) Sealevel history: a dominant control on modern coastal sedimentation styles. *Soc. Econ. Paleontol. Mineral. Ann. Midyear Meet. Abstr.* **1**, 26.

DOMINGUEZ, J.M.L. (1986) Re-evaluation of deltaic sedimentation models in wave-dominated settings, evidence from eastern Brazil (abstract). *Abstracts with Programs, 99th Annual Meeting and Exposition*, Geological Society of America, San Antonio, TX, pp. 586–587.

DOMINGUEZ, J.M.L. (1987) Quaternary sealevel changes and the depositional architecture of beach-ridge strandplains along the east coast of Brazil. PhD dissertation, University of Miami, 288 pp.

DOMINGUEZ, J.M.L., BITTENCOURT, A.C.S.P. & MARTIN, L. (1983) O papel da deriva litoranea de sedimentos arenosos na construção das planicies costeiras associadas as desembocaduras dos rios Sao Francisco (SE/AL), Jequitinhonha (BA), Doce (ES) e Paraiba do Sul (RJ). *Rev. Bras. Geocienc.* **13**, 98–105.

DOMINGUEZ, J.M.L., MARTIN, L. & BITTENCOURT, A.C.S.P. (1987) Sea level history and the Quaternary evolution of river mouth-associated beach-ridge plains along the east/southeast coast of Brazil—a summary. In: *Sea-Level Fluctuation and Coastal Evolution* (Eds Nummedal, D., Pilkey, D.H. & Howard, J.D.). Soc. Econ. Paleontol. Mineral. Spec. Publ. 41, 115–127.

HOBGEN, N. & LUMB, F.E. (1967) *Ocean Wave Statistics*. National Physical Laboratory, Ministry of Technology, London, 263 pp.

KOMAR, P.D. (1976) *Beach Processes and Sedimentation*. Prentice-Hall, New York, 429 pp.

KRAFT, J.C. (1971) Sedimentary facies patterns and geologic history of a Holocene transgression. *Geol. Soc. Am. Bull.* **82**, 2131–2158.

LANESKY, D.E., LOGAN, B.W., BROWN, R.G. & HINE, A.C. (1979) A new approach to portable vibracoring underwater and on land. *J. Sediment. Petrol.* **49**, 654–657.

MARTIN, L. & SUGUIO, K. (1978) Excursion route along the coastline between the town of Cananeia (State of São Paulo) and Guaratiba outlet (State of Rio de Janeiro). In: *1978 International Symposium on Coastal Evolution in the Quaternary*, Spec. Publ. No. 2. Universidade de São Paulo, Brazil, 95 pp.

MARTIN, L., FLEXOR, J.-M., VILAS BOAS, G.S., BITTENCOURT, A.C.S.P. & GUIMARAES, M.M.M. (1979) Courbe de variations du niveau relatif de la mer au cours des 7000 dernières années sur un secteur homogene du littoral bresilien (nord de Salvador, Bahia). In: *Proceedings 1979 International Symposium on Coastal Evolution in the Quaternary* (Eds Suguio, K., Fairchild, R.R., Martin, L. & Flexor, J.-M.) pp. 264–274. Universidade de São Paulo, Brazil.

MARTIN, L., BITTENCOURT, A.C.S.P. & VILAS BOAS, G.S. (1982) Primeira ocorrencia de corais pleistocenicos na costa brasileira; datação do maximo da Penultima Transgressão. *Rev. Cienc. Terra* **1**, 16–17.

MARTIN, L., FLEXOR, J.M., BLITZKOW, D. & SUGUIO, K. (1985) Geoid change indications along the Brazilian coast during the last 7000 years. *Proceedings of the Fifth International Coral Reef Congress*, Tahiti, Vol. 3, pp. 85–90. Antenne Museum-Ephe, French Polynesia.

MELO, U., SUMMERHAYES, C.P. & ELLIS, J.P. (1975) Part IV: Salvador to Vitoria, Southeastern Brazil. In: *Upper Continental Margin Sedimentation Off Brazil* (Eds Milliman, J.D. & Summerhayes, C.P.). *Contrib. Sedimentol.* **4**, 78–116.

REINECK, H.-E. & SINGH, I.B. (1980) *Depositional Sedimentary Environments with Reference to Terrigenous Clastics*. Springer-Verlag, New York, 549 pp.

SCHOFIELD, J.C. (1975) Sea-level fluctuations cause periodic post glacial progradation, South Kaupara barrier, North Island, New Zealand. *N. Z. J. Geol. Geophys.* **18**, 295–316.

SILVA, M.C.P. & DAVIS, G.M. (1983) d'Orbigny's type specimens of Paludestrina (Gastropoda: Prosobranchia) from the southern South America. *Proc. Acad. Nat. Sci. Philadelphia* **135**, 128–146.

STUIVER, M. (1980) Workshop on C-14 data reporting. *Radiocarbon*, **22**, 964–966.

STUIVER, M. & POLACH, H.A. (1977) Reporting of C-14 data. *Radiocarbon* **19**, 355–363.

SUGUIO, K., MARTIN, L. & FLEXOR, J.M. (1979) Sea level fluctuations during the past 6000 years along the coast of the state of São Paulo, Brazil. In: *Earth Rheology, Isostasy and Eustasy* (Ed. Morner, N.A.) pp. 471–486. Wiley, New York.

SUGUIO, K., MARTIN, L. & DOMINGUEZ, J.M.L. (1981) Evolucao do delta do rio Doce (ES) durante o Quaternario—influencia das variaçoes do nivel do mar. *Atas do IV Simposio do Quaternario no Brasil*, Sociedade Brasileira de Geologia, Rio de Janeiro, pp. 93–116.

SWIFT, D.J.P. (1976) Coastal sedimentation. In: *Marine Sediment Transport and Environmental Management* (Eds Stanley, D.J. & Swift, D.J.P.) pp. 255–310. Wiley Interscience, New York.

TASTET, J.P. (1979) L'Holocene du littoral septentrional du Golfe de Guinee. In: *Proceedings, 1979 International Symposium on Coastal Evolution in the Quaternary* (Eds Suguio, K., Fairchild, T.R., Martin, L. & Flexor, J.-M.) pp. 588–606. Universidade de São Paulo, Brazil.

US NAVY (1978) *Marine Climatic Atlas of the World*, Vol. IV, *South Atlantic Ocean*. U.S. Government Printing Office, Washington, DC, 325 pp.

ZENKOVICH, V.P. (1967) *Processes of Coastal Development*. Wiley Interscience, New York, 738 pp.

# Origin and geometry of storm-deposited sand beds in modern sediments of the Texas continental shelf

J. W. SNEDDEN[*,1] and D. NUMMEDAL[†]

[*] *Mobil Exploration and Producing Services, Inc., P.O. Box 650232, Dallas, TX 75265, USA*
[†] *Department of Geology and Geophysics, Louisiana State University, Baton Rouge, LA 70803, USA*

## ABSTRACT

Box cores taken from surficial sediments of the central Texas continental shelf reveal the presence of thin (1–10 cm) discrete beds of sand and silt in an otherwise clayey sedimentary section. The discrete beds are characterized by vertical upward fining grain-size trends, grading from very fine sand to fine silt. Physical sedimentary structures are dominantly planar laminae that locally pass upwards into low-angle inclined lamination. Basal contacts are always sharp, while upper contacts can be either sharp or gradational. Offshore decreases in grain size, thickness and frequency of the discrete sand beds are also observed.

This evidence is compatible with the two hypotheses commonly offered to explain the origin of storm-deposited beds: storm-surge ebb induced turbidity currents or geostrophic storm flows.

Consideration of the driving forces behind shelf sediment transport and mapping of the distribution of one recent storm bed permit further testing of these hypotheses. The along-shelf continuity of a single sand bed, dated as contemporaneous with the passage of Hurricane Carla across the Texas shelf in 1961, is inconsistent with the turbidity current hypothesis. The strike-trending sand body geometry and the seaward fining within the Carla bed imply a sand transport pathway beginning in the shoreface zone and oriented obliquely offshore and along shelf.

Analysis of current and wave measurements from the shelf during the passage of smaller storms, such as Hurricane Allen, suggests that storm flows are geostrophically balanced, driven primarily by the wind-induced hydrostatic pressure gradients. There appears to be a recognizable relationship between the wind field, sea-surface slope, wave propagation, and near-bottom combined-flow kinematics. Building upon this, a sand transport model was developed for Hurricane Carla. This reconstruction indicates that the combined flow present during the storm had sufficient strength and the proper transport kinematics to account for the observed sand body geometry.

## INTRODUCTION

In recent years, a large number of studies of ancient shelf sedimentary sequences have recognized the presence of thin (generally <2 m), discrete beds of sandstone intercalated with shale (see summary by Marsaglia & Klein, 1983; Duke, 1985). These have been termed variously as 'tempestites' (Seilacher, 1982) or 'event beds' (Brandt, 1986), with specific reference to the possibility of formation during storm events. Unfortunately, the number of ancient examples far outnumbers the published reports of possible modern analogues. Another problem is the fact that the best-known modern examples of tempestites come from tidally dominated shelves of the North Sea (Reineck & Singh, 1972; Aigner & Reineck, 1982), or from storm-dominated shelves in high-latitude settings (Nelson, 1982). However, as pointed out by Duke (1985), many of the cited ancient cases come from middle and lower palaeolatitudes.

The Texas continental shelf, in contrast, rep-

[1] Present address: Mobil Research and Development Corporation, Dallas Research Laboratory, P.O. Box 819047, Dallas, TX 75381-9047, USA.

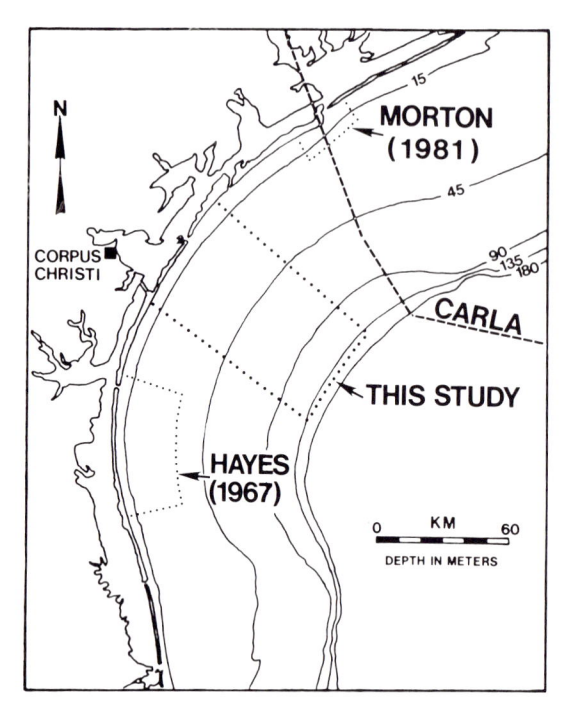

**Fig. 1.** Location of study area and investigation sites of Hayes (1967) and Morton (1981). Dashed line is track of the eye of Hurricane Carla in September of 1961.

resents a low latitude, storm-dominated shelf (Fig. 1). Cores of the modern, near-surface sediments of the shelf contain fine sand and silt in the form of discrete sand beds separated by thicker sections of clay (Fig. 2). These were recognized in short gravity cores taken on the Texas shelf as part of American Petroleum Institute Project 51 (Curray, 1960). However, it was Hayes (1967) who first suggested that the beds may be related to storm passage on the shelf. Hayes (1967) took phleger cores and grab samples in water depths of 18–36 m offshore from central Padre Island in the spring of 1962 and discovered thin (1–9 cm), graded sand deposits in the top of the cores that had not been present in coring surveys prior to the landfall of Hurricane Carla in September of 1961. Carla was an exceptionally large storm, its radius of hurricane-force winds (>33 m s$^{-1}$) stretching nearly 500 km from Brownsville to Port Arthur (Nummedal, 1982). Noting the presence of numerous shore-normal channels across south Padre Island following the storm, Hayes (1967) hypothesized that much of the storm tide produced by Carla returned to the Gulf through these hurricane-cut channels (Fig. 3). He further theorized that these 'storm-surge ebb currents' became turbidity currents upon entering the Gulf of Mexico, transporting shoreline clastics across the shelf.

Morton (1981) later challenged this 'storm-surge ebb' mechanism of sand transport. First, he noted that the storm runoff from Laguna Madre was negligible, most of the excess water passing out of the Brazos–Santiago Pass to the south. Secondly, he observed that most of the washover deposits at Padre Island displayed flood-oriented features, suggesting a rather limited ebb transport through the storm-cut channels. Third, he pointed out that the thickest and most extensive deposits were located offshore from barrier segments with very few storm-cut channels. Citing published studies from other parts of the Gulf of Mexico (e.g. Forristall *et al.*, 1977), Morton (1981) proposed the general mechanism of wind-forcing of bottom currents to explain the erosion, transportation and deposition of the sand in the discrete beds.

It is clear that there is no consensus as to the origin of the discrete sand beds present in near-surface modern sediments of the Texas shelf. Several specific questions remain unanswered with regard to the transport mechanism, the transport pathway and the source of the clastics present in these beds. In addressing these questions, previous studies have concentrated on characteristics of the sand beds themselves (Berryhill *et al.*, 1976, 1977), coastal geomorphology (Hayes, 1967) or data from other parts of the Gulf of Mexico (Morton, 1981). No study conducted in the Texas shelf area has hitherto combined analysis of sedimentological data with actual measurements of the bottom current field from that same area. However, it is clear that any scientific inquiry directed toward these problems must include information on the processes operative upon the shelf as well as the products of those processes.

Studies of sediment transport associated with storm events on other continental shelves have demonstrated the importance of analysis of the wave and current field near the sea-bottom (Lavelle *et al.*, 1978; Adams *et al.*, 1982; Cacchione & Drake, 1982; Wright *et al.*, 1986). Boundary shear stress generated by oscillatory wave motion is often superimposed on the tangential stress associated with the steady unidirectional currents. The two interact in a non-linear fashion to produce a total shear stress that can be greater numerically than the simple sum of the respective stresses (Grant & Madsen, 1979). This wave–current interaction during storm events

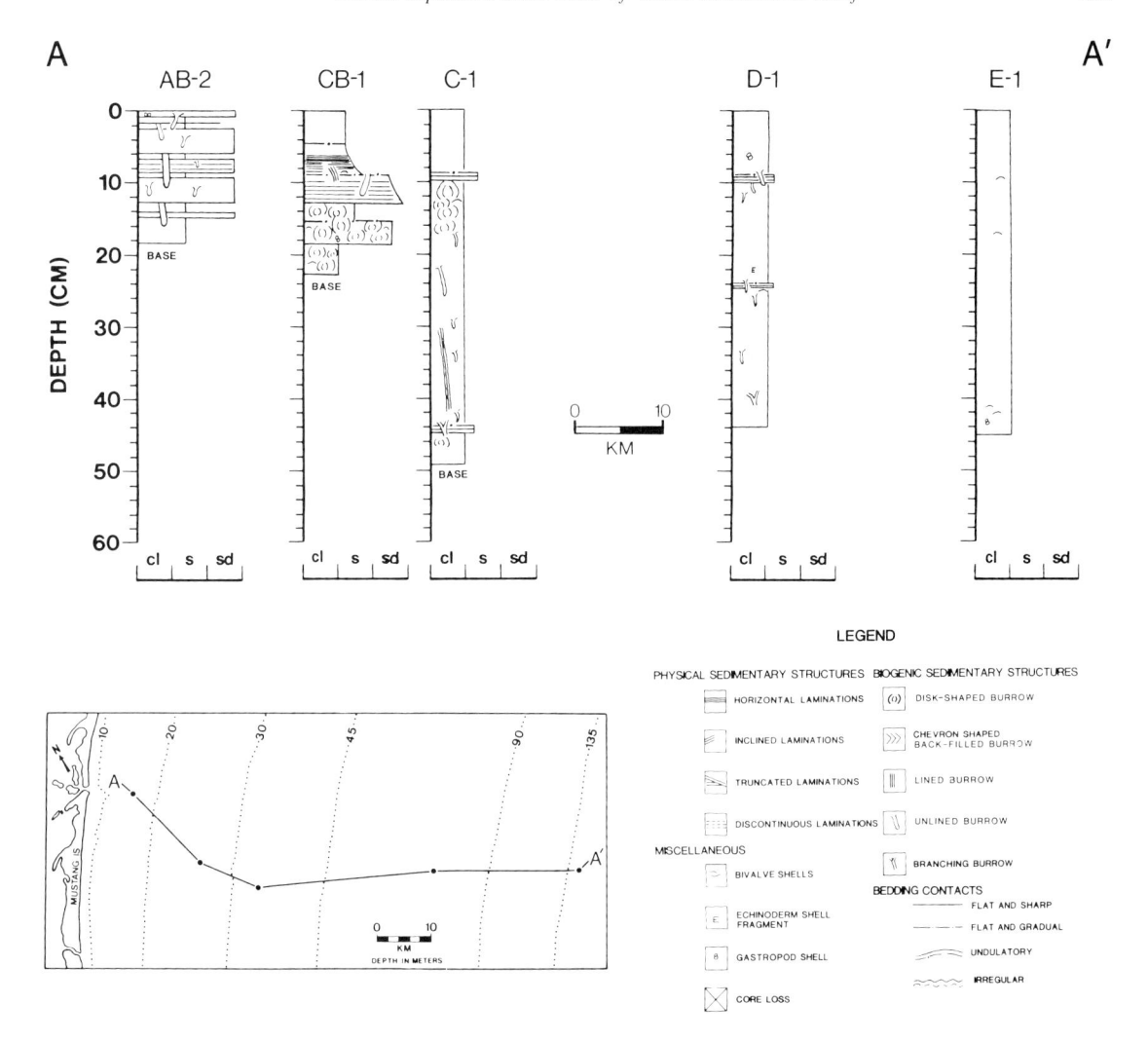

Fig. 2. Cross-shelf box core cross-section showing near-surface modern sediments of the study area. Legend of symbols and core locations are shown in the insets.

has been termed 'combined-flow' (Harms, 1969; Swift et al., 1983; Nøttvedt & Kreisa, 1987). It is the contention of this study that such combined-flow events are responsible for the erosion of coastal sand sources, and the transport and deposition of siliclastic sediments in the central Texas shelf area.

To test these hypotheses, a multi-disciplined approach was used. First, core samples of surface and near-surface modern sediments were obtained from a 50 × 90 km area on the 'central Texas continental shelf', using the terminology of Berryhill et al. (1976, 1977). Sampling was concentrated in this small area offshore from Mustang Island, Texas in order to facilitate direct comparison and correlation of the cores (Fig. 4). This sampling strategy also allowed analysis of along-shelf and cross-shelf patterns in composition, texture and sedimentary structures of the discrete sand beds. Cores were then taken outside this area to facilitate analysis of the lateral continuity of these beds and the large-scale sand body geometry.

The sedimentological data were then compared with measurements of combined flow during fair-weather periods and storm events. Concurrent studies included actual deployment work of current meters in the study area during a summer

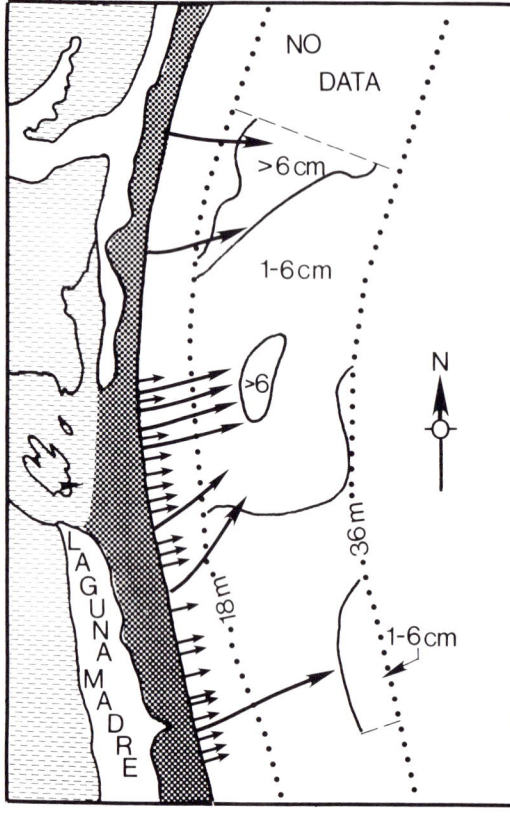

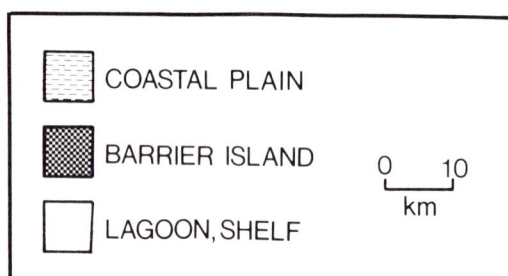

**Fig. 3.** Hypothetical sand transport by storm-surge ebb following Hurricane Carla in 1961. Small arrows represent seaward terminus of hurricane-cut channels and large arrows indicate inferred path of turbidity currents upon shelf. (Modified from Hayes, 1967.)

to winter time period as well as analysis of previously unpublished oceanographic data from one site (Snedden *et al.*, 1988). In addition, observations of winds, tides and surface wave patterns from Hurricane Carla facilitated hindcasting the combined-flow kinematics for this large and important storm (Snedden & Nummedal, 1989).

# SEDIMENTARY CHARACTERISTICS OF THE DISCRETE SAND BEDS

## Methods

Relatively wide, undisturbed core samples of modern sediments in the study area were obtained using a conventional box-coring device mounted on the stern of the R/V *Longhorn*. The box-core samples have dimensions of 20 × 30 × 60 cm. Box cores were collected during four cruises of the R/V *Longhorn* in 1984–1985, with most cores obtained in water depths of 12–30 m (Fig. 4; Table 1). Previous reconnaissance studies (Berryhill *et al.*, 1976, 1977) suggested that this depth range had the greatest amount of variability in the sedimentary characteristics of the discrete sand beds. Once onboard, the box cores were subsampled into plexiglass trays with a similar width but a thickness of 2 cm. The samples were then treated with acetone to prevent further burrowing activity by organisms that were trapped by the coring operation. The plexiglass trays were then radiographed using standard X-ray techniques (e.g. Hamblin, 1962; Coleman, 1966). This permitted identification of biogenic and physical sedimentary structures in the box-core samples. In addition, delineation of discrete sedimentary units was made on the basis of the radiograph, once the X-ray response was calibrated against sand–silt–clay content.

Textural analysis was performed using several different methods. The relative percentages of sand, silt and clay were determined using the sieve-and-pipette technique described by Folk (1968). The size-frequency distribution of silt-sized material was delineated using a Model T Coulter Counter (see Sheldon & Parsons, 1967). Determination of the sand size distribution proved to be somewhat difficult owing to the fine size and the angular nature of many of the grains. Sieve analysis did provide a general breakdown, which was further defined by settling velocity analysis using an automated settling tube (see Anderson *et al.*, 1982). Although the discrete beds will be referred to as 'sand' beds throughout the text, it should be remembered that in some cases these units have a higher percentage of silt than sand.

## Sedimentary structures

Examination of the 23 box cores taken in the central Texas shelf area indicates that the near-surface

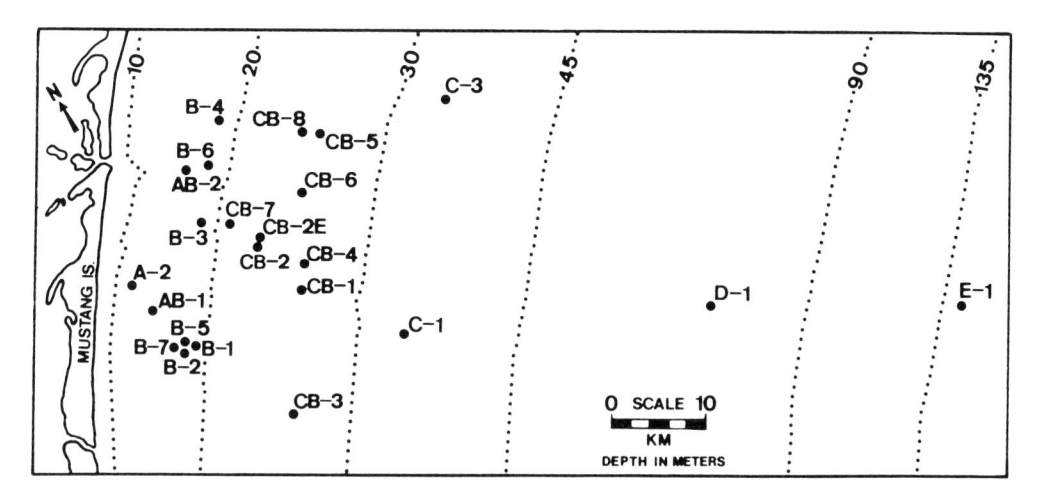

**Fig. 4.** Location of box cores taken in study area. For study area location see Fig. 1.

**Table 1.** Box core locations

| Box core | Water depth (m) | Cruise | Latitude | Longitude |
|---|---|---|---|---|
| E-1 | 140 | July 1984 | 27°12.18′N | 96°23.61′W |
| D-1 | 74 | July 1984 | 27°22.70′N | 96°34.80′W |
| C-1 | 34 | July 1984 | 27°32.90′N | 96°51.70′W |
| C-3 | 34 | December 1984 | 27°43.97′N | 96°42.78′W |
| CB-1 | 26 | July 1984 | 27°37.00′N | 96°56.80′W |
| CB-3 | 26 | November 1984 | 27°34.30′N | 96°59.50′W |
| CB-4 | 26 | November 1984 | 27°39.50′N | 96°56.30′W |
| CB-5 | 25 | December 1984 | 27°45.50′N | 96°50.10′W |
| CB-6 | 25 | December 1984 | 27°42.30′N | 96°54.80′W |
| CB-8 | 24 | December 1984 | 27°46.80′N | 96°50.10′W |
| CB-2 | 24 | July 1984 | 27°40.00′N | 96°58.10′W |
| CB-7 | 21 | December 1984 | 27°41.35′N | 97°00.00′W |
| B-1 | 18 | July 1984 | 27°38.05′N | 97°04.08′W |
| B-2 | 18 | July 1984 | 27°38.05′N | 97°04.08′W |
| B-3 | 18 | July 1984 | 27°42.00′N | 97°01.20′W |
| B-4 | 18 | July 1984 | 27°48.60′N | 96°56.00′W |
| B-5 | 18 | November 1984 | 27°37.81′N | 97°04.12′W |
| B-6 | 18 | November 1984 | 27°46.90′N | 96°57.50′W |
| B-7 | 18 | December 1984 | 27°37.81′N | 97°04.12′W |
| AB-2 | 16 | December 1984 | 27°47.80′N | 96°58.55′W |
| AB-1 | 15 | December 1984 | 27°40.35′N | 97°05.65′W |
| A-2 | 12 | November 1984 | 27°42.52′N | 97°06.20′W |
| LH-1 | 30 | November 1985 | 26°52.97′N | 97°09.36′W |
| LH-4 | 26 | November 1985 | 27°20.11′N | 97°07.96′W |

sedimentary section is composed of a variable number of discrete beds of sand, silt and clay (Fig. 2). Within individual beds there are a variety of physical and biogenic sedimentary structures. Study of the physical structures in the discrete sand beds provides information relevant to the question of the transport mechanism. Analysis of the biogenic structures is germane to the problem of the preservation potential of these sand beds and reconstruction of the initial sand body geometry.

*Physical sedimentary structures*

Horizontal (planar) laminae are the most common form of physical stratification in the discrete sand

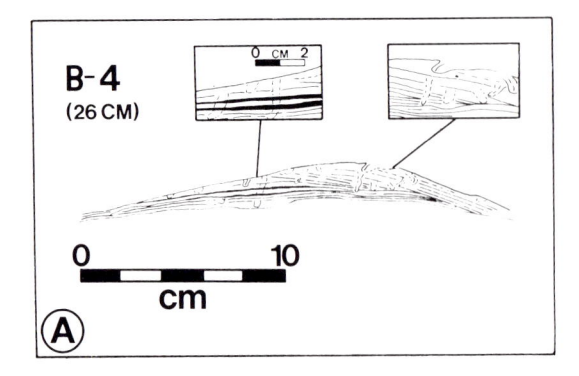

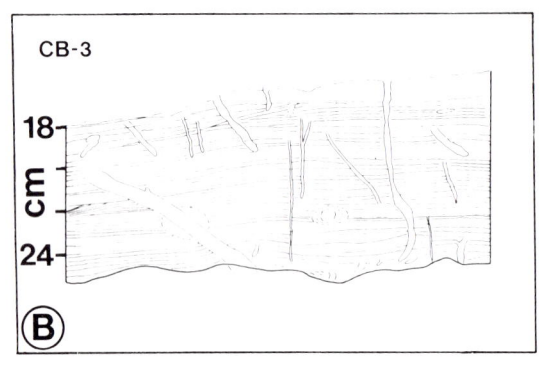

**Fig. 5.** (A) Tracing of an X-ray radiograph of a discrete sand bed at 26 cm core depth in box core B-4 (18 m water depth). Left inset shows horizontal laminae (slightly tilted in coring process) and erosional upper surface of bed. Right inset illustrates the contact between foreset and bottomset laminae. (B) Tracing of discrete sand at 18–25 cm depth in box core CB-3 (26 m water depth).

beds (Fig. 5). These are typically parallel, even, continuous to discontinuous. In a few cases, the discrete sand beds appear massive (unlaminated). The presence of horizontal laminae in graded sequences such as these appears to be common in ancient sequences interpreted to have formed in shelf-depth waters (e.g. Howard, 1972; Hobday & Morton, 1984; Snedden & Jumper, 1990), but the means by which they formed is still unclear. A number of mechanisms have been proposed to explain the occurrence of horizontal laminae in graded sand sequences, including variations in external forcing (Reineck & Singh, 1975), saltation sorting (Kuenen, 1966), depositional sorting (Blow & Bowen, 1980), turbulent bursts and sweeps (Bridge, 1978; Cheel & Middleton, 1986; Cheel, 1990) and the migration of low-amplitude bedforms (Paola et al., 1989). Plane-beds are also common bed configurations found in laboratory flume experiments with unidirectional flow (Simons et al., 1977). They occur at the beginning of lower flow regime (the no movement phase) and upper flow regime stages (Harms et al., 1982). The association of the planar laminae observed in the box cores with sharp lower contacts and the coarser end of the size-frequency distribution in these beds suggests that these laminae formed at higher flow velocities than those characteristic of the lower flow regime (no movement phase).

Recognition of the formation of upper flow regime flat beds in shelf-depth waters has proven to be a source of controversy. Calculation of the Froude number with water depth as the scaling length would dictate that upper flow regime conditions are virtually unattainable with typical shelf current and wave orbital velocities. However, it is possible that a more appropriate scaling length for combined-flow conditions is the thickness of the wave or current-generated boundary layer (C. Adams, pers. comm., 1985). With typical wave boundary layer thicknesses of the order of 10 cm (Komar, 1976), supercritical flow may be possible with bottom flow velocities of 100 cm s$^{-1}$ or greater. Unfortunately, this supposition is hampered by a lack of experimental evidence upon the relationship of bedforms to flow velocity under conditions of combined wave and current flow, high sediment concentrations and large water depths (Komar, 1974, 1985).

Overlying the planar laminae in many of these beds are low-angle inclined laminae. These appear to have tangential contacts with bottomset bedding (Fig. 5A). In controlled flume experiments, Jopling (1965) found that tangential contacts were associated with relatively high stream-flow velocities, fine sediment sizes and a predominance of suspension transport. As will be shown in later sections, the fine size of quartz grains in the discrete sand beds (modal size 68 μm) suggests that suspended load transport is most typical of the fully developed flow (McCave, 1971).

The incipient development of bedforms on the sea-floor is suggested by the bed depicted by Fig. 5A. Although the bed shows evidence of erosion at its upper contact and is possibly form-discordant, bottomset and foreset laminae are apparent. It is possible that the laminae are related to the migration of a flow-transverse bedform, such as a ripple. Development of ripple-like bedforms from initial perturbations during shear flows has been demonstrated both numerically (J.D. Smith, 1970) and experimentally (Mantz, 1978). Early bedform growth appears to be related to a phase lag between topography and the boundary shear stress (Swift & Rice, 1984). However, very high levels of shear-generated turbulence tend to inhibit further development, suggesting that a large portion of bedform growth takes place during current deceleration (Yalin, 1977). Thus, it is possible that the inclined laminae present in the tops of many of the discrete sand beds represent the migration of ripple bedforms during the waning phase of an impulsive flow event (such as a decelerating bottom current or a turbidity current).

On a storm-dominated shelf, such as the central Texas shelf, one would expect to see some evidence of hummocky cross-stratification (HCS). Originally recognized in ancient shelf to shoreface sediments (Campbell, 1966; Harms et al., 1975), HCS is believed to be similar in form and origin to hummocky megaripples found on the inner Atlantic shelf (Swift et al., 1983). Hummocky cross-stratification appears to be a ubiquitous structure in coastal and marine environments, having now been identified in the surf zone, shoreface, inner and outer shelf. However, most agree that it is intimately associated with storm processes (Dott & Bourgeois, 1982; Greenwood & Sherman, 1986; Nøttvedt & Kreisa, 1981; Duke, 1990).

Three of the primary sedimentary structures recognized in the box cores from the central Texas shelf area may be related to HCS and combined-flow bedforms. The inclined laminae mentioned in the preceding section are probably related to the growth of bedforms on the sea-floor during the waning stages of a flow event. Indeed, it is possible that these inclined laminae are part of a larger scale bedform, such as a hummock. Unfortunately, the box core width is at least an order of magnitude smaller than the wavelength of the hummocks recognized in both the modern and ancient. However, some criteria have been formulated for the identification of HCS in narrow cores. One is the presence of low-angle inclined laminae that thicken or 'fan' downwards into a concave-up hollow or swale (Walker, 1985). Downward-fanning laminae appear to be present at 19 cm and 24 cm depth in Fig. 5B.

Another typical feature of HCS is the presence of numerous internal truncation surfaces. Dips on these 'second-order bounding surfaces' (as opposed to the first-order bounding surfaces at the base and top of the sequence) are variable, but usually are less than 15° (Dott & Bourgeosis, 1982). Such a feature is present at 21.8 cm depth in Fig. 5B and it appears to pass laterally into a lamina, a not altogether unusual occurrence (Brenchley, 1985).

Undulating patterns in internal laminations (Fig. 5) are probably further indications of combined-flow influence on the discrete sand beds. Subtle thickening and thinning of the laminae, with amplitudes of a centimetre or less, are similar in form to what Kreisa (1981) recognized in ancient storm-deposited beds and termed 'gentle pinch-and-swell'. Flow-duct experiments have demonstrated that such features can occur when a relatively weak unidirectional current is superimposed upon a stronger oscillatory flow (Arnott & Southard, 1990).

Other than those that might be ascribed to HCS, none of the sedimentary structures present in the discrete sand beds are unique to the transport mechanisms being considered in this study. Plane-bed laminae and ripples are characteristic of deposition from turbidity flows (Bouma, 1962), unidirectional currents and wave-generated oscillatory motion (Reineck & Singh, 1972). Massive, graded beds also can be formed by these processes (Reineck & Singh, 1975; Sanders, 1977). The absence of obvious wave-ripple laminae suggests that either wave orbital motion was negligible or it acted in combination with unidirectional currents. Bed configurations formed under combined-flow conditions can be, under certain conditions, similar to those found in the case of steady unidirectional flow (Harms et al., 1982; Arnott & Southard, 1990).

Although there are some exceptions, the typical bedding sequence in the thicker discrete sand beds consists of, in ascending order, a sharp basal contact, planar laminae (or often a massive, un-laminated division), inclined laminae (often subtle) and a gradational (commonly bioturbated) upper surface (e.g. Fig. 5B). Like the individual sedimentary structures, the sequence of structures is not exclusive to one process or the other. Both Aigner & Reineck (1982) and Nelson (1982) observed sedimentary sequences similar to these in discrete sand beds of modern shelf sediments of the Helgoland Bight (Germany) and Norton Sound (Alaska), respectively, yet proposed different transport mechanisms (offshore flowing pressure-gradient-forced currents versus broad wave-induced liquefaction and storm-surge ebb flow). The most that can be ascertained positively from observation of these vertical sequences is the interpretation that the discrete beds were deposited under conditions of decreasing boundary shear stress.

It should be mentioned that what is described above is most typical of the thicker sand beds that are present in 20–30 m of water. In shallower water (12–20 m), the beds are thinner and invariably display sharp upper contacts, suggesting some post-depositional truncation (Fig. 5A). In fact, in the shallow portion of the study area, in what might be defined as the inner-shelf–shoreface transition (Aigner & Reineck, 1982), the sequence becomes 'amalgamated' and it is difficult to discern individual stratification units (Fig. 6).

### Biogenic sedimentary structures

The majority of the types of burrow structure present in radiographs of box cores taken in the central Texas shelf area have been described previously by Hill (1985) and will not be discussed here. However, analysis of areal trends in bioturbation is relevant to discerning the preservation potential of the sand beds and as a guide in reconstruction of the initial sand body geometry.

In general, the degree of bioturbation decreases in an offshore direction (Fig. 7A). This areal pattern was observed by Hill (1985) and is confirmed by this study. The apparent relationship between the amount of bioturbation and the water depth parallels shore-normal trends in the number and diversity of live macrobenthic organisms present in surficial sediments (Fig. 7B). Grab samples taken across an area just south of the study area show significant decreases in the species density (number of live individuals per sample) and species richness (number of live species per sample) from the shoreline to 30 m (White et al., 1983; Hill, 1985). The two parameters remain approximately constant to 130 m, where they increase slightly. These cross-shelf patterns in biologic parameters have been attributed by Hill (1985) to shore-normal gradients in bottom-water temperature variability and the type of surficial sediment.

Dott (1983) recently suggested that the sand described by Hayes (1967) as a product of Hurricane Carla did not survive the infaunal reworking so typical of the Texas shelf. Indeed analysis of the biomass trends in the central Texas shelf area would lead one to suspect that the preservation potential of a single sand bed is quite low in shelf waters shallower than 20 m. However, the data of Hill (1985) suggest that discrete beds deposited during storms in waters deeper than 20 m could survive the

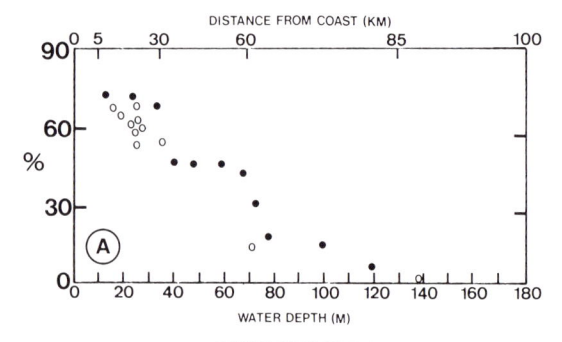

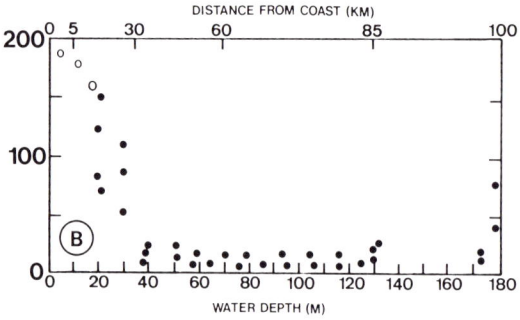

**Fig. 7.** Shore-normal trends in (A) percentage of core bioturbated (solid circles are data of Berryhill et al. (1976) and open circles this study) and (B) number of live individuals per grab sample (solid circles are data of Hill (1985) and open circles data of White et al. (1983)).

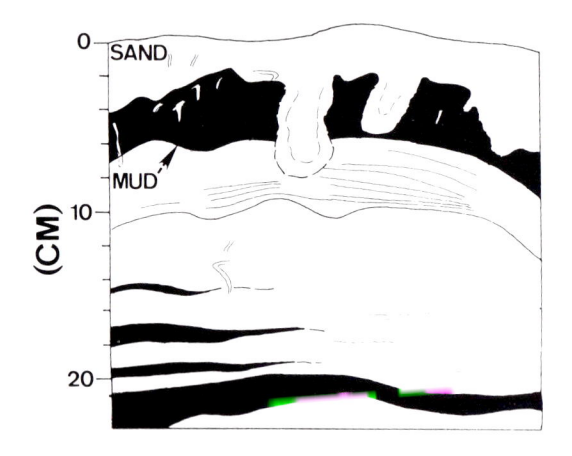

**Fig. 6.** Tracing of an X-ray radiograph of box core AB-1 (water depth 15 m). Note apparent amalgamation of beds from core depth of 7–21 cm.

**Table 2.** Composition of selected sand samples

| Box core | Depth in core (cm) | Quartz | Feldspar | Rock fragments | Carbonate fragments[a] | Other[b] |
|---|---|---|---|---|---|---|
| CB-7 | 9–16 | 87 | 3 | 1 | 8 | 1 |
| CB-1 | 12–13 | 85 | 2 | — | 13 | — |
| CB-8 | 12–13 | 88 | 1 | 1 | 10 | — |
| B-4 | 28–30 | 89 | 1 | 2 | 8 | — |
| Beach[c] | Mustang Island | 92 | 2 | 2 | 5 | — |

[a] Largely shell fragments and occasional whole forams.
[b] Mainly opaque grains.
[c] Sample taken 6 miles southeast of Horace Caldwell Pier, 25 m landward of the berm, 10 cm below surface on 29 June 1984.

destruction by infauna. Supporting evidence for this interpretation is supplied by absolute dating of the box cores and by coring in the study area of Hayes (1967), which will be discussed in later sections.

### Texture and composition

The hydrodynamic properties of detrital grains are a function of their size, shape and density (Anderson *et al.*, 1982). Textural analysis reveals information about the size and shape of the grains, while compositional analysis sheds light on the type of grain and thus its density. Comparison of the textural trends in a vertical and lateral sense also aid in making inferences about temporal and spatial gradients in the competency of a shear flow.

#### Grain composition

The detrital grains in samples from the discrete sand beds are dominantly quartz, paralleling the composition of a representative sample from the beach at Mustang Island, which is shown for comparison (Table 2). These are expected results, considering the provenance and multicyclic nature of the sands present along the Texas coast (Bullard, 1942; Mazzullo & Withers, 1984). In fact, some of the quartz grains show silica overgrowths, which only could have originated during prior history as a framework grain in a cemented sandstone.

#### Grain-size frequency distributions

Figure 8 shows a comparison of the size-frequency distributions of the silt and sand fraction of discrete sand beds in cores from water depths of 24 m and

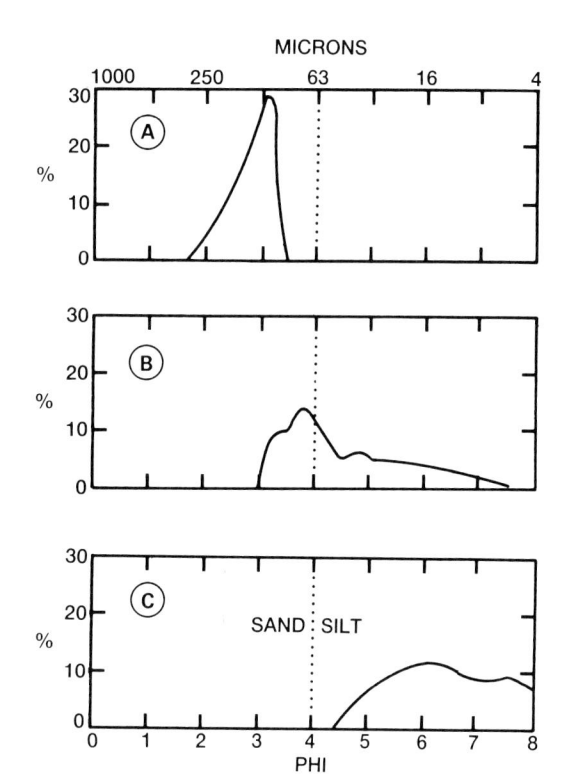

**Fig. 8.** Grain-size frequency distributions of sand and silt from surf zone at (A) Mustang Island, Texas, (B) a discrete sand bed in a box core taken in 24 m of water, and (C) a discrete bed in a box core from 74 m of water in the study area.

74 m, as well as a sample from the surf zone at adjacent Mustang Island. It is apparent that there is a steady depletion of the coarser particles moving into deeper water. This is reflected both in the

shift of the curves toward the fine end of the grain-size spectrum and the transition from the negatively (coarse) skewed distribution of the surf zone samples to the positively skewed shelf samples. Such a pattern has been termed 'progressive sorting' (Russell, 1939). Progressive sorting, the steady depletion of coarse grains down a transport pathway, will result in fine-skewed grain-size frequency distributions regardless of the shape of the input population, and is especially well-pronounced when there is a decrease in the intensity of the flow field down that transport pathway (Swift & Ludwick, 1976). In fact, the beach–shoreface–shelf transect may be regarded as one of the more spectacular examples of the influence of progressive sorting upon grain size frequency distributions (Swift, 1985).

*Grain shape*

As shown by Fig. 8, the modal class of the sand and silt fraction straddles the sand–silt boundary. Grains of this size range tend to be quite angular (Pettijohn et al., 1973), and this was verified by examination of a number of samples from the study area under the electron microscope. This angularity is related to the transport mode. Sand-size grains that are typical of the discrete sand beds are generally transported in suspension (Bagnold, 1966; McCave, 1971). During suspension transport, the grains are insulated from inertial impacts against other grains and there is less tendency for rounding. This conclusion is supported by evidence of delicate features, such as quartz overgrowths, which are preserved upon the surfaces of the grains from the discrete sand beds (Snedden, 1985).

Quartz grain shapes are also known to be a function of source area characteristics and have been quite useful in provenance determinations (e.g. Mazzullo, 1982). Fourier grain-shape analysis of samples from selected discrete sand beds in the study area showed a predominance of grains from what has been referred to as the 'Texas Coast Province' (J. Mazzullo, unpublished data, 1984). This province encompasses the drainage basins of the south Texas coastal plain rivers and is differentiated from the Brazos–Colorado and Rio Grande Province to the northeast and south, respectively. Sands and silts of this type are found in surficial sediments of the shoreface and inner shelf of the study area (Mazzullo & Withers, 1984).

*Vertical grain-size trends*

Settling velocity analysis, which can discriminate grain sizes to a tenth of a phi, reveals that the majority of the variability in sand sizes in a typical vertical sequence occurs within a fraction of a phi interval (Fig. 9). Closely spaced sampling through a discrete sand bed at 9 cm depth in box core CB-1 shows only very subtle upward fining in equivalent sand grain diameters. However, there is a clear increase in clay and silt relative to the proportion of sand. Coulter Counter measurements of the silt sizes also show clear upward fining trends (Fig. 10).

This lack of upward fining in the sand sizes, a

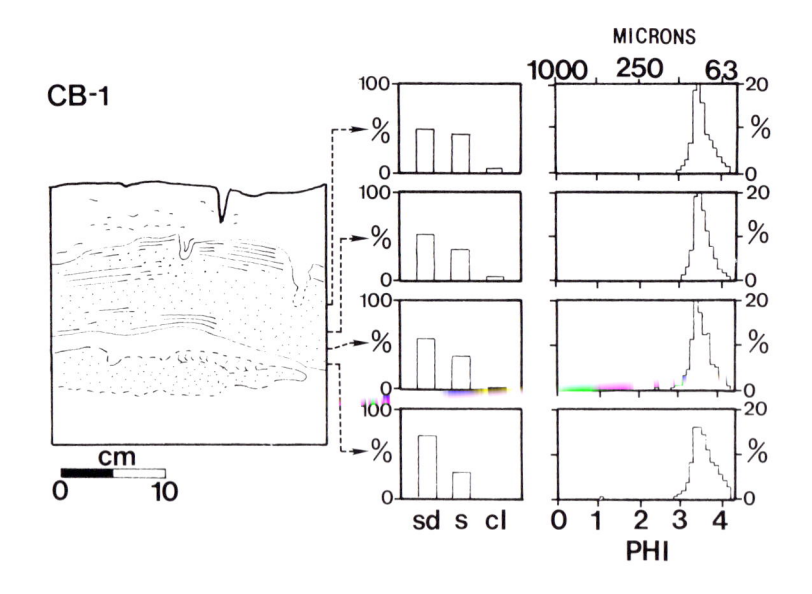

Fig. 9. Vertical trends of sand–silt–clay percentages and equivalent sand size (from settling velocity measurements) in a discrete sand bed of box core CB-1 (26 m water depth).

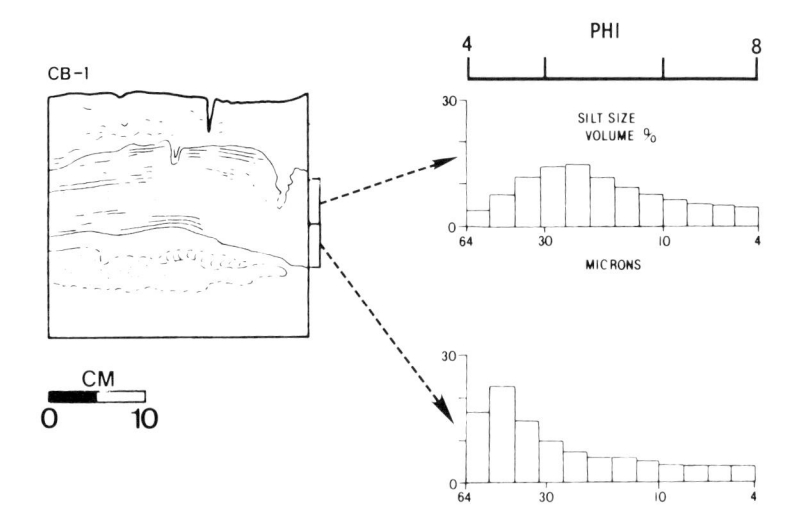

**Fig. 10.** Vertical trend of silt sizes (from Coulter Counter measurements) from two samples of box core CB-1 (26 m water depth).

phenomenon also observed by Berryhill *et al.* (1976, 1977), may be attributed to the phenomenon of progressive sorting. The steady depletion of the coarser particles down a transport pathway will tend to produce a very peaked (leptokurtic) sand-size frequency distribution, especially as the sizes of the deposits approach the sand–silt boundary (Russell, 1939). As the available size range is very narrow, vertical trends in the sand sizes are quite subtle, owing to the lack of variability. However, the availability problem may not be so critical in the silt and clay range, which display quite distinctive upward-fining trends.

The normal (fine) grading in the silt sizes and the upward increases in silt and clay relative to sand, on the other hand, indicate that deposition of the discrete sand beds proceeded as an orderly settling of grains as the levels of shear-generated turbulence decreased. This undoubtedly occurred during the waning energy phase of an impulsive event. This is consistent with the 'temporal acceleration' model for shelf storm deposits proposed by Swift & Rice (1984): (1) erosion and transport occur during flow acceleration; (2) erosion ceases once the sediment-carrying capacity is reached; (3) deposition proceeds as the flow decelerates, turbulence levels fall and the grains settle out of suspension.

*Lateral grain-size trends*

Spatial trends in the grain size present in discrete sand beds are also apparent (Fig. 11). Two points should be mentioned with regard to this figure. First, the samples were taken from a correlative horizon, a bed that absolute dating (next section) suggests was deposited within the time period

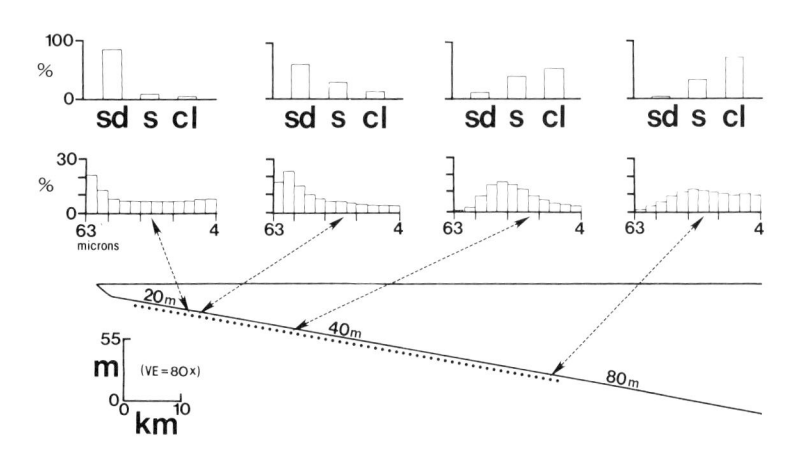

**Fig. 11.** Shore-normal trends of silt sizes and sand–silt–clay ratios within a single correlative horizon across the study area.

1961–1962. This time-equivalence is crucial to the interpretation of the cross-shelf competency of a single-flow event. Secondly, comparison of the grain sizes at each station could be done only within the silt range, as sand-sized material was not present at the 74-m site. Seaward fining is indicated by the shifts in silt sizes and in the sand–silt–clay ratios within this bed. This seaward fining again may be related to progressive sorting, where a deposit becomes increasingly finer down the transport path owing to a gradient of decreasing shear stress and consequent depletion of the coarser grains (Russell, 1939; Swift & Ludwick, 1976).

*Discussion*

Neither the vertical or lateral fining shown by the discrete sand beds are unique to the turbidity current (Stow & Bowen, 1980) or to the current transport mechanism (Swift & Ludwick, 1976). However, Fig. 11 does bring to light some points relevant to questions of source and pathway. The seaward fining is consistent with the concept of a transport pathway oriented at some angle (not necessarily perpendicular) to the trend of the isobaths across the shelf. It also fits with the idea of a coastal source for these clastics. In fact, this seaward fining parallels the size-grading trend shown by the finer grained surficial sediments of this area (Berryhill *et al.*, 1976, 1977). This seaward grading is thought to be a function of the offshore decrease in wave orbital bottom velocities (Curray, 1960). It is possible that the mechanism responsible for the seaward grading in the 'subsurface' sediments also may be depth-sensitive, reflecting an offshore gradient in combined-flow power. In addition, the well-sorted nature of the sand fraction lends credence to the notions of progressive sorting away from a coastal source (cf. Fig. 8).

**Sand body geometry**

Sand body geometry is a direct function of the environment of deposition and the processes that operate within that environment (Potter, 1967). In a known depositional setting, such as the Texas shelf, sand body geometry can be used to infer the transport mechanism and transport pathway.

Interpretation of the large-scale geometry of individual sand beds was not attempted in Hayes (1967) and Morton (1981) as sampling was limited to a small study area. The first regional coring effort on the Texas shelf was conducted by Berryhill *et al.* (1976, 1977). They attempted to reconstruct the morphology of individual sand units by a correlation of their depth below the sea-bed surface. Berryhill's maps indicated that a number of the sand beds were oriented normal to the trend of shelf isobaths and that these beds showed some thickening in areas adjacent to tidal inlets.

However, the analysis of box cores from the central Texas shelf indicates that there are two major problems associated with simply correlating sand beds on the basis of their subsea-floor depth. First, $^{210}$Pb dating of sediment separating sand layers indicates that sediment accumulation rates differ by a factor of five along the Texas shelf (Holmes & Martin, 1977). Secondly, detailed examination of the box cores from the central Texas shelf area suggests that it is impossible to correlate the sand beds in certain areas (such as the inner shelf (10–20 m)) where there is intense bioturbation and a tendency toward amalgamation of the sedimentary sequence.

This is illustrated in Fig. 12, which shows a cross-shelf correlation of the box cores taken in the central Texas shelf study area (location shown in Fig. 13). In box core AB-2, the sequence consists of an amalgamated series of beds with sharp bases and tops. Further seaward in B-6, the core is thoroughly bioturbated. Correlation of individual stratification units is difficult in these cases. However, farther seaward, in water depths of 20 m or more, individual 'event' beds can be recognized and correlated. One bed located in the uppermost 20 cm of the cores is quite distinct. It appears to thin in the offshore direction. Thinning and thickening of this 'Carla' horizon also occurs in shore-parallel cross-sections in the study area (Fig. 14).

There are a number of lines of evidence that suggest that this bed may be the subsurface equivalent of the surficial sand discovered by Hayes (1967) following the passage of Hurricane Carla in September 1961. First, the subsea-floor depth of the bed in the cores is compatible with the sediment accumulation rates that have been calculated for that location in previous studies. For example, box core C-1 was obtained very close to the location of OCS-75, a core that was analysed by Holmes & Martin (1977) using the $^{210}$Pb dating technique (Fig. 15). The $^{210}$Pb procedure has proven to be quite useful in determining sediment accumulation rates in shelf sediments deposited in the last 100 years (e.g. Nittrouer & Sternberg, 1981). The methods

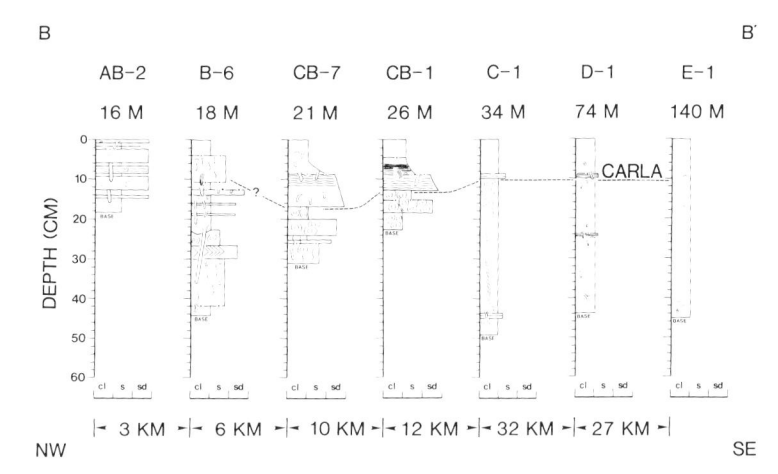

Storm-deposited sand beds of Texas continental shelf 295

**Fig. 12.** Cross-shelf cross-section B–B'.

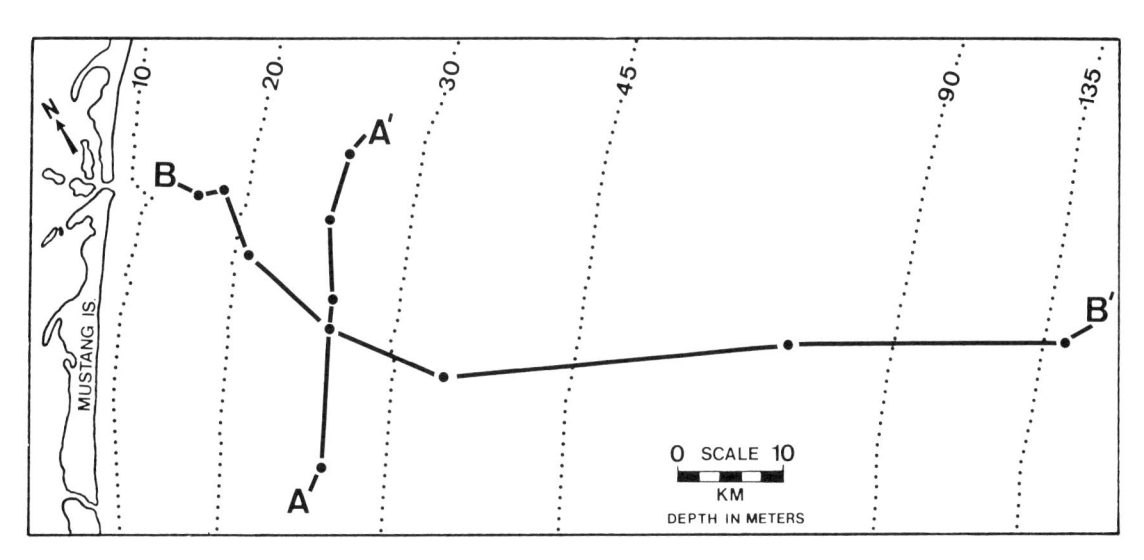

**Fig. 13.** Location of cross-sections A–A' and B–B'.

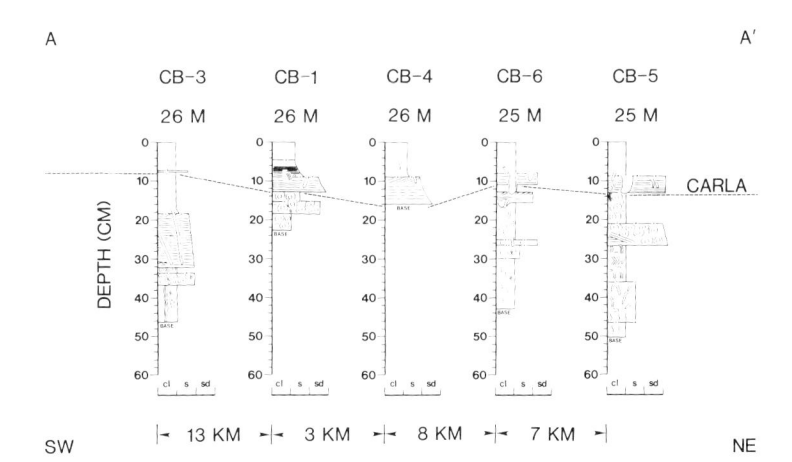

**Fig. 14.** Along-shelf cross-section A–A'.

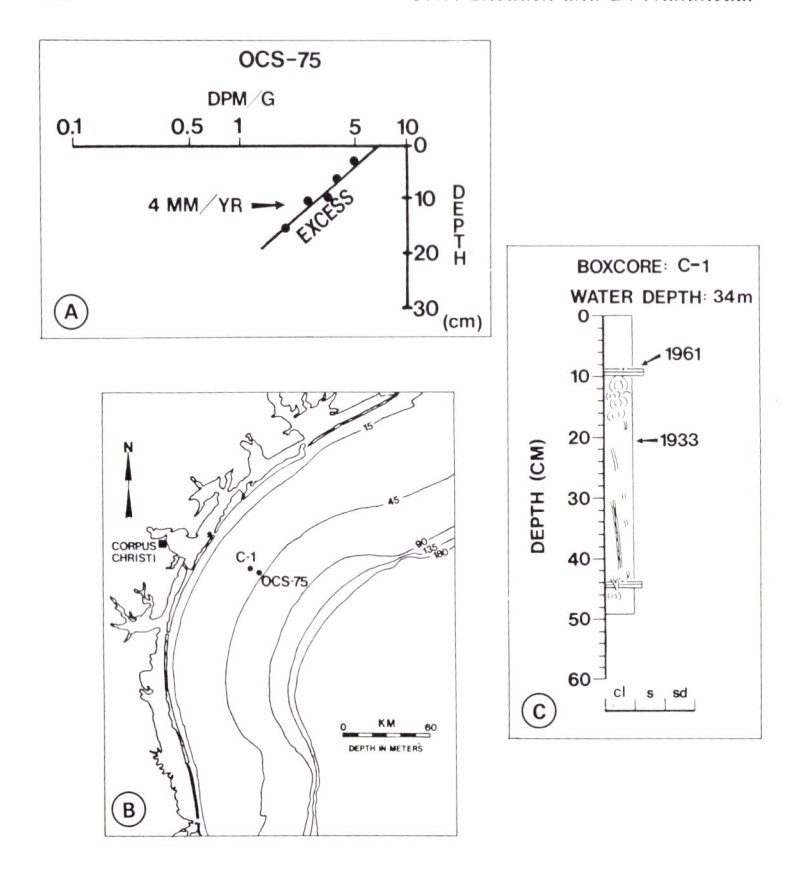

**Fig. 15.** (A) Distribution of excess $^{210}$Pb in OCS-75. (From Holmes & Martin, 1977.) (B) Location of box core C-1 relative to OCS-75. (C) Inferred absolute dates of beds in C-1 using sediment accumulation rates from OCS-75.

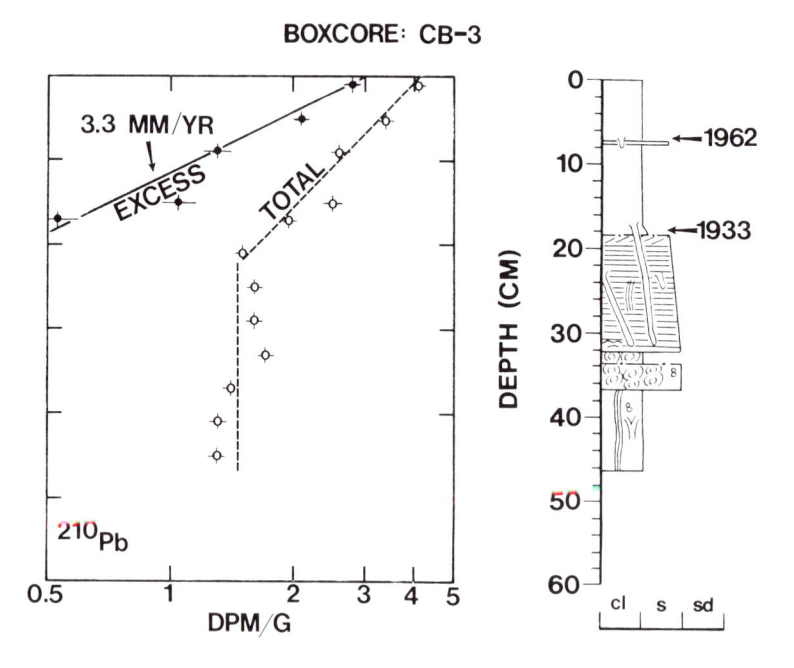

**Fig. 16.** Distribution of total and excess $^{210}$Pb in box core CB-3 (26 m water depth). Absolute dates and sediment accumulation rates from equation given by Faure (1977). (From C. Nittrouer, unpublished data, 1985.)

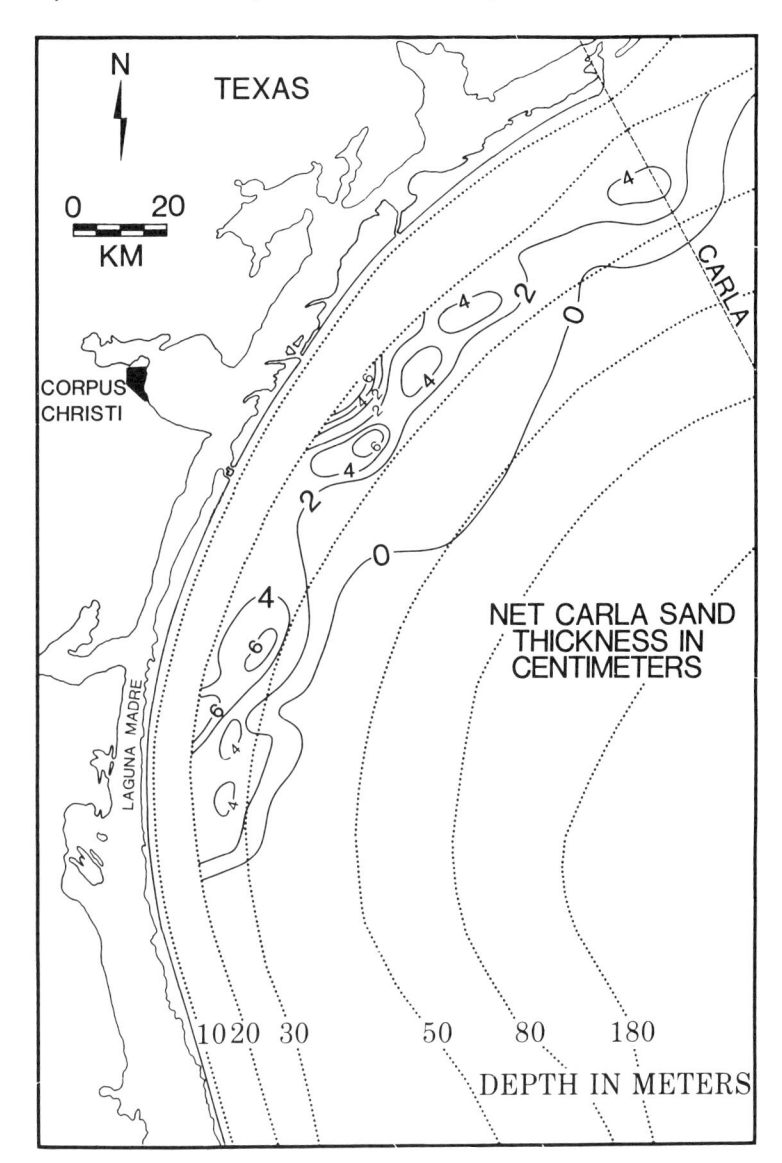

**Fig. 17.** Isopach map of present-day thickness of the Carla sand bed. Contour interval equals 2 cm. Note that the bed is difficult to identify in water depths of less than 20 m owing to pervasive bioturbation of sediments in that area.

used in this procedure are described in Nittrouer *et al.* (1979). Holmes & Martin (1977) found that shelf sediments in this area accumulated at a rate of approximately 4 mm y$^{-1}$. Use of the same rate for box core C-1 indicates that the sand bed in question was deposited about 1961, contemporaneous with the passage of Hurricane Carla.

To further substantiate this hypothesis, a number of cores in shallower water depths were selected for $^{210}$Pb analysis (e.g. Fig. 16). Sediment accumulation rates appear to decrease toward shore and along shelf to the southwest of the central portion of the Texas shelf. The data from these cores appear to support the proposed age of this bed (C. Nittrouer, unpublished data, 1985).

As a further step in establishing the areal distribution of the Carla sand bed, cores were obtained from areas outside the central Texas shelf study area (Fig. 1), including some taken in the investigation site of Hayes (1967). However, these $^{210}$Pb analyses did not supply incontestable evidence for the existence of the Carla bed. This was owing to the presence of a thick surface-mixed layer just above the location of a bed at 8 cm depth, the probable 'Carla bed' recognized by Hayes (1967). The surface-mixed layer, as defined by Nittrouer et al. (1979), is the region where sediments are still under the influence of surface activity (such as sediment reworking by physical and biological processes). It is characterized by high but homogeneous $^{210}$Pb activity. Sediment accumulation rates are calculated only for the zone of logarithmically decreasing $^{210}$Pb activity below the surface-mixed layer. The reason for the presence of such a thick surface-mixed layer in this area (as opposed to the central Texas shelf) is open to question, but may be related to the fairly recent passage of Hurricane Allen across the area in 1980. Further work is required in the south Texas shelf area, perhaps involving other dating procedures such as $^{137}$Cs (e.g. Olsen et al., 1978), to firmly establish the presence of the Carla sand bed there.

With these caveats in mind, it is possible to discern the present-day sand body geometry of the Carla horizon (Fig. 17). This map was prepared by correlation of the box cores used in this study and those of Berryhill et al. (1976, 1977) in the context of sediment accumulation rates calculated for the shelf by Holmes & Martin (1977). These rates were quite consistent with that obtained from $^{210}$Pb analysis of selected cores by C. Nittrouer (unpublished data, 1985–1986). In general, the Carla bed is a strike-trending deposit with smaller scale 'thicks and thins'. These thicks and thins may not be sedimentologically important, as the Carla bed in many places shows evidence of post-depositional truncation, possibly associated with the smaller-scale storms (e.g. Cindy in 1963) that followed Carla. In addition, there is some degree of uncertainty in comparing measurements from three different coring devices (phleger, gravity and box cores) taken over a 22-year period. However, the zero isopach line (the presence or absence of sand) probably closely approximates the depositional limits of the bed. This map does not include a small area on the south side of Hayes (1967) study area, which the writers believe was relict sand from the late Pleistocene Rio Grande delta.

This map further emphasizes the along-shelf continuity of the Carla sand (Fig. 17). This is even more impressive when one considers the large lateral distances (120 km) and small vertical thicknesses (1–9 cm) involved. As mentioned, identification of the Carla sand landward of 20 m was hindered by the intense biological reworking and physical truncation of the sedimentary section. Any transport pathway proposed for the Carla sand must take into account its seaward size-grading, thinning and great along-shelf continuity. A purely shore-normal pathway does not explain the strike continuity and a purely along-shelf pathway is incongruent with the offshore fining and thinning trends. However, a southerly path at an oblique angle to the bathymetric contours is consistent with these observations.

This proposed transport pathway parallels the pathway suggested for surficial sediments by Shideler (1978) and Mazzullo & Withers (1984). However, it should be remembered that the strike-trending geometry of the Carla sand bed is probably a function of both the shore-parallel nature of the transport pathway and the strike-aligned shoreface sand source.

The areal distribution of the Carla sand (Fig. 17) is also incompatible with the storm-surge ebb hypothesis proposed by Hayes (1967). First, there appears to be a lack of any association with shoreline features, such as washover channels. In fact, the greatest thickness and cross-shelf extent is actually offshore from that part of the Texas coast where barrier islands possess a large, continuous foredune ridge (Nummedal, 1982). This large ridge field prevents the formation of numerous hurricane-cut channels, a prerequisite for the mechanism hypothesized by Hayes (1967). Secondly, the sand body is a strike-trending feature, unlike the dip-trending pattern that is typical of deposits of turbidity currents (Berg, 1986).

Thus, the available sedimentological evidence suggests that the turbidity current hypothesis of Hayes (1967) is invalid. These findings agree with the morphological evidence provided by Morton (1981) that the storm-surge ebb flow after Hurricane Carla was not responsible for the discrete sand beds. Unfortunately, the sedimentological data are not sufficient to indicate the specific transport mechanism and the actual transport pathway of sand and silt during Hurricane Carla. This can be revealed only by an understanding of the physical processes that are operative in shelf waters during storm events, as will be discussed in the following section.

## SHELF SAND TRANSPORT PROCESSES

Concurrent with work upon the surficial shelf sediments of the study area was a programme aimed

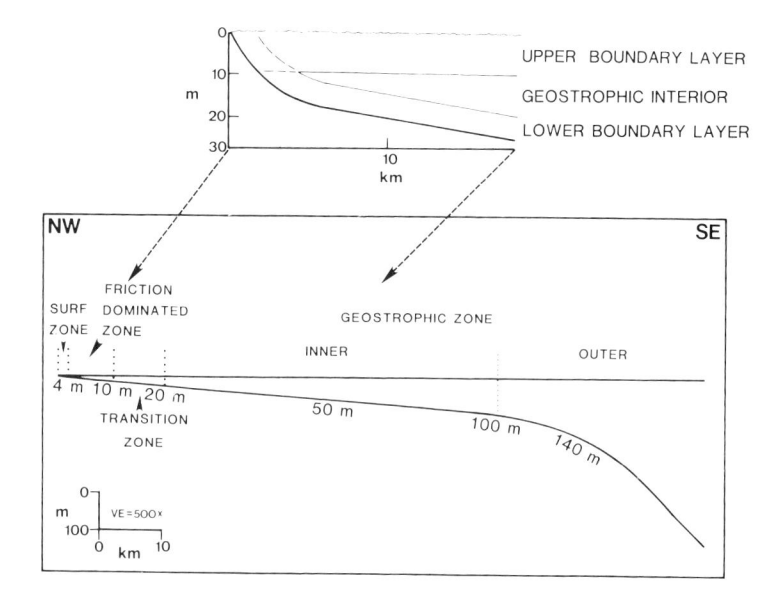

**Fig. 18.** Coastal ocean dynamic zones. Inset shows an expanded view with schematic representation of boundary layers and frictionless (geostrophic) interior described in text. It should be noted that depths and areal distributions are not fixed points in space but depend on the dynamics of the coastal ocean during fair-weather and storm conditions. (From Mooers (1976) and Swift & Niederoda (1985).)

at collection of measurements of the near-bottom flow field during fair-weather periods and small storm events (Snedden et al., 1988). In addition, unpublished data from Hurricane Allen, which made landfall on the Texas Coast in August 1980, were made available to the authors (Snedden, 1985). This information established the finding that there is a causal relationship between the wind field, sea-surface slope, wave propagation and near-bottom combined-flow field on the Texas shelf. This relationship is best understood through a discussion of the forcing mechanisms behind current flow and surface wave kinematics.

## Current flow forcing mechanisms

For many years, oceanographers have concerned themselves with the forcing of current flow in the deep ocean and, more recently, on continental shelves. In the coastal ocean there are many possible forcing mechanisms, including that of the wind, astronomical tides, the barotropic field of mass, the baroclinic field, etc. (Csanady, 1982). Although the shelf response to a large-scale storm, such as Carla, is likely to be quite complex, on the microtidal Texas shelf it is reasonable to assume that the main driving forces are the wind field and the pressure gradient forces that are generated by the interaction of the wind-driven water transport and the coastal boundary. Many previous studies, including both numerical simulations (e.g. Forristall, 1974; J.D. Smith & Long, 1976; Gordon, 1982) and actual measurements during storm events (e.g. Murray, 1970; Beardsley & Butman, 1974; Forristall et al., 1977; N.P. Smith, 1978; Adams et al., 1982; Wright et al., 1986) have demonstrated the dominance of these two forces.

A general qualitative framework for the description of motion on a storm-dominated shelf can be found in Swift et al. (1985). In this system, the shelf and shoreface are divided into four shore-parallel zones, the surf, friction-dominated, transition, and geostrophic zones (Fig. 18). The surf zone is the area where the radiation stress of breaking waves generates littoral along-shore currents. The friction-dominated zone is defined as the area where there is overlap of the upper (wind-influenced) and lower (frictional) boundary layers (Swift & Niedoroda, 1985). During episodes of elevated wind stress the flow takes on the characteristics of a well-stirred, non-rotational 'slab' driven by the wind (Csanady, 1982). In such situations, the majority of the velocity shear is concentrated near the bottom. The friction-dominated zone is generally confined to the shoreface, except during major storm events when it may expand seaward on to the shelf (Swift et al., 1985).

In the geostrophic zone (and to some degree the transition zone), there is significant separation between the surface and bottom boundary layers (Fig. 18). Momentum imparted to the sea surface is not expended directly against the bottom. Instead, motion in the frictionless interior is responsive to

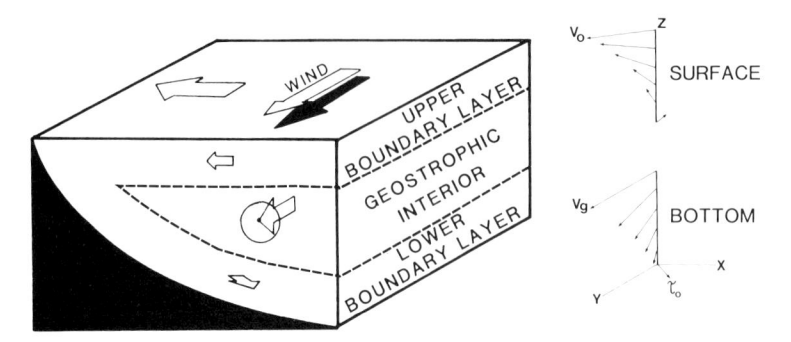

Fig. 19. Idealized diagram of motion in coastal ocean during a downwelling event. Also shown are representations of the distribution of currents in the surface Ekman spiral and bottom Ekman spiral during such an event. $V_0$ is the surface current vector, $V_g$ is the geostrophic velocity vector, and $\tau_0$ is the bottom stress vector. (Modified after Swift *et al.* (1985) and McLellan (1985).)

the sea-surface slope, as manifested in the pressure gradient force. Low frequency (long period) motion in areas distant from the Equator (>10° north or south latitude) tends to be in geostrophic balance with the Coriolis force (a manifestation of the rotation of the Earth) and flow is directed at right angles to the pressure gradient force (in the Northern Hemisphere). This along-shelf flow dominates storm circulation on any continental shelf (Csanady, 1976). At the sea-floor, however, flow departs from this geostrophic balance because of the influence of frictional forces, veering to the left (as one looks in the direction of flow). This veering is particularly prominent during peak flow events (Swift & Niedoroda, 1985). The main difference between the inner and outer portions of the geostrophic zone is that the outer areas can be affected by oceanic circulation, such as western boundary currents and the rings and eddies that spin off those currents (Kirwan *et al.*, 1984).

The transition zone adjoins the friction-dominated and the geostrophic zones (Fig. 18). As the name suggests, it displays aspects of both zones. It can also be the site of interesting flow patterns set up by the convergence or divergence of bottom and surface flow from the adjacent zones (Swift & Niedoroda, 1985).

A qualitative description of the structure of fluid motion during a period of storm-driven coastal downwelling is presented in Swift & Niedoroda (1985). In this case, northerly along-shelf winds blow down a N–S trending coastline (Fig. 19). Surface-water transport in the upper boundary layer is driven to the right of the applied wind stress, up against the coastal boundary. Sea level set-up ensues, and a cross-shelf hydrostatic pressure gradient is formed. The quasi-steady flow in the frictionless interior is in geostrophic balance with Coriolis force and oriented along-shelf. However, in the ageostrophic lower boundary layer, directional veering of flow occurs. Turning is toward the left (as one looks toward the boundary from above) or in this case in the offshore direction (Fig. 19).

## Wave kinematics

Like the steady current flow, wave-induced near-bottom motion in the coastal ocean can be quite complex, particularly during storm events. However, understanding of the surface gravity wave kinematics during hurricanes has been facilitated greatly by the use of synthetic aperture radar (SAR) imagery. King & Shemdin (1978) have summarized the general patterns displayed by the low-frequency waves in different sectors of tropical storm systems. In general, these long-period waves appear to propagate in a fan-shaped pattern outward from an apparent area of optimum wave generation located on the right side of the storm track (Fig. 20). These SAR observations appear to agree with measurements of wave kinematics at fixed locations during hurricanes and tropical storms in the Gulf of Mexico (Forristall *et al.*, 1978, 1980; Forristall & Reece, 1985).

## A model for Hurricane Carla, 1961

One of the reasons for the lack of consensus on the issue of the origin of the Carla sand bed was the absence of data concerning the near-bottom flow field during the passage of Hurricane Carla across the shelf in September 1961. However, there is information upon the wind field, coastal sea-level, and surface gravity waves during the storm. These data were analysed in light of the findings from the study of smaller storms, such as Hurricane Allen (Snedden, 1985) and in the context of the oceanographic principles discussed above. This allowed

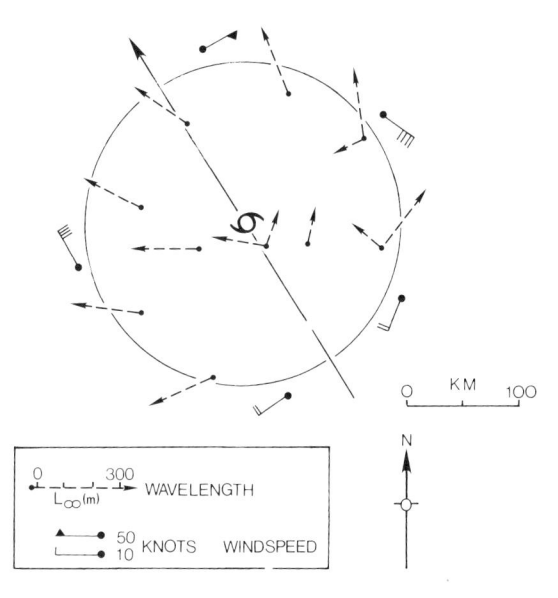

**Fig. 20.** Distribution of low-frequency wave propagation directions and surface winds of Hurricane Gloria in September 1976. Wave propagation directions (orthogonals) are indicated by dashed lines, whose length is proportional to the deep-water wavelength. (Modified from King & Shemdin, 1978.)

for the formulation of a qualitative kinematic model for the kinematics of sand transport during Hurricane Carla, which will be discussed briefly here. For more detailed information on this model (and its assumptions), the reader is referred to Snedden & Nummedal (1989).

Hurricane Carla is considered one of the largest tropical storms in the recorded meteorological history of the Gulf of Mexico (Colon, 1966; Nummedal, 1982). Because of its large size and slow rate of forward progress across the Caribbean and Gulf of Mexico, Carla was able to build an immense storm tide (Harris, 1963). This storm tide was a composite of the astronomical tides, surface waves and, most importantly, the wind-induced sea-level set-up or storm surge. Miyazaki (1965) used a finite-difference scheme in conjunction with actual coastal tide measurements to reconstruct the areal distribution of the storm surge across the shelf during the approach of the hurricane. The model indicates that a substantial cross-shelf sea-surface slope (and hence a pressure-gradient force) had formed at least 24 h prior to hurricane landfall (Fig. 21A). Under such conditions, it would be expected that the shelf flow would approximate the coastal downwelling model (Fig. 19). Surface current flow was likely oriented toward the coast (Fig. 22A) and probably was balanced by along-shelf flow, southwesterly motion in the geostrophic interior and obliquely offshore-trending current flow in the bottom boundary layer (Fig. 22B). As the storm neared the point of landfall, the storm surge became less shore-parallel (Fig. 21B). The along-shelf pressure gradient that developed could have also given rise to a barotropic Kelvin wave, a type of long-period shelf wave that propagates along the coast. However, it is presumed that this long-period decaying wave probably just modulated and locally augmented the downwelling flow already in motion.

Simple calculations of the magnitude of the geostrophic flow generated by these pressure-gradient forces indicates that current speeds could have reached nearly $300\,\mathrm{cm\,s}^{-1}$ (Fig. 23). This exceeds the maximum wind-driven current strength by a factor of six, further emphasizing the importance of the barotropic distribution of mass (and resulting pressure-gradient forces) in driving flow amidst the storm. The peak of current flow occurred just prior to landfall of the storm, in agreement with the measurements from Hurricane Allen and other storm events (Snedden *et al.*, 1988). Thus, significant sand transport must have occurred prior to landfall, not following it as suggested by Hayes (1967).

The reconstructed bottom-current flow pattern shown in Fig. 22B differs considerably from the shore-normal flow patterns proposed by many workers for storm currents (e.g. Hobday & Morton, 1984; Walker, 1985). It should be emphasized that this model is based upon study of the shelf response to Hurricane Allen and to other smaller scale storm events (Snedden, 1985). The along-shelf and ob-

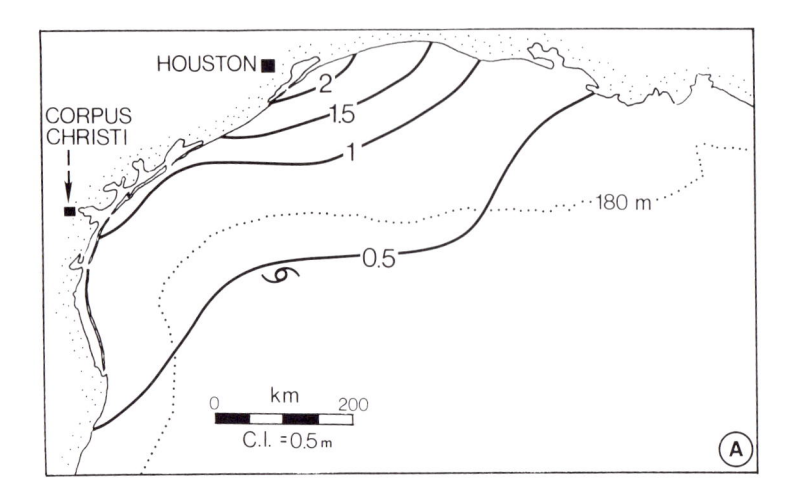

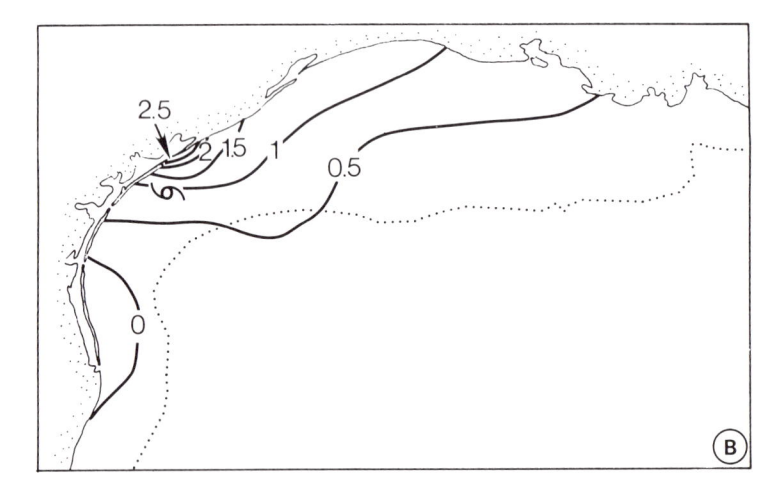

**Fig. 21.** (A) Calculated storm-surge heights, 1800 CST, 10 September 1961, approximately 22 h prior to landfall of Hurricane Carla. (B) 1200 CST, 11 September 1961, 6 h before landfall. (Modified from Miyazaki, 1965.) Position of the eye of the storm is also shown.

liquely offshore-trending, bottom-current motion is similar to the kinematics predicted by many numerical models of storm flow (e.g. Forristall, 1974; Reid et al., 1977) and what has been known to occur on other continental shelves (e.g. J.D. Smith & Long, 1976).

The wave patterns for a translating hurricane (as described by King & Shemdin, 1978; Fig. 20) can be superimposed upon the proposed bottom-current pattern to produce a view of the combined-flow transport. This is illustrated in Fig. 22C using the format of Swift et al. (1985). The observed change in shape of the combined-flow triangles from northeast to southwest along the Texas shelf indicates a change in the transport direction: from offshore transport on the immediate north and south side of the storm track to onshore transport on the storm's southern perimeter. This is owing to the fact that in the north, the steady current vector, $u_c$, augments the offshore bottom stroke of the long-period wave orbitals, $-u_w$, while in the far south it actually augments the onshore stroke, $+u_w$.

The patterns shown in Fig. 22C are applicable to the inner and middle shelf, where calculations for Hurricane Carla indicate that wave and current velocities were subequal (Snedden & Nummedal, 1989). In shallower water, wave-generated shear stresses probably dominated that of the steady current flow, producing a more shore-normal flow (but still asymmetric and oriented in the direction of the steady current motion). Parting lineations, solemarks and combined-flow ripples of ancient shoreface deposits often exhibit an orthogonal

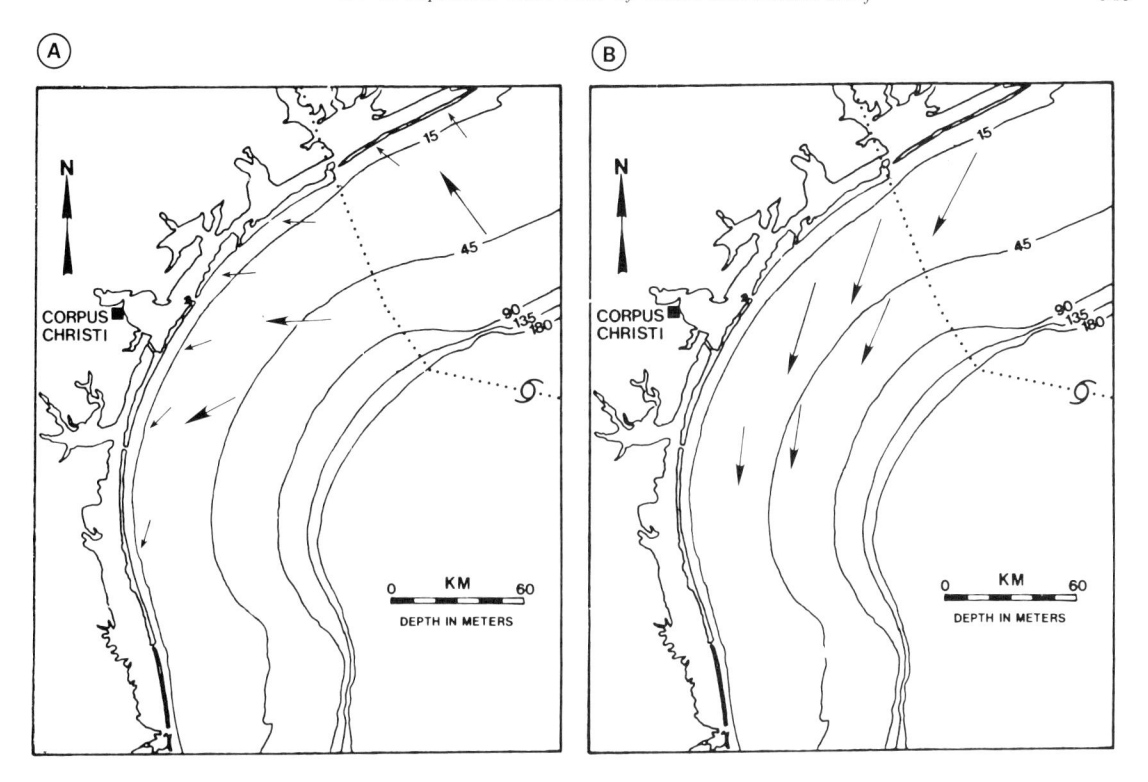

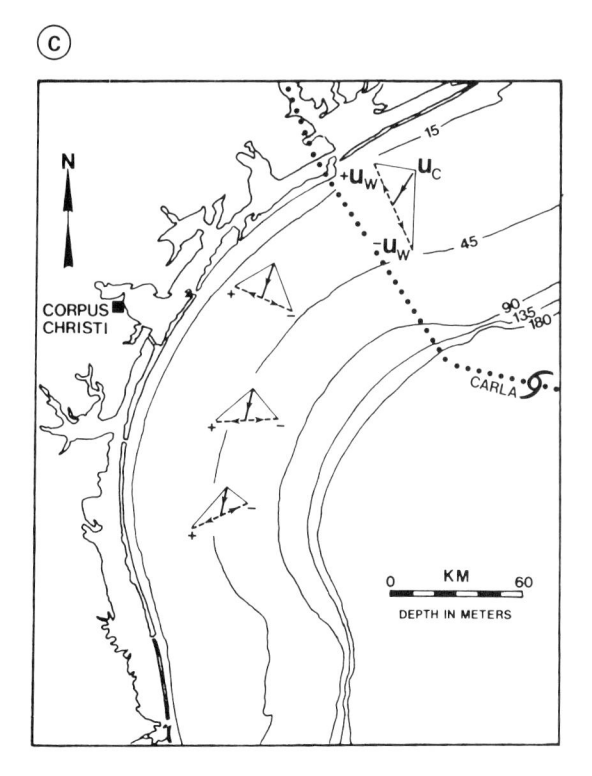

**Fig. 22.** (A) Surface currents (1 m below surface) generated by wind field of Hurricane Carla. Dotted line is the track of the eye of Carla. (B) Near-bottom steady current flow patterns for flow 1 m off bottom. (C) Near-bottom combined-flow transport patterns, where $u_c$ is the steady current flow, $+u_w$ the onshore component of the oscillatory flow, and $-u_w$ the offshore component.

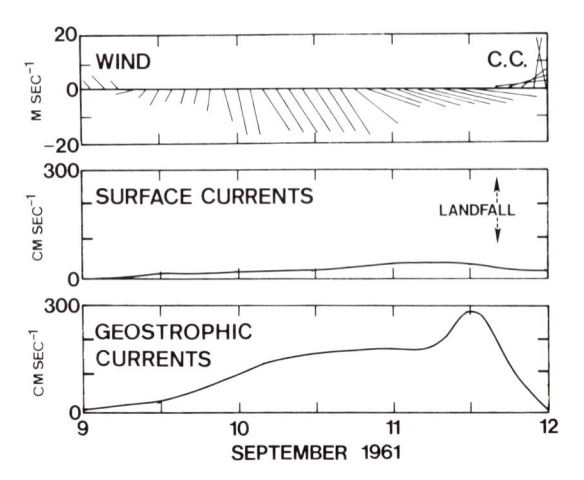

**Fig. 23.** Time series of wind measured at Corpus Christi Naval Air Station, calculated wind-driven surface currents, and calculated geostrophic currents during Hurricane Carla. Orientation of axes for wind, such that vertical axis is the along-shelf ($v$) component, positive to the northeast, and the horizontal axis is the cross-shelf ($u$) component, positive offshore. Surface current velocity based upon Ekman's (1905) empirical expression. Geostrophic calculations use sea surface heights and slopes from numerical model of Miyazaki (1965).

relationship with reconstructed shoreline trends, suggesting a dominance of wave orbital motion in shallow water (Leckie & Krystinik, 1989; Duke, 1990).

In general, the combined-flow transport patterns are consistent with the observed geometry of the Carla sand bed (Fig. 17). The model predicts generally along-shelf and obliquely offshore-trending combined-flow on the shelf immediately adjacent to the storm track, and onshore flow on the storm's southern perimeter. This is compatible with several sedimentological observations. First, the along-shelf continuity and general strike-trending nature of the sand body is a reflection of the along-shelf flow. It is also a reflection of the strike-trending nature of the probable shoreface sand source. The shoreface of the Texas shelf is predominantly sand-rich and is a zone of intensified wave erosion during storms (Swift & Niedoroda, 1985). The transfer of sand from the shoreface to the shelf is expected during downwelling events and has been documented by studies on the Atlantic coastal margin (Niedoroda & Swift, 1981; Swift & Niedoroda, 1985).

The cross-shelf fining trends (Fig. 11) and the offshore decrease in sand bed thickness and frequency (Fig. 12) also can be explained by the obliquely offshore-trending combined-flow. A purely shore-normal flow is not required. In fact, the only flow orientation that explains both the strike-trending sand body distribution on the shelf and the offshore fining is that of along-shelf and obliquely offshore-trending flow.

## SUMMARY AND CONCLUSIONS

The origin of the discrete sand beds found in the modern sediments of the Texas shelf has been a source of controversy in the geological literature for almost 20 years. Specific points of contention involve their clastic source, transport pathway and transport mechanism. These questions were addressed in a two-phased approach involving study of the sedimentary characteristics of the sand beds and a stepwise analysis of the dynamics and kinematics of shelf bottom-current motion during combined-flow events of varying magnitude.

Box cores taken from an area offshore from Mustang Island, Texas were examined to identify physical and biogenic sedimentary structures, texture and composition of discrete beds in the uppermost portion (0–60 cm depth) of the modern sedimentary section of the shelf. The typical discrete sand bed consists of, in ascending order, a sharp basal contact, planar laminations (flat bedding), low-angle inclined laminations, and a gradational upper contact. In the shallowest portion of the shelf, these beds display sharp upper contacts and often are part of a general amalgamated sand sequence.

Individual sand beds appear to be normally graded, although this can be recognized only in the silt and clay fraction, as the sand-size frequency distributions are highly leptokurtic. Offshore decreases in bed thickness and frequency are paralleled by offshore size-grading within correlative sand horizons.

Correlation of a sand bed dated as having been deposited during the period 1961–1962 with surficial sand recognized by Hayes (1967) as a product of Hurricane Carla reveals that individual beds have a tremendous along-shelf continuity. This, combined with recognition of the offshore fining trends, suggests a transport pathway oriented obliquely across the shelf to the southwest.

The strike-trending distribution of a sand bed deposited during the passage of Hurricane Carla in 1961 is not compatible with the storm-surge ebb hypothesis of Hayes (1967). The bed shows little association with shoreline features, such as washover channels, and in fact has its greatest seaward extent offshore from that part of the coast where the well-developed foredune ridges prevent the formation of washover channels by tropical storms.

Previous work on the geostrophic flow generated during the storms of smaller magnitude permitted identification of the primary driving force behind shelf sand transport. Wind-induced hydrostatic pressure gradients are capable of driving storm flows with sufficient magnitude and the proper kinematics to transport sand and silt from coastal sources to be deposited on the shelf. This information facilitated extrapolation to Hurricane Carla, one of the largest storms in recorded meteorological history and one cited frequently as responsible for the deposition of sand in shelf-depth waters. A qualitative kinematic model for combined-flow bottom motion involving along-shelf and obliquely offshore-trending bottom motion on the immediate north and south sides of the storm track, and onshore motion on the storm's southern perimeter is consistent with the distribution of the Carla sand bed across the Texas shelf, as revealed by cores of the shelf sediments.

## ACKNOWLEDGEMENTS

This paper represents the outgrowth of a PhD dissertation by the first author, completed at Louisiana State University. Major funding for this work was provided by ARCO Oil and Gas Company, Cities Service Oil and Gas Company, Marathon Oil and Gas Company, Sohio Petroleum Company, Chevron Oil Field Research Company, The US Minerals Management Survey, and The Louisiana State University Alumni Federation. The authors would like to thank a number of researchers who contributed their expertise to the project by examining samples from the study area with their special methods: Charles Nittrouer, James Mazzullo, Robyn Wright, Stephen Moshier and Curtis Olsen. The authors also would like to acknowledge discussions and interactions with the following individuals: Donald Swift, Charles Adams, Miles Hayes, Robert Morton, William Wiseman, Scott Dinnel, Evans Waddell, Thomas Moslow, Anthony Amos, Henry Berryhill, William Duke and Clyde Moore. The University of Texas is thanked for the use of the R/V *Longhorn*. The authors also extend their thanks to the two anonymous reviewers for constructive suggestions on how to improve this paper.

## REFERENCES

ADAMS, C.E., WELLS, J.T. & COLEMAN, J.M. (1982) Sediment transport on the central Louisiana continental shelf: implications for the developing Atchafalaya River delta. *Contrib. Mar. Sci.* **25**, 23–40.

AIGNER, T. & REINECK, H.E. (1982) Proximity trends in modern storm sands from the Helgoland Bight (North Sea) and their implications for basin analysis. *Senckenbergiana Maritima.* **14**, 183–215.

ANDERSON, J.B., WOLFTEICH, C., WRIGHT, R. & COLE, M.L. (1982) Determination of depositional environments using vertical grain size progressions. *Trans. Gulf Coast Assoc. Geol. Soc.* **32**, 565–79.

ARNOTT, R.W. & SOUTHARD, J.B. (1990) Exploratory flow-duct experiments on combined-flow bed configurations, and some implications for interpreting storm-event stratification. *J. Sediment. Petrol.* **60**, 211–219.

BAGNOLD, R.A. (1966) An approach to the sediment transport problem from general physics. *U.S. Geol. Surv. Prof. Pap.* **422-I**, 40 pp.

BEARDSLEY, R.C. & BUTMAN, B. (1974) Circulation on the New England continental shelf: response to strong winter storms. *Geophys. Res. Lett.* **1**, 181–184.

BERG, R.R. (1986) *Reservoir Sandstones*. Prentice-Hall, New Jersey, 481 pp.

BERRYHILL, H.L., JR., SHIDELER, G.L., HOLMES, C.W., HILL, G.W., BARNES, S.S. & MARTIN, R.G. (1976) *Environmental studies, south Texas outer continental shelf*, Vol. 2, *Geology*. Report PB-251-341, U.S. Department of Commerce, National Technical Information Service, Springfield, VA. 273 pp.

BERRYHILL, H.L., JR., SHIDELER, G.L., HOLMES, C.W., HILL, G.W., BARNES, S.S. & MARTIN, R.G. (1977) *Environmental Studies, South Texas Outer Continental Shelf*, Vol. 3, *Geology*. Report PB-289-144/AS, U.S. Department of Commerce National Technical Information Service, Springfield, VA, 306 pp.

BOUMA, A.H. (1962) *Sedimentology of Some Flysch Deposits: a Graphic Approach to Facies Interpretation*. Elsevier, Amsterdam, 168 pp.

BRANDT, D.S. (1986) Preservation of event beds through time. *Palaios*, **1**, 92–96.

BRENCHLEY, P.J. (1985) Storm influenced sandstone beds. *Mod. Geol.* **9**, 369–396.

BRIDGE, J.S. (1978) Origin of horizontal lamination under turbulent boundary layers. *Sediment. Geol.* **20**, 1–16.

BULLARD, F.M. (1942) Source of beach and river sand on the Gulf coast of Texas. *Geol. Soc. Am. Bull.* **53**, 1021–1044.

CACCHIONE, D.A. & DRAKE, D.E. (1982) Measurements of storm-generated bottom stresses on the continental shelf. *J. Geophys. Res.* **87**, 1952–1960.

CAMPBELL, C.V. (1966) Truncated wave-ripple laminae. *J. Sediment. Petrol.* **36**, 825–828.

CHEEL, R.J. (1990) Horizontal lamination and the sequence of bed phases and stratification under upper flow-regime conditions. *Sedimentology* 37, 517–529.

CHEEL, R.J. & MIDDLETON, G.V. (1986) Horizontal laminae formed under upper flow regime plane bed conditions. *J. Geol.* 94, 489–504.

COLEMAN, J.M. (1966) Ecological changes in a massive freshwater clay sequence. *Trans. Gulf Coast Assoc. Geol. Soc.* 16, 159–173.

COLON, J.A. (1966) Some aspects of hurricane Carla (1961). In: *Hurricane Symposium*, Publ. Am. Soc. Oceanogr. 1, 1–33.

CSANADY, G.T. (1976) Mean circulation in shallow seas. *J. Geophys. Res.* 81, 5389–5399.

CSANADY, G.T. (1982) *Circulation in the Coastal Ocean.* Reidel, London, 280 pp.

CURRAY, J.R. (1960) Sediments and history of Holocene transgression, continental shelf, Northwest Gulf of Mexico. In: *Recent Sediments, Northwest Gulf of Mexico* (Eds Shepard, F.P., Phleger, F.B. & van Andel, T.J.). Am. Assoc. Petrol. Geol. 221–266.

DOTT, R.H. (1983) Episodic sedimentation—How normal is average? How rare is rare? Does it matter? *J. Sediment. Petrol.* 53, 5–23.

DOTT, R.H. & BOURGEOIS, J. (1982) Hummocky stratification: significance of its variable bedding sequences. *Bull. Geol. Soc. Am.* 93, 663–680.

DUKE, W.L. (1985) Hummocky cross-stratification, tropical hurricanes, and intense winter storms. *Sedimentology* 32, 167–194.

DUKE, W.L. (1990) Geostrophic circulation or shallow-marine turbidity currents? The dilemma of paleoflow patterns in storm-dominated prograding shoreline systems. *J. Sediment. Petrol.* 60, 870–883.

EKMAN, V.W. (1905) On the influence of the earth's rotation on ocean currents. *Ark. F. Mat. Astr. Osh. Fysik. K. Sv. Vet. Ak. Stockholm*, 2, 11 pp.

FAURE, G. (1977) *Principles of Isotope Geology.* Wiley, London, 465 pp.

FOLK, R.L. (1968) *Petrology of Sedimentary Rocks.* Hemphills, Austin, TX, 170 pp.

FORRISTALL, G.Z. (1974) Three-dimensional structure of storm-generated currents. *J. Geophys. Res.* 79, 2721–2729.

FORRISTALL, G.Z. & REECE, A.M. (1985) Measurements of wave attenuation due to a soft bottom: the SWAMP experiment. *J. Geophys. Res.* 90, 3367–3380.

FORRISTALL, G.Z., HAMILTON, R.C. & CARDONE, V.J. (1977) Continental shelf currents in tropical storm Delia: observations and theory. *J. Phys. Oceanogr.* 7, 532–546.

FORRISTALL, G.Z., WARD, E.G., CARDONE, V.G. & BORGMANN, L.E. (1978) The directional spectra and kinematics of surface gravity waves in tropical storm Delia. *J. Phys. Oceanogr.* 8, 888–909.

FORRISTALL, G.Z., WARD, E.G. & CARDONE, V.J. (1980) Directional wave spectra and wave kinematics in hurricanes Carmen and Eloise. *Proceedings, 17th International Conference on Coastal Engineering, Sydney, Australia*, pp. 567–585.

GORDON, R.L. (1982) Coastal ocean current response to storm winds. *J. Geophys. Res.* 87, 1939–1951.

GRANT, W.D. & MADSEN, O.S. (1979) Combined wave and current interaction with a rough bottom. *J. Geophys. Res.* 84, 1797–1808.

GREENWOOD, B. & SHERMAN, D.J. (1986) Hummocky cross-stratification in the surf zone: flow parameters and bedding genesis. *Sedimentology*, 33, 33–46.

HAMBLIN, W.K. (1962) X-ray radiography in the study of structures in homogeneous sediments. *J. Sediment. Petrol.* 32, 201–210.

HARMS, J.C. (1969) Hydraulic significance of some sand ripples. *Geol. Soc. Am. Bull.* 80, 363–396.

HARMS, J.C., SOUTHARD, J.B., SPEARING, D.R. & WALKER, R.G. (1975) Depositional environments as interpreted from primary sedimentary structures and stratification sequences. *Soc. Econ. Paleontol. Mineral. Short Course Notes* 2, 161 pp.

HARMS, J.C., SOUTHARD, J.B. & WALKER, R.G. (1982) Structures and sequences in clastic rocks. *Soc. Econ. Paleontol. Mineral. Short Course Notes* 9, 200 pp.

HARRIS, D.L. (1963) Characteristics of the hurricane storm surge. *U.S. Wea. Bur. Techn. Pap.* 48, 100 pp.

HAYES, M.O. (1967) Hurricanes as geologic agents; case studies of Hurricane Carla, 1961 and Cindy, 1963. *Tex. Univ. Bur. Econ. Geol. Rep. Invest.* 61, 56 pp.

HILL, G.W. (1985) Ichnofacies of a modern size-graded shelf, Northwest Gulf of Mexico: In: *Biogenic Structures: Their Use in Interpreting Depositional Environments* (Ed. Curran, H.A.). Soc. Econ. Paleontol. Mineral. Spec. Publ. 35, 195–211.

HOBDAY, D.K. & MORTON, R.A. (1984) Lower Cretaceous shelf storm deposits, Northeast Texas: In: *Siliciclastic Shelf Sediments* (Eds Tillman, R.W. & Siemers, C.T.). Soc. Econ. Paleontol. Mineral. Spec. Publ. 34, 205–214.

HOLMES, C.W. & MARTIN, E.A. (1977) Rates of sedimentation. In: *Environmental Studies, South Texas continental shelf*, Vol. 3, *Geology*. Report PB-289-144/AS, U.S. Department of Commerce, National Technical Information Service, Springfield, VA, 306 pp.

HOWARD, J.D. (1972) Trace fossils as criteria for recognizing shorelines in the stratigraphic record. In: *Recognition of Ancient Sedimentary Environments* (Eds Rigby, J.K. & Hamblin, W.K.) Soc. Econ. Paleontol. Mineral. Spec. Publ. 16, 215–225.

JOPLING, A.V. (1965) Hydraulic factors controlling the shape of laminae in laboratory deltas. *J. Sediment. Petrol.* 35, 777–791.

KING, D.B. & SHEMDIN, O.H. (1978) Radar observations of hurricane wave directions. *Proceedings, 16th International Conference on Coastal Engineering, Hamburg, Germany*, pp. 209–226.

KIRWAN, A.D., JR., MERRELL, W.J., JR., LEWIS, J.K. & WHITAKER, R.E. (1984) Lagrangian observations of an anticyclonic ring in the western Gulf of Mexico. *J. Geophys. Res.* 89, 3417–3424.

KOMAR, P.D. (1974) Oscillatory ripple marks and the evaluation of ancient wave conditions and environments. *J. Sediment. Petrol.* 44, 169–180.

KOMAR, P.D. (1976) The transport of cohesionless sediments on continental shelves. In: *Marine Sediment Transport and Environmental Management* (Eds Stanley, D.J. & Swift, D.J.P.) pp. 107–125. Wiley, New York.

Komar, P.D. (1985) The hydraulic interpretation of turbidites from their grain sizes and sedimentary structures. *Sedimentology* **32**, 395–407.

Kreisa, R.D. (1981) Storm-generated sedimentary structures in subtidal marine facies with examples from the middle and upper Ordovician of southwestern Virginia. *J. Sediment. Petrol.* **51**, 823–848.

Kuenen, P.H. (1966) Experimental turbidite lamination in a circular flume. *J. Geol.* **74**, 523–545.

Lavelle, J.W., Swift, D.J.P., Gadd, P.E., Stubblefield, W.L., Case, F.N., Brashear, H.R. & Haff, K.W. (1978) Fair-weather and storm sand transport on the Long Island inner shelf. *Sedimentology* **25**, 823–842.

Leckie, D.A. & Krystinik, L.F. (1989) Is there evidence for geostrophic currents preserved in the sedimentary record of inner shelf to middle shelf deposits? *J. Sediment. Petrol.* **59**, 862–870.

Mantz, P.A. (1978) Bedforms produced by fine, cohesionless, granular and flakey sediments under subcritical water flows. *Sedimentology* **25**, 83–103.

Marsaglia, K.M. & Klein, G.D. (1983) The paleogeography of Paleozoic and Mesozoic storm depositional systems. *J. Geol.* **91**, 1117–1142.

Mazzullo, J.M. (1982) Grain shape analysis: application to problems of sediment source and transport. *Trans. Gulf Coast Assoc. Geol. Soc.* **32**, 541–554.

Mazzullo, J.M. & Withers, K.D. (1984) Sources, distribution, and mixing of Late Pleistocene and Holocene sands on the south Texas continental shelf. *J. Sediment. Petrol.* **54**, 1319–1334.

McCave, I.N. (1971) Sand waves in the North Sea off the coast of Holland. *Mar. Geol.* **10**, 199–225.

McLellan, H.J. (1965) *Elements of Physical Oceanography*. Pergamon, London, 150 pp.

Miyazaki, M. (1965) A numerical computation of the storm surge of Hurricane Carla 1961 in the Gulf of Mexico. *Oceanogr. Mag.* **17**, 109–140.

Mooers, C.N.K. (1976) Introduction to the physical oceanography and fluid dynamics of continental margins. In: *Marine Sediment Transport and Environmental Management* (Eds Stanley, D.J. & Swift, D.J.P.), pp. 7–22. Wiley, New York.

Morton, R.A. (1981) Formation of storm deposits by wind-forced currents in the Gulf of Mexico and the North sea: In: *Holocene Marine Sedimentation in the North Sea Basin* (Ed. Nio, S.D.). Spec. Publ. Int. Assoc. Sediment. **5**, 385–396.

Murray, S.P. (1970) Bottom currents near the coast during Hurricane Camille. *J. Geophys. Res.* **75**, 4579–4582.

Nelson, C.H. (1982) Modern shallow-water graded sand layers from storm surges, Bering shelf: a mimic of Bouma sequences and turbidite systems. *J. Sediment. Petrol.* **52**, 537–546.

Niedoroda, A.W. & Swift, D.J.P. (1981) Maintenance of the shoreface by wave orbital currents and mean flow: observations from the Long Island Coast. *Geophys. Res. Lett.* **8**, 337–340.

Nittrouer, C.A. & Sternberg, R.W. (1981) The formation of sedimentary strata in an allochtonous shelf environment: the Washington continental shelf. *Mar. Geol.* **42**, 201–232.

Nittrouer, C.A., Sternberg, R.W., Carpenter, R. & Bennett, J.T. (1979) The use of $^{210}$Pb geochronology as a sedimentological tool; application to the Washington continental shelf. *Mar. Geol.* **31**, 297–316.

Nøttvedt, A. & Kreisa, R.D. (1987) Model for combined-flow origin of hummocky cross-stratification. *Geology*, **15**, 357–361.

Nummedal, D. (1982) Hurricane landfalls along the Northwest Gulf Coast. In: *Sedimentary Processes and Environments along the Louisiana–Texas Coast*, Guidebook (Ed. Nummedal, D.). Geol. Soc. Am. Ann. Mtg. 90 pp.

Olsen, C.R., Simpson, H.J., Bopp, R.F., Williams, S.C., Peng, T.H. & Deck, B.L. (1978) A geochemical analysis of the sediments and sedimentation in the Hudson Estuary. *J. Sediment. Petrol.* **48**, 401–418.

Paola, C., Wiele, S.M. & Reinhart, M.A. (1989) Upper-regime parallel lamination as the result of turbulent sediment transport and low-amplitude bedforms. *Sedimentology* **36**, 47–60.

Pettijohn, F.J., Potter, P.E. & Siever, R. (1973) *Sand and Sandstone*. Springer-Verlag, New York, 618 pp.

Potter, P.E. (1967) Sand bodies and sedimentary environments: a review. *Bull., Am. Assoc. Petrol. Geol.* **51**, 337–365.

Reid, R.O., Vastano, A.C., Whitaker, R.E. & Wanstrath, J.J. (1977) Experiments in storm surge simulation: In: *The Sea, Ideas and Observations on Progress in the Study of the Seas* (Eds Goldberg, E.D., McCave, I.N., O'Brien, J.J. & Steele, J.H.) pp. 145–168. Wiley, London.

Reineck, H.E. & Singh, I.B. (1972) Genesis of laminated sand and graded rhythmites in storm sand layers of shelf mud. *Sedimentology* **18**, 123–128.

Reineck, H.E. & Singh, I.B. (1975) *Depositional Sedimentary Environments*. Springer-Verlag, New York, 439 pp.

Russell, R.D. (1939) Effects of transportation on sedimentary particles. In: *Recent Marine Sediments* (Ed. Trask, P.D.) pp. 32–47. American Association of Petroleum Geologists, Tulsa, OK (Dover edn facsimile).

Sanders, J.E. (1977) Primary sedimentary structures formed by turbidity currents and related resedimentation mechanisms: In: *Sedimentary Processes: Hydraulic Interpretation of Primary Sedimentary Structures* (Ed. Middleton, G.V.) Reprint Ser. Soc. Econ. Paleontol. Mineral. **3**, 194–223.

Seilacher, A. (1982) General remarks about event deposits. In: *Cyclic and Event Stratification* (Eds Einsele, G. & Seilacher, A.) pp. 161–173. Springer-Verlag, New York.

Sheldon, R.W. & Parsons, T.R. (1967) *A Practical Manual on the Use of the Coulter Counter in Marine Science*. Coulter Counter Electronics, Inc., Toronto, 66 pp.

Shideler, G.L. (1978) A sediment-dispersal model for the south Texas continental shelf, Northwest Gulf of Mexico. *Mar. Geol.* **26**, 289–313.

Simons, D.B., Richardson, E.V. & Nordin, C.F. (1977) Sedimentary structures generated by flow in alluvial channels. In: *Sedimentary Processes: Hydraulic Interpretation of Primary Sedimentary Structures* (Ed. Middleton, G.V.) Reprint Ser. Soc. Econ. Paleontol.

Mineral. 3, 34–52.

Smith, J.D. (1970) Stability of a sand bed subjected to a shear flow of low Froude number. *J. Geophys. Res.* **75**, 5928–5940.

Smith, J.D. & Long, C.E. (1976) Effect of turning in the benthic boundary layer on continental shelf transport. *Mem. Soc. R. Sci. Leige* **6**, 369–396.

Smith, N.P. (1978) Longshore currents on the fringe of Hurricane Anita. *J. Geophys. Res.* **83**, 6047–6051.

Snedden, J.W. (1985) *Origin and Sedimentary Characteristics of Discrete Sand Beds in Modern Sediments of the Central Texas Shelf.* PhD dissertation, Louisiana State University, 247 pp.

Snedden, J.W. & Jumper, R.S. (1990) Shelf and shoreface reservoirs, Tom Walsh-Owen Field, Texas. In: *Sandstone Petroleum Reservoirs* (Eds Barwis, J.H., MacPherson, J.G. & Studlick, R.J.R.) pp. 415–436. Springer-Verlag, New York.

Snedden, J.W. & Nummedal, D. (1989) Sand transport kinematics on the Texas Continental Shelf during Hurricane Carla, September 1961. In: *Shelf Sedimentation, Shelf Sequences, and Related Hydrocarbon Accumulation* (Eds Morton, R.A. & Nummedal, D.). *Proceedings of 7th Annual GCS-SEPM Research Conference, Corpus Christi*, pp. 63–76.

Snedden, J.W. & Nummedal, D. (1990) Coherence of surf-zone and shelf current flow on the Texas (U.S.A.) coastal margin: implications for interpretation of paleocurrent measurements in ancient coastal sequences. *Sediment. Geol.* **67**, 221–236.

Snedden, J.W., Nummedal, D. & Amos, A.F. (1988) Storm and fair-weather combined-flow on the Central Texas Continental Shelf. *J. Sediment. Petrol.* **58**, 580–595.

Stow, D.A.V. & Bowen, A.J. (1980) A physical model for transport and sorting of fine-grained sediment by turbidity currents. *Sedimentology* **27**, 31–46.

Swift, D.J.P. (1985) Response of the shelf floor to flow: In: *Shelf Sands and Sandstone Reservoirs* (Eds Tillman, R.W., Walker, R.G. & Swift, D.J.P.). Soc. Econ. Paleontol. Mineral. Short Course Notes 13, 135–241.

Swift, D.J.P. & Ludwick, J.C. (1976) Substrate response to hydraulic process; grain size frequency distributions and bedforms. In: *Marine Sediment Transport and Environmental Management* (Eds Stanley, D.J. & Swift, D.J.P.), pp. 159–161. Wiley, New York.

Swift, D.J.P. & Niedoroda, A.W. (1985) Fluid and sediment dynamics on continental shelves. In: *Shelf Sands and Sandstone Reservoirs* (Eds Tillman, R.W., Walker, R.G. & Swift, D.J.P.). Soc. Econ. Paleontol. Mineral. Short Course Notes, 13, 47–135.

Swift, D.J.P. & Rice, D.D. (1984) Sand bodies on muddy shelves; a model for sedimentation in the western interior seaway, North America. In: *Siliciclastic Shelf Sediments* (Eds Tillman, R.W. & Siemers, C.T.) Soc. Econ. Paleontol. Mineral. Spec. Publ. 34, 43–63.

Swift, D.J.P., Figueiredo, A.G., Jr., Freeland, G.L. & Oertel, G.F. (1983) Hummocky cross-stratification and mega-ripples—A geologic double standard?. *J. Sediment. Petrol.* **53**, 1295–1318.

Swift, D.J.P., Niedoroda, A.W., Vincent, C.E. & Hopkins, T.S. (1985) Barrier island evolution, middle Atlantic shelf, U.S.A., Part I: Shoreface dynamics. *Mar. Geol.* **63**, 331–361.

Walker, R.G. (1985) Geological evidence for storm transportation and deposition on ancient shelves. In: *Shelf Sands and Sandstone Reservoirs* (Eds Tillman, R.W., Walker, R.G. & Swift, D.J.P.). Soc. Econ. Paleontol. Mineral. Short Course Notes 13, 243–302.

White, W.A., Calnan, T.R., Morton, R.A., Kimble, R.S., Littleton, T.G., McGowen, J.H., Nance, H.S. & Schmedes, K.E. (1983) Submerged lands of Texas, Corpus Christi Area. *Tex. Univ. Bur. Econ. Geol.* 153 pp.

Wright, L.D., Boon, J.D., Green, M.O. & List, J.H. (1986) Response of the mid-shoreface of the southern Mid-Atlantic Bight to a 'Northeaster'. *Geomarine Lett.* **6**, 153–160.

Yalin, M.A. (1977) *Mechanics of Sediment Transport.* Pergamon, London, 290 pp.

# Model for genesis of shoreface and shelf sandstone sequences, southern Appalachians: palaeoenvironmental reconstruction of an Early Silurian shelf system

S.G. DRIESE, M.W. FISCHER, K.A. EASTHOUSE, G.T. MARKS, A.R. GOGOLA *and* A.E. SCHONER

*Department of Geological Sciences, University of Tennessee, Knoxville, TN 37996-1410, USA*

## ABSTRACT

The geometries and facies distributions of siliciclastic and carbonate sand bodies in Lower Silurian strata of the southern Appalachian Basin of east Tennessee reflect the complex interaction of storm processes within shoreface and shallow-marine shelf environments. The shallower and more proximal siliciclastic shoreface consisted of megaripples and sand waves that migrated northwestward under the influence of offshore-directed storm currents. Within the inner and middle shelf environments, combined storm-wave oscillatory flows and along-shelf geostrophic(?) flows deposited storm sand sheets containing varying scales of hummocky cross-stratification and wave-ripple lamination, which are interbedded with variable proportions of non-bioturbated mudstone. The outer shelf was dominated by non-bioturbated mudstone. A carbonate storm-shelf system existed contemporaneously to the west of the bathymetric basin axis. Storm-generated currents transported coarse skeletal sand eastward into the terrigenous-mud-dominated basin centre resulting in deposition of carbonate storm sand sheets into a 'background' mudstone lithology. Thus, basin-wide facies geometry consists of a central terrigenous-mudstone-dominated facies, into which a regressive, upward-shoaling sandstone wedge and associated facies prograded from the southeast, and carbonate sand sheets contemporaneously extended into the basin from the northwest.

## INTRODUCTION

Lower Silurian sedimentary rocks are widely distributed across the eastern USA, and include a variety of rock types. Within the Appalachian Basin, terrigenous clastic strata exhibit a general pattern of westward thinning and diminishing grain-size, reflecting greater subsidence to the east, as well as primarily eastern source areas (Amsden, 1955). These terrigenous clastic strata represent a clastic wedge associated with the Blountian phase of the Middle Ordovician to Early Silurian Taconic Orogeny (Rodgers, 1953, 1971; Yeakel, 1962; Meckel, 1970), the first of three major Palaeozoic accretionary events to have affected the Appalachian orogen (Williams & Hatcher, 1982, 1983). Palaeoenvironmental interpretations of the laterally extensive, mature quartz sandstone sequences (which constitute a major part of this clastic wedge) have been especially problematic; an excellent example of such a controversial sequence is the Tuscarora Formation (Lower Silurian) of central Pennsylvania, which has been reinterpreted numerous times (see review in Cotter, 1982), most recently as beach (Folk, 1960), meandering fluvial (Yeakel, 1962), braided fluvial (Smith, 1970), and as mixed marine shelf, coastal and fluvial (Cotter, 1982, 1983a).

The difficulty of accurate palaeoenvironmental interpretation of Lower Silurian sequences in the Appalachian region is attributable to:
1 sandstone compositions that are predominantly quartz arenite (Folk, 1960; Hayes, 1974; Sibley

& Blatt, 1976; Schoner, 1985a), thus precluding palaeoenvironmental/provenance interpretations based solely on detrital mineralogy (e.g. Dickinson & Suczek, 1979);

2 an absence of body fossils that might be useful for biostratigraphic and/or palaeoecological interpretation (c.f. Berry & Boucot, 1970); and

3 the variable quality of outcrop exposures of Lower Silurian rocks.

Past workers have relied mainly upon stratigraphy, petrography and regional palaeocurrent analyses to formulate palaeoenvironmental models (e.g. Folk, 1960; Yeakel, 1962; Meckel, 1970; Hayes, 1974; Lampiris, 1975; Whisonant, 1977); clearly these approaches have led to controversy (Tuscarora example cited previously). In this paper we will demonstrate how a more 'process oriented' sedimentological approach has been applied successfully towards a regional palaeoenvironmental interpretation of Lower Silurian siliciclastic and mixed carbonate–siliciclastic strata exposed across the fold-and-thrust belt (Valley and Ridge Province) of east Tennessee.

The primary purpose of our investigation will be to test the applicability of oceanographic shelf transport models based on modern systems (such as those recently summarized by Swift & Niedoroda, 1985) to an analogous Silurian marine shelf system. Our study provides valuable insight into the mechanisms by which sand from coastal environments is transported out to the shelf environment. These depositional processes, in turn, affect the architecture and stratification of the resultant sandstone bodies.

## BACKGROUND INFORMATION

### Tectonic History

Hatcher (1972, 1978) presented a model for the southern Appalachians in which a marginal sea and passive shelf margin developed following a

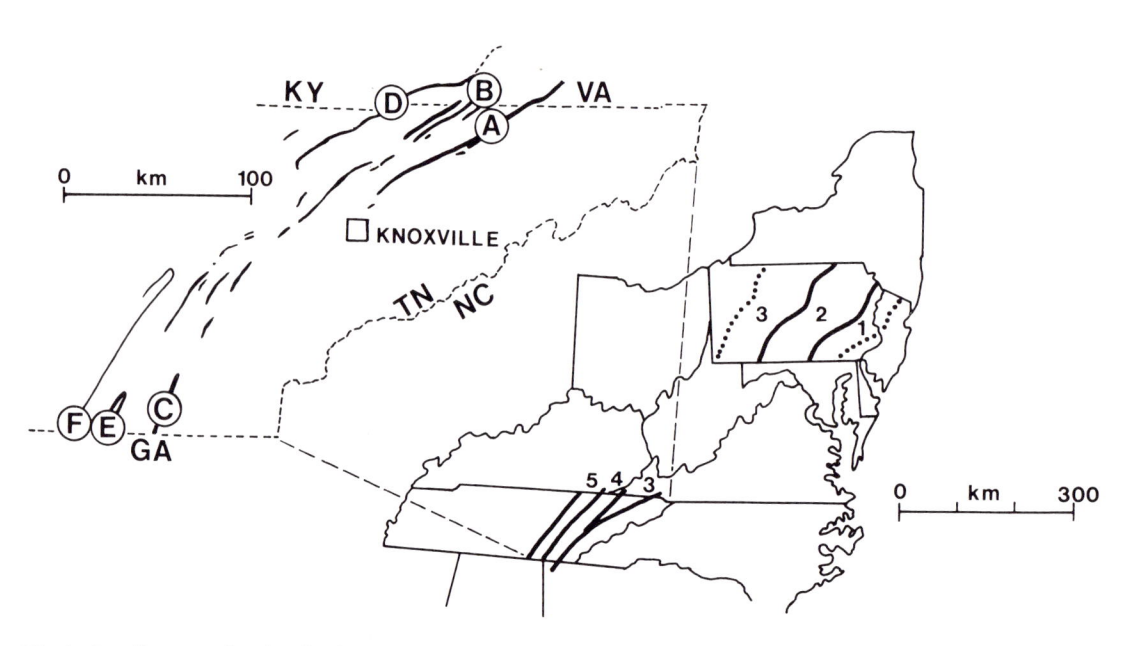

**Fig. 1.** Locality maps showing distribution of Silurian outcrops and generalized palaeoenvironments. Upper left-hand corner: map showing distribution of Silurian outcrops (shaded black) across the fold-and-thrust belt (Valley and Ridge Province) of east Tennessee. Circles indicate outcrop localitites mentioned in text and used in construction of cross-section in Fig. 3A,B. A = Clinch Mountain; B = Powell Mountain; C = Whiteoak Mountain; D = Cumberland Gap; E = Falling Water Creek; F = Sequatchie Valley. Lower right hand corner: map summarizing most recent sedimentological research on Lower Silurian strata in northern Appalachian Basin by Cotter (1982,1983a,b), and in southern Appalachian Basin by Driese and others (Driese et al., 1984, 1986a,b; Schoner, 1985a,b; Schoner & Driese, 1985; Driese, 1985,1986a,b,1987; Easthouse, 1987; Fischer, 1987; Marks, 1987). 1 = alluvial fan complexes; 2 = coastal alluvial plain (braided fluvial); 3 = shoreface and shelf sand wave complexes; 4 = siliciclastic storm shelf; 5 = carbonate storm shelf. Note conspicuous absence of recent data for central Appalachian region. Detailed outcrop locations are given in Appendix 1.

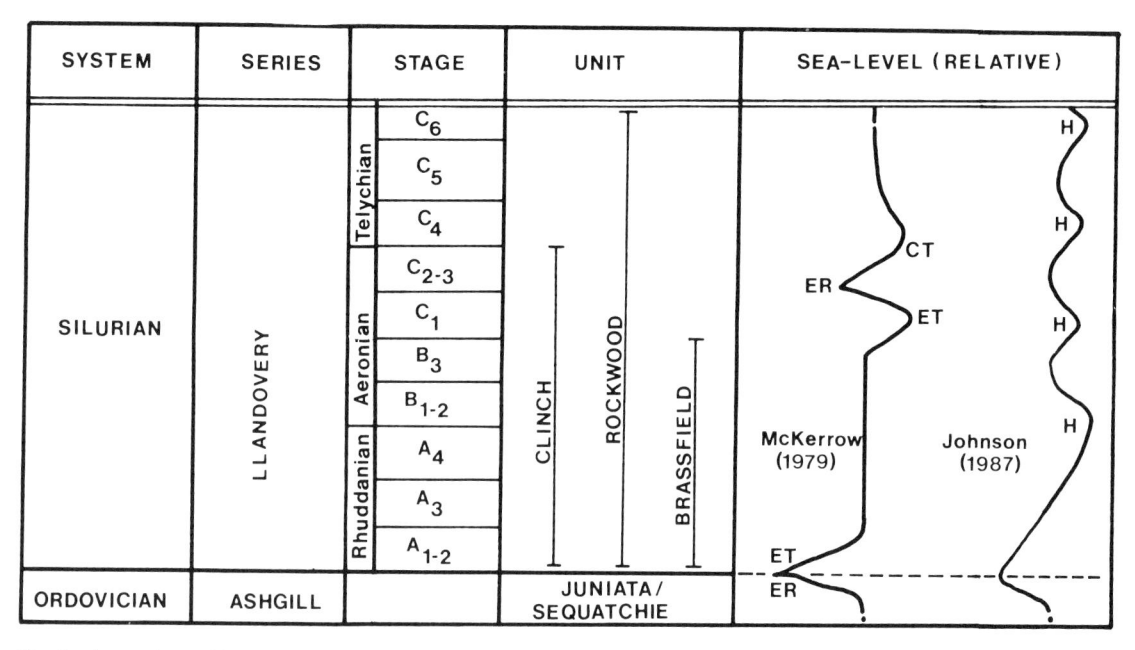

**Fig. 2.** Approximate biostratigraphic correlations of the three principal formational units that comprise the Lower Silurian (Llandoverian) strata exposed across the Valley and Ridge Province of east Tennessee. Stage assignments for the British standard are after Berry & Boucot (1970); the revised nomenclature of Rhuddanian, Aeronian and Telychian is for the Llandovery type district of Wales (Johnson, 1987). Relative sea-level curve on left is from McKerrow (1979); ET = eustatic transgression; CT = continental transgression; ER = eustatic regression. Relative sea-level curve on right is from Johnson *et al.* (1985) and Johnson (1987); H = positions of major sea-level highstands. The Rockwood Formation sections discussed in this paper probably include strata that extend through the entire Llandovery, based on our sparse brachiopod collections (Boucot, pers. comm. to Easthouse, 1986, cited in Easthouse, 1987). All three stratigraphic units overlie Upper Ordovician (Ashgill) rocks, predominantly red beds. Differential preservation of Upper Llandoverian strata is the result of post-Silurian uplift and erosion; Devonian Chattanooga Shale generally overlies Lower Silurian rocks in east Tennessee.

late Precambrian rifting event. By Middle to Late Cambrian time a closing of the marginal sea was initiated by the development of an east-dipping subduction zone along the Inner Piedmont–Blue Ridge crustal block that had rifted away from the continent earlier in late Precambrian time. A major fault system developed (the Hayesville–Fries, later Brevard) that continued to move even into the Silurian, carrying deformed Precambrian and early Palaeozoic sedimentary and Precambrian crystalline basement rocks westward on to the edge of the craton. According to the lithospheric flexure models of Quinlan & Beaumont (1984), emplacement of a 2–4-km-thick thrust load coupled with visco-elastic crustal behaviour would have resulted in the foreland downwarping necessary for deposition of the Silurian clastics preserved in east Tennessee. The mature composition of Silurian sandstone sequences (e.g. Folk, 1960; Hayes, 1974; Schoner, 1985a) in the central and southern Appalachians compares favourably with a recycled orogen interpretation (Dickinson & Suczek, 1979), in which recycled sedimentary rocks are a major component.

## Stratigraphy

Three laterally equivalent lithostratigraphic units comprise the Lower Silurian (Llandoverian) strata exposed in the Valley and Ridge of east Tennessee (Figs 1, 2). The easternmost is the Clinch Sandstone, which is predominantly a mature quartz arenite with only very thin shale beds. Intertonguing with the Clinch Sandstone to the west and south is the Rockwood Formation, a heterolithic succession of varying amounts of interbedded shale, siltstone, sandstone, and sandy limestone, and, locally, sedimentary haematite deposits. The westernmost Lower Silurian lithostratigraphic

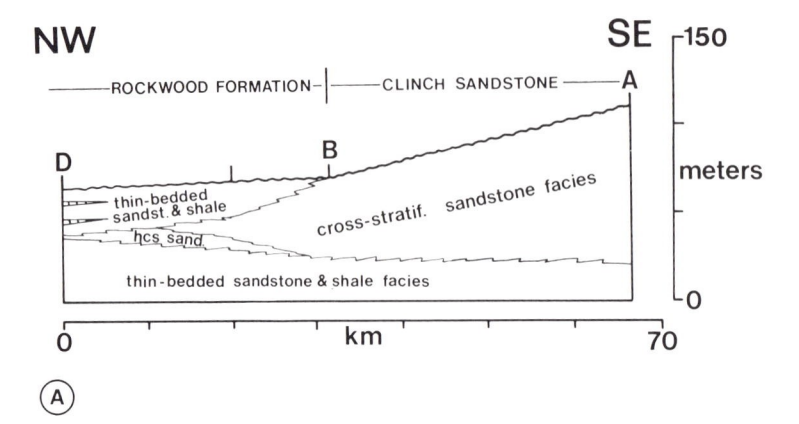

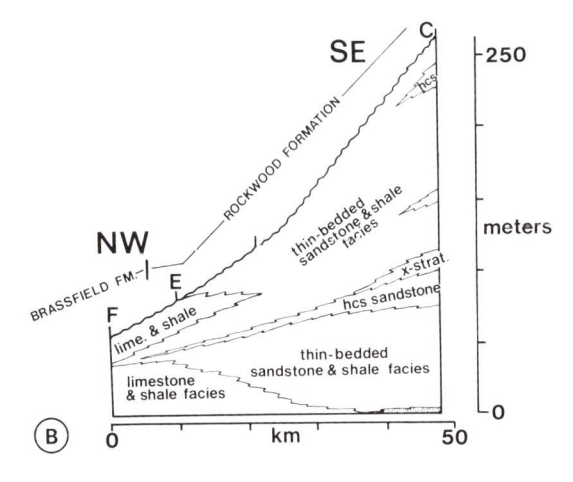

**Fig. 3.** (A) Generalized, palinspastically restored cross-section across upper east Tennessee showing major facies relationships. (B) Generalized, palinspastically restored cross-section across lower east Tennessee showing major facies relationships. Locations of outcrop sections D, B, A, and F, E, C, shown on Fig. 1. See Table 1 for relationships between lithofacies and formal stratigraphic nomenclature. Facies descriptions and interpretations discussed in text. Datum is top of Upper Ordovician in each case.

unit is the Brassfield Formation, a sequence of predominantly interbedded sandy limestone, limestone, dolostone and shale, with only rare sandstone or siltstone beds. All three stratigraphic units in east Tennessee conformably overlie Upper Ordovician (Ashgill) red bed and carbonate sequences (Fig. 2). Pre-Middle Devonian uplift and erosion have resulted in little or no preservation of 'Middle' and Upper Silurian deposits; in most parts of the Tennessee outcrop belt, the Devonian–Mississippian Chattanooga Shale unconformably overlies Lower Silurian strata, although locally a Lower Devonian sandstone unit is present (Dennison & Boucot, 1974).

Age assignments of the Clinch Sandstone are inexact, primarily because it lacks datable body fossils. It probably ranges from the base of the Llandovery to the Llandovery $C_{2-3}$ stage (Fig. 2). Greater confidence may be placed on the age assignments of the more fossiliferous Rockwood and Brassfield Formations (Fig. 2), which are dated primarily on the basis of brachiopod faunas (Berry & Boucot, 1970). All three stratigraphic units are Llandoverian, and were deposited during a major eustatic transgression that followed a eustatic regression at the close of Ashgill time (Fig. 2).

A palinspastically restored stratigraphic cross-section across northeastern Tennessee shows a tongue or wedge of sandstone that has prograded westward over a heterolithic succession, and is locally capped by haematitic strata (Fig. 3A). The character of the sandstone body changes westward, from dominantly high-angle cross-stratified sandstone in the east (Clinch Sandstone), to primarily hummocky cross-stratified and wave-ripple laminated sandstone to the west (Rockwood Formation). The underlying heterolithic succession similarly exhibits a westward change from predominantly bioturbated to non-bioturbated, and rare carbonate beds occur in the westernmost exposures. A second palinspastically restored stratigraphic cross-section across southeastern Tennessee (Fig. 3B) depicts a continuation of the westward-prograding sandstone tongue documented in the first cross-section (Fig. 3A), which pinches out to the west. This sandstone tongue of the Rockwood Formation is sandwiched between two thick heterolithic successions, and intertongues with a carbonate and shale sequence to the extreme west (Brassfield Formation, Fig. 3B). These vertical and lateral facies relationships will be explored in greater detail in a subsequent section.

### Regional palaeoenvironments

Figure 1 summarizes the most recent research on the palaeoenvironments of Lower Silurian (Llandoverian) strata deposited in the Appalachian miogeocline of the eastern USA. Process-oriented research by Cotter (1982, 1983a,b) established the existence of three broad palaeoenvironments across Pennsylvania, which were (from east to west) alluvial fan, coastal alluvial (braided) plain, and marine shelf to shoreface. In east Tennessee, the investigations of Driese and others (e.g. Driese *et al.*, 1984, 1986a,b; Driese, 1985, 1986a,b; Schoner, 1985a,b; Schoner & Driese, 1985; Easthouse & Driese, 1988) serve to define three broad palaeo-environments, one of which we interpret to be similar to the westernmost palaeoenvironment defined by Cotter (1982, 1983a,b); Lower Silurian palaeoenvironments in east Tennessee, which are discussed in this paper, include both shoreface and marine shelf (Fig. 1). Surprisingly lacking are any detailed, process-oriented studies across Maryland, Virginia and West Virginia, although stratigraphic (Hayes, 1974; Lampiris, 1975) and palaeocurrent studies (Whisonant, 1977) seem to corroborate the general patterns depicted in Fig. 1.

### Palaeocurrents

Azimuths obtained from the maximum foreset dip-directions of high-angle cross-strata are shown in Fig. 4A. Most of the data were derived from the Clinch Sandstone, and to a lesser degree from the Rockwood Formation. A few high-angle cross-strata in the Brassfield Formation were used also. The data are essentially unimodal, and show primary transport in a west to northwest direction, which we interpret to be offshore. A minor southeastern (onshore) mode seems weakly developed. These cross-strata dip azimuth patterns resemble those obtained by Yeakel (1962), Meckel (1970) and Whisonant (1977) for Silurian strata in the northern and central Appalachian Basin, which they interpreted as trending down a regional palaeoslope dipping west.

Directional readings obtained from the strike of wave-ripple crests (mainly the form-discordant variety) are depicted in Fig. 4B. Most data were obtained from the rippled upper surfaces of thin sandstone and siltstone beds in the Rockwood Formation; some measurements also were gathered from rippled carbonate grainstone and quartz siltstone beds in the Brassfield Formation. The regional palaeocurrent pattern for wave ripples obtained by us resembles that depicted by Aigner (1985) in his study of the Triassic Muschelkalk of West Germany, i.e. the ripple crests of more proximal (in our case, easterly) deposits are aligned parallel to the inferred palaeoshoreline trend (NE–SW), whereas ripple crests of more distal (in our case, westerly) deposits are oriented normal to the palaeoshoreline (NW–SE). The sedimentological significance of these wave-ripple patterns is discussed in subsequent interpretive sections.

Palaeocurrent data obtained from flute, groove and gutter casts, and from fossil lineations, are plotted in Fig. 4C. Most of the data were obtained from the soles of thin sandstone and siltstone beds in the Rockwood Formation; fewer measurements were derived from carbonate grainstone and quartz siltstone beds in the Brassfield Formation. The orientations of rare flute casts indicate unidirectional palaeoflow to the southwest, which we interpret to parallel the inferred palaeoshoreline trend (NE–SW). Data derived from gutter and groove casts, and fossil lineations, indicate a NE–SW sense of flow. Because these features commonly occur with flute casts (in some cases, on the same bed surface), we infer that the same depositional process produced this whole family of similarly oriented solemarks, namely, southwesterly palaeoflow.

## METHODS OF STUDY

Sixteen outcrop sections and one core (Amoco, Tennessee Division of Geology, D.O.E. Core Hole #7) were measured, described and interpreted for this study. The locations of the sections actually figured or discussed in the text are shown in Fig. 1;

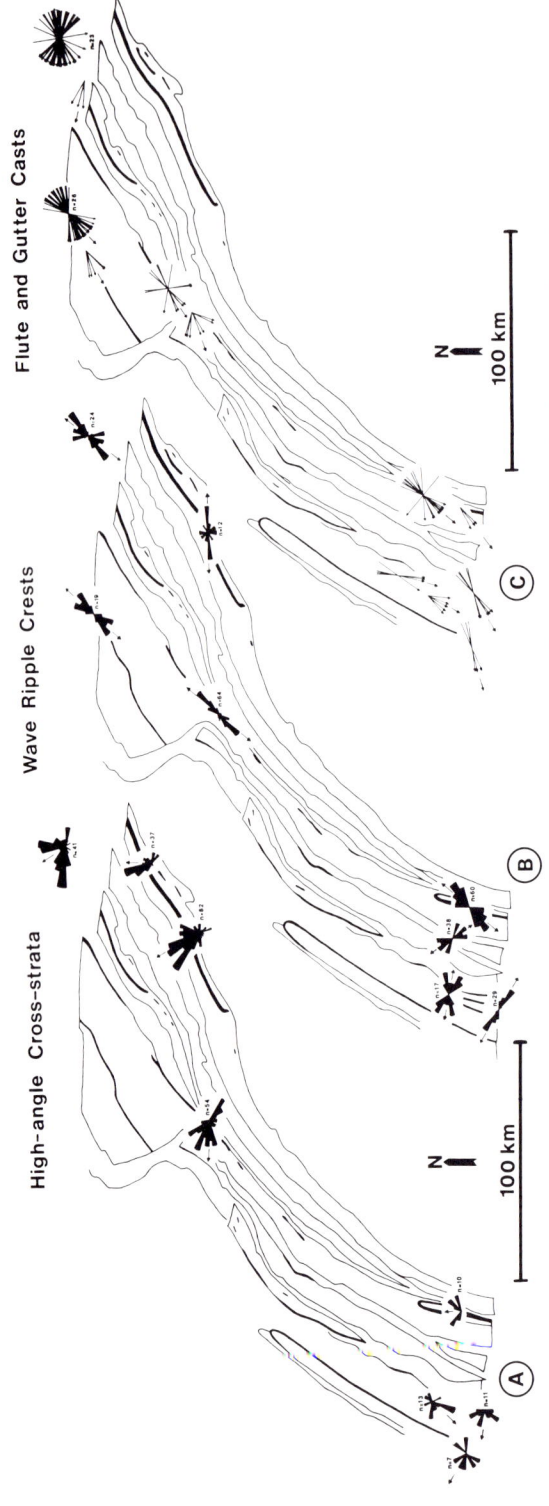

**Fig. 4.** Palaeocurrent data for Lower Silurian rocks, plotted on palinspastic base map of east Tennessee (Roeder & Witherspoon, 1978): (A) dip direction for cross-strata measured from cross-stratified sandstone facies; (B) trends measured from wave-ripple crests in terrigenous and calcareous thin-bedded facies; (C) trends of groove and gutter casts (long arrows) and flute casts (short arrows), measured from terrigenous and calcareous thin-bedded facies. Single arrow associated with each data cluster represents the vector mean for that cluster. Implications of data discussed in text. (Data from Driese, 1985, 1986a,b; Easthouse, 1987; Fischer, 1987; Marks, 1987; Schoner, 1985a,b.)

detailed locations for all sections are given in Appendix I. Outcrop sections were measured using a Jacob's staff and Brunton compass. Primary sedimentary structures were noted and carefully described. A total of approximately 300 hand samples were collected and taken back to the laboratory, where they were slabbed in order to better view sedimentary structures and textures. Over 150 thin-sections were prepared in order to characterize texture and mineralogy. Facies were defined on the basis of recurring associations of sedimentary structures and lithologies.

Palaeocurrent data, after stereonet correction for structural deformation (Ragan, 1973), were plotted on a palinspastic base map (Fig. 4) modified from Roeder & Witherspoon (1978). Care was taken to separate palaeocurrent data from different types of sedimentary structures, which probably recorded the effects of different depositional processes. The vector summation technique (Curray, 1956; Potter & Pettijohn, 1963) was performed using an unpublished computer program written in Basic language by Maher & Dott (1979).

Although the nomenclature for the classification of large-scale bedforms is presently in a state of revision, we have chosen to adopt the nomenclature of Kreisa (1986): (1) *ripple* – height < 5 cm, wavelength <1.5 m; (2) *megaripple*—height 5–60 cm, wavelength 1.5–30 m; and (3) *sand wave*—height >60 cm, wavelength >30 m.

## LITHOFACIES

Four broad lithofacies types are recognized across the outcrop belts of Silurian rocks in east Tennessee. They include: (1) cross-stratified sandstone; (2) hummocky cross-stratified sandstone; (3) thin-bedded sandstone and shale; and (4) thin-bedded limestone and shale (Fig. 3A,B). The relationships between these lithofacies, formal stratigraphic nomenclature, and depositional processes are summarized in Table 1.

### Cross-stratified sandstone facies

*Description*

The cross-stratified sandstone facies (hereafter referred to as the cross-stratified facies) comprises the bulk of the stratigraphic sections in the easternmost Clinch Mountain and Powell Mountain outcrop belts (localities A and B in Figs 1, 3A), which expose the Clinch Sandstone. The facies also constitutes a minor proportion of the Rockwood Formation stratigraphic section, but only in the easternmost Whiteoak Mountain (locality C in Figs 1, 3B) and Lone Mountain belts. The Poor Valley Ridge Sandstone Member of the Clinch Sandstone defined by Miller (1976) is dominated by the cross-stratified facies, in which two major and two minor variants are recognized (Fig. 5A–C).

The first major variant consists of medium to large-scale sets of trough cross-strata up to 1.5 m thick, developed in medium to coarse-grained quartz sandstone (Fig. 6A,D). Trough axes (festoon exposures) are up to 10 m wide. At one locality (Beans Gap), trough sets are internally organized into bundled foresets, in which sandstone (up to tens of centimetres thick) alternates with shale (millimetres to a few centimetres thick) in a generally repetitive pattern (Fig. 6D). These bundled cross-strata show only a weak periodic (tidal?) character (Driese, 1987). Maximum foreset dip-angles rarely exceed 15°. Most bottomset laminae and some foreset laminae show evidence of bioturbation at interfaces between clay drapes and sandstone bundles; trace fossils are dominated by *Arthrophycus*, *Planolites* and *Palaeophycus*. Bottomset laminae typically exhibit tangential (sigmoidal) contacts with the lower bounding surface of the set. Reactivation surfaces or pause planes lacking clay drapes are less common than those with clay drapes. Rarely are complete megaripple bedforms preserved in their entirety (height ranges from 25–75 cm, wavelength from 1–5 m), mantled by a 1–10-cm-thick shale and siltstone drape layer. Small-scale current ripples are oriented either with or opposite to dominant current flow, and occur on megaripple surfaces and along some bundled foresets. Topset laminae were not observed except for the few complete bedforms described previously; most trough sets are truncated sharply along their upper bounding surfaces. Individual cross-sets persist laterally over distances up to 60 m.

The second major variant of the cross-stratified facies consists of medium to very large-scale planar-tabular sets of cross-strata up to 2 m thick, developed in medium to very coarse-grained quartz sandstone (Fig. 6B). Planar sets are composed of centimetre-thick, normally graded, planar-abrupt foresets with abundant reactivation surfaces or pause planes. Rare *Skolithos* and *Diplocraterion* traces descend from the upper surfaces of some sets,

**Table 1.** Comparions between lithofacies defined in text and formally recognized stratigraphic units

| Lithofacies | Description | Process interpretation | Formation(s) |
| --- | --- | --- | --- |
| Cross-stratified sandstone | Small to large-scale, trough or planar-tabular cross-stratified, fine to coarse-grained quartz sandstone interbedded with intensely bioturbated (*Skolithos*, *Diplocraterion*, *Monocraterion*) sandstone. Desiccation cracks and red colouration at top of sand body. Fossil allochems extremely rare | *Shoreface deposition*: megaripples and sand waves migrating offshore (northwest) in response to unidirectional storm flow generated by coastal set-up, in water depths ranging from 5 to 10 m (friction-dominated zone) | Comprises most of the Clinch Sandstone and a small part of the Rockwood Formation section |
| Hummocky cross-stratified sandstone | Medium-scale sets of hummocky cross-stratified, very fine to fine-grained, quartzose to subarkosic sandstone, interbedded with thin lenses and layers of shale and siltstone. *Planolites* and *Palaeophycus* traces locally abundant, as are highly abraded marine fossil allochems | *Inner shelf deposition*: hummocky megaripples generated by combined-flow storm currents consisting of both a unidirectional mean flow and an oscillatory flow, in water depths ranging from 10 to 50 m (more proximal part of geostrophic zone) | Forms major sand bodies within the Rockwood Formation |
| Thin-bedded sandstone and shale | Thin beds of fine to very fine-grained, subarkosic to sublitharenitic sandstone intercalated with grey-green siltstone and fissile to semi-fissile shale. Variable physical sedimentary structures. *Rusophycus*, *Cruziana*, *Planolites*, *Palaeophycus* and *Chondrites* traces, marine fossil allochems locally abundant | *Middle to outer shelf deposition*: small-scale hummocky megaripples, form-discordant wave-ripples, parallel-laminated graded beds generated by combined-flow storm currents. Deep-water waves are along shelf, refracting in shallower water to become shore-parallel, in water depths ranging from 50 to 100 m (more distal part of geostrophic zone). Mean flow is down the basin axis (southwest) | Comprises most of the Rockwood Formation section, lower part of Clinch Sandstone, and part of the Brassfield Formation |
| Thin-bedded limestone and shale | Thin beds of bioclastic or peloidal grainstone to packstone intercalated with grey-green terrigenous siltstone and fissile to semi-fissile shale. Variable physical sedimentary structures. Abundant bioturbation and marine fossil allochems | *Shoreface and shelf deposition*: hydrodynamic processes inferred to be similar to those affecting thin-bedded sandstone and shale facies | Comprises most of the Brassfield Formation section, and parts of the Rockwood Formation |

or along reactivation surfaces between foresets. Shale drape layers are rare between foresets, although shale intraclasts are locally abundant. Winnowed granule and small-pebble lag concentrations are common at the tops of some cross-sets. Foresets are generally inclined at or near the angle of repose (25–33°). Compound cross-sets are common; typically a subsidiary smaller scale cross-stratification is developed within thick, large-scale foresets, with the dip direction of the smaller scale more or less coincident with the larger scale cross-stratification. Some thin (<5 m thick) vertical sequences have a basal large-scale cross-set overlain by progressively smaller scale sets, in some cases accompanied by a slight upward-fining trend. Individual cross-sets persist laterally at least 10–20 m, and possibly more for the largest sets.

Massively bedded, intensely bioturbated to partially cross-stratified sandstone occurs mainly within the upper part of the Poor Valley Ridge Member of the Clinch Sandstone (Fig. 5B) across the Clinch Mountain belt (locality A in Fig. 1), and also at the base of that member (Fig. 5C) along the Powell Mountain belt (locality B in Fig. 1). It is commonly interbedded with the cross-stratified sandstone strata described previously. The deposits typically consist of massive, densely bioturbated, fine to coarse-grained quartz sandstone beds up to 3 m thick (Fig. 6C). Internal stratification features are rarely visible. Robust, centimetre-diameter *Diplocraterion* burrows with well-developed spreite, and millimetre-diameter *Skolithos* tubes, are the most abundant traces (Fig. 6C). Most horizontal bedding surfaces underlain by shale exhibit abundant *Arthrophycus* traces. Rare cross-strata are preserved in the basal parts of some burrowed beds, with burrow density increasing progressively upward. Thick burrowed beds persist laterally for up to 150 m.

Sequences of red sandstone occur only within the Powell Mountain outcrop belt (locality B in Fig. 1), at the very top of the Poor Valley Ridge Sandstone Member of the Clinch Sandstone (Fig. 3A). Haematitic 'ironstone' strata (i.e. haematite-replaced limestone or fossiliferous haematitic sandstone beds) occur in parts of the Rockwood Formation, and are described later as a component of the thin-bedded sandstone and shale facies. These red sandstone deposits are the only strata described in this report that exhibit direct evidence for subaerial exposure. They occur in a stratigraphic position similar to that occupied by the Castanea Member of the Tuscarora Sandstone in Pennsylvania described by Cotter (1982, 1983a,b), and have many lithologic and sedimentological similarities. The deposits consist of maroon red, very fine to fine-grained sandstone beds intercalated with maroon red shaly siltstone. Bed thicknesses range from 1 to 30 cm, and beds are parallel laminated to wave-ripple cross-laminated, or are thoroughly bioturbated. Beds containing ripple cross-laminations exhibit complex interference patterns on their upper surfaces; ripple crests commonly are planed off (flattened), and exhibit a weakly sinous or bifurcating pattern. Associated with the rippled beds are beds containing abundant polygonal desiccation cracks, which exhibit patterns of smaller-scale polygons superimposed within larger-scale polygons; the crack fillings are up to 5 cm deep and are v-shaped in cross-section. Bioturbation is locally very extensive, and some beds of sandstone are so thoroughly burrow-homogenized that no physical sedimentary structures are visible. Trace fossils are dominated by *Skolithos* and *Monocraterion*, with less common *Diplocraterion*.

*Interpretation*

The cross-stratified sandstone facies is interpreted to have been deposited in a storm-dominated shoreface setting, analogous to the modern North American Atlantic shelf. Strong offshore-directed storm currents modified and shaped the shelf sand sheet into a complex system of sand waves and megaripples.

This facies exhibits great similarity to the 'western cross-laminated facies' of the Tuscarora Formation in central Pennsylvania, interpreted by Cotter (1982, 1983a,b) as shelf sand wave deposits. The principal sedimentary structure is high-angle cross-stratification. We interpret our trough cross-stratified variant to have been deposited as three-dimensional megaripples, which migrated offshore under the influence of storm flows. Flow velocities of 80–100 cm s$^{-1}$ are estimated for three-dimensional megaripple bedforms (using the method of Rubin & McCulloch, 1980). Bedform migration was episodic, as evidenced by rare clay drapes and abundant reactivation surfaces, which show only weak evidence for interpretation as tidal bundles (Driese, 1987). Clay drapes and bioturbated siltstone/claystone intervals (up to 50 cm thick) are common, and indicate periods of bedform inactivity (Allen, 1982), perhaps coupled with high suspended-sediment concentrations (McCave, 1970, 1971). Quiescent periods are also suggested by occurrences of infaunal deposit-feeding traces, such as *Planolites* and *Arthrophycus*, along clay drape horizons. Episodic high-energy flows are indicated by deeply incised scour surfaces with up to 1 m of relief, which were probably produced by rapid bedform migration, and by concentrations of shale clasts at the bases of scours.

The planar-tabular cross-stratified variant of this facies is interpreted to have been deposited by the offshore migration of two-dimensional megaripples, as well as by larger-scale sand waves, under the influence of storm flows. The virtual absence of

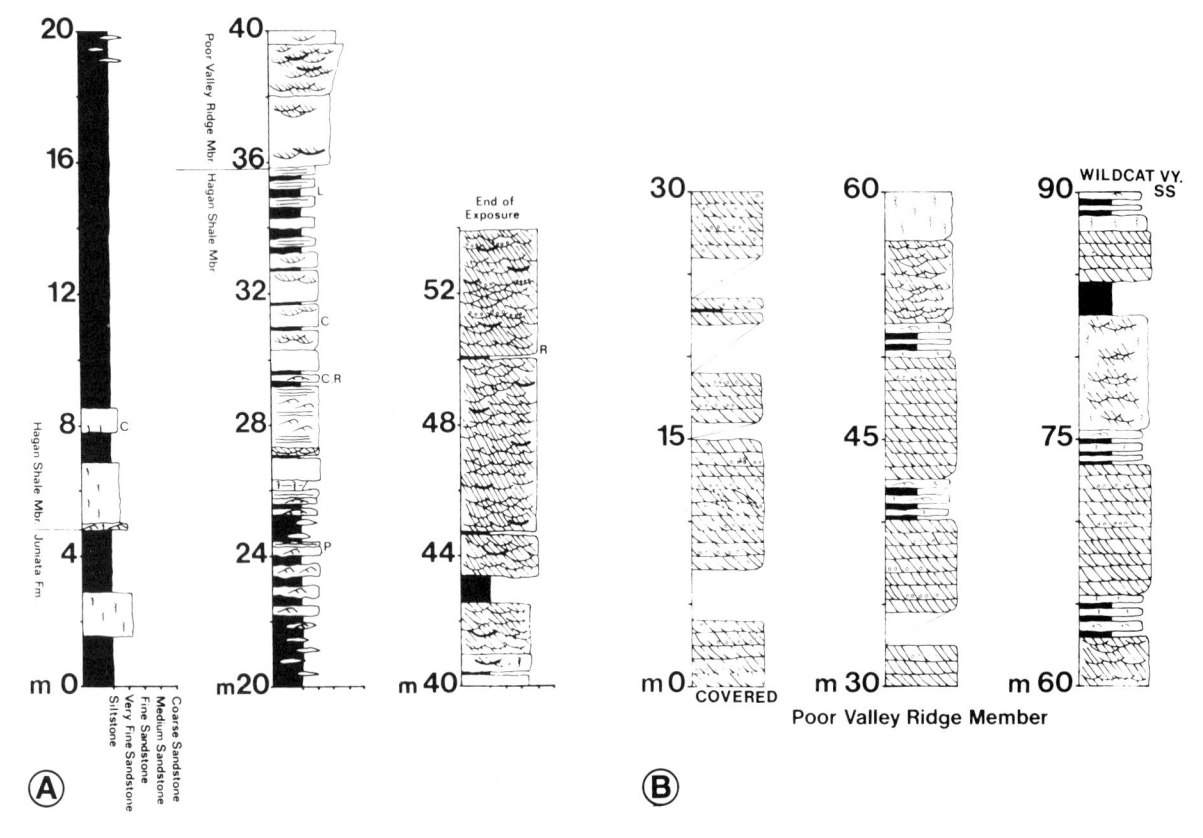

**Fig. 5.** Detailed measured sections of the Clinch Sandstone: (A) Flat Gap locality, Clinch Mountain belt (A in Fig. 1), highlighting transition from predominantly fine-grained Hagan Shale Member to predominantly coarse-grained Poor Valley Ridge Member. (From Fischer, 1987.) (B) Little War Gap locality, Clinch Mountain belt (A in Fig. 1), highlighting repetitious alternation of cross-stratified sandstone and mudstone–siltstone. (From Fischer, 1987.) (C) Outcrop sketch from Powell Mountain belt (Stickleyville, locality B in Fig. 1) showing very large-scale compound cross-stratification interpreted to represent shelf sand-wave deposits. Only major foresets shown; subsidiary medium-scale cross-stratification developed within foresets not shown. Sign post (arrow) is 2.75 m high for scale. Legend shows symbols used in all stratigraphic columns depicted in this paper.

bioturbation and of siltstone/claystone drapes suggests that bedform migration was continuous. The presence of reactivation surfaces in conjunction with a unimodal palaeocurrent pattern (Fig. 4A) does not preclude a possible tidal influence (Klein, 1977), although evidence for tides is weak. Winnowed granule and pebble lags, which cap planar sets, indicate periods of time during which sand was removed more rapidly than was being supplied (storm erosion and reworking of tops of bedforms?) (Levell, 1980). Flow velocities of 70–80 cm s$^{-1}$ are required to have moved the sand as two-dimensional megaripple bedforms (Rubin & McCulloch, 1980).

A shoreface sand sheet interpretation is based on the exclusively subtidal nature of this facies, dominantly unimodal cross-bedding, and the alternation between 5 and 10-m-thick packages of cross-stratified sandstone (which were deposited under the influence of more intense storm flows) and 10-cm to 2-m-thick sequences of parallel-laminated, ripple cross-laminated, and bioturbated sandstone beds that are intercalated with bioturbated claystone and siltstone (probably deposited under waning storm flows or during fair-weather periods) (Fig. 5B,C). Only at locality B (Fig. 5C) was there direct evidence for bedforms on a scale that might be expected from a sand-wave system. At this locality a single cross-set 4.5 m thick was identified that resembled a Class V sand-wave structure of Allen

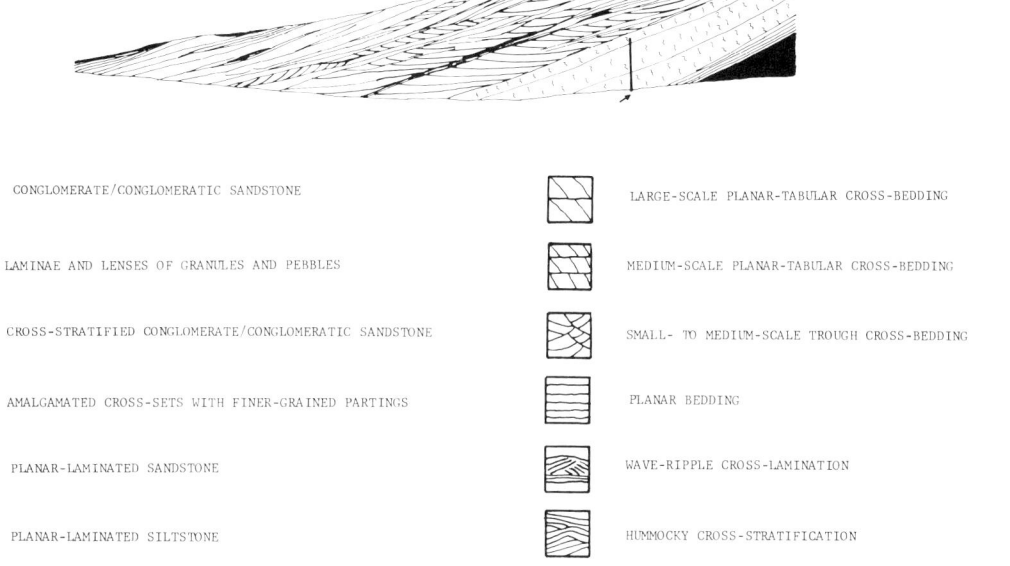

© 

|  |  |
|---|---|
| | CONGLOMERATE/CONGLOMERATIC SANDSTONE |
| | LAMINAE AND LENSES OF GRANULES AND PEBBLES |
| | CROSS-STRATIFIED CONGLOMERATE/CONGLOMERATIC SANDSTONE |
| | AMALGAMATED CROSS-SETS WITH FINER-GRAINED PARTINGS |
| | PLANAR-LAMINATED SANDSTONE |
| | PLANAR-LAMINATED SILTSTONE |
| | SILTY MUDSTONE |
| | INTERLAMINATED MUDSTONE AND FINE-GRAINED SANDSTONE |
| | CROSS-LAMINATED SANDSTONE |
| | AMALGAMATED SANDSTONE CROSS-SETS WITH FINER-GRAINED INTERBEDS/PARTINGS |
| | LARGE-SCALE PLANAR-TABULAR CROSS-BEDDING |
| | MEDIUM-SCALE PLANAR-TABULAR CROSS-BEDDING |
| | SMALL- TO MEDIUM-SCALE TROUGH CROSS-BEDDING |
| | PLANAR BEDDING |
| | WAVE-RIPPLE CROSS-LAMINATION |
| | HUMMOCKY CROSS-STRATIFICATION |
| | CONVOLUTE BEDDING |
| | SYMMETRICAL WAVE-RIPPLE MARKS |
| | SHALE CLASTS |
| | BIOTURBATION |

**Fig. 5.** (continued)

(1980); medium-scale trough cross-sets, tens of centimetres thick and bounded by clay drapes, constitute the individual foresets of this very large-scale set. These medium-scale sets exhibit palaeoflow oblique to the migration direction indicated for the very large-scale set. Furthermore, asymmetrical ripple forms are superimposed across the upper surfaces of medium-scale trough cross-sets, in which both scales of structures are oriented oppositely. Similar large-scale shelf sand bedforms have been reported in the Permian of Arizona (Blakey, 1984; Kreisa, 1986), the Lower Cretaceous of England (Narayan, 1971; Nio, 1976; Allen, 1981a,b, 1982), and the Eocene of Spain (Nio, 1976; H.D. Johnson, 1979; Yang & Nio, 1985).

An abundance of the marine trace fossils *Skolithos* and *Diplocraterion*, such as occurs within the massively bedded, bioturbated sandstone deposits intertonguing with cross-stratified sandstone (Fig. 6C), is normally taken to indicate deposition in the shallow, current-agitated shoreface to foreshore environments of a marine system (Crimes, 1975; Frey, 1975; Rhoads, 1975; Chamberlain, 1978; Seilacher, 1978; Ekdale *et al.*, 1984). The organisms producing these structures were probably vertically burrowing filter-feeders. The extreme density of burrows suggests overall conditions of moderate to low energy, possibly coinciding with reduced sediment influx. Bioturbation was able to continually keep pace with physical sedimentation (Howard, 1978).

Rare preservation of high-angle cross-stratification within bioturbated beds and the common occurrence of a 10–20-cm-thick shale bed capping these deposits, suggest that their occurrences may record temporary low-energy periods during which

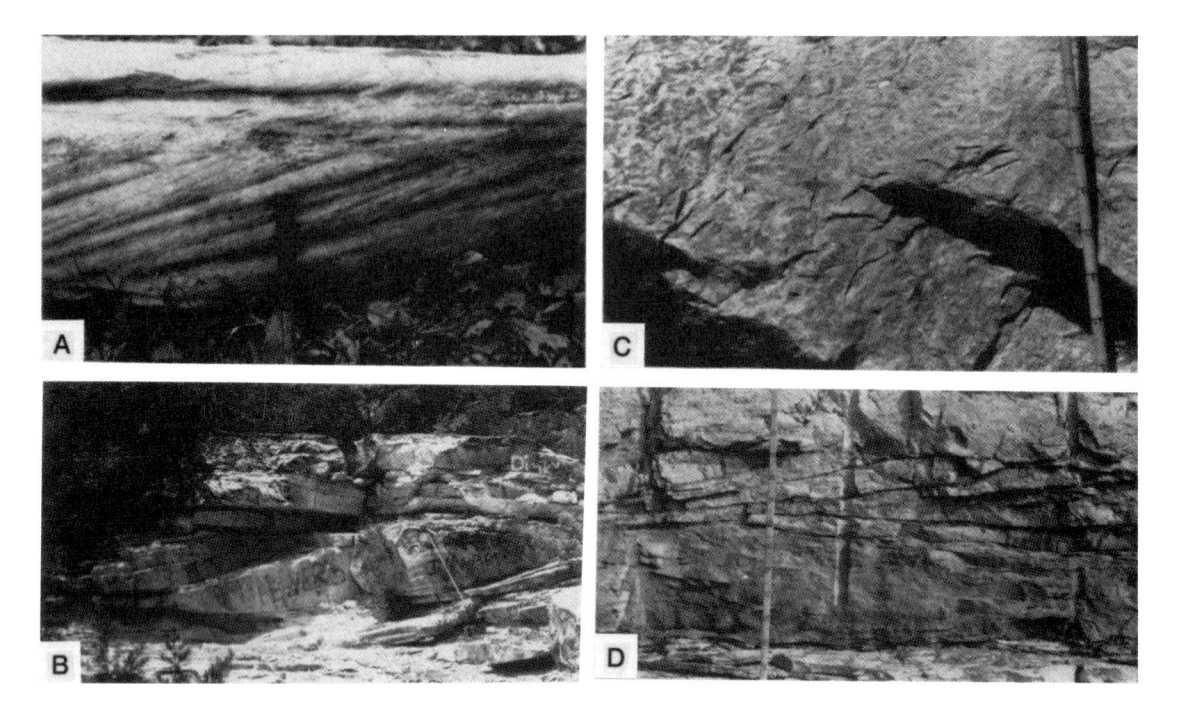

**Fig. 6.** Field photographs of facies occurring in the Clinch Sandstone. (A) Cross-stratified sandstone facies, Clinch Mountain (House Mountain locality)—large-scale, trough cross-set exhibiting compound cross-stratification and reactivation surfaces. Note winnowed gravel–granule lag along upper bounding surface of set. Hammer for scale. (B) Cross-stratified sandstone facies, Clinch Mountain (Little War Gap locality)—very large-scale planar-tabular cross-stratification. Jacob's staff is 1.5 m long. (C) Massive, bioturbated sandstone, Clinch Mountain (Bean Gap locality)—bioturbated sandstone lacking physical sedimentary structures, with trace fossils dominated by *Skolithos* and *Diplocraterion*. View is oblique to bedding surface, 10-cm increments on Jacob's staff. (D) Cross-bedded sandstone facies, Clinch Mountain (Bean Gap locality)—deeply truncated, medium-scale trough cross-set exhibiting bundle structures (tidal?), as described previously by Driese (1987). 10-cm increments on Jacob's staff.

shoreface megaripples and sand waves experienced low activity and became moribund, as in some modern sand banks of the North Sea (Kenyon *et al.*, 1981; Stride *et al.*, 1982). We speculate that possible causes of low bedform activity might include subtle changes in relative sea-level (especially deepening) or changes in hydrodynamic regime (storm tracks, etc.). The great thicknesses of individual burrowed beds (up to 2–3 m thick) suggest rather prolonged periods of low bedform activity.

Rhoads (1967), measuring the actual rate of bioturbation in recent intertidal sediment, showed that the upper 10 cm of sediment can be reworked completely in 1 or 2 months. Ultimately the reworking zone could extend downward 30 cm, although bioturbation to this depth might require several years. Consequently, we do not believe that 2–3 m of sand were first deposited, and then completely reworked; rather, the discontinuous bedding plane surfaces within otherwise massive and homogeneous beds suggest that what appear to be single thick beds actually may be the product of amalgamation of successive thinner beds. Furthermore, deposition of sand was not slow and continuous, but was characterized by intermittent deposition followed by thorough bioturbation of both the most recently deposited bed *and* earlier bioturbated beds (Howard, 1978).

The red sandstone sequences exhibit great lithologic and sedimentological similarity to the Castanea Member of the Tuscarora Formation of Pennsylvania, first described by Swartz (1934) and later interpreted by Cotter (1982, 1983a,b) as the product of deposition on wave-dominated coastal sand flats where energy conditions fluctuated between moderate and low. Both the red sandstone of this report and the Castanea Member occur at the top of predominantly mature, white to tan-coloured

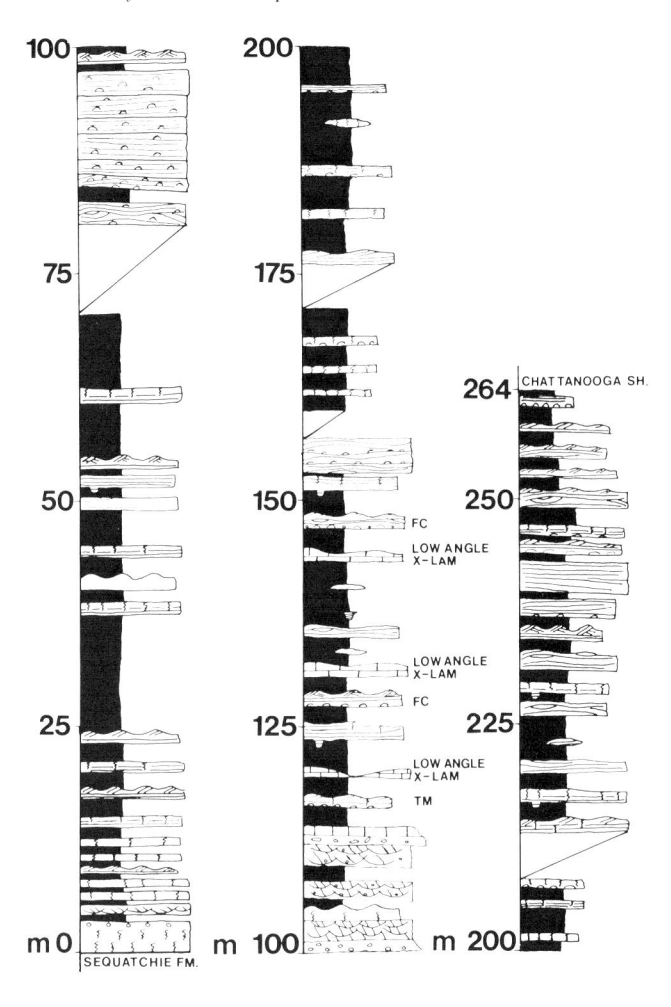

**Fig. 7.** Detailed measured section of the Rockwood Formation (Green Gap, locality C in Fig. 1), showing predominance of thin storm sandstone beds inserted into shelf mudstone (terrigenous thin-bedded facies). The interval from 75 to 100 m represents the hummocky cross-stratified sandstone facies, whereas 100–112 m is the cross-stratified sandstone facies. The vertical sequence terrigenous thin-bedded → hummocky cross-stratified sandstone → cross-stratified sandstone constitutes a progradational, shoaling upward succession. (From Easthouse, 1987.)

quartz arenite successions interpreted as shoreface sand wave and megaripple deposits. Other similar attributes include an abundance of red, haematitic matrix, generally fine grain sizes, rare desiccation cracks, and a high density of *Skolithos* traces.

### Hummocky cross-stratified sandstone facies

#### Description

The hummocky cross-stratified sandstone facies (hereafter referred to as the hummocky facies) occurs mainly within the middle part of the Rockwood Formation (Figs 3B, 7), as at locality C (Fig. 1). It generally consists of medium to very thick-bedded, amalgamated sandstone sequences that form both laterally continuous sheets and discontinuous, tabular to lensoid bodies. This facies is the dominant ridge-forming unit in outcrop, and is both underlain and overlain by the thin-bedded sandstone and shale facies (described below).

The hummocky facies is typified by medium-scale sets of hummocky cross-stratified, very fine to fine-grained sandstone (Fig. 8A,B). Sandstone beds up to 1 m thick are common; most are composed of several stacked hummocky sets, each 10–50 cm thick. The boundaries between individual hummocky sets are accentuated by millimetre to centimetre-thick lags of coarse sand to granule-size quartz grains or bioclastic grains (chiefly fragmented pelmatozoans, bryozoans, brachiopods, and trilobites), or shale lenses and layers 1–10 cm thick. No trace fossils were observed within beds, but *Planolites* and *Palaeophycus* traces are locally abundant at the bases of sandstone beds and associated with thin shale lenses and layers. Most hummocky

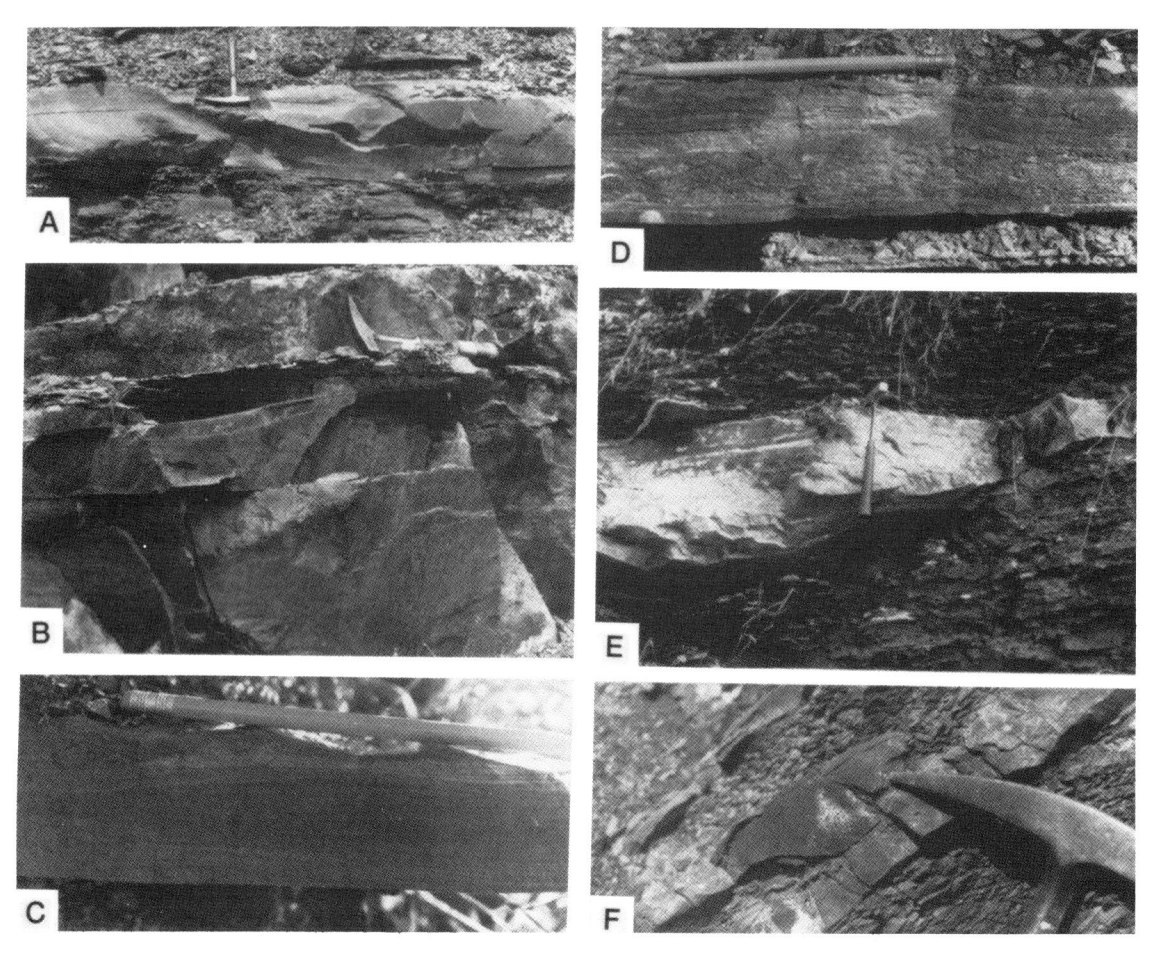

**Fig. 8.** Field photographs of facies occurring within the Rockwood Formation. (A) Hummocky cross-stratified sandstone facies (Bacon Ridge locality)—flat-based, low-angle to hummocky cross-stratified proximal storm sandstone. Hammer for scale. (B) Hummocky cross-stratified sandstone facies (Green Gap locality)—thick, amalgamated hummocky cross-stratification of proximal storm shelf origin. Hammer for scale. (C) Terrigenous thin-bedded facies (Cumberland Gap locality)—sharp-based sandstone bed of distal storm shelf origin, exhibiting micro-hummocky cross-stratification overlain by form-discordant (combined-flow) wave-ripple lamination. Pencil measures 15 cm. (D) Terrigenous thin-bedded facies (Falling Water Creek locality)—sharp-based, mixed carbonate–siliciclastic bed of distal storm shelf origin, exhibiting graded bioclastic lag overlain by parallel laminations. Pencil measures 15 cm. (E) Terrigenous thin-bedded facies (Melton Hill Lake locality)—large-scale sandstone gutter cast. Gutter incised into shale and siltstone. Note basal cross-stratification overlain by parallel laminations. Hammer for scale. (F) Terrigenous thin-bedded facies (Green Gap locality)—small-scale sandstone gutter cast. Gutter incised into sandstone and siltstone/shale. Note laminations in gutter. Hammer for scale.

beds have sharp, flat bases, with rare flute or groove casts. Upper surfaces of beds are always hummocky; antiform crest spacing ranges from 1 to 8 m, but most commonly is 2–5 m. The thickest hummocky beds are laterally persistent for at least 20 m (the scale of most outcrop exposures).

*Interpretation*

The hummocky facies is interpreted to represent deposition in a storm-dominated, inner shelf setting.

Cotter (1983b) described a similar facies in the Rose Hill and Mifflintown Formations ('Middle' Silurian) in central Pennsylvania. Duke (1982) also described hummocky cross-stratified sandstone deposits from the Medina Group (Lower Silurian) in upstate New York and southern Ontario.

Hummocky cross-stratification (HCS) is the diagnostic feature of this facies; HCS is generally taken to indicate deposition on a storm-dominated shelf (Harms *et al.*, 1975; Dott & Bourgeois, 1982;

Walker, 1984; Duke, 1985). Duke (1985) summarized the world-wide stratigraphic distribution of HCS, which is among the most definitive of storm-produced structures. Sandstone beds containing HCS in the Rockwood Formation are generally associated with horizontal laminations and form-discordant wave-ripple laminations. In the continuum of HCS sequences proposed by Dott & Bourgeois (1982), both 'ideal' and amalgamated sequences are represented in the Rockwood Formation. The common interbedding of this facies with the thin-bedded sandstone and shale facies of combined-flow origin is not in conflict with the interpretations of Swift et al. (1983) that HCS is a structure produced by a combination of storm-wave-generated oscillatory flow and a storm-generated, unidirectional mean flow. Swift & Niedoroda (1985) further interpreted HCS to represent storm-flow regimes in which the wave orbital component is high relative to the mean flow component. In the continuum of storm-related structures present along a deepening shelf profile, HCS represents deposition in a more proximal position relative to wave-ripple laminations and graded beds of the distal shelf (Aigner & Reineck, 1982; Aigner, 1985). Shoreward, HCS should grade laterally into the shoreface where high-angle cross-stratification is more abundant (Fig. 3A,B).

The vertical successions of structures observed within Rockwood Formation hummocky beds indicate deposition from waning flow (Brenchley, 1985). The sharp-based soles of the beds indicate an abrupt change from mud to sand (except in the case of amalgamated sequences), the result of initiation of a high velocity, competent flow (Hunter & Clifton, 1982). Erosional structures and lag deposits are common at the bases of beds. Succeeding the basal deposits are planar-laminations, which probably result from combined traction and suspension deposition. The next overlying laminae are hummocky cross-stratified, with slopes of hummocks less than 15°. The hummocky division commonly commences with a first-order erosion surface that cuts into the underlying horizontal division. The erosional hummocky surface is then typically draped with concordant laminae. The laminae generally thicken from domal highs into the troughs, resulting in gradual filling of depressions. The dips of the laminae decrease upward until the hummocky morphology is completely lost. Above the HCS interval in about half the cases there is a planar-laminated division that represents deposition on a flat surface formed when troughs between hummocks have been filled. In instances where flow intensity and sediment aggradation abruptly ended, broadly wavy upper bedding surfaces occur. Finally, form-discordant wave ripples are rarely observed capping HCS beds, but more commonly, non-bioturbated shale and siltstone sharply overlie HCS sandstone beds.

## Thin-bedded sandstone and shale facies

### Description

The thin-bedded sandstone and shale facies (hereafter referred to as the terrigenous thin-bedded facies) comprises the major part of the Rockwood Formation (belts exemplified by localities C, D and E in Figs 1, 3A,B). It both underlies and overlies the Hummocky Facies (Fig. 7), and grades laterally into the thin-bedded limestone and shale facies (Fig. 3B). The Hagan Shale Member of the Clinch Sandstone (Miller, 1976) also is comprised of the terrigenous thin-bedded facies; it occurs within both the Clinch Mountain and Powell Mountain outcrop belts (localities A and B in Figs 1, 3A), stratigraphically underlying the Poor Valley Ridge Sandstone Member. An occurrence of Hagan Shale is summarized in Fig. 4A. In both the Clinch Sandstone and the Rockwood Formation the terrigenous thin-bedded facies is characterized by heterogeneity of lithology.

The terrigenous thin-bedded facies consists of thin beds of fine to very fine-grained, subarkosic to sublitharenitic sandstone intercalated with grey-green siltstone and fissile to semi-fissile shale. The ratio of sandstone beds to siltstone and shale beds ranges from 1:2 to 1:10. Sandstone beds range from 1 to 10 cm in thickness, and exhibit a wide variety of sedimentary structures and sequences (e.g. Fig. 8C–F). About 70–80% of the thin sandstone beds display sharp, flat bases as well as the following vertical sequence, from base to top: flat, sharp base (with or without flute or groove casts) → horizontal laminations → wave-ripple or form-discordant wave-ripple laminations → siltstone or shale (with or without bioturbation). Upper boundaries of these beds are generally sharp, but undulatory, reflecting preservation of ripple forms; diffuse boundaries are less common, and are attributable to bioturbation. Trace fossils are dominated by *Planolites*, *Palaeophycus* and *Chondrites*, which occur on the soles of beds, and along top surfaces; less common are vertical to subvertical, straight tubes exhibiting downward-deflected laminations along the burrow margin. Sharp-based graded beds are also common (Fig. 8D); grain size typically fines upward from fine sand to very fine sand and coarse silt, and a bioclastic lag (commonly haematite-replaced) to 1 cm or so thick defines the base of the bed. Body fossils comprising basal lag deposits include both inarticulate and articulate brachiopods, pelmatozoans, bryozoans, and trilobites, all highly fragmented. Some of the thickest (5–10 cm) sandstone beds display unusually fine-scale hummocky cross-stratification, with wavelengths of 25 cm to 1 m. This type of sedimentary structure (Fig. 8C) has been termed *micro-hummocky cross-stratification* (micro-HCS) by Dott & Bourgeois (1982). Approximately 20–30% of the thin sandstone beds exhibit bases with scours a few centimetres deep. Sandstone gutter casts (see Whittaker, 1973) are common, forming disconnected lenses in cross-section (Fig. 8E,F), but which are sinuous in plan-view. Thin sandstone beds persist laterally for no more than 10 m, and many exhibit a lens-like geometry.

The terrigenous thin-bedded facies within the Hagan Shale Member of the Clinch Sandstone exhibits a few subtle differences when compared with the Rockwood Formation, for there is an overall greater degree of bioturbation. Concomitantly, physical sedimentary structures are less easily observed. Contacts between thin sandstone beds and bounding strata (typically shale and siltstone) tend to be diffuse and less well defined. Similarly, shale strata typically exhibit a blocky to semi-fissile weathering character. Wave-formed structures are less common, and parallel laminations (or no laminations) are more typical. Medium-grained sandstone beds, which are trough cross-stratified, occur locally near the contact with the overlying Poor Valley Ridge Sandstone Member (Fig. 5A). Common trace fossils include *Planolites*, *Palaeophycus* and *Chondrites*, as in the Rockwood Formation, but in addition, *Cruziana* and *Rusophycus* are locally abundant on the soles of thin sandstone beds. Both *Skolithos* and *Monocraterion* occur near the tops of the small-scale trough cross-stratified beds. Body fossils are limited to inarticulate brachiopods (*Lingula*). Clasts of maroon red siltstone and shale, which closely resemble some lithologies found in the underlying Juniata Formation (Upper Ordovician), along with granules and small pebbles of phosphatic composition form concentrations either at the bases or tops of thin sandstone beds.

*Interpretation*

The terrigenous thin-bedded facies represents deposition in a middle to outer shelf setting that was affected by episodic storms.

Individual thin sandstone beds exhibit variable internal stratification features, most of storm origin. The sharp, flat bases common to all beds are indicative of high flow velocities (Hunter & Clifton, 1982); the introduction of layers of coarser sediment into a quiet, muddy environment implies that a strong flow deposited these beds (Hamblin & Walker, 1979). Normally graded beds with bioclastic (fossil) lags at the bases are diagnostic of storm deposits, especially when wave-formed structures occur within the same beds (e.g., Kreisa, 1981; Aigner & Reineck, 1982; Brenchley & Newall, 1982; Walker *et al.*, 1983; Aigner, 1985). Abundant horizontal laminations in these fine to very fine-grained sandstone beds suggest conditions of plane-bed deposition (Mount, 1982), with a minimum shear velocity at the bed of the order of 100 cm s$^{-1}$ (Dott & Bourgeois, 1982). The prevalence of form-discordant wave ripples strongly points to deposition by combined-flow currents (Aigner, 1985). Such combined-flow currents are common during storms, as storm waves interact with the sea-floor and are superimposed on a unidirectional mean flow (Aigner, 1985; Swift & Niedoroda, 1985). Concentrations of mudstone intraclasts at the bases of sandstone beds imply storm scour of the muddy shelf substrate by intense storm flows (Kelling & Mullin, 1975; Walker, 1985). Sporadic occurrences of gutter casts result from intense substrate scour produced by storm-generated helicoidal fluid flow at the bed (Whittaker, 1973; Cotter, 1983b; Aigner, 1985; Walker, 1985); flute casts and associated tool marks also attest to strong unidirectional storm scour. The occurrence of micro-hummocky cross-stratification in Rockwood storm sandstone beds is especially significant. This structure was defined by Dott & Bourgeois (1982) as a smaller scale variant of hummocky cross-stratification, which is a storm-produced structure of combined-flow origin (Swift *et al.*, 1983).

The dominant lithology of the terrigenous thin-bedded facies is fissile to semi-fissile mudstone or shale. The prevalence of fissility, lack of identifiable biogenic structures, and lack of body fossils together imply either oxygen-deficient, anaerobic conditions on the shelf bottom (Rhoads & Morse, 1971; Byers, 1974, 1977; Cluff, 1980; Savrda *et al.*, 1984; Savrda

& Bottjer, 1986), or a substrate unfavourable for colonization by most benthic organisms (generally too 'soupy'; K.R. Walker & Diehl, 1986; Easthouse & Driese, 1988). The substrate argument is perhaps a stronger one in view of the fact that shallow shelves are generally well mixed, owing to frequent wave and current action. An alternative hypothesis could be high shelf productivity overwhelming bottom organisms, thereby producing an anaerobic bottom mud, in spite of the fact that the water column itself was aerobic (Swift, pers. comm., 1987).

Palaeocurrent data from wave-ripple crests (Fig. 4B) suggest onshore to oblique storm wave incidence from the west, northwest, or north. Gutter and flute cast data, in contrast, suggest a strong, unidirectional mean flow to the southwest (Fig. 4C). The common occurrences of sole marks indicating palaeoflow to the southwest (parallel to bathymetric contours) *and* wave-ripple crests oriented NE–SW *within the same beds*, are strong evidence for the combined interaction of a storm-induced along-shelf mean flow (geostrophic current?) interacting with storm-wave-induced oscillatory flow.

## Thin-bedded limestone and shale facies

### Description

The thin-bedded limestone and shale facies (hereafter referred to as the calcareous thin-bedded facies) occurs mainly within the Brassfield Formation (exemplified by locality F in Fig. 1), but also comprises a major part of westernmost exposures of the Rockwood Formation (Fig. 3B). It resembles the terrigenous thin-bedded facies in that it is chiefly composed of shale and siltstone into which thin beds of coarser grained lithologies are inserted. However, unlike the terrigenous thin-bedded facies, the coarser lithologies are not siliciclastic but are carbonate or mixed carbonate–siliciclastic. This facies typically intergrades laterally with the terrigenous thin-bedded facies (Fig. 3B).

The calcareous thin-bedded facies generally consists of beds, 1–30 cm thick, of either trough cross-stratified, parallel laminated, wave-ripple cross-laminated, or bioturbated grainstone and packstone (commonly with an appreciable quartz sand and silt component), which are inserted into semi-fissile grey-green shale and rarer parallel laminated, wave-ripple cross-laminated, or bioturbated quartz siltstone beds. Most carbonate beds have sharp, flat bases but diffuse to irregular tops, owing to extensive bioturbation. Trough cross-stratified beds are generally thickest, and consist of coarse sand, granule and small-pebble sized carbonate clasts dominated by pelmatozoan columnals and plates, ramose bryozoans, whole rugose corals and articulate brachiopods. Many cross-stratified carbonate beds are extensively haematite-replaced (Fig. 9A). Carbonate beds composed of fine to coarse-sand sized grains (chiefly peloids, superficial ooids, quartz sand, and highly abraded bioclasts) exhibit parallel laminations disrupted by vertical 'escape' burrows (Fig. 9C), or either form-discordant wave-ripple laminations or micro-hummocky cross-stratification (Fig. 9B). Bioturbated carbonate beds show diffuse and irregular upper and lower contacts, a lack of primary physical structures, and coarse fossil allochems 'mixed' into a sand sized matrix (Fig. 9D). Some carbonate beds exhibit a crude fining upward trend, in which coarse bioclasts at the base of the bed are succeeded by finer skeletal sand and silt exhibiting parallel laminations or wave-ripple cross-laminations (Fig. 9B).

### Interpretation

The calcareous thin-bedded facies is similar to the terrigenous thin-bedded facies (discussed earlier), exhibiting a wide spectrum of sedimentary structures attributable to combined-flow processes (Fig. 9B,C). Both facies were deposited in shelf settings characterized by low-energy terrigenous mud deposition (post-storm) alternating with high-energy sand deposition during storms. The calcareous thin-bedded facies differs from the terrigenous thin-bedded facies in that coarser beds are carbonate or mixed carbonate–siliciclastic in composition, some beds consist entirely of high-angle cross-strata (Fig. 9A), and the former facies shows greater bioturbation (Fig. 9D).

Several analogous carbonate-and-shale successions of storm-shelf origin have been described recently in the literature; these include the Upper Ordovician Martinsburg Formation of Virginia (Kreisa, 1981), the Triassic Muschelkalk of Germany (Aigner & Reineck, 1982; Aigner, 1985), the 'Middle' Silurian Rose Hill–Mifflintown–Bloomsburg Formations of Pennsylvania (Cotter, 1983b), and the Mississippian Fayetteville Shale–Pitkin Limestone of Arkansas (Handford, 1986). All the following are evidence of combined-flow storm

**Fig. 9.** Field photographs of the Brassfield and westernmost Rockwood Formation. (A) Calcareous thin-bedded facies (Falling Water Creek locality)—high-angle cross-stratification in haematitic, bioclastic lime grainstone of high-energy storm origin. Hammer for scale. (B) Calcareous thin-bedded facies (Falling Water Creek locality)—sharp-based, bioclastic/peloidal lime grainstone of storm shelf origin, exhibiting graded bioclastic base overlain by microhummocky cross-stratification. Pencil measures 15 cm. (C) Calcareous thin-bedded facies (Falling Water Creek locality)—sharp-based, peloidal lime grainstone of storm shelf origin, exhibiting parallel laminations disrupted by 'escape' burrows. Lens cap measures 5.5 cm. (D) Calcareous thin-bedded facies (Anderson Ridge locality)—peloidal lime packstone of storm shelf origin, disrupted by heavy bioturbation. Note relict primary laminations. Lens cap measure 5.5 cm.

deposition of carbonate beds: (1) hummocky or micro-hummocky cross-stratification; (2) sharp-based, normally graded beds consisting of basal bioclastic lags overlain by parallel laminations; (3) gutter or flute casts aligned parallel to palaeoslope; and (4) especially bed sequences of the type: sharp, flat base, with or without flute or groove casts → parallel laminations → form-discordant wave-ripple laminations or micro-hummocky cross-stratification → burrowed silty shale.

Sparse palaeocurrent data from high-angle cross-strata (Fig. 4A) suggest a highly variable pattern of unidirectional flow, which may indicate a complex interplay between storm-induced currents. Wave-ripple crest data (Fig. 4B) suggest storm waves travelling along shelf from northeast to southwest, roughly parallel to the basin axis. This is in contrast to the terrigenous thin-bedded facies to the east, where wave-ripple crests are parallel to palaeo-shoreline and indicate onshore wave approach (Fig. 4B). Aigner (1985) noted a similar change in wave-ripple crest orientation in the Triassic Muschelkalk, which he attributed to the effects of along-shore storm tracks and the Coriolis effect. During Silurian time in east Tennessee, dominantly along-shelf winds blew down the basin axis to the southwest and generated along-shelf wave-trains in offshore water. However, owing to wave refraction associated with shallowing the more nearshore wave-ripple orientations became parallel to the palaeocoastline (Fig. 4B). Rare occurrences of gutter or flute casts whose orientations suggest along-shelf palaeoflow to the southwest (Fig. 4C) reinforce this interpretation of dominantly along-shelf winds.

### Underlying Upper Ordovician strata

It is relevant to briefly describe the Upper Ordovician strata that underlie the Lower Silurian rocks discussed here. They provide evidence for a

Late Ordovician eustatic regression followed by an Early Silurian eustatic transgression (see Fig. 2).

*Juniata Formation*

Within the Clinch Mountain and Powell Mountain outcrop belts (localities A and B in Figs 1, 3A), the Upper Ordovician Juniata Formation lies conformably beneath the Lower Silurian Clinch Sandstone. It consists of maroon red, thin to thick-bedded siltstone to very fine-grained sandstone beds, intercalated with maroon red shale and silty shale. Sandstone beds are either burrow-homogenized, or are parallel laminated to wave-ripple cross-laminated. Shale intraclasts are common at the bases of, and dispersed within, sandstone beds. Desiccation cracks are abundant at the tops of sandstone beds capped by shale strata. Trace fossils are common and include *Skolithos*, *Diplocraterion*, *Planolites* and *Palaeophycus*. Bioturbation is commonly so extensive that individual trace fossils cannot be identified. Rarely are fragments of bryozoans, brachiopods, trilobites and pelmatozoans concentrated near the bases of sandstone beds.

Near the contact with the Hagan Shale Member of the Clinch Sandstone, there is a gradual change from maroon red (Juniata) to grey-green or tan (Hagan). A metre or two beneath the Ordovician–Silurian boundary is a very distinctive, 25–150 cm thick, maroon red shale bed. This bed exhibits peculiar weathering characteristics, which include blocky 'ped' texture (crumbles into polygonal pieces generally 2–7 mm in diameter) and arcuate, slickensided fracture surfaces that resemble 'teepee' structures in carbonate rocks (Driese & Foreman, 1991). Above this distinctive shale bed is a very fine to fine-grained, 20–50 cm thick, completely bioturbated red-brown to tan-yellow sandstone bed. This bed is composed chiefly of quartz sand, but locally includes phosphate grains (sand to small pebbles), chert, maroon red Juniata lithiclasts, and inarticulate brachiopod valves; most of these exotic grains are concentrated both at the base and at the top of the bed (Gogola, 1990). The base of this regionally traceable, distinctive sandstone bed defines the Juniata–Clinch boundary. Phosphate grains and Juniata lithiclasts occur throughout the sandstone beds of the Hagan Shale Member.

*Sequatchie Formation*

Within the Rockwood Formation outcrop belts (localities C, D and E in Figs 1, 3B) the Upper Ordovician Sequatchie Formation lies beneath the Lower Silurian Rockwood Formation. A conspicuous red colour makes the Sequatchie superficially similar to the Juniata; however, it is commonly more calcareous and includes discrete carbonate beds. The maroon red (Sequatchie) to grey (Rockwood) colour change is again transitional. Sedimentary structures resemble those of the Juniata, and desiccation cracks are locally abundant. Again the regionally traceable and distinctive, fine to very fine-grained, basal sandstone bed exists, which defines the Ordovician–Silurian boundary (Gogola, 1990). This 25 cm to 1.5 m thick sandstone bed contains concentrations of phosphate grains, chert, glauconite, maroon red Sequatchie lithiclasts, and fossil grains (brachiopods, bryozoans, pelmatozoans), especially near the base and top of the bed. It is cross-stratified at many localities.

*Evidence for transgression*

That Lower Silurian terrigenous clastic strata were deposited during a eustatic transgression that followed a Late Ordovician regression (Fig. 2) is apparent based on detailed analysis near the boundary. The abundance of desiccation cracks, as well as the suites of sedimentary structures present, suggest locally emergent tidal flat or deltaic depositional conditions for the Upper Ordovician Juniata and Sequatchie Formations (A.M. Thompson, 1970; Milici & Wedow, 1977; J.R. Thompson, 1984; Ghazizadeh, 1985). The distinctive features of the maroon red shale bed, which occurs just below the Ordovician–Silurian boundary, point to a pedogenic origin for this bed, possibly as a Vertisol (Ahmad, 1983; Driese & Foreman, 1991) with gilgai (Knight, 1980). The development of soil peds and slickensided pedogenic fractures typically indicates a seasonal wet–dry climate, in which alternate shrinking and swelling of clays and precipitation and dissolution of evaporites are common (e.g., Allen, 1973; Watts, 1977; Yaalon & Kalmar, 1978; Goldbery, 1982; Gray, 1984; Blodgett, 1985). Thus, there is abundant evidence for both emergence and soil development in the uppermost Ordovician strata just beneath the boundary.

The characteristic of the 'basal sandstone bed', the base of which we interpret as the Ordovician–Silurian boundary, is a lag deposit that records the Early Silurian transgression (Gogola, 1990). The concentrations of phosphate grains, chert grains, maroon red Juniata and Sequatchie lithiclasts,

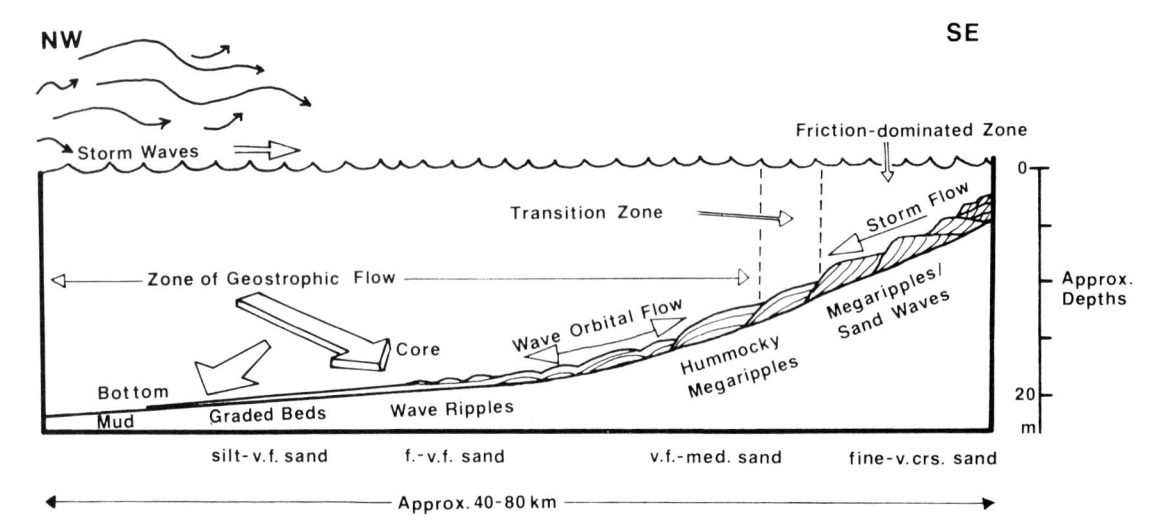

**Fig. 10.** Palaeohydraulic reconstruction of Silurian shelf-bottom showing bedforms and deposits formed in response to storm flow, based on the dynamic zones of the mid-Atlantic shelf (Swift & Niedoroda, 1985). Within friction-dominated zone, storm flows resulted in offshore (northwestward) megaripple and sand-wave migration. Within zone of geostrophic flow, along-shore and onshore storm winds produced coastal set-up, which resulted in along-shelf (southwestward) geostrophic? flow combined with near-bed oscillatory flow. The spectrum of bedform types present reflect grain-size changes produced by progressive sorting and decreasing competence of storm-wave oscillation and geostrophic flow, in an onshore to offshore direction.

quartz granules, glauconite and fossil allochems within the basal sandstone bed suggest initial sediment-starved, subtidal conditions, in which sedimentation rates were very slow and an upwelling component may have existed (Gulbrandsen, 1969; Berner, 1971; Fuller, 1979; Riggs, 1979; Odin & Matter, 1981). The major colour change from maroon red (more-oxidizing conditions) to grey-green (more reducing conditions), more or less coincident with this sandstone bed, reinforces the interpretation of regional transgression. However, as we discuss in a subsequent section, the facies relationships in the sequence indicate that terrigenous influx off adjacent Taconic allochthons soon outpaced sea-level rise yielding an overall progradational pattern for the clastic wedge.

## DISCUSSION

### Shelf dynamics

We have presented evidence for offshore-directed, unidirectional storm flows, offshore and along-shore wave oscillatory (storm) flows, and along-shelf unidirectional (geostrophic?) flows (Fig. 4A–C). The wide spectrum of facies in the Silurian rocks of east Tennessee reflects the response of the shelf substrate to these flows. Our model (Fig. 10) is based mainly on studies of the mid-Atlantic shelf of the USA, which have been summarized recently by Swift (1985) and Swift & Niedoroda (1985). The offshore-oriented cross-strata of the cross-stratified sandstone facies reflect the migration of megaripples and sand waves across a sandy shoreface in response to seaward storm flows, within the *friction-dominated zone*. In this zone, the effect of the Coriolis force is reduced because of the shallow depths and strong turbulence; consequently, currents flow parallel to the direction of the forcing mechanism (Swift & Niedoroda, 1985). Water depths perhaps of the order of 5–10 m seem likely (Fig. 10), based on analogy with the present Atlantic shelf. Storm currents were probably produced by set-up associated with onshore storm winds, whereby the offshore horizontal pressure gradient results in downwelling near the shore and a strong bottom current with an offshore component is developed.

Within palaeowater-depths of 10–100 m(?), a *geostrophic zone* developed (Fig. 10). In this zone, the Coriolis effect is quite significant, currents tend to flow normal to a horizontal pressure gradient associated with coastal set-up or set-down, and flow is parallel to the shore (Swift & Niedoroda, 1985). This zone is characterized by combined flows, in

**Fig. 11.** Model for Silurian shelf in Southern Hemisphere palaeolatitude, showing general distribution of bedforms and sediment types in response to storm-generated flows for a prograding system. Note that southwestward along-shelf, as well as northwestward offshore movement of sand is inferred. Sand source in northern Virginia (vicinity of Massanutten Mountain) is postulated based on maximum pebble size dispersal data of Yeakel (1962), Meckel (1970) and Whisonant (1977).

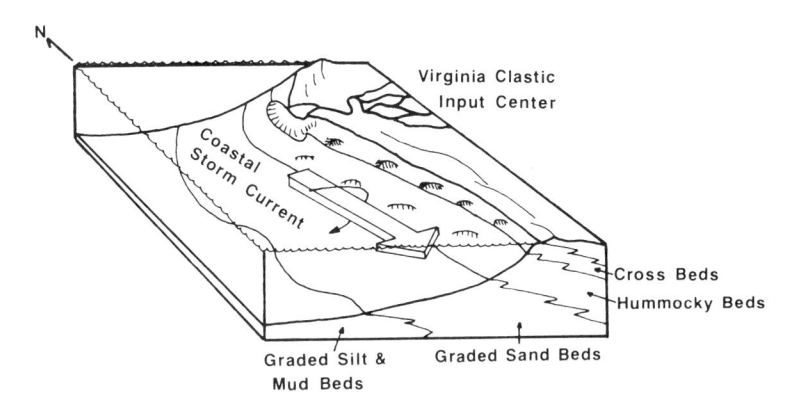

which bedforms are produced by the interaction of wave oscillatory flows and a mean flow component that is unidirectional. The characteristic structures of the hummocky cross-stratified sandstone facies are the product of hummocky megaripples, such as those described by Swift *et al.* (1983) on the Atlantic shelf (Fig. 10). The micro-hummocky cross-stratification strata, wave-ripple laminae, and graded beds of both the terrigenous and the calcareous thin-bedded facies are interpreted as more distal deposits of the geostrophic zone, in which currents overall were weaker (Fig. 10). A *transition zone* is also inferred between the friction-dominated and geostrophic zones, which shifted landward or seaward according to changes in storm intensity and water depth (Fig. 10). The proximal–distal relationships depicted in Fig. 10 are undoubtedly the product of *progressive sorting* as defined by Swift *et al.* (1972), in which the sediment deposited at a given point is a consequence of the hydrodynamic processes occurring at successive points up-current as well as a response to hydrodynamic sorting occurring at that point.

The average slope on the shelf bottom can be estimated by assuming a range of reasonable water depths over the palinspastically restored width of the outcrop belt. Assuming a minimum depth of 10 m and a maximum depth of 100 m for the centre of the basin, gradients of about 0.005–0.010° are computed. These gradients were certainly too low to have initiated or have sustained turbidity flow (see Soegaard & Eriksson, 1985, p. 681), in spite of the arguments presented recently by Walker (1985). Instead, storm currents that developed in response to onshore or alongshore-directed storm winds transported coastal sand across the shelf. Furthermore, along-shore transportation of sand was appreciable, in addition to offshore progradation (Fig. 11). An earlier interpretation of a major clastic input centre in northern Virginia (Massanutten Mountain vicinity) is based on the maximum clast size data presented by Whisonant (1977) for the equivalent Tuscarora Sandstone. Thus, the quartz sand that comprises the Poor Valley Ridge Member of the Clinch Sandstone in Tennessee may have had a more northerly (as opposed to an easterly) source, and was transported southwestward by along-shelf storm-generated flows (Fig. 11).

The reconstructions of Ziegler *et al.* (1977, 1979) indicate that the palaeogeographic setting of the study area was conducive to sedimentation by storm processes. From the given palaeolatitude of about 30° south and the continental configurations, we infer that hurricanes and tropical storms developed at low latitudes and tracked southwestward down the basin axis (Fig. 12). Although a hurricane interpretation is favoured based on palaeolatitude (Duke, 1985), we cannot rule out the possibility of rare, intense mid-latitude winter cyclones tracking northeastward (Fig. 12). With the clockwise rotation of Southern Hemisphere low-pressure centres and a favourable storm track, both hurricanes/tropical storms and winter storms could have produced coastal set-up by the mechanisms described earlier.

## Depositional history

### General

The depositional history of Lower Silurian strata in the southern Appalachian region was primarily controlled by three factors: changes in relative sea-level, changes in tectonic activity and shelf hydro-

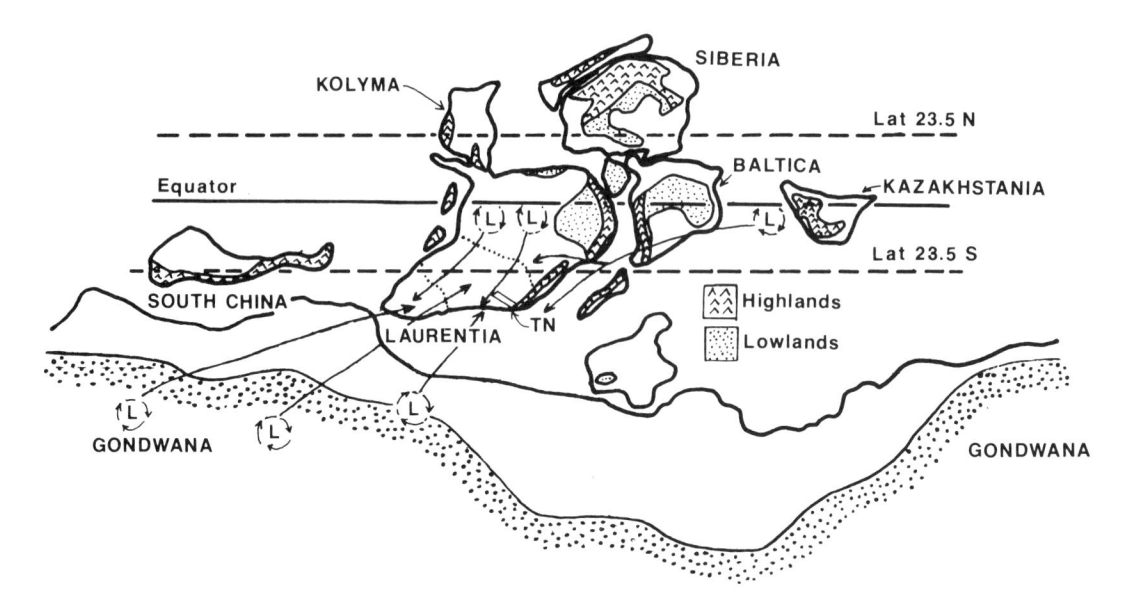

**Fig. 12.** Early Silurian palaeogeographic reconstruction based on Ziegler et al. (1977, 1979) showing position of east Tennessee at about 30° south latitude. We infer storm effects from both low-latitude tropical storms and hurricanes and *possibly* from mid-latitude winter storms as well. Several hypothesized storm tracks (L) are shown.

dynamics. Cotter (1983a,b) discussed Late Ordovician and Early Silurian sea-level fluctuations, so only a brief summary is included here. Convincing evidence exists for an extensive ice sheet in the Gondwana continents in Late Ordovician (Ashgill) time (Fig. 2) (Berry & Boucot, 1973; McKerrow, 1979), which coincided with Juniata and Sequatchie Formation red-bed deposition. These Upper Ordovician red-bed sequences have been interpreted as representing progradational tidal flat and deltaic deposits (A.M. Thompson, 1970; Milici & Wedow, 1977; J.R. Thompson, 1984; Ghazizadeh, 1985). Just prior to Early Silurian (Llandoverian) transgression, soil-forming processes were active, as evidenced by the palaeosols described by Driese & Foreman (1991). Deposition of the Clinch Sandstone and Rockwood Formations began in response to rising sea-level (Fig. 2), caused by the retreat of the Late Ordovician Gondwana glaciers.

Coinciding with this Early Silurian sea-level rise was the emplacement of deformed early Palaeozoic sedimentary, and Precambrian sedimentary and crystalline basement rocks westward on to the eastern edge of the North American craton via transport along the Hayesville–Fries (later the Brevard) fault zone (Hatcher, 1972, 1978). This emplacement of a late Taconic thrust load probably was responsible for the lithospheric flexure that became the Silurian foreland basin in the southern Appalachians (Quinlan & Beaumont, 1984). This tectonically induced subsidence may have contributed, at least partly, to the relative sea-level rise recorded during Early Silurian time (Fig. 2). In addition, the thrust sheets undoubtedly were the major source of the relatively mature quartz sand and granules, which comprise the bulk of the Lower Silurian sandstone section.

The importance of hydrodynamics in the transport of sand on the Clinch–Rockwood shelf was presented above. Sand was transported both offshore (westward and northwestward) and along shore (southwestward) in response to the occurrence of storm winds, which approached the coast primarily from the northwest, north and northeast (Figs 4, 10–12).

*Interpretation*

The depositional history of Lower Silurian strata in the southern Appalachians is summarized in Figs 13, 14. Commencing with rising sea-level during Early Silurian (Llandoverian) time, we postulate a retrograding shoreline dominated by erosional shoreface retreat (c.f. Swift, 1968, 1975; Swift *et al.*, 1971,

**Fig. 13.** Diagrammatic lithofacies profile showing facies relationships from more proximal southeasterly localities (A) to more distal northwesterly localities (F). True stratigraphic thicknesses are not depicted, although the profile is based on actual measured sections (Figs 1, 3A,B, Appendix I); hence it is only partly interpretative. The depositional history for Silurian rocks implied by this figure is discussed in text.

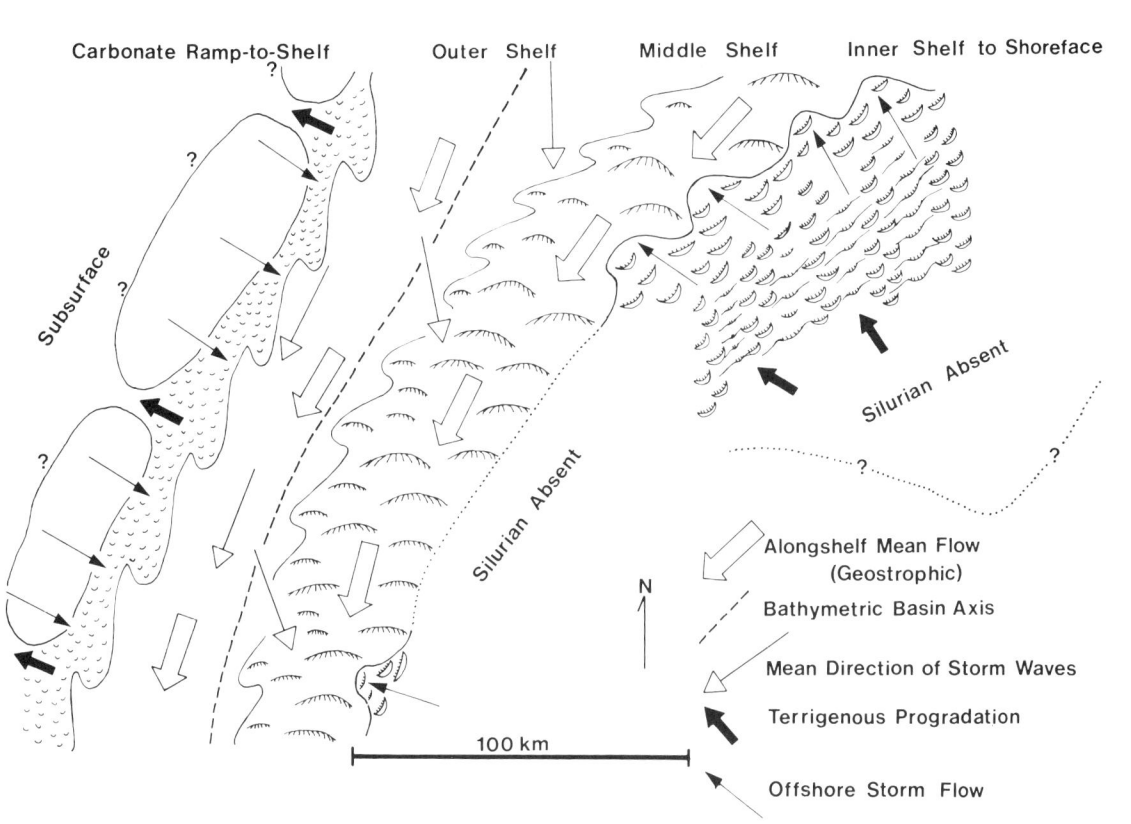

**Fig. 14.** Hypothetical reconstruction of shelf and coastal palaeoenvironments for Lower Silurian rocks presently exposed across the fold-and-thrust belt (Valley and Ridge Province) of east Tennessee. Facies correspond to inferred palaeoenvironments shown in Fig. 13 at time of maximum westward progradation of cross-stratified sandstone. They include: *cross-stratified sandstone facies*—shoreface to inner shelf, megaripple and sand-wave-dominated sand sheet environment; *hummocky cross-stratified sandstone facies*—inner to middle shelf, proximal storm sand sheet environment; *terrigenous thin-bedded facies*—middle to outer shelf, distal storm sand sheet and mud environment; *calcareous thin-bedded facies*—storm-dominated, carbonate bank to shelf environment. Note complex interplay of various types of flows, and their orientations. Explanation in text.

1981; this volume). The Ordovician–Silurian boundary is therefore tentatively identified as a possible *ravinement* surface, and the 'basal' sandstone bed above that surface is a transgressive lag deposit derived from the underlying Upper Ordovician strata (Gogola, 1990).

Late Taconic thrust sheets emplaced against the eastern margin of North America resulted in flexural downwarping and the continued subsidence of the Clinch–Rockwood foreland basin, but more importantly, produced an increase in sediment supply. This quartz sand was of a greater size and maturity than any of Late Ordovician or earliest Silurian time. Major sand transport was both offshore (northwestward) and along shelf (southwestward), in direct response to storm shelf processes (Figs 10, 11). Whereas the earlier sea-level rise was rapid enough to cause shoreline retrogradation, sediment was later introduced at a rate that exceeded sea-level rise, resulting in shoreline progradation and deposition of a clastic wedge (Fig. 13). The shoreface megaripple and sand wave deposits of the cross-stratified sandstone facies represent the most proximal facies preserved in outcrop, except for very local occurrences of wave-dominated coastal sand flat deposits (Fig. 13). The hummocky cross-stratified sandstone facies was situated in a position intermediate between more proximal facies and the distal storm shelf deposits of the terrigenous thin-bedded facies (Fig. 13). During a second transgression, following maximum progradation, extensive iron mineralization occurred within both siliciclastic and carbonate strata as iron was weathered from surrounding sourcelands. Hunter's (1970) and Cotter's (1983b) studies on Silurian strata in the northern and central Appalachian basins indicated that there is a correlation between concentration of iron minerals and major transgressions following sea-level lowstands. Brachiopods identified by A. Boucot (pers. comm. to Easthouse, 1986, cited in Easthouse, 1987) suggest that this second transgression and iron mineralization occurred within the Llandovery $C_{2-3}$ stage, which is in agreement with the relative sea-level curve depicted in Fig. 2. During Llandovery $C_{2-3}$ transgression, continued sea-level rise in combination with diminished clastic input resulted in renewed shoreline retrogradation (Fig. 13), and more distal shelf facies were juxtaposed on top of more proximal deposits. The distribution of palaeoenvironments during maximum progradation is shown in Fig. 14.

The later depositional history of the southern Appalachian Basin is problematical because of variable preservation of Upper Llandoverian strata along an upper unconformity surface (pre-Middle Devonian). Brachiopods collected at locality C (Fig. 1) indicate that strata representing all of the Llandoverian are present at this locality (A. Boucot, pers. comm. to Easthouse, 1986, cited in Easthouse, 1987); facies patterns also suggest that a second progradational event occurred, which was of a lesser magnitude than the first (see Fig. 7, abundance of sandstone strata in the interval 200–264 m, and Easthouse & Driese, 1988). Cotter (1983b) has discussed the post-Llandoverian depositional history of strata preserved in the northern Appalachian Basin, which provides a more complete record.

## SUMMARY AND CONCLUSIONS

The palaeoenvironments and depositional processes for Lower Silurian shelf strata of east Tennessee during maximum progradation are summarized in Fig. 14. Our model, in some respects, resembles one proposed by Cotter (1983b) for 'Middle' Silurian rocks of central Pennsylvania. The entire depositional system can be subdivided broadly into *shelf margin siliciclastics* on the eastern side of the foreland basin, *terrigenous shelf muds* along the basin bathymetric axis and *shelf margin carbonates* on the western margin of the basin (Fig. 14).

Sediment was derived from the east and northeast, primarily from the erosion of thrust sheets consisting of early Palaeozoic sedimentary rocks and/or Precambrian sedimentary and crystalline rocks (i.e. Taconic allochthons). Quartz sand and granules were transported from braided fluvial/alluvial fan systems (not preserved in outcrop) out to shoreface environments by offshore-directed storm flows (Fig. 14). Most of the Poor Valley Ridge Member of the Clinch Sandstone is representative of the shoreface environment (Fig. 5A,B), with upward shoaling to emergent conditions. Sand transport was accomplished chiefly through megaripple and sand wave migration to the northwest (Fig. 4A). Currents in this environment were mainly unidirectional to weakly bidirectional, with little direct evidence for wave oscillatory or along-shelf (geostrophic) flows. We surmise that water depths were in the realm of 5–10 m, and that this regime was entirely friction-dominated (Fig. 10).

Inner and middle shelf deposits consist chiefly of

very fine to fine-grained, submature quartz sand winnowed out of the shoreface environment and deposited by combined flows as storm sand sheets or lenses (Fig. 14). Surface bedforms reflected the interaction of combined flows with a finer grained sediment, and were dominated by hummocky megaripples, micro-hummocks, and form-discordant wave ripples, arranged in a proximal to distal gradient (Fig. 10). Most of the Rockwood Formation is representative of this environment (Fig. 7). Evidence of storm-wave-generated oscillatory flows is abundant. Wave ripples of proximal facies suggest storm wave refraction so as to become nearly parallel to the palaeoshoreline, whereas the wave ripples of distal facies indicate storm waves tracking from the northeast, parallel to the palaeoshoreline (Fig. 4B). A strong along-shelf flow component (possibly geostrophic?) developed, with a sense of flow to the southwest down the basin axis (Fig. 4C, 14). This flow probably originated in response to along-shelf storm winds. Shelf mud layers were deposited on top of storm sand sheets as flow intensities waned. The absence of burrows in shelf mudstone strata suggests that the consistency of the mud substrate was too soupy for most organisms. Alternation of storm and fair-weather conditions resulted in the thin-bedded character of deposits of this environment (Fig. 7). We infer that water depths were approximately 10–50 m, to perhaps as much as 100 m, and that this regime varied from transitional (proximal) to geostrophic-dominated (distal) (Fig. 10).

The outer shelf environment was dominated by terrigenous mud deposition, but during very high-energy storm events deposition of very fine-grained quartz sand and silt and bioclastic carbonate grains occurred. These graded or wave-rippled beds and lenses are interpreted as ultradistal storm layers. Siliciclastic and carbonate components are characteristically mixed because of contribution of a carbonate component from the west, and contribution of a siliciclastic (sand) component from the east (Fig. 14). Parts of the Brassfield and Rockwood Formations (Figs 7, 8) are representative of this environment. Biogenic reworking of mudstone is extremely rare, whereas calcareous lithologies are thoroughly bioturbated. These relationships suggest that substrate consistency ('soupiness' of terrigenous mud versus early cementation of carbonate) was a major control on burrow distributions. Storm currents paralleled the basin axis, and consisted of a southwestward flow (geostrophic?) component and a similarly oriented storm wave oscillatory component (Figs 4B,C, 15). Water depths are estimated to have been 50–100 m.

Outcrop exposures are generally sparse, but we postulate a gently eastward-dipping carbonate bank-to-shelf system on the western margin of the foreland basin (Fig. 14). This system could have developed in association with a migrating peripheral bulge (Nashville Dome?), which existed in response to Taconic thrust loading. Parts of the Brassfield Formation are representative of this environment. Bioclastic sand sheets generated in highly agitated, shallow marine, skeletal sand banks were periodically transported eastward by offshore-directed storm flows into the terrigenous-dominated basin centre. The sheets were then reworked by along-shelf geostrophic flows, along-shelf storm waves and northwesterly storm currents (Figs 4, 15). The abundant bioturbation observed reflects a better substrate condition (firmer) than that of the terrigenous-dominated middle and outer shelf environments. Further studies of this carbonate-dominated portion of the Silurian shelf system, using surface outcrop and subsurface information, should allow more precise interpretations of these rocks.

## ACKNOWLEDGEMENTS

This paper represents a regional synthesis of research conducted both by the senior author, as well as by M.S. thesis students K. Easthouse, M. Fischer, G. Marks, A. Gogola and A. Schoner at the University of Tennessee. Financial support was provided by the donors of the Petroleum Research Fund, through Grant no. 19731-AC2, and by a University of Tennessee Faculty Research and Development Grant, an unrestricted grant from the Appalachian Basin Industrial Associates consortium, the discretionary fund of the Department of Geological Sciences (University of Tennessee), and various student grants-in-aid from Sigma Xi and the southeastern section of the Geological Society of America. J.L. Coleman, Jr. (Amoco Production Company) arranged for the loan of the Avondale core. Discussions in the field with E. Cotter, K.A. Eriksson, R.H. Dott, Jr. and C.W. Byers were extremely helpful. H.E. Clifton, W.L. Duke and R.D. Kreisa reviewed an earlier part of this manuscript (discussions of the Rockwood Formation), which greatly clarified the senior author's understanding of shelf processes. Colleagues K.R.

Walker, T.W. Broadhead and D.J.P. Swift (Old Dominion University), and two anonymous referees reviewed the entire manuscript, and their comments and criticisms are appreciated.

We have summarized many data from unpublished Master's theses. Although most of these are not always referred to specifically in the text, they are listed as Appendix II to aid investigators seeking more local detail. All are available in the libraries of the University of Tennessee, Knoxville.

## APPENDIX I: OUTCROP LOCATIONS

### Clinch Mountain Belt

*Little Moccasin Gap* (36°47'10"N, 82°5'25"W). Clinch Sandstone exposed in roadcuts along US Highway 19, 2 km north of Holston, VA; Brumley, VA 7½' quad.

*Little War Gap* (36°30'19"N, 83°01'22"W). Clinch Sandstone exposed in roadcuts along Tennessee Highway 70 leading to summit of Clinch Mountain, 17.6 km north of Rogersville, TN; Kyles Ford, TN–VA 7½' quad.

*Flat Gap* (36°25'25"N, 83°13'39"W). Clinch Sandstone exposed in roadcuts along Tennessee Highway 31 leading to summit of Clinch Mountain, 8.16 km north of Mooresburg, TN; Lee Valley, TN 7½' quad.

*Bean Gap* (36°21'04"N, 83°23'53"W). Clinch Sandstone exposed in roadcuts along US Highway 25E leading to summit of Clinch Mountain, 2 km southwest of Thorn Hill, TN; Avondale, TN 7½' quad.

*Amoco (Avondale) Core* (36°18'53"N, 83°27'52"W). Core of Clinch Sandstone spudded at elevation of 1120 ft, and includes 465 ft of Silurian section. Located on Doyal Wynn property off Avondale Springs Road, 1.3 km north of Avondale, TN; Avondale, TN 7½' quad.

*House Mountain* (36°06'45"N, 83°45'00"W). Clinch Sandstone exposed in natural bluffs and along jeep trail traversing east side of House Mountain, 1.5 km west of intersection between Idumea and Hogskin Road; John Sevier, TN 7½' quad.

### Powell Mountain Belt

*Stickleyville* (36°42'40"N, 82°52'30"W). Clinch Sandstone exposed in roadcuts along US Highway 58–421 to the summit of Powell Mountain, 2.7 km east of Stickleyville, VA; Duffield, VA and Stickleyville, VA 7½' quads.

*Mulberry Gap* (36°33'55"N, 83°14'50"W). Clinch Sandstone exposed in roadcuts along Tennessee Highway 758, 4.8 km north of Sneedville, TN; Sneedville, TN 7½' quad.

### Wallen Ridge–Lone Mountain–Whiteoak Mountain Belts

*Interstate 75* (36°09'08"N, 84°03'00"W). Rockwood Formation exposed in roadcuts along east side of I-75, 2.2 km southeast of the Norris, TN exit; Norris, TN 7½' quad.

*Clinton Highway* (36°5'18"N, 84°07'45"W). Rockwood Formation exposed in roadcut along west side of US Highway 25W, 1.6 km south of Clinton, TN; Clinton, TN 7½' quad.

*Melton Hill Lake* (36°02'25"N, 84°12'00"W). Rockwood Formation exposed in roadcuts along west side of Melton Lake Drive, 1 km south of Oak Ridge, TN; Clinton, TN 7½' quad.

*Bacon Ridge* (35°47'20"N, 84°33'23"W). Rockwood Formation exposed in roadcuts along Tennessee Highway 72, 12 km southwest of Kingston, TN; Bacon Gap, 7½' quad.

*Green Gap* (35°07'39"N, 85°01'36"W). Rockwood Formation exposed in roadcuts along I-75 to summit of Whiteoak Mountain, 8 km north of Ooltewah, TN; Snow Hill, TN 7½' quad.

### Cumberland Mountain Belt

*Cumberland Gap* (36°35'35"N, 83°39'46"W). Rockwood Formation exposed in roadcut along US Highway 25E and in abandoned railroad tunnel, 0.5 km south of Cumberland Gap, TN; Middlesboro South, TN-KY-VA 7½' quad.

*Falling Water Creek* (35°11'48"N, 85°14'45"W). Rockwood Formation exposed in roadcuts along north side of Falling Water Creek Road, 19.2 km north of Chattanooga, TN; Daisy, TN 7½' quad.

*Tiftonia* (35°01'03"N, 85°23'10"W). Rockwood Formation exposed in roadcuts along I-24 and behind Holiday Inn motel, 1.2 km west of Tiftonia, TN; Wauhatchie, TN 7½' quad.

### Sequatchie Valley Belt

*Anderson Ridge* (35°02'38"N, 85°35'03"W). Brassfield Formation exposed in roadcuts along

US Highway 41-64-72, 4.8 km east of Jasper, TN; Sequatchie, TN 7½′ quad.

## APPENDIX II

These relevant unpublished theses are available at the University of Tennessee, Knoxville, libraries.

EASTHOUSE, K.A. (1987) *The paleoenvironmental analysis of the Lower Silurian Rockwood Formation at Green Gap, Whiteoak Mountain, Southeastern Tennessee*, 218 pp.
FISCHER, M.W. (1987) *Sedimentology of the Clinch Sandstone (Lower Silurian) along Clinch Mountain, Northeastern Tennessee*, 156 pp.
GOGOLA, A.R. (1990) *Depositional and Diagenetic History of the Basal Transgressive Sandstone of the Clinch and Rockwood Formations (Lower Silurian) in Eastern Tennessee*. M.S. Thesis, University of Tennessee, Knoxville, 147 pp.
MARKS, G.T. (1987) *The lithology and paleoenvironmental interpretation of the Clinch Sandstone at Powell Mountain, Southwestern Virginia*, 263 pp.
SCHONER, A.E. (1985) *Sedimentology and the sedimentary petrography of the Lower Silurian Clinch Sandstone at Beans Gap, Grainger County, Tennessee*, 227 pp.
THOMPSON, J.R. (1984) *Depositional environments of the Sequatchie and Rockwood Formations*, 199 pp.

## REFERENCES

AHMAD, N. (1983) Vertisols. In: *Pedogenesis and Soil Taxonomy, II. The Soil Orders* (Eds Wilding, L.P., Smeck, N.E. & Hall, G.F.) pp. 91–133. Elsevier, New York.
AIGNER, T. (1985) *Storm Depositional Systems*. Springer-Verlag, New York, 174 pp.
AIGNER, T. & REINECK, H.E. (1982) Proximity trends in modern storm sands from the Helgoland Bight (North Sea) and their implications for basin analysis. *Senckenbergiana Maritima* **14**, 183–215.
ALLEN, J.R.L. (1973) Compressional structures (patterned ground) in Devonian pedogenic limestones. *Nature (London) Phys. Sci.* **243**, 84–86.
ALLEN, J.R.L. (1980) Sandwaves: a model of origin and internal structure. *Sediment. Geol.* **26**, 281–388.
ALLEN, J.R.L. (1981a) Lower Cretaceous tides revealed by cross-bedding with mud drapes. *Nature* **289**, 579–581.
ALLEN, J.R.L. (1981b) Paleotidal speeds and ranges estimated from cross-bedding sets with mud drapes. *Nature* **293**, 394–396.
ALLEN, J.R.L. (1982) Mud drapes in sand wave deposits: a physical model with application to the Folkestone Beds (early Cretaceous, southeast England). *Philos. Trans. R. Soc. London, Ser. A* **306**, 291–345.
AMSDEN, T.W. (1955) Lithofacies map of the Lower Silurian deposits in central and eastern United States and Canada. *Bull., Am. Assoc. Petrol. Geol.* **39**, 60–74.

BERNER, R.A. (1971) *Principles of Chemical Sedimentology*. McGraw-Hill, New York, 240 pp.
BERRY, W.B.N. & BOUCOT, A.J. (1970) Correlation of the North American Silurian rocks. *Geol. Soc. Am. Spec. Pap.* **102**, 289 pp.
BERRY, W.B.N. & BOUCOT, A.J. (1973) Glacio-eustatic control of Late Ordovician–Early Silurian platform sedimentation and faunal changes. *Geol. Soc. Am. Bull.* **84**, 275–284.
BLAKEY, R.C. (1984) Marine sand-wave complex in the Permian of central Arizona. *J. Sediment. Petrol.* **54**, 29–51.
BLODGETT, R.H. (1985) Palaeovertisols as indicators of climate (abst.). *Bull., Am. Assoc. Petrol. Geol.* **69**, 239 pp.
BRENCHLEY, P.J. (1985) Storm influenced sandstone beds. *Mod. Geol.* **9**, 369–396.
BRENCHLEY, P.J. & NEWALL, G. (1982) Storm influenced inner-shelf sand lobes in the Caradoc (Ordovician) of Shropshire, England. *J. Sediment. Petrol.* **52**, 1257–1269.
BYERS, C.W. (1974) Shale fissility: relation to bioturbations. *Sedimentology* **21**, 479–484.
BYERS, C.W. (1977) Biofacies patterns in euxinic basins: a general model. In: *Deep-water Carbonate Environments* (Eds Cook, H.E. & Enos, P.). Soc. Econ. Paleontol. Mineral. Spec. Publ. 25, 5–17.
CASTON, V.N.D. (1972) Linear sand banks in the southern North Sea. *Sedimentology* **18**, 63–78.
CHAMBERLAIN, C.K. (1978) Recognition of trace fossils in cores. In: *Trace Fossil Concepts* (Ed. Basan, P.B.). Soc. Econ. Paleontol. Mineral. Short Course Notes 5, 133–183.
CLUFF, R.M. (1980) Paleoenvironment of the New Albany Shale Group (Devonian–Mississippian) of Illinois. *J. Sediment. Petrol.* **50**, 767–780.
COTTER, E. (1982) Tuscarora Formation of Pennsylvania. *Fieldtrip Guidebook*. Society of Economic Paleontologists and Mineralogists, Eastern Section, 105 pp.
COTTER, E. (1983a) Shelf, paralic, and fluvial environments and eustatic sea-level fluctuations in the origin of the Tuscarora Formation (Lower Silurian) of central Pennsylvania. *J. Sediment. Petrol.* **53**, 25–49.
COTTER, E. (1983b) Silurian depositional history In: *Silurian Depositional History and Alleghenian Deformation in the Pennsylvania Valley and Ridge (Guidebook)*. 48th Annual Field Conference of Pennsylanian Geologists, Harrisburg, PA, pp. 3–28.
CRIMES, T.P. (1975) The stratigraphical significance of trace fossils. In: *The Study of Trace Fossils* (Ed. Frey, R.W.) pp. 109–130. Springer-Verlag, New York.
CURRAY, J.R. (1956) Analysis of two dimensional orientation data. *J. Geol.* **64**, 117–131.
DENNISON, J.M. & BOUCOT, A.J. (1974) Reference section for Silurian Clinch Sandstone and most nearly complete Devonian section in eastern Tennesee. *Southeastern Geol.* **16**, 79–101.
DICKINSON, W.R. & SUCZEK, C.A. (1979) Plate tectonics and sandstone composition. *Bull., Am. Assoc. Petrol. Geol.* **63**, 2164–2182.
DOTT, R.H. & BOURGEOIS, J. (1982) Hummocky stratification: significance of its variable bedding

sequences. *Geol. Soc. Am. Bull.* **93**, 663–680.

DRIESE, S.G. (1985) Marine shelf storm deposits, Rockwood Formation (Silurian), east Tennessee. *Geol. Soc. Am. Abstr. Prog.* **17**, 89.

DRIESE, S.G. (1986a) Fairweather and storm shelf sequences, Rockwood Formation (Silurian), east Tennessee. *West Virginia Univ. Appalachian Basin Indus. Assoc. Prog.* **10**, 133–175.

DRIESE, S.G. (1986b) Fairweather and storm shelf sequences, Rockwood Formation (Silurian), east Tennessee. *Geol. Soc. Am. Abstr. Prog.* **18**, 218–219.

DRIESE, S.G. (1987) An analysis of large-scale ebb-dominated tidal bedforms: evidence for tidal bundles in the Lower Silurian Clinch Sandstone of east Tennessee. *Southeastern Geol.* **27**, 121–140.

DRIESE, S.G. & FOREMAN, J.L. (1991) Paleopedology and paleoclimate implications of Late Ordovician vertic paleosols, southern Appalachians. *J. Sediment. Petrol.* **62**, in press.

DRIESE, S.G., SCHONER, A.E. & MCDONALD, C.J. (1984) Sedimentology of the Poor Valley Ridge Member of the Clinch Sandstone (Lower Silurian), northeastern Tennessee and southwestern Virginia. *West Virginia Univ. Appalachian Basin Indus. Assoc. Prog.* **6**, 47–75.

DRIESE, S.G., FISCHER, M.W. & EASTHOUSE, K.A. (1986a) Silurian siliciclastic and carbonate storm-shelf sequences, east Tennessee. *Soc. Econ. Paleontol. Mineral. Ann. Midyear Meet. Abstr.* **III**, 31.

DRIESE, S.G., FISCHER, M.W., EASTHOUSE, K.A. & SCHONER, A.E. (1986b) Tidal and storm wave deposition in Lower Silurian shelf sequences, east Tennessee. *Bull., Am. Assoc. Petrol. Geol.* **70**, 583–584.

DUKE, W.L. (1982) The 'type locality' of hummocky cross-stratification: the storm-dominated Silurian Medina Formation in the Niagara Gorge, New York and Ontario. *Proc. Ontario. Petrol. Inst.* **21**, 2.1–2.31.

DUKE, W.L. (1985) Hummocky cross-stratification, tropical hurricanes, and intense winter storms. *Sedimentology* **32**, 167–194.

EASTHOUSE, K.A. (1987) *The paleoenvironmental analysis of the Lower Silurian Rockwood Formation at Green Gap, White Oak Mountain, southeastern Tennessee.* MS Thesis, University of Tennessee, Knoxville, 218 pp.

EASTHOUSE, K.A. & DRIESE, S.G. (1988) Paleobathymetry of a Silurian shelf system: application of proximality trends and trace fossil distributions. *Palaios* **3**, 473–486.

EKDALE, A.A., BROMLEY, R.G. & PEMBERTON, S.G. (1984) Ichnology: the use of trace fossils in sedimentology and stratigraphy. *Soc. Econ. Paleontol. Mineral. Short Course Notes* **15**, 317 pp.

FISCHER, M.W. (1987) *Sedimentology of the Clinch Sandstone (Lower Silurian) along Clinch Mountain, northeastern Tennessee.* MS Thesis, University of Tennessee, Knoxville, 156 pp.

FOLK, R.L. (1960) Petrography and origin of the Tuscarora, Rose Hill, and Keefer Formations, Lower and Middle Silurian of eastern West Virginia. *J. Sediment. Petrol.* **30**, 1–58.

FREY, R.W. (1975) The realm of ichnology, its strengths and limitations. In: *The Study of Trace Fossils* (Ed. Frey, R.W.) pp. 13–38. Springer-Verlag, New York.

FULLER, A.D. (1979) Phosphate occurrences on the western and southern coastal areas and continental shelves of Southern Africa. *Econ. Geol.* **74**, 221–231.

GHAZIZADEH, M. (1985) Depositional environment interpretation of the Juniata Formation at Beans Gap, Clinch Mountain, Tennessee: In: *The Geologic History of the Thorn Hill Paleozoic Section (Cambrian–Mississippian), Eastern Tennessee* (Ed. Walker, K.R.) pp. 93–99. Studies in Geology No. 10, University of Tennessee.

GOGOLA, A.R. (1990) *Depositional and Diagenetic History of the Basal Transgressive Sandstone of the Clinch and Rockwood Formations (Lower Silurian) in Eastern Tennessee.* M.S. Thesis, University of Tennessee, Knoxville, 147 pp.

GOLDBERY, R. (1982) Structural analysis of soil microrelief in paleosols of the Lower Jurassic 'Laterite Derivative Facies' (Mishhor and Ardon Formations) Makhtesh Ramon, Israel. *Sediment. Geol.* **31**, 119–140.

GRAY, M.B. (1984) Slickensided soil fractures—indicators of strain and deformation processes in the Bloomsburg Formation, central Pennsylvania. *Geol. Soc. Am. Abstr. Prog.* **16**, 19.

GULBRANDSEN, R.A. (1969) Physical and chemical factors in the formation of marine apatite. *Econ. Geol.* **69**, 365–382.

HAMBLIN, A.P. & WALKER, R.G. (1979) Storm dominated shallow marine deposits: The Fernie–Kootenay (Jurassic) transition, southern Rocky Mountains. *Can. J. Earth Sci.* **16**, 1673–1690.

HANDFORD, C.R. (1986) Facies and bedding sequences in shelf-storm-deposited carbonates—Fayetteville Shale and Pitakin Limestone (Mississippian), Arkansas. *J. Sediment. Petrol.* **56**, 123–137.

HARMS, J.C., SOUTHARD, J.B., SPEARING, D.R. & WALKER, R.G. (1975) Depositional environments as interpreted from primary sedimentary structures and stratification sequences. *Soc. Econ. Paleontol. Mineral. Short Course Notes.* **2**, 161 pp.

HATCHER, R.D., JR. (1972) Developmental model for the southern Appalachians. *Geol. Soc. Am. Bull.*, **83**, 2735–2760.

HATCHER, R.D., JR. (1978) Tectonics of the western Piedmont and Blue Ridge, southern Appalachians: review and speculation. *Am. J. Sci.* **278**, 276–304.

HAYES, A.W. (1974) *Origin of the Tuscarora Formation (Lower Silurian), southwestern Virginia.* PhD dissertation, Virginia Polytechnic Institute and State University, 162 pp.

HOWARD, J.D. (1978) Sedimentology and trace fossils. In: *Trace Fossil Concepts* (Ed. Basan, P.B.). Soc. Econ. Paleontol. Mineral. Short Course Notes 5, 13–47.

HUNTER, R.E. (1970) Facies of iron sedimentation in Clinton Group, In: *Studies of Appalachian Geology, Central and Southern* (Eds Fisher, G.W., Pettijohn, F.J., Reed, J.C. & Weaver, K.N.) pp. 101–121. Wiley-Interscience, New York.

HUNTER, R.E. & CLIFTON, H.E. (1982) Cyclic deposits and hummocky cross-stratification of probable storm origin in Upper Cretaceous rocks of Cape Sebastian area, southwestern Oregon. *J. Sediment. Petrol.* **52**, 127–144.

JOHNSON, H.D. (1979) Shallow siliciclastic seas. In: *Sedimentary Environments and Facies* (Ed. Reading, H.G.). Elsevier, New York, 207–258.

JOHNSON, M.E. (1987) Extent and bathymetry of North American platform seas in the Early Silurian. *Paleoceanography* **2**, 185–211.

JOHNSON, M.E., RONG, J.-Y. & YANG, X.-C. (1985) Intercontinental correlation by sea-level events in the Early Silurian of North America and China (Yangtze Platform). *Geol. Soc. Am. Bull.* **96**, 1384–1397.

KELLING, G. & MULLIN, P.R. (1975) Graded limestones and limestone–quartzite couplets: possible storm deposits from the Moroccan Carboniferous. *Sediment. Geol.* **13**, 161–190.

KENYON, N.H., BELDERSON, R.H., STRIDE, A.H. & JOHNSON, M.A. (1981) Offshore tidal sand-banks as indicators of net sand transport and as potential deposits. In: *Holocene Marine Sedimentation in the North Sea Basin* (Eds Nio, S.-D., Shuttenhelm, R.T.E. & Van Weering, Tj. C.E.). Spec. Publ. Int. Assoc. Sediment. **5**, 257–268.

KLEIN, G.D.V. (1977) *Clastic Tidal Facies.* Continuing Education Publication Company, Champaign, IL. 149 pp.

KNIGHT, M.J. (1980) Structural analysis and mechanical origins of gilgai at Boorock, Victoria, Australia. *Geoderma* **23**, 245–283.

KREISA, R.D. (1981) Storm generated sedimentary structures in subtidal marine facies with examples from Middle and Upper Ordovician of southwestern Virginia. *J. Sediment. Petrol.* **51**, 823–848.

KREISA, R.D. (1986) Tidal sand wave facies, Rancho Rojo Sandstone (Permian) Arizona: In: *Shelf Sands and Sandstones* (Eds Knight, R.J. & McLean, J.R.). Can. Soc. Petrol. Geol. Mem. **11**, 277–291.

LAMPIRIS, N. (1975) *Stratigraphy of the clastic Silurian rocks of central western Virginia and adjacent West Virginia.* PhD dissertation, Virginia Polytechnic Institute and State University, 85 pp.

LEVELL, B.K. (1980) Evidence for currents associated with waves in Late Precambrian shelf deposits from Finnmark, North Norway. *Sedimentology* **26**, 153–166.

MAHER, L.J. & DOTT, R.H., JR. (1979) *Program for analysis of directional orientation data (for generalized basic).* University of Wisconsin, Madison, unpublished computer program.

MARKS, G.T. (1987) *The lithology and paleoenvironmental interpretation of the Clinch Sandstone at Powell Mountain, southwest Virginia.* MS Thesis, University of Tennessee, Knoxville, 263 pp.

McCAVE, I.N. (1970) Deposition of fine-grained suspended sediment from tidal currents. *J. Geophys. Res.* **75**, 4151–4159.

McCAVE, I.N. (1971) Wave effectiveness at the sea bed and its relationship to bed-forms and deposition of mud. *J. Sediment. Petrol.* **41**, 89–96.

McKERROW, W.S. (1979) Ordovician and Silurian changes in sea level. *J. Geol. Soc. London* **136**, 137–145.

MECKEL, L.D. (1970) Paleozoic alluvial deposition in the central Appalachians: a summary. In: *Studies of Appalachian Geology, Central and Southern* (Eds Fisher, G.W., Pettijohn, F.J., Reed, J.C. & Weaver, K.N.) pp. 49–67. Wiley-Interscience, New York.

MILICI, R.C. & WEDOW, H. (1977) Upper Ordovician and Silurian stratigraphy in Sequatchie Valley and parts of the adjacent Valley and Ridge. *U.S. Geol. Surv. Prof. Pap.* **996**, 38 pp.

MILLER, R.L. (1976) Silurian nomenclature and correlations in southwest Virginia and northeast Tennessee. *U.S. Geol. Surv. Bull.* **1405-H**, 23 pp.

MOUNT, J.F. (1982) Storm-surge-ebb origin of hummocky cross-stratified units of the Andrews Mountain Member, Campito Formation (Lower Cambrian), White-Inyo Mountains, Eastern California. *J. Sediment. Petrol.* **52**(3), 951–958.

NARAYAN, J. (1971) Sedimentary structures in the Lower Greensand of the Weald, England, and Bas-Boulonnais, France. *Sediment. Geol.* **6**, 73–109.

NIO, S.D. (1976) Marine transgressions as a factor in the formation of sandwave complexes. *Geol. Mijnbouw* **55**, 18–40.

ODIN, G.S. & MATTER, A. (1981) De glauconarium origine. *Sedimentology* **28**, 611–641.

POTTER, P.E. & PETTIJOHN, F.J. (1963) *Paleocurrents and Basin Analysis.* Academic Press, New York, 296 pp.

QUINLAN, G.M. & BEAUMONT, C. (1984) Appalachian thrusting, lithospheric flexure, and the Paleozoic stratigraphy of the Eastern Interior of North America. *Can. J. Earth Sci.* **21**, 973–996.

RAGAN, D.M. (1973) *Structural Geology: an Introduction to Geometrical Techniques* (2nd edn). Wiley, New York, 208 pp.

REINECK, H.E. & SINGH, I.B. (1980) *Depositional Sedimentary Environments* (2nd edn). Springer-Verlag, New York, 549 pp.

RHOADS, D.C. (1967) Biogenic reworking of intertidal and subtidal sediments in Barnstable Harbor and Buzzards Bay, Massachusetts. *J. Geol.* **75**, 461–476.

RHOADS, D.C. (1975) The paleoecological and environmental significance of trace fossils. In: *The Study of Trace Fossils* (Ed. Frey, R.W.) pp. 147–160. Springer-Verlag, New York.

RHOADS, D.C. & MORSE, J.W. (1971) Evolutionary and environmental significance of oxygen-deficient marine basins. *Lethaia* **4**, 413–428.

RIGGS, S.R. (1979) Petrology of the Tertiary phosphorite system of Florida. *Econ. Geol.* **74**, 195–220.

RODGERS, J. (1953) Geologic map of east Tennessee with explanatory text. *Tenn. Div. Geol. Bull.* **58**(2), 168 pp.

RODGER, J. (1971) The Taconic orogeny. *Geol. Soc. Am. Bull.* **82**, 1141–1178.

ROEDER, D. & WITHERSPOON, W.D. (1978) Palinspastic map of East Tennessee. *Am. J. Sci.* **278**, 543–550.

RUBIN, D.M. & McCULLOCH, D.S. (1980) Single and superimposed bedforms: a synthesis of San Francisco Bay and flume observations. *Sediment. Geol.* **26**, 207–231.

SAVRDA, C.E. & BOTTJER, A.J. (1986) Trace-fossil model for reconstruction of paleo-oxygenation in bottom waters. *Geology* **14**, 3–6.

SAVRDA, C.E., BOTTJER, D.J. & GORSLINE, D.S. (1984) Development of a comprehensive oxygen-deficient marine biofacies model: evidence from Santa Monica, San Pedro, and Santa Barbara Basins, California continental borderland. *Bull., Am. Assoc. Petrol. Geol.* **68**, 1179–1192.

SCHONER, A.E. (1985a) *The sedimentology and sedimentary petrology of the Lower Silurian Clinch Sandstone at Beans Gap, Grainger County, Tennessee.* MS thesis, University of Tennessee, Knoxville, 227 pp.

SCHONER, A.E. (1985b) The Clinch Sandstone (Lower Silurian) along Highway 25E at Beans Gap. In: *The Geologic History of the Thorn Hill Paleozoic Section (Cambrian–Mississippian), Eastern Tennessee* (Ed.

Walker, K.R.) pp. 100–110. Studies in Geology No. 10, University of Tennessee, Knoxville, TN.

SCHONER, A.E. & DRIESE, S.G. (1985) Sedimentology of the Clinch Sandstone at Beans Gap, Grainger County, Tennessee. *Geol. Soc. Am. Abstr. Prog.* **17**, 134.

SEILACHER, A. (1978) Use of trace fossils for recognizing depositional environments. In: *Trace Fossil Concepts* (Ed. Basan, P.B.). Soc. Econ. Paleontol. Mineral. Short Course Notes 5, 185–201.

SIBLEY, D.F. & BLATT, H. (1976) Intergranular pressure solution and cementation of the Tuscarora orthoquartzite. *J. Sedim. Petrol.* **46**, 881–896.

SMITH, N.D. (1970) The braided stream depositional environment: comparison of the Platte River with some Silurian clastic rocks, north-central Appalachians. *Geol. Soc. Am. Bull.* **81**, 2993–3014.

SOEGAARD, K. & ERIKSSON, K.A. (1985) Evidence of tide, storm and wave interaction on a Precambrian siliciclastic shelf: the 1700 m.y. Ortega Group, New Mexico. *J. Sediment. Petrol.* **55**, 672–684.

STRIDE, A.H. BELDERSON, R.H., KENYON, N.H. & JOHNSON, M.A. (1982) Offshore tidal deposits: sand sheet and sand bank facies. In: *Offshore Tidal Sands* (Ed. Stride, A.H.) pp. 95–125. Chapman & Hall, London.

SWARTZ, F.M. (1934) Silurian sections near Mount Union, central Pennsylvania. *Geol. Soc. Am. Bull.* **45**, 18–134.

SWIFT, D.J.P. (1968) Coastal erosion and transgressive stratigraphy. *J. Geol.* **76**, 444–456.

SWIFT, D.J.P. (1975) Barrier-island genesis: evidence from the central Atlantic Shelf, eastern U.S.A.. *Sediment. Geol.* **14**, 1–43.

SWIFT, D.J.P. (1985) Response of the shelf floor to flow. In: *Shelf Sands and Sandstone Reservoirs* (Eds Tillman, R.W., Swift, D.J.P. & Walker, R.G.). Soc. Econ. Paleontol. Mineral. Short Course Notes 13, 135–241.

SWIFT, D.J.P. & NIEDORODA, R.J. (1985) Fluid and sediment dynamics on continental shelves. In: *Shelf Sands and Sandstone Reservoirs* (Eds Tillman, R.W., Swift, D.J.P. & Walker, R.G.). Soc. Econ. Paleontol. Mineral. Short Course Notes 13, 47–133.

SWIFT, D.J.P., STANLEY, D.J. & CURRAY, J.R. (1971) Relict sediments on continental shelves: a reconsideration. *J. Geol.* **79**, 322–346.

SWIFT, D.J.P. LUDWICK, J.C. & BOEHMER, W.R. (1972) Shelf sediment transport, a probability model. In: *Shelf Sediment Transport: Process and Pattern* (Eds Swift, D.J.P., Duane, D.B. & Pilkey, O.H.) pp. 195–223. Dowden Hutchinson & Ross, Stroudsburg, PA.

SWIFT, D.J.P., YOUNG, R.A., CLARKE, T.L., VINCENT, C.E., NIEDORODA, A. & LESHT, B. (1981) Sediment transport in the Middle Atlantic Bight of North America: synopsis of recent observations. In: *Holocene Marine Sedimentation in the North Sea Basin* (Eds. Nio, S.-D., Shuttenhelm, R.T.E. & van Weering, Tj. C.E.). Spec. Publ. Int. Assoc. Sediment. 5, 361–383.

SWIFT, D.J.P., FIGUEIREDO, A.G., FREELAND, G.J. & OERTEL, G.F. (1983) Hummocky cross-stratification and megaripples: a geological double standard. *J. Sediment. Petrol.* **53**, 1295–1317.

THOMPSON, A.M. (1970) Tidal-flat deposition and early dolomitization in Upper Ordovician rocks of the southern Appalachian Valley and Ridge. *J. Sediment. Petrol.* **40**, 1271–1286.

THOMPSON, J.R. (1984) *Depositional environments of the Sequatchie and Rockwood Formations*. MS thesis, University of Tennessee, Knoxville, 199 pp.

WALKER, K.R. & DIEHL, W.W. (1986) The effect of synsedimentary substrate modification on the composition of paleocommunities: paleoecologic succession revisited. *Palaios* **1**, 65–74.

WALKER, R.G. (Ed.) (1984) Shelf and shallow marine sands. In: *Facies Models*, 2nd edn. Geosci. Can. Reprint Ser. 1, 141–170.

WALKER, R.G. (1985) Geological evidence for storm transportation and deposition on ancient shelves; In: *Shelf Sands and Sandstone Reservoirs* (Eds Tillman, R.W., Swift, D.J.P. & Walker, R.G.). Soc. Econ. Paleontol. Mineral. Short Course Notes 13, 243–302.

WALKER, R.G., DUKE, W.L. & LECKIE, D.A. (1983) Hummocky stratification: significance of its variable bedding sequences: discussion and reply. *Geol. Soc. Am. Bull.*, **94**, 1245–1251.

WATTS, N.L. (1977) Pseudo anticlines and other structures in some calcretes of Botswana and South Africa. *Earth Sci. Processes* **2**, 63–74.

WHISONANT, R.C. (1977) Lower Silurian Tuscarora (Clinch) dispersal patterns in western Virginia. *Geol. Soc. Am. Bull.* **88**, 215–220.

WHITTAKER, J.H.M. (1973) 'Gutter Casts', a new name for scour-and-fill structures: with examples from the Llandoverian of Ringerike and Malmoya, Southern Norway. *Norsk Geol. Tidsskr.* **53**, 403–417.

WILLIAMS, H. & HATCHER, R.D., JR. (1982) Suspect terranes and accretionary history of the Appalachian orogen. *Geology* **10**, 530–536.

WILLIAMS, H. & HATCHER, R.D., JR. (1983) Appalachian suspect terranes, In: *Contributions to the Tectonics and Geophysics of Mountain Chains* (Eds Hatcher, R.D. Jr., Williams, H. & Zietz, I.). Geol. Soc. Am. Mem. **158**, 33–53.

YAALON, D.H. & KALMAR, D. (1978) Dynamics of cracking and swelling clay soils: displacement of skeletal grains, optimum depth of slickensides, and rate of intrapedonic turbation. *Earth Surf. Processes* **3**, 31–42.

YANG, C.S. & NIO, S.D. (1985) The estimation of palaeohydrodynamic processes from subtidal deposits using time series analysis methods. *Sedimentology* **32**, 41–57.

YEAKEL, L.S. JR. (1962) Tuscarora, Juniata and Bald Eagle paleocurrents and paleogeography in the central Appalachians. *Geol. Soc. Am. Bull.* **73**, 1515–1550.

ZIEGLER, A.M., SCOTESE, C.R. MCKERROW, W.S., JOHNSON, M.E. & BAMBACH, R.K. (1979) Paleozoic paleogeography. *Ann. Rev. Earth Planet. Sci.* **7**, 473–502.

ZIEGLER, A.M., HANSEN, K.S., JOHNSON, M.E., KELLY, M.A., SCOTESE, C.R. & VAN DER VOO, R. (1977) Silurian continental distributions, paleogeography, climatology, and biogeography. In: *The Past Distribution of Continents* (Ed. McElhinney, M.W.). Tectonophysics **40**, 13–51.

# Prograding shoreline deposits in the Lower Silurian Medina Group, Ontario and New York: storm- and tide-influenced sedimentation in a shallow epicontinental sea, and the origin of enigmatic shore-normal channels encapsulated by open shallow-marine deposits

W.L. DUKE, P.J. FAWCETT *and* W.C. BRUSSE[1]

*Department of Geosciences, The Pennsylvania State University, University Park, PA 16802, USA*

## ABSTRACT

We have measured 42 exposures of the Lower Silurian Medina Group in New York and Ontario. A thickening and coarsening upward shoaling sequence, 15–25 m thick, occupies most of the upper half of this unit. The lower part of this sequence is dominated by interbedded mudstones and thin wave-rippled and hummocky cross-stratified very fine to fine-grained sandstones. The upper part contains a thick, trough cross-bedded, sheet-like body of fine to medium-grained sandstone. Palaeoflow data, trace and body fossils, and details of stratification strongly suggest that this sequence represents a vertical transition from subtidal, storm-dominated, open epicontinental sea deposits to tide-dominated, wave-influenced shoreline deposits. Offshore bathymetric contours and the straight shoreline were oriented NNE–SSW throughout the study area. The shoreline was an intertidal sand flat dissected by nearly shore-normal estuaries and tidal channels trending NW–SE. These channels were slightly deflected from a shore-normal orientation by strong wave-induced longshore drift directed to the north-northeast along the shoreline.

Superimposed upon the larger, overall shoaling sequence are several smaller coarsening upward sequences, each 1–12 m thick, separated by erosional disconformities (locally, non-depositional paraconformities).

The lateral extent of these small-scale sequences is variable, but most can be traced throughout the study area. Locally, their boundaries are quite subtle; recognition commonly requires very detailed inspection of the outcrop. We infer that these two different scales of superimposed depositional cyclicity were generated by episodic aggradation and shoaling during the episodic west-northwestward progradation of the shoreline. These episodes of progradation were interrupted by rises in relative sea-level, forming the para- and disconformable deepening surfaces between the smaller shoaling sequences.

Large sinuous channels oriented nearly normal to shore (NW–SE) are associated with the erosional boundaries of the small sequences. Typically, these sand-filled channels are both incised into and overlain by mudstones and thin hummocky cross-stratified sandstones deposited offshore in an open, subtidal, storm-dominated setting; however, the presence of the channels just below sequence boundaries places them in the shallowest portion of the smaller sequences. Fossil evidence suggests a restricted marine environment for the fill of these channels. Palaeoflow data reflect reversing flows in these channels; we suggest that they were cut and filled by tidal currents in a shallow subtidal to intertidal setting, during the maximum regressive phase of each minor depositional cycle.

---

[1] Present address: Converse Environmental East, 91 Roseland Ave., Caldwell, NJ 07006, USA.

## INTRODUCTION

The Lower Silurian Medina Group is well exposed in western New York, USA, and southern Ontario, Canada (Fig. 1). In this area, the Medina is composed predominantly of terrigenous mudstones and thin wave-rippled, flat-laminated, and hummocky cross-stratified very fine to fine-grained quartzitic sandstones. These deposits exhibit abundant trace and body fossils from an open, shallow marine environment; we infer offshore, subtidal deposition in a relatively shallow, storm-dominated epicontinental sea. Of particular interest in these rocks are numerous large channels incised decimetres to metres deep into these open marine sediments (e.g. Figs 2, 3). The fill of these channels consists predominantly of fine to medium-grained, trough cross-bedded quartzitic sandstones. Trace and body fossils indicate that these channel deposits also are marine in origin. In many instances, the channels are fully encapsulated by the subtidal, offshore mudstones and storm-deposited sandstones. The origin of channels incised into offshore sediments is, in general, rather enigmatic, both in modern environments and ancient deposits. Thus, determining the origin of these Medina channels has been a major focus of our work. This paper summarizes some results of two ongoing M.S. studies by the junior authors; preliminary findings have been reported previously by Brusse *et al.* (1987, 1988), Duke (1987b), Duke & Brusse (1978), Duke & Fawcett (1987), and Fawcett *et al.* (1988).

For this study, we have measured in detail some 42 Medina exposures along the E–W trending Niagara Escarpment in Ontario and New York (Fig. 1). Roughly half of these exposures are complete or nearly complete sections through the group. The average spacing between these structurally continuous, flat-lying exposures is less than 10 km, permitting confident correlation of lithic units in

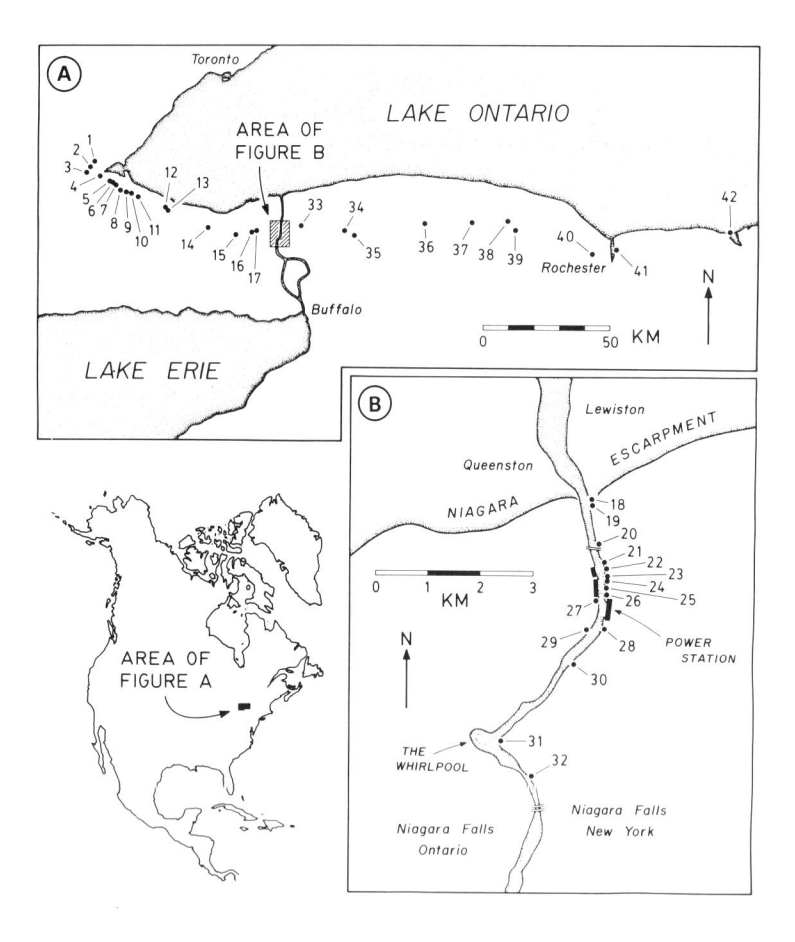

**Fig. 1.** Location maps of exposures measured for this study. (A) Exposures along the Niagara Escarpment. The northward-flowing Niagara River, between exposures 17 and 33, locally forms the border between Canada and the USA. Also indicated are the cities of Toronto (Ontario), Buffalo and Rochester, (New York). (B) Exposures measured within the Niagara Gorge, below the falls. Note locations of bridges and hydroelectric plants. Also indicated are the villages of Queenston (Ontario) and Lewiston (New York), and the twin cities of Niagara Falls.

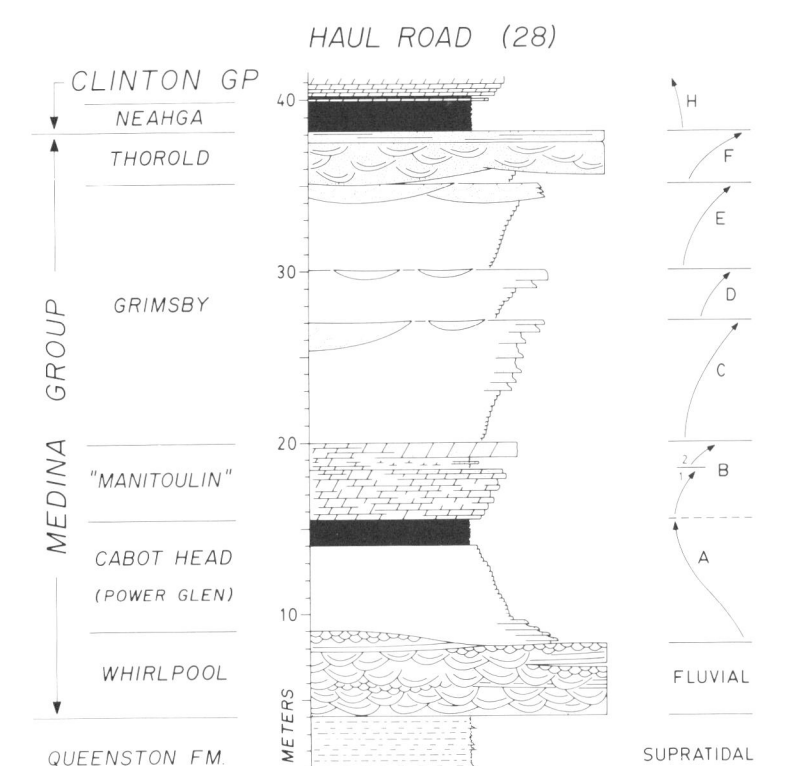

**Fig. 2.** Section measured at Haul Road in the Niagara Gorge (exposure 28, see Fig. 1). Facies symbols are identified in Fig. 4. Lettered arrows at right indicate shallow marine depositional sequences recognized in this study; note that sequence G is absent at this locality. Existing lithostratigraphically defined formations of the Medina Group also are indicated.

**Fig. 3.** A large sandstone-filled channel (outlined by arrowheads) within the overall thickening and coarsening upward Grimsby–Thorold transition at Ball's Falls (exposure 14). For scale, the Thorold sheet sandstone is about 1.5 m thick where indicated on the photograph. Gentle erosional relief is evident at the base of this sheet sandstone, which cuts down gradually to the right. View is to the west-northwest; note inclined lateral accretion surfaces (dipping to the northeast) within both the channel and the basal half of the Thorold. Also note interbedded carbonates and shales of the overlying Clinton Group.

these deposits. Our sections include 15 very closely spaced exposures measured in the nearly continuous outcrops in the Niagara Gorge (Fig. 1B). These superlative natural exposures are among the more extensive in the northeast USA. The N-S trend of the gorge permits examination of lateral facies relationships normal to the dominant outcrop trend, on a scale intermediate between typical surface exposures and typical subsurface drill-core spacings.

## STRATIGRAPHY OF THE MEDINA

The Medina in outcrop has been the focus of a long list of studies by a series of distinguished investigators, including G.K. Gilbert (1899), A.W. Grabau (1901, 1905, 1908, 1909, 1913), E.M. Kindle (1914; Kindle & Taylor, 1913), C. Schuchert (1914), M.Y. Williams (1919), H.L. Fairchild (1901) and G.H. Chadwick (1918, 1935). The stratigraphy of this unit was refined by several additional workers between 1940 and 1975 (see review by Middleton et al., 1987). The first modern sedimentological study of the Medina in outcrop was conducted by I.P. Martini in the mid-1960s (Martini, 1966). More recent studies of the Medina have extended into the structurally continuous subsurface of Ohio (Pees, 1986), and New York and Pennsylvania (Piotrowski, 1981; Laughrey, 1984; Metzger, 1982), where the group grades laterally into the temporally equivalent and physically contiguous strata of the Tuscarora Formation. All available exposures of the basal Whirlpool Sandstone of the Medina have been re-examined recently by G.V. Middleton and colleagues (Middleton et al., 1983, 1987; Middleton & Rutka, 1986).

Despite this long history of interest in the Medina, no consensus has been achieved yet concerning the formal stratigraphy of the unit. The hierarchical rank, boundaries and internal subdivision of the Medina are all subject to debate. Some workers have preferred to rank the Medina as group (e.g., Fisher, 1954; Laughrey, 1984; Brett & Calkin, 1987), whereas others have assigned formational status (e.g., Martini, 1966; Friedman et al., 1982; Duke, 1987b). For the purposes of this paper, we will adhere to the currently accepted formal rank, referring to the Medina as a group.

The formally defined formations of the Medina in the Niagara area are shown in Fig. 2. Note that some workers (e.g. Brett & Calkin, 1987) have preferred to place the Thorold Sandstone with the overlying Clinton shales and carbonates. We prefer to keep the Thorold with the underlying, drawing the Medina–Clinton boundary at the contact between coarse terrigenous sandstones below and grey shales and carbonates above; this division seems to us to be both lithostratigraphically natural and sedimentologically preferable. In contrast, the basal contact of the Medina is not contested; it is drawn at the sharp contact with the inferred supratidal shales and thin, sparsely fossiliferous siltstones and sandstones of the Ordovician Queenston Formation (Middleton, 1987; Middleton et al., 1987).

Wherever one chooses to place the upper boundary of the Medina, the lithostratigraphic formations in this unit (Fig. 2) are the subject of great interest. These formal units are defined largely or entirely by the pronounced colour variation in the Medina. The colours involved (red, white or light grey to green, and dark grey or greenish grey to black) are controlled by variation in the oxidation state of iron-rich cements and the relative abundance of organic matter and finely disseminated pyrite in these rocks. As discussed by Duke (1987b) and Fawcett et al. (1988), much of the colour variation in the Medina is clearly diagenetic in origin, resulting from post-depositional reduction of certain strata by the circulation of low-Eh interstitial fluids. To a large degree, therefore, the existing formations reflect patterns of secondary diagenetic reduction of strata, which originally were oxidized when deposited. The inadequacy of colour variation in the definition of stratigraphic subdivisions of the Medina has been recognized previously by Chadwick (1935) and Armstrong (1953, 1955).

In particular, the Thorold, Grimsby, Cabot Head and Power Glen Formations, as they currently are defined (Fig. 2), do *not* represent genetically significant subdivisions based upon primary lithic attributes of the rocks; in fact, these formations each include a wide range of lithologies, and the colour-based formation boundaries commonly cut across traceable lithic units. In this paper, therefore, we have not utilized the existing formation designations and correlations (which would lead to undesirable complexity and confusion), but have instead established informal correlations based solely upon primary lithic properties. The use of the existing formation names has been avoided herein wherever possible. A formal revision of the stratigraphy of the Medina is beyond the scope of the present text, but will be presented by us elsewhere.

**Table 1.** Summary of recent interpretations of depositional environments in the Medina

*Duke (1982,1985a): outcrop, Ontario and Niagara Gorge*
The basal sandstone (Whirlpool) was not examined in this study. Overlying the Whirlpool, a transgressive sequence (Cabot Head) was interpreted as sandy storm deposits in a shallow mud-dominated sea. The overlying large-scale coarsening upward sequence (Grimsby–Thorold) was interpreted as a prograding shoreline sequence. The inferred shoreline was straight, with a NNE–SSW orientation; progradation was to the west-northwest. Smaller superimposed coarsening upward sequences were interpreted as storm-generated shelf sand ridges on the aggrading basin floor. Channel deposits were interpreted as shore-normal shelf channels incised through the ridges by storm-generated bottom return flows.

*Laughrey (1984): subsurface, northwest Pennsylvania*
The basal sand (Whirlpool), interpreted as a transgressive, sublittoral sheet sand, passes upward to deeper tide and storm-dominated offshore sand bars formed in a shallow mud-dominated sea (Cabot Head). The sands of the upper Medina (Grimsby) exhibit 'barrier bars' with tidal inlets and deltas passing upward to sandy beach ridge, tidal plain, and braided fluvial deposits, sharply overlain by muddy tidal-flat and lagoonal deposits (uppermost Grimsby).

*Leggitt (1985): outcrop, Ontario*
The major dolomitic interval in the Medina (the 'Manitoulin' Member) was shown to contain sharp-based, graded beds of coarse fossil debris and fine-grained, hummocky cross-stratified quartzose sandstones; these were interpreted as offshore storm deposits.

*Lehman (1986): outcrop in vicinity of Rochester, New York*
The entire Medina was interpreted as a single, regressive tidal-flat sequence dissected by tidal channels.

*Lumsden & Pelletier (1969): subsurface, New York*
The Medina was interpreted as a nearshore to subaerial deltaic sequence.

*Martini (1966,1971,1974) and Friedman et al. (1982): outcrop, Ontario and New York*
Various different depositional subsystems were recognized in the inferred tide-influenced deltaic deposits of the Medina. These include open marine prodelta and deltaic outer-fringe deposits, various types of marine bars, beaches, tidal flats, tidal channels, deltaic distributary channels, interdistributary bays and fluvial channels.

*Metzger (1982): subsurface, Chautauqua County, New York*
The basal sand (Whirlpool) was identified as a tide-influenced sublittoral sand sheet, with tidal sand ridges trending NE–SW. Overlying units interpreted as marine shelf, prodelta, distal bar, and tide-dominated deltaic deposits. These deltaic deposits were identified as being deposited in 'elongated dendritic areas that trend nearly north-south'.

*Middleton (1982), Middleton & Rutka (1986) and Middleton et al. (1983,1987): outcrop, Ontario and New York*
The lower portion of the basal sandstone (Whirlpool) was interpreted as the deposit of northwest-flowing braided streams, with the Silurian marine transgression occurring near the top of this unit. Storm deposits were recognized in the transgressive sequence developed in the upper portion of the Whirlpool and the overlying shales and thin sandstones (Cabot Head).

*Pemberton (1979,1987) and Pemberton & Risk (1982): outcrop in vicinity of Hamilton, Ontario*
Based on the examination of trace fossils, the uppermost sandstone beds (Thorold) were interpreted as discrete, offshore, subtidal storm deposits. Normal hemipelagic deposition of muds (containing trace fossils of the *Cruziana* ichnofacies) was interrupted by storm deposition of sand beds, which were colonized by opportunistic organisms (producing an assemblage of trace fossils from the *Skolithos* ichnofacies within sandstones).

*Pees (1986): subsurface, northwest Pennsylvania and northeast Ohio*
The basal sandstone (Whirlpool) was interpreted as being deposited in sublittoral, littoral and subaerial environments. Tidal ridges and off-shore bars were recognized, and transitory (i.e. non-preserved) beaches were inferred.

*Piotrowski (1981): subsurface, northwest Pennsylvania*
The Medina was interpreted as a system of small, channelized deltas with a generalized shoreline trending NE–SW.

## PREVIOUS ENVIRONMENTAL INTERPRETATIONS OF THE MEDINA

Just as the stratigraphy of the Medina remains unsettled, previous environmental interpretations of the unit reflect a wide range of diverse opinions (see summary in Table 1). In a landmark paper from the turn of the century, G.K. Gilbert (1899) discussed the nature and origin of cross-bedding in the Medina. While the latter part of this paper is commonly cited as an early account of the formation of cross-strata by the migration of wave-generated vortex ripples, it has not been generally recognized that the first part of the paper contains the first published description and illustration of the structure now known as hummocky cross-stratification.

In an interpretation that sounds remarkably current, Gilbert considered the structure to be the sedimentary record of giant wave ripples formed by very large, storm-generated surface gravity waves; from this he inferred deposition in a large body of water connected to an open ocean. Gilbert's study thus represents the first inference of a major storm influence in Medina sedimentation. His interpretations soon encountered strong criticism (Fairchild, 1901), and it is now apparent that Gilbert misidentified at least some examples of trough cross-bedding in the Medina as the deposits of 'giant wave ripples'. Regardless of this problem, it is unfortunate that the overall significance of Gilbert's work went unnoticed for almost a century.

The Medina is an important producer of oil and gas and a major source of commercial building materials; as such, it has been the subject of several recent palaeoenvironmental analyses (Table 1). The most interesting recent development has been the reinterpretation of the basal Whirlpool Sandstone of the Medina, a unit traditionally regarded as a transgressive shallow-marine sheet sandstone. Based on a thorough examination of all available outcrop, G.V. Middleton and colleagues concluded that the basal two-thirds of the Whirlpool was deposited in a northwestward-flowing braided fluvial system, with the Silurian marine transgression occurring near the top of the Whirlpool in an interval of transition to fossiliferous shales or carbonates (Middleton, 1982; Middleton & Rutka, 1986; Middleton et al., 1983, 1987). Our own work reported herein has been concerned exclusively with the unequivocally marine strata overlying the basal Whirlpool; limited comments concerning the braided stream interpretation of the lower Whirlpool are presented in a later section of this paper.

Recent studies of the upper strata of the Medina also have produced different, somewhat conflicting interpretations of depositional environments (Table

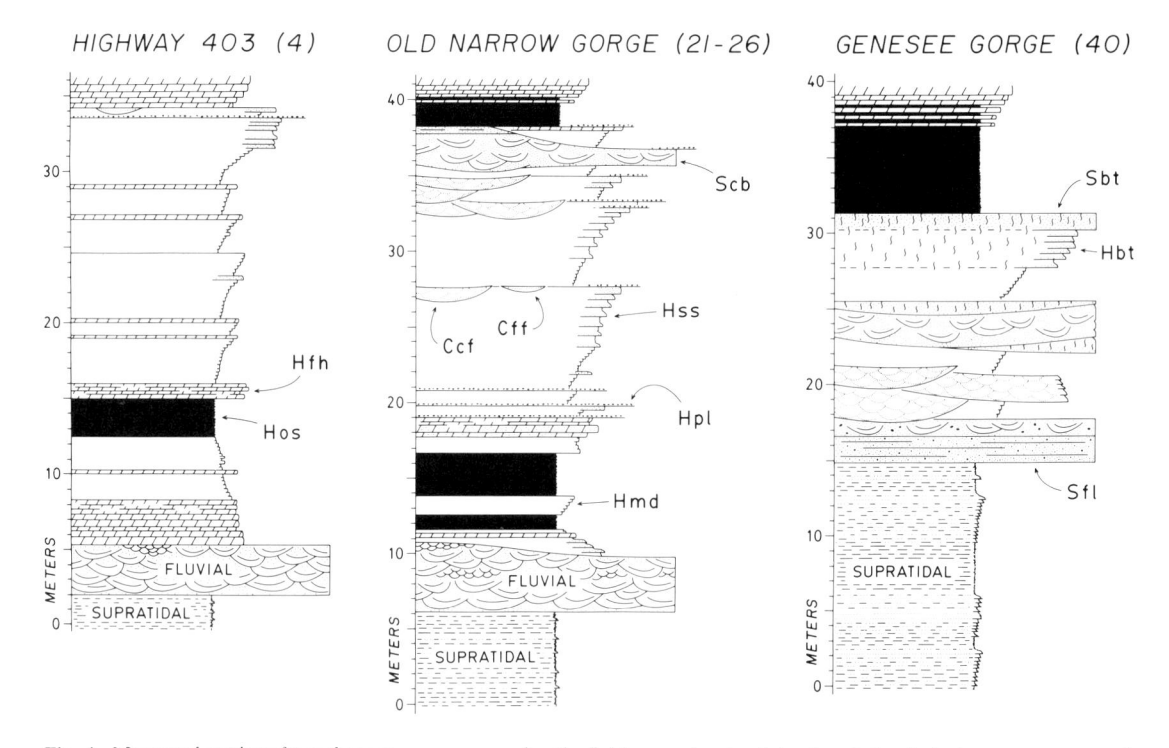

**Fig. 4.** Measured sections from three exposures spanning the field area, showing lateral variation in facies occurrences and vertical facies sequences. Location names are followed by exposure numbers shown in Fig. 1. Symbols of the 11 facies defined in the text are indicated by their respective facies abbreviations. Note that the Clinton carbonates (capping each section), the fluvial deposits of the Whirlpool Formation (in the western and central sections), and the supratidal shales of the Queenston Formation (at the base of each section) do not correspond to facies defined in this study; these units are indicated diagrammatically by a separate set of symbols. The Old Narrow Gorge section represents a synthesis of observations from six closely spaced exposures in the lower Niagara Gorge (exposures 21–26).

**Fig. 5.** Facies Hos (organic shale) and facies Hmd (mudstone with thin sandstones) from the Gully Cut section (exposure 23). Notebook (12 × 19 cm) rests on top of minor sequence A2, composed of facies Hos gradationally overlain by facies Hmd. Note thickening and coarsening upward trend in thin sandstones of facies Hmd. Sharp top of sequence A2 is overlain by facies Hos of sequence A3. Compare with Fig. 19.

1). These interpretations generally reflect two major themes. One suggests that storms dominated sedimentation, at least in Ontario (e.g. Pemberton, 1979, 1987; Duke, 1982, 1985a,b). The other emphasizes the importance of tides, especially in New York (e.g. Martini, 1966, 1971, 1974; Friedman et al., 1982). Data presented herein suggest both a storm and tidal influence in Medina sedimentation. Additionally, the upper Medina shoreline has been interpreted variously as having had an arcuate, deltaic geometry (e.g. Friedman et al., 1982) or a relatively straight configuration (Duke, 1982, 1985a). Our work indicates that the overall geometry of the prograding Medina shoreline was relatively straight, with an approximately NNE–SSW orientation throughout the field area.

Note that several previous studies have inferred extensive tidal flats or fluvial floodplain deposits within the Medina (Table 1). These interpretations have been based in part on reports of desiccation cracks at many stratigraphic levels and geographic locations. Although well-developed desiccation cracks are present locally, most crack-like structures in the Medina are best interpreted as either subaqueous shrinkage cracks or poorly preserved biogenic structures. These features are discussed in detail in the facies descriptions below.

## FACIES DEFINED IN THIS STUDY

For this study, we measured each exposure bed by bed, producing a vertical graphic log for each section at a scale of 1:50 (or greater, where necessary). Physical, chemical, and biological attributes of the rocks were recorded, and palaeoflow data were obtained wherever possible. Detailed location maps for many of these exposures have been presented by Brusse *et al.* (1987).

Herein we have defined 11 facies from our descriptive base; their graphic symbols are indicated in Fig. 4. These facies can be organized conveniently into three facies associations: (1) the host-sediment association includes facies defined from relatively fine-grained, laterally extensive, fossiliferous sediments; (2) the channel association includes facies defined from discrete, sandstone-dominated channel deposits incised into the fine-grained 'host' deposits; (3) the sheet-sandstone association includes facies defined from laterally extensive, sheet-like sandstone bodies. We follow this grouping in the facies descriptions and preliminary interpretations of depositional environments and systems below.

## Description of facies from the host-sediment association

The host-sediment association contains various fine-grained, laterally extensive, fossiliferous deposits. Locally, the channels of the channel association are incised into and fully encapsulated by these deposits (hence the use of the term 'host' in the name of this association). The host-sediment association is dominated by terrigenous shales, mudstones, and thin sandstones, and minor carbonates. Six facies have been defined from these deposits:

### *Facies Hos—organic-rich shale*

This facies consists of very dark grey to greenish grey, organic-rich, fissile shale with finely disseminated pyrite (Fig. 5). Thin calcareous siltstones (a few millimetres to centimetres thick) are present locally. Megascopic trace and body fossils were not observed, but acritarchs and chitinozoans confirm a marine origin (C. Moy, pers. comm., 1988).

### *Facies Hmd—shale and mudstone with minor siltstones and sandstones*

Red, green or light grey shales and mudstones are the dominant lithologies in this facies (Fig. 5). Thin lenticular and continuous coarse siltstones and very fine sandstones (a few centimetres thick) are present; these sharp-based beds are graded or flat-laminated, and rarely are capped with wave-formed interference or symmetrical vortex ripples, or asymmetrical current ripples. These thin coarse beds typically exhibit various trace fossils from the *Chondrites* ichnofacies and a sparse, reworked shallow marine fauna (including brachiopods and bryozoans).

### *Facies Hss—interbedded sandstones and mudstones*

This facies consists of shale or mudstone similar to that of facies Hmd above, interbedded with decimetre-thick beds of very fine to fine-grained sandstone (rarely, medium-grained sandstone). Sandstone beds pinch and swell in thickness laterally; many are lenticular (Fig. 6). They have sharp, erosive bases that commonly display oriented sole marks (scratch, prod, drag and flute marks, and gutter casts; Fig. 7A). Shale rip-up clasts commonly are concentrated above the bases of sandstones.

Very fine to lower fine-grained sandstones in this facies typically are hummocky cross-stratified or flat-laminated; commonly they are capped with wave-formed vortex, interference, or combined-flow ripples, or (rarely) asymmetrical current ripples (Fig. 6). Climbing-ripple cross-lamination is rarely observed. Soft-sediment deformation is fairly common. The much less common upper fine to medium-grained sandstones display flat lamination or medium-scale trough cross-bedding; tops of these beds are sharp.

Occurrences of this facies exhibit trace and body fossils characteristic of an open shallow marine environment. Brachiopods, bryozoans, pelecypods and crinoid ossicles are the most common body fossils; these typically are concentrated within sandstone beds, where locally they are current-aligned. Trace fossils are richly abundant and diverse. S.G. Pemberton has recognized over 20 genera from both the *Cruziana* and *Skolithos* ichnofacies in the deposits of this facies, with elements of the *Cruziana* ichnofacies restricted largely to mudstones and elements of the *Skolithos* ichnofacies confined to the thicker sandstones (Pemberton, 1979, 1987; Pemberton & Risk, 1982; see also Table 1 herein).

Irregular, poorly developed, sandstone-filled mud cracks are present on the bases of sandstones from facies Hss throughout the Medina. These features generally are short in length and rarely branch (Fig. 7A); others are perfectly straight and at least a few decimetres long. We have only rarely observed

**Fig. 6.** Facies Hss (interbedded sandstones and mudstones) forming an overall thickening and coarsening upward interval within sequence E at Albion Falls (exposure 8). (A) Note lateral thickness variation and lenticularity of many of these hummocky cross-stratified sandstone beds, and well-developed wave ripples at lower centre of photograph. Hammer provides scale. (B) Detail of hummocky cross-stratification in sandstone to the right of the hammer in (A).

the complete development of a polygonal pattern. Noting that previous workers had interpreted these features as desiccation cracks, Pemberton (1979) closely examined many examples as part of his investigation of trace fossils in these deposits in exposures in Ontario. He concluded that none of the examples he encountered could be attributed unequivocally to desiccation, suggesting instead a subaqueous origin involving shrinkage resulting from synaresis or biological modification of sediment properties (Pemberton, 1979, pp. 197–199). We concur with this interpretation, and additionally note the unusual occurrence of a single sandstone bed that exhibits such features on both the lower and upper surfaces, suggesting crack formation after burial by an overlying mud layer (Fig. 7B,C).

*Facies Hfh—fossil-hash beds*

This facies (Fig. 8) consists of abundantly fossiliferous carbonates, calcarenites, or carbonate-cemented quartzose sandstones interbedded with thin, fissile shales of variable colour. The coarse beds contain abundant reworked fossil debris (including bryozoans, brachiopods, crinoid ossicles, and rare, large, straight-coned nautiloids). Accumulations of shale rip-up clasts and reworked phosphatic nodules and steinkerns generally are present. Abundant very fine to coarse-grained quartz sand

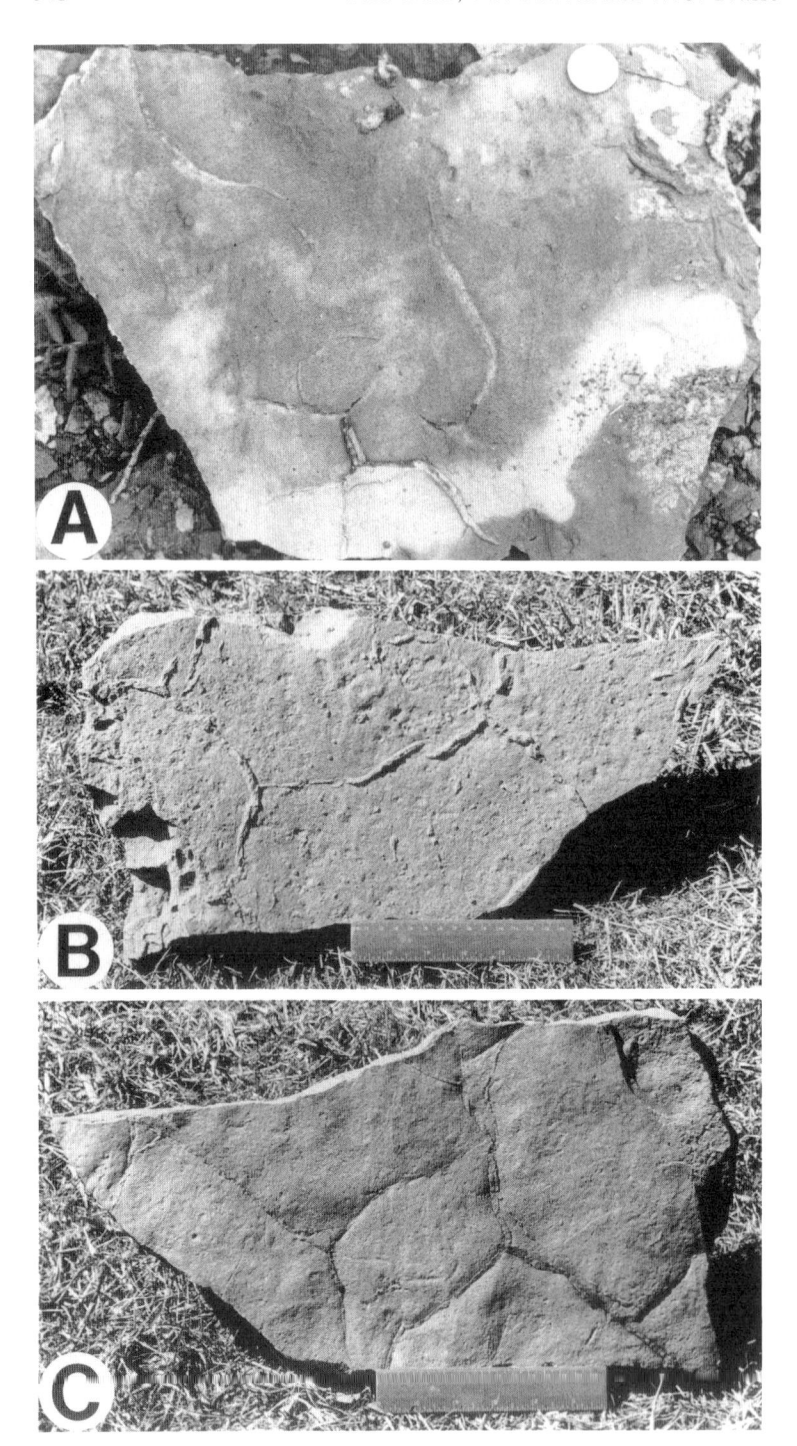

Fig. 7. Inferred subaqueous shrinkage cracks from occurrences of facies Hss. (A) Examples from the base of a sandstone at Sydenham Road (exposure 3). Note delicate scratch marks oriented vertically in this photograph. Coin is 2.4 cm in diameter. (B) Examples from the inferred base of a sandstone collected from float at the Jolley Cut (exposure 5). (C) Examples from the reverse side of the same sandstone shown in (B) above. Although somewhat loaded by compaction, these examples from the inferred top of the sandstone also protrude from the adjacent surface. Scale in (B) and (C) is 15 cm long.

**Fig. 8.** Facies Hfh (fossil-hash beds). Sequence B at Birth of the Falls section (exposure 18). Top of sequence B occurs just above the hammer at the transition to shales (facies Hmd) at base of sequence C. Lighter coloured beds within sequence B are sandy, fossiliferous dolomites; darker beds are dolomitic mudstones. Large dark clasts in dolomite behind hammer iron are reworked phosphatic nodules, steinkerns and skeletal debris.

grains are disseminated through many of the carbonates, which locally grade laterally into carbonate-cemented, fossiliferous quartz sandstones over distances of only a few tens of metres. Hummocky cross-stratification, irregular scour-and-fill structures, and symmetrical vortex ripples are present locally, as are iron oolites, haematite cement, and small grains of pyrite.

*Facies Hpl—erosive-based phosphatic lag deposits*

This facies consists of a laterally extensive, flat to irregularly scoured erosion surface overlain more or less continuously by a thin veneer (millimetres to a few centimetres thick) of reworked phosphatic nodules and steinkerns, coarse phosphatized fossil debris, and phosphatized shale rip-up clasts (Fig. 9). The continuously traceable basal erosion surface commonly steps up and down a few decimetres in the succession over distances of a few tens of metres laterally, thus lying upon different lithologies locally. Relief on the basal erosion surface is quite pronounced in some occurrences, with steep-sided to vertical erosion surfaces developed locally.

The lateral extent of individual occurrences of this facies is somewhat variable. Thus, certain examples can be traced the full length of the extensive Niagara Gorge exposures, a distance of several kilometres; however, the phosphatized debris overlying these examples typically is patchy, being entirely absent locally. Other examples in the gorge can be traced only a few hundred metres; these occurrences grade almost imperceptibly into phosphatic, fossiliferous carbonates of facies Hbc described above. Our regional correlations (presented below) suggest that certain examples can be traced several tens of kilometres. Where traceable over these larger distances, some examples display erosional relief of several metres; such large-scale relief can be observed directly in the continuous Niagara Gorge exposures. Note that extensive occurrences of this facies always occupy special stratigraphic positions, being restricted to the boundaries of coarsening upward sequences (see discussion in a later section).

*Facies Hbt—bioturbated sandstones and mudstones*

Restricted to exposures in the eastern field area, this facies (Fig. 10) is composed of interbedded mudstones and thin sandstones that are totally bioturbated. The trace-fossil assemblage in this facies contrasts markedly with other assemblages from the host-sediment association. Although the density of biogenic structures in these deposits is extremely high, the species diversity is very low: only *Daedalus* and less abundant *Arthropycus* have been recognized. These traces are present in each occurrence of facies Hbt.

**Preliminary interpretation of the host-sediment association**

Environmentally, the deposits of the host-sediment association (other than facies Hbt) clearly represent

**Fig. 9.** Facies Hpl (phosphatic lag deposits). Occurrence capping minor sequence C1 at Birth of the Falls section (exposure 18). Scale in each photograph is 15 cm long. (A) Oblique plan-view showing thin accumulation of phosphatic material, skeletal debris, and shale rip-up clasts overlying sandstone. (B) Cross-section of the same occurrence as (A), at a location a few metres away laterally. Large arrows indicate top of light-coloured lag; small arrows indicate burrows. Burrows penetrate from the base of the lag and are only partially filled with coarse lag material. Note erosional relief at base of lag and low-angle, curved lamination of hummocky cross-stratified sandstone below the lag.

**Fig. 10.** Facies Hbt (bioturbated sandstones and mudstones) within sequence G near the top of the Genesee Gorge section (exposure 40). Note abundant, cross-cutting *Daedalus* burrows.

open shallow-marine deposits, as evidenced by the abundant and diverse body and trace fossils that typify this association. The abundance of mudstone and the dominance of hummocky cross-stratification and various wave-formed ripples in the coarse beds of this association strongly indicate a storm-dominated, subtidal depositional setting, below fair-weather wave base but above storm wave base. This interpretation is supported by an analysis of the mixed assemblage of trace fossils from facies Hss (Pemberton, 1987), indicative of opportunistic invasion of a normally quiet-water setting following the episodic emplacement of storm-deposited sands (see Table 1).

In contrast with the other deposits in this association, the high-density, low-diversity assemblage of trace fossils (*Daedalus* and *Arthrophycus*) in facies Hbt strongly suggests a more proximal setting, in a shallow (possibly intertidal) and restricted (possibly brackish) marine environment (see environmental summaries by Cotter (1983) and Durand (1985)).

### Description of facies from the channel association

In the more extensive exposures examined in this study, individual sandstone-dominated channels are easily recognized by: (1) their limited lateral extent; (2) their consistently sharp, commonly concave-upward erosional basal contacts; (3) their internal structures and structural sequences; and (4) the fining-upward nature of the relatively coarse interbedded sandstones and mudstones filling the channels. In the Niagara Gorge, for example, where dozens of individual channels have been distinguished, numerous exposures exhibit complete transverse sections through one or more large channels (Duke, 1982; Figs 11, 12). Similar exposures are present at a few other localities (e.g. Fig. 3). Channels in less extensive exposures (where complete channel forms cannot be observed) are recognized by their distinctive internal structural sequences.

Herein we have defined two separate facies from the channel association, corresponding to the two major types of channel-fill sequences observed in this study. Although it would be possible to subdivide each type of channel-fill sequence into several separate facies, we have not done so for this paper. The two facies defined below distinguish large, coarse-grained channel bodies from smaller channel bodies with a finer grained fill.

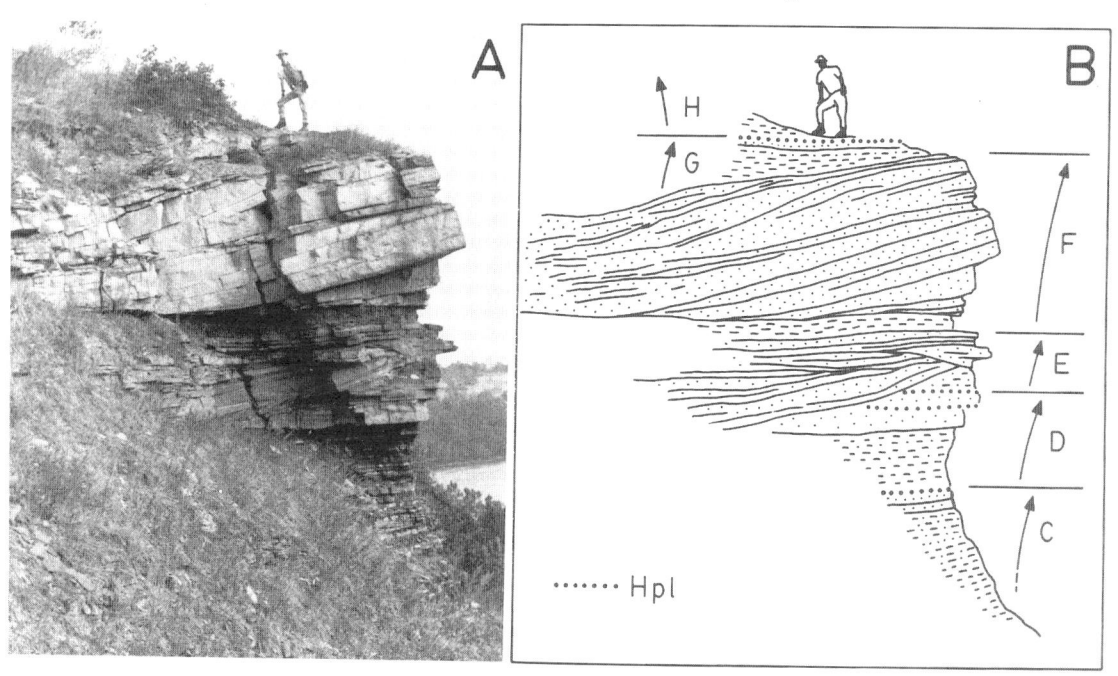

**Fig. 11.** Sequences C–H at the Stack and Drain section (exposure 25). View is toward the south. (A) Note pronounced lateral accretion surfaces (dipping to the northeast) within the light-coloured sheet sandstone capping sequence F. (B) Tracing of photograph (A). Arrows indicate sequences. Note two separate channels (facies Ccf) near top of sequence E at this locality. Compare with view in Fig. 17.

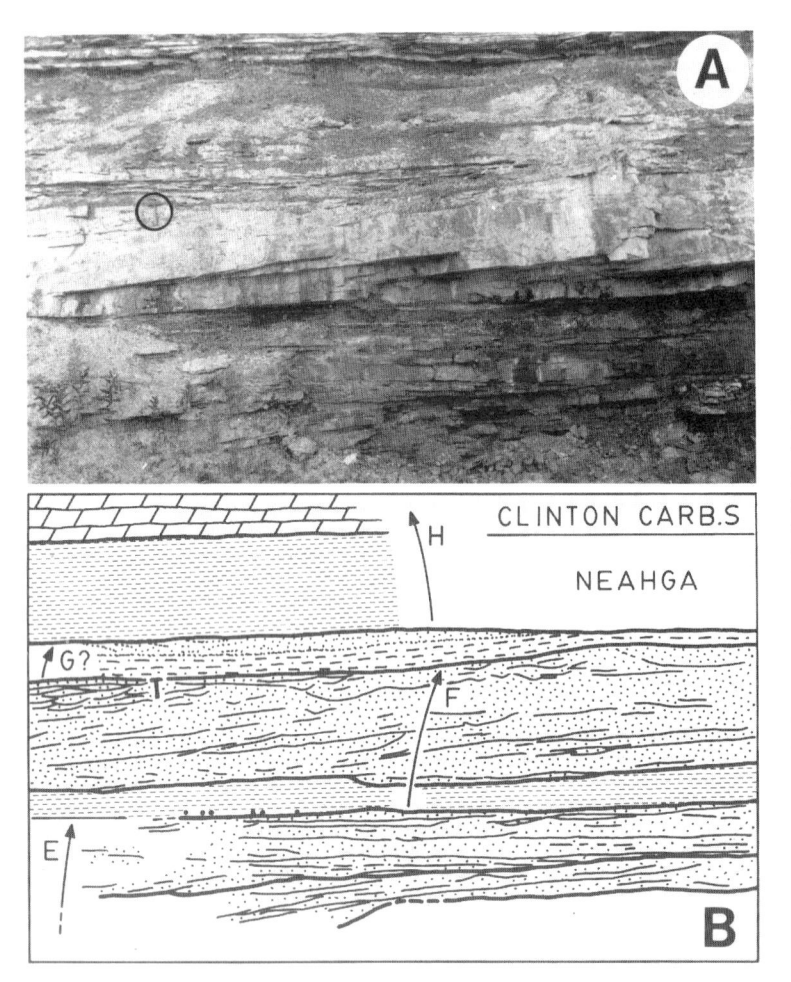

Fig. 12. (A) Photograph and (B) tracing of sequences E, F, G(?) and H at the Sir Adam Beck Power Stations Access Road (exposure 29). Head of hammer (circled) is at contact between sheet sandstone capping sequence F and interval of interbedded sandstones and shales (facies Hss) at base of sequence G. Note that the thin interval identified as the G sequence here may simply represent an unusually fine-grained 'transgressive cap' of the F-sequence sheet sandstone. Clinton dolomites are present at top of photograph. Arrows indicate sequences. Note lateral accretion surfaces dipping to the southeast within the sheet sandstone and to both the southeast and northwest within the coarse channels at top of sequence E. Sequence E is capped by a discontinuous phosphatic lag at this exposure.

*Facies Ccf—coarser grained channel-fill deposits*

This facies (Figs 3, 11, 12) constitutes the fill of most channels. All occurrences possess maximum thicknesses in excess of 1 m. These coarser grained channels have widths ranging from about 10 to 500 m; average width is roughly 50 m. Note that widths were measured normal to channel orientations determined from internal palaeoflow data or the strike of the erosional basal surface measured at inclined channel margins. Depth of channel incision ranges from about 0.5 to 4 m, based upon the maximum preserved thickness of channel-fill deposits (not corrected for compaction). Locally, two or more channels have incised into one another to produce vertically stacked, multistoried channel complexes, smaller examples of which are shown in Figs 11, 12.

A coarse basal lag is present immediately above the erosional surface of about half of the occurrences examined. This lag consists of a thin layer of debris, less than 15 cm thick, which typically includes: (1) abundant fine-grained rip-up clasts derived from underlying marine shales or mudstones; (2) reworked *Lingula* shells; and (3) reworked phosphatic nodules and steinkerns. The bulk of the fill of most occurrences consists of sharp-based, fine to medium-grained sandstones, typically interbedded with thinner shales or mudstones (Figs 11, 12). Sandstones generally are a few decimetres thick; they constitute 80% or more of the thickness of most occurrences. Rarely, sandstones represent almost all the thickness of a channel-fill deposit (Fig. 3). Sandstones typically thin and fine upward in this facies, whereas shale and mudstone partings thicken upward.

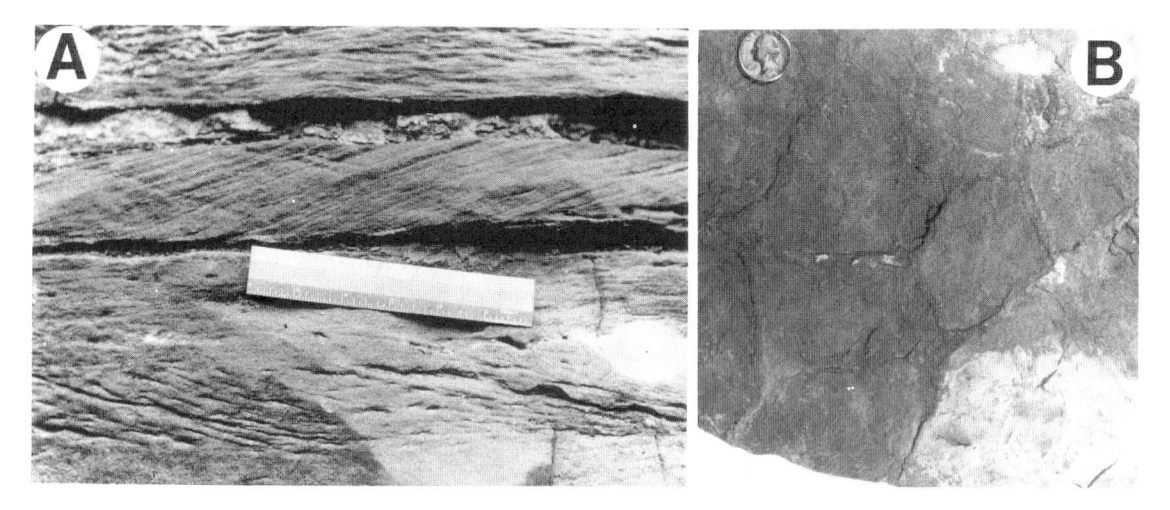

Fig. 13. Examples of structures within facies Ccf. (A) A flow-parallel vertical section from one of several large channels capping sequence D at the Whirlpool Rapids (exposure 32). View is to northeast. Note reversing flow directions in sets of trough cross-bedded sandstone separated by thin shale beds and crudely laminated (flat-laminated and current rippled?) sandstone. Scale is 15 cm long. (B) Sole view of one set of well developed desiccation cracks on the bases of thin sandstones from the top of this same channel occurrence. Coin is 2.4 cm in diameter.

Medium-scale trough cross-bedding (Fig. 13A) is the dominant structure in sandstones, although flat lamination is fairly common, especially in the coarser grained sandstones near the bottoms of channels and the finer grained sandstones near their tops. Trough cross-lamination formed by asymmetrical current ripples also is very common at the tops of channel-fill sequences in this facies; it typically occurs as climbing ripple-drift lamination. Soft-sediment deformation is developed locally. Reworked and abraded *Lingula* shells are common, and horizontal trace fossils (especially *Arthrophycus* and *Planolites*) are present on the bases of many sandstone beds.

Lateral accretion bedding is very well developed in the interbedded fill of many occurrences (Figs 11, 12), and lateral accretion surfaces are common within those channel-fill sequences formed more or less entirely of sandstone (Fig. 3). Where present, lateral accretion structures traverse the entire thickness of a channel-fill sequence (Figs 3, 11, 12). Locally, laterally accreted beds are structureless internally, and they represent the only physical structure present in some instances.

At one location only, in an interval of flat-laminated fine sandstones at the top of an occurrence of this facies, well-developed polygonal dessication cracks were observed *in situ* on the bases of several thin sandstone beds separated by shale partings (Fig. 13B).

*Facies Cff—finer grained channel-fill deposits*

Occurrences of this facies are restricted to (and entirely fill) the smaller, thinner channel bodies in the Medina. These channels generally are only a few decimetres deep and a few metres wide. They are not as common as the larger channels of facies Ccf. Small channels occur either: (1) incised into the tops of the thicker channel-fill sequences; (2) as small, isolated channel bodies shallowly incised into deposits of the host-sediment association; or (3) incised into the tops of thick sandstone bodies of the sheet-sandstone association (see below). *Lingula* and biogenic structures are rarely observed in these deposits. Two end-member types of a gradational spectrum of finer grained deposits filling small channels have been recognized; they are not distinguished from one another in this report. The most common type consists of very fine to fine-grained, flat-laminated sandstones a few centimetres to decimetres thick, interbedded with thin shale or mudstone partings (Fig. 14). Hummocky cross-stratification or wave-formed symmetrical vortex ripples are rarely observed in these sandstones, which typically drape the concave basal erosion surface with little or no evidence of lateral channel

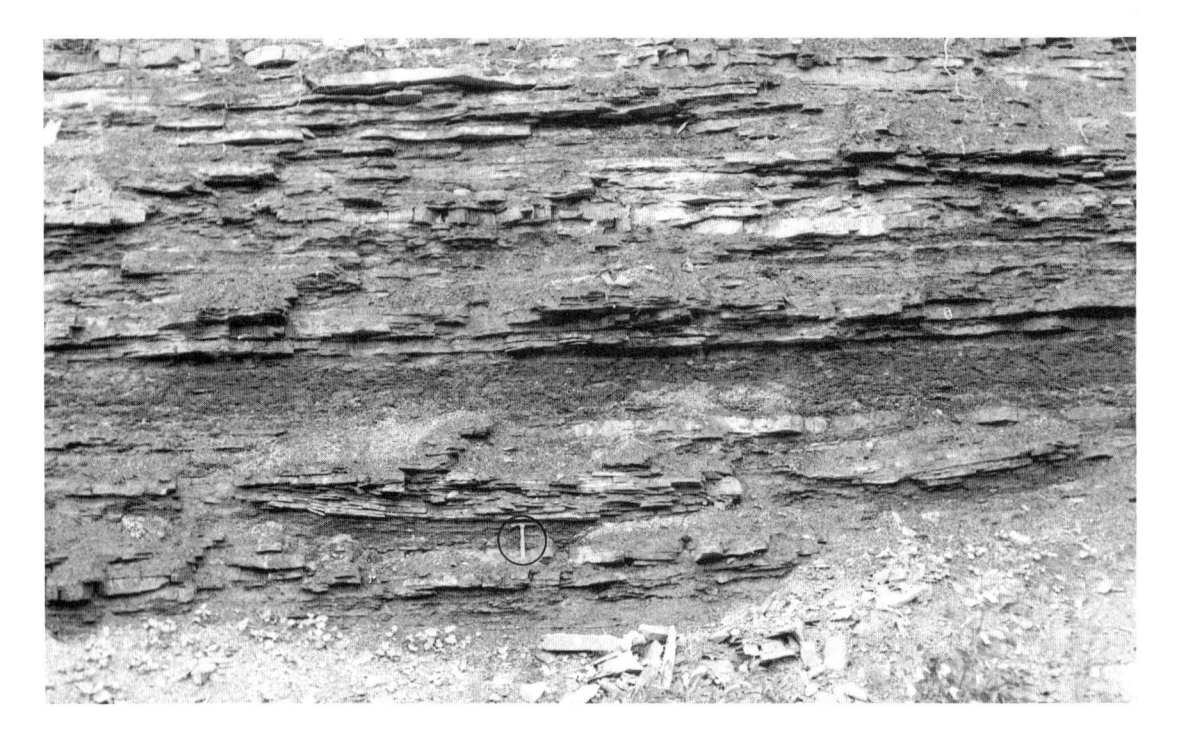

**Fig. 14.** An unusually large occurrence of facies Cff (small, fine-grained channel-fill deposits), just above the hammer, near the top of sequence C at the Sir Adam Beck Power Stations Access Road (exposure 29). Note that this channel displays no evidence of lateral accretion.

migration. A less common type of fill is dominated by mudstone, with very thin beds of very fine to fine-grained sandstone. Locally, lateral accretion surfaces are well developed in some of these small, mud-dominated channel deposits, reflecting varying amounts of lateral channel migration.

**Preliminary interpretation of the channel association**

The limited trace and body fossils of the channel association suggest a restricted, possibly brackish, shallow marine depositional environment; the relative coarseness of these deposits (compared to their 'host' sediments) suggests a more proximal, probably shallower, depositional setting. The abundance of lateral accretion surfaces suggests deposition within sinuous, meandering channels. A more specific interpretation of the depositional system represented by these channels must await a discussion of vertical facies successions and palaeoflow data.

**Description of facies from the sheet-sandstone association**

Occurrences of the sheet-sandstone association represent laterally extensive, shallow marine, coarse to fine-grained sandstone bodies 1–3.5 m thick. In the superlative exposures of the Niagara and Genesee Gorges, individual sheet sandstones are completely continuous over distances up to about 8 km, the maximum lateral extent of outcrop. In contrast with the channel-fill deposits of the channel association, nearly every occurrence of this association is composed entirely of sandstone; very rare, thin shale lenses (a few centimetres thick) are present locally. An individual shale lens generally can be traced no more than a few metres laterally.

Most occurrences of this association are complexly channelled internally (Figs 3, 11, 12). Generally, the dimensions of individual channel forms within sheet sandstones are slightly smaller than (or closely similar to) the dimensions of the discrete channels of facies Ccf at the same locality. A fining upward trend is present at the top of many sheet sandstones.

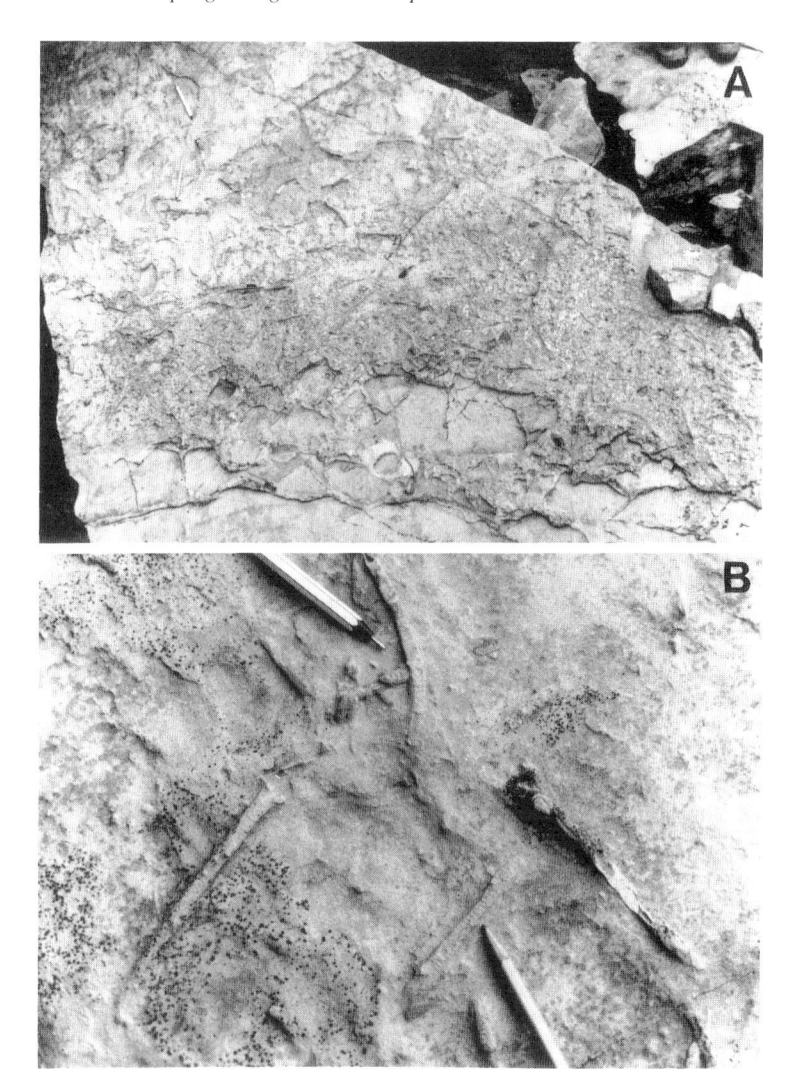

**Fig. 15.** Structures on the base of a fallen block of the sheet sandstone capping sequence F in the Niagara Gorge, between exposures 19 and 20 on the USA side of the river. (A) Sole view showing thin coarse lag of phosphatic fossil debris and shale rip-up clasts. Also note crack-like pattern of structures, especially well developed at bottom of photograph and in the vicinity of the pens at upper left. (B) Detail of structures indicated by pens at upper left of photograph (A). Note annulations that identify these structures as poorly preserved *Arthrophycus* burrows.

Sheet sandstones typically have sharply erosional bases. On the scale of most outcrops, bases commonly display a few decimetres of local erosional relief (Figs 3, 12); however, they more typically are flat (Fig. 11). A lag of shale rip-up clasts and fossil debris, typically highly phosphatic, is commonly present immediately above the basal erosion surface (Fig. 15A). Locally, crack-like structures also are present on the bases of sandstone sheets. Examples are common from the occurrence forming the Thorold Sandstone in exposures in the Niagara Gorge (Fig. 15A), and these may have been interpreted previously by some workers as desiccation cracks. However, close inspection of certain examples reveals that they actually are poorly preserved *Arthrophycus* burrows (Fig. 15B). We suspect that all similar structures from this association are biogenic in origin. Well-developed examples of *Arthrophycus* are present on the bases of sheet sandstones at several localities.

In exposures of restricted lateral extent, sheet-sandstone bodies are recognized solely by their distinctive internal facies sequences. We have defined three facies from the sheet-sandstone association.

*Facies Scb—cross-bedded sheet-sandstone deposits*

Well sorted, medium-grained sandstone (with very rare, lenticular shale breaks described above) is the

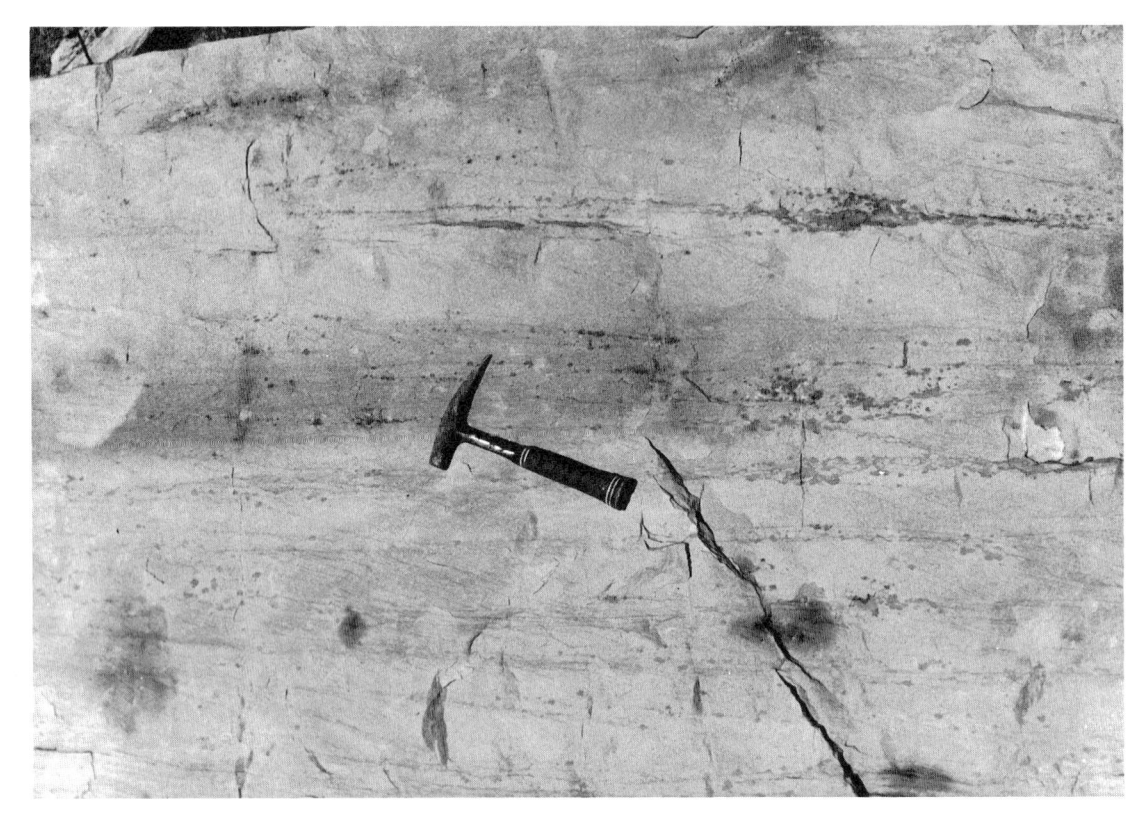

**Fig. 16.** Facies Scb (cross-bedded sheet-sandstone deposits). A flow-parallel section through trough cross-bedding within a fallen block of the sheet sandstone capping sequence F, between exposures 19 and 20 on the USA side of the Niagara River.

dominant lithology of this facies; one occurrence (near the base of the Medina at the Genesee Gorge exposure; Fig. 4) includes coarse-grained, pebbly sandstone. Marine trace fossils, including *Arthrophycus*, *Daedalus* and *Skolithos*, are present in many occurrences. Fragmented *Lingula* shells are found locally.

The sandstone of this facies exhibits medium-scale cross-bedding (Figs 12, 16, 17). This usually is trough cross-bedding, but in some flow-parallel exposures it is not possible to eliminate a planar-tabular geometry. In addition to the dominant cross-bedding, asymmetrical (current) ripple cross-lamination is fairly common; these smaller structures commonly are closely associated with rare mud drapes where the latter are present (Fig. 17). Lateral accretion surfaces and shale rip-up clasts are apparent in many exposures (Figs 3, 11, 12, 17). Locally, silica cementation and an absence of colour variation in these well-sorted deposits makes it difficult to recognize lamination (Fig. 17).

*Facies Sfl—flat-laminated sheet-sandstone deposits*

This facies (Fig. 17) consists of flat-laminated, fine to medium-grained sandstones. Trace fossils similar to those in facies Scb above are present locally. At a few locations, gently undulatory lamination, wave-formed vortex ripples, or hummocky cross-stratification is present in fine-grained sandstone in this facies. An exceptional occurrence at the base of the Medina in the Genesee Gorge exhibits coarse-grained quartzitic and oolitic flat-laminated sandstone (Fig. 4).

*Facies Sbt—bioturbated sheet-sandstone deposits*

This facies consists of intensely bioturbated fine to medium-grained sandstone (Fig. 18). Locally, the intensity of burrowing makes it impossible to recognize specific traces. Typically, however, examples of *Arthrophycus* and *Daedalus* can be discerned; these always occur in very high densities.

**Fig. 17.** Occurrences of facies Scb (cross-bedded sheet-sandstone deposits) and facies Sfl (flat-laminated sheet-sandstone deposits, just visible at top of photo) within sheet sandstone capping sequence F in the Niagara Gorge. View looking north-northwest at the Stack and Drain section (exposure 25). Compare with Fig. 11. Note thin, laterally accreted mud drapes within facies Scb. Close inspection of these mud drapes reveals that the immediately surrounding sandstone is trough cross-laminated. The thicker sandstone beds exhibit faint trough cross-bedding. Three examples of cross-bedding are numbered. Number 1 (just below hammer) and number 2 reflect a strong component of flow to the left (i.e. westward). Number 3 reflects a strong component of flow to the right.

Gently inclined bedding planes (lateral accretion surfaces?) are present locally and are the only physical structures preserved from bioturbation. Note that this facies locally is overlain by an inferred transgression surface, from which many of the biogenic structures penetrate.

**Preliminary interpretation of the sheet-sandstone association**

The overwhelming dominance of relatively coarse sandstone, the sheet-like nature of the occurrences of these sandstone bodies, the limited types of trace and body fossils, the rare presence of vortex ripples and large oolites, and the abundance of trough cross-bedding and lateral accretion surfaces in this association all suggest a continuously energetic, channelized, shallow marine depositional setting, probably in very shallow water near the source of detrital sediments. A shallow shoal, sandy shoreface, or intertidal sand flat exposed to fair-weather waves and tides is indicated; a more detailed interpretation is deferred to a later section.

## LATERAL FACIES DISTRIBUTIONS AND VERTICAL FACIES SUCCESSIONS THROUGHOUT THE STUDY AREA

In the overall coarsening upward interval in the upper half of the Medina (sequences B–F, Figs 2, 4), the lateral distribution of facies varies markedly along the Niagara escarpment. We have noted the following generalizations concerning this variation. (1) Fine-grained, open marine 'host' deposits dominate the western field area; the percentage, thickness, and grain size of sandstones in these deposits generally increase towards the east. (2) Restricted marine host deposits (facies Hbt) are present only in the far eastern field area. (3) Discrete channels are small and rare in the far western field area. Elsewhere, channels are much larger; they increase in abundance towards the east. (4) Sheet sandstones become more abundant and thicker towards the east. (5) Fossiliferous carbonates become less abundant and thinner towards the east.

Based on these observations, the field area may be divided into three gradationally bounded regions characterized by major differences in sedimentation styles. (1) The western region (exposures 1–10, Fig. 1) is dominated by fine-grained host sediments representing open shallow marine sedimentation. Channel deposits are rare, and sheet sandstones are absent. (2) The central region (exposures 11–33) consists of vertical alternations between dominant fine-grained host deposits and subordinate channels and sandstone sheets. (3) The eastern region (ex-

**Fig. 18.** Facies Sbt (bioturbated sheet-sandstone deposits) capping sequence G at Genesee Gorge (exposure 40). Note well-developed *Daedalus* burrows. Thickness of interval shown is about 2 m.

posures 34–42) is dominated by coarse-grained channel deposits and sheet-sandstone bodies.

Representative sections measured in the three different field regions defined above are illustrated in Fig. 4. Note that the vertical facies successions in all but the easternmost of these sections define an overall thinning and fining upward sequence overlying the Whirlpool Sandstone (identified as the A sequence in Fig. 2). This interval is overlain by several thickening and coarsening upward sequences (Fig. 19). The correlation of these depositional sequences is discussed below.

## SEQUENCE CORRELATIONS THROUGHOUT THE STUDY AREA

Figure 20 shows our correlation of sedimentary sequences in the Medina throughout the field area. For convenience, we have shown only nine of our more complete and illustrative sections; the exposures omitted from this diagram are consistent with and supportive of these correlations. The datum employed in these correlations is the Medina–Clinton contact. As shall become evident in a later section of this paper, this surface is more likely to be regionally flat (or at least planar) than is the Queenston–Medina contact.

The correlations of Fig. 20 are straightforward and require few explanatory comments. Little or no vertical grain-size trend is developed in the fossil-hash beds of the B-sequence carbonates traceable throughout the western field region. In the central region, however, the sandy carbonates and shales in the B sequence exhibit two or more very minor sequences of restricted lateral extent (e.g. sequences B1 and B2 in Fig. 2). In the Niagara Gorge exposures, these minor sequences can be traced for only a kilometre or less. Note that we have not attempted to correlate such minor depositional sequences in Fig. 20. The fossiliferous deposits of sequence B grade into a thick siliciclastic sandstone in certain of

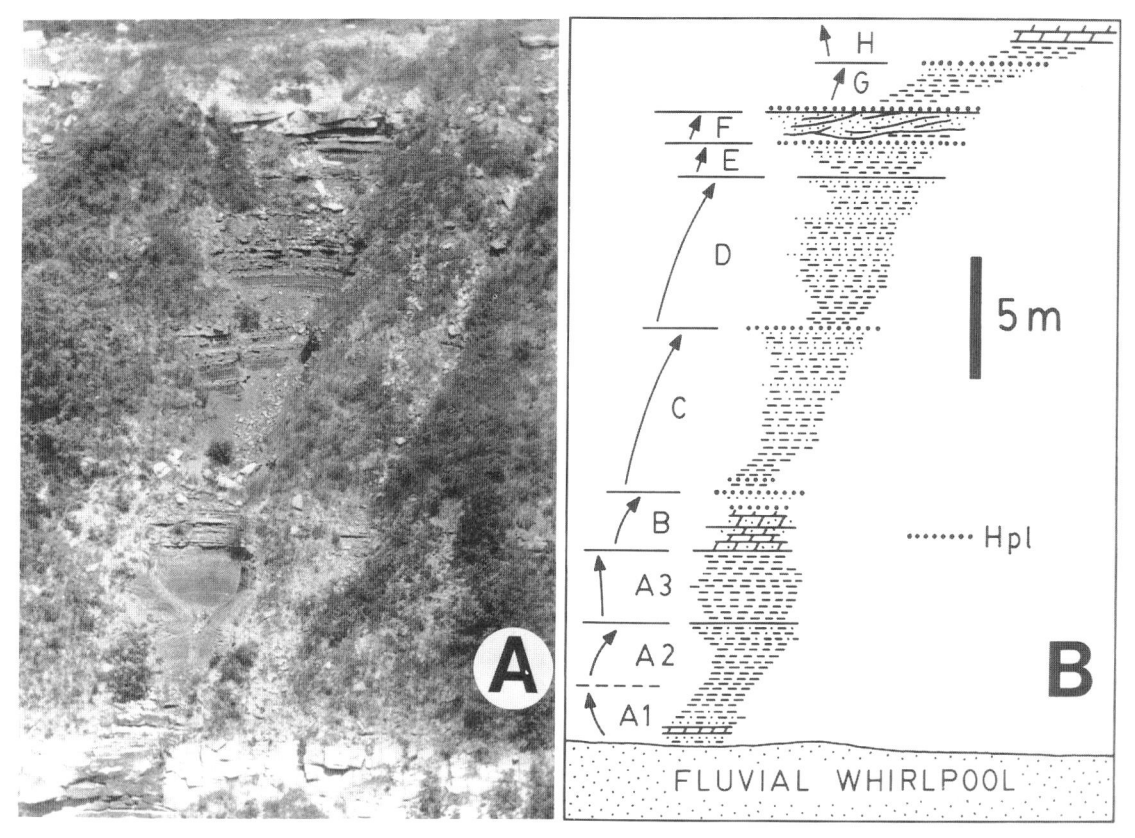

**Fig. 19.** (A) Photograph and (B) tracing of depositional sequences in the Gully Cut section in the Niagara Gorge (exposure 23). This is the only exposure in the gorge where each of the host-sediment cycles may be inspected in the absence of incised channel deposits. Note that both the G and H sequences, including the transition from the Neahga Shale to Clinton carbonates, are exposed in a small rivulet behind the little knoll on the south side of the main outcrop (not visible in photo).

the Niagara Gorge exposures, and throughout the eastern field area, the B sequence consists of a single sheet sandstone.

Note that extensive erosion by channels has locally removed all evidence of certain depositional sequences. In Fig. 20, an example occurs at Ball's Falls (exposure 14), where laterally extensive channel deposits capping sequence D have entirely removed any previously deposited host sediments of that sequence. We suggest that the unusually thick interval of channel deposits at this location reflects the vertical superimposition of sequence D channels incised into the underlying channel deposits of sequence C.

Apart from erosion due to the sequence-capping channels, there is evidence suggesting that the sequence boundaries themselves are erosional, at least locally. This is most clearly evident at boundaries marked by erosive-based phosphatic lags displaying decimetres to metres of erosional relief. Additionally, the truncation of laterally accreted channel-fill deposits and sheet sandstones is apparent at some locally flat sequence boundaries (e.g. at the top of sequences E and F at the Beck Stations Access Road; Fig. 12). Geometries such as these likewise suggest extensive erosion at the contacts between sequences. Where erosion cannot be demonstrated (e.g. in the western field region), sequence boundaries may simply represent extended intervals of non-deposition of terrigenous detritus. This interpretation is supported by the fossil-hash concentrates present locally at many sequence boundaries in the west.

The progressive westward truncation of lithic units and entire depositional sequences at the top of the Medina (Fig. 20) suggests a major discon-

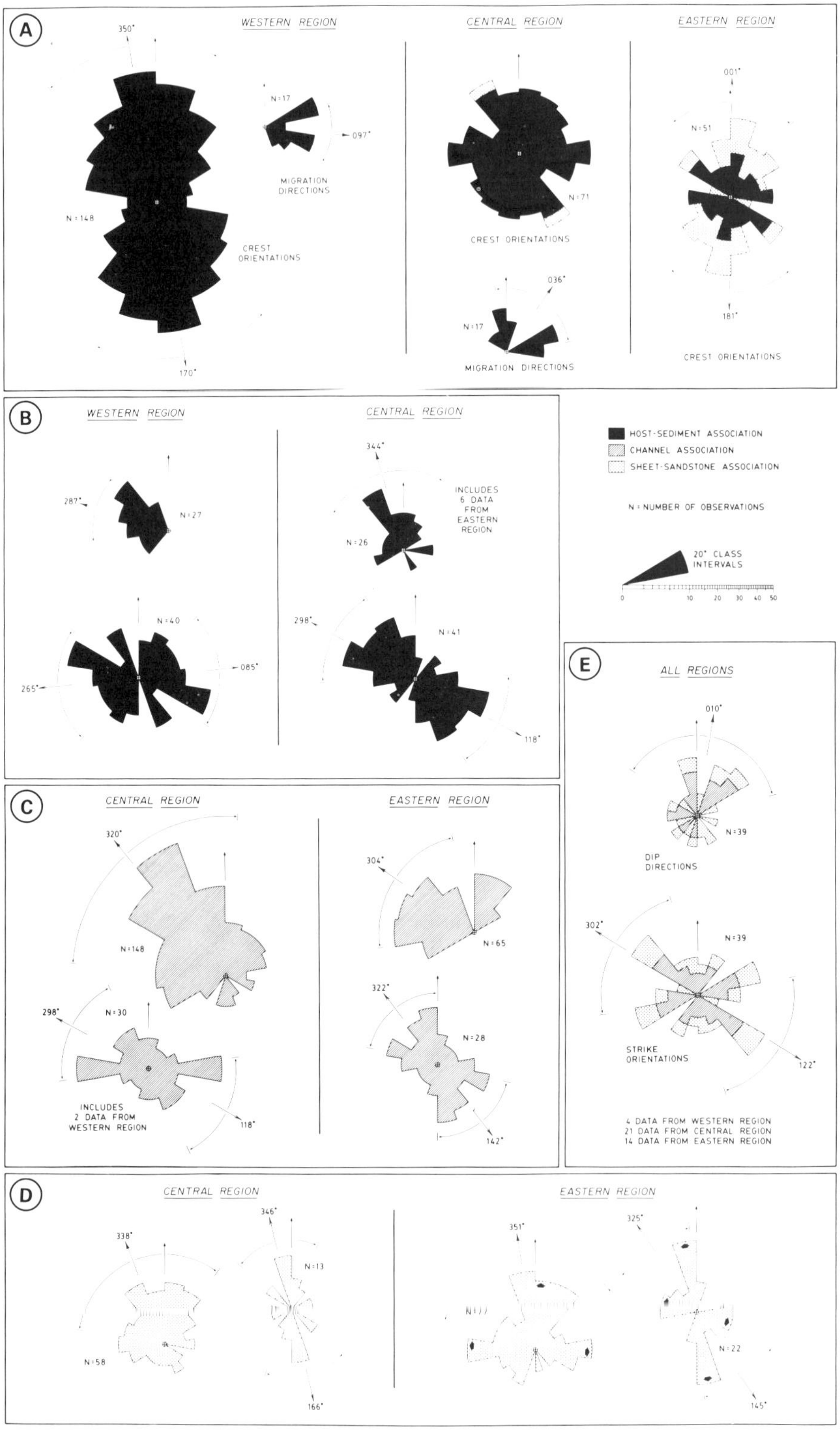
A
WESTERN REGION
350°
N=17
097°
MIGRATION DIRECTIONS
N=148
CREST ORIENTATIONS
170°
CENTRAL REGION
N=71
CREST ORIENTATIONS
036°
N=17
MIGRATION DIRECTIONS
EASTERN REGION
001°
N=51
181°
CREST ORIENTATIONS
B
WESTERN REGION
287°
N=27
N=40
085°
265°
CENTRAL REGION
344°
INCLUDES 6 DATA FROM EASTERN REGION
N=26
298°
N=41
118°
HOST-SEDIMENT ASSOCIATION
CHANNEL ASSOCIATION
SHEET-SANDSTONE ASSOCIATION
N=NUMBER OF OBSERVATIONS
20° CLASS INTERVALS
0 10 20 30 40 50
C
CENTRAL REGION
320°
N=148
298°
N=30
INCLUDES 2 DATA FROM WESTERN REGION
118°
EASTERN REGION
304°
N=65
322°
N=28
142°
E
ALL REGIONS
010°
N=39
DIP DIRECTIONS
302°
N=39
STRIKE ORIENTATIONS
122°
4 DATA FROM WESTERN REGION
21 DATA FROM CENTRAL REGION
14 DATA FROM EASTERN REGION
D
CENTRAL REGION
338°
346°
N=13
N=58
166°
EASTERN REGION
351°
325°
N=22
145°

formable erosion surface capping the group, below the shales and carbonates of the Clinton Group. Although part of this pinch-out may be the result of a westward thinning of sequences, we attribute most of the effect to post-depositional erosion. Across the entire field area, the cumulative amount of erosion at this contact probably is much greater than at the other sequence boundaries described above. Goodman & Brett (1988) have described several additional erosion surfaces within the overlying Clinton Group shales and carbonates in this area. Their work has demonstrated that the westward truncation of depositional sequences continued throughout Clinton deposition.

Note that the Medina is exceptionally thin and coarse-grained (condensed) in the far eastern field area (e.g. at the Genesee Gorge, exposure 40, Fig. 20), and that the basal sandstone there (which contains marine indicators) has been correlated with the sandstones of sequence B in exposures to the west. We therefore agree with previous workers (e.g. Martini, 1966) who suggest that the basal Whirlpool Sandstone (present in the west) is absent in the Genesee area. Subsurface observations south of the outcrop belt form part of the basis for these correlations; these are discussed further in a later section. West of the town of Medina (exposure 36), the Medina Group maintains a remarkably constant thickness.

## SUMMARY OF PALAEOFLOW DATA THROUGHOUT THE STUDY AREA

Figure 21 summarizes approximately 1000 palaeoflow data obtained from the overall coarsening upward interval (sequences B–F) in the three different regions of the study area. Different categories of data are presented for each region; these are discussed below. Note that palaeoflow data from the fining upward marine interval overlying the fluvial Whirlpool (the transgressive A sequence) are not included in this summary, but certain of these data have been discussed previously by Middleton *et al.* (1987). Very few palaeoflow data have been obtained from the largely bioturbated deposits of the G sequence capping the Medina in the eastern field area; consequently, the G sequence is not included in our discussion below.

Wave-generated straight-crested symmetrical vortex ripples are abundant throughout the field area (Fig. 21A). Their fairly consistent N–S crestal orientations may represent either: (1) a consistent direction of wave propagation in the Medina Basin (from the west), with little wave refraction owing to shoaling; or (2) strong wave refraction resulting from a consistent, approximately N–S orientation of bathymetric contours and, probably, the Medina shoreline. Unfortunately, we have no convenient means for calculating the gradient of the Medina Basin floor for a wave-refraction analysis to choose between these possibilities. Nevertheless, we prefer the latter interpretation, which is supported by additional palaeoflow data closely associated with these vortex ripples. These data derive from asymmetrical current ripples, sole marks, gutter casts and rare trough cross-bedding from the host sediments (Fig. 21B). These indicators record flows dominantly towards the west and northwest, presumably in an offshore direction normal (or nearly normal) to both the shoreline and bathymetric contours in the basin. Furthermore, a palaeogeographic reconstruction and detailed consideration of these and other palaeoflow data (presented later in this paper) do not support the idea of wave propagation solely from the west. Note also in Fig. 21A

**Fig. 21.** Summary of all measured palaeoflow data from Medina sequences B–F. Roses (20° class intervals) are presented as equal-area diagrams. Arcs around each resultant vector indicate one standard deviation. 'Western region' refers to exposures 1–10. 'Central region' refers to exposures 11–33. 'Eastern region' refers to exposures 34–42. (A) Palaeoflow data from straight-crested symmetrical vortex ripples. Data are derived from both the host-sediment and sheet-sandstone associations. 'Migration directions' refers to the direction of translation of ripple forms, inferred from foreset dip directions within vortex ripples. Note that no resultant vector is presented for the highly dispersed crestal orientations in the central region. (B) Palaeocurrent indicators from the host-sediment association. Directional indicators mainly include prod marks, asymmetrical current ripples and flute casts. Orientational indicators mainly include scratch and drag marks and gutter casts. (C) Palaeocurrent indicators from the channel association. Directional indicators (mainly trough cross-bedding) also include a small number of asymmetrical current ripples. Orientational indicators mainly include the strike of inclined channel margins and of step-like erosional incisions on channel bases. (D) Palaeocurrent indicators from the sheet-sandstone association. Directional indicators mainly include trough cross-bedding. Orientational indicators mainly include parting and current lineations. See text for discussion of arrows drawn within roses from the eastern region. (E) Attitude of lateral accretion beds and surfaces from both the channel and sheet-sandstone associations. Note that each of these 39 data represents the average of several different attitude readings (usually 5–10) from a single channel-fill sequence.

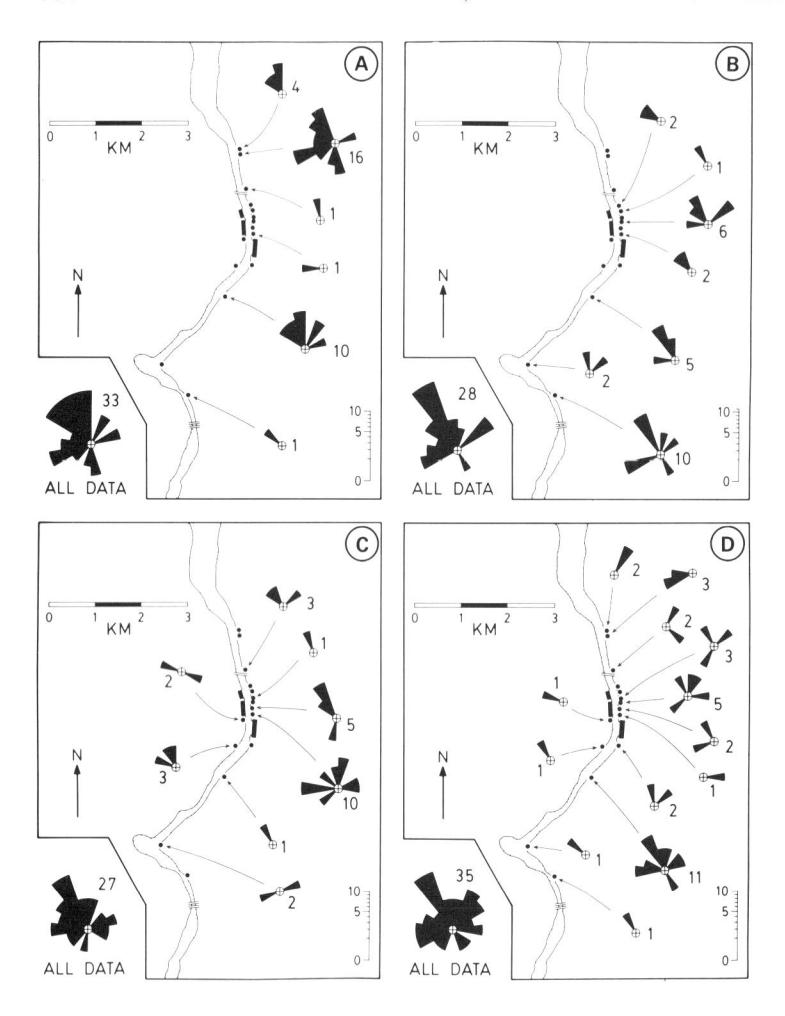

**Fig. 22.** Directional palaeoflow indicators from large isolated channels (facies Ccf) and one sheet sandstone (facies Scb) measured in the various exposures in the Niagara Gorge. Most of these indicators are trough cross-bedding; a few are trough cross-lamination. In each panel, filled circles represent locations of measured sections. Compare with Fig. 1 to determine exposure designations. Number of data are indicated beside each rose. (A) Data from channels capping sequence C. (B) Data from channels capping sequence D. (C) Data from channels capping sequence E. (D) Data from sheet sandstone capping sequence F. Insets show summaries for each set of data. Note that at any particular location, data may have been derived from more than one channel-fill sequence. See text for discussion.

that the migration directions of vortex ripples capping storm-deposited hummocky cross-stratified and flat-laminated sandstones in the western and central field regions were towards the east and northeast, as revealed by foresets preserved within these ripples. We attribute this eastward translation of vortex ripples to onshore-directed sediment transport resulting from the effects of shoaling beneath post-storm swell waves.

Additional support for a straight shoreline is provided by trough cross-bedding and other flow indicators from the discrete channels incised into the host sediments (Fig. 21C). These data also reflect predominantly offshore-directed flows to the northwest; however, the relatively high dispersion of data from the channel association (particularly in the central field area) is noteworthy. Although directed predominantly towards the northwest, these troughs indicate flows that practically box the compass. The origin of this high dispersion is best appreciated by examining data from individual channels. Figure 22 shows the distribution of palaeoflow data from discrete channels cut into the tops of the various sequences present in the overall coarsening upward interval in the Niagara Gorge. The lower of these in-channel deposits (from sequence C; Fig. 22A) shows a pattern of predominantly offshore flows (and minor onshore flows) with a slight dispersion; we attribute this dispersion to local variation in the orientation of the separate channels at the various points of measurement. Moving upward in the succession, channels at the tops of sequences D and E (Fig. 22B,C) reflect increasingly significant flow reversals within individual channels, superimposed upon a similar pattern of local variation in channel orientation. Figure 13A shows a flow-parallel section through a portion of one channel capping sequence D at the Whirlpool Rapids exposure; note the reversal in dip directions in the trough cross-beds interbedded with mudstone at this locality. We therefore attribute the dispersion of palaeoflow data from these channels to minor onshore-directed flow reversals within sinuous channels possessing somewhat variable local orientations. Inspection of the data summaries for each of sequences C, D and E in Fig. 22 shows that the mean flow direction in each system of channels was toward the northeast, providing justification for the inclusion of all channel-derived palaeoflow data in the single set of rose diagrams of Fig. 21C.

Similarly, cross-bedding from the sheet-sandstone association (Fig. 21D) indicates reversing onshore–offshore flows like those from the channel association; note, however, that the dispersion of these data suggests a strong flow component directed approximately northward (i.e. approximately along shore). This is most clearly evident in data derived from the eastern field region (Fig. 21D). Small arrows drawn on the eastern-region roses highlight a trimodal distribution of flow indicators, with bipolar onshore–offshore flows oriented ESE–WNW, and along-shore flows directed approximately northward. The local pattern of flow in the sheet sandstone capping sequence F in the Niagara Gorge is shown in Fig. 22D. This highly variable pattern of local palaeocurrent indicators within the gorge reflects our observation of numerous, complexly incised, rather small channel forms within this sandstone sheet.

Additionally interesting are the orientations of lateral accretion surfaces and beds in 39 individual channel-fill sequences from both the channel association and the sheet-sandstone association (Fig. 21E; note that each data point represents an average of several readings obtained from the deposits of a single channel). Throughout the field area, the strike of these surfaces supports the dominant northwest flow directions derived from trough cross-bedding in channels; curiously, however, the dip directions of a large majority of the lateral accretion surfaces measured are toward the north and northeast. (Note that this northeastward dip direction probably is underrepresented in Fig. 21E: examination of exposures between our measured sections in the Niagara Gorge reveals a large number of channel forms with north and northeast-dipping lateral accretion surfaces, but because these occurrences are not located on our measured sections, they have not been included in our data summaries. By coincidence, the exposures that we did measure apparently exhibit a disproportionately large fraction of those lateral accretion surfaces that dip to the south and southwest.) These data strongly suggest a dominant northeastward migration of shore-normal, sinuous channels along the Medina shelf or shoreline.

## INFERRED DEPOSITIONAL ENVIRONMENTS AND SYSTEMS

We interpret the depositional sequences in the major coarsening upward interval in the upper Medina as resulting from the episodic progradation of a tide-dominated, wave-influenced shoreline into a storm-dominated epicontinental basin. In this interpretation, deposits of the host-sediment association (other than facies Hbt) represent subtidal, offshore muds and storm-deposited sands. Facies Hbt represents more proximal, very shallow subtidal to intertidal-flat deposits composed of a mixture of sand and mud. Occurrences of the channel association represent shallow subtidal to intertidal estuaries and tidal channels along a generally straight shoreline. Occurrences of the sheet-sandstone association represent shoreline-attached, channelized, shallow subtidal to intertidal sand shoals in shallow water continuously exposed to wave activity. Below we detail this preferred interpretation; a discussion of possible alternatives is also presented.

### Preferred interpretation

Our preferred interpretation of the overall coarsening upward interval in the Medina represents in our opinion the most simple and reasonable reconciliation of all our observations. The small-scale coarsening upward sequences in the open marine host sediments were formed by basin-floor aggradation associated with shoreline progradation; the dominance of mudstone and sharp-based hummocky cross-stratified sandstones in these deposits strongly indicate a subtidal origin in a storm-dominated setting below fair-weather wave base, but well above storm wave base. The storm-generated flows that deposited the offshore sands apparently were directed normal to the shoreline, or at least at a very high angle to the shoreline; in some instances, these flows probably were enhanced by ebb-tidal currents or river flood-waters emanating from nearby tidal channels and estuaries (see further discussion below).

The smaller sequences are separated vertically by thin phosphatic lags and fossiliferous carbonates. These represent regionally extensive episodes of transgression interrupting terrigenous sedimentation associated with the advance of the shoreline. Each subsequent sequence was terminated in increasingly shallower water, as evidenced by the vertical increase (in sequences C, D and E) in the number of sequence-capping channels of inferred shallow water origin. Reversing flow directions recorded by cross bedding within individual channels (Fig. 22) are best explained in terms of tidal processes, an inferrence supported by the presence of *Lingula* and *Arthrophycus* within channel-fill sequences. The vertical increase in the number of flood-tidal palaeoflow directions obtained from these channels (Fig. 22) similarly suggests progressive shoaling associated with episodic shoreline advance.

Where channels have incised during progradation into the offshore deposits of the host-sediment association, we suggest that the channel floor was cut to shallow subtidal depths. The presence of well-developed desiccation cracks at the top of one channel-fill sequence (Fig. 13B) indicates a probable intertidal setting for the shallowest portions of these channels.

The sheet sandstone capping sequence F represents the culmination of the overall coarsening upward sequence, and thus likely represents the most shallow deposit in this interval. We interpret it as a very shallow tidal sand shoal. This shoal probably was shoreline-attached; it may have formed at least in part as a sandy intertidal flat (see below).

The dominantly north-northeastward dip directions of lateral accretion surfaces from both the discrete channels and the channelized portions of sheet sandstones (Fig. 21E) suggest that the Medina shoreline was subjected to a strong net longshore drift to the north or northeast, resulting in the alongshore migration of tidal and estuarine channels towards the north or northeast. We are aware of no other interpretation that can account for the predominance of northeasterly dip directions from these lateral accretion surfaces. The presence of a strong along-shore flow component (directed approximately to the north) superimposed upon the ebb-flood palaeocurrent indicators obtained from the sheet-sandstone association (Fig. 21D) supports the lateral accretion dip directions, and further suggests that the inferred longshore drift was very effective in transporting sand in this deposit. This effectiveness implies a very shallow setting, probably above fair-weather wave base and within the intertidal zone (at least in part). Note that we have yet to observe conclusive evidence for subaerial exposure of sheet sandstones in the Medina; this is not surprising, however, considering their dominantly sandy composition and the probable physical and biological erosion and reworking of these sands during post-depositional transgression at the upper sequence boundary.

Additional evidence for wave-induced longshore drift to the north or northeast is derived from inspection of vortex-ripple orientations throughout the field area (Fig. 21A). In the western region, the crests of these ripples are oriented approximately NNW–SSE. In the central region, their orientations are highly scattered (see further discussion below). Orientations are N–S in the eastern area. This progressive clockwise rotation suggests wave approach mainly from the west-southwest, with refraction against an approximately N–S to NE–SW shoreline. This interpretation is consistent with channel orientations throughout the field area (Fig. 21C,E), and would have resulted in the north to northeast longshore drift inferred from other palaeoflow data.

This consistent longshore drift may have been a major source of sediment to the prograding Medina shoreline, with sand being transported along shore from an area south of the outcrop belt. However, many of the abundant and relatively coarse-grained channel deposits present throughout the field area probably were estuaries; we suspect that the rivers feeding into these channels were the dominant sediment source.

Lateral facies variation within each of the smaller coarsening-upward sequences (Fig. 20) is consistent with approximately shore-normal facies variation across the outcrop belt in the inferred depositional system. Thus, sequences become difficult to trace basinward (to the west), where they are characterized by fine-grained sediments with very slight coarsening upward trends. Large tidal channels and transgressive erosion at sequence boundaries appear to be absent in the western field region; however, sequence-bounding carbonates and fossiliferous, phosphatic lags are very well developed, suggesting prolonged periods of transgressive non-deposition of terrigenous detritus in this deeper setting. In contrast, the sheet sandstones present throughout the Medina in the eastern field region represent repeated establishment of intertidal sand shoals during progradation in this shallower, source-proximal area. Certain of the muddy host sediments between the sheet sands in the far eastern region represent open marine deposits, whereas others (i.e. those of facies Hbt) represent restricted, mixed mud and sand deposits from very shallow subtidal to intertidal flats. The inferred nearly shore-normal transect represented by the outcrop belt is consistent with previous environmental interpretations (Table 1), which have emphasized tidal processes in New York and storm processes in Ontario.

The palaeogeography of the Medina coastal system can be refined further by a more detailed consideration of our palaeoflow data. First, the very close similarity in channel-derived palaeocurrent data from the central and eastern field regions (Fig. 21C) strongly suggests that the shoreline orientation in these two regions was nearly identical. This supports our inference that the clockwise rotation of vortex-ripple orientations from west to east (Fig. 21A) resulted from wave refraction, rather than variation in the orientation of the shoreline across the field area. Additional support derives from inspection of the vortex-ripple data from the eastern field region (Fig. 21A). Although the vector-mean orientation of all these data is nearly N–S, those derived solely from the shallow sheet-sandstone association possess a vector mean of 010–190°, reflecting further clockwise refraction owing to shoaling. We believe this orientation most closely approximates the Medina shoreline; thus, our best estimate is that the shoreline was oriented roughly NNE–SSW. This estimate is supported by other data derived from the sheet-sandstone association (especially in the eastern region; Fig. 21D), which reflect onshore–offshore flows precisely normal to this shoreline orientation, and along-shore flows parallel to it.

We further suggest that the high dispersion of vortex-ripple orientations in the central field region (Fig. 21A) reflects a complex pattern of wave-propagation directions resulting from pronounced topographic relief on the intensely channelized seafloor in this region. This conclusion is supported by the bimodal pattern of vortex-ripple migration directions in this region (Fig. 21A). The eastward-directed ripples probably represent shoaling across relatively low-relief portions of subtidal flats, whereas the northward-directed ripples represent shoaling against the northern margins of tidal channels, by waves approaching the shoreline obliquely from the west-southwest. In the eastern region, the dispersion of vortex-ripple orientations derived solely from the host-sediment association (Fig. 21A) similarly reflects channel-controlled propagation directions. In contrast, the relatively smaller scale and highly dispersed orientations of channels in the sheet sandstones does not appear to have strongly affected vortex-ripple orientations in these shallow-water deposits. Similarly, vortex-ripple orientations in the western field region are strongly coherent owing to the absence of large channels in this region.

As a consequence of our estimate of a NNE–SSW shoreline orientation in the Medina, we must conclude that the NW–SE orientations of the isolated channels of the channel association (Fig. 21C,E) reflects their preferred alignment at a high angle to the shoreline, but *not* precisely normal to it. The deflection of these channels to the right (looking seaward) is consistent with (and in fact is required by) our inference of a strong longshore drift to the north-northeast (see further discussion below).

The coast of the German Bay (Helgoland Bight) of the North Sea (Fig. 23) represents a remarkably close modern analogue for the ancient Medina shorelines inferred in this study; coincidentally, even the orientation of the shoreline and channels at this location closely matches the present orientation of equivalent features in the Medina. This modern setting has been studied thoroughly by H.-E. Reineck and colleagues (Reineck et al., 1967, 1968; Gadow & Reineck, 1969; Reineck & Singh, 1975, pp. 315–330 & 355–372, 1980, pp. 394–397; Aigner & Reineck, 1982). Additionally, Ehlers (1988) has presented an enormous survey of geomorphology and dynamics in this area. The description below is summarized from these works.

The maximum tidal range along the portion of the coast illustrated in Fig. 23 is about 3.7 m. Note the dissection of the sandy intertidal flats and subtidal shoals along this coast by numerous shore-normal, straight to sinuous channels of variable dimensions and slightly variable orientations. Smaller channels, too small to map at this scale, dissect the broad intertidal sand flats between the major channels shown. Most of the channels in Fig. 23 are simple tidal channels, but three of them (the Jade, Weser and Elbe) are estuaries. Two of these (the Weser and the Elbe) bring sediment to the shoreline from major German rivers. The thalwegs of all these

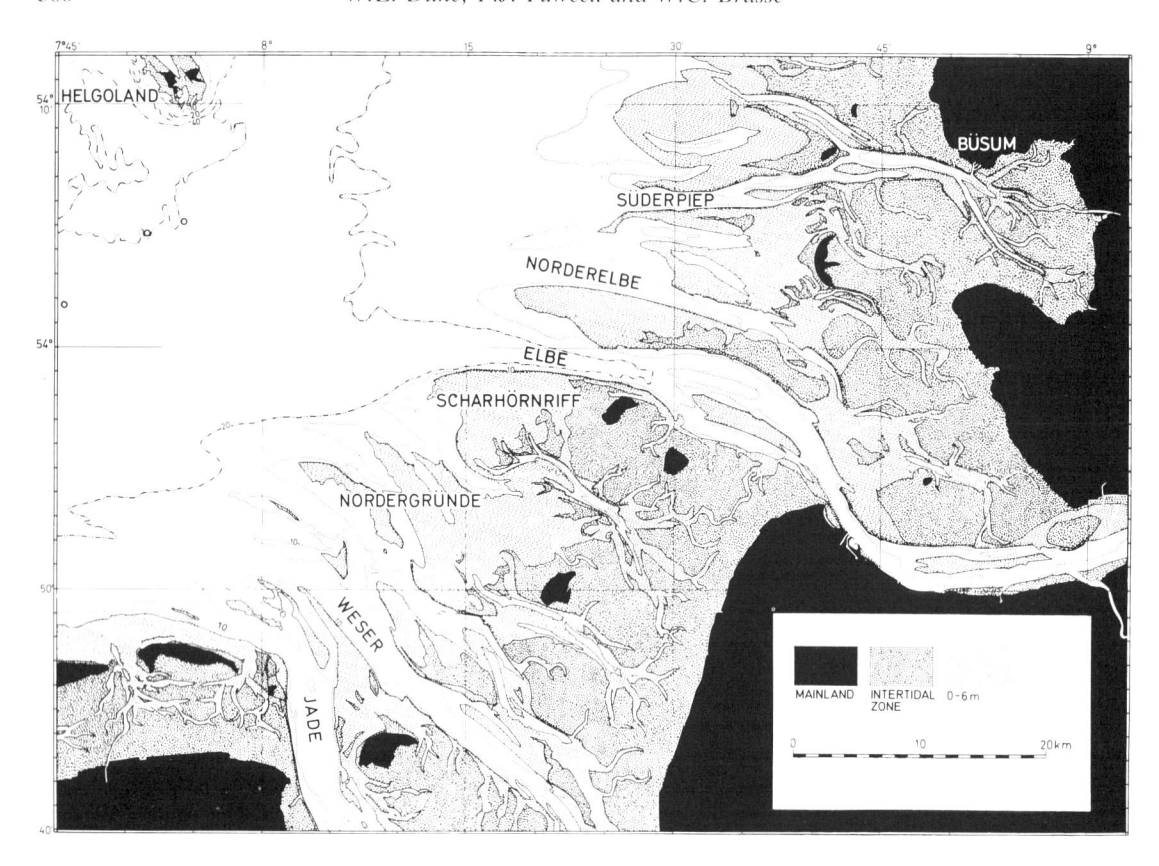

**Fig. 23.** Map of tidal channels and three estuaries (the Jade, Weser and Elbe) along the sandy intertidal flats of the German Bay, North Sea. (From Reineck & Singh (1975, Fig. 463), reprinted by permission of Springer-Verlag.)

large channels are incised to shallow subtidal depths (about 6 m in their upper ends to about 13 m at their seaward mouths; locally, channel depths are as great as 15–20 m). Widths of the major channels increase seaward, ranging from about 1 km on the intertidal flats to about 3 km at their subtidal mouths. These dimensions suggest that the largest of these modern German channels are perhaps two to three times deeper and several times wider than the largest individual Medina channels yet recognized in outcrop; however, they are very similar in size to the multi-story channel complexes recognized locally in the Niagara Gorge. Considering that the modern German channels display multiple, subparallel thalwegs (suggesting that their deposits might resemble our ancient multistory channel complexes), and that the isolated Medina channels have suffered both compaction and transgressive erosion of their upper portions, we suggest that the tidal channel systems in these two depositional settings may have been fairly similar in scale.

The German Bay analogy for the Medina depositional system is further strengthened by the observation that these modern channels are also shifting along shore to the northeast. This effect is most pronounced in the area around Nordergrunde, where the subtidal channels (and their intervening subtidal bars) are shifting to the northeast at an average rate of about 115 m yr$^{-1}$. Note that the channels in this area bend out in a northeasterly direction, in exactly the same manner as we have inferred for the ancient Medina channels.

Thalwegs of the modern tidal channels and estuaries are mantled with medium and coarse quartzitic sands and shelly debris similar to the erosional lags at the bases of the larger Medina channel-fill sequences. The overlying sandy fill within these German channels is in turn mantled by medium-scale three-dimensional ripples (dunes or mega-ripples), the orientations of which reverse with each tidal cycle. Internally, the sands are composed almost entirely of medium-scale trough cross-bedding exhibiting reversing flow directions, with thin layers of shells and flat-laminated fine sand. Locally, muds are dominant in some channels, where sand bypasses the channel floor to a large extent, being washed offshore during storms. These in-channel deposits closely resemble those of facies Ccf.

Offshore from this channelized coastal sand complex, the deeper subtidal channels are incised into muddy basinal sediments and Pleistocene sands. Sharp-based medium to very fine-grained sands interbedded with muds are deposited in subtidal depths seaward of the tidal channels and estuaries; these sands thin and fine basinward, and are flushed into the North Sea by offshore-directed storm-induced coastal downwelling, commonly amplified by ebb-tidal currents or river flood-waters emanating from channel mouths. Box cores from these offshore sand beds display flat lamination, inferred hummocky cross-stratification, and abundant wave-formed ripples (Aigner & Reineck, 1982). These deposits (and their inferred origin) are very similar to those of facies Hss and Hmd. The vertical succession of facies produced by episodic progradation of a coastal system similar to that of the German Bay would be essentially identical to the succession observed in the Medina in outcrop.

We find it difficult to evaluate tidal range along the various Medina shorelines. On the modern Earth, where most coasts currently are undergoing continuing transgression, channelized shorelines without barriers typically are associated with macrotidal areas (Hayes, 1979). There are exceptions to this rule, however (e.g. the portion of the German coast illustrated in Fig. 23 is only mesotidal). High wave energy can likewise inhibit barrier formation (Hayes, 1979). Waves along the ancient Medina coast, for which we infer both a strong storm influence and pronounced fair-weather longshore drift, probably were quite energetic. In any case, the apparent absence of barriers in the Medina may simply reflect its progradational nature, rather than extremely high tidal ranges. The relatively small depth of incision of the Medina channels observed in our work supports a mesotidal origin, whereas the maintenance of inferred estuaries during progradation may suggest a macrotidal origin. Estuaries generally are not maintained during shoreline progradation, but rather act as sediment depocentres from both along-shore currents and fluvial input; nevertheless, small prograding tidal creeks and estuaries are not unknown in the stratigraphic record (e.g. Wright, 1986). Amidst these conflicting observations, we tentatively suggest that the Medina coasts probably were mesotidal.

A palaeogeographic reconstruction of North America during the Early Silurian (Fig. 24) provides a context for the interpretation of sedimentary processes at the tide-dominated, wave-influenced

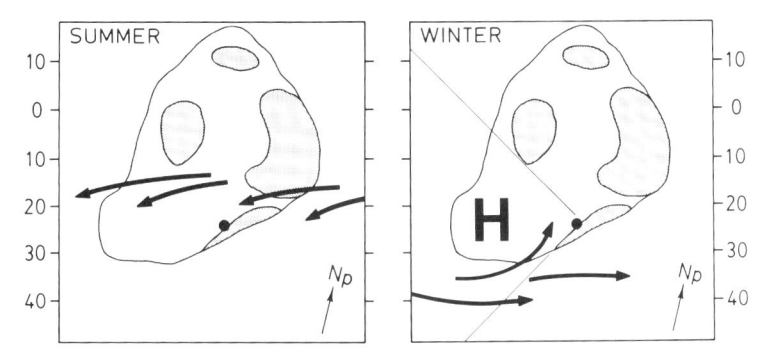

**Fig. 24.** Palaeogeographic reconstruction of North America during the Early Silurian. (After Ziegler *et al.* (1977), Scotese *et al.* (1979), Duke (1982) and Middleton (1987).) In both maps, land areas are densely shaded, shallow seas are lightly shaded, and deep ocean is not shaded. The Medina field area (located at about 25°S) is indicated by the filled circle. Arrows labelled 'Np' indicate the present orientation of true north within the field area. In the southern summer reconstruction, the large arrows represent inferred tracks of Southern Hemisphere tropical storms and hurricanes, which we suggest were the primary generative agents for offshore storm deposits in the Medina. In the southern winter reconstruction, the large arrows represent inferred tracks of Southern Hemisphere mid-latitude storms. Note that rotation of these storms around the high pressure area could have brought them up to the Medina shoreline. The thin straight lines enclose that area of the sea from which the Medina coast was exposed to long-period storm-generated swell waves propagating from the present west and southwest.

Medina shoreline and in the shallow storm-dominated Medina sea. As previously suggested by Duke (1982, 1985a,b, 1987a), the tropical southern palaeolatitude of the field area presumably would have exposed this epicontinental sea to summer tropical storms and hurricanes, either generated locally over this warm epicontinental sea, or advected from the eastern oceanic basin through a narrow seaway to the northeast (Fig. 24).

Fetch distances at the shoreline were much greater to the south and west than to the north; additionally, rotation of surface winds around an inferred high pressure cell west of the field area (Ziegler et al., 1977) would have been counter-clockwise (Fig. 24). Thus, during the winter (and during fair-weather intervals in the summer), the Medina shoreline would have been exposed to long-period swell waves arriving mainly from the present-day south and west (Fig. 24). Longshore drift beneath these waves generally would have been directed to the present-day north-northeast along the Medina shoreline, as is indicated by the palaeoflow data discussed above. Note that Cotter (1983) also has inferred a northeasterly longshore drift for similarly oriented shorelines in the temporally equivalent Tuscarora Formation of Pennsylvania. Dominant wave approach from the west, south, southwest, and north-northwest has been inferred previously for the slightly younger Niagaran reefs of the Great Lakes area (Lowenstam, 1957; Ingels, 1963; Crowley, 1973; Droste & Shaver, 1977). See Middleton (1987) for a further discussion of these Niagaran palaeoflow data.

Given that storm-induced offshore circulation in modern shallow seas generally is geostrophically balanced (see discussion by Swift & Thorne, this volume, pp. 13–15), further comment is needed concerning our inference of storm-deposited sands emplaced offshore by flows directed nearly normal to the Medina shoreline. As seen in Fig. 1, the westernmost exposures examined for this study (exposures 1–3) lie approximately 20 km from exposure 10 (Devil's Punchbowl) and 55 km from exposure 14 (Ball's Falls). These latter two exposures contain the westernmost examples of large, well-developed tidal channels and sheet sandstones correlative with the offshore storm deposits of the westernmost exposures (see Fig. 20). In this area, the angle between the outcrop belt and the inferred shoreline trend is very high; we therefore place the offshore storm deposits in exposures 1–3 a distance of roughly 20–50 km from the adjacent, coeval belt of shallow subtidal and intertidal channels. Thus, we infer that offshore storm flows were directed normal to the shoreline (or at least at a very high angle to the shoreline) for a minimum distance of about 50 km seaward of the low-tide line. The Medina Basin was situated at low latitude (about 25°S; see below), where the value of the Coriolis parameter was relatively small. Additionally, the great lateral extent of the thin depositional sequences in the Medina (Fig. 20) strongly suggests that water depths throughout the study area were nowhere greater than several metres to a few tens of metres; thus, the Coriolis force would have been damped significantly by friction. It therefore seems reasonable to us to suggest that a distance of 50 km was inadequate in this setting to produce a large degree of geostrophic veering of the downwelling flows jetting from the channelized shoreline. Nevertheless, close inspection of palaeocurrent data from the host-sediment association (Fig. 21B) reveals a counter-clockwise shift in these indicators of a few tens of degrees from the central to the western field regions. This progressive rotation to the left (looking seaward) is consistent with a component of geostrophic veering of bottom flows resulting from storm-generated coastal set-up along the Southern Hemisphere Medina shoreline.

## Alternative interpretations of Medina channels

During the course of this investigation, we have entertained numerous different hypotheses for the origin of shallow marine channels in the Medina. Because our interpretation is somewhat different from those previously suggested (see Table 1), we feel compelled to discuss the more reasonable of our alternative hypotheses. These include: (1) fluvial channels cut during rapid falls of sea-level; (2) storm-cut cross-shelf channels; and (3) deltaic distributary channels.

### Fluvial lowstand channels

A possible scenario for channel incision in the offshore sediments of a very shallow epicontinental sea involves a rapid fall of relative sea-level, sub-aerial exposure of offshore sediments, and erosional fluvial incision across these newly exposed deposits. During a subsequent marine transgression, the incised channels would become the focus of transgressive estuarine sedimentation, producing the tidal channel-fill sequences inferred in this study. In such a situation, evidence for the exposure of marine sediments (both between and within channels) could be destroyed by both physical and biological means during the subsequent marine inundation.

While we cannot conclusively eliminate this possible origin for Medina channels, it is noteworthy that we have not yet encountered evidence for subaerial exposure, either at sequence boundaries laterally adjacent to channels, or at the incised bases of the channels themselves. Because this hypothesis involves a more complicated mechanism for channel generation, requiring undocumented sea-level falls, we have rejected it.

## Storm-cut cross-shelf channels

Large sand-filled channels shallowly incised across the inner shelf or in shallow epicontinental seas are a relatively rare feature of both modern and ancient shallow marine deposits; the most likely interpretation of most occurrences involves incision by locally intensified offshore-directed bottom currents associated with storm-induced coastal downwelling (see summary by Cacchione *et al.*, 1984).

The shallow-marine Medina channels have dimensions, spacings and orientations similar to modern storm-cut cross-shelf channels. Palaeoflow data reflect in-channel flows directed predominently offshore. These features of the Medina channels are quite similar to those characteristic of modern storm-cut cross-shelf channels. Most importantly, many of the Medina channels are fully encapsulated by fine-grained sediments deposited in an open, storm-dominated shelf setting. These observations suggest that the Medina channels also may have been incised by storms on the open shelf, as one of us has speculated previously (Duke, 1982, 1985a; see Table 1).

Certain of our more recent observations are not consistent with this inferred origin, however. For instance, our abundant palaeoflow data (Figs 21, 22) strongly suggest that the Medina channels were both sinuous and meandering. Modern cross-shelf channels typically are straight, and there is no evidence to suggest that they migrate laterally. Additionally, the presence of desiccation cracks at the top of one channel-fill sequence represents a major inconsistency. Likewise, reversing palaeoflow directions from trough cross-bedding within several channels (Fig. 22) are difficult to reconcile with an origin resulting from coastal downwelling. Finally, our observation that channels are truncated by minor sequence boundaries has profound implications for their interpretation. The inferred break in the sedimentary record at such a boundary, herein interpreted as resulting from a rise in relative sea-level, leaves open the possibility that the channels were incised in a different (shallower) environmental setting than is represented by the encapsulating deposits. For these reasons, we have rejected the coastal downwelling hypothesis as a general mechanism for channel incision.

This hypothesis should not be dismissed entirely, however, because certain of the very small, non-

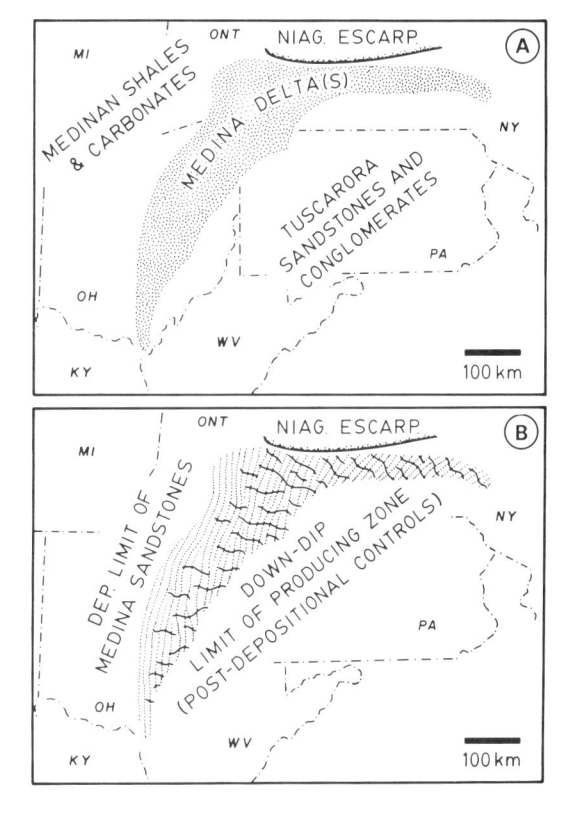

**Fig. 25.** Alternative interpretations of the palaeogeography of the Medina shoreline. (A) The Medina producing region, interpreted as a delta (or deltas) by most previous workers (see Table 1). (After Martini (1971) and Friedman *et al.* (1982).) (B) In the interpretation presented herein, the boundaries of the producing region are controlled in part by various post-depositional effects (see text). Throughout the field area (adjacent to the Niagara Escarpment), the prograding Medina shorelines maintained an approximately NNE–SSW orientation and were dissected by approximately shore-normal tidal channels and estuaries oriented NW–SE.

migratory channels (Facies Cff) below sequence boundaries in the western field region are not conveniently interpreted in terms of tidal processes. Small channels, such as the one capping sequence F at the Highway 403 exposure (Figs 4, 20), cannot be interpreted as the truncated lower portions of distal tidal channels; the distal mouths of these channels should be relatively wide (cf. Fig. 23). Additionally, larger channels and sheet sandstones are absent in the far western field region, leading us to infer that this area was entirely subtidal throughout its depositional history. Thus, the small, distal channels may perhaps have been incised by storm-induced downwelling during maximum regression, when water depths were at a minimum and basin-floor friction was maximized throughout the field area.

*Deltaic distributaries*

Our preferred interpretation of Medina channels is quite similar to the deltaic distributary origin previously suggested by Martini (1966, 1971, 1974), Friedman *et al.* (1982), and several others (see Table 1). The essential difference involves the inferred shape of the shoreline. Our inference of a series of straight shorelines, with bathymetric contours running NNE–SSW on the adjacent northwest-sloping basin floor, contrasts markedly with the previous interpretation involving a broadly curved shoreline.

The deltaic interpretation may have been inspired in part by the broadly arcuate shape of the Medina producing region in the subsurface (Fig. 25A). We suggest that this shape is not especially significant to a depositional interpretation. As indicated in Fig. 25B, the position of the E–W trending updip boundary of the production zone south of Lake Ontario is controlled by the location of the Niagara Escarpment. The position of the entire downdip boundary has been influenced by a variety of post-depositional effects (e.g. structural deformation, fluid-migration paths, diagenesis and cementation). In contrast, the western updip boundary (trending N–S in Ohio and NE–SW to N–S under Lake Erie into Ontario) is controlled by primary lithological variation in these rocks; this variation probably is roughly parallel to palaeobathymetric contours and the ancient shoreline trend. Thus, our inferred NNE–SSW shoreline orientation throughout Medina time probably is approximately applicable to the entire production region. In this regard, however, it is important to note that our analysis cannot eliminate a *nearly* straight shoreline associated with a broad deltaic distributary system elongated by the strong north-northeasterly longshore drift inferred from our palaeoflow data. Perhaps the differences between such a delta and our concept of a straight prograding shoreline dissected by estuaries is more semantic than sedimentological.

### Alternative interpretation of sheet sandstones in the Medina

Thick, sheet-like sandstone bodies in the Medina are best developed in the relatively thin and coarse eastern exposures (Fig. 20). In this area, the Medina can be regarded as a highly condensed deposit (relative to its character further to the west). Because the sheet sandstones commonly are complexly

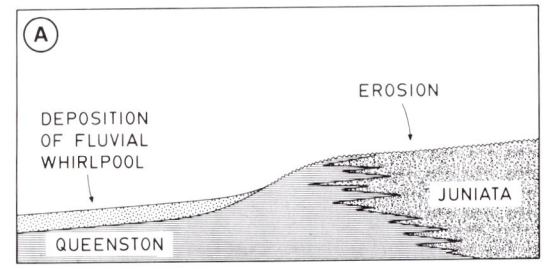

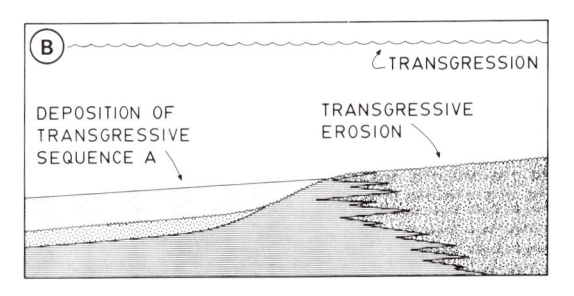

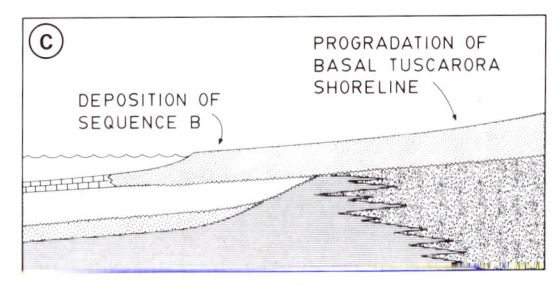

**Fig. 26.** Cartoon illustrating hypothetical depositional relationships between basal Medina and basal Tuscarora strata. No relative scale is implied. See text for explanation.

channelized internally, each might be interpreted as resulting from multiple episodes of reworking by numerous channels incised at the same stratigraphic horizon, during prolonged periods of reduced net sediment aggradation. Thus, the sandstone sheets could represent cannibalization of sediment by sand-dominated channels in a depositional system similar to that represented by the isolated channel-fill sequences, with the finer sediments eroded and bypassed to more distal locations in the basin. In this scenario, deposits of the sheet-sandstone association differ from those of the channel association as the result of variation in long-term sedimentation rates, rather than depth or even proximity to source.

Our work cannot eliminate the possibility of cannibalization as described above; indeed, given the condensation of the Medina in the east, it seems likely that this process operated to influence sedimentation in this area in some undetermined fashion and degree. We nevertheless prefer to interpret the sandstone sheets as shallow subtidal to intertidal sand shoals attached to the shoreline, closer to shore than the isolated channel-fill sequences. Only this latter interpretation offers a plausible explanation of: (1) palaeoflow differences between the two associations, with strong along-shore flow directions recorded in the sheet sandstones but absent in the isolated channel deposits; and (2) the enhanced abundance of *Arthrophycus* and *Daedalus* in sheet sandstones.

## REGIONAL CORRELATIONS AND THE ORIGIN OF DEPOSITIONAL CYCLES IN THE MEDINA

The conceptual basis for our correlations involving the easternmost Medina exposures is derived in part from subsurface work in New York, immediately south of the outcrop belt. In this area, a NE–SW trending subsurface high is developed on the Queenston–Medina contact; the presence of the high in this area has been well documented by oil-industry workers who prefer to retain their anonymity. The subsurface high trends directly towards the Genesee Gorge section (exposure 40). It apparently existed at the surface prior to Medina sedimentation, because the Whirlpool Sandstone is present only to the northwest of the high. Note that several published reports have documented similar stratigraphic relationships along strike to the southwest, in the subsurface of New York and northwestern Pennsylvania (see summary in fig. 7 of Pees, 1986). We suggest that the same stratigraphic relationships observed in the subsurface are apparent in our easternmost exposures. Martini (1966) has previously correlated the basal lithic units in these exposures in much the same way.

The limitation of the Whirlpool to an area northwest of the subsurface high appears to resolve a dilemma of the braided-stream interpretation advanced by G.V. Middleton and colleagues (Table 1). How can this environmental interpretation of the basal Medina be reconciled with the beach environment inferred by Cotter (1983) for the basal Tuscarora Formation in central Pennsylvania? Figure 26 shows a tentative correlation and depositional history of the basal Medina and Tuscarora strata. In panel A, the Whirlpool is deposited in braided streams receiving reworked sands from the Late Ordovician Juniata Formation in central Pennsylvania and the sandy upper portion of the Queenston Formation of New York. This interpretation has been suggested previously by Fisher (1954) and Middleton (1987) to explain the great textural and mineralogical maturity of the Whirlpool. In panel B, marine strata overstep the topographic high southeast of the Whirlpool, forming a thinning and fining upward sequence during transgression (our sequence A). In panel C, renewed influx of coarse sediment from the prograding beach of the basal Tuscarora is deposited as regressive sequence B in the Medina. The erosive base of the B-sequence sandstone (Figs 20, 26) suggests erosional progradation of the Medina B-sequence shoreline during a period of rapid sea-level fall locally. This erosion also may have resulted in the reworking of underlying sandy strata, in a fashion similar to that of the Whirlpool. This may account for the anomalous coarseness of the sheet sandstone in the B sequence and the lack of shale basinward of this sandstone, which grades directly into carbonates in the Niagara region.

In broad terms, our tentative regional correlations are supported by the observation that the oldest marine fossils in the Medina are some of the oldest Silurian fossils in North America (Kilgour & Liberty, 1981). However, detailed subsurface correlations and palaeontological work are needed to determine the origin of the subsurface high and to evaluate the speculative relationships of Fig. 26. Such work is also required before attempting to correlate our Medina sequences with a similar set of sequences observed by Cotter (1983) in the Tuscarora Formation, or with depositional cyclicity developed in other parts of the Appalachian Basin and elsewhere.

In light of the comments above, we believe it is premature at this time to fully evaluate the possible sedimentological, tectonic, climatic, or other controls of the origin of sedimentary sequences observed in the Medina and other Lower Silurian strata. Nevertheless, it is intriguing to speculate about the possibilities. With varying results, numerous authors have attemped to correlate sea-level curves inferred from Lower Silurian strata across North America and around the world. Johnson & Campbell (1980) saw little similarity between Medinan-aged sea-level curves from mid-continental Iowa, the Michigan Basin, and the Medina Group in New York. However, herein we have inferred a bathymetric history for the Medina different from that used by these authors. Our work suggests an overall pattern of transgression (sequence A), major regression (sequences B–F), transgression (base of sequence G), regression (sequence G), followed by renewed transgression at the onset of Clinton time. With the exception of the G-sequence regression, this pattern is similar to the bathymetric history of the Michigan Basin documented by Johnson & Campbell (1980). As these authors suggested, the apparent absence of sea-level fluctuation in cratonic Iowa probably reflects essentially stable global sea-levels during Medinan time, an interpretation generally consistent with the worldwide sea-level correlations of McKerrow (1979). In contrast, the complex history apparently shared by the Michigan and northern Appalachian Basins suggests that both may have been affected by less extensive regional processes (probably tectonic), at least during most of the Medinan Stage. We tentatively attribute this pattern to late Taconic thrusting.

The genesis of the smaller regionally extensive Medina sequences is more difficult to constrain. Dividing the number of these sequences into the length of time represented by the Medinan Stage produces an estimate of less than 900 000 years for the average duration of a single depositional sequence. The high-frequency eustatic sea-level variation recorded in slightly older strata in Ontario (attributed to Late Ordovician glacial cycles) does not appear to have persisted into the Early Silurian (McKerrow, 1979; Petryk, 1981; Middleton, 1987). Thus, the small Medina sequences are best attributed to the effects of one or more regional influences, such as: (1) variation in subsidence rate in the Medina Basin (possibly including episodes of gentle regional uplift) resulting from high-frequency tectonic events in the adjacent Taconic highlands to the southeast, (2) tectonically controlled variation in sedimentation rate at the Medina shoreline; (3) variation in sedimentation rate owing to short-term regional climatic fluctuations; and (4) variation in sedimentation rate resulting from complex system response within the coastal and alluvial plain systems of the Tuscarora to the southeast. Perhaps the more attractive of these alternatives involves the migration of a small peripheral bulge resulting from repeated episodes of late Taconic thrusting, a possibility that lends itself to numerical modelling (e.g. Quinlan & Beamont, 1984).

## SUMMARY AND CONCLUDING REMARKS

Herein we have described spatial facies relationships in exposures of the Lower Silurian Medina Group in New York and Ontario. Most of the upper half of this unit is composed of an overall thickening and coarsening-upward sequence composed primarily of interbedded quartzitic sandstones and mudstones. This sequence varies laterally from about 15 to 25 m in thickness, and is interpreted herein as a prograding shoreline sequence formed in a shallow epicontinental sea. Superimposed on this general coarsening upward trend are thinner (1–12 m), vertically adjacent, laterally extensive coarsening upward sequences separated vertically by non-depositional to erosional disconformities. Physical and biological evidence suggests that these smaller sequences represent episodic shoaling associated with shelf aggradation and shoreline progradation, superimposed upon the larger overall shoaling sequence. Vertical and lateral facies relationships and abundant palaeoflow data suggest that the offshore, subtidal Medina Basin was storm-dominated, whereas the straight Medina shoreline, dissected by numerous nearly shore-normal tidal channels and estuaries, was tide-dominated and strongly wave-influenced. The sandy shoreline was oriented NNE–SSW, as were bathymetric contours on the adjacent basin floor sloping gently to the west-northwest. Strong longshore drift along the shoreline was directed predominantly to the north-northeast, resulting from dominantly oblique wave-approach directions from the west-southwest.

Owing to the episodic progradation of the Medina

shoreline, shallow subtidal to intertidal channels locally are both incised into and overlain by offshore mudstones and thin storm-deposited sandstones. Early in this investigation, we interpreted these rather mysterious sand-filled channels, totally encapsulated by fine-grained offshore sediments, as offshore channels incised by storm-driven coastal downwelling. As we accumulated palaeocurrent data from these channels, however, the presence of subordinate onshore-directed flows within them became increasingly apparent; simultaneously, our more detailed inspection of the outcrops revealed the subtle transgression surfaces that truncate these 'encapsulated' channel deposits. Had we failed to make these observations, we would not have arrived at our present interpretation involving shallow tidal flows. This brief history of our investigation leads us to suggest that perhaps the most important recognition of our study is the necessity for painstakingly detailed inspection of shallow-marine intervals, including a deliberate search for minor disconformities and all available palaeoflow indicators.

## ACKNOWLEDGEMENTS

Principle funding for this work was provided by a Faculty Career Initiation Grant from the Shell Companies Foundation (to W.L. Duke). Additional support was provided by a research assistantship from the Earth Systems Science Center at The Pennsylvania State University (to P.J. Fawcett), an honorarium from the New York State Geological Survey (to W.C. Brusse), and grants from the P.D. Krynine Memorial Fund of the Department of Geosciences at The Pennsylvania State University (to P.J. Fawcett and W.C. Brusse). We are most grateful for this support. Numerous colleagues have shared their insight in the field; these include G.V. Middleton, M.A. Rutka, C.E. Brett, W.M. Goodman, R.G. Walker, A.G. Plint, B.H. Ainsworth, M.J. Risk, S.G. Pemberton, E. Cotter, R.L. Slingerland, and the participants in the 1987 Annual Field Trip of the Eastern Section of the Society of Economic Paleontologists and Mineralogists. C. Moy kindly provided us with her micropaleontological findings. Helpful comments on the preliminary text were provided by the reviewers (G.V. Middleton, S.G. Driese, D.J.P. Swift, and one anonymous person) and by C.E. Brett, W.M. Goodman, R.L. Slingerland, and R.J. Cuffey. Many thanks to all these friends. This paper was written in 1988.

## Note added in proof

Our use of the term *sequence* throughout this paper is strictly generic, and does not correspond with the concept of depositional sequence embodied in modern statements of sequence stratigraphy.

## REFERENCES

AIGNER, T. & REINECK, H. -E. (1982) Proximity trends in modern storm sands from the Helgoland Bight (North Sea) and their implications for basin analysis. *Senckenbergiana Maritima* **14**, 183–215.

ARMSTRONG, H.S. (1953) Geology of part of the Niagara Peninsula of Ontario. In: *Guidebook for Field Trips 4 & 5*. Geological Society of America and Geological Association of Canada, Toronto, Ontario, 20 pp.

ARMSTRONG, H.S. (1955) The Silurian section in the vicinity of Hamilton, Ontario. The Niagara Escarpment of Peninsular Ontario, Canada. In: *Guidebook for the 1955 Annual Field Trip of the Michigan Geological Society*, pp. 44–45. Michigan Geological Society.

BRETT, C.E. & CALKIN P.E. (1987) Niagara Falls and Gorge, New York–Ontario. *Centennial Field Guide*, pp. 97–105. Geological Society of America, Northeastern Section.

BRUSSE, W.C., DUKE, W.L., FAWCETT, P.J., MIDDLETON, G.V., RUTKA, M.A. & SALAS, C.J. (1987) Stop descriptions. In: *Sedimentology, Stratigraphy, and Ichnology of the Lower Silurian Medina Formation in New York and Ontario* (Ed. Duke, W.L.) pp. 96–185. Guidebook for the 1987 Annual Field Trip of the Eastern Section of the Society of Economic Paleontologists and Mineralogists.

BRUSSE, W.C., FAWCETT, P.J. & DUKE, W.L. (1988) Mixed storm- and tide-influenced prograding shoreline deposits in the Lower Silurian Medina Formation, New York and Ontario (abstr.). *Geol. Soc. Am. Abstr. Progr.* **20**, 10.

CACCHIONE, D.A., DRAKE, D.E., GRANT, W.D. & TATE, G.B. (1984) Rippled scour depressions on the inner continental shelf off central California. *J. Sediment. Petrol.* **54**, 1280–1291.

CHADWICK, G.H. (1918) Stratigraphy of the New York Clinton. *Geol. Soc. Am. Bull.* **29**, 327–368.

CHADWICK, G.H. (1935) Geological notes—Thorold Sandstone. *Bull., Am. Assoc. Petrol. Geol.* **19**, 702.

COTTER, E. (1983) Shelf, paralic, and fluvial environments and eustatic sea-level fluctuations in the origin of the Tuscarora Formation (Lower Silurian) of central Pennsylvania. *J. Sediment. Petrol.* **53**, 25–49.

CROWLEY, D.J. (1973) Middle Silurian patch reefs in Gasport Member (Lockport Formation), New York. *Bull., Am. Assoc. Petrol. Geol.* **57**, 283–300.

DROSTE, J.B. SHAVER, R.H. (1977) Synchronization of deposition: Silurian reef-bearing rocks on Wabash Platform with cyclic evaporites of Michigan Basin. In: *Reefs and Evaporites—Concepts and Depositional Models* (Ed. Fisher, J.H.). Am. Assoc. Petrol. Geol. Stud. Geol. 5, 93–109.

DUKE, W.L. (1982) The 'type locality' of hummocky cross-stratification: The storm-dominated Silurian

Medina Formation in the Niagara Gorge, New York and Ontario. *Proc. Ontario Petrol. Inst.* **21**, 2.1–2.31.

DUKE, W.L. (1985a) Prograding shoreline sequence, Silurian Medina Formation, New York and Ontario: storm-dominated sedimentation in a very shallow epicontinental sea (abstr.). *Geol. Soc. Am. Abstr. Progr.* **17**, 16.

DUKE, W.L. (1985b) Hummocky cross-stratification, tropical hurricanes, and intense winter storms. *Sedimentology* **32**, 167–194.

DUKE, W.L. (1987a) Hummocky cross-stratification, tropical hurricanes, and intense winter storms: Reply. *Sedimentology* **34**, 344–359.

DUKE, W.L. (1987b) Revised internal stratigraphy of the Medina Formation in outcrop: an illustration of the inadequacy of color variation as a criterion for lithostratigraphic correlation. In: *Sedimentology, Stratigraphy, and Ichnology of the Lower Silurian Medina Formation in New York and Ontario* (Ed. Duke, W.L.) pp. 16–30. Guidebook for the 1987 Annual Field Trip of the Eastern Section of the Society of Economic Paleontologists and Mineralogists.

DUKE, W.L. & BRUSSE, W.C. (1987) Cyclicity and channels in the upper members of the Medina Formation in the Niagara Gorge. In: *Sedimentology, Stratigraphy, and Ichnology of the Lower Silurian Medina Formation in New York and Ontario* (Ed. Duke, W.L.) pp. 46–65. Guidebook for the 1987 Annual Field Trip of the Eastern Section of the Society of Economic Paleontologists and Mineralogists.

DUKE, W.L. & FAWCETT, P.J. (1987) Depositional environments and regional sedimentation patterns in the upper members of the Medina Formation. In: *Sedimentology, Stratigraphy, and Ichnology of the Lower Silurian Medina Formation in New York and Ontario* (Ed. Duke, W.L.) pp. 81–94. Guidebook for the 1987 Annual Field Trip of the Eastern Section of the Society of Economic Paleontologists and Mineralogists.

DURAND, J. (1985) *Le Gres Armoricain. Sedimentologie—Traces fossiles. Milieux de depot*. Memoires et Documents du Centre Armoricain d'Etude Structurale des Socles, Universite de Rennes, no. 3, 150 pp.

EHLERS, J. (1988) *The Morphodynamics of the Wadden Sea*. Balkema, Rotterdam, 397 pp.

FAIRCHILD, H.L. (1901) Beach structure in Medina Sandstone. *Am. Geol.* **28**, 9–14.

FAWCETT, P.J., BRUSSE, W.C. & DUKE, W.L. (1988) Allostratigraphic correlations indicate the need for a revised member stratigraphy in the Silurian Medina Formation, New York and Ontario (abstr.). *Geol. Soc. Am. Abstr. Progr.* **20**, 17.

FISHER, D.W. (1954) Stratigraphy of the Medina Group, New York and Ontario. *Bull. Am. Assoc. Petrol. Geol.* **38**, 1979–1996.

FRIEDMAN, G.M., SANDERS, J.E. & MARTINI, I.P. (1982) Sedimentary facies: Products of sedimentary environments in a cross section of the classic Appalachian Mountains and adjoining Appalachian Basin in New York and Ontario. In: *Field Excursion Guidebook for Excursion 17A of the 11th International Congress on Sedimentology*, pp. M1–M30. McMaster University, Hamilton, Ontario.

GADOW, S. & REINECK, H.-E. (1969) Ablandiger Sandtransport bei Sturmfluten. *Senckenbergiana Maritima* **1**, 63–78.

GILBERT, G.K. (1899) Ripple-marks and cross-bedding. *Geol. Soc. Am. Bull.* **10**, 135–140.

GOODMAN, W.M. & BRETT, C.E. (1988) Multiple regional unconformities in Silurian strata of the northern Appalachian Basin (abstr.) *Geol. Soc. Am. Abstr. Progr.* **20**, 17.

GRABAU, A.W. (1901) Guide to the geology and paleontology of Niagara Falls and vicinity. *New York State Mus. Bull.* **45**, 1–85.

GRABAU, A.W. (1905) Physical character and history of some New York formations. *Science* **22**, 528–535.

GRABAU, A.W. (1908) A revised classification of the North American Siluric System. *Science* **27**, 622–623.

GRABAU, A.W. (1909) Physical and faunal evolution of North America during Ordovicic, Siluric, and Early Devonic time. *J. Geol.* **17**, 209–252.

GRABAU, A.W. (1913) Early Paleozoic delta deposits of North America. *Geol. Soc. Am. Bull.* **24**, 399–528.

HALL, J. (1843) *Geology of New York, Part 4, Comprising the Survey of the Fourth Geological District*. New York State Geological Survey, Albany, 683 pp.

HAYES, M.O. (1979) Barrier island morphology as a function of tidal and wave regime. In: *Barrier Islands* (Ed. Leatherman, S.P.) pp. 1–27. Academic Press, New York.

INGELS, J.J.C. (1963) Geometry, paleontology, and petrography of Thornton reef complex, Silurian of northeastern Illinois. *Bull. Am. Assoc. Petrol. Geol.* **47**, 405–440.

JOHNSON, M.E. & CAMPBELL, G.T. (1980) Recurrent carbonate environments in the Lower Silurian of northern Michigan and their inter-regional correlation. *J. Paleontol.* **54**, 1041–1057.

KILGOUR, W.J. & LIBERTY, B.A. (1981) Detailed stratigraphy. In: *Colossal cataract—The Geologic History of Niagara Falls* (Ed. Tesmer, I.H.) pp. 94–122. State University of New York Press, Albany, NY.

KINDLE, E.M. (1914) What does the Medina Sandstone of the Niagara section include? *Science* **39**, 915–918.

KINDLE, E.M. & TAYLOR, F.B. (1913) Description of the Niagara folio. *U.S. Geol. Surv. Geol. Atlas, Niagara Folio*, **190**, 25 pp.

LAUGHREY, C.D. (1984) Petrology and reservoir characteristics of the Lower Silurian Medina Group sandstones, Athens and Geneva fields, Crawford County, Pennsylvania. *Pennsylvania Geol. Surv. Mineral Resour. Rep.* **85**, 126 pp.

LEGGITT, S.M. (1985) *A study of the diagenetic history and proposed depositional environment of the Manitoulin Formation in southern Ontario*. Unpublished BA thesis, McMaster University, Hamilton, Ontario, 93 pp.

LEHMAN, B. (1986) *Depositional environments of the Grimsby Sandstone in upstate New York*. Unpublished BS thesis, Hobart and William Smith Colleges, Geneva, NY, 62 pp.

LOWENSTAM, H.A. (1957) Niagaran reefs in the Great Lakes area. In: *Treatise on Marine Ecology and Paleoecology* (Ed. Ladd, H.S.). Geol. Soc. Am. Mem. 67, 215–248.

LUMSDEN, D.N. & PELLETIER, B.R. (1969) Petrology of the Grimsby Sandstone (Lower Silurian) of Ontario and

New York. *J. Sediment. Petrol.* **39**, 521–530.

MARTINI, I.P. (1966) *The sedimentology of the Medina Formation along the Niagara Escarpment (Ontario and New York state)*. Unpublished PhD thesis, McMaster University, Hamilton, Ontario, 420 pp.

MARTINI, I.P. (1971) Regional analysis of sedimentology of Medina Formation (Silurian), Ontario and New York. *Bull. Am. Assoc. Petrol. Geol.* **55**, 1249–1261.

MARTINI, I.P. (1974) Deltaic and shallow marine Lower Silurian sediments of the Niagara Escarpment between Hamilton, Ont., and Rochester, N.Y. A field guide. *Maritime Sediments* **10**, 52–66.

MCKERROW, W.S. (1979) Ordovician and Silurian changes in sea level. *J. Geol. Soc. London* **136**, 137–145.

METZGER, S.L. (1982) Subsurface paleoenvironmental analysis of gas-producing Medina Group (Lower Silurian), Chautauqua County, New York (abstr.). *Bull. Am. Assoc. Petrol. Geol.* **66**, 606.

MIDDLETON, G.V. (1982) *A Brief Guide to the Sedimentology of the Hamilton Area*. Technical Memorandum 82-1, McMaster University, Hamilton, Ontario, 26 pp.

MIDDLETON, G.V. (1987) Geologic setting of the Northern Appalachian Basin during the Early Silurian. In: *Sedimentology, Stratigraphy, and Ichnology of the Lower Silurian Medina Formation in New York and Ontario* (Ed. Duke, W.L.) pp. 1–15. Guidebook for the 1987 Annual Field Trip of the Eastern Section of the Society of Economic Paleontologists and Mineralogists.

MIDDLETON, G.V. & RUTKA, M.A. (1986) The nature of the marine transgression in the Lower Silurian of Ontario (abstr.). *Geol. Soc. Am. Abstr. Progr.* **18**, 55.

MIDDLETON, G.V., SALAS, C.J. & MARTINI, I.P. (1983) Whirlpool Sandstone (Silurian) of southern Ontario, a braided river deposit? (abstr.). *Geol. Soc. Am. Abstr. Progr.* **15**, 128.

MIDDLETON, G.V., RUTKA, M.A. & SALAS, C.J. (1987) Depositional environments in the Whirlpool Sandstone Member of the Medina Formation. In: *Sedimentology, Stratigraphy, and Ichnology of the Lower Silurian Medina Formation in New York and Ontario* (Ed. Duke, W.L.) pp. 31–45. Guidebook for the 1987 Annual Field Trip of the Eastern Section of the Society of Economic Paleontologists and Mineralogists.

PEES, S.T. (1986) Geometry and petroleum geology of the Lower Silurian Whirlpool Formation, portion of NW Pennsylvania and NE Ohio. *Northeastern Geol.* **8**, 171–200.

PEMBERTON, S.G. (1979) *Selected studies in lower Paleozoic ichnology*. Unpublished PhD thesis, McMaster University, Hamilton, Ontario, 516 pp.

PEMBERTON, S.G. (1987). Ichnology of the Thorold Sandstone in the vicinity of Hamilton, Ontario: a Silurian storm-influenced deposit. In: *Sedimentology, Stratigraphy, and Ichnology of the Lower Silurian Medina Formation in New York and Ontario* (Ed. Duke, W.L.) pp. 66–80. Guidebook for the 1987 Annual Field Trip of the Eastern Section of the Society of Economic Paleontologists and Mineralogists.

PEMBERTON, S.G. & RISK, M.J. (1982) Middle Silurian trace fossils in the Hamilton, Ontario region: their identification, abundance, and significance. *Northeastern Geol.* **4**, 98–104.

PETRYK, A.A. (1981) Upper Ordovician glaciation: effects of eustatic fluctuations on the Anticosti platform succession, Quebec. In: *IUGS Subcommission on Silurian Stratigraphy: Ordovician–Silurian Boundary Working Group* (Ed. Lesperance, P.J.), Field Meeting, Anticosti-Gaspe, Quebec, Vol. 2, pp. 81–85.

PIOTROWSKI, R.G. (1981) Geology and natural gas production of the Lower Silurian Medina Group and equivalent rock units in Pennsylvania. *Pennsylvania Geol. Surv. 4th Ser. Mineral Resour. Rep.* **82**, 21 pp.

QUINLAN, G.M. & BEAUMONT, C. (1984) Appalachian thrusting, lithospheric flexure, and the Paleozoic stratigraphy of the Eastern Interior of North America. *Can. J. Earth Sci.* **21**, 973–996.

REINECK, H.-E. & SINGH, I.B. (1975) *Depositional Sedimentary Environments*, 1st edn. Springer-Verlag, Berlin, 439 pp.

REINECK, H.-E. & SINGH, I.B. (1980) *Depositional Sedimentary Environments*, 2nd edn. Springer-Verlag, Berlin, 549 pp.

REINECK, H.-E., GUTMANN, W.F. & HERTWECK, G. (1967) Das Schlickgebiet sudlich Helgoland als Beispiel rezenter Schelfablagerungen. *Senckenbergiana Lethaea* **48**, 219–275.

REINECK, H.-E., DORJES, J., GADOW, S. & HERTWECK, G. (1968) Sedimentologie, Faunenzonierung und Faziesabfolge vor der Ostkuste der inneren Deutschen Bucht. *Senckenbergiana Lethaea* **49**, 261–309.

SCHUCHERT, C. (1914) Medina and Cataract Formations of the Siluric of New York and Ontario. *Geol. Soc. Am. Bull.* **25**, 277–320.

SCOTESE, C.R., BAMBACH, R.K., BARTON, C., VAN DER VOO, R. & ZIEGLER, A.M. (1979) Paleozoic base maps. *J. Geol.* **87**, 217–277.

WILLIAMS, M.Y. (1919) The Silurian geology and faunas of Ontario Peninsula and Manitoulin and adjacent islands. *Geol. Surv. Can. Mem.* **111**, 1–95.

WRIGHT, R. (1986) Cycle stratigraphy as a paleogeographic tool: Point Lookout Sandstone, southeastern San Juan Basin, New Mexico. *Geol. Soc. Am. Bull.* **96**, 331–361.

ZIEGLER, A.M., HANSEN, K.S., JOHNSON, M.E., KELLEY, M.A., SCOTESE, C.R. & VAN DER VOO, R. (1977) Silurian continental distributions, paleogeography, climatology, and biogeography. *Tectonophysics* **40**, 13–51.

# Transgressive Coastal and Shelf Sandstones

# Implications of tide-dominated lagoonal processes on the preservation of buried channels on a sediment-starved continental shelf

G. F. OERTEL*, V. J. HENRY[†] and A. M. FOYLE*

*Department of Oceanography, Old Dominion University, Norfolk, VA 23529, USA; and [†] Department of Geology, Georgia State University, Atlanta, GA 30303, USA

## ABSTRACT

The Holocene sediment cover on outer parts of the USA continental shelf is thin and discontinuous. Much of the Holocene sediment in nearshore areas is reworked or removed by deep shoreface scour during the Holocene transgression. Shallow paralic sand-bodies that form between the basal and ravinement unconformities are partially erased by the retreating Holocene shoreline. In tide-dominated regions of the coast, lagoonal channels are deepened by swift, tidal currents. Deepening further enhances the topographic relief of the antecedent surface. During transgression the topographically higher parts of the surface are most susceptible to removal by shoreface wave-action while the deeper areas are filled and preserved. Tidal enhancement of channels also takes place at ebb tidal-deltas where 'spillover' and distributary channels cut and fill into shallow shoreface deposits.

## INTRODUCTION

Characteristics of buried channels on the inner continental shelf of the southeastern USA reflect a variety of physical conditions during trangression and regression. Hine & Snyder (1985) point out that the record of the Holocene transgression on the middle Atlantic shelf (specifically in the Onslow Bay area, NC) is very thin or missing. The scant record of Holocene deposits makes problematic the understanding of the early Holocene processes and events. The problem is caused by minor sediment input coupled with deep shoreface scour during transgression. As a result, the average vertical distance between ravinement and basal unconformities is small and much of the Holocene record is subject to removal (Belknap & Kraft, 1985; Kraft et al., 1987). Excavation of paralic deposits above the basal unconformity often leaves the deep parts of buried channels as the only depositional record following transgression. As a result, detailed patterns of stream drainage and paths of inlet migration may be difficult to recognize in the stratigraphic record. Thus, following bevelling of the shoreface, deep channels and inlets stand a better chance of preservation than shallow ones.

This paper is a discussion of processes affecting channel development along tide-dominated barrier coastlines (Fig. 1). Tide-dominated coasts are characterized by closely spaced inlets feeding barrier lagoons, with extensive marsh systems dissected by dense networks of tidal channels (Hayden & Dolan, 1979; Oertel, 1987). Strong tidal currents flowing through these channels scour channel floors and maintain topographic relief between marsh surfaces and the channel thalwegs. As marshes expand, decreasing the open-water areas of the lagoon, current speeds increase as the tidal prism is confined into a smaller area. Belknap & Kraft (1985) and Kraft et al. (1987) have found that many of the deeper channels are associated with, and may have inherited their location from, estuaries and palaeo-fluvial valleys under the antecedent lagoon floor.

In contrast with tide-dominated coasts, wave-dominated coasts have long, narrow barrier islands with broad, open-water bays (Hayden & Dolan,

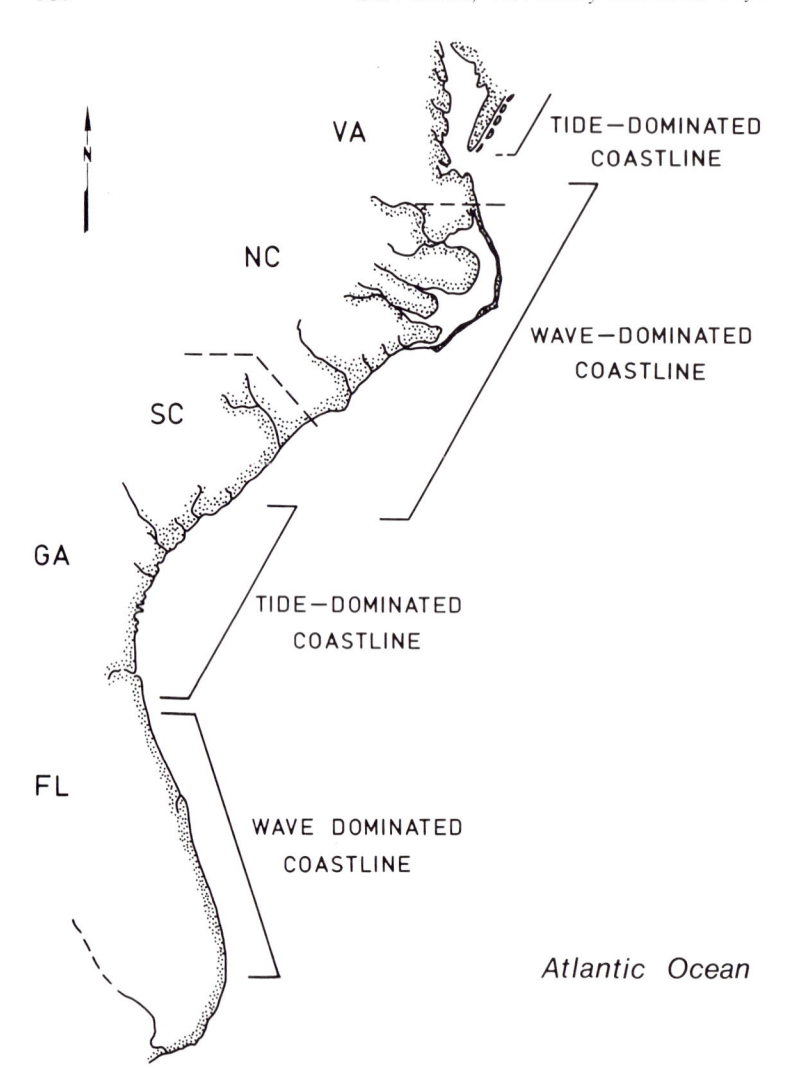

**Fig. 1.** Map of the middle Atlantic and southeastern USA showing the location of tide-dominated and wave-dominated regions.

1979; Oertel, 1987). Adjacent to inlets, the flood and ebb flow fields in the embayments form net flows that result in interior inlet deltas (flood deltas). While tidal channels at inlets and deltas may scour to significant depths, confined inlet flow fields are not distributed far from the inlet (Oertel, 1988). The lack of marshes allows tidal flow to be more uniformly distributed into open bays than into marsh-filled lagoons. Currents in the central parts of open-water lagoons may be influenced to a greater extent by other forces, such as winds and waves.

## SHOREFACE SCOUR

While it is difficult to determine the potential shoreface incision depth, a number of studies suggest a general depth between 16 and 20 m (Swift, 1975; Everts, 1978; Swift et al., 1985). Based on an average of 45 Atlantic and Gulf coast profiles, the average depth between the lower shoreface and the shoreface ramp (Everts, 1978) is approximately −16 m. Field studies by Birkemeier (1984) at Duck, North Carolina suggest that the active part of the shoreface may extend to a depth of −8 to −10 m. Hallermeier (1981) used empirical wave studies and Swift et al. (1985) used dynamic studies to suggest that the active part of the shoreface extends to a depth of −10 to −20 m. Kraft et al. (1987) suggest that shoreface erosion along the Delaware coast reaches −10 m (MSL). While there is much evidence to suggest that typical shorefaces and shoreface ramps extend to −20 m, the lower part of this profile probably is accretional. The erosional and more variable upper part of the shorface determines the depth of the shoreface incision. This is estimated to be the inflection point between the shoreface profile and the shoreface ramp. An average depth for this inflection is approximately −10 m.

At transgressive coasts the preservation of coastal plain and channel traces is dependent inherently upon the depth of the shoreface profile. During transgression the retreating shoreface excavates material located in its path. The trace of coastal-plain channels can be preserved in the rock record only if the initial channel extends deeper than the depth of shoreface incision (Fig. 2). It is shown generally that the depth of shoreface incision is relatively deep on wave-dominated coasts and relatively shallow on tide-dominated coasts (Oertel et al., 1989).

## PARAMETERS AFFECTING INITIAL CHANNEL DEPTH

### Lagoonal channels

Along the coast of Georgia, South Carolina and the Virginian Delmarva Peninsula, large channels on the lagoon floor are primarily inherited from the pre-submerged topography of the mainland (Halsey, 1978; Rhea, 1986). The drainage pattern of these lagoonal channels is dendritic with low-order streams near tidal drainage divides and high-order streams at inlets. Small tidal creeks superimposed on the modern marsh surfaces can be spaced very closely, but are often shallower than mid-tide level. Oertel & Dunstan (1981) found that 80% of the tidal exchange was confined to the larger channels and current speeds were very high. Confinement of the flow to tidal channels in tidal marsh lagoons causes a preferential separation of erosional processes to the channels and accretional processes to marsh surfaces. The confluence of tidal streams is generally the site of the greatest scour in channel floors. Even very low-order tidal streams may cause convergent flows and whirlpools that extend from the channel floor upward to an estuarine front (Oertel & Dunstan, 1981). During transgression, channel position is maintained in association with antecedent drainage patterns. High-velocity currents in tidal channels cause non-deposition or scour. Channels also appear to steepen in response to relatively high sediment advection rates at adjacent marsh surfaces.

Parts of channels with depths greater than the potential depth of shoreface scour have a higher potential for preservation than the parts of channels shallower than that depth. At tide-dominated coasts, the numerous tidal channels in the back-barrier lagoons enhance the probability of deep channel formation that may be preserved following transgression.

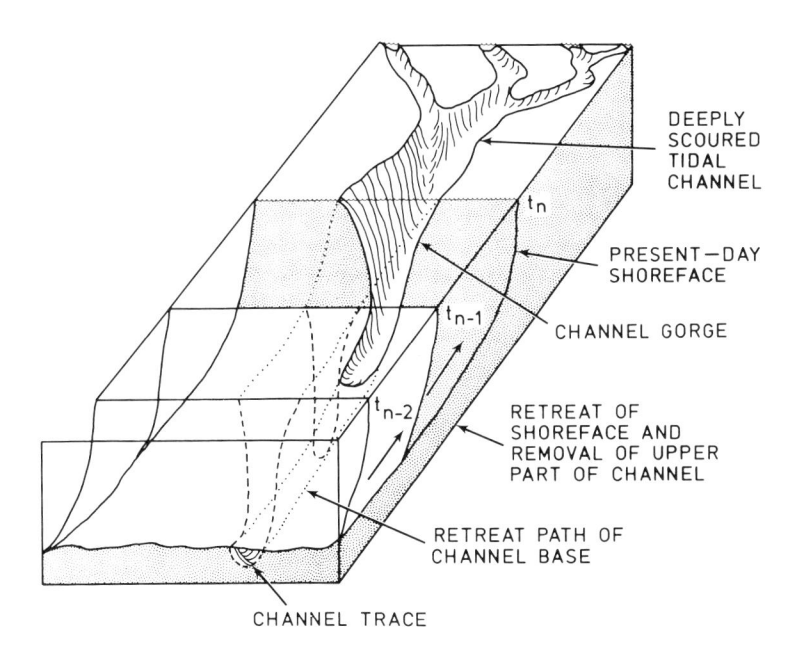

**Fig. 2.** Sketch showing a migrating shoreface bevelling off the upper part of the buried record of a deeply scoured tidal channel.

## Inlet channels

The nature of the interaction between waves and tidal currents is related to channel depth, orientation and channel-fill bedding structure (Oertel, 1975a,b, 1988; Nummedal et al., 1977). Two types of inlet systems (tide-dominated and wave-dominated inlets) are distinguished by relative magnitudes of wave and tidal forces. An extensive literature exists describing sedimentological and morphological characteristics of the deltas associated with these inlet systems (Bruun & Battjes, 1963; Bruun, 1966, 1986; Oertel, 1975a,b; FitzGerald & FitzGerald, 1976; Nummedal et al., 1977; Hubbard et al., 1979), yet the main distinction between types of inlet channels are the depth and stability characteristics.

Wave-dominated coasts of Long Island, New York and North Carolina (Nummedal et al., 1977) have a relatively low density of coastal inlets. Major channels are associated with physiographic drainage basins, whereas lesser channels are associated with migrating inlets and barrier islands. During storms, 'storm cuts' may form by cross-island scour resulting from overwash or 'blowouts' from the lagoon side after the storms pass. The evolution of these storm cuts into inlet channels is determined by tidal currents, which must maintain exchange between the ocean and the lagoon. If the cut is insufficient in size and discharge to flush the influx of littoral sediment then it will close and cease to function as an inlet. Discharge is primarily a function of storm cut access to the lagoonal tidal prism. Since a significant number of barrier lagoons along wave-dominated coasts are 'open water embayments' (Hayden & Dolan, 1979), the probability of significant discharge capture is relatively high. At marsh-filled lagoons, the probability of significant drainage capture is lower because the highly dissected nature of the tidal prism tends to focus toward existing inlets. One criteria for determining whether an inlet is a temporary storm-cut or has developed into a tidal inlet, is whether or not it has developed a delta system with 'net tidal-delta flow' (see Oertel, 1988). Even at wave-dominated coasts with larger 'open-water lagoons', the rate of littoral drift is often high with respect to flushing currents, and storm cuts tend to mend rapidly (Leatherman, 1985; McBride, 1986).

Once an inlet channel is established, littoral sediment transport, littoral currents and coastal currents continue to influence channel sedimentation. Along wave-dominated coasts, substantial longshore currents and drifts produce lateral migration of inlet channels (e.g. Oregon Inlet, North Carolina; Inman & Dolan, 1989). Progressive erosion of the channel primarily is directed laterally toward the downdrift wall and not downward toward the channel floor. As illustrated by Shepard (1960), the base of the channel fill of a migrating inlet is relatively uniform in depth and maximum depth often does not exceed the depth to the base of the adjacent shoreface. Buried inlet retreat paths would probably

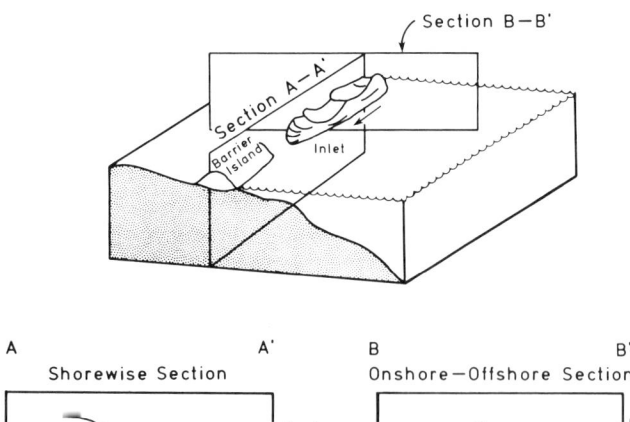

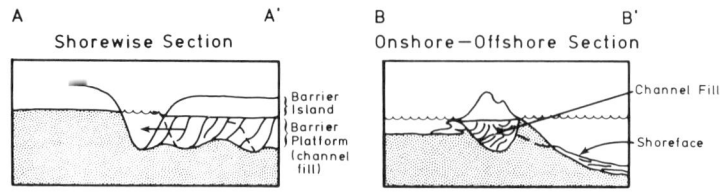

**Fig. 3.** Shore-parallel (A–A') and shore-normal (B–B') sections through a block diagram showing traces left by a laterally migrating inlet.

be removed during landward shoreface translation (Fig. 3). This may be why Hine & Snyder (1985) suggest that large numbers of buried channels on a sediment-starved section of the North Carolina continental shelf are palaeostream valleys.

The tide-dominated coasts of Georgia, South Carolina and Virginia (Nummedal et al., 1977) have inlet systems with numerous tidal channels within barrier lagoons. Only a small percentage of the channels are associated with major river systems. Channels forming the drainage network in these regions are closely associated with topographic lows of antecedent topography. In contrast to the wave-dominated coasts, the tidal prism exchanged through the inlet is relatively large with respect to moderate longshore flows. At many tide-dominated coasts, wave refraction around inlet deltas locally interrupts the unidirectional nature of the longshore current (Oertel, 1975a). Also, seasonally variable directions of wave approach often prevent the development of one 'prevailing' direction of longshore flow. Along the Georgia, South Carolina and Virginia coasts, net southerly directions of longshore flow are produced only by the cumulative effects of storm events on seasonally variable longshore currents. The net longshore currents have a significantly different impact on inlets than prevailing longshore currents. Intermediate-sized inlets migrate at continuous moderate rates depending on the relationship between inlet and longshore currents (Hoyt & Henry, 1967; Oertel, 1975b; FitzGerald & FitzGerald

**Table 1.** Maximum scour depth of inlet gorges on tide-dominated coasts of Georgia (GA), South Carolina (SC) and Virginia (VA)

| Inlet name | Date | Depth (m) |
|---|---|---|
| St Helena Sound, SC | 1965 | 14.3 |
| Port Royal Sound, SC | 1965 | 20.4 |
| Calibogue Sound, SC | 1974 | 16.8 |
| Wassaw Sound, GA | 1975 | 15.8 |
| St Catherines Sound, GA | 1972 | 19.5 |
| Sapelo Sound, GA | 1975 | 27.1 |
| Doboy Sound, GA | 1974 | 17.1 |
| St Simons Sound, GA | 1976 | 23.5 |
| St Andrews Sound, GA | 1976 | 22.6 |
| Wachapreague Inlet, VA | 1984 | 18.0 |
| Quinby Inlet, VA | 1976 | 23.2 |
| Great Machipongo Inlet, VA | 1987 | 26.5 |
| Sand Shoal Inlet, VA | 1984 | 23.2 |
| Average (SD 4.1) | | 20.6 |

**Table 2.** Characteristics of channels from modern lagoons, upper shoreface, lower shoreface (seismic transect from intermediate zone) and inner shelf (seismic transects from the outer zone) of the tide-dominated regions of Georgia and Virginia.

| Location | Channel density (channels per kilometre) |
|---|---|
| Inlets | |
| Georgia | 0.10 |
| Virginia | 0.13 |
| Lagoon channels (>7 m deep) | |
| Georgia | 0.20 |
| Virginia | 0.22 |
| Lagoon channels (>2 m deep) | |
| Georgia | 0.37–1.00 (estimated) |
| Virginia | 0.37–0.75 (estimated) |
| Buried channels | |
| Georgia: | |
| G-1 (upper shoreface) | 2.42 |
| G-2 (upper shoreface) | 2.72 |
| G-3 (lower shoreface) | 1.64 |
| (inner shelf) | 1.42 |
| G-4 (lower shoreface) | 1.13 |
| (inner shelf) | 0.58 |

1976; Oertel, 1988). At larger inlets, inlet tidal currents prevail and the diversion of long-shore currents and resultant migration of the inlets is minor (Oertel, 1975b). At these inlets, deep gorges in the inlet throat generally exceed the depth (approximately 10 m) of local shoreface incision (Table 1).

There are 11 major tidal inlets along the 140 km of tide-dominated Georgia coast. Thus, the inlet density is only about 0.08 inlets per kilometre, or, on average, one major inlet every 12.7 km (Table 2). However, since several of the inlets have multiple entrance channels, 0.10 inlet channels per kilometre (one channel every 10 km) is a more accurate average. The channel density in barrier lagoons is greater. The density of channels with greater than 7 m of cross-sectional relief (along a shore-parallel transect through the modern lagoon) is 0.20 channels per kilometre (one channel every 5 km). When all lagoonal channels with greater than 2 m of cross-sectional relief are included, the modern channel density increases from 0.37 to 1.0 channel per kilometre.

384    G.F. Oertel, V.J. Henry and A.M. Foyle

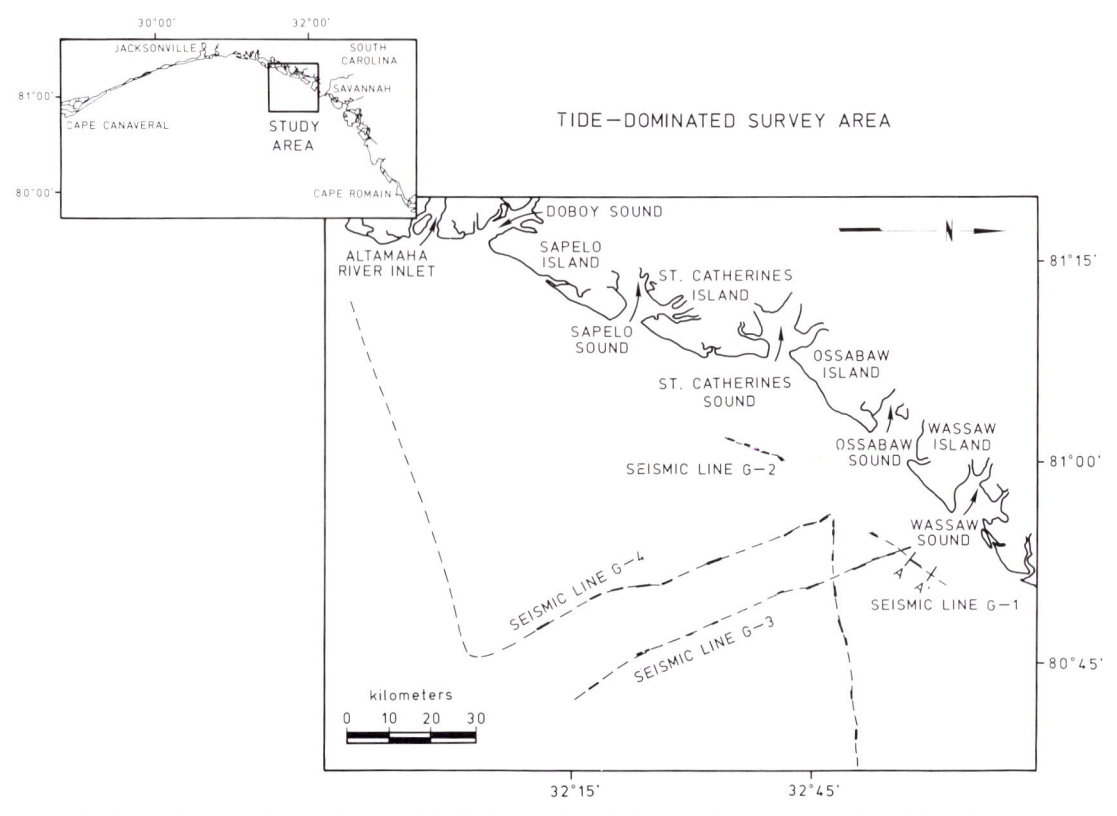

Fig. 4. Study area between Wassaw Sound and St Catherines Sound, Georgia showing the location of the seismic survey lines.

## RESULTS OF SEISMIC PROFILE ANALYSIS

Seismic studies by Hine & Snyder (1985) stongly suggest that many of the buried channels on a section of the wave-dominated North Carolina coast are associated with palaeostream valleys. Belknap & Kraft (1985) have made similar findings on the inner shelf adjacent to the Cape Henlopen headland of Delaware. However, because the relief of antecedent topography along the tide-dominated coasts of Georgia and Virginia is enhanced by swift tidal currents in barrier lagoons and inlets, we hypothesized that remnants of buried shelf channels would reflect the 'enhanced topography' of lagoonal and inlet drainage patterns as well as the palaeostream valleys.

Four seismic survey lines obtained during previous studies (Henry et al., 1981) located 8–12 km seaward of the Georgia shoreface were examined for signatures of buried channels (Fig. 4). The data for this study were obtained using an EG & G 'uniboom' system and a LORAN C navigation system. Two inner lines (G-1 and G-2) were 5–10 km offshore. One line was located parallel to the shoreline of Wassaw Island (G-1) and the other was located adjacent to St Catherines Inlet (G-2). The third line (G-3) was oblique to shore from 46 km seaward of St Catherines Island to approximately 10 km seaward of Wassaw Island. The fourth line (G-4) was 'z shaped' in a zone 10–50 km offshore. Sections of this line were oblique to the shoreline between Altamaha River Inlet and Wassaw Sound.

Seismic sections along the inner shelf (G-1 and G-2) revealed an average of 2.52 channels per kilometre (approximately one channel every 400 m). Channels ranged from approximately 50 m to greater than 1700 m in apparent cross sectional width. Some of the wider channel-traces are believed to have resulted from the migration path of a channel rather than just the 'gorge' of the

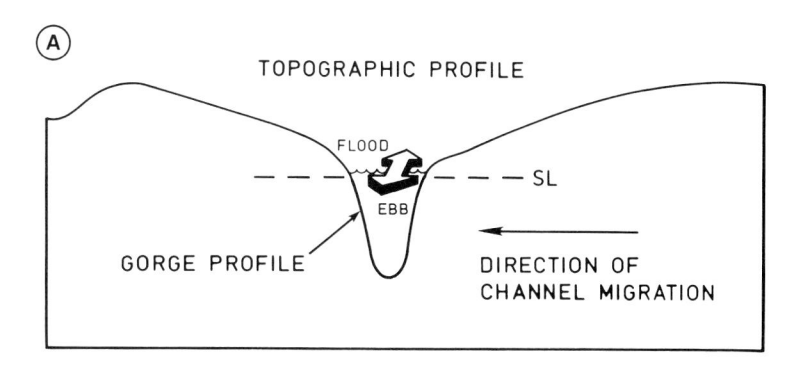

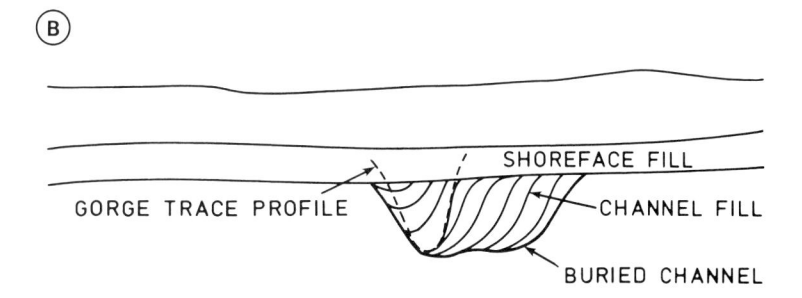

**Fig. 5.** Sketch showing the relationship between (A) the gorge profile of an active channel and (B) the buried 'gorge-trace' and buried channel of a migrating gorge.

active channel (Fig. 5). The question of true versus apparent channel-width is also compounded by the degree of uncertainty in the angular relationships between channel axes and seismic transects (Fig. 6B).

Seismic line G-1 adjacent to Wassaw Island illustrated a total of 37 buried channels, but because of vertical and en echelon stacking and the occurrence of channel structures above and below major reflectors, only 29 were laterally adjacent, forming a channel density of 2.42 channels per kilometre (Table 2). Only 16% of the seismic channels that extended up to the sea-bed penetrated to a depth greater than 10 m (Table 3). The upper layer was not densely channelled and horizontal to low-angle reflectors between channels occupied approximately 21% of the survey line. A large variety of internal structures was found in these channels, with no one predominant type. Twenty-five per cent of the channels had conformable internal structure, 25% were transparent, 25% had inclined reflectors, 12.5% had horizontal relectors and 12.5% illustrated en echelon stacking or some combination of the above types (Fig. 7).

The northern section of G-1 adjacent to Wassaw Sound ebb delta exhibited the largest buried channel complex. The channel trace was 1770 m wide and had a maximum depth of approximately 21 m below the sea-bed (Fig. 8). The undulating bottom reflector of the channel suggests five gorge-cutting events. The internal structure of the channel trace

**Table 3.** Depth characteristics of buried channels from sea-bed and unconformable surfaces

|  | Seismic line |  |  |  |  |  |
| --- | --- | --- | --- | --- | --- | --- |
|  | G-1 | G-2 | G-3 |  | G-4 |  |
|  |  |  | Inner | Outer | Inner | Outer |
| Percentage of buried channels originating at the sea-bed with greater than 10 m relief | 16 | 40 | 11 | 15 | 30 | 51 |
| Average relief of buried channels originating at the sea-bed (m) | 7.9 | 8.3 | 6.5 | 5.5 | 8.6 | 12.3 |
| Average relief of channels buried below a shallow unconformity (m) | 4.6 | 6.5 | — | — | — | — |

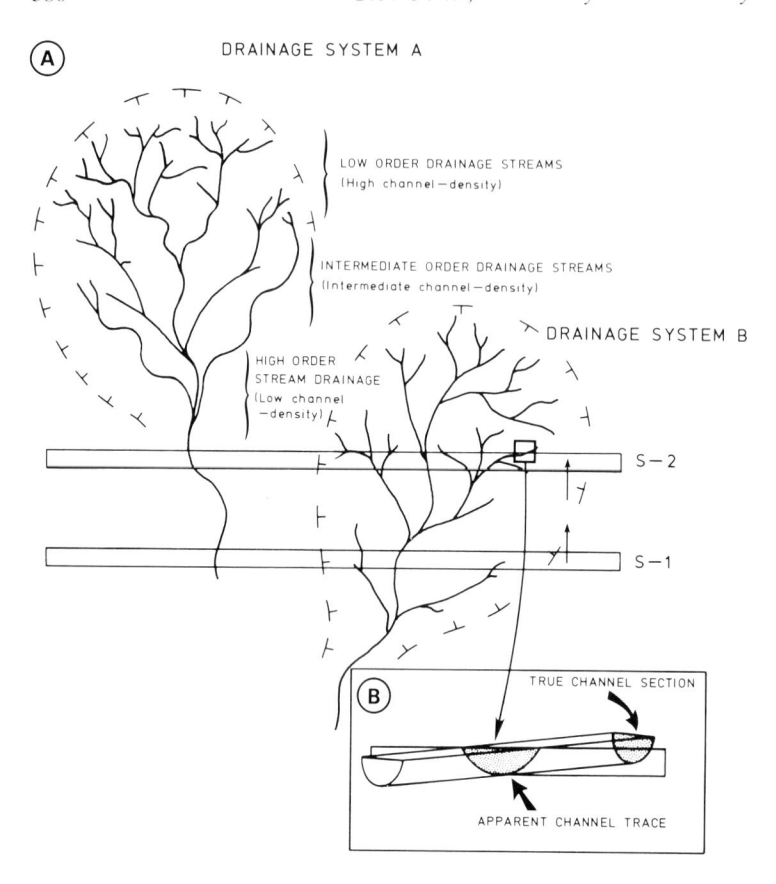

**Fig. 6.** (A) Sketch map showing hypothetical variations in channel-density patterns observed on seismic sections that pass through different parts of an antecedent palaeodrainge system. The density of buried channels is highest on seismic lines that cross the part of a drainage system that primarily has low-order streams. (B) Since seismic sections may cross channel traces at a variety of angles, many channels produce 'apparent' channel sections that are larger than the true section.

is quite complex and contains evidence for multiple stages of channel cut-and-fill. Near the sea-bed, sets of gently inclined beds indicate both northward and southward progradation in different parts of the channel. Some of these reflectors are laterally equivalent to the modern shoals and spillover channels of the modern ebb tidal delta.

The southern part of G-I ran adjacent to the southern part of Wassaw Island and the Ossabaw Sound Inlet. Three relatively large buried channels and a prominent, horizontal, reflector occurred at approximately 10–12m below the sea-bed (Fig. 9). The two buried channels above this reflector appear to correspond to the active tidal channels of Ossabaw Sound. Structures in the lower buried channel are truncated by the deep reflector, suggesting a significant hiatus, possibly the unconformable boundary between the pre-Holocene erosional surface and recent Holocene sedimentation.

The second seismic section (G-2) revealed 25 buried channels seaward of the St Catherines Inlet region. Only 19 channels were laterally adjacent at the surface, because several were stacked through three different horizons. The channel density of laterally adjacent channel structures at the surface was 2.72 channels per kilometre of seismic line. Forty per cent of the surface channel-fills penetrated to greater than 10m below the sea-bed. The average relief of these channels was 8.3m. The deepest gorge trace (−21m sea-bed datum), located 10km seaward of the modern inlet throat of St Catherines Inlet, also was the widest. Two small channel structures in the upper part of the large channel-structure complex are believed to represent cut-and-fill structures of channels and shoals of the modern St Catherines ebb delta (Fig. 10). The large channel structure represents approximately 60% of the entire seismic survey line. The remaining 40% of the survey line contained an almost continuous pattern of small channel structures in the upper part of the section. The gorge traces of the shallow channels were often laterally offset and stacked with intra-truncated scour walls (Fig. 7F). Channel-fill structures (above

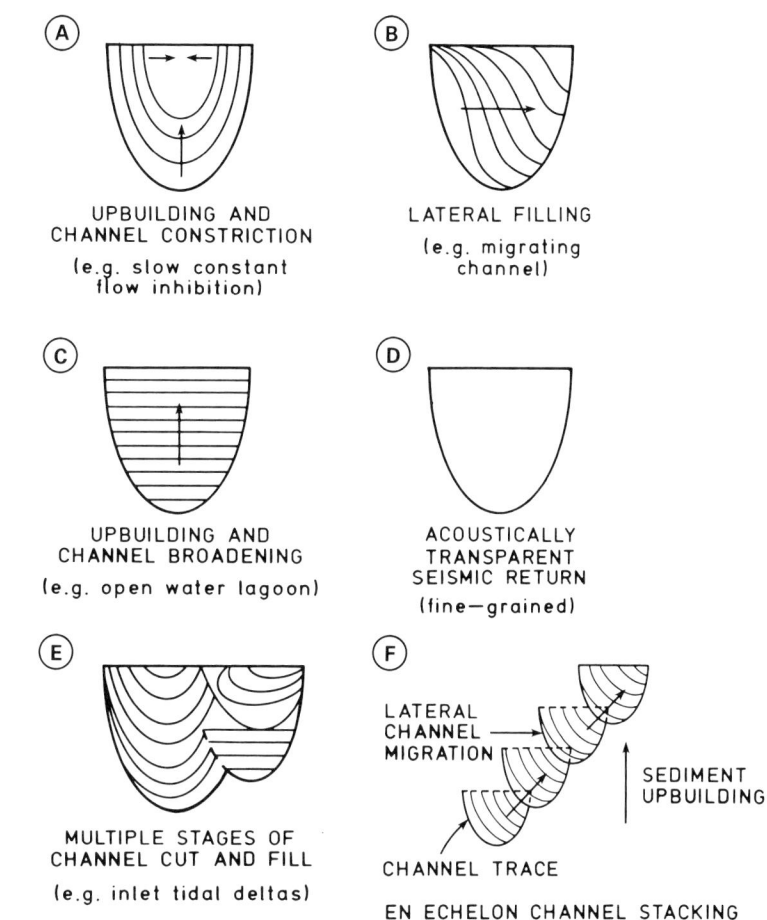

**Fig. 7.** Sketch of channel traces showing the nature of channel fill. (A) Infilling channel with conformable bedding. (B) Laterally migrating channel with inclined bedding. (C) Horizontally bedded channel fill. (D) Acoustically transparent channel fill. (E) Complex channel-fill structures. (F) Laterally offset and stacked channel-fill structures (en echelon channel stacking).

the gorge traces) formed a variety of different patterns. Many beds were conformable with gorge traces suggesting an upbuilding and accreting pattern (Fig. 7). Other fill sequences consisted of inclined foresets, suggesting lateral progradation and channel migration. The two shallow gorge traces with inclined foresets suggest a northward channel migration. Buried channels composed greater than 90% of the seismic line.

The oblique survey line, G-3, between St Catherines Island and Wassaw Island had 45 buried channels. Although the transect was approximately 46 km long, the total shore-parallel distance was only about 30 km. The shore-parallel density of channels was approximately 1.5 channels per kilometre (about one channel every 0.66 km). This was about half the channel density found on the inner shore-parallel survey lines, but 10 times that of modern inlet channels and twice the density of modern lagoon channels. While channel density decreased only slightly from the inner to the outer part of the survey line, the characteristics of the channels varied significantly. The shore-normal distance of the transect ranged from about 46 to 10 km offshore. The inner part of the oblique survey (10–20 km offshore) had an equivalent shore-parallel distance of about 11 km. Channel structures were relatively continuous. Some channels were grouped while others were spaced relatively far apart. Fourteen spaces were found between channels or groups of channels. The arithmetic mean interfluve width was 250 m, which produced a channel-free record for about 20% of this part of the transect. The predominant types of internal bedding structure were gently inclined beds and beds conformable with the channel cut. Inclined bedding did not have a pre-

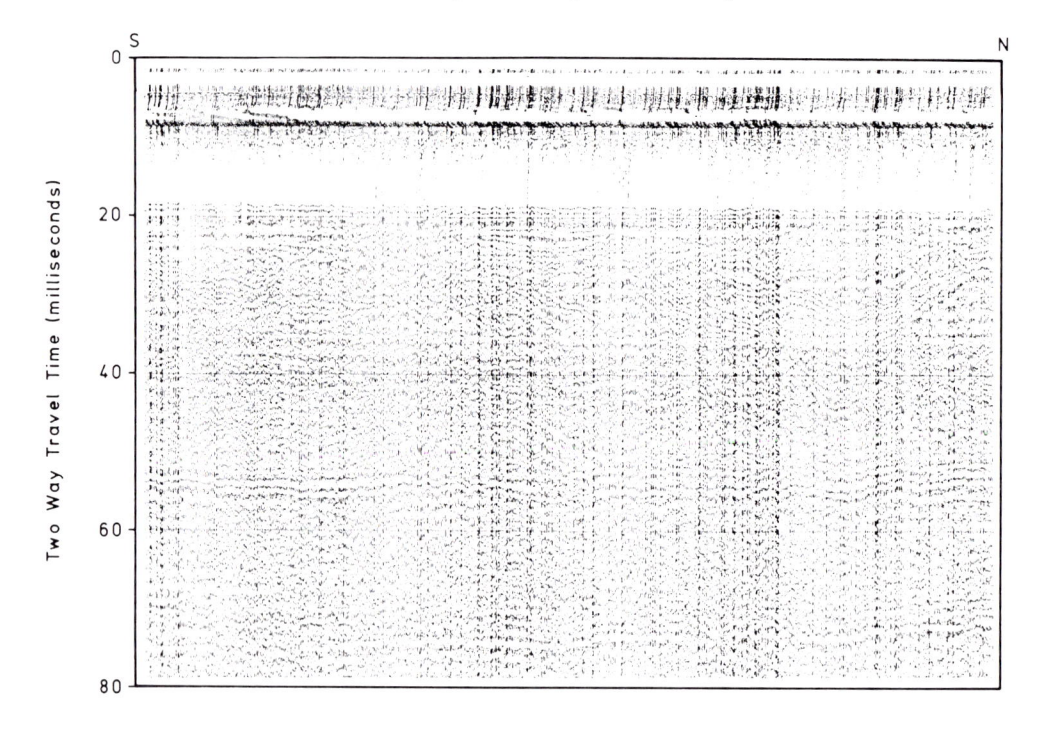

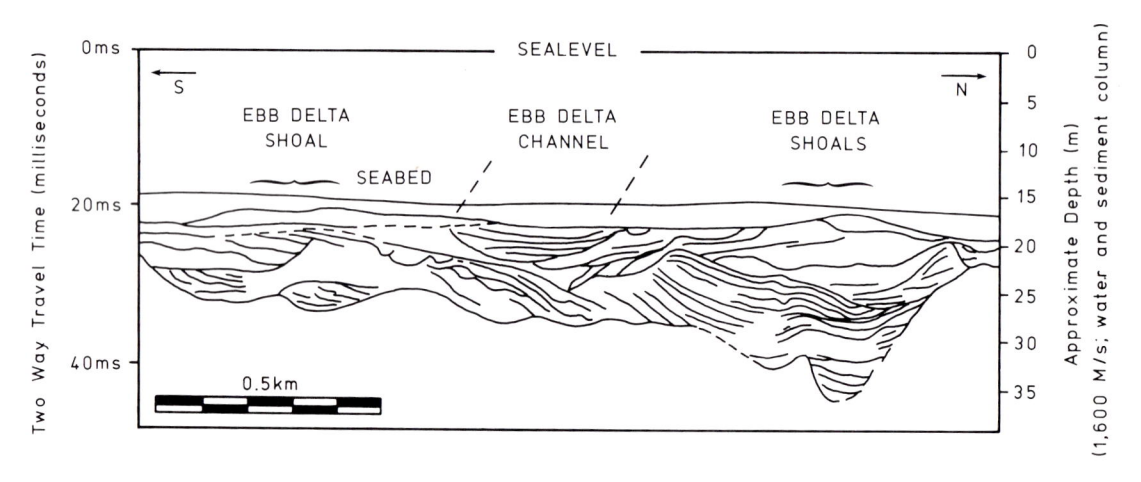

**Fig. 8.** Interpretation of the northern part of seismic line G-1 showing the channel structures adjacent to the entrance of Wassaw Sound, Georgia.

ferred dip direction. Approximately half were inclined in a northerly direction and the other half in a southerly direction. One very broad channel (>2000 m wide) had a lower bounding structure indicating a maximum cut depth of 10 m below the modern sea-bed. This broad channel trace was close and normal to the axis of Ossabaw Sound and is believed to be related to the palaeovalley of the Ogeechee River system and the ebb delta of Ossabaw Sound. Internal structures in most of the buried channels were either conformable or indistinguishable (some transparent). Most of the

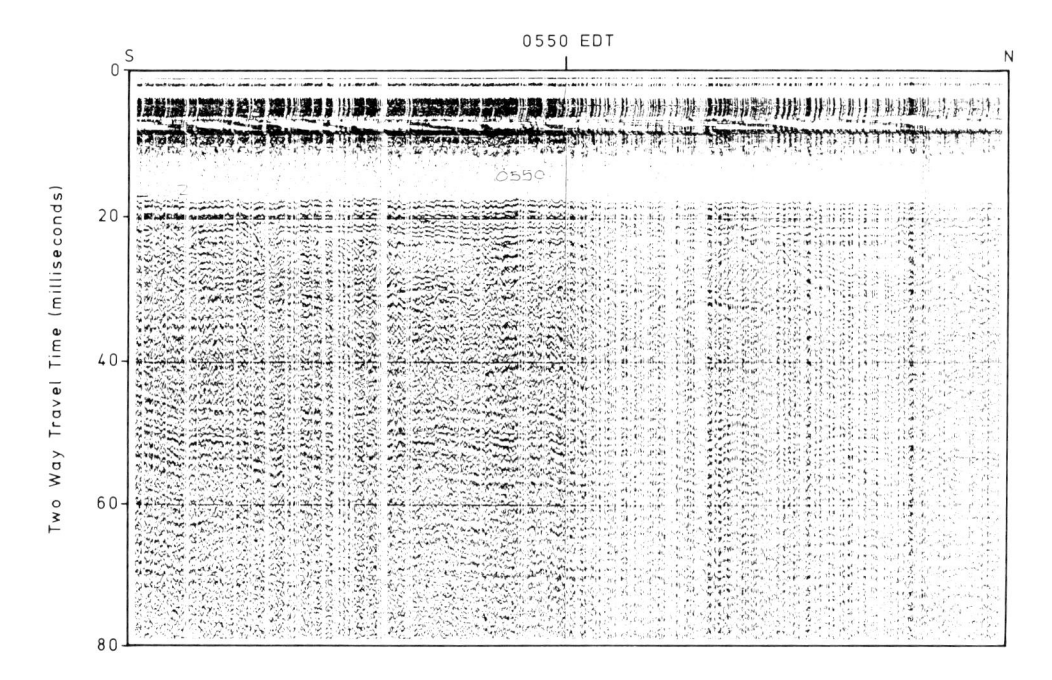

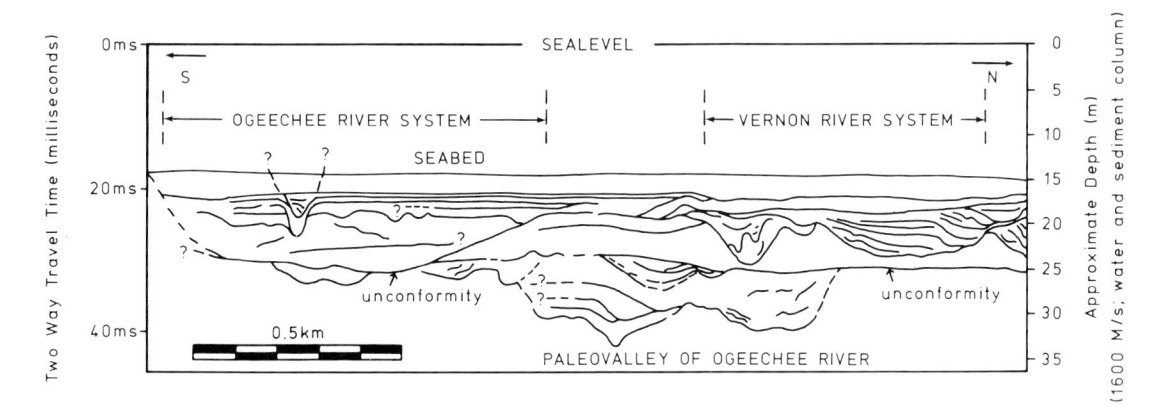

Fig. 9. Interpretation of the southern part of seismic line G-1 showing several palaeostream systems.

channels had lower bounding surfaces that did not extend to great depth. Only 11% of the surface channel-fills penetrated to greater than −10 m. The average relief of the surface channels was about 6.5 m.

The outer part of survey line G-3 (20–46 km offshore) had a different distribution of channel types. Twenty-seven buried channels were present along this part of the survey, yet 80% of the line had low-angle bedding indicative of interfluves.

The average width of the interchannel space was 1 km. Only 15% of the buried channels originating at the sea-bed penetrated to depths greater than −10 m. The average relief of these channels was 5.5 m. Two of the deeper channels displayed relatively complex internal structures with multiple stages of cut and fill. Few en echelon stacking patterns were present and all channels were generally laterally adjacent.

The fourth seismic survey line, G-4, was 'z shaped'

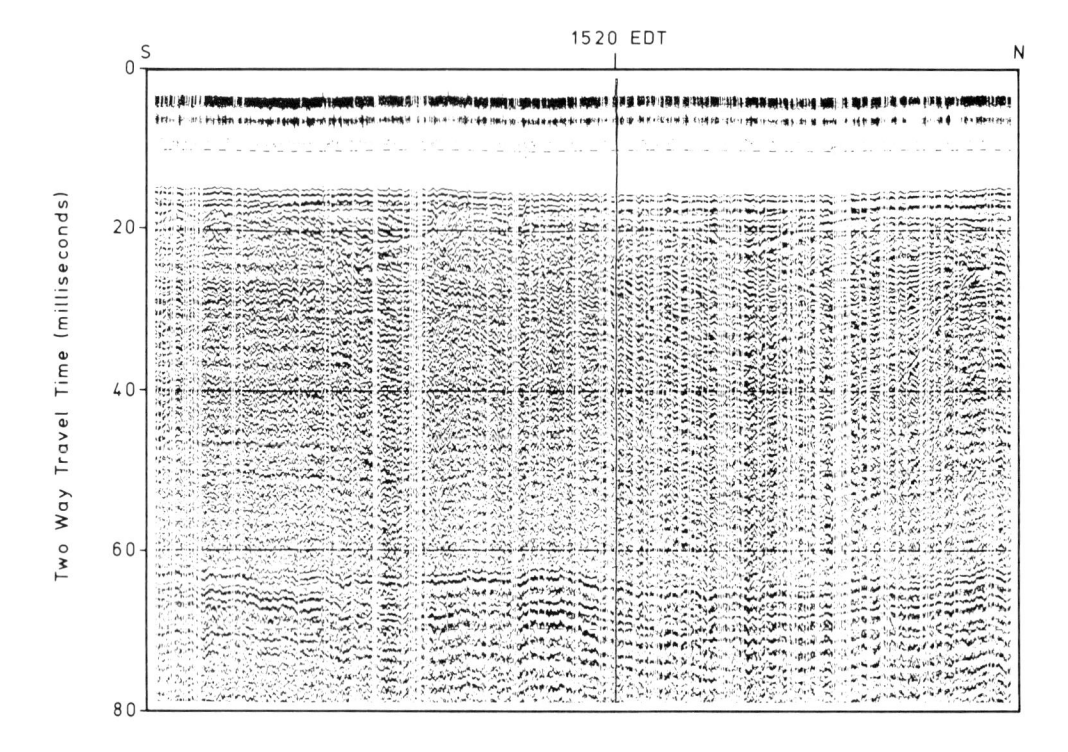

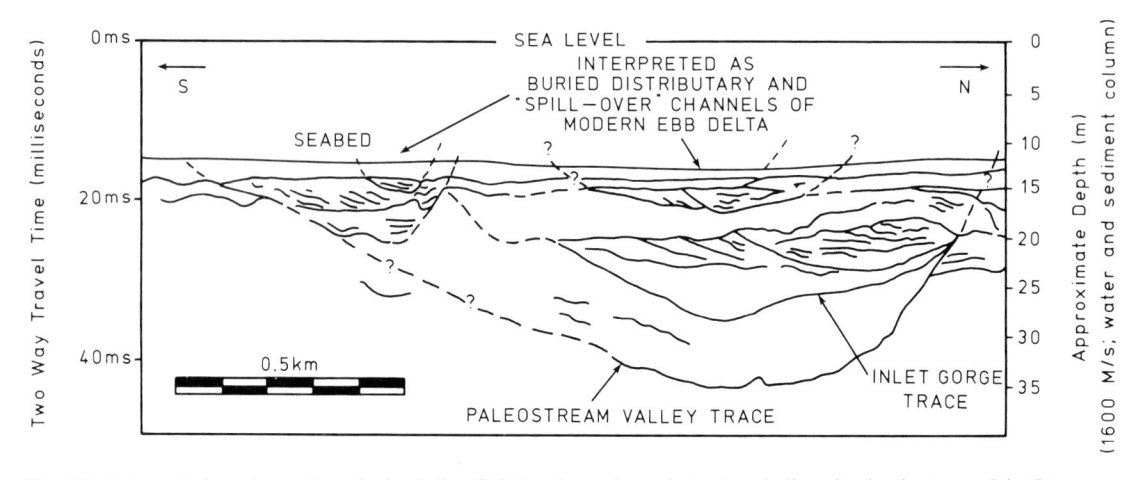

**Fig. 10.** Interpretation of a portion of seismic line G-2 showing a channel structure believed to be the trace of the St Catherines Sound inlet system.

between Altamaha River Inlet and Wassaw Sound Island (Fig. 4). The two seawardmost points on the transect were 46 and 32 km offshore. The two landwardmost points were 7 and 11 km offshore. The total length of the survey was 134 km, however, because of its oblique nature relative to the shoreline, it only represented the equivalent of 87 km of shore-parallel survey. The inner and outer parts of the survey were analysed separately.

The inner part of the line represented an intermediate shore-parallel zone between 12 and 22 km offshore and the outer part was 22–46 km offshore. Eighteen buried channels were discernible in the inner zone, producing a shore-parallel density of 1.13 channels per kilometre. Channel floors were almost uniformly spaced at the three depth horizons. Thirty per cent of the surface channels penetrated to −10 m. The average relief of the channels was 8.6 m. In the outer zone of seismic survey line G-4, 35 channel structures were identified. The equivalent shore-parallel channel density along this part of the survey was lower than any other survey leg (0.58 channels per kilometre). A larger percentage of these outer channels were steeper and deeper than the inner part of the survey (Table 3).

## DISCUSSION

The increase in the number of buried channels with respect to the modern landscape is believed to be related to differences in time-scales. The modern landscape is a 'snap shot' of the landscape in time, while the stratigraphic record represents a composite record of many channel-forming processes through long periods of time. Besides the relict palaeostream valleys, new channels are formed by: (1) the vertical enhancement of antecedent depressions by lagoonal tidal scour; (2) the creation of tidal channels through Holocene lagoonal marshes; (3) the creation of inlet channels associated with inlet circulation and inlet migration; and (4) shoreface channel-development related to the cut and fill of ebb-delta shoals and spillover channels. The density of buried channels on the upper shoreface was about twice as high as the channel density in the modern marsh lagoons and 20 times greater than the density of modern inlet channels.

Initially we hypothesized a seaward decrease in the density of buried channels because of wave bevelling associated with landward shoreface migration. We speculated that during the Holocene transgression the upper part of the sea-bed would have been reworked, destroying shallow buried channels and the upper parts of deep channels. This scenario would favour the preservation of palaeostream valleys and deep, tidally scoured inlet gorges in tide-dominated coastal areas. However, based on the great number of buried channels on the shelf, many of the tidally scoured channels in lagoons may have survived the transgression along with palaeostream valleys and new shoreface channels formed as a result of cut and fill on the distal ends of ebb deltas. Thus, although many of the lower order stream channels are relatively shallow and may be reworked during transgression, the resultant architecture of inner shelf stratigraphy following transgression may move toward an initial enrichment of channel traces.

Much of the apparent patchiness in channel density may not be related to the scour depth of the transgressing shoreface but rather to natural differences in channel density of antecedent drainage basins. Decreases in channel density occur when the coastline migrates from a 'low order' section of a dendritic drainage network (with numerous small tributaries) to a 'high order' section (Fig. 6). These variations in antecedent topography are particularly important in tide-dominated areas where tidal channels are constricted by marshes and tidal currents scour deeply into channel floors.

Channels buried seaward of modern inlets had several characteristics that distinguished them from channels buried seaward of the islands. The density of buried channels adjacent to inlets was relatively high and often illustrated complex channelization related to multiple channels, cut and fill of ebb-delta channels, palaeostream valleys and inlet scour. Adjacent to the islands, buried channels were more regularly spaced and appeared to be cuts of simple channel traces. Internal structures often were acoustically transparent, but some exhibited comfortable and horizontal reflectors (Fig. 7). Along the outer seismic transects the association between channel characteristics and islands or inlets was not as apparent.

## CONCLUSIONS

Processes in a barrier lagoon and on a barrier shoreface determine the degree of channel preservation in continental shelf strata. In wave-dominated areas, shoreface wave bevelling may allow only the preservation of palaeostream valleys and the deeper inlet channels. Traces of these buried channels tend to form shore-normal or dip paths across the shelf (Hine & Snyder, 1985). In tide-dominated areas, many different types of coastal channels may be able to survive a less rigorous shoreface bevelling process.

Tide-dominated inlet systems along the coast of Georgia have well-developed marsh lagoons that confine tidal currents to dendritic systems behind barrier islands. Channelized flow in these lagoons results in deep scour of the inherited antecedent drainage and produces deep channel scars in the basal unconformity. During transgression these channels may be filled with sediment from transgressive barriers, shoreface sediments or inlet shoals. With continued transgression, the erosional portion of the shoreface reworks a portion or all of the buried channels.

Many of the variations in channel density of the modern landscape that are dependant on location within a drainage basin, are inherited in the buried-channel record. Variations in channel density represents a combination of factors including stream-order characteristics of antecedent palaeo-drainage systems, buried palaeovalley systems, new lagoonal channels cut by tidal currents and storms, tidal inlet channels and ebb-delta channels.

In tide-dominated areas, preservation of channel-fill structures is enhanced by two processes. First, the downward bevelling of the shoreface is considerably shallower than in regions with characteristically higher wave-energy regimes. Thus, the depth of reworking by the transgressing shoreface is less and reworking may only bevel off the upper part of buried channels. Second, the low wave-energy regime allows for a slightly thicker shoreface sediment prism than is normally expected at wave-dominated regions. Much of this material is left on the inner shelf forming a reworked carpet over the antecedent surface that was deeply scarred by Holocene tidal-channel scour.

## ACKNOWLEDGEMENTS

The authors express their appreciation to The Skidaway Institute of Oceanography for providing the facilities utilized by the Georgia State University, Coastal and Marine Laboratory and to Old Dominion University for supporting parts of the field and laboratory analysis. Susan Cooke Hoebeke provided graphic support.

## REFERENCES

BELKNAP, D.F. & KRAFT, J.C. (1985) Influence of antecedent geology on stratigraphic preservation potential and evolution of Delaware's barrier systems. *Mar. Geol.* **63**, 235–262.

BIRKEMEIER, W.A. (1984) Time scales of nearshore profile changes. *Proceedings, Coastal Engineering Conference*, Houston, Texas, pp. 1507–1521.

BRUUN, P. (1966) *Tidal Inlet and Littoral Drift*. Universitesforlaget, Norway, 193 pp.

BRUUN, P. (1986) Morphological and navigational aspects of tidal inlets. *J. Coastal Res.* **2**, 123–145.

BRUUN, P. & BATTJES, J. (1963) Tidal inlets and littoral drift. *Proc. Int. Assoc. Hydraulic Res.* **4**, 123–130.

EVERTS, C.H. (1978) *Geometry of Profiles Across Some Inner Continental Shelves of the United States*. TP-78-4, US Army Corps of Engineers, Coastal Engineering Research Center, Fort Belvoir, VA, 92 pp.

FITZGERALD, D.M. & FITZGERALD, S.A. (1976) Sand circulation pattern for Price Inlet, South Carolina. *Proceedings, 15th Conference on Coastal Engineering*, Honolulu, pp. 1868–1880.

HALLERMEIER, R.J. (1981) A profile zonation for seasonal sandy beaches from wave climate. *Coastal Eng.* **4**, 253–277.

HALSEY, S.D. (1978) *Late Quaternary Geologic History and Morphologic Development of the Barrier Island System along the Delmarva Peninsula of the mid-Atlantic Bight*. Unpublished PhD dissertation, University of Delaware, Newark, DL, 592 pp.

HAYDEN, B.P. & DOLAN, R. (1979) Barrier islands, lagoons and marshes. *J. Sediment. Petrol.* **49**, 1061–1072.

HENRY, V.J., JR., MCCREERY, C.J., FOLEY, F.D. & KENDALL, D.R. (1981) Ocean bottom survey of the Georgia Bight. In: *Enviromental Geologic Studies on the Southeastern Atlantic Outer Continental Shelf, 1977–1978* (Ed. Popence, P.) pp. 1–85. US Geological Survey, Open File Report 81-582-A, US Department of the Interior, Reston, Va.

HINE, A.C. & SNYDER, S.W. (1985) Coastal lithosome preservation: evidence from the shoreface and inner continental shelf of Bogue Banks, North Carolina. *Mar. Geol.* **63**, 307–330.

HOYT, J.H. & HENRY, V.J. (1967) Influence of island migration on barrier-island sedimentation. *Geol. Soc. Am. Bull.* **78**, 77–86.

HUBBARD, D.K., OERTEL, G. & NUMMEDAL, D. (1979) The role of waves and tidal currents in the development of tidal-inlet sedimentary structures and sand body geometry. *J. Sediment. Petrol.* **49**, 1073–1092.

INMAN, D.L. & DOLAN, R. (1989) The outer banks of North Carolina: budget of sediment and inlet dynamics along a migrating barrier system. *J. Coast. Res.* **5**, 193–237.

KRAFT, J.C., CHRZASTOWSKI, M.J., BELKNAP, D.F., TOSCANO, M.A. & FLETCHER, C.H., III (1987) The transgressive barrier lagoon coast of Delaware: morphostratigraphy, sedimentary sequences and responses to relative rise in sea level. In: *Sea-level Fluctuation and Coastal Evolution* (Eds Nummedal, D., Pilkey, O.H. & Howard, J.D.). Soc. Econ. Paleontol. Mineral. Spec. Publ. 41, 129–143.

LEATHERMAN, S.P. (1985) Geomorphic and sedimentary analysis of Fire Island, New York. *Mar. Geol.* **63**, 173–195.

McBRIDE, R.A. (1986) *The origin, evolution, orientation and distribution of shoreface-attached sand ridges and their relationship to tidal inlets, North Atlantic shelf, U.S.A.* Unpublished MS thesis, Louisiana State University, 180 pp.

NUMMEDAL, D., OERTEL, G.F., HUBBARD, D.K. & HINE, A.C. (1977) Tidal inlet variability—Cape Hateras to Cape Canaveral. In: *Proceedings Coastal Sediments '77*, American Society for Civil Engineers, Charleston, SC, pp. 543–562.

OERTEL, G.F. (1975a) Ebb tidal deltas of Georgia estuaries. In: *Estuarine Research*, Vol. 2, *Geology and Engineering* (Ed. Cronin, L.E.). Academic Press, New York, 587 pp.

OERTEL, G.F. (1975b) Post-Pleistocene island and inlet adjustment along the Georgia coast. *J. Sediment. Petrol.* **45**, 150–159.

OERTEL, G.F. (1987) Backbarrier and shoreface controls on inlet channel orientation. *Coastal Sediments '87*, American Society of Civil Engineers, New Orleans, LA, pp. 2022–2029.

OERTEL, G.F. (1988) Processes of sediment exchange between tidal inlets, ebb deltas and barrier islands. In: *Hydrodynamics and Sediment Dynamics at Tidal Inlets* (Eds Aubrey, D.G. & Weishar, L.) pp. 297–318. Springer-Verlag, New York.

OERTEL, G.F. & DUNSTAN, W.M. (1981) Suspended sediment distribution and certain aspects of phytoplankton production off Georgia, U.S.A. *Mar. Geol.* **40**, 171–197.

OERTEL, G.F., KEARNEY, M.S., LEATHERMAN, S.P. & WOO, H.J. (1989) Anatomy of a barrier platform: outer Delmarva Peninsula, Virginia. *Mar. Geol.*

RHEA, M.W. (1986) *Comparison of Quaternary shoreline systems in Georgia: morphology, drainage and inferred processes of formation*. Unpublished MS thesis, University of Georgia, Athens, GA, 73 pp.

SHEPARD, F.P. (1960) Gulf coast barriers. In: *Recent Sediments, Northwest Gulf of Mexico*. (Eds Shepard, F.P., Phleger, F. & van Andel, T.). Am. Assoc. Petrol. Geol. 197–220.

SWIFT, D.J.P. (1975) Barrier-island genesis: evidence from the central Atlantic shelf, eastern U.S.A. *Sediment. Geol.* **14**, 1–43.

SWIFT, D.J.P., NIEDERODA, A.W., VINCENT, C.E. & HOPKINS, T.S. (1985) Barrier island evolution, middle Atlantic shelf, U.S.A. Part I: shoreface dynamics. *Mar. Geol.* **63**, 331–361.

# Generation of late Holocene sand ridges on the middle continental shelf of New Jersey, USA— evidence for formation in a mid-shelf setting based on comparisons with a nearshore ridge

J.M. RINE[*], R.W. TILLMAN[†], S.J. CULVER[‡] and D.J.P. SWIFT[§]

[*] *Earth Sciences & Resources Institute, University of South Carolina, Columbia, SC 29208;*
[†] *Consultant, 2121 E. 51st Street, Tulsa, OK 74105, USA;* [‡] *Department of Geological Sciences, Old Dominion University, Norfolk, VA 23508-8520, USA; and* [§] *Department of Oceanography, Old Dominion University, Norfolk, VA 23508-8512, USA*

## ABSTRACT

Data from this study support a new theory that sand ridges on the middle continental shelf of the Middle Atlantic Bight were formed seaward of the nearshore, probably at or near present-day water depths. An implication of this theory is that more sediment transport and deposition is occurring seaward of the nearshore zone than previously hypothesized.

In 1979 a consortium of oil companies and the National Oceanographic and Atmospheric Administration (NOAA) conducted a geological investigation of two linear sand ridges, ridges 1A and 2, on the New Jersey continental shelf. Two separate ridge systems were studied because of differences in their characteristics and because of differences in published theories concerning their origins. Ridge 1A is a nearshore ridge within 4.8 km of the shoreline and in less than 20.3 m of water. Ridge 2 is a mid-shelf ridge, 40 km from shore and in 25–36 m of water.

Three general lithological units are in the sedimentary sequences recovered in vibracores from both ridges 1A and 2. The basal *'non-fossiliferous' mud and sand* unit contains no macrofauna, traces of microfauna, and few burrows within its layers of sand, laminated muds and pebbly sand. The medial *shell-rich poorly sorted sand and mud* unit contains numerous shell fragments and is for the most part bioturbated. The *upper ridge sand* unit generally coarsens upward, is fine to medium-grained, and consists of stacked beds of sand (2.5–70 cm in thickness) that have alternating laminated and non-laminated (bioturbated) layers. The *upper ridge sand* unit is interpreted to be the actively depositing portion of both ridge 1A and ridge 2.

Discriminant analysis of grain-size data and differences in the assemblages of benthic Foraminifera indicate that the upper ridge sand unit of ridge 2 was not deposited in the same setting as ridge 1A. Multiple discriminant analysis, utilizing 50 grain-size parameters, was used to test the validity of the lithostratigraphic units and to determine if the units could be distinguished according to ridges. Within the individual ridges, 97% of the samples in ridge 1A and 95% in ridge 2 were discriminated correctly into lithostratigraphic units. In a comparison of only the upper ridge sand samples from both ridges, 93% of the samples were correctly discriminated according to their respective ridge.

Foraminiferal faunas of the upper ridge sand units in ridges 1A and 2 show considerable differences both in species present and in relative abundances of species that the ridges contain in common. Although both ridges are located within an inner shelf foraminiferal biofacies (shallower than 60 m) the foraminiferal assemblages of ridge 1A are interpreted as representing an open inner shelf environment of normal marine salinity and the assemblages of ridge 2 are interpreted as representing open inner to middle shelf conditions.

The findings of this study support the following model. Time 1, during a period of slow sea-level rise or a stillstand, sand was concentrated by nearshore processes into a barrier/strandplain in the vicinity of the present-day mid-shelf. Time 2, as the rate of sea-level rise increased, the band of concentrated sands was reworked by inner shelf and eventually mid-shelf processes. Time 3 (present day), a majority of the sediment originally deposited in the nearshore has been reworked and redeposited by mid-shelf processes. This mid-shelf deposited sediment comprises the up-to-5.5-m-thick upper ridge sand unit of ridge 2.

## INTRODUCTION

In 1979 a consortium of oil companies conducted a geologically oriented investigation of two linear sand ridges on the New Jersey continental shelf, utilizing vibracores, bathymetric profiles and high-resolution seismic data (Figs 1, 2). The goal of this project was to develop depositional models for shelf-deposited sands to aid in exploring for ancient, hydrocarbon-bearing shelf sandstones in the rock record. Major hydrocarbon reservoirs in sandstones that have been interpreted to be shelfal in origin include the Shannon Sandstone (Tillman & Martinsen, 1984, 1987), the Tocito Sandstone (McCubbin, 1969; Tillman, 1985), and the Cardium Formation (Michaelis & Dixon, 1969), to name but a few. The New Jersey portion of the Middle Atlantic Bight was selected as a site for this investigation because numerous previous studies of the area provided a large data base upon which this study could build. Also, tenable theories as to the origins of the ridges existed in the literature and could be tested.

Data were collected using high-resolution reflection seismic, grab sampling, box coring and vibracoring. Emphasis was placed on sedimentological and palaeontological examination of the vibracores because they obtained cores up to 7 m and generally recovered the entire sedimentary sequence overlying the nonconformity marking the Holocene transgression. These cores furnished data for interpretation of the entire post-transgression history. Seismic data were not included in this paper because the work was done by other investigators and because the 3.5 kHz seismic records could not delineate consistently the obvious lithologic units recovered in the vibracores.

Two separate ridge systems, nearshore ridge 1A and mid-shelf ridge 2, were studied because of differences in their characteristics and because of differences in the theories concerning their origins. Ridge 1A is within 4.8 km of the shoreline, in less than 20 m water depth (Fig. 2), and within a system of shoreface-attached sand ridges that are oriented approximately 30° to the shoreline (Duane et al., 1972). These shoreface-attached ridges are interpreted by Swift (1976) to have originated on the shoreface in response to winter storms. Ridge 2 is approximately 40 km offshore, in 25–35 m water depth, and is subparallel to the present shoreline (Stubblefield et al., 1983, 1984). Presently, two theories attempt to explain the origins of middle shelf ridges. Both of these theories imply that the mid-shelf ridge system contains a core of relict sediment deposited at a lower stand of sea-level. One theory hypothesizes that ridge 2 is a degraded barrier island (Stubblefield et al., 1983, 1984). Another theory is that ridge 2 is a drowned shoreface-attached sand ridge forming part of a shoal retreat massif (Swift, 1973; Swift et al., 1979; Swift et al., 1984). The results from this study substantiate previous theories for origins of the nearshore ridges, but indicates a much more dynamic depositional environment for the mid-shelf than previously hypothesized.

## METHODS

### Coring procedures

Vibracores were obtained with an Alpine Vibracore, which basically consists of a large pneumatic hammering device at the top of a 4 in. (10 cm) inner diameter, 30 ft. long steel pipe containing a plastic tube liner. Penetration through hard layers was accomplished by retrieving the initial vibracore and initiating a second vibracore with a water jetting system that 'jetted' the core tube past the 'resistant' layer. Because a meter used to determine depth of penetration was inoperative, all depths below sediment surface (sea-floor) of jetted cores are estimated and only depths from the tops of these cores are given. A second undesirable result from the lack of a penetration meter was excessive deformation of sedimentary structures in some cored intervals. This deformation resulted in part from extended periods of vibration after the corer reached a resistant layer or point of maximum penetration.

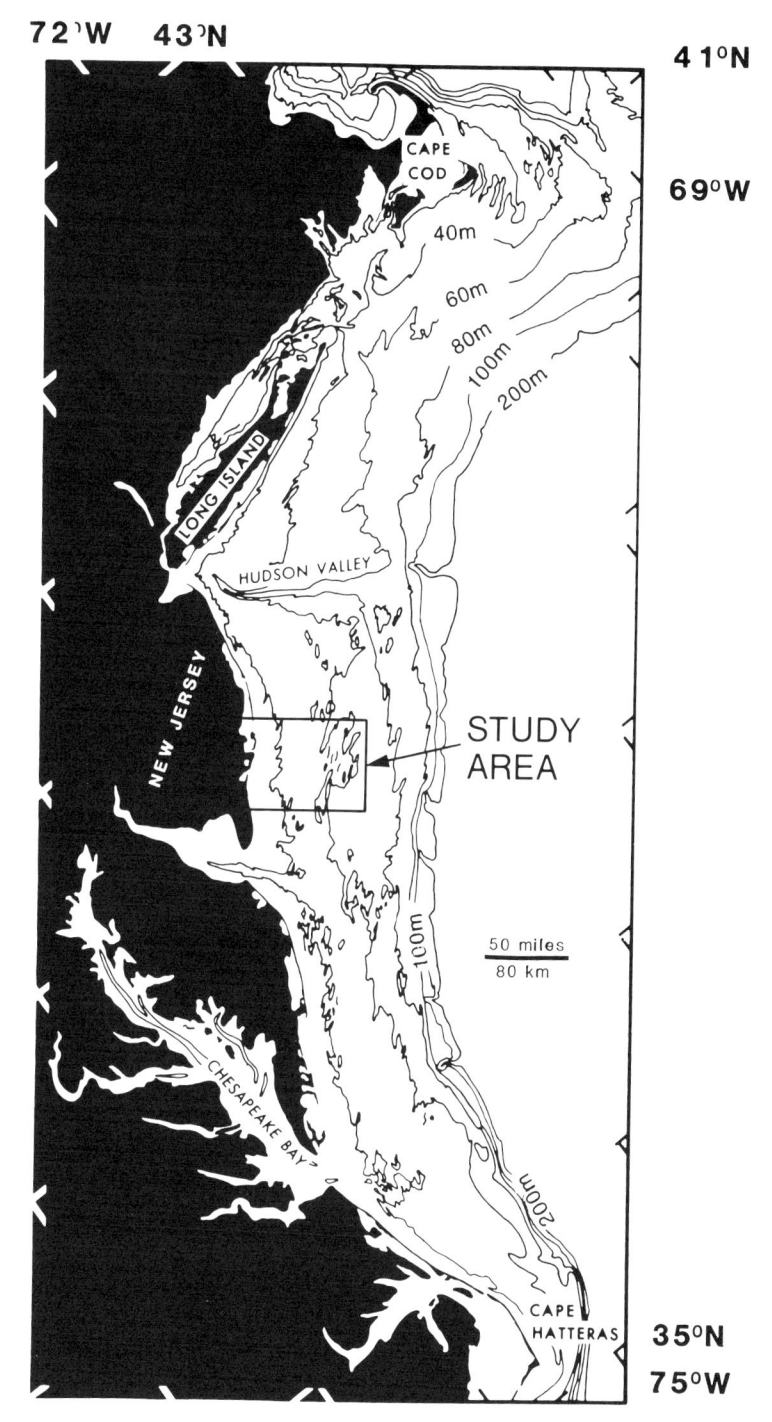

**Fig. 1.** Bathymetric map of the Middle Atlantic shelf, showing location of the study area, which lies between the shoreline and the 60 m isobath.

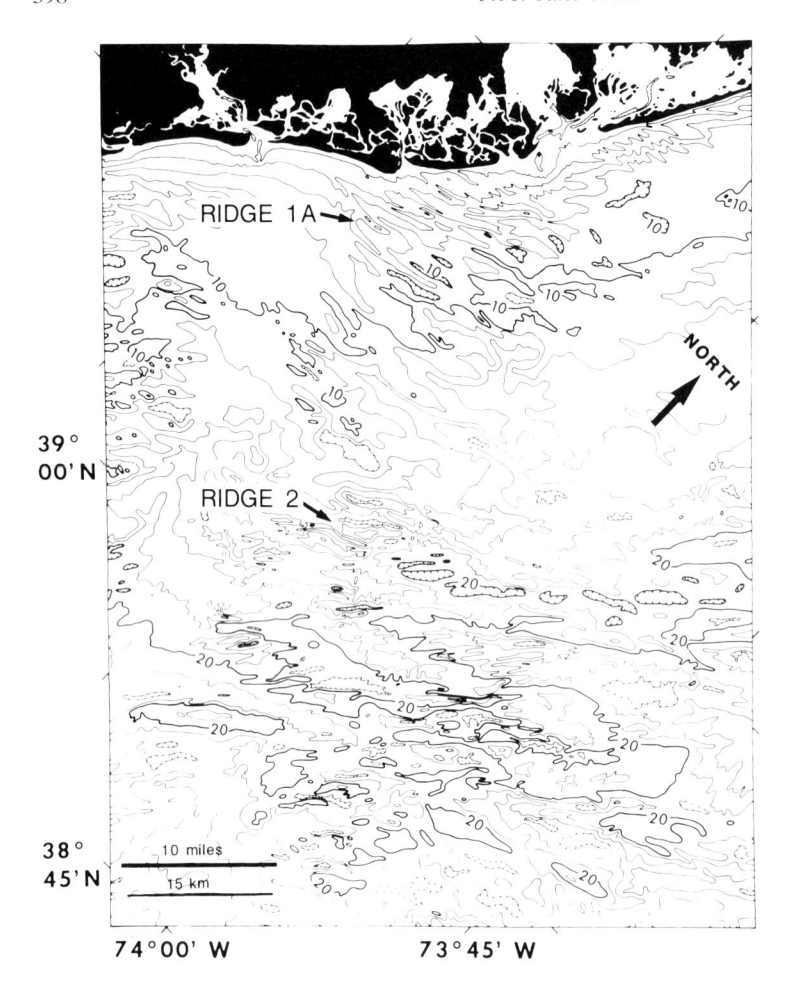

**Fig. 2.** Bathymetric map showing locations of the nearshore, shoreface-attached ridge (ridge 1A) and the mid-shelf ridge (ridge 2) on the New Jersey continental shelf. Contours are in 2 fathom intervals (2 fathoms = 3.66 m).

## Bathymetric profiles

Bathymetric profiles are based on 47 km of 3.5 kHz surveys. Positioning was based on Loran C navigation. A majority of the lines were run normal to the ridge crests with tie lines along the crests and troughs.

## Laboratory procedures

### Core preparation and examination

Whole cores were cut into 0.75 m sections and sliced longitudinally into three segments. A slab 2.0 cm thick from the centre of each core section was X-ray radiographed and photographed immediately after opening. Epoxy peels were also made from the centre slabs. From the remaining core-slab sections, samples were taken for grain-size analysis, micropalaeontological analysis, $^{14}C$ age dating, and porosity and permeability measurements.

Cores were described from epoxy peels, X-ray radiographs and core slabs. Sediment colouration of the core slabs was noted immediately after opening of each core. Sedimentary structures were sketched from epoxy peels and X-ray radiographs. Textural character was determined using sieve and settling tube analyses and a grain-size comparison chart. Only $\frac{1}{4}\phi$ interval sieve data are used for statistical analyses presented in this paper.

Sketches of the cores depict how the cores actually look in epoxy peels and X-ray radiographs. Physical and biogenic structures were noted when observed. Deformation caused by coring procedures was not always accounted for within the written core descriptions because it was difficult to distinguish many forms of coring-induced deformation from sedimentary structures formed by depositional and post-depositional processes (i.e. bioturbation). Some degree of deformation is present in all cores, with the most common deformation feature being the upward bowing of lamina sets. In portions of some cores, vertical amplitudes of the bowing is as much as 13.0 cm. Other types of sediment deformation include vertical flow features and microfaulting. The top several centimetres (10–15 cm) of most cores were disturbed by water sloshing in the core barrel when cores were brought aboard the ship.

When a water jetting technique was utilized to increase core penetration, sediment at the top of the jetted core consistently contained a lag of coarse sand and shell material ('jet-lag'). The contact between the jetting artifact and the relatively undisturbed sediment below was identified in all the jetted cores.

*Grain-size analysis*

Sieving ($\frac{1}{4}\phi$ intervals) was utilized to determine grain-size distributions. The grain-size samples were taken using 1 in. diameter plastic vials, which were pushed into the sediment parallel to bedding. Each vial contained approximately 50 g of sediment. Prior to sieving, samples were removed from the vials, dried, and weighed. A majority of shell material was removed by hand and those samples containing more than 10% mud were wet sieved through a 400 mesh (37 μm) sieve. Weight of the mud was accounted for by reweighing the sample after it was wet sieved and dried and given designated weight percentage values for the 37–4 μm fractions using a cubic spline curve-fit technique (Tillman, pers. comm.).

Weight on individual sieves was obtained using a recording balance that records the cumulative weight as material from each new sieve was added. This information on each sample was processed through two programs termed 'Grain size' and 'Grain factor', and a log-probability cumulative weight percentage graph was plotted for each sample (Fig. 3). From these two programs, 89 variables per sample were calculated. Variables 1–50 are listed in Table 1. Variables 51–89 are a record of weight of material on each sieve and specialized measurements generally more applicable to lithified sediments (A list of variables 51–89 may be obtained by writing to the authors.)

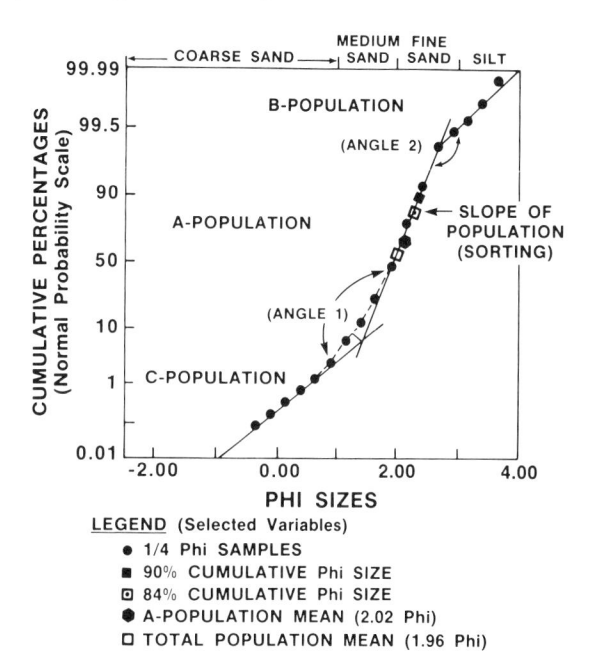

**Fig. 3.** Log-probability cumulative plot showing characteristics of grain size populations. Population A has the steepest slope and lies between angles 1 and 2. The B-population is the fine-grained portion of the plot. Population C is the coarse-grained portion of the plot. Population C may not be present in all samples. Populations are defined in accordance with Moss (1962). Variables computed using plots such as this are given in Table 1.

Based on the work of Moss (1962), Tillman & Reineck (1973), and others, individual data points on each sample are subdivided into one to three grain-size populations. These populations are derived by identifying one to three straight-line segments per cumulative plot (Fig. 3). Once the line segments are drawn (by hand), a variety of points on the segments are digitized and the results of the digitizing and the 'Grain size' results are fed into the 'Grain factor' program in order to calculate the variables for each sample.

The variables utilized in the four discriminant analyses utilized in this study are listed in Table 1. In Table 1 the variables are designated from 1 to 50 on the basis of their importance in the discriminating process. Variables rated as important, in at least three of the four discriminations, listed in order of decreasing ability, are: percentage of sediment in population A, 16% cumulative phi size, mean size, median size, ratio of 0.16 phi to total sample mean, ratio of 0.90 phi to total sample

**Table 1.** List of variables used in statistical analysis of nearshore ridge 1A and mid-shelf ridge 2 units

| | Variable name | Defintion | Ridge 1A units | Ridge 2 units | All ridge 1A and 2 units | Upper ridge sand, ridge 1A and 2 units |
|---|---|---|---|---|---|---|
| 1[a] | $\bar{X}$ Moment | Mean size, total sample | 4 | 9 | 6 | 6 |
| 2 | Sorting (moment) | Standard deviation, total sample | 9 | 18 | 36 | 47 |
| 3 | Skewness (moment) | Skewness, total sample | 40 | 26 | 17 | 45 |
| 4 | Kurtosis (moment) | Kurtosis, total sample | 25 | 42 | 43 | 32 |
| 5 | 1% phi | Size of sand grains below which is included 1% of the sample | 18 | 15 | 28 | 24 |
| 6[a] | 16% phi | Size of sand grains approximately 1 SD from median. Phi size of 16% cumulative | 6 | 14 | 5 | 7 |
| 7[a] | 50% phi | Median, total sample | 8 | 7 | 11 | 4 |
| 8 | 84% phi | Size of sand grains approximately 1 SD from median. Phi size of 84% cumulative | 1 | 49 | 10 | 9 |
| 9 | 90% phi | Size of sand (silt) grains below which is included 90% of total sample | 3 | 31 | 50 | 30 |
| 10 | Percentage greater than 4 phi | Percentage silt + percentage clay | 16 | 22 | 34 | 12 |
| 11 | Percentage greater than 4.75 phi | Percentage finer than smallest sieve | 17 | 13 | 41 | 31 |
| 12 | Modal percentage | Percentage of sand in 1/4 phi size having maximum weight of sediment | 27 | 33 | 29 | 39 |
| 13 | Mode phi size | Phi size of highest histogram peak | 35 | 16 | 21 | 33 |
| 14 | Angle A–B (x10$^{-1}$) | Angle between A and B populations | 47 | 20 | 24 | 23 |
| 15 | Loc. Angle A–B | Phi size at intersection of A and B population line on cumulative curve | 14 | 35 | 37 | 35 |
| 16 | Angle A–C (x10$^{-1}$) | Angle between A and C populations | 34 | 25 | 19 | 50 |
| 17 | Loc. Angle A–C | Phi size at intersection of A and C population lines | 37 | 17 | 12 | 19 |
| 18[a] | Percentage A population (saltation) | Percentage of sediment in A population | 15 | 1 | 2 | 1 |
| 19[a] | Percentage B population (suspension) | Percentage of sediment in B population | 24 | 3 | 7 | 3 |
| 20[a] | Percentage C population (rolling) | Percentage of sediment in C population | 29 | 2 | 4 | 2 |
| 21 | Phi RC-population (rolling) | Phi size range of C population | 10 | 41 | 8 | 49 |
| 22 | Phi RA-population | Phi size range of A population | 19 | 39 | 25 | 34 |
| 23 | Sl C population | Slope on cumulative curve of C population | 39 | 30 | 32 | 40 |
| 24 | Sl A population | Slope on cumulative curve of A population | 32 | 28 | 30 | 8 |
| 25 | Sl B population | Slope on cumulative curve of B population | 41 | 46 | 42 | 16 |
| 26 | A–C mix | Degree of mixing between A and C populations. Measures at angle A–C and in phi units | 36 | 45 | 45 | 13 |
| 27 | A–B mix | Degree of mixing between A and B populations. Measures at angle A–B and in 'phi units' | 49 | 37 | 47 | 44 |
| 28 | F. skew | Skewness value; calculated using Folk equation | 45 | 32 | 49 | 11 |

**Table 1.** (continued)

| | | | | | | |
|---|---|---|---|---|---|---|
| 29 | F. kurt | Kurtosis value; calculated using Folk equation | 48 | 43 | 46 | 37 |
| *C population* | | | | | | |
| 30 | C-mean | Mean | 30 | 44 | 13 | 48 |
| 31 | C-sort | Sorting | 7 | 27 | 14 | 42 |
| 32 | C-skew | Skewness | 28 | 48 | 31 | 38 |
| 33 | C-kurt | Kurtosis | 23 | 34 | 38 | 43 |
| *A population* | | | | | | |
| 34 | A-mean | Mean | 13 | 6 | 35 | 15 |
| 35 | A-sort | Sorting | 22 | 29 | 39 | 14 |
| 36 | A-skew | Skewness | 43 | 38 | 33 | 28 |
| 37 | A-kurt | Kurtosis | 38 | 40 | 40 | 46 |
| *B population* | | | | | | |
| 38 | B-mean | Mean | 26 | 23 | 20 | 29 |
| 39 | B-sort | Sorting | 20 | 47 | 27 | 17 |
| 40 | B-skew | Skewness | 50 | 24 | 44 | 22 |
| 41 | B-kurt | Kurtosis | 46 | 36 | 48 | 20 |
| *Scaled variables* | | | | | | |
| 42 | $0.01P/\bar{X}$ | 0.01 phi/total sample mean | 11 | 5 | 16 | 25 |
| 43[a] | $0.16P/\bar{X}$ | 0.16 phi/total sample mean | 12 | 4 | 9 | 5 |
| 44 | $0.50P/\bar{X}$ | Median/total sample mean | 44 | 21 | 15 | 27 |
| 45[a] | $0.84P/\bar{X}$ | 0.84 phi/total sample mean | 2 | 11 | 1 | 21 |
| 46[a] | $0.90P/\bar{X}$ | 0.90 phi/total sample mean | 5 | 8 | 3 | 18 |
| 47 | $Mode/\bar{X}$ | Mode/total sample mean | 33 | 19 | 23 | 41 |
| 48 | $CMEA/\bar{X}$ | C population/total sample mean | 42 | 50 | 26 | 36 |
| 49 | $AMEA/\bar{X}$ | A population/total sample mean | 31 | 10 | 22 | 10 |
| 50 | $BMEA/\bar{X}$ | B population/total sample mean | 21 | 12 | 18 | 26 |

[a] Variables that are relatively important in at least three of the four discriminations.

mean, ratio of 0.84 phi to total sample mean, percentage of sediment in population B, and percentage of sediment in population C.

### Statistical analysis

Multiple discriminant analysis was performed on the grain-size data from both ridges 1A and 2. All 50 variables listed in Table 1 were utilized for these analyses. Multiple discriminant analysis (Harris & Tillman, pers. comm.) was carried out to test the statistical reliability of lithostratigraphic units recognized on the basis of sedimentary structures and lithology in vibracores. Multiple discriminant analysis of grain-size variables from different units from the New Jersey shelf was found to be a useful technique in discriminating lithostratigraphic units and individual samples within each unit. For a hypothesis to be verified by discriminant analysis a tight clustering of data from each genetic unit is expected. If individual samples are separable on the basis of grain-size variables alone, a stronger intra-unit than inter-unit correlation among individual samples is expected (Klovan & Billings, 1967). $n-1$ discriminant equations were calculated and plotted for each analysis, where $n$ equals the number of lithostratigraphic units in the analysis.

Lithostratigraphic units used in the discriminant analyses are: (1) upper ridge sand, nearshore ridge 1A; (2) upper ridge sand, mid-shelf shelf ridge 2; (3) shell-rich poorly sorted sand and mud, nearshore ridge 1A; (4) shell-rich poorly sorted sand and mud, mid-shelf ridge 2; (5) 'non-fossiliferous' mud and sand, nearshore ridge 1A; (6) 'non-fossiliferous' mud and sand, nearshore ridge 2. Discriminant analysis at a 95% confidence level was carried out on four different groupings of these facies. One analysis correctly discriminated 81% of the samples into the six facies. A second analysis discriminated 93% of the upper ridge sand, nearshore ridge 1A samples as being statistically different from the upper ridge sand unit of the mid-shelf ridge 2.

Third and fourth analyses determined that facies (1), (3), and (5) were distinguishable from one another in 97% of the samples, and that facies (2), (4) and (6) were disguishable from one another in 95% of the samples. Table 1 gives the ranking of importance of the variables used in the discriminant analysis.

### Palaeontological analysis

The microfaunas of 144 samples from 12 core sites, five in ridge 1A and seven in ridge 2, were documented. Each sample, consisting of approximately 50 g dry weight of sediment (primarily sand), was washed over a 63 µm sieve to remove silt and clay and passed through a heavy liquid to concentrate foraminiferal tests. To avoid possible bias in environmental interpretation, all samples were coded upon receipt and interpretations were made prior to consulting any sedimentological or seismic reports. Approximately 300 Foraminifera were randomly picked from each sample.

## BACKGROUND INFORMATION

### Sediments

Over the last half century, sedimentological studies of the continental shelf of the eastern USA have been concerned with sources of sediment and timing of their deposition. Shepard & Cohee (1936) hypothesized that sediment distribution on the Mid-Atlantic shelf of North America reflected Pleistocene depositional patterns because their sampling found little correlation between water depth and grain size (based on 200 samples). More detailed work by Stetson (1938), based on 10 transects from Nova Scotia to Florida, concluded that distribution of sediments on the shelf was 'adjusted' to present sea-level. Much more extensive surveys were undertaken by a Woods Hole research team beginning in the 1960s. These studies used the classification of Emery (1952), in which sediments were classified as biogenic in origin (shell fragments), aeolian or volcanic, chemical precipitates, and relict (from earlier depositional environments). Emery and his colleagues concluded that the Atlantic continental shelf is covered primarily by 'relict' sands with only a thin band (<10 km wide) of 'modern' detrital sediment located in the nearshore (Emery, 1965). More recent studies have shown that this band, which covers much of the nearshore, is just a fair-weather veneer of fine-grained, rip-current fallout (Swift et al., 1972a). This veneer is stripped off during major storms, and the uppermost portion of the underlying relict coastal plain sands is subjected to erosion. At such times, the sands and muds that are eroded may be swept seaward to nourish the leading edge of a transgressive shelf sand-sheet (Swift et al., 1972a).

The advent of high-resolution seismic profiling on the continental shelf added a third dimension to the study of the Atlantic shelf sediments. Based on seismic surveys in the Baltimore Canyon area, Knebel & Spiker (1977) proposed a two-fold shelf stratigraphic scenario, in which a surficial sand sheet, less than 11 000 years old, overlies muddy-sand strata older than 24 000 years BP. This surficial sand sheet has an average thickness of 5–7 m and ranges from less than 1 m to 20 m thick. The two units are divided by a reflector, termed 'R', which they correlated with a reflector traced by McClennen (1973) for 125 km on the outer shelf. Knebel & Spiker (1977) concluded that changes in thickness of the surficial shelf sand-sheet are 'closely related to the bottom morphology and thus to its formation.'

Recent seismic surveys include detailed examinations of the two ridges discussed in this paper (ridges 1A and 2; Figueiredo, 1984; Stubblefield, unpubl. data). Using primarily high-resolution reflection seismic data and some core material, Figueiredo (1984) described the general character and distribution of Holocene, Pleistocene and (uppermost) Tertiary strata of the continental shelf east of Atlantic City, New Jersey. (Correlation of Figueiredo's (1984) results with data presented in this paper will follow in the text.) Stubblefield (unpubl. data) correlated the 'R' reflector of Knebel & Spiker (1977) to a reflector recorded within the study area addressed in this paper.

### Morphology

Morphology of the Atlantic shelf of the eastern USA is complex (Fig. 2) and can be explained best in the context of the relative scales of the features present. Swift et al. (1972a) categorized morphological elements of the central and southern Atlantic shelf into small and large-scale features. Examples of small-scale features present on the shelf are ripples and sand waves (bedforms). The large-scale features are divided into three orders. This study examined cores from 'first order' shoreface-

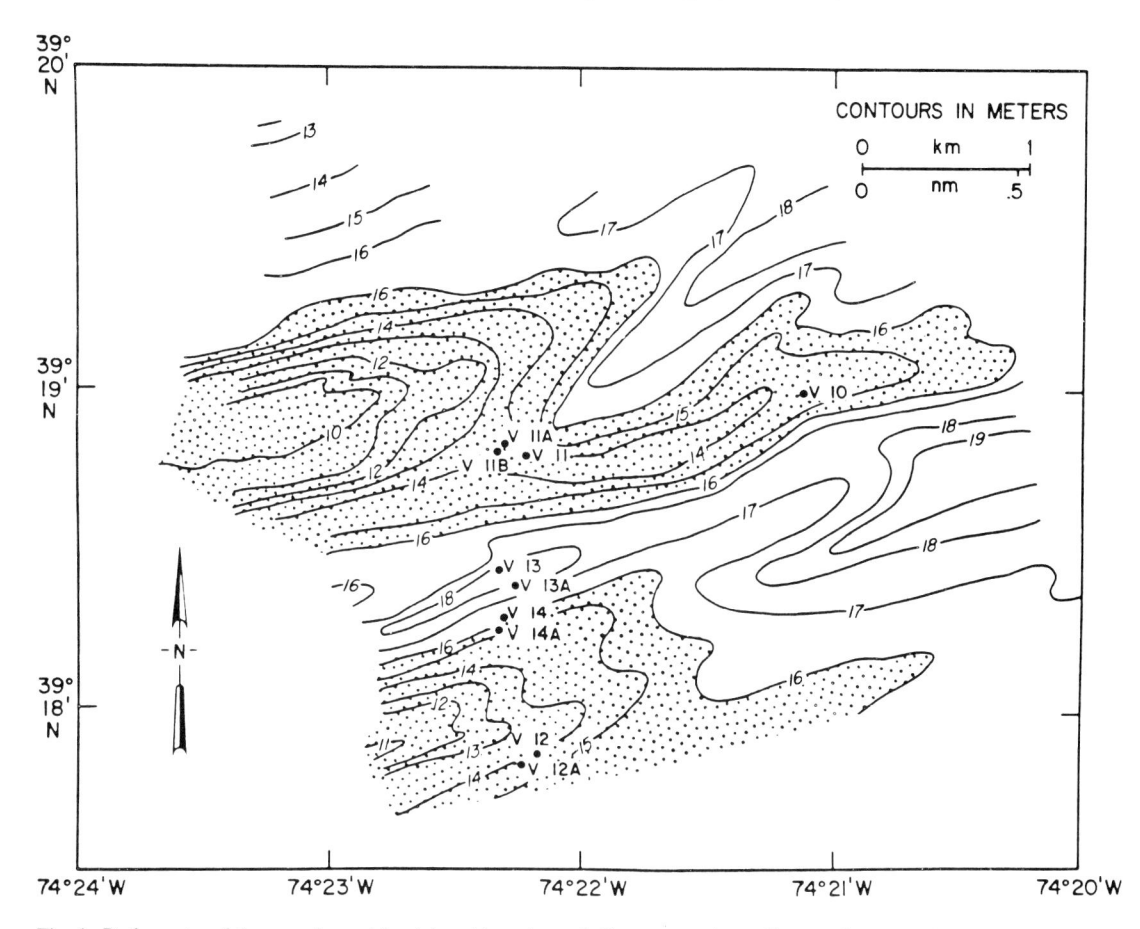

**Fig. 4.** Bathymetry of the nearshore ridge 1A and locations of vibracore stations. Contour interval is in metres. Landward is to the upper left-hand corner of the diagram.

connected (ridge 1A) and mid-shelf (ridge 2) sand ridges. These ridges are both within larger 'second order' ridge fields. The largest, 'third order', features are shoal-retreat massifs, shelf transverse valleys, deltas, scarps, etc. (Swift et al., 1972a).

Stubblefield et al. (1984) described in detail the geometries of the sand ridges on the New Jersey shelf. This description is summarized as follows. Nearshore ridges average 4.2 m in height, 2.7 km in width and 3.4 km in wavelength. Surficial sands on the nearshore ridges grade from coarse-grained sands on the shoreward flank to fine-grained sands on the seaward flank. Mid-shelf ridge heights and widths are slightly greater than the nearshore ridges. Surficial sand-size distribution is somewhat symmetrical with the coarsest sand generally near the upper shoreward flank. Wavelengths of the mid-shelf ridges are less than the nearshore ridge field and average 2.3 km. Orientations of ridges with respect to the present-day shoreline are different in that the nearshore ridges intersect the coastline at 15–30° and the mid-shelf ridges are approximately coast-parallel. Some of the nearshore ridges, such as ridge 1A, are symmetrical to slightly asymmetrical with their seaward side being the steeper side. The mid-shelf ridges, such as ridge 2, are more asymmetrical than the nearshore ridges and are also steepest on their seaward sides.

**Shelf processes**

The continental shelf of New Jersey is a storm-dominated shelf (Duane et al., 1972; Swift & Field, 1981). When discussing shoreface connected ridges, such as ridge 1A, Duane et al. (1972) stated 'storm

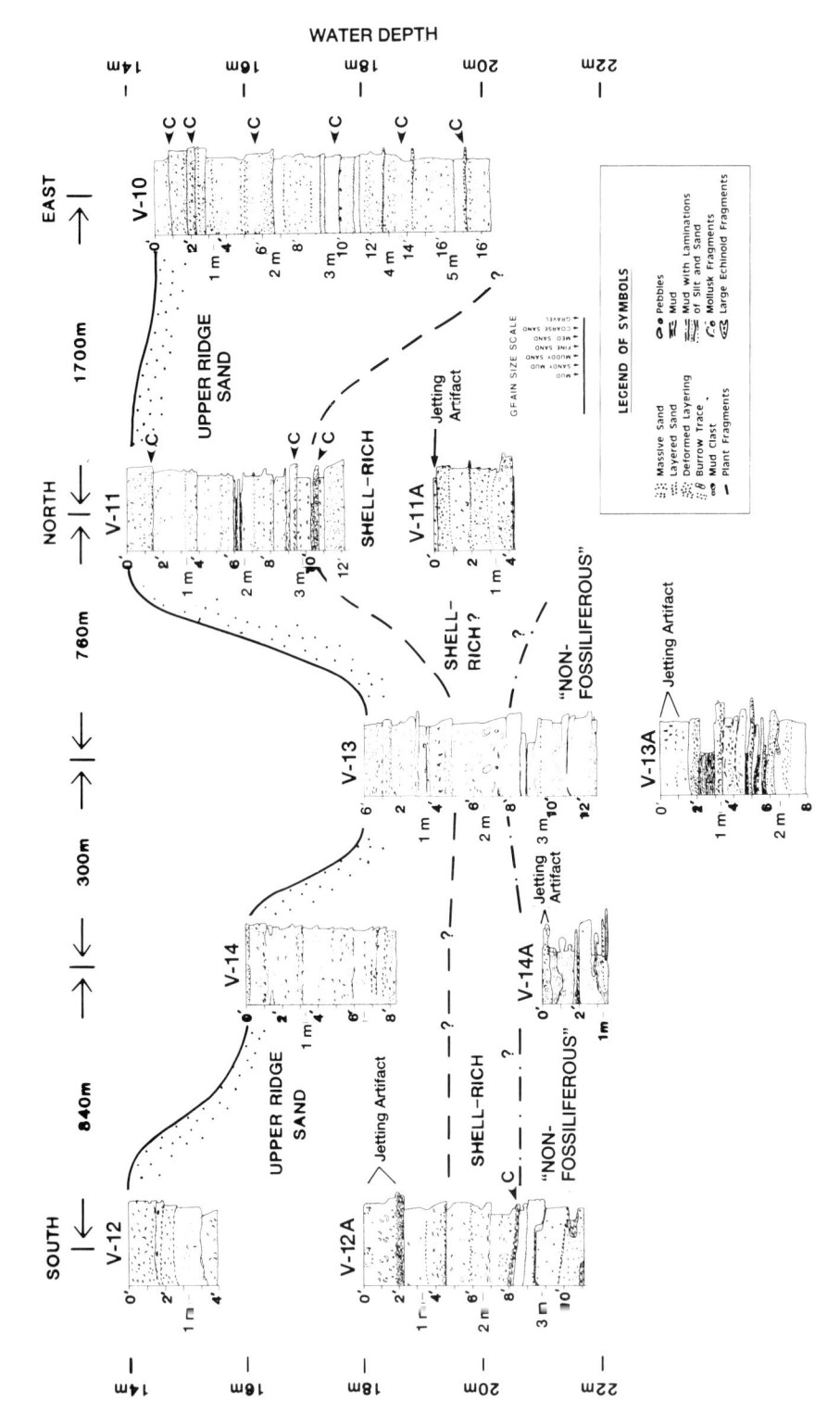

Fig. 5. Composite cross-section of all vibracores from nearshore ridge 1A showing lithologic units within vibracores and unit correlations among vibracores. The tops of the vibracores are positioned on the basis of water depth. The locations of $^{14}C$ analyses are denoted by C. Radiocarbon sample locations and ages are as follows: V 10 (0 ft 11 in to 1 ft 10 in) 4160 ± 110 years BP; V-10 (1 ft 11 in to 2 ft 3 in), fresh shells 140 ± 90 years BP and abraded shells 5975 ± 120 years BP; V-10 (4 ft 11 in to 5 ft 11 in) 4910 ± 145 years BP; V-10 (8 ft 11 in to 11 ft 3 in) 1480 ± 170 years BP; V-10 (12 ft 8 in to 14 ft 3 in) 3460 ± 120 years BP; V-10 (16 ft 11 in to 17 ft 9 in) 1670 ± 110 years BP; V-11 (0 ft to 1 ft 5 in) 3580 ± 100 years BP; V-11 (9 ft 3 in to 9 ft 6 in) 3410 ± 80 years BP; V-11 (10 ft 6 in to 10 ft 8 in) 6130 ± 120 years BP; V-12A (7 ft 10 in t 8 ft 3 in) 6360 ± 90 years BP. See Fig. 4 for location of vibracores.

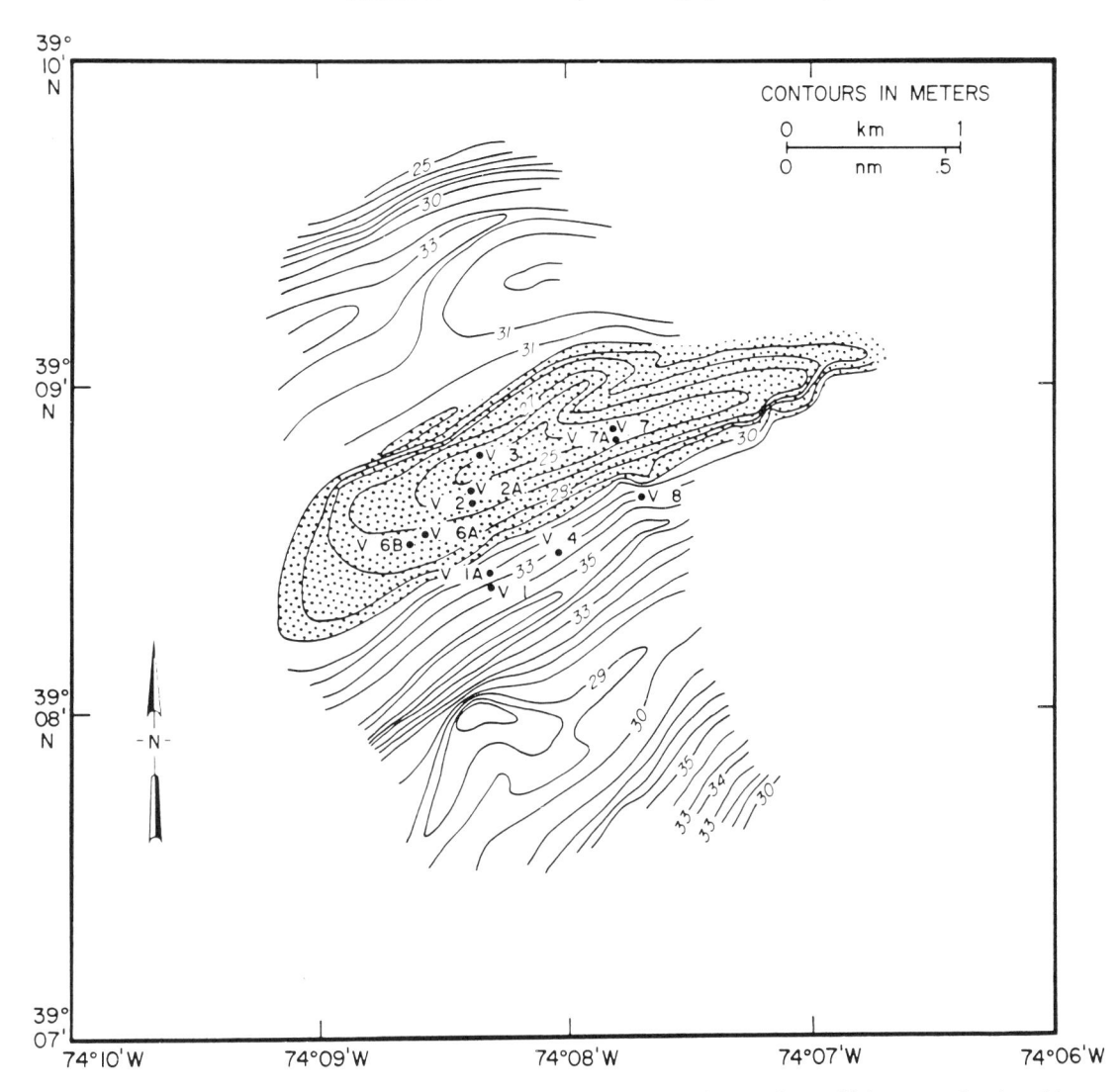

**Fig. 6.** Bathymetry of mid-shelf ridge 2 and locations of vibracore stations. Contour interval is in metres. Landward is to the upper left-hand corner of the diagram.

current and wave trains associated with them are dominant forces shaping the shoals.' The most significant storms on the Atlantic shelf are extratropical (winter) storms. According to Harrison et al. (1967), successive mid-latitude lows passing along the Atlantic seaboard during the winter months generate strong northeast winds. The resulting wind-stress, combined with falling temperatures, destroys water stratification allowing wind-driven currents to affect the bottom. Harrison et al. (1967) also noted that a net southerly drift is established owing to the onshore component of the winter storm winds causing set-up of water along the coastline. Current meter surveys of three sites on the outer New Jersey shelf by Butman et al. (1979) found winter current velocities were much higher than summer velocities. Measurements taken between December 1975 and February 1976 encountered along-shelf currents averaging $15\,\mathrm{cm\,s^{-1}}$, with maxima of $40\,\mathrm{cm\,s^{-1}}$ normally occurring once a month. Measurements taken in June to July found average currents less than $8\,\mathrm{cm\,s^{-1}}$ with the only high velocity current events being of short duration (3–25 h) and attributed to internal wave packets generated by semi-diurnal tides at the shelf edge.

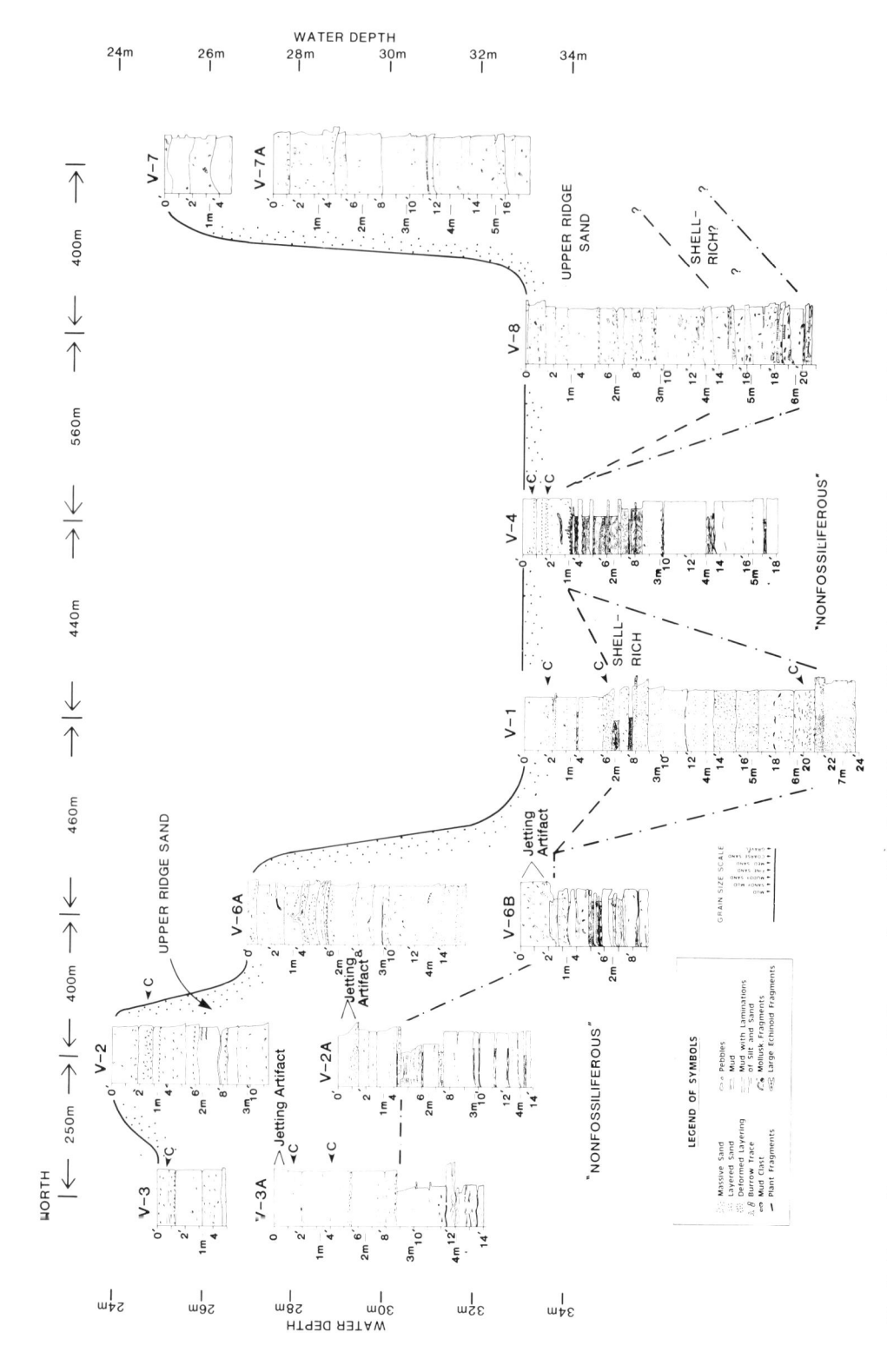

**Fig. 7.** Composite cross-section of all vibracores from mid-shelf ridge 2 showing lithologic units within vibracores and unit correlations between vibracores. The tops of the vibracores are positioned on the basis of water depth. The locations of $^{14}C$ analyses are denoted by C. Radiocarbon sample locations and ages are as follows: V-1 1ft 3in to 2ft 6in) 1155 ± 85 years BP; V-1 (5ft 8in to 6ft 5in) 12 480 ± 155 years BP; V-1 (19ft 3in to 20ft 8in) 13 240 ± 180 years BP; V-2 (1ft 11in to 2ft 11in) 1375 ± 115 years BP; V-3 (0ft to 1ft 4in) 2115 ± 100 years BP; V-3A (0ft 11in to 1ft 11in) 6740 ± 185 years BP; V-3A (3ft 6in to 5ft 6in) 5305 ± 135 years BP; V-4 (0ft to 1ft) 3355 ± 135 years BP; V-4 (1ft to 1ft 10in) 7515 ± 155 years BP. See Fig. 6 for location of vibracores.

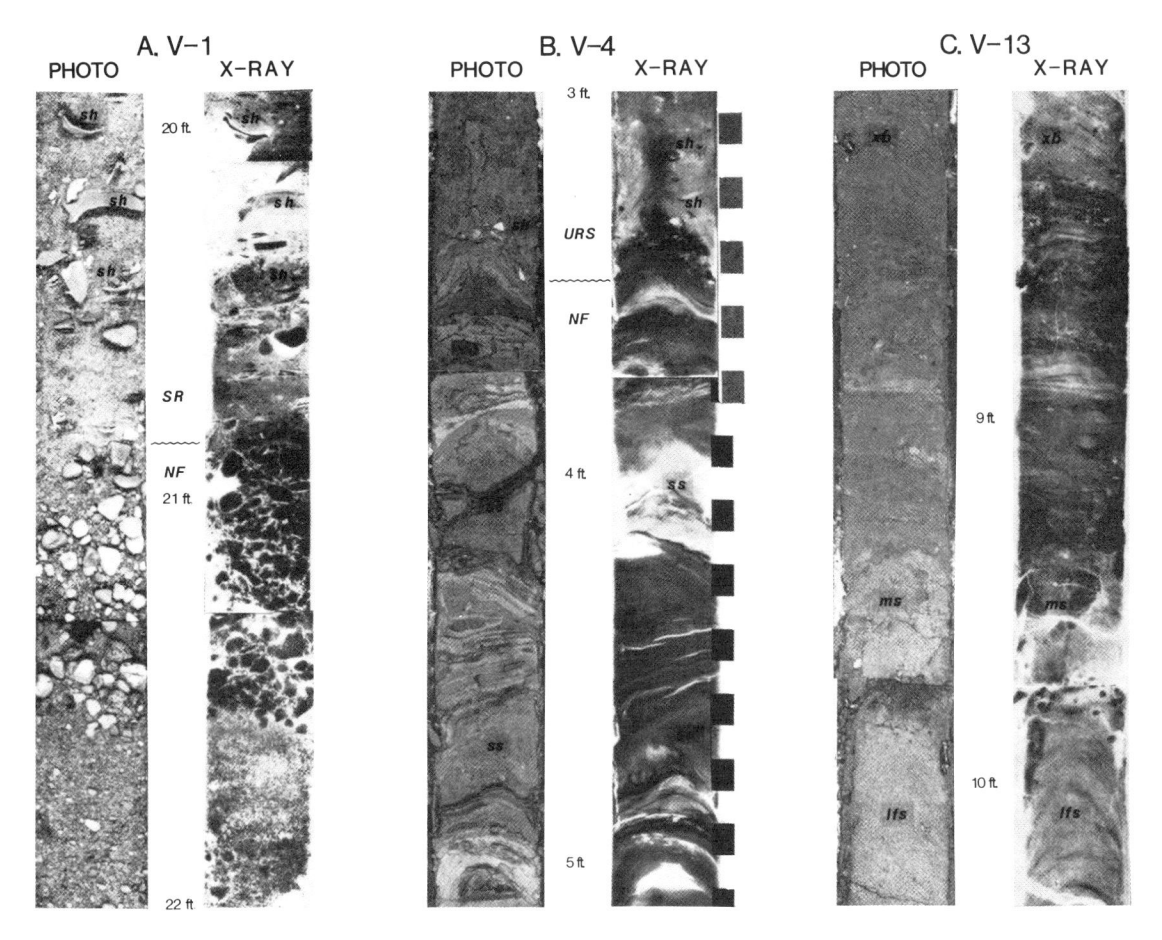

**Fig. 8.** Representative photographs of epoxy-peels and X-ray radiographs of slabs from vibracores that include the non-fossiliferous mud and sand unit. (A) Core interval of V-1 shows sharp contact of shell-rich poorly sorted sand and mud unit (SR) with non-fossiliferous unit (NF) at 20 ft 9 in. The shell-rich unit consists of medium-grained sand with some pebbles and numerous disarticulated bivalve shells (sh). The non-fossiliferous unit contains no macrofossils and consists of a sandy gravel (20 ft 9 in to 21 ft 6 in) overlying a pebble and granule-rich coarse-grained sand. (B) Cored interval of V-4 shows sharp contact between upper ridge sand (URS) and non-fossiliferous (NF) units. The small mud diapir-like structure at 3 ft 3 in is an artifact of coring, as are upward bowed laminations below 3 ft 5 in. Laminations consist of interlayered sand, clayey silt and silty clay. Very thin beds of fine-grained sand (ss) are present within the non-fossiliferous unit, but these sands contain no shell fragments (sh) unlike the fine-grained sands of the upper ridge sand. No lebenspuren was observed in the non-fossiliferous unit of this core. (C) The core interval from the top to 8 ft 6 in of core interval of V-13 consists of cross-bedded (xb) medium-grained sand. Structureless sandy mud (8 ft 6 in to 9 ft 4 in), structureless medium-grained sand (ms) (9 ft 4 in to 9 ft 8 in), and layered fine-grained sand (lfs) (9 ft 8 in to bottom). See Figs 5, 7 for graphic core descriptions.

## GENERAL LITHOSTRATIGRAPHIC UNITS

Three units, based on general lithologic characteristics, are present in both nearshore ridge 1A and mid-shelf ridge 2. These units are designated from bottom to top: 'non-fossiliferous' mud and sand; shell-rich poorly sorted sand and mud; and upper ridge sand. Figueiredo (1984) made seismic correlations between ridges 1 and 2, which showed that the surficial upper ridge sand unit (green unit) was discontinuous between ridges and that the shell-rich and 'non-fossiliferous' units were undifferentiable within a larger seismic unit (red unit) that was correlateable between ridges. Figures 4, 6 show locations of the vibracores on nearshore ridge

**Table 2.** Generalized modern habitats of commonly occurring foraminiferal species on the New Jersey continental shelf and summary of occurrence of commonly occurring foraminiferal species in New Jersey shelf vibracores

| Species | Modern habitat range on New Jersey continental shelf | 'Non-fossiliferous' mud and sand | Shell-rich sand and mud | Ridge 1A upper ridge sand | Ridge 2 upper ridge sand |
|---|---|---|---|---|---|
| *Ammonia beccarii* | Estuary, lagoon, shelf | × | × | × | |
| *Asterigerinata* cf. *mamilla* | Shelf | | × | × | × |
| *Buccella frigida* | Shelf | | × | × | × |
| *Buliminella elegantissima* | Inner shelf | × | | | |
| *Cibicides lobatulus* | Middle shelf | | × | × | × |
| *Elphidium advenum* | Inner shelf | | | × | |
| *E. excavatum* | Estuary, lagoon, shelf | × | × | × | × |
| *E. incertum* | Shelf | × | × | × | × |
| *E. subarcticum* | Shelf | | × | × | |
| *Hanzawaia strattoni* | Middle shelf | | | × | × |
| *Haynesina germanica* | Lagoon, estuary, mudflat, sandflat | × | × | | |
| *Nonionella atlantica* | Middle shelf | | | | × |
| *Pseudopolymorphina novangliae* | Middle shelf | | | | × |
| *Quinqueloculina seminula* | Middle shelf | | × | × | × |
| *Rosalina floridana* | Shelf | | × | × | × |
| *Trochamminina ochracea* | Shelf | | × | | |

1A and mid-shelf ridge 2, respectively. Figures 5, 7 show composite cross-sections through the nearshore and mid-shelf ridges with the approximate vertical distribution of the lithostratigraphic units. Major sedimentary characteristics of these units and foraminferal assemblages are described below.

### 'Non-fossiliferous' mud and sand unit

#### Nearshore ridge 1A

Within the nearshore ridge the 'non-fossiliferous' mud and sand unit is present at the base of cores V-13, V-12A and V-14A (Fig. 5). Macrofossils are absent from this unit, although some foraminiferal tests are present. Sand layers consist of fine to medium-grained sand (220–550 μm mean grain-size range) with some granule to pebble-size grains also present. The sand layers are generally massive appearing or subhorizontally laminated, as in V-13 (Fig. 8C). The sand layers are interlayered with beds of mud and muddy sands. The mud layers are well laminated and contain very few burrow traces (Fig. 8B). Muddy sand layers appear 'churned', and may be bioturbated (over 75% burrowed by volume) or disturbed by coring.

Foraminiferal assemblages in the 'non-fossiliferous' unit of ridge 1A, when present, are generally represented by few specimens and are of low diversity. *Elphidium excavatum* and *Ammonia beccarii* are the most commonly occurring species (Table 2). Plant material is abundant in the samples.

#### Mid-shelf ridge 2

Within the mid-shelf ridge, the 'non-fossiliferous' mud and sand unit is present at the base of cores V-1, V-2A, V-3A, V-4, V-6B and V-8 (Fig. 7). The unit consists of interlayered sand and mud with no observable macrofaunal remains and only rare presence of burrow traces (Fig. 8B). At the base of one core, V-1, the 'non-fossiliferous' unit consists of a very pebbly sand (Figs 7, 8A).

Foraminiferal assemblages in the 'non-fossiliferous' unit of ridge 2, when present, are variable. In vibracore V-1, the few specimens present are of *E. excavatum* and *Elphidium incertum* (Table 2). In V-2A, the samples from the 'non-fossiliferous'

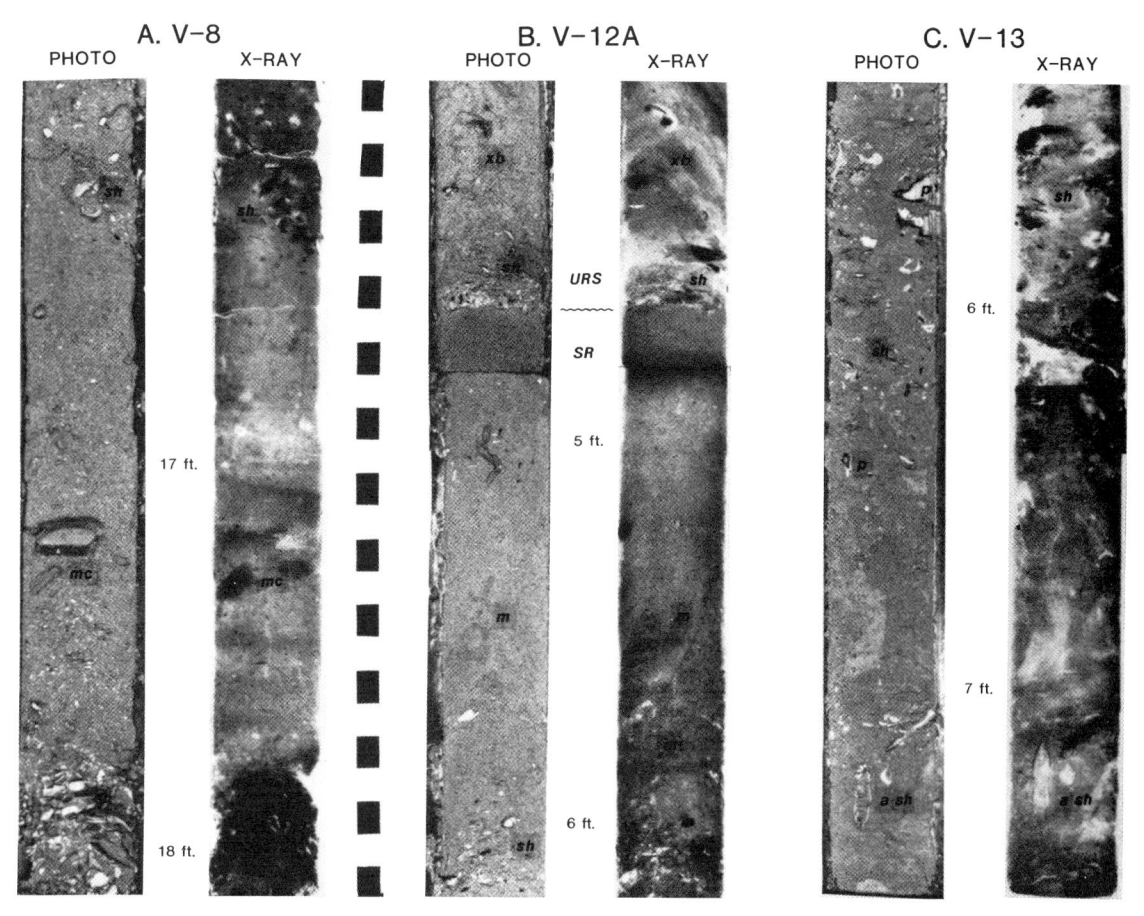

Fig. 9. Representative photographs of epoxy peels and X-ray radiographs of slabs from cored intervals of the shell-rich, poorly sorted sand and mud unit. (A) Cored interval from V-8 shows an absence of physical sedimentary structures (bioturbation?) and abundant shell material (sh). Core interval between the top and 17 ft 2 in consists of fine-grained sand with numerous granules, small mud clasts (mc) and shell fragments. The interval from 17 ft 2 in to the bottom, consists of muddy, poorly sorted, medium-grained sand with some granules and pebbles and abundant mud clasts and shell fragments. (B) Core interval from V-12A contains a sharp contact between the upper ridge sand unit (URS) and the shell-rich unit (SR) at 4 ft 7 in. Some layering (xb) is present in the upper ridge sand unit but none is visible in the underlying shell-rich unit. Mottling (m), visible in X-ray radiograph and epoxy peels, could be the record of extensive burrowing. Both the upper ridge sand and shell-rich units consist of fine to medium-grained sand with abundant shell fragments (sh). (C) Core interval from V-13 consists of a mud-rich, medium-grained sand with abundant shell fragments (sh). Mottling in the X-ray radiograph and epoxy peel probably results from extensive burrowing. Articulated bivalve (a sh), at 7 ft 3 in, is in life-position. Some white patches on the photograph of the epoxy peel (p) are areas where no sediment adhered during the peel-making procedure. See Figs 5, 7 for graphic core descriptions.

unit contain abundant (undifferentiated) plant material. Foraminifera are abundant in only three samples, which are characterized by low diversity assemblages dominated by *E. excavatum* (71–88% of the assemblage) with *Haynesina germanica* a common accessory species (Table 2; 9–21% of the assemblage). Foraminifera, mainly *E. excavatum*, are rare in V-3A, while wood and plant fragments are abundant. Two samples from the 'non-fossil-iferous' unit in V-4 contain assemblages dominated by *H. germanica*, *E. excavatum* and *Buliminella elegantissima*. *Ammonia beccarii* is a common accessory species. V-6B foraminiferal assemblages are of low diversity and are dominatd by *E. excavatum* with *H. germanica* and *A. beccarii* as common accessory species; plant material is abundant. No samples were examined for Foraminifera from the 'non-fossiliferous' unit of V-8.

## Shell-rich poorly sorted sand and mud unit

### Nearshore ridge 1A

Within the nearshore ridge the shell-rich poorly sorted sand and mud unit is present in cores V-11, V-11A, V-12A, V-13 and V-13A (Fig. 5). Sand beds within this unit contain fine to coarse-grained sands (mean grain-size range of 160–800 μm) with shell fragments and granules and/or pebbles. Beds generally fine upward and appear massive except for some apparent cross-bedding in V-13 and sub-horizontal laminations in V-13A (Fig. 5). The massive appearance of most of the sand in this unit is probably the result of bioturbation because recognizeable burrow traces are commonly present. Mud and sandy mud layers consist of laminated silts and clays, and bioturbated sands and muds, as in V-13 (Figs 5, 9).

$^{14}$C dates of bulk shell samples from this unit within the nearshore ridge are 6130 ± 120 years BP (V-11; 3.2–3.3 m) and 6360 ± 90 years BP (V-12A; 2.4–2.5 m). These dates are maximum ages before present because some of the shell material used for dating were stained grey and abraded, indicating that they were probably reworked from older strata or transported for some time before deposition. No age comparison can be made with the underlying unit because of a lack of datable material in the 'non-fossiliferous' unit. However, the sharp contact between these units in cores V-13 and V-12A records either an erosional event or hiatus in deposition.

Foraminiferal assemblages in the shell-rich unit of ridge 1A are dominated by *E. excavatum*. *Ammonia beccarii* is commonly present in V-11, V-11A and V-12. Other common accessory species in V-11 and V-11A are *Asterigerinata* cf. *mamilla*, *Cibicides lobatulus*, *Elphidium subarcticum*, *Rosalina floridana* and *Trochammina ochracea* (Table 2). Foraminiferal assemblages in the shell-rich unit of V-13 and V-13A are quite different from each other, which may indicate the presence of subunits/facies within the shell-rich unit or inaccuracies in the correlation shown in Fig. 5. Two samples in V-13 contain low-diversity faunas dominated by *E. excavatum* (63–64% of the assemblage) with *H. germanica* as a common accessory species (33–35% of the assemblage). Undifferentiated plant material is abundant in these samples. In contrast, V-13A samples are dominated by *E. excavatum* with *Elphidium incertum*, *E. subarcticum*, *Buccella frigida* and *Rosalina florida* as common accessory species (Table 2). *Haynesina germanica* is absent from these higher diversity samples.

### Mid-shelf ridge 2

Within the mid-shelf ridge, the shell-rich poorly sorted sand and mud unit is present only in cores V-1 and V-8 (Fig. 7). Within this unit, mud layers are interbedded with shell-rich, poorly sorted sands (Fig. 9A). $^{14}$C dates from shell material in V-1 give dates of 13 240 ± 180 years BP (V-1; 5.8–6.3 m) and 12 480 ± 155 years BP (V-1; 1.7–2.0 m). The samples used for dating probably include some reworked shells.

The contact with the underlying 'non-fossiliferous' unit in cores V-1 and V-8 is sharp and may record erosion or some break in deposition of unknown duration.

Foraminiferal assemblages in the shell-rich unit of vibracore V-1 are characterized by iron-stained tests. Numbers of individuals and species per sample are lower than in the overlying upper ridge sand unit. Faunal assemblages are dominated by *E. excavatum* with *E. incertum*, *E. subarcticum*, *C. lobatulus* and *Quinqueloculina seminula* as common accessory species (Table 2). In V-8, assemblages are dominated by *E. excavatum* with *A. beccarii* as a common accessory species.

## Upper ridge sand unit

### Nearshore ridge 1A

The upper ridge sand unit is a generally coarsening upward sequence of fine to medium-grained sand (150–400 μm) with numerous stacked beds averaging 20 cm in thickness. Most cores of this unit coarsen upward, such as V-10 (Fig. 5). Only core V-13, which is located on the flank of a ridge, does not coarsen upward (Fig. 5). Beds within the upper ridge sand unit range in thickness from 5 to 45 cm. Bed boundaries are based on changes in texture and/or character of sedimentary structures. A common change in sedimentary structures along a bed boundary is one of laminated sands overlying apparently massive/bioturbated sands (Figs 5, 10A).

Physical sedimentary structures in the upper ridge sand unit are primarily low-angle or sub-horizontal laminations (Fig. 10A, B). Some apparent cross-bedded intervals are present, as within V-10 and V-11 (Fig. 5), but these are rare.

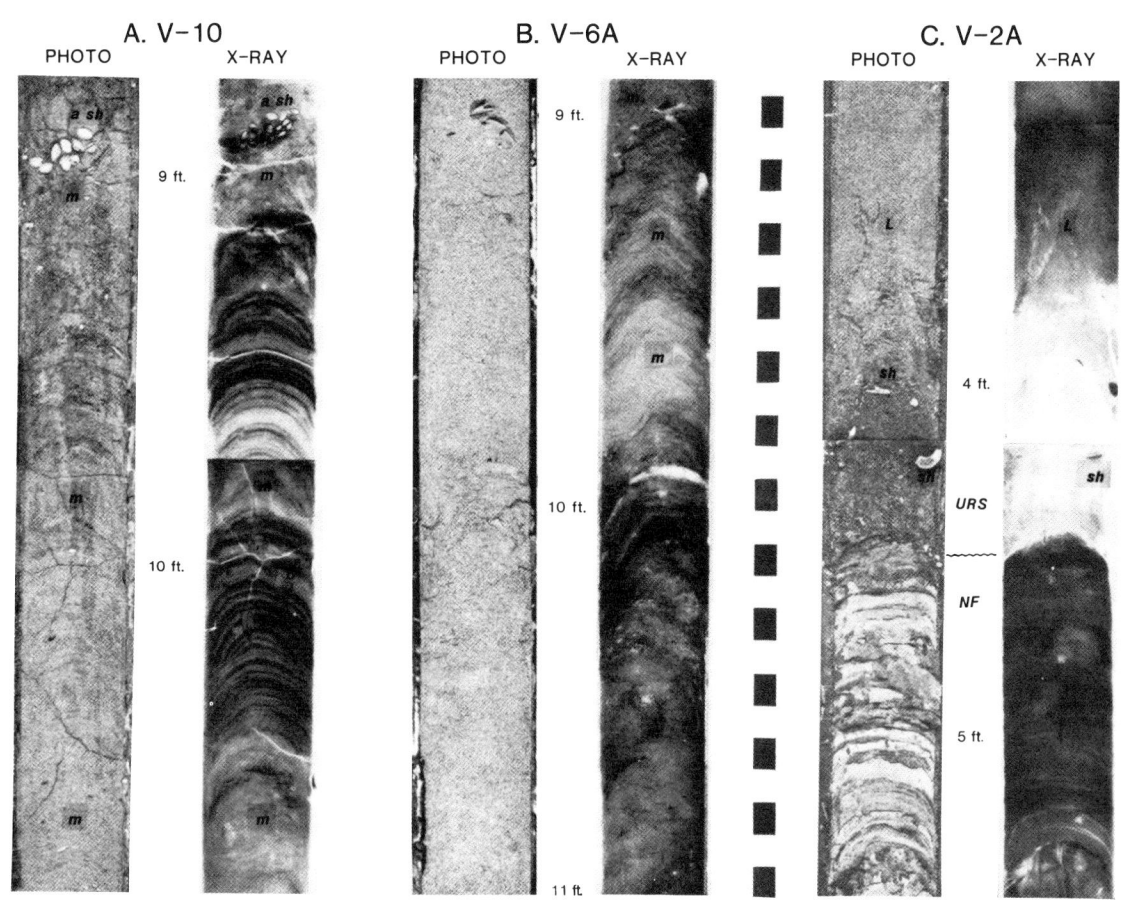

**Fig. 10.** Representative photographs of epoxy-peels and X-ray radiographs of slabs from cores of the upper ridge sand unit. (A) Core interval from vibracore V-10 (mid-shelf ridge) showing laminated to non-laminated character common to the upper ridge sand unit. The poorly layered to non-laminated units are generally mottled (m) owing to bioturbation. The laminations appear parallel to subparallel in spite of the bowing and were probably subhorizontal to horizontal before deformation by coring. Some individual burrow traces (b) are recognizeable. A concentration of juvenile, articulated bivalves (a sh) is at 8 ft. 9 in. The cored interval consists of fine to medium-grained sand. (B) Core interval from V-6A (nearshore ridge) showing the same laminated to non-laminated character present in V-10 (A). In core V-6A, however, the laminations are indistinct and wispy appearing, probably resulting from burrowing by infauna. Mottling (m) present in the X-ray radiographs is the result of extensive burrowing. The cored interval consists of fine to medium-grained sand with only a few shell fragments. (C) Core interval from V-2A shows upper ridge sand (URS) sharply overlying 'non-fossiliferous' unit (NF) at 4 ft 5 in. Upper ridge sand interval consists of vaguely laminated (L), fining upward, coarse to medium-grained sand with some shell material (sh). The underlying 'non-fossiliferous' unit consists of laminations of silty clay and very fine-grained sand. A few thin vertical burrows are present (b; 0.15–1.2 cm diameter; less than 10% by volume).

Almost totally bioturbated intervals and intervals containing distinct burrow traces are common within this unit. Bioturbated and massive appearing intervals increase in abundance with increasing depth of burial. An exception to this is V-13, which is bioturbated throughout the upper ridge sand unit (Fig. 5). Recognizable burrow traces consist of: (1) burrows infilled with mud pellets and shells (V-11); (2) a burrow filled with articulated bivalves in V-10 (Fig. 10A and 5); and (3) 0.3 cm diameter mud-lined burrows and 2.5 cm diameter traces within V-13. The burrow traces are similar to the ichnofaunal types *Ophiomorha* and *Thalassinoides*. $^{14}$C dates of bulk shell material from the upper ridge sand unit in ridge 1A range from 4160 ± 110 years BP to 1480 ± 170 years BP (V-10, Fig. 5). Bulk

**Table 3.** Range of contribution of commonly occurring species in upper ridge sand foraminiferal assemblages

| Species | Ridge 1A Percentage range | Ridge 1A Mean (%) | Ridge 2 Percentage range | Ridge 2 Mean (%) |
|---|---|---|---|---|
| *Ammonia beccarii* | 0–19 | 1.8 | | |
| *Asterigerinata* cf. *mamilla* | 0–7 | 1.8 | 0–9 | 1.9 |
| *Buccella frigida* | 0–4 | 0.9 | 0–5 | 2.4 |
| *Cibicides lobatulus* | 0–4 | 1.7 | 2–46 | 14.1 |
| *Elphidium advenum* | 0–7 | 2.6 | | |
| *E. excavatum* | 76–95 | 87.4 | 30–86 | 56.8 |
| *E. incertum* | 0–5 | 2.1 | 0–13 | 4.7 |
| *E. subarcticum* | 0–4 | 1.0 | | |
| *Hanzawaia strattoni* | 0–3 | 0.9 | 1–30 | 8.3 |
| *Nonionella atlantica* | | | 0–5 | 2.0 |
| *Pseudopolymorphina novangliae* | | | 0–4 | 1.6 |
| *Quinqueloculina seminula* | 0–4 | 0.9 | 0–18 | 4.9 |
| *Rosalina floridana* | 0–5 | 1.5 | 0–36 | 4.8 |

shell samples are an admixture of different aged shells, as revealed by $^{14}$C dating of fresh shells (140 ± 90 years BP) and abraded and worn shells (5975 ± 120 years BP) from the same interval in V-10 (0.6–0.7 m).

Comparison of $^{14}$C dates of bulk shells in the upper ridge sand and the underlying shell-rich units of core V-11 show major differences, which may indicate that the basal contact of the upper ridge sand unit is erosional or records a period of non-deposition. $^{14}$C age dates indicate the shell-rich unit contains shell material that has a calculated age over 2000 years older than material in the upper ridge sand unit. The sharp basal contact of the upper ridge sand unit in V-12A and V-13 also suggests that this boundary is unconformable. This contact is observable seismically with Uniboom (Figueiredo, 1984) and to a lesser degree with 3.5 kHz (Stubblefield, pers. comm.).

Foraminiferal assemblages of the upper ridge sand unit of ridge 1A are dominated by *E. excavatum*, which comprises from 76 to 95% of the assemblage in each sample. *Ammonia beccarii*, *A.* cf. *mamilla*, *B. frigida*, *C. lobatulus*, *Elphidium advenum*, *E. incertum*, *E. subarcticum*, *Hanzawaia strattoni*, *Q. seminula*, and *R. floridana* are the most abundant accessory species (Table 2). Table 3 illustrates the contribution made by each of these taxa to the upper ridge sand unit foraminiferal assemblages in ridge 1A.

*Mid-shelf ridge 2*

The upper ridge sand unit of the mid-shelf ridge is a generally coarsening upward sequence of fine to medium-grained sand (170–400 μm) within stacked, laminated and non-laminated beds. Non-laminated beds are probably bioturbated but core deformation prevents positive identification of burrow dimensions, orientations and types. Cores V-2, V-3A, V-6A, V-7 and V-7A (Figs 7, 10) have interlayered laminated and non-laminated beds. Core V-8 is only laminated in the top 45 cm and is completely bioturbated below (Fig. 7). Thicknesses of beds range from 5 to 70 cm, with most being 30 cm or less. Some of the boundaries between beds are obscured, however, owing to bioturbation and core deformation. Laminated intervals within the cores consist primarily of subhorizontal laminations (Fig. 10). Cross-laminations are present in some cores (Fig. 7, V-8).

$^{14}$C dates of bulk shell material (mixed fresh and abraded shells) from the mid-shelf ridge range from 7515 ± 155 years BP in V-4 (0.3–0.5 m) to 1155 ± 85 years BP in V-1 (0.4–0.8 m; Fig. 7).

Calculated $^{14}$C age dates and the nature of the basal contact of the upper ridge sand unit in ridge 2 indicate that this boundary is erosional or records an unknown period of non-deposition. Comparison of $^{14}$C age dates of bulk shell samples from the upper ridge sand and samples from the shell-rich

**Table 4.** Multiple discriminant analysis on nearshore ridge 1A lithostratigraphic units (number of variables = 50; number of groups = 3; 95% probability)

| Unit | Number of samples | Number correctly discriminated | Percentage correctly discriminated |
|---|---|---|---|
| Upper ridge sand | 102 | 100 | 98 |
| Shell-rich | 27 | 25 | 92 |
| 'Non-fossiliferous | 19 | 19 | 100 |
| Total | 148 | 144 | 97 |

unit in core V-1 (Fig. 7) shows age date differences between units in excess of 11 000 years. The basal contacts of the upper ridge sand unit in cores V-1, V-2A, V-3A, V-4 and V-8 are all sharp (Fig. 7).

*Elphidium excavatum* dominates the foraminiferal assemblages of the upper ridge sand unit in ridge 2, as it does in ridge 1A. This species comprises from 30 to 80% of the assemblage in each sample. *Asterigerinata* cf. *mamilla*, *B. frigida*, *C. lobatulus*, *E. incertum*, *Hanzawaia strattoni*, *Nonionella atlantica*, *Q. seminula* and *R. floridana* are the more abundant accessory species (Table 2). Table 3 lists the contribution made by each of these taxa to the foraminiferal assemblages in the upper ridge sand unit of ridge 2.

### Statistical discrimination of lithostratigraphic units

Discriminant analysis of the grain-size data was carried out to gauge the validity of the units delineated based on visually determined sedimentological features, such as sedimentary structures. As previously mentioned, portions of some cores

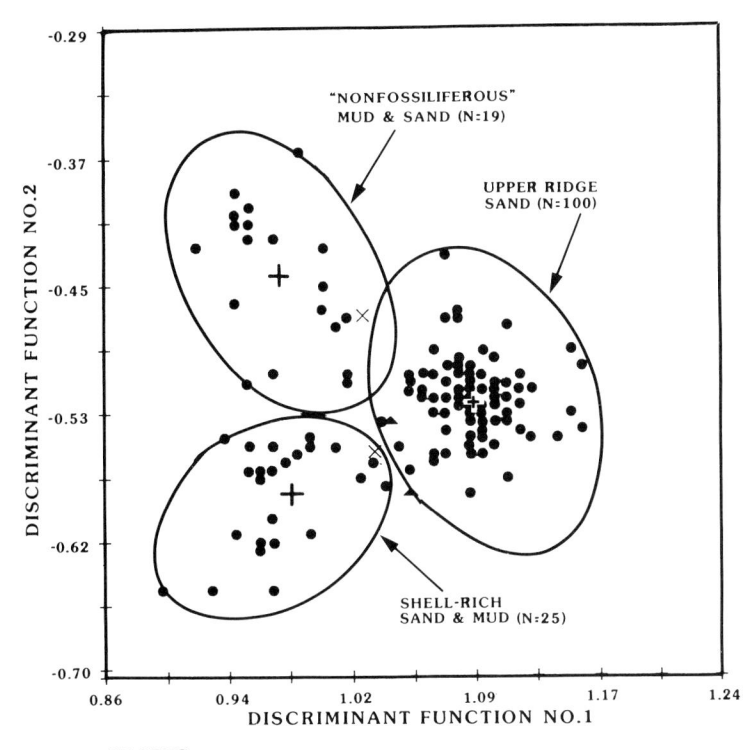

**Fig. 11.** Discriminant functions plot of grain-size data from the three nearshore ridge 1A lithostratigraphic units. Ovals enclose plots of samples that were correctly assigned by discriminant analysis to their 'correct' lithostratigraphic unit, as previously determined by sedimentological and lithologic parameters. The number of samples in individual lithostratigraphic units and the number of samples correctly classified within each unit are shown in Table 4. Ninety-seven per cent of the samples were 'correctly' discriminated using 50 grain-size variables.

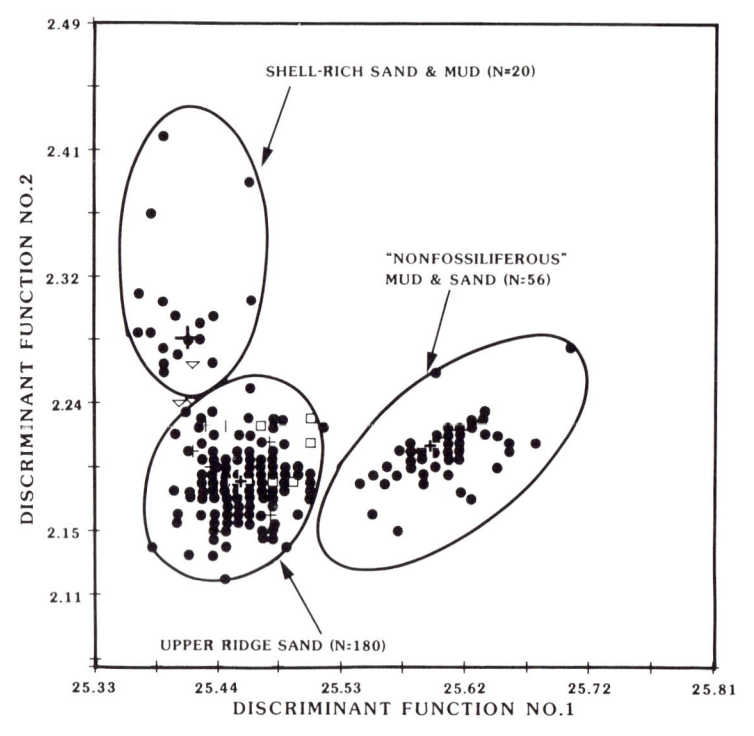

**Fig. 12.** Discriminant functions plot of grain-size data from the three mid-shelf ridge 2 lithostratigraphic units. Ovals enclose plots of samples that were correctly assigned by discriminant analysis to their respective lithostratigraphic unit, as determined independently by sedimentological and lithologic parameters. The number of samples in individual lithostratigraphic units and the number of samples correctly classified within each unit are shown in Table 5. Ninety-five per cent of the samples were 'correctly' discriminated using 50 grain-size variables.

were deformed in varying degrees by the coring process. The deformation however, was not extensive enough to significantly alter the textural characteristics of the cores.

**Table 5.** Multiple discriminant analysis on mid-shelf ridge 2 lithostratigraphic units (number of variables = 50; number of groups = 3; 95% probability)

| Unit | Number of samples | Number correctly discriminated | Percentage correctly discriminated |
|---|---|---|---|
| Upper ridge sand | 183 | 180 | 98 |
| Shell-rich | 26 | 20 | 76 |
| 'Non-fossiliferous | 61 | 56 | 91 |
| Total | 270 | 256 | 95 |

Discriminant analysis of grain-size data results in excellent discrimination among individual samples from the individual ridges on a unit by unit basis. Discriminant analysis of samples from the nearshore ridge 1A correctly discriminated 97% of the samples (144 of 148 samples) into the lithostratigraphic units that had been designated previously on the basis of sedimentary structures and lithology (Table 4). The excellent separation of samples from each of the three nearshore ridge units is shown in Fig. 11. The three lithostratigraphic units within mid-shelf ridge 2 are also successfully discriminated, as shown in a plot of the data in Fig. 12. Of the 270 mid-shelf ridge samples analysed, 256 (95%) were correctly discriminated into the units designated on the basis of sedimentary structures and lithologies (Table 5). Eight of the 14 incorrectly discriminated samples within ridge 2 are from vibracore V-8, which may indicate that the unit boundaries may require some shifting within that core.

## COMPARISON OF NEARSHORE RIDGE 1A AND MID-SHELF RIDGE 2

### General lithostratigraphic units

Both the nearshore and mid-shelf ridges contain similar lithologic units, but differences in the distribution and character of these units between ridges is significant. For example, in both ridges 1A and 2 the upper ridge sand units consist generally of clean sands that coarsen upward; but detailed examination of sedimentological and palaeontological characteristics of these units show that statistically significant differences do exist. These differences are discussed at the end of this section. A more obvious difference between the two ridges is the distribution of the shell-rich lithostratigraphic unit.

### *Unit distribution*

A major difference between the nearshore and the mid-shelf ridges is the distribution of the shell-rich unit. The shell-rich unit is present in all cores from nearshore ridge 1A that penetrate the base of the upper ridge sand unit, but is only present in two cores from mid-shelf ridge 2 (V-1 and V-8). In the nearshore ridge the shell-rich unit occurs between the 'non-fossiliferous' and the upper ridge sand units (Fig. 5). In the mid-shelf ridge the upper ridge sand unit directly overlies the 'non-fossiliferous' unit in cores located near the crest of the ridge and, on the flanks of the ridge, the upper ridge sand unit overlies the shell-rich unit in two of three cores (V-1 and V-8; Figs 6, 7). The patchy distribution of the shell-rich unit within mid-shelf ridge 2 may be the result of infilling of large erosional features, such as the scoured interridge troughs (Stubblefield & Swift, 1976) or relict channels. Channel fills have been delineated in seismic records from the study area (Figueiredo, 1984).

### *Base of upper ridge sand units*

This contact is sharp in both ridges and probably records a period of erosion or non-deposition. No gradational contact or interfingering of units exists in any of the vibracores (Figs 5, 7) and the contact appears to be sharp in seismic sections of the same ridges (Figueiredo, 1984). Consequently, we infer that the upper ridge sand units are not related genetically to the underlying units, although these contacts are located within the relief-forming portion of both ridges (Figs 5, 7). The location of these contacts in the relief-forming portion of the ridges is a result of incising of the interridge area (Stubblefield & Swift, 1976).

### *Foraminiferal assemblages*

Marginal marine conditions are indicated by the foraminiferal faunas present in the cored intervals of the 'non-fossiliferous' units from both nearshore ridge 1A and mid-shelf ridge 2 (Table 2). Foraminifera are poorly represented in the non-fossiliferous unit of ridge 1A. The foraminiferal assemblages, when present, indicate deposition on an inner shelf or estuarine environment, whose closeness to shore is suggested by the presence of abundant plant material. A similar abundance of plant material characterizes the 'non-fossiliferous' unit of ridge 2. Foraminiferal assemblages, when present, vary in composition from core to core in ridge 2 but they all indicate very nearshore conditions. *Ammonia beccarii* and *Buliminella elegantissima* are common in the inner shelf today, while *H. germanica*, a dominant species in some samples, is known to live only in low salinity, often intertidal environments (Table 2; Culver & Banner, 1978).

The shell-rich units in both ridges 1A and 2 contain foraminiferal assemblages that indicate more offshore, open marine conditions than interpreted for the 'non-fossiliferous' units. Foraminiferal assemblages of the shell-rich unit in both ridges 1A and 2 indicate normal salinity, open marine, inner shelf conditions. Faunas are dominated by *E. excavatum* and the typical inner shelf species *A. beccarii* is generally present (Table 2). The only exceptions are two samples from V-13, which contain low-diversity faunas in which the marginal marine *H. germanica* is an abundant accessory species (Table 2).

### *Statistical comparisons*

Discriminant analysis of grain-size parameters from both ridges 1A and 2 resulted in good to fair discrimination (80%) among individual samples on a unit by unit basis for all six lithostratigraphic units. The percentage correctly discriminated for individual facies ranges from a low of 65% for the upper

**Table 6.** Multiple discriminant analysis on nearshore ridge 1A and mid-shelf ridge 2 lithostratigraphic units (number of variables = 50; number of groups = 6; 95% probability)

| Unit | Number of samples | Number correctly discriminated | Percentage correctly discriminated |
|---|---|---|---|
| *Ridge 1A* | | | |
| Upper ridge sand | 102 | 67 | 65 |
| Shell-rich | 27 | 19 | 70 |
| 'Non-fossiliferous' | 19 | 15 | 78 |
| *Ridge 2* | | | |
| Upper ridge sand | 183 | 165 | 90 |
| Shell-rich | 26 | 17 | 65 |
| 'Non-fossiliferous' | 61 | 54 | 88 |
| Total | 418 | 337 | 80 |

ridge sand of ridge 1A and the 'non-fossiliferous' unit of ridge 2 to a high of 90% for the upper ridge sand of ridge 2 (Table 6). In discriminating among the six units, five discriminant equations were found to be statistically viable. Five discriminant scores calculated for each sample were used in 'classifying' each sample. Plots of pairs of discriminant scores of individual samples illustrate the separation among individual units. When all five discriminant scores for each sample are combined into a single equation, mean discriminant scores are determined for each lithostratigraphic unit and individual sample probabilities of unit memberships are calculated (Klovan & Billings, 1967). When two of the five discriminant scores for individual samples are plotted certain facies may be distinguished visually. Figure 13 shows one 'projection' of the grain-size data, which shows nearly discrete groupings of three units (upper ridge sand from ridge 2, 'non-fossiliferous' from ridge 2, and upper ridge sand from ridge 1A). The other three units form dis-

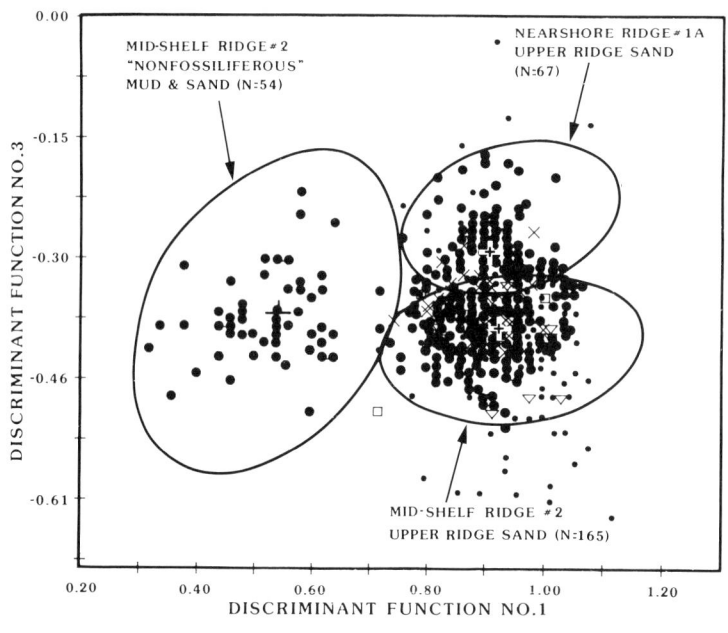

**Fig. 13.** Discriminant functions plot of grain-size data from both nearshore ridge 1A and mid-shelf ridge 2. Ovals enclose plots of three of the six lithostratigraphic units that were correctly discriminated. (Plots of samples from the other three units are represented by small dots. Separation of these units is better shown on $x-y$ plots of other discriminant functions.) Upper ridge sand and 'non-fossiliferous' from ridge 2 and upper ridge sand from ridge 1A are clearly separated on the basis of discriminant functions 1 and 3. The number of samples in individual lithostratigraphic units and the number of samples correctly discriminated within each unit are shown in Table 6. The ovals show only slight overlap indicating that on this particular discriminant score plot, the data points of the three units plot as relatively discrete groups. Eighty per cent of the samples in the six lithostratigraphic units were correctly discriminated using 50 grain-size variables.

Table 7. Comparisons of sedimentary characteristics between upper ridge sand units from nearshore ridge 1A and mid-shelf ridge 2

| Ridge | Mean bed thickness | Average grain size (μm) | Graded beds (%) | Physical structures (%) | Bioturbation (%) | Deformed beds (%)[a] | Number of beds |
|---|---|---|---|---|---|---|---|
| Nearshore ridge 1A | 0.26 m (0.9 ft) | 240 | 63 | 39 | 42 | 19 | 54 |
| Mid-shelf ridge 2 | 0.39 m (1.3 ft) | 220 | 45 | 28 | 54 | 18 | 68 |

[a] Percentage volume of core deformed by coring procedure.

crete groupings when other discriminant functions are used as ordinate and abscissa.

## Comparison of upper ridge sand units

The upper ridge sand units in ridges 1A and 2 are interpreted to be the only 'active' portion of each ridge. As shown above, the shell-rich and 'non-fossiliferous' units are unconformable with the upper ridge sand based on the character of the contacts and variation in radiocarbon dates. Thus, these units are not considered genetically related. Consequently, detailed comparisons between ridges, given below, focus on the upper ridge sand units.

*Sedimentary comparisons*

Table 7 compares the general sedimentological features found in the upper sand ridge units from both ridges 1A and 2. Highlights of those comparisons are as follows. First, units in both the nearshore and mid-shelf ridges consist of stacked beds, with the beds within ridge 2 being slightly thicker. Second, units from both ridges consist of alternating laminated and non-laminated (bioturbated) beds, a sequence similar to that found in ancient and modern (middle to lower) shoreface and inner shelf sand deposits (Howard, 1972; Howard & Reineck, 1981). Laminations within the upper ridge sand unit are generally planar, horizontal to subhorizontal, or dip at a low angle. Inclined or cross-laminated intervals are present but rare. Third, upper ridge sand units from both ridges contain fine to medium-grained sand and generally tend to coarsen upward, but ridge 1A is slightly coarser than ridge 2. Fourth, both ridges are capped by the beds of medium-grained sands (350–400 μm).

*Comparison of radiocarbon dates*

Although not too much emphasis should be placed on radiocarbon dates from bulk shell material, it is interesting to note that the youngest $^{14}$C dates are similar in both ridges 1A and 2. From mid-shelf ridge 2, the youngest date is 1155 ± 85 years BP. (V-1; 0.4–0.8 m; Fig. 7). From nearshore ridge 1A the youngest date is 1480 ± 170 years BP (V-1; 2.7–3.4 m; Fig. 5). This similarity in ages suggests simultaneous deposition of the two ridges or, at least, portions of the two ridges.

The oldest $^{14}$C dates do differ substantially between ridges. The oldest date in the upper ridge sand units of ridge 2 is 7515 ± 155 years BP (V-4; 0.3 to 0.5 m; Fig. 7), whereas the oldest bulk shell date in the upper ridge sand unit of ridge 1A is 4910 ± 145 years BP (V-10; 1.5–1.8 m; Fig. 5). This difference of over 2500 years in oldest ages between ridges probably results, in part, from the presence of older shell material in the surrounding sea-bed that was excavated and redeposited within ridge 2. Because of the gradual rise in sea-level, the mid-shelf ridge has been inundated for a longer period of time than the nearshore ridge and, consequently, the average age of shell remains within the mid-shelf sea-bed is greater than the average age of shells in the sea-bed surrounding the nearshore ridge.

*Comparison of foraminiferal assemblages*

Foraminiferal assemblages of the upper ridge sand unit in ridges 1A and 2 are similar in their species diversity values. Number of species per sample (approximately 300 specimens per sample) range from approximately 10 to 15. They are also similar in that assemblages in both ridges are dominated by *E. excavatum* (see Table 3) and lists of commonly occurring accessory species are similar, although not identical, for both ridges. *Ammonia beccarii*, *E. advenum* and *E. subarcticum*, which are common accessory species in ridge 1A, are extremely rare in ridge 2 and hence not recorded for ridge 2 in Table 3. Similarly, *N. atlantica* and *P. novangliae* are common accessory species in ridge 2 but not in ridge 1A.

Eight species are commonly occurring and abundant in the upper ridge sand of both ridges (Table 3). *Elphidium excavatum* is the overwhelmingly dominant species in ridge 1A assemblages, comprising from 76 to 95% of assemblages (mean contribution of 87%). In ridge 2, the same species only comprises from 30 to 86% of assemblages (mean contribution of 57%). A *t* test shows that the contributions of this species to ridge 1A and 2 assemblages are significantly different at the 90% level. Significant differences also apply to *B. frigida*, *C. lobatulus*, *E. incertum*, *H. strattoni* and *Q. seminula*. The contributions of *A.* cf. *mamilla* and *R. floridana* (Table 3) are not significantly different at the 90% level.

The foraminiferal faunas of the upper ridge sand unit in ridges 1A and 2 show considerable differences both in species present and in relative abundance of species in common. Faunal interpretations are the result of comparisons of modern New Jersey shelf foraminiferal assemblages (e.g. Culver & Buzas, 1980; Miller & Ellison, 1980; Poag *et al.*, 1980) with those recorded in the vibracores from this study.

Foraminiferal assemblages in all samples from the upper ridge sand unit of ridge 1A are interpreted as representing an open inner shelf environment of normal marine salinity. The typical inner shelf species *A. beccarii* is present in most samples, while *C. lobatulus* and *H. strattoni*, which are abundant in the mid-shelf, are present in most samples but in low numbers.

The upper ridge sand unit of ridge 2 contains foraminiferal faunas interpreted as representing open inner to middle shelf conditions. *Ammonia beccarii*, when present, is very rare, while *C. lobatulus*, *H. strattoni* and several other species (Table 3) are more abundant than in ridge 1A.

Both ridge 1A and ridge 2 are located within an inner shelf (<60 m) foraminiferal biofacies dominated by *E. excavatum* (Ellison, 1977; Poag *et al.*, 1980), but the consistent indication is that the entire thickness of upper ridge sand units in both ridges acquired their present faunal characteristics in the position on the shelf that they now occupy. *Ammonia beccarii* reflects the inner shelf aspect of ridge 1A assemblages. In ridge 2 the increased abundance of several species (*C. lobatulus* in particular) reflects the decreasing dominance of *E. excavatum* as a middle shelf (60–100 m) *C. lobatulus* dominated biofacies (Poag *et al.*, 1980) is approached.

*Statistical comparisons*

Discriminant analysis of the textural data from the upper ridge sand units of both ridges 1A and 2 results in good discrimination (93%) between ridges. Of the 285 samples analysed, 266 were correctly discriminated (Table 8).

Many factors could create the textural differences between the upper ridge sand units of ridges 1A and 2, but the most plausible explanation is that the two ridges were deposited in different settings. This conclusion is also supported by the foraminiferal evidence cited above.

## DISCUSSION: ORIGIN OF MID-SHELF RIDGE

### Evolution of model

This section presents a revised model for the origin of the mid-shelf ridge system. The basic premise of the revised model is that the mid-shelf ridges *are not relict features* but were formed in a middle shelf setting. This model initially evolved when sedimentological data from the 1979 cruise indicated that the active portion of the mid-shelf ridge (the upper ridge sand unit) was not 'cored' by a relict deposit and appeared to have been deposited in

Table 8. Multiple discriminant analysis on upper ridge sand units of nearshore ridge 1A and mid-shelf ridge 2 (number of variables = 50; number of groups = 2; 95% probability)

| Unit | Number of samples | Number correctly discriminated | Percentage correctly discriminated |
|---|---|---|---|
| Ridge 1A | 102 | 89 | 87 |
| Ridge 2 | 183 | 177 | 96 |
| Total | 285 | 266 | 93 |

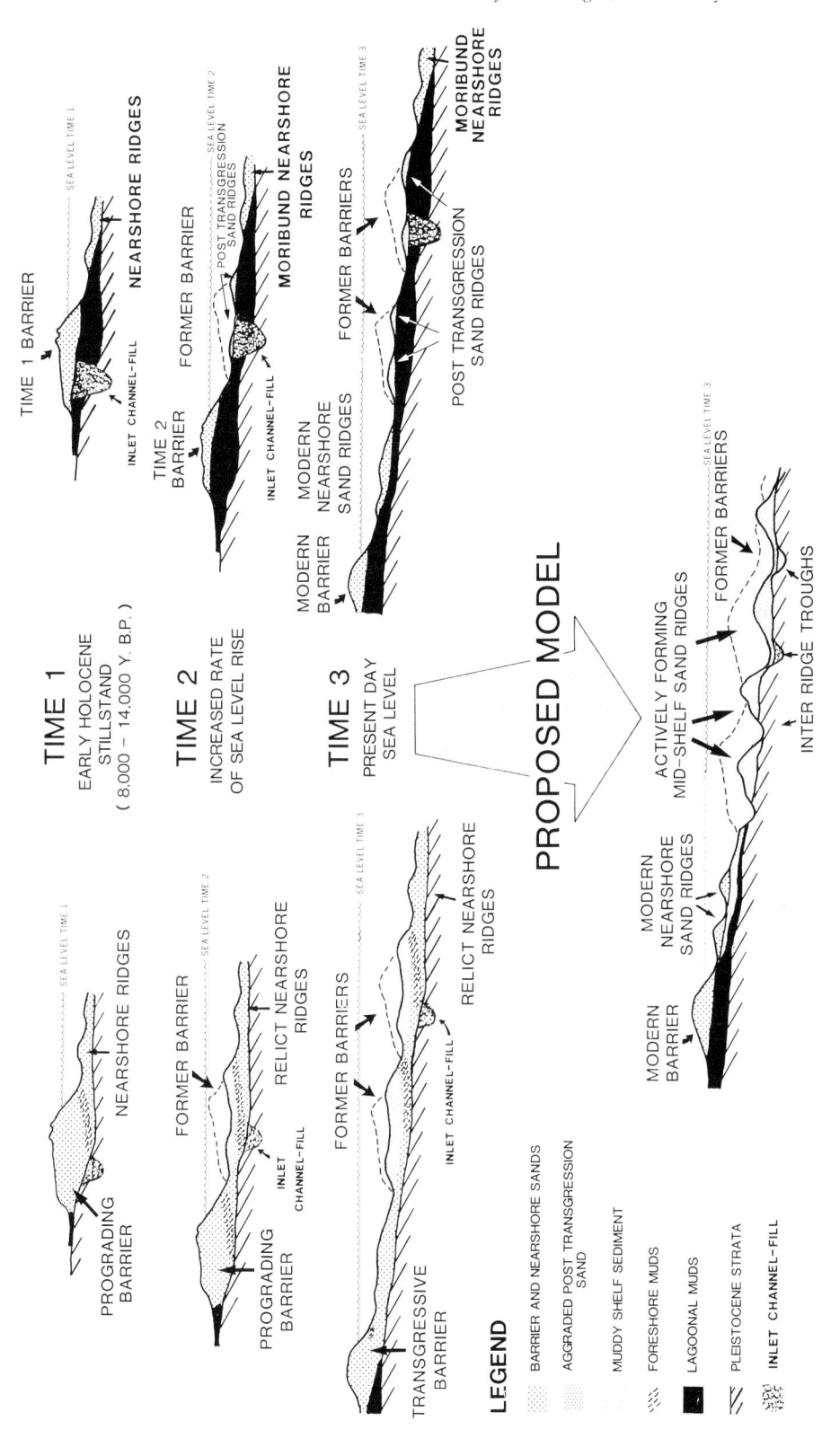

**Fig. 14.** Schematic diagrams comparing the proposed model of mid-shelf sand ridge formation with the models of Stubblefield et al. (1984) and Swift et al. (1973). The proposed model agrees with the theory of Stubblefield et al. (1984) that sands are concentrated in the mid-shelf area during a stillstand, probably as prograding or transgressive barriers and nearshore ridges (Time 1). When the rate of sea-level rise increased (Time 2) these barriers were altered to some unknown degree. The proposed model, however, digresses from the previous models in that it indicates that the actively depositing sand portion of the mid-shelf ridges (upper ridge sand unit) has been deposited at or very near present-day sea-level (Time 3) and the barriers of Stubblefield et al. (1984) and nearshore ridges of Swift et al. (1973) have been completely or almost completely removed.

deeper water than the nearshore ridge. Owing to the degree of coring-induced disturbance in the vibracores, however, the evidence available at that time was not judged to be sufficient to argue against the existing theories (i.e., Swift et al., 1973; Stubblefield et al., 1984), however, information gathered later also supports our initial hypothesis. In 1983, analysis of the foraminiferal assemblages established that the upper ridge sand unit of ridge 2 was deposited within a relatively uniform bathymetric setting that was deeper than nearshore ridge 1A. Finally, statistical analyses of textural data substantiated the validity of both the lithological and palaeontological units established by this study, thus further supporting the new model.

## Summary of revised model

The model proposed herein states that some or all of the mid-shelf ridges on the Middle Atlantic Bight were formed by present-day, mid-shelf processes; and they are not relict features formed during a lower stand of sea-level in a nearshore or inner shelf setting. Major sedimentological evidence supporting this model is that the upper ridge sand unit in mid-shelf ridge 2 may be considered to be a single unit with only gradational changes from top to bottom and no major discontinuities except at the unit base. This observation is supported by foraminiferal data and statistical analyses. If it is assumed that the present top of the upper ridge sand unit is being deposited in a mid-shelf setting (e.g., Stubblefield et al., 1984) and that there are no major discontinuities within the upper ridge sand of ridge 2, then as much as 18 ft (5.5 m) of sand can be attributed to deposition in a mid-shelf setting (Fig. 7; V-7, V-7A). This model is a revised model because it has evolved from previous models proposed by Swift et al. (1973) and Stubblefield et al. (1984). These earlier models and the revised model are graphically presented in Fig. 14.

A precise origin of the sands making up the mid-shelf ridge is not 100% certain, but evidence presented by Stubblefield et al. (1984) suggests that a late Pleistocene to early Holocene sea-level stillstand, between 8000 to 14000 years BP, concentrated sand within a barrier system in the vicinity of the present-day, mid-shelf ridge field. The authors agree, in general, with the Stubblefield et al. (1984) model for concentrating sands on the middle shelf. However, based on the existing data, it is not known whether this early palaeobarrier system was a prograding system, as hypothesized by Stubblefield et al. (1984) or a transgressive barrier system as alluded to by Swift et al. (1973).

The proposed model in Fig. 14 states that the mid-shelf ridges consist predominantly of sand deposited in a mid-shelf setting (the upper ridge sand unit of ridge 2). Based on sedimentological and palaeontological data, and statistical analysis of textural data, the mid-shelf ridges are not 'cored' by the degraded barriers of Stubblefield et al. (1984) nor do they consist of sands deposited as nearshore (earlier deposited) sand ridges as proposed by Swift et al. (1973). The only material that may be a remnant of a stillstand deposit is the shell-rich unit, within the trough adjacent to ridge 2. This unit contains bulk-shell dates within the 8000–14000 year BP period of the stillstand cited by Stubblefield et al. (1983). It is possible that within the mid-shelf, the laterally discontinuous shell-rich unit records inlet channel fills that were deposited in cuts into the underlying 'non-fossiliferous' unit (Fig. 7) and were preserved during subsequent transgression. It has been demonstrated that inlet fills have a high preservation potential within a barrier lithosome (Moslow & Heron, 1978).

## Scenario for development of mid-shelf ridge

Sands were concentrated in the vicinity of the present-day mid-shelf during a sea-level stillstand (Time 1 in Fig. 14). After the stillstand and as the rate of sea-level rise increased, the concentrated sands began to be reworked by inner shelf and eventually mid-shelf processes. The revised model, presented here, proposes that the degree of reworking is much more extensive than that proposed by previous investigators. In the Stubblefield et al. (1984) model, the relict (Time 2 in Fig. 14) nearshore ridges were only negligibly reworked and the former barriers were 'degraded' and capped by post-transgression sands. The Swift et al. (1973, 1984) model also proposed little alteration of the relict nearshore ridges but extensive (but not complete) erosion of the former barriers during a 'step-up' process (Fig. 14, Time 2). As mentioned above, the revised model presented in this paper proposes a greater degree of reworking, but when this reworking began is a matter of conjecture. It probably began between the time of the late Pleistocene to early Holocene stillstand and the time when sea-level reached present-day levels (Time 2 in Fig. 14), as proposed by Stubblefield et al. (1984).

Palaeontological evidence indicates that this re-working continued when ridge 2 was in a mid-shelf setting (post 6000 years BP). Radiocarbon age-dating evidence indicate that the reworking has continued at least to 1200 years BP (based on lowest bulk-shell age date in ridge 2 of 1155 ± 85 years BP from core V-1), and possibly to the present.

## CONCLUSIONS

The major conclusion of this study is that mid-shelf processes on the continental shelf of the Middle Atlantic Bight are more important than previously thought. The model proposed in this study calls for up to 5.5 m of sand deposition within a mid-shelf setting. The implication of this conclusion is that at least some of the shelf ridges on the Atlantic shelf are appropriate modern analogues for ancient shelf sand deposits that are interpreted to have formed far from palaeoshorelines, such as the Cretaceous-aged Shannon Sandstones of the Western Interior Seaway (Tillman & Martinsen, 1984, 1987).

Specific conclusions are as follows:
1 The mid-shelf ridge 2, examined in this study, was formed in a mid-shelf setting by mid-shelf processes.
2 Three main lithologic units are present in both the nearshore ridge and mid-shelf ridge: the 'non-fossiliferous' mud and sand, which was deposited in a marginal marine setting during the late Pleistocene or early to mid-Holocene; the shell-rich, poorly sorted mud and sand, which was deposited as post-transgressive nearshore or shelf sediments; and the upper ridge sand, which is the actively aggrading, ridge-forming unit within both ridges.
3 Sedimentological and textural analyses indicate significant differences between the upper ridge sand units of the nearshore and mid-shelf ridges examined in this study.
4 Foraminiferal assemblages in the upper ridge sand units of nearshore ridge 1A and mid-shelf ridge 2 indicate that these units were deposited in or very near their present shelf setting and environment.

The final conclusion of this study is one that is commonly reached in any analysis of results... 'more work is needed'. More core studies are needed to better document the distribution of lithologic units within the ridges and to clear up discrepancies between the multitude of models proposed in the literature regarding the origins of shelf sand ridges. More studies of physical process are needed over a broader area of the continental shelf. This study has shown that sand ridge formation is not restricted to the inner shelf, consequently, process studies can no longer be focused exclusively on nearshore ridges.

## ACKNOWLEDGEMENTS

The authors thank the companies belonging to the Atlantic Shelf Coring Project consortium for releasing the data presented in this paper. Companies that have participated in the consortium are Atlantic Richfield Oil and Gas (ARCO), Amoco Production Company, Getty Oil Company, Marathon Oil Company, Standard Oil and Gas Company, Cities Service Oil and Gas Corporation, Mobil Oil and Gas Company, and Chevron Oil and Gas Company. The study was a cooperative project, undertaken by the consortium and by the Atlantic Oceanographic and Meteorological Laboratories, National Oceanic and Atmospheric Administration. Steve Conrad ran the statistical programs used in the analysis of the textural data. S.J. Culver thanks J. Goshorn, R. Barringer, W. Mullen, M.A. Buzas, A. Sieben, and R.S. Spencer for their assistance in production of the foraminiferal data. Rod Tillman and Bill Stubblefield acted as co-chief scientists on the 1979 Atlantic Shelf Project cruise, which was carried out on the R.V. *Eastward*. Bill Stubblefield supervised the processing of bathymetry data. Jim Rine supervised the vibracore operations; he was assisted by Chuck Dill and Norman Haskell. Additional scientific crew members included: Chuck Nittrouer, Chuck Siemers, Randi Martinsen, Ted Beaumont and Dick Scott. Fred Mason and Chris Berlin were on-board technicians. The vibracores were processed at the Cities Service Research facility in Tulsa, Oklahoma under the supervision of Jim Rine and Rod Tillman; they were assisted by Fred Mason, Jim Pol, Pierre Kinga and Chris Berlin. Thanks also go to the crew of the R.V. *Eastward* who ably assisted the scientific party. We appreciate the assistance of the formal and informal reviewers.

## REFERENCES

BUTMAN, B., NOBLE, M. & FOLGER, D.W. (1979) Long-term observations of bottom current and bottom sediment movement on the mid-Atlantic continental shelf. *J. Geophys. Res.* **84**(C3), 1187–1205.
CULVER, S.J. & BANNER, F.T. (1978) Foraminiferal assemblages as Flandrian paleoenvironmental

indicators. *Palaeogeogr. Palaeoclimatol. Palaeoecol.* **24**, 53–72.

CULVER, S.J. & BUZAS, M.A. (1980) Distribution of recent benthic foraminifera off the North American Atlantic coast. *Smithsonian Contrib. Mar. Sci.* **6**, 512 pp.

DUANE, D.B., FIELD, M.E., MIESBERGER, E.P., SWIFT, D.J.P. & WILLIAMS, S.J. (1972) Linear shoals on the Atlantic continental shelf, Florida to Long Island. In: *Shelf Sediment Transport: Process and Pattern* (Eds Swift, D.J.P., Duane, D.B. & Pilkey, D.H.) pp. 447–498. Dowden Hutchinson & Ross, Stroudsburg, PA.

ELLISON, R.L. (1977) *Foraminifera of the Outer Continental Shelf of the Middle Atlantic Bight: 1975–1976 Investigations.* Special Report in Applied Marine Science and Ocean Engineering, No. 176, 26 pp. Virginia Institue of Marine Science, Gloucester Point, VA.

EMERY, K.O. (1952) Continental shelf sediments of southern California. *Geol. Soc. Am. Bull.* **63**, 1105–1108.

EMERY, K.O. (1965) Geology of the continental margin off eastern United States. In: *Submarine Geology and Geophysics* (Eds Whittard, W.F. & Bradshaw, R.) pp. 1–20. Butterworths, London.

FIGUEIREDO, JR., A.G. (1984) *Submarine sand ridges: geology and development, New Jersey, U.S.A.* Unpublished PhD dissertation, University of Miami, 409 pp.

HARRISON, W., NORCROSS, J.J., PORE, N.A. & STANLEY, E.M. (1967) *Shelf waters off the Chesapeake Bight.* Environmental Sciences Services Administration, Professional Paper 3, Washington, DC pp. 1–82.

HOWARD, J.D. (1972) Trace fossils as criteria for recognizing shorelines in the stratigraphic record. In: *Recognition of Ancient Sedimentary Environments* (Eds Rigby, J.D. & Hamblin, W.K.). Soc. Econ. Paleontol. Mineral. Spec. Publ. **16**, 215–255.

HOWARD, J.D. & REINECK, H.E. (1981) Depositional facies of high energy beach-to-offshore sequence: comparison with low-energy sequence. *Bull., Am. Assoc. Petrol. Geol.* **65**, 807–830.

KLOVAN, J.E. & BILLINGS, G.K. (1967) Classification of geologic samples by discriminant function analysis. *Bull. Canadian Petrol. Geol.* **15**, 313–330.

KNEBEL, H.J. & SPIKER, E. (1977) Thickness and age of surficial sand sheet, Baltimore Canyon Trough area. *Bull., Am. Assoc. Petrol. Geol.* **61**, 861–871.

MICHAELIS, E.R. & DIXON, G. (1969) Interpretation of depositional processes from sedimentary structures in the Cardium sandstones. *Bull. Can. Petrol. Geol.* **17**, 410–443.

MILLER, D.J. & ELLISON, R.L. (1980) The relationship of Foraminifera and submarine topography on the New Jersey continental shelf. *Geol. Soc. Am. Bull.* **93**, 239–245.

MCCLENNEN, C.E. (1973) Nature and origin of the New Jersey continental shelf topographic ridges and depressions. Unpublished PhD dissertation, University of Rhode Island, Kingston, Rhode Island, 94 pp.

MCCUBBIN, D.G. (1969) Cretaceous strike valley sandstone reservoirs, northwestern New Mexico. *Bull., Am. Assoc. Petrol. Geol.* **53**, 2114–2140.

MOSLOW, T.F. & HERON, JR., S.D. (1978) Relict inlets: preservation and occurrence in the Holocene stratigraphy of southern Core Banks, North Carolina. *J. Sediment. Petrol.* **48**, 1275–1286.

MOSS, A.J. (1962) The physical nature of common sandy and pebbly deposits, 1. *Am. J. Sci.* **260**, 337–373.

POAG, C.W., KNEBEL, H.J. & TODD, R. (1980) Distribution of modern benthic foraminifers on the New Jersey outer continental shelf. *Mar. Micropaleontol.* **5**, 43–69.

SHEPARD, F.P. & COHEE, G.V. (1936) Continental shelf sediments off the Mid-Atlantic states. *Geol. Soc. Am. Bull.* **47**, 441–458.

STETSON, H.C. (1938) The sediments of the continental shelf off the eastern coast of the United States. *MIT WHOI Pap. Phys. Oceanogr. Meteorol.* **V**(4), 48 pp.

STUBBLEFIELD, W.L. & SWIFT, D.J.P. (1976) Ridge development as revealed by subbottom profiles on the central New Jersey shelf. *Mar. Geol.* **20**, 315–334.

STUBBLEFIELD, W.L., KERSEY, D.G. & MCGRAIL, D.W. (1983) Development of middle continental shelf sand ridges: New Jersey. *Bull. Am. Assoc. Petrol. Geol.* **67**, 817–830.

STUBBLEFIELD, W.L., MCGRAIL, D.W. & KERSEY, D.G. (1984) Recognition of transgressive and post-transgressive sand ridges on the New Jersey continental shelf. In: *Siliciclastic Shelf Sediments* (Eds Tillman, D.W. & Siemers, C.T.). Soc. Econ. Paleontol. Mineral. Spec. Publ. **34**, 1–23.

SWIFT, D.J.P. (1973) Delaware Shelf Valley: estuary retreat path, not drowned river valley. *Geol. Soc. Am. Bull.* **84**, 2743–2748.

SWIFT, D.J.P. (1976) Continental shelf sedimentation. In: *Marine Sediment Transport and Environmental Management* (Eds Stanley, D.J. & Swift, D.J.P.) pp. 311–350. Wiley, New York.

SWIFT, D.J.P. & FIELD, M. (1981) Evolution of a classic sand ridge field: Maryland sector, North American inner shelf. *Sedimentol.* **28**, 461–482.

SWIFT, D.J.P., KOFOED, J.W., SAULSBURY, F.P. & SEARS, P. (1972a) Holocene evolution of the shelf surface, south and central Atlantic shelf of North America, In: *Shelf Sediment Transport: Process and Pattern* (Eds Swift, D.J.P., Duane, D.B. & Pilkey, O.H.) pp. 499–574. Dowden Hutchinson & Ross, Stroudsburg, PA.

SWIFT, D.J.P., DUANE, D.B. & MCKINNEY, T.F. (1973) Ridge and swale topography of the Middle Atlantic Bight, North America: secular reponse to the Holocene hydraulic regime. *Mar. Geol.* **15**, 227–247.

SWIFT, D.J.P., SEARS, P.C., BOHLKE, B. & SCOTT, R.B. (1979) Evolution of a shoal retreat massif, North Carolina shelf: Inferences from areal geology. *Mar. Geol.* **27**, 19–42.

SWIFT, D.J.P., MCKINNEY, T.F. & STAHL, L. (1984) Recognition of transgressive and post-transgressive sand ridges on the New Jersey continental shelf: Discussion. In: *Siliclastic Shelf Sediments* (Eds Tillman, R.W. & Siemers, C.T.) Soc. Econ. Paleontol. Mineral. Spec. Publ. **34**, 25–36.

TILLMAN, R.W. (1985) A spectrum of shelf sands and sandstones. In: *Shelf Sands and Sandstone Reservoirs* (Eds Tillman, R.W., Swift, D.J.P. & Walker, R.G.) Soc. Econ. Paleontol. Mineral. Short Course **13**, 1–46.

TILLMAN, R.W. & MARTINSEN, R.S. (1984) The Shannon shelf-ridge sandstone complex, Salt Creek anticline area, Powder River Basin, Wyoming. In: *Siliclastic Shelf Sediments* (Eds Tillman, R.W. & Siemers, C.T.). Soc. Econ. Paleontol. Mineral. Spec. Publ. 34, 85–142.

TILLMAN, R.W. & MARTINSEN, R.S. (1987) Sedimentologic model and production characteristics of Hartzog Draw Field, Wyoming. In: *Reservoir Sedimentology* (Eds Tillman, R.W. & Weber, K.J.). Soc. Econ. Paleontol. Mineral. Spec. Publ. 40, 15–112.

TILLMAN, R.W. & REINECK, H.E. (1973) Discrimination of North Sea sand environments with population-derived grain size parameters. In: *Theme 6, 9th Congress International de Sedimentologie*, Nice, pp. 217–220.

# Sandstones Deposited at Sequence Boundaries

# Evolution of an Upper Cretaceous (Turonian) shelf sandstone ridge: analysis of the Crossfield–Cardium pool, Alberta, Canada

F.F. KRAUSE[1] and D.A. NELSON[1]

*Petroleum Recovery Institute,*
*3512, 33rd Street N.W.,*
*Calgary, Alberta, T2L 2A6, Canada*

## ABSTRACT

The Crossfield Cardium 'A' pool, as delineated in isopachs obtained from 209 wire-line logs, is a body that measures 72 km in length and averages 3.5 km in width, tapering gradually towards the southeast. Mega- and macroscale correlations across the narrow dimension of the reservoir, which are based on electric-logs and cores, reveal further that the sandstone body is lens-like in outline, the eastern edge being steeper than the western edge. Lithofacies associations and distributions indicate that the lens was depositionally controlled and had positive relief over the surrounding sea-floor at the time of its development.

The following lithofacies, identified in 57 cores, characterize the reservoir at meso- and macroscales: L1—massive, laminated and bioturbated, dark grey mudstone; L2—bioturbated, thinly bedded, fine-grained sandstone, and mudstone; L3—thinly bedded, fine to medium-grained sandstone, and mudstone; L4—medium to thick bedded, fine to very coarse-grained sandstone; and L5—granule and small pebble conglomerate. L1 encases and seals the reservoir. L2 occupies the lower half of the reservoir. Here mixing of mudstone and sandstone through bioturbation has affected permeability negatively. L3 segments the reservoir creating barriers to cross-flow. L4 commonly occupies the ridge axis and is the most porous and permeable lithofacies. L5 is found most frequently along the ridge crest and eastern flank. In L4 and L5 diagenetic cements and shale admixtures inhibit permeability.

Ridge evolution has been traced with a series of computer generated log-marker structure surfaces (LMSS). The ridge itself developed on a topographic inhomogeneity and over time grew upwards, eastwards and southwards. This pattern of ridge growth indicates that dominant current directions were ridge-parallel with across-ridge components and would have been directed from the northwest to the southeast. Sediments initially accumulated in a setting where the unidirectional component of flow dominated during deposition. Subsequently, the ridge shoaled to the point where combined wave and current components influenced sedimentation, as indicated by beds with medium scale, low-angle inclined cross-stratification, plane-parallel laminations, and little or no bioturbation. Thick, black, massive, laminated, rippled, and bioturbated mudstones, which enclose the ridge, reflect an over-abundant supply of muds and silts at the time of deposition. These sequences may have formed in response to nearshore mud transport and deposition systems, as observed today along the Suriname coast, South Louisiana shelf, East China and Yellow Seas, and Malabar coast.

---

[1] Present address: Department of Geology and Geophysics, The University of Calgary, Calgary, Alberta T2N 1N4, Canada.

## INTRODUCTION

Following the discovery of the Pembina pool by Mobil in 1953, Shell located the Crossfield pool in 1956 (Nielsen, 1957; Berven, 1966; Javeri, 1966; Nielsen & Porter, 1984). Crossfield (see Figs 1, 2) differed from Pembina in that the producing sandstone was stratigraphically lower than the one from Pembina. However, Crossfield resembled Pembina in that the trap was also thought to be a stratigraphic trap. In the terminology of the day, Crossfield was considered to be a 'shoestring sandstone trap' (Levorsen, 1958). The mental image summoned by the label is quite realistic. Shoestring traps are very long and narrow; their length–width ratios typically are 10:1 to 20:1. Shoestring traps were recognized as 'primary stratigraphic traps' (Levorsen, 1958); that is, traps that owe their origin to depositional history and environment rather than diagenesis, or onlap on to unconformities.

Berven (1966, p. 228) reaffirmed the notion that Crossfield and Garrington (see Fig. 1) were shoestring traps and that they resembled in geometry the shoestring traps described by Bass (1934) and Bass et al. (1937) from southeast Kansas and northeast Oklahoma. These USA examples had been identified by Bass (1934) as offshore bars and compared to the barrier island systems along the present-day New Jersey coast. Berven also attributed an offshore bar origin to Crossfield and Garrington, indicating that they formed during a slow regressive episode and were subsequently drowned by a rapid transgression. It is evident, therefore, that both of these pools were recognized by Berven as *narrow and elongated strandline deposits*; specifically, a barrier island system. It is interesting to note that modern sedimentological studies of the internal geometry and lithofacies of the Kansas and Oklahoma pools conclude that they do not contain strandline deposits as Bass had originally proposed, but represent instead fluvial channel sequences (Walton et al., 1986; Walton, pers. comm., 1986).

The Garrington pool, which is located immediately northeast of Grossfield, has been studied recently by Chakravorty et al. (1978), who adhere to Berven's interpretation of an offshore bar, and Walker (1983a), who suggests instead that the Cardium Formation west of Garrington, between the Caroline and Ricinus pools, forms a 'ragged blanket'. Walker (1983a, p. 24) describes a ragged blanket as a sandstone sheet characterized by non-systematic changes in thickness. Moreover, he affirms:

'... it has been shown that Garrington and Caroline are long, narrow zones of production within sheet sandstones' (Walker, 1984, p. 163).

According to Walker (1983b, p. 211), sandstones in the Garrington–Caroline–Ricinus area were emplaced by turbidity currents into relatively deep water and a long way offshore. Walker's conclusions concerning the geometry and depositional environment of these pools are significantly different from those reached originally by Berven (1966). Unfortunately, Walker (1983a,b,c, 1984) does not shed light on the possible origin of the narrow zones of production. Conceivably, if these narrow zones of production are located within sheet sandstones, but have not been shaped by controls operating at the

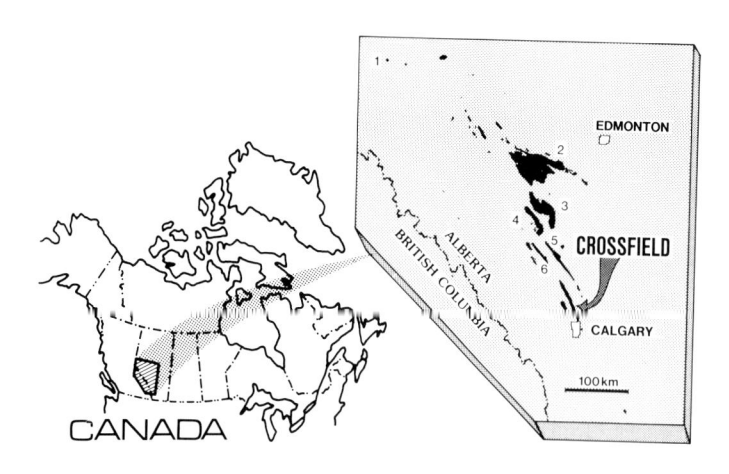

**Fig. 1.** Index map and location of the Crossfield 'A' Pool including other Cardium Formation reservoirs in west-central Alberta. Prominent pools are numbered and include Kawkwa (1), Pembina (2), Willesden Green (3), Ferrier (4), Garrington (5) and Caroline (6). (Pool distribution and location after Wallace-Dudley, 1981.)

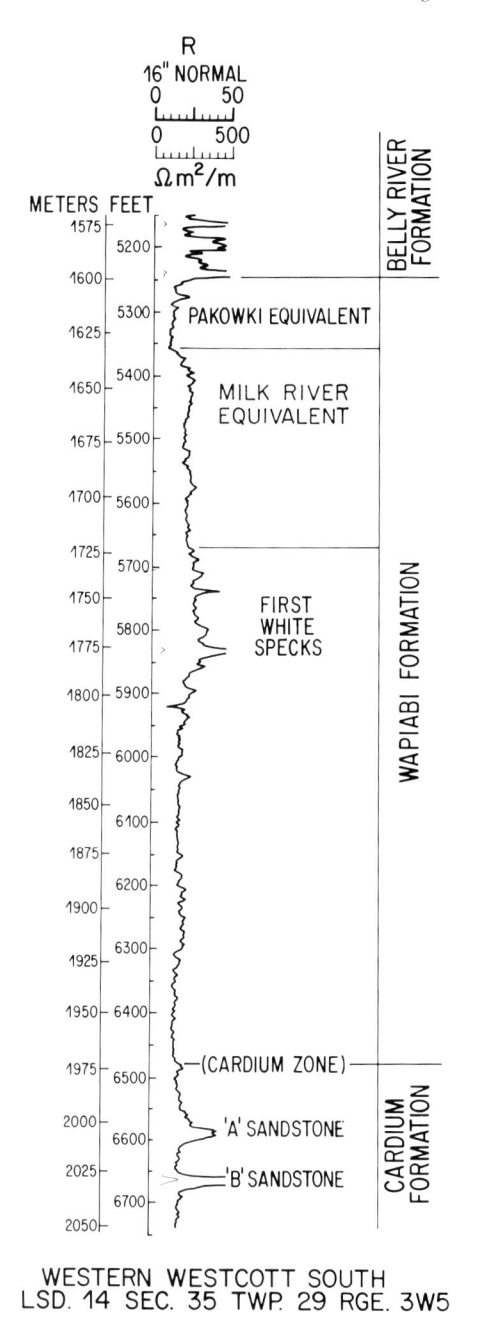

Fig. 2. Upper Cretaceous lithostratigraphy in the Crossfield area. Diagram illustrates SP (spontaneous potential)–R (resistivity) response recorded in well 14-35-29-3W5 of the Crossfield field and rock stratigraphic nomenclature. Note that the rock column consists predominantly of mudstones.

time of deposition, as Berven (1966) implies, then long and narrow pools such as Crossfield may have been moulded by subsequent geological processes, such as diagenesis and/or tectonics.

Obviously, studies that provide answers to the questions raised by these contrasting interpretations are necessary. A clear understanding of the origin of pools like Crossfield has a direct impact on future exploration, development, and production practices in Cardium Formation pools. The objectives of this paper, therefore, will be to:

1 determine the macroscale, or interwell geometry, and megascale, or field-wide view, of the pool (Krause & Collins, 1984; Krause et al., 1987) using all available well data, including all new wells drilled in the immediate vicinity of the field since the studies of Berven (1966) and Javeri (1966) (see Fig. 2);

2 document the internal geometry of the Cardium 'B' sandstone as determined from a continuous transect based on closely spaced cores across the narrow dimension of the pool;

3 interpret the depositional environment and the origin of the Crossfield reservoir in light of our observations.

## EXTERNAL GEOMETRY

Crossfield Field is situated immediately north of the city of Calgary. It is approximately 72 km in length and has an average width of 3.5 km (Figs 1, 3). The field is characterized by a long and very narrow outline, has a length:width ratio of 20:1, and trends NW–SE (azimuth 152°, Fig. 3), as also has been illustrated previously by Berven (1966, his fig. 1) and Javeri (1966, p. 28). In detail, the pool's trend is slightly arcuate, changing 14° from N24°W in the south to N38°W in the north. The arc interior faces west. In plan view, the appearance of the field is that of a long and thin club with the head pointing northward.

### Structure

Structure contours illustrated in Fig. 4 are based on depth picks from electric logs and reflect subsea elevations to the top of the Cardium zone (Krause & Nelson, 1984). Structure contours strike uniformly N–S, except near the ends of the field. In the south they strike N8°E (8°), and in the north they strike N10°W (350°). Our observations in general match

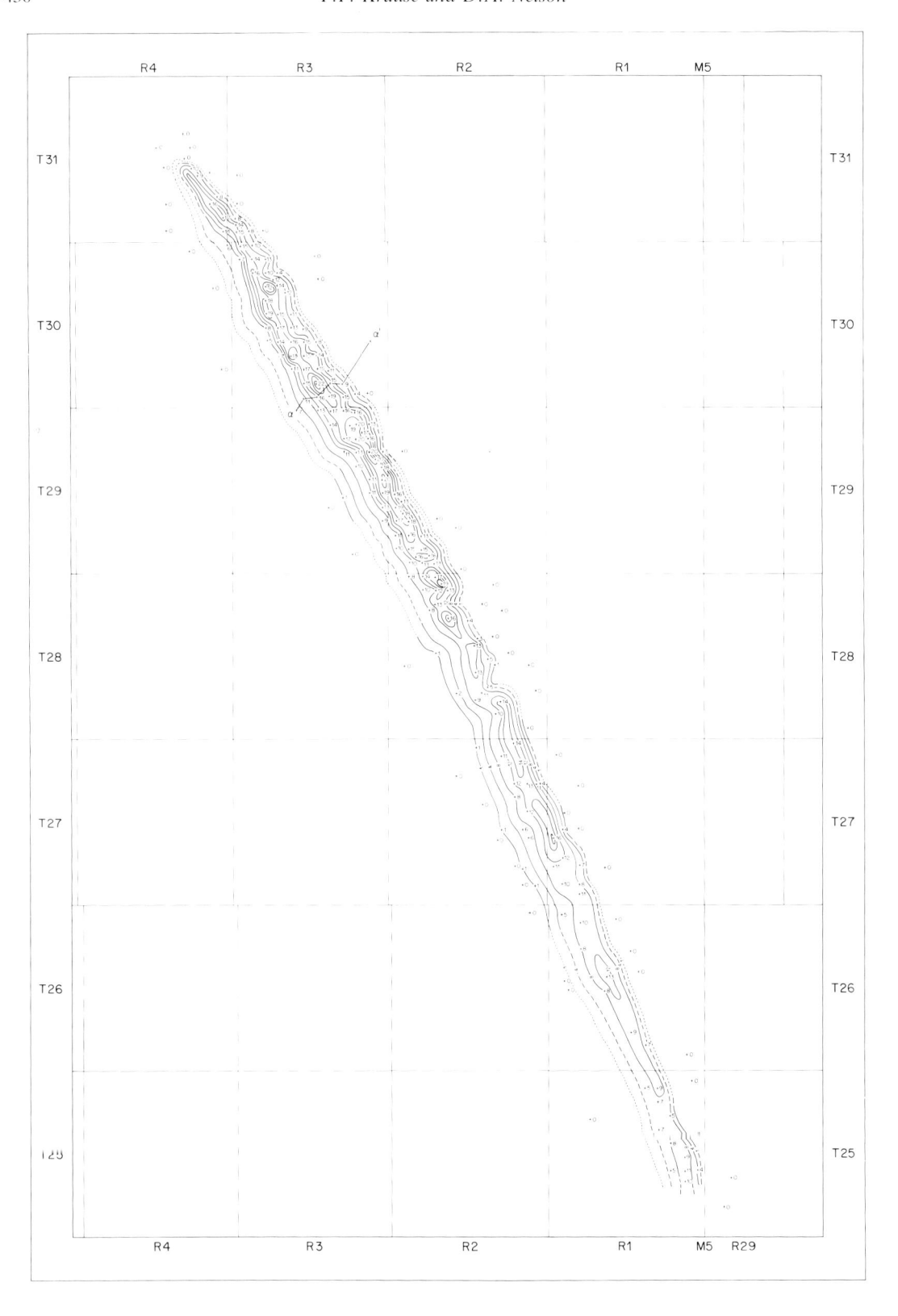

Sand ridges in the Cardium Formation 431

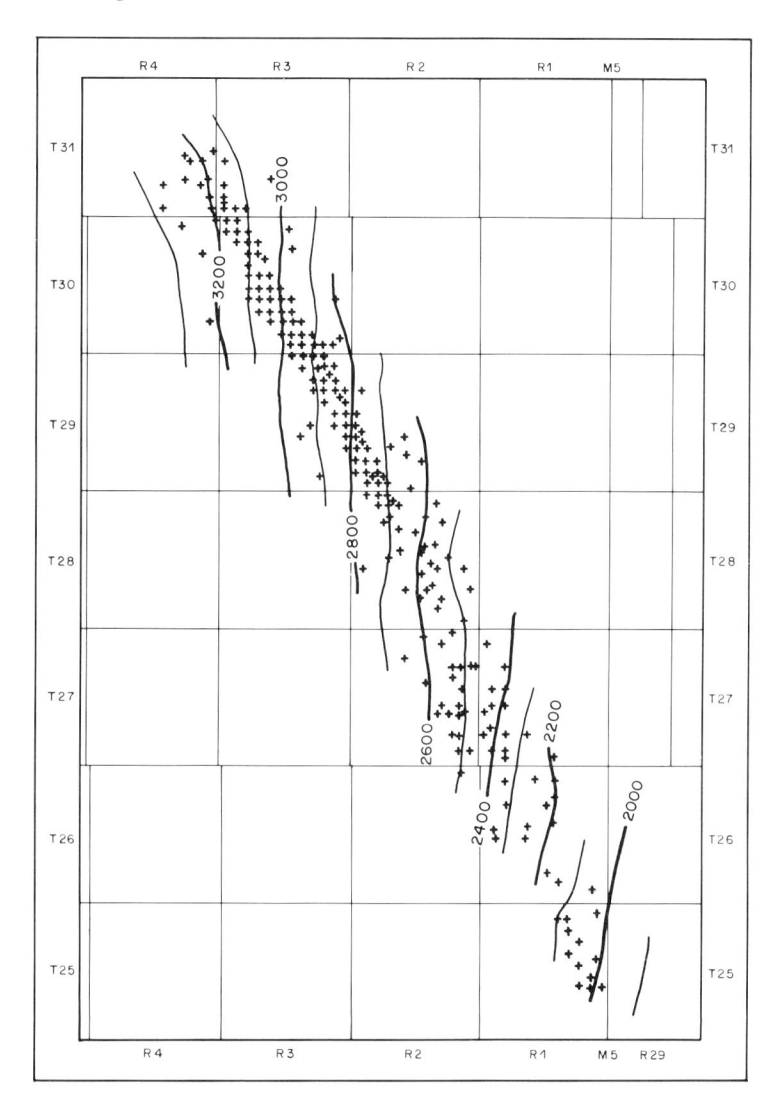

**Fig. 4.** Structure contour map of the Cardium Zone top based on data collected from resistivity logs from wells within and immediately adjacent to the field. Datum is sea-level. Regional westward dip is 0.57° (10 m km$^{-1}$). Scale 1:3 000 000. Note that the Cardium Zone defines a fairly even and constantly dipping surface. Contour intervals in metres.

those published previously by Berven (1966), who also noted N–S structure contour trends. Present-day regional dip is to the west and northwest, and varies from east to west from 0.54° (9.5 m km$^{-1}$) to 0.72° (12.7 m km$^{-1}$).

The pool's proportions and dimensions are clearly defined in the Cardium 'B' sandstone isopach map produced from resistivity logs of 214 wells (Fig. 3). Thickness and width across the pool are variable, however both measures are greatest in the north. Width ranges from 2.5 km in the south to 5 km in the north. Thickness along the pool's length is more variable. However, average sandstone thickness for the 107 wells above a line from Section 25, Town-

**Fig. 3.** Isopach map of the Cardium 'B' sandstone generated from depth picks made on electric logs for wells within and just outside the field. Thickness varies throughout the field along trend. The reservoir interval is thickest in the north and thinner to the south. Location of cross-section a–a', discussed in the text and illustrated in Figs 5, 6, 8, is indicated. Contour intervals are in feet and are variably spaced to reduce contour overlap. Interval span is 3 ft (0.91 m) between 1 and 4 ft (0.30–1.22 m) contours; 4 ft (1.22 m) between 4 and 12 ft (1.22–3.65 m) contours; and 2 ft (0.61 m) between 12 and 20 ft (3.65–6.09 m) contours.

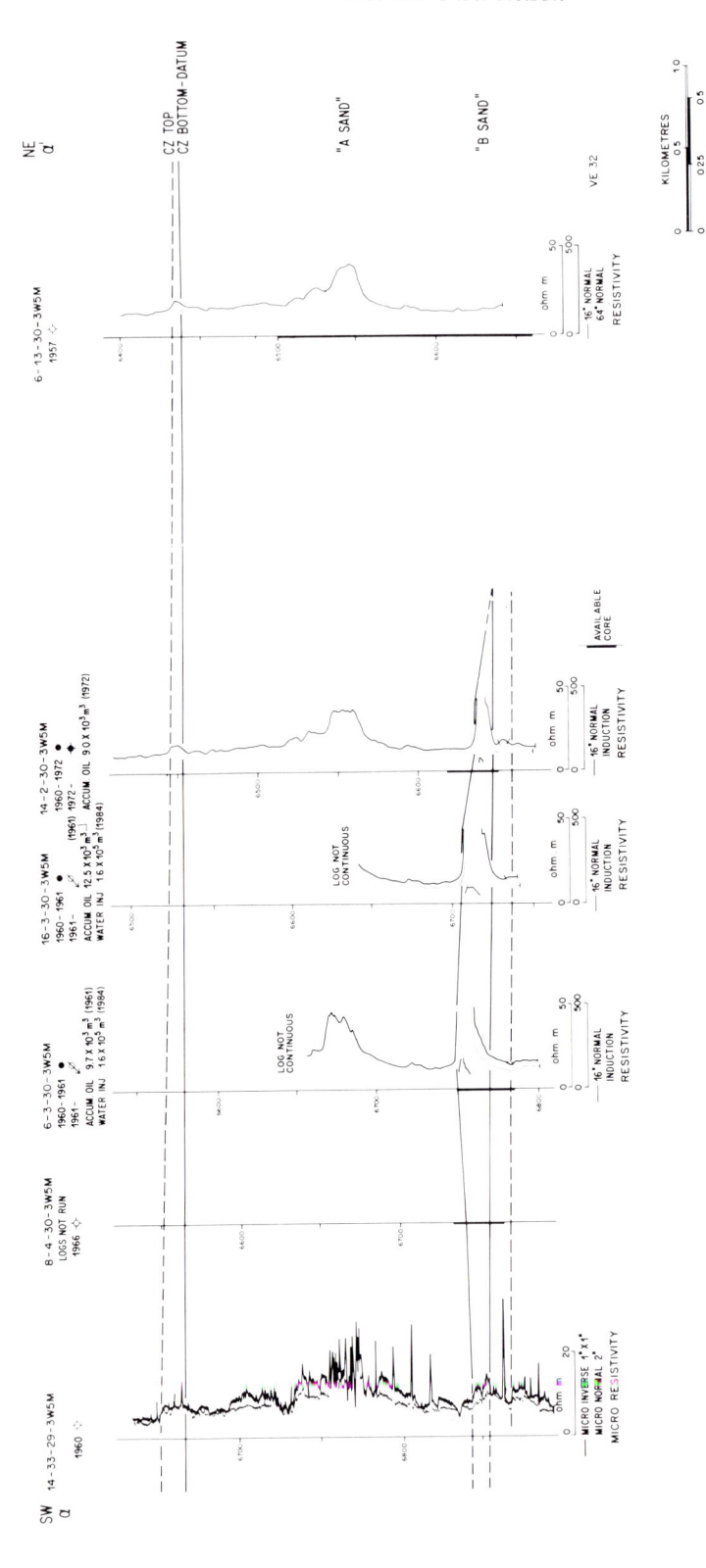

**Fig. 5.** Wire-line log cross-section $a$–$a'$ arranged perpendicular to field trend and based on resistivity logs. A microlog was used for well 14-33-29-3 W5M as this was the only log available. Local datum is the Cardium Zone bottom. The reservoir forms an asymmetrical lense-like feature with positive relief. Note that the Cardium 'B' sandstone is absent in well 6-13-30-3 W5M.

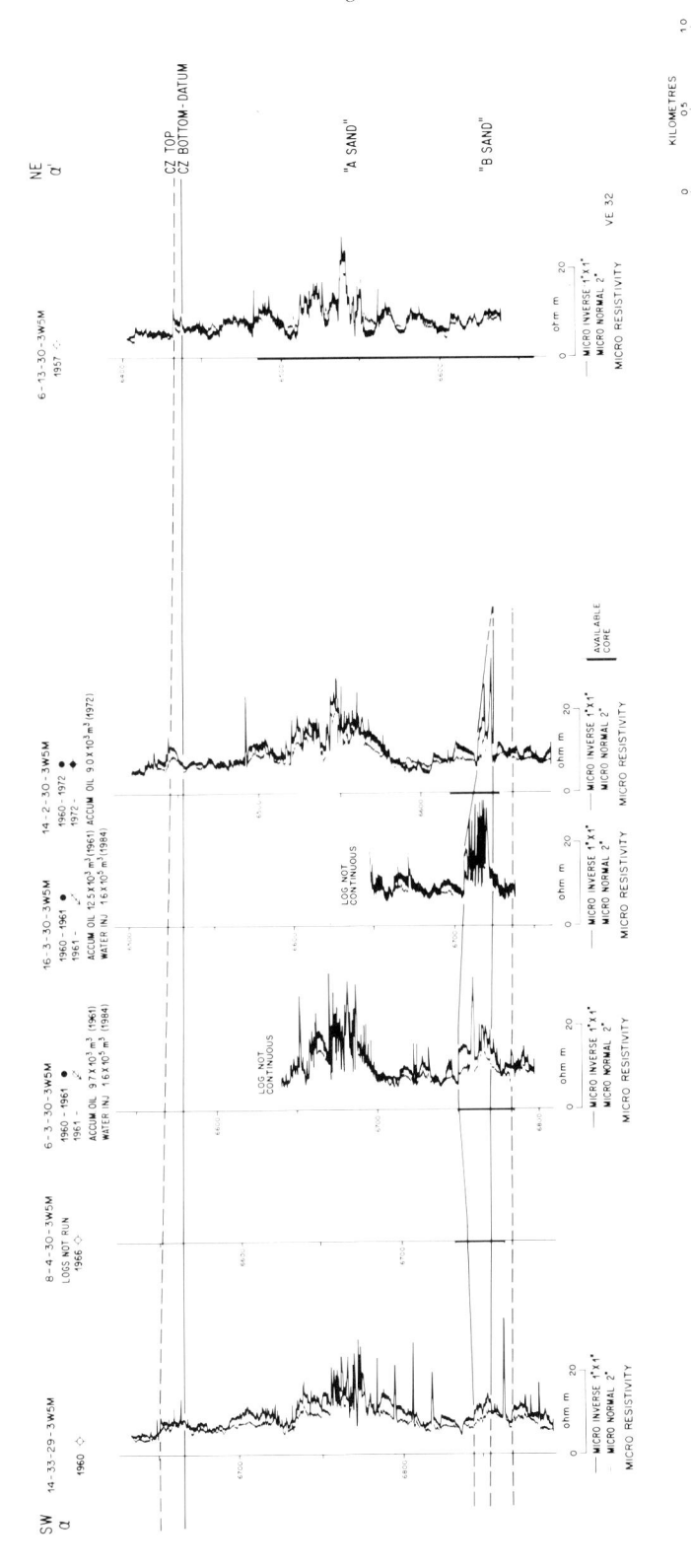

**Fig. 6.** Southwest–northeast wire-line log cross-section based on microresistivity logs. Wells and datum are the same as those illustrated in Fig. 5. Microresistivity logs illustrate the abrupt transition from sandstones to shales along the eastern limit of the field and also indicate that the reservoir itself is complexly interlayered.

ship 25, Range 2W5 is 13.3 ft (4.1 m) and 7.8 ft (2.4 m) in thickness for 52 wells below the same line. The thickness difference between them is 42%. On the other hand, across-trend thickness decreases gradually to the west and rapidly to the east.

**Shape**

The lenticular geometry of the field is illustrated further in the resistivity and microresistivity log cross-sections (Figs 5, 6). Location of these cross-sections is indicated on the isopach map (Fig. 3). This line of section was selected because it has the largest number of cores available in a single transect and includes wells just outside the field boundary. In this fashion, the extent of the Cardium 'B' sandstone is defined. Furthermore, the lateral extent of lithofacies and their relative configuration also can be documented, as is illustrated later.

As observed in the 13 log cross-sections between Townships 28 and 31, the top of the reservoir with respect to the Cardium Zone marker displays positive relief with a steep eastern slope and a gentler western slope (Fig. 7). The profile outline is that of a plano-convex ridge that dips gently to the east. However, lower and upper ridge surfaces are irregular, undulate, and have steeper eastward and gentler southward slopes.

## LITHOFACIES ASSOCIATION

Five main lithofacies characterize the terrigenous clastic deposits of the Cardium 'B' sandstone in this area. They are:

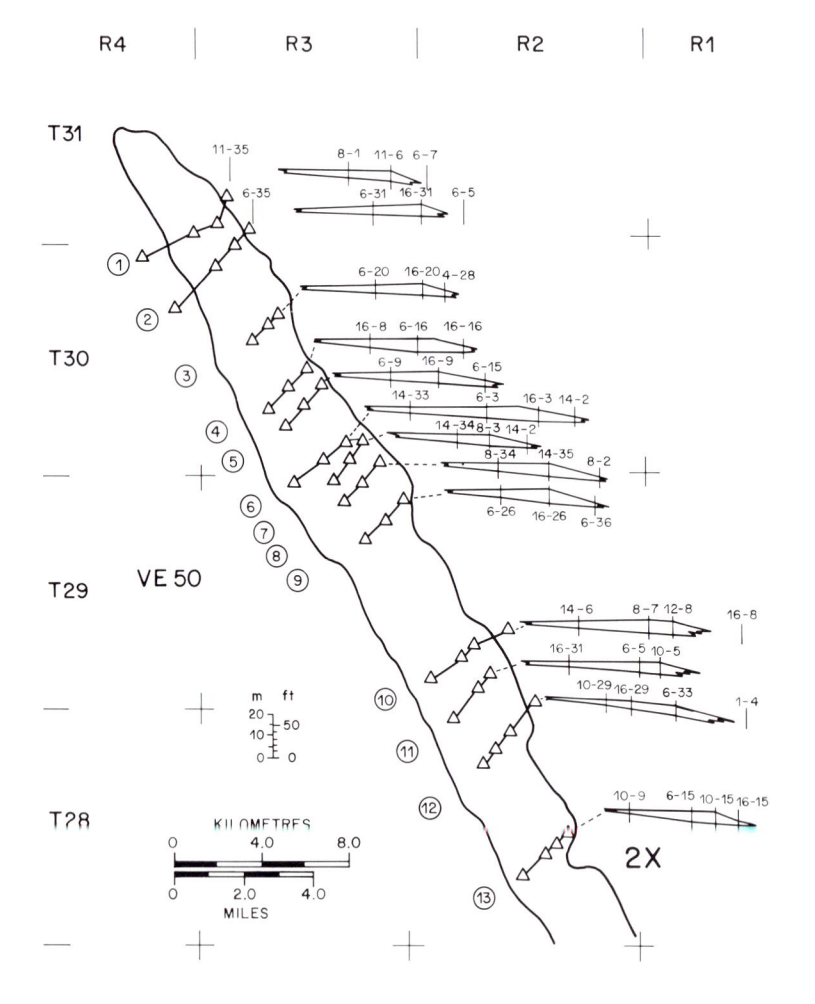

**Fig. 7.** Profile outlines oriented orthogonally to reservoir trend and based on SP (spontaneous potential)–R (resistivity) wire-line logs having the Cardium Zone as datum. Reservoir outlined by the 1 ft (0.30 m) isopach contour. Well locations are indicated by open triangles, and well identifiers are shown on reservoir profiles. The horizontal scale of the profiles is magnified twice.

**Table 1.** Lithofacies identified by various workers in studies based on cores from the Crossfield reservoir and other subsurface localities. Additional Cardium Formation lithofacies schemes are presented in Krause & Nelson (1984)

| THIS STUDY<br>COLLINS et al. 1985 | KRAUSE & NELSON 1984<br>KRAUSE 1983 | PLINT & WALKER 1987*2<br>WALKER 1985*1<br>WALKER 1983C | BERVEN 1966 |
|---|---|---|---|
| **FACIES 1**<br>DARK GREY SHALE<br><30% SANDSTONE:<br>1B - >30% BIOTURBATION<br>1L - LAMINATED, <30% BIOTURBATION<br>1M - MASSIVE, <10% SANDSTONE | **FACIES 1**<br>DARK GREY SHALE | **FACIES 1**<br>MASSIVE DARK MUDSTONES<br><br>**FACIES 2**<br>LAMINATED DARK MUDSTONES | **FACIES 3**<br>DARK BIOTURBATED MUDDY SILTSTONES<br><br>**FACIES 6**<br>SPECKLED GRITTY MUDSTONE | BLACK SHALE |
| **FACIES 2**<br>BIOTURBATED, VERY THIN AND THIN BEDDED SHALE AND FINE TO COARSE GRAINED SANDSTONE,<br>>30% BIOTURBATION | **FACIES 2**<br>BIOTURBATED AND THINLY BEDDED, FINE GRAINED SANDSTONE, SILTSTONE AND SHALE | **FACIES 4**<br>PERVASIVELY BIOTURBATED MUDDY SANDSTONES<br><br>**FACIES 5**<br>BIOTURBATED SANDSTONES | **FACIES 14***1<br>BIOTURBATED SANDSTONES AND MUDSTONES | SANDY SHALE |
| **FACIES 3**<br>VERY THIN AND THIN BEDDED SHALE AND FINE TO COARSE GRAINED SANDSTONE<br><30% BIOTURBATION | **FACIES 3**<br>THINLY BEDDED<br>FINE GRAINED SANDSTONE AND BLACK SHALE,<br><30% BIOTURBATION | **FACIES 15***2<br>INTERBEDDED SANDSTONES AND BLACK MUDSTONES | | SHALY SANDSTONE THINLY BEDDED |
| **FACIES 4**<br>MEDIUM TO THICK BEDDED, FINE TO VERY COARSE GRAINED SANDSTONE WITH <30% CONGLOMERATE | **FACIES 4**<br>MEDIUM TO THICK BEDDED FINE GRAINED SANDSTONE<br>4A - <50% BIOTURBATION OVER LENGTH OF CORE INTERVAL<br>4B - >50% BIOTURBATION OVER LENGTH OF CORE INTERVAL | **FACIES 7**<br>NON BIOTURBATED SANDSTONES<br>**FACIES 9***1<br>GRADED SANDSTONES<br>**FACIES 10***1<br>MASSIVE SANDSTONES | **FACIES 11***1<br>PARALLEL LAMINATED SANDSTONES<br>**FACIES 12***1<br>HUMMOCKY CROSS STRATIFIED SANDSTONES<br>**FACIES 16***2<br>THICK BEDDED, NON-BIOTURBATED SANDSTONE | SANDSTONE MEDIUM TO THICKLY BEDDED |
| **FACIES 5**<br>CONGLOMERATE WITH <30% SANDSTONE<br>5A - CONGLOMERATE WITH SAND MATRIX OR CLAST SUPPORTED FRAMEWORKS WITH <10%SHALE MATRIX<br>5B - CONGLOMERATE WITH SHALE MATRIX OR MUD SUPPORTED | **FACIES 5**<br>COARSE AND PEBBLE CONGLOMERATE THREE SUBFACIES:<br>5A - CONGLOMERATE WITH SAND MATRIX OR CLAST SUPPORTED WITH <10% SHALE MATRIX<br>5B - CONGLOMERATE WITH SHALE OR MUD SIZED SIDERITE, CLAST OR MATRIX SUPPORTED TEXTURE<br>5C - CONGLOMERATE WITH POIKILOTOPIC CALCITE AND COARSE SIDERITE CEMENTS | **FACIES 8**<br>CONGLOMERATE<br>CLAST-SUPPORTED (8A*2)<br>MUD-SUPPORTED (8B*2)<br>THIN GRAVEL LAYERS (8C*2)<br>SAND-SUPPORTED (8D*2) | **FACIES 13***1<br>CONGLOMERATES | CONGLOMERATE WITH SHALE MATRIX |

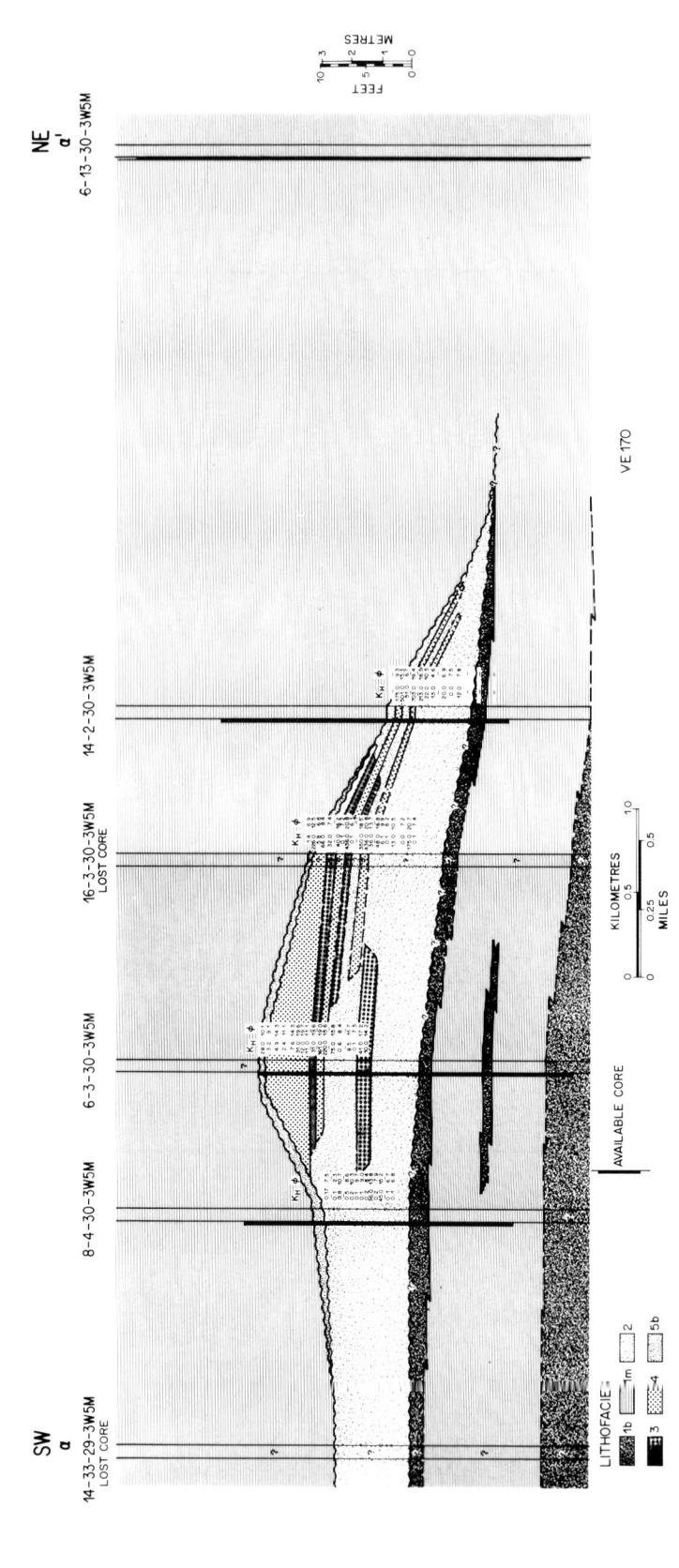

Fig. 8. Southwest–northeast lithofacies cross-section based on cores from wells used in Figs 5, 6. See text for discussion.

1 dark grey mudstone;
2 bioturbated, very thin and thin-bedded mudstone and fine to coarse-grained sandstone.
3 Very thin and thin-bedded mudstone and fine to coarse-grained sandstone.
4 Medium to thick bedded, fine to very coarse-grained sandstone.
5 Conglomerate.

This lithofacies scheme has been used previously by us in studies of the Pembina River Member (Cardium 'A' sandstone) and Cardium Zone Member (Krause, 1982, 1983; Krause & Collins, 1984; Krause & Nelson, 1984, 1987; Collins *et al.*, 1985; Krause *et al.*, 1987). We have introduced minor modifications to make it compatible with evidence gathered in the course of studies of the Cardium 'B' sandstone (Collins *et al.*, 1985). Each lithofacies is recognized by noting the following characteristics: lithology, bedding thickness, grain size, sorting, sedimentary structures, and variation and relationships to other lithofacies. A number of lithofacies also have been identified previously by others; these are summarized in Table 1.

## Lithofacies 1 (L1): dark grey mudstone

Mudrocks of this lithofacies contain variable admixtures of clay, silt, fine-grained sandstone, and infrequently conglomerate. Based predominantly on textural criteria, we recognize three subfacies: 1B, 1L and 1M (Figs 8, 9, 10A, 11H,I, 12A). These subfacies are very similar to Walker's (1983b,c) lithofacies 1, 2, 3 and 6 described by him to the north of Crossfield. As is discussed below, subfacies boundaries are based on the relative proportions of admixed sand and bioturbation or mottling textures contained by the mudstone. Mottling textures may be the product of bioturbation, however, they also may be the product of liquefaction and flowage of mud as is commonly observed along muddy coastlines (Wells & Coleman, 1981; Wells & Kemp, 1984; Rine & Ginsburg, 1985). Bioturbation estimates, where this texture is distinct, have been made visually in accordance with the classification presented by Reineck (1967) and recently adapted by Frey & Pemberton (1985), and Swanson (1981).

*Subfacies 1M (L1M): massive dark grey mudstone*

Subfacies 1M consists of silt and sand-poor mudrocks that appear massive under cursory examination. However, detailed inspection indicates that the massive appearance of this rock may be the result of homogenization of the original sediment by bioturbation activity or rapid accumulation as a result of transport into the area by liquefied flows and slurries. It is easily distinguished from subfacies 1B because mottling or bioturbation are not readily evident, and the sand content is less than 10%. Nevertheless, disrupted and contorted bedding and burrow traces are apparent under X-ray examination. *Gordia* appears to be the most frequently preserved trace and may be the last one formed as it cuts across all previously established mottles. Much less frequently, millimetre to centimetre-thick beds with sand laminae are preserved. These beds may display abrupt bases and gradational tops, or upper and lower boundaries that may be diffuse. Siderite nodules up to 100 mm thick are present.

The distribution and occurrence of this lithofacies is shown in Fig. 8. It is the most abundant rock type illustrated in this figure and, as shown, encases the reservoir.

*Subfacies 1L (L1L): laminated dark grey mudstone*

Subfacies 1L contains 10–30% silt and sand-sized material. It consists of interbedded shales and millimetre to centimetre-thick siltstones and sandstones. All three lithologies preserve millimetre-thick laminae. In addition, siltier and sandier beds often display sharp and erosive bases, graded laminae, ripples and ripple drift cross-lamination, starved ripples, and wavy laminations.

Subfacies 1L is not common in the immediate vicinity of the 'B' sandstone in Crossfield. Although it is not illustrated in cross-sections in this paper, it occurs locally in the vicinity of the reservoir, and is found most frequently in the Crossfield Field area above the 'A' sandstone.

*Subfacies 1B (L1B): bioturbated dark grey mudstone*

This subfacies is similar to subfacies 1L as it also contains 10–30% silt and sand-sized material. It differs from subfacies 1L in that it is more than 30% bioturbated. Where bioturbation has not been complete, partially preserved bedding and sedimentary structures are as in subfacies 1L. Typical trace fossils present are *Chondrites, Planolites, Terebellina, Palaeophycus, Gordia, Nereites* and *Helminthopsis*.

Subfacies 1B underlies the sandstone ridge, as illustrated in Fig. 8. Contacts with underlying and overlying lithofacies are bioturbated. The contact

## LITHOLOGY

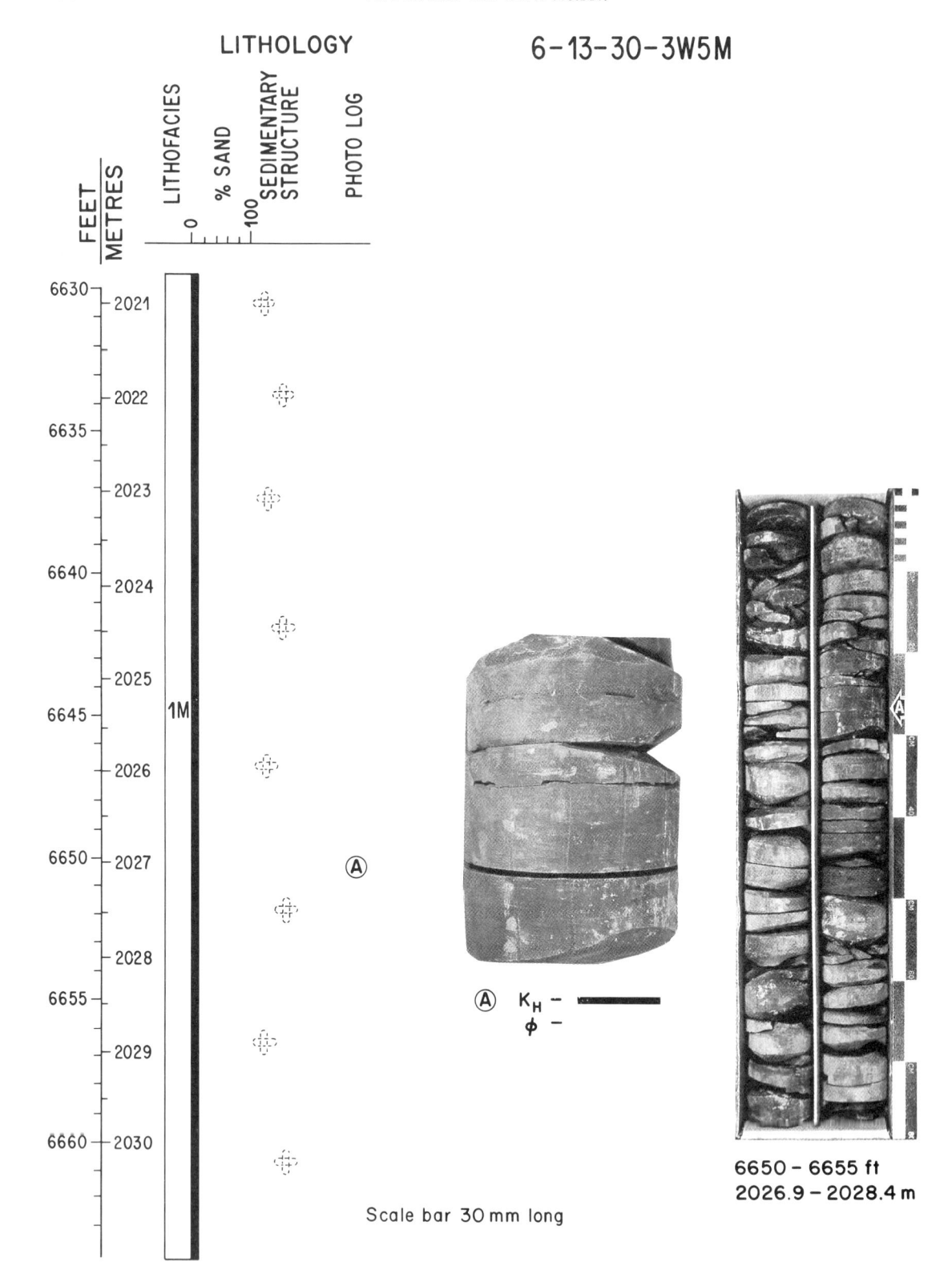

**6-13-30-3W5M**

6650 – 6655 ft
2026.9 – 2028.4 m

Scale bar 30 mm long

with overlying lithofacies 2 is gradational, whereas the contact with underlying subfacies 1M is much more abrupt. This latter behaviour reflects differences in the depositional environment, which at present are incompletely understood. Furthermore, lithofacies 2 characteristically contains coarse sand and granule-sized chert grains, which may be the product of increasing current competence and/or changes in sediment supply.

*Interpretation*

Because the mudrocks described above provide the reservoir's seal and are an integral part of the observed sedimentary sequence, understanding their origin is as important as establishing the depositional environment of the reservoir sandstones and conglomerates. It is clear from our descriptions that the mudrocks enclosing the Crossfield reservoir are not homogeneous and reflect a variety of seemingly conflicting depositional settings. Sedimentary textures preserved in these rocks indicate that the sediments originally accumulated either in areas that normally experience low current energies, or alternatively, in areas in which sediment supply is very high, and sediment lost to erosion following intense periods of scour is rapidly replenished; or possibly in a locale in which current energy typically is high but energy is lost over a fluid mud bottom. The latter behaviour is not unusual, as it has been documented clearly by Wells & Coleman (1981), Wells & Kemp (1984), and Rine & Ginsburg (1985) along the Suriname Coast of northeast South America and beside the Mississippi delta coastline of the USA. Intercalated sandy deposits observed in both lithofacies 1B and 1L preserve scoured bases, current and wave ripples, and graded beds, all indicative of current velocities higher than those required to initiate transport of unconsolidated clay and silt-sized material. In addition, chert granules and pebbles are infrequently interbedded with these deposits. Again, if this material had been transported by currents, higher transport velocities than those needed to initiate movement of unconsolidated mud are required.

Additional signals preserved in these rocks appear to indicate that conditions on or near the sea-floor may have at times been detrimental to the maintenance of normal bottom life. Preservation of shelled epifauna is sparse, mudstones are consistently dark grey or black, siderite of early origin is common, and bioturbation traces are clearly evident only in thin intervals of L1B; elsewhere, L1M with thin intervals of L1L dominates. Typically rocks with similar signals are thought to accumulate in a setting with a density stratified water column with upper and bottom layers separated by a pycnocline (Rhoads & Morse, 1971; Byers, 1977; Savrda *et al.*, 1984). In this setting, the bottom environment is stagnant, oxygen deficient, and typified by low current energies and poor circulation. Although we are not aware of this concept having been proposed previously by others for the Cardium Formation, stratification of the water column in the interior Mesozoic seaway of North America has been invoked by other workers; for example, Stronach

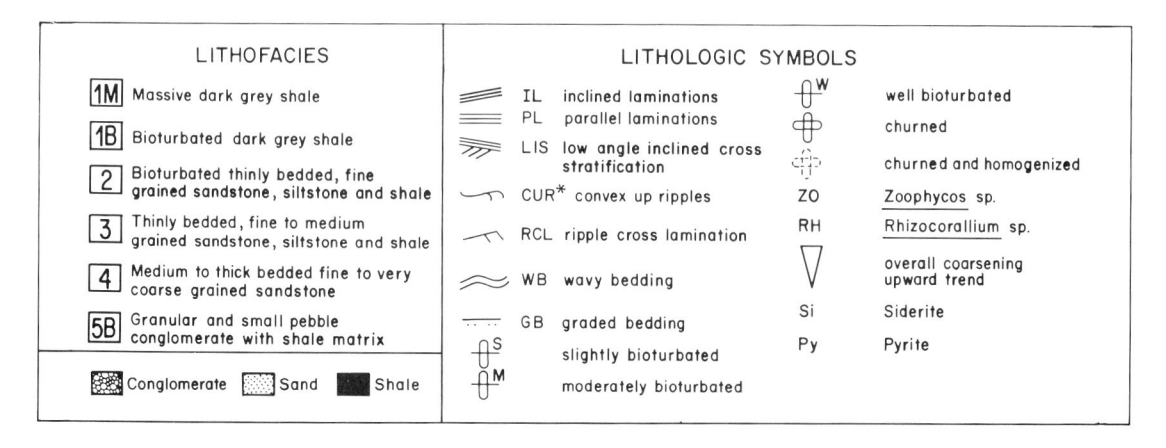

**Fig. 9.** Core and photograph log for well 6-13-30-3 W5M located approximately 3 km east of the sandstone ridge. Note that the entire core is represented by massive dark grey shale (lithofacies 1M). The core legend also applies to Figs 10–12 and Figs 14A, B. *See Krause & Nelson (1984).

(1984), McNeil (1984), Pratt (1984) and Barlow (1986).

Likewise, lithofacies with characteristics similar to L1B, L1L and L1M described by us above have been documented along muddy coastlines and nearshore shelf areas exposed to the open sea (Wells & Coleman, 1981; Rine & Ginsburg, 1985; Nittrouer et al., 1986). Interestingly, in these modern settings high mud concentrations and periodic influxes of fresh and brackish waters preclude the development of normal marine faunas. Thus, lower faunal diversity in this instance is the product of an environment periodically stressed by the introduction of abundant mud and silt, slowly advancing mud banks, and influx of fresh and brackish water plumes where silts and muds are transported by hypo- and hyperpycnal flows. These hypo- and hyperpycnal riverine jets parallel the coastline and project on to the shelf for tens of kilometres before mixing and blending with the surrounding sea water (Muller-Karger et al., 1988; Wright et al., 1988). An extreme example of this condition in modern settings is the fresh and brackish water plume delivered by the Amazon River (Gibbs, 1976; Nittrouer et al., 1986; Muller-Karger et al., 1988). However, this pattern is not restricted to the Amazon shelf and also has been documented elsewhere; for example, on the west shelf of the Mississippi delta (Rezak et al., 1985; Rabalais & Boesch, 1987) and on the Yellow River shelf (Wright et al., 1988).

It is therefore conceivable that the deposits of L1M, L1L and L1B reflect mud and silt accumulations introduced by riverine outflows and dispersed by along-shore current transport in the Late Cretaceous seaway. L1L would have formed in response to resuspension from currents generated during periods of increased wave activity by storms, or as a result of rapid sedimentation from hyperpycnal flows. In addition, L1B and L1M mudstones would have accumulated in more distal positions from riverine outflows than those of L1L, as has been suggested already for similar deposits of the Changjiang delta platform in the East China Sea (Rhoads et al., 1985).

More hospitable, although not necessarily unstressed, conditions are displayed by L1B mudstones, since they preserve the signals associated with a bottom-dwelling benthos. In these deposits it is not uncommon to observe the trace fossils *Chondrites*, *Helminthopsis* and *Nereites*. The former is produced by an endobenthic deposit feeder tolerant of anaerobic bottom conditions (Bromley & Ekdale, 1984; Savrda & Bottjer, 1987). However, trace fossils resembling *Chondrites* also have been noted by Rine & Ginsburg (1985) in the muddy sediments of the Suriname coast. In this latter setting, as already indicated, the bottom is characterized by high concentrations of suspended muds, and the water column often has lowered salinities. Interestingly, *Chondrites* also has been identified by Wightman et al. (1987) in the Lower Cretaceous Upper Mannville deposits of the Lloydminster area of Alberta, which they believe accumulated in brackish waters.

## Lithofacies 2 (L2): bioturbated, thin and very thin-bedded shale and fine to coarse-grained sandstone

This lithofacies (Figs 10B,C,D,E,F, 11E,F,G, 12B,D,F,G) consists of alternating centimetre-thick shale and sandstone interbeds. These beds have been bioturbated extensively, and hydrodynamic imprintings are incompletely preserved. Lithofacies 2 characteristically has greater than 20% shale content and is distinguished from lithofacies 3, which is described later, by being more than 30% bioturbated. Sand-sized material is typically medium to fine grained, although it commonly ranges from very fine to coarse grained. Less frequent is the occurrence of scattered granules and small pebbles (Fig. 12G). Trace fossils are similar to those observed in lithofacies 1B and consist of *Rhizocorallium*, *Teichichnus*, *Chondrites*, *Planolites*, *Zoophycos*, *Terebellina*, *Palaeophycus*, *Helminthopsis* and *Nereites*. Primary sedimentary structures, when preserved, consist of scour-and-fill structures, parallel and ripple cross-laminations, and graded, lenticular and wavy bedding. Clay chips and siderite clasts are commonly associated with these structures. Ovoid siderite nodules and sideritic shale bands also have been noted.

This lithofacies occurs most frequently towards

**Fig. 10.** Core and photograph log for well 8-4-30-3 W5M located west of the ridge crest. Photographs B–F display examples of lithofacies 2 and illustrate the overall coarsening upwards trend of the 'B' sandstone. Also note that bioturbation gradually decreases upwards. Photograph A illustrates the gradational contact between lithofacies 5B and 1M. See Fig. 9 for legend.

## LITHOLOGY

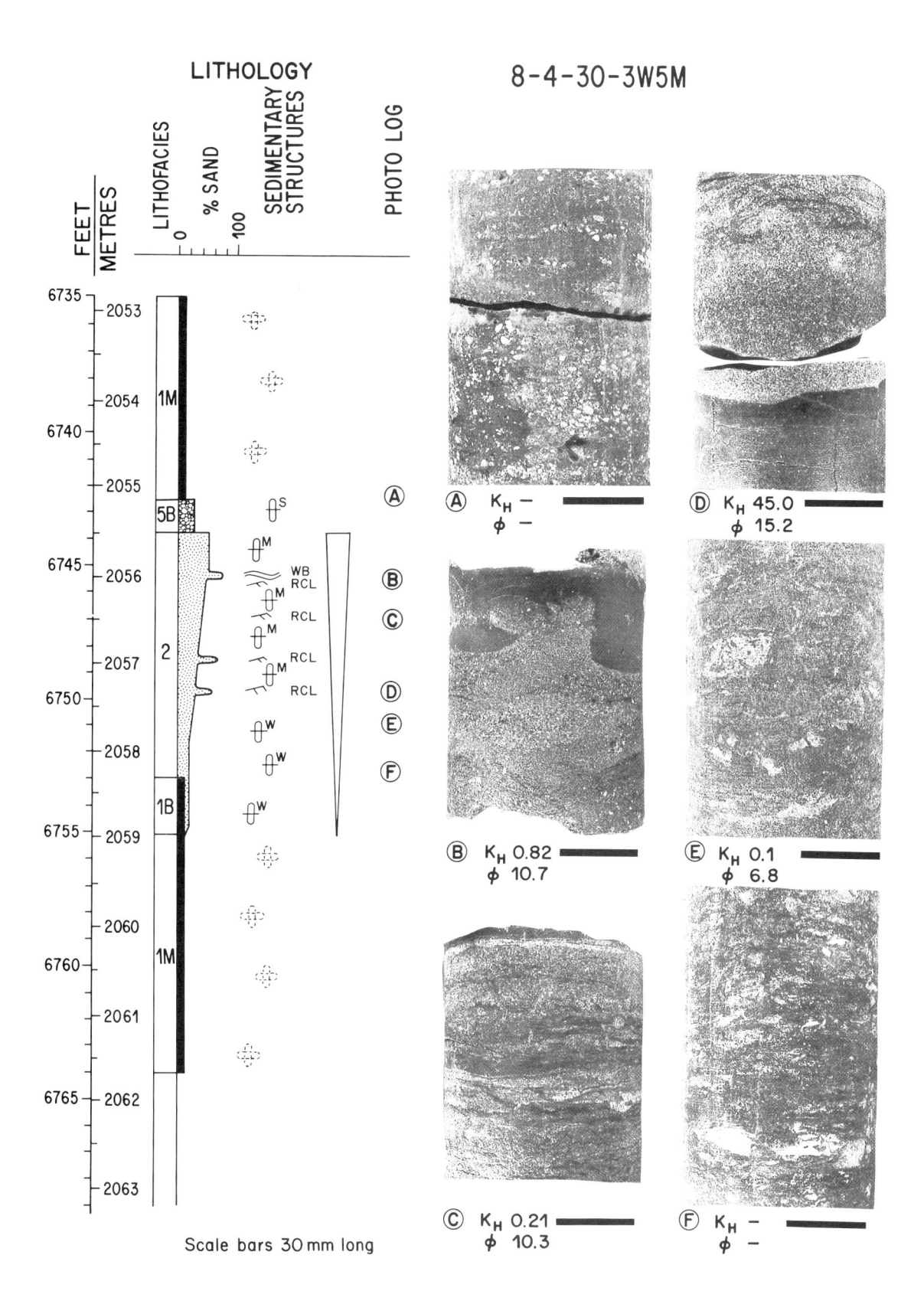

## 8-4-30-3W5M

(A) $K_H$ —
    $\phi$ —

(B) $K_H$ 0.82
    $\phi$ 10.7

(C) $K_H$ 0.21
    $\phi$ 10.3

(D) $K_H$ 45.0
    $\phi$ 15.2

(E) $K_H$ 0.1
    $\phi$ 6.8

(F) $K_H$ —
    $\phi$ —

Scale bars 30 mm long

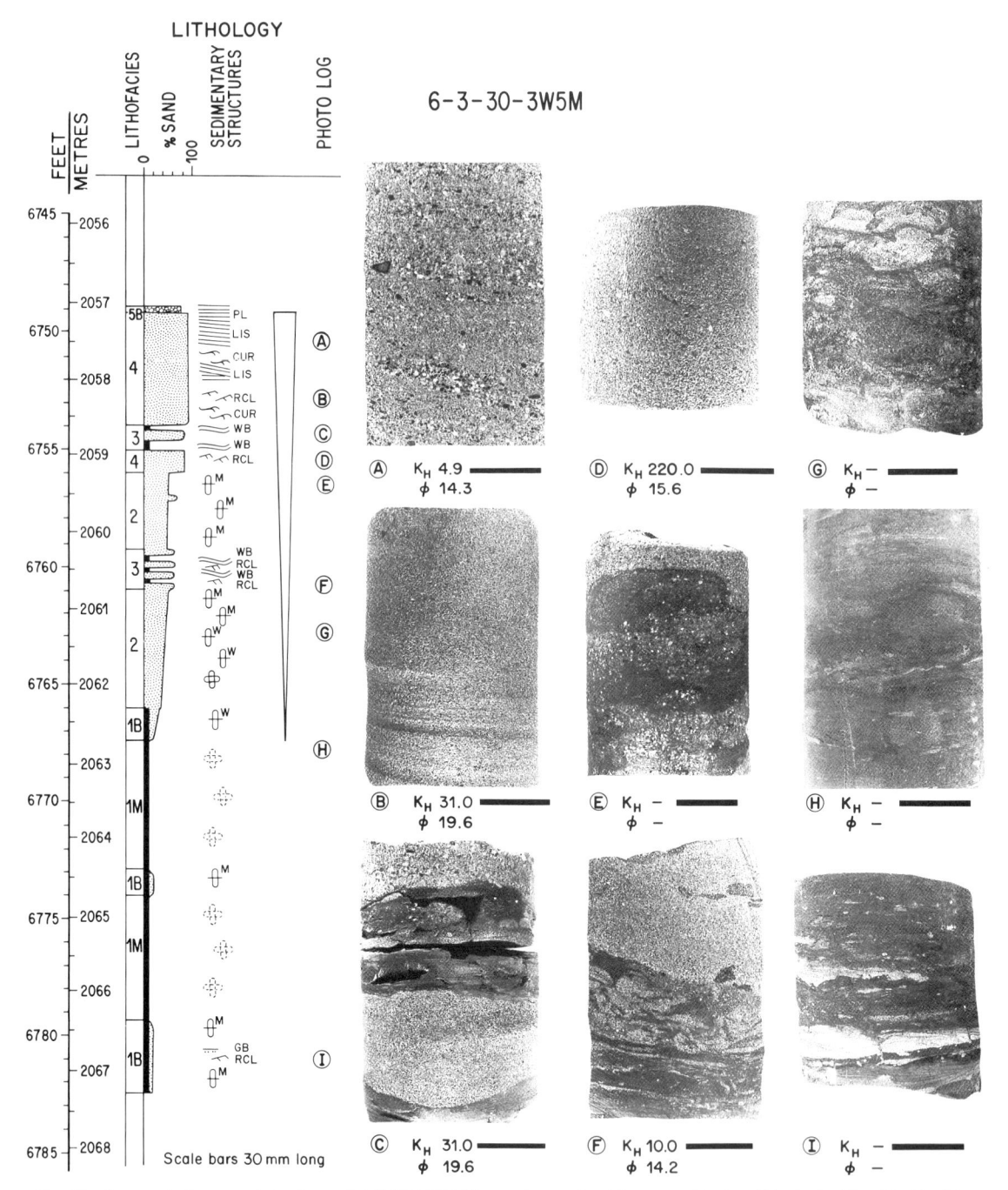

**Fig. 11.** Core and photograph log for well 6-3-30-3 W5M located near the ridge crest. Various lithofacies identified in this core are illustrated in photographs A–I: A, B and D from lithofacies 4; C from lithofacies 3; E, F and G from lithofacies 2; H from lithofacies 1M; and I from lithofacies 1B. See Fig. 9 for legend.

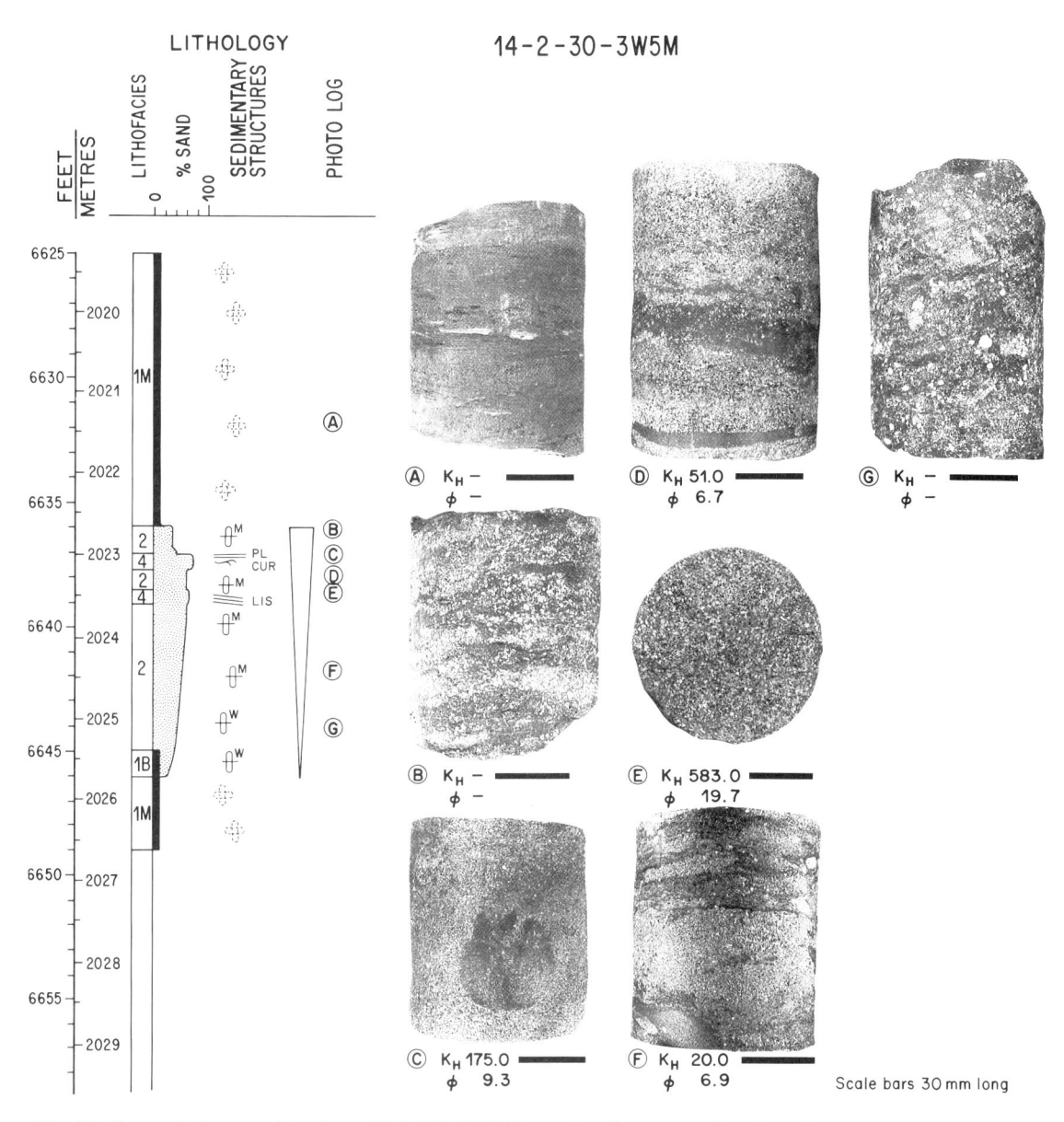

Fig. 12. Core and photograph log for well 14-2-30-3 W5M located on the eastern edge of the ridge. Lithofacies 1M in photograph A; lithofacies 2 in photographs B, D, F and G; and lithofacies 4 in photographs C and E. See Fig. 9 for legend.

the bottom of the reservoir, underlying and interbedded with lithofacies 3 and 4, and overlying lithofacies 1. The distribution of this lithofacies can be appreciated in Fig. 8, where it occupies the lower two-thirds of the Crossfield reservoir. Typically, it is 0.3–7.0 m thick. The transition from underlying L1B to L2 is often characterized by the abrupt appearance of abundant but scattered coarse sand and granule-sized chert particles.

*Interpretation*

Deposits of Lithofacies 2 represent sediments that have been abundantly reworked by benthonic organisms. The rocks preserve most clearly evidence for contrasting energy regimes in the overlying water column during deposition, as is indicated by the physical and biological sedimentary structures preserved. Biological reworking of the sediments would have occurred during periods when the substrate was most stable; that is, during periods of low water agitation and turbulence. Conversely, recurring and frequent physical reworking would have taken place during periods of high agitation and turbulence (Krause & Nelson, 1984). The trace fossil fauna found in these rocks is representative of sublitoral to bathyal marine environments (Crimes, 1975; Frey & Pemberton, 1985).

As indicated earlier, the abrupt appearance of coarser grained material in this lithofacies signals a larger scale environmental change. It is unlikely that this change reflects simply an increase in current competence, since the surrounding matrix resembles that of the underlying lithofacies. In addition these deposits would have been subject to the currents of recurring storms, as it is well established that the Cardium shelf was storm dominated (as discussed by Swagor et al., 1976; Wright & Walker, 1981; Krause, 1982; Krause & Nelson, 1984). Instead, the introduction of coarser grained sediment may reflect readjustments to the basin's margins following a long-term inundation by the sea of coastal areas.

## Lithofacies 3 (L3): thin and very thin-bedded shale and fine to coarse-grained sandstone

Centimetre-thick interbeds of shale, siltstone and fine to coarse-grained sandstone characterize this lithofacies (Fig. 11C). Sandstones with fine to medium grain sizes are dominant, however, an admixture of medium to coarse grains may be observed occasionally. It is easily distinguished from lithofacies 2 as it is less than 30% bioturbated, bed integrity is maintained, and primary sedimentary structures are well preserved. Sedimentary structures commonly observed are: parallel, wave and current ripple cross-laminations; scour-and-fill structures; graded, lenticular, wavy and flaser bedding; and layered bands of siderite and shale clasts. Wave ripples are the structures observed most commonly, but a variety of ripple forms are encountered. Ripple laminae may be unidirectional, opposed, symmetrical, spillover, aggrading with low migration angles, chevron upbuilding, convex and convex bundled.

Lithofacies 3 occurs most frequently interbedded with deposits of lithofacies 4, and ranges in thickness from 0.1 to 1.2 m, as may be observed in Fig. 8.

### Interpretation

Rocks of lithofacies 3 represent deposits similar to those from lithofacies 2, except that they are unbioturbated or only slightly so. If bioturbated, *Chondrites* is often associated with this rock type, as was noted previously by Krause & Nelson (1984) in cores from the Pembina Field. As indicated earlier, the common occurrence of this trace fossil may suggest that bottom conditions were at times oxygen deficient (Bromley & Ekdale, 1984), thus excluding most benthonic organisms from effectively colonizing the sea-floor; or, that the water column was characterized at times by reduced salinities and/or high concentrations of suspended mud. As in lithofacies 2, periods of high and low water turbulence are indicated by couplets of sandstone and shale, respectively. Currents on the bottom were capable of eroding into cohesive beds, as is indicated by common shale-chip conglomerates. Lithological characteristics of the shale chips are similar to those of the undisturbed shale interbeds. Moreover, the bases of the sandstone interbeds are often sharp and erosive. Additional evidence of scour and reworking is abundant and is typified by 'spillover' (Seilacher, 1982), crop-and-fill ripples, 'truncated crest ripples' (truncation of ripple foresets by overlaying shale layers), and flaser bedding. The variety of wave ripple forms observed may be the result of deposition, not only by oscillatory flows, but also by combined flows (Swift et al., 1983).

Preservation of hydrodynamic imprintings in this lithofacies may be the product of local, transitory dysaerobic or anaerobic conditions on the sea-floor, which prevented extensive sea-floor colonization and activity by organisms. Alternatively, they may be the product of increasingly frequent periods of intensified wave activity and rapid regimes of sediment influx.

## Lithofacies 4 (L4): medium to thick bedded, fine to very coarse-grained sandstone

The sandstones of this lithofacies (Figs 11A,B,D, 12C,E) may contain up to 30% of grains coarser than 2 mm. Shale constitutes less than 20% of the rock. Under normal core examination, sedimentary structures in this rock type are sparse, a situation that is aggravated when cores are unslabbed. When

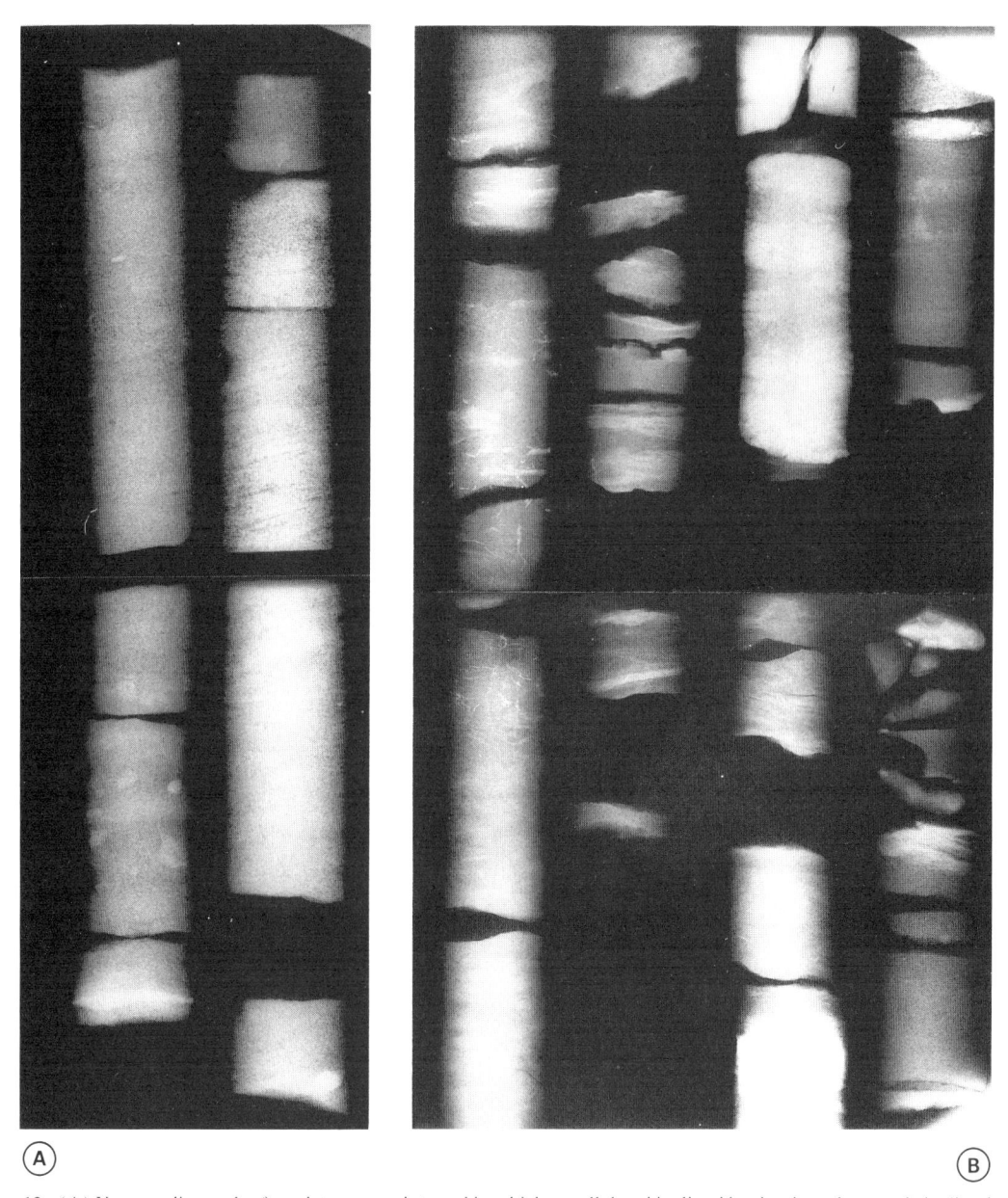

**Fig. 13.** (A) X-ray radiograph of sandstone core interval in which parallel and inclined laminations, low-angle inclined cross-laminations, and ripple cross-laminations are apparent. Brighter intervals along the core highlight siderite-cemented zones. Well 6-3-30-3 W5M. Core interval 6754–6749 ft (2058.6–2057.1 m). Scale bar 6.2 cm long. (B) X-ray radiograph of sandstone core interval displaying bioturbation, vertical and horizontal burrows, *Zoophycos*, wavy and lenticular bedding, ripple cross-laminations, and low-angle inclined cross-stratification. Brighter zones represent high siderite-cement concentrations. Well 6-3-30-3 W5M. Core interval 6763–6754 ft. (2061.4–2058.6 m). Scale bar 6.2 cm long.

these conditions are present, the rock may appear massive (Fig. 11B,D). However, in slabbed cores and under X-ray radiographic examination, the following small and medium-scale sedimentary structures commonly are observed: low-angle inclined cross-stratification (LIS; see Krause & Nelson, 1984), parallel laminations, convex ripples (CUR; see Krause & Nelson, 1984), wave and current

446    F.F. Krause and D.A. Nelson

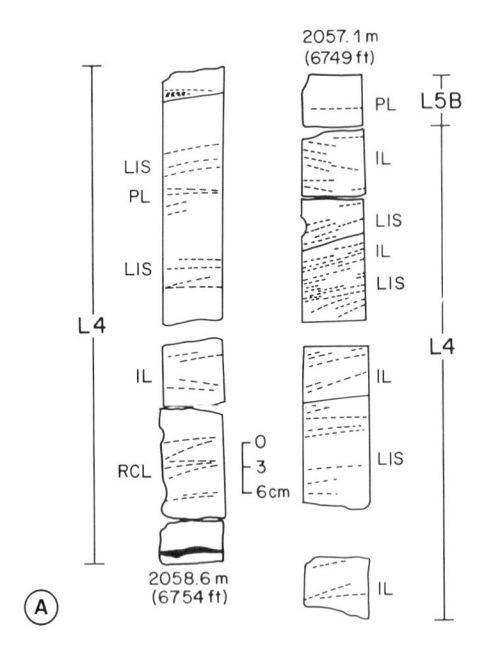

**Fig. 14.** (A,B) Line drawings highlighting sedimentary structures and textures identified on X-ray radiographs shown in Fig. 13A, B. Diagram also depicts the lithofacies identified in this core. Well 6-3-30-3 W5M. See Fig. 9 for legend.

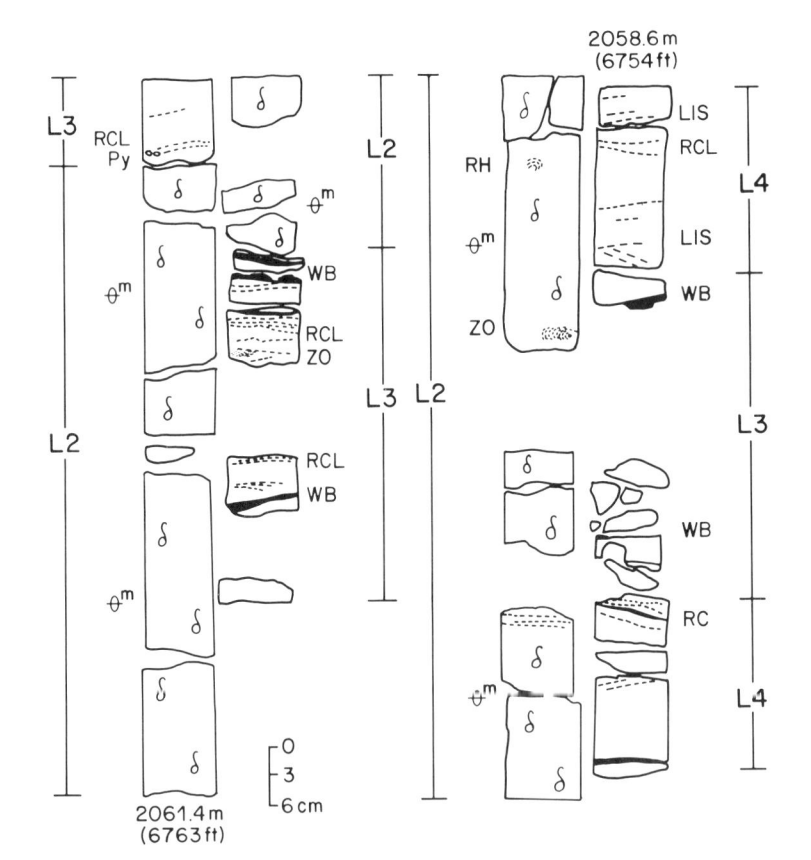

ripple cross-laminations (RCL), trough cross-stratification, climbing ripples, shale rip-up clast conglomerates, reworked nodular siderite clasts, and layers with well-sorted grains (see Figs 13A,B, 14A,B). Burrows and bioturbation textures are also abundant. In addition, *Zoophycos*, *Rhizocorallium* and *Planolites* are the most commonly observed trace fossils.

In addition to the sedimentary structures noted, the rock is extensively cemented by siderite. This pervasive pore-filling material modifies what would otherwise be a rock with excellent hydraulic properties. Although siderite cementation is apparent under hand-lens examination, pore-filling cements are easily identified in X-ray radiographs. Because siderite has a high mass attenuation coefficient to X-rays (Brindley & Brown, 1980), siderite-rich zones will appear in X-ray radiographs as intervals with enhanced contrast (Figs 13A,B; 14A,B).

As illustrated in Fig. 8, L4 occurs along the top of the reservoir and is the dominant lithofacies in the ridge crest. In turn, L4 is overlain by L5A (grain-supported or sandy matrix conglomerate), L5B (shaly conglomerate), or L1M (massive dark grey shale). Contacts with overlying lithofacies are gradational and may appear indistinct because of bioturbation. Contacts with L3 or L2 are sharp and frequently display evidence for erosion. Bioturbated contacts are less common. The thickness of L4, as seen in Figs 8, 11, 12, varies from 0.1 to 1.5 m.

*Interpretation*

The trace fossils identified in this lithofacies are indicative of marine environments that may range from sublitoral to bathyal settings (Seilacher, 1967; Crimes, 1975). Preserved hydrodynamic imprintings indicate that the sediments accumulated in a high-energy setting, higher than that for previous lithofacies. The sedimentary structures most frequently observed and also recorded in X-ray radiographs (Figs 13A,B, 14A,B) are horizontal and low-angle inclined cross-stratification (LIS—may be equivalent in core to HCS, which is easily identified in outcrop, Krause & Nelson, 1984). They would have formed in a setting where the wave orbital component of flow dominated over the unidirectional current component and may reflect plane-bed oscillatory flow conditions and intense combined flows (Harms *et al.*, 1975; Swift *et al.*, 1983; Allen, 1985). The mode of HCS development is at present unknown, although the structure appears to signal sedimentation during storm events. Swift *et al.* (1983) and Allen (1985) have proposed recently a combined wave–current flow origin for this structure. They indicate that storm waves are always accompanied by steady flows, such as geostrophic flows or coastal jets, and that under these conditions HCS may form.

### Lithofacies 5 (L5): conglomerate

Conglomerate clasts of quartz and chert are the most common particles observed. They range in size from granules to medium pebbles, however, the majority of clasts are granule sized. Sand-sized grains are commonly admixed with this lithofacies and constitute less than 70% of the total particles. Sorting ranges from very poor to moderately well sorted. The conglomerate is commonly interbedded with L4, where it occurs in clast-supported and sand-supported interlayers. Conglomerate layers and beds typically are 5–50 mm in thickness.

This lithofacies can be subdivided further into two subfacies, 5A and 5B, based on shale content. Subfacies 5A has less than 10% shale, and 30% of the particles are greater than 2 mm. This subfacies commonly contains sandstone, and forms clast and sandstone matrix-supported frameworks. Siderite cement commonly occludes pores. Sedimentary structures are not readily apparent; their clear identification requires slabbed cores and X-ray radiographs. Subfacies 5B contains more than 10% shale and silt, and conglomerate may represent 5–90% of the particles (Fig. 10A). Subfacies 5B typically is shale-supported, and bioturbation is commonly evident. In contrast to subfacies 5A, subfacies 5B siderite occurs in 50–150 mm thick bands replacing shale and often is nodular.

Conglomerate is distributed patchily, but commonly occurs near the top and flanks of the ridge (Fig. 8). In general, L5 is 0.2–0.3 m thick, however, on occasions greater thicknesses are observed, which may exceed 0.6 m. The distribution of L5, as illustrated in Fig. 10 and observed in many cores throughout the field, suggests that it rests unconformably on other rock types, although scoured and erosional contacts are rarely preserved or easily observed in cores.

*Interpretation*

This lithofacies, in particular L5A, also was moulded by wave and combined flows even though corroborating sedimentary structures are sparse. Evidence for this interpretation is provided by the presence of granule to medium-pebble sized material and interbedding with sandstones of L4. In contrast, L5B may represent reworking of ridge sediments by more infrequent, very high-energy events as water deepened and sedimentation switched to deposition of clay and silt-sized material. Deposition during this deepening event may have returned the area to an accumulation pattern characterized by the deployment of extensive mud banks in nearshore zones. Lithofacies L5B also could be indicative of local influx of muddy, conglomerate-laden density currents as a result of reworking by storm-generated combined wave–current flows. Because we did not observe graded conglomerate beds near the ridge top, and instead observed beds with layered shaly conglomerates in abundance, we presently believe that conglomerates of L5B represent hydrodynamically reworked and sorted ridge sediments amalgamated with shales and silts.

## Lithofacies association, interpretive summary

The lithofacies association that characterizes Crossfield Field is typical of a marine shelf sculptured by storm events. As observed, the lithofacies preserve both current bedding and bioturbation, indicative of a depositional environment influenced by high and low fluid power events. Under these conditions, recurring and frequent physical sediment transport would have taken place during periods of high agitation, increased turbulence and high sediment input. In contrast, biological reworking would have occurred during periods of low water agitation, low sediment suspension, and increased levels of dissolved oxygen in bottom waters.

## RIDGE ARCHITECTURE

Wire-line log correlations illustrated earlier (Figs 5, 6) and various cross-sections through several portions of the field (Fig. 7; Berven, 1966; Collins et al., 1985) show that the reservoir's outline in width cross section is ridge like. However, these wire line log correlations provide little information about the internal distribution of lithofacies and ridge architecture. Correlations utilizing cores in combination with wire-line logs, such as is illustrated in Fig. 8, provide a better although still incomplete view of the reservoir's internal geometry. The view is incomplete because the interpretation of interwell continuity is based on information gathered from two, 110 mm wide cores that are separated by approximately 800 m of uncored rock. In essence, every millimetre of core has to account for 3.6 m of reservoir rock (1:3600). Nevertheless, the cross-section cartoon of the expected lateral lithofacies distribution is based on standard correlation techniques complemented by observations from outcrops of the Cardium Formation, incomplete knowledge of the depositional environment, reservoir fluid production histories, and previous work on other Cardium Formation hydrocarbon pools (e.g. Krause & Collins, 1984; Krause & Nelson, 1987; Krause et al., 1987).

The lithofacies transect illustrated in Fig. 8 is taken almost parallel to dip in the northern and thickest portion of the field. Datum and wells are identical to those illustrated previously in Figs 5, 6. Legend and core descriptions for wells from this cross-section are portrayed in Figs 9, 10–12. Cored well 8-4-30-3W5M was positioned in the transect relative to other cored wells after describing and identifying the cores' lithofacies association because a wire-line log suite was not available. The core intersects both the sandstone ridge and the lower and upper shales. We believe therefore that our correlation is reasonable even though, in this well, we are missing a log suite that includes the top of the Cardium Zone Member (Krause & Nelson, 1984).

As may be appreciated in Fig. 8, laterally and vertically the reservoir's lithofacies are restricted. Eastwards, reservoir lithofacies gradually taper and, over a short distance, thin and disappear into shales of L1M, as is evident between wells 14-2-30-3W5M and 6-13-30-3W5M. The core of the latter well intersects a continuous shale interval (Fig. 9). In contrast, westwards, L2 gradually becomes siltier and shalier. L2 comprises the most common rock type noted in the reservoir and accounts for more than half of the rock volume. In the central and upper portions of the reservoir, L4 is common, where it consists of current bedded, or bioturbated, sandstones and conglomeratic sandstones. In Fig. 8, the thickest development of L4 occurs east of well 6-3-30-3W5. Lithofacies L4 disappears westwards towards well 8-4-30-3W5M. Interestingly, towards the eastern flank of the ridge, interbedded deposits of L2–4 are common. The pattern observed suggests that, towards the eastern flank of the ridge, the three lithofacies are interlayered, upward building and younger layers step progressively backwards. Confirmation of this pattern will require additional cores obtained perhaps during infill drilling programmes. Finally, a thin interval of granular and small-pebble sized material contained within a shale matrix (L5B) discontinuously caps the reservoir.

The pattern that we have observed indicates clearly that the Crossfield reservoir comprises an isolated sandstone body, which is ridge-like in shape. We have not observed a sandstone sheet or blanket-like morphology as had been proposed by Walker (1983a–c, 1984) for the Cardium–Caroline and Cardium–Garrington pools, two narrow reservoirs immediately north of Crossfield. Instead we have confirmed the 'shoestring sandstone' geometry originally proposed by Berven (1966). However, we differ from Berven (1966), Javeri (1966) and also Walker (1983a–c) in the interpretation of the environment of deposition, as discussed previously.

At megascales (Krause et al., 1987) lithofacies distribution patterns similar, but not identical to ours, have been recorded in Upper Cretaceous sandstone ridges in the Woodbine Formation 'C' sandstone of the Kurten Field of south-central Texas (Turner & Conger, 1981, 1984; Phillips & Swift, 1985) and in the Shannon Formation of the Hartzog Draw Field of north-central Wyoming (Hearn et al., 1984, 1986; Gaynor & Swift, 1988).

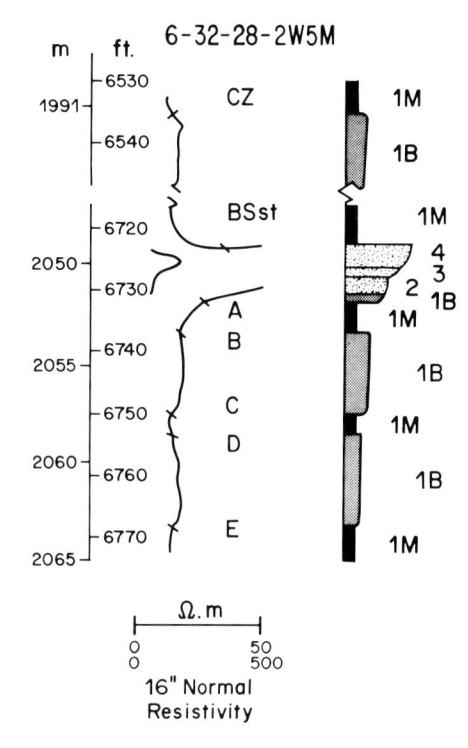

**Fig. 15.** Resistivity log indicating the position of markers A–E in well 6-32-28-2 W5M. Typical lithofacies correlated with the log's signature are shown to illustrate facies changes between markers. Cores and logs were compared throughout the field to determine the particular log/lithofacies response.

## RIDGE EVOLUTION

To characterize the areal geometry and evolution of the ridge we chose several log-markers that are common to the Crossfield area (Figs 15, 16). Significantly, correlations of the marker-defined intervals with cores indicate that the markers chosen reflect major lithofacies changes. Individual intervals may comprise a single lithofacies, or consist of a group of lithofacies that alternate at smaller scales.

**Fig. 16.** East–west dip cross-section, $\beta$–$\beta'$, from the southern end of the field. This transect, based on resistivity logs and the Cardium Zone top as datum, illustrates the trends of the markers that underlie the Cardium 'B' sandstone. Lithosomes 1B and 1M between markers B–C and A–B, respectively, form eastward-thinning sediment wedges. The Cardium 'B' sandstone grew on these wedges forming a ridge with a slightly steeper eastward slope.

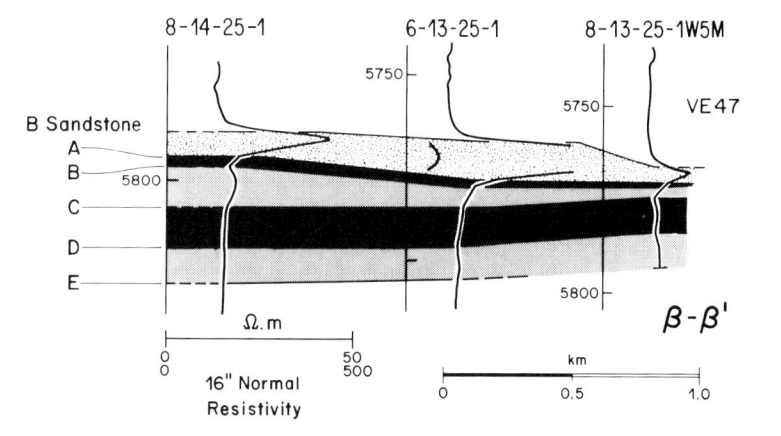

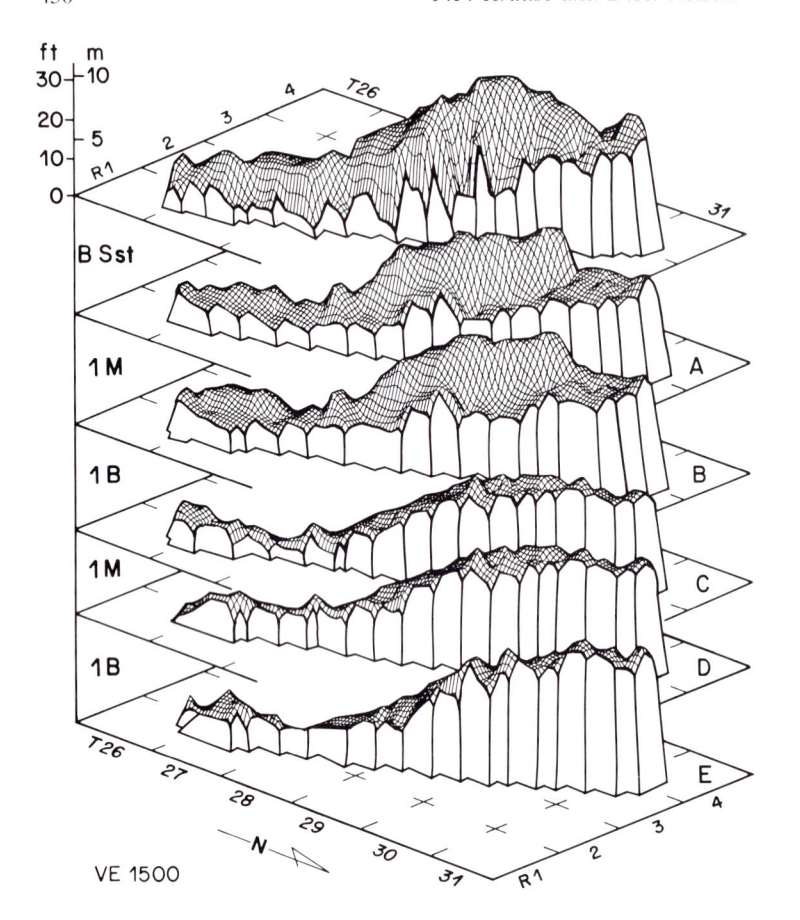

**Fig. 17.** Log marker structure surfaces (LMSS) illustrating topography at the top and below the 'B' sandstone. Log marker structure surfaces A–E correspond to the base of lithofacies intervals shown at left. Datum is Cardium Zone top. View is from the northeast at an angle of 22° above the horizontal. In order to facilitate viewing and avoid overlap, true vertical separation between LMSS has not been maintained.

In the latter instance, the dominant lithofacies was chosen to be representative of the interval. We view these intervals as lithosomes (Wheeler & Mallory, 1956). They represent an individual lithofacies or a grouping of lithofacies that include significant volumes of rock, are distributed over a wide area, and were not deposited over a single, brief instant, but instead reflect multiple sedimentation or erosion and resedimentation events that span varying amounts of time. Thus, time planes may cut through a lithosome. The ridge's lithofacies distribution illustrated in the previous section indicates clearly that the ridge grew episodically. Therefore, time planes will cut through the ridge paralleling depositional slopes. Coarser and finer grained lithofacies may have been juxtaposed in response to local acceleration and deceleration of flows, as proposed by Swift & Rice (1984) in their spatial acceleration model.

A more detailed impression of ridge morphology and development is obtained by examining the log-marker structure surfaces (LMSS) generated from the markers illustrated in Figs 15–18. The thickness of lithosomes 1B and 1M between the lower markers C, D and E is fairly consistent east to west. However, between the upper markers A, B and C, lithosomes 1B and 1M form eastward thinning sediment wedges. It is upon these wedges that the sandstone ridge was initiated. Topography of the markers, as viewed from two directions, is illustrated in three-dimensional diagrams (Figs 17, 18). Log-marker structure surfaces were obtained from isopachs that include 208 wells within the field and its immediate perimeter. Measured thicknesses extended from the Cardium Zone top to the corresponding log-marker.

Log-marker structure surface 'E' has a general southward dip and displays a gently undulating topography that is highest in the north. However, as may be appreciated in Fig. 18, in the northern area

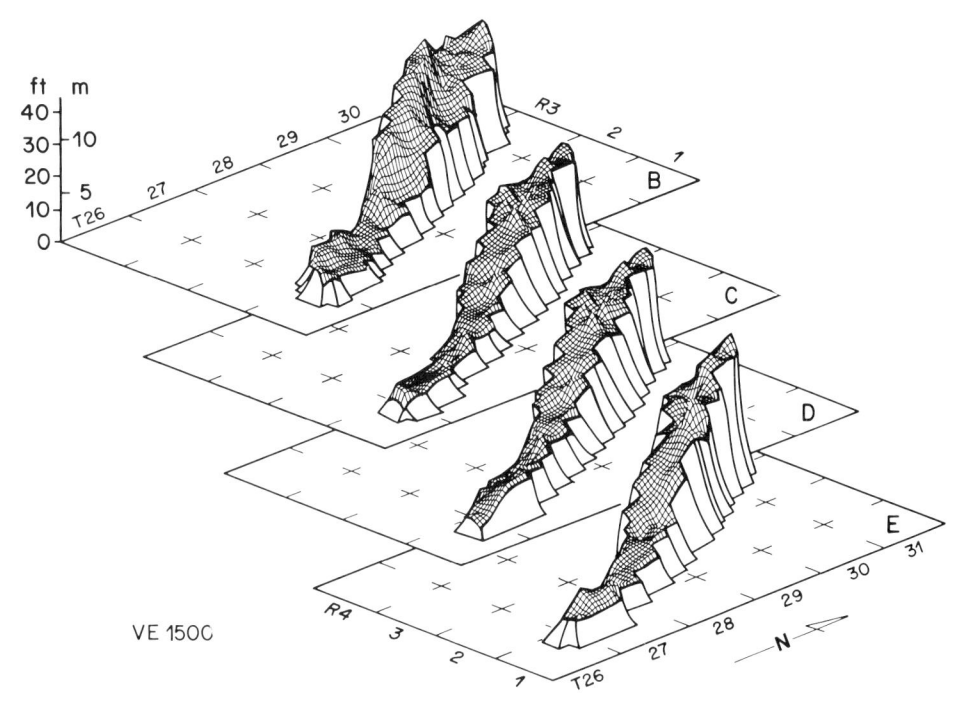

**Fig. 18.** A southeast view of LMSS B–E that underlie the Cardium 'B' sandstone. Angle of view is 22° above horizontal. Distance between markers is not representative of marker spacing on logs.

the surface not only slopes gently southwards but also dips westward. In addition, LMSS 'E' contains small perturbations with dimensions of the order of 4 km, which seemingly trend north and northeast. The topography is much more subdued for LMSS 'D' and 'C'. In both horizons, surfaces undulate gently and slopes are less pronounced than that for LMSS 'E', even though they still maintain more elevated ground in the north. For LMSS 'D', gentle, westward dipping slopes also are present along the eastern margin, whereas westward dipping slopes are found predominantly to the north in LMSS 'C'. Topography changes rapidly between LMSS 'C' and 'B'. It is expressed in LMSS 'B' and 'A' by build-up north of T28 and concomitant slope reversal. Lower and upper LMSS 'A' and 'B-sandstone top' define a ridge-like sandstone body. Topography at the top of the Cardium 'B' sandstone indicates that the ridge grew upwards and extended eastward and southward. In the north, the ridge built on the underlying surface's pre-established high, however, south of a line from Section 25, Township 28, Range 2, ridge deposits infilled a large depression and built upwards.

## DISCUSSION

The reservoir geometry and the described distribution of lithofacies indicates clearly that the Crossfield–Cardium pool formed as a submarine sandstone ridge. The length and width dimensions of the Crossfield ridge, 72 km and 2–5 km, respectively, are comparable to dimensions catalogued by Johnson (1978, fig. 9.40) for modern and ancient submarine storm and tidal ridges. The thickness of 2–4 m is similar to thicknesses recorded by Walker (1983a) for the Caroline and Garrington pools. Comparisons of the Crossfield ridge with the tabulations of modern and ancient submarine ridges reported by Johnson (1978) indicate that Crossfield's thickness is low, but within the reported range.

As determined from analysis of the lithofacies, the ridge that was preserved at Crossfield formed in response to storm flows. Evidence for the latter is retained most clearly by current-bedded sandstones, which most often preserve low-angle inclined stratification (a possible analogue of hummocky cross-stratification) and plane-bed laminations. These sedimentary structures appear to form in response to long-period storm swells and combined-current flow regimes generated during strong atmospheric disturbances (Swift *et al.*, 1983, 1986; Vincent, 1986).

The ridge at Crossfield grew on a muddy bottom, a behaviour that is common to many Jurassic and Cretaceous ridges of the Western Interior of North America, as reported by several authors, for example, Gaynor & Swift (1988), Brenner & Davies (1973), Berg (1975), and Boyles & Scott (1982). Furthermore, the ridge is covered also by muddy siltstones and standstones. Thus the ridge is contained within an envelope of mud. This pattern indicates that the supply of fine-grained material was abundant in this area. In modern settings, abundant muds and silts are typically supplied by riverine plumes. Upon entering the basin, mud and silt plumes are stretched and dispersed slowly alongshore paralleling the coast for tens of kilometres, and in some cases, hundreds of kilometres, as documented by Gibbs (1976), Wells & Coleman (1981), Rine & Ginsburg (1985), Muller-Karger *et al.* (1988), Rezak *et al.* (1985), and Rabalais & Boesch (1987). Not only is material shunted along shore, but hyperpycnal flows are also common near fluvial sources transporting fine-grained sediments on to the shelf, as documented by Wright *et al.* (1988). All of these deposits also are reworked and redistributed by waves, tidal currents and storm flows (Nair, 1976; Wells & Coleman, 1981; McKee *et al.*, 1983; Wells & Kemp, 1984; DeMaster *et al.*, 1985; Rhoads *et al.*, 1985; Kuehl *et al.*, 1986). Crossfield's mudstone envelope may have resulted from processes such as these.

Sedimentary structures on the Guianas shelf (Wells & Coleman, 1981; Rine & Ginsburg, 1985), as well as on the Amazon River shelf (Kuehl *et al.*, 1986; Nittrouer *et al.*, 1986) and the Changjiang delta and East China Sea (Rhoads *et al.*, 1985) resemble the ones we describe from mudstones and siltstones enclosing the Crossfield reservoir sandstones. It is also interesting to note that, in these modern settings, faunas are overwhelmed by vast influxes of very fine-grained materials. This results in a bottom with impoverished faunas, decreased faunal activity and sediment mixing, and presence and colonization by euryptopic organisms (Rhoads *et al.*, 1985). Reduced salinities owing to periodic fresh-water influx may further modulate the environment, as suggested by Pratt (1984). Colonization of Crossfield sediments by hard-shelled benthos may have been restricted by similar processes. In addition, in this setting, any skeletal remains would have been dissolved shortly after burial, reducing the fossil record even further.

The Crossfield ridge would have initiated growth on a transgressive surface, as may be indicated by the appearance of abundant chert granules intermixed with deposits of lithofacies 2 and 1B. Although a scour surface is not apparent in core, the change in grain sizes suggests a rapid and lasting change in transported sediment load. Coarser material may have become available in response to transgressive breaching of nearshore coarser grained stockpiles that accumulated near river mouths and/or were concentrated by wave winnowing. These deposits were flushed and moved to the shelf by episodic storm events. Once initiated, ridge lithofacies and their spatial relationships appear to have been controlled by storm-generated currents, which interacted with the sea-floor and the evolving ridge morphology.

As has been argued convincingly by Swift & Rice (1984) and Gaynor & Swift (1988), storm currents accelerate and decelerate in space as well as in time. As a result, regional velocity gradients control erosion and deposition from place to place on the shelf. Significantly, near topographic irregularities, such as a sand ridge, velocity gradients are pronounced. Near-bottom flows accelerate on the up-current ridge flank and expand and decelerate on the down-current flank (Swift & Field, 1981). This process has been advocated by Gaynor & Swift (1988) to explain the origin of the Shannon sandstone ridges of Wyoming. The ridges would have acted as fractionation mills extracting and concentrating coarser material from the predominantly fine-grained sediment load passing over them during storms. The coarsest material would have been extracted on the up-current flank, and finer material would have been extracted on the down-current flank. Clay and silt drapes would have accumulated as a result of storm current dissipation. It is likely that similar processes operated during the evolution of the Crossfield-Cardium ridge.

## CONCLUSIONS

1 The Crossfield reservoir is a long and narrow sandstone ridge deposit, as was recognized by Berven (1966) and Javeri (1966). The sandstone body trends NW–SE, and plunges less than 1° to the northwest. It is approximately 72 km long, 2.5–5 km

wide, averages 2.4–4.1 m in thickness, and has a length to width ratio of 20:1. It is wider in the north, and thins and narrows southwards. The cross-sectional profile is that of an asymmetrical lens with a plano-convex outline.

2 In cores from the Crossfield pool and immediate vicinity we have recognized five depositional lithofacies: (L1) dark grey mudstone; (L2) bioturbated, thin and very thin-bedded mudstone, and fine to coarse-grained sandstone; (L3) thin and very thin-bedded mudstone, and fine to coarse-grained sandstone; (L4) medium to thick bedded, fine to very coarse-grained sandstone; and (L5) conglomerate. Within the reservoir, the five lithofacies are stacked as an upward-coarsening rock sequence, which, in detail, is typified by lithofacies alternation and repetition. At multiwell scales (megascale—Krause et al., 1987), L1 encases and seals the reservoir, L2 dominates the lower half of the reservoir, L3 commonly segments the middle of the reservoir, L4 occurs predominantly along the reservoir's axis and L5 is found most frequently along the crest and eastern flank. Our observations indicate clearly that the reservoir rock sequence is areally restricted. Along the eastern flank, interbedded deposits of L2, L3 and L4 pinch out into shale. To the west, L2 dominates, gradually becoming siltier and shalier westwards. In addition, intervals of L4 are thicker and more common in the northern half of the field. Towards the south all lithofacies progressively become shalier and beds become thinner. All five lithofacies accumulated in a marine environment as indicated by their trace fossils. We conclude, therefore, that the Crossfield–Cardium reservoir is the deposit of a submarine ridge.

3 The ridge grew on a muddy substrate in response to storm flows. The environment was characterized by an overabundant supply of muds and silts, as is evident from thick mudstone and siltstone sequences enclosing the ridge. This material would have been supplied to the Cardium basin from riverine sources and dispersed along shore or across the shelf by waves, tides, nearshore currents, storm flows, and hyper- and hypopycnal flows. Present-day depositional environments characterized by these patterns have been described from the Guianas coast, Amazon shelf, South Louisiana shelf, East China and Yellow Seas, and Malabar coast (Nair, 1976; Wells & Coleman, 1981; Milliman & Qingming, 1985; Rezak et al., 1985; Rine & Ginsburg, 1985; Nittrouer & DeMaster, 1986).

Ridge evolution, traced by a series of log-marker structure surfaces (LMSS), indicates that the sandstone ridge grew over an irregular, undulating, eastward and southward sloping surface. Ridge sediments would have been transported by currents that flowed from the northwest to the southeast; that is, parallel and perhaps obliquely across the ridge. This behaviour is known from modern environments where ridge-parallel currents with across-ridge components have been recognized on the storm dominated shelf of the Middle Atlantic Bight of the USA by Swift & Field (1981). The surface on which the Crossfield ridge developed was higher in the north than in the south, forming a gentle and protracted undulation from north to south. The ridge grew upwards and eastwards in the north and extended progressively to the south. The end product is a long and narrow ridge dipping gently southwards and having a steeper eastern flanking slope.

## ACKNOWLEDGEMENTS

We thank present and former colleagues at the Petroleum Recovery Institute: G. Addison, I. DeBie, C. DeBree, N. Collins, R. French, M. Hebert, B. Moore, W. Ng, F. Pilles and J. Warner for their assistance with this study; C. Savoie of PetroCanada for his effort in producing the X-ray radiographs; the staff at the I.N. McKinnon and Gallagher libraries and the E.R.C.B. Core Research Centre for their help; J.M. Boyles, G.C. Gaynor and D.J.P. Swift for reviewing an earlier version of the manuscript; and D.J.P. Swift, D.A.W. Keith, G.C. Gaynor, N.N. Rabalais and A.W. Walton for fruitful exchanges of ideas.

## REFERENCES

ALLEN, P.A. (1985) Hummocky cross-stratification is not produced purely under progressive gravity waves. Nature **313**, 562–564.

BARLOW, L.K. (1986) An integrated geochemical and paleoecological approach to petroleum source rock evaluation, Lower Niobrara Formation (Cretaceous), Lyons, Colorado. Mountain Geol. **23**, 107–112.

BASS, N.W. (1934) Origin of Bartlesville shoestring sands, Greenwood and Butler Counties, Kansas. Bull., Am. Assoc. Petrol. Geol. **18**, 1313–1345.

BASS, N.W., LEATHEROCK, C., DILLARD, W.R. & KENNEDY, L.E. (1937) Origin and distribution of Bartlesville and Burbank shoestring oil sands in parts of Oklahoma and Kansas. Bull., Am. Assoc. Petrol. Geol. **21**, 30–66.

BERG, R.R. (1975) Depositional environment of Upper Cretaceous Sussex Sandstone, House Creek Field,

Wyoming. *Bull., Am. Assoc. Petrol. Geol.* **59**, 2099–2110.

BERVEN, R.J. (1966) Cardium sandstone bodies, Crossfield–Garrington area, Alberta. *Bull. Can. Petrol. Geol.* **14**, 208–240.

BOYLES, J.M. & SCOTT, A.J. (1982) A model for migrating shelf-bar sandstones in Upper Mancos Shale (Companion) northwestern Colorado. *Bull., Am. Assoc. Petrol. Geol.* **66**, 491–508.

BRENNER, R.L. & DAVIES, D.K. (1973) Storm-generated coquinoid sandstones—Genesis of high energy marine sediments from the Upper Jurassic of Wyoming and Montana. *Geol. Soc. Am. Bull.* **84**, 1685–1698.

BRINDLEY, G.W. & BROWN, G. (Eds) (1980) *Crystal Structure of Clay Minerals and Their Identification.* Monograph No. 5, Mineralogical Society, London, 495 pp.

BROMLEY, R.G. & EKDALE, A.A. (1984) *Chondrites*: a trace fossil indicator of anoxia in sediments. *Science* **224**, 872–874.

BYERS, C.W. (1977) Biofacies patterns in euxinic basins: a general model. In: *Deep-water Carbonate Environments* (Eds Cook, H.E. & Enos, P.). Soc. Econ. Paleontol. Mineral. Spec. Publ. **25**, 5–17.

CHAKRAVORTY, S.K. BROWN, P.R. & ENDSIN., N. (1978) A review of waterflood performance in Garrington Cardium A and B pools, Unit No. 2. *J. Petrol. Technol.* **30**, 869–874.

COLLINS, H.N., KRAUSE, F.F., FLORES, J.H., HUNTRODS, R.S., HAESSEL, W.W., KIRKEGAARD, M.K., MORRISON, G.R. & WEBSTER, A.R. (1985) *A Preliminary Assessment of the Technical and Economic Feasibility of Enhanced Oil Recovery from the Crossfield Cardium A Pool*, Vols I, II, III & IV. Report for Energy, Mines and Resources by Petroleum Recovery Institute, Computer Modelling Group and Canadian Energy Research Institute, Calgary, Alberta, 500 pp.

CRIMES, T.D. (1975) The stratigraphical significance of trace fossils. In: *The Study of Trace Fossils* (Ed. Frey, R.W.) pp. 109–130. Springer-Verlag, New York.

DEMASTER, D.J., MCKEE, B.A., NITTROUER, C.A., JIANGCHU, Q. & GUODONG, C. (1985) Rates of accumulation and particle reworking based on radiochemical measurements from continental shelf deposits in the East China Sea. In: *Sediment Dynamics of the Changjiang Estuary and the Adjacent East China Sea* (Eds Milliman, J.D. & Quingming, J.). Continental Shelf Res. 4(1/2), 143–158.

FREY, R.W. & PEMBERTON, S.G. (1985) Biogenic structures in outcrops and cores—I. Approaches to ichnology. *Bull. Can. Petrol. Geol.* **33**, 72–115.

GAYNOR, G.C. & SWIFT, D.J.P. (1988) Shannon sandstone depositional model: sand ridge formation on the Campanion Western Interior Seaway. *J. Sediment. Petrol.*, **58**, 868–880.

GIBBS, R.J. (1976) Amazon River sediment transport in the Atlantic Ocean. *Geology* **4**, 45–48.

HARMS, J.C., SOUTHARD, J.B., SPEARING, D.R. & WALKER, R.G. (1975) Depositional environments as interpreted from primary sedimentary structures and stratification sequences. *Soc. Econ. Paleontol. Mineral. Short Course* **2**, 161 pp.

HEARN, C.L., EBANKS, W.J., JR., TYE, R.S. & RANGANATHAN, V. (1984) Geological factors influencing reservoir performance of the Hartzog Draw Field, Wyoming. *J. Petrol. Technol.* **36**, 1335–1344.

HEARN, C.L., HOBSON, J.P. & FOWLER, M.L. (1986) Reservoir characterization for simulation, Hartzog Draw Field, Wyoming. In: *Reservoir Characterization* (Eds Lake, L.W., and Carroll, H.B.). Academic Press, Orlando, FL, 659 pp.

JAVERI, Y.A. (1966) Crossfield Field. In: *Oilfields of Alberta, Supplement* (Ed. Century, J.R.) The Alberta Society of Petroleum Geologists, Calgary, Alberta, 136 pp.

JOHNSON, H.D. (1978) Shallow siliciclastic seas. In: *Sedimentary Environments and Facies* (Ed. Reading, H.G.) pp. 207–258. Elsevier, New York.

KRAUSE, F.F. (1982) Geology of the Pembina Cardium Pool. In: *Depositional Environments and Reservoir Facies in Some Western Canadian Oil and Gas Fields* (Ed. Hopkins, J.C.) University of Calgary Core Conference, Department of Geology and Geophysics, pp. 15–27.

KRAUSE, F.F. (1983) Sedimentology of a tempestite: episodic deposition in the Cardium Formation in the Pembina Oilfield area, west central Alberta, Canada. In: *Sedimentology of Selected Mesozoic Clastic Sequences* (Eds McLean, J.R., & Reinson, G.E.) *Core Expo '83*, Canadian Society of Petroleum Geologists, pp. 43–65.

KRAUSE, F.F. & COLLINS, H.N. (1984) *Pembina Cardium Recovery Efficiency Study: a Geological and Engineering Synthesis.* Petroleum Recovery Institute Report, Calgary, Alberta, 650 pp.

KRAUSE, F.F. & NELSON, D.A. (1984) Storm event sedimentation: lithofacies association in the Cardium Formation, Pembina area, west-central Alberta, Canada. In: *The Mesozoic of Middle North America* (Eds Stott, D.F., and Glass, D.J.). Can. Soc. Petrol. Geol. Mem. 9, 485–511.

KRAUSE, F.F. & NELSON, D.A. (1987) *Geological analysis of a 'shoe-string' sandstone trap. Crossfield-Cardium Pool, Alberta, Canada.* Petroleum Recovery Institute, Report 1987-8, Calgary, Alberta, 43 pp.

KRAUSE, F.F., COLLINS, H.N., NELSON, D.A., MACHEMER, S.D. & FRENCH, P.R. (1987) Multiscale anatomy of a reservoir. Geology of the Pembina-Cardium Pool, west-central Alberta, Canada. *Bull., Am. Assoc. Petrol. Geol.* **71**, 1233–1260.

KUEHL, S.A., NITTROUER, C.A. & DEMASTER, D.J. (1986) Distribution of sedimentary structures in the Amazon subaqueous delta. In: *Sedimentary Processes on the Amazon Continental Shelf* (Eds Niltrouer, C.A. & DeMaster, D.J.). Continental Shelf Res. 6(1/2), 359 pp.

LEVORSEN, A.I. (1958) *Geology of Petroleum.* W.H. Freeman, San Francisco, 703 pp.

MCKEE, B.A., NITTROUER, C.A. & DEMASTER, D.J. (1983) Concepts of sediment deposition and accumulation applied to the continental shelf near the mouth of the Yangtze River. *Geology* **11**, 631–633.

MCNEIL, D.H. (1984) The eastern facies of the Cretaceous System in the Canadian western interior. In: *The Mesozoic of Middle North America* (Eds Stott, D.F., & Glass, D.J.). Can. Soc. Petrol. Geol. Mem. 9, 145–171.

MILLIMAN, J.D. & QINGMING, J. (Eds) (1985) Sediment dynamics on the Changjiang estuary and the adjacent East China Sea. *Continental Shelf Res.* 4(1/2), 251 pp.

MULLER-KARGER, F.E., MCCLAIN, C.R. & RICHARDSON, P.L. (1988) The dispersal of the Amazon's water. *Nature* **333**, 56–59.

NAIR, R.R. (1976) Unique mud banks, Kerala, southwest India. *Bull., Am. Assoc. Petrol. Geol.* **60**, 616–621.

NIELSEN, A.R. (1957) Cardium stratigraphy of the Pembina field. *J. Alberta Soc. Petrol. Geol.* **5**, 64–72.

NIELSEN, A.R. & PORTER, J.W. (1984) Pembina oil field—in retrospect. In: *The Mesozoic of Middle North America* (Eds Stott, D.F. & Glass, D.J.). Can. Soc. Petrol. Geol. Mem. 9, 1–13.

NITTROUER, C.A. & DEMASTER, D.J. (Eds.) (1986) Sedimentary processes on the Amazon continental shelf. *Continental Shelf Res.* **6**(1/2), 359 pp.

NITTROUER, C.A., KUEHL, S.A., DEMASTER, D.J. & KOWSMANN, R.O. (1986) The deltaic nature of Amazon Shelf sedimentation. *Geol. Soc. Am. Bull.* **97**, 444–458.

PHILLIPS, S. & SWIFT, D.J.P. (1985) Shelf sandstones in the Woodbine–Eagle Ford Interval, East Texas: a review of depositional models. In: *Shelf Sands and Sandstone Reservoirs* (Eds Tillman, R.W., Swift, D.J. & Walker, R.G.). Soc. Econ. Paleontol. Mineral. Short Course 13, 503–558.

PLINT, A.G. & WALKER, R.G. (1987) Cardium Formation 8. Facies and environments of the Cardium shore line and coastal plain in the Kakwa Field and adjacent areas, northwestern Alberta. *Bull. Can. Petrol. Geol.* **35**, 48–64.

PRATT, L.M. (1984) Influence of paleoenvironmental factors on preservation of organic matter in Middle Cretaceous Greenhorn Formation, Pueblo, Colorado. *Bull., Am. Assoc. Petrol. Geol.* **68**, 1146–1159.

RABALAIS, N.N. & BOESCH, D.F. (1987) Dominant features and processes of continental shelf environments of the United States. In: *Long-term Environmental Effects of Offshore Oil and Gas Development* (Eds Boesch, N.E. & Rabalais, N.N.) pp. 71–147. Elsevier, New York.

REINECK, H.E. (1967) Parameter von schichtung und bioturbation. *Geol. Rundsch.* **56**, 420–438.

REZAK, R., BRIGHT, T.J. & MCGRAIL, D.W. (1985) *Reefs and Banks of the Northwestern Gulf of Mexico.* Wiley, Toronto, p. 259.

RHOADS, D.C. & MORSE, J.W. (1971) Evolutionary and ecologic significance of oxygen-deficient marine basins. *Lethaia* **4**, 413–438.

RHOADS, D.C., BOESCH, D.F., ZHICAN, T., FENGSHAN, X., LIQIANG, H. & NILSEN, K.J. (1985) Macrobenthos and sedimentary facies on the Changjiang delta platform and adjacent continental shelf, East China Sea. In: *Sediment Dynamics of the Changjiang Estuary and the Adjacent East China Sea* (Eds Milliman, J.D. & Qingming, J.). Continental Shelf Res. 4(1/2), 189–213.

RINE, J.M. & GINSBURG, R.N. (1985) Depositional facies of a mud shoreface in Surinam, South America—a mud analogue to sandy, shallow-marine deposits. *J. Sediment. Petrol.* **55**, 633–652.

SAVRDA, C.E. & BOTTJER, D.J. (1987) The exaerobic zone, a new oxygen deficient marine biofacies. *Nature* **327**, 54–56.

SAVRDA, C.E., BOTTJER, D.J. & GORSLINE, D.S. (1984) Development of a comprehensive oxygen-deficient marine biofacies model. Evidence from Santa Monica, San Pedro, and Santa Barbara Basins, California Continental Borderland. *Bull., Am. Assoc. Petrol. Geol.* **68**, 1179–1192.

SEILACHER, A. (1967) Bathymetry of trace fossils. *Mar. Geol.* **5**, 413–428.

SEILACHER, A. (1982) Distinctive features of sandy tempestites. In: *Cyclic and Event Stratification* (Eds Einsele, G. & Seilacher, A.) pp. 333–349. Springer-Verlag, New York.

STRONACH, N.J. (1984) Depositional environments and cycles in the Jurassic Fernie Formation, southern Canadian Rocky Mountains. In: *The Mesosoic of Middle North America* (Eds Stott, D.F. & Glass, D.J.). Can. Soc. Petrol. Geol. Mem. 9, 43–67.

SWAGOR, N.S., OLIVER, T.A. & JOHNSON, B.A. (1976) Carrot Creek Field, central Alberta. In: *Sedimentology of Selected Clastic Oil and Gas Reservoirs in Alberta* (Ed. Lerand, M.M.), Canadian Society of Petroleum Geologists, 4th Core Conference, pp. 78–95.

SWANSON, R.G. (1981) *Sample Examination Manual.* Am. Assoc. Petrol. Geol., Methods in Exploration Ser., Tulsa, OK, 35 pp.

SWIFT, D.J.P. & FIELD M.E. (1981) Evolution of a classic ridge field, Maryland section, North American Inner Shelf. *Sedimentology* **28**, 461–482.

SWIFT, D.J.P. & RICE, D.D. (1984) Sand bodies on muddy shelves—a model for sedimentation in the Cretaceous Western Interior Seaway, North America. In: *Siliciclastic Shelf Sediments* (Eds Tillman, R.W. & Siemens, C.T.). Soc. Econ. Paleontol. Mineral. Spec. Publ. 34, 43–62.

SWIFT, D.J.P., FIGUEIREDO, A.G., JR., FREELAND, G.L. & OERTEL, G.F. (1983) Hummocky cross-stratification and megaripples. A geological double standard. *J. Sediment. Petrol.* **53**, 1295–1317.

SWIFT, D.J.P., HAN, G. & VINCENT, C.E. (1986) Fluid processes and sea-floor response on a modern storm-dominated shelf: middle Atlantic shelf of North America. Part 1: the storm-current regime. In: *Shelf Sands and Sandstones* (Eds Knight, R.J. & McLean, J.R.). Can. Soc. Petrol. Geol. Mem. 11, 99–119.

TURNER, J.R. & CONGER, S.J. (1981) Environment of deposition and reservoir properties of the Woodbine sandstone at Kurten field, Brazos County, Texas. *Trans. Gulf Coast Assoc. Geol. Soc.* **31**, 213–232.

TURNER, J.R. & CONGER, S.J. (1984) Environment of deposition and reservoir properties of the Woodbine Sandstone at Kurten Field, Brazos Co., Texas. In: *Siliciclastic Shelf Sediments* (Eds Tillman, R.W. & Siemers, C.T.). Soc. Econ. Paleontol. Mineral. Spec. Publ. 34, 215–249.

VINCENT, C.E. (1986) Processes affecting sand transport on a storm-dominated shelf. In: *Shelf Sands and Sandstones* (Eds Knight, R.J. & McLean, J.R.). Can. Soc. Petrol. Geol. Mem. 11, 121–132.

WALKER, R.G. (1983a) Cardium Formation 2. Sand-body geometry and stratigraphy in the Garrington–Caroline–Ricinus area, Alberta—the 'ragged blanket' model. *Bull. Can. Petrol. Geol.* **31**, 14–26.

WALKER, R.G. (1983b) Cardium Formation 1. 'Cardium, a turbidity current deposit' (Beach, 1955): a brief history of ideas. *Bull. Can. Petrol. Geol.* **31**, 205–212.

WALKER, R.G. (1983c) Cardium Formation 3. Sedimentology and stratigraphy in the Garrington–Caroline area, Alberta. *Bull. Can. Petrol. Geol.* **31**, 213–230.

WALKER, R.G. (1984) Cardium Formation 4. Review of the facies and depositional processes in the southern Foothills and Plains, Alberta, Canada. In: *Shelf Sands and Sandstone Reservoirs* (Eds Tillman, R.W., Swift, D.J. & Walker, R.G.). Soc. Econ. Paleontol. Mineral. Short Course 13, 1–49.

WALKER, R.G. (1985) Cardium Formation at Ricinus Field, Alberta: a channel cut and filled by turbidity currents in Cretaceous Western Interior Seaway. *Bull., Am. Assoc. Petrol. Geol.* **69**, 1963–1981.

WALLACE-DUDLEY, K.E. (1981) Oil pools of Western Canada. *Geol. Surv. Can.* Map No. 1559A.

WALTON, A.W., BOUQUET, D.J., EVENSON, R.A., ROFHEART, D.H. & WOODY, M.D. (1986) Characterization of sandstone reservoirs in the Cherokee Group (Pennsylvanian, Desmoinesian) of Southeastern Kansas. In: *Reservoir Characterization* (Eds Lake, L.W. & Carroll, H.B., Jr.) pp. 39–62. Academic Press, Orlando, FL.

WELLS, J.T. & COLEMAN, J.M. (1981) Physical processes and fine-grained sediment dynamics, coast of Surinam, South America. *J. Sediment. Petrol.* **51**, 1053–1068.

WELLS, J.T. & KEMP, G.P. (1984) Interaction of surface waves and cohesive sediments: field observations and geologic significance. In: *Lecture Notes on Coastal and Estuarine Studies*, Vol. 14, *Estuarine Cohesive Sediment Dynamics* (Ed. Mehta, A.J.) pp. 43–65. Springer-Verlag, New York.

WHEELER, H.E. & MALLORY, V.S. (1956) Factors in lithostratigraphy. *Bull., Am. Assoc. Petrol. Geol.* **40**, 2711–2723.

WIGHTMAN, D.M., PEMBERTON, S.G. & SINGH, C. (1987) Depositional modelling of the Upper Mannville (Lower Cretaceous), east central Alberta. Implications for the recognition of brackish water deposits. In: *Reservoir Sedimentology* (Eds Tillman, R.W. & Weber, K.J.). Soc. Econ. Paleontol. Mineral. Spec. Publ. 40, 189–220.

WRIGHT, L.D., WISEMAN, W.J., BORNHOLD, B.D., PRIOR, D.B., SUHAYDA, J.N., KELLER, G.H., YANG, Z.-S. & FAN, Y.B. (1988) Marine dispersal and deposition of Yellow River silts by gravity-driven underflows. *Nature* **332**, 629–632.

WRIGHT, M.E. & WALKER, R.G. (1981) Cardium Formation (Upper Cretaceous) at Seebee, Alberta—storm transported sandstones and conglomerates in shallow marine depositional environments below fair-weather wave-base. *Can. J. Earth Sci.* **18**, 795–809.

# Truncated prograding strandplain or offshore sand body?—sedimentology and geometry of the Cardium (Turonian) sandstone and conglomerate at Willesden Green field, Alberta

D. A. W. KEITH

*Norcen Energy Resources, Calgary, Alberta, T2P 2X7, Canada*

## ABSTRACT

Study of over 1200 well-logs and over 175 cores in and around the Willesden Green field area has allowed recognition in the Cardium Formation of eight facies, mapping of their distribution and large-scale lateral relationships. The Cardium Formation at Willesden Green is an upward-coarsening succession of offshore to lower shoreface mudstones, siltstones and very fine to fine-grained sandstones. The sandstones are informally referred to as the Cardium A sandstone and are unconformably overlain by chert-pebble conglomerates, which are in turn overlain by marine mudstones.

The upper surface of the Cardium A sandstone is undulatory in nature and at Willesden Green forms a convex 'high' with relief of about 30 m. A 'low' separates Willesden Green from the adjacent Ferrier field. Sandstones are restricted to highs whereas siltstones and mudstones typify interfield lows.

The Cardium A sandstones at Willesden Green are subdivided into five upward-coarsening successions termed depositional units, up to three of which are present in any one well. Based on relationships with overlying and underlying geophysical well-log markers, the depositional units demonstrate an eastward offlapping geometry. This offlapping geometry can be demonstrated in Ferrier and other adjacent Cardium fields.

Previous studies have interpreted some Cardium fields as offshore sand bodies. However, in light of the offlapping sandstone geometry within Willesden Green, the relationship with adjacent Cardium fields, and recent studies by others, an alternative interpretation seems appropriate.

The offlapping depositional units may represent lower shoreface to offshore expressions of prograding strandplains developed during falling sea-level. The relatively thin nature of the Cardium A sandstones suggests that the rate of sea-level fall was greater than subsidence and during basinward progradation of the strandplain, subaerial erosion of previously deposited Cardium sandstones would have occurred. During sea-level lowstand and subsequent sea-level rise all remaining upper shoreface, foreshore and subaerial deposits were eroded. Erosion was not uniform, as is evident in the region between the Willesden Green and Ferrier fields, where even lower shoreface sediments were eroded.

Conglomerates that unconformably overly the sandstones may have been introduced to the foreshore during sea-level lowstand. Erosion and reworking of foreshore deposits during stillstand and rising relative sea-level would concentrate the conglomerate. With rising sea-level, southeast-flowing currents accumulated the conglomerate as NW–SE oriented pods in the topographic lows within the upper surface of the Cardium A sandstone.

## INTRODUCTION

Willesden Green field is situated in the Plains of Alberta, east of the folded and thrusted Foothills Disturbed Belt and the Canadian Rocky Mountains (Fig. 1). It has an area of 508 km² and is the second largest Cardium Formation oil–gas field after Pembina. Pembina is the largest conventional oil field in Canada. The reservoir, discovered in 1954, is the Cretaceous (Turonian) Cardium Formation, which occurs at depths of 1600–2100 m within the study area. Cretaceous strata at Willesden Green dip regionally southwest at about 0.65°.

Numerous sedimentological studies have been

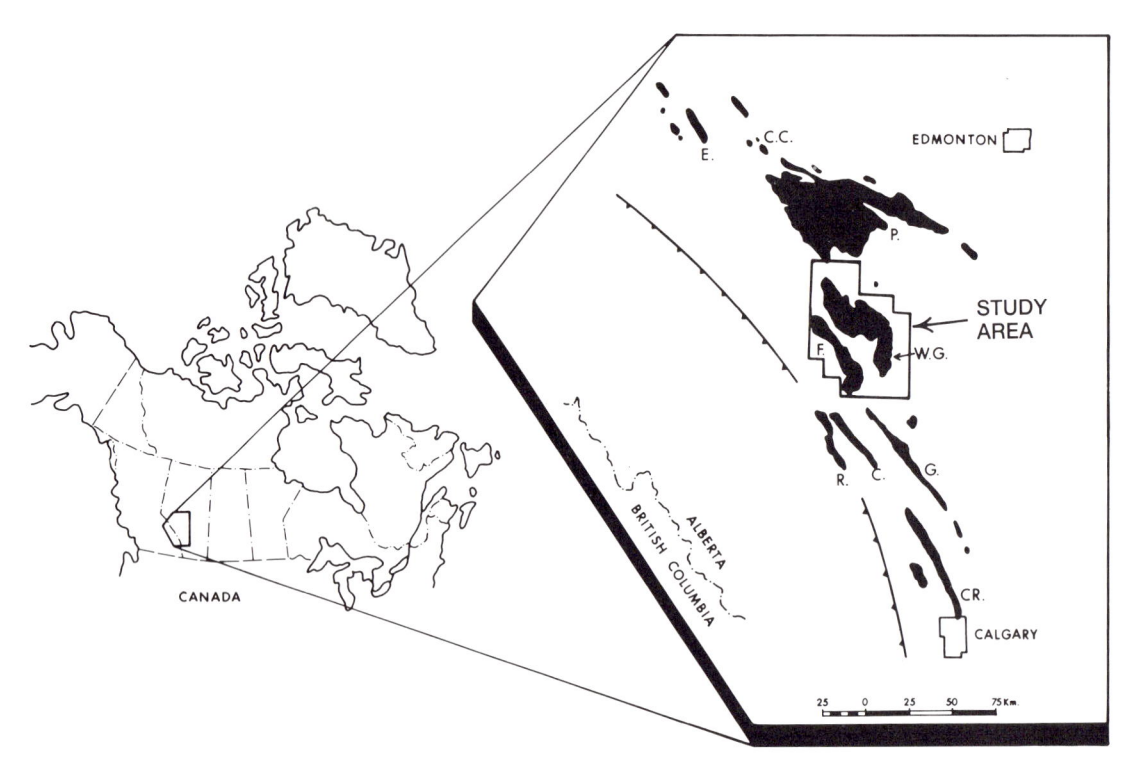

**Fig. 1.** Location of the study area in southwestern Alberta. Locations of Calgary and Edmonton and the approximate eastern limit of the disturbed belt of the Rocky Mountains are shown. Cardium oil and gas fields in the region are shown: Edson (E), Carrot Creek (CC), Pembina (P), Ferrier (F), Willesden Green (WG), Ricinus (R), Caroline (C), Garrington (G), and Crossfield (CR).

conducted on Cardium fields in west- and south-central Alberta: Ferrier, a long and relatively narrow field (Griffith, 1981); Pembina, covering an area of over 3000 km² (Krause & Nelson, 1984; Nielsen & Porter, 1984; Krause *et al.* 1987; and many others); and Caroline and Garrington, both long and narrow (Berven, 1966; Walker, 1983a,b,c).

Willesden Green is somewhat elongate, like the adjacent Ferrier, Caroline and Garrington fields, 65 km in maximum length, but is much wider than these fields, measuring 21 km in maximum width. Thus it is intermediate between the immense Pembina field and the more elongate and narrow Ferrier, Caroline and Garrington fields (Fig. 1).

Previous sedimentological interpretations of these fields have been varied. Cardium sediment transport agents have been seen as turbidity currents and storm-wave dominated currents. Beach (1955) first suggested that the sandstones and conglomerates were turbidity current deposits. Walker (1983a,b,c, 1984, 1985a,b) revived this model in suggesting that deposits at Caroline, Garrington and Ricinus originated by turbidity-current processes in relatively deep water far offshore. Krause (1982, 1983), Krause & Nelson (1984) and Krause *et al.* (1987) suggested episodic offshore transport and reworking related to coastal storm currents in the Pembina field. They referred to the Cardium Formation deposits at Pembina as tempestites (Ager, 1973). More emphasis has been placed recently on the influence of relative sea-level fluctuations to explain geometric field relationships and facies distribution (e.g. Plint *et al.*, 1986).

The incentive for this study has been provided by the large amount of available data on the Willesden Green field, which includes more than 800 logged wells within Willesden Green; about 576 of these have been cored.

The objectives of this paper are: (1) to define the facies and describe the distribution of the coarser grained facies—sandstones and chert-pebble conglomerates; (2) to determine the geometry of the top surface of the coarse-grained facies using overlying and underlying wire-line well-log markers; (3) to relate the distribution of the coarse-grained facies to this surface; (4) to document the internal stratigraphy of the Willesden Green sandstone bodies, and relate the coarse sediment distribution, particularly the conglomerates, to this internal geometry; (5) to determine the regional significance of Willesden Green's internal geometry, and compare it with the adjacent Ferrier field; and (6) to provide an interpretation for the development of these Cardium Formation fields.

## METHODOLOGY

All wire-line well-logs within the study area were examined (over 1400 well-logs). About 31% of the total core within the study area was examined and described with graphic logs to define the facies present within the study area. In an additional 66% of the cored wells not examined, conglomerate thickness values from commercial core analyses were used.

## STRATIGRAPHY

The Cardium Formation is overlain by mudstones and sandstones of the Wapiabi Formation and underlain dominantly by mudstones of the Blackstone or Kaskapau formations. These formations comprise the Upper Cretaceous Alberta Group.

Workers in the Canadian petroleum industry generally refer to the Cardium sandstone at Willesden Green, and in surrounding fields, as the Cardium 'A' sandstone, a practice that is informally perpetuated here. The equivalent sandstone is called the Raven River Member in the Ricinus, Caroline and Garrington fields (Walker, 1983c). Krause & Nelson (1984) refer to the Cardium A sandstone in the Pembina field as the Pembina River Member (Fig. 2).

At Willesden Green the sequence that coarsens upward from mudstones into sandstones is informally referred to as the A sandstone sequence (Fig. 2). The term *sequence* is used in this paper in a generic sense to refer to a succession of facies, and not in a sequence stratigraphic sense as defined by Mitchum *et al.* (1977). The uppermost sandstones within this sequence are truncated by a regionally extensive erosion surface (Krause & Collins, 1984; Keith, 1985; Plint *et al.*, 1986; Krause *et al.*, 1987).

Chert-pebble conglomerates locally lie upon this erosion surface. The overall upward-coarsening sequence from mudstones to conglomerates is referred to as the A sequence. Mudstone facies that lie upon the A sandstone and conglomerates comprise the Cardium Zone Member in Pembina (Krause & Nelson, 1984) and the Dismal Rat Member elsewhere (Plint *et al.*, 1986). Figure 2 shows a composite core litholog from the Willesden Green field compared with similar lithologs from Garrington and Pembina showing the terminology of Krause & Nelson (1984) and Plint *et al.* (1986).

## FACIES DESCRIPTION

Cardium facies recognized by workers in various Cardium fields outside of the study area are summarized by Krause (Table 1, this volume).

In the Willesden Green field, eight facies are recognized within and overlying the A sequence. Two consist of mudstone, three of mixed bioturbated mudstone, siltstone and sandstone, two of dominantly sandstone and one conglomerate facies. Preliminary interpretations relate only to the described sedimentological features and are expanded in a subsequent section.

### Facies 1: massive dark mudstones

These mudstones are dark grey to black, and lack distinct silty or sandy laminae (Fig. 3). Individual trace fossils are not discernible but there is an overall 'stirred or mottled' aspect to the rock, suggesting pervasive bioturbation. Fragments of *Inoceramus* sp. shells are present.

### Facies 2: dark bioturbated muddy siltstones

This facies gradationally overlies facies 1, showing an increase in the percentage of fine-grained sand, up to about 20%. The facies is highly bioturbated (Fig. 4), but there is insufficient sand for the development and recognition of distinct burrow forms. A few thin very fine-grained sandstone laminae are partially preserved or may show soft-sediment deformation.

*Interpretation*

The muddy facies 1 and 2 represent quiet deposition of fines on the basin floor. The absence of sandstone beds and the absence of wave ripples suggest deposition below fair-weather wave base and

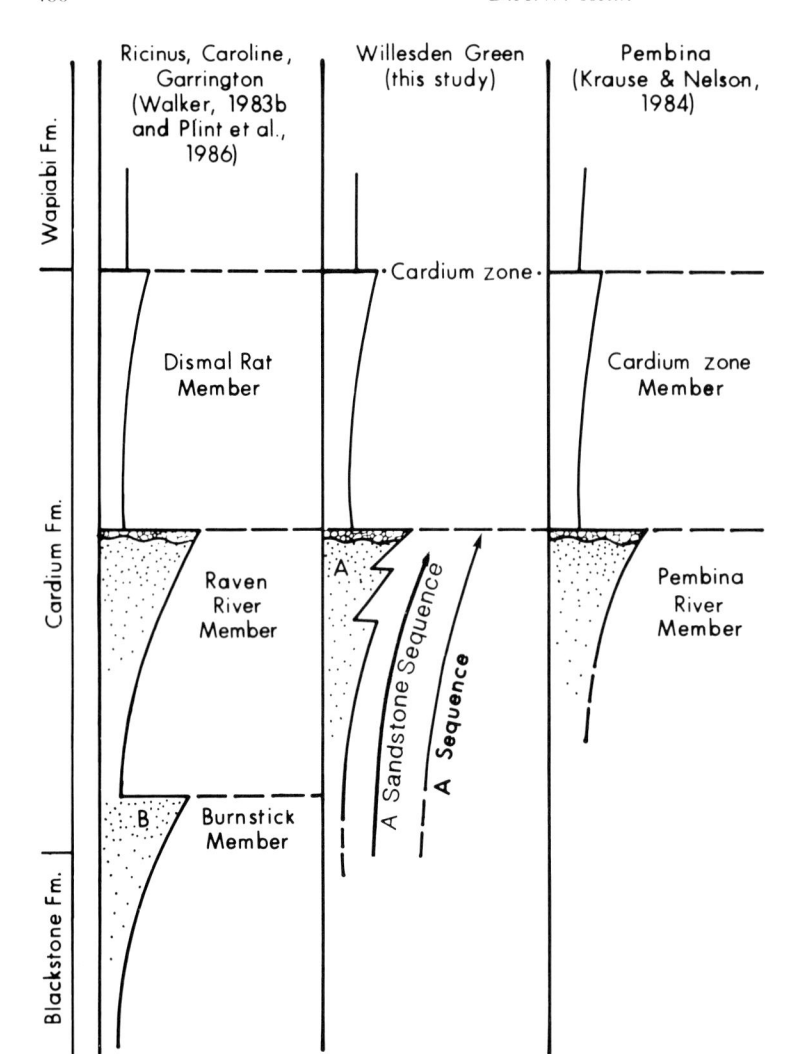

**Fig. 2.** Cardium Formation stratigraphy at Willesden Green and adjacent fields. The figure is not drawn to scale and the coarsening-upward trend of the succession is indicated by the right-hand curvature of the solid line.

possibly even beneath the wave base of the most intense storms. The regional extent of facies 1 (Plint *et al.*, 1986) strongly suggests an extensive, quiet sea-floor removed from any coarse siliciclastic input, and therefore probably relatively deep.

**Facies 3: pervasively bioturbated muddy sandstones**

This facies gradationally overlies facies 2. It contains about 20–50% very fine- to fine-grained sandstone and is characterized by pervasive bioturbation (Fig. 5). Rare fine-grained sandstone beds thicker than about 5 cm may be preserved, but thinner beds are totally bioturbated. Ichnofauna include *Zoophycos*, *Teichichnus*, *Skolithos*, *Planolites* and *Palaeophycus*.

**Facies 4: bioturbated sandstones**

This facies (Fig. 6) gradationally overlies facies 3. Very fine- to fine-grained sandstone comprise 50–90% of the facies, the remainder is composed of siltstone and mudstone (Keith, 1985). With the increased sandstone content relative to facies 1, 2 and 3, there is a larger range of recognizable ichnofauna, including *Zoophycos*, *Rhizocorallium*, *Coneichnus*, *Palaeophycus*, *Rosselia*, *Skolithos*,

**Fig. 3.** Facies 1, massive dark mudstones from the base of the A sandstone sequence. Note absence of lamination and rare patchy occurrence of siltstone (lighter colour). From 15-27-43-7W5, 6024 ft (1836.1 m). Scale in centimetres. Compare with Fig. 13.

**Fig. 4.** Facies 2, dark bioturbated muddy siltstones. Note the churned appearance with only a few partially preserved siltstone and fine-grained sandstone laminae. Some soft-sediment deformation also may be seen (see arrows). This facies is interpreted as episodically deposited silts and very fine-grained sands on a dominantly muddy lower shoreface to inner shelf setting. From 6-18-42-6W5, 6357 ft (1937.6 m). Scale in centimetres. Compare with Figs 3, 13.

*Chrondrites* and *Ophiomorpha*. Sandstone beds thicker than 5 cm are more common in this facies than facies 3. They have sharp bases overlain by horizontal or low-angle inclined lamination. Ripple cross-lamination, of wave origin, commonly caps the sandstone beds. The wave interpretation is based on the symmetrical undulatory laminations with continuity of laminae across the ripple troughs and crests.

*Interpretation*

The pervasively bioturbated muddy sandstones and bioturbated sandstones (facies 3 and 4) show sandstone beds up to 20 cm in thickness with varying degrees of bioturbation. Activity by organisms obliterated most of the primary sedimentary structures. The trace fossil assemblage is characteristic of

**Fig. 5.** Facies 3, pervasively bioturbated muddy sandstones. Note the *Zoophycos* ichnofossil (arrow) and the churned appearance of the sandstones. From 4-9-43-7W5, 1941.8 m. Scale in centimetres.

**Fig. 6.** Facies 4, bioturbated sandstones. Note the thoroughly churned appearance of the sandstones with no preservation of original laminae. From 16-35-42-7W5, 6334 ft (1930.6 m). Scale in centimetres.

the *Cruziana* ichnofacies, indicating environments 'below daily wave base but not storm wave base, to somewhat offshore conditions' (Frey & Pemberton, 1984, p. 193). The Cardium Formation in some outcrop localities is also characterized by the *Cruziana ichnofacies* as described by Pemberton & Frey (1984).

### Facies 5: interbedded sandstones and silty mudstones

Facies 5 commonly occurs above facies 4 and below facies 6. It consists of interbedded sandstones about 3–8 cm thick, alternating with silty mudstones 2–5 cm thick (Fig. 7). Although the proportion of sand is much the same as facies 4, about 50–90%, facies 5 differs in containing only rare isolated burrows with no significant bioturbation. The sandstone beds have sharp bases, are overlain by horizontal or low-angle (less than 10°) lamination. A few beds contain lamination dipping steeper than 15°, in sets up to 15 cm thick. Ripple cross-lamination probably of wave origin, commonly caps the sandstone beds.

### Facies 6: thickly bedded sandstones

This facies forms the main reservoir at Willesden Green. It consists of individual very fine- to fine-

**Fig. 7.** Facies 5, interbedded sandstones and mudstones. A 5-cm lower sandstone and 3-cm upper sandstone bed is separated by a 3-cm mudstone bed. The lower sandstone shows a symmetrical wave ripple profile, interpreted as wave ripples, overlain by subhorizontal sandstone laminae. Note the sharp tops on both sandstone beds and the loading of the upper sandstone bed into the underlying mudstones. These are interpreted as episodically emplaced sands during waning storm flow. From 15-5-43-6W5, 6000 ft (1828.8 m). Scale in centimetres.

**Fig. 8.** Facies 6, thickly bedded sandstones. Note the low-angle inclined laminations truncated and overlain by laminae dipping in the opposite direction. The laminae flatten upward into 5 cm of bioturbated sandstone capped with interlaminated sandstone and mudstone. This is interpreted as hummocky cross-stratified sandstone, the top of which has been bioturbated. From 16-13-42-7W5, 6566 ft (2001.3 m). Scale in centimetres.

grained sandstone beds 10 cm to several metres in thickness, interbedded with very thin mudstone partings (Figs 8, 9). The beds show sharp bases and tops, and contain horizontal or low-angle inclined laminae with dips up to about 15°. There are slight and subtle changes in the dip angles upward through the beds (Fig. 8), and the structure is interpreted as hummocky cross-stratification. Hummocky cross-stratification is the most abundant large-scale sedimentary structure found in this facies. Angle of repose cross-sets more than a few centimetres thick are very rare. The tops of beds commonly show bioturbation or wave-ripple cross-lamination (Figs 8, 9, respectively).

**Fig. 9.** Facies 6, thickly bedded sandstones. Horizontally laminated sandstones truncated by a symmetrically undulating surface (see arrow), which is overlain by 6 cm of wave-rippled sandstone. The upper surface of the sandstone is truncated and capped with mudstone. This is interpreted as an upper plane-bed horizontally laminated sands, the top of which has been reworked into wave ripples, probably during waning storm flow. From 15-5-43-6W5, 6002 ft (1829.4 m). Scale in centimetres.

**Fig. 10.** Facies 7, conglomerate—clast supported. This chert-pebble clast supported conglomerate shows a hint of regular clast grading (5 cm above base to top), although this is rare in the majority of cores examined. Note the angular sandstone clast (2.5 cm by more than 8 cm; see arrow) suggesting erosive conglomerate emplacement on to an at least semi-lithified sandstone. From 10-23-43-9W5, 6589 ft (2008.3 m). Scale in centimetres.

## Interpretation

The interbedded sandstones and mudstones of facies 5 and 6 suggest a normally quiet muddy substrate below fair weather wave base, over which sand was periodically transported. The sharp-based beds suggest sudden emplacement of sand, and the horizontal or low-angle stratification suggests remoulding of the deposit by storm waves, to form hummocky cross-stratification. The identification of hummocky cross-stratification in core is not always easy, but this structure is ubiquitous in Cardium Formation outcrops (M.E. Wright & Walker, 1981; Krause & Nelson, 1984; Duke, 1985), and there seems no reason to doubt that it occurs in the subsurface as well.

The wave ripples that cap the sandstones indicate

**Fig. 11.** Facies 7, conglomerate—pebbly mudstones. In this matrix supported, pebbly mudstone, note the randomly oriented chert pebbles. Note also the disrupted pebble beds (see arrows). From 10-28-38-3W5, 5451 ft (1616.5 m). Width of core about 10 cm.

**Fig. 12.** Facies 7, conglomerate—thin conglomerate stringers. Thin chert-pebble conglomerate beds interbedded with mudstones of facies 8. Note the distinct bases and tops of the lower two pebble beds; the upper pebble bed appears disrupted, possibly the result of burrowing. From 14-22-41-6W5, 6301.5 ft (1920.7 m). Scale in centimetres.

the waning storm and/or fair-weather reworking of the tops of the beds. The absence of medium-scale angle of repose cross-bedding suggests the absence of strong fair-weather (along shore or tidal currents) or storm currents. It could be argued that medium-scale angle of repose cross-bedding would not form in fine- to very fine-grained sands. Instead, sands of this grain size would show abundant current ripple cross-lamination and upper plane-bed horizontal lamination, both of these structures are uncommon in the sandstone facies.

## Facies 7: conglomerates

Conglomerates range in thickness from a few centimetres to over 5 m in thickness. All are composed of well-rounded chert-pebble clasts that range in size from a few millimetres to 7 cm. There are three main types (listed in relative order of importance); clast supported, matrix supported (pebbly mudstones), and thin stringers within mudstones.

### Clast supported conglomerates

These are common in areas of thickest conglomerate accumulation. Where the thickness of individual beds can be determined in core, it is of the order of tens of centimetres. Stratification within beds is generally absent (Fig. 10), and graded bedding or imbrication is extremely rare. The conglomerates are composed of well-rounded chert pebbles, with occasional mudstone rip-up clasts.

*Pebbly mudstones*

This facies (Fig. 11) occurs where the overall conglomerate thickness is less than 1 m. Muddy matrix makes up between 20 and 80% of the volume. There is only rarely stratification, and generally no preferred pebble orientation. Sideritization of the muddy matrix is common.

*Thin conglomerate stringers*

These tend to overlie the sandstone facies and other conglomerate facies, within the lower metre or so of the overlying facies 8 (described below). The stringers range from one pebble diameter in thickness to several cm (Fig. 12). There is no stratification or preferred pebble orientation within the stringers, but graded bedding can be seen in a few beds.

*Interpretation*

The conglomerates pose a special problem in the Cardium and are considered later, after discussion of their distribution and geometric position.

**Facies 8: laminated dark mudstones**

This facies is distinguished from facies 1 by the presence of thin siltstone to very fine-grained sandstone laminae (Fig. 13). The laminae commonly have sharp bases and diffuse tops, with colour grading, light to dark upward, suggesting a grain size grading. Bioturbation is only minor with traces of the pinworm *Gordia* (G. Pemberton, pers. comm., 1984) the only recognizable form. This facies overlies the sandstones and conglomerates in the Willesden Green field and adjacent area. Its significance is discussed below.

*Interpretation*

The deposition of this predominately mudstone facies signals a return to offshore accumulation of fine-grained sediments on the basin floor. The thin

**Fig. 13.** Facies 8, laminated dark mudstones. Note sharp-based silt and fine sandstone laminations, millimetres to 1 cm thick. The laminae are interpreted as episodically emplaced and the occurrence of discontinuous laminae suggests bioturbation. Note that the degree of bioturbation is much less than that of facies 2 or 3 (Figs 4, 5). From 13-11-43-6W5, 5858 ft (1785.5 m). Scale in centimetres.

siltstone to very fine-grained sandstone laminae may represent distal deposits from gravity flows. The lack of significant bioturbation is markedly different than that of facies 1 and 2. The absence of bioturbation in facies 8 may indicate environmental conditions, such as slightly elevated salinity, hindering colonization by bioturbators.

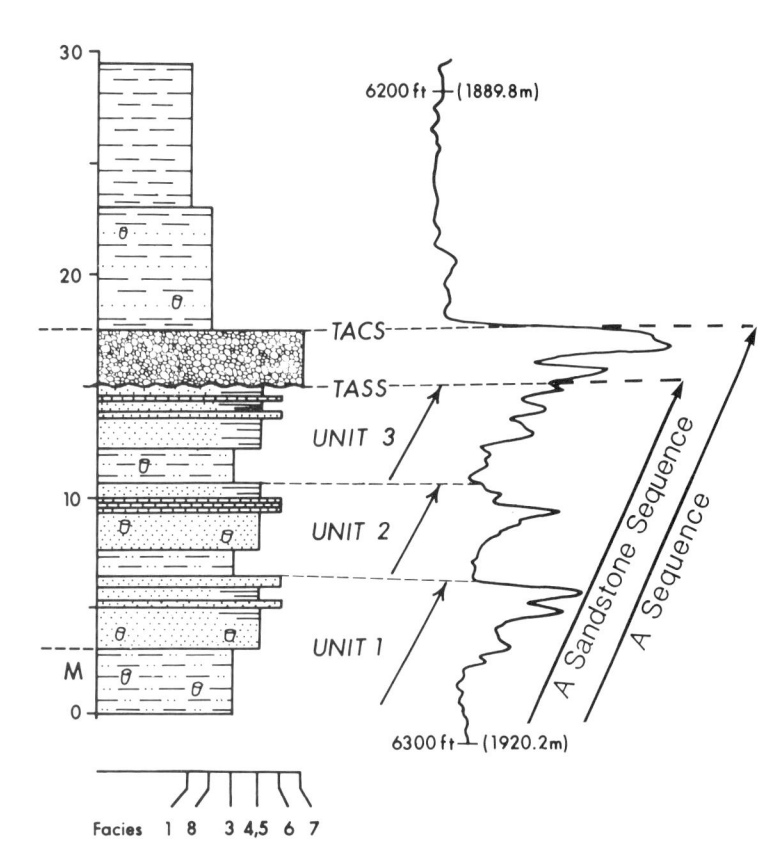

**Fig. 14.** Representative Willesden Green resistivity wire-line well-log and lithology log of the cored interval illustrating characteristic facies sequences and well-log patterns. The Cardium A sandstone sequence is composed of individual upward-coarsening depositional units, which are numbered beginning at the bottom. An unconformity occurs at the 'top of the A sandstone sequence' (TASS). This unconformity is locally overlain by conglomerates. The overall upward-coarsening sequence including the A sandstones and conglomerates is referred to as the A sequence; its top is referred to as the 'top of the A coarse sediments' (TACS).

## VERTICAL FACIES SUCCESSIONS AND THE DEFINITION OF DEPOSITIONAL UNITS

The Cardium Formation deposits in the study area do not form a simple upward progression of facies but, rather, consist of repeated upward-coarsening facies successions, as illustrated in Fig. 14. A resistivity wire-line log and a lithology log of the cored interval, from a representative well in the study area, are shown in Fig. 14. The individual upward-coarsening facies successions that are identifiable in both core and well logs are informally referred to as depositional units. R. Wright (1986) used a similar terminology to describe individual small-scale, asymmetric transgressive–regressive cycles of the overall regressive Point Lookout sandstone of the San Juan Basin, New Mexico, USA.

The depositional units are labelled 1–5, beginning with the stratigraphically lowest. It is shown later that the depositional units offlap towards the northeast, such that a maximum of three depositional units comprise the A sandstone sequence within any individual well.

The base of the A sandstone sequence is comprised of massive dark mudstones (facies 1), and is rarely cored. Stratigraphically lower depositional units coarsen-upwards from facies 1 to facies 2 and gradationally coarsen-upward to facies 4. Sideritized mudstone granules are rarely found at the tops of lower sandstone depositional units. Stratigraphically higher depositional units have facies 3 at their base, gradationally coarsening-upward to facies 5 and/or 6. Thus, stratigraphically higher units generally contain coarser grained sediments than stratigraphically lower depositional units, giving rise to the overall upward-coarsening pattern of the A sandstone sequence. Thicknesses of the facies within the depositional units are extremely variable throughout the study area.

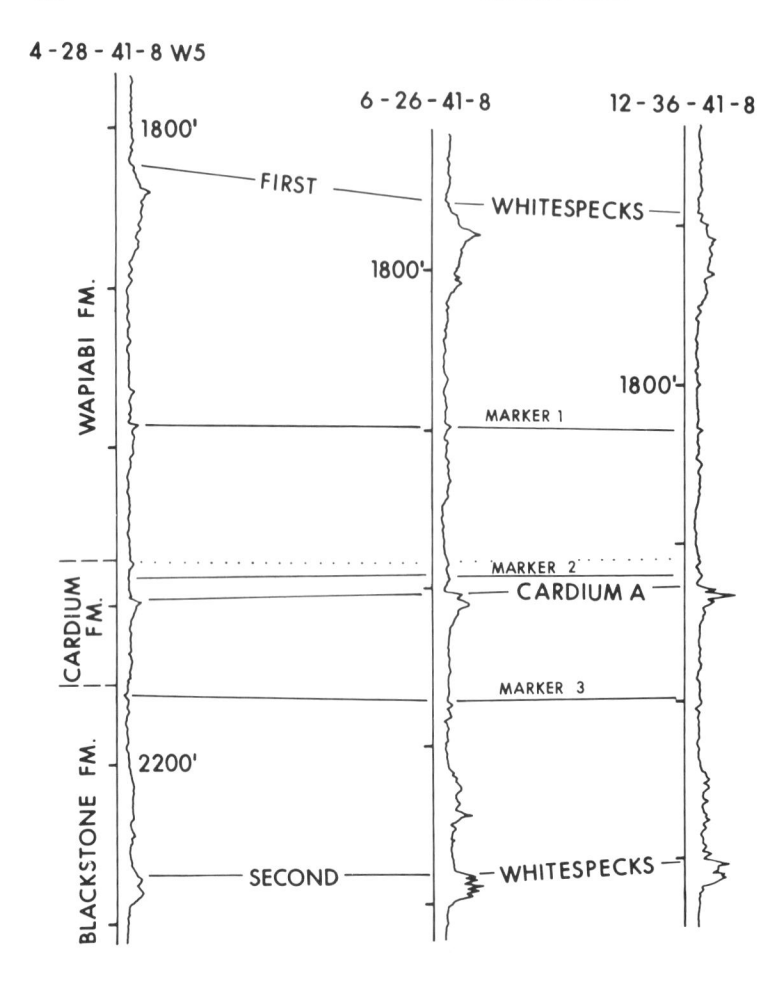

Fig. 15. Stratigraphic cross-section with resistivity wire-line well-logs showing the position of the three datums used in the study. Marker 1 and marker 2 occur above the Cardium A sandstone and marker 3 is situated below the Cardium A sandstone.

The uppermost sandstone within the A sandstone sequence is truncated by an erosion surface, which is referred to as the 'top of the A sandstone surface' (TASS). This regionally extensive surface is locally overlain by chert-pebble conglomerates (facies 7). Its relationship to the overlying conglomerate distribution is investigated below.

The overall upward-coarsening succession, including the A sandstones and locally overlying chert-pebble conglomerates, is referred to as the A sequence. The uppermost surface of the A sequence coarse sediments (i.e. sandstone facies 4, 5 or 6 or conglomerate facies 7) is referred as 'the top of the A sequence coarse sediment' (TACS). The regional geometry of this surface is investigated below. The TACS, whether sandstone or conglomerate, is buried by the laminated mudstones of facies 8. The average thickness of the laminated mudstones in the study area is about 6.3 m (Keith, 1985). This is very similar to the 6.45 m thickness of this facies reported by Walker (1983c) in the Caroline–Garrington Cardium fields, south of the study area and the 7 m average thickness reported by Plint et al. (1986), between the Pembina and Kakwa Cardium fields, northwest of the study area. Massive mudstone (facies 1) overlies the laminated mudstones.

## EXTERNAL GEOMETRY OF WILLESDEN GREEN

Detailed investigation of the TACS in the Willesden Green–Ferrier area reveals a topography on this surface. The topography is determined by the relationship of the TACS to three regionally extensive, high resistivity, wire-line well-log markers. Log marker 1 occurs in mudstones well above the TACS, and marker 3 in mudstones below the A sequence

(Fig. 15). A second marker, marker 2, occurring just above the A sequence is used in cross-sections that are discussed later.

Marker 1 occurs throughout townships 38–44 and ranges 4–10 west of the fifth meridian. Marker 3 can be traced from township 38 to the southern part of township 43. However, in the northern part of the field the Cardium Formation is the drilling target and there is little penetration below the A sequence to marker 3 in this area.

Marker 1 and marker 3 structure maps show no significant deviations from the regional southwesterly dip of Cretaceous strata in the Alberta Basin, suggesting that within the study area, both represent near horizontal surfaces. An isopach of the interval between marker 1 and marker 3 reveals that the two markers are not quite parallel; their separation increases southwestward at about $0.96 \, m \, km^{-1}$ (Keith, 1985).

## Isopach maps

### Marker 1 to TACS

The relative topography on the TACS surface based on the isopach interval between marker 1 and the TACS in 707 wells is shown in Fig. 16 (Keith, 1985, fig. 4.2). To construct Fig. 16, a plane was constructed which parallels marker 1 and passes through the point of maximum separation between the TACS and marker 1 (about 116 m). The inset in Fig. 16 shows the relative position of the TACS above this plane. The point of maximum separation between the TACS and marker 1, which represents the lowest topographic point on this surface, occurs just off the northeastern margin of Ferrier (at 10-31-41-8W5, labelled A, Fig. 16). The point of minimum separation between marker 1 and the TACS (31.7 m), the highest relative topography, occurs within the Willesden Green field (at 6-16-43-8W5, labelled B, Fig. 16). The average slope within the Willesden Green and Ferrier fields between the high and low positions is 1.12°. Between the two fields it is apparent from Fig. 16 that the TACS is relatively low (isopach thick) and is a relative high (isopach thin) within the Willesden Green and Ferrier fields.

It must be emphasized that the construction of Fig. 16 assumes that marker 1 is a near planar surface. As noted above, structure maps constructed for this marker show no significant deviations from the regional structural dip.

### Marker 3 to TACS

As a check on the geometric relationships identified using the upper marker 1 as a datum, a similar map was constructed based on the lower marker, marker 3. The isopach relationship between marker 3 and the TACS was determined using 283 wells (Keith, 1985, fig. 4.4). A very similar topography is shown (Fig. 17). Here, the lowest point (isopach thin) occurs east of Willesden Green (8-13-40-4W5, labelled A, Fig. 17) and the highest point (isopach thick) occurs within the Willesden Green field (6-13-40-5W5, labelled B, Fig. 17). This separation represents a relative topography of about 16 m and a slope of 0.79°.

The locations of the highest and lowest points, and the slopes, differ slightly from those determined in Figure 16. This is because: (1) markers 1 and 3 diverge westward slightly ($0.96 \, m \, km^{-1}$); and (2) Fig. 17 refers only to the southern two-thirds of the area. The topography between the Willesden Green high position and the low between Willesden Green and Ferrier is about 10 m using marker 3. The important point is that the same relative topography is revealed by both maps; Ferrier stands high, there is a relative low between Ferrier and Willesden Green, and Willesden Green stands high. The TACS undulates, as contoured in Figs 16, 17.

## Relationship of sandstone and conglomerate distribution to external geometry

### Sandstone distribution

For the sandstone facies (facies 4–6), thickness values obtained from core data are shown in Fig. 18. These sandstone-thickness isopachs are shown on a topographic surface similar to that of Fig. 16. However, the topographic surface in Fig. 18 is the top of the A sandstone sequence, or TASS (conglomerates have been removed), with respect to marker 1 (see inset in Fig. 18). The sandstone thicknesses were measured in 143 cores and the data points are shown in Keith (1985, Fig. 5.1). There is an even distribution of sandstone-thickness data throughout the field. A linear NW–SE alignment of sandstones may be inferred from the isopachs, with the boundary of the Willesden Green field corresponding closely to the 3 m sandstone isopach. Note the coincidence between topographic highs and regions of thickest sandstone, particularly in the areas identified as A, B and C in Fig. 18. This coincidence

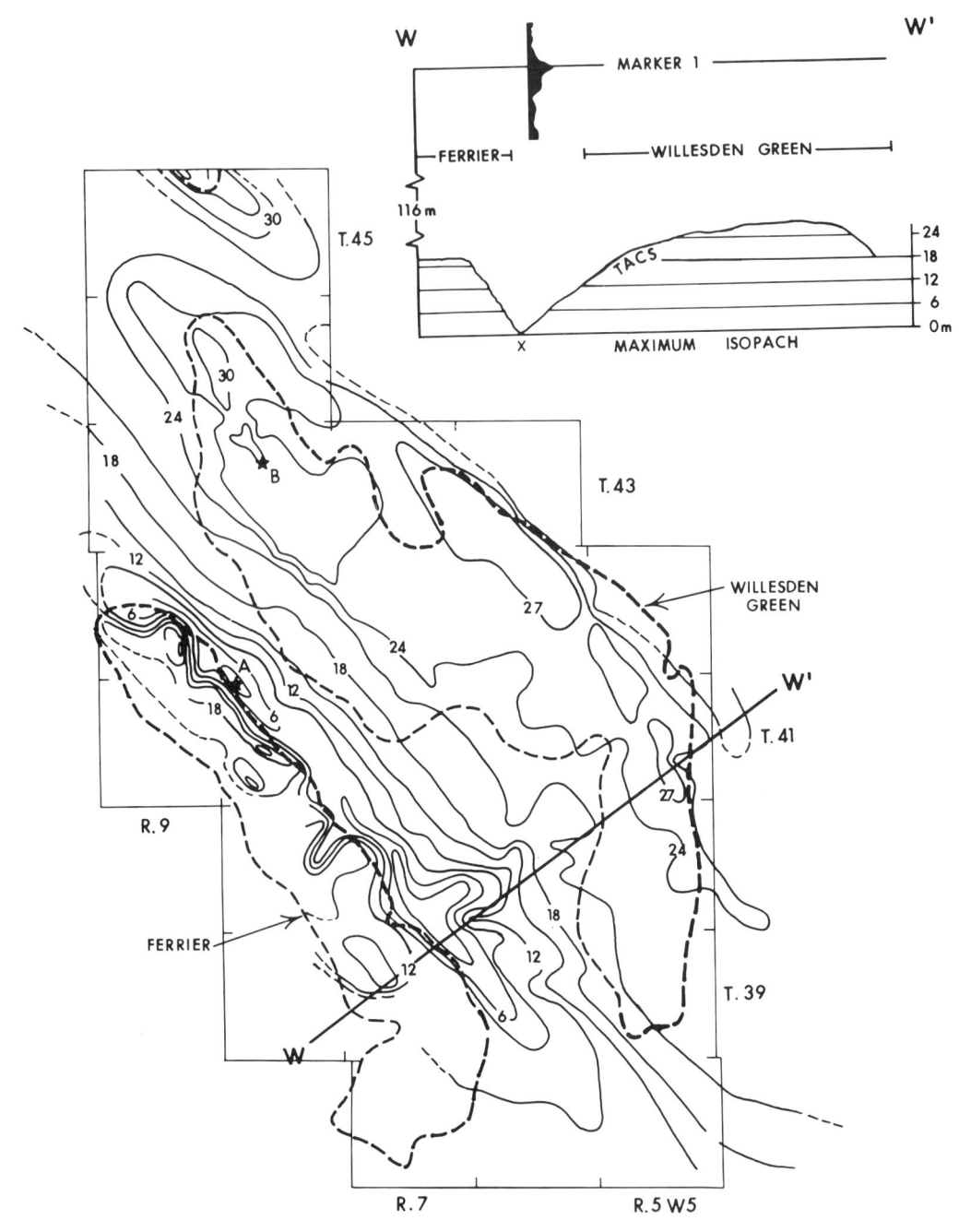

**Fig. 16.** Relative topography on the 'top of the A sequence course sediment' (TACS) based on the isopach of the interval between the TACS and upper marker 1. As cross-section W–W' in the inset illustrates, the interval actually isopached in Fig. 16 is between the TACS and the maximum separation of marker 1 (x, on inset) and the TACS. With respect to upper marker 1, the lowest area on the TACS occurs off the northeastern margin of Ferrier (labelled A) and the highest area on the TACS occurs in northern Willesden Green (labelled B). The field margins of Willesden Green and Ferrier are marked with a dashed line. See text for discussion.

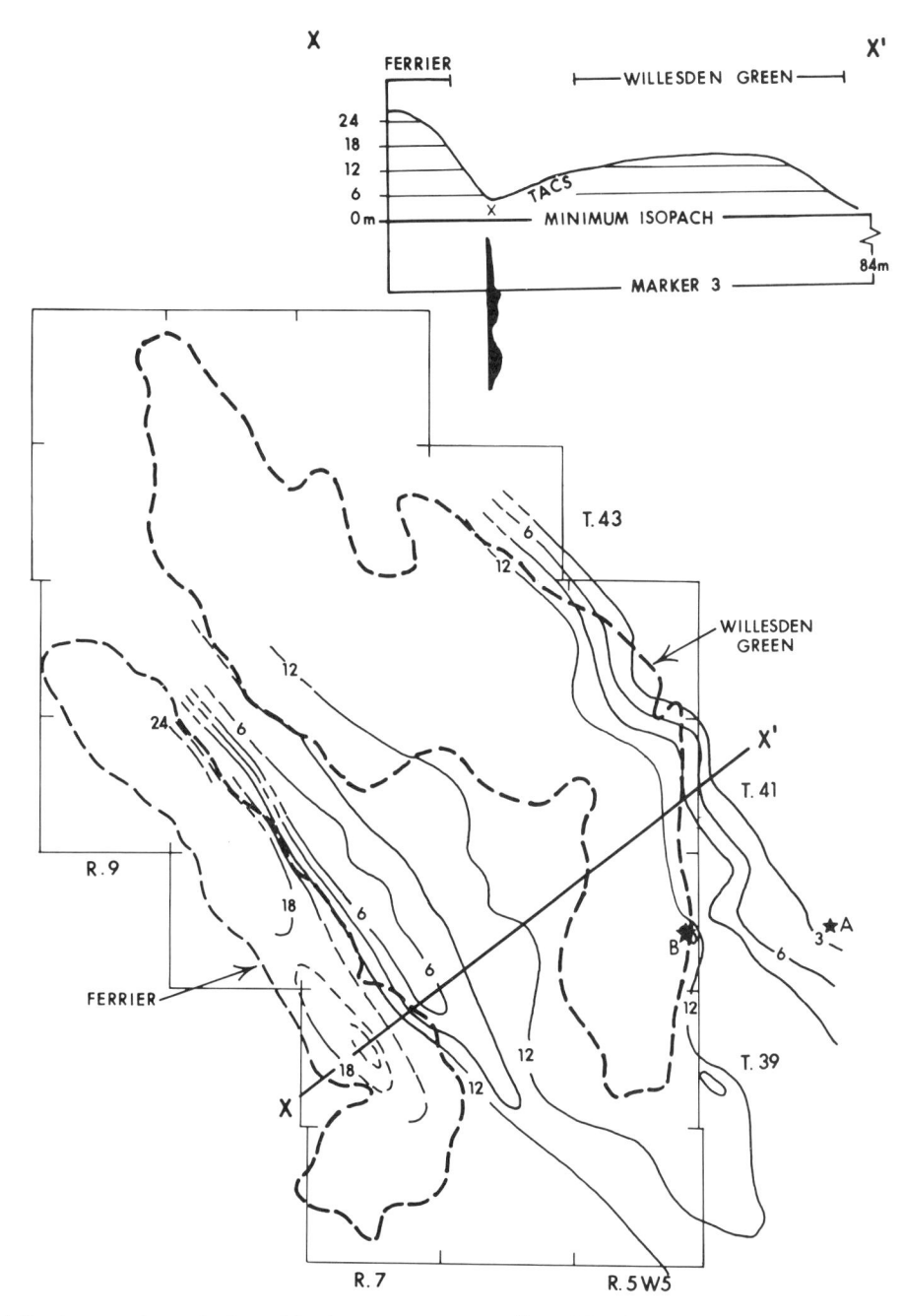

**Fig. 17.** Relative topography on the 'top of the A sequence course sediment' (TACS) based on the isopach of the interval between the TACS and lower marker 3. As cross-section X–X' in the inset illustrates, the interval actually isopached in Fig. 17 is between the TACS and the minimum separation of marker 3 (x, on inset) and the TACS. With respect to lower marker 3, the lowest area on the TACS occurs east of Willesden Green (labelled A) and highest area on the TACS occurs in southeastern Willesden Green (labelled B). The field margins of Willesden Green and Ferrier are marked with a dashed line. See text for discussion.

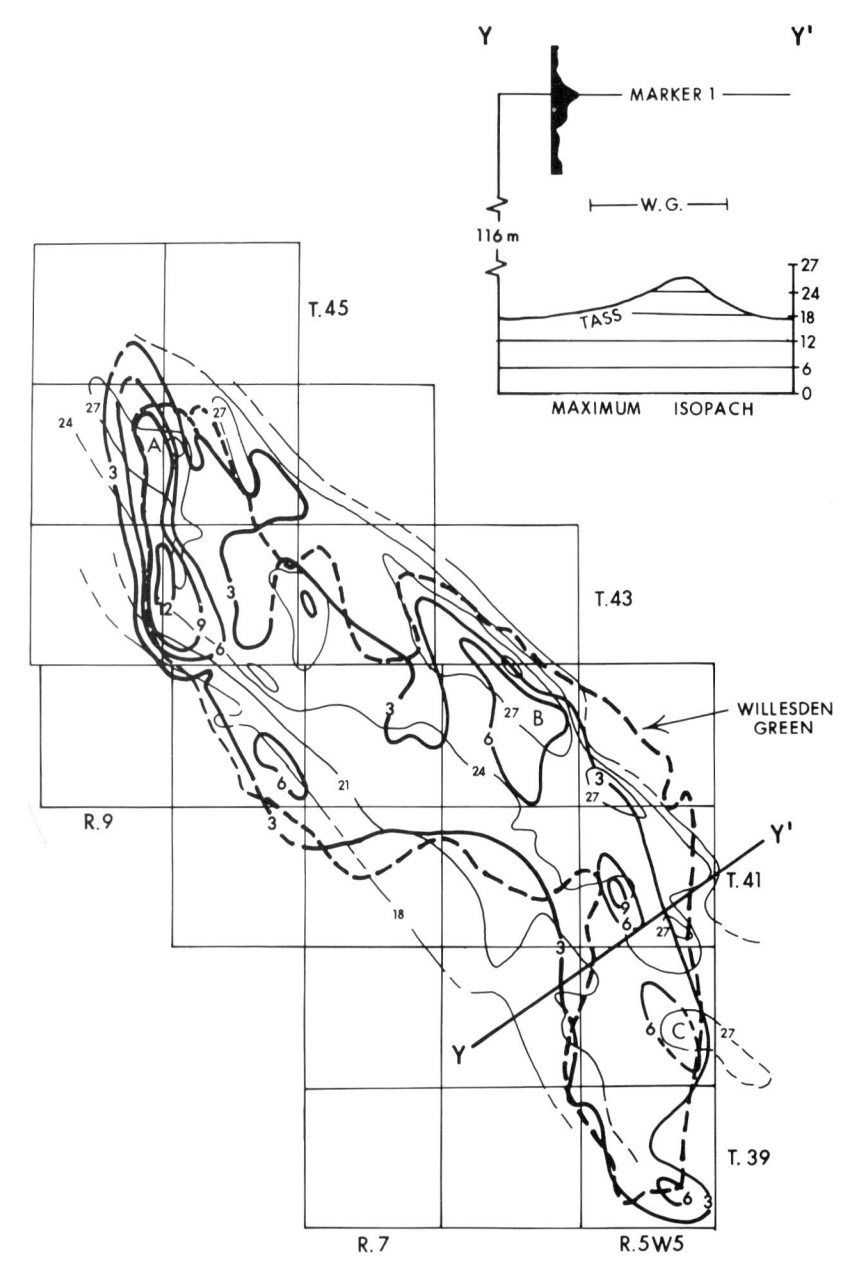

**Fig. 18.** Sandstone facies (facies 4–6) isopach (thick lines) on the relative topography of the 'top of the A sandstone sequence' (TASS). The relative topography of the TASS (thinner lines) was determined in a similar manner to that used in Fig. 16 to determine topography on the TACS (as illustrated in cross-section Y–Y' in the inset), however, here conglomerates that overlie the TASS have been removed. Contour intervals for the TASS and the sandstone are 3 m. Note the coincidence between highs on the TASS and regions of thickest sandstone (see areas labelled A–C). The field margins of Willesden Green and Ferrier are marked with a dashed line. See text for discussion.

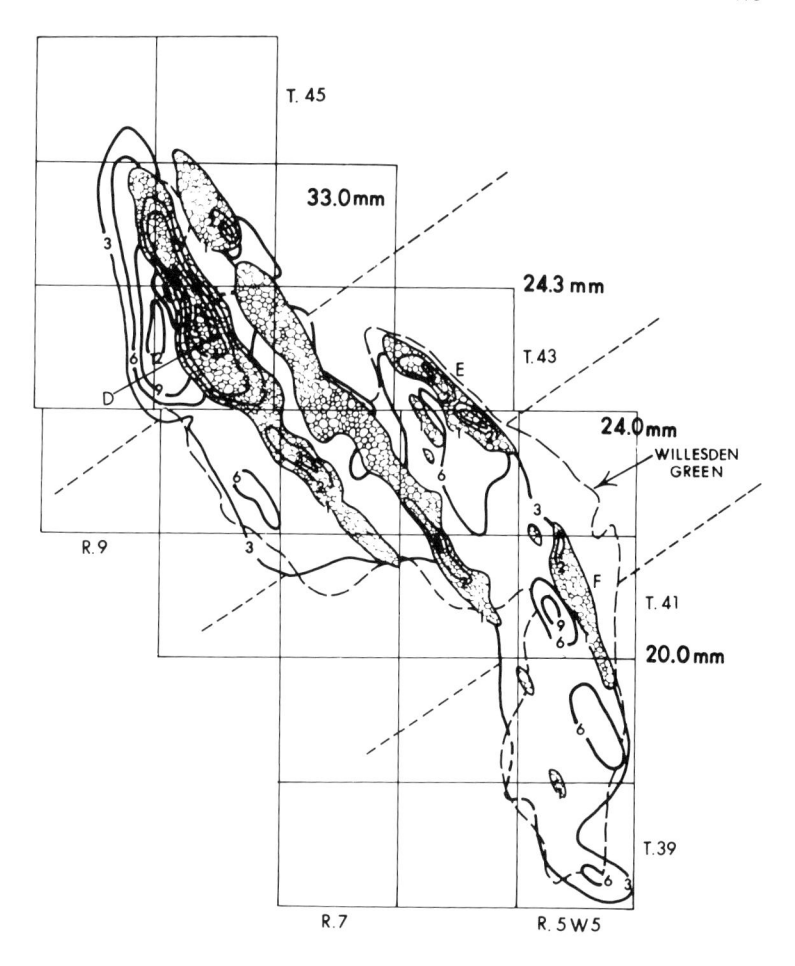

**Fig. 19.** Sandstone facies (facies 4–6) isopach (3 m contour interval) and conglomerate isopach. Conglomerates are highlighted with stippled pattern and are shown with a 1 m contour interval. Note that areas with thick conglomerates tend to occur along the northeast margin of thick sandstones, this is particularly well shown at reference points D–F). Also shown are mean values of chert-pebble long axes (from 160 cores) in each of four approximately equivalently spaced northwest to southeast divisions through Willesden Green. Note the southeastward decrease in mean chert-pebble long axes from 33.0 mm in the northwest to 20.0 mm in southeast Willesden Green. The field margin of Willesden Green is marked with a dashed line.

strengthens the conclusion that the 'highs' determined from isopach relationships of the TACS with upper and lower markers are real, and are composed of the sandstone facies (facies 4–6).

*Conglomerate distribution*

Conglomerate thicknesses were measured in 160 cores. Additional data were collected from 381 wells where commercial core analysis data were available. Good correspondence was observed between conglomerate thicknesses measured in this study and commercial data. The 541 evenly distributed data points are shown in Fig. 19. As is the case in the sandstone distribution, the conglomerates also lie in NW–SE oriented bodies. Maximum thicknesses (greater than 5 m) occur in the northwest part of the field. The conglomerates thin and terminate southeastward.

A plot of maximum chert-pebble diameter from 160 cores shows a progressive decrease southeastward—this is summarized in Fig. 19, where Willesden Green has been divided into four approximately equivalent segments along a NW–SE axis. Mean values of the pebble long axes have been calculated for each segment and range from 33.0 mm in the northwest to 20 mm in the southeast. A similar progressive southeastward decrease in chert-pebble long axes is documented by Griffith (1981, p. 16) in the Ferrier field.

The sandstone contours of Fig. 18 are superimposed on conglomerate thicknesses in Fig. 19. The conglomerate appears to accumulate preferentially along the northeast edge of the thicker sandstones, as shown particularly well at the reference points D–F. This relationship is identical to that documented in other Cardium Formation fields, notably by Swagor *et al.* (1976) and Bergman &

474                                    D.A.W. Keith

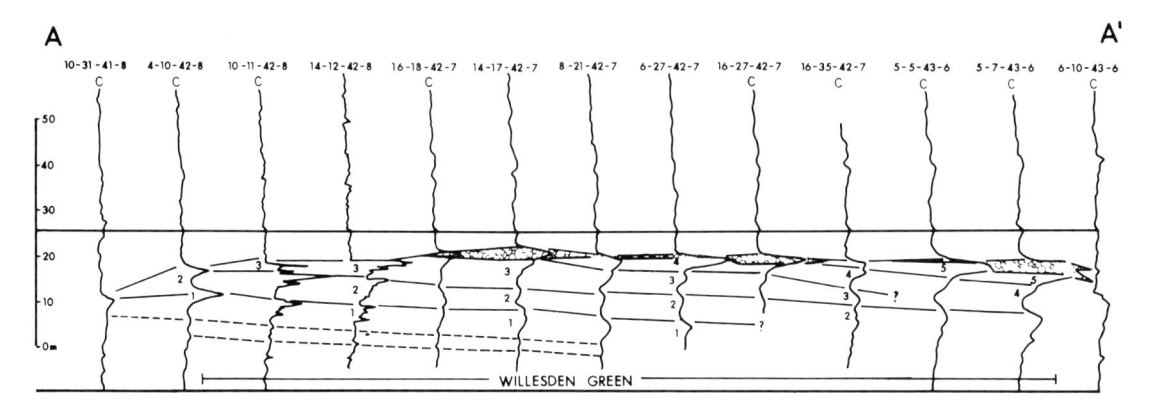

**Fig. 20.** Southwest to northeast stratigraphic cross-section A–A' across northern Willesden Green. See Fig. 22 for location of cross-section, all well locations are west of the fifth meridian. The datum is marker 2 and A sandstone units are numbered from the southeast beginning with 1. Note the successively northeastward offset of offlapping pattern shown by the depositional units. Conglomerates are shown with stippled pattern. Cored wells are indicated by a 'c' below the well location. Vertical exaggeration is about 280×.

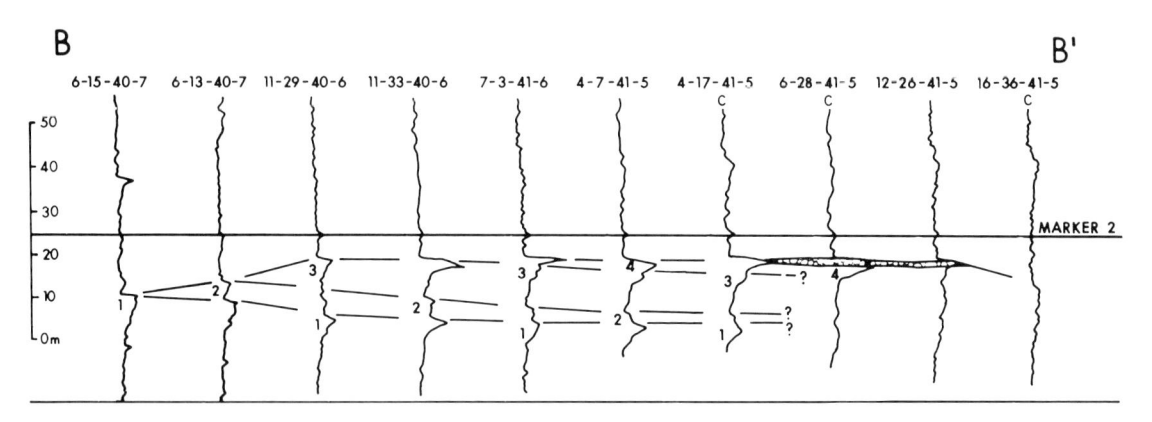

**Fig. 21.** Southwest to northeast stratigraphic cross-section B–B' across southern Willesden Green. See Fig. 22 for location of cross-section, all well locations are west of the fifth meridian. The datum is marker 2 and A sandstone units are numbered from the southeast beginning with 1. Note the successively northeastward offset or offlapping pattern shown by the depositional units. Conglomerates are shown with stippled pattern. Cored wells are indicated by a 'c' below the well location. Vertical exaggeration is about 280×.

Walker (1987) in Carrot Creek; Griffith (1981) in Ferrier; Krause & Collins (1984), and Krause et al. (1987) in Pembina.

## INTERNAL GEOMETRY OF WILLESDEN GREEN

### Willesden Green cross-sections

The detailed internal geometry of Willesden Green has been studied by examination and correlation of all available wire-line well-logs in the field. Individual upward-coarsening depositional units throughout the field were correlated by constructing numerous cross-sections in all orientations. Relationships of the depositional units are particularly well shown in 12 SW–NE cross-sections from Keith (1985, figs 6.1–6.4). These cross-sections are oriented perpendicular to the strike of Willesden Green and its sandstone and conglomerate trends. The 12 cross-sections include a total of 188 well-logs. Marker 2 was used as a datum in these cross-sections since: (1) it represents the transition from mudstone

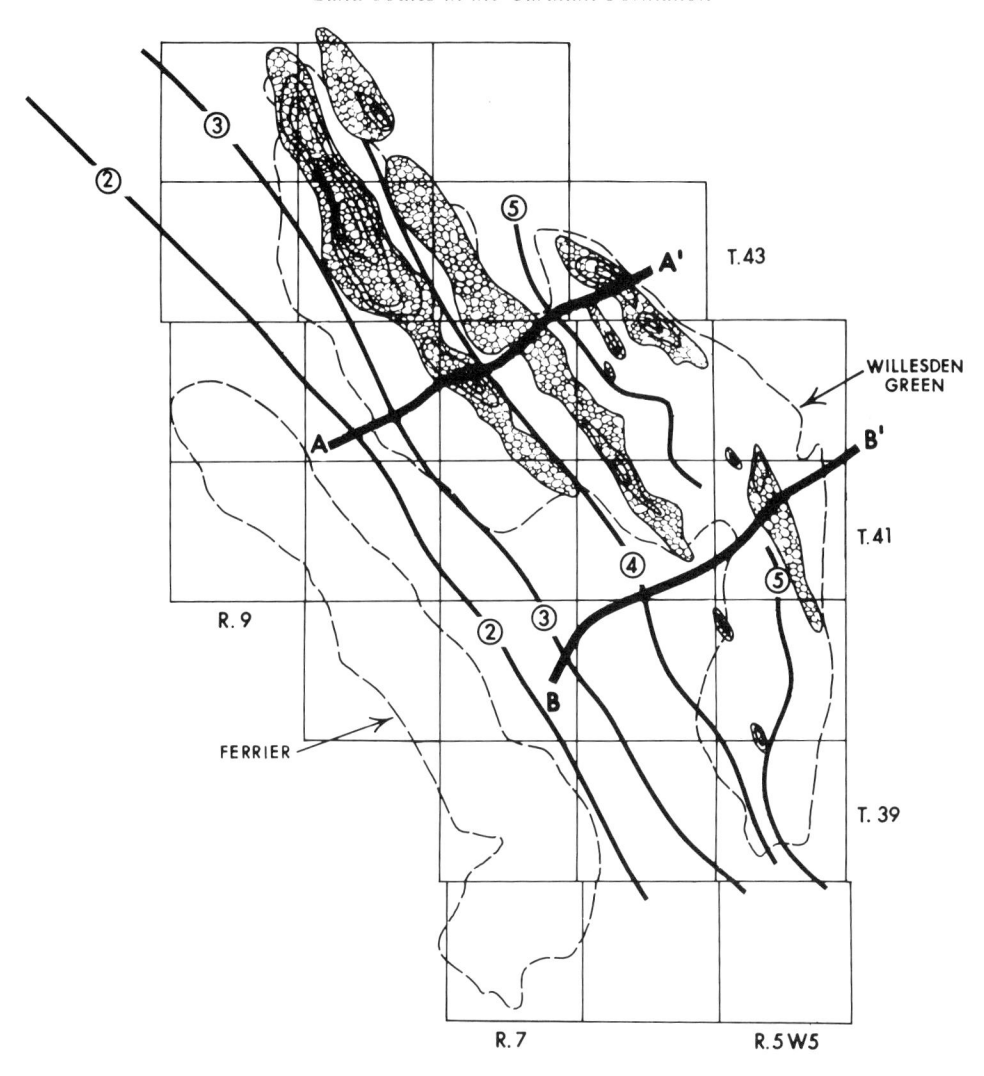

**Fig. 22.** The southwesternmost occurrence or edge of depositional units 2–5 in and adjacent to Willesden Green. Circled numbers indicate which depositional unit the southwest edge represents. Note the NW–SE orientation of depositional units 2–4 and the northern portion depositional unit 5 and the discontinuous nature of depositional unit 5. Also shown is the conglomerate distribution (stippled pattern; 1 m contour interval) and the locations of cross-sections A–A' and B–B'.

facies 8 to mudstone facies 1 and often can be identified in core; (2) within Willesden Green many of the infield wells have expanded scale wire-line logs for limited intervals that include marker 2 but omit markers 1 and 3; and (3) within the study area marker 2 does not show significant relief with respect to upper marker 1 or lower marker 3. Two of the 12 cross-sections from Keith (1985, fig. 6.2, cross-section 6, and fig. 6.4, cross-section 10, respectively) are included here as Figs 20, 21. Conglomerates in the cross-sections are shown by the stippled pattern and cored wells are indicated. Information was extrapolated from adjacent cored wells to wells used in the cross-sections, but which were not cored. Depositional units are labelled on the cross-sections.

*Cross-section A–A'*

The southwestern end of section A–A' (Fig. 20) is located between Willesden Green and Ferrier at 10-31-41-6W5, where there is only a single upward-coarsening sequence (depositional unit 1; for location of section see Fig. 22). To the northeast in 4-10-42-8W5, two upward-coarsening sequences occur (depositional units 1 and 2). The southwestern edge of unit 3 is shown in 10-11-42-8W5; units 1 and 2 are stratigraphically lower relative to marker 2. In 14-12-42-8W5, unit 3 is thicker and units 1 and 2 are stratigraphically lower than in wells 4-10 and 10-11. Conglomerates (about 0.3 m thick) cap unit 3 in 16-18-42-7W5 and units 1 and 2 are lower stratigraphically than western wells, as their resistivity response diminishes thus indicating increasing shale content. In 14-17-42-7W5 about 2 m of conglomerate cap unit 3. This conglomerate occurs in the western band of conglomerates shown in Fig. 21. The southwestern edge of unit 4 occurs in 8-21-42-7W5, where the conglomerates are about 1 m in thickness. In 8-21-42-7W5, unit 3 is stratigraphically lower relative to its position in 14-17, as are units 1 and 2. This sequence continues northeastward to 6-27-42-7W5. At 16-27-42-7W5, about 2 m of conglomerates occur in the position occupied by unit 4 in 6-27-42-7W5. This conglomerate occurs in the central band of conglomerates shown in Fig. 19. Note that although the resistivity wire-line well-log profiles of 6-27-42-7W5 and 16-27-42-7W5 appear similar, core information from 16-27-42-7W5 and adjacent wells confirms that only thin conglomerates are present in 6-27-42-7W5, while thick conglomerates occur in 16-27-42-7W5. The southeasternmost occurrence of unit 5 in cross-section A–A' is present in 16-35-42-7W5, with no conglomerate. In this well, unit 4 is stratigraphically lower than its position in 16-27-42-7W5, unit 3 is also stratigraphically lower and has merged with unit 2. Unit 1 is not recognized. Thin conglomerates cap unit 5 in 5-5-43-6W5, unit 4 is stratigraphically lower than its position in 16-35-42-7W5 and unit 3 is not recognized. Thick conglomerates (about 3 m) cap unit 4 and occur in the position occupied by unit 5 in 16-35-42-7W5. These conglomerates occur along the eastern margin of Willesden Green (Fig. 19). Well 6-10-43-6W5 shows a resistivity response typical of most wells along the eastern margin of Willesden Green. Thin conglomerate stringers are present in the predominantly muddy sediments of 6-10-43-6W5.

## Cross-section B–B'

This cross-section (Fig. 21), located in the southern region of the study area (Fig. 22), shows most clearly the offlapping nature of the units. Its southwesternmost well is located between Willesden Green and Ferrier at 6-15-40-7W5. Unit 1 clearly occupies a progressively stratigraphically lower position toward the northeast along the line of section and shales-out northeast of 4-17-41-5W5. The southeasternmost occurrence of unit 2 is present between Ferrier and Willesden Green at 6-13-40-7W5. As with unit 1, unit 2 also occupies a progressively stratigraphically lower position toward the northeast, becoming difficult to identify in 4-7-41-5W5 and 4-17-41-5W5. In 4-17-41-5W5 unit 2 apparently merges with unit 1. The southwesternmost occurrence of unit 3 occurs in 11-29-40-6W5, again occupying a progressively lower stratigraphical position toward the northeast. Unit 3 terminates between 4-17-41-5W5 and 6-28-41-5W5. Unit 4's southwestern edge occurs at 7-3-41-6W5. Unit 4 ends between 6-28-41-5W5 and 12-26-41-5W5, where a thin conglomerate marks the eastern margin of the Willesden Green field. Unit 5 is absent in cross-section B–B'.

## Discussion

The two cross-sections in Figs 20, 21, along with other cross-sections presented by Keith (1985), show that the Cardium A sandstone succession in the study area consists of five upward-coarsening depositional units. A maximum of three depositional units are generally present in any one well. Based on overlying and underlying well-log markers the depositional units show a northeastern offlapping geometry.

A single upward-coarsening sequence (unit 1) is present between Willesden Green and Ferrier fields. In both cross-sections the most westerly occurrence of unit 2 is adjacent to the western margin of Willesden Green, northeast of the region where unit 1 alone occurs. Units 3–5 progressively offlap toward the northeast. Unit 5 appears to be discontinuous, trending NW–SE in the north, and N–S in southern Willesden Green. The southwest edges of units 2–5 are shown by the thickened solid lines in Fig. 22. The offlapping geometry of the units within Willesden Green are illustrated in a block diagram (Fig. 23), which was constructed with data from Figs 20, 21 and other cross-sections from Keith (1985). As shown in Figs 22, 23 the southwest unit edges parallel each other and the regional trends of both Ferrier and Willesden Green.

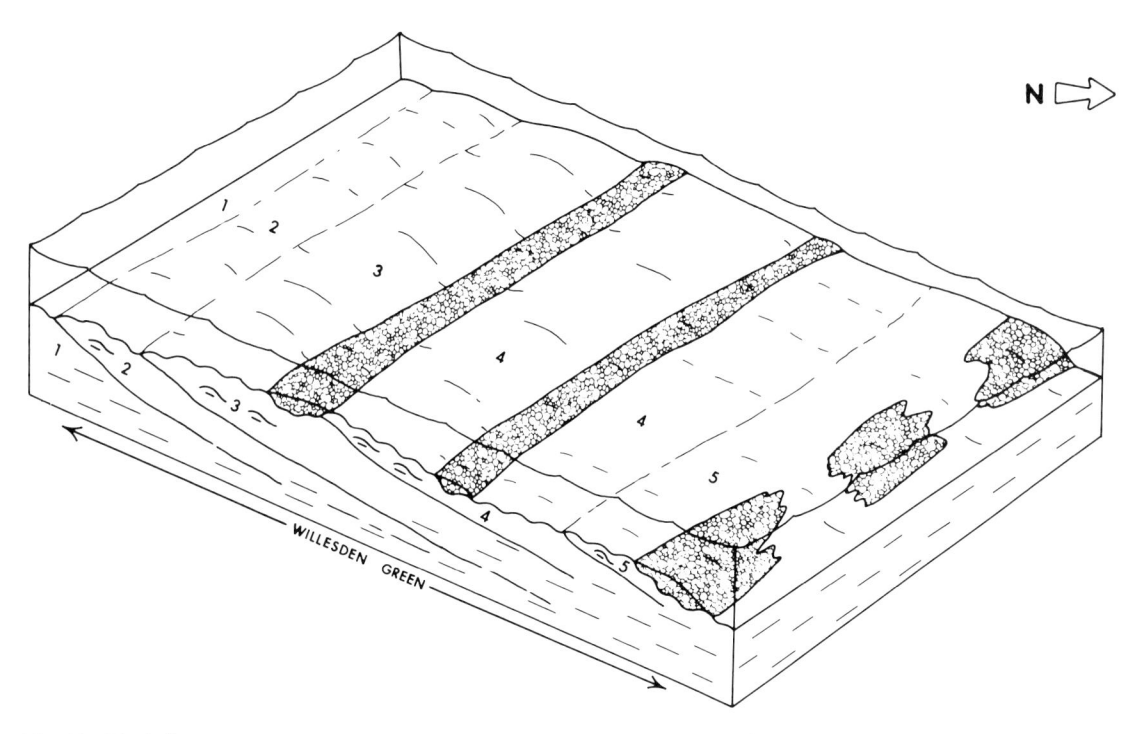

**Fig. 23.** Block diagram illustrating the offlapping relationships of Cardium A sandstone depositional units 1–5 and their relationship to overlying conglomerates (stippled) in and adjacent to Willesden Green. Note that conglomerates occur as southeast-trending bands in topographic lows on the upper surface of the sandstones. Vertical exaggeration is about 280×.

Within any depositional unit, the best sandstone development (facies 4–6) occurs closest to its southwest edge. As the unit occupies a progressively lower stratigraphic position toward the northeast, its resistivity well-log response diminishes, and eventually disappears, indicating higher shale content. In this position the unit is represented only by muddy facies (1 and 2). The 'dip' on a typical unit (e.g. unit 1, Fig. 21, 6-15-40-7W5 to 4-17-41-5W5) is about 0.3° (7 m in 16 km) with respect to marker 2.

## Location of conglomerates

The distribution of the conglomerates in Willesden Green and Ferrier is shown in Fig. 24. The distribution of conglomerates in Ferrier is from Griffith (1981, fig. 9), with the addition of new data from wells drilled after Griffith's study. It is clear from the cross-sections (Figs 20, 21) that the conglomerates post-date all of the depositional units. The conglomerates overlie units 2 and 3 in the northwest region of Willesden Green; unit 4 in a band down the centre of the field, and they overlie unit 5 along the northeast margin of the field. The most westerly band of conglomerate parallels the southwestern edge of unit 4, in areas where unit 3 is lower stratigraphically (relative to its stratigraphic position in a more southwestern location)—this is illustrated in Fig. 20, wells 14-12-42-8W5, 16-18-42-7W5, 14-17-42-7W5 and 8-21-42-7W5. The central band of conglomerate lies half-way between the southwest edges of units 4 and 5, and may be lying in an erosional truncation of the upper portion of unit 4 (see Fig. 20, 6-27-42-7W5, 16-27-42-7W5 and 16-35-42-7W5). The most northeastern band of conglomerate parallels the southwest edge of unit 5. These conglomerates may be filling an erosional truncation of unit 5 along the margin of the field (Figs 20, 22). Alternatively, unit 5 could have thinned at the field margin, with the conglomerate deposited against the thinned-out edge of the sandstone. Thus it appears that areas of conglomerate accumulation are associated with topographic lows on the TASS.

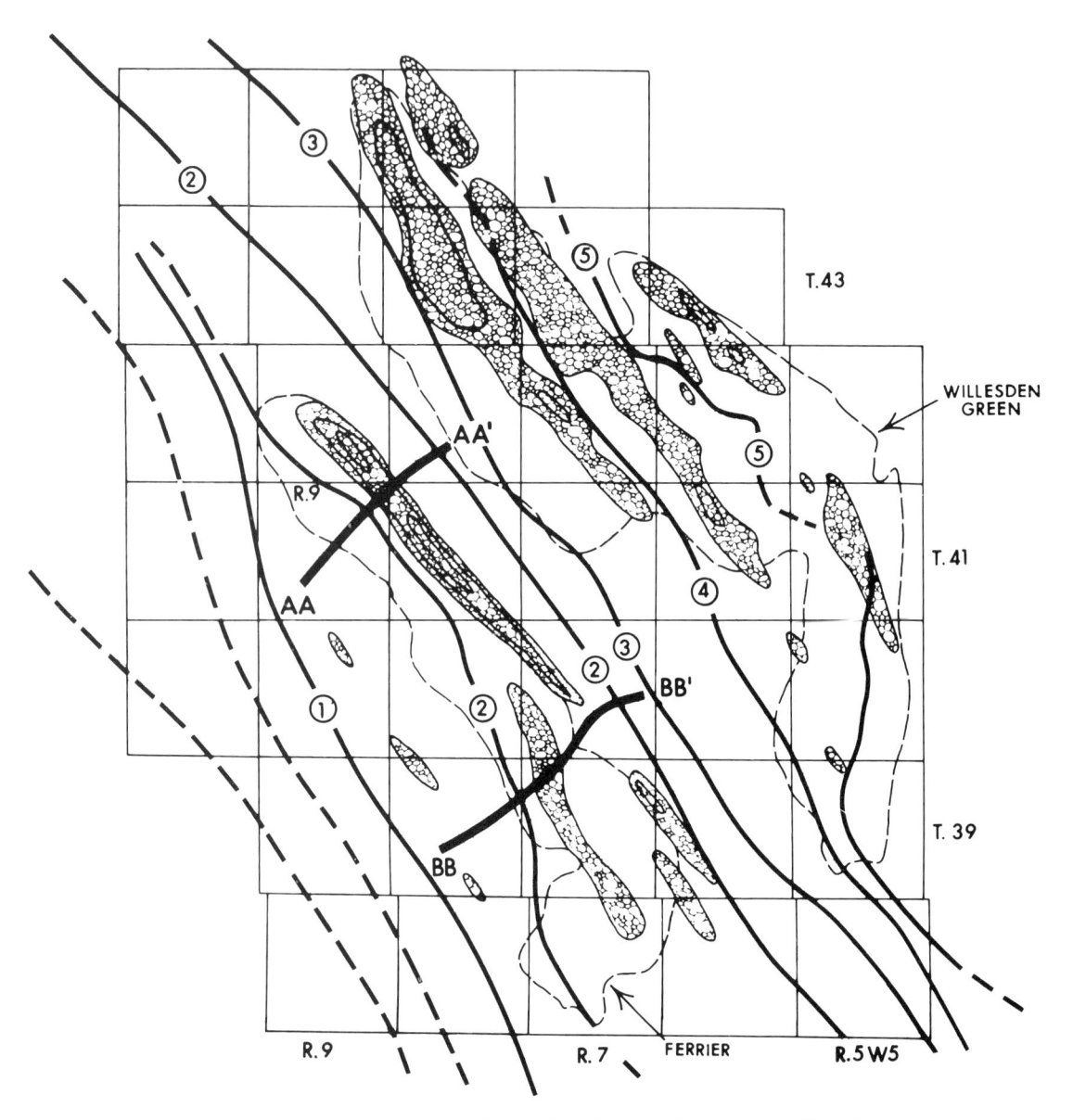

**Fig. 24.** The southwesternmost occurrence or edge of all depositional units in the study area. Circled numbers indicate which depositional unit the southwest edge represents. Two southwest unit edges that occur west of cross-sections presented are indicated by heavy dashed lines. Also shown is the conglomerate distribution (stippled pattern) with a 1 m contour interval and the locations of cross-sections AA–AA' and BB–BB'.

# CORRELATIONS BETWEEN THE WILLESDEN GREEN AND FERRIER FIELDS: THE REGIONAL SIGNIFICANCE OF OFFLAPPING DEPOSITIONAL UNITS

To determine more fully the significance of the offlapping units delineated at Willesden Green, the study area was expanded to the west and south to include areas within and west of the Ferrier field and areas north of the Caroline and Garrington fields. Study of the Ferrier field was thought to be particularly important because any insights with respect to sandstone geometry in Ferrier and its relationship to the offlapping sandstones at Willesden Green might lead to an understanding of the development of the Willesden Green field and its depositional history.

## Ferrier to Willesden Green cross-sections

As shown in Figs 1, 16, 17, the Ferrier field is located to the west of Willesden Green field. Isopach relationship between upper and lower markers with the TACS (Figs 16, 17, respectively), illustrate a topographic low between the Willesden Green and Ferrier fields. Cores in this topographic low show predominantly muddy facies (facies 2 or 3), with rare pebble stringers.

Griffith's work (1981, p. 30) shows multiple Cardium A sandstone bodies at Ferrier. She described them as stacked sandstones and suggested they were emplaced by southerly progradation.

Examination of Griffith's (1981) Ferrier cross-sections and construction of additional cross-sections across Ferrier into Willesden Green (some presented in Keith, 1985) allow correlations between the two fields to be made. Cross-section AA–AA' (Fig. 25) and cross-section BB–BB' (Fig. 26) extend Willesden Green cross-sections A–A' (Fig. 20) and B–B' (Fig. 21) across the Ferrier field.

### Cross-section AA–AA'

The location of this cross-section (Fig. 25) is shown in Fig. 24. The southwesternmost well in section AA–AA' is located west of Ferrier at 8-10-41-9W5. There is a subdued resistivity response at the top of the A sequence, labelled unit 1, in 8-10-41-9W5. Unit 1 is correlated eastward across Ferrier through 4-23-41-9W5, 12-24-41-9W5, 2-25-41-9W5 and 4-31-41-9W5. Correlation of two resistivity log responses situated below unit 1 strengthen overlying correlations. Unit 2 is shown in 12-24-41-9W5. It persists into 2-25-41-9W5, but is absent in 4-31-41-8W5. Conglomerates about 0.3 m thick cap unit 2 in 2-25-41-9W5. East of 2-25-41-9W5, in 4-31-41-8W5, conglomerates thicken to about 4.5 m along the eastern margin of Ferrier. Conglomerates are limited to thin pebble stringers overlying the A sandstone sequence in 10-31-41-8W5. Correlation between 4-31-41-8W5 and 10-31-41-8W5, across the eastern margin of Ferrier, is difficult owing to: (1) a lack of well control; (2) lack of a lower datum (most wells along this margin do not penetrate very far beneath the Cardium Formation sandstones); and (3) post-Cardium Formation thrust faulting along the eastern margin of Ferrier (townships 41 and 42) with offsets in the order of 20 m. Utilizing lower well-log markers and correlation of the pebble stringers of 10-31-41-8W5 into the thick conglomerates of 10-31-41-8W5 indicates that unit 1 is continuous from 4-31-41-8W5 to 10-31-41-8W5. Unit 1 is correlated eastward towards Willesden Green in 4-10-42-8W5 and 10-11-42-8W5, where units 2 and 3, respectively, occur.

### Cross-section BB–BB'

The location of this cross-section (Fig. 26) is shown in Fig. 24. The southwestern well is located west of the Ferrier field at 8-14-39-8W5. The sandstone capping the Cardium A sequence in 8-14-39-8W5, labelled unit 1, is correlated to 2-24-39-8W5 using both upper and lower log markers. A thick sandstone unit, characterized by a more blocky resistivity log response, labelled unit 2, caps the A sandstone sequence in 4-30-39-7W5. Here, thin conglomerates (about 0.3 m) overlie the sandstone. Unit 1 occurs in a stratigraphically lower position in 4-30-39-7W5 than in 2-24-39-8W5. The pattern persists into 4-32-39-7W5, where thick conglomerates (about 1.2 m) cap unit 2. Thin conglomerates in 11-33-39-7W5 cap unit 2 and unit 1 occupies a stratigraphically lower position relative to more southwestern wells. An identical sandstone sequence occurs in 6-9-40-7W5, where conglomerates are absent. At 6-15-40-7W5, between Ferrier and Willesden Green, unit 2 is absent and unit 1 comprises the entire A sequence. Unit 1 is correlated eastward towards Willesden Green in 6-13-40-7W5, where unit 2 is present and unit 1 is stratigraphically lower than its position in more southwestern wells.

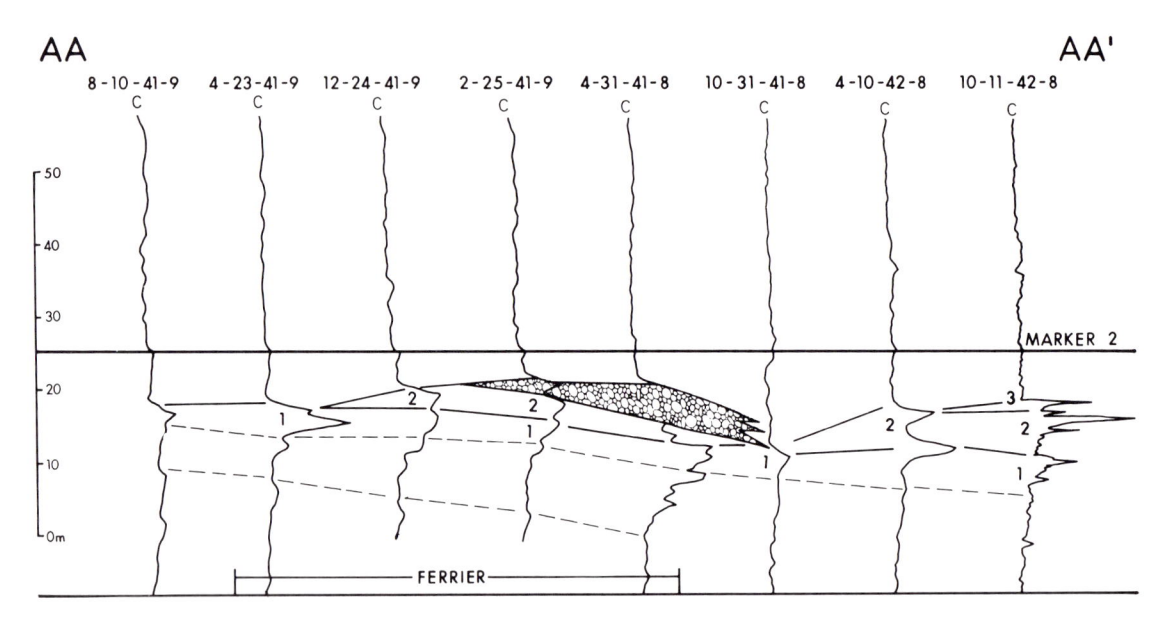

**Fig. 25.** Southwest to northeast stratigraphic cross-section AA–AA' across northern Ferrier and northwestern Willesden Green fields. See Fig. 24 for location of cross-section, all well locations are west of the fifth meridian. This section overlaps section A–A' (Fig. 20). The datum is marker 2 and the A sandstone depositional units are numbered from the southeast beginning with 1. Note the successively northeastward offset or offlapping pattern shown by the depositional units. Conglomerates are shown with stippled pattern. Cored wells are indicated by a 'c' below the well location. Vertical exaggeration is about 280×.

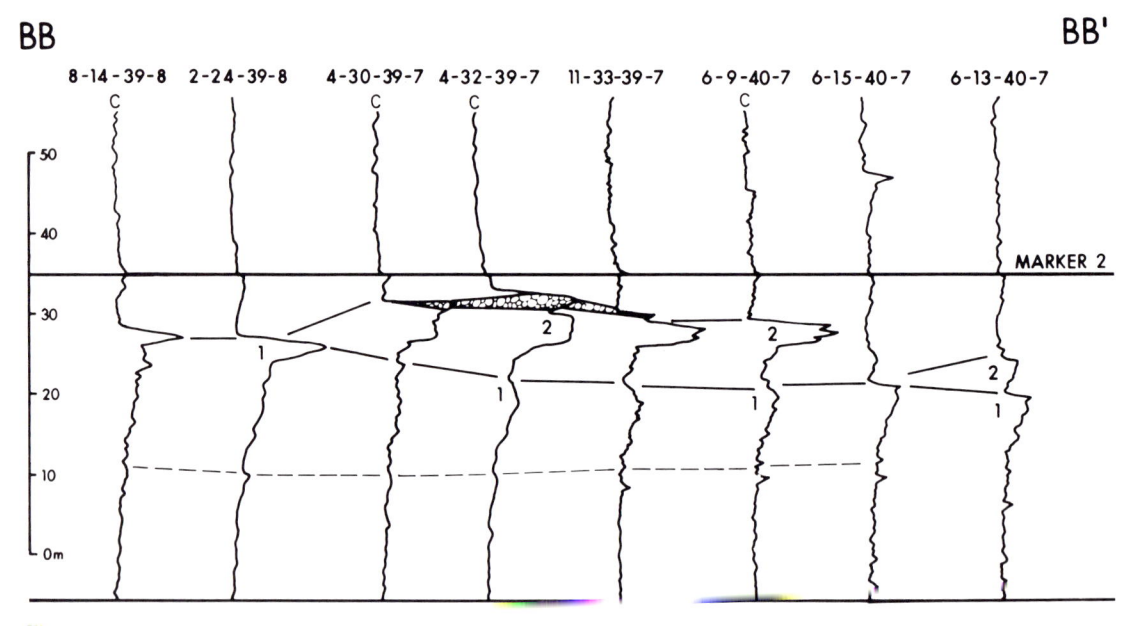

**Fig. 26.** Southwest to northeast stratigraphic cross-section BB–BB' across southern Ferrier and southwestern Willesden Green. See Fig. 24 for location of cross-section, all well locations are west of the fifth meridian. This section overlaps section B–B' (Fig. 20). The datum is marker 2 and A sandstone units are numbered from the southeast beginning with 1. Note the successively northeastward offset or offlapping pattern shown by the depositional units. Conglomerates are shown with stippled pattern. Cored wells are indicated by a 'c' below the well location. Vertical exaggeration is about 280×.

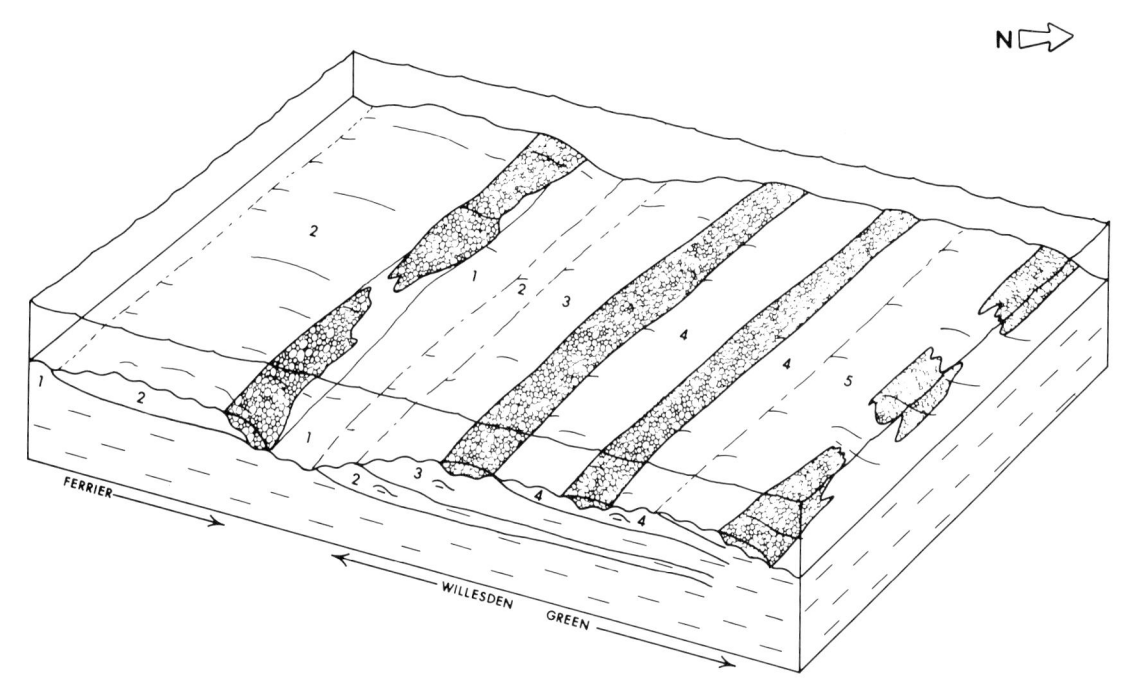

Fig. 27. Block diagram illustrating the offlapping relationships of Cardium A sandstone depositional units 1–5 and their relationship to overlying conglomerates (stippled) in the study area. Note the offlapping relationship of the depositional units throughout the study area, with depositional unit 1 occurring at the top of the A sandstone sequence west of Ferrier, continuous across the deep low between Ferrier and underlying other depositional units in Willesden Green. Note also that conglomerates occur as southeast-trending bands in topographic lows on the upper surface of the sandstones. Data for this block diagram is from Figs 22, 25, 26. Vertical exaggeration is about 280×.

## Discussion

In both cross-sections AA–AA' and BB–BB', the westernmost sandstone, unit 1, comprises the entire Cardium A sequence west of Ferrier. It occupies a progressively lower stratigraphic position northeastward into the Ferrier field, where it is overlain by a second Ferrier sandstone (unit 2). This sand is capped by thick conglomerates. The Ferrier sandstone of unit 2, and the overlying conglomerates, do not continue eastward into Willesden Green. However, the lower sandstone, unit 1, is continuous from Ferrier to Willesden Green. The southwesternmost occurrence or edge of unit 2 in Ferrier is shown in Fig. 24. The NW–SE orientation of the southwest edge of unit 2 in Ferrier is similar to that of the southwest edge orientation of other units in the study area. The relationships of all depositional units in Ferrier and Willesden Green is shown diagrammatically in Fig. 27.

The demonstrable correlation of unit 1 between Ferrier and Willesden Green, and the fact that unit 1 represents the complete Cardium A sandstone sequence to the west of Ferrier (Figs 25, 26) and underlies depositional units 2–5 into Ferrier and Willesden Green, strongly suggest that the offlapping geometry described within Willesden Green also occurs in Ferrier.

## Offlapping sandstones—other evidence

Additional support for a Cardium A sandstone offlapping geometry comes from work in the Pembina field, north of the study area. Here, an offlapping Cardium A sandstone geometry similar to what has been described above occurs and accounts for Cardium Formation fluid distribution trends in Pembina (Kerr, 1980; F.F. Krause, pers. comm., 1987).

## INTERPRETATIONS OF THE WILLESDEN GREEN AND FERRIER FIELDS: DEVELOPMENT OF THE OFFLAPPING DEPOSITIONAL UNITS

Two alternative interpretations of the Willesden Green and Ferrier fields are discussed below, the first, an offshore sand body and the second, which is favoured, an erosionally truncated prograding strandplain.

### Offshore sand bodies

All of the sedimentological information discussed previously suggests that the facies in the study area were deposited in a lower shoreface to inner shelf setting. In addition, no evidence of upper shoreface deposits or subaerial exposure, such as roots, paleosols or *in situ* coals, are found in the study area. Deposits with similar facies in the Cretaceous Western Interior Seaway of North America, which have been interpreted as offshore sand bodies, include the Duffy Mountain Member (Boyles & Scott, 1982); Frontier Formation (Winn *et al.*, 1983); Shannon Sandstone (Spearing, 1979; Tillman & Martinsen, 1984; Gaynor & Swift, 1988); Sussex Sandstone (Berg, 1975; Brenner, 1978; Hobson *et al.* 1982) and the Viking Formation (Koldijk, 1976; Reinsen *et al.*, 1983).

From these comparisons, a conclusion that the Willesden Green and Ferrier Cardium fields represent sandstone bodies deposited some distance from the shoreline may appear reasonable. Sediment could have been transported incrementally to the study area by storm-generated geostrophic flows (Keith, 1984; Walker, 1985a,b). Swift & Niedoroda (1985) and Swift (1985) discuss the importance of geostrophic flow in sand transport on shelves. However, such an interpretation may not adequately account for the depositional-unit offlapping pattern observed in the study area or the apparent eroded nature of the depositional units between Ferrier and Willesden Green.

### Prograding strandplain

The favoured alternative to the offshore development of the Cardium sands in the study area is that the Willesden Green and Ferrier offlapping sandstone units represent lower shoreface to inner shelf toes of an erosionally truncated NW–SE trending, prograding strandplain. The strandplain prograded northeast across the study area, such that all depositional units would have been originally continuous between Ferrier and Willesden Green. With an offlapping sandstone pattern recognized in both Willesden Green and Ferrier it is plausible to suggest that depositional unit 2 in Ferrier is correlative with, and was originally continuous with, depositional unit 2 in Willesden Green. The suggestion is reasonable, based on the offlapping depositional pattern alone. Work by Weimer (1984), Raddysh (1986), Plint *et al.* (1986), Plint & Walker (1987), Bergman & Walker (1987) and others, documenting significant erosion in the Cardium Formation and other similar Cretaceous formations, which is, in part, discussed below, support this suggestion.

The offlapping geometry of the depositional units within the study area is consistent with the geometry of the lower shoreface to inner shelf portions of a prograding shoreline during sea-level fall. As discussed previously, the Cardium A sandstone sequence in the study area shows an upward-coarsening trend, presumably reflecting a sea-level fall during the middle to upper Turonian. A major eustatic sea-level fall has been suggested for this time (Kauffman, 1977; Haq *et al.*, 1987). The relatively thin, fragmentary nature of the Cardium A sandstone sequence in the study area, with only lower shoreface and inner shelf deposits preserved, also suggests that these units where deposited during long-term sea-level fall, where the rate of sea-level fall was greater than subsidence. If subsidence was greater than the rate of sea-level fall, greater preservation of the strandplain deposits would be expected. The prograding depositional pattern recognized in the Willesden Green–Ferrier region as offlapping depositional units deposited during sea-level fall is illustrated in Fig. 28A.

A similar offlapping geometry of progradation strandplain deposits has been documented by R. Wright (1986) in the Point Lookout Sandstone of New Mexico. R. Wright (1986) documented a 30–60 m regressive sandstone complex, the Point Lookout Sandstone in the San Juan Basin of New Mexico, USA. The deposit is composed of 7–15-m asymmetric transgressive–regressive cycles that he termed depositional units, three to five depositional units are present in any typical vertical sequence. These thinner short-term cycles serve as the basic building blocks of the overall regressive Point Lookout Sandstone, in the same manner as the depositional units in the study area are the basic building blocks of the Cardium A sandstone sequence.

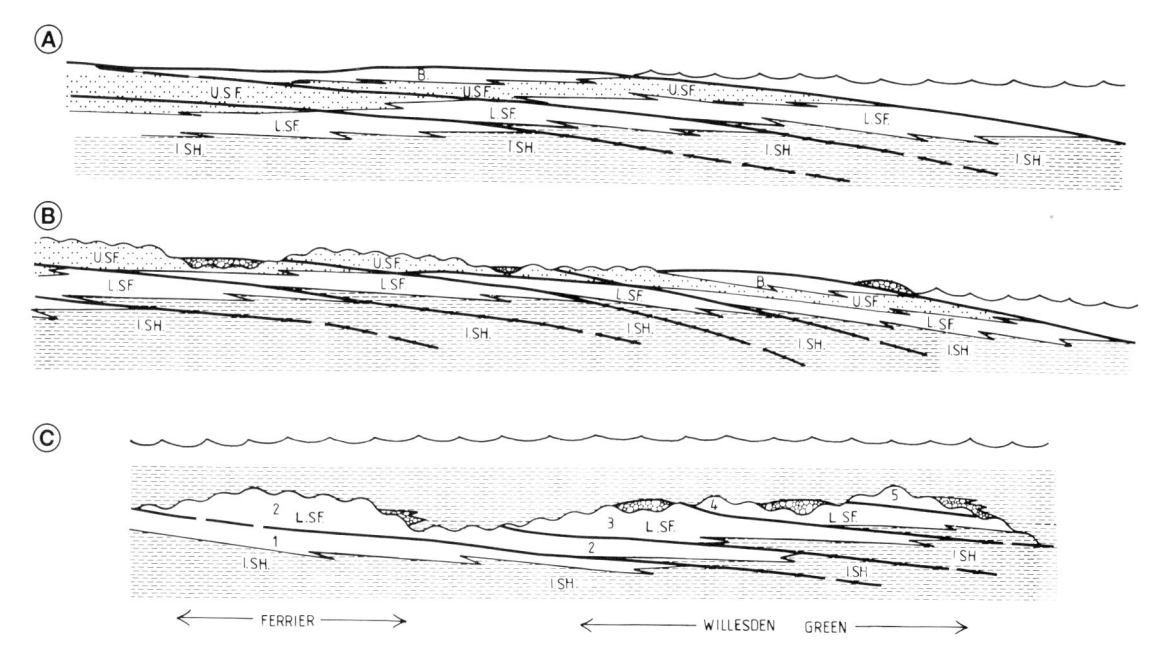

Fig. 28. (A) Diagrammatic SE–NW cross-section through study area during falling sea-level and Cardium shoreline progradation. Locations of beach (B), upper shoreface (USF), lower shoreface (LSF) and inner shelf (ISH) facies are indicated. The heavy solid lines separate individual progradational depositional units. Vertical exaggeration is about 180×. (B) Diagrammatic SE–NW cross-section through study area during sea-level lowstand showing dissection of previously deposited Cardium facies by fluvial erosion. The rivers may have introduced chert pebbles (stippled pattern) to the beach and shoreface during this time. Locations of beach (B), upper shoreface (USF), lower shoreface (LSF) and inner shelf (ISH) facies are indicated. The heavy solid lines separate individual progradational depositional units. Vertical exaggeration is about 180×. (C) Diagrammatic SE–NW cross-section through study area during sea-level rise. Note the erosional truncation of the Cardium facies, caused in part by shoreface retreat, leaving only lower shoreface (LSF) and inner shelf (ISH) facies. Note also the preferential accumulation of conglomerates (stipple pattern) within relative lows on the upper surface of the Cardium sandstone. The heavy solid lines separate the five depositional units that have been recognized in the study area. Shelf mudstones overlie the Cardium sandstones and conglomerates. The approximate locations of Willesden Green and Ferrier are indicated. Vertical exaggeration of this figure (280×) is greater than in Figs 28A,B.

The proposed progradational shoreface could have been nourished by southeasterly longshore flows. Northeastern progradation of the strandplain by the depositional units, which represent 'high-frequency sedimentation cycles', would have been controlled by sea-level, sediment input and subsidence. High-frequency sedimentation cycles, or parasequences, and their relation to progradation strandplain successions have been discussed by numerous authors (Ryer, 1977, 1983; R. Wright, 1986; Swift et al., 1987). It is difficult to suggest what factor or combination of factors was responsible for the depositional unit progradation. Sediment input variations as a result of climatic control has been suggested by a number of workers to account for high-frequency sedimentation cycles. Here, dry periods are represented by shales and the upward-coarsening siliciclastic units representing wetter periods with greater erosion and basinward transport of coarser material (Kauffman, 1985). R. Wright (1986) suggested sediment input fluctuations caused by seasonal rainfall changes were the primary cause of strandplain sedimentation cycles in the Point Lookout Sandstone. Barron et al. (1985) suggested a relationship between Milankovitch-type periodicity in the Earth's orbital elements and climatically controlled central basin rhythmic bedding patterns in the Cretaceous. Other workers, such as Dominguez et al. (this volume), have attributed fluctuations of sand supply to a prograding strandplain to high-frequency sea-level oscillations during a long-term sea-level fall.

deposited Cardium sands, would have occurred by fluvial erosion, as the system prograded basinward. Erosion of previously continuous depositional units between Ferrier and Willesden Green may have been initiated by rivers downcutting during sea-level lowstand. The parallel orientation of the low between Ferrier and Willesden Green to the presumed orientation of the Cardium shoreline is problematic and suggests that other factors, such as structural control, may have influenced the location and depth of this differential erosion.

Chert pebbles of the Cardium conglomerate could have been introduced into the foreshore environment during sea-level lowstand when downcutting rivers could transport coarser material. Tectonic activity in the thrust belt to the west may have provided new sources of coarse material at this time. Tectonic activity has been suggested as the control of conglomerate sources for the Kakwa Cardium field located northwest of the study area (Plint & Walker, 1987). The chert pebbles may have been concentrated in the upper shoreface and been bypassed by finer grained sediments (Plint & Walker, 1987). The lack of conglomerate below the unconformity, or any significant sandstone development above the unconformity does suggest that pebbles were not fully introduced to the system until after rising sea-level encroached upon the gravel beaches.

Any record of progradation of the strandplain deposits would have been eroded during subsequent sea-level rise (Fig. 28C). As discussed previously, the upper surface of the Cardium A sandstone (TASS) has been truncated by a regionally extensive unconformity. This surface marks a transgressive surface of erosion developed while the shoreface retreated westward during stillstand and subsequent transgression. Combined flow (wave- and wind-generated currents) during storms would have eroded any evidence of subaerial exposure, the foreshore and upper shoreface deposits within the study area, leaving only lower shoreface and inner shelf deposits. Bergman & Walker (1987) have documented erosion at a correlative horizon in the Cardium Carrot Creek field. They attribute much of the erosion to shoreface retreat by wave-scour during stillstand prior to transgression. A series of lows in the upper surface of the Cardium sandstones or TASS may have been developed or enhanced during shoreface retreat. Initial erosion, as in the case of the low between Ferrier and Willesden Green, may have begun subaerially during falling

and lowstand in sea-level. The lows are oriented NW–SE, parallel to the strike of Willesden Green and Ferrier and the presumed orientation of the shoreline.

Sea-level rise and retreat of the shoreface would result in a redistribution of the chert pebbles by southeastward longshore currents. Evidence of southeastward directed currents controlling conglomerate distribution is the thickness and greater abundance of conglomerates (as with sandstone) in the northwest of Willesden Green and the decrease in pebble size to the southeast. The chert pebbles accumulated within the NW–SE lows on the TASS. The conglomerates are therefore suggested to represent transgressive lag deposits infilling previously developed lows on the Cardium A sandstones.

## CONCLUSIONS

The regressive Cardium A sandstone sequence in the study area comprises five individual upward-coarsening deposition units that consist of facies reflecting lower shoreface to inner shelf depositional environments. The depositional units offlap each other towards the northeast across the Ferrier and Willesden Green fields and are interpreted as having formed the toes of an eroded, prograding strandplain. The Cardium shoreline may have prograded across the entire study area, but later erosional truncation precludes better definition. Cardium sandstones were deposited during a period of falling sea-level, at a rate that exceeded the rate of subsidence, and accounts, in part, for the thin, fragmentary nature of the Cardium A sandstone sequence. During falling sea-level, and at sea-level lowstand, incisement by rivers would have eroded previously deposited strata. Chert pebbles may have been supplied to the Cardium shoreline during the sea-level lowstand by the more competent downcutting rivers, and tectonic activity to the west may have provided coarse-sediment source areas.

With subsequent sea-level rise, the shoreface retreated to the southwest and chert pebbles that had accumulated along the Cardium shoreline were reworked by the transgressing sea. Several NW–SE oriented lows on the upper surface of the Cardium A sandstones may have been developed or enhanced during shoreface retreat. During sea-level rise the chert-pebble lag infilled the lows on the A sandstone surface. Continued sea-level rise resulted in deposition of muds that blanketed Cardium sandstones and conglomerates.

It cannot be overstated that the Cardium sandstones that we see preserved in the Willesden Green and Ferrier fields clearly demonstrate a northeast offlapping geometry, and any interpretation of the depositional history of the Cardium Willesden Green and Ferrier fields must account for this geometry.

## ACKNOWLEDGEMENTS

I would like to thank Donald Swift, Rod Tillman, William Duke and Federico Krause who provided suggestions that greatly improved the manuscript. Peter McCabe, George Dix, Grant Mossop, and Shaun O'Connell provided suggestions that improved an earlier version of the manuscript. Madelaine Keith drafted all figures. This study began as an MSc thesis at McMaster University. Initial funding was provided by a National Science and Engineering Research operating grant to R.G. Walker. I would like to thank Roger Walker for his supervision during the early stages of this study and Federico Krause for his continuing encouragement and support. I would also like to thank J. Walter Keith, Anadarko Petroleum Company of Canada Limited, Home Oil Company Limited, and Shell Canada Resources Limited who provided data used early in the study. The study was continued with the support of Gulf Canada Corporation. Access to word-processesing facilities and final figure preparation was provided by the Alberta Research Council.

## REFERENCES

Ager, D.V. (1973) Storm deposits in the Jurassic of the Moroccan High Atlas. *Palaeogeogr. Palaeoclimatol. Palaeoecol.* **15**, 83–93.

Barron, E.J., Arthur, M.A. & Kauffman, E.G. (1985) Cretaceous rhythmic bedding sequences: a plausible link between orbital variations and climate. *Earth Planet. Sci. Lett.* **72**, 327–340.

Beach, F.K. (1955) Cardium, a turbidity current deposit. *J. Alberta Soc. Petrol. Geol.* **3**, 123–125.

Berg, R.R. (1975) Depositional environment of Upper Cretaceous Sussex Sandstone, House Creek field, Wyoming. *Bull., Am. Assoc. Petrol. Geol.* **59**, 2099–2110.

Bergman, K.M. & Walker R.G. (1987) The importance of sea-level fluctuations in the formation of linear conglomerate bodies, Carrot Creek Member of the Cardium Formation, Cretaceous Interior Seaway, Alberta, Canada. *J. Sediment. Petrol.* **57**, 651–665.

Berven, R.J. (1966) Cardium sandstone bodies, Crossfield–Garrington area, Alberta. *Bull. Can. Petrol. Geol.* **14**, 208–240.

Boyles, J.M. & Scott, A.D. (1982) A model for migrating shelf-bar sandstones in Upper Mancos Shale (Campanian), northwest Colorado. *Bull., Am. Assoc. Petrol. Geol.* **66**, 491–508.

Brenner, R.L. (1978) Sussex Sandstone of Wyoming—example of Cretaceous offshore sedimetation. *Bull., Am. Assoc. Petrol. Geol.* **62**, 181–200.

Duke, W.L. (1985) *Sedimentology of the Upper Cretaceous (Turonian) Cardium Formation in outcrop in southern Alberta.* Unpublished PhD thesis, McMaster University, Hamilton.

Frey, R.W. & Pemberton, S.G. (1985) Biogenic structures in outcrops and cores—I. Approaches to ichnology. *Bull. Can. Petrol. Geol.* **33**, 72–115.

Gaynor, G.C. & Swift, D.J.P. (1988) Shannon Sandstone depositional model: Sand ridge dymanics on the Campanian Western Interior Shelf. *J. Sediment. Petrol.* **58**, 868–880.

Griffith, L.A. (1981) *Depositional environment and conglomerate diagenesis of the Cardium Formation, Ferrier Field, Alberta.* Unpublished MSc thesis, University of Calgary.

Haq, B.U., Hardenbol, J. & Vail, P.R. (1987) Chronology of fluctuating sea-levels since the Triassic. *Science* **235**, 1156–1167.

Hobson, J.P., Jr, Fowler, M.L. & Beaumont, E.A. (1982) Depositional and statistical exploration models, Upper Cretaceous offshore sandstone complex, Sussex Member, House Creek Field, Wyoming. *Bull., Am. Assoc. Petrol. Geol.* **66**, 689–707.

Kauffman, E.G. (1977) Geological and biological overview—Western Interior Cretaceous basin. *Mountain Geol.* **14**, 75–99.

Kauffman, E.G. (1985) Cretaceous evolution of the Western Interior Basin of the United States. In: *Fine-grained Deposits and Biofacies of the Cretaceous Western Interior Seaway: Evidence of Cyclic Sedimentary Processes* (Eds Pratt, L.A., Kauffman, E.G. & Zelt, F.B.), Guidebook for Society of Economic Paleontologists and Mineralogists Fieldtrip, 15 August 1985.

Keith, D.A.W. (1984) Cardium, Willesden Green, West-Central Alberta: sedimentology and stratigraphy of a shelf sandstone body (abstr.). *Program and Abstracts, Canadian Society of Petroleum Geologists/Canadian Institute of Mineralogy National Convention*, Calgary, p. 273.

Keith, D.A.W. (1985) *Sedimentology of the Cardium Formation (Upper Cretaceous) Willesden Green Field, Alberta.* Unpublished MSc thesis, McMaster University, Hamilton.

Kerr, W.E. (1980) A geological explanation for the variation in fluid properties across the Pembina Cardium Field. *J. Can. Petrol. Technol.* **19**, 76–84.

Koldijk, W.S. (1976) Gilby Viking B: a storm deposit. In: *The Sedimentology of Selected Clastic Oil and Gas Reservoirs in Alberta* (Ed. Lerand, M.M.) pp. 62–77. Canadian Society of Petroleum Geologists, Calgary.

Krause, F.F. (1982) Geology of the Pembina Cardium Pool. In: *Depositional Environments and Reservoir*

Facies in Some Western Canadian Oil and Gas Fields (Ed. Hopkins, J.C.) pp. 15–72. University of Calgary Core Conference.

KRAUSE, F.F. (1983) Sedimentology of a tempestite: episodic deposition in the Cardium Formation in the Pembina Oilfield area, west central Alberta, Canada. In: *Sedimentology of Selected Mesozoic Clastic Sequences* (Eds McLean, J.R. & Reinson, G.E.), *Core Expo '83*, pp. 43–65. Canadian Society of Petroleum Geologists, Calgary.

KRAUSE, F.F. & COLLINS, H.N. (1984) *Pembina Cardium Recovery Efficiency Study: a Geological and Engineering Synthesis*. Vols I & II (Appendices). Petroleum Recovery Institute, Calgary, Alberta, 300 pp.

KRAUSE, F.F. & NELSON, D.A. (1984) Storm event sedimentation: facies association in the Cardium Formation, Pembina area, west-central Alberta, Canada. In: *The Mesozoic of Middle North America* (Eds Stott, D.F. & Glass, D.J.). Can. Soc. Petrol. Geol. Mem. 9, 485–511.

KRAUSE, F.F., COLLINS, H.N., NELSON, D.A., MACHEMER, S.D. & FRENCH, P.R. (1987) Multiscale anatomy of a reservoir: geological characterization of the Pembina–Cardium pool, west-central, Canada. *Bull., Am. Assoc. Petrol. Geol.* **71**, 1233–1260.

MITCHUM, R.M., VAIL, P.R. & THOMPSON, S. (1977) Seismic stratigraphy and global changes of sea-level. Part 2. The depositional sequence as a basic unit for stratigraphic analysis. In: *Seismic Stratigraphy—Applications to Hydrocarbon Exploration* (Ed. Payton, C.E.). Mem., Am. Assoc. Petrol. Geol. 26, 53–62.

NIELSON, A.R. & PORTER, J.W. (1984) Pembina oil field—in retrospect. In: *The Mesozoic of Middle North America* (Eds Stott, D.F. & Glass, D.J.). Can. Soc. Petrol. Geol. Mem. 9, 1–13.

PEMBERTON, S.G. & FREY, R.W. (1984) Ichnology of storm-influenced shallow marine sequence: Cardium Formation (Upper Cretaceous) at Seebe, Alberta. In: *The Mesozoic of Middle North America* (Eds Stott, D.F. & Glass, D.J.). Can. Soc. Petrol. Geol. Mem. 9, 281–304.

PLINT, A.G. & WALKER, R.G. (1987) Cardium Formation 8. Facies and environments of the Cardium shore line and coastal plain in the Kakwa Field and adjacent areas, northwestern Alberta. *Bull. Can. Petrol. Geol.* **35**, 48–64.

PLINT, A.G., WALKER, R.G. & BERGMAN, K.M. (1986) Cardium Formation 6. Stratigraphic framework of the Cardium in the subsurface. *Bull. Can. Petrol. Geol.* **34**, 213–225.

RADDYSH, H.K. (1986) *Sedimentology of the Viking Formation at Gilby A and B Fields, Alberta*. Unpublished BSc thesis, McMaster University, Hamilton.

REINSEN, G.E., FOSCOLOS, A.E. & POWELL, T.G. (1983) Comparison of Viking Sandstone Sequences, Joffre and Caroline Fields. In: *Sedimentology of Selected Mesozoic Clastic Sequences* (Eds McLean, J.R. & Reinson, G.E.) pp. 101–117. Canadian Society of Petroleum Geologists, Calgary.

RYER, T.A. (1977) Patterns of Cretaceous shallow-marine sedimentation, Coalville and Rockport areas, Utah, *Geol. Soc. Am. Bull.* **88**, 177–188.

RYER, T.A. (1983) Transgressive–regressive cycles and the occurrence of coal in some Upper Cretaceous strata of Utah. *Geology* **11**, 201–210.

SPEARING, D.R. (1976) Upper Cretaceous Shannon Sandstone: an off-shore shallow marine sand body. *Wyoming Geological Association, 28th Annual Filed Conference, Guidebook*, pp. 65–72.

SWAGER, N.S., OLIVER, T.A. & JOHNSON, B.A. (1976) Carrot Creek field, central Alberta. In: *Sedimentology of Selected Clastic Oil and Gas Reservoirs in Alberta* (Ed. Lerand, M.M. ). Canadian Society of Petroleum Geologists, Calgary, 125 pp.

SWIFT, D.J.P. (1985) Response of the shelf floor to flow. In: *Shelf Sands and Sandstone Reservoirs* (Eds Tillman, R.W., Swift, D.J.P. & Walker, R.G.). Soc. Econ. Paleontol. Mineral. Short Course Notes 13, 353–402.

SWIFT, D.J.P. & NIEDORODA, A. (1985) Fluid and sediment dynamics on contentianl shelves. In: *Shelf Sands and Sandstone Reservoirs* (Eds Tillman, R.W., Swift, D.J.P. & Walker, R.G.). Soc. Econ. Paleonotol. Mineral. Short Course Notes, 13, 47–133.

SWIFT, D.J.P., HUDELSON, P.M., BRENNER, R.L. & THOMPSON, P. (1987) Shelf construction in a foreland basin: storm beds, shelf sandbodies and shelf-slope depositional sequences in the Upper Cretaceous Mesaverde Group, Book Cliffs, Utah. *Sedimentology* **34**, 423–457.

TILLMAN, R.W. & MARTINSEN, R.S. (1984) The Shannon shelf–ridge sandstone complex, Salt Creek Anticline area, Powder River Basin, Wyoming. In: *Siliciclastic Shelf Sediments* (Eds Tillman, R.W. & Siemers, C.T.). Soc. Econ. Paleontol. Mineral. Spec. Publ. 34, 82–142.

WALKER, R.G. (1983a) Cardium Formation 1. 'Cardium a turbidity current deposit' (Beach, 1955): a brief history of ideas. *Bull. Can. Petrol. Geol.* **31**, 205–212.

WALKER, R.G. (1983b) Cardium Formation 2. Sand-body geometry and stratigraphy in the Caroline–Garrington–Ricinus area, Alberta—The 'ragged blanket' model. *Bull. Can. Petrol. Geol.* **31**, 14–26.

WALKER, R.G. (1983c) Cardium Formation 3. Sedimentology and stratigraphy in the Caroline–Garrington area, Alberta. *Bull. Can. Petrol. Geol.* **31**, 213–230.

WALKER, R.G. (1984) Shelf and shallow marine sands. In: *Facies Models*: 2nd edn (Ed. Walker, R.G.). Geosci. Can. Reprint Ser. 1, 141–170.

WALKER, R.G. (1985a) Cardium Formation at Ricinus field, Alberta: a channel cut and filled by turbidity currents in Cretaceous Western Interior seaway. *Bull., Am. Assoc. Petrol. Geol.* **69**, 1963–1981.

WALKER, R.G. (1985b) Upper Cretaceous (Turonian) Cardium Formation, southern foothills and plains, Alberta. In: *Shelf Sands and Sandstone Reservoirs* (Eds Tillman, R.W., Swift, D.J.P. & Walker, R.G.). Soc. Econ. Paleontol. Mineral. Short Course Notes 13, 353–402.

WEIMER, R.J. (1984) Relation of unconformities, tectonics and sea-level changes, Cretaceous Western Interior, U.S.A. In: *Interregional Unconformities and Hydrocarbon Accumulation* (Ed. Schlee, J.S.). Mem., Am. Assoc. Petrol. Geol. 36, 7–35.

WINN, R.D., JR., STONECIPHER, S.A. & BISHOP, M.G. (1983) Depositional environments and diagenesis of offshore sandridges, Frontier Field, Wyoming. *Mountain Geol.* **20**, 41–58.

WRIGHT, M.E. & WALKER, R.G. (1981) Cardium Formation (U. Cretaceous) at Seebe, Alberta—storm transported sandstones and conglomerates in shallow marine depositional environments below fair-weather wave base. *Can. J. Earth Sci.* **18**, 795–809.

WRIGHT, R. (1986) Cycle stratigraphy as a paleogeographic tool: Point Lookout Sandstone, southeastern San Juan basin, New Mexico. *Geol. Soc. Am. Bull.* **96**, 661–173.

# Facies of the lower Ferron Sandstone and Blue Gate Shale Members of the Mancos Shale: lowstand and early transgressive facies architecture

P.E. RIEMERSMA[1] *and* M.A. CHAN

*Department of Geology and Geophysics, University of Utah, Salt Lake City, UT 84112-1183, USA*

## ABSTRACT

Marine sandstones of the Upper Cretaceous lower Ferron Sandstone Member are abruptly overlain by dark grey marine shale of the Blue Gate Shale Member of the Mancos Shale Formation in east-central Utah. Virtually continuous outcrop exposures along the flanks of the San Rafael Swell allow detailed facies description and correlation of an unconformity between the lower Ferron Sandstone and the Blue Gate Shale.

Four distinct facies are recognized in the lower Ferron Sandstone. Three of these, the bioturbated siltstone, hummocky cross-stratified sandstone, and amalgamated hummocky cross-stratified sandstone facies of the lower Ferron Sandstone, form a shallowing upwards (regressive) sequence. This sequence records southeastward progradation of a storm and wave-dominated shoreline during relative sea-level fall. The fourth facies, the bioturbated sandstone facies, was deposited during early relative sea-level rise and rests above an unconformity capping the coarsening upward sequence.

The unconformity above regressive facies of the lower Ferron Sandstone is interpreted to represent a type 2 unconformity with no evidence of subaerial exposure or incision by channels. This unconformity was probably formed by submarine processes just below fair-weather wave base, with minor truncation of underlying strata. The unconformity (termed the lower Ferron sequence boundary) appears to be correlative to Upper Cretaceous erosion surfaces at the top of the Codell Sandstone Member of the Carlisle Shale in Colorado, near the base of the Wall Creek (Frontier) Sandstone in Wyoming and is at the base of the Semilla Sandstone in New Mexico. It represents a sequence boundary marking a regional relative sea-level fall.

A thin, trough cross-bedded, poorly-sorted sandstone unconformably overlies offshore bioturbated siltstones of the lower Ferron Sandstone. This pebbly and coarse-grained fishtooth sandstone rests above the sequence boundary. The sandstone is a transgressive lag probably deposited during maximum sea-level lowstand and reworked during the subsequent transgression.

Dark grey laminated mudstones and thin siltstones of the laminated shale facies of the lowermost Blue Gate Shale overlie the lower Ferron Sandstone. These sediments were deposited during a rapid relative sea-level rise and possibly oxygen deficient marine conditions. Thin, sandy calcarenite beds occur at the top of the sequence (calcarenite facies). Ammonites from the calcarenite interval identify the lowermost Blue Gate Shale as correlative to the Juana Lopez Member in New Mexico and Colorado.

Shelf sandstones of the lower Ferron Sandstone record rapid progradation of the shoreface during sea-level fall. Facies of the lower Ferron Sandstone form a thin sequence of fine to very fine-grained sandstones deposited by storm processes. The bioturbated sandstone facies of the lower Ferron Sandstone records relative rising sea-level. The lower Blue Gate Shale was probably deposited during a period of reduced clastic deposition, and possibly oxygen deficient conditions during early sea-level rise.

---

[1] Present address: 203 David Dr., Apt B2 Bryn Mawr, PA 19010, USA.

## INTRODUCTION

The Ferron Sandstone and Blue Gate Shale Members of the Mancos Shale Formation were deposited along the western edge of the Western Interior Cretaceous Seaway. The Ferron Sandstone was deposited as an eastward thinning clastic wedge of strandplain, deltaic and shoreface sediments in middle and late Turonian time. Cessation of Ferron Sandstone clastic deposition is recorded by the overlying transgressive marine Blue Gate Shale Member.

An objective of this study is to identify and characterize the sequence boundary above the lower Ferron Sandstone and document the facies succession underlying and overlying this unconformity both perpendicular and parallel to depositional strike. Careful analysis of sequence boundaries is important in correlating regional unconformities and evaluating regional sea-level changes. Facies analysis of the lower Ferron also will serve as a model of deposition during sea-level lowstand.

The Ferron Sandstone can be divided informally into upper and lower members based on age, stratigraphic relationships, and facies analysis. Lower Ferron outcrops examined in this study in Castle Valley are north of the town of Moore, near Dry Wash (Fig. 1). The lower Ferron Sandstone fines to scattered siltstone lentils east of Emery, Utah (Thompson et al., 1986). The upper Ferron Sandstone is exposed in southern Castle Valley east of highway 10 and pinches out north of the town of Ferron into an interval of scattered shelf sandstones (Fig. 1). Davis (1954) interpreted the source area for the sandstone in north Castle Valley as being to the northwest, whereas the source area for the southern sequence (upper Ferron in this report) lay to the southwest. Subsurface cross-sections and outcrop studies indicate that the delta-front sandstone that forms the basal unit of the upper Ferron Sandstone in southernmost Castle Valley is equivalent to the lower Ferron Sandstone (Ryer & McPhillips, 1983). The lower Ferron Sandstone is separated from the overlying upper Ferron Sandstone at all measured sections in this study by dark grey laminated shale of the lowermost Blue Gate Shale.

The lower Ferron Sandstone was informally referred to as the Farnham unit by Katich in 1951. Cotter (1975) completed the first detailed stratigraphic and depositional interpretation of the lower Ferron Sandstone in Castle Valley. As part of our study, 20 sections were measured along both the west and east flanks of the San Rafael Swell in Utah. Sections are sequentially numbered from west to east (see Fig. 1).

A calcareous interval in the lowermost Blue Gate Shale is correlative to the Juana Lopez Formation, which is recognized as a regional marker interval (Molenaar, 1975). The underlying lower Ferron Sandstone is likely correlative to the Codell Sandstone Member in Colorado and the Semilla Sandstone Member in New Mexico (Glenister & Kauffman, 1985).

This study will describe and interpret facies of the lower Ferron Sandstone and Blue Gate Shale. Analysis of the lower Ferron sequence boundary and regional time correlative unconformities will follow, culminating in reconstruction of the depositional history of regressive and transgressive facies.

## LOWER FERRON SANDSTONE FACIES

The lower Ferron Sandstone can be divided into four facies (Fig. 2), in ascending order: a bioturbated siltstone facies; an interbedded hummocky cross-stratified sandstone and bioturbated siltstone facies (hummocky sandstone facies); an amalgamated hummocky cross-stratified sandstone facies (amalgamated sandstone facies); and a bioturbated sandstone facies. The first three facies form a coarsening upward sequence. The lower Ferron outcrop orientation is approximately parallel and perpendicular to depositional strike.

### Bioturbated siltstone facies

*Description*

The bioturbated siltstone facies is characterized by widespread burrowing, an increase in grain size and decrease in burrowing intensity upward, a massive weathering profile and large septarian concretions. It is commonly 8–10 m thick, has a gradational lower contact with the Tununk Shale, and occurs below hummocky cross-stratified beds (Fig. 2). The bioturbated siltstone facies coarsens upward from a mottled light and dark grey highly bioturbated (85%) shaly siltstone to a grey to light brown bioturbated (70%) sandy siltstone (<10% very fine-

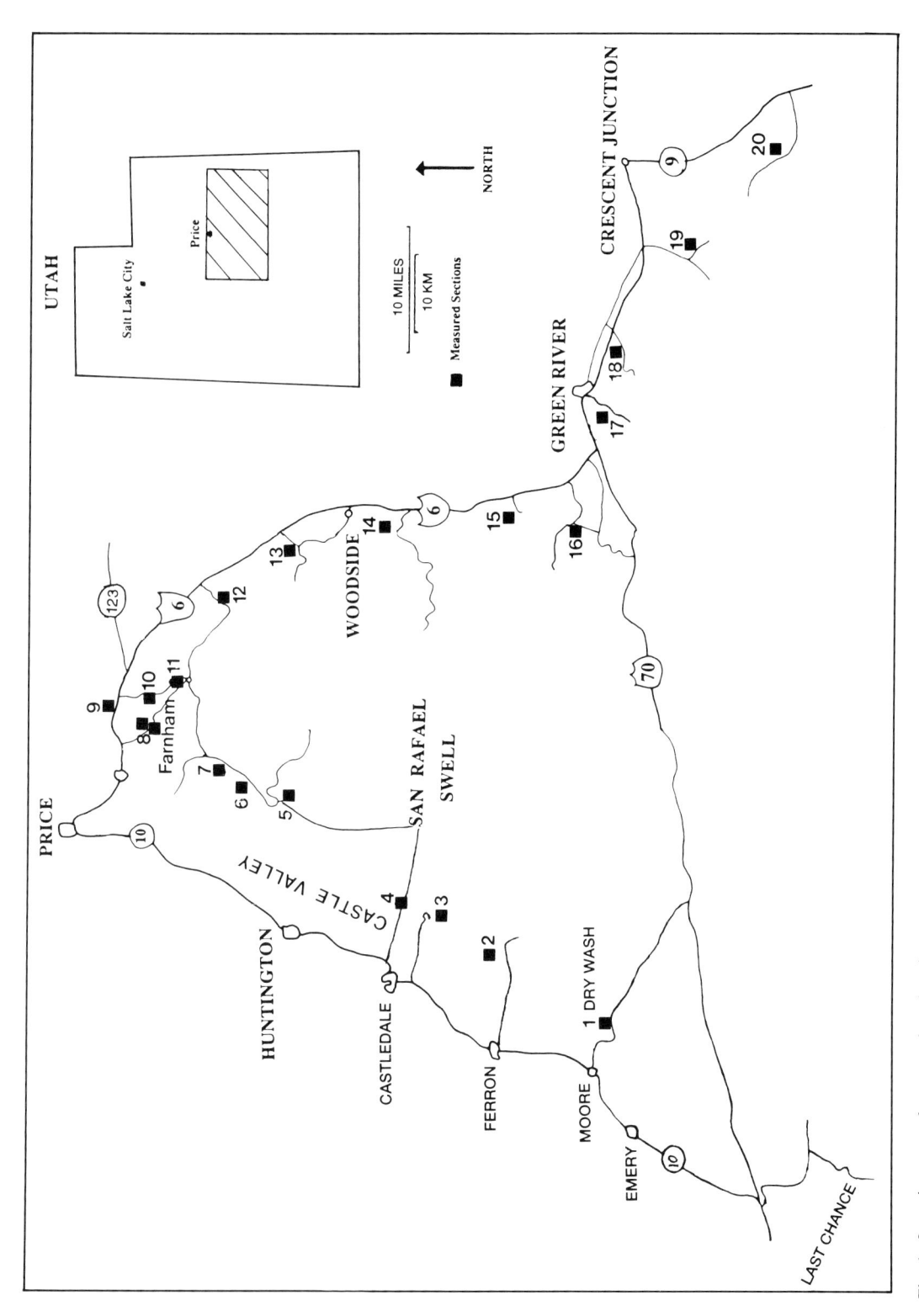

Fig. 1. Location map of measured sections in Utah.

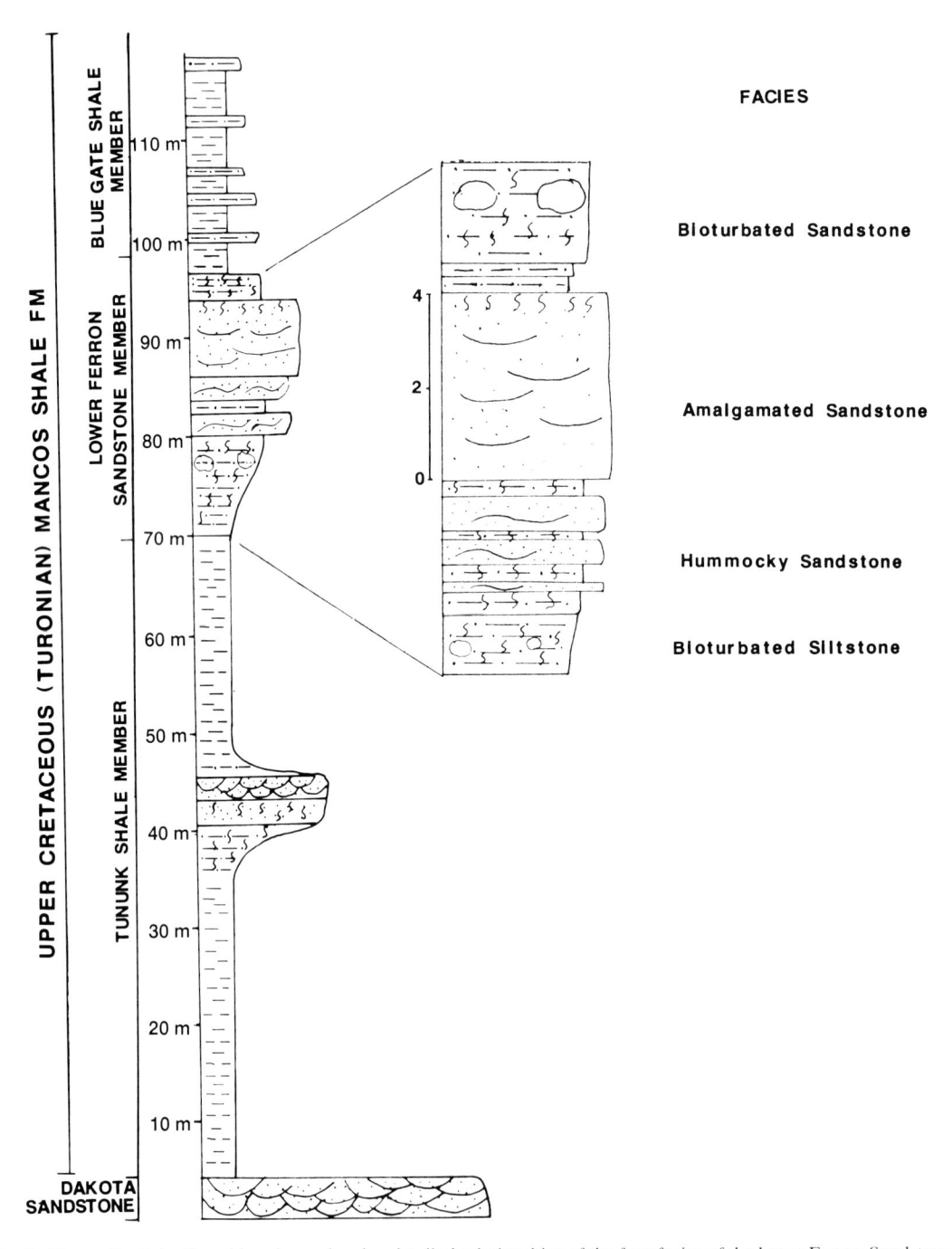

Fig. 2. Generalized stratigraphic column showing detailed relationships of the four facies of the lower Ferron Sandstone.

grained sand). The degree of burrowing decreases slightly upward and has obscured any previous bedding or sedimentary structures. Vertical to oblique *Teichichnus* burrows are visible in the uppermost (2-3 m) sandy portions, whereas few distinct burrow outlines are visible in the underlying mottled shaly siltstone.

This facies is best exposed approximately along depositional strike in a SW-NE orientation in north Castle Valley. It appears to thin and decrease in maximum grain size (sandy siltstone to shaly siltstone) to the southeast where it is exposed along the east side of the San Rafael Swell.

A spectacular septarian concretion horizon 3 m thick within the uppermost bioturbated siltstone facies can be traced laterally 64 km in Castle Valley (section 2 to section 12). Septarian concretions range from 10 cm to 300 cm in diameter and average an impressive 150 cm. Occasional isolated small fossiliferous concretions (<25 cm) also are found within the facies.

*Interpretation*

The bioturbated siltstone facies is interpreted to have been deposited below fair-weather wave base and probably below storm wave base, based on the evidence of: (1) widespread burrowing; (2) low sand content; and (3) position stratigraphically below hummocky sandstone beds. This facies also may have been deposited above storm wave base, with storm deposits completely burrowed, obscuring erosive surfaces. Inadequate sand supply offshore may have prohibited formation of wave bedforms.

The slight coarsening upward trend within the siltstone may reflect shallowing water conditions. This trend may also reflect increased deposition of suspended silt and fine sand from distal shoreface events deposited seaward of the lower shoreface. Biological conditions were optimal for complete homogenization of the sediment by deposit feeders.

**Interbedded hummocky cross-stratified sandstone and bioturbated siltstone facies**

*Description*

The interbedded hummocky cross-stratified sandstone and bioturbated siltstone facies (referred to as hummocky sandstone facies) is 2-4 m thick and contains hummocky beds separated by bioturbated sandy siltstone intervals with few relict sedimentary structures preserved (Fig. 2). The lower boundary of the hummocky sandstone facies is placed at the first hummocky sandstone bed above the bioturbated siltstone facies and the Tununk Shale. Hummocky sandstones vary in thickness from 5 to 150 cm. Sandstone beds thicken and thin laterally and may be almost completely bioturbated. Hummocky sandstone beds have sharp bases, may contain several internal truncation surfaces (especially in thick beds) and have wave-rippled or burrowed (*Ophiomorpha*) tops. Thompson *et al.* (1986) have locally identified trace fossils of *Asterosoma, Planolites* and *Rhizocorallium* within sandstone beds. No scour marks are observed at the base of beds. The hummocky sandstone facies is present in north Castle Valley and along the east San Rafael Swell. Gradationally overlying this facies near Farnham (sections 8 and 9) (Figs 2, 3, 4) is an amalgamated hummocky cross-stratified sandstone and elsewhere a bioturbated shaly siltstone.

*Interpretation*

The presence of hummocky sandstones suggests deposition above storm-weather wave base but below fair-weather wave base (Walker, 1979). Truncation surfaces, common within thick hummocky sandstone beds, record superimposed storm events (Dott & Bourgeois, 1982). Bioturbated siltstones represent periods of mud deposition following or during storms, with later reworking of the sediment by organisms. The hummocky sandstone facies reflects an increase in sand supply and dominance of physical wave processes over biological reworking of the sediment. This facies preserves a record of rapid, episodic (high fluid energy) sedimentation (hummocky sandstones) interpreted to occur in the lower shoreface.

**Amalgamated hummocky-cross stratified sandstone facies**

*Description*

The amalgamated hummocky cross-stratified (amalgamated sandstone) facies is 4-5 m thick at maximum and is not present at all locations (Fig. 3). This facies consists of a massive, fine-grained (150 μm) sandstone containing broad swales in gently dipping (<15°) laminations (Fig. 5).

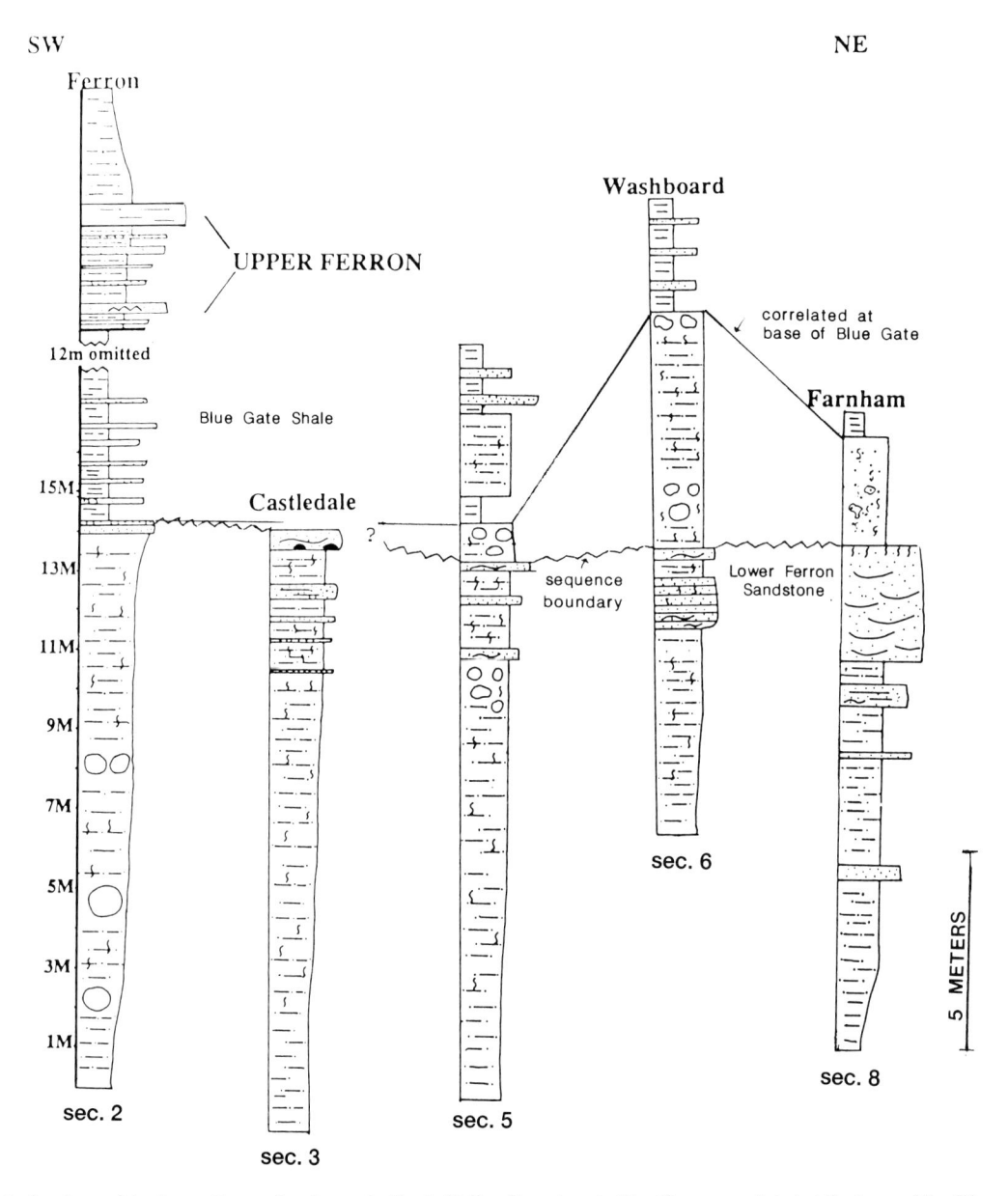

**Fig. 3.** Sections of the lower Ferron Sandstone in Castle Valley (locations in Fig. 1) are correlated at the base of the Blue Gate Shale and at an unconformity (lower Ferron sequence boundary). For key see Fig. 4.

Numerous truncation surfaces and rare shale lenses several centimetres thick produce amalgamated bedding. The original beds may have been 20–30 cm thick but the bedding surfaces have been reworked and obscured. The amalgamated sandstone facies is only exposed in an extreme northwestern outcrop belt near Farnham (Fig. 1). It overlies 2–3 m of thick hummocky sandstones separated by thin bioturbated siltstones (Fig. 3). No unconformity or erosional contact at the base of the amalgamated sandstone facies was observed in outcrop. The basal erosive surface of many swales is locally outlined by a shell lag (*Inoceramus*). Rare *Cylindrichus* occur within the unit and the upper 30 cm of the facies is moderately burrowed by *Ophiomorpha*.

The amalgamated sandstone facies is different-

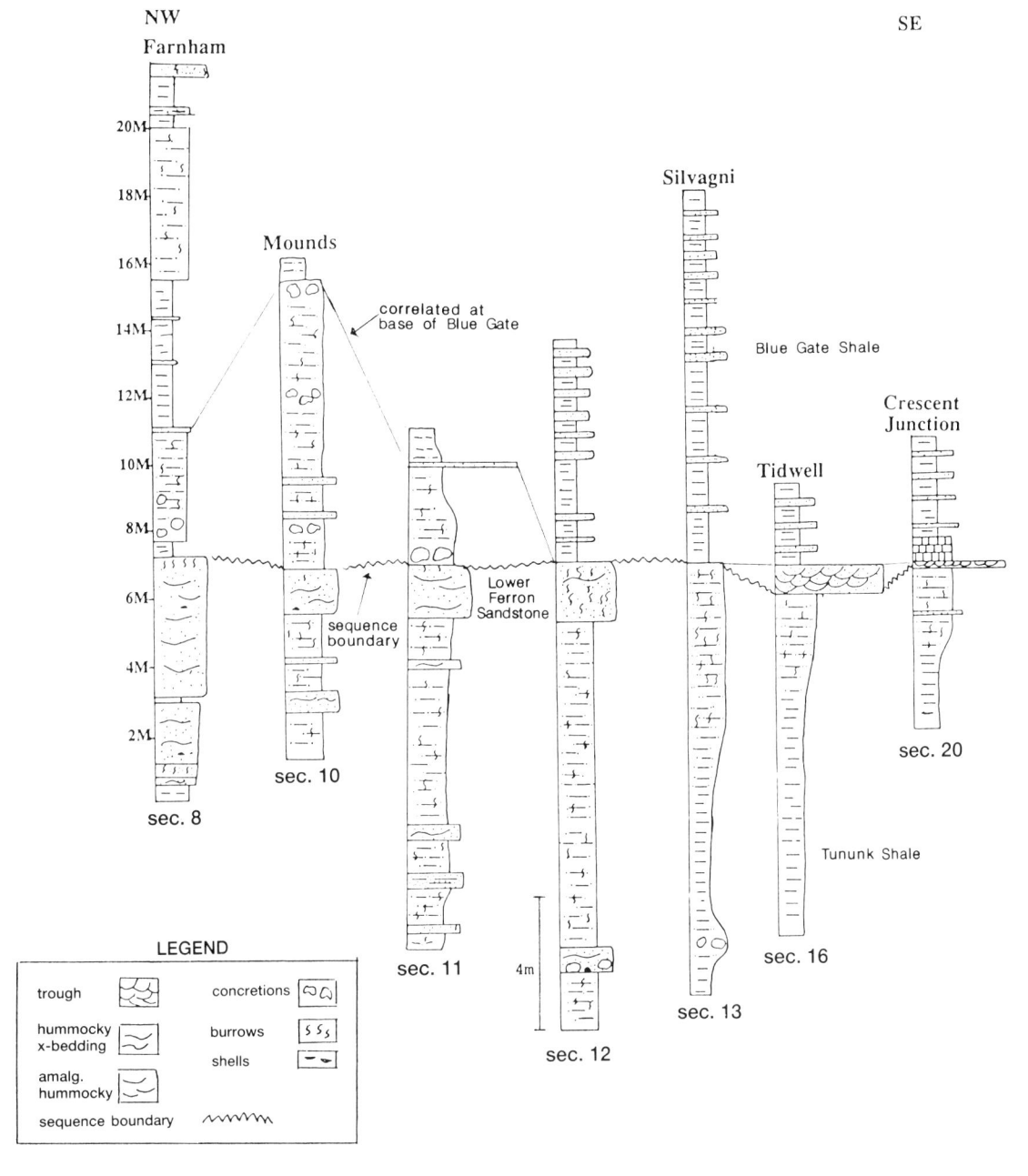

**Fig. 4.** Sections along the east flank of the San Rafael Swell are correlated at the base of the Blue Gate Shale. The sequence boundary lies above hummocky sandstones and below a thin coarse-grained sandstone (sections 16 and 20). Hummocky sandstones become increasingly bioturbated to the southeast.

iated from the hummocky sandstone facies by: (1) an absence of antiformal (hummocky) laminations; (2) a lack of burrowed or mudstone intervals; and (3) an increase in the number of truncation surfaces. Swales in amalgamated hummocky cross-strata are differentiated from trough cross-bedding as: (1)

Fig. 5. Amalgamated sandstone facies (4–5 m thick) contains broad swales in gently dipping lamination. Sandstone appears massive (owing to amalgamation of bedding) and fine grained. Trough cross-strata or extensive burrowing is absent.

swales are wider and shallower than trough cross-bedding (length of swales varies from 1–2.5 m and typical height is 30 cm); (2) dip of amalgamated laminations do not exceed 15° in contrast to the angle of repose lamination in troughs; and (3) amalgamated hummocky cross-strata geometry remains the same regardless of vertical outcrop orientation.

The amalgamated sandstone facies is similar to the shallow broad swales and low-angle subparallel lamination of swaley cross-stratification as described by Leckie & Walker (1982). Amalgamated sandstones may contain more hummocky laminations than swaley cross-strata but the essential difference between the two in our opinion has not been resolved fully.

*Interpretation*

Numerous truncation surfaces within the sandstone and the resulting amalgamation of bedding suggest that alternating erosion and redeposition events were common. Amalgamation of successive hummocky sandstone beds provides evidence of frequent storm events and storm deposition above storm wave base (Dott & Bourgeois, 1982). The homogeneous fine grain size reflects efficient sorting (winnowing) of suspended sand during transport, or a fine-grained source of sand. Shell lags common at the base of swales may have been swept offshore during storms and deposited at the erosive base of beds or winnowed during subsequent storms events.

The position of the amalgamated sandstone facies above the hummocky sandstone facies and similarity of swale geometry and grain size suggest a common origin for both facies. The amalgamated sandstone facies appears to be a high-energy variant of hummocky cross-stratification (Dott & Bourgeois, 1982) in which the swales are preferentially preserved and hummocks are eroded, resulting in numerous truncation surfaces (Dott & Bourgeois, 1982). Shell lags are useful in distinguishing significant first-order boundary surfaces (Dott & Bourgeois, 1982).

It is difficult to determine whether the amalgamated sandstone was deposited within fair-weather wave base. The facies probably preserves or records only episodic high-energy storm conditions and not fair-weather activities. Other descriptions of amalgamated hummocky cross-strata or swaley cross-strata infer deposition above fair-weather wave base with erosion or reworking of all fair-weather current and wave deposits during storms (Bourgeois, 1980). Amalgamated sandstones 40 m thick have been described in the transgressive Cretaceous Cape Sebastian Sandstone in Oregon (Bourgeois, 1980). Amalgamated sandstones with few burrowed zones in the Cape Sebastian Sandstone imply 'water depths at which storms of average intensity erode the bottom to a depth that would destroy evidence of everyday faunal activity' (Bourgeois, 1980). Other fair-weather deposits, such as mudstone beds, burrows and trough cross-strata, were likely reworked during intense storms (Dott & Bourgeois, 1982).

Bioturbated very fine-grained sandstones, siltstones and rare thin (10 cm) shale beds directly overlie amalgamated sandstones in the lower Ferron Sandstone and are inferred to record a relative rise in sea-level. As a result of the processes associated with the sea-level rise, the successive unit to be expected in a normal progradational sequence either was never deposited or was eroded from above the lower Ferron amalgamated sandstone facies.

Other investigators interpret amalgamated and swaly cross-stratification to be deposited in the upper shoreface (Tillman, 1986). A regressive shoreface sequence in the Lower Cretaceous Lower Gates Formation contains swaley cross-strata underlain by amalgamated hummocky cross-stratified sandstones and overlain by parallel-laminated beach sandstones (Leckie & Walker, 1982). Tillman (1986) concludes that 'in a shallowing sequence swaley cross-strata is always shallower than hummocky cross-strata and deeper than beach (foreshore) laminations.' He envisions swaley cross-strata to be an upper shoreface indicator.

The amalgamated sandstone facies probably represents a higher or a more sustained moderate fluid energy environment than underlying interbedded hummocky sandstones and bioturbated siltstones. The water depth of this facies is more difficult to determine. We consider it unlikely that the amalgamated sandstone facies was deposited above fair-weather wave base without preserving isolated angle of repose cross-strata or localized bioturbated zones within the sandstone. This facies probably represents what we informally term a middle shoreface, an environment which is just below fair-weather wave base and is heavily influenced by storm events.

**Bioturbated sandstone facies**

*Description*

Bioturbated, very fine-grained sandstones or shaley siltstones within the bioturbated sandstone facies vary in thickness from 2 to 6 m and overlie hummocky (sections 3–7, 10 and 11) and amalgamated (sections 8 and 9) sandstones in north Castle Valley (Figs 2, 3, 4). The contact of the bioturbated sandstone facies with underlying hummocky sandstones is planer and marks a sharp increase in burrowing and decrease in grain size. This contact represents a significant unconformity (the Ferron sequence boundary), which can be correlated throughout the study area.

At Farnham (sections 8 and 9), 3 m of light brown and grey, very fine-grained sandstone beds, 15–30 cm thick, appear wave-rippled. At this locality physical structures have not been obliterated completely by burrowing organisms (*Teichichnus*). Elsewhere, the facies is a heavily bioturbated (85%) mottled light and dark grey shaley siltstone. A large (2 m) septarian concretion horizon occurs within this facies. Scattered plant fragments are also contained within the sandstones and siltstones.

The bioturbated sandstone facies is laterally continuous in north Castle Valley but is not present on the east flank of the San Rafael Swell (south of section 11), probably owing to depositional thinning down dip and erosion during the transgression. The upper boundary of the facies is in sharp, planar contact with overlying fissile dark grey shale of the Blue Gate Shale.

*Interpretation*

The bioturbated sandstone facies records an increase in mud deposition and biological reworking of the sediment, and a decrease or cessation of storm-dominated sedimentation (amalgamated and hummocky sandstone facies). Slower sedimentation rates allowed burrowing organisms sufficient time to homogenize the sediment. Wave ripples at the most landward outcrop near Farnham suggest wave influence. This facies is interpreted to overlie an unconformity with very low relief that was formed during relative sea-level fall. Bioturbated sandstones and siltstones were deposited during sea-level stillstand and early relative sea-level rise. During the initial transgression, offshore sedimentation on the shelf was minimal as sediment was probably trapped landward of the shoreline. However, erosive retreat of the shoreface landward probably supplied some sand and silt to the adjacent inner shelf. Relative sea-level rise reduced the frequency, magnitude and effectiveness of storm and wave energies at the shelf bottom. The sharp upper contact of the bioturbated sandstone facies with the Blue Gate Shale separates periods of burrowing and sand deposition from mudstone deposition. The supply of very fine-grained sand and silt apparently decreased above this contact as relative sea-level rise continued.

The bioturbated sandstone facies is a transgressive facies of the lower Ferron Sandstone that overlies a significant regional unconformity (sequence boundary), and separates storm-dominated sandstone facies of the regressive lower Ferron from dark grey mudstones of the lower Blue Gate Shale.

## LOWER FERRON SANDSTONE—DISTRIBUTION OF FACIES

The bioturbated siltstone facies, hummocky sandstone facies, and amalgamated sandstone facies comprise a coarsening upward sequence deposited during a relative drop in sea-level. The distribution of these three lower Ferron Sandstone facies records a southeastward progradation of the shoreline. Outcrop exposures are roughly parallel to depositional strike in Castle Valley and down depositional dip along the east flank of the San Rafael Swell.

Outcrops of the lower Ferron Sandstone in north Castle Valley south of Farnham (sections 3–7) consist of hummocky sandstones underlain and overlain by bioturbated siltstones (bioturbated siltstone facies and bioturbated sandstone facies, respectively) (Fig. 3). Thickness of hummocky sandstone beds decreases to the southwest (1.5 m to 0.3 m) and are no longer present at section 2. The hummocky sandstone horizon can be traced continuously a distance of 32 km (sections 3–7). The amalgamated sandstone can be traced 8 km in a NE–SW direction without a significant change in thickness, supporting the interpretation that the line of section is parallel to the lower Ferron shoreline. The bioturbated sandstone facies becomes increasingly bioturbated and silty as it is traced south of Farnham. Near Ferron (section 2), the absence of hummocky sandstones makes it indistinguishable from the bioturbated siltstone facies (Fig. 3).

Outcrop exposures along the east San Rafael Swell contain hummocky cross-stratified beds that decrease in frequency and thickness to the southeast. The percentage bioturbation increases and grain size decreases, yielding a heavily bioturbated lower Ferron sandstone (section 12) and siltstone (section 13) (Figs 4, 6). As the amalgamated sandstone facies is traced 5 km south of Farnham in an offshore direction, it is gradually replaced by hummocky and wave-rippled sandstone beds. An increase in burrowing (especially at the top of beds), decrease in number of truncation surfaces and appearance of hummocky laminations accompanies the change from dominantly amalgamated sandstone beds to discrete hummocky sandstone beds. Three kilometres south of Farnham, the amalgamated sandstone has decreased in thickness from 4 m to 1.2 m of fine-grained sandstone. Few shell coquinas outline the base of swales.

**Fig. 6.** Dark grey shale of the laminated shale facies of the Blue Gate Shale (section 13) overlies grey bioturbated siltstones and silty shales correlative to the lower Ferron Sandstone. Note abrupt change in lithology at the planar contact (unconformity) between the Blue Gate and the lower Ferron Sandstone (arrow).

## COARSE SANDSTONE TRANSGRESSIVE LAG

A thin, pebbly, coarse-grained poorly sorted sandstone rests above siltstones of the lower Ferron Sandstone. It is distinct from lower Ferron sandstone deposits described previously and it rests above a significant unconformity separating the lower Ferron Sandstone from the dark grey shale of the Blue Gate Shale. Examination of the distribution, composition, sedimentary structures and thickness of this sandstone may help in evaluation of mechanisms responsible for creating regional and local unconformities.

### Description

The medium to coarse-grained poorly sorted pebbly sandstone contains locally abundant fish and shark teeth and large clay clasts up to 8 cm in diameter. Sandstones are often discontinuous and commonly 5–10 cm thick, with a maximum thickness of 50 cm observed at one locality (section 16) (Fig. 4). The sandstone contains variable distributions, abundances and sizes of chert pebbles (Fig. 7). Pebbles are: (1) randomly distributed through thick cross-bedded intervals; (2) concentrated at the base of beds and cross-bedded sets; or (3) concentrated as a pebble lag at the top of the sandstone unit. Composition of the sandstone reflects varying amounts of winnowing and concentration of biogenic material and chert pebbles. Thin transgressive lag sandstone beds (<5 cm) commonly contain a pebble lag at the base, some rare *Inoceramus* fragments, and are more poorly sorted than thicker beds, probably reflecting more extensive winnowing.

Trough cross-bedding is the dominant sedimentary structure in sandstone beds greater than 5 cm thick. Thin sandstones from 2 to 5 cm are commonly wave and current rippled. Tabular cross-bed sets range from 5 to 50 cm thick. At the thickest exposure (section 16) (Fig. 4), cross-bed set thickness decreases upward and the uppermost centimetres of the sandstone are reworked into wave ripples. The vertical sedimentary structure sequences record gradual decreases in depositional fluid energies. Palaeocurrent measurements ($n = 41$) from trough cross-bedding show a dominant south-southeast current direction. Occurrences of burrows are rare.

The northernmost outcrop of the coarse-grained sandstone is at one isolated locality (section 11), where a thin (2–5 cm) pebbly sandstone overlies the transgressive bioturbated sandstone facies of the lower Ferron Sandstone. The southernmost outcrop examined is south of Crescent Junction (section 20). The thickest exposure of the sandstone (section 16) can be traced 7 km northwest (section 15) to its gradual pinchout into a pebbly interval. This pebbly contact separates silty shales and siltstones of the Tunuk Shale (and correlative lower Ferron) and fissile shale of the lower Blue Gate. The transgres-

**Fig. 7.** Chert pebbles and biogenic debris concentrated at the top of the poorly sorted, coarse-grained, trough cross-bedded, lag sandstone (near section 16). This sandstone is interpreted to overlie an unconformity termed the lower Ferron sequence boundary.

sive lag, though areally widespread, is not very thick. These sandstones (with a maximum thickness of 1.5 m) can be described best as thin extensive sand sheets that do not appear to have been reworked into offshore sand ridges.

The lower contact of the sandstone is sharp and planar with no evidence of significant scour or channelling into underlying deposits. The upper surface of the sandstone exhibits local scour and erosional relief of up to 0.5 m. Rare, metre-wide medium-grained, lenticular, sandstone channel fills occur 3 m below the main coarse-grained sandstone.

*Interpretation*

Poorly sorted coarse-grained sandstones were likely deposited during relative sea-level lowstand as shallowing increased wave and current effectiveness and transported coarse material offshore. There is no evidence of subaerial exposure or broad channelling at the base of the transgressive lag to suggest incision and subaerial exposure of the middle and outer shelf. Submarine processes were likely responsible for transport or redistribution of the lag material.

Concentration of biogenic debris is most likely to occur by winnowing of shale and reworking of the sandstone during the initial rise in sea-level. Wave-rippled tops and a decrease in set thickness of the coarse-grained sandstone at section 16 are interpreted to represent a decrease in current energy as the transgression continued.

The source of coarse-grained sand and pebbles in the transgressive lag is difficult to determine. One possibility is that coarse sediment was eroded from fluvial stream gravels (unexposed in outcrop) as base level dropped during maximum sea level lowstand, then was transported far offshore during storms.

## LOWER BLUE GATE SHALE MEMBER FACIES

The lower Blue Gate Shale Member contains in its lower portion a laminated shale facies exposed along the west and east flanks of the San Rafael Swell and an upper calcarenite facies restricted to the eastern flank of the San Rafael Swell. The laminated shale facies is characterized by its dark grey colour and thinly bedded (fissile) or laminated structure. The calcarenite facies contains thin fossiliferous calcareous beds. These facies are lithologically similar and are correlative to the regionally widespread Juana Lopez Member in New Mexico and Colorado. The laminated shale facies unconformably overlies deposits of the lower Ferron Sandstone and correlative shales of the uppermost Tunk Shale. The fissile shale of the Blue Gate Shale Member (where it does not overlie sandstones and siltstones of the lower Ferron) is easily differentiated from the underlying light grey, silty Tunk Shale Member by its dark grey (almost black) colour, fissility and decreased silt content (Figs 6, 8).

**Fig. 8.** Laminated shale facies contains dark grey fissile shale with thin interbedded siltstones (section 12). Thickness and frequency of occurrence of siltstones increase up section. Characteristic dark grey colour and fissility distinguish this facies from silty grey shale of the Tunk Shale or overlying upper Blue Gate Shale.

### Laminated shale facies

*Description*

The laminated shale facies is 15 m thick and contains dark grey fissile mudstones interbedded with thin

(1–15 cm) siltstones. This facies in eastern exposures (Fig. 4) rests in sharp contact with underlying: (1) bioturbated and hummocky sandstones (section 12); (2) bioturbated siltstones (section 13) (Fig. 6); (3) silty shales probably correlative to the uppermost lower Ferron (sections 14 and 15); and (3) thin pebbly coarse sandstones and pebble stringers. This facies in north Castle Valley (sections 2–11) (Fig. 3) overlies several metres of brown massive calcareous sandy siltstones and very fine-grained sandstones included in the bioturbated sandstone facies of the lower Ferron Sandstone.

Dark grey shale is platy, thinly laminated and contains numerous thin (1–4 cm) bentonites. Thin siltstones and fine-grained sandstones are interbedded with fissile shale and make up 10–15% of the facies (Fig. 8). Siltstones and sandstones are 1–8 cm thick, increase in thickness and frequency up section, and are laterally continuous over 0.5 km.

Sandstones are commonly massive to parallel laminated and contain wave-rippled tops, flaser bedding, climbing ripples and sparse horizontal burrowing. Sandstones typically fine upward, but some wave-rippled sandstones have a thin (1–3 cm) medium-grained top. Sharp but planar bases of beds have locally abundant groove marks with a bidirectional N–S orientation. Flaser bedding in thin fine-grained sandstones and siltstones represent alternating deposition of sand and mud. Partial Bouma sequences, climbing ripples and groove-marked bases suggest probable rapid deposition from a turbulent decelerating flow.

The laminated shale facies is continuous along the west and east flanks of the San Rafael Swell. Within the laminated shale facies at Farnham in north Castle Valley is a 6 m, dark grey bioturbated (70%) shaley siltstone interval. Fissile mudstone of the laminated shale facies forms the majority of eastern exposures of the lower Blue Gate Shale (Fig. 4).

*Interpretation*

The fissility and fine lamination of the shale suggests an absence of burrowing, as burrowing would have disrupted the bedding. The dark grey colour, in contrast to surrounding grey marine shale, probably represents an increased organic carbon content. The hypothesized increased carbon and absence of burrowing organisms suggests that the bottom of the water column may have been oxygen deficient. These conditions may have been present only at the sediment–water interface. Sparse horizontal burrowing within siltstones is interpreted to represent a brief oxygenated period resulting from sediment gravity flows bringing in oxygenated waters, allowing a short-lived benthic community to develop (Byers, 1977).

Palaeocurrents, as determined from abundant groove marks and climbing ripples, suggest a southerly flow. Scour-marked bases, climbing ripples and partial Bouma sequences suggest thin siltstones and sandstones were deposited as distal turbidites or storm debris flows. Wave-rippled and coarser grained tops of siltstones and sandstones suggest that wave-dominated conditions existed, if not during, at least following deposition of these thin, extensive deposits.

A regional unconformity, herein proposed to be the lower Ferron sequence boundary, separates the laminated shale facies from underlying bioturbated siltstones, and coarse and fine-grained sandstones. Fissile black shale, similar to the laminated shale facies, seems frequently to occur close to or at the base of transgressive marine sequences (Hallam & Bradshaw, 1979). Restricted and dampened current and wave-generated activities during initial sea-level rise (Schalanger & Jenkins, 1976) and a shallow, broad shelf with a low oceanward gradient (Hallam, 1981) could inhibit biological activity and increase preservation of organic matter. Increase in thickness and frequency of siltstones beds and shell coquinas upward within the lower Blue Gate Shale could have resulted from an increased rate of sediment supply to the shoreface during continued sea-level rise.

**Calcarenite facies**

*Description*

The upper contact of the laminated shale facies is poorly exposed in outcrop and is gradational with a thin interval of calcareous beds. The calcarenite facies (3 m thick) is composed of fissile dark grey shale with thin calcarenite beds 1–2 cm thick composed almost entirely of *Inoceramus* fragments. The calcarenite facies is restricted to the east flank of the San Rafael Swell and has not been observed north of section 12. This facies is not well-exposed at all sections, however, calcarenite float at the top of most sections is common. This calcareous interval, only 3 m thick, is termed a calcarenite facies because of its recognition as a regional marker bed. At section 15 a sandy calcarenite 3 cm thick contains *Inoceramus dimidius* and *Prionocyclus wyomingenesis* (ammonite) fragments. An upper calcareous bed contains a coquina of well-preserved *Scaphites*. These fossils provide an Upper Turonian age for this interval, and are correlatable with the Juana Lopez Member in Colorado and New Mexico (Molenaar, 1975).

At one complete exposure (section 15) the gradational upper boundary of the calcarenite facies is defined as the first presence of a massive brown mudstone (1 m thick), signifying a change in shale colour and lamination from underlying dark grey fissile shale.

*Interpretation*

Thin calcarenite beds cap a coarsening upwards sequence and probably reflect an increase in current and wave energies during early sea-level rise. The thickest calcarenite beds in the Juana Lopez Member in New Mexico also occur at the top of the unit (Dane *et al.*, 1966). Abundant *Inoceramus* and local ammonite coquinas suggest that extensive winnowing and sufficient wave and current energies existed to form a lag of shell material. Carbonate deposition was likely enhanced by a virtual cessation of clastic deposition. Brown mudstones and grey silty shales of the Mancos Shale Formation above the calcarenite facies record increased clastic deposition, the end of oxygen deficient conditions (return to benthic activity), and signal a return to more typical marine shale deposition.

An early and late Juana Lopez age has been described previously for the laminated shale facies and calcarenite facies, respectively (Molenaar, 1975). Molenaar collected *Prionocyclus macombi*, *I. dimidius*, and *Lopha lugubris* (identified by W.A. Cobban) within 15 m of interbedded fissile black shale and calcareous siltstone near measured section 20 (Fig. 4) in this study. These fauna were given an early Juana Lopez age. Samples of *P. wyomingensis*, *I. dimidius*, *Baculites*, and *Scaphites* were collected by Molenaar from a sandy calcarenite bed at the top of the interval. This calcarenite unit can be traced from east of Farnham to the Grand Junction area.

## INITIAL DISCUSSION OF THE UNCONFORMITY

We suggest that an unconformity lies above the amalgamated sandstone facies and hummocky sandstone facies in north Castle Valley and above bioturbated siltstones and silty shales correlative to the lower Ferron Sandstone along the east San Rafael Swell (Figs 3, 4). The unconformity is identified and defined by a rapid vertical change in marine facies and the presence of a thin coarse-grained pebbly sandstone above this surface at sections 15–20. The unconformity, termed the lower Ferron sequence boundary, separates lowstand sea-level deposits from transgressive deposits. Recognition of subtle unconformities is essential in reconstructing the depositional history of many shallow-marine deposits. Well-exposed outcrops allow this unconformity to be traced above shoreface sandstones to offshore basinal shales. The unconformity separates (in a vertical succession) (Fig. 4): (1) amalgamated hummocky sandstone facies and bioturbated sandstone facies (sections 8 and 9); (2) hummocky sandstone facies and the Blue Gate Shale (section 12); (3) bioturbated siltstone facies and the Blue Gate Shale (section 13); and in distal sections is overlain by a thin coarse-grained pebbly sandstone (sections 15–20). The unconformity is planar with no erosional relief or truncation of underlying strata visible in outcrop. Coarse-grained sandstones and conglomerates in offshore basinal sequences in the Western Interior Cretaceous Seaway have been used to help identify unconformities (Plint & Walker, 1987).

It is difficult to determine if the lateral change from hummocky sandstones to bioturbated siltstones along the east flank of the San Rafael Swell is the result of depositional thinning in an offshore direction or erosional truncation by the lower Ferron unconformity. Because no current-generated features or coarse lag deposits occur at the top of regressive lower Ferron sandstones, and event marker beds (such as bentonite layers) have not been correlated precisely between sections, it is not possible to estimate the amount of erosional truncation of underlying strata or erosional relief on the unconformity. Recent studies of unconformities in the Western Interior Cretaceous Seaway in Canada illustrate significant regional erosion and relief of 20–40 m on a subtle and generally planar unconformity in outcrop (Bergman & Walker, 1987; Plint & Walker, 1987).

Detailed correlation of electric logs of the early Santonian Bad Heart Formation and surrounding marine mudstone formations in Alberta indicates at least 38 m of erosional relief on an unconformity overlying shoreface sandstones of the Bad Heart Formation (Plint & Walker, 1987). Subsurface correlations indicate truncation of well-log markers and thinning of the Bad Heart Sandstone, which in turn allows identification of significant regional erosion surfaces. Recognition of subtle erosional truncations such as these in the lower Ferron outcrops is very difficult. As in the lower Ferron Sandstone, angle-of-repose, medium-scale cross-bedding is absent in the Bad Heart Sandstone. Like the lower Ferron, the Bad Heart Sandstone is also abruptly overlain by a thick sequence of marine mudstone, however, in contrast, a thin conglomerate veneer is occasionally present above the Bad Heart Sandstone. In the lower Ferron, a thin, extensive, transgressive lag rests above siltstones correlative to the uppermost Tununk Shale and to the distal lower Ferron Sandstone.

## REGIONAL STRATIGRAPHIC RELATIONSHIPS

The unconformity (sequence boundary) at the top of the lower Ferron Sandstone can be correlated with the transgressive unconformity at the top of the Codell Sandstone Member of the Carlisle Shale in central and eastern Colorado. It is probably correlative to the lower contact of the Semilla Sandstone Member of the Mancos Shale Formation in New Mexico. The laminated shale and calcarenite facies of the Blue Gate Shale are similar and time-correlative to shales and calcarenites of Juana Lopez age in Colorado and New Mexico.

Time-correlative facies in Colorado are similar in lithology and structure to facies of the lower Ferron Sandstone and Blue Gate Shale. Offshore strata equivalent in age to the Ferron Sandstone near Pueblo, Colorado include the Blue Hill Shale Member and the upper middle Turonian sandstone of the Codell Sandstone Member (Scott, 1969). The Blue Hill Member of the Carlisle Shale gradationally underlies the Codell Sandstone. The uppermost 2.8 m of the Blue Hill is a highly bioturbated brown-grey sandy shale (Glenister & Kauffman, 1985), similar to bioturbated shaly siltstones underlying the lower Ferron sequence boundary along the east flank of the San Rafael Swell. The Codell Sandstone contains hummocky cross-laminated fine-grained sandstones (storm deposits) changing to trough cross-bedded middle to upper shoreface sandstones at the top of the section (Glenister & Kauffman, 1985). Disconformably overlying the Codell Sandstone is a shale informally termed the Upper Carlisle Shale Member. Near Pueblo, the Carlisle is a maximum of 51 cm thick with the lowermost 10 cm consisting of dark grey thinly laminated shale similar to the laminated shale facies of the Blue Gate Shale.

In the San Juan Basin of New Mexico and southern Colorado, the discontinuous offshore Semilla Sandstone Member is time equivalent to the Blue Hill Shale Member and the Codell Sandstone (Dane et al., 1968). At several localities in the San Juan Basin, where the Semilla achieves its greatest thickness, it is interpreted to represent sand ridges (La Fon, 1981). However, the Semilla appears to be expressed more commonly as a <1 m thick, medium to coarse-grained trough cross-bedded sandstone. This sandstone has been interpreted to overlie a sequence boundary (D. Nummedal, pers. comm., 1986) and appears similar and possibly time-correlative to the transgressive lag overlying the lower Ferron sequence boundary.

The lithology and fauna of the laminated shale and calcarenite facies is similar to what Weimer (1982), Hook (1980), and Glenister & Kauffman (1985) described for the Juana Lopez Member in Colorado and New Mexico. Shared characteristics include the thinly bedded structure of dark grey shale, increased frequency of thin siltstones upward and fossiliferous calcarenite beds at the top of the sequence.

Weimer (1982) identified and described the Juana Lopez Member in southwestern Colorado near Ridgeway. The Juana Lopez is 12.2 m thick and consists of fissile non-calcareous shale interbedded with thin siltstone layers. Siltstones are thinly bedded, rippled, and have scoured bases and sharp upper and lower contacts. The upper 6.1 m contains shell fragments of bivalves and ammonites, with fossils increasing in abundance upward. Ammonites found in the uppermost 30 cm of the Member include *Scaphites warreni* and *P. wyomingenesis*, with *P. macombi* found in float. The bivalve *Lopha lugubris* is found throughout the section. The gradational basal contact is placed at the first prominent siltstone and the top contact is placed at the top of the bioclastic layer. The Juana Lopez at this locality is underlain by 100 m of fissile, black, non-calcareous shale of the Mancos Shale.

In New Mexico, the thickness of the type section of the Juana Lopez Member in the east San Juan Basin is 32.2 m thick and the lower and upper contacts are conformable and gradational. The Juana Lopez consists of thin, highly fossiliferous calcarenites interbedded with dark grey fissile non-calcareous shale (Hook, 1980). The calcarenites are generally composed of *Inoceramus* prisms. The upper 3 m of the member form a ledge and the thickest calcarenites. This upper section contains *I. dimidius*, *S. warreni*, *P. wyomingenesis* and *Prionocyclus novimexicanus*. *Prionocyclus macombi* are found in the lower 3 m of the section.

At Pueblo, Colorado the Juana Lopez Member is highly discontinuous and is bounded by disconformities. It is represented by rounded lenses and cobbles of hard calcarenite that locally form lenticular beds (Glenister & Kauffman, 1985).

Identification of other unconformities underlying the Juana Lopez at the top of the Codell Sandstone in Colorado and below coarse lag deposits of the Semilla in New Mexico, support identification of the lower Ferron sequence boundary as a regional unconformity.

## DEPOSITIONAL HISTORY

The depositional history of the lower Ferron Sandstone and Blue Gate Shale is reconstructed from facies descriptions and correlation of outcrops parallel and perpendicular to depositional strike. The geometry and character of facies of the lower Ferron Sandstone and Blue Gate Shale were strongly influenced by relative sea-level changes. Thin regressive sandstones of the lower Ferron were deposited during relative sea-level fall, while transgressive facies of the lower Ferron and Blue Gate Shale record relative sea-level rise.

The lower Ferron Sandstone gradationally overlies silty prodelta shales of the Tununk Shale. The bioturbated siltstone, hummocky sandstone and amalgamated sandstone facies of the lower Ferron Sandstone form a shallowing upwards shoreface sequence (Fig. 9). The vertical succession of facies is increasingly dominated by episodic storm events, as evidenced by an increase in truncation surfaces within hummocky sandstones, increase in grain size and sorting, and decrease in burrowing. Outcrops of the lower Ferron Sandstone preserve storm and wave-dominated, fine-grained, middle and lower shoreface sediments interpreted to have been deposited as falling sea-level increased the frequency and magnitude of storm deposition. No fluvial or shoreline facies are exposed. Lateral facies relationships support a NE—SW orientation of the lower Ferron shoreface with an increase in water depth (and a progradation of sediments) to the southeast (Fig. 8). The source area for sediments was to the northwest, with the amalgamated sandstone facies being the most proximal landward deposits exposed. The lateral continuity of the relatively thin lower Ferron shoreface sequence and absence of shoreline facies suggest that the rate of progradation seaward may have been rapid. Sediments originally deposited in the (unexposed) upper shoreface were probably redeposited offshore as falling sea-level increased erosion at the base of the shoreface. Rapid sea-level fall resulted in more progradation than aggradation of sediments.

Deposition of storm-dominated facies of the lower Ferron ceased as continued sea-level fall resulted in non-deposition and formation of an unconformity. The unconformity (sequence boundary) above regressive facies of the lower Ferron was probably formed during maximum sea-level lowstand, with reworking or erosion of this surface during the subsequent transgression (Fig. 9).

Sea-level drop does not appear to have been rapid enough to subaerially expose the shelf in the study area. The unconformity is planar in outcrop with no erosional truncations, channelling or overlying non-marine facies. The lower Ferron sequence boundary was probably formed by storm processes with minor truncation and erosion of underlying facies.

Thin coarse-grained sandstones were deposited offshore, following or coincident with formation of the lower Ferron sequence boundary (Fig. 9). These lag sandstones are the coarsest sandstones observed in the study area and are seaward of lower Ferron shoreface sandstones. Coarse-grained sand and pebbles in the transgressive lag may have been eroded from (unexposed) coarse-grained shoreline deposits during sea-level lowstand, when waves were likely to be particularly effective in transport of sediment offshore. Another hypothesis is that as relative sea-level fall continued, a low oceanward gradient was formed. Fluvial streams would then prograde over a thin shoreline sequence far offshore, depositing a thin mantle of coarse sediment. Because of the low offshore gradient, significant incision of the shelf would not occur. Alternatively, the transgressive lag may have been deposited during early relative sea-level rise as erosional shoreface retreat eroded coarse-grained nearshore deposits and transported them offshore.

Though the origin of the lag is uncertain, reworking and winnowing of the sandstone lag by current and wave activity concentrated biogenic material. Conditions favoured formation of a thin

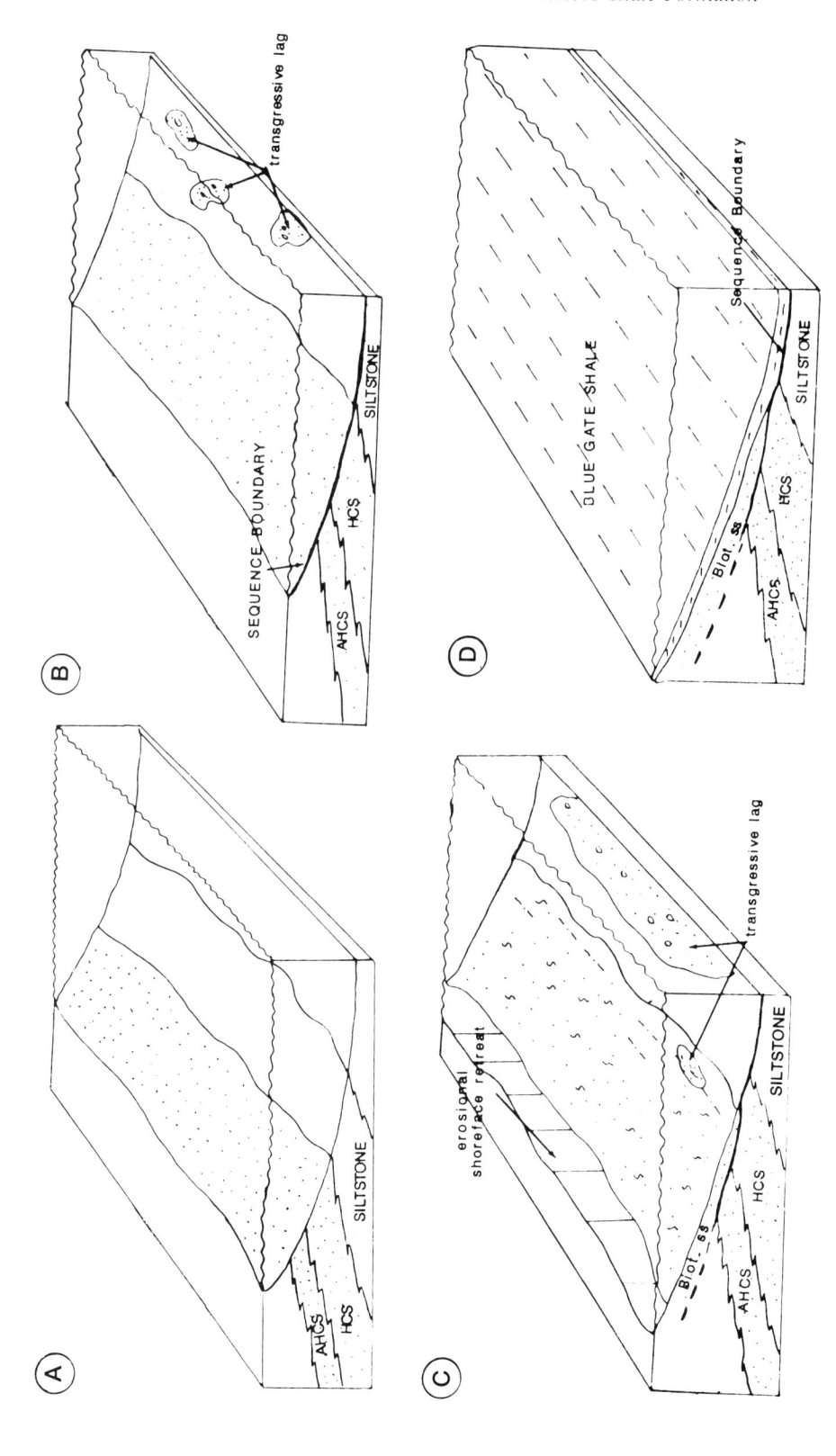

**Fig. 9.** Depositional history of the lower Ferron Sandstone and Blue Gate Shale. Rapid progradation (southeast) of lower Ferron Sandstone facies (A) with formation of the sequence boundary during sea-level lowstand (B). Transgression results in erosional shoreface retreat and deposition of bioturbated sandstone facies (C). Laminated shale facies of lower Blue Gate Shale is deposited as sea-level rise continues (D). AHCS = amalgamated sandstone facies; HCS = hummocky sandstone facies; Siltstone = bioturbated siltstone facies; biot. ss = bioturbated sandstone facies.

sheet of transgressive lag, deposited extensively offshore but not present everywhere, and not reworked into sand ridge deposits.

The bioturbated sandstone facies of the lower Ferron Sandstone is interpreted to have been deposited during sea level stillstand and early relative sea level rise. The bioturbated sandstone facies was probably derived from erosion of (unexposed) landward upper or middle shoreface sediments during erosional shoreface retreat (Fig. 9). Sand was deposited in the inner shelf just seaward of the shoreline. Very fine-grained sand was restricted to the most landward outcrop with increased silt and shale proportions offshore. The bioturbated sandstone facies is a likely time-correlative to distal coarse-grained sandstones. Pinchout of the bioturbated sandstone facies southeast of section 11 probably represents the seaward limit of the shoreline. Deposition offshore was slow and restricted, as much sediment was trapped landward of the shoreline. Eventually silt and sand-sized sediment supply decreased, and burrowing organisms thoroughly bioturbated the sediment of the bioturbated sandstone facies.

As the transgression continued, the laminated and calcarenite facies of the lower Blue Gate Shale were deposited during a period of low clastic influx and possibly dysaerobic conditions hostile to benthic organisms.

## SEQUENCE STRATIGRAPHIC CONCEPTS

A change in sea-level is influenced by the interaction of rates of sedimentation, subsidence and eustatic sea-level change. Any change in sea-level ultimately affects and controls the base level and accommodation potential of sediments and the distribution of facies within sequences (Haq *et al.*, 1987). Sequence stratigraphy emphasizes identification of key surfaces and unconformities or ' sequence boundaries that represent or record significant changes in base level owing to relative sea-level change. Facies in this study can be examined using the Vail formalism of sequence terminology as described by Haq *et al.* (1987).

### Regressive half-sequence, lower Ferron Sandstone

During a relative sea-level fall, base level is lowered and shelf sequences may record a shoaling upward sequence, termed the regressive half-sequence (Haq *et al.*, 1987). At maximum sea-level fall, storms may effectively transport coarse sediment far offshore. The lower Ferron Sandstone reflects a regressive interval deposited during sea-level fall. Outcrops of the lower Ferron Sandstone preserve storm and wave-dominated fine-grained middle and outer shoreface facies. A decrease in thickness of the lower Ferron Sandstone and an increase in bioturbation south along the east San Rafael Swell may reflect depositional thinning downdip or truncation of overlying sediments by the overlying sequence boundary.

### Sequence boundary

Identification, dating and correlation of sequence boundaries and correlative conformities on a regional and global scale will help define the origin of relative sea-level changes in relation to global eustatic sea-level changes or regional and local tectonic and sedimentological changes.

Type 1 unconformities are produced by a rapid sea-level fall extensive enough to expose subaerially the entire shelf. Prominent erosional truncations result (Haq *et al.*, 1987). When the sea-level fall is less extensive and slower, the entire shelf may not be exposed, resulting in a less prominent type 2 unconformity.

A sequence boundary is proposed to lie directly above the amalgamated hummocky sandstone facies (where present), directly above uppermost hummocky sandstone beds and directly below the coarse-grained transgressive lag (Figs 4, 10). The absence of any overlying non-marine facies, rooted zones and channelling associated with this surface support its interpretation as a type 2 unconformity. This sequence boundary was probably formed by submarine processes during maximum relative sea-level fall, with winnowing and redistribution of lag material following early sea-level rise. It is proposed that formation of the sequence boundary was just below fair-weather wave base, such that only minor truncation of underlying strata occurred. This unconformity appears regionally and is correlative to disconformities above the Codell Sandstone and below the Semilla Sandstone.

### Transgressive surface

A transgressive surface at the base of the bioturbated sandstone facies in Castle Valley and elsewhere at the base of the Blue Gate Shale is coincident with the sequence boundary. This surface

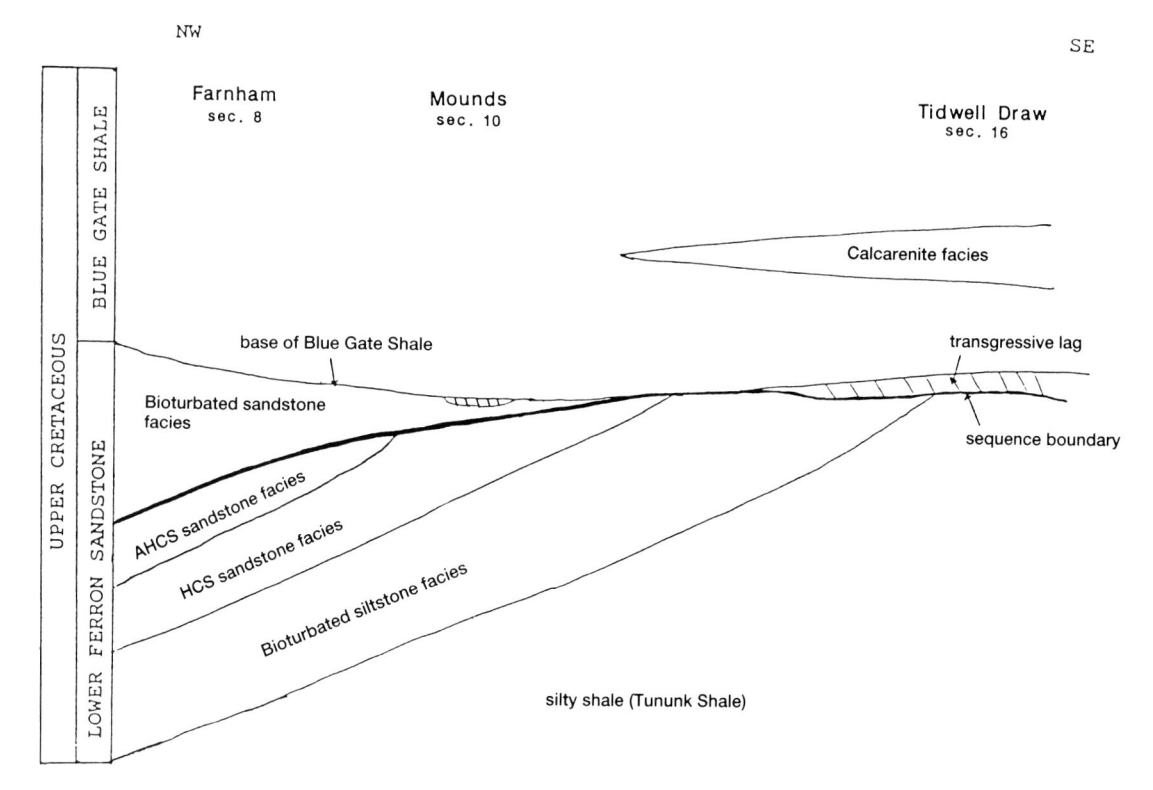

**Fig. 10.** Conceptual sketch (not to scale) of downdip facies along the east flank of the San Rafael Swell. The sequence boundary overlies regressive facies of the lower Ferron Sandstone. The base of the Blue Gate Shale overlies the bioturbated sandstone facies (transgressive lower Ferron) and a thin transgressive lag.

is interpreted to have formed during relative sea-level rise. In Castle Valley this surface marks the cessation of hummocky sandstone deposition. At sections 16–20, along the east San Rafael Swell, the base of the transgressive lag marks this surface. A transgressive surface of dual origin may form in type 2 unconformities. A nearshore ravinement surface, formed as erosional shoreface retreat progressively climbs landward, may be paired with an offshore erosional surface (Nummedal et al., 1987) associated with increased submarine winnowing. The transgressive surface in this study may be described best as an offshore erosional surface that may be paired with an unexposed ravinement surface in a landward direction.

## Transgressive half-sequence, lower Ferron Sandstone

Following maximum sea-level lowstand, rising sea-level chokes fluvial sedimentation. Transport of sediment on to the shelf is thereby decreased. Transgressive facies are deposited as waves and currents winnow and rework upper portions of the lowstand deposits during the early part of sea-level rise. The transgressive sequence of the lower Ferron Sandstone, the bioturbated sandstone facies, is only exposed as several metres of bioturbated sandstone and siltstone and is interpreted to have formed during a period of low sedimentation and widespread burrowing and effective winnowing of mud. The transgressive sequence is continuously exposed in north Castle Valley, roughly parallel to depositional strike. Deposits seaward of section 12 along the east San Rafael Swell were either not deposited or were eroded during the transgression. Near Green River, Utah, coarse-grained sand, pebbles and sharks teeth were reworked and concentrated by offshore currents and winnowing of shale.

## Transgressive Blue Gate Shale and condensed sequence

At maximum rate of sea-level rise, clastic influx has effectively ceased and carbonate deposition is enhanced. Thin limestone and calcarenites deposited at this time are termed a 'condensed sequence' and represent long intervals of slow deposition. A condensed sequence is globally synchronous but may differ slightly in age with local changes in subsidence and sedimentation (Haq et al., 1987).

The laminated shale facies was deposited during low clastic influx and possibly dysaerobic marine conditions. Calcarenites of Juana Lopez age in the study area reflect enhancement of carbonate deposition during maximum rate of sea-level rise and are termed a 'condensed' interval correlative over a large area of the Western Interior Seaway. Continued transgression resulted in deposition of grey blocky silty shale of the Blue Gate Shale and summoned a return to open-marine conditions. Continued sea-level rise resulted in sufficient space to accommodate thick prograding delta deposits of the upper Ferron Sandstone in south Castle Valley, and shale of the Mancos Shale Formation along the east San Rafael Swell.

## SUMMARY AND CONCLUSIONS

### Lowstand and transgressive shelf sands facies architecture

This study has documented the depositional architecture of the lower Ferron Sandstone and Blue Gate Shale Members of the Mancos Shale Formation in east-central Utah. The lower Ferron Sandstone is interpreted to have been deposited during falling sea level and is an example of shoreface and offshore sedimentation during sea level lowstand conditions.

The thin (2–6 m), laterally continuous sequence of storm-dominated sandstones of the lower Ferron Sandstone may be characteristic of shelf sandstones deposited along wave-dominated coastlines during sea-level fall. The effect of subsidence during deposition of the lower Ferron Sandstone was mitigated by falling sea-level, resulting in rapid progradation and little aggradation of shoreface sandstones. An absence of shoreline facies in the lower Ferron and little evidence for erosion of such deposits also supports rapid progradation of the shoreface.

The lower Ferron Sandstone contrasts with shoreface deposits of the upper Ferron Sandstone Member exposed in southern Castle Valley. The upper Ferron Sandstone in the Emery area consists of five major cycles of deltaic sedimentation resulting in a number of stacked progradational sequences (Ryer, 1981). Delta front sandstones vary from 15 to 25 m in thickness (Ryer, 1981). Sandstones of the upper Ferron Sandstone were built into a subsiding basin during a period of general, relative sea-level rise that provided accommodation space to accumulate and preserve a thick sequence of sediment. Seaward growth of these sandstones was relatively slow as the ratio of metres aggraded to metres prograded was relatively high.

Marine sedimentation during sea-level fall was probably a cycle of erosion and redeposition by storms. Erosion at the foot of the lower Ferron shoreface likely redeposited sand offshore on the inner shelf as hummocky sandstones. These sandstones probably did not have a high preservation potential, as continued sea-level fall would result in their erosion and redeposition farther offshore.

The lower Ferron sequence boundary is a subtle yet significant unconformity that separates regressive and transgressive deposits. The sequence boundary, curiously, does not appear in outcrop to have been a surface of significant erosion. However, similar erosion surfaces in the Bad Heart Sandstone in Alberta truncate well-log markers and show significant erosional relief. Though this surface is interpreted to have formed during maximum sea-level lowstand, it is coincident with a transgressive surface, and the two are indistinguishable.

Recognition of thin conglomeratic and pebbly intervals in offshore marine deposits of the lower Ferron has proven useful in identifying breaks in otherwise shaly deposition. Thin coarse-grained transgressive lag sandstones are extensive and appear to have been winnowed and redistributed during the transgression. How these sandstones were transported offshore remains speculative. The preferred interpretation is that coarse sand and pebbles were transported offshore during sea-level lowstand by storms. Alternative interpretations involve erosion of overlying coarse grained sediments during erosional shoreface retreat as sea-level rose. Outcrop observations seem to support submarine erosion and formation of the sequence boundary with erosional shoreface retreat possible in more landward facies.

Transgressive deposits of the lower Ferron Sandstone are thin and heavily bioturbated. Sandstone and siltstones of the bioturbated sandstone facies were likely eroded from the shoreline and shoreface during sea-level stillstand and early sea-level rise and deposited in the adjacent inner shelf. Early transgressive deposition in the lower Blue Gate Shale is dominated by (probably organic rich) mudstone, time-correlative and lithologically similar to the Juana Lopez Member. Conditions at the sediment–water interface were likely oxygen deficient over a large region, prohibiting benthic activities.

## ACKNOWLEDGEMENTS

This paper forms part of an MS thesis written at the University of Utah, under the supervision of Marjorie A. Chan.

Fieldwork was supported and partially funded by Atlantic Richfield Co., Marathon Oil Co., and a grant from the American Association of Petroleum Geologists. D.J.P. Swift, R. Tillman and an anonymous reviewer are thanked for their review of the manuscript and constructive suggestions. The senior author greatly appreciates the support and encouragement from D.J.P. Swift.

## REFERENCES

BERGMAN, K.M. & WALKER, R.G. (1987) The importance of sea-level fluctuations in the formation of linear conglomerate bodies; Carrot Creek Member of the Cardium Formation, Cretaceous Western Interior Seaway, Alberta, Canada. *J. Sediment. Petrol.* **57**(4), 651–665.

BOURGEOIS, J. (1980) A transgressive shelf sequence exhibiting hummocky stratification: the Cape Sebastian Sandstone (Upper Cretaceous), southwestern Oregon. *J. Sediment. Petrol.* **50**, 681–702.

BYERS, C.W. (1977) Biofacies patterns in euxinic basins: a general model. In: *Deep-water Carbonate Environments* (Eds Cook, H.E. & Enos, P.). Soc. Econ. Paleontol. Mineral. Spec. Publ. 25, 5–17.

COTTER, E. (1975) Late Cretaceous sedimentation in a low energy coastal zone: the Ferron Sandstone of Utah. *J. Sediment. Petrol.* **45**, 669–685.

DANE, C.H., KAUFFMAN, E.G. & COBBAN, W.A. (1968) Semilla Sandstone, a new member of the Mancos Shale in the southeastern part of the San Juan Basin, New Mexico. *U.S. Geol. Surv. Bull.* **1254-F**, 21 pp.

DANE, C.H., COBBAN, W.A. & KAUFFMAN, E.G. (1966) Stratigraphy and regional relationships of a reference section for the Juana Lopez Member, Mancos Shale, in the San Juan Basin, New Mexico. *U.S. Geol. Surv. Bull.* **1224-H**, 15 pp.

DAVIS, L.J. (1954) Stratigraphy of the Ferron Sandstone. *Intermountain Association of Petroleum Geologists, Fifth Annual Field Conference Guidebook*, pp. 55–58.

DOTT, R.H. & BOURGEOIS J. (1982) Hummocky stratification: significance of its variable bedding sequences. *Bull. Geol. Soc. Am.* **93**, 663–680.

GLENISTER, L.M. & KAUFFMAN, E.G. (1985) High resolution stratigraphy and depositional history of the Greenhorn regressive cyclothem, Rock Canyon Anticline, Pueblo, Colorado. *Golden, Colorado Field Trip no. 9*, pp. 170–183, Soc. Econ. Paleontol. Mineral. Field Trip Guidebook No. 4. Ann. Midyear Meet.

HALLAM A. (1981) Facies Interpretation and the Stratigraphic Record. W.H. Freeman, San Francisco, CA, 291 pp.

HALLAM, A. & BRADSHAW, M.J. (1979) Bituminous shales and oolitic ironstones as indicators of transgression and regressions. *J. Geol. Soc. London* **136**, 157–164.

HAQ, B.U., HARDENBOL, J. & VAIL, P. (1987) Chronology of fluctuating sea levels since the Triassic. *Science* **235**, 1156–1167.

HOOK, S.C. (1980) Reinterpretation of type section of Juana Lopez Member of Mancos Shale, *N.M. Geol.* **2**(2), 17–22.

LA FON, N.A. (1981) Offshore bar deposits of Semilla Sandstone Member of Mancos Shale (Upper Cretaceous), San Juan Basin, New Mexico. *Bull., Am. Assoc. Petrol. Geol.* **65**, 706–721.

LECKIE, D.A. & WALKER, R.G. (1982) Storm- and tide-dominated shorelines in Cretaceous Moosebar–Lower Gates interval outcrop equivalents of deep basin gas trap in western Canada. *Bull., Am. Assoc. Petrol. Geol.* **66**(2), 138–157.

MOLENAAR, C.M. (1975) Some notes on the Upper Cretaceous stratigraphy of the Paradox Basin. In: *Four Corners Geological Society Guidebook, 8th Field Conference, Canyonlands*, pp. 191–192.

NUMMEDAL, D. & SWIFT, D.J.P. (1987) Transgressive stratigraphy at sequence-bounding unconformities: some principles derived from Holocene and Cretaceous examples. In: *Sea-Level Fluctuation and Coastal Evolution* (Eds Nummedal, D., Pilkey, D.H. & Howard, J.D.). Soc. Econ. Paleontol. Mineral. Spec. Publ. 41.

PLINT, A.G. & WALKER, R.G. (1987) Morphology and origin of an erosional surface cut into the Bad Heart Formation during major sea-level change, Santonian of west-central Alberta, Canada. *J. Sediment. Petrol.* **57**(4), 639–650.

RYER, T.A. (1981) Deltaic coals of the Ferron Sandstone Member of the Mancos Shale. Predictive model for Cretaceous coal-bearing strata of Western Interior. *Bull., Am. Assoc. Petrol. Geol.* **65**, 2323–2340.

RYER, T.A. & MCPHILLIPS, M. (1983) Early Late Cretaceous paleogeography of east-central Utah. In: *Mesozoic paleogeography of west-central United States*. Society of Economic Paleontologists and Mineralogists, Rocky Mountain Section, West-Central United States Paleogeography Symposium.

SCHLANGER, S.O. & JENKYNS, H.C. (1976) Cretaceous oceanic anoxic events: causes and consequences. *Geol. Mijnbouw* **55**, 179–184.

SCOTT, G.R. (1969) General and engineering geology of the northern part of Pueblo, Colorado. *U.S. Geol. Surv. Bull.* **1262**, 131 pp.

THOMPSON, S.L., OSSIAN, C.R. & SCOTT, A.J. (1986) Lithofacies, inferred processes, and log response characteristics of shelf and shoreface sandstones, Ferron Sandstone, central Utah. In: *Modern and Ancient Shelf Clastics* (Eds Moslow, T.F. & Rhodes, E.G.). Soc. Econ. Paleontol. Mineral. Core Workshop 9, pp. 325–363.

TILLMAN, R.W. (1986) Swaley cross-stratification and associated features, Upper Cretaceous Western Interior Seaway of United States (abstr.) *Bull., Am. Assoc. Petrol. Geol.* **70**(5), 656.

WALKER, R.G. (1979) Facies models 7, shallow marine sands. In; *Facies Models* (Ed. Walker, R.G.). Geosci. Can. Reprint Ser. **1**, 75–89.

# Index

References to figures appear in *italic type*.
References to tables appear in **bold type**.

A sandstone sequence 459, *467*, 467, 479
　upward-coarsening pattern 467
A sequence 459, *467*, 481
Abrolhas Bank 261
acceleration zones 24, 27
accommodation, subaerial 221, 222, 229
accommodation/supply ratio 18–19, 181, 182, 183–4, 195–6, 235–7, 248
　changes controlled by R 241, *242*, 243
　rapid increase in 228
　and sequence architecture 227–34
accretionary wedges 6
accumulation rate 8, 22, 38, 39–40, 42, 43, 64, 79, 84, 98, 104, 117
　decreased 75
　determination of 294, 297–8
　dynamic 9, 21, 25, 78
　high 96
　Mesaverde Formation 72
　and preserved bed thickness 65
　*see also* sedimentation rates
accumulation zones 143, 233
Adaville Formation 238
Adriatic shelf 116
advection, along-shelf 137
aerial photos, use of, Doce strandplain 260–1, *262*
Alabama margin 161
Alberta Basin 174–5
Alberta Group 459
allochemical particles 105
allocyclic forcing 184
allostratigraphic units 156
alluvial fan (fill) 194
alluvial fan geometric systems tract 247
alluvial fan systems 332
alluvial fans 93, 94
　Tertiary 261, 264
alluvial ridges 264
Altamaha River Inlet 384
amalgamated beds 100
　hummocky 112
amalgamated facies 167
amalgamated sand and gravel facies 139
amalgamated sand lithofacies 97, 98, 128, 143, 170, 195, 212, 216, 234
　cross-stratified/micro-cross laminated 131
　lagoonal channel 131
　proximal 129, 130, 145
amalgamated sandstone lithofacies 109–10, 112, 115, 172, 493–7, 498, 502, 504
amalgamated sandstone sequences *312*, 321
amalgamated shale lithofacies 115
amalgamated washover beds 128
amalgamation depth, critical 67
amalgamation limit 101

amalgamation surface 110
Amazon River mouth 107, 115
Amazon River plume 440
Amazon shelf 9, 212, 453
　condensed sedimentation 213
　highstand sedimentation *218*
　prograding mud sheet 219
　sedimentary structures 452
*Ammonia beccarii* 408, 410, 412, 415, 418
ammonite coquinas 502
*Anachis obesa* 276
*Anomalocardia brasiliana* 269, 270, 271, 272
　radiocarbon dating of 276
Appalachian Basin 309, 330, 372
Appalachians, southern
　depositional history 329–32, 330, *331*, 332
　palaeoenvironments and palaeocurrents 313, *331*
　stratigraphy 311–13
　tectonic history 310–11
Aquitaine shelf *160*, 181
　early Holocene parasequences 161
Argentine coast 161, 164
Argentine shelf 181
*Arthrophycus* 315, 317, *349*, 351, 353, 355, 356, 364, 371
*Asterigerinata* cf. *mamilla* 410, 412, 413, 418
*Asterosoma* 493
Atchafalaya River 166
Atlantic coast, USA, relative sea-level rising 259–60
Atlantic shelf 127, 216
　Cape Hatteras 133
　eastern, USA
　　morphology 402
　　sediment studies 402
　Georgia Bight 124
　highstand preservation potential on 246–7
　middle, Pleistocene deposits 247
　Middle Atlantic Bight 137, 161–2, *162*, *163*, 164
　　formation of sand ridges 395–421, 453
　Middle Atlantic shelf 181
　modern, thalweg deposits 206
　New England shelf *142*, 143, 212, 233
　New Jersey shelf 161–2, 181, *211*, 396–421
　North American 25, 27
　North Carolina shelf and coast 161–2, 384
　southeastern USA
　　affected by transgression and regression 379
　　scour depth of inlet gorges **383**
　　seismic profile analysis 384–91
　USA 211, 212
Atlantic shoreline, modern, modification of migrating barriers 120–1
autocyclic forcing 184
autocyclic switching 228–9, 237
autocyclicity 157–8, 183
　potential role of 181–2

511

avulsion 158
  river 165, 166, 184, 199, *200*

back-barrier depositional systems 145, 241, 247
back-barrier depositional systems tract **106**, 209, 210
back-barrier flooding *120*, 120
back-barrier sediment wedge 247
back-barrier wedge (fill) 210, 233
back-step fill *205*, 241
  fluvial 210
back-step shelf wedge (fill) 210–12
back-step shelf wedge systems tract 233
back-step systems tracts 241, 243
back-step wedge deposits 237, 243
back-step wedge (fill) 194
back-step wedge (fill) deposits 133, *134*, 206, 207
back-step wedge (fill) systems tract *205*
back-step wedge geometric systems tracts 202, 204–12, 231–2, 234, 247
  transgression, Vail and Galloway 204–6
back-step wedge systems tracts 241, 243
  comparison with modern transgressive deposits 206–12
  development of 233
back-step–offlap couplet 217
back-stepping valley fill 233
backshore beds 128
*Baculites* 502
Bad Heart Formation 502–3, 508
Balize subdelta 166
barchan sand waves 141
barrier coastlines, channel development along 379
barrier formation, causes of 119–20
barrier island coast 118
barrier island systems 428
barrier islands 16, 382
  degraded 396, 420
  dispersal systems *127–8*, 127–8
  migration of 209
barrier lagoons 379
barrier migration 120
  sand budget *126*
  sand dispersal 125–7
barrier overstep 123, *124*, 127
barrier retreat 120–3
  continuous 120, 121, *122*
  cyclic 123, *125*
  discontinuous modes 123
  marsh–lagoon mode 121, *122*, *123*
  open lagoon mode 121, *122*, *123*
barrier roll-over 16, 120, *121*
barrier shorefaces, erosional retreat 126
barrier step-up *124*
barrier–island–lagoonal systems 260, 279
  formed by drowning 264, *265*
barrier–lagoon system 123
barrier–lagoon systems tract 205
barrier–lagoon–estuary systems 118–31
barrier–overwash deposits 240
bars
  offshore 428
  river-mouth 15
base level changes 36–7
basin fan (canyon-cut) onlap surface 230–1, 236, 237
  formation of 227–9

basin fan (fill) 194, 198–9
basin fan (fill) subsystems 202–4, 231
basin fan (fill) systems tract 228
basin fan geometric systems tract 247
  comparison with modern lowstand deposits 198–204
  models and observation 198
basin fan initiation 228
basin fan systems tract 198–9, 201
  formation of 236
  prograding complex 212
  seismic identification 198–9
basin fan-prograding complex sequences, autocyclic 228–9
basin infill, controls on style of 6–9
basin sedimentation 197
basin-floor aggradation 363
basin-floor fan 198, 202–3, 235, 236, 237
  development of 230–1
basins
  cratonic 6
  deposition on floor of 459–60
  evolution of 92
    and textural maturity 93
  foreland 5, 6, 168, 170, 243–4
  subsiding 48–9, *50*, 233, 508
bayline 221
bayline advance, strandplain and deltas 223
bayline migration 221
bayline and shoreline, relative movements of 223
beach prism 128
beach ridges 114, 165
beach systems 128
beach–dune–washover fan depositional complex 172
beach-ridge deposits, Pleistocene 279
beach-ridge sets, cusp-shaped 266, 267
beach-ridge terraces 264
  Pleistocene 268, 270, *272*
Beaufort shelf, glaciomarine sediment wedge 212
bed roughness 81, 82
bed sets
  upward-coarsening 179
  upward-fining 183
bed-shear approximation, maximum 82
bed-shear velocity 82, 83
bedding
  current 448, 451
  flaser 444, 501
  graded 440, 444, 466
  inclined 387
  parallel 269
  wavy 440, 444
bedform migration 96, 317–18, 328, 332
bedforms 402
  combined-flow 289
  incipient development of *288*, 288–9
  shelf sand 318–19
bedload armouring 60, 61–79, 79–84
bedload convergences 142
bedload layer 62
  thickness of 62, 83
bedload partings 137, 141
beds
  graded 439
  hummocky 95, 96

stacked 410, 412, 417
transgressive 78
Bengal shelf 116, 143
bentonite 501
  episodic 172
bentonitic markers 178
bidirectional downlap 198, 203, 230
Big Horn Basin 172, 241
biogenic debris, concentration of 499, 500
bioturbated mud facies 115, 128, 195, 212, 216
bioturbation 109, 290, 294, 324, 333, 408, 410, 437, 440, 448
  burrow traces 411
  and the reworking zone 320
  *see also* mudstone, bioturbated; sandstone bioturbated; siltstone bioturbated
biozones, and condensed sections 213
Blackhawk Formation 96, 112
Blackstone Formation 459
Bloomsburg Formation 325
Blue Gate Shale Member (Mancos Shale Formation) 490, 497, 502, 503
  condensed sequences 508
Blue Hill Shale Member (Carlisle Shale) 503
body fossils 340, 346, 351
  as basal lags 324
bottom currents 328
  flow patterns 301–2, *303*
  offshore-directed 369
  storm-generated *21*, 61
  wind-forcing of 284
bottom flows 15
  geostrophic veering of 368
bottom friction, effect of 54
bottom wave power 117
Bouma sequences 96
  partial 501
Bouma structures 145
boundary layer, laminar 84
boundary roughness 81
boundary shear stress 284
bounding surfaces 98, 110, 227, 235, 248
  *see also* sequence boundaries
*Brachidootes exustus* 269
Brassfield Formation 312, 313, 325, *326*, 333
Brazilian coast 161
  east
    microtidal 261
    physiographic units 261
    relative sea-level, falling 260
break-point bars 114, 126, 128
Brevard fault system 311, 330
Bridge Creek Limestone *174*, 180
  high-frequency condensed sequences 173–4
British Isles, sediment dispersal pathways 137, *139*, 141
Bruun model 39
  and shoreface retreat 16–17
Bruun's rule 277–9
*Buccella frigida* 410, 412, 413, 418
*Buliminella elegantissima* 409, 415
buoyant spreading 115
burial, permanent 17, 19, 21, 22, 61, 100
buried channel complexes 385–6
buried channels, as palaeostream valleys 383

butterfly effect 182
bypass surface, marine 78
bypassing 11, 71, 93, 115
  coastal 96
  differential 27
  down-shelf 13, 15, 17, 117
  of fines 233
  of gravel 178, 240
  of mud 161
  of suspended sediment 142
  in transgressive shelf regimes 133
  upper-slope 199
  *see also* river mouth bypassing; shoreface bypassing
bypassing ratio 210

$^{14}$C dating
  bulk shell material 411–12, 412–13, 417
  shell fragments 410
Cabot Head Formation 342
calcarenite 503–4, 506, 508
  lower Blue Gate Shale 500, 501–2
Calvert Formation 166, 167, 168
Caminada strandplain 126
Canadian Shield 91
canyon formation 199, 228, 236
canyon-cut 227
canyons, modern 212
Cape Sebastian Sandstone 496
carbonate bank-to-shelf systems 333
carbonate beds 313, 325–6, *326*
  evidence for combined-flow deposition 325–6
  shelf margin *331*, 332
carbonate sand 262
carbonates
  bioclastic 333
  deposition of 92, 502
  fossiliferous 364
  sequence-bounding 364
Cardium A sandstone 459
  lithofacies associations 459–66, 474–7
Cardium B sandstone *430*, 431
  lithofacies associations 434–48
Cardium Formation 396, 457–85
  erosion in 482, 483–4
  lithofacies schemes **435**, 459
  repeated upward-coarsening facies successions 467–8
  sand bodies in 457–85
  sand ridges in 427–53
  *see also* Cardium A sandstone; Cardium B sandstone
Cardium shoreline 484
Cardium Zone Member 437, 448, 459
Cardium zone top 429, *431*, *449*
Carla horizon 294, *297*, 297, 298
  thinning and thickening 294, 298
Carlisle Shale 503
Caroline Field 458, 468, 479
Caroline pool 428, 449, 451
Carrot Creek Field 484
Castanea Member (Tuscarora Sandstone) 317, 320–1
Castlegate Sandstone 240–1
Chandeleur Islands 166
Changjiang delta platform, East China Sea 440
channel association, Medina Group 351–4, 357, 363
channel complexes, stacked *351*, 352, 359

channel cut-and-fill, multiple stages 386
channel density 385, 386
  combination of factors 392
  decreases in *386*, 391
  patchiness in 391
  patterns variation *386*
  shore-parallel 391
  upper shoreface 391
  and variation in channel characteristics 387, 389
channel deposits, buried, preservation of 379–92
channel depth, initial 381–3
  inlet channels 382–3
  lagoonal channels 381
channel development, shoreface 391
channel fill 340, 351, 387
  coarser-grained *341*, *351*, *352*, 352–3
  finer-grained 353–4
  nature of *387*
  sandstone 500
channel incision, depth of 352
channel preservation, determining processes 391
channel sequences, fluvial 428
channel stacking 385, 386, *387*
channel switching 266
  allocyclic 267
channel traces, migration path of channel 384
channel width, true vs. apparent 385
channels
  formation of 391
  incised 340, *341*, 364, 365–6, 367, 373
  internal structures 385
  migrating *387*
  potential for preservation 381
  sandstone-dominated 351
  sinuous, meandering 354, 363, 369
  small, incised by storm-induced downwelling 370
  tidal 129, 131, 363, *381*
  truncated by minor sequence boundaries 359, 369
chaos theory 35
chaotic behaviour 182–3, 184
chenier plains 114, 223
chert grains 439, 452
chert pebbles/clasts 447, 465, 484, 499
  redistributed 484
Chesapeake Bay 166, *167*
Chincoteague shoal, sediment dispersal in vicinity of *140*
*Chione cancellata* 269, 271, 272
*Chondrites* 323, 324, 438, 440, 444, 461
*Chondrites* ichnofacies 346
Choptank Formation 166, 167
*Cibicides lobatulus* 410, 412, 413, 418
clastic sediments 92
  supply rate 195–6
clastic source 294
clastic wedge geometry 238, *239*
clastic wedges 309
  deposition of *331*, 332
  Ferron Sandstone 490
  model, $Q$-dominated 178
  multiple wedges 168
clay chips 440
clay clasts 499
climate

and the Atlantic shelf 405
and strandplain sedimentation cycles 483
climatic cycles 7, 114, 174, 182
Clinch Sandstone 311, 312, 315, *318–19*, *320*, 330
  palaeocurrent data 313, *314*
clinoform structures
  expanded 172, *173*
  internal 116
Clinton Group *341*, 342, 361
coarsening-up cycles 172, 174–5, 179
coarsening-up units 159, 167
coarsening-upward sequences 167, 172, 410, 412, 415, 417, 502
  Cardium A sandstone 453, 459, 467–8, 474–6, 484
  lower Ferron Sandstone 498
coastal compartment orientation, North American Atlantic Shelf 16, 121–2, 137
coastal engineering, approach to shallow-water regime 52–5
coastal equilibria 223
coastal equilibrium profiles
  $Q$-dependent case 39–41, 46, 113
  $R$-dependent case 37–9, 45, *47–51*, 76, 113, 224
  transgressive shape 78, 119–23
coastal evolution, during highstand and sea-level fall 221
coastal ocean dynamic zones *299*, 299–300
coastal plain 224
  and accommodation/supply ratio 225–6, 229
coastal plain system, sediment transport efficiency 225–6
coastal progradation 12–19, 223
coastal retreat 16
coastal-plain–slope unconformity 193, 198, 199, 247
coasts
  microtidal 261
  sand-free 117
  tide-dominated 379, 381, 384
  transgressive
    fine sediment source 127
    mesotidal and macrotidal 119, 367
    microtidal 118–19
    rocky 119
    sediment dispersal 125–7
  wave-dominated 379–80
Codell Sandstone Member (Carlisle Shale) 490, 503, 506
Cody Shale 172, 181, 239
coffee-rock 268, 270
cohesion, and sediment transport 80–1
Columbia River plume 143, *144*
combined flow 284–5, 304, 323, 328–9, 484
  conditions for 288
  deposition of carbonate beds 325–6
  intense 447
  interaction with fine-grained sediment 333
combined-flow power, decrease in 294
combined-flow processes 325, *326*
combined-flow transport patterns
  and Carla sand bed 304
  near-bottom *303*
compositional maturity 92
condensed intervals, and maximum flooding surface 213–17
condensed sections 98, 100, 103, 130, 145, 160, 213, 246
  back-barrier 167

distal 104, 105, 117, 234
  inundites 110
  and preservation filter 21–3
  proximal 234
  transgressive setting 103–4
condensed sequences
  Blue Gate Shale 508
  high-frequency 173–4
*Coneichnus* 460
conformities, correlative 157
conglomerate 445, 459, 502
  Cardium A sandstone 479, 484
  Cardium B sandstone *438, 441*, 447–8
  Willesden Green Field *464*, 465–6, 468, 473–4, 476, 477
conglomerate stringers *465, 466*, 476, 479
conjugate systems 102–3, 105, 112, 127–8, 209
  stacked 145
Connecticut shelf valley 212
contacts
  occluded 145
  serial 210
continuity principle, and shelf flows 24
coquina lenses 173
*Corbula swiftiana* 269, 272
Corinth Gulf 229
Coriolis force 300, 328, 368
*Crassostrea rhizophorae* 274
*Crepidula plana* 269
Cretaceous Western Interior Basin
  depositional sequences and parasequences 168–78
  *Q*- and *R*-dominated sedimentation patterns 238–44
  repeated flooding of 93
  ¥-dependent models 241, *242*, 243
Cretaceous Western Interior Seaway 490, 508
cross-bedding 269, 275, 276, 410
  tabular 499
  trough 346, 353, 356, 357, 361, 367, 499
    supporting a straight shoreline 362
  unimodal 318
cross-lamination 412, 440, 444
  asymmerical ripple 356
  climbing-ripple 346
  current-ripple 465
  ripple 109, 461, 462
  wave and current ripple (RCL) 445
  wave-ripple 327
cross-sets
  compound 316–17
  trough 318–19
cross-strata
  formation of 343
  high-angle *325*
    giving palaeocurrent data 313, *314*
    and storm-induced currents *314, 326*
  planar-tabular sets 315, *320*
  trough 315
cross-strata sets 109–10
cross-stratification 131, 312, 325
  high-angle *312*, 319, 323
  hummocky (HCS) 71, 72, 82, 289, 312, 321–3, 351, 353, 367, 447, 463, 464
    early interpretation 343–4

  generation of 323
  low-angle inclined (LIS) 445, 447
  micro-hummocky (micro-HCS) 324, 325, *326*, 329
  swaley 495–7
  trough 112, 445
  wave-ripple 463
Crossfield Field
  external geometry 429–34
    shape 434
    structure 429–31
  internal geometry *432–3, 438*, 448–9
Crossfield pool 428
  lithofacies 434–48, 453
  a submarine sandstone ridge 451
Crossfield reservoir 449, 452
Crossfield ridge
  contained within mud envelope 452
  formed in response to storm flows 451
  growth on muddy substrate 452, 453
  growth on a transgressive surface 452
*Cruziana* 324
*Cruziana* ichnofacies 346, 461–2
current bedding 448, 451
current climate, Cretaceous interior seaway 71, *72*
current flow
  forcing mechanisms 299–300
  peak 301
current ripples 315, 361, 439, 444
  asymmetrical 353
current velocities, New Jersey shelf 405
currents
  along-shelf 405
  combined-flow 324
  cross-shelf 71
  ebb-tidal 363–4
  littoral 15, 16, 100, 112, 128, 130
  longshore, seasonally variable 383
  nearshore 453
  rip 114
  and shelf sediment transport 226
  tidal 379
  unidirectional 284
  *see also* longshore drift; storm currents
cut and fill, multiple stages of 389
cut-off percentage 65–6, 73, 79, 85, 98, 100
cycle boundaries 192–3, 213
cycles
  first-order 157
  second-order 157, 244
  third-order 191, 244
  fourth-order 157, 183
  fifth-order 157, 183
  glacio-eustatic 246
  sea-level 161
  transgressive–regressive 247
cyclic variation, in regime variables 191
cyclicity 174, 184
  depositional 175
  Mesaverde Formation 68, 76, 79
*Cyclostremiscus caraboensis* 269, 272, 273
*Cyclostremiscus plana* 269
cyclothems, Upper Cretaceous Western Interior Basin 168

*Cylindrichus* 495
*Cyrtopleura costata* 271, 273, 276

*Daedalus* 349, 351, 356, 371
Dakota Sandstone 118
Damietta prodelta plume deposits 111–12
debris sheets 216
deceleration sheets 117–18, 143, 217
deformation
   coring-induced 399, 412, 414
   soft-sediment 346, 459
Delaware Bay, back-stepping fill 207, *209*
Delaware estuary, mouth shoal *132*
Delaware shelf valley complex *134*
Delaware-Maryland shelf, stratigraphy *208*
delivered grain size 6
delta complexes
   Mississippi delta 166
   shelf-edge 231
delta front 41–2
delta lobes, Mississippi 165–6
delta mouth bar systems 107–10
   lithofacies *99*, 109–10
delta mouth bars 103
delta systems, shelf-edge 162
delta-strandplain unit 162
deltaic deposits 107, 160
deltaic systems
   shelf-margin 201
   and tidal inlets 282
deltas 93, 96, 107, 114
   avoidance of term 279–80
   back-barrier tidal 16
   bayhead 223
   cuspate *113*, 113–14
     becoming retreating cuspate forelands 133, *134*
   ebb,
     Ossabaw Sound 388
     St Catherines 386–7, *390*
     Wassaw Sound 385–6
   ebb-tidal 119, *129*, 129
     dispersal systems 129, 130
   flood 380
   flood-tidal 119, *129*
     building of 130–1
   intra-lagoonal 266, 270, 279
   lowstand, conceptual model *200*
   micro-birdfoot 240
   oceanic *164*, 223
   offlap wedge *216*, 218
   platform 41, 201, 218, 235
     prograding 198, 237
   platform to shelf-edge transition 227–8
   prograding 508
     equilibrium shoreface profile 226
   shelf *200*
     creating downlap and onlap 201
   shelf margin *200*
   shelf-edge 41, 190, 204, 212, 235, 237
     building marine terrace-scarps 218–19
     lowstand 247
   shelf-phase 201
   tidal 107
   wave-dominated 112

density currents, conglomerate-laden 448
density flows 8
denudation rates
   regional 7
   sub-aerial 6
Denver Basin 240
depocentres 156, 236
   basinward movement of 234
   secondary 219, 234
   shelf-edge deltaic 246
deposition 69
   above storm wave base 493
   back-barrier *208*
   deltaic 327
   fan to wedge-shaped 201
   inner shelf or estuarine 415
   middle shoreface 496–7
   and near-bottom sediment concentration 142
   offlap fill 219
   offlap wedge 229, 236, 237
   and point-source input 117, 231
   prodelta plume 77
   rapid, from turbulent decelerating flow 501
   rhythmic 173, 174
   shallow-marine 351, 354
     channelized and energetic 357
   shelf-margin 46–7
   subtidal, storm-dominated 351
   transgressive 194
     modern, comparison with back-step wedge systems tracts 206–12
   upper shoreface 109–10
   washover fan 128
   wedge to fan-shaped transition 198–9
   *see also* depositional regimes; depositional systems; sedimentation
deposition-bypassing ratio 9–10, 27
depositional agents 7
depositional break in slope 46, 48
depositional complexes 157
depositional environments 5–6, 101, 145, 165
depositional patterns
   problems with existing models 158–9
   scales of **157**
depositional regimes 6–9, 35, 51, 90, 118–19, 154, 170
   accommodation balancing supply 181
   allochthonous 17, 41–4, 52, 190
   autochthonous 17–19, 137
   shelf 17–19
depositional sequences 142, 155
   basinward-stepping 172
   as episodes of coastal onlap 213
   Galloway model 154–7, 192–3, 247–8
   hierarchical 183
   regressive, remoulding into transgressive 181
   Vail model 39, *45*, 45, 46–7, 52, *154*, *192*, 192, 193, 247
   vs. Galloway model 154–7
depositional shoreline break (shelf edge) 39, 46, 213, 218
depositional systems 23–4, 60, 101–5, 158, *170*, 191
   accommodation-dominated 161
   back-step wedge 237
   basinward shifts 170, 172
   contacts between 102–3

ebb-delta–flood-delta–tidal-channel 129–31
and lithofacies patterns 103–5
prodelta plume 110–12, 191
regressive 17, 106–18, 142, 145, 160, 167
shallow-marine, classes of 105–43
shelf 79
and source diastem 101–2
source-sink model 209
spatial arrangements 145
as three-dimensional assemblages of facies 156
transgressive 145, 160, 167
barrier–lagoon–estuary systems 118–31
shoreface–shelf systems 131–43, 172, 174, 212, 216, 234
transgressive 18, 145, 160, 167
depositional systems complexes, Nayarit strandplain 165
depositional systems tracts 60, 105, **106**, 106, 145–6, 154–5, 158–9, *170*, 183
bounded by flooding surfaces 159
regressive 191
Vailian 192
vs. geometric systems tracts 191–2
depositional units, Cardium Formation 467–8, 474–6
desiccation cracks 317, 327, 345, 353, 364, 369
diagenesis 342
Diamond Shoals *135*
diastems 101–2, 105, 145, 166, 167, 170, 209, 241
surf-cut 110, 170, 172
system boundary 170
thalweg 110, 170
transgressive 175
*see also* marine erosion surface; ravinement surfaces
diffusive mixing 40
Dilco Coal Member 178
*Diplocraterion* 315, 319, *320*, 327
*Diplocraterion* burrows 317
disconformities 247, 372, 503, 506
Dismal Rat Member (Cardium Formation) 459
dispersal systems 7–8, 19, 27, 90, 101–2, 129, 145
accommodation-dominated *102*, 102
advective 25, 26, 137
back-barrier 127, *128–9*
basic lithofacies 98–9
deltaic 107
and depositional systems 90–2
diffusive 25, *26*
dune-aeolian-flat 128
elements of 24
and facies assemblages 23–5
as a hierarchy of subsystems 91
littoral, variations in 133
regressive
immature 94
mature 96
supply-dominated 101–2, *102*
transgressive 96, 118, 125–6, 127, *128–9*
transgressive shelves, autochthonous 145
Doce River 266
distributaries abandoned 266–7
drowning of river mouth 274
Doce River strandplain 259–80
effects of submergence phase II 266
geomorphological units *263*, 264
lagoonal deposits 268–74
littoral zone deposits 275–7
regional setting 261–2
shoreline, submergence phase I 264, 266
two-dimensional evolutionary scheme 264–8
doublets 160
down-slope creep 195
downlap, bidirectional 198, 203, 230
downwarping, foreland 311, 332
downwelling *14*, 15, 100, 114, 328, 369
amplified 367
fluid motion during 300
storm-driven 373
downwelling flows 301
geostrophic veering of 368
drainage divides 221, 222
drainage systems, rejuvenated 168, 225
drapes 195, 215, 323, 353, 452
clay 315, 317, 318
mud 356, *357*
shale 316
drift-divergence nodes 121
Drumcliff Shell Bed 168
Duffy Mountain Member 482
dune–aeolian–flat dispersal system 128

East China Sea 216, 219
East China Shelf 116, 219
East Greenland Basin 91
ebb-channel–flood-channel pairs *128*, 133, 137
ebb delta systems *129*, 129
ebb–flood-channel systems *129*, 129–30, *132*, 133
and tidal sand ridges 137
Ebro delta 9, 201
prograding slope model 202
echinoderm fragments 277
effective accommodation potential 248
effective sediment accommodation potential 196
effective sediment supply rate 195–6, 248
elevator tectonics 222
*Elphidium advenum* 412, 418
*Elphidium excavatum* 408, 409, 410, 412, 413, 415, 418
*Elphidium incertum* 408, 410, 412, 413, 418
*Elphidium subarcticum* 410, 412, 418
equilibrium
Davisian and Penckian models 7
dynamic 36, 154
steady state 36
theory 35
equilibrium depth, wave-determined 8, 9, 27, 219
equilibrium index 11, 179, 182
equilibrium point 221, 223–4
equilibrium profiles 34, 36–41, *45–51*, 224–5
equilibrium shelf, concept of 9–12, 60
equilibrium surfaces 179, 182
classes of 11, 36–41
responsive 234
erosion 359
fluvial 484
marine 232, 233, 241, 243
post-depositional 361
of the shoreline 279
slope 199
subaerial 193

erosion surfaces 105, 191, 277, *278*, 349
   disconformable 359, 361
   regional 503
   subaerial 156
      basal 240
   'top of the A sandstone surface' (TASS) 468
erosional source environment 172
estuaries 16, *132*, 365
   circulation in 127
   disequilibrium 13, 118–19
   equilibrium 13, 106, 119
   mesotidal 131
   mouths moving landward 133, 161
   as sediment depocentres 367
   sediment-trapping 16, 27
   subtidal to intertidal 363
estuarine fill, backstepping 210
eustacy 158, 168, 182, 191, 248
eustatic cycles 194
eustatic sea-level curve 47, *52*
event bed thickness 62–4, 64–7, 79, 97–8
   Mesaverde Formation 69–79
   and storm climate 79
event beds 20
   formation of 60–7, 154
   preservation of, Mesaverde Formation 72
   *see also* storm beds
event intensity, and return period power 97–8
event strata 20, 60, 61, 82, 96–101, 145, 170
   problems in interpretation 60
   *see also* depositional sequences; depositional systems; lithofacies
event-scale processes *see* strata formation
expanded sections 101, 145, 160, 246
extreme values, classical theory of 53–4
extreme-event deposits 100

facies 60, 69, 156
   assigned according to degree of condensation 143, 145
   transgressive, mixed 216
facies assemblages, and shelf dispersal systems 23–5
facies differentiation 25, 50, 154, 166, 190
facies patterns
   accommodation dominated 26–7
   autochthonous 26–7, 96, 137, 139
   shelf 19–27
   supply-dominated 25–6
   *see also* lithofacies
facies relationships, east Tenessee *312*, 312–13
facies sequences (Galloway) 157
facies successions, upward-coarsening 174–5, *175*
fair-weather deposits 100
fair-weather flow 20, 60
fan systems, modern 203
fan-aprons, submarine 229
fan-delta bars 107
fan deltas 93, 94, 114
   coalesced 229
fans, continental, modern 204
Farnham unit *see* lower Ferron Sandstone Member
fault systems 311, 330
faulting, listric 118
Fayetville Shale 325
Ferrier Field 458

   correlations with Willesden Green field 479–81
Ferrier sandstone unit 481
Ferron Sandstone Member (Mancos Shale Formation) 490
   *see also* lower Ferron Sandstone Member
fetch 61
   Cretaceous interior seaway 70–1
   and the Medina shoreline 368
fill deposits 194, 206, 212, 219
   fluvial 210
   *see also* channel fill
fine-sediment dispersal 115
   lagoons and estuaries 127
fining trends, cross-shelf 304
fining-upward sequences, storm beds 69
fining-upwards beds 410
fining-upwards trend 354–5
firm grounds 214
fissility 324–5, 346, 500, 501
flaser bedding 444, 501
flats, back-barrier 16
flexural moat 197
flood deposits, central bar mouth 110
flood-water, river 363–4
flooding, passive 123
flooding surfaces 156, 157, 158, 167, 172, *174*, 178, 179, 183, 202
   evolving 118
flow conditions, plane-bed oscillatory 447
flow regimes, combined-current 451
flow reversal 362–3, 364, 369
flows
   along-shelf 300, 304
   bottom 15, 368
   geostrophic 15, 301, *304*, 305, 447, 482
   high-energy, episodic 317
   hypo- and hyperpycnal 440, 452, 453
   liquified 437
   longshore 483
   near-bottom 452
   onshore 304
   onshore and offshore directed 362–3
   oscillatory 81, 333
   spatial acceleration model 450
   storm-generated 15, 100, 328, *331*, 332, 363–4, 368
   stratified 84
   supercritical 288
   *see also* longshore drift; storm flows
fluid power 8, 34
flute casts 313, *314*, 322, 324
fluvial entrenchment 199
fluvial entrenchment surfaces 233, 235, 236, 240, 241, 243, 247
   formation of 229–30, *230*
fluvial plain 224
   and accommodation/supply ratio 224–5, 229
fluvial profile
   regrading of 221, 222
   *see also* river gradients
fluvial sedimentation, timing of (Posamentier) 221
   fluvial morphodynamics 221–2
   $Q$-dependency, modern highstand coasts 223–4
foraminiferal assemblages, New Jersey shelf sand ridges 408–9, 410, 412, 413, 415, 417–18, 421

foredune ridge 298
foreland basins 5
　complex tectonics 243–4
　geometry of 168, 170
　subsidence in 6
foreland thrusts, in- vs. out-of-sequence 244, 248
foresets, bundled 315, *320*
form drag 81
fossil debris 355
　reworked 347
fossil-hash beds 347, *349*, 349, 358, 359
friction-dominated zone 299, *328*, 328
frictional drag 107, *108*, 133
frictional forces, influence of 300
Frontier Formation 238, 482
　unconformities 240
furrows, erosional 141

Gallup Sandstone 176, 180, 181
Gallup–Tocito parasequences 178
Gammon Ferruginous Member 241
Garrington Field 458, 468, 479
Garrington pool 428, 449, 451
geoid, deformation of 7
geometric systems tracts 106, 159, *170*, 183, 227, 235, 247
　terminology 194–5
　vs. depositional systems tracts 191–2
geomorphic cycle 34
　in the marine environment 34
Georges Bank 96, 137
Georges Bank Shoals *142*, 143
Georgia Bight, USA southern Atlantic shelf 124
Georgia (USA), seismic profile analysis on Atlantic shelf 384–91
　Little St Simons Island–Tybee Island 390–1
　oblique survey line 387–9
　sections along inner shelf 384–8
geostrophic adjustment 8
geostrophic flows 15, 305, 447
　and pressure gradient forces 301, *304*
　storm-generated 482
geostrophic zone 299–300, *328*, 328–9
German Bight, North Sea 124, 143
glacio-eustacy 182
glacio-eustatic cycles 7, 161
*Gordia* 437, 438, 466
gorge profile, active channel 385
gorge traces
　buried 385
　St Catherines Sound inlet system 386–7, *390*
grade, concept of 34
graded sand sequences, horizontal laminae in 288
graded stream profile 34, 36–7
　and regulation of sediment input 37
gradient reduction deposits 195
grain composition, discrete sand beds 291
grain shape 292
grain size 98, 100, 190
　distinguishing proximal and distal shoreface lithofacies 100
　down-current changes in 141
　fining upwards 324
　and progressive sorting 19–20, 23
　regional controls of, tectonism vs. base level changes 92

　and system bias 98
grain-size analysis, shelf sand ridges 399, 401, 415
grain-size data
　Damietta prodelta plume 112
　shelf sand ridges, multiple discriminant analysis 401–2, 413–15, *416*
grain-size distribution 62, *63*, 64
grain-size frequency distributions 291–2
grain-size profiles *94*, 96
grain-size range, outer shelf 101
grain-size trends
　lateral 293–4
　vertical 292–3
grains, angularity of 292
grainstone, bioturbated 325
gravel 240
　basal, transgressive 139
gravel waves 141
gravity flows 8, 501
　distal deposits from 466
　downslope 236
gravity remobilization 198
Greenhorn Formation 168, 173
Grimsby Formation 342
Grimsby–Thorold transition *341*
groove casts 322
groove marks 501
Guianas shelf, sedimentary structure 452
Gulf Coast Basin, Tertiary subsidence vs. sedimentation rates 182
Gulf of Mexico
　barrier retreat 123
　northern coastal plain, coastal-plain–slope unconformity 199
　northwest 116–17, 248
　　depositional framework 115, *116*, 117–18
　　Pleistocene retrogressive system 201
　　stratigraphy of 246
　relative sea-level, rising 259–60
　shelf, highstand preservation on 244, *245*, 246
　Texas shelf 9, *116*, 204, 212, 216–17, 283–305
　upper continental slope 199, 201
　*see also* Mississippi delta; Mississippi fan
Gumbel distribution 61, 79, 93
gutter casts 313, *314*, 361
　formation of 324
　sandstone *322*, 324


Hagan Shale Member (Clinch Sandstone) *314*, 323, 324, 327
half-sequences, lower Ferron Sandstone
　regressive 506
　transgressive 507–8
*Hanzawaia strattoni* 412, 413, 418
hardgrounds 173–4
Hartzog Draw Field 449
Hayesville–Fries fault system 311, 330
*Haynesina germanica* 409, 410, 415
HCS *see* cross-stratification, hummocky
headland retreat 126
headland–barrier system 125, 223
*Heleobia* sp 269, 270, 271, 272, 273, 274
Helgoland Bight *see* North Sea, Helgoland Bight

*Helminthopsis* 438, 440
hiatuses 156, 157
high-frequency sequences 158, 179, 183, 184
    accommodation-dominated 181
    Point Lookout Sandstone 170, 172
    regressive 172
    upward-fining 161
highstand, Vail and Galloway 212–13
highstand coasts, modern, $Q$-dependency of 223–4
highstand deposits 78, 194, 196–7
highstand preservation potential 161
    index of 196–7, 244
    variations in 244–7
highstand systems tracts 47, 49, *51*, 156, 173, 191, *192*
    progradational parasequence stacking 213
    Woodbinean sequence 175, *176*
hinge zone width 49, *51*
host–sediment association, Medina Group 346–51, 357, 361, 363, 365, 367
hummocks and swales 109
Hurricane Allen 298
Hurricane Carla 284
    model for 300–4
hurricanes, translating, wave patterns for 302
hydraulic climate 93
hydrocarbon reservoirs 396
hydrostatic pressure gradients, wind-induced 305

ice sheet, Gondwanaland, and red bed formation 330
*Iellina* sp. 269, 271, 272
incised valley fill 240
incision surface, fluvial 199, *200*
Indus fan 203–4
inference, and the sedimentary record 4
inflection point, shoreface profile and shoreface ramp 381
infra-gravity waves 134
inlet channels 382–3, 391
inlet gorges, preservation of 391
inlet scour 130
inlet systems
    tidal channels within barrier lagoons 383
    tide-dominated 391–2
        and wave-dominated 382
inlets
    backfilled 130
    mesotidal 119
    migrating 120, *382*, 382, 383
    retreating landward 131
    tidal 124, 126, 130
inner shelf, storm-dominated 322–3
*Inoceramus* 459, 495, 499, 501
*Inoceramus dimidius* 501, 502, 504
interfingering contact 210
interfingering systems 103, 129–30, 133, 145
interstratified sand and mud lithofacies 97, 98, 99, 112, 131, 195, 212
interstratified sandstone and shale facies 115, 170, 172
intraclasts, shale 327
inundites 110, 127
Irish Sea 143
isostatic balance 37
isostatic equilibrium profile 37

jets
    coastal 447
    river-mouth, flood-stage *14*, 42
    riverine, hypo- and hyperpycnal 440, 452, 453
Juana Lopez Member 490, 500, 502, 503–4, 508, 509
Juniata Formation 324, 327, 330
Juniata–Clinch boundary 327

Kakwa Field 468
Kaskapau Formation 459
Kimmeridge Clay Formation 228
Korea Strait *210*
Kuparuk River Formation 116, 172
Kurten Field *176*, 176, *177*, 449

LaFourche subdelta 126, 166
lagoonal channels, drainage pattern of 381
lagoonal deposits 210
    Doce strandplain 268–74
        depositional histories 270–1, 273–4
        sedimentary facies 268–70, 271–3
lagoonal infilling 270–1, 273
lagoons 16
    back-barrier 221
    elongate 266, 273
        final closure 273–4
    fossil 223
    marsh-filled 121, 122
    open 121
lags *22*, 22, 96, 139, 323, 503
    basal
        coarse 352
        fossiliferous 324
    bioclastic 321, 324
    coarse 504
    granule and pebble 316, 318
    pebble 499
    phosphatic 359, 364
        erosive-based 349, *350*
        fossiliferous 364
    quartz grains 321
    rip-up clasts and fossil debris 355
    sandstone *500*, 504
    shell 495, 496
    transgressive 332, 504, 506
        coarse sandstone 499–500, 508
Laguna Madre, storm run-off 284
lakes, elongate 274
laminae
    bottomset 315
    downward-fanning 289
    horizontal (planar) 287–8
    low-angle inclined 288, 289
    parallel 324
    plane-bed 289
    ripple 444
    sand 437
laminar friction factor 82
laminated mud lithofacies 97, 98, 99, 109, 112, 115, 142, 195, 212, 234
    Louisiana-Texas shelf 117, 217
laminated shale lithofacies 172
    distal 115

laminations
  flat 353, 367
  horizontal 324, 410
  low-angle inclined 461, 462
  parallel 325, 327, 445
  planar 323
  plane-bed 451
  shelf sand ridges 412
  upper sand ridge units 417
Lance Formation 238
landslides, submarine 8
lateral accretion bedding *341, 351, 352*, 353
lateral accretion surfaces *341, 351, 352*, 356, *357, 357*, 363
levee-channel and overbank deposits 203–4
levees, subaqueous *108*, 109
lever point, position of 277–8, *278*
Lewis Shale 239
limestone 172, **316**, 325–6
  basin-centre 168
  *see also* Niobrara Formation (limestone)
*Lingula* shells 352, 353, 356, 364
lithofacies
  barrier–lagoon *130*
  barrier–lagoon–estuary systems 127–31
  delta mouth bar systems 99, 109–10
  in depositional systems 5, 103–5
  distal, regressive shelves 115–18
  inner shelf 118
  interstratified 129
  prodelta plume 112
  proximal
    regressive shelves 115
    stacking of 105
    transgressive shelf systems 139, 141–2
  regressive 191
  stacked *see* magnafacies
  towards a genetic classification 96–9
lithofacies assemblage, transgressive shelf depositional system 141–2
lithofacies changes 20–3
lithofacies differentiation 90, 92
  on a prograding shelf 99–101
lithofacies patterns, and depositional systems 103–5
lithologies 103
lithosomes 60
  classification *145*, 146
  Crossfield area 449–50
lithospheric flexure 311, 330
littoral drift 121
  abundant supply of *126*, 223
littoral energy fence *12*, 13
littoral transport capacity, loss of 133
littoral zone deposits, Doce River strandplain 275–7
littoral zone sequences, and Bruun's rule 277–9
lobe switching 158
log-marker structure surfaces (LMSS) *450*, 450–1, *451*, 453
longshore drift
  and channel migration 382
  Doce River strandplain 266, 267
  and Medina shoreline 364, 365, 368, 372
longshore drift divergence 266
longshore sediment supply, and Bruun's rule 277–8

*Lopha lugubris* 502, 503
Louisiana shelf 110, 117, *159*, *215*, 247
  condensed sections 216
  hiatal surface 215
  lowstand facies tract *200*
  parasequences 160–1
  regime conditions 41–4, 52
  transgressive shoreface-shelf system 216
Louisiana–Texas margin 161
Louisiana–Texas shelf, mud plume 117, 217
lower Blue Gate Shale Member (Mancos Shale Formation)
  early transgressive deposition 509
  facies 500–2
lower Ferron Sandstone Member (Mancos Shale Formation) 490, 500, 501
  deposited during falling sea-level 508
  facies 490, 493–8
lower Ferron sequence boundary 497, 501, 502–3, 508
Lower Gates Formation, swaley cross-stratification 497
lower Mesaverde Formation 241
  depositional systems architecture 172
lower shoreface environment 271
  shoaling waves zone 276
lowstand coastal plain river development, non-Vailian 199
lowstand deposits 77, 77–8, 176, 194, 508
  basin fan (fill) subsystems 202–4
  in the offlap wedge systems tract 201–2
  seismic recognition of basin fan (fill) geometry 198–9
lowstand sea-level
  deposition during 490, 500
  and formation of sequence boundaries 504
  and lower Ferron sequence boundary 502
lowstand systems tracts 49, *51*, 156, 194, 197, 198, 240–1
  Eaglefordian sequence 175, *176*
lowstand tracts 191, *192*
lowstand wedge 46, 47
lowstands, glacial 198
*Lunarca ovalis* 271, 273, 276

*Macoma constricta* 269
magnafacies *170*, 172, 176
mainland–beach detachment *120*, 120, 123
Malabar coast 453
Mancos Shale Formation 168, 172, 181, 508
  Blue Gate Shale, depositional history 504–6
  lower Blue Gate Shale, lithofacies 500–2
  coarse sandstone transgressive lag 499–500
  lower Ferron Sandstone,
    depositional history 504–6
    lithofacies 490, *492*, 493–8
  regional stratigraphic relationships 503–4
  sequence stratigraphic concepts 506–8
margins
  continental
    evolution of 9–10, 27
    prograding 93
    prograding complex *204*
  convergent 5, 6, 93
  divergent 93
  passive 5, *5*, 6–7, 37, 310–11
    Atlantic-type *46*, 46

depositional styles *245*
  hinged *5*, 222
marine erosion surface 206, 210, 237, 247
  formation of 232–3
marine terraces 211
Markov process *18*, 19, 91
marsh lagoons 382, 391–2
marshes, fringing, as fine sediment source 127
Marshybank Formation 175
Martinsburg Formation 325
Maryland Miocene 166–8, 180, 181
mass wasting 205
maximum flooding surface **106**, 156, 157, 173, 175, 178, 193, 205, 206, 209, 210, 236, 237
  and condensed intervals 213–17
  formation of 234
  Parker Creek Bone Bed 168
Medina Basin
  latitude of 368
  storm-dominated 372
Medina channels
  as deltaic distributaries 370
  as fluvial lowstand channels 368–9
  preferred interpretation 363–8
  as storm-cut cross-shelf channels 369–70
Medina Group 322, 340–73
  coarsening upward interval, preferred explanation 363–8
  condensed deposit 361, 370
  control of colour variation in 342
  depositional cycles, regional correlations and origin of 371–2
  facies 345–57
    channel association 351–4
    host–sediment association 346–51
    sheet–sandstone association 354–7
  inferred depositional environments and systems 363–71
  palaeoflow data summary 361–3
  previous environmental interpretations **343**, 343–5
  sequence correlations 359–61
  stratigraphy 342
Medina shoreline 361–2, 363, 364, 365, 368, 370
  episodic progradation 372–3
  tide-dominated 372
Medina–Clinton boundary 342, 358
megaripples 315, 317, 319–20, 328
  hummocky 329, 333
  migration of 130, 317–18
Mesaverde Formation *63*, 64, 112, *114*, *171*, 172, 181, 238, 240
  character of 68, *69*
  grain-size data 71–2
  interpreted geological history *71*, 74–6
  strata-forming processes and regime sedimentation 67–9
Mesaverde Group 168
mica, in shoreface sediments 276–7
Michigan Basin 372
micro-HCS *see* cross-stratification, micro-hummocky
micro-resistivity logs, Crossfield Field *433*
mid-shelf processes, importance of 420, 421
Middle Wilcox Yoakum Canyon 228

Mifflintown Formation 322, 325
Milankovich cycles 7
Milankovich mechanism 182, 184
Milankovich periodicity, and rhythmic bedding patterns 483
minimum bed thickness 98, 100, 104, 117
Mississippi delta 8, 41–4, 109, 126, 165, 223, 440
  depositional scarp *216*, 218
  and northwest Gulf of Mexico shelf, depositional framework 115, *116*
  subdeltas *165*, 165–6, 223
Mississippi dispersal system 91
Mississippi fan 199, 203–4
  Pleistocene, bidirectional downlap 203
  transition from offlap wedge to basin fan systems tract 201
Mississippi River 6, 20, 41, 91
*Monocraterion* 324
morphodynamics 105
  barrier–lagoon–estuary systems 119–25
  delta mouth bars 107–8
  prodelta plume 110–12
  strand plain and regressive shoreface-shelf systems 112–14
  transgressive
    storm-dominated 133–7
    tidal 137
mottling textures 437, 459, 490, 497
mouth bar systems 170
mud 262
  laminated 269, 271
  plastic 269, 273
  prodeltaic 279
  sandy 269
  subtidal, offshore 363
  terrigenous *331*, 332, 333
mud cracks, sandstone-filled 346–7, *348*
mud deposits 101, 118, 128, 276, 333
  plume-like 143
  Texas shelf 215
  transgressive 142–3, 212, 233
mud plumes 117, 217
mud sheets 216–17, 219
mud-flat coasts 114
mudbanks 448
Muddy Sandstone 238, 240
mudrocks, providing reservoir seal 437, 439–40
mudstone 340, 346, 372, 373, 453, 459, 460, 463, *464*, 467
  bioturbated *349*, 350
  Cardium B sandstone 437–40, *441*, *442*, *443*
  fissile to semifissile 324–5
  laminated 466, 468
  massive 468
  pebbly *465*, 466
  shelf 101
  silty 462
mudstone granules, sideritized 467
mudstreams, advective 115, 117
  nearshore 112
*Mulinia cleryana* 271, *272*, 273, *274*, 276
multiple discriminant analysis
  sand ridge grain-size data 401–2
  sand ridge unit **413**, **416**, **418**

multiple-event beds 66, 100, 101
  amalgamated 100
  thin 69
Muschelkalk 325, 326
Muskiki Formation transgressive parasequences 174–5
Mustang Island, Texas 284–5, 291, 304

Nantucket shoals 137, 143
Nanushuk-Torok Group 229
Navarro Group 112
Nayarit coast/strandplain *164*, 165, 223, 260, 280
Nayarit shelf *26*
Nayarit shelf plume 117–18
near bottom motion, coastal ocean 300–1
*Nereites* 438, 440
Netherlands coast, parasequences 161, 164
Netherlands shelf 181
New England shelf (southern) *142*, 143
  mud patch 212, 233
New Jersey shelf 161–2, *211*, *245*
  comparison of nearshore and mid-shelf ridges 415–18
  current velocities 405
  non-fossiliferous mud and sand unit 401, *406*, *407*, 408–9, 415, 421
  shell-rich poorly sorted sand and mud unit *404*, *406*, 410, 415, 421
  upper ridge sand units *404*, *406*, 410–13, *411*, 417–18, 421
New Mexican shelf 181
Niagara Gorge *340*, 342, 349
Niagara River *340*
Niger Delta 20, 94
Niger Shelf 94, 143
Nile Prodelta Shelf 110–11
Nile River, Damietta mouth 115
Niobrara Formation (Limestone) 168, 172, 173, 239
nodules, phosphatic reworked 352
non-linear systems theory, implications of 182–3
*Nonionella atlantica* 413, 418
Norfolk banks *138*
North America, during the Silurian 367, 367–8
North Carolina coast, buried channels associated with palaeostream valleys 384
North Carolina shelf 161–2
North Sea 211, 216
  Frisian and German Bights 124, 143
  Helgoland Bight 289
    modern analogue, Medina shorelines 365–7
  northeastern, back-step fill 212
  northern, Late Juraassic basin fans 228
  Southern Bight 137
Norton Sound, Alaska 289
  modern shelf 80

*Odostomia* sp. 271
offlap 484
  Cardium A depositional units 481
    development of 482–4
  Willesden Green Field depositional units 476–7
offlap apron 213
offlap components 156, 191, 213, 243
offlap wedge (fill) 194
offlap wedge geometric systems tract 247
  comparison with modern highstand deposits 213–19

  models and observations 212–19
offlap wedge progradation 210, 247
offlap wedge systems tracts 213, 235, 241, 243
  development of 230, 234
  lowstand deposits within 201–2
  stacked, Ebro delta 202
offlap wedges, stacked 202
offshore back-step fills 212, 231
  formation of 233–4
offshore circulation, storm-induced 368
offshore veering 15
Ogeechee River system, palaeovalley of 388
*Olivella* sp. 271, 273, 276
ommission surface, burrowed 214
onlap
  coastal
    basinward shift 198
    downward shift in 193, 227
  diagnostic of transgressive deposits 142
onlap components 156, 191, 205–6
onlap surface, canyon-cut 227–9, 230–1, 236, 237
oolites 357
*Ophiomorpha* 411, 416, 493, 495
orbital forcing 174, 183, 184
organic debris 276
oscillatory flows 81
  storm-wave-generated 333
Ossabaw Sound, active tidal channels 386
overwash frequency rate 127
oxygen deficiency 439, 444, 501, 509

*P. novangliae* 418
packstone, bioturbated 325
Padre Island, Texas continental shelf 284
palaeoenvironments, Appalachian miogeocline *310*, 313
palaeostream valleys, preservation of 391
*Palaeophycus* 315, 321, 323, 324, 327, 438, 440, 460
paracycles 49–50, 52, 157
parasequence architecture *180*, 181
  varying with accommodation/supply ratio and preservation index 183
parasequence boundaries, and parasequence set boundaries 156
parasequence boundary truncations *154*, 156
parasequence sets 146, 156, 183, 192
parasequences 5, 106, 112, 142, 156, 158–9, *170*, 183, 191
  back-stepping 183, 191, 204–5, 232, 233
    transgressive 181
  coarsening upwards 158
  in coastal settings 184
  collapsed 183
    and expanded *180*, 181
  compound 162, 164
  Cretaceous Western Interior Basin 168–78
  depositional regime control 181
  formation of
    driven by Milankovich periodicity 182
    Vailian theory 179
  internal structure of 179–81
  Louisiana shelf 160–1
  Maryland Miocene 167–8, 181
  modern highstand shelves 164–6
  origins of 181–3

outer Aquitaine shelf 161
progradational *171*, *173*, 213
in regressive settings 170-4
shallow-marine 172
stacked 159, 181, 183, 191, 213
transgressive settings 174-8
modern shelves 161-4, *162*, *163*
upward-fining 181
paratruncation 118, 156
Parker Creek Bone Bed 167-8
Parkman Sandstone Member, Mesaverde Formation 172
parvafacies *170*, 172
pause planes 315
$^{210}$Pb dating 294, *296*, 297-8
peat 270, 271, 273, 274
'ped' texture 327
Pembina Field 457, 458, 468, 481
Pembina pool 428
Pembina River Member (Cardium A sandstone) 437, 459
peneplanation 225
peripheral bulge 197
migration of 333, 372
*Phacoides pectinatus* 269
phosphatic nodules, reworked 347, 349
Pierre Shale 168, 172, 241
Pitkin Limestone 325
planar beds 275
planar structure 275, 276
plane-bed deposition 324
*Planolites* 315, 317, 321, 323, 324, 327, 353, 438, 440, 447, 460, 493
plant material 408, 409, 410
Plum Point Member (Calvert Formation) 166, 168
plumes 143
buoyant 107, *108*, 109
Columbia River 143, *144*
muddy bottom 117, 217
point bars 131
Point Lookout Sandstone 112, *169*, 176, 181, 240, 467
offlaping progradation strandplain deposits 482
regressive parasequences 170, 172
strandplain sedimentation cycles 483
*Polinices* sp. 276
ponding deposits 195, 212
Poor Valley Ridge Sandstone Member (Clinch Sandstone) 315, 317, *318*, 323, 324, 329
shoreface environment 332
Portuguese continental margin, prograding complex *204*
Powder River Basin 168, *171*, 240, 241
Power Glen Formation 342
preservation, relative 6, 64-7, 71-2, 79, 84-5, 97-101
preservation filter 21-3, 27, 64-5, 98
*see also* reworking ratio
preservation index *see* stratigraphic preservation index
preservation potential 84, 100, 101, 181
storm-generated sand beds 290-1
*see also* highstand preservation potential
preservation potential index 196
preservation ratio 21-3, 181
pressure gradient 299
pressure-gradient force 301
primary structures, and textural maturity 96
*Prionocyclus macombi* 502, 503, 504

*Prionocyclus novimexicanus* 504
*Prionocyclus wyomingenesis* 502, 503
prodelta plume depostional system 191
prodeltas, intra-lagoonal 269, 270, 274
prograding complex 202, 204, 236
and autocyclic processes 237
development of 231
progressive sorting *18*, 19-20, 27, 28, 50, 61, 92, *95*, 100, 105, 145, 294, *328*, 329
beach-shoreface-shelf transect 292
efficiency of 21-2, 25, 26
episodic 23
and event beds 20
and facies differentiation *21*
and grain size 292-3
in mature regressive settings 94
role of grain size 23
and sediment pathways 141
trends *95*, 96
progressive sorting systems, efficient 96
*Pseudaspidoceras* internal moulds 173-4

quartz clasts 447
quartz grain shapes, and source area characteristics 292
quartz sand 332, 333, 347, 349
quartz sandstone, coarse-grained 315, *320*
Queenston Formation 342
Queenston-Medina contact 358, 371
*Quinqueloculina seminula* 410, 412, 413, 418

radiocarbon dating 268-74, 421
bulk shell material 417
'ragged blanket', Cardium Formation 428
ramp surface 234
Raven River Member 459
ravinement process 127, 243
destruction of proximal deposits 128
importance of 247
ravinement surfaces 102, 104, 164, 166, 167, 172, 176, 180, 206, 209-10, 211, 232, 233, 236, 237, 240, 241, 247, 332, 507
formation of 15-17, *16*, 120-3, 133, 232
and the lever point 277
Mesaverde Formation 240
ravinement unconformity 128
re-entrainment 19
probability of 92
reactivation surfaces 315, 318
red beds 327
formation of 330
reflector geometry 146
regime canals 34
regime concepts, and understanding of stratigraphy 148-9
regime conditions
Louisiana Prodelta Shelf 41-4, 52
varying 155
regime models 5, 9-12, 34
*Q*-dependent 243, 244
*R*-dependent 241, *242*, 243
evaluation of 221-4
use of for sequence interpretation 235-7
¥-dependent 227-34, 241, *242*, 243

regime oscillation 181
regime parameter 141, 196
regime ratio 18, 90, 101, 106, 142
  and bottom grain size 139, 141
  cyclic oscillation of 17–19, 179
regime sedimentation 11–19, 34, 60–8, 90, 154
  characteristics of *10*
  coastal equilibrium profiles,
    *Q*-dependent case 39–41
    *R*-dependent case 37–9, 78
  Cretaceous Western Interior Basin 168, 170, 238–44
  graded stream profile 36–7
  isostatic equilibrium profile 37
    simulations 46–50
  Mesaverde Formation 68–73
  terminology 195–7
regime theory 4, 35, 60, 247, 248
  application of 4–5, 35–50
regime topographic profiles *see* regime sedimentation; topographic profiles
regime variables 90, 146, 158, 190, 191, 195, 248
  controlling sedimentation processes 219–27
  cyclic variation in 191
  *see also* variables, geohistorical
regimes 51, 190
  accommodation-dominated *17*, 26–7, 118, 131, 133
    progressive sorting 20
  allochthonous *17*, 17, 28, 40, 50, 105
    role of *Q*-dependent profiles 106, 113
  autochthonous *17*, 17, 27, 50
    and coastal profile model 39
    and *R*-dependent coastal equilibria 190
  with intruding ocean currents 131, 133
  mixed autochthonous-allochthonous 142
  oscillations of 181
  *Q*- or *R*-dominated 197
  sedimentary *17*, 60
  storm-dominated 8, 131
  supply-dominated *17*, 106–7, 113–18
    progressive sorting 19–20
  tide-dominated 8, 131–2
  transgressive 131, 133
regional marker beds 501
regression 76, 77, 106
  eustatic 312, 326–7
  mid-Holocene, Atlantic shelf 121, 166–8
  Silurian 372
rejuvenation 93, 225
  drainage systems 168, 225
reservoirs 19, 23
resistivity logs *449*
  Crossfield Field 431, *432*
resuspension 61–4, 100, 112, 134
  by wave orbital currents 96
  damping turbulence 82
resuspension events 22
return period 97, 100
return period intercept 64, 73–4, 101
return period power 64, 66, *67*, 73–4, 98
  lower shoreface and inner shelf 101
reworking 392, 443, 493, 494, 507
  biological 298, 448, 497
  depositional 64

  and development of mid-shelf ridges 420
  physical 444
  and redistribution 450, 452
  of ridge sediments 447–8
reworking depth 100
reworking ratio 21–2, *22*, 26, 67, 75–6, 79, 93, 98, 101, 104–5, 110, 143, 145, 179
  high 100, 115, 117, 139, 141, 234
  low 96, 101
  and preserved bed thickness 65
*Rhizocorallium* 440, 447, 460, 493
Rhône delta 201, *202*, 202, 210
Rhône margin 161
rhythmic deposition 173, 174
Ricinus pool 428
ridge and swale systems 133, *135*, 137
Rio Grande depocentre *245*
rip currents 114
rip-up clasts
  fine-grained 352
  phosphatized 349, *350*
  shale 346, 347, 355, 356
ripple cross-laminations 109, 461, 462
ripples 289, 315, 346, 402
  climbing 501
  convex 445
  crop-and-fill 444
  wave-formed 351, 367
  *see also* current ripples; ripple cross-lamination; vortex ripples; wave ripples
river gradients
  differing 225–6
  response to changes in relative sea-level 36–7, 225
river-mouth bypassing 13, *14*, *15*, 17, 28, 40, 42, 106, 181, 190
  distal *15*
  proximal *14*
river mouths
  deltaic and estuarine 13
  drowning of 274, 279
river valleys
  submerging 13
  transgressive flooding of 206–7
rivers, graded 34, 221
rock terraces 6
Rockwood Formation 311, *312*, 312, 313, 315, *321*, 321, 323, 324, *326*, 330, 333
  deposition from waning flow 323
  hummocky cross-stratification in 323
  palaeocurrent data 313, *314*
Rocky Mountain foreland, in- and out-of-sequence thrusts 244
*Rosalina floridana* 410, 412, 413, 418
Rose Hill Formation 322, 325
*Rosselia* 460
rough turbulent friction factor 81–2
*Rusophycus* 324


*S. schumoi* 272
Sabine Bank area, transgressive deposits 216
Sable Island Bank 141
St Catherines Island 384

St Catherines Sound inlet system 386–7, *390*
salinity, reduced 444, 452
Salisbury embayment, Chesapeake Bay 247
San Diego County, California, continental shelf, stacked aggradational wedges 210–11
San Juan Basin 168, 170, 172, 176, 178, 503
San Rafael Swell 490, 493, 497, 498, 501, 503, 506
sand 273, 408, 410
   medium-grained 275, 412
   micaceous 276
   mid-shelf ridge 420
   muddy 269
   off-shore transport of 15
   quartzose 262
   relict 402
   shelf-deposited, depositional models for 396
   storm-deposited 351, 363, 368
   transgressive 277
   and the transgressive sand sheet 126
   very fine-grained 275–6
   well-sorted 275
sand beds
   frequency of, and progressive sorting 101
   storm-deposited
     sedimentary characteristics 286–98
     Texas continental shelf 283–305
     *see also* event beds; tempestites
   storm-generated, typical bedding sequence 289
sand bodies 105
   Cardium Formation 457–85
   composite 105
   Mancos Shale Formation 489–509
   offshore 482
sand body geometry 294–8
sand limit 101
sand patches 142
sand ribbons 141
sand ridge complexes, tidal 119
sand ridge fields *see* sand ridge systems
sand ridge systems
   shore-face attached 396
   transgressive 133–4, *136*, 176, 178, 179, 403
sand ridges 96, 133–4, 137, 176, 178, 181
   Cardium Formation 427–53
     ridge architecture 448–9
     ridge evolution 449–51, 453
   Huthnance stability model 134, 137
   mid-shelf, formation of 420–1
   New Jersey shelf 395–421
     comparison of nearshore and mid-shelf ridges 415–18
     geometries of 403
     lithostratigraphic units 407–15
     origin of mid-shelf ridge 418–21
   as sediment dispersal systems 137, 139
   and sedimentation paracycles 176
   shore-connected 403, 405
   submarine, storm and tidal 451
   tidal 137
sand sheets
   bioclastic 333
   inner shelf 262
   middle Atlantic Bight 133

Nayarit sandplain 165
shelf, transgressive 402
shoreface 318–21
storm 333
surficial 402
transgressive
   patchy *136*, 212
   ridged 161
sand shoals 363, 364, 371
   intertidal 365
sand source, shoreface 304
sand spits 125
   coastwise progradation of 273–4
   formation of 266
   progradation of 279
sand storage 126, 127, 128
   subtidal 119
sand supply, fluctuations in 483
sand transport
   along-coast 329
   by longshore drift 364
   during Hurricane Carla 300–1
   shelf, driving force behind 305
   and storm shelf processes *328*, *329*, 332
   storm-surge ebb, after Hurricane Carla (hypothetical) *286*
sand transport processes, shelf 298–304
sand wave deposits, shelf 317
sand wave fields 141
sand waves 315, 317, 320, 402
sandstone 373, 447, 452, 459, 479, 501, 508
   amalgamated 170
   basal, Silurian transgression 327–8
   bioturbated 176, 315, 319, 349, *350*, 448, 460–2, *463*, 497–8, 499, 502, 506, 507, 509
     muddy 460
   conglomeratic 448, 502
   cross-stratified 21, 315, 315–21, 316
   current-bedded 451
   deltaic 175
   fine to coarse-grained *438*, 440, *441*, *442*, *443*, 443–4, *445*, *446*, 447
   fine-grained 321, *322*, 327, 346
     proximal 115
   fining-upwards 501
   flat-laminated 353, *354*
   hummocky 504, 508
   hummocky cross-stratified 321–3, 329, 346, *492*, 493–7, 498, 502, 503
   inlet 131
   lenticular 110, 112, 176, 346, *347*
   lowstand 105
   marine 94
   massive-bedded 317
   offlapping 481
   quartz arenite 309–10
   quartzitic 340, 372
   red 317, 320–1
   regressive 504
   sheet, coarse-grained 133
   siliclastic 358–9
   stacked 479
   strandplain 96, 112, 240

coarsening upwards 115
subarkosic-litharenitic 323
thickly bedded 462–5
thin-bedded **316**, *321*, 323–5
transgressive, autochthonous 96
trough cross-stratified 112
upper shoreface 503
upward-coarsening 176, 178
*see also* sheet sandstone
sandstone beds, mouth bar lithofacies 109
sandstone-shale couplets 444
*Scaphites* 502
*Scaphites warreni* 503, 504
scarps
  constructional 115–18
    prograding 41–2
  depositional 41–2, 116–18, *216*, 218
  prograding 218–19
  shelf 123
scour 444, 500
  of antecedent drainage 392
  cross-island 382
  shoreface 380–1
  storm 324
  tidal 379, 391
scour surfaces, incised 317
scour-and-fill structures 440, 444
sea-level
  changes in, influences on and effects of 506
  eustatic 194
    cyclic variation in 76
    effect on stratal geometries 49, *51*
    falling 221, *222*
    Holocene rise 227
    variation in 7, 37, 47, *52*, 76, 93, 158, 212–13
  falls in 76–7, 78, 166, 504
  glacio-eustatic,
    increase in 161
    Quaternary fluctuations 158
  high-frequency oscillations, effects of 279
  Pleistocene changes 246
  regional (diastrophic) changes in 7
  relative
    fall in 178, 239–40, 506
    fluctuations in 458
    rise in 176, 228, 259–60, 497, 508
  relative change 37, 190, 504
    cyclic 39, 241
    rate of 7, 34, 38, 162, 164, 236
  relative sea-level curves 262
  rise in 133
    and Doce strandplain 264, 266–7
  in Vailian theory 158
sea-level cycles
  glacio-eustatic 182
  higher order *see* paracycles
  Paleogene 246
sea-level oscillations 7, 52
section condensation *see* condensed sections
sediment accommodation potential *238*
sediment accumulation 37, 38–9, 190
sediment accumulation rate *see* accumulation rate
sediment column, single-event beds in 66

sediment concentration 42–3
sediment continuity equation 11–12
sediment diffusivity 82
sediment dispersal
  delta mouths to flanking strandplains *113*, 113–14
  transgressive coasts 125–7
sediment dispersal pathways 141
  on the British Isles shelf 137, *139*, 141
  southern New England shelf 143
sediment dispersion, diffusive 190
sediment input rate 6–7, 9, 34, 38, 39, 190
sediment load, fine-grained 24
sediment prisms 52, 93
  aggrading 9
sediment recirculation 127
sediment settling velocity 81
sediment slumping 195
sediment starvation 17–18, 72, 112
  outer shelf 143
sediment supply 235–6, 244, 248
  high 47, 48, 52
  northwest Gulf of Mexico 246
  and subsidence 76, 243–4
sediment supply index 9, 21, 25, 78
sediment transport 7–9, 118–19, 195
  advective 27–8, 40, 42
  and bedload armouring 61–4, 79–84
  diffusive 24–5, 40, 107, *144*
  ebb- and flood-dominated 109
  efficiency of 225–6
  multiple systems 27
  physical 448
  shelf 226
sediment transport pathways 304
  Carla sand 298
  and seaward fining 294
sediment transport rate 7–9, 34, 190, 226–7, 235, 236
  increase in 227–8
sediment trapping 16, 210
sediment yield 6
sedimentary sequences, amalgamation of 294, *295*
sedimentary structures
  Amazon River shelf 452
  biogenic 290–1, 304
  high-energy setting 447
  physical 287–90, 304, 410
  primary 440, 444
  Surinam shelf 452
  Texas shelf sand beds 286–91
sedimentary textures, Cardium Formation mudrocks 439
sedimentation
  allochthonous 106–18, 195
  autochthonous 118–31, 132, 161, 195
  by storm processes 329, *330*
  condensed, time-transgressive 214
  continental shelf, topographic control of 24
  cyclic 160, 161, 508
  lowstand 160, 508
  of Medina Group 345, 372
  multiple events 450
  point source 231
  *Q*-dominated 197, 235–6, 240–1, 244, 246
  *R*-dominated 197, 236–7, 240, 244

relative to subsidence 168, 170
shelf 51, 60
    model for 12–19, 50, 79, 90
    transgressive estuarine 368–9
    *see also* deposition; fluvial sedimentation
sedimentation cycles 76
    high-frequency 483
        *see also* parasequences
sedimentation rates *22*, 38, 101
    central Atlantic margin 246–7
    high 118
    slower 497
sediments
    relict 13, 139, 141, 215, 234, 262, 396, 402
    riverborne 266, 267, 268, 277
    suspended 127
seismic reflectors 156
Semilla Sandstone Member 490, 503, 506
septarian concretions 490, 493, 497
Sequatchie Formation 327, 330
sequence architecture 191, 247, 248
    and accommodation/supply ratio 227–34
    model and observed compared (Western Interior Seaway) 243–4
    systems tracts 197–219
sequence boundaries 156, 490, 506
    erosional 359
    high-order 78
    lower Ferron 497, 501, 502–3
    Vailian 193
        type 1 199, 201, 204, 229
        type 2 229
    vs. cycle boundaries 192
sequence development, and rate of clastic sediment supply *235*
sequence formation, Cretaceous Western Interior Basin 44, 76, 238
sequence hierarchies 157–8
sequence stratigraphy 60
    correlation, using continuous reflectors of condensed sections 214–15
    terminology **155**, 191–3
sequences
    second-order 168, 244, 248
    third-order 77, 183, 191, 244
        Maryland Miocene 167–8
    fourth-order 166, 178, 179–80, 157
    fifth-order *77*, 164–6, 157
    composite 277
    hierarchical 168
    imbricated 170
    small-scale 77, 167, 172
    stacked 247, 508
        asymmetric 241
    upward-fining 131
    *see also* cycles; high-frequency sequences
serial systems 103, 128, 131, 145
set-up 24, 54, 405
    and on-shore storm winds 328
    sea-level 301
    wave-induced 109
shale *321*, 323–6, 346, 437–8, 440, 444, 448
    bioturbated 440, *441*, *442*, 443–4

estuarine 131
fissile 499, 502, 503
laminated 500–1, 508
organic-rich *345*, 346
prodelta 504
red, pedogenic origin for 327
semi-fissile 325
thin-bedded *442*, 444
transgressive 504
shale chips/clasts 444
shale lenses 494
shale prisms 238–9
shale–silt sequences, upward-coarsening 172
Shannon Sandstone 241, 396, 449, 482
Shannon sandstone ridges 452
shear stress
    bed 42, 43
    bottom, peak 82
    boundary 61
    wave-current 61–2
    wave-generated 302
sheet sands
    coarse-grained 141
    fine-grained 116, 141–2
sheet sandstone 132, 170, 354
    bioturbated 356–7, *358*
    cross-bedded *352*, 355–6, *356*, *357*
    flat-laminated 356
    representing sediment cannibalization 371
    as subtidal to intertidal sand shoals 371
    tidal sand shoal 364
sheet–sandstone association, Medina Group 354–7, 357, 363, 365
shelf currents/flows 15, 24, 62
    and stratification 20–3
shelf deposits, inner and middle shelf 332–3
shelf dynamics, Silurian 328–9
shelf edge, Abrolhas Bank 261
shelf gradient, high 48, 52
shelf incision 213
shelf processes 403, 405
shelf regrading 127
shelf sandstone sequences, Silurian, genesis of 309–33
shelf surface
    evolution of 9–10
    regime response 226–7
shelf systems
    regressive 103, 112–18
    transgressive 209
        distal lithofacies 142–3
        lithofacies on *104*
        proximal lithofacies 139, 141–2
shelf systems tract 205
shelf valleys 133, 161
shelf–margin systems tracts 47, 156, 231
shelf–margin tracts 191, *192*
shelf–valley complexes 133, *134*, 206
shell beds 167, 180
shell coquinas 498, 501, 502
shell debris 130
shell-rich unit, New Jersy shelf ridges *404*, 410, 415
shelves
    accommodation-dominated 26–7

allochthonous 26, 115
    role of $Q$-dependent profile 39–41, 45, 106, 113
autochthonous 25, 26–7, 41, 105
    role of $R$-dependent profiles 37–9, 45, 78, 113, 143, 161–4, 181, 236–7, 241
continental,
    active sedimentation, offlap wedge geometry 213
    evolution of 5–6
    structural settings 5
    distal prodelta 8–9
    epicontinental 6
    marine, and storm events 448
    middle to outer, episodic storms 324–5
    modern, transgression 6
    muddy 142
    pericontinental 6
    Quaternary, systems and sequence patterns 160–6
    storm-dominated 403, 405, 444, 453
        motion on 299
    supply-dominated
        facies differentiation 25–6
        progressive sorting in 23
    transgressive 27
        modern 212
        sediment dispersal systems on 137, 139
Ship Shoal 166
shoal retreat massifs 133, *134*, *135*, 137, 396, 403
    Middle Atlantic Shelf *140*
    ridged 137
shoaling 362
    episodic 372
    sea-floor 107
shoaling upward trend, Mesaverde Formation *71*, 72
shoaling upwards sequences 506
shoals
    ebb-delta, cut and fill of 391
    estuary-mouth 133
    *see also* sand shoals
shoestring sandstone traps 428, 449
shoreface
    constructional 125
    migrating *381*
    regime response 226
    storm-dominated 317
shoreface bypassing 13, *14*, 15–17, *17*, 17–18, 27, 39, 181
shoreface deposits 22–3
    mouthbar flanks 109–10
    upper Ferron Sandstone Member 508
shoreface erosion 6, 9
    releasing fines 142
shoreface incision depth 380–1
shoreface profile 38, 381
    translation of 277, *278*, 279
shoreface retreat *232*, 509
    and accommodation/supply ratio 233
    and development of Cardium A sandstone 484
    erosional 16–17, *38*, 39, 78, 96, 102, 119–20, 133, 143, 162, 209, 330, *331*, 497
    ebb delta front 130
    and lows in Cardium sandstone upper surface 484
shoreface sequence, shallowing-upwards 504
shoreface–shelf systems **106**, 112–18, 131–43, 172, 174, 212, 216, 234

shoreline 18–19, 38, 221, 223–4
    advancing 279
    behaviour 13
    erosion of 279
    migration 38, 40
    prograding 221, 223, 332, 363, 498
    retrograding 330, *331*, 332
    transgressive **119**
shoreline deposits, prograding, Lower Silurian 340–73
siderite 439
siderite cementation 447
siderite clasts 440, 445
siliclastics, shelf margin *331*, 332
silt 333
silt trapping 223
siltstone 313, 323, 346, 444, 452, 453, 460, 500–1
    bioturbated 490, *492*, 493, 495, 498, 502, 504
    Cardium B sandstone 437–40, *441*, *442*, *443*
    Juana Lopez Member 503
    muddy 459, *461*
    quartz 325
    shaley 497
single-event beds 66, 79
sinks 19, 23
    partial 90–1, *91*, 92
*Skolithos* 315, 317, 319, *320*, 324, 327, 356, 460
*Skolithos* ichnofacies 346
slickenside fracture surfaces 327
slope complexes, prograding, recognition of 204
slope depositional processes 204
slope fans 198, 202, 235–6, 237
    development of 231
    recognition of 203–4
slope gradient, high 48, *49*, 52
slope ratio 39
*Solariorbis* sp. 271
sole marks 325, 361
sorting efficiency 92–3, 96
sorting episodes 20
sorting profiles, marine dispersal systems 94
source diastems *see* diastems
source environments 90, 127, 145
sources 19, 23
South Louisiana shelf 453
southeastern African shelf 115
Spanish Mediterranean shelf 143, *214*
    modern surface of non-deposition 213–14
*Sphenia antilhensis* 276
stacking patterns
    progradational 176
    retrogradational 203
steady-state sedimentation, theory of *see* regime theory
steinkerns, reworked 347, 349, 352
stillstand, deposition during 497, 506
storm beaches 114
storm beds
    amalgamated 79
    deposition of *21*
    generation of *21*, 60–4, 79
        limitation of theory 67
    in Mesaverde Formation 68–79
    mud tops, preservation of 101
    preservation of 64–7, 79

theory of 84–5
*see also* event beds; event strata
storm currents 8, 15, 403, 405, 452, 458
   action of 42
   and sand transport 329
storm cuts 382
storm debris flows 501
storm deposition 324, 496, 503
storm events 493, 496
   episodic 452, 504
   extreme 81
storm flows 15, 328
   episodic 100
   offshore 368
   offshore-directed *331*, 332
storm layers, ultradistal 333
storm return period 61, 62, *63*, 64, 65, 84
   of smallest storms 66
storm strata *see* event strata
storm swells, long-period 451
storm tides 301
   from Hurricane Carla 284
storm waves 447
   work of 464
storm winds, along-shelf 333
storm surge 301
storm-surge ebb currents 284, *286*
storm-surge ebb hypothesis 298
storm-surge heights, prior to Hurricane Carla *302*
strandplain and regressive shoreface-shelf systems 112–18, *169*, 170
strandplains 121, 176, 223, 240
   beach-ridge 260
   Doce River 259–80
   forward growth 126
   Nayarit *164*, 165
   prograding 482–4
   *Q*- or *R*-dependent coastal profiles 113
strata formation
   by fluid dynamic processes 21–2
   and depositional facies 97, 143
   in the shallow-marine environment 60–7
stratal bundling 68
stratal patterns 103
stratification 5, 60
   concepts, and problem of facies differentiation 92, 99–101
   flood 110
   low-angle inclined 451
   in the Mesaverde Formation 67–79
   and shelf flow 20–3
   storm-generated 60, 82
   vertical patterns 23
stratification parameters 64–7, 79, 100
stratification pattern 20–3, 27, 96–7
   changing to seaward 100
   and variations in stratal condensation 143
stratigraphic columns, synthetic 64
stratigraphic cycles, high-order 76
stratigraphic preservation index 21–2, 183, 196–7
stratigraphic sequences
   Galloway model *193*, 193
   genetic 156, 157

stratigraphy 35
   basin-margin, Cretaceous Western Interior Basin 168, 170
   science of 4
stress, tangential 284
*Strigilla carnaria* 271
subsidence 5, 6, 16, 37, 508
   Appalachian Basin 309
   basin 48–9, *50*, 233, 332, 372, 508
   by crustal loading 6
   compaction-induced 7
   Cretaceous Western Interior Basin 76, 168, 238, *239*
   due to subduction 6
   episodic 244
   flexural 197
   limiting power of 221
   and sediment supply 243–4
   tectonic 233, 330
substrate consistency, and burrow distribution 325, 333
supercritical flow 288
superimposed systems 103, 129, 145, 170
   delta mouth bar 107–10
surf zone 275, 299
   as an eroding source environment 91
   seaward migration of 110
surf zone beds 22–3
surface-mixed layer 297–8
Surinam Coast, muddy sediments of 440, 453
suspension deposits 109
suspension transport 288
   and grain shape 292
Sussex Sandstone 482
swales 109, 120, 493, 495–6, *496*, 498
swamps
   freshwater 264, 270, 271, 273, 274
   mangrove 264, 268–74, 269, 270, 279
swash action 275
swash aggradation 128
swash bars 130
systems, occluded 103
systems differentiation 183, 184
systems tracts 146, 156, 157, 158, 159, 179, 183, 191
   and cyclic variation in regime variables 191
   timing of 248

TACS ('top of the A sequence coarse sediment') 468–7, 479
   topography on 468–9, *470*, *471*
*Tagelus plebeius* 269
TASS ('top of the A sandstone surface') 468, 477
   relative topography of *472*
   truncation of 484
Teapot Sandstone 240
tectonic loading 168
tectonic tilting 233, 237
tectonism 92, 93, 484
*Teichichnus* 440, 460, 497
*Teichichnus* burrows 493
*Tellina* sp. 272
tempestites 127, 142
   mud 101, 109
Tennessee, east
   Lower Silurian palaeocurrent data 313, *314*

Silurian lithofacies 315–26
  compared with recognized stratigraphic units **316**
  cross-stratified sandstone facies 315–21, *331*, *332*
  hummocky cross-stratified sandstone facies 321–3, *331*, *332*
  thin-bedded limestone and shale facies 325–6, *331*
  thin-bedded sandstone and shale facies *321*, 323–5, *331*, *332*
Silurian sedimentation by storm processes 329, *330*
Tennessee (east), Valley and Ridge Province 310, 331
  biostratigraphic correlations *311*
*Terebellina* 438, 440
Texas shelf 9
  central, mud sheet interpretations 216–17
  prograding complex *116*, 204, 212
  storm-deposited sand beds 283–305
textural gradients *17*, 18
textural maturity 145
  and basin evolution 93
  and the dispersal system 92–3
  and grain-size profiles 94–6
  and primary structures 96
  and topographic profiles 96
  vs. compositional maturity 92
*Thalassinoides* 411
thalwegs
  Helgoland Bight channels 365–7
  river 223
Thermaikos Plateau 213
Thorold Sandstone 342, 355
Thorold sheet sandstone *341*
thrust faulting, post-Cardium Formation 479
thrust sheets, Taconic 332
  as quartz sand source 330
thrusting, Taconic 372
thrusts, foreland 244, 248
tidal bundling 131
tidal channels 363
  deeply scoured *381*
  ebb-dominated, dispersal system 131
  flood-dominated 129
  migrating 129
tidal creeks, and marsh surface 381
tidal delta–tidal channel complexes 105–6
tidal flats 327
tidal flow, in lagoonal channels 381
tidal flushing 107
tidal inlets 124
  as sand source 126
  short life of 130
tidal mixing 109
tidal scour 379
  lagoonal 391
tidalites 127, 142
tides, effects on river mouths 109
Tiger shoal 215–16
Tocito Sandstone 396
  transgressive sequences and sand ridge deposits 176, 178
topographic evolution, as a diffusive process 40–1
topographic fill sedimentation 219
  Holocene 212
  terminology 195

topographic profiles 35, 36–41, 232
  control by accommodation/supply ratio 224–7, 248
  formation of fluvial entrenchment surface 229–30, *230*
  Louisiana prodelta shelf *44*
  $R$-dependent 37–8, 45, 232
  steady-state 51–2
  and textural maturity 96
  *see also* coastal equilibrium profiles
topographic surface 35
topography, antecedent, importance of variations in *386*, 391
Torrivio sandstone 178
trace fossils 315, 323, 327, 340, 346, 351, 353, 356, 440, 461–2, 493
  indicating a marine environment 447
transgression 76, 77
  eustatic 312, 326–7
    evidence for 327–8
  Holocene 396
  and preservation of buried channel deposits 392
  Silurian 344, 372
transgressive surfaces 75, 112, 156, 184, 236–7, 241, 357, 484, 506–7
  formation of 231–2
  stepped *180*, 181
  and the turnaround event 206, 232
transgressive systems tracts 47, 49, *51*, 106, 156, 173, 175, 204–5
transgressive tracts 191, *192*
transition zone *328*, 329
  flow patterns in 300
transport divergence zones 143
transport system 7–8
transport velocity, gradients of 43
Trinity River, incised valley fill 206
Trinity Shoal 166, 215–16
*Trochammina ochracea* 410
truncation
  at top of Medina Group 359, 361
  post-depositional, Carla bed 298
  trough sets 315
truncation surfaces 276, 493, 494, 498, 504
  internal 289
Tununk Shale 490, 493, 499, 500, 503, 504
turbid zone, nearshore 127
turbidites, distal 501
turbidity current deposit hypothesis 298
turbidity currents 284, 294, 428
  deposits 458
turbidity maximum 127
turbulence 444
  dampling of 82
Tuscarora Formation 309, 317, 329, 368, 371–2

unconformities 156, 193, 198, 228, 499–500
  coastal plain-slope 193, 198, 199, 247
  drowning 228
  planar 504
  regional 484, 498, 501, 502–3
  sequence-bounding 221
  subaerial 157
  Teapot Sandstone 240
  transgressive 503

type 1 506
type 2 507
Vailian 158
see also sequence boundaries
uplift 241
upper Carlisle Shale Member 503
upper Ferron Sandstone 490, 508
upper flow regime conditions 288
upper Mississippi canyon 199
upper shoreface, eroding, as source environment 127
upper shoreface environment 275

valley fill deposits 233
  and the basal conformity 206–7, 209
valleys, incised 240
variables
  geohistorical 6–9, 34, 35, 36, 46
    control on simulated stratal geometry patterns 47–50
veering
  due to frictional forces 300
  geostrophic 368
  offshore 15
velocity gradients, regional 452
velocity shear 299
vibracores 263, 274, 396, *404, 406–7*, 408–14, *409, 411*
  core preparation and examination 398–9
  coring procedures 396
Viginia Bank Massif 141
Viking Formation 482
Virginia shelf 161–2
vortex ripple migration *360–1*, 362, 365
vortex ripples 343, 346, 353, 356, 357
  crestal orientations, Medina Basin *360*, 361, 364
  and wave refraction 365

Walther's law 100, 101
Wapiabi Formation 459
washover chutes 128
washover deposits, Padre Island 284
washover fan systems 128
washover fans 120
washover systems 102
washovers 270
Wassaw Island 384, 385
Wassaw Sound ebb delta 385–6
water column, densely stratified 439
water depth 64, 66, 79
water stratification, destruction of 405
water-sediment interface, aggrading 5
wave base 8, 9, 27, 35
wave bevelling *386, 391, 392*
wave breaking, effect of 54–5

wave climate 64, 67, 70, 226
  conditions 61
  Cretaceous interior seaway 70–1
  and the Doce strandplain 261
wave energy, variation in 8
wave forecasting 61
wave height 53–4, 55, 61, 71
  extreme 61, 100
  maximum 62
wave kinematics 300
wave motion, oscillatory 284
wave orbital action 81–2, 226
wave orbital motion *12*, 100, 302, 304
  high-frequency 24
wave propagation directions, Medina coast *360–1*, 365
wave ripples 323, 439, 444, 464–5, 497, 499
  form-discordant 324, 325, 333
  and HCS 343
wave set-up 54
wave trains 114, 405
  prevailing 266
wave-ripple crests 325
  giving palaeocurrent data 313, *314*
wave–tidal current interaction 382
waves
  shoaling 113–14
  storm 447, 464
  swell, long-period 368
Whirlpool Sandstone (Medina Group) 342, 361, *370*, 371
  reinterpretation of 344
Willesden Green Field 457–85
  external geometry 468–74
    conglomerate distribution 473–4
    sandstone distribution 469, *472*, 473
  facies description 459–66
  and Ferrier Field
    correlations between 479–81
    interpretations of 482–4
  internal geometry 474–7
  stratigraphy 459
Willesden Green reservoir 462
wind field 299
winds, along-shelf 326, 333
Woodbine Formation 112, 449
  transgressive parasequences and sand ridge deposits 175–6, *177, 178*

Yangtze River 8
Yellow River and shelf 8, 440
Yellow Sea 269

zones of convergence 137
*Zoophycos* 440, 447, 460